Encyclopedia o

Ian P. Stolerman • Lawrence H. Price
Editors

Encyclopedia of Psychopharmacology

Second Edition

Volume 2

M–Z

With 216 Figures and 147 Tables

Editors
Ian P. Stolerman
Section of
Behavioural Pharmacology
Institute of Psychiatry,
Psychology and Neuroscience
King's College London
London, UK

Lawrence H. Price
Department of Psychiatry and Human Behavior
The Warren Alpert Medical School of
Brown University
Butler Hospital
Providence, RI, USA

ISBN 978-3-642-36171-5 ISBN 978-3-642-36172-2 (eBook)
ISBN 978-3-642-36173-9 (print and electronic bundle)
DOI 10.1007/978-3-642-36172-2
Springer Singapore Heidelberg New York Dordrecht London

Library of Congress Control Number: 2015930031

© Springer-Verlag Berlin Heidelberg 2010, 2015
This work is subject to copyright. All rights are reserved by the Publisher, whether the whole or part of the material is concerned, specifically the rights of translation, reprinting, reuse of illustrations, recitation, broadcasting, reproduction on microfilms or in any other physical way, and transmission or information storage and retrieval, electronic adaptation, computer software, or by similar or dissimilar methodology now known or hereafter developed. Exempted from this legal reservation are brief excerpts in connection with reviews or scholarly analysis or material supplied specifically for the purpose of being entered and executed on a computer system, for exclusive use by the purchaser of the work. Duplication of this publication or parts thereof is permitted only under the provisions of the Copyright Law of the Publisher's location, in its current version, and permission for use must always be obtained from Springer. Permissions for use may be obtained through RightsLink at the Copyright Clearance Center. Violations are liable to prosecution under the respective Copyright Law.
The use of general descriptive names, registered names, trademarks, service marks, etc., in this publication does not imply, even in the absence of a specific statement, that such names are exempt from the relevant protective laws and regulations and therefore free for general use.
While the advice and information in this book are believed to be true and accurate at the date of publication, neither the authors nor the editors nor the publisher can accept any legal responsibility for any errors or omissions that may be made. The publisher makes no warranty, express or implied, with respect to the material contained herein.

Printed on acid-free paper

Springer is part of Springer Science+Business Media (www.springer.com)

Preface to the Second Edition

This second edition of the *Encyclopedia of Psychopharmacology* is intended to reflect the many advances that have occurred in the field since the publication of the first edition in 2010. In the preclinical arena, this includes not only the substantial progress that has been made in the psychopharmacology of complex psychological and behavioral processes, such as cognition and drug self-administration, but also the introduction and refinement of novel methods and areas of inquiry, such as optogenetics and ultrasonic vocalization. At the clinical level, the release of the fifth edition of the *Diagnostic and Statistical Manual of Mental Disorders* (DSM-5) by the American Psychiatric Association signals the first major change in psychiatric diagnostic taxonomy since the publication of DSM-IV in 1994. Whatever one's opinion as to the wisdom, necessity, or execution of that change (and contributors to the *Encyclopedia* reflect the full range of diversity in this regard), there can be no disputing the need for an up-to-date compendium of psychopharmacology to fully take it into account. It should be noted that some of the definitions given are based on those in the DSM or other standard sources of reference; we therefore acknowledge the role of such important works in the development of the *Encyclopedia*.

All of the 252 essays and approximately 1,150 brief definitions of the first edition have been reviewed and, as needed, revised to accommodate new findings and understanding or even just to improve readability. In addition, 32 new essays and about 10 new brief definitions have been added in order to more fully address topics that were either insufficiently covered in the first edition or have since emerged as important in their own right. Both old and new contributors were enlisted to accomplish this massive undertaking that has increased the size of the *Encyclopedia* by more than 30%. The panel of section editors was also expanded so that it now includes a member based in Asia and an expert to coordinate and develop the entries on substance abuse and dependence. Readers should be aware that there are areas in any active scientific field where experts do not agree with each other; our policy has not required authors to adhere strictly to the views of the section editors or the editors in chief; rather, we have aimed to allow scope for expert contributors to express their own views on the field and have encouraged them to indicate in their contributions the areas where controversy exists.

We extend our thanks, again, to the many staff members at Springer for their unflagging support of this project; to our exceptional section editors, whose careful oversight, detailed suggestions, and substantive essays so profoundly shaped the final product; and, finally, to our many contributing authors, who labored with such dedication to distill the broad and deep knowledge they have of their subjects into entries which, we hope, will be readily accessible and efficiently informative to the nonspecialist readers of this work.

Institute of Psychiatry, Ian P. Stolerman
Psychology and Neuroscience
King's College London
London, UK

Butler Hospital Lawrence H. Price
Brown University
Providence, RI, USA
March 2015

Preface to the First Edition

Psychopharmacology is the study of the effects of psychoactive drugs on the functioning of the central nervous system at all levels of analysis, thus embracing cognition, behavior, psychological states, neurophysiology, neurochemistry, gene expression, and molecular biology. It includes, as integral parts of its domain, the medicinal and social uses of the substances, the interaction of environmental and genetic factors with the actions of psychoactive drugs, their use for probing the functionality of the central nervous system, and drug dependence and abuse.

The modern science of psychopharmacology was developed in the second half of the twentieth century as a consequence of innovative "blue-skies" laboratory research and the parallel recognition of the specific therapeutic effects of a new generation of drugs that brought about the so-called psychopharmacological revolution. This *Encyclopedia of Psychopharmacology* attempts to bring together much of this knowledge that has been acquired in the last 50 years in a format that makes it accessible to a diverse range of nonspecialists and students. It facilitates the dissemination of what has been learned to a broader population of researchers, clinicians, and advanced students. The level of the entries is typically midway between articles for informed laypeople and reviews for specialized biomedical experts. The diversity of disciplinary areas of which psychopharmacology is comprised means that it is a useful educational resource when specialists need to extend their role into areas beyond their primary fields of expertise. People in related fields, students, teachers, and laypeople will benefit from the information on the most recent developments of psychopharmacology.

The scope of the product is defined by the definition above and by the scope of the journal *Psychopharmacology*, a premier peer-reviewed journal for publication of original research and expert reviews. The aim is to provide detailed information on psychopharmacology and its subdisciplines, such as clinical psychopharmacology, molecular neuropsychopharmacology, behavioral pharmacology in laboratory animals, and human experimental psychopharmacology. Nevertheless, the interdisciplinary nature of the field has engendered very different ideas about the meaning of the word "psychopharmacology," and there is probably no definition that could meet with universal and permanent acclaim. It is like the proverbial story of the blind men and the elephant. Some see psychopharmacology primarily as the scientific study of the psychological and behavioral effects of substances in normal subjects. Others regard it as mainly concerned with the development and evaluation of pharmacotherapies for psychiatric states. Yet a further important group of scientists have been engaged in defining the actions of psychoactive drugs across the huge range of disciplines of which neuroscience is composed including, among others, neurochemistry, neurophysiology, and molecular biology. It is hoped that the project will help to shed light on the interconnections between the different parts of the elephant, showing how the effects of drugs within the intact conscious, behaving organism are related to their properties at the levels of molecules, cells, and neural systems.

The wide-ranging entries in the *Encyclopedia of Psychopharmacology* are written by leading experts, including basic and clinical scientists in academia and industry. The entries fall into several main categories. Target essays include descriptions of fundamental psychological and biological processes that are influenced by psychoactive drugs; here the emphasis is at the behavioral and psychological level, although important functions in the neuropharmacological and psychosocial domains are included. Targets at the molecular and cellular levels constitute the primary domain of the *Encyclopedia of Molecular Pharmacology* (Offermanns and Rosenthal, Eds. Springer 2008) that may be seen as a parallel endeavor. Those entries named according to psychiatric disorders describe the role of pharmacotherapy in treatment, with reference to differential therapies for subclasses of a disorder, with cross-references to essays on the drugs used. Conditions where drug treatment is not normally used are excluded except for substance use disorders, where the main characteristics of all the disorders are pertinent even though there is no pharmacotherapy for some of them. Drugs are the stock-in-trade of psychopharmacologists and therefore many entries are named according to drugs or drug classes. These essays review their general pharmacology, with an emphasis on psychotropic and other central nervous system effects; their neuropharmacological mechanisms of action are described in relation to receptor or enzymatic targets, key events that occur downstream from the receptor, and brain regions, with cross-references to the psychiatric states in which the drugs are used most often. Other essays focus upon the key methods used in the field, describing the main features of the techniques and outlining their roles in psychopharmacology, the types of information obtained, and why they are needed; the advantages and limitations of a technique may also be summarized.

The essays are complemented by more than 1,150 short definitions; essays and definitions cross-reference each other and other relevant entries. Entries appear in an alphabetical sequence that makes the printed encyclopedia easy to use. Users of the online version benefit from hyperlinks that correspond to the cross-references in the printed version.

I thank the many members of the publisher's staff who have participated in this project and have supported it so ably. Throughout the enterprise I have also enjoyed the support of an outstanding team of Field Editors, all of whom have sustained internationally recognized records of scholarly activity in psychopharmacology. The team includes individuals based in pharmaceutical industry as well as in academia, reflecting the frequent and often essential collaborations between these sectors. It would not have been possible to produce the *Encyclopedia of Psychopharmacology* without their guidance on its content and their assistance with the selection of authors and reviews of the submitted entries. I also thank the hundreds of individual authors whose exceptional work forms the substance of the product and who have given so generously of their time and expertise.

Institute of Psychiatry Ian P. Stolerman
King's College London
London, UK
August 2009

Editors-in-Chief

Ian Stolerman, B.Pharm., Ph.D. Emeritus Professor of Behavioural Pharmacology, Institute of Psychiatry, Psychology and Neuroscience, King's College London, London, UK
ian.stolerman@kcl.ac.uk

Lawrence H. Price, M.D. Professor of Psychiatry and Human Behavior, The Warren Alpert Medical School of Brown University, Providence, RI, USA

President and Chief Operating Officer, Butler Hospital, Providence, RI, USA
Lawrence_Price_MD@brown.edu

Section Editors

Section: Behavioral Pharmacology in Laboratory Animals

Klaus A. Miczek, Ph.D. Moses Hunt Professor of Psychology, Psychiatry, Pharmacology and Neuroscience, Tufts University, Medford, MA, USA
klaus.miczek@tufts.edu

Trevor W. Robbins, Ph.D., FRS Professor of Cognitive Neuroscience, Department of Psychology, University of Cambridge, Cambridge, UK
twr2@cam.ac.uk

Section: Clinical Psychopharmacology

Malcolm Lader, D.Sc., M.D., Ph.D. Emeritus Professor of Clinical Psychopharmacology, Institute of Psychiatry, Psychology and Neuroscience, King's College London, London, UK
malcolm.lader@kcl.ac.uk

Christopher J. McDougle, M.D. Director, Lurie Center for Autism, Lexington, MA, USA

Professor of Psychiatry and Pediatrics, Massachusetts General Hospital, Lexington, MA, USA

Nancy Lurie Marks Professor in the Field of Autism, Harvard Medical School, Boston, MA, USA
cmcdougle@partners.org, CMCDOUGLE@mgh.harvard.edu

Seiya Miyamoto, M.D., Ph.D. Director of Schizophrenia Treatment Center, Department of Neuropsychiatry, St. Marianna University School of Medicine, Kawasaki-shi, Kanagawa, Japan
s2miya@marianna-u.ac.jp

Section: Human Experimental Psychopharmacology

Harriet de Wit, Ph.D. Professor, Human Behavioral Pharmacology Laboratory, Department of Psychiatry and Behavioral Neuroscience, University of Chicago, Chicago, IL, USA
hdew@uchicago.edu

Theodora Duka, M.D., Ph.D. Professor of Experimental Psychology, School of Life Sciences, University of Sussex, Falmer, Brighton, Sussex, UK
t.duka@sussex.ac.uk

Section: Molecular Neuropsychopharmacology
Daniel Hoyer, Ph.D., D.Sc., FBPhS Chair and Head, Department of Pharmacology and Therapeutics, School of Biomedical Sciences, Faculty of Medicine, Dentistry and Health Sciences, The University of Melbourne, Parkville, VIC, Australia
d.hoyer@unimelb.edu.au

Section: Preclinical Psychopharmacology
Thomas Steckler, M.D. Pharma R&D Quality & Compliance, Janssen Research and Development, Beerse, Belgium
tsteckle@its.jnj.com

Section: Substance Use Disorders
Stephen T. Higgins, Ph.D. Professor of Psychiatry and Psychology, Department of Psychiatry, University of Vermont, Burlington, VT, USA
stephen.higgins@uvm.edu

Contributors

Chadi G. Abdallah Department of Psychiatry, Yale University School of Medicine, New Haven, CT, USA

Clinical Neuroscience Division, National Center for PTSD, West Haven, CT, USA

Alfonso Abizaid Department of Neuroscience, Carleton University, Ottawa, ON, Canada

Francisco Aboitiz Departamento de Psiquiatría, Facultad de Medicina, and Centro Interdisciplinario de Neurociencia, Pontificia Universidad Católica de Chile, Santiago, Chile

Anthony Absalom Anesthesiology Department, University Medical Center Groningen, University of Groningen, Groningen, The Netherlands

Ram Adapa Division of Anaesthesia, School of Clinical Medicine, Cambridge University Hospitals NHS Foundation Trust, Cambridge, UK

Albert Adell Department of Neurochemistry and Neuropharmacology, Instituto de Investigaciones Biomédicas de Barcelona, Consejo Superior de Investigaciones Científicas (CSIC), Barcelona, Spain

Institut d'Investigacions Biomèdiques August Pi i Sunyer (IDIBAPS)

Centro de Investigación Biomédica en Red de Salud Mental (CIBERSAM), Spain

Irene Adelt Department of Psychiatry and Psychotherapy, LKH Klagenfurt, Klagenfurt, Austria

Anders Ågmo Department of Psychology, University of Tromsø, Tromsø, Norway

Yesne Alici Department of Psychiatry and Behavioral Sciences, Memorial Sloan-Kettering Cancer Center, New York, NY, USA

Christer Allgulander Department of Clinical Neuroscience, Division of Psychiatry, Karolinska University Hospital, Karolinska Institutet, Stockholm, Sweden

Osborne F. X. Almeida NeuroAdaptations Group, Max Planck Institute of Psychiatry, Munich, Germany

Shimon Amir Department of Psychology, Center for Studies in Behavioral Neurobiology, Concordia University, Montreal, QC, Canada

Stephan G. Anagnostaras Department of Psychology and Program in Neurosciences, University of California, San Diego, La Jolla, CA, USA

Rodrigo Andrade Department of Pharmacology, School of Medicine, Wayne State University, Detroit, MI, USA

Per E. Andrén Biomolecular Imaging and Proteomics, National Laboratory for Mass Spectrometry Imaging, Department of Pharmaceutical Biosciences, Uppsala University, Uppsala, Sweden

Anne M. Andrews Semel Institute for Neuroscience and Human Behavior, Hatos Center for Neuropharmacology, California NanoSystems Institute, University of California, Los Angeles, Los Angeles, CA, USA

Raymond F. Anton Department of Psychiatry and Behavioral Sciences, Alcohol Research Center, Medical University of South Carolina, Charleston, SC, USA

Francesc Artigas Department of Neurochemistry and Neuropharmacology, Instituto de Investigaciones Biomédicas de Barcelona, Consejo Superior de Investigaciones Científicas (CSIC), Barcelona, Spain

Institut d'Investigacions Biomèdiques August Pi i Sunyer (IDIBAPS)

Centro de Investigación Biomédica en Red de Salud Mental (CIBERSAM), Spain

Vera Astreika Cleveland, OH, USA

John Atack Translational Drug Discovery Group, School of Life Sciences, University of Sussex, Beerse, Belgium

Aldo Badiani Department of Physiology and Pharmacology, Sapienza University of Rome, Rome, Italy

Ahmed S. BaHammam University Sleep Disorders Center, College of Medicine, National Plan for Science and Technology, King Saud University, Riyadh, Saudi Arabia

Glen B. Baker Neurochemical Research Unit and Bebensee Schizophrenia Research Unit, Department of Psychiatry, University of Alberta, Edmonton, AB, Canada

Robert L. Balster Department of Pharmacology and Toxicology, School of Medicine, Virginia Commonwealth University, Richmond, VA, USA

Christof Baltes Varian Medical Systems Imaging Laboratory GmbH, Dattwil, Switzerland

Tomek J. Banasikowski Centre for Neuroscience Studies, Queen's University, Kingston, ON, Canada

Segev Barak School of Psychological Sciences and the Sagol School of Neuroscience, Tel-Aviv University, Tel-Aviv, Israel

Michael T. Bardo Center for Drug Abuse Research Translation (CDART), College of Arts & Sciences, University of Kentucky, Lexington, KY, USA

Andrea Bari Department of Neurosciences, Neuroscience Institute, Medical University of South Carolina, Charleston, SC, USA

Thomas R. E. Barnes Department of Medicine, Imperial College, London, UK

James E. Barrett Department of Pharmacology and Physiology, Drexel University College of Medicine, Philadelphia, PA, USA

Lucie Bartova Department of Psychiatry and Psychotherapy, Division of Biological Psychiatry, Medical University of Vienna, Vienna, Austria

Pierre Baumann Centre de Neurosciences psychiatriques, Département de Psychiatrie, DP-CHUV, Prilly-Lausanne, Switzerland

Jill B. Becker Psychology Department and Molecular & Behavioral Neuroscience Institute, University of Michigan, Ann Arbor, MI, USA

Robert H. Belmaker Beersheba Mental Health Center, Beersheba, Israel

Catherine Belzung UFR Sciences et Techniques, Université François Rabelais, Tours, France

Fabrizio Benedetti Department of Neuroscience, University of Turin Medical School, and National Institute of Neuroscience, Turin, Italy

Fernando Benetti Memory Center, Brain Institute, Pontifical Catholic University of Rio Grande do Sul, Porto Alegre, RS, Brazil

Richard J. Beninger Department of Psychology, Centre for Neuroscience Studies, Queen's University, Kingston, ON, Canada

Anne Berghöfer Institute for Social Medicine, Epidemiology and Health Economics, Charité - Universitätsmedizin Berlin, Berlin, Germany

Leandro José Bertoglio Departamento de Farmacologia, CCB, Universidade Federal de Santa Catarina, Florianópolis, SC, Brazil

Daniel Bertrand Department of Neuroscience, HiQScreen Sàrl, Geneva, Switzerland

João M. Bessa Life and Health Science Research Institute (ICVS), School of Health Sciences, University of Minho, Braga, Portugal

ICVS/3B's – PT Government Associate Laboratory, Braga/Guimarães, Portugal

Rick A. Bevins Department of Psychology, University of Nebraska-Lincoln, Lincoln, NE, USA

Saurabha Bhatnagar Department of Physical Medicine and Rehabilitation, Massachusetts General Hospital, Spaulding Rehabilitation Hospital, Brigham and Women's Hospital, and Harvard Medical School, Charlestown, MA, USA

Warren K. Bickel Addiction Recovery Research Center, Virginia Tech Carilion Research Institute, Roanoke, VA, USA

Michel Billiard Department of Neurology, Gui de Chauliac Hospital, Montpellier, France

István Bitter Department of Psychiatry and Psychotherapy, Semmelweis University, Budapest, Hungary

Lisiane Bizarro Instituto de Psicologia, Universidade Federal Do Rio Grande do Sul, Porto Alegre, RS, Brazil

Katherine A. Blackwell Department of Psychiatry, Yale School of Medicine, New Haven, CT, USA

Kelly Blankenship Department of Psychiatry, Indiana University School of Medicine, Indianapolis, IN, USA

Michael Bloch Yale OCD Research Clinic, New Haven, CT, USA

Alyson J. Bond Addictions Department, Institute of Psychiatry, King's College London, London, UK

Heleen B. M. Boos Department of Psychiatry, Rudolf Magnus Institute of Neuroscience, University Medical Center, Utrecht, CX, The Netherlands

Jean-Philippe Boulenger Department of Adult Psychiatry, Hôpital la Colombiere CHU Montpellier, University Montpellier 1. Inserm U 888, Montpellier, France

Chase H. Bourke Department of Psychiatry and Behavioral Sciences, Laboratory of Neuropsychopharmacology, Emory University School of Medicine, Atlanta, GA, USA

C. M. Bradshaw Division of Psychiatry, Psychopharmacology Section, University of Nottingham, Nottingham, UK

Marc N. Branch Department of Psychology, University of Florida, Gainesville, FL, USA

William Breitbart Department of Psychiatry and Behavioral Sciences, Memorial Sloan-Kettering Cancer Center, New York, NY, USA

David A. Brent Western Psychiatric Institute and Clinic, University of Pittsburgh School of Medicine, Pittsburgh, PA, USA

Holden D. Brown Brains On-Line, LLC, South San Francisco, CA, USA

Verity J. Brown School of Psychology and Neuroscience, University of St Andrews, Scotland, UK

Marcy J. Bubar Center for Addiction Research, University of Texas Medical Branch, Galveston, TX, USA

Alan J. Budney Department of Psychiatry, Geisel School of Medicine at Dartmouth, DH Addiction Treatment and Research Program, Lebanon, NH, USA

Catalin V. Buhusi Utah State University, North Logan, UT, USA

David B. Bylund Department of Pharmacology and Experimental Neuroscience, University of Nebraska Medical Center, Omaha, NE, USA

Martine Cador Team Neuropsychopharmacology of addiction, UMR CNRS 5287, University of Bordeaux, Bordeaux, France

Wiepke Cahn Department of Psychiatry, Rudolf Magnus Institute of Neuroscience, University Medical Center, Utrecht, CX, The Netherlands

Stéphanie Caillé Team Neuropsychopharmacology of addiction, UMR CNRS 5287, University of Bordeaux, Bordeaux, France

Mary E. Cain Department of Psychology, Kansas State University, Manhattan, KS, USA

Paul D. Callaghan ANSTO LifeSciences, Australian Nuclear Science and Technology Organisation (ANSTO), Kirrawee DC, NSW, Australia

Delphine Capdevielle French National Institute of Health and Medical Research, INSERM U888, La Colombière Hospital, Montpellier, France

Stephanie A. Carmack Department of Psychology and Program in Neurosciences, University of California, San Diego, La Jolla, CA, USA

Antonio Pádua Carobrez Departamento de Farmacologia, CCB, Universidade Federal de Santa Catarina, Florianópolis, SC, Brazil

Linda L. Carpenter Department of Psychiatry and Human Behavior, Butler Hospital, Brown University, Providence, RI, USA

Ximena Carrasco Servicio de Neurología y Psiquiatría infantil, Hospital Luis Calvo Mackenna Facultad de Medicina, Universidad de Chile, Santiago, Chile

Marilyn E. Carroll Department of Psychiatry and Neuroscience, Medical School, University of Minnesota, Minneapolis, MN, USA

Helen J. Cassaday School of Psychology, University of Nottingham, Nottingham, UK

F. Xavier Castellanos New York University Child Study Center, Nathan Kline Institute for Psychiatric Research, New York, NY, USA

Samuel R. Chamberlain Department of Psychiatry, University of Cambridge, Addenbrooke's Hospital, Cambridge, UK

Cambridge and Peterborough NHS Foundation Trust, Cambridge, UK

R. Andrew Chambers Department of Psychiatry, IU Neuroscience Center, Indiana University School of Medicine, Indianapolis, IN, USA

Lucy G. Chastain Department of Psychiatry and Behavioral Sciences, Emory University School of Medicine, Atlanta, GA, USA

Darren R. Christensen Faculty of Health Sciences, University of Lethbridge, Lethbridge, AB, Canada

MacDonald J. Christie Discipline of Pharmacology, The University of Sydney, NSW, Australia

Arthur Christopoulos Drug Discovery Biology, Monash Institute of Pharmaceutical Sciences, Monash University, Parkville, Melbourne, VIC, Australia

Yogita Chudasama Department of Psychology, McGill University, Montreal, QC, Canada

Andrea Cipriani Department of Psychiatry, Medical Sciences Division, University of Oxford, Oxford, UK

Leslie Citrome Department of Psychiatry and Behavioral Sciences, New York Medical College, Valhalla, NY, USA

Paul B. S. Clarke Department of Pharmacology and Therapeutics, McGill University, Montréal, QC, Canada

Emil F. Coccaro Department of Psychiatry, University of Chicago, Chicago, IL, USA

Caroline Cohen Therapeutic Strategic Unit Aging, Chilly-Mazarin, France

Maria Isabel Colado Departamento de Farmacologia, Facultad de Medicina, Universidad Complutense, Madrid, Spain

Francis C. Colpaert SCEA-CBC, Puylaurens, France

Roshan Cools Department of Psychiatry, Donders Institute for Brain, Cognition and Behaviour, Centre for Cognitive Neuroimaging, Radboud University Medical Center, Nijmegen, The Netherlands

Christoph U. Correll Department of Psychiatry, The Zucker Hillside Hospital, Glen Oaks, NY, USA

Pietro Cottone Boston University School of Medicine, Boston, MA, USA

Ronald L. Cowan Departments of Psychiatry and Radiology and Radiological Sciences, Psychiatric Neuroimaging Program, Vanderbilt Addiction Center, Vanderbilt University School of Medicine, Nashville, TN, USA

Philip J. Cowen Department of Psychiatry, Warneford Hospital, University of Oxford, Oxford, Oxfordshire, UK

John Cryan Department of Anatomy & Neuroscience, University College Cork, Cork, Ireland

A. Claudio Cuello Department of Pharmacology, McGill University, Montreal, QC, Canada

Christopher L. Cunningham Department of Behavioral Neuroscience and Portland Alcohol Research Center, Oregon Health & Science University, Portland, OR, USA

Kathryn A. Cunningham Center for Addiction Research and Department of Pharmacology and Toxicology, University of Texas Medical Branch, Galveston, TX, USA

H. Valerie Curran Department of Psychology, Clinical Psychopharmacology Unit, University College London, London, UK

Megan M. Dahmen Via Christi Regional Medical Center – Good Shepherd Campus, Wichita, KS, USA

Lynette C. Daws Department of Physiology and Pharmacology, University of Texas Health Science Center at San Antonio, San Antonio, TX, USA

Peter Paul De Deyn Laboratory of Neurochemistry and Behaviour, Institute Born-Bunge, University of Antwerp, Wilrijk, Antwerp, Belgium

E. Ronald de Kloet Department of Endocrinology and Metabolism, Leiden Academic Center for Drug Research and Leiden University Medical Center, Leiden, RA, The Netherlands

Luis de Lecea Department of Psychiatry and Behavioral Sciences, Stanford University School of Medicine, Palo Alto, CA, USA

R. H. de Rijk Department of Psychiatry, Leiden University Medical Center, Leiden, RA, The Netherlands

Marlene Dobkin de Rios Department of Psychiatry and Human Behavior, University of California, Irvine, CA, USA

M. M. de Rosa Almeida Instituto de Psicologia (UFRGS), Laboratório de Psicologia, Neurociência e Comportamento (LPNeC), Porto Alegre, RS, Brazil

Mischa De Rover Cognitive Psychology Unit, Institute of Psychology, Leiden University, Leiden, AK, The Netherlands

Harriet de Wit Department of Psychiatry and Behavioral Neuroscience, Human Behavioral Pharmacology Laboratory, University of Chicago, Chicago, IL, USA

Pedro L. Delgado Department of Psychiatry, College of Medicine, University of Arkansas for Medical Sciences (UAMS), Little Rock, AR, USA

Psychiatric Research Institute, University of Arkansas for Medical Sciences (UAMS), Little Rock, AR, USA

Richard A. Depue Department of Human Development, Laboratory of Neurobiology of Personality, College of Human Ecology, Cornell University, Ithaca, NY, USA

Giuseppe Di Giovanni Department of Physiology and Biochemistry, Laboratory for the Study of Neurological Disorders, University of Malta, Msida, Malta

Vincenzo Di Matteo Istituto di Ricerche Farmacologiche "Mario Negri", Santa Maria Imbaro, Chieti, Italy

Anthony H. Dickenson Neuroscience, Physiology and Pharmacology, University College London, London, UK

Anthony Dickinson Department of Psychology, University of Cambridge, Cambridge, UK

Wilhelmus H. I. M. (Pim) Drinkenburg Janssen Research & Development, A Division of Janssen Pharmaceutica NV, Beerse, Belgium

Maryana Duchcherer Department of Psychiatry, University of Alberta, Edmonton, AB, Canada

Theodora Duka Behavioural and Clinical Neuroscience, School of Psychology, University of Sussex, Falmer, Brighton, UK

Stephen B. Dunnett School of Biosciences, Cardiff University, Wales, Cardiff, UK

Andrea R. Durrant Research and Psychiatry Departments, Ezrath Nashim - Herzog Memorial Hospital, Jerusalem, Israel

Linda A. Dykstra Behavioral Neuroscience Program, Department of Psychology, University of North Carolina at Chapel Hill, Chapel Hill, NC, USA

Nicole Edgar Department of Psychiatry, University of Pittsburgh, Pittsburgh, PA, USA

Center for Neuroscience, University of Pittsburgh, Pittsburgh, PA, USA

Bart Ellenbroek School of Psychology, Victoria University of Wellington, Wellington, New Zealand

Robin Emsley Department of Psychiatry, Faculty of Medicine and Health Sciences, University of Stellenbosch, Tygerberg, Cape Town, South Africa

Jacques Epelbaum Faculté de Médecine, Centre de Psychiatrie and Neuroscience, UMR 894, Université Paris Descartes, Paris, France

C. Neill Epperson Departments of Psychiatry and Obstetrics/Gynecology, Penn Center for Women's Behavioral Wellness, Perelman School of Medicine at the University of Pennsylvania, Philadelphia, PA, USA

Craig A. Erickson Department of Psychiatry, University of Cincinnati School of Medicine, Cincinnati, OH, USA

Edzard Ernst Complementary Medicine Group, Peninsula Medical School, University of Exeter, Exeter, UK

Ennio Esposito Consorzio Mario Negri Sud, Laboratory of Neurophysiology, Santa Maria Imbaro, Chieti, Italy

Brian A. Fallon New York State Psychiatric Institute, Columbia University, New York, NY, USA

Maurizio Fava Depression Clinical and Research Program (DCRP), Harvard Medical School, Massachusetts General Hospital, Boston, MA, USA

David Feifel Department of Psychiatry, University of California, San Diego, CA, USA

Carlo Ferrarese Department of Surgery and Interdisciplinary Medicine, Neurology Unit, S. Gerardo Hospital, University of Milano-Bicocca, Monza, MB, Italy

Matt Field Department of Psychological Sciences, University of Liverpool, Liverpool, UK

Robert L. Findling Division of Child and Adolescent Psychiatry, Kennedy Krieger Institute, Johns Hopkins University, Baltimore, MD, USA

Naomi A. Fineberg Department of Psychiatry, NHS Foundation Trust, Queen Elizabeth II Hospital, Welwyn Garden City, Hertfordshire, UK

Gabriele Fischer Department of Public Health, Medical University of Vienna, Vienna, Austria

Shelly B. Flagel Department of Psychiatry, Molecular and Behavioral Neuroscience Institute, University of Michigan, Ann Arbor, MI, USA

W. Wolfgang Fleischhacker Department of Psychiatry and Psychotherapy, Medical University Innsbruck, University Klinik für Biologische Psychiatrie, Innsbruck, Austria

Peter J. Flor Department of Systemic and Molecular Neuroendocrinology, University of Regensburg, Regensburg, Bavaria, Germany

Stan B. Floresco Department of Psychology, University of British Columbia, Vancouver, Canada

Richard W. Foltin The New York State Psychiatric Institute, College of Physicians and Surgeons of Columbia University, New York, NY, USA

Ariadna Forray Department of Psychiatry, Yale School of Medicine, New Haven, CT, USA

Stephen C. Fowler Department of Pharmacology and Toxicology, Schieffelbusch Institute for Life Span Studies, University of Kansas, Lawrence, KS, USA

Christine A. Franco Department of Psychiatry, Yale University School of Medicine, New Haven, CT, USA

Ingmar H. A. Franken Faculty of Social Sciences, Institute of Psychology, Erasmus University Rotterdam, Rotterdam, The Netherlands

Joseph H. Friedman Movement Disorders, Butler Hospital and Brown University, Providence, RI, USA

Kim Fromme Department of Psychology, The University of Texas at Austin, Austin, TX, USA

Cristiane Furini Memory Center, Brain Institute, Pontifical Catholic University of Rio Grande do Sul, Porto Alegre, RS, Brazil

Kristina G. Gaud Department of Psychiatry and Human Behavior, Butler Hospital, Brown University, Providence, RI, USA

John Geddes Warneford Hospital, University of Oxford, Oxford, Oxfordshire, UK

Greg A. Gerhardt Department of Anatomy and Neurobiology, Morris K. Udall Parkinson's Disease Research Center of Excellence, Center for Microelectrode Technology, University of Kentucky Medical Center, Lexington, KY, USA

Mark A. Geyer Department of Psychiatry, University of California San Diego, La Jolla, CA, USA

Helen E. Gibson Department of Molecular Pharmacology, Physiology and Biotechnology, Providence, RI, USA

Edwin Glueck Department of Psychology and Program in Behavioral Neuroscience, Western Washington University, Bellingham, WA, USA

Guy M. Goodwin University Department of Psychiatry, Warneford Hospital, Headington, Oxford, UK

Meghan M. Grady Neuroscience Education Institute, San Diego, CA, USA

Frederico Guilherme Graeff Institute of Neuroscience and Behavior – INeC, Ribeirão Preto, SP, Brazil

Jon E. Grant Department of Psychiatry and Behavioral Neuroscience, Pritzker School of Medicine, University of Chicago, Chicago, IL, USA

A. Richard Green School of Life Sciences, Queen's Medical Centre, University of Nottingham, Nottingham, UK

Guy Griebel Translational Sciences Unit, Chilly-Mazarin, France

Jeffrey W. Grimm Department of Psychology and Program in Behavioral Neuroscience, Western Washington University, Bellingham, WA, USA

Charles S. Grob Department of Psychiatry, Harbor/UCLA Medical Center, Torrance, CA, USA

Anna I. Guerdjikova Research Institute, Lindner Center of HOPE, Mason, OH, USA

Department of Psychiatry & Behavioral Neuroscience, University of Cincinnati College of Medicine, Cincinnati, OH, USA

Jason C. G. Halford Psychological Sciences, Institute of Psychology, Health and Society, University of Liverpool, Liverpool, UK

John H. Halpern Division of Alcohol and Drug Abuse, McLean Hospital, Harvard Medical School, The Laboratory for Integrative Psychiatry, Belmont, MA, USA

Liisa Hantsoo Departments of Psychiatry and Obstetrics/Gynecology, Penn Center for Women's Behavioral Wellness, Perelman School of Medicine at the University of Pennsylvania, Philadelphia, PA, USA

Kelli J. Harding New York State Psychiatric Institute, Columbia University Medical Center, New York, NY, USA

Oliver Hardt Centre for Cognitive and Neural Systems, The University of Edinburgh, Edinburgh, UK

Ben J. Harrison Melbourne Neuropsychiatry Centre, Department of Psychiatry and Melbourne Health, The University of Melbourne, Carlton, Melbourne, VIC, Australia

Jaanus Harro Division of Neuropsychopharmacology, Department of Psychology, Estonian Centre of Behavioural and Health Sciences, University of Tartu, Tartu, Estonia

John A. Harvey Department of Pharmacology and Physiology, Drexel University College of Medicine, Philadelphia, PA, USA

Victoria L. Harvey Neuroscience, Physiology and Pharmacology, University College London, London, UK

Sarah H. Heil Department of Psychiatry and Psychology, Vermont Center on Behavior and Health, University of Vermont, Burlington, VT, USA

David J. Hellerstein Department of Psychiatry, Columbia University, New York, NY, USA

Uriel Heresco-Levy Research and Psychiatry Departments, Ezrath Nashim – Herzog Memorial Hospital, Jerusalem, Israel

Psychiatry Departments, Hadassah Medical School, Hebrew University, Jerusalem, Israel

Giovanni Hernandez Department of Physiology, Faculty of Medicine, Université de Montréal, Montréal, QC, Canada

Lieve Heylen Neuroscience Department, Johnson & Johnson Pharmaceutical Research and Development, Program Management Office, Beerse, Belgium

Charles J. Heyser Department of Neurosciences, University of California, San Diego, La Jolla, CA, USA

Christoph Hiemke Department of Psychiatry and Psychotherapy, University of Mainz, Mainz, Rhineland-Palatinate, Germany

Stephen T. Higgins Department of Psychiatry and Psychology, Vermont Center on Behavior and Health, University of Vermont, Burlington, VT, USA

Stephen J. Hill Faculty of Medicine & Health Sciences, School of Life Sciences, The University of Nottingham, Nottingham, UK

Cecilia J. Hillard Department of Pharmacology and Toxicology and Neuroscience Research Center, Medical College of Wisconsin, Milwaukee, WI, USA

Ian Hindmarch University of Surrey, Guildford, UK

Alex Hofer Biological Psychiatry Division, Department of Psychiatry and Psychotherapy, Medical University Innsbruck, Innsbruck, Austria

Lee Hogarth School of Psychology, University of Exeter, Exeter, UK

Andrew Holt Department of Pharmacology, University of Alberta, Edmonton, AB, Canada

Sabine M. Hölter Helmholtz Zentrum München, German Research Center for Environmental Health (GmbH), Institute of Developmental Genetics, München, Germany

Cyril Höschl Prague Psychiatric Centre & 3rd Medical Faculty, Charles University, Prague, Czech Republic

Daniel Hoyer Department of Pharmacology and Therapeutics, School of Biomedical Sciences, Faculty of Medicine, Dentistry and Health Sciences, The University of Melbourne, Parkville, VIC, Australia

Pedro E. Huertas Division of Alcohol and Drug Abuse, McLean Hospital, Harvard Medical School, The Laboratory for Integrative Psychiatry, Belmont, MA, USA

Raymond S. Hurst Forum Pharmaceuticals, Watertown, MA, USA

Samuel B. Hutton Department of Psychology, University of Sussex, Falmer, Brighton, UK

Roberto William Invernizzi Department of Neuroscience, Mario Negri Institute for Pharmacological Research, Milan, Italy

Anthony R. Isles Institute of Psychological Medicine and Clinical Neurosciences, and MRC Centre for Neuropsychiatric Genetics and Genomics, School of Medicine, Cardiff University, Cardiff, UK

Ivan Izquierdo Memory Center, Brain Institute, Pontifical Catholic University of Rio Grande do Sul, Porto Alegre, RS, Brazil

Anne Jackson School of Pharmacy and Biomolecular Sciences, University of Brighton, Brighton, East Sussex, UK

Héctor Jantos Department of Pharmacology and Therapeutics, School of Medicine Clinics Hospital, Montevideo, Uruguay

Bankole A. Johnson Department of Psychiatry, School of Medicine, University of Maryland, Baltimore, MD, USA

Susan Jones Department of Physiology, Development and Neuroscience, University of Cambridge, Cambridge, UK

Eileen M. Joyce Institute of Neurology, University College London, London, UK

Peter W. Kalivas Department of Neurosciences, Medical University of South Carolina, Charleston, SC, USA

Shitij Kapur Institute of Psychiatry, King's College London, London, UK

Siegfried Kasper Department of Psychiatry and Psychotherapy, Medical University of Vienna, Vienna, Austria

Sidney H. Kennedy Department of Psychiatry, University Health Network, University of Toronto, Toronto, ON, Canada

Robert Kessler Neurochemical Brain Imaging and PET Neurotracer Development, University of Alabama School of Medicine, Birmingham, AL, USA

Falk Kiefer Central Institute of Mental Health, University of Heidelberg, Mannheim, Germany

Deborah R. Kim Departments of Psychiatry and Obstetrics/Gynecology, Penn Center for Women's Behavioral Wellness, Perelman School of Medicine at the University of Pennsylvania, Philadelphia, PA, USA

Grasielle Clotildes Kincheski Departamento de Farmacologia, CCB, Universidade Federal de Santa Catarina, Florianópolis, SC, Brazil

Becky Kinkead Department of Psychiatry and Behavioral Sciences, Emory University School of Medicine, Atlanta, GA, USA

Tim C. Kirkham School of Psychology, University of Liverpool, Liverpool, UK

Clemens Kirschbaum Department of Psychology, Dresden University of Technology, Dresden, Germany

Stephen J. Kohut Neurobiology Program, Alcohol and Drug Abuse Research Center McLean Hospital, Harvard Medical School, Belmont, MA, US

Scott H. Kollins Department of Psychiatry & Behavioral Sciences, Department of Psychology & Neuroscience, Duke University School of Medicine, Durham, NC, USA

John H. Krystal Department of Psychiatry, Yale University School of Medicine, New Haven, CT, USA

Clinical Neuroscience Division, National Center for PTSD, West Haven, CT, USA

Michael J. Kuhar Division of Neuroscience, Yerkes National Primate Research Centre of Emory University, Atlanta, GA, USA

F. Hoffmann La Roche Pharmaceutical Research and Early Development Department, Roche Innovation Center, Basel, Switzerland

Malcolm Lader Institute of Psychiatry, Psychology and Neuroscience, King's College London, London, UK

Adriaan A. Lammertsma Nuclear Medicine and PET Research, VU University Medical Centre, Amsterdam, The Netherlands

James D. Lane Department of Psychiatry and Behavioral Sciences, Duke University School of Medicine, Durham, NC, USA

Xavier Langlois Neuroscience Department, Johnson & Johnson Pharmaceutical Research and Development, Program Management Office, Beerse, Belgium

Hilde Lavreysen Janssen Research and Development, A Division of Janssen Pharmaceutica, Beerse, Belgium

Matthew E. Layton Program of Excellence in Addictions Research, Washington State University-Spokane, Spokane, WA, USA

James F. Leckman Sterling Hall of Medicine Child Study Centre, Yale University School of Medicine, New Haven, CT, USA

Björn Lemmer Institute of Experimental and Clinical Pharmacology and Toxicology, Ruprecht-Karls-University of Heidelberg, Mannheim, Germany

Brian E. Leonard Emeritus Professor of Pharmacology, Pharmacology Department, National University of Ireland, Galway, Ireland

Mark G. LeSage Department of Medicine, Minneapolis Medical Research Foundation and University of Minnesota, Minneapolis, MN, USA

Stefan Leucht Klinik fur Psychiatrie und Psychotherapie der TU-Munchen, Klinikum rechts der Isar, Munich, Bavaria, Germany

Josee E. Leysen Nuclear Medicine and PET Research, VU University Medical Centre, Amsterdam, The Netherlands

Marco Leyton Department of Psychiatry, McGill University, Montreal, QC, Canada

Joachim Liepert Department of Neurorehabilitation, Kliniken Schmieder, Allensbach, Germany

Janka Lincoln Department of Psychiatry, Kansas University School of Medicine, Wichita, KS, USA

Craig W. Lindsley Departments of Pharmacology & Chemistry, Vanderbilt University Medical Center, Vanderbilt Center for Neuroscience Drug Discovery, Nashville, TN, USA

Juan J. López-Ibor Institute of Psychiatry and Mental Health, San Carlos Hospital, Complutense University, Madrid, Spain

Maria-Inés López-Ibor Department of Psychiatry and Medical Psychology, Complutense University, Madrid, Spain

Irwin Lucki Department of Psychiatry, University of Pennsylvania, Philadelphia, PA, USA

Linda Lundström Pharmaceuticals Division, Hoffmann-La Roche Ltd, CNS Discovery Functional Neuroscience, Basel, Switzerland

Fadi T. Maalouf Western Psychiatric Institute and Clinic, University of Pittsburgh School of Medicine, Pittsburgh, PA, USA

Kai MacDonald Department of Psychiatry, University of California, San Diego, CA, USA

Angus Mackay Advisory Board on the Registration of Homoeopathic Products, Medicines and Healthcare Products Regulatory Agency, London, UK

Husseini K. Manji Johnson and Johnson Pharmaceutical Research and Development, Titusville, NJ, USA

Karl Mann Central Institute of Mental Health, University of Heidelberg, Mannheim, Germany

John R. Mantsch Department of Biomedical Sciences, Marquette University, Milwaukee, WI, USA

Josef Marksteiner Department of Psychiatry and Psychotherapy, LKH Klagenfurt, Klagenfurt, Austria

Andreas Marneros Klinik und Poliklinik für Psychiatrie, Psychotherapie und Psychosomatik, Martin Luther Universität Halle-Wittenberg, Halle, Germany

Barbara J. Mason Committee on the Neurobiology of Addictive Disorders, Pearson Center for Alcoholism and Addiction Research, The Scripps Research Institute, La Jolla, CA, USA

Dmitriy Matveychuk Neurochemical Research Unit and Bebensee Schizophrenia Research Unit, Department of Psychiatry, University of Alberta, Edmonton, AB, Canada

R. Hamish McAllister-Williams Academic Psychiatry, Institute of Neuroscience, Wolfson Research Centre, Newcastle University, Newcastle upon Tyne, UK

Silvia Gatti McArthur Pharmaceuticals Division, Hoffmann-La Roche Ltd, CNS Discovery Functional Neuroscience, Basel, Switzerland

Christopher J. McDougle Lurie Center for Autism and Department of Psychiatry, MassGeneral Hospital for Children, Lexington, MA, USA

Nancy Lurie Marks Professor in the Field of Autism, Harvard Medical School, Boston, MA, USA

Susan L. McElroy Research Institute, Lindner Center of HOPE, Mason, OH, USA

Department of Psychiatry & Behavioral Neuroscience, University of Cincinnati College of Medicine, Cincinnati, OH, USA

Patrick D. McGorry Department of Psychiatry, Orygen Youth Health Research Centre, Centre for Youth Mental Health, University of Melbourne, Melbourne, VIC, Australia

Iain S. McGregor School of Psychology, University of Sydney, Sydney, NSW, Australia

Warren H. Meck Department of Psychology and Neuroscience, Duke University, Durham, NC, USA

Mitul A. Mehta Centre for Neuroimaging Sciences, Institute of Psychiatry at King's College London, London, UK

David K. Menon Department of Anaesthesia, School of Clinical Medicine, University of Cambridge, Cambridge, UK

Klaus A. Miczek Tufts University, Medford, MA, USA

David S. Middlemas Department of Pharmacology, Kirksville College of Osteopathic Medicine, A. T. Still University of Health Sciences, Kirksville, MO, USA

Paola V. Migues McGill University, Montreal, QC, Canada

Michele Stanislaw Milella Department of Psychiatry, McGill University, Montreal, QC, Canada

Michael Minzenberg Department of Psychiatry, Imaging Research Center, University of California, Davis Medical Centre, Sacramento, CA, USA

Nicholas Mitchell Department of Psychiatry, University of Alberta, Edmonton, AB, Canada

Suzanne H. Mitchell Department of Behavioral Neuroscience, L470, Oregon Health & Science University, Portland, OR, USA

Nobumi Miyake Schizophrenia Treatment Center, Department of Neuropsychiatry, St. Marianna University School of Medicine, Kawasaki, Kanagawa, Japan

Seiya Miyamoto Schizophrenia Treatment Center, Department of Neuropsychiatry, St. Marianna University School of Medicine, Kawasaki, Kanagawa, Japan

Tooru Mizuno Department of Physiology, University of Manitoba, Winnipeg, MB, Canada

Michel Le Moal Neurocentre Magendie Inserm U862, Université de Bordeaux, Bordeaux, France

Hanns Möhler Institute of Pharmacology and Neuroscience Center, University of Zurich and Swiss Federal Institute of Technology (ETH), Zurich, Switzerland

Daniel Monti Department of Pharmacology and Therapeutics, School of Medicine, Clinics Hospital, Montevideo, Uruguay

Jaime M. Monti Department of Pharmacology and Therapeutics, School of Medicine, Clinics Hospital, Montevideo, Uruguay

Sharon Morein-Zamir Department of Psychiatry, University of Cambridge, Cambridge, UK

Micaela Morelli Department of Biomedical Sciences, Section of Neuropsychopharmacology, University of Cagliari, Cagliari, Italy

Center of Excellence for Neurobiology of Dependence, University of Cagliari, Cagliari, Italy

Section of Cagliari, National Research Council (CNR), Neuroscience Institute, Cagliari, Italy

National Institute of Neuroscience (INN), University of Cagliari, Cagliari, Italy

Celia J. A. Morgan Department of Psychology, Clinical Psychopharmacology Unit, University College London, London, UK

Michael M. Morgan Department of Psychology, Washington State University Vancouver, Vancouver, WA, USA

Sarah Morgan Pharmacovigilance Risk Management, Vigilance and Risk Management of Medicines, Medicines and Healthcare Products Regulatory Agency, London, UK

A. Leslie Morrow Departments of Psychiatry and Pharmacology, Bowles Center for Alcohol Studies, University of North Carolina School of Medicine, Chapel Hill, NC, USA

Johannes Mosbacher Research & Development, F. Hoffmann – La Roche Ltd, Basel, Switzerland

Darrell D. Mousseau Cell Signalling Laboratory, Department of Psychiatry, University of Saskatchewan, Saskatoon, SK, Canada

Ronald F. Mucha Department of Psychology, University of Würzburg, Würzburg, Bavaria, Germany

Thomas Mueggler Hoffmann-La Roche Ltd, DTA Neuroscience, Basel, Switzerland

Karen E. Murray Department of Psychiatry and Behavioral Sciences, Laboratory of Neuropsychopharmacology, Emory University School of Medicine, Atlanta, GA, USA

Jociane C. Myskiw Memory Center, Brain Institute, Pontifical Catholic University of Rio Grande do Sul, Porto Alegre, RS, Brazil

Karim Nader McGill University, Montreal, QC, Canada

Sunila G. Nair Harborview Medical Center, Seattle, WA, USA

Kazuyuki Nakagome National Center of Neurology and Psychiatry, Kodaira, Tokyo, Japan

Behrouz Namdari Department of Psychiatry & Behavioral Sciences, Duke University School of Medicine, Durham, NC, USA

Susan Napier Department of Medicinal Chemistry, Schering-Plough Corporation, Newhouse, Lanarkshire, UK

S. Stevens Negus Department of Pharmacology and Toxicology, Virginia Commonwealth University, Richmond, VA, USA

J. Craig Nelson Department of Psychiatry, University of California San Francisco, San Francisco, CA, USA

Charles B. Nemeroff Department of Psychiatry and Behavioral Sciences, University of Miami, Leonard M. Miller School of Medicine, Miami, FL, USA

Paolo Nencini Department of Psychiatry, McGill University, Montreal, QC, Canada

Jelena Nesic School of Life Sciences, University of Sussex, Brighton, East Sussex, UK

Eric J. Nestler Fishberg Department of Neuroscience, Mount Sinai School of Medicine, New York, NY, USA

Inga D. Neumann Department of Systemic and Molecular Neuroendocrinology, University of Regensburg, Regensburg, Bavaria, Germany

Paul A. Newhouse Department of Psychiatry, Vanderbilt University School of Medicine, Nashville, TN, USA

David E. Nichols Chemical Biology and Medicinal Chemistry, Eshelman School of Pharmacy, University of North Carolina, Chapel Hill, IN, USA

Sander Nieuwenhuis Cognitive Psychology Unit, Institute of Psychology, Leiden University, Leiden, AK, The Netherlands

David Nutt Hammersmith Hospital, Imperial College London, London, UK

Brian L. Odlaug Department of Public Health, Faculty of Health and Medical Sciences, University of Copenhagen, Copenhagen, Denmark

Berend Olivier Department of Psychopharmacology, Utrecht University, Utrecht, The Netherlands

Kieran O'Malley Child Psychiatry, Charlemont Clinic, Dublin, Ireland

Søren Dinesen Østergaard Depression Clinical and Research Program (DCRP), Harvard Medical School, Massachusetts General Hospital, Boston, MA, USA

Sven Ove Ögren Department of Neuroscience, Karolinska Institutet, Stockholm, Sweden

Michael J. Owens Department of Psychiatry and Behavioral Sciences, Laboratory of Neuropsychopharmacology, Emory University School of Medicine, Atlanta, GA, USA

Ashwini Padhi Community Mental Health Team, Birmingham and Solihull Mental Health Foundation Trust, Solihull, West Midlands, UK

Matthew I. Palmatier Department of Psychology, East Tennessee State University, Johnson City, TN, USA

Subhash C. Pandey Department of Psychiatry, University of Illinois at Chicago and Jesse Brown VA Medical Center, Chicago, IL, USA

Christos Pantelis Department of Psychiatry, Melbourne Neuropsychiatry Centre, Sunshine Hospital, The University of Melbourne, St Albans, VIC, Australia

George I. Papakostas Depression Clinical and Research Program (DCRP), Harvard Medical School, Massachusetts General Hospital, Boston, MA, USA

Department of Psychiatry, Massachusetts General Hospital, Harvard Medical School, Boston, MA, USA

Sarah Parylak The Scripps Research Institute, University of California San Diego, La Jolla, CA, USA

Torsten Passie Department of Psychiatry, Social Psychiatry and Psychotherapy, Hannover Medical School, Hannover, Germany

Amee B. Patel G.V. (Sonny) Montgomery VA Medical Center, Jackson, MS, USA

Darrel J. Pemberton Division of Janssen Pharmaceutica NV, Johnson & Johnson Pharmaceutical Research and Development, Beerse, Belgium

Paul R. Pentel Department of Medicine, Hennepin County Medical Center, Minneapolis, MN, USA

Jorge Perez-Parada Department of Psychiatry, University of Alberta, Edmonton, AB, Canada

Lukas Pezawas Department of Psychiatry and Psychotherapy, Division of Biological Psychiatry, Medical University of Vienna, Vienna, Austria

Noah S. Philip Department of Psychiatry and Human Behavior, Butler Hospital, Brown University, Providence, RI, USA

Melanie M. Pina Department of Behavioral Neuroscience and Portland Alcohol Research Center, Oregon Health & Science University, Portland, OR, USA

Sakire Pogun Center for Brain Research, Ege University, Bornova, Izmir, Turkey

Institute on Drug Abuse, Toxicology and Pharmaceutical Science, Ege University, Bornova, Izmir, Turkey

Laura C. Politte Departments of Psychiatry and Pediatrics, Lurie Center for Autism, Massachusetts General Hospital, Lexington, MA, USA

Patrizia Porcu Neuroscience Institute, National Research Council of Italy (CNR), Cagliari, Italy

David J. Posey Indianapolis, IN, USA

H. D. Postma Department of Psychiatry, Rudolf Magnus Institute of Neuroscience, University Medical Center, Utrecht, CX, The Netherlands

Marc N. Potenza Department of Psychiatry, Yale University School of Medicine, New Haven, CT, USA

Jack Price Department of Neuroscience, Institute of Psychiatry, King's College London, London, UK

Lawrence H. Price The Warren Alpert Medical School of Brown University and Butler Hospital, Providence, RI, USA

Jos Prickaerts Department of Psychiatry and Neuropsychology, University of Maastricht, Maastricht, MD, The Netherlands

Jorge A. Quiroz Child and Adult Psychiatry, Neuroscience Translational Medicine, Roche, Nutley, NJ, USA

Ashley D. Radomski Neurochemical Research Unit and Bebensee Schizophrenia Research Unit, Department of Psychiatry, University of Alberta, Edmonton, AB, Canada

Michael E. Ragozzino Department of Psychology, University of Illinois at Chicago, Chicago, IL, USA

Gary Remington Department of Psychiatry, Centre for Addiction and Mental Health, University of Toronto, Toronto, ON, Canada

Kenneth J. Rhodes Discovery Neurobiology, Biogen Idec, Cambridge, MA, USA

Peter Riederer University Hospital, Center for Mental Health, Clinic and Policlinic for Psychiatry, Psychosomatics and Psychotherapy, University Wuerzburg, Wuerzburg, Germany

Anthony L. Riley Department of Psychology, American University, Washington, DC, USA

Sakina J. Rizvi Department of Psychiatry, Departments of Pharmaceutical Sciences and Neuroscience, University Health Network, University of Toronto, Toronto, ON, Canada

Trevor W. Robbins Department of Psychology, University of Cambridge, Cambridge, UK

Angela Roberts Department of Physiology, Development and Neuroscience, University of Cambridge, Cambridge, UK

David C. S. Roberts Department of Physiology and Pharmacology, Wake Forest University Health Sciences, Winston-Salem, NC, USA

Terry E. Robinson Department of Psychology, University of Michigan, Ann Arbor, MI, USA

Jonathan P. Roiser Institute of Cognitive Neuroscience, University College London, London, UK

John M. Roll Program of Excellence in Addictions Research, Washington State University-Spokane, Spokane, WA, USA

Hans Rollema Rollema Biomedical Consulting, Mystic, CT, USA

Markus Rudin Molecular Imaging and Functional Pharmacology, Institute for Biomedical Engineering, University and ETH Zurich, Zurich, Switzerland

Institute for Pharmacology and Toxicology, University of Zurich, HCI D426, Zurich, Switzerland

Jennifer R. Sage Department of Psychology and Program in Neurosciences, University of California, San Diego, La Jolla, CA, USA

Barbara Jacquelyn Sahakian Department of Psychiatry, University of Cambridge, Addenbrooke's Hospital, Cambridge, UK

Behavioural and Clinical Neuroscience Institute, University of Cambridge, Cambridge, UK

Gessica Sala Department of Surgery and Interdisciplinary Medicine, Neurology Unit, S Gerardo Hospital, University of Milano-Bicocca, Monza, MB, Italy

Melita Salkovic-Petrisic Department of Pharmacology, University of Zagreb School of Medicine, Zagreb, Croatia

Frank Sams-Dodd Willingsford Limited, Southampton, UK

Peter A. Santi Department of Otolaryngology, University of Minnesota, Minneapolis, SE, USA

Martin Sarter Department of Psychology, University of Michigan, Ann Arbor, MI, USA

Lawrence David Scahill Marcus Autism Center, Children's Healthcare of Atlanta, and Department of Pediatrics, Emory University School of Medicine, Atlanta, GA, USA

Jean-Michel Scherrmann Department of Pharmaceutical Sciences, Inserm U1144, Faculty of Pharmacy, University Paris Descartes, Paris, France

Angelika Schmitt University Hospital, Center for Mental Health, Clinic and Policlinic for Psychiatry, Psychosomatics and Psychotherapy, University Wuerzburg, Wuerzburg, Germany

Tomasz Schneider School of Medicine, Pharmacy and Health, Durham University, Oxford, UK

Michael D. Scofield Department of Neurosciences, Medical University of South Carolina, Charleston, SC, USA

Robert Taylor Segraves Department of Psychiatry, Case Western Reserve University School of Medicine, Cleveland, OH, USA

Pandi-Perumal Seithikurippu Ratnas Center for Healthful Behavior Change (CHBC), Division of Health and Behavior, Department of Population Health, New York University Medical Center, Clinical & Translational Research Institute, New York, NY, USA

Dana E. Selley Department of Pharmacology and Toxicology, Virginia Commonwealth University, Richmond, VA, USA

Marianne Seney Department of Psychiatry, University of Pittsburgh, Pittsburgh, PA, USA

Center for Neuroscience, University of Pittsburgh, Pittsburgh, PA, USA

Patrick M. Sexton Drug Discovery Biology, Monash Institute of Pharmaceutical Sciences, Monash University, Parkville, Melbourne, VIC, Australia

Yavin Shaham Behavioral Neuroscience Branch, IRP/NIDA/NIH, Baltimore, MD, USA

Colin M. Shapiro Department of Psychiatry, Toronto Western Hospital, University Health Network, University of Toronto, Sleep and Alertness Clinic Toronto, Toronto, ON, Canada

David V. Sheehan Depression and Anxiety Disorders Research Institute, University of South Florida College of Medicine, Tampa, FL, USA

Kathy H. Sheehan Depression and Anxiety Disorders Research Institute, University of South Florida College of Medicine, Tampa, FL, USA

Peter Shizgal Center for Studies in Behavioural Neurobiology (Groupe de recherche en neurobiologie comportementale) and Department of Psychology, Concordia University, Montréal, QC, Canada

Etienne Sibille Department of Psychiatry, University of Pittsburgh, Pittsburgh, PA, USA

Center for Neuroscience, University of Pittsburgh, Pittsburgh, PA, USA

Michelle M. Sidor Department of Psychiatry, University of Pittsburgh School of Medicine, Pittsburgh, PA, USA

Stacey C. Sigmon Department of Psychiatry and Psychology, Vermont Center on Behavior and Health, University of Vermont, Burlington, VT, USA

Nicola Simola Department of Biomedical Sciences, Section of Neuropsychopharmacology, University of Cagliari, Cagliari, Italy

Laura J. Sim-Selley Department of Pharmacology and Toxicology, Virginia Commonwealth University, Richmond, VA, USA

Samuel G. Siris Hofstra North Shore LIJ School of Medicine, The Zucker Hillside Hospital, Glen Oaks, NY, USA

Mark Slifstein Department of Psychiatry, Columbia University and New York State Psychiatric Institute, New York, NY, USA

Nuno Sousa Life and Health Science Research Institute (ICVS), School of Health Sciences, University of Minho, Braga, Portugal

ICVS/3B's – PT Government Associate Laboratory, Braga/Guimarães, Portugal

Linda P. Spear Department of Psychology, Binghamton University, Binghamton, NY, USA

Edoardo Spina Department of Clinical and Experimental Medicine and Pharmacology, Policlinico Universitorio, Unit of Pharmacology, University of Messina, Messina, Italy

Will Spooren Pharmaceuticals Division, Hoffmann-La Roche Ltd, CNS Discovery Functional Neuroscience, Basel, Switzerland

Stephen M. Stahl University of California, San Diego, CA, USA

University of Cambridge, Cambridge, UK

Thomas Steckler Pharma R&D Quality & Compliance, Janssen Research and Development, Beerse, Belgium

Dan J. Stein Department of Psychiatry, University of Cape Town, Cape Town, South Africa

Gregory D. Stewart Drug Discovery Biology, Monash Institute of Pharmaceutical Sciences, Monash University, Parkville, Melbourne, VIC, Australia

Oliver Stiedl Department of Functional Genomics and Department of Molecular and Cellular Neurobiology, Center for Neurogenomics and Cognitive Research, VU University Amsterdam, Amsterdam, The Netherlands

Kimberly A. Stigler Department of Psychiatry, Christian Sarkine Autism Treatment Center, James Whitcomb Riley Hospital for Children, Indiana University School of Medicine, Indianapolis, IN, USA

Luis Stinus Team Neuropsychopharmacology of addiction, UMR CNRS 5287, University of Bordeaux, Bordeaux, France

Maxine L. Stitzer Psychiatry and Behavioral Sciences, Johns Hopkins Bayview Medical Center, Baltimore, MD, USA

Ian P. Stolerman Section of Behavioural Pharmacology, Institute of Psychiatry, Psychology and Neuroscience, King's College London, London, UK

James J. Strain Department of Psychiatry, Icahn School of Medicine at Mount Sinai, New York, NY, USA

Tomiki Sumiyoshi Department of Clinical Research Promotion, National Center of Neurology and Psychiatry, Tokyo, Japan

Joji Suzuki Department of Psychiatry, Brigham and Women's Hospital, Addiction Psychiatry Service, Harvard Medical School, Boston, MA, USA

Takefumi Suzuki Department of Neuropsychiatry, Keio University School of Medicine, Tokyo, Japan

Department of Psychiatry, Inokashira Hospital, Tokyo, Japan

Per Svenningsson Department of Physiology and Pharmacology, Karolinska Institute, Hasselt University, Stockholm, Sweden

David S. Tait School of Psychology and Neuroscience, University of St Andrews, Scotland, UK

John Talpos Neuroscience Therapeutic Area, Janssen Research and Development, Beerse, Belgium

Sophie Tambour Department of Behavioral Neuroscience, University of Liège, Liège, Belgium

Tiffany Thomas Division of Child and Adolescent Psychiatry, University Hospitals Case Medical Center, Cleveland, OH, USA

Murray R. Thompson School of Psychology, University of Sydney, Sydney, NSW, Australia

Fiona Thomson Neuroscience Discovery, Merck Research Laboratories, West Point, PA, USA

Sara Tomlinson Neurochemical Research Unit and Bebensee Schizophrenia Research Unit, Department of Psychiatry, University of Alberta, Edmonton, AB, Canada

Duc A. Tran Department of Physical Medicine and Rehabilitation, Massachusetts General Hospital, Spaulding Rehabilitation Hospital, Brigham and Women's Hospital, and Harvard Medical School, Charlestown, MA, USA

Lucio Tremolizzo Department of Surgery and Interdisciplinary Medicine, Neurology Unit, S. Gerardo Hospital, University of Milano-Bicocca, Monza, MB, Italy

Gustavo Turecki McGill Group for Suicide Studies, McGill University, Montreal, QC, Canada

Marc Turiault Geneva Business News, Geneva, Switzerland

Peter J. Tyrer Department of Medicine, Imperial College, London, UK

Stephen Tyrer Campus for Ageing and Vitality, Wolfson Research Centre, Institute of Neuroscience, Newcastle University, Newcastle upon Tyne, Tyne and Wear, UK

Thomas M. Tzschentke Grünenthal GmbH, Preclinical Research and Development, Department of Pharmacology, Aachen, Germany

Hiroyuki Uchida Department of Neuropsychiatry, Keio University School of Medicine, Tokyo, Japan

Annemarie Unger Department of Psychiatry and Psychotherapy, Medical University of Vienna, Vienna, Austria

Joachim D. Uys Department of Neurosciences, Medical University of South Carolina, Charleston, SC, USA

Tayfun Uzbay Department of Medical Pharmacology, Faculty of Medicine, Psychopharmacology Research Unit, Gulhane Military Medical Academy, Etlik, Ankara, Turkey

Dietrich van Calker Division of Psychopharmacotherapy, Department of Psychiatry and Psychotherapy, University of Freiburg, Freiburg, Germany

Debby Van Dam Laboratory of Neurochemistry and Behaviour, Institute Born-Bunge, University of Antwerp, Wilrijk, Antwerp, Belgium

Ryan G. Vandrey Johns Hopkins University School of Medicine, Baltimore, MD, USA

Peter Verhaert Department of Biotechnology, Netherlands Proteomics Centre, Delft University of Technology, Delft, Netherlands

Joris C. Verster Division of Pharmacology, Utrecht University, Utrecht Institute for Pharmaceutical Sciences, Utrecht, The Netherlands

Centre for Human Psychopharmacology, Swinburne University of Technology, Melbourne, VIC, Australia

Cécile Viollet Faculté de Médecine, Centre de Psychiatrie and Neuroscience, UMR 894, Université Paris Descartes, Paris, France

Jan Volavka Department of Psychiatry, School of Medicine, New York University, New York, NY, USA

Marie-Louise G. Wadenberg Department of Natural Sciences, Linnaeus University, Kalmar, Sweden

Sharon L. Walsh Department of Behavioral Sciences, Center on Drug and Alcohol Research, College of Medicine, University of Kentucky, Lexington, KY, USA

Johannes Wancata Department of Psychiatry and Psychotherapy, Medical University of Vienna, Vienna, Austria

Elizabeth C. Warburton School of Physiology and Pharmacology, University of Bristol, Bristol, UK

Melissa R. Warden Department of Neurobiology and Behavior, Cornell University, Ithaca, NY, USA

Koichiro Watanabe Department of Neuropsychiatry, Kyorin University School of Medicine, Tokyo, Japan

Ina Weiner School of Psychological Sciences and the Sagol School of Neuroscience, Tel-Aviv University, Tel-Aviv, Israel

Galen R. Wenger Department of Pharmacology and Toxicology, University of Arkansas for Medical Sciences, Little Rock, AR, USA

Tara L. White Department of Community Health, Laboratory of Affective Neuroscience, Center for Alcohol and Addiction Studies, Brown University, Providence, RI, USA

Heather Wilkins Department of Psychiatry, Clinical Neuroscience Research Unit, University of Vermont College of Medicine, Burlington, VT, USA

Lawrence S. Wilkinson Behavioural Genetics Group, School of Psychology and MRC Centre for Neuropsychiatric Genetics and Genomics, School of Medicine, Cardiff University, Cardiff, UK

Paul Willner Department of Psychology, School of Human Sciences, Swansea University, Swansea, UK

Gail Winger Department of Pharmacology, University of Michigan, Ann Arbor, MI, USA

James Winslow NIMH Non-Human Primate Research Core, National Institute of Mental Health, National Institutes of Health, Bethesda, MD, USA

Roy A. Wise Behavioral Neuroscience Branch, NIDA IRP, Baltimore, MD, USA

Jeffrey M. Witkin Lilly Res Labs, Eli Lilly & Co., Indianapolis, IN, USA

Kim Wolff Institute of Pharmaceutical Science, King's College London, London, UK

James H. Woods Department of Pharmacology, University of Michigan, Ann Arbor, MI, USA

Jennifer Wright Department of Pharmacology and Therapeutics, McGill University, Montréal, QC, Canada

Huaiyu Yang Department of Psychiatry, Massachusetts General Hospital, Harvard Medical School, Boston, MA, USA

Gorkem Yararbas Institute on Drug Abuse, Toxicology and Pharmaceutical Science, Ege University, Bornova, Izmir, Turkey

Kimberly A. Yonkers Departments of Psychiatry, Obstetrics, Gynecology and Reproductive Sciences and the School of Epidemiology and Public Health, Yale School of Medicine, New Haven, CT, USA

Alice M. Young Department of Psychological Sciences, Texas Tech University, Office of the Vice President for Research, Lubbock, TX, USA

Andrew Young School of Psychology, University of Leicester, Leicester, UK

Simon N. Young Department of Psychiatry, McGill University, Montréal, QC, Canada

Shuang Yu NeuroAdaptations Group, Max Planck Institute of Psychiatry, Munich, Germany

Alison R. Yung Institute of Brain, Behaviour and Mental Health, University of Manchester, Manchester, UK

Ross Zafonte Department of Physical Medicine and Rehabilitation, Massachusetts General Hospital, Spaulding Rehabilitation Hospital, Brigham and Women's Hospital, and Harvard Medical School, Charlestown, MA, USA

David H. Zald Department of Psychology and Psychiatry, Vanderbilt University, Nashville, TN, USA

Helio Zangrossi Jr. Department of Pharmacology, School of Medicine of Ribeirão Preto, University of São Paulo, Ribeirão Preto, SP, Brazil

Joseph Zohar Division of Psychiatry, Chaim Sheba Medical Center, Tel Hashomer Ramat Gan, Israel

Eric P. Zorrilla The Scripps Research Institute, University of California San Diego, La Jolla, CA, USA

Magnetic Resonance Imaging (Functional)

Ben J. Harrison[1] and Christos Pantelis[2]
[1]Melbourne Neuropsychiatry Centre, Department of Psychiatry and Melbourne Health, The University of Melbourne, Carlton, Melbourne, VIC, Australia
[2]Department of Psychiatry, Melbourne Neuropsychiatry Centre, Sunshine Hospital, The University of Melbourne, St Albans, VIC, Australia

Synonyms

fMRI; Functional magnetic resonance imaging

Definition

Functional magnetic resonance imaging (fMRI) is a specialized form of MRI and a modern neuroimaging technique that is typically used for investigating brain activity in animals and humans. Most fMRI experiments measure blood-oxygenation-level-dependent (BOLD) contrast – an endogenous hemodynamic signal that reflects blood-oxygenation changes linked to neuronal activity. BOLD fMRI is thus an *indirect* or surrogate measure of neuronal function. Because intracranial recordings of local field potentials better predict changes in BOLD signal amplitude than multiunit activity, it is generally considered to reflect synaptic input and local processing in neuronal ensembles as opposed to neuronal spiking activity per se. In conventional applications, BOLD fMRI has a temporal resolution in the order of seconds (1–3 s) and a spatial resolution in the order of millimeters (cubes of tissue 3–5 mm on each side) when covering the whole brain. Largely due to its noninvasive nature and good spatiotemporal resolution, but also through growing understanding of the biophysical basis of the BOLD signal, as well as advances in the acquisition, design, and statistical analysis of brain mapping experiments, BOLD fMRI has been a principal research tool in human cognitive neuroscience since the mid-1990s.

Principles and Role in Psychopharmacology

Blood circulation and energy metabolism are closely linked to neuronal synaptic activity in the brain – an observation first suggested by nineteenth-century researchers. fMRI, and specifically BOLD contrast fMRI (BOLD fMRI), is a modern neuroimaging technique that exploits the fact that such processes, particularly blood flow and blood oxygenation, are regionally coupled to changes in neuronal activity levels.

© Springer-Verlag Berlin Heidelberg 2015
I.P. Stolerman, L.H. Price (eds.), *Encyclopedia of Psychopharmacology*,
DOI 10.1007/978-3-642-36172-2

Background Principles

Active neurons consume oxygen that is carried by hemoglobin in capillary red blood cells. Under periods of increased neuronal demand for oxygen (⇑ oxygen utilization), there is an accompanying increase in cerebral blood flow (CBF) to active brain areas, or functional hyperemia. Central to understanding BOLD fMRI is that, during this period of increased blood flow, blood oxygenation increases more than is necessary to satisfy the increased neuronal demand for oxygen, leading to local changes in the relative concentration of oxygenated and deoxygenated blood, as well as local cerebral blood volume.

Magnetic resonance imaging (MRI), extending the principle of nuclear magnetic resonance (NMR), has the capacity to measure physiological changes associated with increased (or decreased) neuronal activity, such as measurements of tissue perfusion, blood volume, and blood oxygenation. To understand these measurements, it is necessary to have basic knowledge of the physical, biophysical, and engineering principles of MRI and its functional variants such as BOLD fMRI. For this, several comprehensive introductory texts are available (inc. Buxton 2002; Huettel et al. 2004; Jezzard et al. 2003).

Two early discoveries of special relevance to BOLD fMRI are (1) that deoxygenated hemoglobin is paramagnetic and (2) that there is an oxygenation dependence of the transverse relaxation time of water protons in whole blood at high magnetic field strengths (1.5 or greater). This led Ogawa et al. (1990) to investigate whether altering blood oxygenation levels would influence the visibility of blood vessels on T_2^*-weighted MR images. When increasing the relative concentration of deoxygenated hemoglobin in blood, they observed reduced T_2^*-weighted signal intensity in local vasculature on gradient-echo images (GE). Ogawa and colleagues went on to suggest that their observation of "BOLD contrast" could potentially be used to investigate neuronal activity, albeit indirectly, through changes in blood flow and tissue oxygenation.

The first BOLD fMRI studies of the brain in humans were reported in 1992 and involved sensory-related activation of the visual and motor cortices (Fig. 1). This work confirmed that MRI could be used to investigate regional changes in brain activity, similar to functional brain mapping studies undertaken at the time with PET imaging. Since these initial studies, the growth of BOLD fMRI in neuroscience applications has been extraordinary. While BOLD fMRI initially provided a noninvasive and improved brain mapping alternative to PET imaging, it has since taken on its own unique role in cognitive neuroscience research, as well as having a variety of clinical and commercial applications. According to a recent estimate, over 19,000 peer-reviewed articles are returned from an ISI/Web of Science search with the keyword terms "fMRI" or "functional MRI" or "functional magnetic resonance imaging," where the rate of

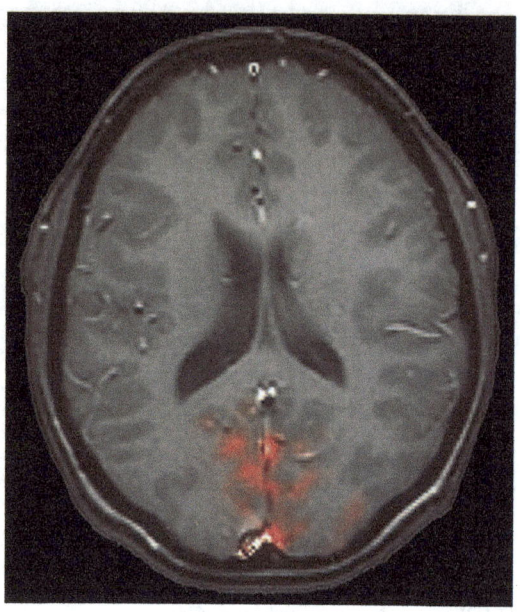

Magnetic Resonance Imaging (Functional), Fig. 1 An early blood-oxygenation-level-dependent (*BOLD*) fMRI study of visual cortex activation in a single human subject. This study was performed in October 1992 at the Magnetic Resonance Centre of Pedralbes in Barcelona, Spain (gradient-echo sequence at 1.5 T, GE Signa), single-slice acquisition, 96 × 64 pixel matrix; round surface coil; TR = 7 s. The subject was stimulated with an 8 Hz visual flicker in a blocked-design experiment that compared four blocks of visual stimulation alternating with four blocks of darkness. Eight images were acquired per block (image courtesy of J. Pujol)

fMRI-related publications has risen from a total of four papers in 1992 to eight papers per day by 2007 (Logothetis 2008). Of the total number of reports, approximately 43 % investigated functional localization and/or anatomy associated with specific stimuli or tasks (sensory, cognitive, and emotional); 22 % were "region of interest" (ROI) studies examining the physiological properties of distinct brain regions; 8 % were related to neuropsychology; 5 % were on the properties of the fMRI signal; and the remaining work was related to various topics including plasticity, drug action, experimental design, and analysis methods. For a specific overview of fMRI applications in clinical neuroscience, including pharmacological fMRI (ph-fMRI), see Matthews et al. (2006).

Spatial and Temporal Resolution

Spatial resolution in fMRI experiments is defined by voxel size – 3D rectangular prisms (volume elements) that form the basic unit of measurement in MR images. In whole brain fMRI studies, voxels will typically have a resolution of 3–5 mm (on a side), which is determined by the field of view, matrix size, and slice thickness of the imaged volume. Reducing the size of voxels generally comes at the risk of decreasing signal-to-noise (given the various noise sources in BOLD fMRI; see below) and increasing acquisition times, especially in whole brain studies. Larger voxels, on the other hand, may contain to a larger extent signal contributions from distinct tissue types or regions, known as "partial volume effects." Determining appropriate voxel size is therefore a trade-off with respect to spatial coverage, resolution, and acquisition time in fMRI studies.

Ultimately, the spatial resolution of BOLD fMRI is constrained by the specificity of the brain's vascular system, which can be said to define the technique's *functional resolution*. This refers to the anatomical coincidence between hemodynamic and neuronal activities, which varies in the brain depending on characteristics of the local microvasculature (e.g., density and architecture). BOLD fMRI is sensitive to signal changes in the capillary bed (<10 μm in diameter) and the venous side of the circulation, including venules and larger draining veins (100 μm to mm in diameter), since arteries and arterioles are close to full saturation and contain no deoxygenated blood. In the case of large vessels, signal changes can be displaced up to several millimeters from the activated site and may obscure the localization of smaller magnitude changes, such as those in the capillary bed supplying active neurons. Higher spatial specificity is therefore advantageous in BOLD fMRI experiments. To this end, advanced acquisition sequences have been developed to emphasize or de-emphasize vascular components of the BOLD signal that are distant from neuronal activity, while fMRI experiments can also be designed to better localize hemodynamic responses to specific regions of interest (see example in Fig. 2). The experiments that have compared BOLD fMRI and intracranial microelectrode recordings has shown that the BOLD signal is a robust (linear) predictor of neuronal activity only when considered at the supra-millimeter scale (3–6 mm^2).

Temporal resolution in BOLD fMRI experiments is usually defined by repetition time (TR), or the time taken to acquire one image of the brain or specified volume (number of slices) of interest. In conventional experiments, TR may range from 1 to 3 s, giving such measurements an intermediate level of temporal resolution in-between electrophysiological (ms) and positron emission tomography (PET) imaging (tens of seconds) techniques. Like its spatial resolution, temporal resolution in BOLD fMRI is constrained by technical and physiological factors. In the former case, a major contribution to its current success has been the development of novel acquisition schemes, notably echo-planar imaging (EPI), that permit rapid functional imaging of T_2*-weighted images of the whole brain. Currently, BOLD fMRI with segmented GE-EPI can acquire single slices at a sampling rate of less than 100 ms. Depending on the experimental design, reducing TR (⇑ temporal resolution) will improve the statistical estimation of the BOLD hemodynamic response to a certain extent, although this presents a trade-off between

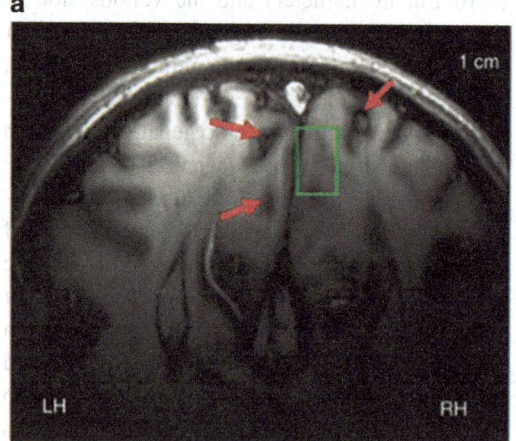 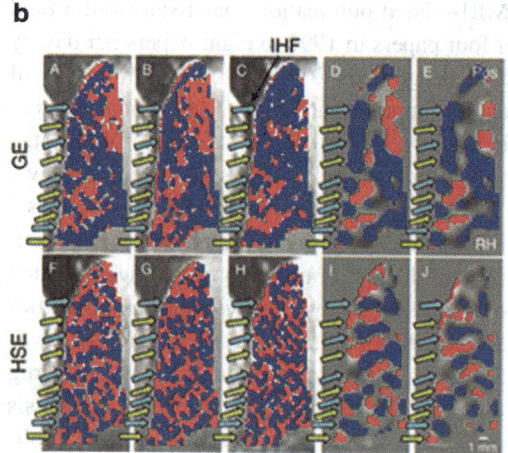

Magnetic Resonance Imaging (Functional), Fig. 2 BOLD fMRI of ocular dominance (*OD*) columns in the human visual cortex: vertical neuronal columns that respond preferentially to visual stimuli presented to one eye rather than to stimuli presented to the other eye. fMRI studies of OD are often presented to showcase the level of spatial specificity that can be achieved with advanced fMRI techniques. In humans, OD columns are separated by approximately 1 mm. *Left panel* (**a**): The imaging slice from a single subject selected in a study by Yacoub et al. (2007) that permitted a resolution of 0.25×0.25 mm^2 in plane for a slice thickness of 3 mm. *Right panel* (**b**): Differential functional OD maps depicting increased activity for left eye stimulation (*blue*) and right eye stimulation (*red*) for this subject across distinct sessions (*A, B, C & F, G, H*) and different filtered averages (*D, E* and *I, J*). The upper and lower rows show maps obtained using gradient-echo (*GE*) and hahn spin-echo (*HSE*) fMRI, respectively. Both approaches reproduced the expected OD columns, although with increased specificity seen with the HSE method due to its enhanced ability to suppress the influence of large vessels. *Pos* posterior, *RH* right hemisphere, *IHF* interhemispheric fissure (Reproduced with permission from Yacoub et al. 2007 © 2007 Elsevier Inc.)

image quality and spatial coverage. Ultimately, temporal resolution in BOLD fMRI is constrained by the slower nature of the hemodynamic response to neuronal activity, whose onset lags behind the timing of actual neuronal events by 4–6 s (see below). Despite these absolute timing differences, well-designed experiments have been able to discriminate the *relative timing* of BOLD signals between different stimuli or brain regions within a few hundred milliseconds (reviewed in Chap. 8, Huettel et al. 2004).

Characteristics and Generation of the BOLD Response

As introduced above, BOLD signal is inversely proportional to the concentration of deoxygenated hemoglobin, which is influenced by local changes in three physiological parameters: cerebral blood volume (CBV), CBF, and the cerebral metabolic rate of oxygen consumption (CMRO2). Signal increases reported in BOLD fMRI experiments are related to the fact that neuronal activity increases regional CBF and glucose utilization (CMRglu) to a larger extent than CMRO2. The net effect of neuronal excitation is therefore to decrease the concentration of deoxygenated hemoglobin ("deoxyhemoglobin washout"; Brown et al. 2007), which in turn increases BOLD signal strength. It is now understood that the characteristic BOLD signal changes observed in fMRI studies reflects the summation of these competing events (CBF, CMRO2, and CBV), resulting in a complex response function that is controlled by several parameters (Buxton 2002). In other words, the BOLD signal does not reflect a single physiological process, but rather represents the combined effects of CBF, CBV, and CMRO2 (Fig. 3).

The BOLD response to a short duration event or single stimulus has a canonical hemodynamic waveform shape, which is often described as consisting of (1) a fast response lasting 1–2 s ("initial dip") in which there is a small decrease in BOLD signal amplitude; (2) a larger amplitude

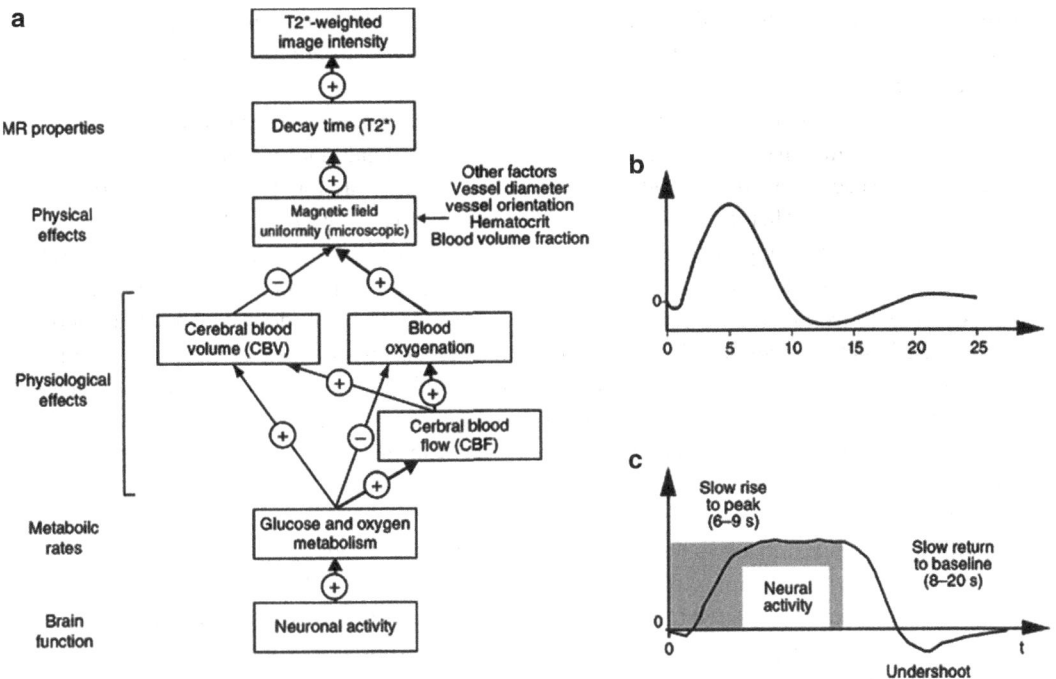

Magnetic Resonance Imaging (Functional), Fig. 3 *Left panel* (**a**): Measuring signal changes in fMRI experiments depicted as a complex multistage process, beginning with neuronal activity and ending with BOLD signal measurement as a property of the MRI scanner and pulse sequence. This figure presents a schematic illustration of interactions in the formation of the BOLD signal. *Positive/negative arrows* indicate positive/negative correlations between the parameters. The right pathway (*bold arrows*) is the most significant effect in most BOLD fMRI. *Right panels* (**b**) *and* (**c**): Simplified schematic representation of the BOLD hemodynamic response waveform to a short duration stimulus (**b**) and to a block of multiple consecutive stimuli (**c**) (parts of this figure reproduced with permission from http://www.eecs.umich.edu/~dnoll)

hyperemia associated with the inflow of oxygenated blood, which peaks at approximately 4–6 s after stimulus presentation; and (3) a refractory period lasting 6–12 s where the signal undershoots the baseline due to the combination of reduced regional CBF and increased CBV ("post-stimulus undershoot"). The second phase of hyperemia (or hyperoxic phase) is the common focus for detecting increases in brain activity measured by BOLD fMRI. The so-called "initial dip" is suspected to result from early oxygenation changes (oxygen extraction) localized to capillaries and has been argued as more closely related to neuronal activity than the ensuing hyperemia. However, this observation remains controversial and is not reliably detected in BOLD fMRI studies. Like the BOLD response to a single event, multiple repetitions of the same stimuli in blocks (see below) will see the BOLD response rise to a steady plateau and decline once the block ends, although there are variations to this rule, such as an initial overshoot, slow increasing and decreasing ramps, or an undershoot at the end of the stimulus.

Neurovascular Coupling and Neuronal Correlates of BOLD

The process by which neural activity influences the hemodynamic properties of the surrounding vasculature (principally arterioles) is referred to as neurovascular coupling, although the mechanism(s) responsible for this is not fully understood. One hypothesis regarding a principal signaling route is that a feedforward pathway involving neuronal-glial interactions after neurotransmitter release stimulates regional CBF. Astrocytes play a crucial role in neurotransmitter recycling, using energy reliant on glycolysis (nonoxidative glucose metabolism) to clear extracellular glutamate and convert it to glutamine after neuronal firing. Increased glycolysis in astrocytes is suspected to trigger intracellular events that couple glutamate cycling rate to the production of vasoactive agents, including nitric oxide and eicosanoids. Therefore, according to this view, neurovascular coupling is mediated by neuronal signaling mechanisms via glial pathways, as opposed to signaling mechanisms of an energy deficit in neurons per se. This view supports, in part, the notion that glycolysis is relevant to the detection of BOLD activity changes and in explaining the apparent mismatch between CBF, CMRglu, and CMRO2 during evoked brain activity (Raichle and Mintun 2006).

Detailed biophysical models have also been proposed to explain the complex shape of the hemodynamic response observed in BOLD fMRI studies, accounting for the changes in CBF, CBV, and CMRO2 that accompany increased neuronal activity. The most prominent is the "balloon model" of Buxton and colleagues. According to this work, the apparent discrepancy between CBF and CMRO2 results from how oxygen is supplied to neurons related to its poor diffusion in brain tissue. That is, blood flow must increase more than oxygen consumption to maintain tissue-oxygen gradients supporting oxygen delivery to tissue because its extraction (by passive diffusion) from blood is less efficient at higher flow rates (Buxton 2002). Evidence favoring this model versus the former hypothesis (and vice versa) can be found in expanded form in Buxton et al. (2004) and Raichle and Mintun (2006), respectively.

Regardless of the precise cause(s) of the physiological changes that give rise to the BOLD signal, evidence has been marshaled in support of a close relationship between evoked hemodynamic and neuronal activity changes. Notably, in the work of Logothetis et al. (2001), which compared BOLD fMRI and intracranial electrophysiological measurements recorded simultaneously in monkeys, BOLD signal was found to be spatially well localized and scaled with neuronal activity. Specifically, these authors reported that the amplitude of the BOLD signal was better correlated with recordings of local field potentials rather than multiunit activity (Logothetis et al. 2001). That is, BOLD signal better reflects the weighted average of synchronized activity of the input signals into a neuronal ensemble than their spiking (action potential) activities. This suggests that BOLD signal changes, primarily, reflect input and integrative processes rather than output (communicative) activity. However, there remains some debate about the contributions of different types of neuronal activity to the BOLD signal (local field potentials vs. spiking activity), as the former will be correlated with the latter in many instances (Raichle and Mintun 2006).

Experimental Design

In a conventional fMRI experiment, time series of T_2^*-weighted images are acquired while subjects are exposed to a specific stimulus or set of stimuli ("task-on") that is systematically varied with respect to a "control-off" condition, typically in the context of a serial or cognitive subtraction or factorial design. The goal of this approach is to evoke significant changes in blood flow and oxygenation within a given region or network associated with the "task-on" state that will modulate BOLD signal intensity about its mean value. The duration of these stimulus presentations or epochs must be tailored to the dynamics of the hemodynamic response and will be repeated multiple times to establish sufficient contrast and functional signal to noise ratios for the mapping of "activation" responses. In practice, the magnitude of task-related changes in fMRI studies is small (up to ± 5 % but usually less) in comparison to the total image intensity and variability across time due to various sources of physical (MR system) and physiological noise. Careful experimental design and the use of postprocessing methods for maximizing the detection of activation in the BOLD time series is therefore a critical feature of fMRI studies (Chaps. 8–13, Huettel et al. 2004).

One common approach is to take advantage of the summed signal as a way of minimizing the influence of noise in fMRI experiments (Fig. 4). The idea here is that the BOLD response summed

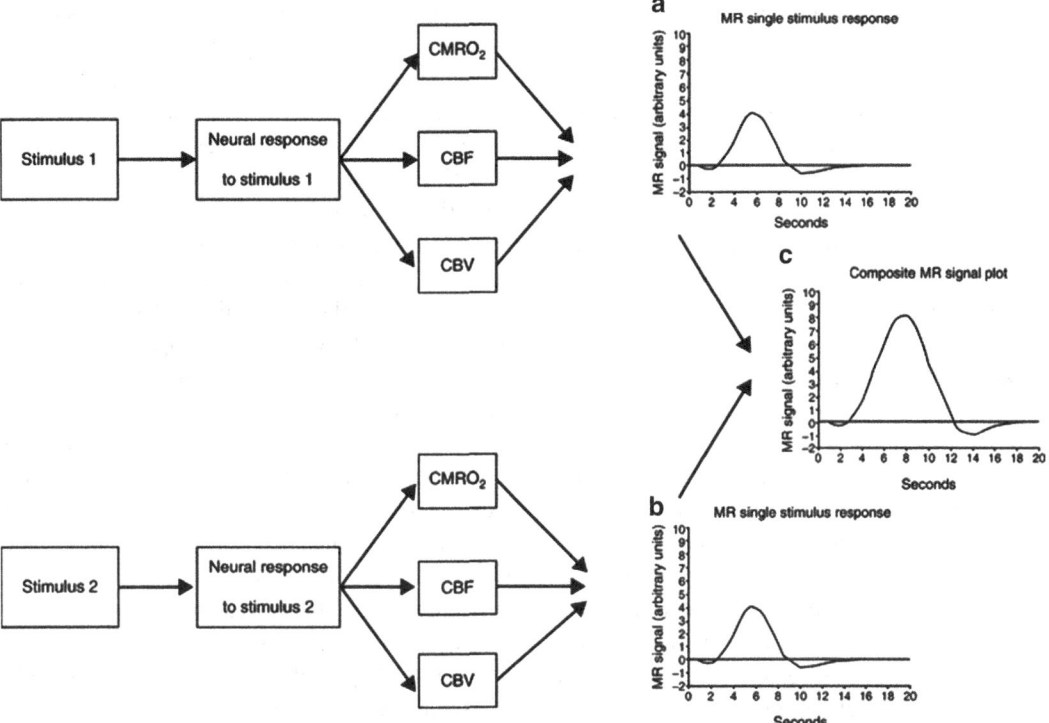

Magnetic Resonance Imaging (Functional), Fig. 4 Linearity of the hemodynamic response: events leading from the presentation of two stimuli to the generation of a summed BOLD signal. To a certain degree, the BOLD response to successive neural events can be predicted from the summed responses (superposition) to single neural events given an appropriate time delay between them. The figure assumes that stimulus 1 is presented 1 s before stimulus 2 and each stimulus evokes a response that alters CMRO2, cerebral blood flow, and CBV. The net effect of these physiological changes leads to the BOLD response for individual stimuli (**a**, **b**). The observed BOLD response (**c**) is a composite of the unobserved BOLD responses to the single stimuli (reproduced with permission from Brown et al. 2007. © 2007 Springer Netherlands)

over several trials will reduce the influence of random noise sources as a result of averaging (Brown et al. 2007). Blocked designs that present the same class of stimuli on multiple occasions seek to capitalize on this strategy. In turn, this involves selecting the correct number and timing of stimuli to occur within a block, the duration of the block itself, and its number of repetitions, as well as the number of different block types to be included in a single acquisition for later comparison. Overall, block designs are powerful in terms of detecting significant sustained (steady-state) activation in fMRI studies but are generally poor estimators of the time course of the regional hemodynamic response to neuronal events because of their reliance on linear summation of individual responses.

Event-related designs are a second common approach in fMRI experiments and involve the presentation of specific stimuli as short duration events in order to detect transient associated changes in neuronal activity. With this approach, each event is separated temporally by an interval ranging from a few to tens of seconds and typically in a random order of predefined range. Investigators typically assume a canonical shape to the hemodynamic response to each stimulus presented and model it as a weighted sum to consecutive stimuli – although this linearity assumption may not hold, especially for the early phase of the hemodynamic response. Compared to blocked designs, event-related designs are superior in investigating the shape of regional hemodynamic responses and to compare features such as amplitude or relative timing differences between events. Event-related designs also allow for the investigation of BOLD responses sorted by response types, for instance, comparing correct versus incorrect or fast versus slow responses. By comparison, their detection power is relatively poor with respect to blocked designs due to the fewer number of events that can be presented and averaged in a single experimental run.

Analysis of BOLD fMRI

It was previously stated that the BOLD fMRI time series is influenced by a number of sources of physical and physiological noise. In the former case, this includes system noise that causes fluctuations in MR signal (e.g., signal drift) due to magnetic field inhomogeneities and other factors. In the latter case, this includes gross head motion artifacts, motion related to the cardiac (beat-to-beat) and respiratory (breath-to-breath) cycles, as well as slow variations in respiratory rate and volume, which change the pressure of arterial CO_2 – a potent vasodilator. Awareness of these various noise sources in BOLD fMRI studies has led to a range of methods to reduce or mitigate their influence, which continue to be improved upon and refined.

Preprocessing of the raw fMRI time series, prior to statistical analysis, generally has two main goals: firstly, to reduce unwanted or uninteresting variability from data and secondly, to prepare data for statistical analysis and inference given that many statistical tests applied in fMRI studies make assumptions that are met through such preprocessing steps. In practice, this involves modifying the raw data in a series of steps often including *image realignment*, to correct and to diagnose head motion artifact in a time series; *image normalization*, to transform data from different subjects into a common neuroanatomical space; *spatial smoothing*, to filter (or blur) the data to reduce spatial noise and to improve its normality for statistical parametric tests; and *temporal smoothing*, to remove low- or high-frequency noise sources, such as mentioned above.

There are a growing number of ways to perform statistical analysis in BOLD fMRI experiments. This has been assisted greatly by the development of publicly available neuroimaging analysis software packages, such as Statistical Parametric Mapping (http://www.fil.ion.ucl.ac.uk/spm/), FMRIB Software Library (http://www.fmrib.ox.ac.uk/fsl/FSL), and Analysis of Functional Neuro Images (http://afni.nimh.nih.gov/). The majority of fMRI studies to date have adopted a conventional voxel-based mapping approach based on extensions of the general linear model for time-series analysis. The basic premise behind such approaches is that the observed fMRI data can be accounted for by a combination of several experimental (or model) parameters and uncorrelated (or independently distributed) noise. Given the high number of statistical tests performed (voxel by voxel), some correction factor for multiple comparisons will generally be applied, leading to the generation of statistically threshold "activation" maps related to the experiment at hand. This may be performed for the whole brain or specific regions of interest.

Other techniques, based on multivariate analysis techniques, can also be used to investigate which brain areas are "activated" by a task or a stimulus in fMRI studies. These techniques, as opposed to the general linear model approach, are data driven and therefore do not require the specification of experimental models a priori. Another important distinction between this class of statistical tests and the former is that they are sensitive for testing not only "where" activation occurs in a given experimental context but also how different regions or networks of regions may interact or show interdependence in their activities over time (Fig. 5). Such relationships have been characterized as representing distinct forms of brain-functional connectivity, which has become a topic of specific interest with BOLD fMRI in recent years.

Major Strengths of the Method

- Is a safe, noninvasive, highly repeatable, and widely available technique for measuring changes in brain activity in vivo

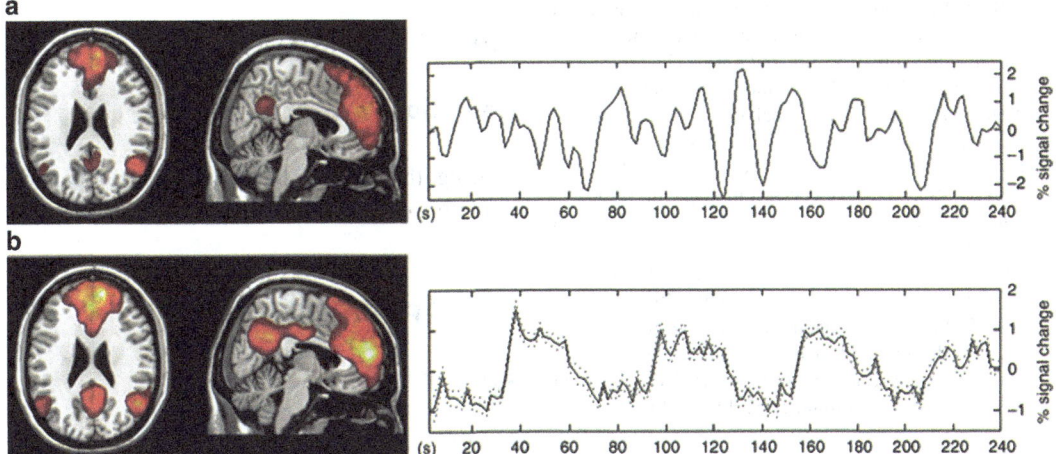

Magnetic Resonance Imaging (Functional), Fig. 5 Global functional connectivity of a large-scale and distributed brain network characterized from two distinct task states using independent component analysis (*ICA*). *Top panel* (**a**): Correlated fluctuations of the BOLD signal among regions of the so-called default-mode network in a group of healthy subjects scanned at rest (NB: time-course plot is of a single subject). *Bottom panel* (**b**): Correlated fluctuations of the BOLD signal among "default mode network" regions in the same group of subjects performing a moral dilemma task (NB: time course plot is of the group mean). The task consisted of four alternating 30 s control (C) and moral dilemma (D) condition blocks (CDCDCDCD) (modified from Harrison et al. (2008) Proc Natl Acad Sci USA 105:9781–9789. © 2008 by the National Academy of Sciences of the USA)

- Has superior spatial resolution compared to other human neuroimaging techniques
- Affords high flexibility in experimental design and data modeling

Major Limitations of the Method

- Measures neuronal activity indirectly via changes in blood oxygenation levels
- Has a temporal resolution in the order of seconds due to the nature of the hemodynamic response
- Is susceptible to influences of non-neural changes in the body

Cross-References

- ▶ Cognitive Neuroscience
- ▶ Cognitive Subtraction
- ▶ Echo-Planar Imaging
- ▶ Gradient-Echo Images
- ▶ Local Field Potentials
- ▶ Neuroimaging
- ▶ Nuclear Magnetic Resonance
- ▶ Time Series

References

Brown GG, Perthen JE, Liu TT, Buxton RB (2007) A primer on functional magnetic resonance imaging. Neuropsychol Rev 17:107–125

Buxton RB (2002) Introduction to functional magnetic resonance imaging: principles and techniques. Cambridge University Press, Cambridge

Buxton RB, Uludag K, Dubowitz DJ, Liu TT (2004) Modeling the hemodynamic response to brain activation. Neuroimage 23:S220–S233

Harrison BJ et al. (2008) Consistency and functional specialization in the default mode brain network. Proc Natl Acad Sci USA 105:9781–9789

Huettel SA, Song AW, McCarthy G (2004) Functional magnetic resonance imaging. Sinauer Associates, Sunderland

Jezzard P, Matthews PM, Smith SM (2003) Functional MRI: an introduction to methods. Oxford University Press, New York

Logothetis NK (2008) What we can do and what we cannot do with fMRI. Nature 453:869–879

Logothetis NK, Pauls J, Augath M, Trinath T, Oeltermann A (2001) Neurophysiological investigation of the basis of the fMRI signal. Nature 412:150–157

Matthews PM, Honey GD, Bullmore ET (2006) Application of fMRI in translational medicine and clinical practice. Nat Rev Neurosci 7:732–744

Ogawa S, Lee TM, Nayak AS, Glynn P (1990) Oxygenation-sensitive contrast in magnetic resonance image of rodent brain at high magnetic fields. Magn Reson Med 14:68–78

Raichle ME, Mintun MA (2006) Brain work and brain imaging. Annu Rev Neurosci 29:449–476

Yacoub E, Shmuel A, Logothetis NK, Ugurbil K (2007) Robust detection of ocular dominance columns in humans using Hahn Spin Echo BOLD functional MRI at 7 Tesla. Neuroimage 37:1161–1177

Major Tranquilizer

Synonyms

Antipsychotic; Neuroleptics

Definition

A medication used in the treatment of psychotic disorders of any type with a particular emphasis on inducing sedation. The term is less used than its synonyms and is poorly characterized in pharmacological terms. It tends to be used less by psychiatrists than by nonspecialists.

Cross-References

▶ Antipsychotic Drugs
▶ First-Generation Antipsychotics
▶ Second- and Third-Generation Antipsychotics

Mania

Definition

A state in which the individual experiences euphoria, irritability, lack of sleep, and increased drives, often to the point of poor judgment.

MAO-B Inhibitor

Definition

A drug that blocks the action of monoamine oxidase type B.

Mass Spectrometry

Synonyms

Mass spectroscopy

Definition

Mass spectrometry is an analytical technique that measures the molecular mass of an analyte. A mass spectrometer typically consists of three modules: (1) an ion source in which the sample is vaporized and ionized to make it analyzable; (2) a mass analyzer, which separates the sample components on the basis of their mass (more precisely their mass-to-charge ratios); (3) and a detector, which provides data for calculating the abundance of each ion present.

Cross-References

▶ Electrospray Ionization
▶ Imaging Mass Spectrometry
▶ Matrix-Assisted Laser Desorption Ionization
▶ Metabolomics
▶ Neuropeptidomics
▶ Posttranslational Modification
▶ Proteomics
▶ Two-Dimensional Gel Electrophoresis

Matching Law

Definition

A ▶ behavioral allocation function that relates the rate at which an operant response is performed to the rate of reward delivery. The generalized matching law incorporates additional independent variables, such as the strength, amount, imminence, and likelihood of reward.

Maternal Deprivation Model

Definition

An animal model in which young rats are separated from their mother for a single period of 24 h. The optimal day for separation is postnatal day 9. These animals develop a large number of schizophrenia-like phenomena in adulthood. Interestingly, most of these phenomena occur after puberty in accordance with the clinical literature on schizophrenia.

Cross-References

▶ Schizophrenia: Animal Models
▶ Simulation Models

MATRICS

Synonyms

Measurement and Treatment Research to Improve Cognition in Schizophrenia

Definition

This is an initiative of the National Institutes of Health in the USA to enhance the methodology for assessing cognitive impairment in schizophrenia, using neuropsychological tests, for the purpose of clinical trials. This initiative has also boosted interest in cognitive assessment in experimental animals in order to evaluate putative cognitive enhancing compounds.

Matrix-Assisted Laser Desorption Ionization

Synonyms

Laser desorption ionization; MALDI

Definition

Formation of gas-phase ions from molecules that are present in a solid or liquid matrix that is irradiated with a pulsed laser.

Cross-References

▶ Imaging Mass Spectrometry
▶ Mass Spectrometry
▶ Metabolomics
▶ Neuropeptidomics
▶ Proteomics
▶ Two-dimensional Gel Electrophoresis

Mecamylamine

Definition

A nicotinic antagonist that is well absorbed from the gastrointestinal tract and crosses the blood-brain barrier.

Medazepam

Definition

Medazepam is a benzodiazepine that has anxiolytic, sedative, and anticonvulsant properties. It has a very long duration of action, mainly due to a long ▶ elimination half-life (36–200 h) and conversion to active metabolites including the benzodiazepines ▶ diazepam and *N*-desmethylmedazepam, both of which also are long acting and also have long-acting metabolites. Like most similar compounds, medazepam is subject to ▶ tolerance, ▶ dependence, and ▶ abuse.

Cross-References

- Anxiolytics
- Benzodiazepines

Medial Forebrain Bundle

Definition

The MFB is a complex bundle of axons coming from the basal olfactory regions, the periamygdaloid region, and the septal nuclei and passing to the lateral hypothalamus with some carrying on into the tegmentum. It contains both ascending and descending fibers. It is commonly accepted that the MFB is part of the reward system involved in the integration of reward and pleasure. Electrical stimulation of the MFB can serve as a positive reinforcer and is believed by some experts to cause sensations of pleasure.

Medial Prefrontal Cortex

Synonyms

mPFC

Definition

A brain structure located in the ▶ prefrontal cortex that receives dopamine input from the midbrain. It has been implicated in many cognitive functions including drug reward, ▶ behavioral inhibition, and stress reactivity.

Medication Adherence

Koichiro Watanabe[1] and Takefumi Suzuki[2,3]
[1]Department of Neuropsychiatry, Kyorin University School of Medicine, Tokyo, Japan
[2]Department of Neuropsychiatry, Keio University School of Medicine, Tokyo, Japan
[3]Department of Psychiatry, Inokashira Hospital, Tokyo, Japan

Synonyms

Adherence; Compliance; Concordance; Obedience; Observance

Definition

Defining adherence in a medical context is not an easy task. The participants at the World Health Organization (WHO) Adherence meeting in June 2001 concluded that defining adherence as "the extent to which the patient follows medical instructions" was a helpful starting point. The concept has evolved to the extent to which a person's behavior – taking medication, following a diet, and/or executing lifestyle changes – corresponds with agreed recommendations from a healthcare provider (http://apps.who.int/medicinedocs/en/d/Js4883e/6.html). Suboptimal adherence to professional instructions, including regularly taking medications, represents a significant clinical problem and a serious public health concern. It is therefore of great importance to accurately identify predictive factors of adherence. This is especially relevant in chronic disorders and conditions, where poor adherence to medications in the long run is expected to result in worse outcomes (Fenton et al. 1997). The same holds true for major psychiatric illnesses; nonadherence or poor/partial adherence can have devastating consequences and is a significant barrier to improvement and recovery for patients with serious mental illnesses (SMIs) such as schizophrenia and bipolar disorder (Velligan et al. 2003).

Current Concepts and State of Knowledge

History

The term "compliance" has been widely used since the 1950s to define the extent of a patient's concordance with a doctor's instructions among those with SMIs. However, the term "compliance" intrinsically has a compulsory and forcible nuance which can be stigmatizing for patients to obey doctor's decisions. From the perspective of autonomy of patients, the term "adherence" has been preferred since the 1990s, exemplified by the pharmacological treatment of HIV/AIDS. "Adherence" has now been regarded to represent the patient's own values/beliefs and decisions that are indispensable for any medical treatment or intervention.

Adherence in the Real World

Treatment adherence in real-world settings has been disappointing and problematic. For instance, maintenance treatment with antipsychotic medications plays a critical role in relapse prevention of schizophrenia. Nevertheless, Velligan et al. (2006) reported that, in the first 3 months after discharge, only 40 % of patients with schizophrenia were still adherent based on pill counts (defined as taking at least 80 % of the prescribed doses). Weiden et al. (1995) likewise indicated that as many as 75 % of patients with schizophrenia become nonadherent within 2 years after hospital discharge. Further, Lacro et al. (2002) reviewed 39 studies published between 1980 and 2000 that included data on the prevalence of medication nonadherence in patients with schizophrenia. The authors found the average rates of nonadherence ranging from 20 % to 89 %, depending on the methodology and patient population (e.g., recently hospitalized versus more stable ambulatory patients), with the mean rate of 41.2 % (range: 20.2–55.6 %) in the studies with better methodology.

Definitions of medication adherence varied greatly, hampering comparison across the studies and limiting the ability to draw firm conclusions on this issue. Also, it is critical to point out that adherence is not a black-or-white issue; partial adherence (or taking only a part of medications intentionally or unintentionally, or taking them too much) is a clinical problem in the real world. Velligan et al. (2006) found in various studies that the percentages ranged from ingesting 50 % to 90 % of prescribed medication, while categorical classifications were expressed as not taking any medication vs. taking nearly every dose.

In evaluating adherence, blood levels of psychotropic medications (and therapeutic and toxic ranges) are available for some agents such as lithium, valproate, haloperidol, clozapine, and clonazepam, but are not readily available for many other drugs. It is important to note that pharmacokinetic parameters of some agents, and hence plasma levels, are greatly influenced by such various factors as metabolizing enzymes including cytochrome P450 systems. Some studies have used a pharmacy-based measure or the medication possession ratio (MPR) to define the levels of adherence. MPRs are calculated by dividing the number of days' supply of medication the patient received by the number of days' supply they needed to receive to take their medication continuously as prescribed (Velligan et al. 2006). For example, Valenstein et al. (2002) evaluated MPR in a context of admission rate in patients with schizophrenia and found that those with 80–110 % of MPR had the lowest admission rate; the rate increased among those whose MPRs were lower and higher.

Medication event monitoring (MEM), in which pill bottle caps record time and date whenever the bottle is opened, may be considered as the "gold standard" for adherence monitoring. However, even MEM can yield hard-to-assess results because of high rates of missing data (Velligan et al. 2006). Therefore, it is proposed that all studies of adherence should include at least two measurements of adherence, of which one should be a direct measure (Velligan et al. 2006). In fact, Jonsdottir et al. (2010) investigated four different measures of adherence, including one direct measure of serum concentration of medications, finding that while outpatients with schizophrenia and bipolar disorder had relatively good adherence to medications in their sample, their self-report scores overestimated adherence. Recently, adherence has been conceptualized to be composed of two separate components. The first part is "persistence," defined as continuously refilling prescriptions in accordance with the duration of therapy as medically indicated. The second component of adherence is "compliance." Even if patients with schizophrenia persist with their treatment (i.e., continue to show up at the clinic and fill the prescriptions), they may not necessarily be compliant (i.e., do not take their drug in accordance with the prescribed dosage and schedule) (Cooper et al. 2007). While similar rates of persistence and compliance, 40 % and 50 % in 1 year, respectively, have been reported in schizophrenia, symptomatology, insight into (or awareness of) the illness, adverse effects, and psychosocial circumstances all need to be taken into thoughtful account in judging adherence.

Factors Influencing Adherence

Any intervention improve adherence must be tailored to the needs and illness-related demands of each individual patient and caregiver. To accomplish this, health systems and providers need to be vigilant for not only accurately assessing adherence but also the factors that influence it (WHO 2003). Factors that have been identified to influence adherence often conceptually overlap or theoretically influence each other but are heuristically differentiated into those related to the patient, the patient's environment, the treating clinician, and the treatment itself (Fleischhacker et al. 2003). In this context, the most consistently implicated issues in patients with schizophrenia are suboptimal insight and a lack of therapeutic alliance (Fenton et al. 1997; Lacro et al. 2002; Jónsdóttir et al. 2013). Other important clinical correlates are negative subjective attitudes toward medications, previous nonadherence, shorter illness duration for first-episode patients, substance abuse (Jónsdóttir et al. 2013), and inadequate discharge planning as well as suboptimal aftercare environments. In bipolar disorder, attitudes toward illness and health beliefs are related to nonadherence.

Other important factors include substance and alcohol abuse and a comorbid personality disorder.

Psychotropic agents are associated with many adverse effects, some of which may be stigmatizing to patients. They range from metabolic disturbances, sexual dysfunction, motor and anticholinergic side effects, to dysphoric experiences such as dizziness and impaired cognition. Side effects, including weight gain and negative subjective feelings such as poor well-being, have also been suggested to be related to nonadherence, although there are some contradictory findings. Experts evaluated which side effects had the greatest influence upon adherence; weight gain, oversedation, akathisia, and sexual dysfunction were found to be the offenders (Velligan et al. 2009).

Neurocognitive impairment can be noted both in schizophrenia and bipolar disorder and is identified as an important predictor of functional outcome in the real world. While it is theoretically possible that those with more impairments in cognition are worse in terms of adherence, the results are mixed on the relationships between neurocognitive dysfunction and nonadherence in schizophrenia and in bipolar disorder as well, likely due to differences in the nature and magnitude of cognitive impairments in the populations under study.

Strategies to Improve Adherence

Osterberg and Blaschke (2005) have suggested several strategies to improve adherence. Successful approaches include a combination of educational activities (involving both patients and their families), cognitive-oriented interventions, and the periodic use of reinforcement techniques. Educational approaches appear to be most effective when they are combined with behavioral techniques and supportive services. Patient-tailored strategies using reinforcements with a wide variety of techniques, such as monetary rewards or vouchers, frequent contact with the patient, and other types of personalized reminders, may also have a role.

The relative effectiveness of second-generation antipsychotics (SGAs) versus first-generation antipsychotics (FGA_S) is a matter of debate. SGAs generally have fewer motor side effects than FGAs; consequently, their use may result in lower rates of discontinuation and better adherence. Still SGAs have many troublesome (non-motor) adverse effects and are a heterogeneous group with each SGA having different characteristics. At the very least, they are more costly than older agents. The possibility of better adherence is far from definite and differential effectiveness with SGAs has been questioned in large pragmatic studies such as Clinical Antipsychotic Trials of Intervention Effectiveness (CATIE), Cost Utility of the Latest Antipsychotics in Severe Schizophrenia (CUtLASS), and European First-Episode Schizophrenia Trial (EUFEST); more comparative studies are necessary.

Long-acting (depot) antipsychotic agents (LAIs) are often the treatment of choice for patients with schizophrenia who have trouble adhering to a regimen of oral agents. The recent development of SGA LAIs has the potential to improve adherence, since these agents claim to combine better efficacy and tolerability of the SGA agents through improved drug delivery technologies. However, this theoretical possibility of better adherence and less worsening in the long run has not been unequivocally proven to date, with different study designs yielding somewhat different results (e.g., randomized comparisons versus mirror image investigations).

It is a common clinical practice to prescribe psychotropic medications so that they are taken multiple times a day. While this convention appears, at least in part, to reflect the biological half-life of the compound, a complicated medication regimen is an obvious hurdle for adherence among patients with SMIs. Active marketing of longer acting (or extended release) compounds of the short-acting parent drug appears to be a reflection of such a concern. Further, it is relevant to be aware of the possibility that the burden of adherence in patients with somatic complications necessitating medical treatment can be even more problematic. Every effort should be made to make the medication regimen as simple as possible.

Finally, in order to improve medication adherence, it is firstly important to establish a good therapeutic alliance with patients. Treatment plans should be formulated and implemented using shared decision-making techniques respecting patients' needs, values, beliefs, and desires in an easy-to-understand manner. It is also critically important to capture patients' impressions of the medications. For this purpose, some rating scales such as the Drug Attitude Inventory may be of clinical use. Other user-friendly clinical scales include the Brief Adherence Rating Scale and the Brief Evaluation of Medication Influences and Beliefs for antipsychotics and the Antidepressants Adherence Scale for antidepressants. The importance of this treatment process cannot be overemphasized, especially in psychiatry.

Conclusion

To conclude, physicians should pay close attention to medication adherence in patients with SMIs. Physicians also need to be aware of the gravity of nonadherence in psychiatric practice and should not underestimate the consequences of this common problem. The starting point would be to honestly ask patients about their impressions or opinions of their medications and treatments, respect and reflect their willingness as clinically appropriate, keep an eye upon the factors that may hamper adherence, and try to address obstacles to acceptable adherence, all of which can be achievable through a good therapeutic alliance.

Cross-References

- ▶ Akathisia
- ▶ Bipolar Disorder
- ▶ Clinical Antipsychotic Trials of Intervention Effectiveness Study
- ▶ Clonazepam
- ▶ Clozapine
- ▶ CUtLASS
- ▶ Cytochrome P450
- ▶ EUFEST
- ▶ First-Generation Antipsychotics

- ▶ Haloperidol
- ▶ Lithium
- ▶ Schizophrenia
- ▶ Second-generation antipsychotics
- ▶ Valproic Acid

References

Cooper D, Moisan J, Grégoire JP (2007) Adherence to atypical antipsychotic treatment among newly treated patients: a population-based study in schizophrenia. J Clin Psychiatry 68(6):818–825

Fenton WS, Blyler CR, Heinssen RK (1997) Determinants of medication compliance in schizophrenia: empirical and clinical findings. Schizophr Bull 23:637–651

Fleischhacker WW, Oehl MA, Hummer M (2003) Factors influencing compliance in schizophrenia patients. J Clin Psychiatry 64(Suppl 16):10–13

Jónsdóttir H, Opjordsmoen S, Birkenaes AB et al (2010) Medication adherence in outpatients with severe mental disorders: relation between self-reports and serum level. J Clin Psychopharmacol 30(2):169–175

Jónsdóttir H, Opjordsmoen S, Birkenaes AB et al (2013) Predictors of medication adherence in patients with schizophrenia and bipolar disorder. Acta Psychiatr Scand 127(1):23–33

Lacro JP, Dunn LB, Dolder CR et al (2002) Prevalence of and risk factors for medication nonadherence in patients with schizophrenia: a comprehensive review of recent literature. J Clin Psychiatry 63(10):892–909

Osterberg L, Blaschke T (2005) Adherence to medication. N Engl J Med 353(5):487–497

Valenstein M, Copeland LA, Blow FC et al (2002) Pharmacy data identify poorly adherent patients with schizophrenia at increased risk for admission. Med Care 40(8):630–639

Velligan DI, Lam F, Ereshefsky L et al (2003) Psychopharmacology: perspectives on medication adherence and atypical antipsychotic medications. Psychiatr Serv 54(5):665–667

Velligan DI, Lam YW, Glahn DC et al (2006) Defining and assessing adherence to oral antipsychotics: a review of the literature. Schizophr Bull 32(4):724–742

Velligan DI, Weiden PJ, Sajatovic M et al (2009) Expert consensus panel on adherence problems in serious and persistent mental illness. The expert consensus guideline series: adherence problems in patients with serious and persistent mental illness. J Clin Psychiatry 70(Suppl 4):1–46

Weiden P, Rapkin B, Zygmunt A et al (1995) Postdischarge medication compliance of inpatients converted from an oral to a depot neuroleptic regimen. Psychiatr Serv 46(10):1049–1054

WHO (2003) Adherence to long-term therapies – evidence for action. http://www.who.int/chp/knowledge/publications/adherence_introduction.pdf

Medicine

Synonyms

Drug; Human medicinal product

Definition

Any substance or combination of substances that may be used in or administered to human beings either with a view to restoring, correcting, or modifying physiological functions by exerting a pharmacological, immunological, or metabolic action or to making a medical diagnosis.

Megakaryocytes

Definition

The megakaryocyte is a nucleated cell originating from the bone marrow and responsible for the production of platelets (thrombocytes), which are necessary for the process of hemostasis. The cytoplasm, just like the platelets that bud off from it, contains alpha-granules and dense bodies.

Melanin-Concentrating Hormone

Definition

Lateral hypothalamic peptide involved in the regulation of feeding-motivated behaviors including food intake. Like hypocretin/orexin, cells producing this peptide are located in the lateral hypothalamic region, but melanin-concentrating hormone (MCH)-producing cells do not co-express hypocretin/orexin.

Melatonin

Definition

Melatonin is a neurotransmitter that participates in the regulation of the sleep/wake cycle. It is produced endogenously and also available over the counter as an effective hypnotic for sleep onset.

Memantine

Definition

Memantine is used as an anti-dementia drug with some efficacy in the treatment of moderate to severe ▶ Alzheimer's disease. It reduces glutamatergic neurotransmission by acting as a low-affinity NMDA-receptor antagonist, and this action is thought to underlie its protective effects against neuronal ▶ excitotoxicity. In addition to its anti-glutamate action, memantine also acts as a noncompetitive antagonist at serotonin 5-HT$_3$ and nicotinic acetylcholine receptors and as an agonist at the dopamine D2 receptor; the role of these actions in the anti-dementia properties of the drug is unknown. Adverse side effects include confusion, dizziness, drowsiness, headache, insomnia or sleepiness, agitation, and hallucinations. In addition to Alzheimer's disease, memantine is currently being tested as a potential treatment for a number of other disorders. Preclinical studies as well as the clinical absence of withdrawal symptoms suggest that this drug has a low abuse potential.

Membrane Potential

Synonyms

Membrane voltage

Definition

A cell's membrane potential is the voltage difference across the plasma membrane that is present in all living cells.

Cross-References

▶ Intracellular Recording

Memorial Delirium Assessment Scale

Synonyms

MDAS

Definition

The Memorial Delirium Assessment Scale (MDAS) is a 10-item, 4-point clinician-rated scale (possible range, 0–30) designed to diagnose and quantify the severity of delirium, validated in hospitalized patients with advanced cancer and AIDS. Items included in the MDAS reflect the diagnostic criteria for delirium in the DSM-IV, as well as symptoms of delirium from earlier or alternative classification systems (e.g., DSM-III, DSM-III-R, ICD-9). The MDAS is both a good delirium diagnostic screening tool and a reliable tool for assessing delirium severity in patients with advanced disease. Scale items assess disturbances in arousal and level of consciousness, as well as in several areas of cognitive functioning (memory, attention, orientation, disturbances in thinking) and psychomotor activity. A cutoff score of 13 is diagnostic of delirium. The MDAS is designed to be administered repeatedly within the same day, in order to allow for objective measurement of changes in delirium severity in response to medical changes or clinical interventions. The MDAS has advantages over other delirium tools in that it is both a diagnostic and a severity measure that is ideal for repeated assessments and for use in treatment intervention trials.

Memory

Definition

Memory refers to the ability to recover information about past events or knowledge, or the process of recovering information about past events or knowledge, as well as cognitive reconstruction. The brain engages in a remarkable reshuffling process in an attempt to extract what is general and what is particular about each passing moment.

Memory may be divided into short-term (also known as working or recent memory) and long-term memory. Short-term memory recovers memories of recent events, while long-term memory is concerned with recalling the more distant past.

Cross-References

▶ Long-Term Potentiation and Memory
▶ Short-Term and Working Memory in Animals
▶ Short-Term and Working Memory in Humans

Memory-Storage Effect

Definition

Memory-storage effect refers to the gradual effect of a drug to alter the long-term memory of an estimated duration of time that is dependent on the translation of the clock reading into memory. In the Peak Internal (PI) procedure, a memory effect is observed as a gradual horizontal shift in the response function in PI trials following drug administration that is proportional to the estimated duration (see Meck 1996).

Cross-References

▶ Timing Behavior

References

Meck WH (1996) Neuropharmacology of timing and time perception. Cogn Brain Res 3:227–242

Meprobamate

Synonyms

Miltown

Definition

Meprobamate is an obsolete sedative with a medium duration of action that resembles the barbiturates used in the treatment of ▶ anxiety. It has largely been replaced by the benzodiazepines. Unwanted effects include sedation, headaches, paradoxical excitement, confusion, cognitive and psychomotor impairment, and confusion in the elderly. Interaction with alcohol can be hazardous. It depresses respiration and is toxic in overdose. Long-term use can induce dependence with severe withdrawal reactions. Recreational use and abuse can occur: it is a scheduled substance.

Cross-References

▶ Barbiturates
▶ Minor Tranquilizer

Meptazinol

Definition

Meptazinol is an opioid analgesic for use in moderate to severe pain. It is most commonly used to treat pain in obstetrics (childbirth). As a partial μ-opioid receptor agonist, its mixed agonist/antagonist activity results in a lower risk of dependence and abuse in comparison to full μ-agonists such as ▶ morphine. Meptazinol exhibits not only a short onset of action but also a shorter duration of action relative to other opioids.

Cross-References

- Addiction
- Analgesics
- Dependence
- Opioids
- Pain and Psychopharmacology
- Tolerance

Mesolimbic System

Definition

Brain pathway that is dominated by dopamine projections branching from the ventral tegmental area to the subcortical ventral striatum (nucleus accumbens) and to the prefrontal cortex.

Mesotelencephalic Dopamine Reward Systems

Definition

The mesotelencephalic dopamine reward system has three components, the nigrostriatal, mesolimbic, and mesocortical pathways, consisting of cell bodies in the substantia nigra and ventral tegmental area that project to a number of regions including the nucleus accumbens, amygdala, striatum, and prefrontal cortex. These areas of the brain are strongly implicated in reward-related learning.

Meta-analysis

Definition

A statistical technique for combining data from independent but methodologically similar studies to answer related hypotheses and to estimate an overall effect across all the studies. Examples of its use include clinical trials and studies of behavioral and psychiatric genetics.

Metabolism

Synonyms

Biotransformation

Definition

In pharmacology metabolism is the irreversible transformation of drugs into chemically different substances that are called metabolites.

Cross-References

- Absorption
- Distribution
- Excretion
- Liberation
- Pharmacokinetics

Metabolomics

Definition

The study of the complete set of small-molecule metabolites, such as metabolic intermediates, hormones, and other signaling molecules, and secondary metabolites, found within a biological sample.

Cross-References

- Electrospray Ionization
- Imaging Mass Spectrometry

- ▶ Mass Spectrometry
- ▶ Matrix-Assisted Laser Desorption Ionization
- ▶ Neuropeptidomics
- ▶ Posttranslational Modification

Metabotropic Glutamate Receptor

Synonyms

mGluR

Definition

G-protein-coupled receptor for which glutamate is the endogenous ligand.

Meta-phenylenediamine

Synonyms

m-PD

Definition

All microelectrode array sites used in neurophysiological studies are electroplated with meta-phenylenediamine (m-PD) by applying a potential of +0.5 V to the Pt sites versus a silver/silver chloride (Ag/AgCl) reference electrode (Bioanalytical Systems, RE-5) in a deoxygenated 0.05 M phosphate buffered saline (PBS, pH 7.1–7.4) with 5.0 mM mPD. The mPD forms a size-exclusion layer over the sites, blocking DA, ascorbic acid (AA), DOPAC, and other electroactive compounds.

Meta-plasticity

Definition

Some forms of neuronal plasticity affect Long Term Potentiation (LTP) and Long Term Depression (LTD), which are already forms of plasticity. Therefore, the term "meta-plasticity" was introduced.

Methadone

Definition

Methadone is an opioid racemate drug that acts as an agonist at the μ opiate receptor. It is the best-studied substance available for opioid maintenance in terms of clinical effectiveness in reducing illicit opioid consumption; reducing high-risk behaviors, such as needle sharing; and increasing rates of treatment retention.

Cross-References

▶ Opioid Use Disorder and Its Treatment

Methamphetamine

Synonyms

Desoxyephedrine; Methylamphetamine; N-Methylamphetamine

Definition

Methamphetamine is a ▶ psychostimulant and sympathomimetic drug that has a blood ▶ half-life of 9–15 h in humans. The primary metabolite of methamphetamine is ▶ amphetamine, a chemical that itself is a potent psychostimulant. Methamphetamine is clinically available for the treatment of obesity, ▶ narcolepsy, and in some cases ADHD. Methamphetamine is a highly potent drug of abuse, with its illicit use reaching epidemic proportions in several Western countries including North American, Asian, and Pacific regions. Chronic exposure to large doses of methamphetamine can lead to schizophrenia-like psychosis and neurotoxic degeneration of dopaminergic neurons.

Cross-References

- ▶ Addiction
- ▶ Adolescence and Responses to Drugs
- ▶ Amphetamine
- ▶ Attention-Deficit/Hyperactivity Disorder
- ▶ Dependence
- ▶ Dopamine
- ▶ Half-Life
- ▶ Narcolepsy
- ▶ Neurotoxicity
- ▶ Sensitization to Drugs

Methergoline

Synonyms

Metergoline; Metergoline phenylmethyl ester; ((8beta)-1,6-Dimethylergolin-8-yl)methyl]carbamic acid phenylmethyl ester

Definition

Methergoline is a synthetic compound that acts as a nonselective ▶ partial agonist at serotonin (5-hydroxytryptamine) receptors. It is potent at some subtypes of both 5-HT$_1$ and 5-HT$_2$ receptors but has little action at 5-HT$_3$ receptors. It also acts as a ▶ dopamine agonist.

Cross-References

▶ Drug Discrimination

Methylenedioxymethamphetamine (MDMA)

Iain S. McGregor[1], Paul D. Callaghan[2] and Murray R. Thompson[1]
[1]School of Psychology, University of Sydney, Sydney, NSW, Australia
[2]ANSTO LifeSciences, Australian Nuclear Science and Technology Organisation (ANSTO), Kirrawee DC, NSW, Australia

Synonyms

3,4-Methylenedioxymethamphetamine; E; Eccie; Ecstasy; Hug drug; Love drug; Love hormone; XTC

Definition

MDMA is a popular recreational drug that is renowned for its ability to produce euphoria and unique prosocial effects. It is the best known and most commonly used member of the family of phenythylamines (substitutes for amphetamines) that are sometimes called entactogens, empathogens, or the MDxx class of drugs. MDMA has multiple neurochemical effects, the most prominent of which is to promote the release of serotonin via an action on the serotonin transporter (SERT). The prosocial effects of MDMA have recently been linked to the release of the neuropeptide oxytocin. High doses of MDMA can cause long-term depletion of serotonin in the brains of laboratory animals, but whether this also occurs in humans and whether this leads to associated psychopathology such as depression and cognitive impairment remain unclear.

Pharmacological Properties

History

MDMA was first synthesized in 1912 by the company E. Merck. Although commonly thought to have been designed as an appetite suppressant, the original patent bears no record of this and simply states that MDMA was deemed to contain primary constituents for therapeutically active compounds. The first reported pharmacological study involving MDMA occurred in 1927 although basic toxicology studies were not undertaken until the 1950s. Further studies at the University of Michigan, supported by the US Army, reported LD_{50} values for five different species, with the lowest LD_{50} value found in dogs and the highest in mice.

The first systematic use of MDMA was as an adjunct to insight-oriented psychotherapy, with administration of MDMA producing an easily controllable altered state of consciousness with positive emotional and sensual overtones. The colloquial term for MDMA changed from "Empathy" as was used by therapists in the 1970s to "Ecstasy," emphasizing the drug's euphoric effects. Heavy media attention in 1985 sensationalized Ecstasy's euphoric effects and caused a surge in recreational use.

In 1986, MDMA became a schedule 1 drug in the United States, deemed not to possess any recognized therapeutic value. By the 1990s, Ecstasy had become strongly linked to the club and rave culture, with its use by groups of young people attending all-night dance parties where vigorous dancing occurred to highly repetitive and hypnotic "techno" music. The popularity of MDMA has continued to grow to the point where it is now one of the most widely used illicit drugs in the world.

Mechanisms of Action

MDMA is a ring-substituted amphetamine, with a methylenedioxy group attached to the aromatic ring of amphetamine. It has multiple, complex pharmacological actions. The most important property is to potently release serotonin (5-HT) from axon terminals into the synapse and to inhibit 5-HT reuptake. To a lesser extent, MDMA also releases dopamine, noradrenaline, and acetylcholine. MDMA reverses the action of the SERT causing 5-HT stores from the neuron to be pumped into the synapse. An additional related action is to block the reuptake of 5-HT ("▶ Uptake"), which further increases synaptic 5-HT concentrations. Pretreatment with SERT ligands including SSRIs (e.g., fluoxetine) prevents MDMA-induced 5-HT release in brain slices and in vivo. In addition to these effects on 5-HT efflux, MDMA-mediated inhibition of monoamine oxidase prevents the breakdown of 5-HT and other neurotransmitters such as dopamine, further contributing to elevated monoamine levels. A further effect of MDMA is to inhibit tryptophan hydroxylase (the rate-limiting enzyme for 5-HT synthesis). This effect may contribute to depletion of 5-HT stores in the days following MDMA use.

MDMA possesses two stereoisomers: $(-)$-MDMA has a higher affinity for postsynaptic 5-HT receptors while $(+)$-MDMA has a higher affinity for the SERT. The two isomers differ in their behavioral effects in rhesus monkeys and subjective effects in humans. These two isomers also differ in the rate in which they are metabolized across individuals which may result in large interindividual differences in the overall response to MDMA.

MDMA also binds to various 5-HT receptors with moderate to high affinity. Receptor-binding studies indicate that MDMA possesses a high affinity for the 5-HT_2 family of receptors and a moderate affinity for 5-HT_1-type receptors. Activation of 5-HT_{1A} receptors largely acts to inhibit serotonergic cell firing although the resultant inhibitory effects on 5-HT release are overridden through MDMA-induced effects at the SERT in forebrain regions.

MDMA acts to increase synaptic dopamine levels, but these increases are generally smaller than the increases in 5-HT in any given region. MDMA-induced dopamine release may involve both an indirect 5-HT_{2A} receptor mechanism and a direct action on the dopamine transporter (DAT). Dopamine levels are also augmented by an action of MDMA on the vesicular monoamine transporter 2 (VMAT2) causing dopamine efflux from vesicular stores via carrier-mediated exchange.

MDMA also causes a significant release of norepinephrine via an interaction with the norepinephrine transporter (NET). Acetylcholine release also occurs in the prefrontal cortex and dorsal hippocampus following MDMA, and there is also the involvement of GABA, glutamate, nitrergic, and sigma (σ1) systems. MDMA causes major endocrine changes ("▶ Neuroendocrine Markers for Drug Action") including an increase in plasma oxytocin, vasopressin, cortisol, and prolactin.

Pharmacokinetics

Human pharmacokinetic studies show that MDMA's distinctive pharmacodynamic effects occur at doses of 1 mg/kg or above with peak MDMA serum concentrations observed 2 h post-administration, coinciding with peak psychological effects. MDMA has nonlinear pharmacokinetics, with increasing doses resulting in unpredictable tissue/blood concentrations of parent drug (de la Torre et al. 2012). Like most other psychoactive drugs, MDMA is primarily metabolized by the liver via cytochrome P450 family of enzymes, with the CYP2D6 isozyme involved in the primary pathway (of which poor, normal, and extensive metabolizers are seen in the general population due to genetic polymorphisms). MDMA has a very complex metabolic pathway in comparison to other amphetamine analogs, and this may explain the sometimes complex and unpredictable relationship between Ecstasy tablet intake and acute effects of the drug. Single doses of MDMA inhibit CYP2D6 function within 2 h in humans, for a period of up to 10 days, converting the subject to poor metabolizer status (also affecting other pharmaceuticals metabolized by this pathway). This is relevant to understanding adverse effects as the primary metabolites of MDMA in humans, HHMA and HMMA, which are readily broken down in the body to orthoquinones, highly reactive compounds that may lead to free radical-induced brain injury. Recent studies indicate that genetic polymorphisms of the 5-HT transporter and catechol-O-methyltransferase also contribute to the pharmacokinetics and adverse pharmacodynamic effects of MDMA in humans (Pardo-Lozano et al. 2012).

Neurotoxicity

Exposure to relatively high doses of MDMA can cause a long-lasting reduction in brain monoamine levels in a variety of animal species. While rats and primates show a primary reduction in brain 5-HT, mice show primary reductions in brain dopamine. Reductions in SERT density in the cortical, limbic, and striatal regions have also been reported in many studies with rats and primates. Given that the SERT protein is primarily located in 5-HT axons, MDMA-induced axotomy has been invoked as the primary reason for this effect and has been confirmed in some histological studies. Abnormal 5-HT axonal immunoreactivity has been seen in primates 7 years post-MDMA treatment. However, these patterns of findings do not confirm a neurotoxic effect in the classic sense. Gliosis is not typically observed following MDMA administration, nor is there any damage to serotonergic cell bodies. The widely discussed notion of MDMA-induced neurotoxicity therefore remains controversial (Baumann et al. 2007).

In addition to global SERT changes, alterations in the density of specific 5-HT receptor subpopulations can be seen following MDMA. Significant reductions in 5-HT$_{2A}$ receptor density in the cortical, striatal, thalamic, and hypothalamic regions have been reported in rats months after MDMA treatment, although opposite findings on 5-HT$_{2A}$ receptor density have been reported in some human studies. 5-HT$_{1B}$ receptor density was reduced in MDMA-treated rats in the globus pallidus, hippocampus, and medial thalamus but increased in the nucleus accumbens and lateral septum.

High ambient temperatures at the time of dosing, typical of the dance parties where MDMA is often taken, may exacerbate MDMA-induced 5-HT depletion. A neuroprotective effect of coadministered drugs (e.g., haloperidol, ketanserin, pentobarbitone, and various antioxidants) may result from an induction of hypothermia or by preventing the hyperthermic effects of MDMA. However, some drugs (e.g., cannabinoids) are protective independently of their body temperature effects.

Some human studies have found that Ecstasy users differ from controls on a range of measures related to 5-HT, including a reduction of cerebrospinal 5-HIAA levels and a blunted neuroendocrine responses to serotonergic ligands. Decreased global and regional SERT density in Ecstasy users has been reported in some PET imaging and SPECT imaging studies although these are generally modest effects and may recover with abstinence from the drug.

Positive Effects in Humans

MDMA induces a positive mood state in humans along with increased energy and euphoria, typical of amphetamine and its derivatives. However, MDMA users also report a unique sense of intimacy and empathy coupled with an increased feeling of closeness to others ("▶ Entactogen") that is not always typical of amphetamine ("▶ Social Behavior"). In addition, MDMA users also report mild hallucinogen-like enhancement of perceptions and sensations with augmented responses to touch and music. Unlike amphetamines, MDMA appears to have relatively low abuse potential in humans, perhaps due to rapid tolerance developing to the positive effects with repeated use.

SSRIs attenuate many of the acute psychological effects of MDMA in humans, consistent with a primary action of MDMA on SERT. SSRIs also reduce MDMA-induced heart-rate changes. Other studies showed that MDMA-induced perceptual changes and emotional excitation are partially mediated by postsynaptic 5-HT$_{2A}$ receptors since these effects can be attenuated by ketanserin (Liechti and Vollenweider 2001).

The positive acute effects of MDMA in humans may involve other neurochemical systems. Thus, the antipsychotic drug haloperidol partially antagonized the positive and mania-like mood states induced by MDMA. In a recent laboratory study, the increased feeling of sociability after MDMA was associated with increased plasma levels of oxytocin in human subjects (Dumont et al. 2009).

Effects in Laboratory Animals

The acute effects of MDMA have been investigated in a diverse range of laboratory animal species. A key consideration in utilizing animal models is in establishing appropriate species-equivalent dosing levels to model human MDMA use (Green et al. 2009). This issue is still far from resolved with many animal studies using MDMA dose regimes that are in the extreme range. Species-specific pharmacokinetics also complicates the picture.

MDMA has amphetamine-like sympathomimetic effects, increasing blood pressure and heart rate. It exerts a powerful influence on body temperature, with the direction of change (hyperthermia or hypothermia) dependent upon the ambient temperature of the environment (Green et al. 2003). The hyperthermic response to MDMA appears in part to be reliant upon the mitochondrial uncoupling protein 3 (UCP-3) acting in striated myocytes. MDMA also produces peripheral vasoconstriction, further preventing heat loss.

Behaviorally, MDMA causes amphetamine-like hyperactivity and locomotor sensitization in rodent species. Intravenous self-administration of MDMA is seen in mice, rats, and nonhuman primates although rates are significantly less than that of other abused drugs such as the psychostimulants cocaine and methamphetamine. Self-administration of MDMA in rats is increased at high ambient temperatures, and this may be in part due to augmentation of MDMA-stimulated increases in dopamine and neuronal activation in reward-relevant brain regions. Rats will also show a conditioned place preference to MDMA, an effect that involves dopamine, opioid, and endocannabinoid systems.

In line with its characteristic prosocial effects in humans, MDMA reduces aggression and increases social interaction in rodents. In the social interaction test, rats spend increased times in adjacent contact following acute MDMA treatment, and this effect is also augmented at high ambient temperatures. This prosocial effect of MDMA is reduced by oxytocin antagonists and is mimicked by the 5-HT$_{1A}$ agonist 8-OH-DPAT (McGregor et al. 2008).

MDMA-Associated Hazards and Psychopathology

Hyperthermia and other components of the serotonin syndrome are the main acute hazards facing human MDMA users, particularly when the drug is taken in high doses and in hot environments. Despite considerable media attention, lethal effects of MDMA (taken alone) appear comparatively rare. However, combining MDMA with other serotonergic drugs (e.g., monoamine oxidase inhibitors) can be extremely dangerous due to the possibility of serotonin syndrome. Other problems for users relate to the fact that Ecstasy tablets do not always contain MDMA, with a wide range of adulterants reported in analytical studies.

Acute adverse psychological effects are occasionally reported with MDMA, most commonly anxiety and paranoia. A greater research focus, however, has been on possible lasting adverse psychological effects of MDMA use, effects that might be associated with serotonin depletion. In various studies, Ecstasy use has been linked to anxiety, depression, and mild cognitive impairment. However, many of these studies have inherent methodological problems. For example, Ecstasy users typically use other substances and coincident heavy cannabis use is a particularly troublesome confound in studies probing cognitive impairment after MDMA. There is also evidence that people with a preexisting childhood tendency toward anxiety and depression are more likely to become Ecstasy users, providing an additional confound. There is therefore a need for prospective longitudinal studies to control for premorbid psychiatric and cognitive problems in assessing MDMA-related harms. An example of this is the recent Netherlands XTC Toxicity (NeXT) study. This has uncovered subtle abnormalities in brain function in a sample of young persons taking MDMA for the first few times (de Win et al. 2008).

Preclinical studies are also important in addressing the issue of whether MDMA exposure has lasting adverse consequences. Consistent, lasting adverse effects have been reported in a number of behavioral tests in rodents pretreated with MDMA. These include increased anxiety as assessed in the emergence test and the elevated plus maze, increased depressive-like symptoms in the Porsolt test ("▶ Depression: Animal Models"), and impaired novel object recognition and spatial memory. The social interaction test has been found to be particularly sensitive to detecting lasting adverse effects of MDMA in rodents, with decreased social behavior detected even months after low-dose MDMA exposure. Many of the above long-term effects are seen with low-dose regimes of MDMA that do not deplete brain 5-HT. As yet, unspecified neuroadaptations in nonserotonergic brain systems may therefore underlie these lasting adverse effects (McGregor et al. 2008).

Therapeutic Uses

Despite concerns relating to the neurotoxicity and possible psychopathology associated with MDMA use, a number of reputable scientists have called for further study of the use of MDMA as a therapeutic for anxiety and depression and relationship issues. This marks something of a return to the original use of MDMA as a tool for assisting interpersonal psychotherapy in the 1970s and 1980s. The Multidisciplinary Association for Psychedelic Studies (MAPS) (http://www.maps.org/mdma/) is currently sponsoring small clinical trials of MDMA in several countries for the treatment of traumatic stress disorder and is also sponsoring a study of MDMA for alleviation of anxiety linked to terminal cancer.

Conclusions

MDMA is a controversial drug with a unique and complex pharmacology. No other drug, with the possible exception of GHB ("▶ Sodium Oxybate"), has the capacity to produce such marked facilitatory effects on social behavior in humans and other animal species. It is therefore encouraging to see that recent psychopharmacological studies of MDMA have started to focus on the positive prosocial effects of the drug in humans (Bedi et al. 2009; Dumont et al. 2009) and not just on its possible adverse effects and neurotoxicity.

Despite a plethora of human and animal studies spanning more than two decades, experts cannot appear to reach a consensus on the relative harms associated with MDMA use: some claim MDMA is largely innocuous (Nutt 2009) while others proclaim its dangers (Parrott 2002). Fortunately, our overall knowledge of MDMA psychopharmacology continues to grow, as research studies involving both human Ecstasy users and laboratory animals given MDMA evolve in their sophistication, scope, and power. Perhaps given another decade of research, a greater consensus will emerge, and we will understand not only how MDMA acts in the brain to produce "chemical love" but also whether this is a good or a bad thing for the health of the individual.

Cross-References

▶ Amphetamine
▶ Anxiety: Animal Models
▶ Depression
▶ Dopamine Transporter
▶ Hypothermia
▶ Neurotoxicity
▶ Serotonin Syndrome
▶ Serotonin Transporter
▶ Social Behavior

References

Baumann MH, Wang X, Rothman RB (2007) 3, 4-Methylenedioxymethamphetamine (MDMA) neurotoxicity in rats: a reappraisal of past and present findings. Psychopharmacology (Berl) 189:407–424

Bedi G, Phan KL, Angstadt M, de Wit H (2009) Effects of MDMA on sociability and neural response to social threat and social reward. Psychopharmacology (Berl) 207:73–83

de la Torre R, Yubero-Lahoz S, Pardo-Lozano R, Farré M (2012) MDMA, Methamphetamine and CYP2D6 pharmacogenetics: what is clinically relevant? Front Genet 235(3):1–8

de Win MM, Jager G, Booij J, Reneman L, Schilt T, Lavini C, Olabarriaga SD, den Heeten GJ, van den Brink W (2008) Sustained effects of ecstasy on the human brain: a prospective neuroimaging study in novel users. Brain 131:2936–2945

Dumont GJ, Sweep FC, van der Steen R, Hermsen R, Donders AR, Touw DJ, van Gerven JM, Buitelaar JK, Verkes RJ (2009) Increased oxytocin concentrations and prosocial feelings in humans after ecstasy (3, 4-methylenedioxymethamphetamine) administration. Soc Neurosci 4:359–366

Green AR, Mechan AO, Elliott JM, O'Shea E, Colado MI (2003) The pharmacology and clinical pharmacology of 3, 4-methylenedioxymethamphetamine (MDMA, "ecstasy"). Pharmacol Rev 55:463–508

Green AR, Gabrielsson J, Marsden CA, Fone KC (2009) MDMA: on the translation from rodent to human dosing. Psychopharmacology (Berl) 204:375–378

Liechti ME, Vollenweider FX (2001) Which neuroreceptors mediate the subjective effects of MDMA in humans? A summary of mechanistic studies. Hum Psychopharmacol 16:589–598

McGregor IS, Callaghan PD, Hunt GE (2008) From ultrasocial to antisocial: a role for oxytocin in the acute reinforcing effects and long-term adverse consequences of drug use? Br J Pharmacol 154:358–368

Nutt DJ (2009) Equasy – an overlooked addiction with implications for the current debate on drug harms. J Psychopharmacol 23:3–5

Pardo-Lozano R, Farré M, Yubero-Lahoz S, O'Mathúna B, Torrens M, Mustata C, Pérez-Mañá C, Langohr K, Cuyàs E, Carbó M, de la Torre R (2012) Clinical pharmacology of 3,4-Methylenedioxymethamphetamine (MDMA, "Ecstasy"): the influence of gender and genetics (CYP2D6, COMT, 5-HTT). PLoS One 7(10): e47599

Parrott AC (2002) Recreational ecstasy/MDMA, the serotonin syndrome, and serotonergic neurotoxicity. Pharmacol Biochem Behav 71:837–844

Methylphenidate and Related Compounds

Behrouz Namdari[1] and Scott H. Kollins[2]
[1]Department of Psychiatry & Behavioral Sciences, Duke University School of Medicine, Durham, NC, USA
[2]Department of Psychiatry & Behavioral Sciences, Department of Psychology & Neuroscience, Duke University School of Medicine, Durham, NC, USA

Synonyms

Dexmethylphenidate; Dex-MPH; Dextromethylphenidate; D-Methylphenidate; D-MPH; Methylphenidate; MPH; Ritalin

Definition

Methylphenidate and related compounds are psychostimulant medications that were initially created to treat barbiturate-induced comas and which are now used mainly for the treatment of attention-deficit hyperactivity disorder.

Pharmacological Properties

Background

Methylphenidate received its first Food and Drug Administration (FDA) indication for hyperactivity and was patented in 1955 by Ciba pharmaceuticals (which later was integrated into the Novartis Corporation). Currently, MPH is indicated for the treatment of attention-deficit hyperactivity disorder (ADHD) for adults, adolescents, and children (above the age of five) as well as for narcolepsy (Novartis 2013). MPH is most commonly prescribed for the treatment of ADHD, and it is the most commonly prescribed medication for this illness. MPH exhibits robust effect sizes up to 1.3 compared to placebo and is found to reduce the core symptoms of ADHD in approximately 70 % of child, adolescent, and adult patients to whom it is prescribed. MPH also has shown positive results in many studies of off-label uses including treatment of refractory depression, cancer-related pain, neurocognitive deficits associated with pediatric cancer survival, HIV-related cognitive impairment, and traumatic brain injury. Since its introduction, MPH prescriptions have been steadily increasing with a rapid incline in the 1990s as ADHD awareness and public acceptance increased. Recently, there has been growing concern over the recreational use and diversion of MPH particularly among college students (Challman and Lipsky 2000; Volkow et al. 2002). This entry provides an overview of MPH, including a review of mechanism of action, physiological effects, and potential for abuse.

Chemistry

Methylphenidate is a piperidine-derived molecule which possesses two points of chirality. The original formulation of MPH included the optical isomers D, L threo-methylphenidate and D, L erythro-methylphenidate. The erythro-methylphenidate isomers were later linked to cardiac side effects and were found to lack any central nervous system activity. As such, erythro-methylphenidate was subsequently eliminated from all available preparations. Of the two threo isomers, the d isomer was found to be the chirally active molecule in humans. Consequently, pure D-threo-methylphenidate (dexmethylphenidate or D-methylphenidate) is twice as potent as racemic MPH and can be taken at half the dose. Both racemic and pure D-threo-methylphenidate preparations are available (Ritalin and Focalin, respectively), but the lower-dose requirements of pure D-threo-methylphenidate may theoretically result in less drug-drug interactions and metabolism requirements (Challman and Lipsky 2000).

Pharmacokinetics

Oral methylphenidate is absorbed through the gut and transdermal MPH is absorbed through the skin. The different preparations have varying half-lives which are directly correlated with the speed of their dispersion from its vehicle, while peak plasma concentration is inversely correlated with the speed of drug release. MPH is approximately 15 % plasma protein bound. In racemic formulations, the inactive L-enantiomer has minimal effect on the pharmacokinetics of the active d enantiomer (Capp et al. 2005).

Metabolism

Methylphenidate is metabolized primarily by hepatic (and peripheral) esterases and not through oxidative pathways (p450 system). MPH is rapidly excreted through the urine after de-esterification with less than 10 % cleared unchanged. After 48 h, approximately 90 % of a single dose of radiolabeled MPH is cleared from the body (Capp et al. 2005). Plasma levels of MPH are similar between the genders after taking the same dose although plasma levels of the inactive metabolites are higher in women. Currently, there is no information regarding methylphenidate dose modifications for individuals with renal or hepatic impairment. Methylphenidate can inhibit the metabolism of Coumadin, anticonvulsants, and tricyclic antidepressants. Precaution and dose adjustments of these medications should be considered when MPH is initiated, discontinued, or titrated (Novartis 2013).

Mechanism of Action

Methylphenidate is a central nervous system stimulant, and its full mechanism of action is currently unknown. On a synaptic level, studies have shown that MPH blocks the dopamine (DA) and the norepinephrine (NE) transporters (along with weakly blocking the serotonin transporter). These transporters run along the presynaptic neuron's axonal cell membrane and are responsible for the reuptake of the neurotransmitters back into the presynaptic neuron for re-storage or degradation. When it is taken, MPH reduces the clearance of DA and NE in the synaptic cleft which allows the neurotransmitters more time to interact with postsynaptic receptors (D1, D4, and Alpha 2A receptors). Ultimately, the presynaptic neuron is able to impart its DA and NE "message" at a higher intensity when MPH is present (Challman and Lipsky 2000). These "messages" are believed to enhance task-specific neuronal signaling and suppress background noise. Thus, MPH leads to improved attention to salient, task-related stimuli and reduces distractibility from other sources (Volkow et al. 2002).

Location of Action

Rat studies have shown methylphenidate activity in the striatum, nucleus accumbens, olfactory tubercle, and prefrontal cortex. In humans, microdialysis studies have shown in vivo increases of intra-synaptic dopamine in the striatum and nucleus accumbens when MPH is taken (Froehlich et al. 2010). Positron emission tomography (PET) studies have also shown that the highest concentration of MPH can be found in the striatum, and, at therapeutic dosages, MPH occupies more than half of the brain's dopamine transporter (Volkow et al. 2002).

Multi-neurotransmitter Hypothesis

Given its direct effects on dopamine neurotransmission and heavy activity in DA-rich brain structures, DA was originally implicated as the neurotransmitter responsible for MPH's effect. The DA hypothesis was further reinforced by genetic studies which indicated differential response to MPH based on allelic variation in genes coding for several receptors involved in dopamine neurotransmission. To evaluate part of the dopamine hypothesis, knockout mice were developed that lacked the gene for the DA transporter. These mice showed higher baseline activity and learning delays which was expected if DA impairment is responsible for ADHD symptoms. However, when the knockout mice were given MPH, they showed a decrease in hyperactivity which was perplexing given the absence of the DAT where MPH was presumed to have its effect. This study, along with others, led to the conclusion that there are likely other targets responsible for MPH's central nervous system activity. The involvement of norepinephrine and serotonin was established following additional pharmacological and genetic studies (Froehlich et al. 2010). Another possible mechanism of action for MPH's central nervous system effects is a modification of brain-derived neurotrophic factor (BDNF) levels. A recent small noncontrolled study measured BDNF levels before treatment and after 6 weeks of treatment with MPH. The study showed significant increases in BDNF after 6 weeks of treatment when compared to the baseline measure which correlated with improvements in ADHD symptoms (Amiri et al. 2013).

Abuse

Like cocaine (a commonly abused psychostimulant), methylphenidate has the potential for euphoria and reinforcement. Euphoria can be produced when MPH's concentration is high enough to occupy greater than 60 % of the dopamine transporter. Interestingly, therapeutic dosages of MPH often reach this mark without any reinforcing effects. It is believed that oral MPH does not routinely produce euphoria because of its slow rate of absorption which mimics tonic DA neuron firing. On the other hand, intravenous (IV) or intranasal (IN) MPH absorption is quick and bypasses first-pass metabolism yielding a rapid peak in concentration. Consequently, when MPH is taken as IV or IN, its concentration rise mimics rapid (phasic) shifts in DA neuron firing which can cause euphoria and lead to abuse. Available oral preparation of MPH can be dissolved into a liquid and taken IV or crushed into a powder and taken IN. This poses difficulty when prescribing MPH since it could be readily taken recreationally or diverted (Swanson and Volkow 2003).

Physiological Effects

Central Nervous System

The quality and intensity of methylphenidate's effect differ between individuals. Generally, MPH's central nervous system (CNS) benefits include a decreased sense of fatigue, a mild euphoria, and an increase in mental alertness. Conversely, MPH use may result in worsening of anxiety, tension, and agitation. MPH use may lead to treatment emergent psychotic/manic symptoms or may switch someone with bipolar disorder to into a manic/hypomanic state. Additionally, individuals with motor tics or Tourette's disorder may have worsening of their involuntary movements, while others may develop them. MPH may also reduce the seizure threshold which is particularly worrisome in someone with epilepsy or other seizure disorders. Headache (16–53 %), insomnia (13.4–46 %), and diminished appetite (10–41 %) are the most frequently experienced negative CNS effects of MPH. Often, these side effects resolve as plasma levels of the drug decrease or if the medication is discontinued altogether (Godfrey 2009; Novartis 2013).

Peripheral

Cardiovascular Although methylphenidate is regarded as a psychostimulant, it can produce similarly activating effects peripherally by inducing the sympathetic nervous system. As such, there is a concern for sudden death, stroke, cardiac infarction, and arrhythmias (problematic heart rate or rhythm) in individuals with known cardiac disease (including conduction defects, cardiomyopathy, coronary heart disease, and heart rhythm abnormalities). Methylphenidate can also affect the circulatory system by increasing the heart rate by about 3–6 beats per minute and blood pressure by about 2–4 mmHG. These elevations are unlikely to be clinically significant in the short term but may yield substantial consequence if left untreated or in those with preexisting cardiac disease. Some individuals may have dramatic changes in their cardiac measures when taking MPH which may require discontinuation or additional treatment. Heart rate and blood pressures alterations often can be treated successfully with antihypertensive medications. Methylphenidate should not be prescribed with monoamine oxidase inhibitors or pressor against due to its effects on blood pressure (Godfrey 2009).

Growth Methylphenidate has been linked to growth suppression in children. It is unclear why this occurs. On average, children (ages 7–10) who took MPH regularly (7 days a week throughout the year) were 2 cm shorter and 2.7 kg lighter than non-medicated controls in one three-year study. Data is currently mixed if children regain these deficits when the medications are discontinued (Faraone et al. 2008).

Peripheral Vasculature Methylphenidate is associated with sympathetic-driven peripheral vasculature vasoconstriction. This can lead to varying degrees of painful and/or cold fingers and toes. In some, the peripheral vasculopathy can precipitate Raynaud's phenomenon or skin ulceration (Novartis 2013).

Vision Methylphenidate can lead to sympathetic changes in vision. Some individuals can have dry eyes, blurred vision, mydriasis, or difficulties with visual accommodation. Methylphenidate can also worsen or cause glaucoma (Novartis 2013).

Pregnancy and Lactation Methylphenidate has a pregnancy category of C. Several rat and rabbit pregnancy studies using doses larger than those equivalent to human therapeutic doses showed the potential for teratogenic effects. Rabbit newborns had an increased rate of spina bifida when pregnant mothers were given 40 times the maximum recommended human dose (MRHD), and rat newborns had skeletal variations when pregnant mothers were given 7 times the MRHD. In a separate rat study, offspring of mothers who were given 4 times the MRHD during pregnancy and lactation had decreased body weights. No formal studies have been performed in humans in pregnancy, and it is unknown if methylphenidate is excreted in human milk. Consequently, precaution is advised during these time periods (Novartis 2013).

Methylphenidate Versus Amphetamine

Amphetamine and related molecules is the second most commonly prescribed psychostimulant for ADHD after methylphenidate. It too blocks the dopamine and norepinephrine transporters (as well as weakly blocking serotonin transporter) resulting in an increase in neurotransmitters in the synaptic cleft. Amphetamine differs from MPH in that it also binds to reserpine-sensitive vesicles in the presynaptic cell which allows dopamine to leak out into the synapse. Consequently, amphetamine's effect is partially independent from presynaptic neuronal firing, while MPH's effect is not. In general, racemic amphetamine has twice the potency compared to racemic MPH and requires half the milligram equivalents when prescribed (Challman and Lipsky 2000). In placebo-controlled trials for ADHD, the effect size for amphetamine is comparable to MPH, and head-to-head studies have found no significant difference among the two in efficacy (Brown et al. 2005). Amphetamine carries a similar concern for diversion and abuse as methylphenidate.

Cross-References

▶ Action Potentials
▶ Amphetamine
▶ Attention-Deficit and Disruptive Behavior Disorders
▶ Dopamine
▶ Narcolepsy
▶ Norepinephrine
▶ Serotonin
▶ Substance Abuse

References

Amiri A, Parizi GT, Kousha M, Saadat F, Modabbernia MJ, Najafi K, Roushan ZA (2013) Changes in plasma BDNF levels induced by methylphenidate in children with ADHD. Prog Neuropsychopharmacol Biol Psychiatry. doi:10.1016/j.pnpbp.2013.07.018

Brown RT, Amler RW, Freeman WS, Perrin JM, Stein MT, Feldman HM, Wolraich ML (2005) Treatment of attention-deficit/hyperactivity disorder: overview of the evidence. Pediatrics 115(6):e749–e757. doi:10.1542/peds.2004-2560

Capp PK, Pearl PL, Conlon C (2005) Methylphenidate HCl: therapy for attention deficit hyperactivity disorder. Expert Rev Neurother 5(3):325–331. doi:10.1586/14737175.5.3.325

Challman TD, Lipsky JJ (2000) Methylphenidate: its pharmacology and uses. Mayo Clin Proc 75(7):711–721. doi:10.4065/75.7.711

Faraone SV, Biederman J, Morley CP, Spencer TJ (2008) Effect of stimulants on height and weight: a review of the literature. J Am Acad Child Adolesc Psychiatry 47(9):994–1009. doi:10.1097/CHI.0b013e31817eOea7

Froehlich TE, McGough JJ, Stein MA (2010) Progress and promise of attention-deficit hyperactivity disorder pharmacogenetics. CNS Drugs 24(2):99–117. doi:10.2165/11530290-000000000-00000

Godfrey J (2009) Safety of therapeutic methylphenidate in adults: a systematic review of the evidence. J Psychopharmacol 23(2):194–205. doi:10.1177/0269881108089809

Novartis (2013) Ritalin hydrochloride package insert. East Hanover

Swanson JM, Volkow ND (2003) Serum and brain concentrations of methylphenidate: implications for use and abuse. Neurosci Biobehav Rev 27(7):615–621

Volkow ND, Fowler JS, Wang GJ, Ding YS, Gatley SJ (2002) Role of dopamine in the therapeutic and reinforcing effects of methylphenidate in humans: results from imaging studies. Eur Neuropsychopharmacol 12(6):557–566

Mexazolam

Definition

Mexazolam is a benzodiazepine derivative that has anxiolytic, anticonvulsant, hypnotic, sedative, amnesic, and muscle-relaxant properties.

Cross-References

▶ Anxiolytics
▶ Benzodiazepines

Mianserin

Definition

Mianserin is a tetracyclic second-generation antidepressant with combined serotonergic-noradrenergic mechanism of action. It increases serotonergic (5-HT) and noradrenergic (NA) neurotransmission by acting as an antagonist mainly at 5-HT$_2$ and α2 presynaptic and somatodendritic auto- and hetero-receptors. This drug also has a strong antihistaminic effect, but, unlike the ▶ tricyclic antidepressants, it has almost no anticholinergic and cardiotoxic properties. In addition to its antidepressant effects, mianserin also has anxiolytic, sedative-hypnotic, antiemetic, and appetite-enhancing effects. Clinical effects of mianserin usually become noticeable after 1–3 weeks of treatment. Common side effects include dizziness, blurred vision, drowsiness, weight gain, dry mouth, and constipation, while more serious adverse reactions may include hypomania, fainting, seizures, and hematological problems. As with other antidepressants, abrupt or rapid discontinuation of mianserin therapy may induce withdrawal effects, such as rebound depression, anxiety, panic attacks, anorexia, and insomnia.

Cross-References

▶ SNRI Antidepressants

Microdialysis

Albert Adell and Francesc Artigas
Department of Neurochemistry and Neuropharmacology, Instituto de Investigaciones Biomédicas de Barcelona, Consejo Superior de Investigaciones Científicas (CSIC), Barcelona, Spain
Institut d'Investigacions Biomèdiques August Pi i Sunyer (IDIBAPS)
Centro de Investigación Biomédica en Red de Salud Mental (CIBERSAM), Spain

Synonyms

Brain microdialysis; Intracerebral microdialysis

Definition

Brain microdialysis is a sampling technique developed to study the concentration of chemicals (mainly neurotransmitters and their metabolites) in the extracellular compartment of the brain by means of implanting a small tubing equipped with a dialysis membrane. During the last decades, the necessity to measure the release of neurotransmitters in vivo in the central nervous system (CNS) has prompted the development of innovative techniques for sampling the extracellular fluid in the brain of experimental animals. Historically, one of the methods that evolved for this purpose was the push-pull perfusion which involved the stereotaxic insertion of a push-pull cannula into a selected area of the brain. Being an open flow system, push-pull perfusion allowed a direct contact of perfusion fluid with brain tissue, which often caused tissue damage, microbial and blood contamination, etc. To circumvent such drawbacks, a semipermeable membrane was attached to the cannula tip, and this device was called dialysis bag or dialytrode. This was soon replaced by a more straightforward approach, the intracerebral dialysis, in which the dialysis bag was substituted by a hollow fiber, the dialysis membrane (Ungerstedt 1984).

Principles and Role in Psychopharmacology

The term dialysis refers to the passage of small molecules through a semipermeable membrane, a process driven by a concentration gradient. The endogenous substances diffuse out of the extracellular fluid into the perfusion medium. In the same way, exogenous compounds can be infused locally through the dialysis probe and reach the brain compartment. Unlike push-pull perfusion, dialysis is based on a closed flow system. Therefore, only a single perfusion pump is needed. The probe is constantly perfused with a physiological solution at a low flow rate (typically <2 µL/min), and perfusate samples are then collected for further analysis. Due to its relative ease of use, microdialysis has become the technique of choice for the in vivo analysis of neurotransmitters in the extracellular compartment of the brain of experimental animals. Its use has been fundamental in the identification of the mechanism of action of numerous psychoactive drugs. In particular, microdialysis has enabled to clarify neuronal elements (neurotransmitters, receptors) and brain networks affected by two of the most important drug classes in psychiatry: antidepressants and antipsychotics (Artigas and Adell 2007).

The low concentration of endogenous neurotransmitters in the extracellular brain space has been one of the main difficulties associated with the dialysis technique. The development of highly sensitive high-performance liquid chromatographic (HPLC) methods has made possible the increasing use of the microdialysis technique for the in vivo analysis of nanomolar concentrations of neurotransmitters and their metabolites (usually at higher concentrations) in the brain. Capillary electrophoresis has also been successfully applied to the analysis of amino acids and amines. However, the need to sample for relatively long periods of time (typically >10 min) is one of the main drawbacks of microdialysis, compared with other in vivo techniques assessing brain function, such as electrophysiology and electrochemical techniques.

The dialysis membrane constitutes a real barrier between the perfusion fluid and the interstitial brain space, which usually excludes the transport of large molecules that may interfere with the substances of interest in the analytical procedure. Furthermore, enzymes that could cause a breakdown of the neuroactive compounds are also prevented from being picked up by the dialysate fluid. The implant and functioning of microdialysis probes may cause tissue reactions ranging from an excessive washout of neurotransmitters and metabolites if flow rates are too high to glial reaction surrounding the probe that may act as an actual barrier for the passage of components from the extracellular brain space to the inner part of the microdialysis probe. These aspects need to be examined in detail while establishing the experimental protocols. In particular, the effect of flow rate and duration of experiments need to be carefully assessed.

Methodology

Microdialysis probes are implanted stereotaxically in the brain of anesthetized animals. The coordinates for rat or mouse brain are usually taken from the corresponding atlas (Paxinos and Watson 2005; Franklin and Paxinos 1997). This allows a theoretical precision of 0.1 mm in the placement of microdialysis probes, although the actual precision is impaired due to individual variations in the size and shape of the brain of experimental animals.

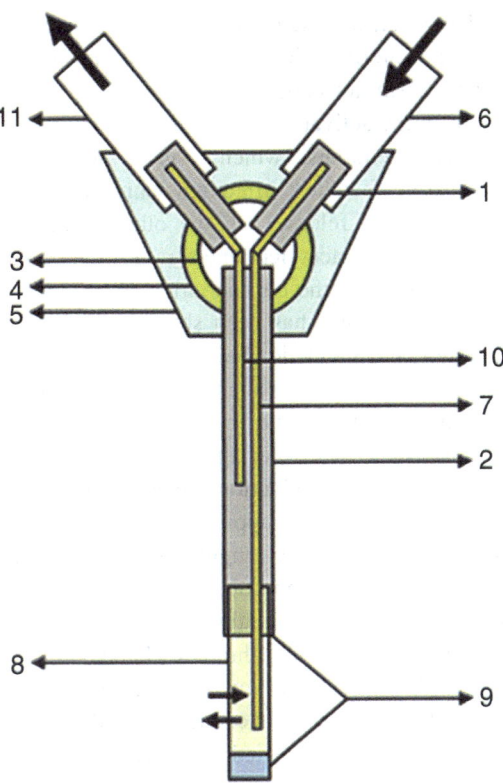

Microdialysis, Fig. 1 Schematic representation of a concentric microdialysis probe made up of the following components (see text for details): 27-gauge stainless steel tubing (*1*), 25-gauge stainless steel tubing (*2*), epoxy resin (*3* and *9*), dental cement (*4*), hotmelt adhesive (*5*), polyethylene tubings (*6* and *11*), fused silica capillary tubings (*7* and *10*), and dialysis membrane (*8*). *Small arrows* indicate the interchange process through dialysis membrane. *c* indicates the direction of perfusion fluid

Once the dialysis probe has been positioned in the area of the brain to be studied, flushing with artificial cerebrospinal fluid (CSF) is recommended in order to check the integrity of the membrane. Then, the probe is secured to the skull with anchor screws and dental cement.

Construction of a Dialysis Probe

At present, the type of dialysis probe most commonly used has a concentric structure (Fig. 1). This sort of probe has been used in a wealth of experimental research because it is well suited for reaching deep structures and/or small nuclei of the brain. A detailed description of the materials and suppliers can be found in Adell and Artigas (1998). Briefly, the body of the probe is made up of 20-mm-long 25-gauge stainless steel tubing. The inflow and outflow tubes threaded through the 25-gauge tubing consist of fused silica capillary tubing of 0.11 mm OD, 0.04 mm ID. The upper exposed end of fused silica tubings is inserted into a 7-mm piece of 27-gauge (0.41 mm OD, 0.20 mm ID) stainless steel tubing. The junction of the 27- and 25-gauge stainless steel tubings is sealed with epoxy glue and covered with dental cement to harden the assembly. Sampling of monoaminergic neurotransmitters usually requires a dialysis membrane consisting of a regenerated cellulose hollow fiber (0.17 mm OD, 0.15 mm ID), with a molecular weight cutoff of 6,000 Da, which is placed over the protruding lower portion of the inlet-fused silica tubing and glued with epoxy resin to the inside surface of the 25-gauge stainless steel tubing. The tip of the hollow fiber is sealed also with epoxy glue. The length of the dialysis membrane exposed to the tissue varies according to the brain area to be examined. Finally, the 27-gauge protective steel tubes are friction fitted with 20-mm lengths of polyethylene tubing of 0.61 mm OD, 0.28 mm ID to facilitate the connection of the probe to the perfusion pump and outflow line. These polyethylene and steel tubes are secured together with hotmelt glue.

Perfusion Fluids

One of the crucial aspects in microdialysis studies is that the composition of the perfusion medium must be physiological, i.e., isotonic with respect to that of the interstitial space. However, several fluids are being used currently that differ in their electrolytic composition/concentration (reviewed by Benveniste and Hüttemeier 1990). With little variation, the fluids used to perfuse dialysis probes are those derived from Krebs-Ringer solutions or artificial CSF. Typically, the concentration of Ca^{2+} ions in the perfusion fluid may vary from the physiological 1.2–3.3 mM. Since Ca^{2+} ions are essential for the process of exocytosis, some authors have used higher concentrations to stimulate transmitter release.

Although the buffering capacity of the extracellular fluid for some cations such as K^+ and Ca^{2+} is high, differences in the ionic composition of the perfusion and interstitial fluids may alter the responsiveness of neurons.

Other perfusing solutions also contain glucose to prevent its depletion from the interstitial space produced by the continuous drainage of perfusion. Glucose addition to the perfusion fluid also provides the essential nutrient for neurons to cope with the cell damage and disruption of the blood-brain barrier caused by probe implantation. However, the concentration of lactate, pyruvate, aspartate, and glutamate in rat cortical dialysates is similar when the perfusion fluid contains 0 or 3 mM glucose, which suggests that the presence of glucose does not play a critical role in neuronal metabolism during microdialysis experiments. In addition, it should be kept in mind that the presence of glucose may favor bacterial growth in the perfusion fluid, thus altering the extraction of neurotransmitters.

One common problem inherent to most microdialysis studies is the very low concentration of neurotransmitters in the dialysate fluid caused by efficient mechanisms of removal from interstitial space, such as reuptake or enzymatic degradation. To circumvent this complication, an uptake blocker or an inhibitor of enzymatic breakdown is included in the perfusion medium. In the absence of such agents, the extracellular concentration of a transmitter reflects the balance between the processes of release and inactivation. However, in the presence of such agents, the release component is amplified, and this has to be taken into consideration while interpreting dialysis results. For example, the use of an uptake blocker may allow to detect changes in the extracellular level of transmitters that otherwise may be overlooked. The most common compounds added to the perfusion fluid are uptake inhibitors such as citalopram for serotonin, nomifensine for dopamine, desipramine for norepinephrine, and physostigmine or neostigmine to block the enzymatic degradation of acetylcholine.

However, the use of such agents can complicate the interpretation of results, as the higher neurotransmitter concentration they induce may result in the activation of terminal autoreceptors in nerve endings, which usually results in a negative feedback affecting neurotransmitter synthesis and release. On the other hand, the addition of acetylcholinesterase inhibitors, which increases the concentration of acetylcholine in dialysates, has been shown to influence markedly the interaction between cholinergic and dopaminergic brain systems. Therefore, these modified perfusion fluids must be used with caution, as they may alter the function of drugs whose mechanism of action is examined.

The choice of an appropriate flow rate for the perfusion (usually ranging between 0.1 and 2 µL/min) is an important practical point for several reasons. The relative recovery of neurochemical compounds through dialysis membranes declines as the flow rate increases. High flow rates generate a concentration gradient, and compounds can be carried away from the extracellular space with the subsequent enrichment of dialysate samples so that the absolute recovery per time unit is enhanced. However, such a washing effect may reduce the tone on terminal autoreceptors and, therefore, alter the dynamics of the release process.

For all of the above reasons, low flow rates are preferred in order to approach ideal dialysis conditions and maximize the recovery of transmitter substances from the interstitial space. Flow rates of 0.25–3 µL/min provide a sample volume of 5–60 µL per 20-min fraction, which can be collected in plastic microvials and is easy to handle for further analysis by HPLC procedures.

Quantitative Aspects

Different attempts have been made to calculate the extracellular concentration of neurotransmitters from its amount in dialysate samples. The simplest method is to calibrate the probes for in vitro recovery and correct dialysate concentration for that value. To this aim, the probes are immersed in a beaker filled with a known concentration of the substances of interest dissolved in the perfusion fluid. The ratio between the concentration of the substance in the probe effluent and in the medium provides the recovery value of such a substance. The term "absolute recovery" refers to the total amount of a compound that passes into the perfusion fluid per unit of time, whereas "relative recovery" describes the concentration of a compound in the dialysate relative to that in the perfusion medium expressed as a percentage value. The flow rate of the perfusion fluid is inversely related to the relative recovery, and the magnitude of absolute recovery is limited by perfusion flow rate. It is important to note herein that the absolute recovery is proportional to the concentration of the substance outside the dialysate, whereas the relative recovery is not. In addition, the calibration in vitro of the probes depends on the temperature and physicochemical stability of the compounds to be analyzed. The validity of these calibration procedures is based on the assumption that the conditions in vitro and in vivo are similar. However, the brain interstitial space is a more complex matrix, and the tortuosity of the diffusion created by cell membranes and the drainage of endogenous compounds induced by the continuous perfusion are factors that must be taken into consideration.

Several refined mathematical models have been described to determine the actual concentration of transmitters in the extracellular compartment of the brain, yet they are usually too complex to be used routinely in neuropsychopharmacology research. A more practical approach was reported in which the dialysate concentration of a neurotransmitter is measured at different flow rates and extrapolated to a flow rate of zero (Justice 1993). With no net flow, the dialysate is in equilibrium with the extracellular fluid. Therefore, the level found at zero flow should represent the actual in vivo concentration of the transmitter.

Finally, it should be considered that, for many applications, the knowledge of the in vivo concentration of a transmitter is not necessary. Instead, the change with respect to baseline level is what actually matters and is to be related to the mechanism of action of drugs.

Statistical Analysis of Data

The working hypothesis usually tested in microdialysis experiments is whether a physiological or pharmacological manipulation affects the concentration of a transmitter in dialysate samples. Typically, the experimental approach consists in the collection of several pretreatment samples until stable baseline values are attained. Then, drugs are administered or animals are subjected to certain procedures (e.g., stress, forced motor activity, behavioral manipulation, etc.) and a number of posttreatment samples are collected. These temporal data series are often analyzed by means of analysis of variance (ANOVA) for repeated measures followed by appropriate post hoc tests to compare pre- and posttreatment periods. In more complex experiments (e.g., when assessing the effects of different drug doses or in various areas of the brain), the use of two-way ANOVA for repeated measures is better suited, with dose (or region) as the independent factor and time (or treatment) as the dependent variable. AUCs of the posttreatment periods can be also calculated as an integrated estimate of drug action and compared by means of one- or two-way ANOVA.

Neural Origin of Transmitters in Dialysate Samples

In order to determine the neural origin of neurotransmitter efflux measured in dialysis experiments, several specific criteria must be fulfilled. First, basal transmitter release from nerve terminals has to be impulse dependent. This is usually assessed by the addition of the sodium channel blocker tetrodotoxin, which impairs the release of the transmitter. A second requirement for the neuronal origin of a putative transmitter is its disappearance or decrease from dialysate, when Ca^{2+} is omitted from the perfusion medium. The basis for such an action is that the impulse-dependent release of a transmitter by exocytosis is dependent on the availability of extracellular Ca^{2+} (Augustine et al. 1987). Finally, the ability of elevated concentrations of K^+ to depolarize neural structures and stimulate the output of transmitters has been taken as an additional criterion for their neural provenance.

Working Practices

As detailed in the preceding sections, microdialysis is only a sampling procedure. The combination with appropriate analytical techniques has made it possible to monitor changes in the concentration of small molecules in the interstitial space of the CNS. In addition, researchers have devised a number of different experimental approaches to exploit the capabilities of this technique.

Local Administration of Chemicals and Drugs

Due to the ability of the microdialysis membrane to allow the passage of small molecules in both directions, microdialysis probes have been used to deliver chemicals in restricted areas of the brain by reverse dialysis. Ions or pharmaceutical agents known to affect neural function can be dissolved in the perfusion fluid and delivered to the brain structures of interest, provided that the molecular weight cutoff of the membrane is appropriate. Changes in the concentration of transmitters can thus be monitored locally or distally, by means of a second dialysis probe implanted in an area anatomically or functionally related to the brain structure in which the first probe is located (see section "Construction of a Dialysis Probe").

When appropriate amounts of chemical agents or drugs have to be dissolved in the perfusion fluid, it is important to check that they do not alter pH or osmolarity of the fluid. Usually, once stable baseline values are obtained, the standard dialysis fluid is replaced by one containing the compound(s) of interest. This procedure may be particularly useful when examining the effects of substances with a poor penetration into the brain or when assessing regional differences in the effects of drugs.

Quantitative effects of drugs in vivo can be estimated through ED_{50} values calculated after local application of drugs. The ED_{50} values obtained in this manner are by no means comparable to those obtained using cell cultures, membranes, synaptosomes, or other in vitro preparations. Several factors account for these differences, including (1) the recovery of the dialysis membrane, usually much lower than 100 %; (2) the diffusion of chemicals within the brain once they have crossed the dialysis membrane and the tortuosity of the neural tissue; (3) the continuous drainage of applied drugs by the CSF; and (4) the unspecific binding to cell membranes, particularly of lipophilic molecules. All these factors contribute to reduce the actual concentration of drugs reaching the active sites in the brain, dramatically. In contrast, in vitro drug affinities for receptors/transporters are calculated under almost ideal conditions, i.e., with enriched preparations and unlimited access of the chemicals to their cellular targets, generally at equilibrium and under nondegrading conditions.

Systemic Administration of Drugs

Experiments involving the systemic administration of drugs constitute the vast majority of the applications of microdialysis. In such experiments, however, appropriate controls must be carried out, because the procedure of drug delivery or the vehicle used may change transmitter function due to the associated stress or the sensitivity of some neuronal groups to sensory stimuli. The changes in the concentration of transmitters in an area of the brain after systemic administration of drugs do not necessarily parallel those found after their local application. In general, when drugs are applied locally, larger concentrations are needed to reach effects similar to those obtained after systemic application. This possibly reflects a better distribution of drugs administered systemically through the diffuse network of brain capillaries. Another factor frequently ignored is the fact that the diffusion of a chemical agent delivered by reverse dialysis is limited to a small portion of the brain tissue surrounding the dialysis probe. In contrast, the changes in the extracellular concentration of the transmitter in the same area after systemic administration of a drug result from an integrated response of the whole CNS, i.e., local and transynaptic effects.

Dual Probe Approaches

Experiments carried out with two or more probes implanted in the same animal present two distinct advantages. First, such an approach allows to reduce the number of animals used in a single experiment, and, second, it is ideal to examine functional interactions between different brain areas. This latter asset was first employed for dopamine and serotonin systems to study how the release in terminal areas is regulated by the activity at the level of cell bodies. This was possible because the cell bodies of those neuronal systems are tightly packed in the midbrain, substantia nigra, ventral tegmental area, and raphe nuclei. Therefore, the local application of drugs known to interact with receptors or transporters located on these monoaminergic cell bodies induces changes of the release of the transmitter in projection areas. Dual probe microdialysis studies have been extremely helpful for the study of the functional connections between different brain areas and the transmitter/receptors involved (Adell and Artigas 1991; Santiago et al. 1991).

Coupling to Electrical Stimulation

Similar to the experiments described in the previous section, electrical stimulation coupled with microdialysis in distal areas has been used to assess the existence of functional connections between brain areas. This is usually achieved by inserting an electrode in an area containing the cell bodies of a certain transmitter system and a dialysis probe in the corresponding projection areas. For instance, the electrical stimulation of the substantia nigra or the raphe nuclei results in an enhanced release of dopamine and serotonin, respectively, in projection areas. Similar experimental procedures combining electrical stimulation and microdialysis have been used to study the modulatory role of prefrontal cortex on dopaminergic and cholinergic activity in subcortical structures such as the dorsal striatum or the nucleus accumbens.

Microdialysis, Table 1 Advantages and limitations of the microdialysis technique

Advantages	Easy to use routinely
	Easy manufacture of probes
	No enzymatic degradation of transmitters in samples
	Coupling to chemical methods of analysis (HPLC, mass spectrometry, capillary electrophoresis, etc.)
	Possibility of local administration of drugs
	Possibility of concurrent determination of drugs after systemic administration
	Dual probe approaches
	Possibility of concurrent recording of electrical activity
	Concurrent study of behavior in freely moving animals
Limitations	Invasive procedure, causes neuronal death and reactive gliosis
	Limited spatial resolution
	Limited temporal resolution
	Analytical difficulties with some transmitters
	Low membrane recoveries with high molecular weight compounds

Advantages and Limitations of Microdialysis

Since its first applications, microdialysis has become increasingly popular to study brain function. The use of alternative in vivo procedures such as push-pull perfusion or voltammetry has remained constant or even declined during last years. A comparison between microdialysis and voltammetry reveals that microdialysis is applicable to most types of small molecules, whereas the use of voltammetry is limited to easily oxidizable compounds such as catecholamines and serotonin. Moreover, microdialysis appears to be simple to use on a routine basis and can easily be applied to study freely moving animals.

Table 1 summarizes some of the advantages and limitations of microdialysis. Certainly, microdialysis is by no means a definitive method for the assessment of the active transmitter concentrations in the brain. Yet, it has a number of advantages over its predecessor, the push-pull perfusion, which has led to its more widespread use. The main limitations of microdialysis are the size of the probes and the tissue damage caused by their insertion. For certain applications, size may not be a problem (e.g., to assess the effects of drugs in large brain regions). However, the study of physiologically or pharmacologically induced changes of transmitters in small nuclei may pose some constraint because a larger proportion of neurons are damaged.

Finally, the low amount of certain neurotransmitters in brain dialysates makes it necessary to collect samples every 20 or 30 min, a time scale which is far from that of neuronal events. This may not be a problem in pharmacological studies because most drugs reach peak levels at a time compatible with the usual periods of sampling of 20 or 30 min. This enables to follow up drug-induced transmitter changes. However, microdialysis may not be suitable for the study of the effects of neuronal stimulation on transmitter release at a physiological time scale. Recent advances in the detection of very low concentration of certain transmitters with capillary electrophoresis have permitted a considerable shortening of the sampling periods. Yet, this is still far from the scale at which neuronal excitation or inhibition is associated to the release of a transmitter. It is hoped that future methodological and technical developments will overcome some of these limitations.

Cross-References

▶ Antidepressants
▶ Antipsychotic Drugs

References

Adell A, Artigas F (1991) Differential effects of clomipramine given locally or systemically on extracellular 5-hydroxytryptamine in raphe nuclei and frontal cortex. An in vivo brain microdialysis study. Naunyn-Schmiedebergs Arch Pharmacol 343:237–244
Adell A, Artigas F (1998) In vivo brain microdialysis: principles and applications. In: Boulton AA, Baker GB, Bateson AN (eds) In vivo neuromethods. Humana Press, Totowa, pp 1–33
Artigas F, Adell A (2007) The use of brain microdialysis in antidepressant drug research. In: Westerink BHC, Cremers TIFH (eds) Handbook of microdialysis: methods, applications and clinical aspects. Elsevier/Academic, Amsterdam, pp 527–543
Augustine GJ, Chanton MP, Smith SJ (1987) Calcium action in synaptic transmitter release. Ann Rev Neurosci 10:633–693
Benveniste H, Hüttemeier PC (1990) Microdialysis-theory and application. Prog Neurobiol 35:195–215
Franklin KBJ, Paxinos G (1997) The mouse brain in stereotaxic coordinates. Academic, San Diego
Justice JB Jr (1993) Quantitative microdialysis of neurotransmitters. J Neurosci Meth 48:263–276
Paxinos G, Watson C (2005) The rat brain in stereotaxic coordinates. Elsevier/Academic, Amsterdam
Santiago M, Rollema H, De Vries JB, Westerink BHC (1991) Acute effects of intranigral application of MPP^+ on nigral and bilateral striatal release of dopamine simultaneously recorded by microdialysis. Brain Res 538:226–230
Ungerstedt U (1984) Measurement of neurotransmitter release by intracranial dialysis. In: Marsden CA (ed) Measurement of neurotransmitter release in vivo. Wiley, Chichester, pp 81–105

Microelectrode Arrays

Synonyms

MEA

Definition

Ceramic-based multisite microelectrode arrays (MEAs; 15×330 μm or 20×150 μm recording sites) with platinum recording sites and polyimide insulation that have been recently described (Hascup et al. 2007). The triangular design of the microelectrodes yields a microelectrode array ranging in thickness from 37 to 125 μm and an overall length of 8–10 mm. These arrays are wire bonded to printed circuit board holders that adapt MEAs for measures in brain slices and studies in anesthetized and awake mice, rats, and nonhuman primates. More than 20 varieties of geometries ranging from Pt 4 to 16 recording sites have been designed.

Microiontophoresis and Related Methods

Roberto William Invernizzi[1], Ennio Esposito[2], Vincenzo Di Matteo[3] and Giuseppe Di Giovanni[4]
[1]Department of Neuroscience, Mario Negri Institute for Pharmacological Research, Milan, Italy
[2]Consorzio Mario Negri Sud, Laboratory of Neurophysiology, Santa Maria Imbaro, Chieti, Italy
[3]Istituto di Ricerche Farmacologiche "Mario Negri", Santa Maria Imbaro, Chieti, Italy
[4]Department of Physiology and Biochemistry, Laboratory for the Study of Neurological Disorders, University of Malta, Msida, Malta

Synonyms

Iontophoresis

Definition

The term microiontophoresis is derived from the ancient Greek term *phoretikos*, which refers to the production or induction of movement. Microiontophoresis is a technique with which drugs and other ionized particles can be ejected in very small amounts from solutions contained in glass micropipettes. This ejection is accomplished by applying a voltage across the micropipette and causing the electrode to become polarized. Ionized particles in solution migrate in the applied field and will be ejected from the tip as they carry the current into the tissue. This technique is widely used to determine the effects of various substances on firing parameters of both central and peripheral neurons and muscles. In investigating the phenomenon of synaptic transmission at the neuromuscular junction, during the 1950s, this technique became very popular. A technique appropriate for the study of synaptic pharmacology was first realized by Nastuk (1953) and was later developed by del Castillo and Katz (1955), and it consisted essentially of the microiontophoretic method, i.e., movement of charged particles produced by an electrical current, restricted to a micropipette with a tip diameter of the order of 1 μm. Thus, solutions of acetylcholine chloride were used, and by passing a suitable current to this solution, acetylcholine ions could be ejected from the 1 μm orifice onto a correspondingly localized area of subsynaptic membrane at the neuromuscular junction. Later, Curtis and his colleagues adopted this technique for studying the mammalian central nervous system (CNS) (Curtis and Eccles 1958b). The experiments of Curtis and coworkers, however, involved an important modification of the original method, in that this group used multibarrel micropipettes. In the production of these, several lengths of tubing are fused together and then pulled so as to produce a single collective tip, but with each barrel having its own orifice. Multibarrel micropipettes are usually composed of five to seven barrels (Fig. 1). Usually, the central barrel is the recording electrode, whereas the other side barrels contain drug solutions (Fig. 1). As the drug molecules would tend to diffuse from solution in the pipette tip into the extracellular environment, it is necessary to apply a small current to reduce that efflux. This is known as a "holding" or "retaining" current (Fig. 2). It is also a usual practice to include a barrel containing sodium chloride solution, which can be used to control the effects of the current itself. This may be done either by periodically passing through the control barrel the same current used for drug ejection or by passing continuously a current adequate to cancel out the instantaneous sum of ejecting and retaining currents passing through the drug-containing barrels. This is known as "current balancing." The use of microiontophoresis is suitable for any ionized molecule, but nonionized compounds can be ejected by the closely related variant "electro-osmosis," which is attributable to the presence of an electrical "double layer" within the barrel tip. When an aqueous solution is in contact with glass, negative ions are tightly adsorbed on the glass surface, leaving the bulk of solution

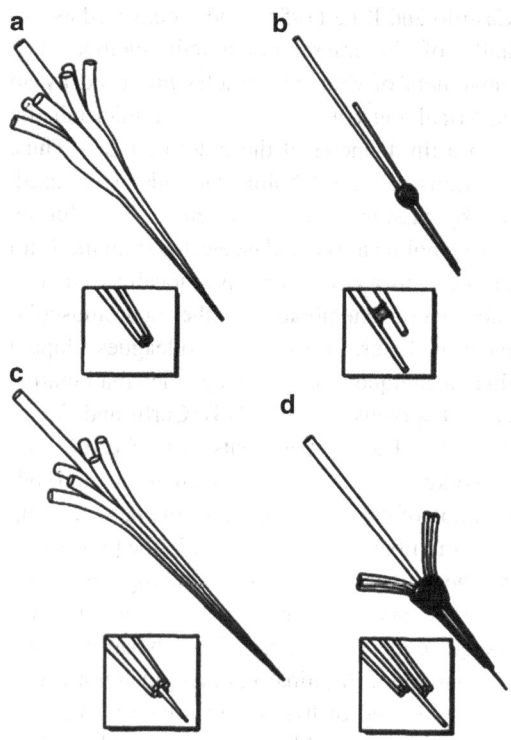

Microiontophoresis and Related Methods, Fig. 1 Examples of different types of multibarrel micropipette assemblies used in microiontophoretic experiments. (**a**) Standard 7-barrel assembly in most common use, introduced first by Curtis. (**b**) Twin, or parallel micropipette. (**c**) Coaxial assembly. (**d**) Staggered tip multibarrel (From Hicks 1984)

One advantage of micropressure ejection is that it can be applicable to all compounds; however, it is not devoid of problems and artifacts and is unlikely to replace microiontophoresis as a microapplication method.

Principles and Role in Psychopharmacology

General Principles

Each barrel of a micropipette assembly to be used for drug ejection is filled with a solution of the ionized compound, and the solution is connected to the iontophoresis machine by a suitable lead, which is in contact with the drug solutions. The establishment of a potential difference between the drug solution and the medium surrounding the barrel tip will then cause the movement of ions through the solution and out of the pipette tip (Fig. 2). A chief advantage of the microiontophoretic method is that it is possible to examine the effects of drugs on single neurons in vivo without affecting the whole nervous system or other physiological responses, such as those that may occur when drugs are administered systemically (Aghajanian 1972). If a voltage is applied to a solution, ions and charged molecules will migrate toward and away from the source of the imposed electrical field depending on the sign of their net charge. This phenomenon is the fundamental principle of microiontophoresis: the desired charged particles are ejected from the mouth of one barrel of a multipipette assembly by appropriately charging the interior of that barrel (Fig. 2). An outward current will cause the "ejection" of positively charged ions, and an inward current flow, the ejection of negatively charged particles. If the pipette assembly is positioned close to a neuron, so that the recordings of its activity can be made through another electrolyte-filled barrel, drugs may be ejected, and their pharmacological effects are inferred by the resulting changes in the rate and/or firing pattern.

The Transport Number and the T_{50} Value

An important technical consideration for experiments employing microiontophoresis is the

carrying a net positive charge. The passage of positive (or outward) current then causes the ejection of a small volume of solution containing the compound of interest (Fig. 2). It should be noted, however, that this mechanism has nothing to do with the osmotic pressure of a solution or the establishment of any osmotic gradient. The term electro-osmosis derives simply from the fact that the driving force is the movement of the *solvent*, not the solute, just as the case of osmotic movements across a semipermeable membrane. An alternative method of applying both ionized and nonionized compounds from micropipettes is the use of pressure. A suitable source of pressure, usually a cylinder of compressed gas, is connected to the open end of a micropipette barrel. Pressure usually up to 20 lb per square inch (p.s.i.) will eject fluid from a 1 μm pipette tip.

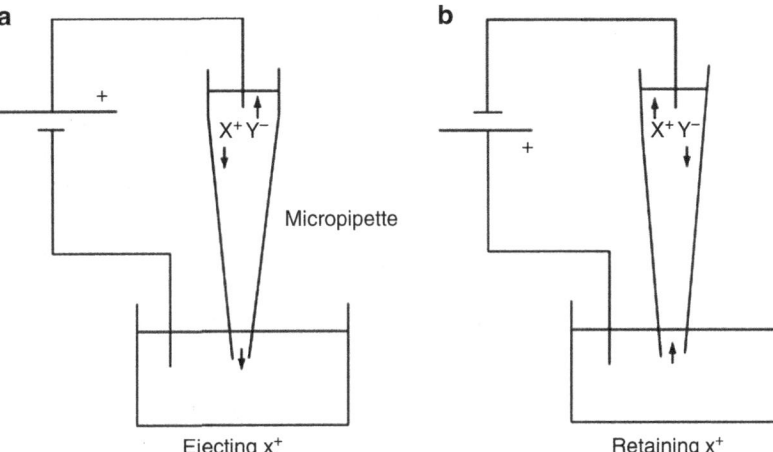

Microiontophoresis and Related Methods, Fig. 2 Schematic diagram of a micropipette that contains a salt X^+Y^-, showing the direction of current necessary to eject (**a**) and retain (**b**) the ion X^+ (From Hicks 1984)

transport number. The transport number is a measure of the amount of drug released from the micropipette by iontophoretic expulsion, and it is important, because it helps one to evaluate dose-response relations between different compounds, and it can also provide some indication of the absolute potency of compounds. The transport number varies for individual compounds and is based on the interaction of the following variables: their solubility, the extent of their dissociation in solution, their polarity, and the nature of the external medium into which the drugs are administered. The transport number may be formally described by the following equation:

$$n = R_i Z F i^{-1},$$

where

n = apparent transport number of the drug ion,
Z = valency,
F = Faraday's constant, in Coulombs
i = intensity of ejecting current, in nanoamperes, and
R_i = rate of microiontophoretic release (which is equivalent to total release minus the sum of the rate of steady-state spontaneous release and where applicable, the release due to electroosmosis).

During microiontophoresis, the total number of ions transported is related in a direct manner to the amount of current applied to the solution, according to Faraday's law. However, only a certain proportion of the charge imposed is carried by the ion species of interest. This value, which is "n," the transport number, is not constant for a given material but will vary not only from pipette to pipette but also, to a lesser extent, between different barrels of the same micropipette assembly containing identical solutions. Despite these inconsistencies, it remains valid that under steady-state conditions, drug release from micropipettes conforms to Faraday's law: the amount of drug released is proportional to the magnitude of current passed (Hicks 1984).

Another important parameter to consider when interpreting microiontophoretic data is the T_{50} value, which is the time taken for a response to reach 50 % of its maximum (Fig. 3). The basis for this procedure is the hypothesis that each individual response to an agonist may be considered as a cumulative dose-response relationship reflecting the gradual increase in tissue concentration of drug during the ejection period. If a series of such responses are obtained, reaching the same maximum amplitude, they can be readily characterized by the T_{50} value (Fig. 3). Moreover, T_{50} value is easier to measure accurately than a response size. Thus, it can be very difficult to obtain reproducible graded response amplitudes to some very potent compounds such as amino acids. Any changes in firing rate during a response, which ideally should be of the plateau

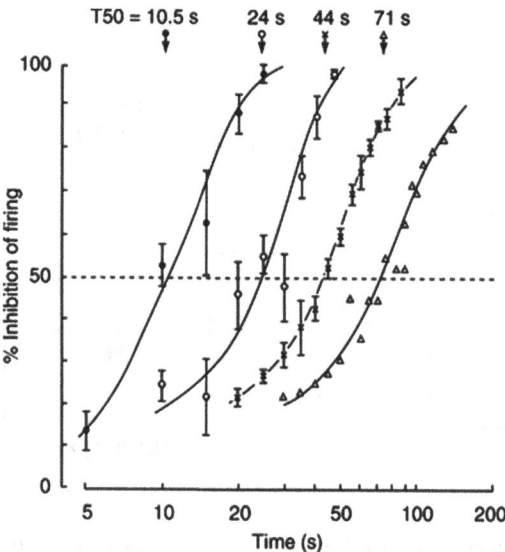

Microiontophoresis and Related Methods, Fig. 3 Time courses of inhibition of neuronal firing following microiontophoretic application of GABA with four different currents, 20 nA (*filled circle*), 10 nA (o), 5 nA (×), and 2 nA (*open triangle*). Each curve was obtained from the same neuron at a depth of 957 μm in the *middle* suprasylvian gyrus of the cat cortex. The neuron was driven by continuous microiontophoretic application of L-glutamate (20 nA). Each of the points for 20, 10, and 5 nA applications of GABA is the mean ± SEM of three values obtained from three separate applications of the same current of GABA. The values of T_{50} shown are the times taken to achieve 50 % inhibition of neuronal firing (From Hill and Simmonds (1973) Br J Pharmacol 48:1–11)

variety, may further complicate any assessment of the response size, whereas in the determination of T_{50} values, all responses increase to the same maximal level, which may be 100 % inhibition or a clear maximal plateau of excitation tending toward overdepolarization.

Microiontophoresis in the Central Nervous System

It is now more than 50 years since Curtis and Eccles (1958a, b) first employed the technique of microiontophoresis in the CNS. Microiontophoresis has provided so far a great contribution in the identification of the central effects of neurotransmitters, including glutamate, aspartate, γ-aminobutyric acid (GABA), noradrenaline, serotonin, dopamine, and a variety of neuropeptides (enkephalins, cholecystokinin, neurotensin, tachykinins). Microiontophoresis also allows the histological confirmation of the sites of electrophysiological recordings, and the neuroanatomical determination of pathways by applying dyes, markers, and materials, which are carried by axonal transport for tracing fiber tracts. Alterations in neuronal sensitivity due to the influence of anesthetic compounds have been monitored when pharmacological agonists have been tested using microiontophoresis.

Applications

The largest number of studies has been concerned with the central nervous system. These studies have yielded information on (1) the qualitative sensitivity of neurons to putative neurotransmitters and drugs, (2) quantitative estimates of variations and sensitivity in different CNS regions or of different cell types and the following lesions or the administration of drugs, (3) the pharmacology of transmitter receptors, (4) the effects of modifier of putative transmitter effects (antagonistic or enhancing substances) on synaptic transmission, and (5) the mechanisms and ionic conductances underlying transmitter effects.

Excitatory Amino Acids

Some of the earliest iontophoretic studies demonstrated marked excitatory activity of several simple dicarboxylic acids, including L-glutamic and L-aspartic acids (Curtis and Watkins 1960). Responses to some of these amino acids, especially glutamate and aspartate, terminate rapidly when an ejecting iontophoretic current is switched off. It is unclear to what extent this is due to the kinetics of iontophoresis or reflects the presence of rapid and efficient uptake processes. Some authors have reported long-lasting changes of cortical neuronal firing following iontophoresis of glutamate sufficient to at least double the resting firing rate. The development of a series of potent amino acid analogs with very high agonist potency led to the discovery of NMDA and non-NMDA glutamate receptors. This discovery was strengthened by additional findings that phosphonate analogs of amino acids, such as 2-amino-5-phosphonovaleric acid (AP-5) blocked the effects of NMDA, but not of quisqualic and kainic acids.

Inhibitory Amino Acids

Both glycine and GABA act as potent inhibitors of neuronal activity in the CNS, usually causing hyperpolarization associated with increased membrane conductance to chloride. Glycine is selectively antagonized by strychnine, whereas the effects of GABA are blocked by picrotoxin and bicuculline. Microiontophoretic experiments showing potentiation of the inhibitory effects of GABA by benzodiazepines were among the earlier experimental evidence for the modulatory action of these drugs on $GABA_A$ receptors (Gallager 1978).

Acetylcholine

Almost every region of the brain has been examined for its sensitivity to iontophoretically applied cholinergic agents. Most of the earlier work in vivo was concerned primarily with establishing the direction of responses to cholinomimetics and whether the effects involved muscarinic or nicotinic receptors. Many studies examined only cells encountered randomly in a particular brain region, but others have often succeeded in relating the direction of responses to cholinomimetics with some specific function. In the cerebral cortex deep pyramidal tract, cells are excited by acetylcholine. Several authors have also described an inhibitory action of acetylcholine, largely muscarinic in nature, in more superficial levels of the cortex and an excitatory action, which appears to have a predominant nicotinic pharmacology, in the same superficial layers. Some authors have shown that acetylcholine enhances the stimulus-evoked responses of visually driven cortical units, without affecting the overall excitability of the cell. Thus, orientation and direction specificity of neurons is preserved and increased relative to the nonpreferred responses. This phenomenon is reminiscent of the effects of some amines, which can also increase the signal-to-noise ratio by potentiating evoked activity and suppressing background. Recently, Di Giovanni and Shi (2009) showed the lack of effect of scopolamine, an antagonist of muscarinic receptors on SNc dopaminergic neurons. Surprisingly, scopolamine microiontophoretically applied did not cause any significant change in either the firing rate or pattern of the spontaneously active dopamine neurons.

Noradrenaline

Early microiontophoretic studies have shown that noradrenaline would cause a depression of neuronal firing in the cat cerebral cortex, and a large number of experiments have revealed similar responses in most areas of the CNS. This inhibition often seems to involve a voltage-dependent hyperpolarization accompanied by an increased membrane resistance, although a decreased membrane resistance was found on neurons of the locus coeruleus in slice preparation in vitro. The biochemical basis of this hyperpolarization has been the subject of much argument. Although it was originally suggested that they may be mediated by an increase in the intracellular concentration of cyclic AMP, some group failed to reproduce these findings. Overt excitatory effects of noradrenaline have also been observed in many areas of the CNS. Neuronal responses to iontophoretic application of noradrenaline, apparently excitatory as well as inhibitory, can be enhanced by antidepressants. However, this potentiation can occur even after the loss of most amine-containing terminals, and it may be restricted to certain layers of the cortex. The pharmacology of responses to iontophoretically applied noradrenaline has been extensively studied. Some authors have postulated that, in the neocortex, excitatory responses to noradrenaline are mediated by α_1-adrenergic receptors, whereas inhibitory responses occur through β-adrenergic receptors. Activation of α_2-adrenergic receptors does also elicit inhibitory responses.

Dopamine

Dopamine was first tested iontophoretically in the cerebral cortex, where profound suppression of spontaneous cell firing was observed. This action has been confirmed by several authors, although excitatory effects have also been reported. Much attention has been centered on the effects of dopamine in the neostriatum where its action is usually inhibitory in the caudate nucleus. Bunney and Aghajanian (1976) have performed a laminar analysis of amine responses in the rat cerebral cortex. They found that neurons in layers II and III, which receive a dense noradrenergic projection, were more sensitive to noradrenaline than dopamine, whereas the reverse pattern was noted in layers V and VI, which receives a greater dopamine-containing projection. These authors also reported that desipramine, a selective inhibitor of noradrenaline reuptake, would enhance noradrenaline responses in layers II and III, but not in deeper layers, while benztropine-enhanced dopamine responses only in layers V and VI. Dopamine receptors are present not only on innervated cells but also on the dopaminergic neurons themselves: the so-called autoreceptors. Activation of such receptors by dopamine or apomorphine causes marked inhibition of cell firing, and these effects are blocked by neuroleptic drugs. Microiontophoretic studies of dopamine response pharmacology have mostly proved consistent with behavioral and neurochemical work. Phenothiazines, for example, block dopamine but not noradrenaline responses in the cerebral cortex and the striatum. Iontophoretically applied α-flupenthixol can also block the effects of dopamine, although intravenously administered α-flupenthixol or pimozide did not modify neuronal responses to iontophoretic dopamine.

Serotonin

There is an extensive scientific literature regarding the effects of microiontophoretically applied serotonin on different areas of the central nervous system. Indeed, the microiontophoretic technique contributed substantially to the elucidation of the physiology and pharmacology of the central serotonergic system. Thus, an important factor controlling the activity of central serotonergic neurons is neuronal feedback inhibition. This is thought to be a homeostatic response, which, under physiological conditions, acts to compensate for increases in synaptic availability of serotonin. Thus, as the concentration of serotonin increases in the brain, the activity of central serotonergic neurons correspondingly decreases. The mechanism underlying this feedback regulation is both local or intrinsic to the raphe region (where serotonergic cell bodies are located) and through a feedback loop from postsynaptic target neurons. Serotonin released in the raphe region from dendrites and possibly from axon terminals appears to inhibit serotonergic neurons by activating somatodendritic autoreceptors, which produces hyperpolarization of the cell membrane via an increase in potassium conductance. Historically, the first drug reported to exert a preferential action on the 5-HT autoreceptor was LSD (lysergic acid diethylamide) applied microiontophoretically on the dorsal raphe nucleus of rats. Subsequently, several other hallucinogenic indoleamines, notably 5-MeODMT (5-methoxy-N,N-dimethyltryptamine), were found to share this property with LSD. Since that time, several highly selective 5-HT$_{1A}$ agonist compounds such as 8-OH-DPAT have been synthesized and shown to suppress the firing of serotonergic neurons with potencies comparable with, or even greater than, that of LSD. On the basis of electrophysiological data, the serotonin autoreceptor has been characterized as the 5-HT$_{1A}$ subtype. Microiontophoretic technique also contributed to characterize the action of serotonin agonists and antagonists and to elucidate the physiological role of serotonin receptor subtypes such as 5-HT$_{1B}$, 5-HT$_{2A}$, and 5-HT$_{2C}$. As regards the 5-HT$_{2C}$, it was found that this receptor subtype exerts a tonic inhibitory influence on the activity of dopamine-containing neurons in the substantia nigra pars compacta and the ventral tegmental area. Apparently, this inhibitory effect is mediated through the activation of nondopaminergic (presumably GABA-ergic) neurons in the substantia nigra pars reticulata. Thus, it was recently shown that microiontophoretic application of 5-HT$_{2C}$ receptor agonists stimulates the basal activity of nondopaminergic (presumably GABA-ergic) neurons in the substantia nigra pars reticulata (Invernizzi et al. 2007) (Fig. 4). By using microiontophoresis, it was also found that serotonin exerts a tonic inhibitory influence on the activity of noradrenergic neurons in the locus coeruleus.

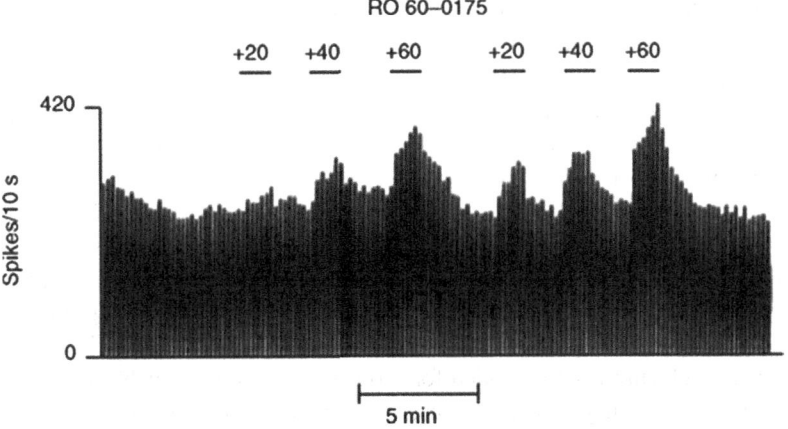

Microiontophoresis and Related Methods, Fig. 4 Representative rate histogram showing the effect of the selective 5-HT$_{2C}$ receptor agonist RO 60-0175 on the basal activity of a nondopaminergic (presumably GABA-ergic neuron) of the rat substantia nigra pars reticulata. Microiontophoretic application of RO 60175 causes excitation of basal neuronal activity, which is proportional to the amount of the ejecting current applied (numbers above each bar in nA) (From Invernizzi et al. 2007)

Opiates and Opioids

Microiontophoresis has proved exceedingly valuable for opiate system studies, since it allows the testing of discrete units activated by noxious or nonnoxious stimuli in the same preparation. In most such studies, the applied opiates have depressed noxious stimulus-evoked activity, although usually in parallel with the effects on spontaneous or chemically induced firing. Microiontophoresis has also been proved as a popular means for comparing qualitatively opiate responses in normal and opiate-tolerant animals. Thus, inhibitory responses to morphine were encountered less frequently in the neocortex of morphine-tolerant rats than in controls. It was shown that iontophoretically applied naloxone would elicit a large increase of firing in the locus coeruleus noradrenergic neurons in morphine-tolerant rats, presumably as a correlate of the withdrawal phenomenon in such animals. Also, opioid peptides have been tested iontophoretically in many regions of the central nervous system. Opioid peptides were found to excite hippocampal neurons; however, these effects were apparently mediated through an indirect action on transmitter release or to a naloxone-sensitive depression of local inhibitory interneurons.

Peptides

Microiontophoretic or pressure ejection has been used to apply a wide range of endogenous and synthetic peptides to neurons in vivo and in vitro. However, partly because of the lack of selective antagonists, there has been little progress in relating the observed responses to a physiological role, and as a result, attention has been concentrated on the mechanism of the observed responses, and potential interactions with neurotransmitters. Substance P, for example, appears to interact selectively with acetylcholine. Microiontophoretic substance P has also been found to enhance the response of spinal cord neurons to noxious stimulation but not innocuous ones, in some cases leading to the occurrence of responses in initially unresponsive units. Some excitatory effects of substance P can be mimicked by capsaicin, also applied iontophoretically. It was also reported that the excitatory effect of substance P on noradrenalin-containing neurons in the locus coeruleus is blocked by the selective antagonist [D-Pro^2, D-Trp^{7,9}] substance P. Thyrotropin-releasing hormone (TRH) has been found to enhance the excitatory effects of acetylcholine on cortical neurons, with no effects on resting firing rate. Somatostatin exerts a potent excitatory effect on hippocampal neurons.

Cholecystokinin (CCK) and neurotensin are also frequently excitatory, while angiotensin has excitant properties, which appear to be restricted to the subfornical organ and related structures. However, it is important to point out that peptides present special problems for microiontophoresis. Larger molecules tend to be adsorbed on to charged surfaces, which include the internal wall of a micropipette tip. Some peptides may also undergo denaturation or degradation during iontophoretic experiments. This problem may be exacerbated if very high currents are applied for long periods of time through high resistance tips, in that any change of local temperature may have a major impact on the stability of a peptide.

Nitric Oxide

Nitric oxide (NO) has been linked to many regulatory functions in mammalian cells. Studies of NO release are hampered by the short half-life of the molecule. Therefore, the study of nitrergic manipulation in the CNS has benefit of microiontophoresis techniques. Compelling evidence shows that different NO synthase inhibitors, NO donors or NO/cGMP pathway modulators can be administered by microiontophoresis in discrete nuclei of the CNS and are capable of influencing neuronal discharge and useful for revealing the physiological role of this gaseous neurotransmitter in the brain. For instance in the striatum, iontophoresis of 3-morpholinosydnonimine hydrochloride (SIN 1), a nitric oxide donor, produced reproducible, current-dependent inhibition of glutamate-induced excitation. Conversely, microiontophoretic application of *N*-omega-nitro-L-arginine methyl ester (L-NAME), an inhibitor of nitric oxide synthase, produced clear and reproducible excitation of glutamate evoked firing. To evaluate the involvement of cyclic guanosine monophosphate (cGMP) in the electrophysiological effects produced by the NO donor, the effects of methylene blue, an inhibitor of guanylyl cyclase, microiontophoretically applied on the responses of neurons to SIN 1 were tested. The effect of SIN 1 was significantly reduced during continuous iontophoretic administration (50 nA) of methylene blue (Di Giovanni et al. 2003).

Microiontophoresis Coupled with Other Techniques

Microiontophoresis has been extensively used coupled to neuronal recording in vivo and in vitro. The single unit neuronal activity is generally recorded by one of the capillary of the iontophoretic electrode in vivo. In addition single electrode microiontophoresis has been used to eject a compound in the proximity of a specific neural site in vitro recording (Fig. 5).

Microiontophoresis has been also coupled to other techniques such as fast-scan cyclic voltammetry. This approach was first developed by Millar and coworkers who used fast-scan cyclic voltammetry and single-unit recording at a single carbon-fiber electrode. They combined these techniques with iontophoresis to compare the response of striatal medium spiny neurons (MSNs) to locally applied dopamine and endogenous dopamine released by electrical stimulation. Despite the utility of this approach, it has been used only rarely in behaving animals. Recently, Kenan and colleagues evaluated this approach in awake animals to probe the effects of dopamine in the nucleus accumbens (NAc), a striatal subregion involved in reward processing, on evoked dopamine release and the electrical activity of MSNs.

Advantages and Disadvantages of Microiontophoresis

The original microiontophoretic technique was developed for answering questions concerned with synaptic transmission and the neuromuscular junction. Using this preparation, it is a simple matter to microscopically examine the muscle fiber being studied, to determine the distance of the micropipette from the tissue, and to have ready access to known synaptic inputs. These advantages are not valid for the CNS. Nevertheless, with some further precautions and considerations, the technique has been used successfully in the CNS for about 50 years. It is important to consider other potentially confounding technical factors limiting the utility of microiontophoresis, as it is used in central investigations. Of primary concern is the site of drug administration relative to cell soma, where the strongest depolarizing or

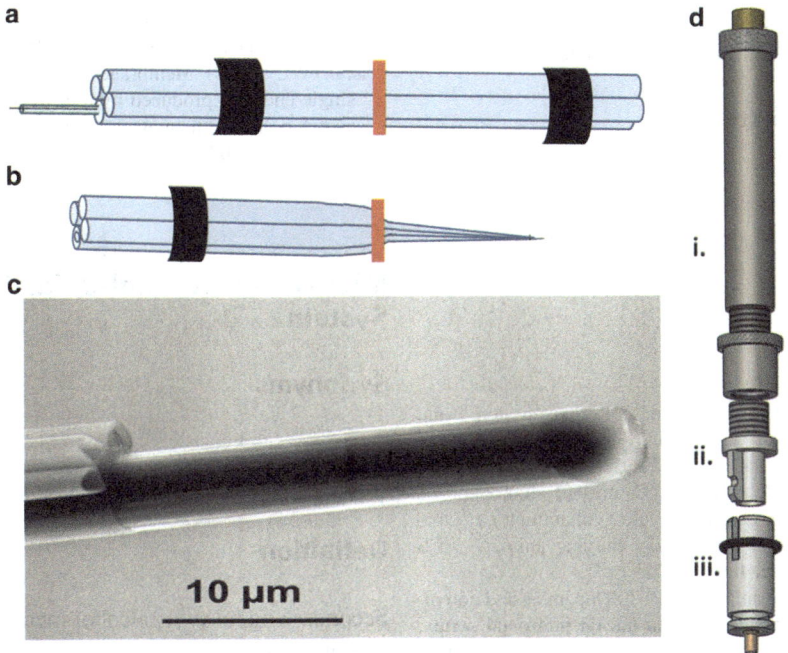

Microiontophoresis and Related Methods, Fig. 5 Modifications to probe construction and hardware for iontophoresis in freely moving animals. (**a**) Four-barrel perfused pipette ready for pulling with heat shrink on either end where the glass will contact the chucks of the puller. A smaller capillary containing a T-650 carbon fiber is loaded into one barrel. (**b**) One of the two electrodes created from the two-pull electrode making technique. The *vertical* bar indicates the midpoint of the pipet before pulling. (**c**) Environmental scanning electron micrograph of the probe tip showing the glass iontophoresis barrels and carbon fiber. (**d**) Hardware used to attach and lower iontophoresis probe into the head of a freely moving animal. (*i*) Commercially available Biela manipulator from Crist Instrument Co. (Part# 3-MMB-3D). (*ii*) Custom-machined adapter that allows use of Biela manipulator with a guide cannula from Bioanalytical Systems, Inc. fabricated by UNC Physics Machine Shop (Chapel Hill, NC). (*iii*) Commercially available guide cannula from Bioanalytical Systems. Reprinted with permission from Belle AM, Owesson-White C, Herr NR, Carelli RM, Wightman RM (2013). Controlled iontophoresis coupled with fast-scan cyclic voltammetry/electrophysiology in awake, freely moving animals. ACS Chem Neurosci. 4(5):761–771. Copyright (2013) American Chemical Society

hyperpolarizing influences are manifested, and the dendritic field, where synaptic influences are normally expressed and where antagonists of transmitters must accumulate to modify transsynaptic excitations. Another consideration for central investigations also concerns the spatial distribution of drugs in the CNS. Since the CNS is densely packed with cells, microiontophoretically administered compounds cannot affect single neurons in isolation. This must be kept in mind when interpreting the data.

Conclusions

Microiontophoresis, an experimental technique introduced more than 50 years ago, prompted a great impetus to the study of the physiology and pharmacology of the central nervous system. By mimicking the synaptic function, it provided a crucial step in establishing the physiological role of most neurotransmitters, including amines, amino acids, and neuropeptides. Although it is now considered by some as a "classical" neurophysiologic approach to the study of central nervous system, it is likely that it will still contribute substantially to the progress of neuroscience.

Cross-References

▶ Antidepressants
▶ Excitatory Amino Acids and Their Antagonists
▶ Extracellular Recording

- Hallucinogens
- Inhibitory Amino Acids and Their Receptor Ligands
- Intracellular Recording
- Neurotensin
- Opioids
- Somatostatin
- Tachykinins

References

Aghajanian GK (1972) LSD and CNS transmission. Annu Rev Pharmacol 12:157–168

Belle AM, Owesson-White C, Herr NR, Carelli RM, Wightman RM (2013) Controlled iontophoresis coupled with fast-scan cyclic voltammetry/electrophysiology in awake, freely moving animals. ACS Chem Neurosci 4(5):761–71

Bunney BS, Aghajanian GK (1976) Dopamine and norepinephrine innervated cells in the rat prefrontal cortex: pharmacological differentiation using microiontophoretic techniques. Life Sci 19:1783–1789

Curtis DR, Eccles RM (1958a) The excitation of Renshaw cells by pharmacological agents applied electrophoretically. J Physiol 141:435–445

Curtis DR, Eccles RM (1958b) The effect of diffusional barriers upon the pharmacology of cells within the central nervous system. J Physiol 141:446–463

Curtis DR, Watkins JC (1960) The excitation and depression of spinal neurones by structurally related amino acids. J Neurochem 6:117–141

del Castillo J, Katz B (1955) On the localization of acetylcholine receptors. J Physiol 128:157–181

Di Giovanni G, Ferraro G, Sardo P, Galati S, Esposito E, La Grutta V (2003) Nitric oxide modulates striatal neuronal activity via soluble guanylyl cyclase: an in vivo microiontophoretic study in rats. Synapse 48:100–107

Di Giovanni G, Shi W-X (2009) Effects of Scopolamine on Dopamine Neurons in the Substantia Nigra: Role of the Pedunculopontine Tegmental Nucleus. SYNAPSE 63(8):673–680

Gallager DW (1978) Benzodiazepines: potentiation of a GABA inhibitory response in the dorsal raphe nucleus. Eur J Pharmacol 49:133–143

Hicks TP (1984) The history of development of microiontophoresis in experimental neurobiology. Prog Neurobiol 22:185–240

Hill RG, Simmonds MA (1973) A method for comparing the potencies of -aminobutyric acid antagonists on single cortical neurones using micro-iontophoretic techniques. Br J Pharmacol 48(1):1–11

Invernizzi RW, Pierucci M, Calcagno E, Di Giovanni G, Di Matteo V, Benigno A, Esposito E (2007) Selective activation of 5-HT_{2C} receptors stimulates GABA-ergic function in the rat substantia nigra pars reticulata: a combined in vivo electrophysiological and neurochemical study. Neuroscience 144:1523–1535

Nastuk WL (1953) Membrane potential changes at a single endplate produced by transitory application of acetylcholine with an electrically controlled microjet. Fed Proc 12:102

Microsomal Ethanol-Oxidizing System

Synonyms

MEOS

Definition

Secondary pathway of alcohol metabolism in the liver which contributes to the conversion of alcohol into acetaldehyde. This process is upregulated following chronic alcohol use, resulting in metabolic tolerance.

Midazolam

Definition

Midazolam is an ultrashort-acting benzodiazepine derivative with potent anxiolytic, amnestic, hypnotic, anticonvulsant, skeletal muscle relaxant, and sedative properties. It has an elimination half-life of about 2 h. It is used in some countries for the short-term treatment of insomnia and in many countries as a premedication before surgery. Intravenous midazolam is indicated for procedural sedation (often in combination with an opioid, such as fentanyl), for preoperative sedation, for the induction of general anesthesia, and for sedation of ventilated patients in critical care units.

Cross-References

- Benzodiazepines

Mild Cognitive Impairment

Josef Marksteiner and Irene Adelt
Department of Psychiatry and Psychotherapy,
LKH Klagenfurt, Klagenfurt, Austria

Definition

Mild cognitive impairment (MCI) is a relatively recent term. It is used to describe individuals who have cognitive impairments beyond those expected for their age and education, but that do not interfere significantly with their daily activities (Petersen et al. 1999). The criteria for MCI are shown in the next section. It is considered to be a transitional stage between normal aging and various types of dementia. Although MCI can present with a variety of symptoms, when memory loss is the predominant symptom, it is termed "amnestic MCI" and is frequently seen as a risk factor for Alzheimer's disease (AD) (Morris et al. 2001). Individuals who have impairments in cognitive domains other than memory are classified as nonamnestic single- or multiple-domain MCI. The subtype of MCI may influence rates of progression to dementia and has a major influence on subsequent type of dementia diagnosis (Yaffe et al. 2006). Studies clearly suggest that MCI patients tend to progress to probable Alzheimer's disease at a significantly higher rate than healthy individuals of the same age. It is important that people with cognitive impairment are diagnosed as early as possible, so that they can benefit from therapeutic interventions.

Criteria for MCI

The criteria for MCI are those previously proposed by Petersen et al. (1999):

- Presence of a subjective memory complaint
- Preserved general intellectual functioning
- Demonstration of a memory impairment by cognitive testing
- Intact ability to perform activities of daily living
- Absence of dementia

The revised MCI criteria are those proposed by the Stockholm Group consensus:

- Presence of a cognitive complaint from either the subject or a family member
- Absence of dementia
- Change from normal functioning
- Decline in any area of cognitive functioning
- Preserved overall general functioning but possibly with increasing difficulty in the performance of activities of daily living

Mild Neurocognitive Disorder

The fifth edition of the American Psychiatric Association's (APA) Diagnostic and Statistical Manual (DSM-5) (American Psychiatric Association 2013) replaces the classic MCI construct with a new entity, mild neurocognitive disorder (mild NCD), which was previously placed in the appendix of DSM-IV-TR (American Psychiatric Association 2000). The APA's Fact Sheet for DSM-5 mild NCD describes the new category as providing an opportunity for early detection and treatment of cognitive decline before patients' deficits become more pronounced and progress to major neurocognitive disorder (dementia) or other debilitating conditions. Its inclusion in the manual is intended to help clinicians develop effective treatment plans as well as encourage researchers to evaluate diagnostic criteria and potential therapies (APA Fact Sheet 2013). The mild NCD category seems to attempt a more emphatic difference from normal aging by requiring a level of cognitive decline that requires compensatory strategies and accommodations to help maintain independence and perform activities of daily living. To be diagnosed with this disorder, there must be changes that impact cognitive functioning (APA Fact Sheet 2013). Mild NCD is primarily subtyped according to the etiological/pathological entities underlying the cognitive decline. To establish the diagnosis of mild NCD, an updated listing of cognitive domains, including complex attention, executive function, learning and memory, language, perceptual motor, or social cognition, is needed (American Psychiatric Association 2013).

Role of Pharmacotherapy

Epidemiology and Risk Factors

People with MCI are more likely to develop Alzheimer's or other dementias than are those without cognitive impairment. In fact, about half the people with MCI will progress to Alzheimer's disease within 5 years. In the general population, the prevalence of MCI was estimated to be about 3 % (Ritchie et al. 2001). In the same population, a prevalence rate of about 18 % was reported for age-associated cognitive decline, which is a similar concept to MCI but is related to normal cognitive aging processes rather than incipient dementia. Most of the studies described an increase in prevalence of MCI with age. A higher prevalence rate of MCI has been found in females.

Vascular diseases were identified as risk factors for MCI in some studies. Patients, particularly in general hospitals, represent a high-risk group for MCI, since risk factors like cardiovascular diseases are quite common (Bickel et al. 2006). MCI is also positively associated with stroke and peripheral arterial obstructive disease.

Depressive symptoms are highly prevalent among elderly MCI subjects and in cognitively normal elderly individuals. These symptoms are associated with an increased risk of developing MCI (Ravaglia et al. 2008). Depressive symptoms may also increase the conversion rate from MCI to dementia (Barnes et al. 2006). A synergistic interaction between apolipoprotein E genotype (epsilon3/epsilon4 or epsilon4/epsilon4) and depression was reported with regard to the incidence of MCI (Geda et al. 2006). Association between MCI and apoE allele 4 status largely depends on MCI subtype. For Alzheimer's disease, the apoE allele 4 status is probably the best-established risk factor.

Symptoms

How do the memory difficulties in MCI differ from those of normal aging? Numerous studies have examined the cognitive performance of patients with MCI. In general, information on cognitive performance over time is essential for the definition of MCI to substantiate a worsening over time. Neuropsychological examinations have demonstrated that, in general, these patients are impaired on tests of memory compared with age-matched healthy individuals. Forgetting things that are usually remembered may be a first symptom of MCI patients. MCI may encompass deficiencies in any or all of the following categories: language; visuospatial ability, for example, placement of things in time and space becomes more difficult; executive function, for example, decision-making becomes more challenging; and episodic memory, for example, what happened yesterday. A general recommendation for individuals concerned about their memory would be to discuss these concerns with their physician.

MCI patients may report distinct difficulties in other areas of cognition, such as naming objects or people and complex planning tasks. These symptoms are comparable to but less severe than the neuropsychological deficits found for Alzheimer's disease, especially for amnestic MCI. A careful interview may reveal that the MCI patient has mild difficulties with daily activities. These problems should be confirmed by a significant other.

People with MCI may also experience:

- Depression
- Irritability
- Anxiety
- Aggression
- Apathy

Diagnosis

A great deal of research has focused upon techniques to try to improve ways of identifying people with MCI. The diagnosis of MCI requires considerable clinical judgment, which consists of a comprehensive clinical assessment, including clinical observation, laboratory examinations, neuropsychological assessment, and neuroimaging. These examinations are also performed to rule out any alternate diagnosis which may lead to cognitive impairment. A similar assessment is usually performed for the diagnosis of Alzheimer's disease or other types of dementia.

As part of the physical examination, a neurological examination checks for signs of Parkinson's disease, strokes, tumors, or other medical conditions that can impair memory as well as physical function.

Laboratory Tests
Simple blood tests can rule out physical problems that can affect memory, such as vitamin B12 deficiency or an underactive thyroid gland.

Neuroimaging
There is emerging evidence that magnetic resonance imaging (MRI) can detect deterioration, including progressive loss of gray matter in the brain, from MCI to Alzheimer's disease. MRI-based volumetric measurements of medial temporal lobe structures can discriminate between normal elderly control subjects and patients with Alzheimer's disease. The extent of medial temporal lobe atrophy distinguishes probable Alzheimer's disease and amnestic MCI from healthy subjects. Neuroimaging also helps to monitor disease progression from a diagnosis of MCI to different stages of AD. Furthermore, MRI or CT is necessary to rule out the possibility of a tumor, evidence of stroke, or bleeding, which could also cause some of the symptoms seen in MCI patients.

Biomarker
Currently, a major focus of research is on biomarkers in Alzheimer's disease and in MCI patients. A promising area of research for biochemical diagnosis of AD and mixed forms of dementia is the analysis of cerebrospinal fluid (CSF).

Biological markers can serve as early diagnostic indicators and as markers of preclinical pathological change. This approach is likely to take on an increasingly important role in the diagnosis and monitoring of AD and probably also in MCI. Some biomarkers are currently in the process of implementation as outcome variables. In a large multicenter study, the CSF biomarkers Aβ42, T-tau, and P-tau could predict with good accuracy which MCI patients developed AD, which replicates finding earlier smaller studies (Mattsson et al. 2009). Sensitivity, specificity, and ease of use are the most important factors that ultimately define the usefulness of a biomarker for diagnosis.

Neuropathology
MCI often causes the same types of brain changes seen in Alzheimer's disease or other forms of dementia. The difference between MCI and other types of dementia lies in the extent of these changes. The available data suggest that MCI is associated with the early stages of the neuropathological changes that are found in the lesions of AD, including the accumulation of neuritic plaques, neurofibrillary tangles, synaptic and neurotransmitter associated deficits, and significant neuronal cell death. There is evidence suggesting that even amnestic MCI patients may not meet the neuropathological criteria for Alzheimer's disease and that these patients may be in a transitional stage of evolving Alzheimer's disease.

Treatments and Drugs
In general, there is no cure for MCI. There are several medications, as well as many nonpharmacological approaches, that can potentially improve MCI-related symptoms. Treatment of coexisting conditions, such as high blood pressure or depression, may help reduce cognitive problems. As MCI may represent a prodromal state to clinical Alzheimer's disease, treatments proposed for Alzheimer's disease, such as antioxidants and cholinesterase inhibitors, have also been tested for MCI. Patients with MCI are frequently treated with "off-label" cholinesterase inhibitors and memantine, as well as other possible cognition-enhancing drugs.

Acetylcholinesterase Inhibitors (AChEIs) and Memantine
Randomized, placebo-controlled trials examining the therapeutic value of acetylcholinesterase inhibitors (AChEIs) have been performed during the last several years. All of them have had negative outcomes with regard to the primary outcome parameter, which was to prevent the conversion from MCI to Alzheimer's disease. Negative results have also been obtained with steroidal and anti-inflammatory compounds and antioxidants. As these compounds are thought to target the early steps in the pathophysiological cascade of dementia, the negative results

are disappointing. So far, no disease-modifying drugs with proven efficacy are available for the treatment of MCI patients. For example, rivastigmine showed no benefit in delaying the progression to Alzheimer's disease or on cognitive function for individuals with MCI (Feldman et al. 2007). Donepezil showed only minor, short-term benefits. During the first year of a 3-year study, the rate of progression from MCI to Alzheimer's was significantly lower in the people who took donepezil (Petersen et al. 2005). However, that difference disappeared by the end of the study. A recent study showed that depression is predictive of progression from amnestic MCI to Alzheimer's disease, and treatment with donepezil delayed progression to AD among depressed subjects with amnestic MCI (Lu et al. 2009). Similarly, galantamine does not change the conversion rate from MCI to Alzheimer's disease, but may increase the risk of sudden death from heart attacks and strokes when used in people who have MCI (Winblad et al. 2008). However, additional clinical trials are in progress to determine if any medications will prevent or delay the rate of progression from MCI to dementia.

High Blood Pressure Drugs

People who have MCI are also more likely to have problems with the blood vessels inside their brains. High blood pressure can worsen these problems and cause memory difficulties. Therefore, antihypertensive drugs are under investigation to determine whether they can reduce the conversion rate from MCI to dementia. Because of other medical complications, it is essential to keep blood pressure at normal levels.

Antidepressants

Depression is common in people who have MCI, and depression itself can cause memory problems. Treating depression may help to improve memory. However, studies to date in this area have been of short duration, and it is not clear whether there is a longer-term benefit associated with antidepressant treatment. Further studies are needed to investigate the role of antidepressants in depressed MCI patients, especially to ascertain whether these drugs can reduce the conversion rate from MCI to dementia.

Antioxidants

The antioxidant vitamin E may help to protect brain cells from the oxidative stress that appears to play a role in dementia, but it works no better than placebo in relieving symptoms or delaying the progression of MCI. Ginkgo appears to improve memory and concentration in older adults with no major memory problems, but it is still uncertain if ginkgo can reduce the memory problems associated with MCI.

Physical Activity

Observational studies have shown that physical activity reduces the risk of cognitive decline. In a randomized controlled trial of a 24-week physical activity intervention, participants in the intervention group improved, suggesting that a 6-month program of physical activity provided a modest improvement in cognition over an 18-month follow-up period (Lautenschlager et al. 2008). In general, however, it is still controversial as to whether physical activities can prevent or reverse MCI. Nevertheless, physical activity can be part of a healthy lifestyle for older people with or without MCI.

Conclusion

The concept of MCI is currently of high interest. A number of studies have been dealing with different aspects of MCI with regard to diagnosis and therapeutic strategies. Up to now, no drug is licensed for the indication of MCI. However, there is hope that new compounds may be efficacious in the treatment of MCI to reduce its conversion rate to dementia.

References

American Psychiatric Association (2000) Diagnostic and statistical manual of mental disorders, 4th edn, text revision. American Psychiatric Publishing, Washington, DC

American Psychiatric Association (2013) Diagnostic and statistical manual of mental disorders, 5th edn. American Psychiatric Publishing, Washington, DC

APA Fact Sheet (2013) Mild neurocognitive disorder. http://www.dsm5.org/Documents/Mild Neurocognitive Disorder Fact Sheet.pdf

Barnes DE, Alexopoulos GS, Lopez OL, Williamson JD, Yaffe K (2006) Depressive symptoms, vascular disease, and mild cognitive impairment: findings from the Cardiovascular Health Study. Arch Gen Psychiatry 63:273–279

Bickel H, Mosch E, Seigerschmidt E, Siemen M, Forstl H (2006) Prevalence and persistence of mild cognitive impairment among elderly patients in general hospitals. Dement Geriatr Cogn Disord 21:242–250

Feldman HH, Ferris S, Winblad B et al (2007) Effect of rivastigmine on delay to diagnosis of Alzheimer's disease from mild cognitive impairment: the InDDEx study. Lancet Neurol 6:501–512

Geda YE, Knopman DS, Mrazek DA, Jicha GA, Smith GE, Negash S, Boeve BF, Ivnik RJ, Petersen RC, Pankratz VS, Rocca WA (2006) Depression, apolipoprotein E genotype, and the incidence of mild cognitive impairment: a prospective cohort study. Arch Neurol 63:435–440

Lautenschlager NT, Cox KL, Flicker L, Foster JK, van Bockxmeer FM, Xiao J, Greenop KR, Almeida OP (2008) Effect of physical activity on cognitive function in older adults at risk for Alzheimer disease: a randomized trial. JAMA 300:1027–1037

Lu PH, Edland SD, Teng E, Tingus K, Petersen RC, Cummings JL (2009) Donepezil delays progression to AD in MCI subjects with depressive symptoms. Neurology 72:2115–2121

Mattsson N, Zetterberg H, Hansson O, Andreasen N, Parnetti L, Jonsson M, Herukka SK, van der Flier WM, Blankenstein MA, Ewers M, Rich K, Kaiser E, Verbeek M, Tsolaki M, Mulugeta E, Rosen E, Aarsland D, Visser PJ, Schroder J, Marcusson J, de Leon M, Hampel H, Scheltens P, Pirttila T, Wallin A, Jonhagen ME, Minthon L, Winblad B, Blennow K (2009) CSF biomarkers and incipient Alzheimer disease in patients with mild cognitive impairment. JAMA 302:385–393

Morris JC, Storandt M, Miller JP, McKeel DW, Price JL, Rubin EH, Berg L (2001) Mild cognitive impairment represents early-stage Alzheimer disease. Arch Neurol 58:397–405

Petersen RC, Smith GE, Waring SC, Ivnik RJ, Tangalos EG, Kokmen E (1999) Mild cognitive impairment: clinical characterization and outcome. Arch Neurol 56:303–308

Petersen RC, Thomas RG, Grundman M, Bennett D, Doody R, Ferris S, Galasko D, Jin S, Kaye J, Levey A, Pfeiffer E, Sano M, van Dyck CH, Thal LJ (2005) Vitamin E and donepezil for the treatment of mild cognitive impairment. N Engl J Med 352:2379–2388

Ravaglia G, Forti P, Lucicesare A, Rietti E, Pisacane N, Mariani E, Dalmonte E (2008) Prevalent depressive symptoms as a risk factor for conversion to mild cognitive impairment in an elderly Italian cohort. Am J Geriatr Psychiatry 16:834–843

Ritchie K, Artero S, Touchon J (2001) Classification criteria for mild cognitive impairment: a population-based validation study. Neurology 56:37–42

Winblad B, Gauthier S, Scinto L, Feldman H, Wilcock GK, Truyen L, Mayorga AJ, Wang D, Brashear HR, Nye JS (2008) Safety and efficacy of galantamine in subjects with mild cognitive impairment. Neurology 70:2024–2035

Yaffe K, Petersen RC, Lindquist K, Kramer J, Miller B (2006) Subtype of mild cognitive impairment and progression to dementia and death. Dement Geriatr Cogn Disord 22:312–319

Milnacipran

Synonyms

Milnacipran hydrochloride

Definition

Milnacipran, under the brand name Ixel, became available in the late 1990s in Europe and several other countries for the treatment of major depressive disorder. It was approved in 2009 for the treatment of fibromyalgia in the USA.

Minor Tranquilizer

Synonyms

Antianxiety medication; Anxiolytic; Sedative; Tranquilizer

Definition

A medication used in the treatment of anxiety disorders and also as a nonspecific sedative. It is a more precise term for the drugs used to lessen anxiety than its synonyms. The distinction from sedative is not a clear one. Minor tranquilizers lie on a continuum, from mild antianxiety effects, through more pronounced effects and induction of sleep and relaxation, to anesthesia of increasing depth.

Cross-References

▶ Anxiolytics
▶ Generalized Anxiety Disorder

Mirtazapine

Synonyms

Remeron; Zispin

Definition

Mirtazapine is the first atypical antidepressant with noradrenergic and specific serotonergic receptor antagonist properties (NaSSa); it was introduced in 1994 and has fewer serotonergic, anticholinergic, and antiadrenergic side effects than ▶ tricyclic antidepressants but maintains comparable effectiveness. It also has beneficial effects on symptoms of anxiety and sleep disturbances. Its most common side effects are increased appetite and subsequent weight gain, drowsiness, and dizziness.

Cross-References

▶ Antidepressants

Mismatch Negativity

Synonyms

Mismatch field; MMN

Definition

Mismatch negativity (MMN) is similar to the P300 in that it is an event-related potential component seen when subjects are presented with a rare or "deviant" stimulus in a train of frequent or regular stimuli. It can be found using stimuli in a variety of sensory modalities but has been most researched using auditory or visual stimuli. The deviant stimuli can vary from the frequent ones in any way, for example, tone, pitch, and loudness. Unlike the P300, subjects do not need to be paying conscious attention to the stimuli. MMN is a negative potential observed over fronto-central scalp with a typical latency of 150–250 ms after the onset of the deviant stimulus.

Cross-References

▶ Event-Related Potential

Misoprostol

Synonyms

Cytotec

Definition

Misoprostol is an FDA-approved drug for the prevention of nonsteroidal anti-inflammatory drug-induced gastric ulcers. Misoprostol is also widely used by obstetricians and gynecologists for the induction of labor, therapeutic abortion, and the early termination of pregnancy. Chemically, misoprostol is a synthetic prostaglandin E1 analogue. The most commonly reported adverse effects of misoprostol are diarrhea, abdominal pain, nausea, and headache.

Cross-References

▶ Autism: Animal Models

Mitochondrial Complex Chain

Definition

Mitochondria provide the energy source of cells by synthesis of adenosine triphosphate (ATP).

A series of enzymes and cofactors (complexes I–V) in the inner membrane of the mitochondrion oxidize carbohydrates to release electrons that drive the proton pumps necessary to provide the energy for ATP synthesis. A range of drugs and toxins act on these enzyme complexes to disrupt or enhance energy production via the mitochondrial electron transport chain, affecting cell metabolism and ultimately cell survival.

Mobile Phase

Definition

In high-pressure liquid chromatography (HPLC) the mobile phase is the buffer that is pumped through the chromatographic column. All chemicals used to prepare the buffer should be of at least HPLC grade, and they should be made up using ultrapure, HPLC-grade water. In addition, the buffer should be thoroughly filtered which is best achieved by vacuum filtration. This has the benefit that it also removes dissolved gas in the mobile phase that otherwise can come out of solution causing minute air bubbles to form. These get lodged in the column, with resultant adverse effects on the separation, or pass through to the detector, where they make irregular baseline noise, and rapid spikes.

Cross-References

▶ High-Pressure Liquid Chromatography

Moclobemide

Synonyms

Aurorix; Manerix

Definition

Moclobemide is an antidepressant that inhibits reversibly and preferentially MAO-A. It was introduced in 1977 but was not approved by the FDA in the USA. It is mainly used in the treatment of major depression and social anxiety, and it is claimed to have a favorable side-effect profile but with equal effectiveness to ▶ tricyclic antidepressants and SSRIs within 1 week of treatment. It is rapidly absorbed and has a relatively short ▶ half-life, but the CNS effects persist for many hours, and it is considered safe.

Cross-References

▶ Antidepressants
▶ Monoamine Oxidase Inhibitors

Modafinil

Michael Minzenberg
Department of Psychiatry, Imaging Research Center, University of California, Davis Medical Centre, Sacramento, CA, USA

Synonyms

2-[(diphenylmethyl)sulfinyl]acetamide

Definition

Modafinil is a non-amphetamine psychostimulant currently FDA-approved for the treatment of sleepiness in narcolepsy and shift-work sleep disorder.

Pharmacological Properties

Modafinil (brand name Provigil) is a racemate, with the two enantiomers being approximately equipotent in behavioral effects, but different in pharmacokinetic profile Robertson and Hellriegel (2003). The R-enantiomer (armodafinil) reaches higher plasma concentrations than the racemic form between 6 and 14 h after administration, with a longer duration of wake-promoting activity in healthy adults. Modafinil is readily absorbed after single or multiple oral doses, reaching peak plasma concentrations 2–4 h after administration. The presence of food in the gastrointestinal tract can slow the rate of absorption but does not affect the total extent of absorption. Steady-state plasma concentrations are achieved between 2 and 4 days with repeated dosing. It is highly lipophilic, and approximately 60 % bound to plasma proteins, primarily albumin. Major pharmacokinetic parameters are independent of doses in the range of 200–600 mg/day. The major circulating metabolites, modafinil acid and modafinil sulfone, do not exert any significant activity in the brain or the periphery. The elimination half-life is approximately 12–15 h, and single daily dosing is adequate and common in clinical practice. Elimination occurs primarily in the liver, via amide hydrolysis and, to a lesser extent, by cytochrome P450-mediated oxidation. Excretion occurs in the urine, with less than 10 % of the oral dose excreted as the unchanged drug. Elimination is slow in the elderly and in individuals with hepatic or renal impairment. Some drug-drug interactions are apparent with modafinil. In vitro, modafinil exerts a reversible inhibition of CYP 2C19, a smaller but concentration-dependent induction of CYP 1A2, 2B6, and 3A4, and a suppression of 2C9 activity. There are significant interactions of modafinil with ethinylestradiol and triazolam, though not with methylphenidate, dextroamphetamine, or warfarin.

Neurochemical Effects of Modafinil

Effects of Modafinil on Catecholamine Systems

Modafinil is structurally unrelated to amphetamine, with a differing profile of pharmacological and behavioral effects. While the in vitro potency of modafinil in binding both the dopamine transporter (DAT) and norepinephrine transporter (NET) is low relative to methylphenidate, bupropion, and benztropine, modafinil shows DAT occupancy comparable to methylphenidate at clinically relevant doses (Madras et al. 2006).

Modafinil has a complex profile of effects on central dopamine (DA) and norepinephrine (NE) systems, lacking many neurochemical and behavioral effects observed with amphetamine administration Minzenberg and Carter (2008). For instance, in contrast to amphetamine, modafinil does not significantly affect DA release or turnover; it shows negligible effects on cerebral cortical blood flow and shows different patterns of metabolic activation compared to amphetamine; it does not produce behavioral stereotypies or rebound hypersomnia; and in healthy humans, modafinil has effects on the resting EEG that are distinct from amphetamine. Nevertheless, a parenteral administration of modafinil does raise extracellular DA levels in the rat prefrontal cortex (de Saint et al. 2001), and in the caudate nucleus of narcoleptic dogs, though only minimally in the rat hypothalamus (de Saint et al. 2001). It causes a modest increase in DA in the accumbens after intraperitoneal doses up to 300 mg/kg. In rat brain slices, modafinil inhibits the activity of ventral tegmental area DA neurons, an effect that appears to be mediated by D2 receptors. This suggests that modafinil inhibits DA reuptake, leading to DA cell body autoreceptor activation, to diminish DA cell firing. Modafinil's effects on wakefulness are abolished in DAT knockout mice. In a rodent drug discrimination paradigm, modafinil partially generalizes to a cocaine-like stimulus; in addition, modafinil's effects on activity levels in mice are modestly attenuated by D1 receptor antagonism, though not by D2 antagonism. Modafinil reduces blood prolactin levels in humans, without effects on blood growth hormone or thyroid-stimulating hormone. Overall, these findings suggest that modafinil's effects on arousal and behavioral activity are at least partly mediated by synaptic DA, but in a manner differing from that of amphetamine, and possibly favoring corticostriatal over subcortical limbic circuits.

Modafinil also elevates extracellular NE levels in prefrontal cortex (along with DA) and hypothalamus. Pretreatment with α-adrenergic receptor antagonists diminishes modafinil-induced increases in arousal and activity in rats and monkeys. However, modafinil does not reduce cataplexy in dogs or humans with narcolepsy, a feature that is similar to other DAT inhibitors, and in contrast to α1B receptor agonists and NET inhibitors. In addition, pretreatment with low doses of the α2 antagonist yohimbine potentiates modafinil-induced wakefulness and activity, whereas higher doses attenuate the activity increases. This biphasic response to yohimbine suggests that low doses may preferentially block the inhibitory terminal α2 autoreceptor to enhance NE release and thus augment postsynaptic adrenergic receptor activation by modafinil, whereas higher doses also block postsynaptic α2 receptors, attenuating modafinil effects. These findings make it likely that postsynaptic α2 receptors mediate some of the behavioral effects of modafinil. Importantly, modafinil also augments pupillary dilation parameters in a manner consistent with phasic activity of locus coeruleus neurons. This effect may also be mediated through α2 receptor activation; in this case, those receptors (autoreceptors) are located on locus coeruleus cell bodies. Modest attenuation of modafinil-induced arousal and activity has also been observed after pretreatment with the β-blocker propranolol, suggesting that postsynaptic β receptors also mediate these modafinil effects.

Taken together, these varied findings suggest that modafinil may potentiate both DA and NE neurotransmission. It appears likely that the elevations in extracellular NE observed after modafinil are responsible for the majority of the adrenergic receptor-mediated effects, which may involve α2, α1, and β receptors. D1 and D2 receptors probably also mediate modafinil's effects on cognition and behavior. In addition, however, Wisor and Eriksson (2005) have proposed that the elevated synaptic DA resulting from DAT inhibition may lead to DA activation of adrenergic receptors. There remains the possibility that enhanced DA in the prefrontal cortex results from competition with increased NE levels for binding to the NET, which plays an important role in terminating DA action in the prefrontal cortex. DA has an affinity for cloned mouse α1B receptors that is on the same order of magnitude as NE, and DA can activate adrenergic receptors in various brain regions. These observations suggest a mechanism whereby the modafinil inhibition of DAT inhibition may be related to adrenergic receptor-mediated behavioral effects.

Effects of Modafinil on GABA, Glutamate, and Serotonin Systems

Modafinil also has consistent effects on central glutamate and gamma aminobutyric acid (GABA) neurotransmitter systems. The regional effects on extracellular glutamate occur at ascending doses in this order: thalamus = hypothalamus < striatum = hippocampus. Glutamate levels in the globus pallidus and substantia nigra are unchanged after the highest doses administered. These effects on glutamate may interact with adrenergic mechanisms.

Modafinil causes a dose-dependent decrease in extracellular GABA. These regional GABA effects occur at ascending doses in this order: cortex < striatum/pallidum = hypothalamus < thalamus = hippocampus = substantia nigra = nucleus accumbens.

The effects on extracellular GABA may be mediated by modafinil's effects on other neurotransmitter systems. Cortical GABA effects require intact catecholamine neurons. In addition, modafinil elevates extracellular serotonin (5HT) in the frontal cortex, central nucleus of the amygdala, and dorsal raphe nucleus, but minimally in the hypothalamus. Modafinil and the 5HT reuptake inhibitors fluoxetine, paroxetine, and imipramine mutually enhance the effects of each other on elevations in cortical 5HT. Taken together, this literature suggests that modafinil effects on GABA are at least partly mediated by 5HT. Ultimately, modafinil's effects on GABA may be mediated by adrenergic effects on 5HT activity.

Effects of Modafinil on Orexin and Histamine Systems

The clinical efficacy of modafinil in narcolepsy, a condition characterized by deficient orexin (hypocretin) in the brain, suggests that modafinil may have clinically relevant effects on this neurochemical system. Modafinil does activate orexin cells in the perifornical area of mice and rats. However, modafinil induces wakefulness more potently in orexin knockout mice than in wild-type mice, with similar patterns of Fos-immunoreactivity. In addition, modafinil does not bind to the orexin 1 receptor and retains effects on both extracellular striatal DA and wake-promoting activity in orexin 2 receptor-deficient narcoleptic dogs. Therefore, modafinil's effects on arousal do not appear to be mediated through the orexin system, and the precise role of orexin in the cognitive and clinical effects of modafinil remains unknown. Modafinil also elevates extracellular histamine (HA) in the anterior hypothalamus. However, a direct injection of modafinil into the tuberomammillary nucleus (the site of HA cell bodies) does not affect HA release. Given the multiple effects on catecholamines, 5HT, and GABA described earlier for modafinil, it appears likely that modafinil's effects on HA are mediated by one or more of these other neurotransmitter systems.

Effects of Modafinil on Cognition

Studies in rodents indicate that modafinil can improve working memory performance in a dose- and delay-dependent manner and that the processing of contextual cues is also enhanced with modafinil Minzenberg and Carter (2008). These effects may be augmented with sustained dosing regimens. In healthy humans (with or without sleep deprivation), working memory, recognition memory, sustained attention, and other tasks dependent on cognitive control (and on function of the prefrontal cortex) are enhanced with modafinil Minzenberg and Carter (2008); Minzenberg et al. (2008). Some evidence suggests that the magnitude of modafinil's effects in healthy adults may depend on underlying cognitive abilities. Those with high general intellectual abilities, or high performance in specific cognitive domains, appear to exhibit less improvement after modafinil, suggesting that these individuals already experience optimal levels/patterns of catecholamine activity in the modulation of cognition. Among psychiatric populations, there is now consistent evidence that modafinil (in well-tolerated dosing regimens) improves attention and response inhibition in children and adolescents with attention-deficit/hyperactivity disorder (ADHD) Turner et al. (2004). These improvements in cognition may form the basis for clinical efficacy in ADHD. Among adult psychiatric patients, modafinil improves several cognitive functions dependent on the prefrontal cortex in schizophrenia, major depression, and adult ADHD, with some null findings reported in schizophrenia. However, these studies have significant limitations evident in their design. The range of clinical samples and cognitive functions that are subjected to empirical study with modafinil is expected to expand in the future.

Clinical Effects of Modafinil

Modafinil has consistently shown efficacy in measures of alertness in narcolepsy and shift-work sleep disorder in randomized, double-blind placebo-controlled studies. In these studies, modafinil has shown efficacy in open-label extension phases extending for as long as 136 weeks, and it has been well tolerated, with no evidence of significant adverse events or abuse. Modafinil has also been evaluated for the treatment of fatigue and sedation in a number of other neurological and medical conditions, including multiple sclerosis, idiopathic Parkinson's disease, chronic fatigue syndrome, polio, HIV infection, dementias, obstructive sleep apnea, postanesthetic sedation, and fibromyalgia, with generally favorable but somewhat mixed results (Ballon and Feifel 2006).

In samples of adult psychiatric patients, two studies of patients with major depression have found significant improvements in mood symptoms on modafinil compared to placebo Fava et al. (2005), and modafinil has been associated with greater rates of abstinence in cocaine-dependent adults. It also shows clinical efficacy in adults with ADHD Greenhill et al. (2006). In contrast, adjunctive modafinil has shown modest and inconsistent efficacy for symptoms of schizophrenia, though these studies have been plagued by small sample sizes and other methodological limitations.

In childhood/adolescent ADHD, modafinil improved parent, teacher, and clinician ratings of ADHD symptoms in several short-term (4–9 weeks), randomized, double-blind, placebo-controlled trials, at mean doses ranging from 195 to 368 mg daily.

Throughout these clinical trials, modafinil has been well tolerated. However, case reports have appeared describing significant adverse events in routine clinical use of modafinil, including the exacerbation of psychosis, acute mania, clozapine toxicity, premature ventricular contractions, irritability, and verbal aggression. Nonetheless, these events have not been observed at a significant rate in modafinil-treated patients compared to placebo-treated patients in clinical trials, and no serious (e.g., life-threatening) sequelae have ensued in these reported isolated cases. Modafinil also appears to have a relatively low potential for abuse, which may be a function of its pharmacodynamic profile and/or its physical properties, being insoluble in water and unstable at high temperatures, which minimizes its bioavailability upon smoking or intravenous use. Nonetheless, careful clinical judgment should be exercised in the decision to initiate therapy with modafinil, with particular attention to both its side effect profile and its potential for drug-drug interactions.

Cross-References

▶ Attention-Deficit and Disruptive Behavior Disorders
▶ Drug Interactions

References

Ballon JS, Feifel D (2006) A systematic review of modafinil: potential clinical uses and mechanisms of action. J Clin Psychiatry 67:554–566

de Saint HZ, Orosco M, Rouch C, Blanc G, Nicolaidis S (2001) Variations in extracellular monoamines in the prefrontal cortex and medial hypothalamus after modafinil administration: a microdialysis study in rats. Neuroreport 12:3533–3537

Fava M, Thase ME, DeBattista C (2005) A multicenter, placebo-controlled study of modafinil augmentation in partial responders to selective serotonin reuptake inhibitors with persistent fatigue and sleepiness. J Clin Psychiatry 66:85–93

Greenhill LL, Biederman J, Boellner SW, Rugino TA, Sangal RB, Earl CQ et al (2006) A randomized, double-blind, placebo-controlled study of modafinil film-coated tablets in children and adolescents with attention-deficit/hyperactivity disorder. J Am Acad Child Adol Psychiatry 45:503–511

Madras BK, Xie Z, Lin Z, Jassen A, Panas H, Lynch L et al (2006) Modafinil occupies dopamine and norepinephrine transporters in vivo and modulates the transporters and trace amine activity in vitro. J Pharmacol Exp Ther 319:561–569

Minzenberg MJ, Carter CS (2008) Modafinil: a review of neurochemical actions and effects on cognition. Neuropsychopharmacology 33(7):1477–1502

Minzenberg MJ, Watrous AJ, Yoon JH, Ursu S, Carter CS (2008) Modafinil Shifts human locus coeruleus to low-tonic, high-phasic activity during functional MRI. Science 322(5908):1700–1702

Robertson P Jr, Hellriegel ET (2003) Clinical pharmacokinetic profile of modafinil. Clin Pharmacokinet 42:123–137

Turner DC, Clark L, Pomarol-Clotet E, McKenna P, Robbins TW, Sahakian BJ (2004) Modafinil improves cognition and attentional set shifting in patients with chronic schizophrenia. Neuropsychopharmacology 29:1363–1373

Wisor JP, Eriksson KS (2005) Dopaminergic-adrenergic interactions in the wake promoting mechanism of modafinil. Neuroscience 132:1027–1034

Molindone

Definition

Molindone, a primarily dopamine D2 blocking dehydroindolone antipsychotic with an ▶ elimination half-life of 6.5 h, is mainly metabolized by 2D6 CYP450 isoenzymes. It is normally regarded as a first-generation antipsychotic.

Cross-References

▶ First-Generation Antipsychotics

Monoamine Oxidase Inhibitors

Andrew Holt[1], Dmitriy Matveychuk[2], Darrell D. Mousseau[3] and Glen B. Baker[2]
[1]Department of Pharmacology, University of Alberta, Edmonton, AB, Canada
[2]Neurochemical Research Unit and Bebensee Schizophrenia Research Unit, Department of Psychiatry, University of Alberta, Edmonton, AB, Canada
[3]Cell Signalling Laboratory, Department of Psychiatry, University of Saskatchewan, Saskatoon, SK, Canada

Definition

Monoamine oxidase inhibitors (MAOIs) inhibit the enzyme MAO, of which there are two major isoforms, MAO-A and MAO-B.

Pharmacological Properties

History

MAOIs were originally developed as antidepressants because of their ability to restore brain levels of the biogenic amines noradrenaline and 5-hydroxytryptamine (serotonin), both of which are thought to be functionally deficient in depression. Yet, several of these drugs have also proven useful in the treatment of panic disorder, social anxiety disorder, eating disorders, pain syndromes, Parkinson's disease, and Alzheimer's disease. MAOIs are no longer first-line drugs for most depressive or anxiety disorders, but remain important agents when first- or second-line drugs prove to be ineffective or intolerable; they are especially effective in atypical depression and depression associated with anxiety, panic, or phobias (Kennedy et al. 2009). The MAOIs may be classified according to selectivity (for either the MAO-A or MAO-B isoform) and reversibility (reversible or irreversible) (see Kennedy et al. 2009 and Stahl and Felker 2008 for details about properties and distribution of MAO-A and MAO-B). Noradrenaline and serotonin are preferred substrates for MAO-A, while β-phenylethylamine and benzylamine are preferred substrates for MAO-B. Selective inhibitors of MAO-A and MAO-B include clorgyline and L-deprenyl, respectively. Figure 1 and Table 1 list several MAOIs, some of which are currently available clinically, and others that are used only as research tools or that are in preclinical development.

Side Effects

While the inhibition of MAO-A is required for antidepressant activity, clinically used nonselective inhibitors of MAO reduce the degree of inactivation of dietary sympathomimetic amines, such as tyramine, by MAO-A in the gut wall and by MAO-A and MAO-B in the liver. When the inhibitor acts irreversibly, the substantial increase in systemic tyramine stimulates the release of noradrenaline from sympathetic varicosities in the vascular wall. This also occurs with clorgyline, despite its lack of effect upon hepatic MAO-B. As a consequence of MAO-A inhibition, noradrenaline accumulates and produces a series of symptoms, usually starting with headaches and, if not dealt with appropriately, potentially culminating in a hypertensive crisis. This adverse effect is called the "cheese effect", so named because it was originally observed in patients taking MAOIs who had ingested foods such as aged cheeses that have high tyramine content. This food-drug interaction can occur with some of the MAOIs used currently in the clinic, including tranylcypromine, phenelzine, and isocarboxazid (all of which act as irreversible inhibitors of both MAO-A and MAO-B). Thus, when patients receive a prescription for such MAOIs, they should be informed by their physician and pharmacist of the foods and beverages that should be avoided while on these medications (Stahl and Felker 2008).

Concern about these food-drug interactions led to the development of selective MAO-B inhibitors such as L-deprenyl (selegiline) as putative antidepressants. However, L-deprenyl did not exhibit antidepressant effects until selectivity was lost at higher doses that also caused irreversible MAO-A inhibition. Nonetheless, L-deprenyl

Monoamine Oxidase Inhibitors, Fig. 1 Some monoamine oxidase inhibitors and their structures

Monoamine Oxidase Inhibitors, Table 1 Some monoamine oxidase inhibitors and their actions (Table modified from Kennedy et al. 2009)

Generic name	Reversible/irreversible selectivity
Iproniazid	Irreversible, nonselective
Isoniazid	Irreversible, nonselective
Phenelzine	Irreversible, nonselective
Isocarboxazid	Irreversible, nonselective
Tranylcypromine	Irreversible, nonselective
Clorgyline	Irreversible, MAO-A selective
Pargyline	Irreversible, MAO-B selective
Selegiline	Irreversible, MAO-B selective
Moclobemide	Reversible, MAO-A selective
Brofaromine	Reversible, MAO-A selective
Lazabemide	Very slowly reversible, MAO-B selective
Mofegiline	Irreversible, MAO-B selective
Milacemide	Slowly reversible, MAO-B selective
Toloxatone	Reversible, MAO-A selective
Befloxatone	Reversible, MAO-A selective
Pirlindole	Reversible, MAO-A selective
Rasagiline	Irreversible, MAO-B selective
Ladostigil	Irreversible, MAO-B selective
PF 9601 N	Irreversible, MAO-B selective
Aliphatic N-propargylamines	Irreversible, MAO-B selective

is used in the treatment of Parkinson's disease and has been demonstrated to have neuroprotective properties in a large number of neurotoxicity tests in vivo and in vitro (Maruyama and Naoi 2013; Youdim 2013). In fact, the neuroprotective actions of this drug have stimulated considerable research into related drugs as potential neuroprotective agents (rasagiline is an example of productive research in this area), though such effects may be mediated through mechanisms distinct from MAO inhibition (vide infra). Transdermal L-deprenyl has recently been reported to be effective as an antidepressant (Culpepper and Kovalick 2008; Stahl and Felker 2008); under these conditions, the drug is delivered directly into the systemic circulation, thereby avoiding extensive first-pass metabolism and reaching the brain in sufficiently high concentrations to inhibit both MAO-A and MAO-B.

In a further effort to develop antidepressants that would not precipitate the "cheese effect", reversible inhibitors of MAO-A (RIMAs) were developed. Moclobemide, a competitive RIMA available clinically for several years, retains antidepressant properties, but the risk of a hypertensive crisis is reduced since dietary tyramine can still compete with moclobemide for the active site on MAO-A. RIMAs have a further advantage over irreversible inhibitors in that after cessation of treatment, enzyme activity increases as inhibitors are cleared from the body and recovery is usually complete in 2–5 days. This is in sharp contrast to a period of approximately 2 weeks that is required before synthesis of new enzyme restores MAO activity to normal levels following cessation of treatment with irreversible inhibitors. This is an important consideration when shifting a patient to a drug regimen that would be contraindicated for concomitant use with a MAOI (e.g., a selective serotonin reuptake inhibitor [SSRI] that, in combination with MAO inhibition, would result in a significant and dangerous elevation in levels of serotonin).

A hypertensive crisis is a feared adverse effect associated with irreversible MAOIs, though this is usually the result of a food-drug or drug-drug interaction (Gillman 2011). Paradoxically, orthostatic hypotension is a more common adverse cardiovascular effect with MAOIs. Serotonin syndrome may result if a MAOI is given with another drug that also increases the availability of serotonin (see Table 2). Insomnia can be a problem with tranylcypromine (a MAOI with a structure similar to that of amphetamine), and the irreversible MAOIs can produce weight gain, peripheral edema, and sexual dysfunction (Kennedy et al. 2009). A discontinuation syndrome (arousal, mood disturbances, and somatic symptoms) may occur with phenelzine or tranylcypromine if treatment is discontinued abruptly.

Drug-Drug Interactions

In addition to the food-drug interaction mentioned earlier, drug-drug interactions must be taken into consideration when prescribing

Monoamine Oxidase Inhibitors, Table 2 Potential drug-drug interactions involving monoamine oxidase inhibitors (Table modified from Kennedy et al. (2009) and Stahl and Felker (2008))

Interacting drugs	Examples	Possible result
Drugs that stimulate the release of or inhibit the reuptake of noradrenaline at sympathetic neurons; decongestants	Amphetamines, methylphenidate, ephedrine, phenylephrine, phenylpropanolamine, pseudoephedrine, oxymetazoline, some antidepressants, tramadol, sibutramine, phentermine	Hypertension
Drugs metabolized by monoamine oxidase	Phenylephrine (oral), sumatriptan, citalopram	Hypertension, increased serum levels of sumatriptan, citalopram
Drugs that inhibit serotonin reuptake	SSRIs, clomipramine, imipramine, meperidine, dextromethorphan, propoxyphene, venlafaxine, chlorpheniramine, brompheniramine, tramadol	Serotonin syndrome, confusion, agitation, hypomania, sweating, myoclonus, fever, coma, possible fatality
Serotonin agonists	Sumatriptan	Serotonin syndrome
β-Blockers		Increased hypotension, bradycardia
Oral hypoglycemics		Increased hypoglycemic affects

MAOIs; some of these may be life-threatening. These potential drug-drug interactions are summarized in Table 2.

Metabolism

Most of the MAOIs are metabolized extensively and various cytochrome P450 (CYP) enzymes are involved in this metabolism (see Kennedy et al. 2009). Interestingly, phenelzine is metabolized, in part, through oxidation by MAO.

Beyond Inhibition of MAO

The MAOIs are multifaceted drugs and have been reported to bind to a wide variety of other enzymes, receptor systems, and uptake pumps, thereby contributing to their therapeutic and/or adverse effects. Depending on the MAOI involved, interactions with the following have been reported: other amine oxidases; various transaminases, decarboxylases, dehydrogenases, and cytochromes P450 (CYPs); biogenic amine receptors and transporters; imidazoline binding sites; and sigma receptors (Holt et al. 2004; Gillman 2011). Several MAOIs also cause marked increases in brain levels of trace amines such as β-phenylethylamine and tryptamine, both of which can affect the normal function of classical neurotransmitter amines such as noradrenaline, dopamine, and serotonin. There has been extensive interest in recent years in the possible neuroprotective effects of MAOIs and their potential for the treatment of neurodegenerative disorders. For example, the MAO-B inhibitors L-deprenyl and rasagiline are neuroprotective and are used in Parkinson's disease and, to a lesser extent, as therapeutic agents in Alzheimer's disease (Youdim 2013). In many cases, these neuroprotective actions appear to be independent of the inhibition of MAO. L-Deprenyl and rasagiline have been reported to prevent the initiation of apoptotic cascades by upregulating the anti-apoptotic protein Bcl-2, downregulating pro-apoptotic proteins such as BAD and BAX, and preventing the activation and nuclear translocation of glyceraldehyde-3-phosphate dehydrogenase (Mousseau and Baker 2012; Song et al. 2013). L-Deprenyl has also been shown to increase levels of brain-derived neurotrophic factor (BDNF) in the cerebrospinal fluid (CSF) of patients with Parkinson's disease, whereas rasagiline increases expression of glial cell line-derived neurotrophic factor (GDNF) in the CSF of nonhuman primates (Maruyama and Naoi 2013). In addition, both L-deprenyl and rasagiline modulate glutamatergic receptor activity in the rat hippocampus in vitro (Song et al. 2013). Preliminary reports indicate that L-deprenyl might be useful for the treatment of negative symptoms in schizophrenia. Phenelzine has been shown to provide neuroprotection in animal models of stroke (global ischemia) and traumatic brain injury and to improve neurological and behavioral symptoms in an animal model of multiple sclerosis (experimental autoimmune encephalomyelitis) (Song et al. 2013; Musgrave et al. 2011). Phenelzine's contribution to neuroprotection might rely on actions as diverse as the inhibition of GABA transaminase (GABA-T) and elevation of brain GABA, the elevation of brain ornithine, and the sequestration of toxic aldehydes such as 3-aminopropanal, acrolein, formaldehyde, and 4-hydroxy-2-nonenal (Song et al. 2013). These highly reactive aldehydes can covalently modify cellular proteins, nucleic acids, and aminophospholipids; a toxic increase in these aldehydes is thought to contribute to the development and/or exacerbation of numerous neurodegenerative disorders. Furthermore, phenelzine has been reported to protect neurons and astrocytes against formaldehyde-induced toxicity and to increase BDNF expression in the rat frontal cortex and whole brain. Tranylcypromine, another irreversible and nonselective MAOI, has been shown to increase the expression of messenger RNA (mRNA) for BDNF and cAMP response binding protein (CREB) in the rat brain, events that can contribute to neurogenesis, and to increase the expression of the anti-apoptotic Bcl-2 and Bcl-XL proteins in several brain regions (Song et al. 2013). Although unrelated to MAO, primary amine oxidase (PrAO, previously called semicarbazide-sensitive amine oxidase [SSAO]) is inhibited by some MAOIs (e.g., phenelzine). PrAO has been the subject of extensive research in the areas of inflammation, neuropsychiatry, and cardiovascular complications of diabetes in recent years (Jiang et al. 2008). PrAO is responsible for the production of the toxic aldehydes formaldehyde and methylglyoxal and has been reported to have increased activity in Alzheimer's disease. In addition, PrAO has been found to be co-localized with cerebrovascular β-amyloid deposits in Alzheimer's disease patients (Jiang et al. 2008).

The RIMA moclobemide has been reported to have antiParkinsonian activity, neuroprotective effects in a model of cerebral ischemia, and antinociceptive actions in an animal model of neuropathic pain. Clorgyline (an irreversible inhibitor of MAO-A), like L-deprenyl and rasagiline, contains an N-propargyl moiety and has been reported to be neuroprotective in vitro and in vivo (Song et al. 2013), albeit at doses lower than are required to inhibit MAO. Increased activity and expression of MAO has been reported in Alzheimer's disease, suggesting that MAOIs should be investigated more thoroughly as adjunctive drugs in this neuropsychiatric disorder. Ladostigil, a drug which combines the activity of rasagiline with anticholinesterase properties, has been reported to contribute to increased antioxidant activity, induction of BDNF and GDNF mRNA, regulation of anti-apoptotic Bcl-2 proteins, and stabilization of the mitochondrial membrane potential (Song et al. 2013). Ladostigil is currently in phase II clinical trials for the treatment of Alzheimer's disease (Youdim 2013), while related drugs that also possess iron-chelating properties or inhibitory potency versus glutamate release are in preclinical development.

Summary

Although MAOIs are not currently used extensively for treatment of mood and anxiety disorders because of clinically relevant food-drug and drug-drug interactions, they continue to have an important place in treatment of psychiatric and neurological disorders and represent exciting leads for the development of future drugs with potential neuroprotective activity.

Acknowledgments The authors are grateful to CIHR (MOP86712 [GBB] and MOP37895 [AH]), the University of Alberta DUP program (GBB), the CRC/CFI programs, the Alzheimer Society of Saskatchewan, the Saskatchewan Health Research Foundation (DDM) and Donald R. and Nancy Romanow Cranston for funding. DM is the recipient of a QEII scholarship.

Cross-References

- Antidepressants
- Brain-Derived Neurotrophic Factor
- Depression
- Neuroprotection
- Panic Disorder
- Social Anxiety Disorder
- Trace Amines

References

Culpepper L, Kovalick LJ (2008) A review of the literature on the selegiline transdermal system: an effective and well-tolerated monoamine oxidase inhibitor for the treatment of depression. Prim Care Companion J Clin Psychiatry 10:25–30

Gillman PK (2011) Advances pertaining to the pharmacology and interactions of irreversible nonselective monoamine oxidase inhibitors. J Clin Psychopharmacol 31:66–74

Holt A, Berry MD, Boulton AA (2004) On the binding of monoamine oxidase inhibitors to some sites distinct from the MAO active site, and effects thereby elicited. Neurotoxicology 25:251–266

Jiang ZJ, Richardson JS, Yu PH (2008) The contribution of cerebral vascular semicarbazide-sensitive amine oxidase to cerebral amyloid angiopathy in Alzheimer's disease. Neuropathol Appl Neurobiol 34:194–204

Kennedy SH, Holt A, Baker GB (2009) Monoamine oxidase inhibitors. In: Sadock BJ, Sadock VA, Ruiz P (eds) Comprehensive textbook of psychiatry, 9th edn. Lippincott Williams & Wilkins, New York, pp 3154–3164

Maruyama W, Naoi M (2013) "70th Birthday Professor Riederer" Induction of glial cell line-derived and brain-derived neurotrophic factors by rasagiline and (−)deprenyl: a way to a disease-modifying therapy? J Neural Transm 120:83–89

Mousseau DD, Baker GB (2012) Recent developments in the regulation of monoamine oxidase form and function: is the current model restricting our understanding of the breadth of contribution of monoamine oxidase to brain [dys]function? Curr Top Med Chem 12:2163–2176

Musgrave T, Benson C, Wong G, Browne I, Tenorio G, Rauw G, Baker GB, Kerr BJ (2011) The MAO inhibitor phenelzine improves functional outcomes in mice with experimental autoimmune encephalomyelitis (EAE). Brain Behav Immun 25:1677–1688

Song MS, Matveychuk D, MacKenzie EM, Duchcherer M, Mousseau DD, Baker GB (2013) An update on amine oxidase inhibitors: multifaceted drugs. Prog Neuropsychopharmacol Biol Psychiatry 44:118–124

Stahl SM, Felker A (2008) Monoamine oxidase inhibitors: a modern guide to an unrequited class of antidepressants. CNS Spectr 13:855–870

Youdim MB (2013) Multi target neuroprotective and neurorestorative anti-Parkinson and anti-Alzheimer drugs ladostigil and m30 derived from rasagiline. Exp Neurobiol 22:1–10

Monoamines

Definition

Monoamines (so-called because they have one organic substituent attached to the nitrogen atom) include serotonin, norepinephrine, and dopamine, all of which are neurotransmitters that are important in the pathophysiology and treatment of psychiatric disorders. Monoamines are subdivided into catecholamines and indoleamines.

Mood Disorders

Synonyms

Affective disorders

Definition

Depressive, manic, and hypomanic disorders, such as major depression, mania, hypomania, bipolar disorder, dysthymia, and cyclothymia.

Mood Stabilizers

Guy M. Goodwin
University Department of Psychiatry, Warneford Hospital, Headington, Oxford, UK

Synonyms

Long-term treatments for bipolar disorder

Definition

Mood stabilizers are pragmatically defined by their clinical efficacy in bipolar disorder. Bipolar disorder is a complex condition, as it is expressed as episodic periods of contrasting mood disturbance – mania and depression – and its long-term or maintenance treatment must prevent new episodes of both. Any medicine that achieves this can be said to be a mood stabilizer.

Pharmacological Properties

History

The first mood stabilizer was lithium. It was discovered over 60 years ago by guided serendipity. Lithium salts of urea were found by the Australian doctor John Cade to be sedative in animals. He had reasoned that urea itself was an active component, but realized that, in fact, lithium was unexpectedly tranquilizing. Immediate trials in patients with mania suggested acute efficacy, and subsequent experience showed that lithium could markedly modify the course of bipolar disorder (then called manic depression) in the long term. The effects were anecdotally so dramatic for some extremely disabled patients that a group of psychiatrists and scientists, led most notably by Mogens Schou, quickly became self-proclaimed lithium enthusiasts and, by the 1960s, had influenced practice in many parts of the world. The history of lithium's acceptance was interrupted by highly vocal criticism of the methodology of its early adopters. While the critics were right about the methodology, they were wrong about lithium (which they proclaimed on equally little evidence to be a dangerous yet inactive placebo). Subsequent trials have repeatedly shown that lithium is an effective medicine for the prevention of relapse (especially to the manic pole) in bipolar disorder (Geddes et al. 2004). When the illness starts with mania and tends to relapse to the manic pole, lithium can produce remarkable mood stability. However, such success is seen in only about 30 % of patients with the severe form of the disorder. Therefore, for many patients, other mood stabilizers are required to meet their unmet needs.

Valproate and carbamazepine are also often described as mood stabilizers, although the evidence is weaker than for lithium. Both also tend to be most active against mania.

There is good evidence for antipsychotics in mood stabilization (Malhi et al. 2005): quetiapine and lurasidone act against both poles of the illness. It also turns out that other agents can be effective in long-term treatment against, for example, the manic pole of the illness, without having an important impact on depression (e.g., aripiprazole, olanzapine, and risperidone), and vice versa (e.g., lamotrigine). Some authors have argued for extending the term mood stabilizer to include medicines with all these effects. Even more loosely, there was at one time a tendency to extend the term mood stabilizer to include anticonvulsants, effectively by extrapolation from the examples of valproate and carbamazepine. In the case of topiramate and gabapentin, subsequent clinical trials proved to be negative. There seems little advantage to a more liberal definition except for marketing the drugs.

Mechanism of Action

It follows from the definition that there are no pharmacological properties that define mood stabilizers as a class. Moreover, our understanding of the neurobiology of mania and depression remains highly provisional. However, the fact of clinical efficacy has often preceded an understanding of mechanism and has been a major driver to further research. Thus, the individual medicines effective in stabilizing mood have a variety of effects, defined for the most part in animal experiments.

Lithium inhibits the phosphoinositide second-messenger system in the brain and peripheral tissues. This may be the basis for its therapeutic (and nontherapeutic) effects, although it has not been investigated as fully as might be expected. There are also some effects on monoamine function in the brain, which have weak parallels with better-defined psychotropic drugs like antidepressants and antipsychotics. In recent years, interest has been directed to downstream cellular changes in transcription factors and other molecular pathways that may mediate neuroprotection. Lithium has some actions in common with valproate in these cellular models (Harwood 2011).

The net actions of the anticonvulsants are variously pro-GABAergic and antiglutamatergic. The efficacy of lamotrigine in bipolar depression is of particular interest given its lack of neurotoxicity at effective doses and selective effects on membrane polarization and glutamate release. Glutamate is increasingly implicated in the neuronal circuits that are believed to regulate mood, and there is an evolving pharmacology targeted on the glutamate system. Whether this next generation of molecules will be effective is still an open question.

The antipsychotics that can prevent long-term relapse in bipolar disorder have a common action in blocking dopamine receptors. However, most have additional actions on serotonergic function in the brain and, in the specific case of quetiapine, an active metabolite that appears to block noradrenergic reuptake, a property shared with a number of antidepressants. Whether these effects, in addition to dopamine blockade, are important remains an open question.

Pharmacokinetics

Lithium provides an unusual example for psychopharmacology of a drug whose plasma levels are routinely monitored in clinical practice. Such monitoring is necessary because the drug has a narrow therapeutic index (also known as therapeutic ratio): in other words, the difference between a blood level that is just effective and that which can poison the patient is relatively small. For efficacy, 0.5 mmol/L is regarded as the minimum effective level in plasma taken 18 h after the last dose of lithium. Levels over 1.5 mmol/L are potentially toxic and definitely not recommended. The most common well-tolerated level is around 0.7 mmol/L, although slightly higher levels (0.8–1.0 mmol/L) are often recommended. Choice of level should always be informed by any symptoms the patient describes. A variety of adverse effects can limit doses to well below 1.0 mmol/L.

Although not routinely used to guide treatment, levels of valproate and carbamazepine are often available because of their use in epilepsy. A valproate level between 50 and 125 µg/mL has been associated with acute response in mania. Importantly, because combined treatments are so often necessary in bipolar patients, carbamazepine promotes enzyme induction and can lower the levels of a range of other agents (including oral contraceptives); hence, higher doses of comedication may often be necessary. Conversely, valproate approximately doubles the availability of lamotrigine, thus halving the dose of that drug required for efficacy.

In the elderly, it is common for the required dosage to be substantially less than that used in younger people. Side effects are an important guide, as ever, to what the dosage should be. The highest *well-tolerated* dose (whatever the actual figure) is usually the best choice in bipolar disorder.

Tolerability and Safety

In general, mood-stabilizing medicines have to be reasonably well tolerated because patients are expected to take them week in and week out for many years – often indefinitely. There is a tendency for doctors to underestimate the adverse impact of medicines on their patients. The most important adverse subjective effects of psychotropic drugs that are used to treat bipolar disorder tend to be tiredness, sedation, and weight gain. In addition, lithium can produce tremor, increased urine volumes, and thyroid dysfunction. Attention to minimizing these problems (by optimizing doses) is essential for good adherence to prescribed medicines.

Weight gain was formerly seen as a largely cosmetic problem. It is now realized to be a much more important health issue, because obesity combined with lack of exercise and smoking is associated with the so-called metabolic syndrome. This is a composite term for biochemical, blood pressure, and weight indices associated with older age and higher body mass index. It is the prelude to diabetes, coronary heart disease, and stroke. Several antipsychotics used to treat bipolar disorder, including clozapine, olanzapine, and quetiapine, are particularly associated with increased weight gain and the risk of dyslipidemia, hypercholesterolemia, and elevated glucose. In an increasingly obese population, this is a growing concern that requires active prevention wherever possible and the treatment of risk factors.

In pregnancy, there is a risk of teratogenicity from several of the medicines used as mood stabilizers. Thus, especially in the first 3 months of fetal development, drugs may interfere with the formation of the most complex organs. The neural tube and heart are especially vulnerable. Lowest risks appear to be associated with the antipsychotics. Higher teratogenic risks are associated with lithium and especially the anticonvulsants (valproate > carbamazepine > lamotrigine).

Decisions about the use of medication in pregnancy by women with bipolar disorder are always difficult. Sudden discontinuation (Goodwin 1994) or switching medicines risks destabilizing mood and precipitating relapse of the bipolar disorder. Treatment of an acute episode in a pregnant woman may both be highly stressful and require much higher doses of psychotropic drugs than would be required for prophylaxis.

Treatment issues also arise postpartum. Childbirth greatly increases the risk of relapse in patients with bipolar disorder in the weeks and months after delivery. Bipolar women with a previous history of a severe postpartum episode (puerperal psychosis) and bipolar women with a family history of puerperal psychosis will have a >50 % risk of severe relapse. Medication given to prevent this outcome will often appear in breast milk and have potential consequences for the neonate.

Conclusion

The term mood stabilizer is in common use, even though it imprecisely describes what such medicines do and certainly does not define a meaningful pharmacological action or even set of actions. The strictest definition, proposed by Bauer and Mitchner (2004), is prophylaxis and the prevention of recurrence *and* evidence of short-term efficacy for both poles of the illness. Lithium may just meet the criterion despite its relative weakness against depression; quetiapine and lurasidone certainly do, but clinical experience with these agents is less. Therefore, for any agent to be called a mood stabilizer, it is probably necessary to know how it performs in comparison to lithium in the long term. It is also necessary to gauge its relative effects against the manic and depressive poles of the illness.

Medicines that reduce the risk of relapse to mania and/or depression in bipolar patients can contribute to mood stability, whatever we call them. Some contemporary uses of the term are overinclusive. However, mood stability is a goal that patients, their families, and attending doctors share. If a medicine helps, then doctors are more likely to prescribe it and patients are more likely to take it. In other words, the term mood stabilizer is chosen to be comforting and persuading. This does not detract from the scientific challenge to understand the mechanisms underlying mood stabilization and how to improve treatment.

Cross-References

▶ Anticonvulsants
▶ Antidepressants
▶ Bipolar Disorder
▶ Classification of Psychoactive Drugs
▶ Glutamate
▶ Monoamines

References

Bauer MS, Mitchner L. What is a "mood stabilizer"? An evidence-based response. Am J Psychiatry. 2004;161(1):3–18.

Geddes JR, Burgess S, Hawton K, Jamison K, Goodwin GM. Long-term lithium therapy for bipolar disorder: systematic review and meta-analysis of randomised controlled trials. Am J Psychiatry. 2004;161:217–22.

Goodwin GM. The recurrence of mania after lithium withdrawal: implications for the use of lithium in the treatment of bipolar affective disorder. Br J Psychiatry. 1994;164:149–52.

Goodwin GM. Evidence-based guidelines for treating bipolar disorder: revised second edition – recommendations from the British Association for Psychopharmacology. J Psychopharmacol. 2009;23:346–88.

Harwood AJ, Agam G. Search for a common mechanism of mood stabilisers. Biochem Pharmacol. 2003;66:179–89.

Harwood AJ, Prolyl Oligopeptidase, Inositol Phosphate Signalling and Lithium Sensitivity CNS Neurol Disord Drug Targets. 2011;10:333–339.

Malhi GS, Berk M, Bourin M, Ivanovski B, Dodd S, Lagopoulos J, et al. Atypical mood stabilisers: a 'typical' role for atypical antipsychotics. Acta Psychtr Scand. 2005;111 Suppl 426:29–38.

Moperone

Definition

Moperone is a first-generation ▶ (typical) antipsychotic drug that belongs to the ▶ butyrophenone class; it is approved in Japan for the treatment of ▶ schizophrenia. It has higher antagonist affinity for D_2 than $5\text{-}HT_{2A}$ receptors. It also has high binding affinity for sigma receptors. It can induce extrapyramidal motor side effects, insomnia, and thirst, but generally displays low toxicity.

Cross-References

▶ Butyrophenones
▶ Extrapyramidal Motor Side Effects
▶ First-Generation Antipsychotics

Morphine

Definition

Morphine is a highly potent opiate analgesic drug. It is the principal active ingredient in opium that is derived from the opium poppy, *Papaver somniferum*. It is considered to be the prototypical ▶ μ-opioid agonist. It acts directly on the μ-opioid receptors to relieve pain. Morphine has a high potential for addiction; tolerance and both physical and psychological dependence develop rapidly.

Cross-References

▶ Addiction
▶ Analgesics
▶ Dependence
▶ Diamorphine

Morris Water Maze

Synonyms

Morris water navigation task; Water maze

Definition

The Morris water maze was developed by Richard Morris (1984) and is used to assess spatial learning in rats and mice. The apparatus comprises a pool of varying diameter, typically 1.2–2 m (smaller for mice) and depth of approximately 60 cm which is filled with opaque (e.g., milky) water and contains a hidden platform. The animal is placed in the pool at different starting points and swims to the hidden platform (there are many variations of this to assess, e.g., working memory, daily learning, delayed match to sample). There are no intramaze cues and the dominant strategy in relocating the hidden platform is thought to be truly spatial. The main measure for the water maze is the latency to find the platform. In order to control for search strategies, other measures can be taken, including time spent in each quadrant during the main trials and probe trials (where the platform is removed) and analyses of the path length. The advantages of the Morris water maze are the rapid acquisition of the task; the ability to distinguish between learning and performance, markers of motivation, and motor ability (e.g., swim speed); and the innate motivation of rats to want to find the platform without being distressed.

Cross-References

▶ Spatial Learning in Animals

References

Morris R (1984) Developments of a water-maze procedure for studying spatial learning in the rat. J Neurosci Methods 11(1):47–60

Mosapramine

Synonyms

Y-516

Definition

Mosapramine is a first-generation ▶ (typical) antipsychotic drug that belongs to the iminodibenzyl class; it is approved in Japan for the treatment of schizophrenia. It is a potent dopamine antagonist with high affinity for D_2, D_3, and D_4 receptors, but lower affinity for 5-HT$_{2A}$ receptors. It can induce ▶ extrapyramidal motor side effects and drowsiness, but generally displays low toxicity.

Cross-References

▶ Extrapyramidal Motor Side Effects
▶ First-Generation Antipsychotics
▶ Schizophrenia

Motor Activity and Stereotypy

Stephen C. Fowler
Department of Pharmacology and Toxicology, Schieffelbusch Institute for Life Span Studies, University of Kansas, Lawrence, KS, USA

Synonyms

Exploratory behavior; Locomotor activity; Repetitious behavior; Repetitive behavior; Spontaneous activity

Definition

Motor activity generally refers to a laboratory animal's horizontal movements within an enclosure that permits the use of a variety of methods for quantifying such movements. The term is rarely used in the context of laboratory animals exercising in running wheels or on treadmills.

Stereotypy refers to abnormally repetitive behavior that reflects dysfunction of the nervous system induced by drugs (especially those that elevate brain dopamine levels), brain lesions, genetic mutations, or environmental circumstances (e.g., wild animals in captivity). Depending on the context in which the term is used, stereotypy or stereotyped behavior may include specific features of locomotor activity (e.g., moving predominantly in the same direction along the inside perimeter of an enclosure) or may be restricted to repetitive behaviors that primarily occur in the absence of locomotion (e.g., rhythmic head movements, known as focused stereotypy; gnawing on the wire-mesh floor of a cage, known as oral stereotypy).

Impact of Psychoactive Drugs

The locomotor activity assay is one of the most frequently used procedures in the initial exploration of a drug's putative psychoactive effects. Typically, a mouse or rat is administered a fixed dose of a drug and is then placed in an enclosure with a flat floor large enough to enable free movement of several body lengths in any direction in the horizontal plane. A human observer or an instrument is used to quantify the distance an animal moves. Other species-typical rodent behaviors, such as grooming or rearing, may also be recorded depending on the objective of the research, the recording method, and other experimental conditions. Comparisons of a variety of techniques for measuring locomotor activity can be found in Fowler et al. (2001). The success of locomotor activity as a drug assay depends largely on the fact that most types of mice and rats spontaneously engage in "exploratory" behaviors, when placed in environments new to them. As it moves from place to place in the new environment, a mouse or rat uses its sensory organs and information-processing endowments to learn the available species-relevant attributes of the environment. Importantly, the study of drug effects on spontaneously expressed behavior is economical because the experimenter does not need to impose any preconditions on the animals (e.g., neither food restriction nor explicit training in a specific task, etc.) in order to assure that easily measurable amounts of species-typical behavior will occur in placebo-treated animals ("controls"). The behavioral output of controls then affords a baseline condition with which the effects of drug treatments can be compared. Depending on type of drug, drug dose, amount of experience in the environment, amount of experience with the drug, duration of the observation period, route of drug administration, lighting conditions, time of day, and many additional variables, the drug treatment may increase, decrease, or have no effect on measures of locomotor activity compared to control performance. When interpreting the results of these kinds of experiments, one should keep in mind the fact that a priori suppositions about a drug's "stimulant" or "depressant" activity may not be confirmed in a straightforward manner. For example, the drug, pentobarbital, which in clinical terms is described as a "sedative-hypnotic," increases locomotor activity in rodents at low doses and decreases or abolishes locomotor activity at higher doses as an anesthetic dose level is neared. The psychomotor stimulant drug, D-amphetamine sulfate, also increases locomotor activity in rats at doses around 1.0 mg/kg, but at doses around 5.0 mg/kg of amphetamine produces a focused stereotypy syndrome characterized by the absence of locomotion and the concurrent expression of rapid rhythmic head movements. In research situations where a priori information about a drug is lacking, a relatively wide range of doses should be used in locomotor activity assays to detect lower-dose increases and higher-dose decreases in locomotor activity.

Amphetamine-Induced Locomotor Activity: Illustrative Data

Figure 1 provides data illustrating increased locomotor activity induced by amphetamine. The measure of locomotor activity was the distance traveled during six 10-min time blocks in a 1-h recording session. Distance traveled was calculated from the rats' center of force x-y coordinates as the rats moved on the 28 × 28-cm load plate of a force-plate actometer (Fowler et al. 2001). The saline group exhibited rapid habituation during the first 40 min, and the distance traveled remained low for the remainder of the hour. Compared with the saline group, the amphetamine-treated rats showed significantly more locomotor activity at all but during the first 10-min interval, the standard errors of the mean (sem) overlapped. The curvature in the amphetamine plot between block 2 and block 6 is likely the result of changes in the brain level of amphetamine. Brain microdialysis studies have shown that intraperitoneally administered D amphetamine sulfate reaches a peak level at about 22–26 min after injection and is eliminated with a half-life of about 40–45 min. During the sixth time block (see Fig. 1), the amphetamine group exhibited more distance traveled than the control group in every time block except the first, suggesting that amphetamine continued to induce abnormally high levels of locomotion 1 h after treatment. Given that the time-related diminution of locomotion in the control group is a representative example of habituation, one can

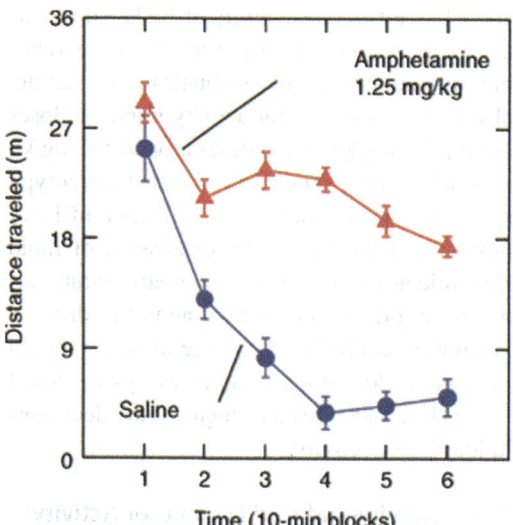

Motor Activity and Stereotypy, Fig. 1 Effect of D-amphetamine sulfate on locomotor activity in two separate groups ($n = 8$) of male Sprague Dawley rats. The amphetamine was injected intraperitoneally a few seconds before the rats were individually placed in a dark force-plate actometer for a 1-h recording session. Data are based on the rats' first exposures to the apparatus. The saline group exhibited habituation during the first 40 min, and distance traveled remained low for the remainder of the hour. Compared to the saline group, the amphetamine-treated rats showed hyperactivity throughout the hour. Over the entire hour, the saline group mean distance traveled was 59.71 m (s.e.m. 7.23 m), and the amphetamine-treated group traveled 133.51 m (s.e.m. 6.59 m)

hypothesize that amphetamine, at least in part, produced elevated levels of locomotor activity by interfering with the habituation process. A well-designed experiment with cocaine, a psychomotor stimulant with many pharmacological effects similar to those of amphetamine, supports this hypothesis (Carey et al. 2003). Thus, the increased locomotor activity induced by a variety of CNS-active drugs is not simply a "motor effect," because the multiple sites of action of such drugs reside in multiple anatomical loci that perform information processing, associative, and motivational functions, as well as motor functions.

Amphetamine-Induced Focused Stereotypy

In male Sprague Dawley rats, the first experience with a 2.5 mg/kg, ip, dose of amphetamine will induce an increase in locomotor activity in a large majority of the subjects compared to the untreated controls (see Fig. 2). However, when given repeatedly, the same dose of amphetamine predominantly evokes, a syndrome characterized by an absence of locomotion accompanied by rapid head movements that have been described as "sniffing" and/or "head bobbing." This syndrome is focused stereotypy or the "stationary phase" of the amphetamine response, and it has been observed in observation arenas as large as 3.0×3.5 m (Schiorring 1971).

The focused stereotypy score in Fig. 2 was calculated by combining a quantitative measure of spatial confinement with the variance of vertical force variation within the boundaries of the space used (see Fowler et al. 2007a for details). This score is low for a sleeping animal (high-spatial confinement but low-force variance because of a lack of "in-place' movements). The focused stereotypy score is also low when spatial confinement is low (movements are dispersed across the floor), despite the presence of high-force variance associated with ambulation. When spatial confinement is pronounced and force variance is also high, a high focused stereotypy score is obtained. In Fig. 2, the switch from first-dose locomotor stimulation to tenth-dose substantial focused stereotypy (Fig. 2, panel b) can be appreciated by comparing the distance traveled data of injection 1 (panel a, triangles) with the same measure for injection 10 (panel a, squares), during the second 30 min of the session. Correspondingly, during the same time period, the focused stereotypy scores after injection 1 were near zero but averaged higher than one after injection 10. This change in response topography from locomotion to stationarity and focused stereotypy is the result of a sensitization process, whereby rats become more sensitive to amphetamine or other psychomotor stimulants as repeated dosing ensues. After injection 10, distance traveled (see Fig. 2, panel a, squares) did not drop to no-drug levels because three rats did not make a complete switch to focused stereotypy and continued to display substantial locomotor activation. However, all but one of the eight rats after the tenth injection showed a *decrease* in

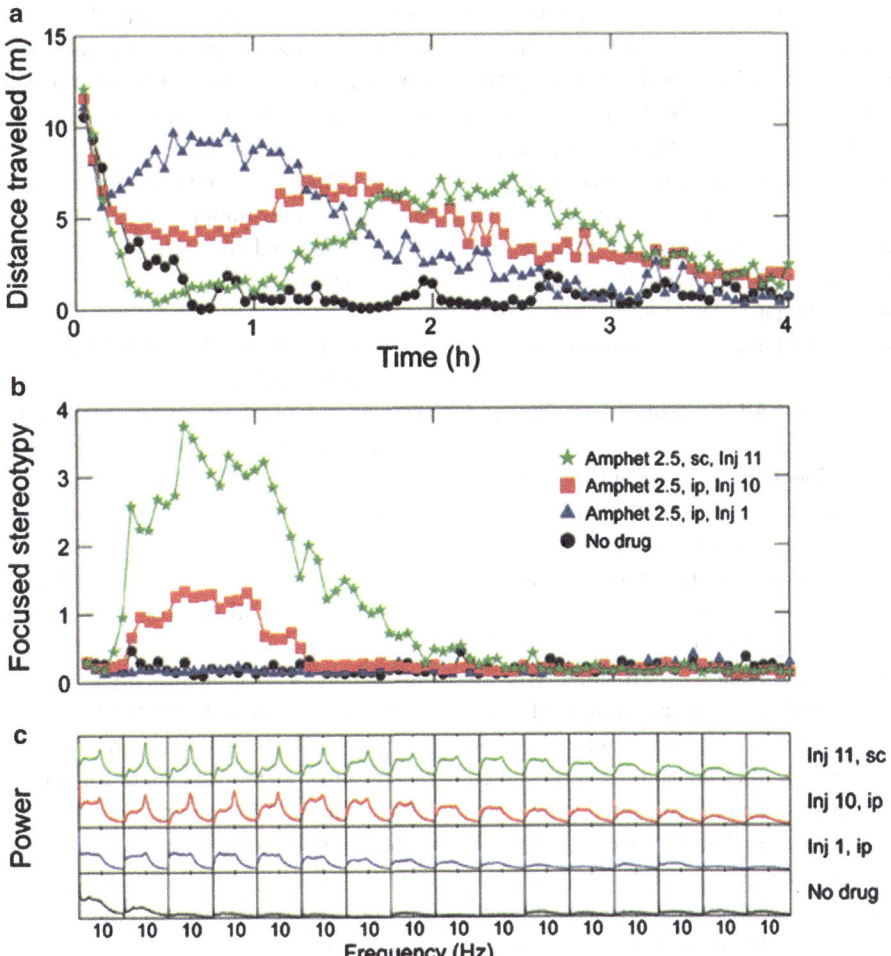

Motor Activity and Stereotypy, Fig. 2 Measures from a force-plate actometer showing how the behavioral effects of D-amphetamine sulfate at 2.5 mg/kg depends on time after treatment, amount of experience with the drug, and the route of administration. After experiencing a 4-h habituation session (*black circles*, panels **a** and **b**), in subsequent sessions, eight male Sprague Dawley rats received injections (*Inj*) of amphetamine eleven times, 3–4 days apart. The first 10 injections were given ip, and the 11th injection (*green stars*) was administered sc in the same volume and dose as the previous ip injections. Panel (**a**) shows distance traveled for successive 3-min periods in the 4-h recording session. In panel (**b**) are plotted, also in 3-min intervals, group mean focused stereotypy scores for the indicated treatment conditions. Panel (**c**) shows the group mean power spectra of the vertical force variations over 15-min periods for the same treatments. The distinctive spectral peaks near 10 Hz in the *red* and *green* power spectra reflect the rhythm of the head movements of focused stereotypy

distance traveled compared to injection 1. The aberrant rat actually exhibited evidence of sensitization of locomotor response, as suggested by its 67.9 % *increase* in distance traveled between injection 1 and 10. Two important points can be made from these observations: (1) individual differences in response to drugs are to be expected, especially when genetically heterogeneous outbred strains of rats are used, and (2) when amphetamine or similar drugs are under study, at doses near the threshold for expression of focused stereotypy, the observed behavioral effect may be in opposite directions, with a subset of rats showing increased locomotor activity and another subset exhibiting locomotor suppression. Another important point is that a univariate (i.e., single-dependent-variable method such as distance traveled) approach to behavioral pharmacology with indirect-acting dopamine agonists can easily lead to erroneous interpretations of a drug's effects. For example, if one did not have a stereotypy score (Fig. 2, panel b) and relied only on the distance traveled information (Fig. 2, panel a), one may erroneously conclude that, between injection 1 and injection 10, tolerance (decrease in distance traveled) had occurred instead of sensitization.

Effect of Route of Administration on Focused Stereotypy

An 11th injection of 2.5-mg/kg amphetamine was given, but this injection was by the subcutaneous route (sc). Figure 2 (panels a, b, stars) shows that changing the route of administration from ip to sc intensified and lengthened the expression of focused stereotypy. Gentry et al. (2004) also found more pronounced focused stereotypy for 3.0-mg/kg sc methamphetamine compared to the same dose given ip. The group mean increase in focused stereotypy after the sc injection resulted from (1) a recruitment of the three previously low-stereotypy rats to full stereotypy and (2) all five rats that were already expressing focused stereotypy had higher scores after sc amphetamine treatment compared to the previous ip treatment with the same dose. Pharmacokinetic studies (Gentry et al. 2004) have shown that sc compared with ip amphetamine in male Sprague Dawley rats reaches a lower peak blood concentration, takes longer to reach its peak, and is eliminated more slowly (i.e., longer duration of action). Thus, the pharmacokinetic data indicating a longer duration of action for sc compared to ip administration of amphetamine are consistent with the behavioral data shown in Fig. 2 (in panels a and b, compare behavioral measures for injections 10 and 11). However, neither the reported pharmacokinetic time-to-peak nor peak level achieved is in accord with the behavioral data. In Fig. 2 (panel b), focused stereotypy scores began to rise earlier in the session for the sc injection than for the ip injection (opposite the pharmacokinetic later rise in concentration), and the sc-related focused stereotypy scores reached a substantially higher level than those seen for the ip injection (again, opposite the lower pharmacokinetic peak for sc dosing). While no explanation is yet available for this difference in the stereotypy-evoking efficacy of the ip and sc injection methods, the empirical data show that route of administration can substantially influence the focused-stereotypy-inducing effects of amphetamine.

Rhythmicity of Head Movements During Focused Stereotypy

During the expression of amphetamine-induced focused stereotypy, not only are the head movements observably repetitive they are also strongly rhythmic (Fowler et al. 2001; Fowler et al. 2007a). Application of signal-processing techniques, such as power spectral analysis, to a rat's variations in vertical force recorded with a force-plate actometer shows that the head movements of focused stereotypy have a tightly regulated rhythm near 10 Hz (Hz = cycles/s). Panel c in Fig. 2 presents group mean power spectra that were calculated for each 15-min interval in the 4-h session. Once the expression of focused stereotypy begins in male Sprague Dawley rats, the near 10-Hz rhythm can be continuously sustained for an hour or more (see Fig. 2, panel c, top row of functions, frames 2, 3, 4, and 5 from left to right). The near 10-Hz rhythm has been recorded in rats separately treated with four different indirect-acting dopamine agonists: amphetamine, nomifensine, methylphenidate, and cocaine. The precision of rhythm regulation in rats expressing focused stereotypy rivals or exceeds that of consummatory licking (7.0 Hz), hindlimb scratching behind the ear (8.0 Hz), within-bout vibrissal whisking (e.g., 7.0 Hz), or grooming (forepaws/face/head: 7.2 Hz; flank licking: 3.6 Hz). The existence of these species-typical rhythmic reflexes invite the conjecture that the head movement rhythm of amphetamine focused stereotypy is also reflexive. Both the atypical antipsychotic drug, clozapine, and the α-1 noradrenergic antagonist, prazosin, have been shown to slow the head movement rhythm of amphetamine-induced focused stereotypy without abolishing the overall syndrome (Fowler et al. 2007a). The frank rhythmicity of amphetamine-induced head movements adds a new measurable dimension to the focused stereotypy response, and experimental analyses of rhythm production and modulation may provide insights into the neurotransmitter systems and neuroanatomical loci that mediate drug-induced stereotypy.

Locomotor Activity and Stereotypy in Drug Self-Administration Experiments

Measurements of locomotor activity and stereotypy in rodents have been used in drug self-administration research in at least two different ways. One way exemplified by the work of Piazza et al. (1989) used the locomotor activity response to a novel environment (i.e., a 170-cm perimeter, 10-cm-wide circular track with photobeams to detect the rats' locomotion) to analyze individual differences in rats' susceptibility to acquiring amphetamine self-administration later in another environment. Piazza et al. (1989) found that rats with a relatively low level of locomotor response to novelty did not acquire the amphetamine self-administration nose-poke response, while those rats exhibiting higher amount of locomotor activity acquired the response. This rodent-based experiment was one of the first to suggest that behavioral response to novelty and susceptibility to drug-abuse-like behaviors are correlated. A second way that locomotor activity and stereotypy have been combined with self-administration research is illustrated by an experiment that used a force-plate actometer as the floor of a self-administration chamber where five lever presses by rats resulted in an intravenous cocaine infusion in a 24-h "binge" session (Fowler et al. 2007b). The work provided measures of rotational behavior ("circling in the same direction"), distance traveled, power spectra of the vertical force of movements (as in panel c of Fig. 2), and focused stereotypy in order to determine whether or not self-administered cocaine produced the same kind of unconditioned behaviors that are elicited by noncontingently administered bolus doses of cocaine. The results showed that all six rats displayed behaviors typical of noncontingently administered bolus doses. The average duration of active cocaine intake was 14.7 h (± 1.8 h). During this time, rats rotated in the same direction an average of 1206.5 (± 372.7) times, traveled an average of 914.6 m (± 65.9 m), and exhibited a power spectral peak near 10 Hz during 32.0 % (± 8.4 %) of the active binge duration. These data show that locomotor activity and focused stereotypy are major behavioral manifestations of self-administered cocaine and raise the possibility that unconditioned behaviors evoked by indirect-acting dopamine agonists contribute to the reinforcing effects of psychomotor stimulant drugs.

Differences in Expression of Stimulant-Induced Behaviors in Rats and Mice

Although it is difficult to offer accurate generalizations that are valid for all strains of rats or mice, mice tend to move more than rats, and mice tend to have shallower locomotor activity habituation functions than rats. Response topographies of stimulant-induced stereotypies of rats and mice have similarities and differences. Generally, amphetamine and other indirect-acting dopamine agonists, in sufficient doses, evoke some degree of spatial confinement in both rats and mice. Also, after very high doses of these drugs, most types of laboratory rats and mice exhibit self-injury (usually biting the dorsal wrist area sufficiently to cause bleeding and exposure of subcutaneous structures). Despite these similarities, the topography of the focused stereotypies in rats and mice can be very different in appearance. For example, while rats exhibit a complete and enduring arrest of locomotion during focused stereotypy, mice tend to retain a substantial locomotor component in their stereotypy response. Typically, mice observed in force plate actometers do not display the narrow-band head-movement rhythm, characteristic of the rat response to a 5.0-mg/kg dose of amphetamine. Different stereotypy topographies in mice and rats make it difficult if not impossible to devise a rating scale that can be used to quantify stereotypies equally well in both species. Comparing stereotypies in rats and mice is further hindered by the heterogeneity among inbred strains of mice in their responses to amphetamine and other psychomotor stimulants. For example,

BALB/cJ mice show a low-dose *decrease* in their locomotor activity response to 1.0 mg/kg amphetamine, and at a higher dose of 10.0 mg/kg, BALB/cJ mice exhibit vertical leaping ("popping") behavior. C57BL/6J, DBA/2J, 129SvJ, and C3H/HeJ mice exhibit neither low-dose suppression of locomotor activity nor higher-dose vertical leaping behavior.

Neurobehavioral Interpretations and Clinical Importance of Stereotypy

Stereotyped behaviors are generally thought to reflect the dysfunction of the central nervous system (Robbins et al. 1990; Teitelbaum et al. 1990). Persons with mental retardation, autism, schizophrenia, Tourette's syndrome, Huntington's disease, or obsessive-compulsive disorder (OCD) often exhibit excessively repetitious and apparently purposeless behaviors as do persons who have used excessive amounts of amphetamine over multi-day periods. Substantial clinical and experimental evidence points to dopamine as the brain neurotransmitter with a major role in the expression of stereotyped behavior in the aforementioned clinical abnormalities. In rodent studies of the behavioral effects of psychomotor stimulants, links have been established between dopaminergically innervated subcortical structures and expression of increased locomotor activation and stereotypy induction. Dopamine agonists are thought to induce locomotor activity by acting on the nucleus accumbens, while actions of the drugs on the caudate/putamen are believed to be critical for the expression of focused stereotypy. In the clinic, dopamine receptor-blocking drugs (e.g., the antipsychotic drug haloperidol) can be effective in suppressing the spontaneously occurring stereotypies, such as hand flapping, in persons with autism. In the laboratory, most, but not all, antipsychotic drugs can block the expression of amphetamine-induced focused stereotypy. Although a massive amount of empirical evidence implicates a role for dopamine in the evocation of locomotor activity and the expression of stereotyped behavior, much remains to be discovered about dopamine's interaction with a multiplicity of other neurotransmitters (e.g., glutamate, GABA, acetylcholine, serotonin, neuropeptides) and their receptors in the nucleus accumbens, caudate/putamen, and elsewhere in the brain. Given the large number of drug targets located in these brain regions, the measurement of pharmacologically induced locomotor activity and stereotypy is likely to continue to be important in the conduct of psychopharmacology research for many decades to come.

Acknowledgment This work was supported by MH043429 and HD002528.

Cross-References

▶ Aminergic Hypothesis for Schizophrenia
▶ Amphetamine
▶ Antipsychotic Drugs
▶ Cocaine
▶ Habituation
▶ Methylphenidate and Related Compounds
▶ Open-Field Test
▶ Phenotyping of Behavioral Characteristics
▶ Self-Administration of Drugs
▶ Sensitization to Drugs

References

Carey RJ, DePalma G, Damianopoulos E (2003) Cocaine-conditioned behavioral effects: a role for habituation processes. Pharmacol Biochem Behav 74:701–712.

Fowler SC, Birkestrand BR, Chen R, Moss SJ, Vorontsova E, Wang G, Zarcone TJ (2001) A force-plate actometer for quantitating rodent behaviors: illustrative data on locomotion, rotation, spatial patterning, stereotypies and tremor. J Neurosci Methods 107:107–124.

Fowler SC, Pinkston JW, Vorontsova E (2007a) Clozapine and prazosin slow the rhythm of head movements during focused stereotypy induced by d-amphetamine in rats. Psychopharmacology (Berl) 192:219–230.

Fowler SC, Covington HE, Miczek K (2007b) Stereotyped and complex motor routines expressed during cocaine self-administration: results from a twenty-four hour binge of unlimited cocaine access in rats. Psychopharmacology (Berl) 192:465–478.

Gentry WB, Ghafoor AU, Wessinger WD, Laurenzana EM, Hendrickson HP, Owens SM (2004) (+)-Methamphetamine-induced spontaneous behavior in rats depends on route of (+)METH administration. Pharmacol Biochem Behav 79:751–760.

Piazza PV, Deminière JM, Le Moal M, Simon H (1989) Factors that predict individual vulnerability to amphetamine self-administration. Science 245:1511–1513.

Robbins TW, Mittleman G, O'Brien J, Winn P (1990) The neuropsychological significance of stereotypy induced by stimulant drugs. In: Cooper SJ, Dourish CT (eds) Neurobiology of stereotyped behaviour. Clarendon, Oxford, pp 25–63.

Schiorring E (1971) Amphetamine induced selective stimulation of certain behaviour items with concurrent inhibition of others in an open-field test with rats. Behaviour 39:1–17.

Teitelbaum P, Pellis SM, DeVietti TL (1990) Disintegration into stereotypy induced by drugs or brain damage: a microdescriptive behavioural analysis. In: Cooper SJ, Dourish CT (eds) Neurobiology of stereotyped behaviour. Clarendon, Oxford, pp 169–199.

Motor Memory

Definition

A motor memory develops by repetition of movements that the individual is already able to perform. Repetition is supposed to improve motor skills and to enhance some degree of automation. Presumably, ▶ long-term potentiation is involved in the induction of a motor memory.

Movement Disorders Induced by Medications

Thomas R. E. Barnes
Department of Medicine, Imperial College, London, UK

Synonyms

Antipsychotic-induced movement disorders; Drug-induced motor syndromes; EPS; Extrapyramidal side effects

Definition

Discussion of medication-induced movement disorder generally refers to the side effects of antipsychotic medication which affect motor behavior. However, other drugs, such as lithium, valproate, antidepressants, and psychostimulants like amphetamine, can cause movement problems, principally tremor. Patients developing antipsychotic drug-induced movement disorders exhibit a range of neurological phenomena: dyskinesia; dystonia; parkinsonian features including bradykinesia, tremor, and rigidity; and restless movements as part of antipsychotic-induced akathisia. While some motor syndromes such as parkinsonism, acute dystonia, and acute akathisia are more common during acute drug treatment, others such as tardive dyskinesia, tardive dystonia, and chronic akathisia are more common with long-term treatment. The motor phenomena exhibited by those developing these disorders can cause functional impairment and be socially stigmatizing; and parkinsonism, akathisia, and dystonia also have mental manifestations that can be unpleasant and distressing (Owens 1999). Further, the disorders can confound clinical assessment of the psychotic illness. For example, features of parkinsonism such as bradykinesia overlap phenomenologically with symptoms of depression and negative symptoms, while akathisia may be misdiagnosed as anxiety or an exacerbation of psychotic symptoms.

Role of Pharmacotherapy

The four main diagnostic categories of extrapyramidal side effects (EPS) associated with antipsychotic medication are parkinsonism, akathisia, dystonia, and tardive dyskinesia. The pathophysiology of these movement disorders involves the dopamine D2 receptor-blocking properties of antipsychotic drugs. However, it should be noted that both parkinsonism and dyskinesia are observed in antipsychotic-naive patients with first-episode psychoses, suggesting that such movements reflect a neuro-dysfunction intrinsic to the pathophysiology of schizophrenia (Pappa and Dazzan 2009). While EPS were relatively common with the first-generation antipsychotics (FGAs: such as chlorpromazine, fluphenazine, haloperidol, and trifluoperazine), one of the main claims for second-generation antipsychotics (SGAs: such as aripiprazole, clozapine, olanzapine, quetiapine, and risperidone) has been a lower risk of developing such disorders, although the individual SGAs vary in their liability for EPS burden. Clinical trials and meta-analyses (Leucht et al. 2003, 2013) suggest that while SGAs have a lower liability for acute EPS and tardive dyskinesia when compared with haloperidol, even at low dosage, the evidence that this is the case in relation to other FGAs in moderate dosage is less convincing. Further, the common use of high dose or combined SGAs, or combined SGAs and FGAs, in clinical practice may compromise any such advantage. Parkinsonism and akathisia remain major problems despite the widespread use of SGAs. With regard to tardive dyskinesia, the evidence suggests a lower risk with SGAs (Correll and Schenk 2008), but prospective, longitudinal studies of SGAs as monotherapy are required to quantify the risks of tardive dyskinesia with particular drugs.

Parkinsonism

Antipsychotic-induced parkinsonism (sometimes called "pseudoparkinsonism") usually occurs within days of beginning antipsychotic treatment or after a dosage increase. It comprises a triad of bradykinesia, rigidity, and tremor. Bradykinesia is probably the core feature and manifests itself as difficulty with the initiation of movements and slowness and interruption of the normal flow of movement. When tested by passive movement of the limbs, muscle rigidity may be revealed as being of the lead-pipe (i.e., stiffness that is uniform throughout the range of movement) or cogwheel (i.e., a ratchet-like resistance) type. Other clinical signs include a mask-like expression, lack of spontaneous gesture, and a reduction in the normal arm swing when walking. Subjectively, patients experience slowed thinking ("bradyphrenia"), fatigue, weakness and stiffness, and sometimes apathy and diminished interest and initiative. There is some evidence that the development of parkinsonism may indicate an increased risk of developing tardive dyskinesia later.

While drug-induced parkinsonism mimics idiopathic Parkinson's disease, it is rather more an akinetic rigid syndrome, with the classical resting tremor being relatively uncommon. Nevertheless, the coincidental onset of idiopathic Parkinson's disease should be borne in mind as a differential diagnosis, and if this is suspected because of the nature of the clinical presentation, or because the signs and symptoms prove to be persistent after antipsychotic discontinuation, a neurology referral may be indicated.

Treatment options include reduction of the dose of the causal, or suspected, antipsychotic or switching to another with evidence for a lower risk of parkinsonism. Prescription of an antimuscarinic agent (also called anticholinergic or antiparkinsonian) may be useful (Barnes and McPhillips 1996). The anticholinergic agents most commonly used are as follows:

Benzatropine (benztropine) mesylate – 0.5–6 mg/day by mouth in one to two divided doses; can be a sedative, so if one dose is greater, administered at bedtime. Tablets are not available in the United Kingdom.

Trihexyphenidyl hydrochloride (benzhexol hydrochloride) – 1–2 mg daily, increased gradually; usual maintenance dose 5–15 mg daily in three to four divided doses, to a maximum of 20 mg daily.

Orphenadrine hydrochloride – 150 mg a day by mouth in divided doses initially, titrated slowly upward if necessary. Usual dose range is 150–300 mg daily in divided doses.

Procyclidine hydrochloride – 2.5–5 mg by mouth up to three times a day initially, titrated slowly upward if necessary. Usual maximum is 30 mg a day in divided doses.

Biperiden – 2 mg by mouth, one to two times a day.

Anticholinergic drug prescription should be regularly reviewed, partly because parkinsonism can wane spontaneously, and after 3 months or so there may no longer be a need for such treatment, and partly because antimuscarinic drugs are associated with their own unwanted effects such as blurred vision, headaches, dry mouth, increased heart rate, difficulty in urinating, and constipation, as well as confusion and disorientation, inability to concentrate, and memory impairment. The elderly are at greater risk of anticholinergic side effects.

Akathisia

Akathisia is characterized by a subjective feeling of inner restlessness and objectively by increased restless movement (Barnes 1992). These movements are not dyskinetic (involuntary, repetitive movements) but rather resemble normal patterns of restless movement. Most typically, they involve the legs, for example, walking on the spot, pacing around, or shuffling and tramping of the legs when sitting. When required to sit or stand still, patients can experience a mounting sense of tension and a compulsive desire to move in an attempt to gain some respite. The clinical significance of akathisia relates not only to the subjective dysphoria and unease experienced but also to its adverse influence on medication adherence and its status, like parkinsonism, as a possible risk factor for tardive dyskinesia.

Treatment options include reduction of the dose of the causal antipsychotic or switching to another with evidence for a lower risk of akathisia. Low-dose propranolol (initially 20 mg bd) or other lipophilic beta-blockers, an anticholinergic drug (e.g., procyclidine; see above), or a benzodiazepine (e.g., diazepam 2–4 mg tds) may be helpful. The evidence base for such interventions is limited, though strongest for propranolol or other lipophilic beta-blockers (Miller and Fleischhacker 2000). Note that the beta-blockers are contraindicated in those with asthma or peripheral vascular disease. Taylor et al. (2007) also cite evidence for a possible reduction in symptoms with low-dose clonazepam, diphenhydramine (an antihistamine), and $5HT_2$ antagonists such as cyproheptadine, mirtazapine, trazodone, and mianserin (Laoutidis and Luckhaus 2013).

Dystonia

Dystonia is defined as sustained muscle contractions causing twisting and repetitive movements or abnormal postures, which can be painful. Acute dystonia usually occurs early in drug treatment and is short-lived, although it can be distressing and frightening. It can also occur as an antipsychotic drug-withdrawal phenomenon. Tardive dystonia occurs late in the treatment, tends to be persistent, and in severe cases can be disabling and disfiguring. The neck, jaw, and tongue are the most common sites to be affected by the muscle spasms, but trunk and limbs can also be involved.

With regard to *treatment options for acute dystonia*, the intervention of choice is an anticholinergic drug (Barnes and McPhillips 1996), for example, procyclidine (see above for oral administration if symptoms are mild. Also by intramuscular or intravenous injection as an emergency treatment for acute drug-induced dystonia, 5–10 mg), benzatropine (see above for oral administration if symptoms are mild, but this drug may also be given intramuscularly or even intravenously as an emergency treatment for acute drug-induced dystonia: 1–2 mg repeated if symptoms reappear to a maximum of 6 mg daily), orphenadrine (see above for oral administration if symptoms are mild), or trihexyphenidyl hydrochloride (see above for oral administration). Patients may be given an oral antimuscarinic drug to be taken prn ("as required") should there be any signs of recurrence within the next few days. Other drug treatments advocated include antihistamines, such as diphenhydramine, and benzodiazepines, such as diazepam (Owens 1999).

With regard to *treatment options for tardive dystonia*, clinicians should first consider the differential diagnosis, including idiopathic torsion dystonia or secondary dystonia associated with conditions such as Huntington's disease or Wilson's disease. Further, there is some overlap with the features of tardive dyskinesia, with which tardive dystonia may coexist. The most common phenomena are sustained, forced, involuntary closing of the eyelids (blepharospasm), twisting of the neck to one side (torticollis) or drawing the head back (retrocollis), and involvement of the laryngeal and pharyngeal muscles affecting speech and swallowing.

The condition can be hard to treat. Withdrawal of antipsychotic medication is not usually a realistic clinical option for people with an established psychotic illness. Switching to an antipsychotic with a lower liability for EPS may be helpful in a proportion of cases over time, and the best evidence is for clozapine. There are also several specific drug treatments that may be beneficial:

Dopamine-depleting agents ("▶ Tetrabenazine," "▶ Reserpine") are used in low dose and have a reasonable evidence base to support their efficacy. They have a range of potentially unpleasant side effects including drowsiness, parkinsonism, depression, and orthostatic hypotension, although tetrabenazine has a lower side-effect burden than reserpine.

There is evidence for benefit with anticholinergic agents, sometimes in high dosage, although the relevant studies have yielded mixed results, and use of these drugs runs the risk of exacerbating any co-existing tardive dyskinesia. Benzodiazepines (such as clonazepam), which have muscle relaxant properties, are also commonly used.

For focal tardive dystonia which has not responded to standard measures, local injection of botulinum A toxin into the affected muscle by a neurologist with experience of the technique may be effective.

Tardive Dyskinesia

Tardive dyskinesia is a syndrome of abnormal involuntary movements, repetitive and stereotypic in nature and commonly referred to as choreiform (i.e., rapid, jerky). The condition usually appears late in the course of treatment and is partly related to advancing age and possibly the psychotic illness for which the antipsychotic medication is prescribed. Tardive dyskinesia most commonly affects the orofacial muscles; characteristically a combination of movements is observed, including chewing, tongue twisting and protrusion, and lip smacking and puckering. But virtually all parts of the body can be involved including the trunk and limbs and respiratory muscles. The condition is most pronounced when patients are aroused and tends to ease during states of relaxation. The abnormal involuntary movements of tardive dyskinesia can result in considerable social and physical disability, although patients are usually unaware of them.

Treatment options include reducing the dosage of the antipsychotic or switching to an SGA with evidence for a low liability for tardive dyskinesia, particularly if the patient has been receiving an FGA. Limited data from small studies do not provide convincing evidence of the value of these approaches (Soares-Weiser and Rathbone 2005), but clozapine is probably the antipsychotic most likely to diminish dyskinetic movements in patients with existing tardive dyskinesia. Discontinuing antimuscarinic agents may also be a worthwhile therapeutic option, given the evidence that such drugs can worsen tardive dyskinesia, but the available evidence does not allow for any confident statement on the likely effects of such a strategy.

A range of other potential anti-dyskinetics have been tested (Owens 1999; Taylor et al. 2007; Bhidayasiri et al. 2013), including dopamine depleters (e.g., tetrabenazine, reserpine, oxypertine), cholinomimetic agents (e.g., choline, lecithin, deanol), GABA agonists (e.g., sodium valproate, gamma-vinyl GABA), calcium channel blockers (e.g., diltiazem, verapamil), and vitamin E (alpha-tocopherol). But none has the strength of evidence for efficacy or data on adverse effects that would allow for a clinical recommendation as a treatment for tardive dyskinesia (Soares and McGrath 1999).

Cross-References

▶ Antipsychotic Drugs
▶ First-Generation Antipsychotics

References

Barnes TRE (1992) Neuromuscular effects of neuroleptics: akathisia. In: Lieberman J, Kane JM (eds) Adverse effects of psychotropic drugs. Guilford Publications, New York, pp 201–217

Barnes TRE, McPhillips MA (1996) Antipsychotic-induced extrapyramidal symptoms: role of anticholinergic drugs in treatment. CNS Drugs 6:315–330

Bhidayasiri R, Fahn S, Weiner WJ, Gronseth GS, Sullivan KL, Zesiewicz TA, American Academy of Neurology (2013) Evidence-based guideline: treatment of tardive syndromes: report of the Guideline Development Subcommittee of the American Academy of Neurology. Neurology 81(5):463–469

Correll CU, Schenk EM (2008) Tardive dyskinesia and new antipsychotics. Curr Opin Psychiatry 21:151–156

Laoutidis ZG, Luckhaus C (2013) 5-HT2A receptor antagonists for the treatment of neuroleptic-induced akathisia: a systematic review and meta-analysis. Int J Neuropsychopharmacol 29:1–10

Leucht S, Wahlbeck K, Hamann J, Kissling W (2003) New generation antipsychotics versus low-potency conventional antipsychotics: a systematic review and meta-analysis. Lancet 361:1581–1589

Leucht S, Cipriani A, Spineli L, Mavridis D, Orey D, Richter F, Samara M, Barbui C, Engel RR, Geddes JR, Kissling W, Stapf MP, Lässig B, Salanti G, Davis JM (2013) Comparative efficacy and tolerability of 15 antipsychotic drugs in schizophrenia: a multiple-treatments meta-analysis. Lancet 382(9896):951–962

Miller CH, Fleischhacker WW (2000) Managing antipsychotic-induced acute and chronic akathisia. Drug Saf 22:73–81

Owens DGC (1999) A guide to the extrapyramidal side effects of antipsychotic drugs. Cambridge University Press, Cambridge

Pappa S, Dazzan P (2009) Spontaneous movement disorders in antipsychotic-naïve patients with first-episode psychoses: a systematic review. Psychol Med 39:1065–1076

Soares KV, McGrath JJ (1999) The treatment of tardive dyskinesia – a systematic review and meta-analysis. Schizophr Res 39:1–16

Soares-Weiser K, Rathbone J (2005) Neuroleptic reduction and/or cessation and neuroleptics as specific treatments for tardive dyskinesia. Cochrane Database Syst Rev (3); Art. No. CD000459. doi:10.1002/14651858.CD000459.pub2

Taylor D, Paton C, Kerwin R (2007) The Maudsley prescribing guidelines, 9th edn. Informa Healthcare, London

MR Image Analysis

Definition

Unlike, for example, X-ray or computerized tomography (CT), MR (magnetic resonance) images usually do not provide absolute values of MR contrast parameters, as such measurements are associated with long acquisition times that are not acceptable in clinical applications. For this reason, MR quantification is mostly based on relative measurements in comparison to a defined reference such as baseline data acquired under normal or resting conditions or data derived from a healthy control group.

In functional MRI, small and relative signal changes caused by the hemodynamic response to changes in neuronal activity are assessed. Sophisticated analysis methods are essential to detect these small changes, ranging from approaches that incorporate prior knowledge from the stimulation paradigm applied to solely data-driven or exploratory methods.

Cross-References

▶ BOLD Contrast
▶ Functional Magnetic Resonance Imaging
▶ Stimulation Paradigm

Multibarrel Micropipette

Definition

An assembly of glass micropipettes usually fused together and terminating in a common tip; used for the concomitant recording of neuronal activity and the application of transmitters, drugs, or other compounds of interest.

Multimeric Protein Complex

Definition

A complex formed by several proteins.

Multiunit Activity

Synonyms

Multiple-unit spiking activity

Definition

The electrophysiologically recorded multiunit activity (MUA) is thought to represent the average spiking of small neuronal populations close to the vicinity of the placed microelectrode. It is obtained by band-pass filtering the recorded signal in a frequency range of 400 to a few thousand Hz.

Cross-References

▶ Magnetic Resonance Imaging (Functional)

Mu-Opioid Agonists

Synonyms

Morphine-like compounds

Definition

These are drugs acting selectively on the mu receptors of the endogenous opioid system. Examples are morphine, fentanyl, and sufentanil.

Cross-References

▶ Endogenous Opioid
▶ Opioids

μ-Receptor

Definition

The term "*μ-opioid peptide receptor*" (μ from morphine) represents the ▶ G-protein-coupled receptor that responds selectively to the majority of clinically useful ▶ opioid drugs. It is usually named the μ-receptor or MOR. It is expressed in areas of the nervous system that mediate therapeutic and adverse effects of most opioid drugs. The MOR protein is produced by a single gene. Several mRNA splice variants are known to exist and produce receptor proteins that display different properties when expressed in cells. When activated, the MOR predominantly transduces actions via inhibitory G proteins. The direct electrophysiological consequences of MOR activation are usually inhibitory.

μPET

A μPET machine is a relatively high-resolution positron emission tomography imaging device for the noninvasive assessment of small animals, for example, to study receptor occupancy (with specific radioligands) or functional activity (with radiolabeled ▶ deoxyglucose) of the brain, to evaluate animal models of psychiatric or neurological disorders, or to develop novel radiotracers for use in man.

Cross-References

▶ Deoxyglucose
▶ Positron Emission Tomography (PET) Imaging

Muscarine

Definition

A poisonous substance that is found in certain types of mushrooms. The substance mimics the actions of ▶ acetylcholine at muscarinic acetylcholine receptors.

Muscarinic Cholinergic Receptor Agonists and Antagonists

Michael E. Ragozzino[1] and Holden D. Brown[2]
[1]Department of Psychology, University of Illinois at Chicago, Chicago, IL, USA
[2]Brains On-Line, LLC, South San Francisco, CA, USA

Synonyms

Muscarinic agonists; Muscarinic antagonists

Definition

Muscarinic acetylcholine receptors represent one of the two classes of receptors that mediate the action of acetylcholine in the nervous system and other body parts. Muscarinic acetylcholine receptors were so named because of their greater sensitivity to muscarine toxin over nicotine. Muscarinic agonists and antagonists are agents that activate or block, respectively, muscarinic acetylcholine receptors at an orthosteric or allosteric site.

Pharmacological Properties

Muscarinic cholinergic receptors are G-protein-coupled receptors that are ubiquitously expressed in the central nervous system. There are different muscarinic receptor subtypes, referred to as M_1–M_5, when a receptor subtype is described based on pharmacology and m_1–m_5 when based on its molecular properties. The M_1, M_3, and M_5 muscarinic receptor subtypes are coupled to $G_{q/11}$ proteins that activate phospholipase-C resulting in the mobilization of intracellular calcium. The M_2 and M_4 muscarinic receptor subtypes are linked to $G_{i/o}$ proteins with activation of these receptors inhibiting adenylate cyclase and decreasing intracellular cyclic AMP. Antibodies specific to the muscarinic acetylcholine receptor proteins indicate that the m_1, m_2, and m_4 receptors are most abundant in the brain. The m_1 receptor is most concentrated in the neocortex, hippocampus, striatum, and amygdala and is found to be postsynaptic. The m_2 receptor is most abundant in the basal forebrain, thalamus, neocortex, and striatum. These receptors are located on cholinergic terminals, but are also located postsynaptically. The m_4 receptor has its highest density in the striatum but also in the neocortex and hippocampus. These receptors are found postsynaptically, as well as presynaptically, on cholinergic neurons. The m_3 and m_5 receptors are sparser in the brain with the m_3 receptor most common in the neocortex, thalamus, and hippocampus, while the highest density of m_5 receptors is located in the substantia nigra. It appears these receptors are present postsynaptically.

Orthosteric and Allosteric Sites

The orthosteric site refers to the binding site of an endogenous agonist on its receptor protein. This is also the binding site for competitive antagonists and inverse agonists. For muscarinic receptors, there is proposed to be a high sequence conservation of the orthosteric site across receptor subtypes that has limited the development of selective, competitive muscarinic agonists and antagonists for muscarinic receptor subtypes.

An allosteric site refers to a distinct site from the endogenous agonist binding site that modulates receptor activity by producing a conformational change in the receptor.

Muscarinic Agonists and Antagonists

The identification of different muscarinic receptor subtypes and varied regional expression has led to interest in developing muscarinic receptor subtype agonists and antagonists. Described below are the effects of muscarinic agonists and antagonists that have preference for one type of

Muscarinic Cholinergic Receptor Agonists and Antagonists, Fig. 1 Chemical structures of muscarinic M_1, M_2, M_4, and M_5 agonists and antagonists

muscarinic acetylcholine receptor over the other subtypes. The focus is on M_1-, M_2-, M_4-, and M_5-preferring drugs as most developed compounds preferentially act at these muscarinic receptor subtypes (see Fig. 1).

M_1 Muscarinic Receptor-Preferring Drugs

There has been considerable effort in the development of pharmacological agents that selectively act at the M_1 muscarinic acetylcholine receptor, particularly agonists. This is due in large part to studies indicating that M_1 muscarinic acetylcholine receptors are altered in Alzheimer's disease (Langmead et al. 2008). The disease is marked by cognitive deficits, and thus there has been an interest in developing selective M_1 muscarinic agonists to alleviate these deficits. Developing highly selective M_1 muscarinic receptor agonists has been a challenge as noted above. Partial M_1 receptor agonists has been developed that have a higher selectivity for the M_1 muscarinic receptor at the orthosteric site than full receptor agonists (Foster et al. 2012; Ragozzino et al. 2012). These include drugs such as CDD-102A and AF267B, which have both shown to improve learning, as well as working and long-term memory in rodents (Foster et al. 2012; Ragozzino et al. 2012).

In addition to the development of partial M_1 muscarinic agonists, different M_1 muscarinic receptor allosteric agonists have been developed (Langmead et al. 2008). These drugs are also referred to as M_1 positive allosteric modulators. These positive allosteric modulators have an advantage over current competitive agonists that act at the orthosteric site because of great selectivity. However, because they modulate the actions of acetylcholine at the receptor, they may have limited benefits as treatments if there is a considerable loss of endogenous cholinergic signaling (Foster et al. 2012). Although similar to partial M_1 muscarinic orthosteric receptor agonists, M_1 muscarinic allosteric agonists, such as benzylquinolone carboxylic acid, attenuate a working memory deficit induced by the nonspecific, muscarinic cholinergic receptor antagonist, scopolamine (Chambon et al. 2012).

Some of the strongest evidence suggesting that M_1 muscarinic receptors support learning and memory comes from experiments examining the effects of M_1 muscarinic-preferring antagonists in animal models (Tzavos et al. 2004). For example, infusions of the M_1-preferring antagonist pirenzepine into specific brain regions of the rodent impair learning or memory. More recently, VU0255035 has been identified as a M_1 muscarinic antagonist at the orthosteric site (Sheffler et al. 2009). This drug has not been characterized as well as pirenzepine, but appears to have mild effects on learning, which could be advantageous in certain treatment conditions, e.g., epilepsy. Another approach to study muscarinic receptor activity is through the use of snake toxins that bind to specific muscarinic receptor subtypes. Muscarinic toxin 7 (MT-7) is one such compound that exhibits greater selectivity for the M_1 muscarinic receptor over other subtypes. Because MT-7 acts as a more selective M_1 muscarinic receptor antagonist compared to that of pirenzepine, it has been used to study the role of M_1 muscarinic receptors in learning. One experiment demonstrated that injections of MT-7 into the rodent dorsomedial striatum do not affect initial learning of a spatial discrimination, but specifically impair spatial reversal learning (McCool et al. 2008). Thus, the results from blockade of M_1 muscarinic receptors indicate that this muscarinic receptor subtype is important for learning, memory, and behavioral flexibility. Taken together, several experiments indicate that blockade of M_1 muscarinic acetylcholine receptors in various brain areas impairs learning and memory, indicating that M_1 muscarinic acetylcholine receptors may support several forms of learning and memory.

M_2 Muscarinic Receptor-Preferring Drugs

Another pharmacological approach to modify brain cholinergic activity has been through M_2 muscarinic acetylcholine receptors. Some M_2 muscarinic acetylcholine receptors are found to be heteroreceptors in different brain regions. However, many M_2 receptors act as autoreceptors providing negative feedback at cholinergic terminals. The localization of M_2 muscarinic acetylcholine receptors on cholinergic neuron terminals has led to the examination of M_2-preferring antagonists on brain acetylcholine activity and cognitive function. In particular, M_2-preferring antagonists enhance brain acetylcholine efflux as measured by in vivo microdialysis (Ragozzino et al. 2009). Furthermore, administration of M_2-preferring antagonists such as Sch 72788, BIBN99, AF-DX 116, or methoctramine given either systemically or centrally improves memory consolidation as well as working memory on a variety of tasks. To provide more direct evidence that changes in acetylcholine output are related to learning, a study demonstrated that infusion of the M_2-preferring agonist, oxotremorine sesquifumarate, into the dorsomedial striatum simultaneously blocked a behaviorally induced increase in striatal acetylcholine output and impaired reversal learning. These effects were reversed by the M_2-preferring antagonist, AF-DX 116. Thus, blockade of M_2 muscarinic acetylcholine receptors can provide a mechanism for modulating brain acetylcholine release and cognitive functioning (Ragozzino et al. 2009).

Degeneration of cholinergic neurons is one of the hallmarks in Alzheimer's disease. Because

M_2 muscarinic acetylcholine receptor antagonists may enhance brain cholinergic levels, this type of treatment has been proposed to be used in Alzheimer's disease (Langmead et al. 2008). A number of compounds have been developed that exhibit significant selectivity for the M_2 muscarinic acetylcholine receptor. One of the serious drawbacks about using a M_2 muscarinic antagonist as a drug therapy is that M_2 muscarinic receptors are located in the heart where they slow heart rate. Thus, treatment with a M_2 muscarinic antagonist can lead to tachycardia.

Another hallmark of Alzheimer's disease is the formation of amyloid plaques in the brain. The plaques are principally composed of amyloid-β peptides, and accumulation of these peptides is considered to be a primary factor in the progression of the disease. The amyloid-β peptide is formed from amyloid precursor protein. The precursor protein is processed by either a nonamyloidogenic pathway (α-secretase) or an amyloidogenic pathway (β-secretase). The α-amyloid precursor protein may actually be neurotrophic and neuroprotective (Langmead et al. 2008). There is some evidence that M_1 muscarinic receptor agonists facilitate nonamyloidogenic processing of the amyloid precursor protein and thus may alter the progression of the disease. One study found that chronic treatment with the partial M_1 muscarinic receptor agonist, AF267B, attenuated learning and memory deficits, as well as reduced hippocampal amyloid-β peptide levels in a mouse model of Alzheimer's disease (Foster et al. 2012). Thus, treatment with partial M_1 muscarinic receptors agonists may be beneficial to alleviating cognitive deficits and reducing disease progression.

M_4 Muscarinic Receptor-Preferring Drugs

The M_4 muscarinic acetylcholine receptor has gained significant interest in developing selective agents as novel treatments in Parkinson's disease. Despite interest in the M_4 muscarinic acetylcholine receptor, there is a relative lack of compounds that are highly selective for this muscarinic receptor subtype. Comparable to therapeutic generation for the M_1 muscarinic receptor, development of positive allosteric modulators for the M_4 muscarinic receptor holds promise in generating novel treatments for various disorders and diseases.

PTAC and BuTAC are two compounds that are M_4-preferring agonists. As with many other muscarinic agonists, these drugs do not display a strong selectivity for the M_4 muscarinic acetylcholine receptor (Langmead et al. 2008). Therefore, there is a real possibility of producing unwanted side effects with such treatments.

In contrast, there are various M_4-specific positive allosteric modulators that have shown high selectivity and potency for the M_4 muscarinic receptor (Foster et al. 2012). These positive allosteric modulators hold promise for future therapies such as in schizophrenia. VU0152100 and LY2033298 are two positive allosteric modulators at the M_4 muscarinic receptor. Both of these treatments have shown to attenuate amphetamine-induced hyperactivity (Foster et al. 2012).

In contrast to these M_4-preferring agonists, tropicamide is a drug that acts as a M_4-preferring antagonist. Interest in the M_4 muscarinic acetylcholine receptor related to Parkinson's disease is related to striatal dopaminergic-cholinergic interactions. Because of reduced striatal dopamine activity in Parkinson's disease, treatment with a M_4 muscarinic acetylcholine receptor antagonist may have benefits in enhancing dopaminergic transmission and reducing symptoms in the disease. In support of this idea, tropicamide, a muscarinic acetylcholine receptor antagonist with moderate binding selectivity for the M_4 muscarinic acetylcholine receptor subtype, suppresses tremulous jaw movements in rats, a model of Parkinson's disease, without significant impairment on memory tasks (Betz et al. 2007).

M_5 Muscarinic Receptor-Preferring Drugs

As described above, M_5 muscarinic receptors appear to be restricted to midbrain dopamine neurons. Although the generation of M_5 receptor knockout mice has increased our understanding of this receptor, the development of M_5 muscarinic-preferring drugs has been quite limited. A M_5 muscarinic receptor-preferring positive allosteric modulator, VU0238429, has been developed and shown to modulate dopaminergic transmission in brain slices (Foster et al. 2012). Further development is needed to develop compounds that can be used in vivo.

Conclusions

Muscarinic acetylcholine receptors are found throughout the central nervous system. There are various neurodegenerative and psychiatric disorders that exhibit abnormalities in muscarinic acetylcholine receptor function. These various diseases and disorders often exhibit altered muscarinic acetylcholine receptor function for specific muscarinic receptor subtypes. Thus, the development of compounds that selectively modulate the different muscarinic receptor subtypes will provide an opportunity for novel treatments that can reduce severe impairments in cognitive or motor functioning.

Cross-References

▶ Allosteric Site
▶ Alzheimer's Disease
▶ Muscarine
▶ Orthosteric Site
▶ Parkinson's Disease

References

Betz AJ, McLaughlin PJ, Burgos M, Weber SM, Salamone JD (2007) The muscarinic receptor antagonist tropicamide suppresses tremulous jaw movements in a rodent model of parkinsonian tremor: possible role of M4 receptors. Psychopharmacology (Berl) 194:347–359

Chambon C, Jatzke C, Wegener N, Gravius A, Danysz W (2012) Using cholinergic M_1 receptor positive allosteric modulators to improve memory via enhancement of brain cholinergic communication. Eur J Pharmacol 697:73–80

Fisher A (2012) Cholinergic modulation of amyloid precursor protein processing with emphasis on M1 muscarinic receptor: perspectives and challenges in treatment of Alzheimer's disease. J Neurochem 120(Suppl 1):22–33

Foster DJ, Jones CK, Conn PJ (2012) Emerging approaches for treatment of schizophrenia: modulation of cholinergic signaling. Discov Med 14:413–420

Langmead CJ, Watson J, Reavill C (2008) Muscarinic acetylcholine receptors as CNS drug targets. Pharmacol Ther 117:232–243

McCool MF, Patel S, Talati R, Ragozzino ME (2008) Differential involvement of M1-type and M4-type muscarinic cholinergic receptors in the dorsomedial striatum in task switching. Neurobiol Learn Mem 89:114–124

Ragozzino ME, Mohler EG, Prior M, Palencia CA, Rozman S (2009) Acetylcholine activity in selective striatal regions supports behavioral flexibility. Neurobiol Learn Mem 91:13–22

Ragozzino ME, Artis S, Singh A, Twose TM, Beck JE, Messer WS Jr (2012) The selective M_1 muscarinic cholinergic agonist CDD-0102A enhances working memory and cognitive flexibility. J Pharmacol Exp Ther 340:588–594

Sheffler DJ, Williams R, Bridges TM, Xiang Z, Kane AS, Byun NE, Jadhav S, Mock MM, Zheng F, Lewis LM, Jones CK, Niswender CM, Weaver CD, Lindsley CW, Conn PJ (2009) A novel selective muscarinic acetylcholine receptor subtype 1 antagonist reduces seizures without impairing hippocampus-dependent learning. Mol Pharmacol 76(2):356–368

Tzavos A, Jih J, Ragozzino ME (2004) Differential effects of M1 uscarinic receptor blockade and nicotinic receptor blockade in the dorsomedial striatum on response reversal learning. Behav Brain Res 154:245–253

Myelination

Definition

A process by which axonal trees of major projection neurons (i.e., neurons that project to distant brain regions) are sheathed in myelin, a fatty membrane produced by surrounding neuroglial cells. This myelin sheathing insulates the propagation of action-potentials, the electrical signals that convey information from the cell body to the axonal synaptic terminals, from nonspecific dissipation in the neuropil. Myelin sheathing reduces the energy requirements while increasing the speed of long-range information relays in the brain.

N

Nafion

Definition

A perfluorinated ion-exchange resin that allows the passage of cations (such as biogenic amines) and precludes the passage of anions (such as metabolites of biogenic amines and ascorbic acid).

Naloxone

Definition

Naloxone is one of the best-known opioid receptor antagonists that shows high affinity for μ-opioid receptors in the CNS and as result of blockade of these receptors often produces rapid onset of withdrawal symptoms in opiate-dependent subjects. Naloxone also blocks with a lower affinity, at κ- and δ-opioid receptors. Its potency when administered orally is very low. It is used clinically to counteract the effects of opioid overdose, for example, heroin or morphine overdose, where it is most commonly injected intravenously.

Cross-References

- ▶ Naltrexone
- ▶ Opioid Antagonist

Naltrexone

Synonyms

ReVia; Vivitrol (injectable)

Definition

Naltrexone is an orally active opioid receptor antagonist and is known to bind to all three opioid receptors (mu, delta, kappa) as a function of the dose administered. Approved in oral form (ReVia®) in 1994 for the treatment of alcohol dependence, naltrexone appears most effective in reducing heavy drinking and is believed to act by blocking some of the reinforcing properties of alcohol. Problems with compliance led to the development of a long-acting (30 days) injectable form of naltrexone (Vivitrol®), approved in 2006.

Narcolepsy

Definition

A chronic disorder of sleep characterized by excessive daytime sleepiness and sudden onset of sleep, without the normal transition through lower levels of arousal, and with an

© Springer-Verlag Berlin Heidelberg 2015
I.P. Stolerman, L.H. Price (eds.), *Encyclopedia of Psychopharmacology*,
DOI 10.1007/978-3-642-36172-2

altered circadian pattern. Individuals with narcolepsy often experience cataplexy, which refers to sudden attacks of muscular weakness, often brought on by strong emotional states.

Narcotics Prison Farm

Synonyms

Federal medical center, Lexington; Lexington narcotics farm

Definition

The US Narcotics Prison Farm was established in 1935 in Lexington, Kentucky, as a prison hospital for drug addicts and was operated by the US Federal Bureau of Prisons and Public Health Service. Inmates were either committed by courts to serve sentences at the farm or voluntarily admitted themselves. Patients/prisoners received treatment for their substance abuse problem and were given new employment skills by working on the farm or on related jobs. The Narcotics Prison Farm was home to the ▶ Addiction Research Center, a significant research center focused on the study of drug abuse and dependence, particularly opioid dependence, using the inmate population. Research of this type on prisoners is rarely conducted today because of ethical concerns, although prisons remain important locations for the study of treatment programs.

Cross-References

- ▶ Abuse Liability Evaluation
- ▶ Addiction Research Center
- ▶ Opioid Use Disorder and Its Treatment
- ▶ Sedative, Hypnotic, and Anxiolytic Dependence

NARI Antidepressants

Megan M. Dahmen[1] and Janka Lincoln[2]
[1]Via Christi Regional Medical Center – Good Shepherd Campus, Wichita, KS, USA
[2]Department of Psychiatry, Kansas University School of Medicine, Wichita, KS, USA

Synonyms

Noradrenergic reuptake inhibitors; Selective noradrenergic reuptake inhibitors; Selective norepinephrine reuptake inhibitors; SNRIs

Definition

Selective noradrenergic reuptake inhibitors (selective NRIs) are a group of drugs that exert action primarily by the inhibition of norepinephrine reuptake at the norepinephrine transporter protein. The affinity of a drug for such a receptor can be expressed mathematically as the dissociation constant (K_i). To be considered as a selective NRI, the binding affinity of the most occupied receptor, in this case the norepinephrine reuptake pump, must be at least tenfold higher than the binding affinity of the next most occupied receptor. Parenthetically, the lower the K_i value, the higher the affinity of the drug for the receptor. Ki values for the selective NRIs are compared in Table 1.

Pharmacological Properties

Agents that meet the criteria for selective NRIs include reboxetine, atomoxetine, viloxazine, and several secondary amine tricyclic antidepressants (TCAs) (e.g., desipramine). The history, the pharmacokinetics, and the efficacy of reboxetine, atomoxetine, and viloxazine will be described together, as they are most similar in mechanism of action, having little affinity for receptors other than norepinephrine. Secondary amine TCAs and bupropion, a dual dopamine and norepinephrine

NARI Antidepressants, Table 1 Ki values in nM and reference studies (Adapted from http://pdsp.med.unc.edu/)

	5HT-T	5HT1A	Alpha-1	M1	D2	H1	NET
Reboxetine	107 Millan et al. (2001)	>10,000 Millan et al. (2001)	>10,000 Millan et al. (2001)		>10,000 Millan et al. (2001)		15.8 Millan et al. (2001)
Atomoxetine	77 Bymaster et al. (2002)	>1,000 Bymaster et al. (2002)		>1,000 Bymaster et al. (2002)	>1,000 Bymaster et al. (2002)	>1,000 Bymaster et al. (2002)	5 Bymaster et al. (2002)
Viloxazine	17,300 Tatsumi et al. (1997)						155 Tatsumi et al. (1997)
Nortriptyline	279 Owens et al. (1997)	294 Cusack et al. (1994)	55 Cusack et al. (1994)	40 Stanton, et al. (1993)	2,570 Cusack et al. (1994)	6.3 Cusack et al. (1994)	1.8 Owens et al. (1997)
Desipramine	163 Owens et al. (1997)	6,400 Cusack et al. (1994)	100 Cusack et al. (1994)	110 Stanton et al. (1994)	3,500 Cusack et al. (1993)	60 Cusack et al. (1994)	0.63 Owens et al. (1997)
Protriptyline	19.6 Tatsumi et al. (1997)						1.41 Tatsumi et al. (1997)
Maprotiline	5,800 Tatsumi et al. (1997)					0.79 Kanba and Richelson (1984)	11.1 Tatsumi et al. (1997)
Bupropion	9,100 Tatsumi et al. (1997)	>35,000 Cusack et al. (1994)	4,200 Cusack et al. (1994)	>35,000 Stanton et al. (1993)	DT 520 Tatsumi et al. (1997)	11,800 Cusack et al. (1994)	52,000 Tatsumi et al. (1997)

5HT-T serotonin transporter (i.e., uptake pump), $5HT_{1A}$ serotonin 1A receptor, Alpha 1 adrenergic alpha 1 receptor, M1 cholinergic muscarinic receptor, D_2 dopamine 2 receptor, H_1 histaminic 1 receptor, NET norepinephrine reuptake transporter, DT dopamine reuptake pump

reuptake inhibitor, will be discussed separately in this essay. Safety and tolerability will be described for the selective NRI class as a whole.

Reboxetine, Atomoxetine, and Viloxazine

History
Reboxetine holds the distinction of being the first truly selective NRI, and it has been used in the treatment of clinical depression, anxiety disorders, and attention deficit disorders. It is licensed for use in several countries but is unavailable in the USA (Fleishaker 2000). Reboxetine has been denied approval by the U.S. Food and Drug Administration (FDA) multiple times (Page 2003). Viloxazine has also been unable to gain FDA approval. As first reported in 1976 by Lippman and Pugsley, viloxazine was found to inhibit norepinephrine reuptake in the hearts of rats and mice, but unlike previously discovered agents such as imipramine, did not have similar actions in brain tissues. Since that time, more research has been conducted suggesting possible benefits in conditions including depression, narcolepsy, nocturnal enuresis in children, and alcoholism. Another selective NRI, atomoxetine, is the first nonstimulant drug approved for the management of symptoms secondary to attention deficit hyperactivity disorder (ADHD). Initially, studies were conducted to evaluate the efficacy of atomoxetine in depression, but these studies were stopped in 1990 as the manufacturer was focusing efforts on a more promising agent for the same indication, fluoxetine. Interestingly, the drug was originally named tomoxetine but was changed to atomoxetine after the FDA raised concerns about possible confusion with the previously marketed drug tamoxifen, an agent used in the treatment of breast cancer.

Mechanisms of Action
Antidepressants modulate various neurotransmitter systems involved in mood. Reboxetine selectively inhibits the neuronal reuptake of

norepinephrine, which is believed to mediate its antidepressive effects. In contrast to secondary amine TCAs (desipramine) and maprotiline, reboxetine lacks affinity for alpha 1, histaminergic, and muscarinic receptors and has extremely limited serotonergic properties.

Atomoxetine is closely related in structure to reboxetine. Administration of atomoxetine results in an increase of norepinephrine in the prefrontal cortex, an area of the brain associated with attention and memory. An increase of norepinephrine at this site is thought to correlate with improvement in target symptoms of ADHD.

Pharmacokinetics

Reboxetine is metabolized by cytochrome P450 (CYP) 3A4. Inhibitors of CYP 3A4 could potentially increase reboxetine plasma levels, while potent CYP3 A4 inducers will reduce the plasma concentrations of reboxetine. Reboxetine undergoes hydroxylation, oxidative dealkylation, and oxidation of both the parent compound and its metabolites, with subsequent renal elimination. Patients with hepatic impairment or renal impairment (defined as a creatinine clearance of <50 mL/min) should be started on initial doses of 2 mg twice daily, with maximum daily doses of 6 mg.

The metabolism of atomoxetine also warrants discussion, as the half-life is largely dependent on CYP 2D6. For extensive metabolizers the reported half-life is approximately 5 h, as compared to poor metabolizers (roughly 7 % of Caucasians and 2 % of African Americans), where the half-life could be prolonged up to 20 h (Caballero and Nahata 2003). Dose adjustment may be warranted when using atomoxetine in combination with CYP 2D6 inhibitors. As reboxetine, atomoxetine, and viloxazine have only weak serotonergic activity, these agents can be safely used in combination with selective serotonin reuptake inhibitors (SSRIs) with virtually no overlap in mechanism of action, although metabolic drug interactions may occur.

Efficacy

Reboxetine has not been granted approval in the USA due to a lack of compelling evidence of efficacy. A meta-analysis of four short-term (4–8 week), double-blind, multicenter trials was published in 2002. The Hamilton Depression Rating Scale was used as the primary assessment tool. In these studies, improvements were observed in psychomotor retardation, cognitive disturbance, anxiety, and insomnia. Interestingly, depressive symptoms were not discussed, leaving the reader to question the efficacy of reboxetine in the treatment of mood symptoms. Reboxetine in total daily doses of 8–10 mg may be effective in the management of depressive symptoms (Ferguson et al. 2002).

Atomoxetine should be considered as a treatment option in ADHD patients who do not tolerate stimulants, especially if insomnia and impact on growth are areas of concern. When the administration of atomoxetine is started, it can take 2–4 weeks before a therapeutic effect is appreciated, a timeframe that is comparable to other selective NRIs. This latency period could be viewed as a disadvantage of atomoxetine, as the effects of stimulants, such as methylphenidate, occur within 1–3 h of achieving an effective dose. Atomoxetine does not have abuse potential and is not labeled in the USA as a controlled substance. These features make it a particularly appealing choice over stimulants in the treatment of patients who have a history of substance abuse disorders.

Few advantages of viloxazine over other selective NRIs have been reported, although viloxazine may be a less cardiotoxic alternative to imipramine.

Tricyclic Antidepressants (TCAs)

History

Several TCAs are considered to be selective NRIs, including nortriptyline, lofepramine with its active metabolite desipramine, protriptyline, and the tetracyclic agent maprotiline (Preskorn 2009). TCAs were developed from phenothiazines to be used as possible sedatives, antihistamines, analgesics, and antiparkinsonian drugs in the 1950s (Brunton et al. 2006). Imipramine, the first TCA developed, was tested in 500 patients with various psychiatric disorders. Of the patients

NARI Antidepressants, Table 2 TCA receptor binding profile (Adapted from Preskorn 2009)

Receptor	Effect associated with blockade	
Histamine 1 (H1)	Sedation	Increase of appetite and weight
	Antipruritic effect	
Muscarinic receptor	Dry mouth	Urinary retention
	Sinus tachycardia	Memory impairment
	Constipation	
Serotonin 5HT2 uptake pump	Antidepressant	Loose stools
	Nausea	Insomnia
		Anorgasmia
Serotonin HT2C	Antianxiety	Appetite decrease
	Decrease of motor restlessness	
Serotonin HT3	Decreases nausea	Decreases vomiting
Fast Na + channels	Delayed repolarization leading to arrhythmia, seizures, delirium	

who were tried on imipramine, only depressed patients with psychomotor retardation treated daily for 1–6 weeks showed improvement in symptoms (Domino 1999).

Mechanism of Action

Unlike reboxetine, atomoxetine, and viloxazine, TCAs bind to a variety of receptors in addition to the norepinephrine reuptake pump, including serotonin reuptake pumps, muscarinic acetylcholine receptors, histamine 1 receptors, norepinephrine alpha 1 receptors, and fast Na channels (Table 2). Activity at these receptor sites is more prominent with tertiary than secondary amine TCAs and is responsible for efficacy as well as side effects.

Pharmacokinetics

Pharmacokinetics of TCAs that are considered to be selective NRIs is similar. They are well absorbed from the gastrointestinal tract independently of food. First-pass metabolism accounts for 40–50 % of the total metabolism that occurs in the intestine and the liver.

The main pharmacokinetic drug-drug interactions take place on plasma proteins involved in the distribution of the medication and during hepatic metabolism involving enzymes CYP 450. TCAs have a high affinity for plasma proteins and share this trait with other medications such as phenytoin, valproic acid, aspirin, and warfarin. If used concomitantly with these agents, displacement from the plasma proteins can occur, thus increasing the unbound form to toxic levels, although this mechanism is more theoretical than actual. CYP 450 enzymes involved in the metabolism of TCAs include 2D6, 2C19, 1 A2, and 3A3-4, with the 2D6 enzyme pathway identified as the rate-limiting step in the elimination of TCAs. Inducers and inhibitors of the enzymes aforementioned can increase or decrease the metabolism of TCAs, causing either inadequate levels with poor therapeutic response or increased levels leading to toxicity (Preskorn et al. 2004).

Efficacy

There are multiple double-blind placebo-controlled studies showing the effectiveness of TCAs in treating acute depression. The therapeutic response usually starts with the improvement of somatic symptoms. The main indication for protriptyline (dose 15–60 mg/day), nortriptyline (75–150 mg/day), desipramine (75–200 mg/day), and maprotiline (75–225 mg/day) is the treatment of depression in adults, but nortriptyline is also used as a treatment for neuropathic pain, as a nocturnal enuresis, and as a second-line treatment in attention deficit hyperactivity disorder in children, adolescents, and adults. Currently, lofepramine is not approved for use in the USA, but is used for the treatment of depression and anxiety in Europe.

TCA serum levels can be used as a clinical guide to achieving adequate therapeutic response. TCAs have a narrow therapeutic index, and therapeutic drug monitoring (TDM) can facilitate safe use of TCAs (Janicak et al. 2001). The optimal therapeutic plasma concentrations of desipramine and nortriptyline are 50–150 ng/ml.

Bupropion

History

Bupropion has a long and complex history. First discovered in 1969 by Mehta, bupropion was granted approval as an antidepressant by the FDA in 1985 under the trade name Wellbutrin. Prior to marketing, subjects involved in clinical trials were reported to experience increased risk of seizures. At the time of approval, doses up to 900 mg daily were used. Subsequently, marketing was halted and further research revealed that the risk of seizures was dose-dependent. Bupropion was released in 1989 with a lower maximum daily dose and more strict contraindications for use. Since that time, bupropion has become widely prescribed for depression and has also been granted indications for smoking cessation and, as an extended-release form, for seasonal affective disorder.

Mechanisms of Action

Bupropion is not a true selective NRI as it exhibits more potent reuptake of dopamine than norepinephrine. Bupropion has only weak serotonergic activity (Table 1).

Pharmacokinetics

Currently, bupropion is available in four dosage forms: an immediate-release tablet intended to be taken thrice daily, a sustained-release tablet to be taken twice daily, and two extended-release forms (one a hydrochloride salt and one a hydrobromide) to be taken once daily. Caution should be used when bupropion is used concomitantly with drugs dependent on CYP 2D6 for clearance, as doses ≥ 300 mg of bupropion can cause appreciable inhibition of drugs metabolized by this CYP enzyme.

Efficacy

Bupropion has proven to be effective for the treatment of depressive symptoms as monotherapy. It is also used off-label as an adjunctive antidepressant, particularly in patients who have only partially responded to SSRIs. Additionally, bupropion has been evaluated in the management of symptoms associated with ADHD, but to date does not have an indication for this disorder.

NARI Antidepressants, Table 3 Noradrenergic pathways (Adapted from Stahl 2004)

Beginning of the pathway	Projecting to	Control of
Locus coeruleus	Frontal cortex	Attention
	Alpha-2 receptor	Concentration
		Cognition
		Libido increase
Locus coeruleus	Frontal cortex	Mood
	Beta-1 receptor	
Locus coeruleus	Limbic cortex	Energy
		Fatigue
		Emotions
		Psychomotor agitation and retardation
Locus coeruleus	Cerebellum	Tremors
Locus coeruleus	Brainstem	Blood pressure
Locus coeruleus	Spine	Pain modulation
	Alpha-2	
Spinal sympathetic neurons	Heart	Heart rate
	Beta-1 receptor	
Spinal sympathetic neurons	Urinary bladder	Bladder emptying
	Alpha 1	

Selective NRIs: Safety and Tolerability

The side effect profiles of reboxetine, atomoxetine, and viloxazine can be attributed to the unwanted effects of norepinephrine, as activity at other receptors is limited. The concept of selectivity can thus be a bit deceiving. While an agent such as reboxetine is selective for the neurotransmitter norepinephrine, its activity is nonselective in terms of distribution, contributing to effects not only in the brain but elsewhere in the body as well (Table 3). In the limbic system, the stimulation of noradrenergic receptors has been correlated with agitation. Acute stimulation of norepinephrine receptors in the brainstem and spinal cord can contribute to elevations in blood pressure, although the extent of these elevations does not appear to reach clinical relevance in most cases. Such elevations have been reported with atomoxetine. In most cases, the side effects that are present with selective NRIs lessen over time. The most frequent side effects of reboxetine include dry mouth, constipation, urinary retention, blurred vision, headache, drowsiness, dizziness, excessive sweating, and insomnia. When looking at these side effects, one might associate dry mouth, constipation, urinary retention, and blurred vision with anticholinergic activity, but as was already mentioned, reboxetine does not directly block muscarinic cholinergic receptors. Instead, anticholinergic-like effects are the result of norepinephrine receptor stimulation in the sympathetic nervous system, which in turn causes a reduction of parasympathetic cholinergic tone, frequently referred to as sympathomimetic effects. This is observed clinically in the fact that reboxetine, atomoxetine, and viloxazine are generally better tolerated than tertiary amine TCAs, which elicit full effects at these receptor sites. In addition to anticholinergic effects, TCAs are linked to lethal cardiac arrhythmias, seizures, and neurotoxicity associated with elevated serum TCA levels. Toxic levels are generally accepted as >500 ng/ml for desipramine and nortriptyline.

Selective NRIs should be used cautiously in several patient populations. The potential of antidepressants to lower the seizure threshold is a class effect, and as such, all drugs in this class should be used with caution in patients with seizure disorders. Bupropion is contraindicated in patients who have epilepsy or concomitant medical conditions that may lower the seizure threshold, including the active discontinuation of alcohol or benzodiazepines and eating disorders. Daily doses greater than 450 mg of bupropion are associated with increased incidence of seizure.

As a class, selective NRIs depend on hepatic function for proper metabolism. Caution should be used, and dosage adjustments may be warranted, in hepatic insufficiency. In 2004, the FDA mandated that a warning be added to the atomoxetine drug label following two reports in which patients with no documented liver insufficiencies presented with elevated bilirubin levels and hepatic enzymes.

Selective NRIs are contraindicated with monoamine oxidase inhibitors (MAOIs). Concurrent use can result in increased catecholamine concentrations, which present, clinically, as hypertensive crisis, confusion, and seizures. Reboxetine has the ability to block the neuronal uptake of tyramine and, as a result, would theoretically be protective against hypertensive crisis secondary to the dietary ingestion of tyramine. As the threshold and duration of these effects are unknown, the combination of reboxetine and MAOIs should be avoided.

The effects of selective NRIs have not been extensively studied in pregnancy or breastfeeding, and therefore, these drugs should be used during pregnancy only when it is determined that the benefits outweigh the potential risks. Currently, maprotiline is the only selective NRI that is classified as pregnancy category B. Bupropion was previously classified as pregnancy category B, but this labeling was changed by the FDA to category C. All other selective NRIs are category C, with the exception of nortriptyline, which is classified as category D due to its increased risk of teratogenicity, and therefore should be avoided in pregnancy.

In October 2004, the FDA mandated the addition of a black box warning to the product information labels of antidepressants and selective NRIs disclosing a risk for suicide in children and adolescents, and later extended the warning to include young adults. Close monitoring for suicidal ideation or changes in behaviors is warranted in patients who are started on therapy, particularly during the first few months of therapy or following dosage changes. Along with the warning, manufacturers in the USA must provide Patient Medication Guides that are to be given to the patient with all prescriptions for antidepressants.

Conclusion

In summary, selective NRIs can be divided into two main categories: those that have primary activity only on norepinephrine receptors, such as reboxetine, atomoxetine, and viloxazine and those that have significant activity at other receptor sites in addition to norepinephrine, including secondary amine TCAs, maprotiline, and bupropion. With respect to the first group, the effects of norepinephrine activity alone result in increased alertness and concentration. For this reason, atomoxetine is beneficial in treating symptoms associated with attention deficit disorders. Agents with exclusive norepinephrine activity, such as reboxetine and viloxazine, have not been overwhelmingly advantageous in the treatment of depressive symptoms. When looking at the second group, it appears that activity at other receptor sites, including serotonin and dopamine, may be beneficial in the treatment of depressive symptoms. This interpretation is supported by the fact that the TCAs and bupropion result in more robust clinical response when used to treat depressive illness.

Cross-References

▶ Anticholinergic Side Effects
▶ Antidepressants
▶ Anxiety

References

Brunton LL, Lazo JS, Parker KL (eds) (2006) Drug therapy of depression and anxiety disorders. In: Goodman and Gilman's -the pharmacological basis of therapeutics, Chap 17. 11th edn. McGraw-Hill, New York, p 431

Bymaster FP, Katner JS, Nelson DL et al (2002) Atomoxetine increases extracellular levels of norepinephrine and dopamine in prefrontal cortex of rat: a potential mechanism for efficacy in attention deficit/hyperactivity disorder. Neuropsychopharmacology 27(5):699–711

Caballero J, Nahata MC (2003) Atomoxetine hydrochloride for the treatment of attention deficit hyperactivity disorder. Clin Ther 25(12):3065–3083

Cusack B, Nelson A, Richelson E (1994) Binding of antidepressants to human brain receptors: focus on newer generation compounds. Psychopharmacology (Berl) 114(4):559–565

Domino EF (1999) History of modern psychopharmacology: a personal view with an emphasis on antidepressants. Psychosom Med 61:591–598

Ferguson JM, Mendels J, Schwartz GE (2002) Effects of reboxetine on Hamilton depression rating scale factors from randomized, placebo-controlled trials in major depression. Int Clin Psychopharmacol 17:45–51

Fleishaker JC (2000) Clinical pharmacokinetics of Reboxetine, a selective norepinephrine reuptake inhibitor for the treatment of patients with depression. Clin Pharmacokinet 39(6):413–427

Janicak P, Davis JM, Preskorn SH, Ayd FJ Jr (2001) Principles and practice of psychopharmacotherapy, 3rd edn. Lippincott Williams & Wilkins, Philadelphia, pp 268–274

Kanba S, Richelson E (1984) Histamine H1 receptors in human brain labelled with [3H]doxepin. Brain Res 304(1):1–7

Millan MJ, Gobert A, Lejeune F et al (2001) S33005, a novel ligand at both serotonin and norepinephrine transporters: I. Receptor binding, electrophysiological, and neurochemical profile in comparison with venlafaxine, reboxetine, citalopram, and clomipramine. J Pharmacol Exp Ther 298(2):565–580

Owens MJ, Morgan WN, Plott SJ et al (1997) Neurotransmitter receptor and transporter binding profile of antidepressants and their metabolites. J Pharmacol Exp Ther 283(3):1305–1322

Page ME (2003) The promise and pitfalls of reboxetine. CNS Drug Rev 9(4):327–342

Preskorn SH (2004) Reboxetine: a norepinephrine selective reuptake pump inhibitor. J Psychiatr Pract 10(1):57–63

Preskorn SH (2009) Outpatient management of depression, 3rd edn. Professional Communications, Caddo, pp 77–126

Preskorn SH, Feighner JP, Stanga C, Ross R (2004) Antidepressants: past, present and future (Handbook of experimental pharmacology), vol 157. Springer, New York, pp 48–52

Stahl SM (2004) Essential psychopharmacology – neuroscientific basis and practical applications, 2nd edn. Cambridge University Press, Cambridge, pp 162–167

Stanton T, Bolden-Watson C, Cusack B et al (1993) Antagonism of the five cloned human muscarinic cholinergic receptors expressed in CHO-K1 cells by antidepressants and antihistamines. Biochem Pharmacol 45(11):2352–2354

Tatsumi M, Groshan K, Blakely RD et al (1997) Pharmacologic profile of antidepressants and related compounds at human monoamine transporters. Eur J Pharmacol 340(2–3):249–258

N-Back Test

Definition

A much-used test of cognition in which subjects are presented with a series of stimuli (e.g., spatial locations, visual objects, letters, etc.) and required to decide whether the current stimulus is the same as the stimulus seen n trials back. In variants of this test, subjects have to respond to the current trial with the stimulus presented n (e.g., two) trials back.

Necrosis

Definition

A term used to define the mechanism by which cells die due to the degradative action of enzymes. Necrosis follows an initial causative factor such as ischemia, a neurotoxin, or injury.

Cross-References

▶ Apoptosis

Negative Reinforcement

Definition

Like positive reinforcement, negative reinforcement increases the likelihood that a behavior associated with it will be continued. A negative reinforcer is an aversive event whose *removal* following an operant response increases the frequency with which that operant occurs. Negative reinforcers can range from uncomfortable physical sensations to actions causing severe physical distress. Taking a drug to relieve the withdrawal distress is, arguably, an example of negative reinforcement. If a person's ▶ withdrawal syndrome (stimulus) goes away after taking the drug (behavior), then it is likely that the person will seek out the drug as soon as the first withdrawal discomfort appears in the future.

Negative Symptoms Syndrome

Synonyms

Deficit symptoms syndrome

Definition

A syndrome which involves a deficiency of a number of mental features or capacities which would be expected to be present in a healthy individual. These features and capacities include the ability to sustain an adequate level of attention during tasks, activities, or social encounters; a full range of genuine, appropriately responsive, and appropriately sustained affect; an adequate quantity of spontaneous speech which contains objective content and which is delivered without disruptive delays in either initiating or sustaining the speech; an appropriate level of attention and caring concerning grooming and personal hygiene; an appropriate level of motivation and persistence concerning work, school, or other productive activities; an appropriate level of energy for initiating and sustaining activities and capacity for self-starting and self-direction; an appropriate level of interest and investment in, and pleasure derived from, enjoyable or recreational activities; and an age-appropriate interest in, and capacity for, friendship, cooperation, and interpersonal closeness or intimacy.

Cross-References

▶ Depressive Disorder and Schizophrenia

Nemonapride

Definition

Nemonapride is a first-generation ▶ (typical) antipsychotic drug that belongs to the benzamide class; it is approved in Japan for the treatment of schizophrenia. It is a potent dopamine antagonist with high affinity for D_2, D_3, and D_4 receptors. In addition, it is a potent $5-HT_{1A}$ receptor agonist and has relatively high affinity for sigma receptors, with little affinity for $5-HT_{2A}$ receptors. It can induce ▶ extrapyramidal motor side effects and has a propensity to elevate prolactin secretion, but generally displays low toxicity.

Cross-References

▶ Extrapyramidal Motor Side Effects
▶ First-Generation Antipsychotics
▶ Schizophrenia

Neologisms

Definition

Creation of new words not corresponding to linguistic conventions.

Neonatal Abstinence Syndrome

Definition

Expression of drug withdrawal behavior in neonates of drug-addicted mothers, a syndrome best characterized in neonates exposed prenatally to opiates such as heroin or methadone.

Nerve Growth Factor

A. Claudio Cuello
Department of Pharmacology, McGill University, Montreal, QC, Canada

Synonyms

NGF

Definition

Nerve growth factor (NGF) is a protein primarily responsible for the differentiation and survival of sympathetic and spinal cord primary sensory neurons in the peripheral nervous system and for the differentiation and survival of cholinergic neurons of the basal forebrain during the development of the nervous system and responsible for the maintenance of their neuronal phenotype after maturity. NGF has other broader effects over endocrine and inflammatory mechanisms.

Pharmacological Properties

Preamble

NGF was the first molecule identified as having well-defined trophic actions over cells of the nervous system. Following the discovery of NGF, a plethora of other proteins with differential trophic effects over PNS or CNS neurons have also been identified. The discovery of NGF introduced the concept that neuronal survival, differentiation, and growth are controlled by trophic factors and initiated a major research field in the neurosciences and neuropharmacology.

History

The discovery and identification of NGF is one of the most fascinating chapters in the history of neuroscience. The immediate roots of this discovery can be found in the pioneering investigations of Rita Levi-Montalcini and Viktor Hamburger who demonstrated that the grafting of sarcoma tumors in chick embryos elicited a remarkable hypertrophic growth of spinal cord primary sensory and sympathetic ganglia with an ingrowth of the corresponding nerves into the sarcoma (Levi-Montalcini 1987). The possibility that a "diffusible factor" was responsible for such a remarkable neurotrophic response was consecutively investigated in vitro by Rita Levi-Montalcini who, while in Rio de Janeiro, demonstrated that the culture media of sarcoma tissue induced a remarkable "halo" of radially outgrowing neurites emerging from the ganglia. These powerful microscopic images have remained iconic, given these dramatic effects and their significance. Stanley Cohen and, later, Rita Levi-Montalcini serendipitously identified similar trophic effects with material extracted from snake venom glands and from rodent submaxillary glands. These highly purified preparations allowed them to establish the proteinaceous nature of NGF and to generate antibodies for immunoneutralization and biological characterization. Later, these preparations allowed Angeletti and Brashaw to establish NGF's amino acid sequence. Because of these pioneering findings, the Nobel Prize in Medicine was granted in 1986 to both Rita Levi-Montalcini and Stanley Cohen. A family of NGF-like molecules (the neurotrophins) and their receptors were consequently identified after these seminal contributions.

Neurobiology and Mechanisms of Action

It is currently known that NGF is a member of the so-called neurotrophin family of trophic factors which is composed of NGF, brain-derived neurotrophic factor (BDNF), and the neurotrophins, NT3 and NT4/5. These proteins have a high degree of homology with NGF and their tertiary structure reveals antiparallel strands and four hairpin loops. Three disulfide bridges known as a "cysteine knot" maintain the 3D configuration. The neurotrophins can be differentiated from each other principally by the amino acid sequence of their hairpin loops, which apparently grants them their biological specificity (Skaper 2008). The neurotrophins have subunit molecular masses in the range of 12–14 kDa and are found as noncovalently linked homodimers of approximately 26 kDa. The 3D molecular structure of NGF and its relative homology with other members of the neurotrophin family are represented in Fig. 1.

Three main high-affinity receptor tyrosine kinases have been identified and referred to as TrkA, TrkB, and TrkC, responding preferentially to NGF, BDNF, and NT3, respectively, and NT4 to NT4. While NGF acts fairly narrowly on TrkA, some overlap regarding receptor responses can be expected of the rest of the neurotrophins (Chao 2003; Kaplan and Miller 2000; Skaper 2008), as illustrated in Fig 2.

These receptors are mainly responsible for the diverse and specific neurotrophic responses elicited by the neurotrophins over defined groups of CNS and PNS neurons. A ubiquitous and less discriminating "co-receptor," p75 neurotrophin receptor (NTR), appears to act cooperatively with Trk receptors in eliciting neurotrophic actions. However, in certain conditions, in the absence of Trk receptors, $p75^{NTR}$ are thought to mediate cell death via apoptotic mechanisms, acting cooperatively with the sortilin receptor (Nykjaer et al. 2004), a receptor for neurotensin, which is also present in NGF-rich CNS regions.

Both TrkA and $p75^{NTR}$ receptors are nowadays considered to be high-affinity receptors for NGF. NGF is generated by a variety of tissues and cell types. In the mature CNS, it is mainly produced by neurons located in target regions for NGF-dependent neurons (target-derived trophic support) where their axonal branches and presynaptic boutons terminate. NGF is synthesized as a larger precursor molecule, preproNGF, whose signal peptide is cleaved in the endoplasmic reticulum to produce proNGF. ProNGF is a secreted molecule and is the most abundant molecular species of NGF in the adult CNS. It is the preferred ligand of the $p75^{NTR}$/sortilin complex. It can also bind TrkA, but with weaker affinity than mature NGF (Fahnestock et al. 2004). The mature form of NGF (mNGF), on the other hand, is most likely generated extracellularly and binds more tightly to TrkA. The biological activities of proNGF and NGF are determined by the relative levels of TrkA and $p75^{NTR}$ (Masoudi et al. 2009). The best-characterized biological actions of mNGF are its neurotrophic effects. These are achieved by mNGF binding to monomeric or homodimeric TrkA receptors and also to hetero-TrkA–$p75^{NTR}$ complexes; dimeric complexes result in higher ligand–receptor affinity (Fig. 3).

Synthesis, Release, and Metabolism of NGF

As indicated earlier, mature and biologically active NGF is the immediate product of a larger precursor protein, proNGF. Western blot analysis reveals that the predominant form of NGF in the adult CNS is proNGF (Fahnestock et al. 2001). It has been proposed that proNGF can be cleaved intracellularly by the action of furin converting proNGF into NGF. However, the degree to which the "conversion" of proNGF into mNGF takes place intra- or extracellularly is debatable. Studies with transfected cell lines have shown that a furin-resistant form of proNGF can be released into the extracellular space and induce cell death via the activation of $p75^{NTR}$ (Lee et al. 2001), possibly when TrkA levels are low. Cell biology studies indicate that the secretion of NGF is activity-dependent. Ex vivo studies with superfused cerebral cortex tissue of adult rats

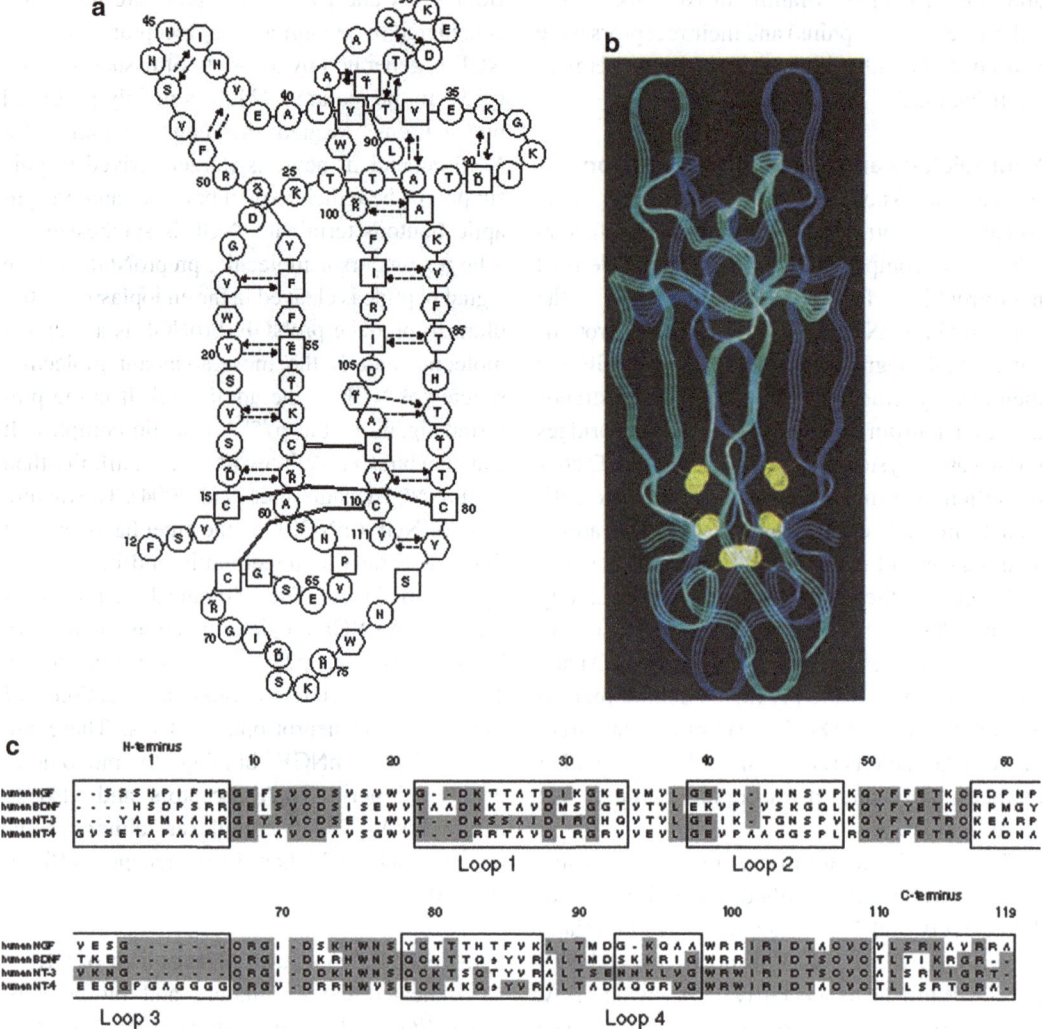

Nerve Growth Factor, Fig. 1 (a) Schematic representation of the β-NGF monomer subunit highlighting structurally important residues. *Bold*, absolutely conserved residues for all NGF and NGF-related sequences. *Squares*, buried residues in the β-NGF subunit with a relative side chain. The hexagons represent residues involved in the dimer interface as identified using computer imaging techniques. The side chain of the residue participates in a hydrogen bond with either a side-chain or a main-chain atom. The main-chain hydrogen bonds involved in the β-sheet structure are displayed as *arrows*, pointing in the direction of the donor acceptor. The cysteine knot is shown as *three solid lines* linking six cysteine residues near the *bottom* of the molecule. (From McDonald et al. (1991) Nature; with author's and publisher's permission). (b) Ribbon representation of the β-NGF dimer. The *cyan* and *dark blue* ribbons each represent a subunit. The Sγ atom for each half-cystine residue is also shown drawn as a *sphere*. (From McDonald et al. (1991) Nature; with author's and publisher's permission). (c) Structure-based sequence alignment of the neurotrophins. Numbering refers to the sequence of mature human NGF. Positions are numbered from the first residue in each neurotrophin. Note that, because of differences in the lengths of the N-termini of the different neurotrophins, homologous positions in different molecules do not have equivalent numbering. Conserved residues are *shaded* and low sequence homology (*boxes*) and sequence of hairpin loops are represented. *Dashes* indicate gaps introduced for the sake of alignment (From Skaper 2008. CNS Neurol Disord Drug Targets; with author's and publisher's permission)

Nerve Growth Factor, Fig. 2 Neurotrophins and their receptors. The neurotrophins display specific interactions with the three Trk receptors: NGF binds TrkA, BDNF and NT4 bind TrkB, and NT3 binds TrkC. In some cellular contexts, NT3 can also activate TrkA and TrkB albeit with less efficiency. All neurotrophins bind to and activate $p75^{NTR}$. CR1–CR4, cysteine-rich motifs; C1/C2, cysteine-rich clusters; LRR1–3, leucine-rich repeats; Ig1/Ig2, immunoglobulin-like domains (From Skaper 2008. Drug Targets. CNS Neurol Disord; with author's and publisher's permission)

have revealed that proNGF is the molecular form released in an activity-dependent manner, following depolarization or neurotransmitter stimulation (Bruno and Cuello 2006). These studies also indicate that in the adult and fully differentiated nervous system, the conversion of proNGF occurs in the extracellular space by the coordinated action of plasminogen and tissue plasminogen activator (tPA) producing plasmin which converts proNGF into mNGF (Bruno and Cuello 2006). Mature NGF and corresponding activated TrkA receptor are internalized in endosomal vesicles, which are transported retrogradely by axonal processes from nerve terminals to perinuclear regions of neurons, where the vesicles continue signaling via the Erk–CREB pathway (Grimes et al. 1996). These studies also indicate that the metalloproteinase MMP9 rapidly degrades and inactivates any remaining mNGF, which is not bound to the cognate receptor (TrkA) and rapidly internalized.

The above metabolic pathway involving plasmin for the maturation of NGF and MMP9 for its degradation has been pharmacologically validated in vivo, showing that the inhibition of tPA results in the brain accumulation of proNGF and, conversely, the inhibition of MMP9 in the accumulation of mNGF (Bruno and Cuello 2006). Figure 4 illustrates the activity-dependent release of proNGF and its consequent conversion to mature NGF, binding to its cognate receptors and its eventual degradation in the extracellular space as well as the protease cascade and endogenous inhibitors.

The best-defined and most dramatic actions of the endogenously generated NGF are illustrated during embryonic stages and the early postnatal period. In brief, in in vitro conditions, the deprivation of NGF support leads to cell death of NGF-dependent embryonic neurons, typically small-size spinal cord primary sensory and sympathetic neurons. The early studies on the deprivation of NGF trophic support were performed by immunoneutralization with anti-NGF polyclonal antibodies, and more recent studies were done with NGF knockout (KO) or NGF-deficient animal models. Mice lacking NGF or TrkA (KO models) do not survive beyond weeks after birth. The phenotype of NGF(−/−) or TrkA(−/−) mice is one of dramatic loss of NGF-dependent spinal cord sensory neurons and sympathetic neurons, but with a lesser effect on NGF-dependent forebrain neurons. However, mice carrying a single NGF allele have shown marked atrophy of forebrain

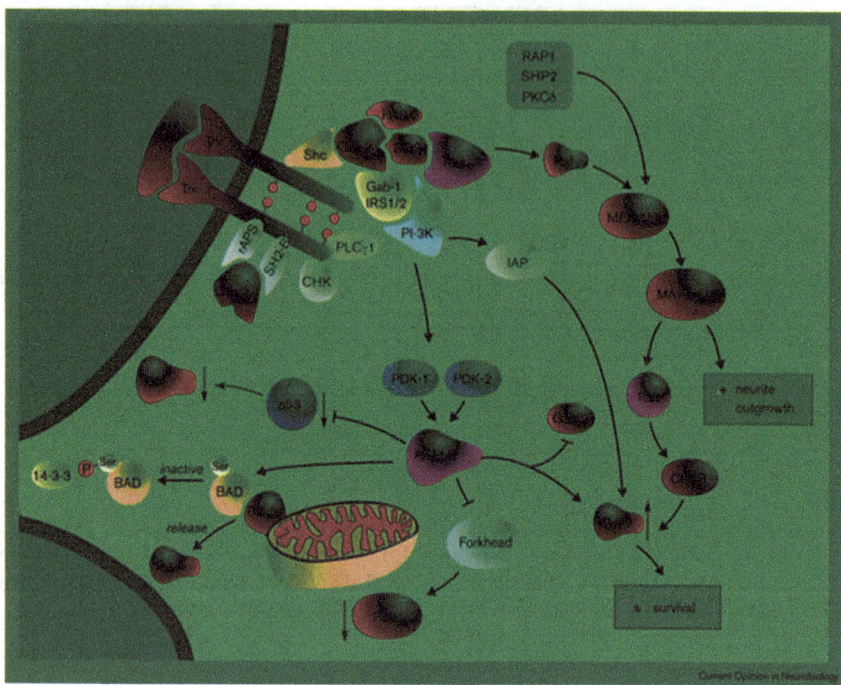

Nerve Growth Factor, Fig. 3 Trk signaling pathways regulating survival and neurite growth in neuronal cells. Neurotrophin (*NT*) binding to Trk stimulates receptor transphosphorylation, resulting in the recruitment of a series of signaling proteins to docking sites on the receptor. These proteins include Shc, which activates Ras through Grb-2 and SOS; FRS-2, rAPS, SH2-B, and CHK, which participate in activating MAPK; and PLCγ1 and CHK bind to phosphorylated Tyr 785. MAPK activity is also regulated through Raf, Rap 1, SHP-2, and PKCδ. The MEK and MAPK pathway is thought to regulate neurite growth and survival. Trk activates Pl-3 K through the RAS and the Gab-1/IRS-1/IRS-2 family of adapter proteins. Pl-3 K activity stimulates the activities of PDK2, which in turn activate Akt. The targets of Pl-3 K/Akt anti-apoptotic activity, including BAD, forkhead, GSK-3, Bcl-2 IAP, and the p53 pathway involved in cell death (From Kaplan and Miller 2000. Curr Opin Neurobiol; with author's and publisher's permission)

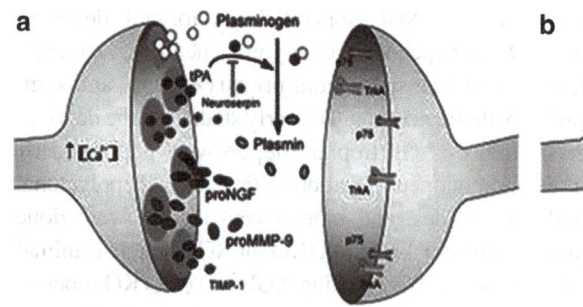 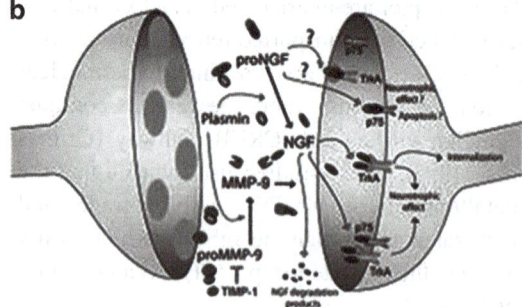

Nerve Growth Factor, Fig. 4 Schematic representations of events leading to proNGF conversion into mNGF and its degradation. Neuronally stored proNGF, plasminogen, tPA, neuroserpin, proMMP9, and TIMP-1 would be released into the extracellular space upon neuronal stimulation. Released tPA would induce the conversion of plasminogen to plasmin, where its activity is tightly regulated by secreted neuroserpin. The generated plasmin would convert proNGF into mature NGF and activate proMMP9 into active MMP9. Mature NGF would interact with its cognate receptors (TrkA and p75^{NTR}) or suffer degradation by activated MMP9 (From Bruno and Cuello 2006. PNAS; with author's and publisher's permission)

cholinergic neurons of the nucleus basalis and medial septum. The most accepted view is that during development, the presence of NGF in defined target areas will attract axonal growth to the sites of termination (synapses in the CNS) and secure the eventual survival of NGF-dependent neurons (Sofroniew et al. 2001).

Many of the concepts derived from investigations on embryonic tissue have been applied to the adult nervous system. Thus, the concept of "target-derived" NGF support of CNS neurons in the adult brain was readily accepted. It was shown early on that axotomy of the septal–hippocampal cholinergic pathway in the adult brain resulted in the loss of the corresponding NGF-sensitive cholinergic neurons. Their recovery by the application of NGF was interpreted as an indication of a similar NGF dependency for neuronal survival in the adult and fully differentiated CNS. However, the substantial excitotoxic destruction of the target tissue (hippocampus), while sparing their axonal input, resulted in atrophy, but not cell death, of the NGF-dependent cholinergic neurons of the medial septum, unequivocally demonstrating that NGF in the adult is key for the maintenance of neuronal phenotype, but not survival (Sofroniew et al. 2001). A similar NGF dependency for the phenotypic characteristics of cholinergic neurons of the nucleus basalis has also been amply documented (Cuello 1994). Further confirmation of the essential role of NGF in the maintenance of the CNS cholinergic phenotype in the mature brain is provided by the evidence that blocking "endogenous" NGF function through the application of either antibodies mapping out the ligand (NGF) or NGF mimetic peptides with antagonist actions over the receptor (TrkA) in the cerebral cortex of adult rats results in the loss of preexisting cortical cholinergic synapses (Debeir et al. 1999). Such an observation indicates that the "steady-state" number of CNS cholinergic synapses is dependent on a continuous supply of endogenous NGF, in line with the classical Hebbian concept that brain activity unleashes a growth-dependent synaptic efficacy. While the evidence for a target-derived action of NGF is strong, it is not clear whether endogenous NGF can also act in a paracrine fashion. However, it has been shown in both rodents and primates that exogenously applied NGF can also act in a paracrine fashion (Tuszynski et al. 2005), a concept (see below) which might have therapeutic applications.

NGF Pharmacology

Both NGF purified from submaxillary glands and genetically engineered recombinant forms of NGF have been successfully applied in a large variety of in vitro and in vivo experimental models. Both forms have been shown to be equally effective. In most in vitro models, NGF displays trophic effects at concentrations as low as 10^{-13} M. In the adult CNS, the recovery of target-deprived or lesioned NGF-dependent neurons can be achieved with the application of doses as low as 1 μg/kg/day typically administered in the cerebrovascular space for period of a week or longer. The drawback of the application of the NGF protein resides in its broad actions over a large variety of target cells and tissues in the CNS and periphery, its rapid proteolytic degradation, and the difficulty with which it crosses the blood–brain barrier (BBB) for potential CNS reparative applications.

A number of strategies have been explored to develop smaller-molecular-weight compounds with NGF agonist or antagonistic actions. The initial efforts have involved the testing of NGF diverse fragments, then the synthesis of cyclic peptides mimicking NGF loops, and, lately, the use of dimeric peptide-mimetic molecules. The majority of these molecules displayed some TrkA antagonistic actions in either in vitro or in vivo conditions (Longo et al. 2007; Skaper 2008). A recently developed dimeric NGF mimetope named D3 has been shown to display similar efficacy to NGF in recovering the CNS cholinergic phenotype and correcting memory impairments of aged and cognitively impaired rats (Bruno et al. 2004). Whether analogous low-molecular-weight compounds are capable of crossing the BBB will find therapeutic applications remains to be established.

NGF in Health and Disease

The multifaceted aspects of NGF have been repeatedly signaled by Rita Levi-Montalcini, Luigi Aloe, and collaborators, and they would indicate the possible involvement of NGF in a multitude of functions beyond the classical neurotrophic effects. Two areas of potential therapeutic applications have been given the most attention. These are the potential reparative effects of NGF in Alzheimer's disease (AD) and the control of the NGF proinflammatory/pronociceptive effects in the periphery.

It is well established that the experimental, exogenous application of NGF in the cerebrovascular space can rescue the cholinergic phenotype and memory functions in the axotomy model of the septal–hippocampal pathway and in the model of retrograde degeneration of nucleus basalis neurons following large stroke-type lesions of the cerebral cortex (Cuello 1994). Furthermore, CNS cholinergic axonal sprouting and synaptogenesis have resulted from the application of NGF in rodent and primate animal models (Cuello 1994). Likewise, it has been shown that NGF is capable of reversing the well-known age-related CNS cholinergic atrophy. Since acetylcholine is known to play a key role in higher CNS functions such as attention, learning, and memory and since this transmitter system is the most vulnerable to AD neuropathology, it was naturally considered early on as a possible trophic factor for reparative therapy for this neurodegenerative disease. The first therapeutic attempts in AD were made by the intracerebral application of highly purified murine NGF. Despite anecdotal cognitive improvements, these trials were interrupted as the patients developed an unacceptable weight loss and pain. More recently, application of NGF has been restricted to the forebrain in order to overcome these undesirable effects. This has been done in the knowledge that NGF, besides acting as a "target-derived" neurotrophin, also displays paracrine trophic effects. Thus, autologous fibroblasts genetically modified to express human NGF have been grafted in the area of the nucleus basalis in a limited number of AD sufferers (Tuszynski et al. 2005). The drawback of this approach is the invasive nature of the procedure. The development of small-molecular-weight TrkA agonists capable of crossing the BBB and eliciting neuroregeneration without the activation of pain receptors or provoking loss of body weight would be a desirable objective.

AD pathology occurs concurrently with a marked atrophy of NGF-dependent basal forebrain cholinergic neurons. However, in AD, there are no signs of a diminished synthesis of NGF, but on the contrary, an elevation of the NGF precursor, proNGF, has been consistently reported (Fahnestock et al. 2001). This creates the paradoxical availability of an abundance of NGF precursor along with cholinergic atrophy. This can now be explained by the finding of an altered metabolism of NGF in AD such that the conversion of proNGF to mNGF is impaired and the degradation of mNGF exacerbated. Such circumstances could offer a conceptual framework for a less invasive pharmacological strategy to correct the NGF dysmetabolism observed in AD.

As noted by Levi-Montalcini and Aloe and observed by others, NGF is capable of recruiting/activating mast cells and inducing an inflammatory response in animal models and in arthritic tissue in humans (Watson et al. 2008). Furthermore, there is good evidence that NGF is elevated in a variety of severe pain conditions in humans. These include arthritis, chronic cystitis, and pancreatitis. The application of NGF or the overexpression of NGF in transgenic animal models results in hyperalgesia and even allodynia. The production and release of NGF is stimulated by a variety of cytokines and inflammatory reagents present in damaged or inflamed tissues. In turn, NGF acts in a proinflammatory fashion by recruiting mast cells and facilitating mast cell degranulation. The pronociceptive effects of NGF are thought to be elicited mainly by the stimulation of TrkA receptors present in small-diameter spinal cord primary sensory afferents of types Aδ and C. These fibers contain and release either substance P or calcitonin gene-related peptide (CGRP) or both. NGF stimulates the production and release of these pronociceptive peptides orthodromically in the spinal cord (first CNS pain signal) and antidromically in the periphery, facilitating further plasma extravasation and mast cell degranulation (Watson et al. 2008). NGF modulates a number of ligand- and voltage-gated ion channels participating in nociception. In particular, NGF enhances the synthesis, membrane expression, and density of the ligand-gated transient receptor potential cation channel, vanilloid subfamily member 1 (TRPV1). This receptor responds to capsaicin and a number of noxious stimuli and agents released by injured tissue such as bradykinin. It also sensitizes spinal cord primary sensory fibers to nociceptive stimuli. In consequence, NGF has "feedforward" properties in enhancing and prolonging both inflammation and pain. A number of pain-controlling strategies aimed at blocking NGF actions in the periphery have been explored. Thus, chimeric molecules with dimeric TrkA domains fused to the constant region (Fc) of human IgG domains have been shown to be effective in controlling pain and inflammation in the carrageenan-inflamed rat hind limb model, when injected locally. However, the large molecular size and immunological reactions are potential drawbacks for the clinical application of this molecule. More recently, a recombinant and refolded protein with the amino acid sequence of the d5 domain of TrkA has been shown to be equally effective in a variety of preclinical models. The clinical application of such an approach has not, as yet, been explored. Antibodies against NGF are known to control nociceptive effects. The development of a monoclonal antibody capable of reducing pain and inflammation in a number of preclinical models including bone fractures has encouraged the humanization of these antibodies for further, ongoing, clinical trials with some apparent success (Watson et al. 2008). Small-molecular-weight molecules acting as NGF antagonists on TrkA receptors may also find an application on this front (Skaper 2008).

Cross-References

▶ Brain-Derived Neurotrophic Factor
▶ Mild Cognitive Impairment
▶ Neuroprotection

References

Bruno MA, Cuello AC (2006) Activity-dependent release of precursor nerve growth factor, conversion to mature nerve growth factor, and its degradation by a protease cascade. Proc Natl Acad Sci U S A 103:6735–6740

Bruno MA, Clarke PB, Seltzer A, Quirion R, Burgess K, Cuello AC, Saragovi HU (2004) Long-lasting rescue of age-associated deficits in cognition and the CNS cholinergic phenotype by a partial agonist peptidomimetic ligand of TrkA. J Neurosci 24:8009–8018

Chao MV (2003) Neurotrophins and their receptors: a convergence point for many signalling pathways. Nat Rev Neurosci 4:299–309

Cuello AC (1994) Trophic factor therapy in the adult CNS: remodelling of injured basalo-cortical neurons. Prog Brain Res 100:213–221

Debeir T, Saragovi HU, Cuello AC (1999) A nerve growth factor mimetic TrkA agonist causes withdrawal of cortical cholinergic boutons in the adult rat. Proc Natl Acad Sci U S A 97:4067–4072

Fahnestock M, Michalski B, Xu B, Coughlin MD (2001) The precursor pro-nerve growth factor is the predominant form of nerve growth factor in brain and is increased in Alzheimer's disease. Mol Cell Neurosci 18:210–220

Fahnestock M, Yu G, Michalski B, Mathew S, Colquhoun A, Ross GM, Coughlin MD (2004) The nerve growth factor precursor proNGF exhibits neurotrophic activity but is less active than mature nerve growth factor. J Neurochem 89:581–592

Grimes ML, Zhou J, Beattie EC, Yuen EC, Hall DE, Valletta JS, Topp KS, LaVail JH, Bunnett NW, Mobley WC (1996) Endocytosis of activated TrkA: evidence that nerve growth factor induces formation of signaling endosomes. J Neurosci 16:7950–7964

Kaplan DR, Miller FD (2000) Neurotrophin signal transduction in the nervous system. Curr Opin Neurobiol 10:381–391

Lee R, Kermani P, Teng KK, Hempstead BL (2001) Regulation of cell survival by secreted proneurotrophins. Science 294:1945–1948

Levi-Montalcini R (1987) The nerve growth factor 35 years later. Science 237:1154–1162

Longo FM, Yang T, Knowles JK, Xie Y, Moore LA, Massa SM (2007) Small molecule neurotrophin receptor ligands: novel strategies for targeting Alzheimer's disease mechanisms. Curr Alzheimer Res 4:503–506

Masoudi R, Ioannou MS, Coughlin MD, Pagadala P, Neet KE, Clewes O, Allen SJ, Dawbarn D, Fahnestock M (2009) Biological activity of nerve growth factor precursor is dependent upon relative levels of its receptors. J Biol Chem 284:18424–18433

McDonald NQ et al (1991) New protein fold revealed by a 2.3-A resolution crystal structure of nerve growth factor. Nature 354:411–414

Nykjaer A, Lee R, Teng KK, Jansen P, Madsen P, Nielsen MS, Jacobsen C, Kliemannel M, Schwarz E, Willnow

TE, Hempstead BL, Petersen CM (2004) Sortilin is essential for proNGF-induced neuronal cell death. Nature 427:843–848

Skaper SD (2008) The biology of neurotrophins, signalling pathways, and functional peptide mimetics of neurotrophins and their receptors. CNS Neurol Disord Drug Targets 7:46–62

Sofroniew MV, Howe CL, Mobley WC (2001) Nerve growth factor signaling, neuroprotection, and neural repair. Annu Rev Neurosci 24:1217–1281

Tuszynski MH, Thal L, Pay M, Salmon DP, HS U, Bakay R, Patel P, Blesch A, Vahlsing HL, Ho G, Tong G, Potkin SG, Fallon J, Hansen L, Mufson EJ, Kordower JH, Gall C, Conner J (2005) A phase 1 clinical trial of nerve growth factor gene therapy for Alzheimer disease. Nat Med 11:551–555

Watson JJ, Allen SJ, Dawbarn D (2008) Targeting nerve growth factor in pain: what is the therapeutic potential? BioDrugs 22:349–359

Neuroactive Steroids

A. Leslie Morrow[1] and Patrizia Porcu[2]
[1]Departments of Psychiatry and Pharmacology, Bowles Center for Alcohol Studies, University of North Carolina School of Medicine, Chapel Hill, NC, USA
[2]Neuroscience Institute, National Research Council of Italy (CNR), Cagliari, Italy

Definition

This entry delineates the pharmacological, physiological, pathological, and therapeutic relevance of neuroactive steroids. These steroids are endogenous or synthetic compounds that can cross the blood–brain barrier and rapidly alter neuronal excitability via membrane receptors. We focus on neuroactive steroids that have specific binding sites on γ-aminobutyric acid type A ($GABA_A$) receptors which directly influence both synaptic and extrasynaptic transmission and indirectly alter many physiological processes including the hypothalamic–pituitary–adrenal (HPA) axis function, inflammation, and neuroregeneration. Neuroactive steroid levels are modulated by stress, ovarian cycle, pregnancy, as well as numerous pharmacological agents. We summarize the sites of neuroactive steroid actions, the systemic and molecular consequences of these actions, and the potential therapeutic relevance of their effects for neuropsychiatric diseases.

Pharmacological Properties

Neuroactive steroids are endogenous neuromodulators that influence brain processes in fundamental ways that affect mood, behavior, and all organ systems controlled by brain function. Pioneering work from Baulieu and collaborators in the early 1980s demonstrated the persistence of substantial amounts of pregnenolone, progesterone, dehydroepiandrosterone (DHEA), and their sulfate metabolites in the brain of adrenalectomized/gonadectomized animals. This suggests that the brain was capable of local synthesis of these steroids that were termed neurosteroids (Baulieu 1998). Subsequent studies have shown that neurons are able to locally synthesize neurosteroids de novo from cholesterol. The biosynthetic pathway for neuroactive steroids is shown in Fig. 1.

The classical action of steroid hormones is to regulate transcriptional activity and protein biosynthesis over minutes to hours via interaction with nuclear steroid receptors. However, neuroactive steroids rapidly (milliseconds to seconds) alter neuronal excitability by binding to membrane receptors in the plasma membrane (Paul and Purdy 1992). Neuroactive steroids may include any steroid that acts on membrane receptors in brain to alter physiological properties of neurons. These steroids include glucocorticoids, estrogens, progesterone, DHEA, as well as the 3α,5α-reduced metabolites of progesterone and deoxycorticosterone that are the primary focus of this entry.

GABAergic Neuroactive Steroids

The 3α,5α- and 3α,5β-reduced metabolites of progesterone, deoxycorticosterone, DHEA, and testosterone enhance GABAergic transmission and produce inhibitory neurobehavioral effects (see Morrow 2007, for review). The most studied are the progesterone metabolite

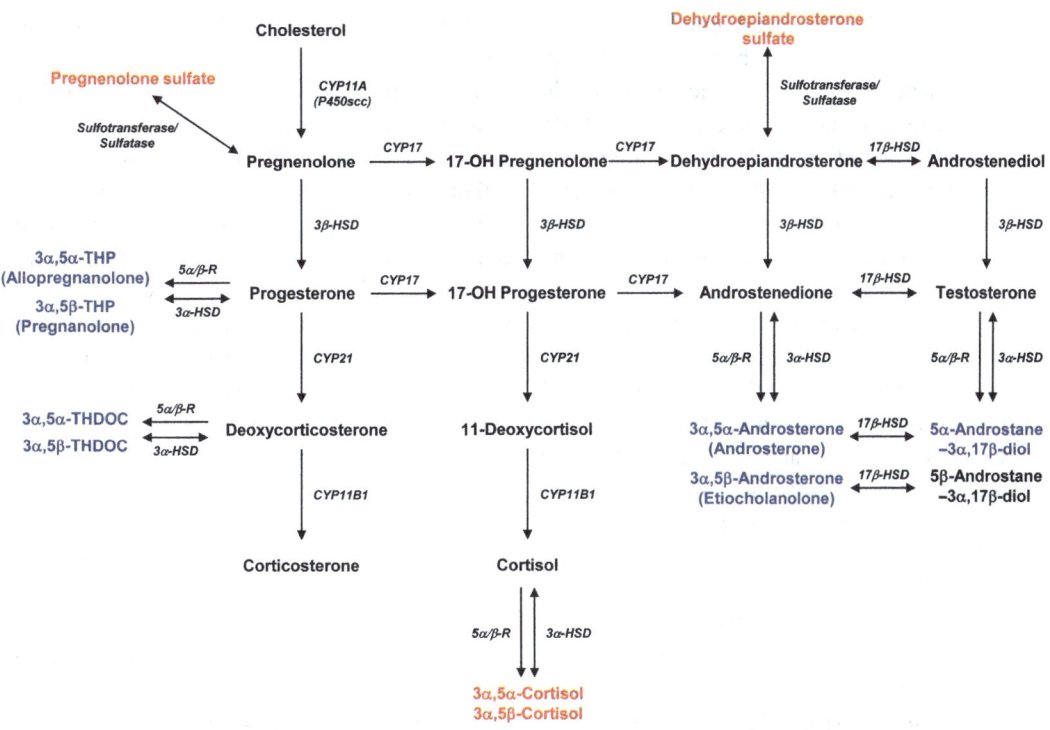

Neuroactive Steroids, Fig. 1 Biosynthetic pathway for neuroactive steroids. Neuroactive steroids with inhibitory activity on neurons are shown in *blue*, while neuroactive steroids with excitatory activity on neurons are shown in *red*. $3\alpha,5\alpha$-THP ($3\alpha,5\alpha$)-3-hydroxypregnan-20-one; $3\alpha,5\beta$-THP ($3\alpha,5\beta$)-3-hydroxypregnan-20-one; $3\alpha,5\alpha$-THDOC ($3\alpha,5\alpha$)-3,21-dihydroxypregnan-20-one; $3\alpha,5\beta$-THDOC ($3\alpha,5\beta$)-3,21-dihydroxypregnan-20-one; *3β-HSD* 3β-hydroxysteroid dehydrogenase; *5α/β-R* 5α/5β-reductase; *3α-HSD* 3α-hydroxysteroid dehydrogenase; *17β-HSD* 17β-hydroxysteroid dehydrogenase

($3\alpha,5\alpha$)-3-hydroxypregnan-20-one ($3\alpha,5\alpha$-THP or allopregnanolone) and the deoxycorticosterone metabolite ($3\alpha,5\alpha$)-3,21-dihydroxypregnan-20-one ($3\alpha,5\alpha$-THDOC or allotetrahydrodeoxycorticosterone). The demonstration of the GABAergic activity of these endogenous compounds (Majewska et al. 1986) initiated decades of further studies, including recognition of their nanomolar potency at GABA$_A$ receptors. In contrast, the excitatory neuroactive steroids include the sulfated derivatives of pregnenolone and DHEA, as well as the $3\alpha,5\alpha$- and $3\alpha,5\beta$-reduced metabolites of cortisol. These latter steroids are weak inhibitors of the GABA$_A$ receptors, while both pregnenolone sulfate and DHEA sulfate also act as weak *N*-methyl-D-aspartate (NMDA) receptor agonists (micromolar potency).

Pharmacological Activity

Systemic administration of GABAergic neuroactive steroids exerts a variety of pharmacological activities. In the 1940s, Hans Selye reported the sedative, anesthetic, and anticonvulsant actions of progesterone. However, only in the mid-1980s it was discovered that these properties are due to its GABAergic metabolites (Paul and Purdy 1992). Likewise, $3\alpha,5\alpha$-THP, $3\alpha,5\alpha$-THDOC, and their precursor progesterone have potent sleep-inducing properties in rats, and $3\alpha,5\beta$-THP (pregnanolone) has soporific and anesthetic properties in humans. Moreover, progesterone ameliorates the symptoms of catamenial epilepsy, a seizure disorder associated with the cyclical variations of steroid hormones during the menstrual cycle.

Progesterone and its metabolites, $3\alpha,5\alpha$-THP and $3\alpha,5\alpha$-THDOC, have anxiolytic, antidepressant-like, and analgesic properties in animal and human studies. $3\alpha,5\alpha$-THP, $3\alpha,5\alpha$-THDOC, and $3\alpha,5\beta$-THP have anxiolytic properties in several animal models of anxiety-like behavior. In addition to the GABAergic neuroactive steroids, lower doses of pregnenolone sulfate are also anxiolytic, while higher doses are anxiogenic, likely due to a combination of mechanisms, including metabolism to progesterone and its derivatives. Likewise, DHEA and DHEA sulfate also have anxiolytic activity in mice. Progesterone and $3\alpha,5\alpha$-THP have antidepressant-like properties in the Porsolt swimming test and the tail suspension test in rats. Systemic administration of $3\alpha,5\alpha$-THP, $3\alpha,5\alpha$-THDOC, and progesterone also induces analgesic effects in inflammatory and neuropathic pain models in rodents. Furthermore, progestins and androgens, respectively, activate and inhibit feminine sexual behavior of rodents via their GABAergic metabolites $3\alpha,5\alpha$-THP and the 5α-reduced testosterone metabolite $(3\alpha,5\alpha,17\beta)$-androstane-3,17-diol.

Neuroactive steroids, like various drugs of abuse, can have rewarding properties too (see Biggio and Purdy 2001, for review). $3\alpha,5\alpha$-THP and $(3\alpha,5\alpha,17\beta)$-androstane-3,17-diol have reinforcing properties via $GABA_A$ receptor activation. In addition, $3\alpha,5\alpha$-THP and other GABA agonist neuroactive steroids modulate ethanol intake in animal models of ethanol self-administration, increasing consumption when administered at low doses or when modulating low levels of ethanol intake and reducing consumption when administered at high doses or for modulation of high levels of ethanol intake. Furthermore, ethanol-dependent rats show enhanced $GABA_A$ receptor sensitivity to $3\alpha,5\alpha$-THP and $3\alpha,5\alpha$-THDOC and are sensitized to the anticonvulsant actions of neuroactive steroids, suggesting this class of compounds may be therapeutic during ethanol withdrawal. Indeed, neuroactive steroid therapy may have advantages over benzodiazepine therapy since benzodiazepines exhibit cross-tolerance with ethanol.

In the past decade, studies have shown that $3\alpha,5\alpha$-THP has neuroprotective, neurotrophic, and antiapoptotic effects (see Schumacher et al. and Mellon in Morrow 2007 and Brinton 2013). $3\alpha,5\alpha$-THP reduces brain swelling and inflammatory cytokines after traumatic brain injury, suggesting a role for neuroactive steroids in the inflammation process. In addition, progesterone and $3\alpha,5\alpha$-THP have neuroprotective effects after traumatic brain injury, spinal cord injury, or cerebral ischemia and promote remyelination after peripheral nerve damage and in murine models of multiple sclerosis. Estradiol has also been shown to promote neurogenesis, while DHEA sulfate and $3\alpha,5\beta$-THP hemisuccinate are neuroprotective in experimental models of ischemia and stroke, respectively. Furthermore, brain $3\alpha,5\alpha$-THP concentrations are decreased in humans with Alzheimer's disease as well as in mouse models of this disease, and $3\alpha,5\alpha$-THP administration to these mice reverses the neurogenic deficits and promotes neuronal regeneration (Brinton 2013).

Mechanisms and Sites of Action

The sites of action of both excitatory and inhibitory neuroactive steroids are depicted in Fig. 2. The inhibitory actions of the GABAergic neuroactive steroids are mediated by synaptic and extrasynaptic $GABA_A$ receptors. These steroids interact with $GABA_A$ receptors via specific binding sites on the α subunits that allosterically modulate binding to GABA and benzodiazepine recognition sites (Hosie et al. 2006). $GABA_A$ receptor function is enhanced by potentiation of GABA-mediated Cl^- conductance as well as direct stimulation of Cl^- conductance. $GABA_A$ receptors appear to have multiple neurosteroid recognition sites that likely reflect distinct recognition sites on $GABA_A$ receptor subtypes. Neuroactive steroids modulate both synaptic and extrasynaptic $GABA_A$ receptors with lower potency at synaptic receptors that contain $\gamma 2$ subunits and higher potency at extrasynaptic receptors that contain δ subunits (see Belelli and Lambert 2005). These steroids also have modulatory actions at serotonin type 3 receptors, neuronal nicotinic acetylcholine receptors, and

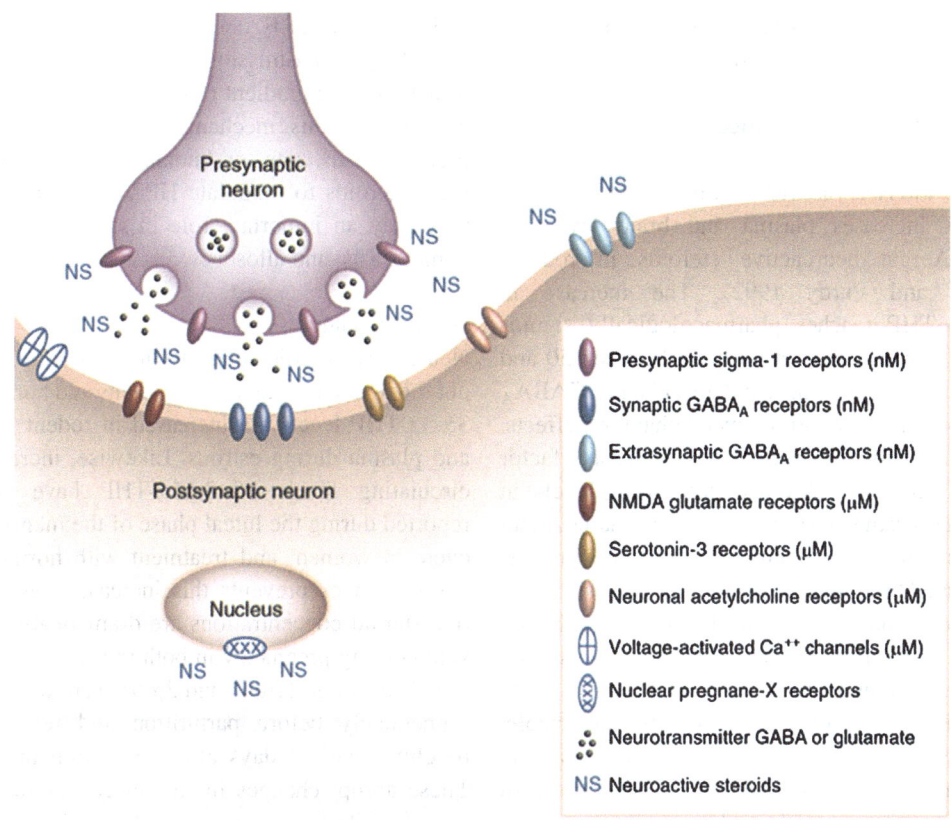

Neuroactive Steroids, Fig. 2 Sites of action of inhibitory and excitatory neuroactive steroids. The inhibitory actions of neuroactive steroids are mediated by synaptic and extrasynaptic γ-aminobutyric acid type A (GABA$_A$) receptors. The excitatory actions of sulfated steroids such as pregnenolone or dehydroepiandrosterone (DHEA) sulfate are partially mediated by direct, but low-potency, activation of N-methyl-D-aspartate (NMDA) receptors and inhibition of GABA$_A$ receptors. Pregnenolone sulfate also potently inhibits GABA or glutamate release at nerve terminals, through presynaptic sigma-1 receptors. Neuroactive steroids have modulatory actions at serotonin type 3 receptors, neuronal nicotinic acetylcholine receptors, and voltage-activated calcium channels, albeit with micromolar potency at these sites. Another site of action includes the nuclear pregnane-X receptor

voltage-activated calcium channels, albeit with micromolar potency (see Rupprecht 2003).

Studies of the structural requirements for neuroactive steroid activity at GABA$_A$ receptors include 3α reduction and 5α/5β reduction of the A ring as well as hydroxylation of C21. The 5β-reduced metabolites of progesterone and deoxycorticosterone, 3α,5β-THP and 3α,5β-THDOC, are equipotent modulators of GABAergic transmission. Humans synthesize these 5β-reduced neuroactive steroids and the concentrations of 3α,5β-THP are physiologically relevant and are comparable with those of 3α,5α-THP in plasma and cerebrospinal fluid. In addition, 3α,5α-reduced cortisol and 3α,5β-reduced cortisol have antagonist properties at both GABA and neurosteroid recognition sites of GABA$_A$ receptors, and these compounds are the most abundant metabolites of cortisol in human urine.

The excitatory actions of sulfated steroids such as pregnenolone or DHEA sulfate are partially mediated by direct, but low-potency, interactions with NMDA receptors (Fig. 2). Pregnenolone sulfate also potently inhibits GABA or glutamate release at different neuronal sites, effects that are blocked by sigma-1 receptor antagonists. Another site of action includes the nuclear pregnane-X receptor that regulates P450 enzymes and steroid hormone levels across many tissues, but there are no effects of the excitatory or inhibitory neuroactive steroids on nuclear progesterone or glucocorticoid receptors.

Physiological Significance

Stress and HPA Axis Activation
Stress increases plasma and brain levels of GABAergic neuroactive steroids in rodents (Paul and Purdy 1992). The increase in $3\alpha,5\alpha$-THP reaches pharmacologically significant concentrations in rat brain between 50 and 100 nM that are sufficient to enhance $GABA_A$ receptor activity and produce behavioral effects. Similarly, stress or corticotropin-releasing factor (CRF) infusion elevates $3\alpha,5\alpha$-THP levels in human plasma. The increase in plasma neuroactive steroid levels elicited by stressful stimuli is mediated by activation of the HPA axis, since it is absent in adrenalectomized animals. Although adrenalectomized animals exhibit no circulating concentrations of $3\alpha,5\alpha$-THP and $3\alpha,5\alpha$-THDOC, brain levels are still detectable, suggesting that brain synthesis may play an important role in neuroactive steroid actions. In addition, anxiogenic drugs that inhibit GABAergic transmission, activate the HPA axis, and produce stress also increase brain and plasma $3\alpha,5\alpha$-THP levels.

Activation of the HPA axis in response to acute stress increases the release of CRF from the hypothalamus that stimulates the release of adrenocorticotropic hormone (ACTH) from the pituitary, which, in turn, stimulates the adrenal cortex to release glucocorticoids and the GABAergic neuroactive steroids. Glucocorticoids provide negative feedback upon the hypothalamus and pituitary. Likewise, GABAergic neuroactive steroids inhibit CRF production and release, ACTH release, and subsequent corticosterone levels following stress in rats but activate the HPA axis in C57BL/6J mice. In this mouse, neuroactive steroids promote the physiological response to stress through $GABA_A$ δ subunit-containing receptors located on CRF neurons in the paraventricular nucleus (PVN) of the hypothalamus. Under basal conditions, neuroactive steroids act in the PVN to decrease HPA axis activity, but following stress, there is an activation of CRF neurons due to excitatory GABAergic transmission caused by the collapse of the chloride gradient (Sarkar et al. 2011). It is unknown if this mechanism extends to other mouse strains, rats, or humans. The ability of these steroids to modulate HPA axis activation may play an important role in stress response, homeostasis, and allostasis.

Ovarian Cycling
Neuroactive steroid concentrations vary throughout the ovarian cycle in both rodents and humans. $3\alpha,5\alpha$-THP levels are increased in rodent brain and plasma during estrous. Likewise, increased circulating levels of $3\alpha,5\alpha$-THP have been reported during the luteal phase of the menstrual cycle in women, and treatment with hormonal contraceptives prevents this increase. Neuroactive steroid concentrations are dramatically elevated during pregnancy in both rats and women. Levels of progesterone and $3\alpha,5\alpha$-THP decrease immediately before parturition and return to baseline levels 2 days after parturition in rats. These abrupt changes in steroid concentrations may contribute to postpartum depressive symptoms. Levels of $3\alpha,5\alpha$-THP remain low throughout development and increase immediately before puberty in both humans and rodents. Moreover, $3\alpha,5\alpha$-THP levels decline during menopause in women (see Smith et al. in Morrow 2007).

Modulation by Pharmacological Agents
Numerous psychoactive drugs increase the brain and plasma concentrations of GABAergic neuroactive steroids and their precursors in rats (Biggio and Purdy 2001). For instance, systemic administration of ethanol increases brain and plasma concentrations of $3\alpha,5\alpha$-THP, which contributes to ethanol actions in rats (Morrow 2007). Likewise, acute administration of psychotropic drugs like caffeine, nicotine, morphine, $\Delta 9$-tetrahydrocannabinol, and γ-hydroxybutyric acid increases the brain and plasma concentrations of $3\alpha,5\alpha$-THP in rats.

In contrast, chronic ethanol consumption or chronic administration of nicotine or morphine to rats does not increase neuroactive steroids.

However, cerebrocortical 3α,5α-THP levels are increased following abstinence from nicotine or morphine in rats. Moreover, ethanol dependence induces adaptations in HPA axis function, with blunted neuroactive steroid responses to stress or to ethanol challenge in both animals and humans. Further studies are needed to determine the effects of chronic drug exposure on neuroactive steroids or whether humans exhibit adaptations under conditions of drug abuse.

Neuroactive steroid concentrations are increased by several drugs that are used to treat mood disorders, and it is hypothesized that this increase may contribute to their therapeutic efficacy (see Girdler et al. in Morrow 2007). Fluoxetine, reboxetine, venlafaxine, or imipramine increase the cerebrocortical and plasma concentrations of progesterone, 3α,5α-THP, and 3α,5α-THDOC in rats. Moreover, chronic treatment with fluoxetine increases cerebrospinal fluid 3α,5α-THP concentrations in depressed patients. Treatment with other antidepressants, like amitriptyline, clomipramine, viloxazine, nortriptyline, mirtazapine, and lithium, also induced an increase in circulating 3α,5α-THP level that was associated with improved symptomatology in depressed patients. Other drugs that are used to treat mood disorders, including lithium and carbamazepine, also increase the brain levels of 3α,5α-THP in rats.

The atypical antipsychotic drugs, clozapine and olanzapine, increase the brain and plasma concentrations of pregnenolone, progesterone, and their metabolites, 3α,5α-THP and 3α,5α-THDOC, as well as the serum levels of corticosterone. The increase in neuroactive steroids is prevented by adrenalectomy, suggesting the involvement of the HPA axis. In contrast, the typical antipsychotics, risperidone and haloperidol, induce a modest increase in serum progesterone and corticosterone levels, but do not alter brain 3α,5α-THP. Pregnenolone is elevated in the postmortem brain of patients with schizophrenia, though it is unknown if this is secondary to the disease or antipsychotic treatment. Nonetheless, elevations of pregnenolone and 3α,5α-THP may contribute to the therapeutic efficacy of clozapine and olanzapine in patients with schizophrenia.

Regulation of $GABA_A$ Receptors

Changes in neuroactive steroid levels that occur in vivo under physiological conditions such as puberty, ovarian cycle, pregnancy, or lactation are associated with changes in $GABA_A$ receptor function and expression (see Smith et al. in Morrow 2007). Likewise, the chronic administration of progesterone, 3α,5α-THP, or hormonal contraceptives to rats also alters $GABA_A$ receptor function and expression. The onset of puberty is associated with abrupt increases in 3α,5α-THP levels and a marked increase in α4 and δ subunit expression in the mouse. Furthermore, administration of 3α,5α-THP, which is normally anxiolytic, increases anxiety in pubertal female mice, perhaps via decreased outward currents at the highly expressed α4β δ $GABA_A$ receptors. The diestrus phase of the mouse ovarian cycle is accompanied by elevated levels of progesterone and 3α,5α-THP, as well as increased expression of the δ subunit and decreased expression of the γ2 subunit of the $GABA_A$ receptor, with consequent increase in tonic inhibition and decreased seizure susceptibility and anxiety.

The regulation of $GABA_A$ receptor subunit expression during pregnancy in rodents is complicated by differential changes across different brain regions. However, the increase in neuroactive steroid levels observed during pregnancy appears to be consistently associated with a decrease in γ2 and δ subunit expression. Moreover, δ subunit knockout mice exhibit anxiety- and depression-like maternal behavior during the postpartum period, suggesting that δ subunit plasticity may be relevant to the development of postpartum depression (see Smith et al., in Morrow 2007).

Pharmacological treatment with steroids also has the ability to alter $GABA_A$ receptor subunit expression. Thus, the administration of ethinylestradiol and levonorgestrel, two of the synthetic steroids most frequently used in hormonal contraceptives, decreased pregnenolone, progesterone, and 3α,5α-THP concentrations, increased the expression of $GABA_A$ receptor γ2 subunit, and increased anxiety-like behavior in female rats. Likewise, progesterone or 3α,5α-THP administration to female rats increased expression of α4 and δ subunits and decreased expression of the α1 subunit in the hippocampus, but the effects were no longer observed after 4–5 days, suggesting a compensatory mechanism to counteract the effects of continued steroid exposure on $GABA_A$ receptors. Furthermore, steroid withdrawal from long-term exposure of both progesterone and 3α,5α-THP results in marked increases in α4 and δ subunit expression in the hippocampus with subsequent changes in receptor function, sensitivity to benzodiazepines, and increased anxiety and seizure susceptibility (see Smith et al., in Morrow 2007). Finally, neonatal administration of β-estradiol-3-benzoate induces a marked and persistent decrease in the cerebrocortical concentrations of 3α,5α-THP in adult female rats, which is associated with compensatory changes in $GABA_A$ receptors.

Pathological Significance

Neuroactive steroid concentrations are altered in various diseases including depression, premenstrual dysphoric disorder, alcoholism, and schizophrenia (Morrow 2007). However, the significance of these alterations remains a matter of speculation. Because neuroactive steroids contribute to regulation of the HPA axis response to stress and are modulated by drugs used as medications for neuropsychiatric disease, neuroactive steroids may play a homeostatic role that both reflects and influences the allostatic state of the person and contributes to the wide range of symptomatology associated with these disease states.

Patients with major depression have elevated cortisol levels, hypersecretion of CRF, and suppression of feedback mechanisms marked by blunted dexamethasone suppression of cortisol levels. Some neuroactive steroid concentrations are decreased in patients with major depression as well as in animal models of depression, and the administration of antidepressant-specific serotonergic reuptake inhibitors increases these neuroactive steroids in patients and in rodent brain and plasma (Rupprecht 2003). This increase might be mediated by a direct effect of specific serotonergic reuptake inhibitors on the biosynthetic enzyme 3α-hydroxysteroid dehydrogenase that produces GABAergic neuroactive steroids or by increased serotonin neurotransmission that stimulates the release of CRF to activate the HPA axis. The acute anxiolytic and antidepressant-like effects of neuroactive steroids, demonstrated in rodent models, may be important for prevention and recovery from major depression. Hence, neuroactive steroids may contribute to the therapeutic efficacy of antidepressant medications by contributing to GABAergic inhibition, modulation of the HPA axis, as well as the unknown mechanisms that underlie their acute antidepressant-like activity.

Neuroactive steroid levels are also altered in premenstrual dysphoric disorder, although the literature is controversial with reports of increases, decreases, and no change in 3α,5α-THP plasma levels (see Girdler in Morrow 2007). Differences in analytic methods, diagnostic criteria, or presence of other comorbid psychiatric disorders might account for these discrepancies. Premenstrual dysphoric disorder results in dysregulation of the HPA axis and sympathetic nervous system responses to stress, with a blunted 3α,5α-THP response to stress. Women with a history of depression, regardless of premenstrual dysphoric disorder symptoms, also had reduced basal neuroactive steroid levels and a blunted 3α,5α-THP response to stress. All this experimental evidence emphasizes the important link between HPA axis function and neuroactive steroid levels in the maintenance of homeostasis and healthy brain function.

Alterations in HPA axis responsiveness are found in alcoholism during drinking and abstinence (see Morrow in Morrow 2007). ACTH and cortisol secretion increased during ethanol intoxication and acute alcohol withdrawal, but attenuated responsiveness of the HPA axis is found in both drinking and abstinent alcohol-dependent patients. Alcohol-dependent patients have low cortisol and 11-deoxycortisol basal levels, a reduced cortisol response to exogenous ACTH challenge, and attenuated ACTH and cortisol responses after pituitary stimulation by ovine or human CRF. Altered cortisol and ACTH responses to ovine CRF and naloxone have also been found in sons of alcoholics. The levels of GABAergic neuroactive steroids have not been studied under these conditions but are likely to be impacted since both human and animal studies show that neuroactive steroid responses to stress or HPA axis activation are blunted when ACTH or glucocorticoid responses are blunted. Basal levels of $3\alpha,5\alpha$-THP are reduced during ethanol withdrawal in humans, when circulating cortisol levels are elevated. Furthermore, abstinent alcoholics show a blunted pregnenolone sulfate response to adrenal stimulation and a delayed deoxycorticosterone response to CRF challenge, supporting the idea that blunting of the HPA axis also impacts the neuroactive steroid responses to stress in alcoholics.

The levels of the neuroactive steroid precursors pregnenolone and DHEA as well as $3\alpha,5\alpha$-THP have been studied in postmortem brain of patients with schizophrenia. Pregnenolone and $3\alpha,5\alpha$-THP were present in human postmortem brain tissue at considerably higher concentrations than typically observed in serum or plasma. Pregnenolone and DHEA levels were higher in both posterior cingulate and parietal cortex of subjects with schizophrenia versus control subjects. $3\alpha,5\alpha$-THP levels tended to be decreased in parietal cortex in subjects with schizophrenia compared with controls. In addition, neuroactive steroid induction represents a potential mechanism contributing to the efficacy of clozapine and olanzapine, both of which increase plasma pregnenolone and $3\alpha,5\alpha$-THP levels in rodents. Clinical evidence suggests antipsychotic properties for neuroactive steroids. Progesterone administration ameliorates symptoms of postpartum psychosis in women. Furthermore, adjunct treatment with pregnenolone or 17β-estradiol improves schizophrenia symptomatology in both men and women. Most recently, two independent studies show that pregnenolone improves cognition and negative symptoms in patients with schizophrenia. These findings suggest that neuroactive steroids are candidate modulators of schizophrenia pathophysiology and/or therapeutics.

Progesterone and $3\alpha,5\alpha$-THP promote the viability of neurons in the brain and spinal cord. Neuroprotective effects have been documented in animal models of traumatic brain injury, experimentally induced ischemia, spinal cord lesions, multiple sclerosis, Alzheimer's disease, and Niemann–Pick type C disease (see Schumacher et al. and Mellon, in Morrow 2007; Brinton 2013). Decreased brain concentrations of $3\alpha,5\alpha$-THP have been found in humans with multiple sclerosis and with Alzheimer's disease. Indeed, progesterone had remarkable efficacy in a clinical study of traumatic brain injury where it dramatically reduced mortality and increased recovery of function. This is a new area of avid investigation that will likely lead to a better understanding of the role of neuroactive steroids in health and disease.

Therapeutic Potential of Neuroactive Steroids

The preceding sections have alluded to the therapeutic potential of GABAergic neuroactive steroids and their precursors for behavioral disorders that involve anxiety, dysregulation of the HPA axis, cognitive impairment, and brain inflammation. Many neuropsychiatric disorders exhibit these attributes that may contribute to severity of disease or recovery under standard treatment protocols. Therefore, neuroactive steroids might be considered as primary or adjunctive therapy that supports normal brain function and cognition, normalization of HPA axis function, and the reduction of inflammatory processes that contribute to neuronal dysfunction. Since brain inflammation is implicated in depressive disorders, alcoholism, brain lesions, traumatic brain injury, and possibly cognitive disorders such as schizophrenia and Alzheimer's disease, the use of neuroactive steroids or precursors may ameliorate this component of the pathology and thereby support recovery or disease management.

Few neuroactive steroid compounds are under development for neurological or psychiatric disease, despite the data reviewed here. No GABAergic neuroactive steroids are presently approved for clinical use in any disease; therefore, compounds are not readily available for clinical testing. The precursor pregnenolone is commercially available as a nutritional supplement, and therefore, little impetus exists for clinical trials sponsored by the pharmaceutical industry. The GABAergic neuroactive steroids could also have abuse or dependence potential, and this factor increases the risk of untoward effects – a deterrent to investment in the development of new compounds. Despite these limitations, clinical trials are underway for traumatic brain injury, epilepsy, schizophrenia, and anxiety and depressive disorders. Indeed, the stimulation of steroidogenesis by direct activation of the 18-kDa cholesterol translocator protein has recently been shown to have therapeutic potential for anxiety disorders, without the side effects concomitant with the use of benzodiazepine anxiolytics. Further studies are needed to evaluate the therapeutic potential of neuroactive steroids.

Cross-References

▶ Alcohol Abuse and Dependence
▶ Anticonvulsants
▶ Antidepressants
▶ Antipsychotic Drugs
▶ Premenstrual Dysphoric Mood Disorder

References

Baulieu EE (1998) Neurosteroids: a novel function of the brain. Psychoneuroendocrinology 23:963–987

Belelli D, Lambert JJ (2005) Neurosteroids: endogenous regulators of the $GABA_A$ receptor. Nat Rev Neurosci 6:565–575

Biggio G, Purdy RH (2001) Neurosteroids and brain function, vol 46, International review of neurobiology. Academic, New York

Brinton RD (2013) Neurosteroids as regenerative agents in the brain: therapeutic implications. Nat Rev Endocrinol 9:241–250

Hosie AM, Wilkins ME, da Silva HM, Smart TG (2006) Endogenous neurosteroids regulate $GABA_A$ receptors through two discrete transmembrane sites. Nature 444:486–489

Majewska MD, Harrison NL, Schwartz RD, Barker JL, Paul SM (1986) Steroid hormone metabolites are barbiturate-like modulators of the GABA receptor. Science 232:1004–1007

Morrow AL (2007) Special edition on neurosteroids. Pharmacol Ther 116:1–172

Paul SM, Purdy RH (1992) Neuroactive steroids. FASEB J 6:2311–2322

Rupprecht R (2003) Neuroactive steroids: mechanisms of action and neuropsychopharmacological properties. Psychoneuroendocrinology 28:139–168

Sarkar J, Wakefield S, MacKenzie G, Moss SJ, Maguire J (2011) Neurosteroidogenesis is required for the physiological response to stress: role of neurosteroid-sensitive $GABA_A$ receptors. J Neurosci 31:18198–18210

Neurobiological

Definition

Neurobiology is the study of the nervous system and the organization of parts (neurons, glia, and nerve fibers) into circuits that process information and mediate behavior.

Neurodegeneration

Definition

The progressive neuronal loss caused by cell death.

Cross-References

▶ Apoptosis
▶ Neuroprotectants: Novel Approaches for Dementias

Neurodegeneration and Its Prevention

Melita Salkovic-Petrisic[1], Angelika Schmitt[2] and Peter Riederer[2]
[1]Department of Pharmacology, University of Zagreb School of Medicine, Zagreb, Croatia
[2]University Hospital, Center for Mental Health, Clinic and Policlinic for Psychiatry, Psychosomatics and Psychotherapy, University Wuerzburg, Wuerzburg, Germany

Definition

"Neurodegeneration" is a common expression, and most frequently, it is not defined in manuscripts. An important question is how we define aging, and it is often difficult to differentiate between age-related processes and neurodegenerative ones. This includes neuronal systems that are vulnerable to aging but also to degenerative pathology, and this may happen in certain brain regions while others are spared. Degeneration (in one system only or in several) thus may start with only a functional disturbance triggered by dose-related genetic aberrations or by environmental factors.

This entry focuses on Alzheimer's disease, a classical neurodegenerative disorder for which aging is a major risk factor but with significant differences between the pathology of Alzheimer's disease and the physiological aging process. Thus, we regard this entry as a "teaching example." But in addition, Alzheimer's disease presents many pathological features and mechanism that are unspecific in the sense that they are common to many neurodegenerative disorders, for example, Parkinson's disease. We present many pathological processes which are a general basis for "neurodegeneration," such as loss of adenosine triphosphate (ATP), disturbances of the respiratory chain activity, disturbances of energy metabolism, apoptosis, and reactive oxygen species (ROS).

Impact of Psychoactive Drugs

Among all dementias, Alzheimer's disease (AD) is the most frequent in old age. In origin, AD can be divided into two different forms:

1. A very small proportion of 1 % or less of all Alzheimer cases is caused by missense mutations in presenilins 1 and 2 on chromosomes 14 and 1, respectively, or in the amyloid precursor protein (APP) gene on chromosome 21, leading to autosomal dominant familial AD with an early onset. These currently known three types of mutations are involved in both the overexpression and abnormal cleavage of APP, and increased formation of its derivative amyloid peptides (e.g., Aβ1-40, 1–42, 1–43) being the inevitable starting point for the disease process of familial AD in the form of the cause-effect principle (β-amyloid cascade hypothesis).

2. In contrast, millions of people worldwide suffer from sporadic AD, and no such mutations were found as yet for this dementia form. Among susceptibility genes that may contribute to the development of sporadic AD, the most important and powerful risk is an allelic abnormality of the apolipoprotein E (APOE) 4 gene on chromosome 19 involved in the lipid transport in the brain and in the metabolism of the plasma membrane constituent cholesterol. The list of confirmed risk genes has recently been extended to at least nine genes that have been considered to carry less powerful risk than APOEε4 polymorphism (http://www.alzgene.org/; reviewed by Bertram et al. 2010). Morphologically, AD is characterized by cell loss in susceptible brain regions, extracellular deposition of β-amyloid containing neuritic plaques, and neurofibrillary tangles mainly composed of hyperphosphorylated tau protein. Aggregated proteins must be successfully removed or degraded and neurons are particularly vulnerable to disruption of interactions among the autophagic and endocytic pathways important for this degradation, especially as the brain ages. Such a disruption may occur at different stages of the autophagy pathway and have different implications for pathogenesis and therapy of late-onset disorders like AD (reviewed by Nixon 2013).

Therefore, besides the above risk factors, aging has been found to be the main risk factor for AD. A multitude of inherent variations in fundamental metabolic processes are set into motion at the cellular, molecular, and genetic levels which result from functional imbalances of regulative systems of which the functionally most important ones include:

- Energy production (reduced) and energy turnover (increased)
- Insulin action (reduced) and cortisol action (increased)
- Acetylcholine action (reduced)
- Noradrenaline action (increased), indicative of an increased sympathetic tone, in brain areas due to environmental stress, which in

addition causes hypercortisolemia via chronic disturbance of the hypothalamic-pituitary-adrenal axis
- Formation of ROS (increased) and capacity of their degradation (reduced)
- Loss of membrane lipids and shift of unsaturated fatty acids in membranes in favor of saturated fatty acids
- Dysregulation of intracellular pH
- Shift of gene expression profile from anabolic site (reduced) to catabolic site (increased)

These variations and shifts may indicate an uncoupling of synchronization of biological systems that may correspond to increased entropy, that is, increased disturbances of a biological system. Smaller additional internal or external events, even one that is ineffective in itself, may change biological and/or biophysical properties of the aging brain. Such events may shift a system in a stepwise manner to increased entropy/decreased criticality ending in a catastrophic reaction – disease (Hoyer and Plaschke 2004).

Physiologically, the mature mammalian brain uses the nutrient glucoseal most exclusively to ensure its structure and function. From the glycolytic glucose metabolite fructose-6-phosphate (F-6-P), UDP-N-acetylglucosamine(UDP-GlcNAc) is formed and used for protein O-glycosylation, for example, for tau protein (Hart 1997). The metabolite acetyl-CoA serves as the source of the generation of (1) the neurotransmitter acetylcholine for learning and memory processes and (2) the membrane constituent cholesterol in the 3-hydroxy-3 methylglutaryl-CoA cycle. Finally, the energy-rich compound ATP is formed which is essential for most cellular and molecular activities. A hierarchy of ATP-utilizing processes has been proposed in the following order: protein synthesis > RNA/DNA synthesis > Na^+ cycling > Ca^{2+} cycling > protonleak. The position in this hierarchy may be determined by the sensitivity of each process to changes in energy charge. Highly ATP-dependent processes are:

- Sorting, folding, transport, and degradation of proteins
- Maintenance of pH 6 in the endoplasmic reticulum/Golgi apparatus
- Heat shock protein-guided transport across the latter compartments
- Axonal transport of proteins
- Regulation of the conformational state of insulin-degrading enzyme
- Maintenance of intracellular/extracellular ion homeostasis
- Maintenance of biophysical membrane properties
- Maintenance of a membrane potential in neurons
- Regulation of synaptic membrane composition
- Maintenance of synaptic transmission
- Control of the metabolism of both APP and tau protein

β-amyloid deposits as a hallmark of AD tell us they are important. However, in contrast to familial AD, it has yet to be proven that faulty metabolism of APP and the increased formation of the derivative β-amyloid contribute to the generation of sporadic AD. Recent evidence has been provided that an insulin-resistant brain state (IRBS) may play a pivotal role in the generation of the latter form of dementia (Grünblatt et al. 2007; Riederer and Hoyer 2006; Salkovic-Petrisic et al. 2006). Several age-related candidates may contribute to the causation of an IRBS:

- Increased concentration of cortisol that may cause neuronal insulin receptor desensitization by inhibiting phosphorylation of its tyrosine residues
- Increased activity and metabolism of noradrenaline and increased concentration of cAMP that lead to the decrease of the receptor's tyrosine kinase activity
- Increased concentration of ROS (the uncharged species hydrogen peroxide) that may persistently inhibit the activity of phosphotyrosine phosphatase, thus inhibiting its dephosphorylation and rendering the insulin receptor ineffective
- Increased concentration of long-chain fatty acids that may reduce insulin's receptor binding

As shown in Fig. 1, IRBS may induce diverse abnormalities in cellular and molecular brain

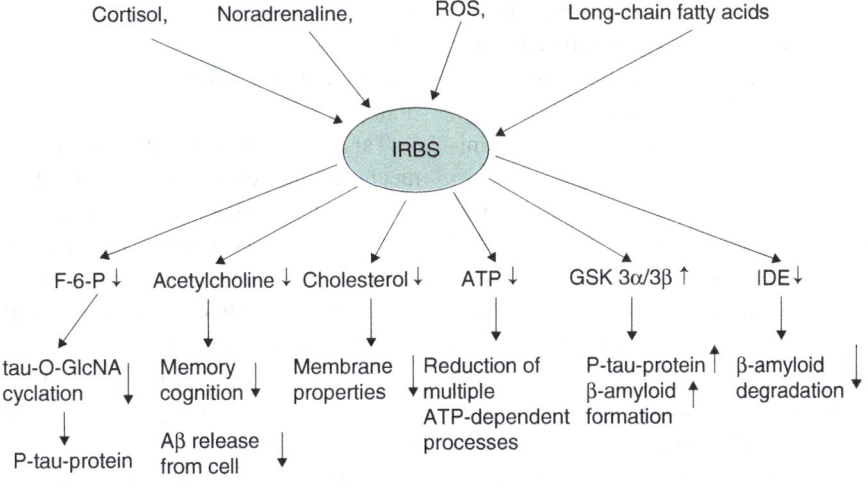

Neurodegeneration and Its Prevention, Fig. 1 Candidates which may induce an insulin-resistant brain state (*IRBS*) and a survey of diverse effects thereafter

metabolism (Hoyer 2004), among which are the reduction of glucose metabolism including the compound F-6-P, acetyl-CoA, and ATP. F-6-P decrease favors tau hyperphosphorylation; the fall of acetyl-CoA diminishes the formation of both acetylcholine and cholesterol. The reduction of the former decreases the cellular release of APP via its muscarinic receptors, likewise, learning, memory, and cognitive capacities decline. The reduction of cholesterol in the cell membrane changes its properties and may, thus, damage receptor function. The deficit in ATP has a cascade-like effect on diverse ATP-dependent cellular and molecular processes (see above) which may damage cell functions and may jeopardize the survival of the cell. The compromised insulin signaling downstream of the insulin receptor may activate glycogen synthase kinase-3α/β which may result in hyperphosphorylated tau protein and increased production of β-amyloid (Hoyer 2004). The reduced insulin signal may downregulate insulin-degrading enzyme (IDE) resulting in a reduced capacity to degrade β-amyloid and, thus, favoring its cellular accumulation.

In contrast to the hereditary form of AD, sporadic AD may be considered to be a complex and self-propagating metabolic brain disease with predominating abnormalities in oxidative/energy metabolism, including decreases in the formation of both acetylcholine and cholesterol finally ending in the formation of neuritic plaques and neurofibrillary tangles. Therefore, a "rational" therapy may be based upon several different strategies.

In line with that idea, novel approaches to the treatment of neurodegenerative disorders including AD are enhancing the efficacy of autophagy by targeting the stages that are specifically disrupted, among which stimulation of the induction of autophagy has lately received the greatest attention (Nixon 2013). Compounds have been tested both in animal models and in AD clinical trials that are efficacious in enhancing autophagy either by inhibition of the autophagosome formation via mTOR inhibition or via AMPK activation; these compounds include inhibitors of mTOR kinase, a main regulator of autophagy induction, e.g., curcumin, resveratrol, latrepirdine (also known as dimebolin or Dimebon), and those that act via other mTOR-independent pathways, e.g., activation of 5′AMP-activated protein kinase, AMPK, or metformin (Nixon 2013). Unfortunately, all have failed in clinical trials indicating that a lot more research needs to be done including establishment of clinically reliable and useful biomarkers of autophagy to assess the therapeutic potential of this strategy. However, the success of autophagy induction intervention in AD may depend on first relieving the block of lysosomal clearance or stimulating autophagy induction very early in the course of disease before the generation of such a block, and approach of enhancing the lysosome efficiency which has been found successful in preclinical AD studies so far (e.g., cathepsin activators) needs to be proven in clinical trials as well (Nixon 2013).

According to all available information about the pathology of sporadic AD, it is evident that multiple drug regimens are necessary in order to come closer to a "disease-modifying" treatment strategy. Therefore, it is necessary to develop multifunctional drugs. While the development of anti-Alzheimer (dementia) drugs involves all currently known pathological principles and is driven by globally acting industry as well as by smaller and start-up companies, the concept of developing multifunctional drugs is still one followed by relatively few firms. Worldwide, enormous capacity focuses development onto new substances to inhibit acetylcholine esterase or to block glutamatergic N-methyl-D-aspartate (NMDA) receptor subunits. Currently, several drug companies are developing novel compounds as acetylcholinesterase inhibitors and also various types of glutamate receptor antagonist. Other strategies are related to nicotinic and muscarinic receptor subtype modulation. These are not expected to be disease-modifying strategies. The therapeutic potency, side effects, and adverse reaction profile of receptor subtype-specific nicotinic and muscarinic drugs still have to be evaluated. There is profound knowledge about the β-amyloid pathology. Based on this, many companies try to develop (1) protein aggregation inhibitors, (2) β- and γ-secretase inhibitors and γ-secretase modulators, or (3) vaccination strategies against Aβ-induced plaques. For the latter, caution is suggested if the data from Holmes et al. (2008) are taken into consideration. These postmortem human brain studies performed on AD patients who have died after vaccination with AN 1792 give evidence that reduction of plaque load is not correlated with any cognitive improvement. However, it has been suggested that vaccination at advanced stages of AD comes too late, although recent data question whether that bapineuzumab (monoclonal anti-Aβ antibody) has therapeutic activity also in mild-to-moderate AD regardless of the APOE4 + or APOE4 − status. Therefore, early vaccination strategies are envisaged as mentioned below.

Targeting of β-amyloid in the development of mechanism-based treatments for AD has obviously been associated with difficulties, and current doubts concerning its causative role in sporadic AD seem to grow, suggesting that it may not be as attractive a target as previously thought. In line with that idea, and since the finding that tau protein plays one of the central roles in AD neurodegeneration, interest in tau-targeted treatments is growing. The first tau aggregates form in a few nerve cells in discrete brain areas and would seem to become self-propagating and spread to distant brain regions almost in a prion-like manner (Spillantini and Goedert 2013). The pathway leading from soluble and monomeric to hyperphosphorylated, insoluble, and filamentous tau is central to AD pathology, and this has led to the development of new therapeutic strategies tested in preclinical and a few also in clinical studies, respectively: (1) inhibition of tau aggregation, e.g., inhibitory polyphenols, methylene blue; (2) inhibition of tau phosphorylation which depends on the balance between the activities of tau kinases and tau phosphatises, e.g., GSK-3 inhibitors like lithium and tideglusib, or PP2A activators like metformin and sodium selenate; (3) reductions of tau level, e.g., increased tau degradation by the autophagy-lysosome system; (4) tau immunization with antibodies against phosphorylated or non-phosphorylated tau (passive immunization); (5) anti-inflammatory treatments as microglial activation has been shown to promote the hyperphosphorylation of tau; and (6) microtubule stabilization as hyperphosphorylated tau reduces its ability to interact with microtubules and tau aggregation destabilizes microtubules, e.g., epothilone D, (reviewed by Spillantini and Goedert 2013). Current pharmacotherapeutic strategies for sporadic AD consist of minimizing the central acetylcholine deficit by means of acetylcholinesterase inhibitors and by an NMDA receptor antagonist. In addition to this standard therapy, other proposed options for interfering with pathological mechanisms are antioxidants to defeat the action of ROS (vitamins A, E, and C or Ginkgo biloba); these treatment options have not yet been approved. All these therapeutic interventions together may temporarily improve the quality of life of AD patients, but they are unable to reduce or halt the progression of this devastating disorder. Antioxidative drug developments are of more general interest as oxidative stress is proven in all neurodegenerative disorders. The same holds true for anti-inflammatory drug developments.

In line with the widespread loss of neurons and synapses that occurs in AD and attempted stem cell-based therapies that have aimed to replace missing or defective cells in other neurodegenerative disorders (e.g., Parkinson's disease), recent research has been focused also on such a therapeutic strategy in AD. However, although transplantation of neuronal stem cells can improve cognition, reduce neuronal loss, and enhance synaptic plasticity in animal AD models, the mechanisms that mediate these effects seem to involve neuroprotection and trophic support rather than neuronal replacement (reviewed by Chen and Blurton-Jones 2012). In addition to these indirect beneficial effects, transplantation of neuronal stem cells could also provide a powerful tool to deliver therapeutic proteins to the damaged regions of the brain which remains to be explored in future studies.

Considering the neuronal loss in neurodegenerative disorders and AD in particular, it has to be mentioned that recent data indicate that some drugs might have therapeutic potential in relation to their effects on adult neurogenesis, the birth of new neurons in the adult brain. Drugs used to treat AD such as acetylcholinesterase inhibitors and NMDA receptor antagonists were shown to stimulate neurogenesis in in vitro and in vivo animal studies (Jin et al. 2006; Namba et al. 2009; Chen et al. 2010). Interestingly, the long-acting glucagon-like peptide 1 (GLP-1) agonist liraglutide, which is on the market as a therapeutic to treat diabetes mellitus type 2, is shown to increase adult hippocampal neurogenesis besides preventing memory impairments as well as degenerative processes in the $APP_{swe}/PS1_{\Delta E9}$ transgenic mouse model for AD (McClean et al. 2011). Additionally, neurotrophic growth factors (e.g., BDNF and NGF) and hematopoietic growth factors (e.g., VEGF) and Ginkgo biloba extracts were also shown to promote the generation of new neurons in the adult brain. Considering these beneficial effects on synaptic changes and BDNF downregulation which occurs early in the progression of AD, drugs which induce the upregulation of BDNF might also represent a novel therapeutic strategy, suggested as a "synaptic repair" therapy for neurodegenerative diseases that targets pathophysiology rather than pathogenesis (Lu et al. 2013).

One aspect that, at least in our minds (see above), deserves more attention is influencing the "glucose metabolism." Namely, a growing body of evidence convincingly suggests an association between diabetes mellitus type 2 and AD in that one disease increases the risk of the other. Recent research has suggested the development of insulin dysfunction as a candidate mechanism in AD, involving the effects of insulin on cerebral glucose metabolism, amyloid-β, vascular functions, lipid metabolism, and inflammation/oxidative stress; different nodes of the insulin network could be altered for a subgroup of AD patients (Craft et al. 2013). Since diabetes type 2 and AD overlap in some pathophysiological features, it has been proposed that agents that may show efficacy in one disease may be useful against the other. Drug-related interactions are given by (1) α- and β-(PPAR) γ-agonists, (2) glycogen synthase kinase (GSK)-3-α and -β-inhibitors, as well as (3) inhibitors of advanced-glycation-end (AGE) product biosynthesis or inhibitors of AGE receptors. It is to hope that part of such research will come from antidiabetic research and respective drug developments. A new line of research in the field of neuroprotection in AD condition has recently become attractive, based on the preclinical evidence that activation of GLP-1 receptors in the brain represents a strategy worth assessing for the treatment of sporadic AD. As mentioned earlier in the text, GLP-1 mimetics have been recently registered for the treatment of diabetes mellitus type 2. Growth-promoting/neuritogenic effects of

GLP-1R stimulation in the brain result from crosstalk between the PI-3 K and extracellular signal-regulated mitogen-activated protein kinase (ERK) pathway, both being also important elements downstream the two IR signaling pathways. Clinical trials have recently started to investigate the neuroprotective effects of long-acting GLP-1 analogues in an AD population (reviewed by Hölscher 2013). Since all these antidiabetic drugs target insulin resistance, it does not come as a surprise that influencing the "IRBS" seems now to be a plausible mode of action for treating the cause of sporadic AD (Grünblatt etal. 2007; Riederer and Hoyer 2006; Salkovic-Petrisic et al. 2006). Multiple mechanisms of actions that these antidiabetic drugs may have or share could be additional value in the therapeutic AD strategies being currently developed. For example, metformin is a strong AMPK activator (induction of autophagy) but acts also as protein phosphatase 2A/PP2A/activator (inhibition of tau hyperphosphorylation) and is an important regulator of glucose metabolism and homeostasis (decrement in insulin resistance), indicating that multifunctional-drug approach seems to already exist and could be a good one.

Beside this "retrograde" approach where already designed drugs have been afterwards found to show additional mechanisms, a promising "anterograde" approach introduces tailored multifunctional drugs, designed to have particular moieties with a desirable activity. Representative example of such approach comes from the development of iron-chelating neuroprotective-neurorescue drugs where the drug molecule has been designed to possess the neuroprotective N-propargyl moiety of the anti-Parkinsonian drug (monoamine oxidase-B inhibitor – rasagiline) and the antioxidant-iron chelator moiety of the 8-hydroxyquinoline derivative of the iron chelator (VK28), a combination which has recently demonstrated a therapeutic potential in neurodegenerative processes in vitro and in transgenic mice AD models in vivo (Kupershmidt et al. 2012). Therefore, the therapeutic concept for the treatment of neurodegenerative diseases has changed from superiority of "silver bullet" agents targeting one function over "dirty drugs" targeting more than one function to giving the priority to the latter ones as can be seen also from the change in their names to "multifunctional" (or multimodal) drugs. Unfortunately, with cognition and the present level of knowledge, we are still in dark as to which multiple functions should be combined for the most efficient AD disease-modifying option. An alternative, more conventional, strategy is to use mixtures of different drugs each of which acts on just one of the proposed targets.

There is still a debate about cholesterol, hypercholesterolemia (as measured in plasma/serum), and the use of anti-hypercholesterolemia drugs, statins. As we have pointed out earlier (Hoyer and Riederer 2007), it is necessary to distinguish between the peripheral and central cholesterol-related pathology. There is no question that the treatment of peripheral hypercholesterolemia is useful to reduce and avoid, for example, cardiovascular disturbances. Chronic increase of plasma/serum cholesterol may also disturb the functioning of brain capillaries and thus may provoke β-amyloid pathology within the capillaries' epithelium. If so, "peripheral" anti-hypercholesterolemia treatment by peripherally acting statins may be useful. Experimental proof for this working hypothesis is, however, still lacking.

As peripheral cholesterol does not pass the blood-brain barrier, knowledge about soluble and membrane-bound cholesterol in brain regions involved in the pathology of sporadic AD is speculative. While there is evidence for (1) reduced cholesterol in membranes undergoing degeneration and (2) disturbed membrane fluidity based on an imbalance of the cholesterol/fatty acid ratio, there are no data available to judge the role of soluble neuronal or extra-neuronal cholesterol. In case of such lack of knowledge, it is not justified to use centrally acting statins, as they may lead to worsening of AD if used in a chronic treatment design. In line with this are all the negative outcomes of prospective clinical studies using statins for sporadic AD (Hoyer and Riederer 2007). However, all current pharmacotherapeutic actions are too late to be effective as "disease-modifying" or even neuroprotective or neuro-restorative strategies. Therefore, development of biomarkers in order to detect early phases of pathological development, for example, in the population of "mild cognitively impaired patients" or even earlier on the basis of "health-control checkups," is as important as drug development. In fact, there is current effort by both basic and clinical research and industrial interest to get this dual concept into a reality.

In addition to the drug-based disease-modifying therapeutic approaches, a growing body of evidence suggests that some risk factors that contribute to the development of late-onset dementias and AD are modifiable, to mention only those like physical inactivity, depression, diabetes mellitus type 2, and insulin resistance. Blood pressure, diabetes, and plasma lipids/cholesterol are of importance especially in middle-aged persons and must be controlled regularly. However, carefully planned clinical trials need to confirm (if possible at all) their disease-modifying activity. Considering already performed clinical trials which tended to be preventive ones but failed, a question has arisen – have AD prevention trials been too short, too late, too small, or too narrow? Recently therapeutic strategies in AD population have made a great switch from "AD treatment" to "AD prevention," with the latter gaining more and more attention. In line with the AD staging, it has been proposed that different strategies of AD prevention and treatment should be chosen for different disease stages, e.g., at the preclinical stages of AD, the onset of disease should be prevented by controlling the initial pathogenic factor, such as insulin resistance; at pre-dementia stage, the disease progression should be stopped by applying different combinations of preventive and curative drugs; and at dementia stage, a "cocktail therapy" of drugs targeting multiple pathogenic mechanisms should be used. However, different biomarkers are necessary to distinguish various stages of AD. The concept of AD prevention has moved forward in line with the preclinical stages of non-demented subjects, stage in which there are no abnormal biomarkers, and stage of suspected non-amyloidal pathology in which biomarkers for amyloid are normal but biomarkers of neurodegeneration are abnormal. Temporal ordering of the biomarkers defines their roles in future therapeutic strategies, but their predictive value and responsiveness to change should be carefully evaluated, e.g., markers of amyloid deposition in PET studies may reflect the trajectories of the different underlying pathology, and its presence alone is not sufficient to produce cognitive decline, and therefore, they are not necessarily markers of AD (Liang et al. 2013). The potential to predict at a time when subjects with early-onset familial AD are still asymptomatic but who will or may (on the basis of biomarkers for the late-onset sporadic AD) go on to develop AD, combined with new techniques to monitor disease progression prior to symptom onset, has added attraction to presymptomatic clinical trials, which are also considered to be secondary to prevention studies (Liang et al. 2013). Several AD secondary prevention trials that are currently in the planning stages or have begun enrolling subjects are focused on two different populations in which the efficacy of the passive immunization approach (monoclonal antibodies against β-amyloid, solanezumab, and crenezumab) is being tested. One secondary prevention trial is targeting asymptomatic individuals with biomarker evidence of AD pathology who will be treated with solanezumab for 3 years, and the other preventive trial is targeting presymptomatic individuals with dominant mutations or other genetic risk factors for AD who will begin receiving crenezumab prior to the expected age of onset. Hopefully, the results of these clinical trials will shed a new light on the prevention of neurodegeneration in AD at this early stage which could be considered to be a real disease-modifying strategy.

Conclusions

Sporadic AD is a spectrum disorder released by multiple triggers and caused by multiple pathologies. The richness of the field and its potential can be seen from the large number of possible targets: multiple targets could be treated with multifunctional drugs. Thus, a "rational" therapy could be based upon different subtype-tailored and disease stage-tailored strategies, respectively, which at the current level of knowledge may include different combinations of the following:

(1) acetylcholinesterase inhibitors, (2) glutamate N-methyl-D-aspartate (NMDA) receptor channel antagonists, (3) NMDA receptor subunit inhibitors, (4) nicotinic receptor modulators, (5) muscarinic receptor modulators, (6) ß-amyloid protein aggregation inhibitors, (7) ß- and gamma-secretase inhibitors, (8) gamma-secretase modulators, (9) alpha-secretase stimulants, (10) vaccination strategies against Aß-induced plaques, (11) inhibitors of tau protein aggregation, (12) inhibitors of tau protein phosphorylation, (13) tau immunization with antibodies against phosphorylated or non-phosphorylated tau (passive immunization), (14) microtubule stabilization, (15) anti-inflammatory strategies, (16) antioxidants, (17) enhancement of autophagy by mTOR-dependent or independent modulators, (18) improvement of lysosomal clearance, (19) stem cell-based therapies, (20) "synaptic repair" therapies, (21) strategies based on deep-brain stimulation, (22) AD prevention strategies in preclinical stages of the disease, (23) gamma- and ß-(PPAR) agonists, (24) GSK-3-alpha and ß inhibitors, (25) inhibitors of AGE product biosynthesis, (26) inhibitors of AGE receptors, (27) activators of GLP-1 receptors, and (28) anti-hypercholesterolemia treatments. Thus, the treatment of various subtypes of AD and disease stages including a combination of specific pathology-related strategies and the development of multifunctional drugs is envisaged.

Acknowledgments Professor Siegfried Hoyer is to be acknowledged as the coauthor of the first edition of this paper which laid the ground for those that have followed.

Cross-References

▶ Acetylcholinesterase and Cognitive Enhancement
▶ Dementias and Other Amnestic Disorders
▶ Muscarinic Cholinergic Receptor Agonists and Antagonists
▶ Neuroprotectants: Novel Approaches for Dementias
▶ Neuroprotection
▶ Vaccines and Drug-Specific Antibodies

References

Bertram L, Lill CM, Tanzi RE (2010) The genetics of Alzheimer disease: back to the future. Neuron 68:270–281

Chen WW, Blurton-Jones M (2012) Concise review: Can stem cells be used to treat or model Alzheimer's disease? Stem Cells 30:2612–2618

Chen TF, Huang RF, Lin SE, Lu JF, Tang MC, Chiu MJ (2010) Folic Acid potentiates the effect of memantine on spatial learning and neuronal protection in an Alzheimer's disease transgenic model. J Alzheimers Dis 20:607–615

Craft S, Cholerton B, Baker LD (2013) Insulin and Alzheimer's disease: untangling the web. J Alzheimers Dis 33 (Suppl 1):S263–275

Grünblatt E, Salkovic-Petrisic M, Osmanovic J, Riederer P, Hoyer S (2007) Brain insulin system dysfunction in streptozotocin intracerebroventricularly treated rats generates hyperphosphorylated tau protein. J Neurochem 101:757–770

Hart GW (1997) Dynamic O-linked glycosylation of nuclear and cytoskeletal proteins. Annu Rev Biochem 66:315–335

Holmes C, Boche D, Wilkinson D, Yadegarfar G, Hopkins V, Bayer A, Jones RW, Bullock R, Love S, Neal JW, Zotova E, Nicoll JA (2008) Long-term effects of Abeta42 immunisation in Alzheimer's disease: follow-up of a randomised, placebo-controlled phase I trial. Lancet 372:216–223

Hölscher C (2013) Central effects of GLP-1: new opportunities for treatments of neurodegenerative diseases. J Endocrinol [Epub ahead of print]. doi:10.1530/JOE-13-0221

Hoyer S (2004) Glucose metabolism and insulin receptor signal transduction in Alzheimer disease. Eur J Pharmacol 490:115–125

Hoyer S, Plaschke K (2004) The aging brain – the burden of life. In: Herdegen T, Delgado Garcia J (eds) Brain damage and repair: from molecular research to clinical therapy. Kluwer, Dordrecht, pp 1–22

Hoyer S, Riederer P (2007) Alzheimer disease – no target for statin treatment. A mini review. Neurochem Res 32(4–5):695–706

Jin K, Xie L, Mao XO, Greenberg DA (2006) Alzheimer's disease drugs promote neurogenesis. Brain Res 1085:183–188

Kupershmidt L, Amit T, Bar-Am O, Weinreb O, Youdim MB (2012) Multi-target, neuroprotective and neurorestorative M30 improves cognitive impairment and reduces Alzheimer's-like neuropathology and age-related alterations in mice. Mol Neurobiol 46:217–220

Liang Y, Ryan NS, Schott JM, Fox NC (2013) Imaging the onset and progression of Alzheimer's disease: implications for prevention trials. J Alzheimers Dis 33(Suppl 1):S305–S312

Lu B, Nagappan G, Guan X, Nathan PJ, Wren P (2013) BDNF-based synaptic repair as a disease-modifying strategy for neurodegenerative diseases. Nat Rev Neurosci 14:401–416

McClean PL, Parthsarathy V, Faivre E, Holscher C (2011) The diabetes drug liraglutide prevents degenerative processes in a mouse model of Alzheimer's disease. J Neurosci 31:6587–6594

Namba T, Maekawa M, Yuasa S, Kohsaka S, Uchino S (2009) The Alzheimer's disease drug memantine increases the number of radial glia-like progenitor cells in adult hippocampus. Glia 57:1082–1090

Nixon RA (2013) The role of autophagy in neurodegenerative disease. Nat Med 19:983–997

Riederer P, Hoyer S (2006) From benefit to damage. Glutamate and advanced glycation end products in Alzheimer brain. J Neural Transm 13(11):1671–1677

Salkovic-Petrisic M, Tribl F, Schmidt M, Hoyer S, Riederer P (2006) Alzheimer-like changes in protein kinase B and glycogen synthase kinase-3 in rat frontal cortex and hippocampus after damage to the insulin signalling pathway. J Neurochem 96:1005–1015

Spillantini MG, Goedert M (2013) Tau pathology and neurodegeneration. Lancet Neural 12:609–622

Neurodevelopmental Hypothesis

Definition

The neurodevelopmental hypothesis proposes that ▶ schizophrenia results from abnormalities in neuronal connectivity which arise from fetal life onwards but are not expressed until the onset of illness.

Neuroendocrine Markers for Drug Action

R. H. de Rijk[1] and E. Ronald de Kloet[2]
[1]Department of Psychiatry, Leiden University Medical Center, Leiden, RA, The Netherlands
[2]Department of Endocrinology and Metabolism, Leiden Academic Center for Drug Research and Leiden University Medical Center, Leiden, RA, The Netherlands

Definition

The hypothalamic-pituitary-adrenal (HPA) axis is a neuroendocrine system that coordinates adaptation of body and brain function to environmental demands, i.e., daily- and sleep-related events and stressors (de Kloet et al. 2005). The hormones of the HPA axis operate in a feedforward cascade from hypothalamic corticotropin-releasing factor (CRF) and vasopressin via pituitary adrenocorticotropin hormone (ACTH) to adrenocortical secretion of glucocorticoids in the circulation. Glucocorticoids in turn feedback on the brain and pituitary to shut down the activated HPA axis. In this chapter the mechanism underlying HPA axis functioning is first described and subsequently the different tests being used to evaluate HPA axis reactivity. Then, specific psychiatric disorders characterized by dysregulation of the HPA axis are discussed with reference to genetic and early life susceptibility factors. The chapter concludes with an analysis of HPA axis responses that may serve as predictors for the efficacy of psychoactive drugs in humans.

HPA Axis Regulation

CRF and its co-secretagogue vasopressin (AVP) organize the behavioral, sympathetic, and neuroendocrine response to daily- and sleep-related events and stressors (de Kloet et al. 2005). The neuroendocrine response proceeds via the hormones of the HPA axis, e.g., hypothalamic CRF and AVP, pituitary ACTH, and adrenal glucocorticoids. CRF, AVP, and pro-opiomelanocortin (POMC) peptides also have elaborate neuropeptide networks in the limbic-midbrain which modulate the processing of circadian and stressful information (Herman et al. 2003). Additionally, many neurotransmitter systems influence dynamically at several levels the activity of the HPA axis. These include among others GABA, glutamate, monoamines, neuropeptides, and gas-based messengers. More recently, the endocannabinoid system has been proposed to exert an inhibitory action on HPA axis activation.

Under basal conditions the hormones of the HPA axis are released in hourly pulses (Lightman and Conway-Campbell 2010). The magnitude of these pulses changes during day and night with the largest amplitude at the start of the active period, which is in the morning for man and late afternoon for rodents. Awakening triggers an additional distinct HPA response. The pulsatile HPA axis pattern shows gender differences. The frequency of the pulses alters during chronic stressful or inflammatory conditions. The ultradian rhythm becomes disordered during Cushing and Addison's disease, during chronic stressful and inflammatory conditions, and during the aging process. The pulse is generated because of the closed feedback organization underlying self-regulation of HPA axis activity. Any system that operates with feedback delay will oscillate, hence the origin of pulsatility. The ACTH pulses are amplified on the adrenal level. Frequency encoding is a common mechanism in information processing by hormonal systems, and evidence is accumulating that resilience in target tissues depends on pulsatile exposure to glucocorticoids.

Superimposed on the ultradian rhythm is the response to stressors. In fact, it has been shown in rats that the magnitude of the HPA axis response depends on the phase of the rhythm: corticosterone responses were larger when a stressor is experienced at the ascending rather than at the descending phase. The onset, duration, and magnitude of the HPA axis response is affected by the previous experience, and, in particular, early life events are potent stimuli capable to program the reactivity of the axis to stressors in later life. The most profound psychological stressors are characterized by loss of control, no information, and no prediction of upcoming events combined with a sense of fear and uncertainty. This condition results in a profound and long-lasting activation of the HPA axis (de Kloet et al. 2005; Joëls et al. 2012).

Receptors

The actions of the various HPA axis hormones are mediated by receptors. CRF has a high affinity to the CRF_1 receptors and a lower affinity to the CRF_2 receptors, which have actually their own privileged ligands urocortin II and III. CRF_1 receptors are on pituitary corticotrophs and have a discrete localization in the limbic-midbrain circuitry, such as in the amygdala, where CRF promotes emotional arousal. Activation of the CRF_2 receptor ligands produces complimentary responses. There is evidence that CRF_1 and CRF_2 receptors mediate different phases of the stress response from activation to later adaptations. CRF_1 antagonists have been developed which suppress fearful and emotional responses in animals and humans (Ising et al. 2007). These antagonists potentially represent a new generation of anxiolytic and/or antidepressant agents that have an action mechanism within the stress system (Ising et al. 2007; Steckler 2010).

AVP binds to V_{1A} receptors regulating the pressor response, to V_2 receptors mediating its antidiuretic action in the kidney, and to V_{1B} receptors in pituitary corticotrophs inducing ACTH release. These V_1 and V_2 receptors are also localized in discrete brain regions. The oxytocin (OT) receptor which is localized in the central amygdala, ventromedial nucleus, ventral subiculum, and olfactory tubercle also displays high affinity to AVP. The neuropeptides AVP and OT act potently in fear conditioning paradigms and have an important modulatory function in psychosocial aspects of behavior (Donaldson and Young 2008).

The glucocorticoids cortisol (man) and corticosterone (man and rodent) readily penetrate the brain and bind to two types of nuclear receptors: high affinity mineralocorticoid receptors (NR3C2 or MR) and lower affinity glucocorticoid receptors (NR3C1 or GR), which are colocalized in high density in the limbic system. These nuclear receptors also have lower affinity membrane variants that can mediate rapid actions of glucocorticoids in the brain. The action of glucocorticoids mediated by the two types of receptors operates in complimentary fashion. The low and high affinity MR are involved in appraisal processes and emotional expressions and involved in the onset of the stress response, while the termination of the stress response is facilitated via the GR types (negative feedback) and the way the stressful experience is handled and stored in the memory for future use. Analogs of cortisol that activate selectively MR and GR are being developed for treatment of anxiety disorders and depression (de Kloet et al. 2005; Joëls et al. 2012).

An organism is healthy and resilient when the HPA axis is readily activated as long as it is also turned off efficiently. If coping with stress fails HPA axis, regulation is compromised, CRF is usually hyperactive, and stress-induced glucocorticoid secretion can be inadequate or excessive and prolonged. Psychoactive drugs that interact with the stress neurocircuitry in the limbic-midbrain have the potential to modulate HPA axis regulation, either by direct interaction with the "core" of the HPA axis or indirectly by modulation of afferent pathways. Recovery of HPA axis responsiveness and pulsatility by, e.g., antidepressants, precedes recovery from depression (Holsboer 2000).

Testing of HPA Axis Reactivity

Cortisol, the human's principal glucocorticoid, binds in plasma to serum albumin and with higher affinity to corticosteroid-binding globulin or transcortin. Hence, total cortisol can be measured as index for adrenocortical output; free cortisol is considered to be the biologically active fraction. Currently, the free cortisol concentration is often determined in saliva, which can be obtained by a much less stressful procedure than blood via venipuncture. In 24 h urine cortisol and 17-ketosteroids (17-OHCS) are measured and are an index for daily production. Cortisol levels in hair segments can be used as a retrospective calendar of HPA activity during specific time periods preceding sample collection (Meyer and Novak 2012).

Basal values of cortisol are often reported, but these levels must be judged with care. In single measurements, hourly pulses of the hormone may produce variable data unless temporal patterns are measured. Moreover, the amplitude of the cortisol pulse displays a strong circadian rhythm producing high levels of the hormone during awakening and low levels from usually 11.00 p.m. to 5.00 a.m. Finally, 30 min after awakening a sharp rise in cortisol is observed on top of the circadian rhythm. This rise is called the cortisol awakening response (CAR).

Chemical challenges of the HPA axis date back to the 1970s, when administration of 1 mg dexamethasone was used to test the negative feedback capacity of the HPA axis; this is the so-called dexamethasone suppression test (DST). Glucocorticoid receptors (GR) in the pituitary corticotrophs are probably the primary target of dexamethasone, which at these low dosages is hampered to penetrate the brain by multidrug resistance P-glycoprotein located in the blood-brain barrier. Lower doses of dexamethasone of 0.5 mg and even 0.25 mg can detect more subtle differences in negative feedback function.

An extension of the DST, the combined dexamethasone-CRF challenge, was developed as a pharmacological test of reactivity of the HPA axis, further amplifying differences in dexamethasone sensitivity. The principle of the test is that the individuals receive dexamethasone at 23.00 p.m., and then the next afternoon the pituitary-adrenal response to a CRF injection is measured. The CRF in this test enhances escape from dexamethasone suppression. Several other pharmacological challenges of the HPA axis have been developed and applied. To test the integrity of the pituitary gland, ovine CRF or AVP was administrated to depressed patients. Furthermore, ACTH challenges are being performed often using high (250 μg)- or low (1 μg)-dose Synacthen (synthetic ACTH), to test synthesis and secretion of corticosteroids by the adrenal cortex (Holsboer 2000).

Physical challenges of the HPA axis may involve, for example, strenuous exercise for 20 min on a treadmill under controlled conditions of O_2 usage and CO_2 production. This condition results in a profound activation of the HPA axis. Such conditions in which energy metabolism is challenged usually are accompanied by a profound increase in adrenal sensitivity to ACTH.

Psychological challenges characterized by lack of control and lack of social support trigger the largest ACTH and cortisol response. The Trier Social Stress Test (TSST) is a well-established psychosocial stress test (Dickerson and Kemeny 2004). This test involves a public speaking and mental arithmetic task which needs to be performed in front of an audience and camera and results in a large increase in ACTH, cortisol, and sympathetic activity (Kirschbaum et al. 1993).

Hypo- and Hypercortisolism States

Health and resilience are characterized by a reactive HPA axis that is readily turned on and off. Several disease states have been associated with changes in basal and activated HPA axis responses. Thus, *hyper* cortisolemic states usually enhance the risk for infection because of suppression of the immune system. Examples are Cushing syndrome, melancholic depression, panic disorder, anorexia nervosa, sleep deprivation, addiction, and malnutrition. Alternatively, *hypo* cortisolism enhances the risk for inflammatory disorders because the cortisol action is insufficient to restrain the initial stress reactions. Examples are Addison's disease, atypical depression, chronic fatigue syndrome, fibromyalgia, post-traumatic stress disorder (PTSD), and autoimmune diseases (Chrousos and Gold 1992).

Common functional genetic variants that are associated with changes in HPA axis reactivity have been described (DeRijk 2009). Most evidence for specific genetic components is provided by single nucleotide polymorphisms (SNPs) in the MR and GR genes found to modulate both negative feedback and stress-induced activity of the HPA axis. Importantly, these SNPs in the MR and GR genes also associate with aspects of depression, indicating their importance as vulnerability factors. Alternatively, a trait defect might be acquired as a result of unfavorable environmental circumstances, particularly during early life (Daskalakis et al. 2013). These so-called programming effects cause changes in the expression of MR and GR with consequences for HPA axis reactivity, neuronal activity, and behavior. In addition, also gender and aging have been found to modulate HPA axis reactivity.

Depression

In depressed patients, basal plasma ACTH and cortisol have been found to be elevated (Holsboer 2000). Also urinary-free cortisol, saliva-free cortisol, or cerebrospinal fluid cortisol levels are increased. Escape from dexamethasone suppression was observed in a subgroup of depressed patients suggesting resistance to corticosteroid action. Hypercortisolism in depression is further supported by the enlarged adrenal glands of the patients. This indicates enhanced adrenal sensitivity to ACTH stimulation. The combined dexamethasone suppression-CRF stimulation test (Dex-CRF test) appears to be the most sensitive tool to detect depression-related changes in HPA axis reactivity. Moreover, several studies reported that normalization of the HPA axis precedes or parallels the clinical response to antidepressant treatment (Binder et al. 2009).

Post-traumatic Stress Disorder

Core features of HPA axis changes in PTSD include low basal cortisol secretion and enhanced negative feedback control of the HPA axis in the face of high sympathetic and emotional reactivity (Yehuda 2002). The enhanced negative feedback was found using low-dose dexamethasone (0.25 mg) or metyrapone tests. Blunted ACTH responses to CRF stimulation are explained by downregulated CRF_1 receptors, possibly as a result of sustained, increased endogenous CRF levels. However, findings have not been consistent. Differences could involve disease stages, gender, genetic background, or type of trauma among others. Using the combined Dex-CRF test did not reveal HPA axis abnormalities in PTSD patients when compared to trauma controls (also exposed to trauma but without PTSD). However, PTSD patients with a comorbid depression showed an attenuated ACTH response compared to PTSD patients without comorbid depression. This indicates the presence of PTSD subgroups with different HPA axis regulations.

Psychoactive Drugs that Modulate HPA Axis Reactivity

Almost all drugs that change central neuronal processes initially modulate basal and stress-induced HPA axis activity, because they acutely change homeostasis (Table 1). Here, drugs that affect parts of the regulatory mechanisms of the HPA axis and that are used in clinical practice are described.

Neuroendocrine Markers for Drug Action, Table 1 Drug effects on HPA axis activity

Drugs				HPA axis					
				Acute effects				Prolonged effects	
System involved	Indication	Drugs	Action	Basal cortisol		Reactivity		Basal cortisol	Reactivity
Biogenic monoamines									
NE 5-HT DA	Psychostimulant	Amphetamine	Reuptake inhibition	No effects					
	Antidepressant	TCA	Reuptake inhibition	Increase				Decrease	Normalization
	Antidepressant	TCA/SSRI	Reuptake inhibition	Increase				Decrease	Normalization
	Antidepressant	Mirtazapine	NaSSA, antagonist of 5-HT$_2$ and H$_1$	Decrease					
	Antidepressant	Moclobemide	MAO-inhibitor	Decrease					
	Antidepressant	Nefazodone	Antagonist 5-HT$_{1a}$ and α1	Small increase				Decrease	
	ADHD	Methylphenidate	Reuptake inhibition, agonist 5-HT$_{1A}$ and $_{2B}$	Increase					
NE	Hypertension, migraine	Clonidine	α2 agonist	Decrease					
	Nasal congestion	Methoxamine	α1 agonist	Increase					
	Hypertension	Prazosin	α1 antagonist			Decrease			
	Hypertension	Doxazosin	α2 antagonist	Decrease					
	Hypertension	Propranolol	β2 antagonist			Increase			
	Heart failure	Nebivolol	β1 antagonist	No effect		No effect			
	Depression	Carvedilol	α1/β antagonist	Increase					
	Hypertension	Yohimbine	α2 antagonist	Increase		Increase			
	Hypertension	Reserpine	Vesicle depletion	Decrease					
5-HT	Anxiety	Buspirone	5-HT$_{1A}$ agonist	Increase		Increase			
	Anxiety	Gepirone	5-HT$_{1A}$ and $_{2A}$ agonist	Increase		Increase			
DA	Schizophrenia	Haloperidol	Antagonist	No effect					
	Schizophrenia	Olanzapine	5-HT$_{2C}$ antagonist	Decrease		Decrease			
	Schizophrenia	Quetiapine	DA NE 5-HT antagonism	Decrease					
	Schizophrenia	Clozapine	DA NE 5-HT Ach H antagonism	Decrease		Decrease			

(continued)

Neuroendocrine Markers for Drug Action, Table 1 (continued)

Drugs				HPA axis				
				Acute effects		Prolonged effects		
System involved	Indication	Drugs	Action	Basal cortisol	Reactivity	Basal cortisol	Reactivity
	Schizophrenia	Risperidone	DA 5-HT antagonism	Small decrease			
	Schizophrenia	Sulpiride	D_2 antagonism		Decrease		
	Parkinson	Levodopa	D_2 agonist	Decrease	Decrease	Decrease	Decrease
	Parkinson	Pramipexole	D_2 agonist	Unclear			
Acetylcholine	Motion sickness, anesthesia	Nicotine	nAch agonist	Increase	Increase	Small effects	Attenuation
	Motion sickness, anesthesia	Scopolamine	mAch agonist				
	Peptic ulcers	Pirenzepine	mAch agonist				
	Myasthenia gravis	Physostigmine	Cholinesterase inhibitor	Enhanced			
	Surgery	Neostigmine	Cholinesterase inhibitor	Decreased			
Amino acids							
GABA	Anxiety	Alprazolam	Agonist		Decrease		
	Anxiety	Pivagabine (PVG)	Agonist		Decrease		
	Anxiety	Diazepam	Agonist			Decrease	
	Anxiety	Oxazepam	Agonist			Decrease	
	Anesthesia	Propofol	Agonist	No effect			
	Surgery trauma	Etomidate	Agonist	Decrease			
	Anesthesia	Thiopental	Agonist barbiturate	Decrease			
	Overdoses Benzodiazepine	Flumazenil	Antagonist	Decrease			
Glutamate	Anesthesia	Ketamine	NMDA antagonist	Increase			
	Amyotrophic lateral sclerosis	Riluzole	Blocks release, antagonist	No effect	No effect		
Neurotransmitter release	Anesthesia	Halothane	Reduced transmitter release and action	Increase			
	Anesthesia	Isoflurane	Reduced transmitter release and action	Increase			
	Anesthesia	Sevoflurane	Reduced transmitter release and action	Increase			

Category	Indication	Drug	Mechanism				
	Anesthesia	Droperidol + opioid	Dopamine antagonist				Increase
	Inflammation	Aspirin	Inhibition of PG synthesis				No effect
Neuropeptides							
CRH							
	Inflammation	NBI-34041	CRHR1 antagonist			Decrease	
Corticosteroids	Inflammation	Dexamethasone	GR agonist		Decrease	Decrease	Decrease
	Psychotic depression, Cushing disease	Mifepristone	GR antagonist		Large increase	Decrease	Large increase
	Salt wasting	Fludrocortisone	MR agonist			Decrease	Decrease
	Hypertension	Spironolactone	MR antagonist			Increase	Increase
Cannabinoids	Marijuana	Tetrahydrocannabinol (THC)				Increase	
Opioids	Pain	Codeine				Decrease	
	Pain	Cocaine	Reuptake inhibition			Increase	
	Anesthesia, pain	Remifentanil	Agonist			No effect	
	Addiction	Naloxone	Mu antagonist			Increase	
	Addiction	Methadone	Antagonist			Increase	
Ion channels	Bipolar epilepsia	Valproate	Inhibition GABA breakdown				Enhanced
	Bipolar epilepsia	Carbamazepine	Inhibition sodium channels			Enhanced	Enhanced
	Bipolar	Lithium				Enhanced	Enhanced

Benzodiazepines. Benzodiazepines are a major first line of treatment in anxiety disorders. However, little is known of their effect on HPA axis reactivity in man. Administrations of 1 mg alprazolam strongly inhibit psycho(social) stress-induced increases in ACTH and cortisol in males. Other GABA agonists such as Pivagabine (PVG), a hydrophobic 4-aminobutyric acid derivative (7 days treatment, 900 mg, twice a day), could inhibit ACTH and cortisol responses following a psychosocial stressor. Basal levels of cortisol are decreased following administration of codeine (an opiate), diazepam, or oxazepam (benzodiazepines). Prolonged use of benzodiazepines results in profound suppression of the basal and stress-related HPA activity, and discontinuation of these drugs results in rebound activation. Flumazenil, which has been proposed to exert GABA antagonistic properties, did show however a decrease in plasma ACTH and cortisol, following administration. Valproate, used to treat bipolar disorders and epilepsy, has several modes of action including inhibition of GABA breakdown. Administration leads to enhanced HPA axis responsiveness. Taken together, drugs that stimulate GABAergic transmission decrease the activity of the HPA axis. Carbamazepine, acting on voltage-dependent sodium channels, is used for mood stabilization in bipolar disorder and in epilepsy. This drug enhances both basal as well as ACTH and cortisol concentrations as outcomes of the Dex-CRF test following prolonged treatment. Also lithium did enhance Dex-CRF responses but also increased morning cortisol levels in the DST (1.5 mg).

Cholinergic drugs. These agents are mainly used to control blood pressure (hypertension glaucoma) and myasthenia gravis and are used during anesthesia. Furthermore, they are suggested to improve clinical signs of Alzheimer's disease. Physostigmine, an acetylcholine esterase inhibitor used in myasthenia gravis, did not change basal ACTH and cortisol secretion per se, excluding direct effects on basal HPA secretion under these conditions. However, eating-induced ACTH and cortisol responses were further increased. Also acute administration of physostigmine increased plasma cortisol lasting for several hours. Neostigmine (a parasympathomimetic, acetylcholine esterase inhibitor) given epidurally significantly reduced the plasma levels of cortisol in the early surgical period. Ectothiopate (parasympathomimetic, irreversible acetylcholine esterase inhibitor), used as eyedrops in glaucoma, has not been tested on HPA axis parameters. Scopolamine, used to treat motion sickness or together with atropine in anesthesia, is a muscarinic antagonist but did not have an effect on the HPA axis. Pirenzepine, a muscarinic 1 receptor antagonist used in peptic ulcers, enhanced the circadian differences of basal cortisol levels. Rivastigmine, tacrine, donepezil, and galantamine (all cholinesterase inhibitors) used in Alzheimer's diseases have not been tested on HPA axis parameters, and no clear data are available. Nicotine was found to be a strong activator of the HPA axis. In habitual smokers, a few cigarettes activate the HPA axis, but only small changes of basal HPA axis activity are observed, with an attenuated responsiveness of the HPA axis to psychological stress but not to injection of CRF. Etomidate, used in trauma settings to intubate patients, is short acting and activates GABA R and possibly also mACh receptors. It lowers plasma cortisol levels because of inhibition of the 11β hydroxylase blocking the conversion of deoxycortisol to cortisol.

Biogenic monoamines and antidepressants. Monoamines clearly activate the HPA axis. Modulators of monoamines are used for the regulation of blood pressure and in psychiatric disorders. Oral administration of 5-HTP, the direct precursor of serotonin (5-HT) induces fast dose-dependent increases in circulating ACTH and cortisol. Also administration of the 5-HT-releasing agent D-fenfluramine or the piperazine-based 5-HT receptor agonist mCPP increases acutely ACTH and cortisol responses. Buspirone, a $5-HT_{1A}$ agonist used to treat anxiety disorder, induces ACTH and cortisol responses following acute challenges. Gepirone, a partial $5-HT_{1A}$ agonist having anxiolytic properties, also resulted in large increases in plasma cortisol.

Noradrenaline (NE) or serotonin (5-HT) reuptake-inhibiting antidepressants such as reboxetine or citalopram acutely stimulate cortisol and ACTH release in healthy volunteers. Mirtazapine, a noradrenergic and specific serotonergic antidepressant (NaSSA), acutely inhibits the ACTH and cortisol release, probably due to its antagonism at central 5-HT$_2$ and/or 5-HT$_1$ receptors. The classic TCAs, such as desipramine, imipramine, and clomipramine, also induce an acute increase in ACTH and cortisol levels. Following longer treatment with most reuptake inhibitors, including SSRIs and noradrenergic specific reuptake inhibitors, a gradual normalization of the hyperactive HPA axis is found in depressed patients (Binder et al. 2009). These chronic effects probably depend, at least in part, on the increased expression of MR and GR, leading to a normalization of HPA axis reactivity. Nefazodone, a 5-HT$_{2A}$ and α_1 adrenergic antagonist also inhibiting NE and 5-HT reuptake, increased slightly cortisol levels following administration in healthy subjects. Moclobemide, a selective MAO-A inhibitor mostly used in refractory depression, slightly decreases cortisol levels upon acute administration.

Methylphenidate is a norepinephrine and dopamine reuptake inhibitor with affinity for the serotonergic 5-HT$_{1A}$, and 5-HT$_{2B}$ is used in ADHD as a psychostimulant. Acute administration induces increases in cortisol, while prolonged treatment with methylphenidate did not seem to affect basal cortisol or cortisol levels during exercise. Amphetamine increases levels of norepinephrine, serotonin, and dopamine in the brain by actions on the transporter and synaptic vesicles. However, acute amphetamine administration up to 20 mg did not activate the HPA axis. Cocaine is a dopamine, norepinephrine, and a serotonin reuptake inhibitor, and acute administration induces an increase in plasma ACTH and cortisol.

Clonidine is an α2-adrenergic agonist used in hypertension and migraine, and its acute administration resulted in a strong decrease of the activity of the HPA axis, with low ACTH and cortisol levels. Methoxamine, an α1 agonist used for nasal congestion, enters the CNS following peripheral administration. ACTH and cortisol increased early during methoxamine infusion, and ACTH returned to baseline promptly after the infusion ceased.

Prazosin, an α1 antagonist used as an antihypertensive, lowers challenge-induced activation of the HPA axis. Yohimbine is an α2-adrenergic antagonist acting on presynaptic terminals that inhibit (nor)adrenaline release. Acute administration of yohimbine results in increases in CSF noradrenaline levels leading to increases in plasma ACTH and cortisol. Following a serotonergic challenge, coadministration of yohimbine further increases cortisol values. Doxazosin, an α2 antagonist for treatment of severe hypertension, resulted in a short-lasting decrease of plasma cortisol. For propranolol, a β receptor antagonist, mixed results have been reported ranging from no effects (psychosocial stress) to increased cortisol responses (exercise). Prolonged administration of carvedilol, having α1 and β receptor antagonistic properties, leads to a noticeable increase in serum cortisol levels. Nebivolol, a β1 selective blocker with vasodilator properties, did not seem to affect basal cortisol or during exercise.

Reserpine, which depletes stores of NE by inhibiting vesicular uptake, decreased basal excretion of urinary 17-OHCS significantly in Cushing patients and caused normalized suppression of plasma cortisol by 1 mg dexamethasone.

Glutamate. Ketamine is a clinical available NMDA receptor antagonist also used as an anesthetic, because of its effects on opioid receptors at higher dosages. Furthermore, ketamine is found to reduce the symptoms of opiate withdrawal, suggesting its potential role in the treatment of addiction. Acute administration of ketamine (e.g., 0.5 mg/kg) induces increases in plasma cortisol. The antiglutamatergic drug riluzole did not have any effects upon HPA axis activity under baseline and stress-induced conditions in elderly subjects.

Dopaminergic drugs. These drugs are typically used to treat Parkinson's disease, while antagonists of the dopamine system are used to treat psychotic symptoms, which often occur in schizophrenia and depression.

Haloperidol, a dopamine antagonist used in schizophrenia, did not lower basal cortisol levels in control subjects. Among the atypical antipsychotic medication, olanzapine, an antagonist of several 5-HT receptors, the histamine H_1 receptor, and the dopamine (D_2) receptor, lowers basal cortisol levels at night in healthy male controls following a standard 8-day treatment. Similar effects on cortisol have been observed in patients with schizophrenia. Treatment of patients with olanzapine resulted in a blunted ACTH and cortisol responses following an mCPP challenge. Also quetiapine, showing dopaminergic, adrenergic, and serotonergic receptor antagonism, lowers nocturnal cortisol output in healthy subjects. Clozapine, used for treatment-resistant schizophrenia, shows antagonist properties for different subtypes of dopaminergic, adrenergic, cholinergic, and histaminergic receptors. Basal cortisol levels decrease slightly during clozapine treatment. Clozapine might also inhibit mCPP-induced cortisol responses. Risperidone, acting at D_2 and serotonergic ($5-HT_{2A}$) receptors, induced a modest decrease of cortisol levels following treatment. Sulpiride, a selective antagonist at postsynaptic D_2 receptors, suppresses increases in plasma cortisol levels induced by repetitive blood sampling at 180–240 min, compared with the response to placebo.

Levodopa, first-line treatment for Parkinson's disease, decreases plasma cortisol after acute administration. Also during physical exercise which increases cortisol, Levodopa resulted in a decrease of plasma cortisol in patients with Parkinson's disease. However, in children, administration of levodopa in combination with propranolol increased plasma cortisol. Pramipexole, a D_2 receptor agonist used in Parkinson's disease, does not have a clear effect on cortisol levels. In conclusion, antagonists of dopaminergic transmission seem to inhibit HPA axis reactivity.

Anesthetic and sedative drugs. This class of drugs generally inhibits excitatory channels (e.g., glutamate) and facilitates inhibitory channels (e.g., GABA). Ether induces a profound activation of the HPA axis with high levels of ACTH and cortisol. Halothane also induces an increase in plasma ACTH following administration without surgery in healthy volunteers. General anesthesia maintained with an isoflurane-nitrous oxide-oxygen mixture or sevoflurane is followed by a sharp increase in plasma cortisol. The combination of droperidol, a dopamine D_2 antagonist, with an opioid analgesic induced plasma cortisol elevations. In contrast, a mixture of propofol (acting at $GABA_A$ and maybe glycine and endocannabinoids) and remifentanil (mu-opiate receptor agonist) did not activate the HPA axis and even blocked HPA axis responses during surgery. However, under anesthesia with propofol, synthetic ACTH was still capable of inducing a profound cortisol plasma level elevation. Thiopental, a barbiturate acting on GABA receptors, also decreased plasma levels of cortisol during anesthesia and surgery. Other analgesic drugs such as aspirin did not affect basal or stress-induced HPA axis responses.

Opioids. Endogenous opioids, such as POMC-derived endorphins and the enkephalins, exert inhibitory influences on HPA axis activity; see remifentanil (mu-opiate receptor agonist) above. Naloxone, an mu-opiate receptor antagonist, increases HPA axis activity by blocking the inhibitory tone of opiates on hypothalamic CRF secretion. However, the acute administration of methadone, an opioid agonist, results in a profound short-lasting activation of the HPA axis.

Cannabinoids. Acute administration of high doses of tetrahydrocannabinol (THC) induces release of ACTH and cortisol, while low doses of THC dampen stress-induced HPA axis activity by targeting the hypothalamus, particularly in regular users. Endogenous cannabinoids (endocannabinoids) were found to limit activation of the HPA axis. This inhibitory action of endocannabinoids is thought to mediate the rapid feedback action of glucocorticoids by attenuating the release of the excitatory neurotransmitter glutamate that activates the CRH neurons. In addition, endocannabinoids also enhance the inhibitory influence of projections from the prefrontal cortex. The very high doses of exogenous cannabinoids probably act via ascending aminergic pathways. The differential effects of cannabinoids depend on the dose and the context (administered or endogenous) the compound operates (Hill and Tasker 2012).

CRF antagonists. CRF is central in the activation of the HPA axis, and increased central release of the peptide is thought to play a role in the pathogenesis of several stress-related disorders. The CRF_1 receptor antagonist decreased psychosocial stress-induced ACTH and cortisol responses, but did not affect basal and circadian changes in HPA axis activity. The antagonist also did not interfere with HPA axis stimulation by CRF injection. This indicates that the effect of the CRF antagonist is primarily in the brain to modulate behavioral appraisal rather than that it interferes directly with stress-induced ACTH release on the level of the pituitary. CRF_1 receptor antagonists (e.g., NBI-34041) are considered promising next-generation drugs for the treatment of depression and anxiety disorders (Ising et al. 2007; Steckler 2010).

Corticosteroids. Dexamethasone has been used to test negative feedback of the HPA axis. To test for Cushing's disease, dosages of 4 mg are used which in all healthy subjects results in suppression of morning ACTH and cortisol levels. Lower dosages such as 1.5, 1, 0.5, or 0.25 mg test for negative feedback at the level of the pituitary gland for more subtle changes, as observed in patients suffering from depression or PTSD. Also hydrocortisone (cortisol) has been used to test HPA axis negative feedback.

Glucocorticoid antagonists. Mifepristone (RU 486) is a compound with progesterone as well as cortisol-blocking activities. Within a day of treatment (e.g., 200 mg/day), increases in plasma ACTH and cortisol are observed, as a result of loss of the feedback action of endogenous glucocorticoids. The compensatory increase in ACTH also resulted in increases in deoxycortisol, progesterone, and androgens. During prolonged treatment a resetting of the HPA axis at higher levels occurred, showing an increase in the amplitude of circadian rhythmicity.

Mineralocorticoids. Fludrocortisone, a potent MR agonist that promotes the efficacy of SSRI's, exerts also a significant inhibition of ACTH and cortisol after acute administration (0.5 mg p.o.) and therefore better can be qualified as a potent mixed agonist for MR and GR. Spironolactone, an MR antagonist, induces acute increases in plasma ACTH and cortisol, which supports a function of the MR in inhibiting the HPA axis under basal conditions.

Conclusions

Stress-induced HPA axis activation is blocked by corticosteroids as well as GABA, opioid, and endocannabinoid agonists. HPA axis activity is induced by glutaminergic, noradrenergic, and serotonergic agonists. The pharmacokinetic characteristics of the drugs need to be taken into account with respect to their penetration through the blood-brain barrier and their site of action in the complex neural circuitry underlying HPA axis activation.

Prolonged use of drugs may result in adaptation of the HPA axis leading to *hypo-* or *hyper* cortisolemic states. Alternatively, antidepressants such as SSRIs and TCAs and, to some extent, benzodiazepines may normalize an overactive HPA axis. Hence, drugs may affect a wide range of peripheral and central functions through their action on the HPA axis.

These actions exerted by the drugs may be either directly targeted to the core of the HPA axis – the adrenal, pituitary, or PVN – or the drugs may act indirectly on circuits involved in the processing of stressful information underlying emotional, motivational, and/or cognitive processes. Irrespective of the site and mode of action of the drugs, it is often difficult to disentangle to what extent primary or secondary psychotropic effects proceed through the HPA axis.

Cross-References

- ▶ Addictive Disorder: Animal Models
- ▶ Aminergic Hypotheses for Depression
- ▶ Analgesics
- ▶ Anticonvulsants
- ▶ Antidepressants
- ▶ Anti-Parkinson Drugs
- ▶ Antipsychotic Drugs
- ▶ Anxiety: Animal Models
- ▶ Arginine-Vasopressin
- ▶ Benzodiazepines

- Beta-Adrenoceptor Antagonists
- Cannabinoids and Endocannabinoids
- Cocaine
- Corticosteroid Receptors
- Corticotropin-Releasing Factor
- Depression: Animal Models
- Glucocorticoid Hormones
- Opioids
- SSRIs and Related Compounds
- Stress: Influence on Relapse to Substance Use
- Trier Social Stress Test

References

Binder EB, Kunzel HE, Nickel T, Kern N, Pfennig A, Majer M, Uhr M, Ising M, Holsboer F (2009) HPA-axis regulation at in-patient admission is associated with antidepressant therapy outcome in male but not in female depressed patients. Psychoneuroendocrinology 34(1):99–109

Chrousos GP, Gold PW (1992) The concepts of stress and stress system disorders. JAMA 267(9):1244–1252

Daskalakis NP, Bagot RC, Parker KJ, Vinkers CH, de Kloet ER (2013) The three-hit concept of vulnerability and resilience: toward understanding adaptation to early-life adversity outcome. Psychoneuroendocrinology 38(9):1858–1873

de Kloet ER, Joëls M, Holsboer F (2005) Stress and the brain: from adaptation to disease. Nat Rev Neurosci 6(6):463–475

DeRijk RH (2009) Single nucleotide polymorphisms related to HPA axis reactivity. Neuroimmunomodulation 16:340–352

Dickerson SS, Kemeny ME (2004) Acute stressors and cortisol responses: a theoretical integration and synthesis of laboratory research. Psychol Bull 130(3):355–391

Donaldson ZR, Young LJ (2008) Oxytocin, vasopressin and the neurogenetics of sociality. Science 322:900–904

Herman JP, Figueiredo H, Mueller NK, Ulrich-Lai Y, Ostrander MM, Choi DC, Cullinan WE (2003) Central mechanisms of stress integration: hierarchical circuitry controlling hypothalamo-pituitary-adrenocortical responsiveness. Front Neuroendocrinol 24(3):151–180

Hill MN, Tasker JG (2012) Endocannabinoid signaling, glucocorticoid-mediated negative feedback and regulation of the HPA axis. Neuroscience 204:5–16

Holsboer F (2000) The corticosteroid receptor hypothesis of depression. Neuropsychopharmacology 23:477–501

Ising M, Zimmermann US, Kunzel HE, Uhr M, Foster AC, Learned-Coughlin SM, Holsboer F, Grigoriadis DE (2007) High-affinity CRF1 receptor antagonist NBI-34041: preclinical and clinical data suggest safety and efficacy in attenuating elevated stress response. Neuropsychopharmacology 32(9):1941–1949

Joëls M, Sarabdjitsingh RA, Karst H (2012) Unraveling the time domains of corticosteroid hormone influences on brain activity: rapid, slow and chronic modes. Pharmacol Rev 64:901–938

Kirschbaum C, Pirke KM, Hellhammer DH (1993) The 'Trier Social Stress Test' – a tool for investigating psychobiological stress responses in a laboratory setting. Neuropsychobiology 28(1–2):76–81

Lightman SL, Conway-Campbell BL (2010) The crucial role of pulsatile activity of the HPA axis for continuous dynamic equilibrium. Nat Rev Neurosci 11(10):710–718

Meyer JS, Novak MA (2012) Hair cortisol: a novel biomarker of hypothalamic-pituitary-adrenocortical activity. Endocrinology 153(9):4120–4127

Steckler T (2010) Developing small nonpeptidergic drugs for the treatment of anxiety disorders: is the challenge still ahead? Curr Top Behav Neurosci 2:415–428

Yehuda R (2002) Post traumatic stress disorder. N Engl J Med 346(2):108–114

Neuroethics

Definition

Neuroethics is the study of the ethical, legal, and social questions that arise when scientific findings about the brain are carried into medical practice, legal interpretations, and health and social policy. Neuroethics is a subfield within the broader domain of bioethics, which encompasses the ethical and moral implications of all biological and medical advances. Neuroethics was established to address the rapid developments within cognitive neuroscience and neuropsychiatry and addresses findings relating specifically to the sciences of the mind encompassing the central nervous system and the underlying brain mechanisms of human behavior.

In response to the advances in cognitive neuroscience and neuropsychiatry and their increasing potential for broader application in the "real world," and in part to the use of pharmacological cognitive enhancers in healthy individuals, the Neuroethics Society (www.neuroethicssociety.org) was established.

Cross-References

▶ Ethical Issues in Human Psychopharmacology
▶ Pharmaceutical Cognitive Enhancers: Neuroscience and Society

Neurofibrillary Tangles

Definition

Intraneuronal pathological fibers consisting of neurofilament and neurotubules based on hyperphosphorylation of cytoplasmic tau protein.

Neurogenesis

João M. Bessa[1,2], Osborne F. X. Almeida[3] and Nuno Sousa[1,2]
[1]Life and Health Science Research Institute (ICVS), School of Health Sciences, University of Minho, Braga, Portugal
[2]ICVS/3B's – PT Government Associate Laboratory, Braga/Guimarães, Portugal
[3]Neuroadaptations Group, Max Planck Institute of Psychiatry, Munich, Germany

Synonyms

Adult neurogenesis; Postnatal neurogenesis

Definition

Neurogenesis refers to the production of new neurons in the brain; it is a complex process that begins with the division of a precursor cell and ends with the formation of a fully differentiated, functioning neuron.

Impact of Psychoactive Drugs

Neurogenesis
Postnatal cell proliferation in the brain of mammals has been occasionally reported during the first half of the twentieth century; however, this phenomenon only gained momentum in the 1960s (Altman 1962). By then, it was not clearly established if these newly born cells could develop into functional neurons, and the potential importance of the original observations was not fully appreciated. In the 1990s, however, a clear-cut demonstration of neurogenesis in the adult brains of various species (from rodents to humans) was provided (Manganas et al. 2007); it was later shown that after maturation and differentiation, newly generated neurons are functionally integrated into preexisting neuronal circuits (most notably in the hippocampus). In the subventricular zone, neurogenesis gives rise to neurons that migrate through the rostral migratory stream and which are integrated as interneurons in the olfactory bulb.

How Is Neurogenesis Measured?
Early studies used [^{3}H]-thymidine, which incorporates into replicating DNA during the S-phase of the cell cycle, to label dividing cells by autoradiography. An important technical improvement was the introduction of the synthetic thymidine analog BrdU (5-bromo-3-deoxyuridine) that substitutes for thymidine in the newly synthesized DNA of proliferating cells. BrdU incorporated into DNA can then be easily visualized with immunocytochemical techniques using specific antibodies against BrdU. This technique allows quantitative analysis of proliferation, migration, differentiation, and survival of newborn cells by varying the time interval between the pulse administration of BrdU and the sacrifice of animals. Other specific antigenic or histochemical markers are usually combined with BrdU immunocytochemistry to reveal cell phenotypes, differentiation state, and survival. While BrdU labeling is the most commonly used method for studying adult neurogenesis, other markers, such as proliferating nuclear antigen (PCNA) and Ki-67, may substitute or complement analysis by BrdU labeling.

Neurogenesis, Depression, and Antidepressant Treatments
Several thousand new cells are added to the adult rodent hippocampus each month, although the

rate of constitutive neurogenesis declines with age. At the same time, however, there is a parallel loss of cells in this brain region through the process of apoptosis; therefore, neuronal turnover is likely to be of greater relevance than either neurogenesis or apoptosis alone.

While early studies in songbirds suggested a functional role for adult neurogenesis in seasonal song learning, the functional importance of this process in mammals remains unclear (Shors et al. 2002). However, it is a highly regulated process through the actions of hormones, growth factors, neurotransmitters, and environmental factors. In particular, glucocorticoids (e.g., cortisol) exert a strong negative influence on neurogenesis, an effect that may explain the marked reduction in hippocampal granule cell proliferation after exposure to stress; on the other hand, antidepressants stimulate hippocampal neurogenesis and can reverse the effects of stress. Since many depressed subjects display glucocorticoid hypersecretion (as a result of HPA axis dysregulation), a link between high levels of circulating adrenal steroids with reduced neurogenesis and depression is plausible. This view has been supported by observations of hippocampal atrophy in depressed patients (Sheline et al. 2003).

In contrast to adrenal steroids and stress, antidepressant drugs (including tricyclics and selective serotonin reuptake inhibitors SSRIs), as well as electroconvulsive therapy, lead to increased neurogenesis in the hippocampus in association with decreased depression-like behavior in animal models (Malberg et al. 2000). Altered monoaminergic transmission provides the best mechanistic explanation for these changes; two neurotransmitters of particular relevance are noradrenaline (NA) and serotonin (5-HT), the latter being a major regulator of adult neural cell proliferation. In adult rats, d,l-fenfluramine, which releases 5-HT throughout the central nervous system, has a significant stimulatory effect on hippocampal dentate granule cell division, whereas reductions of 5-HT by lesioning 5-HT neurons with 5,7-dihydroxytryptamine (5,7-DHT) or the inhibition of 5-HT synthesis with parachlorophenylalanine (PCPA) results in long-term disruption of the proliferative capacity of the hippocampus. The neurogenic actions of 5-HT are mediated primarily by 5-HT_{1A} receptors but also by 5-HT_{2A} and 5-HT_{2C} receptors. Additionally, the recently introduced antidepressant, agomelatine, which is a mixed MT1/MT2 melatonin receptor agonist and $5\text{-HT}_{2B/2C}$ receptor antagonist, also increases hippocampal neurogenesis and can reverse the anti-neuroproliferative effects of prenatal stress. Likewise, stress-induced reductions in neurogenesis can be reversed with corticotropin-releasing factor type I receptor (CRF-R1) and arginine-vasopressin V1b receptor (AVP-1bR) antagonists, consistent with their ability to suppress adrenocortical secretion. However, the antidepressant clinical efficacy of these drugs remains to be established. Finally, the neurokinin type 1 receptor (NK1R) antagonists, which have been shown to increase neurogenesis in animal models, have failed to reveal significant antidepressant efficacy in clinical studies. Besides the antidepressants mentioned above, the atypical antidepressant tianeptine and the mood stabilizer lithium have also been found to increase proliferation and survival of new neurons in the dentate gyrus.

Together, the observations described, especially those derived from validated animal models of depression ("▶ Depression: Animal Models"), form the backbone of the notion that the therapeutic actions of antidepressants in patients depend on the generation of new hippocampal neurons. Irradiation of the hippocampus of naive ("nondepressed") mice abolishes the ability of antidepressants to relieve signs of anxiety behavior (Santarelli et al. 2003); anxiety is one behavioral element commonly seen in depressed patients but does not fully reflect the depressed state. However, other experiments in which hippocampal neurogenesis was blocked by chemical means clearly indicate that a wide range of antidepressants can improve depression-like behavior in an appropriate model of the disease (rats exposed to a chronic mild stress paradigm) even in the absence of ongoing neurogenesis (Bessa et al. 2009). However, recent evidences using these models suggest that newborn hippocampal neurons are necessary for the long-term behavioral effects of antidepressants (Mateus-Pinheiro et al. 2013).

If Not Neurogenesis, Then What?

Although the exact mechanisms by which stress and glucocorticoids on one hand, and antidepressants and serotonin on the other, affect neurogenesis have not been completely elucidated, there is evidence that implicates changes in the expression of cAMP response element binding (CREB) protein and, in turn, of brain-derived neurotrophic factor (BDNF). BDNF plays key roles in the survival and guidance of neurons during development and is required for the survival and normal functioning of neurons in the adult brain. Stress triggers a decrease in brain levels of BDNF, thus potentially compromising neuroplasticity and neuronal development and survival. Since all current antidepressant treatments (drugs and electroconvulsive therapy) upregulate CREB and BDNF expression, it can be inferred that the cAMP-CREB cascade and BDNF are common targets of both stress (through elevated glucocorticoid secretion) and antidepressant treatments, with the opposing stimuli acting to influence neuroplasticity.

To gain an insight into how stress and antidepressants act, it is important to note that their effects are not restricted to the regulation of adult neurogenesis. Exposure to chronic stress induces significant atrophy, debranching, and synaptic reorganization of neurons in all divisions of the hippocampus, changes that negatively correlate with hippocampus-dependent cognitive functions such as learning and memory. Further, the changes in dendritic and synaptic plasticity correlate with decreased expression of neural cell adhesion molecule (NCAM) and synaptic-related proteins such as synapsin 1 and are reversible with a variety of antidepressant drugs. Importantly, some of the stress-induced alterations in hippocampal plasticity propagate to other brain regions; for example, chronic stress interferes with dendritic and synaptic structure of pyramidal neurons in the prefrontal cortex in association with the onset of deficits in working memory and behavioral flexibility. In parallel, other brain regions show signs of hyperactivity, consistent with the consensual view that the cluster of depressive symptoms results from a "push-and-pull" interplay between different brain regions. For example, anhedonia, a core symptom of depression that is characterized by decreased interest in pleasurable activities or the inability to experience pleasure, has been associated with a dysfunction of the brain reward pathway. Dopamine (DA) released from cell bodies in the ventral tegmental area (VTA) at terminals in the nucleus accumbens (NAc) plays a central role in this pathway, and dysfunctional or suboptimal DA transmission has been implicated in the pathophysiology of depression. Altered dopaminergic neurotransmission in the VTA-NAc pathway modulates depressive-like behavior in animals, and antidepressants influence dopaminergic activity in the VTA and its targets, including the NAc. Consistent with this, antidepressants can reverse stress-induced neuroplastic changes in the NAc (Bessa et al. 2013).

Neurogenesis, Schizophrenia, and Antipsychotics

Psychopharmacological interest in neurogenesis extends beyond depression and the actions of antidepressant drugs to other psychiatric disorders, as recently evidenced by studies suggesting a role of disturbed neurogenesis in the pathogenesis of schizophrenia (Reif et al. 2006) and its modulation by antipsychotic drugs. Patients with schizophrenia show reduced volumes of the frontal lobes as well as of the medial temporal lobes, in particular of the hippocampus, with the dentate gyrus displaying a reduced number of cells with the potential to proliferate. Further, preclinical studies indicate that chronic antipsychotic treatment can stimulate neurogenesis both in the subgranular zone of the hippocampal dentate gyrus and in the subventricular zone. Neurogenesis in the subventricular zone is stimulated by the first-generation antipsychotic drug ("▶ First-Generation Antipsychotics") haloperidol, a strong D_2 receptor antagonist. In contrast, the second-generation antipsychotics such as olanzapine, risperidone, clozapine, and quetiapine induce neurogenesis mainly in the hippocampus (Newton and Duman 2007), thus resembling the actions of antidepressants. Interestingly, the second-generation drugs are poorer D_2 receptor antagonists, as compared to haloperidol, but have affinity for 5-HT receptors. Since treatment with atypical antipsychotics improves neurocognitive function, it seems likely that hippocampal neurogenesis may play a role in the amelioration of the cognitive symptoms in schizophrenia.

Conclusions

Neurogenesis occurs during immediate postnatal life, right through adulthood, although at an ever-diminishing rate. However, the relevance of neurogenesis in the actions of psychoactive drugs and the etiopathophysiology of psychiatric illnesses remains uncertain. Evidence that neurogenesis occurs in parallel with the apoptotic loss of neurons suggests that neuronal turnover and accompanying changes in neuroplasticity may be more important than neurogenesis per se. Although hippocampal neurogenesis does not appear to play a role in the early mood-improving actions of antidepressants, neurogenesis may be essential for the sustained behavioral effects of these drugs and might serve to rectify behaviors in associated behavioral domains such as anxiety and cognition.

Cross-References

- Antidepressants
- Antipsychotic Drugs
- Apoptosis
- Arginine-Vasopressin
- Brain-Derived Neurotrophic Factor
- Corticotropin-Releasing Factor
- Depression: Animal Models
- First-Generation Antipsychotics
- Lithium
- Schizophrenia
- Second-Generation Antipsychotics
- SSRIs and Related Compounds
- Synaptic Plasticity

References

Altman J (1962) Are new neurons formed in the brains of adult mammals? Science 135:1128–1129

Bessa JM, Ferreira D, Melo I, Marques F, Cerqueira JJ, Palha JA, Almeida OFX, Sousa N (2009) The mood-improving actions of antidepressants do not depend on neurogenesis but are associated with neuronal remodeling. Mol Psychiatry 14:764–773

Bessa JM, Morais M, Marques F, Pinto L, Palha JA, Almeida OFX, Sousa N (2013) Stress-induced anhedonia is associated with hypertrophy of medium spiny neurons of the nucleus accumbens. Transl Psychiatry 4(3):e266

Malberg JE, Eisch AJ, Nestler EJ, Duman RS (2000) Chronic antidepressant treatment increases neurogenesis in adult rat hippocampus. J Neurosci 20:9104–9110

Manganas LN, Zhang X, Li Y, Hazel RD, Smith SD, Wagshul ME, Henn F, Benveniste H, Djuric PM, Enikolopov G, Maletic-Savatic M (2007) Magnetic resonance spectroscopy identifies neural progenitor cells in the live human brain. Science 318:980–985

Mateus-Pinheiro A, Pinto L, Bessa JM, Morais M, Alves ND, Monteiro S, Patrício P, Almeida OFX, Sousa N (2013) Sustained remission from depressive-like behavior depends on hippocampal neurogenesis. Transl Psychiatry 3:e210, 15

Newton SS, Duman RS (2007) Neurogenic actions of atypical antipsychotic drugs and therapeutic implications. CNS Drugs 21:715–725

Reif A, Fritzen S, Finger M, Strobel A, Lauer M, Schmitt A, Lesch KP (2006) Neural stem cell proliferation is decreased in schizophrenia, but not in depression. Mol Psychiatry 11:514–522

Santarelli L, Saxe M, Gross C, Surget A, Battaglia F, Dulawa S, Weisstaub N, Lee J, Duman R, Arancio O, Belzung C, Hen R (2003) Requirement of hippocampal neurogenesis for the behavioral effects of antidepressants. Science 301:805–809

Sheline YI, Gado MH, Kraemer HC (2003) Untreated depression and hippocampal volume loss. Am J Psychiatry 160:1516–1518

Shors TJ, Townsend DA, Zhao M, Kozorovitskiy Y, Gould E (2002) Neurogenesis may relate to some but not all types of hippocampal-dependent learning. Hippocampus 12:578–584

Neuroimaging

Synonyms

Brain imaging; Brain mapping

Definition

Neuroimaging is a family of techniques used for obtaining images of the structure or the function of the human brain. Neuroimaging studies investigate structural and functional brain maturation in health and in diseases of the brain, such as schizophrenia, bipolar disorder, depression, attention deficit hyperactivity disorder (ADHD), drug dependence, and autism. These studies often aim to find genetic and environmental factors which might affect brain structure and function over time. Through additions to diagnostic radiology, neuroimaging has broadened to become a distinct field in neuroscience. The neuroimaging techniques that are currently used in research include single-photon emission computerized tomography (SPECT), positron emission tomography (PET), and distinct forms of magnetic resonance imaging (MRI), including structural and functional MRI, MR spectroscopy (MRS), and diffusion tensor imaging (DTI).

Cross-References

▶ Magnetic Resonance Imaging (Functional)
▶ Positron Emission Tomography (PET) Imaging
▶ SPECT Imaging

Neuroinformatics

Definition

Biophysical rules and mechanisms by which neuronal networks gather, represent, store, integrate, transform, and implement information for behavioral output. Understanding information representation and computational processing as a biophysical product of neural tissue is an emerging field of neuroscience that relies on a non-traditional integration of biological, mathematical, engineering, and computer science fields. Aims of this area of research are to define the information-processing and learning and memory capabilities and limits of biological neural networks and to characterize how physical mechanisms that integrate gene-environment interactions result in changes in neural information processing.

Neurokinin A

Synonyms

Neurokinin 2; Neurokinin-α; Neuromedin L; NK2; NKA; Substance K

Definition

NKA is a neuropeptide from the tachykinin family similar in structure to SP and generated by alternative splicing from the same preprotachykinin gene as SP. NKA is mostly expressed in smooth muscles and in discrete areas of the central nervous system implied in mood disorders. NKA binds and activates the NK2 receptor.

Neurokinin B

Synonyms

Neurokinin 3; Neuromedin K; NK3; NKB

Definition

NKB is a ten amino acid neuropeptide from the tachykinin family generated by alternative splicing from the same preprotachykinin gene as substance P. It is present in smooth muscles and widely expressed in the central nervous system. NKB binds and activates the NK3 receptor.

Neuroleptic Malignant Syndrome

Synonyms

NMS

Definition

A severe, potentially lethal adverse effect of antipsychotic treatment of unclear pathophysiology. Features include muscle rigidity, fever, altered consciousness, and autonomic symptoms such as dysregulations of blood pressure and respiration. Most patients also show elevations of creatine phosphokinase levels. NMS represents a psychiatric emergency. Treatment, next to intensive care measures, includes dopaminergic drugs such as bromocriptine as well as dantrolene.

Cross-References

▶ Antipsychotic Drugs

Neuropeptide Y

Definition

Orexigenic peptide produced primarily by cells in the hypothalamic arcuate nucleus. This peptide potently increases appetite and decreases energy expenditure via its actions in the hypothalamus. In the hippocampus and cortex, neuropeptide Y is associated with antiepileptic effects.

Neuropeptides

Definition

Neuropeptides are substances with a peptide structure that are synthetized in nervous tissue and used there as messenger molecules. Some neuropeptides fulfill all criteria for a neurotransmitter and may have additional roles as neuromodulators or growth factors.

Neuropeptidomics

Synonyms

Peptidomics of the brain

Definition

Neuropeptidomics is a technological approach for detailed analyses of endogenous peptides from the brain.

Cross-References

▶ Electrospray Ionization
▶ Imaging Mass Spectrometry
▶ Mass Spectrometry
▶ Matrix-Assisted Laser Desorption Ionization
▶ Posttranslational Modification

Neuroplasticity

Definition

A general term referring to changes in information content and processing within neural networks, instantiated by a large variety of biophysical changes of constituent neurons embracing molecular to cellular to higher scales of observation. Changes in neuronal gene-expression correspond to phenotypic changes, including, but not limited to, alterations in neuronal excitability, neurotransmitter release and receptor densities, location and number of axodendritic synaptic contacts, and branching morphology of axons and dendrites. Net up or down changes in neuron-to-neuron synaptic strength are characterized electrophysiologically by long-term potentiation (LTP) and long-term depression (LTD). These changes underlie an inherent structural and functional flexibility (i.e., plasticity) of neurons and neurocircuits brought on by a combination, and interaction, of developmentally timed patterns of gene expression and environmental stimuli. Incorporating all the physical processes and learning and memory capabilities of brain systems, neuroplasticity seeks to achieve the most adaptive mapping of behavioral programming on present and future conditions of the external world.

Neuroprotectants: Novel Approaches for Dementias

Kenneth J. Rhodes
Discovery Neurobiology, Biogen Idec, Cambridge, MA, USA

Definition

Neuroprotection can be defined as a pharmacological activity that promotes the survival and function of neurons. Drugs that offer neuroprotection may be used to slow or prevent loss of brain function in a neurodegenerative disease.

Pharmacological Properties

Dementia can be defined as a loss of memory function accompanied by impairments in other cognitive domains, including language, decision making, object recognition, spatial navigation, and other functions (American Psychiatric Association 2000). In order for a patient to be classified as demented, their impairments in memory and other cognitive abilities must be sufficient to affect routine aspects of daily life. Many types of dementia are caused by progressive, age-related neurochemical and neurodegenerative changes in areas of the brain required to support memory and other cognitive functions. As a result, efforts in drug discovery and development have focused on the identification of neuroprotective drugs as a means to reduce or prevent neurodegeneration and thereby slow or reverse cognitive impairment and dementia. Although many diseases of the central nervous system are accompanied by dementia, by far the most prevalent and devastating cause of dementia worldwide is Alzheimer's disease (AD). Because the incidence and prevalence of AD are increasing rapidly as the world's population ages, strategies for preventing or reversing the neurodegenerative changes of AD have become areas of intense focus for academic and industry research and will be the focus of this brief review.

The pathophysiology of AD has been the subject of intensive study for the past several decades. Tremendous progress in our understanding of disease mechanisms has been stimulated by the integration of genetics, histopathology, biochemistry, and animal models. As a comprehensive review of the scientific literature on AD pathophysiology is well beyond the scope of this encyclopedia, this review will focus on neuroprotective strategies centered on the role of amyloid-β (Aβ), a 38–43 amino acid peptide fragment derived from the amyloid precursor protein (APP), and its contributions to the cognitive impairment and neurodegeneration of AD. Aβ is the core molecular component of the neuritic plaque, one of the hallmark neuropathological features of AD. The production and accumulation of Aβ is thought to contribute directly to defects in neuronal communication, amyloid plaque deposition, neurodegeneration, and cognitive impairment in AD.

The Aβ peptide is generated as a result of a two-step enzymatic cleavage of transmembrane APP. The first enzymatic cleavage, which generates the amino-terminus of Aβ and liberates a 99-amino acid C-terminal APP fragment (C-99), is termed β-site APP cleavage enzyme (β-secretase, BACE). The second enzymatic cleavage, which cuts within the C-99 fragment and generates the carboxyl-termini of Aβ, is mediated by a multi-subunit, integral membrane aspartyl protease termed γ-secretase (GS).

In AD patients and normal healthy adults, there are two predominant forms of the Aβ peptide: Aβ40 and Aβ42. Aβ42 is less soluble than Aβ40 and rapidly and spontaneously assembles into soluble dimmers, fibrils, and insoluble higher molecular weight aggregates, which likely deposit in brain tissue as amyloid plaques. Mutations in APP that enhance cleavage by BACE are associated with increased Aβ production and early onset, familial AD. Mutations of GS or APP that enhance the production of Aβ, and specifically increase the production of Aβ42, are also associated with early onset, familial AD. Recently, dimers and higher molecular weight multimers of Aβ42, produced in vitro or isolated from AD brain tissue, have been shown to directly impair synaptic function and cause neuronal death in vitro (Shankar et al. 2008). These dimers and multimers isolated from human AD brain have also been shown to impair memory processes when infused directly into the brains of experimental animals. Together, there is now a wealth of preclinical data to indicate that the cascade of events, triggered by the formation of Aβ42, evokes detrimental changes in brain physiology leading to clinical symptoms of dementia as well as progressive neuropathology and neurodegeneration in AD.

The precise mechanisms whereby dimers and multimers (hereafter termed oligomers) of Aβ cause neurotoxicity are not well understood. Proposed mechanisms include the activation of specific cell surface receptors, including the direct or allosteric modulation of ion channels. These channels, when stimulated, are thought to allow the entry of sodium and calcium ions into the neuron, thereby activating proteases, kinases, and other signaling molecules. The signaling events evoked by the binding of Aβ oligomers to the neuronal surface may lead to the removal of neurotransmitter receptors from synapses, thereby impairing neuronal communication, or may trigger processes that lead to neuronal injury or death (Kamenetz et al. 2003).

The primary goals of neuroprotective strategies in AD are to prevent the formation of Aβ peptides or increase their rate of removal from the brain. To this end, several companies are pursuing strategies to inhibit or modulate BACE or GS. Of these two strategies, GS inhibitors (GSIs) have progressed furthest into clinical development, with potent and selective GSIs (LY-450139, BMS-788163) currently in later stages of clinical development (see Harrison et al. 2004). GSIs potently and dramatically inhibit GS activity and thereby reduce the production of Aβ peptides by blocking the cleavage of the C-99 fragment of APP. GSIs have been shown to improve cognitive function in transgenic mice that overexpress mutant human APP. However, these compounds typically suffer from dose-limiting toxicities that may stem from inhibiting the cleavage of GS substrates other than APP, such as Notch. Notch is a cell surface receptor that, upon cleavage by GS, releases an intracellular C-terminal fragment that regulates a gene expression program critical for cell fate decisions in the gastrointestinal tract and in the immune system. A crucial component of GSI development is finding dose levels or a dose regimen that provides sufficient GS inhibition to reduce Aβ formation and AD pathology without causing intolerable digestive or immunological toxicity. Currently there is little evidence to indicate that GSIs reduce Aβ-mediated toxicity in vitro or in transgenic models of AD in vivo. Nevertheless, there is a hypothesis that GSIs, by their ability to halt Aβ production, will prevent progressive neurodegeneration and preserve cognitive function in AD patients. This hypothesis is currently being evaluated in large, multinational clinical trials.

BACE inhibitors have also been the focus of intense drug discovery activity (see Ghosh et al. 2008). On the basis of data from BACE1 knockout mice, a deletion of the BACE1 gene nearly completely suppresses Aβ production in vivo. Over the last several years, many pharmaceutical companies and academic laboratories have been pursuing BACE inhibitors as potential disease-modifying therapies for AD. Although many potent and selective BACE inhibitors have been identified, these compounds typically suffer from poor oral absorption or are substrates for transport molecules that efficiently clear drugs out of the central nervous system, resulting in insufficient concentrations of drug in the brain to effectively inhibit BACE. Thus far, CTS-21166 has progressed into early phases of clinical development and has shown pharmacological activity by reducing plasma and CSF Aβ levels. Additional clinical testing is required to evaluate the ability of this or other BACE inhibitors to ameliorate the cognitive decline and neurodegeneration in AD.

Another strategy to provide neuroprotection from toxicity mediated by Aβ is to neutralize or clear Aβ peptides once they are formed or deposited in amyloid plaques. This approach stems from the observation that transgenic animals expressing mutant human APP, which develop plaque-like Aβ deposits with increasing age, show a reduced accumulation of Aβ and a reduced plaque density following immunization with aggregates of Aβ or when they are treated systemically with antibodies that recognize Aβ peptides. This observation stimulated active and passive immunotherapy approaches targeting Aβ removal in AD patients. A clinical trial of active vaccination with aggregated Aβ42 (AN1792) was halted due to development of aseptic meningoencephalitis in some trial participants, but long-term follow-up of trial subjects revealed several important findings (Vellas et al. 2009). First, patients who developed strong antibody titers to Aβ42 aggregates, and specifically antibodies that recognized aggregated Aβ in plaques, showed a reduced rate of cognitive decline as compared to patients who did not develop such antibodies (Hock et al. 2003). Moreover, vaccine responders in the AN1792 study, followed up to 4 years, showed significantly reduced cognitive decline. Second, when their brains were examined after death, patients in the AN1792 study who developed sustained anti-Aβ titers showed unexpectedly low densities of Aβ plaques (Vellas et al. 2009). Together, these findings suggest that antibodies to Aβ can affect amyloid pathology in the AD brain and reduce the severity of the clinical course in AD. Whether this strategy reduced the overall rate of neurodegeneration, as evidenced by the preservation of brain gray matter volume, remains to be determined.

Although the AN1792 trial was halted, strong interest remains in exploring passive immunotherapy using monoclonal antibodies that target specific amino acid sequences within the Aβ peptide (see Nitsch and Hock 2005). In experimental animals, passive immunotherapy with anti-Aβ antibodies dramatically reduces amyloid pathology, improves behavioral performance, and seems to prevent neurodegeneration. Several humanized anti-Aβ antibodies, including Elan/Wyeth's bapineuzumab and Eli Lilly and Company's solanezumab, are now in Phase III clinical trials as disease-modifying therapies for AD. The outcome of these large, multinational trials will provide incredibly valuable information about the role of Aβ in neurodegenerative processes and cognitive impairment in AD.

Another approach to AD targeting Aβ involves the modulation of GS activity with the goal of selectively reducing the formation of Aβ42. Compounds that selectively reduce Aβ42 are called GS modulators (GSMs) because they reduce the generation of Aβ42 without affecting the total amount of Aβ that is produced. In contrast to GSIs, which reduce the production of all Aβ isoforms (and inhibit cleavage of all GS substrates), GSMs typically reduce the production of Aβ42, have no effect on Aβ40, and increase production of nonpathogenic Aβ38 roughly in proportion to the decrease in Aβ42. Although it is presumed that GSMs work by binding to an allosteric site on the GS enzyme complex, the binding site for GSMs and their mechanism of action have not been clearly elucidated. One potential advantage of GSMs over GSIs is that GSMs may avoid the toxicities associated with GS inhibition. GSMs alter the ratios of cleavage products that result from GS activity but do not inhibit GS activity completely. In the case of Notch, for example, although the site of Notch cleavage may be altered by a GSM, Notch C-terminal fragment is still generated and its ability to signal is not affected.

The first prototype GSM to enter large-scale clinical development was R-flurbiprofen (Flurizan). Although this molecule has very weak GSM activity in vitro ($IC_{50} > 150$ μM) and penetrates the brain poorly, it was shown to reduce Aβ42 production in experimental animals and was advanced into clinical development (see Wilcock et al. 2008). Although a large Phase III clinical trial of R-flurbiprofen failed to demonstrate efficacy, considerable interest remains in developing more potent GSMs that achieve brain tissue concentrations sufficient for the effective and selective reduction of Aβ42. Several companies and academic laboratories are currently working to achieve this goal.

Other approaches to neuroprotection in dementia that are currently under active investigation in clinical trials are the disruption of Aβ fibrils (e.g., PBT2 – a chelator of divalent metal ions that seems to disrupt Aβ production and fibril formation) and the inhibition of excitotoxicity by blocking excitatory amino acid receptors (e.g., memantine, Dimebon), preventing neuronal apoptosis by preserving mitochondrial function (e.g., Dimebon, otherwise called latrepirdine or Pf-01913539). There is generally a limited amount of published preclinical data from animal models to substantiate the activity of these novel agents as neuroprotective therapies in AD. Results and commentary about the latest research and clinical trial findings in Alzheimer's disease can be found at www.alzforum.org.

Future Directions

Exciting avenues for future research on neuroprotectants for dementia and AD center on the identification of a cell surface receptor or receptors for Aβ oligomers and an elucidation of the signaling events that link Aβ binding to neuronal membranes with altered activity and morphology of synapses (Kamenetz et al. 2003; Shankar et al. 2008). A better understanding of these pathways may allow for the generation of therapies that specifically target the impaired synaptic function and early structural defects of AD. It seems likely that the modulation of these fundamental processes in synaptic transmission and memory may also provide novel therapeutic approaches to preserving cognitive function across many types of dementias.

References

American Psychiatric Association (2000) Diagnostic and statistical manual of mental disorders, 4th edn, text revision. American Psychiatric Association, Washington, DC

Ghosh AK, Gemma S, Tang J (2008) Beta-secretase as a therapeutic target for Alzheimer's disease. Neurotherapeutics 5(3):399–408

Harrison T, Churcher I, Beher D (2004) Gamma-secretase as a target for drug intervention in Alzheimer's disease. Curr Opin Drug Discov Dev 5:709–719

Hock C, Konietzko U, Streffer JR, Tracy J, Signorell A, Müller-Tillmanns B, Lemke U, Henke K, Moritz E, Garcia E, Wollmer MA, Umbricht D, de Quervain DJ, Hofmann M, Maddalena A, Papassotiropoulos A, Nitsch RM (2003) Antibodies against beta-amyloid slow cognitive decline in Alzheimer's disease. Neuron 38(4):547–554

Kamenetz F, Tomita T, Hsieh H, Seabrook G, Borchelt D, Iwatsubo T, Sisodia S, Malinow R (2003) APP processing and synaptic function. Neuron 37(6):925–937

Nitsch RM, Hock C (2005) Targeting beta-amyloid pathology in Alzheimer's disease with Abeta immunotherapy. Neurotherapeutics 5(3):415–420

Shankar GM, Li S, Mehta TH, Garcia-Munoz A, Shepardson NE, Smith I, Brett FM, Farrell MA, Rowan MJ, Lemere CA, Regan CM, Walsh DM, Sabatini BL, Selkoe DJ (2008) Amyloid-beta protein dimers isolated directly from Alzheimer's brains impair synaptic plasticity and memory. Nat Med 14:837–842

Vellas B, Black R, Thal LJ, Fox NC, Daniels M, McLennan G, Tompkins C, Leibman C, Pomfret M, Grundman M, AN1792 (QS-21)-251 Study Team (2009) Long-term follow-up of patients immunized with AN1792: reduced functional decline in antibody responders. Curr Alzheimer Res 6:144–151

Wilcock GK, Black SE, Hendrix SB, Zavitz KH, Swabb EA, Laughlin MA, Tarenflurbil Phase II Study investigators (2008) Efficacy and safety of tarenflurbil in mild to moderate Alzheimer's disease: a randomised phase II trial. Lancet Neurol 7:483–493

Neuroprotection

A. Richard Green[1] and Maria Isabel Colado[2]
[1]School of Life Sciences, Queen's Medical Centre, University of Nottingham, Nottingham, UK
[2]Departamento de Farmacologia, Facultad de Medicina, Universidad Complutense, Madrid, Spain

Definition

The term neuroprotection is used to refer to either an endogenous mechanism within the central nervous system which protects neurons from cell death or, more commonly, the action of a drug to prevent cell death following the action of either a neurotoxic drug or the consequences of brain injury.

Current Concepts and State of Knowledge

Mechanisms Involved in Cell Death

Neurons can die by one of two distinct mechanisms. One is programmed cell death or apoptosis, the other is necrosis. Both types of cell death have been reported to occur following a variety of insults to the brain (Higgins et al. 2010). Insults to the integrity of the brain can be chemical, for example, neurotoxins, or traumatic, for example, stroke. Both chemical and traumatic insults result in neurodegeneration ("▶ Neurodegeneration and Its Prevention").

Neuroprotection Following Stroke and Traumatic Brain Injury

Stroke
Tissue damage following a stroke can be mitigated either by restoring blood flow with a thrombolytic ("clot busting") drug or by interfering with the biochemical changes which occur following the ischemic insult and thereby minimizing the damage. At present only the thrombolytic drug recombinant tissue plasminogen activator or rt-PA is in clinical use for the acute treatment of stroke (Green 2008).

Any compound which protects the brain from cell death following an acute ischemic stroke is known as a neuroprotective agent or neuroprotectant, and its mechanism of action is generally based on trying to interfere with the biochemical changes which occur in the brain following an acute ischemic insult (Green 2008). The biochemical chain of events following ischemia is often referred to as the ischemic cascade. The initial change that occurs during ischemia is a pathological release of glutamate, an excitatory amino acid ("▶ Excitatory Amino Acids and Their Antagonists"). This release produces further changes including depolarization of the cells, the production of free radicals, excitotoxicity, and cell death. It should also be noted that reperfusion, while mitigating neurodegeneration, also results in the production of free radicals and is thus subject to studies examining whether additional treatment with a neuroprotective agent might be advantageous.

Neuroprotection can be determined in various ways in animal models of acute stroke. Primarily, neuroprotection is measured by histological evidence that the size of the damaged tissue (infarct area) is smaller. This can be achieved by staining tissue or counting intact cell populations. Such studies should be followed up by measuring the degree of motor impairment in living animals. Clinically, the use of scales which quantify the degree of motor impairment is the generally accepted outcome measure.

Since the initial event following the onset of cerebral ischemia is glutamate release, considerable efforts have been made to provide neuroprotection by producing drugs that are antagonists at the glutamate receptor. Antagonists of both the glutamate NMDA (N-methyl-D-aspartate) receptor and the AMPA (α-amino-3-hydroxy-5-methyl-4-isoxazolepropionic acid) receptor have been examined in animal models of acute stroke. Some of these have also been investigated in clinical trials with stroke patients. Glutamate antagonist compounds include the NMDA-receptor antagonists dizocilpine, aptiganel, and selfotel. Clinical studies have also been conducted with the NMDA glycine-site antagonist gavestinel, the NMDA polyamine-site antagonist eliprodil, and the NMDA channel blocker magnesium. AMPA receptor antagonists such as 2,3-dioxo-6-nitro-1,2,3,4-tetrahydrobenzo[f]quinoxaline-7-sulfonamide (NBQX) have also been examined but less actively pursued, primarily because of safety concerns. All these compounds were effective in providing neuroprotection in animal models of stroke. However all failed in clinical trials, either because the trial had to be terminated early because of safety issues or because the compound failed to demonstrate efficacy (Green 2008).

Another approach to providing neuroprotection in stroke has been to administer compounds which increased the function of GABA, an inhibitory amino acid neurotransmitter ("▶ GABA"). Such approaches have included administration of the benzodiazepine ("▶ Benzodiazepines") diazepam or clomethiazole, a drug which increases GABA function at its receptor. Again, despite clear evidence for efficacy in animal models of acute stroke, both drugs failed to provide neuroprotection in patients as indicated by a decrease in stroke-induced disability.

Since a major consequence of both ischemia and reperfusion is excessive free-radical production, attempts have been made to provide neuroprotection in stroke as well as other degenerative conditions such as Parkinson's disease ("▶ Anti-Parkinson Drugs") and Alzheimer's disease ("▶ Dementias and Other Amnesic Disorders") by the use of compounds that scavenge or trap free radicals. In stroke research such compounds included ebselen, a selenium compound with glutathione peroxidase-like activity; the lazaroid compound tirilazad; and edaravone, a hydroxyl radical scavenger. Only edaravone has been claimed to be of clinical benefit but, at present, its use is restricted to Japan. One compound that showed considerable preclinical promise was the nitrone-derived compound disodium 2,4-disulphophenyl-N-tert-butylnitrone (NXY-059). However, it failed to demonstrate efficacy in late phase clinical trials.

The failure of compounds acting on the ischemic cascade to provide neuroprotection in stroke patients has led to an increase in approaches which interfere with the role of cytokines in cerebral ischemia. Both tumor necrosis factor alpha (TNFα) and interleukin-1 beta (IL-1β) have been shown to be involved in the inflammatory response to brain injury, and TNFα and IL-1β antibodies have been shown to be neuroprotective in experimental stroke studies in animals. A recombinant IL-1 receptor antagonist is being examined clinically. Due to the possible problems in producing and administering protein-based compounds, research is also being undertaken to produce non-peptide compounds (Green 2008).

A recent individual animal meta-analysis demonstrated unequivocal evidence of the efficacy of NXY-059 in preclinical models of stroke (Bath et al. 2009). In addition there was general agreement that the clinical trials on the compound had been well conducted. Consequently the failure of this particular compound in the clinic resulted in considerable discussion in the scientific press as to the reasons for the failure of this drug and every other experimental neuroprotective drug developed to demonstrate successful translation to clinical efficacy. This problem of translation has resulted in most pharmaceutical companies terminating their stroke research, and rather little work on neuroprotective drug approaches to stroke is now occurring (Minnerup et al. 2012).

Traumatic Brain Injury

Traumatic brain injury (or TBI) is the term used to describe injury to the brain caused by mechanically induced damage such as that resulting from impact to the skull induced by a car accident or by a firearm. The pathological changes seen include those which occur following stroke but also additional changes such as bleeding (hemorrhage) and mechanical damage to cerebral tissue due to impact. Several of the compounds developed to treat acute ischemic stroke have also been examined in TBI. Compounds such as aptiginal, the excitatory amino acid antagonist and IL-1β antibodies, have been shown to provide modest benefit in animal models of TBI, but no compound has yet proved to have efficacy in human TBI.

Neuroprotection and Neurotoxic Amphetamines

3,4-Methylenedioxymethamphetamine (MDMA or "ecstasy") is an amphetamine and a commonly used recreational drug, often ingested at warm and crowded dance clubs and raves. There is substantial evidence that the compound produces long-lasting neurotoxic changes in the brains of experimental animals. In the rat, MDMA, given as either a single high dose or several lower doses over a relatively short period of time, induces a long-term loss of 5-hydroxytryptamine (5-HT or serotonin) nerve terminals in several regions of the brain. This effect is reflected by a substantial decrease in the activity of tryptophan hydroxylase, the first (and rate limiting) enzyme responsible for the synthesis of 5-HT, and in the concentration of 5-HT and of 5-HT and its metabolite, 5-hydroxyindole acetic acid (5-HIAA). There is also a reduction in the number of 5-HT transporters and a decrease in the immunoreactivity of 5-HT axons in several brain areas (Green et al. 2003).

In contrast to its effects in rats, MDMA is generally accepted to behave as a selective dopamine neurotoxin in mice, inducing a long-term decrease in the concentration of dopamine and its metabolites, a reduction in the number of dopamine transporters, and a decrease in the immunoreactivity of tyrosine hydroxylase in striatum and substantia nigra (Granado et al. 2011). Neurotoxicity in mice is evident following repeated and higher doses of MDMA than those administered to rats.

Another substituted amphetamine, methamphetamine (METH), is also known to induce neurotoxic loss of dopamine axon terminals in the striatum and cell body in the substantia nigra of mice. The fact that direct administration of MDMA and METH into the brain does not produce toxicity suggests that the parent compounds are not responsible for the toxicity and that toxicity probably results from the effects of metabolic products (Esteban et al. 2001). This interpretation is now supported by recent studies on the cathinone compound mephedrone which shares many of the pharmacological properties of MDMA but which does not have free radical producing catechol or quinone metabolites and does not induce neurodegeneration in experimental animals.

The mechanisms producing this neurodegeneration are not totally understood at present, but evidence indicates that several factors play an essential role: acute hyperthermia, monoaminergic transporters, an oxidative stress process, strong microglia activation, and the release of pro-inflammatory cytokines such as IL-1β and TNFα (Torres et al. 2011). Some of these potential mechanisms are also almost certainly involved in ischemia-induced neurodegeneration.

There have been a variety of compounds which, when given concurrently with MDMA or METH, have been shown to provide protection. However, recent studies have indicated that a significant number of such compounds, including several with clear efficacy as neuroprotective agents in animal models of acute ischemic stroke, only prevent amphetamine-induced damage either because they prevent the acute hyperthermia which follows amphetamine administration (e.g., antipsychotics such as haloperidol) or because they induce hypothermia. Compounds in this latter class include barbiturates such as pentobarbitone and NMDA antagonists such as dizocilpine. Hyperthermia is one of the major features of acute MDMA toxicity in both rodents and humans, and its presence has been shown to markedly enhance the neurotoxic damage induced by MDMA administration (Green et al. 2003).

In contrast, there are some other compounds that are able to protect against MDMA-induced damage through some specific neurochemical mechanism and not merely by decreasing body temperature. The 5-HT uptake inhibitors ("► SSRIs and Related Compounds") fluoxetine and fluvoxamine, coadministered with MDMA, completely prevent the long-term loss of 5-HT concentration without altering the hyperthermia following MDMA. Both compounds inhibit the 5-HT transporter and could be blocking the entry of a toxic metabolite of MDMA into the 5-HT nerve terminal. Monoaminergic transporters also appear to be involved in MDMA-induced neurotoxicity in mice since 1-[2-[bis-(4-fluorophenyl) methoxy]ethyl]-4-(3-phenylpropyl) piperazine dihydrochloride (GBR 12909), a selective dopamine uptake inhibitor, also provided protection against long-term dopamine depletion induced by MDMA and METH without altering body temperature.

The role of free radicals in the damage induced by MDMA and METH has been demonstrated by several different approaches. α-Phenyl-N-tert-butyl nitrone (PBN) is a hydroxyl radical trapping agent that partially prevents the neuronal damage induced by MDMA on 5-HT nerve terminals as a result of its free-radical scavenging activity. PBN, at a dose that did not modify hyperthermia, was shown to prevent MDMA-induced hydroxyl radical formation. Other free-radical scavenging drugs (sodium ascorbate, L-cysteine) have also been found to protect against MDMA-induced damage. Supporting the existence of an oxidative stress process is the fact that the antioxidant alpha-lipoic acid prevented the long-term loss of 5-HT in the striatum but did not alter the hyperthermia induced by MDMA. In addition to hydroxyl radicals, nitrogen reactive species have also been shown to be involved in MDMA-induced neurotoxicity. In mice, there is also evidence for a key role of oxygen and nitrogen reactive species in the MDMA-induced neurotoxicity on dopamine neurons. Pretreatment with nitric oxide synthase (NOS) inhibitors S-methylthiocitrulline (S-MTC), N-(4-(2-((3-chlorophenylmethyl) amino)-ethyl) phenyl) 2-thiophene carboxamidine (AR-R17477AR) or 3-bromo-7-nitroindazole provides significant neuroprotection against the long-lasting MDMA-induced dopamine depletion suggesting that MDMA leads to radicals that combine with nitric oxide to produce peroxynitrites. Peroxynitrites are also formed following neurotoxic doses of METH, and NOS inhibitors attenuate the dopamine damage induced by METH (Sanchez et al. 2003).

Microglial activation is also emerging as an important element of the MDMA and METH neurotoxic cascade. Microglia are considered to be the resident immune cells of the brain and are activated in response to brain injury, leading to the secretion of a variety of cytotoxic factors such as cytokines, prostaglandins, and reactive oxygen/nitrogen species, many of which have been involved in amphetamine-induced neurotoxicity. On the other hand, microglia may also initiate tissue repair and regeneration through secretion of growth and neurotrophic factors, thereby exerting a beneficial neuroprotective role. MDMA and METH cause a prompt and transient increase in microglial activation that seems not to be related to increased rectal temperature. There have been several attempts to implicate microgliosis in amphetamine-induced neurotoxicity, but results have not always been successful. Thus, while dizocilpine and dextromethorphan prevent microgliosis following METH and provide neuroprotection against the loss of dopamine nerve terminals, some cannabinoid ("▶ Cannabinoids") CB2 receptor agonists do not prevent MDMA neurotoxicity. These data suggest that microglial activation occurs in response to amphetamine-induced neuronal damage but does not cause it. MDMA and METH also induces an increase in the brain levels of proinflammatory cytokines such as IL-1β and TNFα, respectively. Further studies are required in order to establish the specific role of these cytokines in amphetamine neurotoxicity and therefore develop appropriate approaches for providing neuroprotection.

A recent study has shown that repeated administration of low-dose METH disrupts the integrity of the blood–brain barrier (BBB). Immediately after METH treatment, there is a reduction in striatal laminin expression and increased IgG immunoreactivity colocalizing with areas of greater metalloproteinase activity (Urrutia et al. 2013). Administration of the JNK1/2 inhibitor, 2H-dibenzo[cd,g]indazol-6-one (SP600125), prevented all these effects and attenuated the increased expression of p-JNK1/2 induced by the drug, indicating that changes in BBB structure induced by METH are mediated by the JNK pathway. Additional studies are required in order to determine the role of an increased blood–brain-barrier permeability in inducing the subsequent dopaminergic neurodegeneration caused by the drug.

Hypothermia and Neuroprotection

There is a substantial body of evidence that hypothermia produces neuroprotection, not only in animal models of stroke but also when neurodegeneration has been produced by administration of neurotoxins such as methamphetamine and MDMA. This has meant that all experimental studies on neuroprotection must both monitor body temperature (and ideally also brain temperature) and must also ensure that body temperature is also adjusted where necessary to attenuate any body temperature changes. Only when this is done can the experimenter be confident that the drug under investigation is producing neuroprotection by interfering with pathological processes involved in neurodegeneration rather than merely lowering body temperature. This is particularly important in rodent studies as many administered compounds have been reported to lower body temperature though nonspecific mechanisms.

Lowering body (or brain) temperature deliberately by the use of specific drugs or by cooling techniques such as exposing the head or body to low temperature using ice baths or cooling helmets is being investigated clinically. It is the only approved neuroprotective approach in persons who have suffered global cerebral ischemia due to cardiac arrest. Both the severity of the hypothermia and the speed of temperature decrease have to be controlled closely for benefit to occur.

The generally accepted explanation for the efficacy of hypothermia as a neuroprotectant is that free-radical formation induced either by the ischemic insult or by the administration of the neurotoxin is markedly attenuated during hypothermia.

Cross-References

▶ Anti-Parkinson Drugs
▶ Apoptosis
▶ Barbiturates
▶ Benzodiazepines
▶ Cannabinoids and Endocannabinoids
▶ Dementias and Other Amnestic Disorders
▶ Excitatory Amino Acids and Their Antagonists
▶ GABA
▶ Methylenedioxymethamphetamine (MDMA)
▶ Neurodegeneration and Its Prevention
▶ Neurotoxins
▶ SSRIs and Related Compounds

References

Bath PMW, Gray LJ, Bath AJG, Buchan AM, Miyata T, Green AR (2009) Effects of NXY 059 in experimental stroke: an individual animal meta-analysis. Br J Pharmacol 157:1157–1171

Esteban E, O'Shea E, Camarero J, Sanchez V, Green AR, Colado MI (2001) 3,4-methylenedioxymethamphetamine induces monoamine release but not toxicity when administered centrally at a concentration occurring following a peripherally injected neurotoxic dose. Psychopharmacology (Berl) 154:251–260

Granado N, Ares-Santos S, Oliva I, O'Shea E, Martin ED, Colado MI, Moratalla R (2011) Dopamine D2-receptor knockout mice are protected against dopaminergic neurotoxicity induced by methamphetamine or MDMA. Neurobiol Dis 42:391–403

Green AR (2008) Pharmacological approaches to acute ischaemic stroke: reperfusion certainly, neuroprotection possibly. Br J Pharmacol 153: S325–S338

Green AR, Mechan AO, Elliott JM, O'Shea E, Colado MI (2003) The pharmacology and clinical pharmacology of 3, 4-methylenedioxymethamphetamine (MDMA, "ecstasy"). Pharmacol Rev 55:463–508

Higgins GC, Beart PM, Shin YS, Chen MJ, Cheung NS, Nagley P (2010) Oxidative stress: emerging mitochondrial and cellular themes and variations in neuronal injury. J Alzheimers Dis 20(Suppl 2):S453–S473

Minnerup J, Sutherland BA, Buchan AM, Kleinschnitz C (2012) Neuroprotection for stroke: current status and future perspectives. Int J Mol Sci 13:11753–11772

Sanchez V, Zeini M, Camarero J, O'Shea E, Bosca L, Green AR, Colado MI (2003) The nNOS inhibitor, AR-R17477AR, prevents the loss of NF68 immunoreactivity induced by methamphetamine in the mouse striatum. J Neurochem 85:515–524

Torres E, Gutierrez-Lopez MD, Mayado A, Rubio A, O'Shea E, Colado MI (2011) Changes in interleukin-1 signal modulators induced by 3,4-methylenedioxymethamphetamine (MDMA): regulation by CB2 receptors and implications for neurotoxicity. J Neuroinflammation 8:53

Urrutia A, Rubio-Araiz A, Gutierrez-Lopez MD, El Ali A, Hermann DM, O'Shea E, Colado MI (2013) A study on the effect of JNK inhibitor, SP600125, on the disruption of blood-brain barrier induced by methamphetamine. Neurobiol Dis 50:49–58

Neuroprotective Agent

Synonyms

Neuroprotectant; Neuroprotective drug

Definition

A drug or novel pharmacological agent which when administered after a cerebral trauma or neurotoxic compound attenuates the neuronal cell damage induced by the trauma or neurotoxin and thereby lessens the subsequent functional problems induced by the cerebral trauma.

Neuropsychiatric Disorders

Synonyms

Mental disorders; Psychiatric disorders

Definition

Neuropsychiatric disorders are any illness with a psychological origin manifested either in symptoms of emotional distress or in abnormal behavior. Neuropsychiatric disorders (psychiatric disorders) have been classified according to criteria defined in the *Diagnostic and Statistical Manual of Mental Disorders* (APA – American Psychiatric Association: Diagnostic and Statistical Manual of Mental Disorders – DSM-5, 2013) into major (clinical) depression, bipolar disorder, schizophrenia, anxiety disorders and attention-deficit hyperactivity disorder, substance use disorder, and other categories and subcategories. Due to the absence of sharp boundaries to and high occurrence as comorbidities in neurological disorders (i.e., stroke, Alzheimer's disease), an integrative view of psychiatric and neurological disorders is currently discussed.

Neurotensin

Lucy G. Chastain[1], Becky Kinkead[1] and Charles B. Nemeroff[2]
[1]Department of Psychiatry and Behavioral Sciences, Emory University School of Medicine, Atlanta, GA, USA
[2]Department of Psychiatry and Behavioral Sciences, University of Miami, Leonard M. Miller School of Medicine, Miami, FL, USA

Definition

Neurotensin (NT) is a tridecapeptide that functions both as a hormone and a neurotransmitter. It regulates intestinal motility, feeding behavior, nociception, thermoregulation, and anterior pituitary hormone secretion. In the central nervous system (CNS), NT modulates classic monoamine neurotransmitter systems, most notably dopamine (DA). Because of its modulation of DA circuits, NT has been implicated in the pathophysiology of schizophrenia and drug abuse and may represent a potential therapeutic target for treatment of these disorders.

Introduction

Neurotensin (NT) is a neuropeptide that was first isolated from bovine hypothalamus by Carraway and Leeman in 1973. The amino acid sequence of NT is N-Glu-Leu-Tyr-Glu-Asn-Lys-Pro-Arg-Arg-Pro-Tyr-Ile-Leu-C. The NT gene is highly conserved between species and encodes a 169–170 amino acid precursor protein containing both the tridecapeptide NT and a closely related hexapeptide, neuromedin N (NN), which are cleaved from the precursor protein after translation. Based on sequence homology, other structurally related endogenous peptides that bind to NT receptors have been identified, including NN, xenin, xenopsin, LANT-6, contulakin-G, and kinetensin, demonstrating the existence of a family of NT-related peptides (Boules et al. 2007; Kinkead and Nemeroff 2006).

NT functions as a neurotransmitter and hormone and is present in both the peripheral and central nervous system (CNS), as well as in the gastrointestinal tract, particularly the intestine. It has been implicated in a variety of physiological processes including gastric motility, vasoactivity, appetite, nociception, thermoregulation, and anterior pituitary hormone secretion. In the CNS of rodents, NT cell bodies are located in the hypothalamus, hippocampus, amygdala, bed nucleus of the stria terminalis (BNST), lateral septum, substantia nigra (SN), ventral tegmental area (VTA), olfactory tubercles, striatum, basal forebrain, and periaqueductal gray (PAG) (see Binder et al. 2001a for neuroanatomy review). In neurons, NT is stored in presynaptic vesicles and its release is Ca^{2+}-dependent. NT neurotransmission is terminated by cleavage of the peptide by peptidases.

NT Receptors

There are four known NT receptors (Kinkead and Nemeroff 2006 for review). The NT_1 and NT_2 receptors are G protein-coupled receptors with the classic seven transmembrane-spanning regions. NT_2 is a receptor with low affinity for NT that also binds the histamine H1 receptor antagonist levocabastine. The NT_1 receptor is a levocabastine-insensitive receptor with high affinity for NT. Two other identified receptors, NT_3 (sortilin) and NT_4 (SorLA/LR11), are members of the family of Vps10p domain receptors with high affinity for NT. In contrast to NT_1 and NT_2, the NT_3 and NT_4 receptors are type I amino acid receptors with a single transmembrane-spanning region. The majority of NT_3 and NT_4 receptors are found intracellularly and have been theorized to play a role in intracellular sorting processes. NT_3 has also been posited to play a role in cell death, and NT_4 may play a role in terminating NT function.

NT Receptor Agonists and Antagonists

NT is easily degraded by peptidases and must be injected directly into the brain in order to exert CNS effects. Thus, over the years, many groups targeting brain NT receptors for drug development have sought a compound that could be delivered peripherally (Boules et al. 2013). The last six amino acids on the NT C-terminus carry the full biological activity of NT, and several NT analogs that can be delivered systemically have been developed, including PD149163, NT66L, NT67L, NT69L, and NT77L. To date, all identified NT receptor agonists are modified peptide analogs of NT(8–13). Preclinical studies have suggested NT agonists may be useful as novel analgesic medications and as antipsychotic drugs (Boules et al. 2013); these studies are discussed below.

Several small molecule NT receptor antagonists have also been identified; SR 48692 and SR 142948A are the best characterized (Kinkead and Nemeroff 2006). Both these antagonists cross the blood-brain barrier and possess nanomolar affinity for the NT_1 receptor. Although SR 48692 has a lower affinity for NT_2 than NT_1 receptors, neither SR 48692 nor SR 142948A discriminates well between the NT_1 and NT_2 receptors, and there is some evidence that SR 48692 binds to the NT_3 receptor as well. SR 142948A and SR 48692 do not block all NT-mediated effects, indicating that there are pharmacological subtypes of NT receptors at which the available antagonists have low affinity. Whether these two NT receptor antagonists interact with the NT_4 receptor is yet to be determined.

Functional Role of NT in the CNS

One of the most studied effects of NT in the CNS is its modulation of DA neurotransmission. NT differentially modulates DAergic activity in the mesocortical, mesolimbic, and nigrostriatal pathways (for review, see Binder et al. 2001a). In general, the effect of NT is to decrease DA

D_2-type receptor-mediated effects with the overall effect on cell firing and behavior depending on the brain region. This effect is mediated in part by NT activation of the NT_1 receptor, leading to allosteric receptor/receptor interactions between the activated NT receptor and DA D_2-type receptors and subsequent functional antagonism of the D_2 receptor. Within the midbrain where NT_1 receptors are found on DA cell bodies, NT functions to increase DAergic cell activity by attenuating D_2 receptor-mediated autoinhibition. The behavioral effects of increased NT neurotransmission in the midbrain resemble those of peripherally administered DA agonists. In the nucleus accumbens (NAcc), NT_1-D2 receptor colocalization takes place primarily on postsynaptic GABAergic neurons. Injection of NT into the NAcc leads to desensitization of the D2 receptor, increased accumbal GABA release, and decreased accumbal DA levels, effects resembling those of peripherally administered antipsychotic drugs. The net effect of increased accumbal NT neurotransmission is thus inhibition of the NAcc and its terminal regions, which may result in reduced behavioral effects of DA agonists.

In addition, NT plays a key role in regulating hormone secretion (Rostène and Alexander 1997; Geisler et al. 2006; Boules et al. 2013). As noted above, NT is found in high concentrations in the hypothalamus. Central injection of NT in rats stimulates the hypothalamic-pituitary-adrenal (HPA) axis, increasing plasma levels of corticosterone and adrenocorticotropic hormone (ACTH). Blocking NT neurotransmission with NT receptor antagonists does not affect baseline levels of corticosterone but suppresses stress-induced increases in this hormone, suggesting endogenous NT mediates stress-induced activation of the HPA axis. NT also stimulates prolactin secretion at the level of the anterior pituitary gland.

In conjunction with other peptides (including leptin, galanin, neuropeptide Y, and others), NT also modulates appetite and feeding behavior and is an important regulator of the central feeding circuit. Activation of the NT system decreases feeding in rats, an effect which is mediated via the NT_1 receptor and which is downstream of leptin activation. Central injections of leptin increase hypothalamic NT mRNA expression as well as reduce food intake and decrease body weight in rats. In addition, NT receptor antagonist administration and NT_1 gene knockout block leptin-associated appetite suppression. Thus, the NT system, in particular the NT_1 receptor, may be a potential novel target for the development of anti-obesity drugs.

NT also plays an important role in modulating nociception (Dobner 2006; Mustain et al. 2011). NT fibers and cell bodies are found in multiple brain regions involved in pain transmission and modulation, including the PAG, the rostroventral medulla (RVM), and the dorsal horn of the spinal cord. NT dose-dependently modulates nociception. When low doses of NT are injected into the RVM, the result is an enhanced nociceptive response and hyperalgesia. When higher doses of NT are injected into the RVM, the result is antinociception, which is not mediated by endogenous opioids. Thus, NT modulates nociception via differences in intensity of signaling. Blocking NT neurotransmission via NT receptor antagonist administration in rats and NT gene knockout in mice significantly attenuates baseline nociception as well as stress-induced antinociception (Dobner 2006). These studies suggest endogenous NT facilitates both nociception and stress-induced antinociception. Stress-induced antinociception is likely mediated via NT activation of the NT_2 receptor, because mice lacking the NT_2 gene exhibit deficits in stress-induced antinociception. In preclinical studies, subcutaneous and topical administration of NT receptor agonists produced antinociceptive effects, suggesting these drugs may be useful as analgesics.

Role in Psychiatric Disorders

Both clinical and preclinical studies have implicated NT in the pathophysiology of schizophrenia, and it has been suggested that NT may function as an endogenous antipsychotic agent or neuroleptic (Cáceda et al. 2006). Decreased concentrations of NT are found in the cerebrospinal fluid (CSF) of a subset of schizophrenic patients, and NT levels normalize following effective treatment with antipsychotic drugs (Cáceda et al. 2006). Mice lacking the NT gene exhibit deficits in sensorimotor gating similar to those reported in schizophrenic patients. Mice lacking NT_1 receptors show alterations in striatal DAergic activity resembling mesolimbic system dysfunction observed in schizophrenics (Boules et al. 2013). In addition, central injection of NT and viral vector overexpression of NT_1 receptors in the NAcc have many of the same behavioral effects as systemic administration of antipsychotic drugs (Cáceda et al. 2006). Based on these findings, it has been proposed that decreased NT neurotransmission may be an underlying factor in the pathophysiology of schizophrenia and NT receptor agonists may represent a novel treatment for schizophrenia. Animal studies evaluating the therapeutic potential for NT receptor agonists have shown that several of these drugs produce behavioral effects predictive of antipsychotic drug efficacy (Boules et al. 2013).

In addition to the possibility that NT receptor agonists may represent a novel class of antipsychotic drugs, evidence suggests that modulation of NT neurotransmission may be involved in the mechanism of action of antipsychotic drugs (Binder et al. 2001b). To date, all clinically effective antipsychotic drugs, none of which bind to NT receptors, have been shown to increase NT neurotransmission (mRNA expression, NT peptide concentrations, or NT release) in the NAcc. This has led to the hypothesis that increased NT release in the NAcc may mediate some of the therapeutic effects of antipsychotic drugs. In fact, endogenous NT neurotransmission appears to mediate the behavioral effects of some, but not all, antipsychotic drugs (Kinkead and Nemeroff 2006; Cáceda et al. 2006). Pretreatment with an NT receptor antagonist blocks the behavioral effects of the typical antipsychotic drug haloperidol in two animal models of sensorimotor gating, prepulse inhibition (PPI) of the acoustic startle response, and latent inhibition. As mentioned previously, mice lacking the NT gene exhibit deficits in PPI (similar to those seen in schizophrenia) that are restored by the antipsychotic drug clozapine, but not haloperidol, quetiapine, or olanzapine. These results suggest that increased endogenous NT is critically involved in the mechanism of action of a subset of antipsychotic drugs. In combination with the clinical data indicating that there is a subset of schizophrenic patients with deficits in NT neurotransmission, these findings raise the possibility of preselecting an effective antipsychotic drug for this subgroup of patients, thereby dramatically improving patient outcomes and reducing the risk of inadequate treatment.

NT has also been suggested to play a role in drug abuse. Drug abuse is characterized by a loss of control leading to compulsive drug use, despite obvious detrimental consequences. It has been theorized to be, in part, due to dysregulation of the mesocorticolimbic DA system, which mediates reward processing and learning. Because of its close association with the mesocorticolimbic DA system, NT is implicated in mechanisms underlying reward and addiction (Cáceda et al. 2006; Dobner 2005; Boules et al. 2013). As demonstrated by studies utilizing central injection of the peptide, NT has rewarding properties in the VTA and modulates the behavioral effects of psychomotor stimulants in the NAcc. In addition, NT/NN mRNA expression and NT peptide concentrations are increased in the NAcc, striatum, and prefrontal cortex following acute cocaine or amphetamine administration. Chronic administration of NT receptor antagonists diminishes the hyperlocomotor effects of acute cocaine, but not acute amphetamine. In contrast, supra-activation of NT systems following intra-NAcc injection of NT, peripheral administration of NT receptor agonists, and NT_1 receptor overexpression in the NAcc significantly reduces amphetamine-induced hyperlocomotion. Likewise, systemic administration of NT receptor agonists diminishes the acute hyperlocomotor effects of amphetamine and cocaine. Therefore, increased NTergic neurotransmission counteracts the acute locomotor effects of these drugs.

Much evidence implicates the endogenous NT system in sensitization to drugs, particularly in sensitization to psychomotor stimulants. In humans, the repeated use of psychomotor stimulants can result in long-lasting alterations in the response to subsequent drug use, including the emergence of drug craving. In rats, this phenomenon, termed behavioral sensitization, is considered a useful animal model for drug craving. Repeated intra-VTA or ICV injection of NT leads to behavioral and chemical sensitization. Perhaps, most intriguing is the finding that pretreatment with an NT receptor antagonist diminishes locomotor sensitization to amphetamine and cocaine. From these data, it can be inferred that NT neurotransmission is involved in the initiation of behavioral sensitization to these psychomotor stimulants. There is also some evidence that the role of NT in sensitization can be generalized to other drug classes including alcohol, nicotine, and NMDA receptor antagonists (Cáceda et al. 2006). Based on these findings, it is possible that genetic vulnerability within the NT system may contribute to drug addiction and provide a rationale for the development of compounds modifying NT neurotransmission in the treatment of drug addiction.

Disclosures

Lucy G. Chastain: NIH funding
Becky Kinkead: NIH funding
Charles B Nemeroff: NIH funding
Scientific Advisory Board: American Foundation for Suicide Prevention (AFSP), CeNeRx BioPharma (2012), National Alliance for Research on Schizophrenia and Depression (NARSAD), Xhale, PharmaNeuroBoost (2012), Anxiety Disorders Association of America (ADAA), Skyland Trail
Stockholder or Equity: CeNeRx BioPharma, PharmaNeuroBoost, Revaax Pharma, Xhale
Consulting: Xhale, Takeda, SK Pharma, Shire, Roche, Lilly, Allergan, Mitsubishi Tanabe Pharma Development America, Taisho Pharmaceutical Inc., Lundbeck
Board of Directors: AFSP, NovaDel (2011), Skyland Trail, Gratitude America, ADAA
Patents: Method and devices for transdermal delivery of lithium (*US 6,375,990B1*), Method of assessing antidepressant drug therapy via transport inhibition of monoamine neurotransmitters by ex vivo assay (*US 7,148,027B2*)
Other Financial Interests: CeNeRx BioPharma, PharmaNeuroBo
Income sources or equity of $10,000 or more: PharmaNeuroBoost, CeNeRx BioPharma, NovaDel Pharma, Reevax Pharma, American Psychiatric Publishing, Xhale
Honoraria: Various
Royalties: Various
Expert Witness: Various

Cross-References

▶ Alcohol
▶ Analgesics
▶ Antipsychotic Drugs
▶ Appetite Suppressants
▶ Blood–Brain Barrier
▶ Cocaine
▶ Latent Inhibition
▶ Leptin
▶ Nicotine
▶ Opioids
▶ Prepulse Inhibition
▶ Schizophrenia
▶ Schizophrenia: Animal Models

References

Binder EB, Kinkead B, Owens MJ, Nemeroff CB (2001a) Neurotensin and dopamine interactions. Pharmacol Rev 53:453–486
Binder EB, Kinkead B, Owens MJ, Nemeroff CB (2001b) The role of neurotensin in the pathophysiology of schizophrenia and the mechanism of action of antipsychotic drugs. Biol Psychiatry 50:856–872
Boules M, Shaw A, Fredrickson P, Richelson E (2007) Neurotensin agonists: potential in the treatment of schizophrenia. CNS Drugs 21(1):13–23
Boules M, Li Z, Smith K, Fredrickson P, Richelson E (2013) Diverse roles of neurotensin agonists in the central nervous system. Front Endocrinol 4(36):1–16
Cáceda R, Kinkead B, Nemeroff CB (2006) Neurotensin: role in psychiatric and neurological diseases. Peptides 27:2385–2404
Dobner PR (2005) Multitasking with neurotensin in the central nervous system. Cell Mol Life Sci 62:1946–1963

Dobner PR (2006) Neurotensin and pain modulation. Peptides 27:2405–2414

Geisler S, Bérod A, Zahm DS, Rostène W (2006) Brain neurotensin, psychostimulants, and stress – emphasis on neuroanatomical substrates. Peptides 27:2364–2384

Kinkead B, Nemeroff CB (2006) Novel treatments of schizophrenia: targeting the neurotensin system. CNS Neurol Disord Drug Targets 5:205–218

Mustain CW, Rychahou PW, Evers BM (2011) The role of neurotensin in physiologic and pathologic processes. Curr Opin Endocrinol Diabetes Obes 18:75–82

Rostène WH, Alexander MJ (1997) Neurotensin and neuroendocrine regulation. Front Neuroendocrinol 18:115–173

Neurotoxicity

Definition

A phenomenon in which a substance (▶ neurotoxin), condition, or state alters the normal activity or structure of the nervous system such that damage to nervous tissue occurs. Neurotoxicity can result from exposure to certain drug therapies as well as from certain ▶ drugs of abuse.

Neurotoxicity and Schizophrenia

Heleen B. M. Boos, H. D. Postma and Wiepke Cahn
Department of Psychiatry, Rudolf Magnus Institute of Neuroscience, University Medical Center, Utrecht, CX, The Netherlands

Definition

Neurotoxicity is the tendency of substances (also called neurotoxins), conditions, or states to alter the normal activity of the nervous system. This can eventually disrupt or even kill neurons (key cells that transfer and process signals in the brain and parts of the nervous system). Neurotoxicity can result from exposure to drug therapies and certain kinds of drug use or abuse.

The "neurotoxicity theory" of schizophrenia states that psychosis, especially when untreated, is neurotoxic to the brain and that brain changes are an inherent feature of the neurobiological disease process in schizophrenia. Although the pathophysiology of schizophrenia is still unknown, it is known that antipsychotics (dopamine receptor blockers) reduce symptoms of schizophrenia, but that recreational drugs like cocaine and amphetamine increase dopamine in the brain and may induce psychosis.

The term *neurotoxic* is used to describe a substance, condition, or state that damages the nervous system and/or brain, usually by killing neurons. The term is generally used to describe a condition or substance that has been shown to result in observable physical damage.

Current Concepts and State of Knowledge

Introduction

Although Kraepelin (1919) suggested that dementia praecox ("▶ Schizophrenia") was a chronic, deteriorating psychotic disorder, evidence was lacking to prove this; in postmortem studies, few or no abnormalities were found in the brains of patients with schizophrenia. In the 1960s and 1970s, schizophrenia was thought by some not to be a brain disease at all. These clinicians instead considered problematic family interactions as a cause. Poor mother-child interactions, in particular, were thought to cause or worsen schizophrenia.

This began to change in the mid-1970s, when new in vivo neuroimaging techniques, such as computerized tomography (CT), revealed that patients with schizophrenia had enlarged ventricles when compared with age-matched controls (Johnstone et al. 1976). Interestingly, even 50 years earlier, the first evidence of brain ventricular enlargement had already been provided by a pneumoencephalography (PEG) study in a small sample of schizophrenia patients (Jacobi and Winkler 1927).

Because of the lack of progression of cerebral ventricular enlargement in initial studies, the neurodevelopmental hypothesis began to emerge, according to which schizophrenia is posited to result from abnormalities in neuronal connectivity, which arise during fetal life but are not expressed until the onset of illness. Despite evidence in favor of the neurodevelopmental hypothesis, such as delayed developmental milestones, there are some substantial symptoms of schizophrenia that cannot be adequately explained by this model. The apparent progression of clinical aspects of the syndrome in some patients, including deterioration, dilapidation, and treatment resistance, suggests that schizophrenia is a progressive illness. Furthermore, recent longitudinal magnetic resonance imaging (MRI) studies in patients with schizophrenia have shown that through the course of the illness, brain volume declines progressively. It remains unclear what underlying neuropathological mechanisms are causing these progressive brain volume changes in schizophrenia, but it is thought that psychosis itself could be neurotoxic.

This essay will briefly describe the scientific evidence in favor of schizophrenia being a progressive brain disease. It will then discuss the neurotoxicity hypothesis of schizophrenia and the concept of neuroprotection.

Neuroimaging Studies

In 1976, the new in vivo neuroimaging technique CT was first used in schizophrenia research by Johnstone et al. (1976). In this study, patients with schizophrenia showed enlarged ventricle volumes when compared with age-related healthy control subjects. In the last two decades, numerous studies have been conducted using MRI. In 2000, a cross-sectional meta-analysis (Wright et al. 2000) convincingly showed that brain volume changes are present in schizophrenia. Lateral ventricle volume was found to be increased (16 %), while cerebral volume was reduced (2 %). The latter was primarily attributed to a decrease in gray matter volume (2 %). A small but significant reduction was also found in white matter volume (1 %). The improved quality of the MRI scans made it also possible to manually delineate specific brain areas of interest. Regional pathology findings indicate larger reductions in temporal and frontal lobes, especially in medial temporal structures ("▶ Hippocampus" and "▶ Amygdala"). The finding of reduction in frontal and medial temporal areas of the brain in patients with schizophrenia has been corroborated by using a voxel-based morphometry approach.

It has long been argued that the brain volume changes found in schizophrenia are at least partly caused by antipsychotic medication. Indeed, in the early stages of schizophrenia, progressive decreases in gray matter (Cahn et al. 2002) and frontal lobe volume (Gur et al. 1998; Madsen et al. 1999) have been found to correlate with the amount of antipsychotic medication taken. Those patients who were prescribed the highest doses of antipsychotic medication also had the greatest progressive decreases in brain volumes. Nevertheless, this brain volume decrease might not be a direct effect of the medication, as those who are prescribed the highest doses of antipsychotic medication are generally the most severely ill patients. Increases and decreases in brain volumes depend on the type of antipsychotic medication. Basal ganglia volumes decrease on typical antipsychotic medication and increase (or normalize) on atypical antipsychotics (Hakos et al. 1994; Scheepers et al. 2001). Recent longitudinal MRI studies have shown that olanzapine and clozapine actually attenuate brain tissue loss in schizophrenia, whereas typical antipsychotics do not (DeLisi 2008; Lieberman et al. 2005; van Haren et al. 2008).

In the last decade, many studies have focused their attention on investigating the effects of the first psychotic episode on the brain. Studying the early phase of the illness is advantageous because the confounding effects of chronicity and long-term medication can be excluded. MRI studies in antipsychotic-naïve patients appear to show a relative paucity of brain abnormalities, which stands in marked contrast with findings in more chronically ill schizophrenia patients. Several explanations can be offered to elucidate the discrepancy in brain abnormalities between those patients who are chronically ill and those who are in the early phase of the illness. Medication might increase brain abnormalities and could contribute to these brain volume changes. Finding few brain abnormalities in antipsychotic-naïve patients with schizophrenia could also be the result of a selection bias favoring the inclusion of patients who have a less severe form of schizophrenia. Finally, of course, progression of the illness itself may lead to an increase of brain abnormalities.

Indeed, there is a growing body of evidence that brain abnormalities become greater in schizophrenia over the course of the illness. Various reviews of longitudinal MRI studies in patients with first-episode schizophrenia conclude that there is increased loss of brain volume, especially gray matter, over time, particularly in the frontotemporal cortical areas, and that there is sulcal and ventricular expansion. In schizophrenia, there is a 3 % decrease in whole brain volume compared with healthy control subjects, with a 0.5 % decrease per year, which is consistent with findings in postmortem studies in schizophrenia (Hulshoff Pol and Kahn 2008). Although changes in brain volume over time are reported in both first-episode patients and chronic patients with schizophrenia, the magnitude of change in first-episode patients (e.g., −1.2 % in 1 year for whole brain volume) suggests that these brain volume reductions are particularly prominent during the first years of illness (Cahn et al. 2002).

Nevertheless, progressive brain volume reductions are only relevant if they are associated with the clinical characteristics and outcome in schizophrenia. The most consistent finding of longitudinal MRI studies in first-episode and chronic schizophrenia is the relationship between reduced brain volume (gray matter decrements and ventricular increments) and poor outcome. Psychotic symptoms have also been examined in relation to brain volume loss over time. A recent MRI study investigated the relationship between psychosis and brain volume change in first-episode patients with schizophrenia over the first 5 years of illness. Associations between gray matter volume loss, lateral and third ventricle volume increase, and longer duration of psychosis were found. Total duration of psychotic symptoms was further associated with greater decreases in total brain and cerebellar volume (Cahn et al. 2009). Other MRI studies, which examined smaller brain structures, found reduced volumes of the medial temporal lobe, superior temporal gyrus, and hippocampus in patients with psychotic symptoms. Furthermore, a long duration of untreated psychosis (DUP) is associated with poor clinical and social outcomes. Various research groups have now found that patients with a longer DUP have more decreased gray matter volume than patients with a shorter DUP (Lappin et al. 2006). These findings suggest that brain volume loss over time could be attributable to the "toxic" effects of the psychotic state.

Besides psychosis, treated or untreated, there are other factors that could be neurotoxic in schizophrenia, such as cannabis use and stress. About 28–50 % of patients with schizophrenia use cannabis. Clinically, patients who use cannabis have more positive (but not negative) symptoms, an earlier disease onset, and an increased number of psychotic episodes, when compared with patients who do not use cannabis. Rais et al. (2008) found significantly more decrease in brain volume in patients using cannabis when compared with nonusing patients over a 5-year period.

At present, there is only indirect evidence that life events might affect brain volumes in schizophrenia. This includes findings that lower gray and white matter volumes in schizophrenia are associated with dysregulated dopaminergic-/noradrenergic-mediated stress responses.

Neurotoxicity

The neurotoxicity theory of schizophrenia states that psychosis, especially when untreated, is neurotoxic to the brain and that brain changes are an inherent feature of the neurobiological disease process. As mentioned previously, MRI studies show volume reductions over time, particularly of gray matter. Nevertheless, the brains of schizophrenia patients do not reveal characteristic histopathology, as do other neurodegenerative diseases. Moreover, postmortem studies have not found evidence of neuronal injury or degeneration. A neurodegenerative process in the brain usually accompanies loss of neuronal cells and microglial cells (a type of glial cells that are the resident macrophages of the brain).

In schizophrenia, a lack of microglial cells has been found. Some researchers have postulated that neuronal cell death occurs in schizophrenia, but that this reflects apoptosis, programmed cell death, rather than necrosis, in which microglial cells phagocytose or "eat" the injured cells and leave scar tissue at the place where the necrotic cell used to be. Various intracellular and extracellular events, like increased glutamate stimulation, can induce a programmed apoptotic cascade which results in cell destruction. This apoptotic cascade could then produce synaptic loss and synaptic remodeling and would compromise cell function and alter brain morphology (Lieberman 1999).

Postmortem studies in schizophrenia have also reported reduced dendritic spines, a measure of the amount of synaptic contacts between neurons. Furthermore, they have found smaller dendritic arbors on the pyramidal cells of the cortex, damage to myelinated fiber tracts, and increased neuronal density because of reduced neuropil, the synaptic syncytium between neurons where synaptic connections are formed between branches of axons and dendrites (Davis et al. 2003). This decreased interneuronal neuropil could cause functional and anatomic hypoconnectivity in a schizophrenic brain and would explain the decreased cortical volumes as seen on MRI (Davis et al. 2003).

A higher concentration of neurotransmitters in the brain could lead to excitotoxicity, resulting in cell death and, in turn, brain volume reduction. Although the pathophysiology of schizophrenia is still unknown, it is known that antipsychotics, which are dopamine receptor blockers, reduce symptoms of schizophrenia and that recreational drugs like cocaine and amphetamine increase dopamine in the brain and may induce psychosis. It is thought that in schizophrenia there is an increase in dopamine in the mesolimbic system. This is the so-called dopamine hypothesis of schizophrenia. Nevertheless, this hypothesis only explains the positive symptoms and does not explain the negative symptoms and cognitive problems seen in schizophrenia. Phencyclidine (PCP), an N-methyl-D-aspartate (NMDA) receptor (a glutamate receptor) antagonist, disinhibits excitatory glutamatergic pathways, which can result in neuronal damage. PCP induces symptoms more similar to the broader array of those seen in schizophrenia. The glutamate hypothesis of schizophrenia postulates that there is hypofunction of the NMDA receptor in the schizophrenic brain. The reduced activity of the NMDA receptor affects the glutamate concentration but also influences other neurotransmitter systems in the brain, such as dopamine and GABA.

The NMDA receptor is an ionotropic excitatory receptor with glutamate as its ligand. Binding of glutamate causes an influx of Na^+ and Ca^{2+} and, thereby, postsynaptic membrane depolarization. When there is a reduced amount of NMDA receptors, the excess glutamate stays in the synaptic cleft and increases the stimulation of other ionotropic receptors, like α-amino-3-hydroxy-5-methyl-4-isoxazole propionic acid (AMPA) and kainite receptors. This overstimulation, also called excitotoxicity, leads to the dysregulation of Ca^{2+} homeostasis and causes oxidative stress and, thereby, apoptosis (Deutsch et al. 2001).

Antagonism, or reduced activity, of the NMDA receptor, and thus a hypofunctioning of glutamate signaling, may also result in altered dopamine concentrations. As a consequence, prefrontal D_1 receptors might be hypostimulated, which could lead to the negative and cognitive symptoms of schizophrenia. An ensuing episodic hyperactivity of the mesolimbic dopamine system could lead to positive symptoms (Jarskog et al. 2007).

NMDA receptors are also present at the GABAergic inhibitory interneurons within the cortex. Glutamate activation normally leads to the release of GABA to inhibit glutamatergic neurons and the release of glutamate. With reduced NMDA receptor activity, a decreased amount of GABA will be available to inhibit glutamate activity and thereby result in the heightened activity of glutamatergic neurons. GABAergic inhibition is of great importance in critical circuits of normal brain function and could be the cause of the cognitive symptoms in patients (Reynolds et al. 2004).

Neuroprotection

Neuroprotection refers to treatment that helps to maintain the functional integrity of the brain in response to neurobiological stress. Neuroprotection is already a rapidly advancing concept in the treatment of neurological disorders. It is also increasingly viewed as a possible therapeutic approach for psychiatric disorders, to improve loss of function or prevent neurodegeneration from occurring. The most common treatment for schizophrenia is antipsychotic medication. Almost all patients with schizophrenia receive antipsychotic medication during their illness; therefore, it is not clear whether the progressive brain changes occurring in the brains of the patients are due to the illness itself (untreated psychosis) or perhaps due to the use of antipsychotic medication. In other words, it is unclear whether antipsychotics themselves are neurotoxic or neuroprotective to the brain (Hulshoff Pol and Kahn 2008).

Increases and decreases in brain volume appear to vary according to the type of antipsychotic medication. Basal ganglia volumes decrease on typical antipsychotics and increase or normalize on atypical agents. Recent longitudinal MRI studies have shown that olanzapine and clozapine actually attenuate brain tissue loss in schizophrenia, whereas typical antipsychotics do not. It has been suggested that antipsychotic drugs, specifically the atypicals, have an effect on synaptic remodeling and neurogenesis and thereby ameliorate the pathophysiology of schizophrenia (Lieberman et al. 2008). The progressive brain loss seen in patients who are treated with haloperidol and other typical antipsychotics could be due to the fact that typical antipsychotics are not neuroprotective, thus allowing the progressive brain loss of the disease to continue increase despite treatment.

It has been suggested that physical exercise, psychoeducation, and cognitive therapy might be neuroprotective in schizophrenia, but very limited research in this area has been done so far. Further studies are needed to examine the possible neuroprotective effects of medication and psychosocial treatments in this debilitating disorder.

Cross-References

▶ Aminergic Hypothesis for Schizophrenia
▶ Antipsychotic Drugs
▶ Antipsychotic Medication: Future Prospects
▶ Apoptosis
▶ Classification of Psychoactive Drugs
▶ First-Generation Antipsychotics
▶ Movement Disorders Induced by Medications
▶ Neurodegeneration and Its Prevention
▶ Neurogenesis
▶ Neurotoxins
▶ Schizophrenia
▶ Second- and Third-Generation Antipsychotics

References

Cahn W, Hulshoff Pol HE, Lems EB, van Haren NE, Schnack HG, van der Linden JA, Schothorst PF, van Engeland H, Kahn RS (2002) Brain volume changes in first-episode schizophrenia: a 1-year follow-up study. Arch Gen Psychiatry 59(11):1002–1010

Cahn W, Rais M, Stigter FP, van Haren NE, Caspers E, Hulshoff Pol HE, Xu Z, Schnack HG, Kahn RS (2009) Psychosis and brain volume changes during the first five years of schizophrenia. Eur Neuropsychopharmacol 19(2):147–151

Davis KL, Stewart DG, Friedman JI, Buchsbaum M, Harvey PD, Hof PR, Buxbaum J, Haroutunian V (2003) White matter changes in schizophrenia: evidence for myelin-related dysfunction. Arch Gen Psychiatry 60(5):443–456

DeLisi LE (2008) The concept of progressive brain change in schizophrenia: implications for understanding schizophrenia. Schizophr Bull 34(2):312–321

Deutsch SI, Rosse RB, Schwartz BL, Mastropaolo J (2001) A revised excitotoxic hypothesis of schizophrenia: therapeutic implications. Clin Neuropharmacol 24(1):43–49

Gur RE, Cowell PE, Turetsky BI et al (1998) A follow-up magnetic resonance imaging study of schizophrenia. Relationship of neuroanatomical changes to clinical and neurobehavioral measures. Arch Gen Psychiatry 55(2):145–152

Hakos MH, Lieberman JA, Bilder RM et al (1994) Increase in caudate nuclei volumes of first-episode schizophrenic patients taking antipsychotic drugs. Am J Psychiatry 151(10):1430–1436

Hulshoff Pol HE, Kahn RS (2008) What happens after the first episode? A review of progressive brain changes in chronically ill patients with schizophrenia. Schizophr Bull 34(2):354–366

Jacobi W, Winkler H (1927) Encephalographische studien an chronische schizophrenen. Arch Psychiatr Nervenkr 81:299–332

Jarskog LF, Miyamoto S, Lieberman JA (2007) Schizophrenia: new pathological insights and therapies. Annu Rev Med 58:49–61

Johnstone EC, Crow TJ, Frith CD, Husband J, Kreel L (1976) Cerebral ventricular size and cognitive impairment in chronic schizophrenia. Lancet 2(7992):924–926

Kraepelin E (1919) Dementia praecox and paraphrenia. E&S Livingstone, Edinburgh

Lappin JM, Morgan K, Morgan C, Hutchison G, Chitnis X, Suckling J, Fearon P, McGuire PK, Jones PB, Leff J, Murray RM, Dazzan P (2006) Gray matter abnormalities associated with duration of untreated psychosis. Schizophr Res 83(2–3):145–153

Lieberman JA (1999) Is schizophrenia a neurodegenerative disorder? A clinical and neurobiological perspective. Biol Psychiatry 46(6):729–739

Lieberman JA, Tollefson GD, Charles C et al (2005) Antipsychotic drug effects on brain morphology in firstepisode psychosis. Arch Gen Psychiatry 62(4):361–370

Lieberman JA, Bymaster FP, Meltzer HY, Deutch AY, Duncan GE, Marx CE, Aprille JR, Dwyer DS, Li XM, Mahadik SP, Duman RS, Porter JH, Modica-Napolitano JS, Newton SS, Csernansky JG (2008) Antipsychotic drugs: comparison in animal models of efficacy, neurotransmitter regulation, and neuroprotection. Pharmacol Rev 60(3):358–403

Madsen AL, Karle A, Rubin P et al (1999) Progressive atrophy of the frontal lobes in first-episode schizophrenia: interaction with clinical course and neuroleptic treatment. Acta Psychiatr Scand 100(5):367–374

Rais M, Cahn W, Van HN, Schnack H, Caspers E, Hulshoff Pol PH, Kahn R (2008) Excessive brain volume loss over time in cannabis-using first-episode schizophrenia patients. Am J Psychiatry 165(4):490–496

Reynolds GP, Abdul-Monim Z, Neill JC, Zhang ZJ (2004) Calcium binding protein markers of GABA deficits in schizophrenia – postmortem studies and animal models. Neurotox Res 6(1):57–61

Scheepers FE, de Wied CC, Hulshoff Pol HE et al (2001) The effect of clozapine on caudate nucleus volume in schizophrenic patients previously treated with typical antipsychotics. Neuropsychopharmacology 24(1):47–54

van Haren NE, Hulshoff Pol HE, Schnack HG et al (2008) Focal gray matter changes in schizophrenia across the course of the illness: a 5-year follow-up study. Neuropsychopharmacology 18(4):312–315

Wright IC, Rabe-Hesketh S, Woodruff PW, David AS, Murray RM, Bullmore ET (2000) Meta-analysis of regional brain volumes in schizophrenia. Am J Psychiatry 157(1):16–25

Neurotoxins

Stephen B. Dunnett
School of Biosciences, Cardiff University, Wales, Cardiff, UK

Synonyms

Catecholamine toxins; Excitotoxins; Immunotoxins; Metabolic toxins; Plant toxins

Definition

Neurotoxins are poisons that disable or kill neurons.

Current Concepts and State of Knowledge

Some neurotoxins have been discovered as synthetic chemicals, whether intended for neuronal targeting (e.g., ▶ 6-hydroxydopamine, MPTP) or as side effects of other applications (e.g., paraquat and other pesticides). Others are signaling molecules of normal cells but delivered in excess (e.g., nitric oxide, ▶ glutamate). Nevertheless, the most powerful neurotoxins come from the natural world, having evolved for use by animals and plants both for predation and for defense, such as snake, insect, and spider venoms or plant toxins. Neurotoxicology is a wide field in its own right, seeking to understand the pharmacology and mechanisms of neuronal action of toxins in order to promote safety, whether in the testing of manufactured products or protection of the environment, in humans and animals alike. Neurotoxins have also, however, emerged over the last four decades as a powerful tool in experimental neuroscience, and this is the focus of this entry. In particular, I briefly consider the development of a range of molecules that allow selective targeting, disruption, and death of specific populations of neuronal cells in the CNS. These tools can then be used to make lesions for experimental analysis of the normal function of the targeted neurons, to generate effective and efficient animal models of human diseases, to study mechanisms of neurodegeneration, and to develop novel therapeutics.

The utility of particular toxins relates to multiple factors, including *specificity* for the particular target population alone; *validity* to reproduce the specific neuropathology, and ideally the pathogenic process, involved in the target disease; *reliability* to yield consistent toxicity from one application to the next; and *practicality* for safe, simple, efficient, and cost-effective handling and use within the laboratory environment.

Monoamine Neurotoxins

6-Hydroxydopamine (6-OHDA)

6-OHDA is structurally related to dopamine (DA) and is selectively incorporated via active uptake channels into catecholamine (DA and noradrenaline, NA) neurons, where it is metabolized to produce toxic products that cause cell death. When injected systemically, 6-OHDA will induce ▶ sympathectomy, but will not cross the ▶ blood-brain barrier, so direct administration into the brain is required to induce lesions in the central neurons. The precise stereotaxic placement of the injection will determine which population of central catecholamine neurons are targeted. Thus, injection into the nigrostriatal pathway is effective in producing an effective lesion of central DA neurons of the ventral mesencephalon (substantia nigra and ▶ ventral tegmental area) resulting in widespread DA denervation of the neostriatum, ventral striatum, and associated cortical and limbic projections and provides a widely used animal model of ▶ Parkinson's disease (PD). In adult animals, bilateral 6-OHDA lesions, whether made by intraventricular or nigrostriatal injection, are associated with a profound akinetic syndrome and failure to engage in any voluntary behaviors results in ▶ catalepsy, aphagia, and adipsia, making these animals extremely difficult to maintain in good health and suitable for experimental analysis. By contrast, unilateral lesions produce a marked motor asymmetry involving neglect of contralateral space, postural and response bias to the ipsilateral side, and marked head-to-tail turning response ("rotation") when activated by an arousing stimulus or stimulant drug. Rotation in unilateral lesioned animals has been very widely used as a simple, reliable, quantitative functional test of experimental therapeutics – including neuroprotective agents, trophic factors, cell therapies, and novel pharmaceutics – targeted at DA system function, in particular with respect to PD, and also covering applications in schizophrenia, ADHD, and addiction.

Centrally administered 6-OHDA will affect all catecholamine neurons and their projections within the vicinity of the spread of the injection. To some extent, the lesions can be made selective to an individual population of neurons by the judicious placement of the injection, but, for example, nigrostriatal placement will inevitably also affect collateral NA projections in the medial forebrain bundle. If selective targeting of DA neurons is required, NA neurons can be protected by pretreatment i.p. with the selective uptake inhibitor desipramine. Conversely, selective NA depletion can be achieved by use of a selective NA toxin, such as DSP-4 (see below), or by 6-OHDA injection caudal to the ventral mesencephalon where the ascending DA neurons are located. Moreover, the forebrain projections of brainstem NA cell groups to hypothalamic, limbic, and cortical targets separate in dorsal and ventral bundles that can again be selectively disrupted by differential placement. Such manipulations were important in the pioneering studies that first distinguished the specific substrates for reward and motivational systems in the brain.

Dihydroxytryptamines (DHTs)

5,6-DHT and 5,7-DHT are structurally related to ▶ serotonin (5-hydroxytryptamine, 5-HT) and cause death of these neurons by a similar process of active uptake and intraneuronal cytotoxicity. Thus, injection into the cerebral ventricles or into the ascending fiber pathways in the medial forebrain bundle can produce relatively extensive depletion of forebrain serotonin. This has proved useful in studies of the role of 5-HT systems in the ▶ hippocampus and in particular in the interdependence and interaction of serotonergic and cholinergic afferents in regulating hippocampal function, both physiologically (such as in maintenance of the theta rhythm) and in studies of the central substrates of learning and memory. However, DHTs do not have the same degree of potency and selectivity as 6-OHDA. Both 5,6-DHT and 5,7-DHT cross-react with catecholamine neurons, and greater attention needs to be paid to selective blockade of uptake into both DA and NA neurons and to selective placement into the target terminal areas such as the hippocampus.

DSP-4

An alternative catecholamine neurotoxin is DSP-4 (N-2-chloroethyl-N-ethyl-2-bromobenzylamine) which, following peripheral administration, appears to have a relatively selective toxicity against NA neurons and nerve terminals, in particular the cortical and hippocampal projections of the locus coeruleus. In spite of the ease of administration and the relative selectivity for NA neurons within the brain, the toxin has been reported to also produce collateral damage in DA and 5-HT systems and lasting changes within the periphery as well as center, which may be overcome by direct central injection into circumscribed brain regions. Moreover, the precise mechanisms of DSP-4 toxicity remain ambiguous. This neurotoxin may nevertheless prove useful in combination with other toxins, such as 5,7-DHT, in comparing the interaction of NA and complementary 5-HT projections in the forebrain on the profiles of anatomical compensation and behavioral and electrophysiological changes within the target nuclei, in particular in the hippocampus or neocortex.

Other Models of Parkinson's Disease (PD)

1-Methyl-4-Phenyl-Tetrahydropyridine (MPTP)

MPTP was discovered as a by-product of a failed illegal synthesis of meperidine analogs, when it caused a profound parkinsonian syndrome in drug abusers who self-administered the substance. The peripheral administration of the drug in monkeys and mice induces major mesencephalic DA cell loss and a similarly marked PD-like motor syndrome. As in the idiopathic disease, MPTP-induced PD is responsive to L-DOPA.

MPTP is not itself the toxic agent. Rather it is metabolized by monoamine oxidase (MAO-B) to the active form, MPP+, which is taken up into cells via active DA uptake channels, and acts on mitochondria to cause cell death via mechanisms believed to involve reduced oxidative phosphorylation, lipid peroxidation, and disturbance of calcium homeostasis. MAO inhibitors such as deprenyl inhibit toxicity. Interestingly, the drug has little effect in rats which is believed to be due to the fact that the dominant isoform of MAO in this species, MAO-A, does not provide an equivalent substrate for MPTP conversion to MPP+, whereas if MPP+ itself is injected into the rat brain, it is as toxic in rats as it is in mice and monkeys.

MPTP continues to be widely used to model PD in mice and monkeys. Animals exhibit an acute parkinsonian syndrome that recovers rapidly after a single injection, and they need to be treated chronically to achieve stable bilateral cell loss, associated with profound parkinsonian debility, including ▶ akinesia, aphagia, and adipsia. As a consequence, unilateral lesions that retain the animals in better health are more suitable for long-term studies of reparative and neuroprotective therapeutics and can be achieved in monkeys by unilateral infusion of the toxin into the ascending carotid artery, thereby restricting drug distribution to just the ipsilateral hemisphere.

Industrial and Agricultural Chemicals

A variety of other toxins – including paraquat, rotenone, other industrial pyridines, and agricultural fertilizers – have been identified that appear to induce preferential toxicity against DA neurons after peripheral administration. This may relate to these neurons being particularly sensitive to oxidative stress and associated sensitivity of nigral neurons to iron toxicity. Indeed DA agonists such as ▶ methamphetamine and DA itself can be toxic to DA neurons both in vitro and in vivo when administered in high concentration. Although heavily investigated in terms of seeking to understand mechanisms of pathogenesis in PD, none of these toxins have been found to be sufficiently reliable, consistent, and specific to provide a practical alternative to 6-OHDA and MPTP.

A variety of other pharmaceutical agents are associated with extensive disruption of dopamine metabolism, such as ▶ reserpine and α-methyl-para-tyrosine. These drugs offer reversible tools to study acute dopamine depletion and were widely used in particular in early studies of experimental Parkinsonism. Repeated high doses of ▶ amphetamines (in particular methamphetamine) can also induce partial cell loss in the substantia nigra.

Ubiquitin-Proteasome System (UPS) Inhibitors
The UPS is the major system of the cell for tagging and digesting damaged, misfolded, and aberrant proteins. PD has been associated with impairment in proteolytic activity of the 20/26S proteasomes in the substantia nigra, leading to the investigation of UPS blockade as a model of PD. Thus, the injections of UPS inhibitors such as lactastatin, epoxomicin, or synthetic inhibitor peptides into the substantia nigra have been reported to produce not just selective DA depletion but also protein aggregation in nigral neurons and formation of Lewy body-like protein inclusions in the cells, characteristic of the major neuropathological hallmark of the human disease. Although this remains a plausible and attractive model of PD, there continues considerable debate about the reproducibility of the model, which seems to depend on quite precise lesion parameters to reproduce the specific pathology.

Excitotoxins

Excitatory Amino Acids (EAAs)
The excitotoxins are a class of neurotoxins with the common feature of being glutamate receptor (GluR) agonists. The first discovered EAAs with neurotoxic potential in the brain were monosodium glutamate (active after peripheral administration) and kainic acid (requiring central injection). Their common action as GluR agonists led to the "excitotoxicity hypothesis," viz., that at a sufficient dose, such EAAs induce a prolonged depolarization of glutamate receptive neurons, resulting in their cell death through some process of overactivation or overstimulation. Subsequent studies have identified a cascade of changes following GluR depolarization producing a "cycle of toxicity," which includes sodium influx, osmotic imbalance and cell lysis, calcium influx, activation of the mitochondrial energy chain and oxidative stress, inflammatory response, induction of specific cell death gene expression pathways, and other metabolic changes that combine to cause neuronal death via both necrotic and apoptotic mechanisms.

Kainic Acid (KA)
EAAs such as kainic acid offered a distinct advantage over older lesion methods (such as aspiration, electrolytic, and radiofrequency lesions) in permitting localized induction of cell loss while sparing axons of passage. This allowed for the first time making selective lesions in areas traversed by en passant fiber bundles, such as in the hypothalamus and neostriatum in rodents, and permitted important studies for the first time of the relative role of discrete hypothalamic nuclei vs. ascending brainstem regulatory systems in the central control of essential motivational and reward-related processes and early studies of behavioral anatomical and neurochemical changes in the striatum to produce the first usable animal model of ▶ Huntington's disease (HD).

Nevertheless, the excitatory potential of kainic acid was also associated with a marked potential for inducing sustained neuronal firing originating from the lesion focus, but inducing epileptogenic activity throughout the brain, overt seizures, and cell loss in susceptible neuronal populations in remote areas of the brain, in particular in the ▶ hippocampus and piriform lobe.

Receptor Subclass Selectivity
The profile of EAA-induced hyperactivity and cell toxicity depends on the match between the profile of glutamate receptor distribution on the

different populations of neurons in the target area and the GluR selectivity of the agonist. EAAs with excitotoxic properties have been identified binding to all major classes of GluR, notably the kainate, AMPA, and NMDA selective subclasses of ionotropic receptors, and metabotropic receptors sensitive to quisqualate. Some EAAs are relatively selective for a single receptor subclass, such as kainate, AMPA, and NMDA themselves, whereas others have mixed receptor targets, such as ibotenic acid.

With different cell populations in a nucleus or brain area expressing different receptors, judicious selection of toxin can provide different profiles of cell death. Thus, a comparison of alternative toxins following injection into the nucleus basalis magnocellularis (NBM) identified quisqualic acid and AMPA as the toxins of choice for the (relative) selective targeting of the cholinergic neurons of the basal forebrain-cortical projection system in comparison to the damage in non-cholinergic neurons produced by ibotenic acid, NMDA, and kainate.

Targeting NMDA receptors using NMDA itself has emerged as the neurotoxin of choice in the ▶ hippocampus and cortex, and NMDA has provided a potent toxin for demonstrating the effects of hippocampal lesions on a range of learning phenomena. Moreover, the NMDA receptor antagonist AP-5 has been particularly influential in mapping parallel effects of hippocampal NA denervation disrupting behavioral and electrophysiological function, supporting the hypothesis that hippocampal LTP both in vivo and in slice preparations provides a valid cellular substrate of the neural processes underlying learning and memory within hippocampal circuits.

More recently, quinolinic acid has emerged as the EAA of choice within the striatum. After intrastriatal injection, quinolinic acid is relatively selective for the medium spiny (▶ DARPP-32-positive) projection neurons of the striatum with relative sparing of the large ACh and medium NADPH-diaphorase-positive and other aspiny interneurons. It has proved of additional interest as an animal model of HD because this molecule is found in the normal brain as an endogenous product in one of the two main pathways of tryptophan metabolism, and the lesion has been widely used to explore the development of novel cellular, neurotrophic, and neuroprotective therapeutics for the human disease.

Nevertheless, receptor distribution is not the sole determinant of selectivity. Alternative excitotoxins also differ in (1) epileptogenic potential, as a consequence of which kainic acid is seldom used today for in vivo lesion studies; (2) lipophilicity, as a consequence of which ibotenic acid shows much greater diffusion through myelinated fiber tracts causing, e.g., more damage in deep layers of the cortex after striatal injection than does, say, quinolinic acid; and (3) inflammatory activity that can cause the demyelination of passing axons and reduce the fiber-sparing benefits originally claimed for the EAAs. Thus, the selection of an appropriate excitotoxin may be suggested by theoretical considerations but will require systematic empirical validation before use within any new model application.

Metabolic Toxins

3-Nitropropionic Acid (3-NP)

3-NP is a plant and fungal toxin associated with a toxic dystonic syndrome in man and has its action as an inhibitor of succinate dehydrogenase, a component of mitochondrial complex II energy metabolism. In the 1990s, 3NP was optimistically adopted as a particularly efficient metabolic toxin since it appeared to provide selective striatal lesions even after peripheral administration. For reasons that remain poorly understood, i.p. injections produce selective destruction of the medium spiny projection neurons of the neostriatum, with relative sparing of the striatal interneurons. This seemed to reproduce the profile of cell loss in HD and was adopted in several labs as a simple and efficient method to reproduce the human pathology. However, the toxin has a number of practical difficulties for experimental application. Standardized dosing can produce very great variability of toxicity with some animals exhibiting very extensive nonspecific lesions and large necrotic holes in the striatum, whereas other identically treated animals have no detectable pathology at all. In order to achieve consistent and relatively selective striatal lesions, it is necessary to administer very many small injections over a regular and extended time period, typically daily over weeks or even months, combined with daily testing with a functional readout that will allow setting a criterion for the cessation of dosing each animal at a comparable stage. The only effective protocols for achieving reliable and reproducible lesions that can be used to then test experimental therapeutics involve a major investment in time and resources that largely offsets the original promise of efficiency offered by the peripheral route of administration.

Other Mitochondrial Toxins
Although 3-NP may be relatively unreliable in its toxicity and specificity, a variety of other toxins similarly affect the mitochondrial energy chain, including malonic acid (MA, malonate), which inhibits the same mitochondrial complex II enzymes as 3-NP, and aminooxyacetic acid (AOAA). Unlike 3-NP, these toxins do not cross the blood-brain barrier and so require central injection. Thus, when injected into the striatum, both MA and AOAA induce effective striatal lesions. These toxins have been described as inducing secondary or "indirect" excitotoxicity to neurons in the area of injection, since the profile of lesion toxicity is similar to that induced by GluR agonists, they potentiate the effects of low doses of the latter EAAs, and their toxicity is partially blocked by the NMDA antagonist MK-801. These metabolic toxins have again been of particular interest as another model of HD since they are associated with a similar cellular impairment in metabolic function as seen in the human disease and have been argued to model a final common pathway in cell death. Moreover, toxicity is age dependent and selective for medium spiny neurons with relative sparing of striatal interneurons. However, in contrast to the human disease, malonate appears to have a toxic effect also on DA afferents to the striatum and can be more variable in its effects than the consistent results obtained with the classic EAA excitotoxins.

Immunotoxins

Cholinergic Neurotoxins
The selective targeting of cholinergic neurons for a long time relied on the selective placement of excitotoxins into the circumscribed cell body nuclei in the medial septum and the NBM. This achieved effective lesions of the target neurons and extensive cholinergic deafferentation of the targets of cholinergic projections in the cortex and hippocampus, respectively, but could never be demonstrated to be specific; co-localized non-cholinergic neurons and parallel projections were typically equally affected (e.g., damage to the parvalbumin-positive ▶ GABA neurons of the medial septum projecting in parallel to the hippocampus). A claim was made for cholinergic selectivity for the ethylcholine mustard aziridinium ion, AF-64A, but this has not been well supported. In the hands of most people, AF-64A produces cavitation at the site of injection and extensive nonspecific lesions. Greater specificity has subsequently been demonstrated for the immunotoxin 192-IgG ▶ saporin.

Immunotoxins
Following a similar logic to the familiar neuroanatomical use of antibodies to label specific target molecules on cells in a range of sensitive immunohistochemistry techniques, immunotoxins involve conjugating an immunoglobulin antibody (or immunoglobulin fragment) to target cell surface receptors or channels with a cytotoxin that will then induce death of the target cell. The trick is to select targets such as neurotrophic factor receptors, which will internalize the ligand, thereby allowing the toxin to be transported into the cell for intracellular concentration, often with active transport back to the cell body where its primary toxic action takes effect. A range of antibody fragments and cytotoxins have now been developed for a range of application, the first of which to receive major attention was 192-IgG saporin.

192-IgG Saporin
192-IgG saporin was the first used and most extensively explored of the burgeoning range of

immunotoxins. The 192-IgG immunoglobulin recognizes the p^{75} low affinity NGF receptor which is widely expressed on cholinergic neurons in the forebrain. After central injection, 192-IgG saporin binds to the NGF receptor, achieving selective internalization preferentially in cholinergic neurons, retrograde transport of the cytotoxin saporin to the cell body, and consequent death of the cell. Although the preparation and stability of the toxin and precise parameters of injection are important, 192-IgG saporin has proved the most successful technique for selective lesion of septal and basal forebrain neurons sparing co-localized cells and has been widely used to confirm that many of the electrophysiological and behavioral changes in learning and memory function associated with excitotoxic lesions of the medial septum and NBM are indeed attributable to the selective depletion of the cholinergic innervations of the hippocampus and cortex, respectively.

Other Immunotoxins

More recently, other immunotoxins have been developed using a similar conjugate strategy to develop tools for simple and effective lesion of NA neurons by targeting the synthetic enzyme dopamine-ß-hydroxylase, DA neurons by targeting the DA transporter, striatal neurons by targeting the substance P receptor, and selective lesions of basal forebrain neurons receiving glutamatergic projections by conjugating saporin to the NMDA receptor. Indeed, the selectivity provided by immunotoxins allows a variety of previously intractable experimental issues to be addressed, such as the relative contributions of striosome and matrix compartments of the striatum by targeting the striosome projection neurons with a saporin-antibody complex recognizing the μ-opiate receptor.

Conclusions

The last three decades have provided a wide range of novel neurotoxins that allow selective disruption and death of targeted populations of neurons within the brain. Targets may be selected by anatomical location, morphological cell type, specific neurotransmitter (via selective uptake channel), pharmacological and immunological targeting of identified receptors, or disruption of subcellular metabolic cell processes. These then provide powerful tools for making selective manipulations as the independent variable in physiological, pharmacological, and behavioral analyses of normal and abnormal brain function in health and disease. Nevertheless, while most available toxins are effective in reproducing particular profiles of cell loss and features of cellular and subcellular pathology of the corresponding human disease, they frequently do not reproduce the neuropathogenic process accurately, and the lesions are typically acute rather than slowly progressive as is characteristic of most human neurodegenerative diseases. Consequently, neurotoxin-based lesions can provide effective models to study functional organization in the nervous system and have value for evaluating symptomatic and reparative approaches to treatment. However, such lesions are less suitable for the experimental analysis of neuroprotective processes and for the development of therapeutic strategies to alter or reverse the progressive course of disease. For such applications, alternative genetic models are increasingly considered preferable.

Acknowledgments I thank Dr. Simon Brooks and Dr. Emma Lane for their helpful comments on the manuscript. I acknowledge the ongoing support for our own studies from the UK Medical Research Council, the Parkinson's Disease Society, the High Q Foundation, and the FP6 and FP7 research programs of the European Union.

Cross-References

- ▶ Anti-Parkinson Drugs
- ▶ Apoptosis
- ▶ Blood–Brain Barrier
- ▶ Dementias: Animal Models
- ▶ Excitatory Amino Acids and their Antagonists
- ▶ 6-hydroxydopamine
- ▶ Long-Term Potentiation and Memory

References

Balázs R, Bridges RJ, Cotman CW (2006) Excitatory amino acid transmission in health and disease. Oxford University Press, Oxford/New York

Dunnett SB (2005) Motor functions of the nigrostriatal dopamine system: studies of lesions and behaviour. In: Dunnett SB, Bentivoglio M, Björklund A, Hökfelt T (eds) Dopamine, vol 21, Handbook of chemical neuroanatomy. Elsevier, Amsterdam, pp 235–299

Dunnett SB, Björklund A (1999) Parkinson's disease: prospects for novel restorative and neuroprotective treatments. Nature 399(Suppl):A32–A39

Feldman RS, Meyer JS, Quenzer LF (1997) Principles of neuropsychopharmacology. Sinauer, Sunderland

Iversen LL, Iversen SD, Bloom FE, Roth RH (2009) Introduction to neuropsychopharmacology. Oxford University Press, Oxford/New York

Sanberg PR, Nishino H, Borlongan CV (2000) Mitochondrial inhibitors and neurodegenerative disorders. Humana press, Totowa

Schwarting RKW, Huston JP (1996a) The unilateral 6-hydroxydopamine lesion model in behavioral brain research. Analysis of functional deficits, recovery and treatments. Prog Neurobiol 50:275–331

Schwarting RKW, Huston JP (1996b) Unilateral 6-hydroxydopamine lesions of meso-striatal dopamine neurons and their physiological sequelae. Prog Neurobiol 49:215–266

Neurotransmitter

Definition

Neurotransmitters are chemical substances that relay, amplify, and modulate signals between one neuron and another or between a neuron and another cell. Typically, neurotransmitters are packaged into vesicles that cluster beneath the membrane on the presynaptic side of a synapse and are released into the synaptic cleft, where they bind to receptors in the membrane on the postsynaptic cell. Release of neurotransmitters usually follows arrival of an action potential at the synapse but may follow graded electrical potentials. Low level "baseline" release may also occur without electrical stimulation.

Neurotransmitter Transporters

Definition

Proteins that function to take up neurotransmitters from the extracellular milieu into cells such as neurons and glia. These transporters are targets for many psychotherapeutic drugs (e.g., Fluoxetine, Prozac©) and sites of action of abused drugs (e.g., cocaine).

Neurotrophic Factors

Definition

A neurotrophic factor is a neuropeptide that regulates the growth, differentiation, and survival of certain neurons in the peripheral and central nervous systems.

Cross-References

▶ Brain-Derived Neurotrophic Factor

Neurovascular Unit

Definition

In addition to the capillary endothelial cells, the site of anatomical blood-brain barrier, neurons, and nonneuronal cells such as pericytes, astrocytes, and microglia together constitute a functional unit, often referred to as a neurovascular unit.

Cross-References

▶ Blood–Brain Barrier

Neutral Antagonists

Definition

A ligand that binds to a receptor does not increase or decrease cellular activity, but can block the actions of both agonists and ▶ inverse agonists.

Cross-References

▶ Inverse Agonists

Nicotinamide Adenine Dinucleotide

Synonyms

NAD

Definition

Coenzyme utilized by alcohol dehydrogenase (ADH) to convert alcohol into acetaldehyde; the rate-limiting factor of ADH-related metabolic tolerance.

Nicotine

Paul B. S. Clarke
Department of Pharmacology and Therapeutics, McGill University, Montréal, QC, Canada

Synonyms

(−)-1-Methyl-2-(3-pyridyl)pyrrolidine

Definition

Nicotine is by far the most extensively studied chemical in tobacco smoke. This entry summarizes its pharmacokinetics, mechanisms of action, and behavioral effects in humans and animals. The contributions of nicotine to tobacco addiction are highlighted, as currently understood from the perspective of animal and human studies. One such contribution may reflect nicotine's ability to enhance the impact of other (sensory or chemical) reinforcers that are associated with tobacco smoke.

Pharmacological Properties

Pharmacokinetics

Nicotine is a tertiary amine and weak base, such that it is more than 50 % ionized at physiological pH. In its nonionized form, nicotine tends to pass rapidly through membranes. For example, when nicotine reaches the brain via the carotid arteries, it is swiftly taken up and then released slowly back into the bloodstream. Animal studies have shown that the brain can maintain a threefold higher nicotine concentration than plasma, although how much of the drug is free to act on receptors in the brain is unclear.

Nicotine is extensively metabolized by cytochrome P450 and other hepatic enzymes, giving rise to numerous metabolites, principally cotinine ("▶ Cotinine"). The plasma elimination half-life of nicotine is around 2 h in humans but with notable interindividual differences reflecting genetic variation. Elimination is faster in rats (half-life approx. 1 h) and more especially in mice (6–7 min). Metabolic pathways are also species dependent; for example, the main P450 enzyme responsible for nicotine-cotinine conversion differs between humans and rats (CYP2A and CYP2B1/2, respectively).

Peripherally formed nicotine metabolites appear to have only minor pharmacological effects. However, in animals and potentially in humans, nicotine is also metabolized within the brain. One metabolite formed locally within the rat brain, nornicotine, is a weak nicotinic agonist ("▶ Nicotinic Agonists and Antagonists") but may accumulate with repeated nicotine dosing.

It was long believed that the act of cigarette smoking extracts nicotine rapidly from the lungs, such that each puff would promptly deliver a highly concentrated bolus to the brain. However, empirical evidence now reveals that arterial levels of nicotine rise quite gradually, peaking only 15–30 s after each puff. This finding has major implications for our understanding of nicotine reinforcement.

Mechanisms of Action

Nicotine exerts almost all of its known actions via nicotinic acetylcholine receptors (nAChRs). Each nicotinic receptor comprises five protein subunits, arranged around a water-filled channel. The receptor is normally closed but opens upon binding of the agonist. Channel opening allows the passage of ions, notably Na^+ and/or Ca^{++}. These positively charged species rapidly enter the cell, causing depolarization. This depolarization in turn stimulates the opening of voltage-gated ion channels, with a consequent cascade of intracellular signaling events. Hence, at the cellular level, stimulation of nAChRs can trigger contraction (skeletal muscle), action potential generation, and modulation of transmitter or hormone release.

Nicotinic receptors are widely expressed by neurons in the brain, spinal cord, and autonomic ganglia and by primary sensory neurons. Indeed, there is scarcely a nucleus in the brain that does not express nicotinic receptors. Although most CNS nAChRs are located in or on neurons, some appear associated with glial cells. Nicotinic AChRs also occur on other nonneuronal cells, for example, in the skeletal muscle, lung, skin, immune cells, and vascular endothelium.

Neuronally expressed nicotinic receptors are highly heterogeneous (Hurst et al. 2013). This multiplicity partly reflects the diversity of protein subunits which can combine to form a receptor. Each type of nAChR subunit is encoded by a different gene, and neuronal nAChRs are formed from combinations of α2–α7 and β2–β4 subunits. Theoretically, then, a vast number of nAChR subtypes might exist, but the real number appears far smaller, with perhaps one dozen accounting for most pharmacological actions of nicotine. Subtypes of nAChR are named according to their constituent subunits, with the stoichiometry indicated if known (e.g., α4β2 or α4$_2$β2$_3$). By convention, an asterisk indicates the possible presence of additional subunits (e.g., α4β2*). Most nAChR subtypes comprise combinations of both α and β subunits, but α7 subunits can additionally form homo-oligomeric receptors (i.e., α7$_5$). Individual neurons are capable of expressing more than one nAChR subtype, each of which may contain several different types of subunit.

Nicotinic AChR subtypes differ in several important respects (Hurst et al. 2013). First, each subtype possesses a unique anatomical pattern of expression. Second, nAChR subtypes differ in their channel properties. For example, α7-containing receptors are unusually permeable to Ca^{++} relative to Na^+ ions, and this property has important implications for transmitter release and other intracellular functions mediated by Ca^{++}. Third, nAChR subtypes differ in their sensitivity to nicotinic agonists and antagonists and in their propensity to desensitize and resensitize. Fourth, some receptor subtypes proliferate in response to chronic agonist administration, although the functional consequences of such upregulation are not well understood.

Nicotinic receptors transition through three main functional states: resting, activated, and desensitized. In the absence of agonist, the receptor-associated channel is predominantly closed. Binding of agonist leads to channel opening, an extremely rapid transition typically occurring within milliseconds. The activated receptor may subsequently enter a desensitized state in which the channel is again closed. With time, the desensitized receptor assumes the resting, activatable state. Certain nAChR subtypes desensitize rapidly (e.g., α7 in less than 1 s), whereas others desensitize slowly if at all. Although desensitization was once thought to require high agonist concentrations, it is now known that even very low (nanomolar) concentrations of nicotine can desensitize some nAChR subtypes, without significant receptor activation. This is potentially significant, since in habitual smokers, plasma nicotine levels remain in this range even after a night of abstinence. With prolonged agonist exposure, nicotinic receptors can enter a more persistent desensitized state, termed inactivation, lasting several days (Wu and Lukas 2011).

It is not at all clear to what extent cigarette smoking stimulates versus desensitizes neuronal nAChRs. The answer will surely depend on the nAChR subtype, previous nicotine exposure, and the temporal pattern and amount of nicotine intake. Most animal models related to tobacco smoking employ relatively brief nicotine exposures which are unlikely to mimic the complexity of tobacco smoking.

Behavioral Effects of Nicotine in Humans and Animals

Nicotine exerts a plethora of acute behavioral effects, which is unsurprising given that nAChRs are very widely expressed in the brain and elsewhere. Certain behavioral effects depend not only on the dose administered but also on recent or remote drug history; some effects undergo transient or persistent tolerance, whereas others show sensitization or else remain largely stable across repeated tests.

Comparison of behavioral effects in humans versus animals is hindered by several factors. First, behavioral procedures are usually quite different. Second, most human studies are conducted in tobacco smokers; these individuals often have many years of drug experience. Third, very few animal studies have attempted to mimic the complex pharmacokinetics associated with tobacco smoking. Instead, nicotine is typically administered as an acute subcutaneous or intraperitoneal "depot" injection, in doses providing plasma levels higher than those found in most habitual smokers (Matta et al. 2007). The use of relatively high doses in animals leads to numerous behavioral effects, not all relevant to tobacco smoking. Behavioral responses to nicotine that are most related to reinforcement are described in a later section.

In drug-naïve rats, nicotine can produce marked behavioral disruption, associated with prostration and ataxia. However, upon repeated dosing, persistent tolerance develops, such that a generalized stimulant effect emerges in tests of locomotor stimulation and operant responding. This transition is often viewed in terms of behavioral sensitization ("▶ Sensitization to Drugs"), although it is at least partly due to tolerance to the initial depressant effect. The behavioral-activating effect of nicotine appears quite general, since operant responding is also increased during time-out periods when reinforcers are unavailable and also in tasks where low rates of responding are preferentially reinforced.

Nicotine's status as a mild psychostimulant drug is supported by several additional findings. For example, like amphetamine and similar drugs, nicotine improves sustained attention and produces an amphetamine-like discriminative stimulus ("▶ Drug Discrimination") (i.e., drug cue) in nicotine-experienced rats (Smith and Stolerman 2009). Pharmacological and lesion studies that have been conducted in animals further suggest that like psychostimulants, nicotine can increase locomotor activity and serve as a reinforcer via a mechanism dependent on mesolimbic dopaminergic transmission (see below).

Many other behavioral effects of systemic nicotine administration have been reported in rodents or human subjects or both. These include antinociception ("▶ Analgesics"), improved cognition ▶ Cognitive Enhancers, reduced or increased anxiety ("▶ Anxiety: Animal Models"), attenuated aggression, reduced food intake ("▶ Appetite Suppressants"), and conditioned taste aversion ("▶ Conditioned Taste Aversions").

Antinociceptive effects of nicotine have been observed in several animal pain models ("▶ Antinociception Test Methods"), and the underlying neuronal and receptor mechanisms have been partially elucidated. Pain-suppressing effects of tobacco smoking have also been demonstrated in human subjects; limited evidence suggests that nicotine is at least partly responsible.

Sustained attention is improved by nicotine in both humans and animals (Levin 2013; "▶ Attention" and "▶ Rodent Tests of Cognition"). Improvements have been documented in rats, using several procedures including the five-choice serial reaction time task. In this procedure, major differences between rat strains were observed. Interestingly, attentional enhancement by nicotine was not fully mimicked by amphetamine, suggesting divergent mechanisms. Attentional processing is also increased in human subjects receiving nicotine via skin patches. Enhancement has been found not only in abstinent smokers but also in nonsmokers. Patients suffering from schizophrenia or attentional deficit hyperactivity disorder have also shown improvements in attention.

Learning and/or memory is improved by nicotine in a wide variety of vertebrate species and procedures. In rodents, nicotine tends to enhance these cognitive aspects more strongly than attention, whereas the reverse has been found in humans. It is unclear whether this divergence represents a species difference or reflects the particular behavioral tests used in rodents versus humans. In rats, nicotine-associated improvements in working memory have been extensively studied using the eight-arm radial maze ("▶ Short-Term and Working Memory in Animals"). Here, several nicotinic agonists have proven effective, and nicotine is also effective when given chronically. Mechanisms of memory improvement have been partially elucidated in terms of nAChR subtypes, transmitter systems, and brain structures (Levin 2013).

Although in most animal and human studies, subjects are exposed to nicotine during acquisition and test sessions, a few reports employing posttrial administration suggest that nicotine can also improve memory consolidation. Nicotine has had beneficial effects in rodents with impaired learning or memory due to brain lesions. However, nicotine does not appear to improve memory in Alzheimer's disease ("▶ Dementias and Other Amnestic Disorders") or schizophrenic patients ("▶ Schizophrenia"), despite improvements in attentional processing.

Tests of anxiety in rodents have provided clear evidence of both anxiolytic and anxiogenic effects, with dose and baseline anxiety acting as modulating variables ("▶ Anxiety: Animal Models"). Genetic deletion of different nAChR subunits suggests that multiple nAChR subtypes modulate anxiety levels in complex ways. Rodents undergoing nicotine withdrawal show signs of anxiety, and in smokers, tobacco withdrawal-related anxiety is alleviated by nicotine replacement.

Aggressive behavior is reduced by acute systemic administration of nicotine in several species, including rats and human subjects ("▶ Aggression"). In some studies, the effect is demonstrably not due to motor impairment. Irritability and aggressiveness commonly occur during smoking cessation and are alleviated by nicotine replacement therapy especially in subjects with high trait hostility. Mechanisms underlying these effects of nicotine remain unexplored.

Tobacco smoking reduces the body weight set point, and weight gain is a common consequence of cessation. Many smokers, especially young women, regard body weight gain as a deterrent to quitting. However, smokers tend to eat as much as nonsmokers, and acute administration of nicotine appears not to reduce hunger or caloric intake in hungry smokers, suggesting that metabolic factors may be critical instead. Nicotine deprivation does not appear to contribute much to post-cessation weight gain, since the latter is only mildly inhibited by nicotine replacement therapy. In rodents, nicotine not only exerts complex effects on peripheral energy metabolism but also suppresses appetite. The anorexigenic effect of nicotine likely arises from peripheral and central actions, especially within the lateral hypothalamus.

Conditioned taste aversion is readily produced by nicotine in some strains of mice and in adult, but not in peri-adolescent, rats. The effect is due to a central action of nicotine, and it appears dependent on dopaminergic transmission in the nucleus accumbens.

Nicotine provides a recognizable and reliable discriminative stimulus (cue), in humans, monkeys, and rodents ("▶ Drug Discrimination"). Typically, nicotine has been administered by nasal spray (humans) or by systemic injection (animals). In rats, the nicotine cue has been identified as central in origin, likely involving multiple nicotinic receptor subtypes, transmitter pathways, and brain regions (Smith and Stolerman 2009). The multiplex nature of the nicotine cue, together with the potential for redundancy across different neural pathways, has hindered mechanistic dissection. As mentioned, the nicotine cue has psychomotor stimulant-like properties in rodents, with contributions from ascending dopaminergic projections as well as α4β2 nAChRs. However, the nicotine cue is not critically dependent upon mesolimbic dopamine transmission in rats, unlike the drug's reinforcing and locomotor stimulant effects (Smith and Stolerman 2009). Candidate trigger sites have been identified in medial prefrontal cortex and hippocampus.

Subjective Effects of Nicotine in Human Subjects

Nicotine acutely elicits multiple and complex subjective effects, including changes in mood, alertness, and anxiety. These effects depend on a host of factors, such as personality, smoking status, degree of abstinence, situational context (e.g., stress), passive versus self-administration, other drug use, and dose. Nicotine has been administered intravenously or via nasal spray in most studies. Stronger subjective responses have been seen in nonsmokers than in smokers. Generally, low to moderate doses tend to improve mood in smokers, especially during abstinence. However, dysphoria is commonly encountered as well, particularly at higher doses. Nicotine increases subjective arousal in smokers but reduces it in subjects who have never smoked; it can also make both types of individual feel less relaxed. A number of studies have suggested that nicotine can exert euphoric, "head-rush" effects resembling cocaine. However, such findings have little relevance to tobacco smoking, since they were obtained in known substance abusers given rapid (10-s) intravenous infusions of nicotine in doses equivalent to one or two cigarettes (i.e., 0.01–0.04 mg/kg). Puff-size doses of nicotine (e.g., 0.1 mg/infusion IV), in contrast, have only mild subjective effects.

Nicotine as a Contributor to Tobacco Addiction

In many individuals, cigarette smoking represents an addiction: it is a compulsive behavior, unaided quitting is rare, and relapse is common. Nicotine has been accorded a leading role in tobacco addiction, mostly notably in the 1988 US Surgeon General's Report. This document concluded that "nicotine is the drug in tobacco that causes addiction" ("▶ Nicotine Dependence and Its Treatment"). The main arguments that have been forwarded in support of this conclusion are as follows. Nicotine is consumed in ways that avoid the many pyrolysis products found in tobacco smoke (e.g., snuff, chewing tobacco). The tobacco withdrawal syndrome can largely be attributed to nicotine withdrawal, since most symptoms are countered by nicotine replacement therapy, and a nicotine withdrawal syndrome can be produced reliably in animals. Nicotine replacement therapy doubles the chance of quitting. In the absence of tobacco, nicotine can serve as a positive reinforcer in human and animal self-administration studies. Nicotine shares important characteristics with other drugs of abuse such as amphetamine (e.g., cue properties, sensitization, release of mesolimbic dopamine, facilitation of brain stimulation reward). However, as discussed below, several of these statements are subject to important qualifications, and consequently, a more nuanced view of nicotine's role in tobacco addiction is beginning to emerge.

Animal Models Related to Nicotine Reward

"Reward" is a multifaceted concept, and the reward-related effects of nicotine are commonly studied in the context of intravenous self-administration, conditioned place preference, or the drug's ability to enhance other reinforcers. Most such studies are performed using rats and increasingly mice.

Intravenous self-administration ("▶ Self-Administration of Drugs"). Intravenous infusions of nicotine can serve as a primary reinforcer in several mammalian species including humans (Le Foll and Goldberg 2008). However, the great majority of studies have employed rates of drug delivery and doses outside the range experienced by smokers. In particular, nicotine has nearly always been given as a rapid (e.g., 1-s) bolus, whereas after a cigarette puff, the drug is released only slowly into the circulation, with arterial levels peaking after some 15–30 s. Furthermore, smokers receive only 1–2 μg/kg per puff (Matta et al. 2007), and although there are isolated reports of self-administration at doses as low as 3 μg/kg, most animal studies continue to use 15–30 μg/kg per infusion. To date, studies of brain mechanisms related to nicotine self-administration have relied on this conventional fast-infusion high-dose procedure.

Conditioned place preference ("▶ Conditioned Place Preference and Aversion") has been observed after acute nicotine administration in rats and mice (Le Foll and Goldberg 2008) but is often neither as reliable nor as large as that seen with drugs such as morphine and psychostimulants.

The reinforcement-enhancing effect of nicotine refers to an ability to heighten the impact of other reinforcers. It was first described in rats engaged in intracranial self-stimulation and has now been extended to a wide range of unconditioned (sensory, gustatory, and pharmacological) and conditioned reinforcers. This phenomenon also complicates the interpretation of intravenous self-administration studies performed in animals. This is because in a typical experiment, nicotine infusions are programmed to coincide with presentation of a cue light which is somewhat reinforcing in its own right; nicotine makes this sensory reinforcer more reinforcing (Caggiula et al. 2009). Reinforcement-enhancing effects of nicotine have also been reported in human subjects and could potentially contribute to tobacco addiction (see below).

Nicotine Withdrawal in Humans and Animals

Many of the signs and symptoms of tobacco withdrawal ("▶ Withdrawal Syndromes") can be alleviated by nicotine replacement therapy, giving rise to the notion that tobacco withdrawal is principally due to *nicotine* abstinence. However, several caveats are in order (Hughes 2007). For example, even in placebo-controlled trials, nicotine patch and nicotine gum can produce detectable cues, suggesting that some nicotine-treated subjects may *expect* to obtain withdrawal relief. It has also been argued that some tobacco withdrawal symptoms are nonspecific to nicotine but would also be seen upon loss of other reinforcers. Another puzzling feature of tobacco withdrawal is that it is not readily precipitated by administration of the centrally active nicotinic antagonist mecamylamine.

Rodents readily express signs of spontaneous or antagonist-precipitated nicotine withdrawal after chronic exposure (Malin and Goyarzu 2009). In most studies, however, nicotine is delivered via subcutaneous osmotic minipumps and reaches constant 24 h exposure levels beyond the range of all but the heaviest smokers. For example, the standard minipump infusion dose used in adult rats is 3 mg/kg/day, which tends to provide nicotine plasma levels in excess of 50 ng/ml. By way of comparison, a typical smoker (15 cigarettes/day) attains between-cigarette "trough" levels of no more than around 20 ng/ml.

Both spontaneous and nAChR antagonist-precipitated types of withdrawal are manifested by "somatic" and "affective" signs. Somatic signs include behaviors (e.g., ptosis, writhing and grasping) and pharmacological features reminiscent of opiate withdrawal, although there is little evidence of opioid involvement in tobacco withdrawal signs in humans. "Affective" signs are manifested by elevations in brain-simulation reward thresholds and conditioned place aversion. Somatic signs derive at least partly from inhibition of central nAChRs; hence they are also termed "somatically expressed signs" (Malin and Goyarzu 2009). Affective signs, in contrast, exclusively reflect central nAChR function. Somatic and affective signs are also distinguishable in terms of nAChR subtype mediation.

Nicotine in Relation to Nicotinic Cholinergic Transmission

Nicotinic AChRs represent an important target for the neurotransmitter acetylcholine (ACh). Although nAChRs and ACh are both widely expressed in the brain, evidence for nicotinic cholinergic transmission in the brain has only emerged in the past 25 years. For several reasons, it is not known how many nAChRs actually participate in cholinergic transmission. First, many previous attempts to localize nAChRs using immunocytochemistry are now considered unreliable because of doubts about antibody specificity. Second, some evidence suggests that ACh may be capable of acting not only as a synaptic transmitter but also via volume (extrasynaptic) transmission. Third, α7-containing nAChRs may also be activated by endogenous levels of *choline*. To date, there has been little attempt to integrate knowledge about nicotinic cholinergic transmission with the known behavioral effects of nicotine.

Nicotine or Nicotine-Plus?

Nicotine, in its pure form, is only mildly rewarding in standard animal models (intravenous self-administration and conditioned place preference). Most if not all human studies also concur with this conclusion. How, then, might nicotine help maintain tobacco addiction? Several explanations have been proposed. First, standard behavioral procedures may not capture the full reinforcing potential of this drug in animals. For example, the relevant studies tend to be of short duration, lasting at most a few weeks. This limitation may be important because it is known that in the case of cocaine, rats become much more motivated to seek the drug after they have had several months of drug exposure. Alternatively, if cigarette smokers use nicotine as a cognitive tool or for emotional support, these aspects would not be modeled in widely used animal tests such as intravenous self-administration.

Even if nicotine serves as only a weak *primary* reinforcer in humans, it may still play an important role by potentiating other reinforcers associated with tobacco smoking. Foremost here might be sensory cues. Cigarette smoke is detected not only visually but also in the mouth, nose, throat, and airways, partly via activation of nicotinic receptors that are located on sensory nerve terminals. This sensory barrage gives rise to sight, taste, aroma, and irritation, which in turn contributes to smoking satisfaction in habitual smokers (Rose 2008). Although smoke-related sensory stimuli may serve to some degree as primary reinforcers, they would appear particularly well placed to acquire conditioned-reinforcing properties. Significantly, nicotine enhances the impact of both unconditioned and conditioned reinforcers, according to animal studies (above).

Tobacco smoke comprises a complex mixture of thousands of chemicals, and only a tiny percentage have been analyzed pharmacologically or tested for possible behavioral effects. One such compound is the nicotinic agonist nornicotine, which is much less prevalent than nicotine and is only weakly self-administered by rats. Acetaldehyde, in contrast, is only 50 % less prevalent than nicotine in mainstream smoke, and although it is not detectably reinforcing at smoking-comparable doses in rats, it can synergize with high-dose bolus infusions of nicotine to support intravenous self-administration. Tobacco smoke also contains identified MAO-inhibiting chemicals ("▶ Monoamine Oxidase Inhibitors"), and smokers possess reduced levels of MAO-A and MAO-B isoforms, both in the CNS and periphery. Monoamine neurotransmitters, and dopamine in particular, are metabolized by MAO, giving rise to the suggestion that MAO inhibition enhances neurochemical actions of nicotine that reinforce tobacco smoking. Rats that are treated with MAO-inhibiting drugs can become much more motivated to self-administer intravenous nicotine, but two caveats are in order. First, this facilitatory effect has been reported with levels of MAO inhibition that far exceed that experienced by smokers, and second, evidence indicates that MAO inhibition is at best only partly responsible for the behavioral effects observed.

Cross-References

▶ Aggression
▶ Analgesics

- ▶ Antinociception Test Methods
- ▶ Anxiety: Animal Models
- ▶ Appetite Suppressants
- ▶ Attention
- ▶ Conditioned Place Preference and Aversion
- ▶ Conditioned Taste Aversions
- ▶ Cotinine
- ▶ Dementias and Other Amnestic Disorders
- ▶ Drug Discrimination
- ▶ Monoamine Oxidase Inhibitors
- ▶ Nicotine Dependence and Its Treatment
- ▶ Nicotinic Agonists and Antagonists
- ▶ Rodent Tests of Cognition
- ▶ Schizophrenia
- ▶ Self-Administration of Drugs
- ▶ Sensitization to Drugs
- ▶ Short-Term and Working Memory in Animals
- ▶ Withdrawal Syndromes

References

Caggiula AR, Donny EC, Palmatier MI, Liu X, Chaudhri N, Sved AF (2009) The role of nicotine in smoking: a dual-reinforcement model. Nebr Symp Motiv 55:91–109

Hughes JR (2007) Effects of abstinence from tobacco: etiology, animal models, epidemiology, and significance: a subjective review. Nicotine Tob Res 9:329–339

Hurst R, Rollema H, Bertrand D (2013) Nicotinic acetylcholine receptors: from basic science to therapeutics. Pharmacol Ther 137:22–54

Le Foll B, Goldberg SR (2008) Effects of nicotine in experimental animals and humans: an update on addictive properties. Handb Exp Pharmacol 192:335–367

Levin ED (2013) Complex relationships of nicotinic receptor actions and cognitive functions. Biochem Pharmacol 86:1145–1152

Malin DH, Goyarzu P (2009) Rodent models of nicotine withdrawal syndrome. Handb Exp Pharmacol 192:401–434

Matta SG, Balfour DJ, Benowitz NL, Boyd RT, Buccafusco JJ, Caggiula AR, Craig CR, Collins AC, Damaj MI, Donny EC, Gardiner PS, Grady SR, Heberlein U, Leonard SS, Levin ED, Lukas RJ, Markou A, Marks MJ, McCallum SE, Parameswaran N, Perkins KA, Picciotto MR, Quik M, Rose JE, Rothenfluh A, Schafer WR, Stolerman IP, Tyndale RF, Wehner JM, Zirger JM (2007) Guidelines on nicotine dose selection for in vivo research. Psychopharmacology (Berl) 190:269–319

Rose JE (2008) Disrupting nicotine reinforcement: from cigarette to brain. Ann N Y Acad Sci 1141:233–256

Smith JW, Stolerman IP (2009) Recognising nicotine: the neurobiological basis of nicotine discrimination. Handb Exp Pharmacol 192:295–333

Wu J, Lukas RJ (2011) Naturally-expressed nicotinic acetylcholine receptor subtypes. Biochem Pharmacol 82:800–807

Nicotine Dependence and Its Treatment

Maxine L. Stitzer
Psychiatry and Behavioral Sciences, Johns Hopkins Bayview Medical Center, Baltimore, MD, USA

Synonyms

Smoking cessation

Definition

Dependence and abuse are part of a diagnostic system (DSM-5) designed to classify the nature and extent of problems substance use disorders that people have in controlling their use of chemical substances. The concepts of dependence and abuse can apply to any drug that people use on a regular basis, including tobacco, alcohol, marijuana, opiates, and stimulants. Since nicotine is thought to be the main chemical that makes tobacco substance use disorder use addictive, the terms nicotine dependence and tobacco dependence are often used synonymously. Features of dependence include excessive time spent acquiring, using, and recovering from the use of drugs; using larger amounts of them and for longer periods of time than intended; developing tolerance, having difficulty in stopping them (repeated unsuccessful quit attempts); experiencing craving for the drug, having withdrawal symptoms when the use of the drug is stopped; interference in social, recreational, or work activities; and continued use despite knowledge of adverse consequences. Not all these criteria apply to cigarette smoking, but many of the key criteria (e.g., continued use despite knowledge of adverse consequences; withdrawal symptoms, and difficulty quitting with repeated attempts) do apply.

Who Smokes and Who Quits?

It is not difficult for people to start smoking, since tobacco cigarettes worldwide are a legal, readily available, and often heavily promoted product. Smoking prevalence may still be rising in many countries, but in the US, rates have steadily declined over the past 50 years with growing recognition of the health risks associated with smoking, as well as increasingly stringent environmental restrictions and the rising cost of cigarettes. Today, the smoking rate in the US general population stands at about 19 % with smoking concentrated in lower socioeconomic strata and in individuals with other mental health and substance abuse problems. With much lower than historical rates of smoking, this is still a remarkably high prevalence rate for something that is known to be a very serious behavioral risk factor for premature morbidity and mortality.

Interestingly, at least among US smokers, about 70 % at any given time claim that they want to quit smoking. Furthermore, many of these smokers, perhaps 25–30 %, do attempt to quit each year. The problem is that the quit attempt for many is short-lived with a return to smoking, usually within the first week or two after quitting, being the ultimate outcome. This return to smoking following a quit attempt is called relapse. Scientists have been trying to understand when, why, and how people relapse during a quit attempt to get clues to help smokers succeed in quitting permanently.

Relapse Circumstances

Smokers have been asked to report on the circumstances of their return to smoking both in retrospective and real-time surveys, and the factors surrounding relapse have been well documented (Shiffman et al. 1996). Withdrawal symptoms are a clinical reality when smokers try to quit and are a factor contributing to relapse. These symptoms include irritability, anxiety, restlessness, and difficulty concentrating, and they last for about 2–3 weeks after quitting. Interestingly, withdrawal symptom intensity has not been clearly linked to smoking cessation outcomes, but craving levels do appear to be important. Specifically, higher craving levels at any given point of time are associated with poorer outcomes at later points of time, whether the timeframe examined is early or later in the quit attempt. Intensity of postquit craving may reflect the smoker's level of dependence on cigarettes. Alleviation of withdrawal symptoms can make people feel less irritable and more comfortable after quitting; at the same time, a reduction in craving may be especially important for improving quit success.

A another important factor associated with relapse is negative affect. While relapse can occur in positive mood states, people frequently report going back to smoking when they feel negative emotions including anger, frustration, stress, and even boredom. Emotions may serve as a strong cue for smoking, particularly if people have previously used smoking to enhance positive emotions and/or to feel better in the presence of negative mood states. Thus, any intervention that can help people avoid or improve postquit negative mood states may be beneficial.

Drinking alcohol increases the likelihood of a relapse. Not only is alcohol a cue for smoking, it may directly enhance the pleasant effects of smoking and promote continued smoking through behavioral disinhibition. Due to this interaction between alcohol and cigarettes, smokers are often advised to avoid alcohol during their quit attempt. Finally, self-confidence in being able to stay away from cigarettes is often associated with better outcomes in a quit attempt. Self-confidence may be influenced by many factors, but encouragement and support from people in the social network of a smoker who is trying to quit may be helpful for boosting self-confidence.

All relapses must by definition begin with the first inhalation of tobacco smoke following the start of a quit attempt. This initial reexposure, whether it happens within the first 24 h after the quit attempt or after several weeks or months, is called a slip or lapse. Theoretically, the smoker who slips by taking several puffs or even smoking a whole cigarette could just stop and return to abstinence. However, research shows that slips indicate a poor prognosis and are associated with long-term failure in the quit attempt. People who have an initial slip or lapse within the first 2 weeks after they quit would have only a 10 % chance of succeeding in their quit attempt (vs. a 90 % chance of failing), whereas those who do not smoke at all in the first 2 weeks have a much better chance (about 50 %) of long-term success (Kenford et al. 1994). Thus, it would be beneficial if treatments could prevent or reduce the negative impact of smoking slips during a quit attempt.

Successful treatments must simultaneously address all the factors that contribute to relapse, which include withdrawal symptoms and cravings, response to environmental and internal emotional cues, and the negative impact of smoking slips. There are currently two things that smokers can do to improve their chances of quitting. One is to use an approved medication and the other is to use behavioral counseling support for smoking cessation. Many smokers quit on their own with no help from medicines or therapy. It is estimated that the success rate for unaided quit attempts is about 5 % on a given try. In contrast, when modern treatments for smoking cessation are tested, long-term (e.g., 6–12 months) success rates may be as high as 30 % (Fiore et al. 2008).

Medications: Nicotine Replacement Products

There are several places to find reviews of medications for smoking cessation (Fiore et al. 2008; Foulds et al. 2006; Le Foll and George 2007; Nides 2008). The first medications approved for smoking cessation were the nicotine replacement therapies (NRT). These products come in five different forms: patch, gum, lozenge, inhaler, and nasal spray. All the products deliver pure nicotine to the body to provide a short-term substitute for cigarettes when the smoker is trying to quit, and these can lower the intensity of withdrawal symptoms and cravings. Further, all the NRT improve smoking cessation outcomes, approximately doubling the chances of a successful quit when compared with placebo medication. The NRT products differ primarily in the route and speed at which nicotine is absorbed, as well as in their convenience and ease of use (see Table 1). Nicotine delivered in medications is much safer than nicotine delivered by smoking tobacco. This is because all the toxins and carcinogens delivered in smoked tobacco are eliminated, while nicotine itself is not cancer causing. Nicotine does stimulate heart rate and blood

Nicotine Dependence and Its Treatment, Table 1 Pros and cons of smoking cessation therapies

Treatment	Pros	Cons
Medications		
Nicotine patch	Easy to use (applied once daily); available OTC; provides steady nicotine levels	No smoker control
Nicotine gum	Smoker controls dose and timing; available OTC	Acceptability/ effort of use; difficult to use with dentures
Nicotine lozenge	Smoker controls dose and timing; available OTC	Side effects include nausea, hiccups, heartburn
Nicotine inhaler	Mimics hand-to-mouth behavior of smoking; smoker controls dose and timing	Needs prescription; intensive puffing required to get adequate dose
Nicotine nasal spray	Rapid onset effect; smoker controls dose and timing	Needs prescription; side effects include burning nose, runny eyes, sneezing
Bupropion (Zyban)	Twice per day pill	Needs prescription; side effects include insomnia and dry mouth; contraindicated with seizures, head trauma, eating disorders
Varenicline (Chantix)	Twice per day pill; few medical contraindications to use	Needs prescription; side effects include nausea and sleep disturbance; possible suicidal thoughts

(continued)

Nicotine Dependence and Its Treatment, Table 1 (continued)

Treatment	Pros	Cons
Behavioral support		
Group	Tips for successful quitting; social support from leader and group members	Effort required to find and use groups; location and scheduling barriers
Individual	Intense, tailored help	Hard to find
Telephone (quit lines)	Easy to use; provides guidance and social support	Therapy may be less intense and less structured than in-person methods
Internet	Easy to use; provides good information and suggestions	Less social support than in-person help; quality may vary across sites
Self-help booklets	Easy to use; provides good information and suggestions	Not shown to be efficacious in boosting quit rates

pressure, but it can be used safely even by people with heart disease, because it is safer than smoking cigarettes (Benowitz and Gourlay 1997).

Nicotine can be absorbed from any surface of the body; this includes any part of the skin, mouth, and nasal membranes as well as the surface of the lung, as in smoking. (Nicotine is not well absorbed by the stomach owing to the acid content of that organ.) The various NRT formulations have taken advantage of this absorption versatility to deliver nicotine by different routes.

Nicotine patches are perhaps the most widely used method of nicotine delivery. They are very convenient to use, being applied and changed only once daily, and may be used for 16 or 24 h per day. Patches are the only formulation that delivers steady levels of nicotine throughout the day, and this is their advantage. Patches are meant to be used for 8–12 weeks after quitting to suppress withdrawal symptoms and cravings. Most patches deliver about 21 mg/day of nicotine. The dose content of patches can be tapered during the later weeks of use (e.g., to 14 then 7 mg/day), to gradually wean away from nicotine, or the patches can be stopped abruptly with no ill effects. Side effects are few and include skin irritation and disturbed sleep. Nicotine patches can be obtained either by prescription or over the counter (OTC).

With all the remaining nicotine replacement products, the number and timing of doses is controlled by the smoker. This is both the advantage and downside of these products. The advantage is that the products can be used "as needed" when difficult situations arise, in which the individual may be tempted to smoke. The disadvantage is that people sometimes fail to use the medications in adequate amounts to gain the full benefits of nicotine replacement. Gum, for example, should be used at about 9 pieces per day for full effect, but many people use less. Gum and lozenge are very similar in that they both deliver nicotine through the mouth membranes; lozenge may be easier and more socially acceptable to use since it does not require chewing. Both are available OTC. Both gum and lozenge come in 2 mg and 4 mg doses. The higher dose is recommended for heavier smokers (e.g., more than 1 pack per day).

The nicotine inhaler also delivers nicotine across mouth mucous membranes. This product requires a hand-to-mouth behavior that some smokers find comforting since it mimics smoking behavior. The disadvantage of the inhaler is that intensive puffing is required to obtain adequate doses of nicotine, and there may be initial side effects including throat burning, watery eyes and nose, and coughing. The inhaler is only available by prescription. For all NRT taken by mouth, acidic drinks such as coffee and juice should be avoided for 15 min before product use, since these can reduce nicotine absorption.

Nicotine nasal spray delivers nicotine through nasal mucous membranes. This product delivers a relatively high dose with rapid absorption that most closely mimics nicotine delivery from cigarettes. Side effects include burning, runny nose and eyes, and sneezing, but these effects subside with continued use. The spray is only available by prescription and is recommended for heavier smokers.

It is important to note that the use of NRT product combinations has recently been shown to further improve the chances of success beyond those obtained with the use of a single medication. In particular, combined use of patch plus a short-acting smoker-controlled medication such as gum or nasal spray can produce better cessation outcomes than either medication alone (Fiore et al. 2008). This may be because the advantages of steady nicotine replacement levels and smoker-controlled dosing are combined.

Medications: Non-Nicotine Products

In addition to all the NRT, there are also two non-nicotine medications that have been approved by the FDA for smoking cessation based on evidence of efficacy. Bupropion (Zyban) is an antidepressant medication whose benefits for smoking cessation were discovered by chance during its use as an antidepressant. Subsequent research has shown that bupropion doubles quit rates when compared with a placebo control and appears to reduce both withdrawal symptoms and craving. Interestingly, the medication is equally effective for smokers with and without a history of depression, so it can be used by any smoker. Bupropion comes in 150 mg pills and is taken twice daily starting 1–2 weeks prior to the quit date; it needs to be obtained by prescription. There are some limitations on who can take it, related to medical conditions. For example, people with a history of head trauma, seizures, or eating disorders should not take this medicine (Fiore et al. 2008). Side effects are primarily insomnia and dry mouth. Even though this medication works in people with and without a history of depression, smokers who are concerned about negative affect may want to try this medication.

Varenicline (Chantix) is the most recent medication approved for smoking cessation. This drug, a partial agonist, was specifically developed as a smoking cessation aid. It attaches to the receptor responsible for nicotine effects in the brain, acting like a substitute to reduce withdrawal symptoms and cravings. However, there is also some evidence that varenicline can reduce the reinforcing effects (e.g., satisfaction) of nicotine when smokers have a postquit slip or lapse, thus potentially addressing this important relapse risk factor (West et al. 2008). The research on which FDA approval of this drug is based showed that varenicline efficacy is better than that of both placebo and of bupropion and that the medication can triple the chances of a successful quit attempt when compared with placebo (Gonzales et al. 2006; Jorenby et al. 2006). Varenicline must be obtained by prescription; the only medical contraindication is severe kidney impairment. Side effects are mainly nausea and disturbed sleep, but concerns have also been raised about increased suicidal thoughts in people taking this medication.

Table 1 summarizes the available smoking cessation medications and highlights their pros and cons. Smokers generally choose a smoking cessation product based on convenience and availability, the advice of their doctor, and their own personal experience with the use of these products. Because smokers generally require several attempts before they can quit for good, there is an opportunity to try different medications and see which ones work best for each individual.

The medications listed earlier are broadly recommended for use by all types of smokers, but there are two specific groups for whom the medications may not be recommended (Fiore et al. 2008). Research with adolescent smokers (under 18 years of age) has shown that counseling therapy is beneficial but has not shown that medications (NRT or bupropion) improve the chances of successful quitting. Therefore, medications are not currently recommended for adolescent smokers. Pregnant women are the second group where recommendations must be qualified. In this case, there has been only limited research conducted with medications because of safety concerns. Nicotine most likely does have adverse effects on the fetus, which is why it is important for pregnant women to stop smoking. However, they should try to stop with behavior therapy alone, if possible. If this does not work, pregnant smokers should consult with their doctor about the use of smoking cessation medications.

Behavior Therapies

Smokers trying to quit will benefit from using a behavior therapy program in addition to medications. Behavior therapy is frequently available through hospitals or community agencies and should consist of individual or group counseling sessions offered both before and after the target quit day. In these programs, smokers are taught how to prepare for a quit attempt and what to expect when they quit. Relapse prevention skills are also stressed with interactive discussion of how to handle difficult situations, in which the newly abstinent smoker may be tempted to light up a cigarette. Research has shown that behavior therapy can improve the chances of successful quitting through problem solving and social support, and that in general, the more therapy people get, the better their outcome is likely to be. Specifically, 4–8 or more in-person therapy sessions have been shown in research to be optimal for improving outcomes (Fiore et al. 2008).

More recently, there is evidence that therapy delivered over the telephone via quit lines is efficacious (Fiore et al. 2008). This is an encouraging finding, since some people find it difficult or inconvenient to go to an in-person therapy program, and some may be reluctant to engage in face-to-face therapy. Ideally, quit lines should offer a structured series of calls following the initial contact, with calls initiated by the therapist who provides guidance and support during the quit attempt comparable with what would be found in an individual or group therapy program. There are also Internet programs now available that can help smokers quit. The quality of these programs may vary. An ideal program would include personalized help with specific issues and problems, monitoring and feedback on personal progress, and a chance to interact with others who are trying to quit.

In summary, there are several ways that smokers can improve their chances of success in smoking cessation, but the most important principle is that a combination of medication and behavior therapy has been shown in research to be the optimal approach for best outcomes. There are seven FDA approved medications available either by prescription or OTC, and behavior therapy is more available than ever with the advent of telephone and Internet-based counseling. Smokers attempting to quit are advised to take full advantage of these resources.

Cross-References

▶ Nicotine
▶ Nicotinic Agonists and Antagonists
▶ Withdrawal Syndromes

References

Benowitz NL, Gourlay SG (1997) Cardiovascular toxicity of nicotine: implications for nicotine replacement therapy. J Am Coll Cardiol 29:1422–1431

Fiore MC, Jaen CR, Baker TB, Bailey WC, Benowitz NL et al (2008) Treating tobacco use and dependence clinical practice guideline. 2008 update. US Government Printing Office, Washington, DC; ISSN 1530–6402. ISBN 978-1-58763-351-5

Foulds J, Steinberg MB, Williams JM, Ziedonis DM (2006) Developments in pharmacotherapy for tobacco dependence: past, present and future. Drug Alcohol Rev 25:59–71

Gonzales D, Rennard SI, Nides M, Oncken C, Azoulay S, Billing CB, Watsky EJ, Gong J, Williams KE, Reeves KR (2006) Varenicline, an $\partial 4\beta 2$ nicotinic acetylcholine receptor partial agonist, vs sustained-release bupropion and placebo for smoking cessation. JAMA 296:47–55

Jorenby DE, Hays JT, Rigotti NA, Azoulay S, Watsky EJ, Williams KE, Billing CB, Gong J, Reeves KR (2006) Efficacy of varenicline, an $\partial 4\beta 2$ nicotinic acetylcholine receptor partial agonist, vs placebo or sustained-release bupropion for smoking cessation. JAMA 296:56–63

Kenford SL, Fiore MC, Jorenby DE, Smith SS, Wetter D, Baker TB (1994) Predicting smoking cessation. Who will quit with and without the nicotine patch. J Am Med Assoc 271:589–594

Le Foll B, George TP (2007) Treatment of tobacco dependence: integrating recent progress into practice. Can Med Assoc J 177:1373–1380

Nides M (2008) Update on pharmacological options for smoking cessation treatment. Am J Med 121:S20–S31

Shiffman S, Paty JA, Gnys M, Kassel JA, Hickcox M (1996) First lapses to smoking: within-subjects analysis of real-time reports. J Consult Clin Psychol 64:366–379

West R, Baker CL, Cappelleri JC, Bushmakin AG (2008) Effect of varenicline and bupropion SR on craving, nicotine withdrawal symptoms, and rewarding effects of smoking during a quit attempt. Psychopharmacology (Berl) 197:371–377

Nicotinic Acetylcholine Receptor Subtypes

Definition

Receptors formed by different combinations of subunits.

Nicotinic Agonists and Antagonists

Hans Rollema[1], Daniel Bertrand[2] and Raymond S. Hurst[3]
[1]Rollema Biomedical Consulting, Mystic, CT, USA
[2]Department of Neuroscience, HiQScreen Sàrl, Geneva, Switzerland
[3]Forum Pharmaceuticals, Watertown, MA, USA

Synonyms

Nicotinic acetylcholine receptor ligands; Nicotinics

Definition

The wide range of pharmacological effects of nicotine and other nicotinic natural products, such as epibatidine, anabasine, lobeline, and cytisine, together with significant advances in the knowledge of the role and function of nicotinic acetylcholine receptors (nAChRs; "▶ Nicotinic Receptor"), continues to stimulate interest in the potential of nAChR antagonists, agonists, partial agonists, and modulators as treatments for central nervous system (CNS) disorders. Focused efforts to design selective nAChR ligands as new therapies have been ongoing for three decades, but to date, only two synthetic nAChR ligands have been approved for clinical use – the nAChR antagonist mecamylamine for hypertension in 1950 and the nAChR partial agonist varenicline for smoking cessation aid in 2006. Evidence is accumulating for the involvement of nAChRs in several CNS disorders in addition to nicotine dependence, including alcohol use disorder, depression, schizophrenia, attention-deficit/hyperactivity disorder (ADHD), neurodegenerative diseases, pain, and inflammation. Advances have been made in the design and development of subtype selective compounds targeting these disorders.

Current Concepts and State of Knowledge

Pharmacological Properties of Neuronal Nicotinic Acetylcholine Receptors

The name of Doctor Jean Nicot de Villemain, who introduced tobacco to the French royal court in 1561 and demonstrated its therapeutic properties by curing the queen's migraine, was immortalized in pharmacology when nicotine, the natural alkaloid in the tobacco plant, was named after him. Only centuries later, with the discovery of a class of receptors for the neurotransmitter acetylcholine (ACh) that are specifically activated by nicotine, the nicotinic acetylcholine receptors (nAChRs), we are beginning to understand the details of nicotine's effects on the central nervous system (CNS), on the peripheral nervous system, and on non-neuronal cells.

In this update of "Nicotinic Agonists and Antagonists" (Rollema et al. 2010), we briefly describe the current knowledge of nAChRs and discuss nicotinic ligands with therapeutic potential for various indications.

Structure and Subunits

nAChRs belong to a superfamily of ligand-gated ion channels characterized by a conserved sequence in the N-terminal domain flanked by linked cysteines. These highly conserved membrane proteins are pentameric structures, composed of different subunits. To date, eight α (2–7, 9, and 10) and three β (2, 3, and 4) subunits have been identified in the mammalian CNS, while three additional subunits have been identified in muscle receptors that are termed γ, δ, and ε. nAChRs consist of five identical subunits (homomeric) or of two or more different subunits (heteromeric). The α7 subunit forms homomeric receptors, a major component of the nAChRs in the CNS. Heteromeric receptors formed by α4 and β2 subtypes are the other principal receptor types in the mammalian brain. Certain subunit combinations are more widely expressed than others, while some are restricted to well-defined neuronal pathways. Since nAChRs can contain multiple α and β subunit types and the subunit composition varies in different species, insight in receptor composition and localization, especially in the human brain, is critically important.

The subunits in nAChRs are arranged in a doughnut-like manner forming a pore in their center, the ion channel. They have an extracellular ligand-binding site with an N- and C-terminal and four membrane-spanning domains (Fig. 1). Three loops in the N-terminal of the α subunit form the major component of the ligand-binding site and are positioned next to three loops of the adjacent subunit that form the remainder of the binding site. In this specific arrangement, both subunits interact with the ligand and, therefore, determine the binding properties.

Functional Activity

Functionally, nAChRs modulate the flow of ions across the cell membrane under the control of ACh, the natural endogenous nAChR ligand. When ACh is released from presynaptic cholinergic axon terminals, it binds to the extracellular binding domain of the receptor and controls channel opening. The subsequent net influx of cations through the channel pore depolarizes the cell membrane and elevates intracellular calcium, thereby influencing the activity of the target neurons. Nicotinic compounds can have therapeutic effects by acting as extracellular signaling molecules that can mimic, attenuate, or potentiate the action of ACh.

The simplest representation of nAChR ligand binding is a bimolecular reaction: the ligand binds and is subsequently released with a single association and dissociation constant, so that ligand affinity is characterized by a binding constant that is the equilibrium between the ON and OFF rates. Binding affinities provide however little or no information on the conformational states in which the ligand stabilizes the receptor, since agonists, which transition the nAChR to the open active state, are indistinguishable from antagonists, which stabilize the nAChR in a closed conformation. Functional assays are needed to classify ligands according to their functional efficacy and pharmacological effects: full agonists are as efficacious as ACh in activating the receptor; partial agonists are less efficacious than ACh; and antagonists inhibit receptor activation by ACh.

Importantly, functional efficacy or intrinsic activity (the maximal amount of receptor activation) is not correlated with functional potency (the concentration of ligand required to cause half maximal activation of the receptor (Fig. 2a). An additional complexity, common to all ligand-gated channels, is that sustained presence of an agonist progressively stabilizes the receptor in a desensitized, nonconducting state. Thus, exposure to an agonist is expected to cause transient activation, followed by sustained

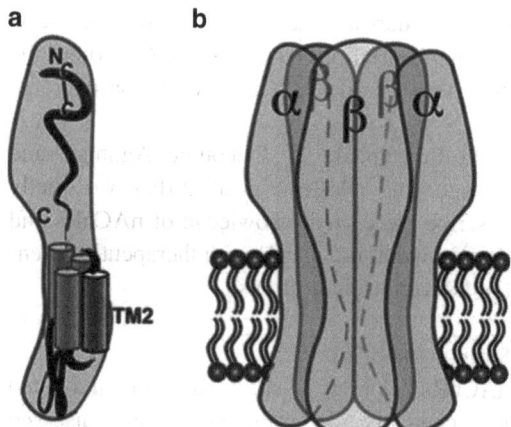

Nicotinic Agonists and Antagonists, Fig. 1 Schematic representation of the nAChR and the ACh binding site. (a) Representation of a single subunit formed of α-helices (cylinders), spanning the membrane four times with the N- and C-termini facing the extracellular domain. Note the position of the second transmembrane domain (TM2) that is facing the ionic pore. (b) Schematic drawing of an nAChR inserted into the lipid bilayer. The ligand-binding site lies at the interface between two adjacent α and β subunits (From Rollema et al. 2010)

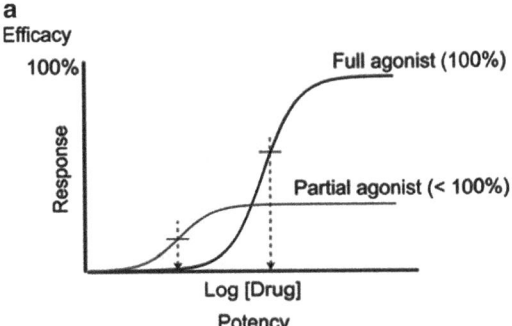

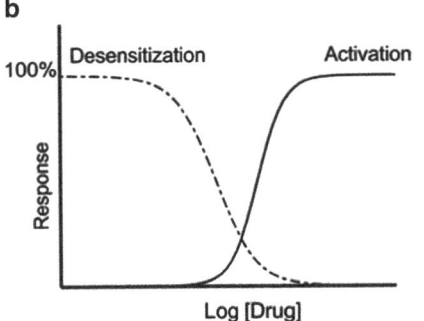

Nicotinic Agonists and Antagonists, Fig. 2 Concentration-dependent nAChR activation and desensitization. (**a**) Typical concentration activation curves for a full agonist (100 % efficacy) and a partial agonist (<100 % efficacy). Note that the potency of the compound (EC_{50}) does not correlate with its efficacy (relative activation vs. ACh). (**b**) Concentration relationships for activation and desensitization are represented on the same graph. Note that desensitization occurs at lower concentrations when compared with that of activation, indicating that sustained exposure to an agonist will mainly cause receptor desensitization (From Rollema et al. 2010)

inhibition due to desensitization, which typically occurs at concentrations one or two orders of magnitude lesser than those required for activation. Ligands display therefore a higher affinity for receptors in their desensitized state than in their active state, which is easily seen when plotting activation and desensitization profiles in one graph (Fig. 2b). Since radioligand binding data are acquired during prolonged incubation with the ligand to reach equilibrium (>0.5 h), agonist-binding affinities correlate best with receptor desensitization potencies.

Transition from the closed state to the open state reflects changes in the three-dimensional structure of the receptor. Though the simplest model involves only one transition between the resting and active states, evidence points toward the existence of an intermediate state, thought to be closed and termed the "flip state" (Fig. 3). The complex receptor structure allows high-affinity binding at sites other than the ligand-binding ("▶ Orthosteric") site. Molecules that bind at these so-called allosteric sites can either increase (positive allosteric modulators) or decrease (negative allosteric modulators) the evoked current and thereby affect nAChR function. For comprehensive reviews on the structure, properties, and functions of nAChRs, see, e.g., Albuquerque et al. 2009, Taly et al. 2009, and Hurst et al. 2013.

nAChR Upregulation

Repeated exposure to nicotine results in an increase or "upregulation" in surface nAChRs, which has been shown to occur in laboratory animals and in smokers. Several mechanisms may underlie functional receptor upregulation, e.g., decreased turnover of cell-surface receptors, increased transitions of receptors from low to high sensitivity state, increased synthesis of new proteins, increased receptor trafficking to the cell surface, nicotine acting as a chemical "chaperone," and facilitating assembly of the nAChR and trafficking to the cell surface (review Govind et al. 2012). Numerous studies have shown that nAChR agonists, partial agonists, and antagonists can upregulate different nAChR subtypes to the same extent as nicotine. nAChR upregulation might have therapeutic relevance, especially if it occurs at such low concentrations that the net result of upregulation and nAChRs desensitization is a gain of function.

nAChR Antagonists, Agonists, and Modulators

Given the diversity of physiological roles of specific nAChR subtypes and their relationship with disease states, the possibility of subtly modulating the function of these receptors with selective ligands has fueled interest in developing agonists, partial agonists, antagonists, and allosteric modulators as potential pharmacotherapeutic agents for a number of CNS disorders. A schematic representation of the major nAChRs subtypes that

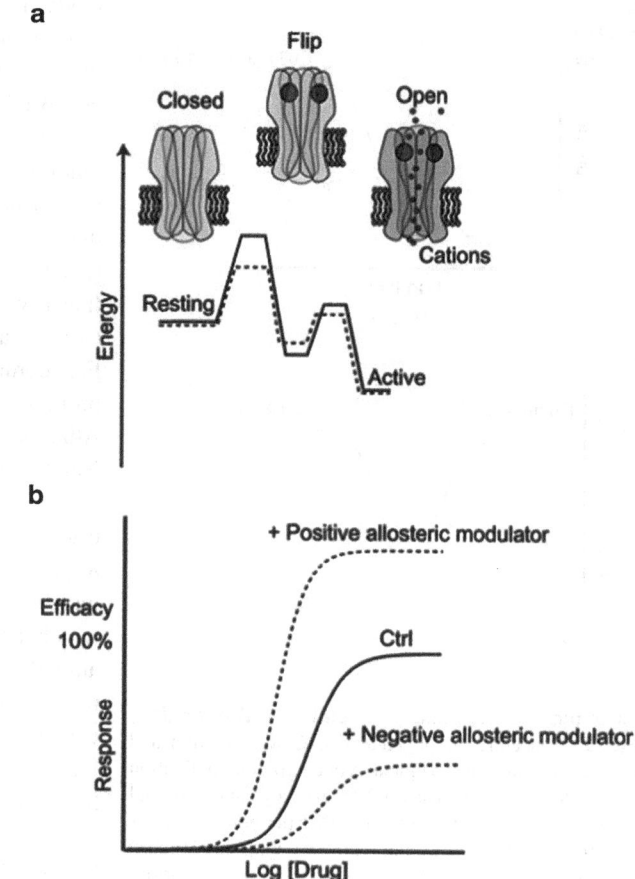

Nicotinic Agonists and Antagonists, Fig. 3 Activation and modulation of the nAChRs (**a**) Schematic representation of the energy barriers between the resting (*closed*) and the active (*open*) state. The intermediate state, also referred to as the "flip state," has been shown to play an important role in partial agonism. (**b**) Effects of positive (+) or negative (−) allosteric modulators on the concentration activation profiles. Positive allosteric modulators cause an enhancement of the agonist-evoked response and potency and increase the apparent cooperativity (slope of the curve) compared to the control (agonist only)-evoked response (Ctrl) (From Rollema et al. 2010)

are most relevant as CNS targets for therapeutic intervention, as well as peripheral nAChR subtypes that could mediate adverse events, is given in Fig. 4.

In drug discovery, the selectivity of novel molecules for a target nAChR subtype is a key parameter that is assessed preclinically through binding and functional studies using transgenic animal models and nAChR ligands that have well-established selectivity for one or more nAChR subtypes. The chemical tools that are most commonly used as selective agonists or antagonists in in vivo pharmacological or in in vitro receptor studies are shown in Fig. 5.

Since often the same subtypes and comparable mechanisms underlie different disorders, ligands targeting an individual nAChR subtype may be beneficial for more than one disease. On the other hand, the physiological role of individual nAChR subtypes is highly complex and far from being completely understood, also because disease state and ligand exposure can substantially alter the number and type of nAChRs expressed in various brain regions. Given these complexities, it is not surprising that results from animal models do not always translate to the clinic and that several developmental candidates with a promising preclinical profile have been discontinued because of the lack of efficacy or an insufficient therapeutic window. Unfortunately, the actual reason for discontinuation of clinical studies with novel compounds is often not known or not disclosed, making it difficult to evaluate if the proposed mechanism at a target nAChR subtype for a certain indication was appropriately tested in the clinic.

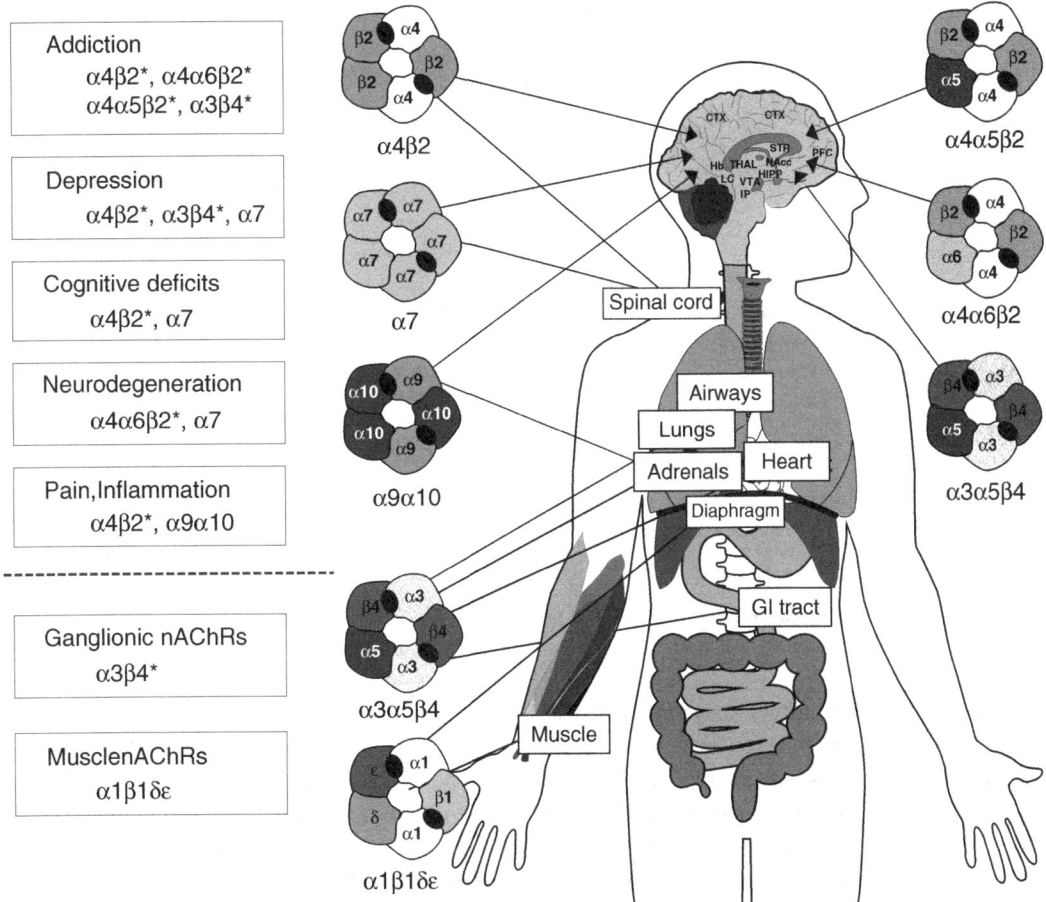

Nicotinic Agonists and Antagonists, Fig. 4 Schematic overview of the major central nAChR subtypes that are targets for treatment of the CNS disorders discussed in this chapter. Examples of CNS areas that are modulated by nAChRs are the cortex (CTX), prefrontal cortex (PFC), hippocampus (HIPP), thalamus (THAL), locus coeruleus (LC), ventral tegmental area (VTA), nucleus accumbens (NAcc), striatum (STR), habenula (Hb), interpeduncular nucleus (IP), and spinal cord. Medicinal chemistry efforts targeting CNS disorders attempt to minimize activity at peripheral nAChR subtypes such as α1βδε (muscle) and α3β4* (ganglionic, chromaffin cells in adrenal glands) that control a variety of critical physiological functions. An asterisk in the heteromeric nAChR subtypes indicates that other subunits may be present. Note that the figure is not inclusive of all nAChR subtypes expressed in the body and of all areas that contain nAChRs; additional receptor subtypes or areas not shown could either represent therapeutic targets or mediate adverse events (Adapted from Rollema et al. 2010)

Therapeutic Potential of Targeting Nicotinic Receptors

The following sections discuss nAChR ligands that target certain nAChR subtypes and are or have been in clinical development for various therapeutic indications.

Addiction

Various nAChR subtypes modulate the release of neurotransmitters and mediate neuronal events that are associated with the development and maintenance of drug dependence. Therefore, nAChR ligands are of interest as potential treatments of drug addictions, in particular

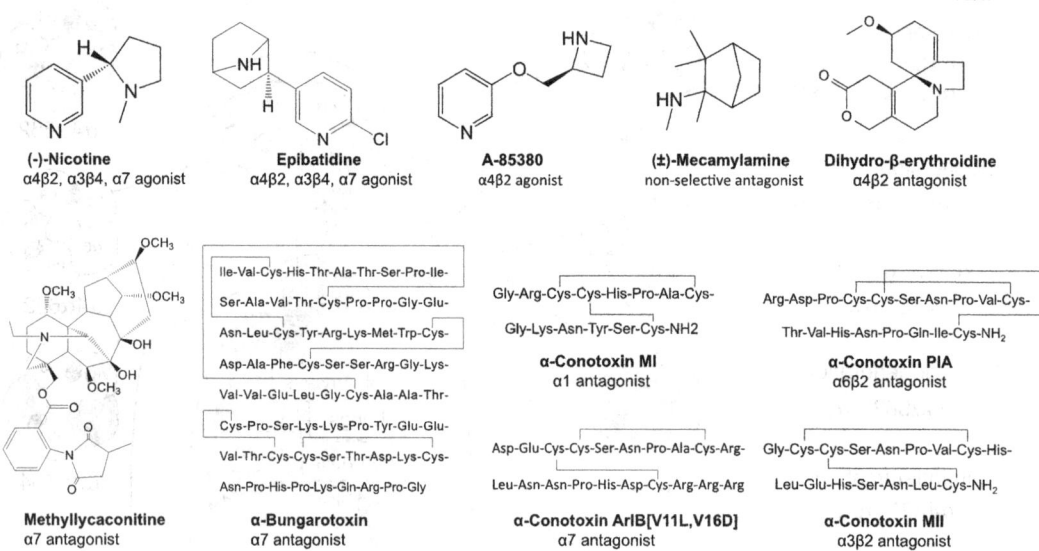

Nicotinic Agonists and Antagonists, Fig. 5 nAChR ligands used as pharmacological tools

compounds that will affect mesolimbic dopamine, a key mediator of the rewarding and addictive effects of drugs of abuse (reviews by Changeux 2010; D'Souza and Markou 2011; Hurst et al. 2013). Varenicline is so far the only nAChR ligand, besides nicotine itself in the form of nicotine replacement therapy (NRT), which has been approved as a drug for smoking cessation, and ongoing nAChR-based addiction research is aimed at ligands that are potent and selective at α4β2* or other nAChR subtypes (Fig. 6). Though the main focus of drug discovery efforts is on nicotine dependence, a limited number of nicotinic compounds are also being explored as potential treatments for dependence to drugs of abuse other than nicotine, such as alcohol, cocaine and amphetamine.

Nicotine Dependence ("▶ Nicotine Dependence and Its Treatment") is a chronic and recurring disorder, since it is extremely difficult to quit smoking, because of the rewarding effects of nicotine and the withdrawal symptoms that are almost immediately relieved by taking nicotine again and lead to relapse. The rewarding effects of nicotine are mediated via the interactions of inhaled nicotine with high-affinity nAChRs, such as α4β2, α4α6β2*, or α4α5β2* nAChRs, resulting in repeated pulsatile increases in mesolimbic dopamine release. Other nAChR subtypes, e.g., α3β4*, are thought to play a role in the pronounced withdrawal symptoms, such as anxiety, depression, and insomnia, which are experienced during abstinence.

Given the serious health consequences of smoking, pharmacotherapy of nicotine dependence is highly cost-effective, as it improves long-term abstinence. This can be achieved with the first two approved smoking cessation aids, NRT and bupropion, that have modest end-of-treatment quit rates, with odds ratios ≤2. NRT replaces nicotine from smoking with "safe" nicotine, i.e., without the toxic and carcinogenic components from tobacco smoke. Nicotine-containing gums, patches, or inhalers provide a constant delivery of nicotine that enters the brain and reduces craving and withdrawal symptoms during a quit attempt. Bupropion is an antidepressant that was found to help smokers quit and subsequently marketed as a smoking cessation aid. Its activity is likely mediated by the dopamine and norepinephrine reuptake inhibitory properties, but the weak nAChR antagonist properties of bupropion, and especially of its active metabolite hydroxybupropion, may contribute to its efficacy.

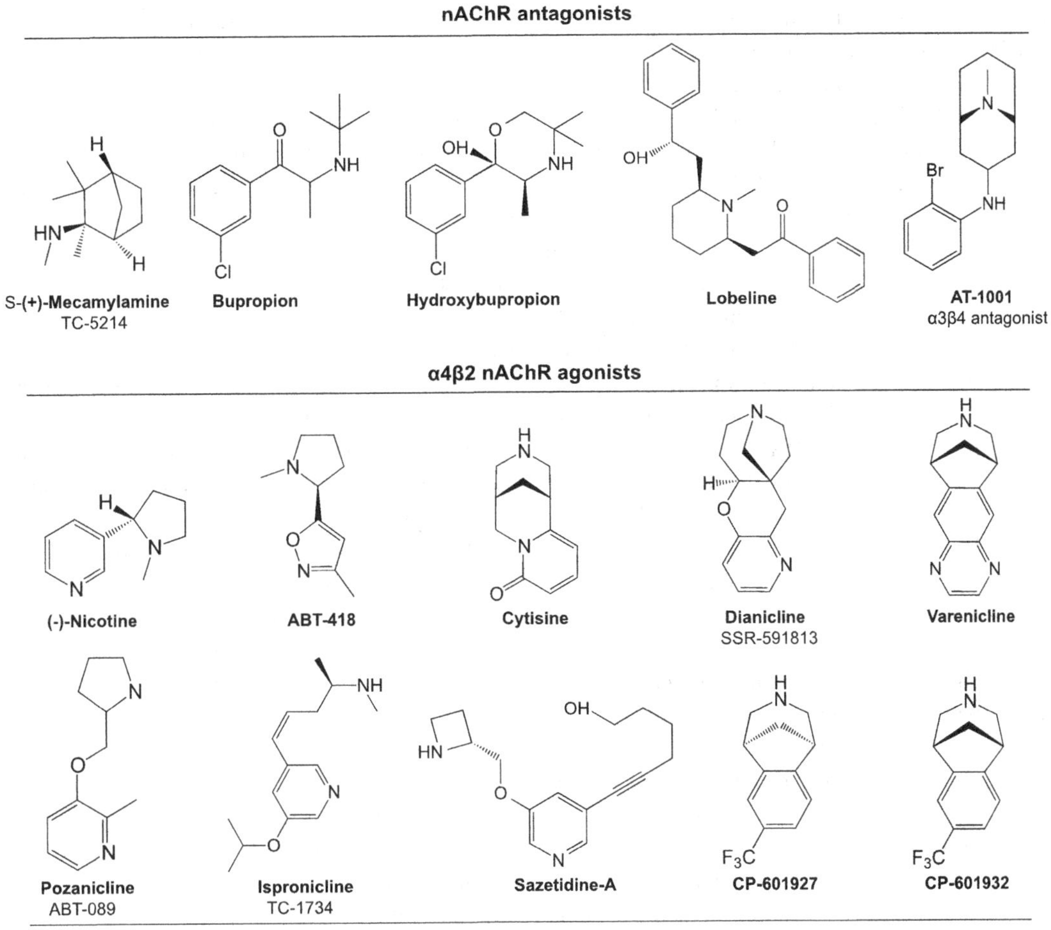

Nicotinic Agonists and Antagonists, Fig. 6 Compounds discussed under "Addiction" and "Depression"

Treatment strategies that would not only provide nicotine-like reinforcement to relieve craving when abstinent, but also attenuate the rewarding effects of nicotine during relapse should theoretically improve efficacy. Such a dual action can be achieved with partial agonists that will exert a mild nicotine-like effect in the absence of nicotine and thus relieve craving, but will also compete with inhaled nicotine and prevent nicotine binding to the nAChRs, reducing its reinforcing effects. This concept was supported by the activity of α4β2 nAChR partial agonists. The natural alkaloid cytisine has been used as a smoking cessation aid in Eastern Europe since the 1950s and was later found to be an α4β2 nAChR partial agonist. The α4β2 partial agonist ABT-418 was the first synthetic partial agonist tested for smoking cessation and had positive effects in a small Phase 2 trial, while the selective α4β2 partial agonist varenicline, shown to have high efficacy as a smoking cessation aid, was marketed in 2006. Two related α4β2 nAChR partial agonists, CP-601927 and CP-601932, were less efficacious than varenicline in a Phase 2 trial and not further developed. Other α4β2 nAChR ligands, e.g., the partial agonist dianicline and the natural alkaloid lobeline, were also examined in clinical trials, but had no significant effect on smoking behavior.

Further advances in nAChR-based treatments of nicotine dependence could come from compounds that are selective for nAChR subtypes that recently have been shown to play an important role in modulating mesolimbic dopamine release (α6* nAChRs), in withdrawal and reward (α3β4* nAChRs) and in the aversive effects of nicotine (α5* nAChRs). Varenicline is a potent partial agonist at α4α6β2* nAChRs, which likely contributes to its clinical efficacy, while the α4α6β2* partial agonist pozanicline (ABT-089), currently being studied for ADHD (see "Cognitive Deficits"), may also have potential as a smoking cessation aid. Interest in α3β4* and α3β4α5* subtypes is based on their high expression in the habenula and interpeduncular nucleus, which play a key role in nicotine self-administration and reward, as well as on the finding that polymorphisms in the gene cluster CHRNA5/A3/B4 are closely associated with the risk for heavy smoking and inability to quit. Recent preclinical studies showed that AT-1001, an α3β4 selective partial agonist/antagonist, is a potent inhibitor of nicotine self-administration in rats without affecting mesolimbic dopamine release, suggesting that α3β4* ligands could represent smoking cessation agents with a novel mechanism of action.

Alcohol Use Disorder. Common cholinergic mechanisms may play a role in nicotine dependence and alcohol use disorder (AUD) that are highly comorbid diseases. The CHRNA5/A3/B4 cluster (see above) has been associated with alcohol dependence, and particular nAChR subunits are thought to mediate the rewarding effects of alcohol. There is a great need for new AUD treatments with better efficacies and side-effect profiles than those currently available ("▶ Disulfiram"; "▶ Acamprosate"; "▶ Naltrexone"), but nAChRs have not received much attention as potential therapeutic targets. However, the demonstrated efficacy and safety of the α4β2 partial agonist varenicline for smoking cessation, together with preliminary findings that the nAChR antagonist mecamylamine reduces the stimulant and euphoric effects of alcohol, has prompted studies on α4β2 partial agonists or antagonists as treatments for AUD. Preclinical studies have shown that varenicline and the mixed α4β2/α3β4 partial agonist CP-601932 selectively reduce alcohol seeking and consumption in rodent models. A recent large, multisite double-blind trial found that varenicline treatment was well tolerated and significantly reduced alcohol consumption and craving in smokers and nonsmokers with alcohol dependence, suggesting that varenicline could be a viable option for the treatment of AUD. Given the uncertainty over which nAChR subtypes mediate the effects of alcohol, studies on compounds with different nAChR subtype selectivity might provide clues for the potential of other nAChRs as targets for alcohol dependence treatments.

Depression

The cholinergic-adrenergic theory of depression states that over-activation of nAChRs by ACh contributes to the development and exacerbation of depressive symptoms (reviewed by Philip et al. 2010; Mineur and Picciotto 2010; Hurst et al. 2013). This hypothesis is supported by several preclinical studies demonstrating that compounds that block or reduce nAChR activity have antidepressant-like activity in murine models, while clinical trials showed that the nonselective nAChR open-channel blocker mecamylamine improves mood and anxiety in patients with Tourette's syndrome and enhances the response to antidepressants in treatment-resistant depressed patients. These findings reinforced the idea that reducing the activity of nAChRs that play a role in depression results in antidepressant effects. However, as described below, more recent placebo-controlled clinical trials with nAChR ligands failed to meet outcome measures of antidepressant activity.

A reduction in signaling of nAChR subtypes can be achieved either by receptor blockade with selective antagonists or by reducing receptor function with low efficacy partial agonists and agonists that promote receptor desensitization. As discussed above, agonists and partial agonists are several orders of magnitude more potent in desensitizing than in activating nAChRs and could therefore have antidepressant activity at the low concentrations that are associated with receptor desensitization. Negative allosteric modulators are also of interest as potential antidepressants, since these compounds may attenuate the effect of ACh without having an effect on the nAChR of its own, which might be associated with improved selectivity compared to orthosteric antagonists and agonists. Nonselective nAChR antagonists will block any nAChR subtype thought to be involved in depression but can at the same time obviously interfere with the physiological functioning of off-target nAChRs. Selectively reducing the activity of a nAChR subtype that is part of pathways that regulate mood is thus expected to result in better antidepressant efficacy and/or improved side-effect profiles. While pharmacological studies support the putative role of nAChRs in depression, the precise subtypes and underlying mechanisms remain to be determined. Based on the pharmacological activity of nAChR subtype selective ligands, on phenotypes of transgenic animals and on the expression pattern of nAChR subtypes in brain areas involved in mood and stress (e.g., nucleus accumbens, prefrontal cortex, amygdala, hippocampus, habenula, interpeduncular nuclei), the α4β2, α7, and α3β4 nAChR subtypes are considered promising targets for novel nicotinic antidepressants. Given that over-activation of nAChRs is central to the cholinergic theory of depression, most drug discovery efforts are aimed at developing nAChR ligands that inhibit or block nAChR activity.

A large number of nonselective and β2- or α7-selective antagonists, and selective α4β2 partial agonists (Fig. 6), have antidepressant-like activity in the forced swim and tail suspension tests, the preclinical animal models used to assess antidepressant-like activity. While preclinical data show that nAChR partial agonists and antagonists may have potential as antidepressants per se, the focus is on antidepressant-augmentation strategies, since several compounds act in a synergistic manner when co-administered with selective serotonin reuptake inhibitors ("▶ SSRI") or tricyclic antidepressants ("▶ TCA") in preclinical models. Therefore, clinical studies are investigating the use of nAChR ligands as adjunct therapy with classical antidepressants in treatment-resistant patients, and data are available for (±)-mecamylamine, S-(+)-mecamylamine (TC-5214), varenicline, and CP-601927. Bupropion should also be mentioned here, since as a norepinephrine and dopamine reuptake inhibitor, it has obviously antidepressant activity, but its noncompetitive nAChR antagonist properties may also contribute to the antidepressant activity.

Two small studies had initially shown that the nAChR channel blocker (±)-mecamylamine improved the response to antidepressants, but subsequent four large trials conducted with the S-(+)-enantiomer of mecamylamine (TC-5214) did not meet the primary endpoints, and further development of S-(+)-mecamylamine has been halted. Likewise, the α4β2-selective partial agonist, CP-601927, which has antidepressant-like activity in murine models, did not meet the primary endpoint in a large Phase 2 augmentation strategy for treatment-resistant depression. Although the α4β2 partial agonist varenicline, which potentiates the effect of the antidepressant sertraline in mice, significantly improved depression scores in smokers on a stable antidepressant regimen, this was a small, open-label study, and large, placebo-controlled trials would be needed to confirm if varenicline can indeed enhance the antidepressant response.

In summary, while available preclinical data support both the putative role of nAChRs in depression and the antidepressant potential of nAChR antagonists and partial agonists, the results of large placebo-controlled clinical trials have so far been disappointing. The lack of effect of S-(+)-mecamylamine and CP-601927 as adjunct therapies in treatment-resistant patients could be due to several factors, and one plausible explanation is that the extent of nAChR inactivation by both compounds is insufficient to cause a clinical effect. This was proposed in a study showing that the portion of central α4β2 nAChRs inactivated by unbound human brain concentrations of mecamylamine or CP-610927, after clinical doses used in augmentation studies, is less than 25 %. Interestingly, the recommended dose of varenicline for smoking cessation was predicted to inactivate >70 % of α4β2 nAChRs (Weber et al. 2013). Clearly, further large blinded trials with selective nAChR antagonists, agonists, or allosteric modulators that are capable to cause robust reductions in nAChR signaling of a targeted nAChR subtype are needed to clarify this important issue and demonstrate the validity of the nAChR approach to depression treatment.

Cognitive Deficits

The central cholinergic system plays an important role in cognitive processes and nicotine has been shown to improve information processing, attention, and memory in both laboratory animals and humans. The high prevalence of smoking in individuals with mental disorders has led to the speculation that nicotine use may be a form of self-medication to ameliorate some of the cognitive symptoms ("▶ Cognitive Enhancers"; "▶ Cognitive Enhancers: Novel Approaches") associated with disorders such as schizophrenia and ADHD. Beyond the adverse health effects of tobacco use, nicotine is not well suited as a therapeutic agent owing to its pharmacokinetic properties and poor toleration, especially among nonsmokers. Therefore, a number of selective nAChR ligands are being investigated as potential therapies for psychiatric and neurodegenerative disorders (reviews: Hajós and Rogers 2010; Wallace and Porter 2011; Hurst et al. 2013). The largest effort has centered on the most prevalent among brain nAChRs, the $\alpha 4\beta 2$ and $\alpha 7$ nAChRs, since these subtypes have not only been associated with psychiatric and neurodegenerative disorders, but also to cognition and attention, based on the effects of selective ligands on cognitive function in preclinical animal models ("▶ Rodent Tests of Cognition"). The following sections address the major psychiatric and neurodegenerative disease areas and discuss the nicotinic ligands that have been proposed and/or tested for clinical benefits (Fig. 7).

Alzheimer's Disease. Alzheimer's disease (AD) is characterized by the accumulation of neuritic plaques containing amyloid-beta (Aβ), neurofibrillary tangles, neuronal loss, and progressive deterioration of memory and cognitive function. Loss of cholinergic neurons and diminished cholinergic signaling contribute to the declining cognitive function with disease progression. It has been well established that nAChR density decreases with disease progression, and several studies found that either Aβ1-42 or Aβ1-40 can directly interact with both $\alpha 7$ and $\alpha 4\beta 2$ nAChRs. These specific interactions between nAChRs and very low concentrations of Aβ have been proposed to play an important role in the regulation of synaptic plasticity, but the precise consequences are not fully understood. It has been reported that this interaction can either inhibit or activate the nAChR or facilitate internalization of Aβ. Beyond the specific interactions with Aβ, activation of nAChRs has also been shown to protect neurons and other cells from a variety of toxic insults in vitro and in vivo, which has been linked to $\alpha 7$ and $\alpha 4\beta 2$ subtypes. The potential for cognitive enhancement and neuroprotection makes $\alpha 7$ and $\alpha 4\beta 2$ nAChRs attractive drug targets for the treatment of AD, and several selective compounds that have advanced to clinical trials are discussed below.

The development of several early drug candidates that are efficacious in preclinical cognition models has been discontinued, because of either lack of efficacy, insufficient therapeutic windows, or other reasons. For instance, initial studies in healthy volunteers with GTS-21 (DMXB-A), an anabaseine derivative and the first $\alpha 7$ nAChR agonist to undergo clinical testing for AD, indicated positive effects on several memory tasks. However, because of its suboptimal pharmacokinetic properties, low efficacy at $\alpha 7$ nAChRs, and antagonism at $\alpha 4\beta 2$ nAChRs, drug discovery efforts have been focused on $\alpha 7$ nAChR agonists with improved drug-like properties. MEM3454 (R3487) is a potent $\alpha 7$ partial agonist that enhances cognition and auditory gating in rodents and improves working memory and quality of episodic secondary memory scores in a Phase 2 trial for mild to moderate AD, but development in AD patients was discontinued for unknown reasons. ABT-107, a potent and high-efficacy $\alpha 7$ nAChR agonist that improved cognitive performance in rats, was discontinued in Phase 1. The development of the $\alpha 7$ agonist, ABT-126, which had cognitive benefits in Phase 2 studies in patients with mild to moderate AD, was halted because of insufficient efficacy, but is still being investigated for cognitive impairment associated with schizophrenia (see below).

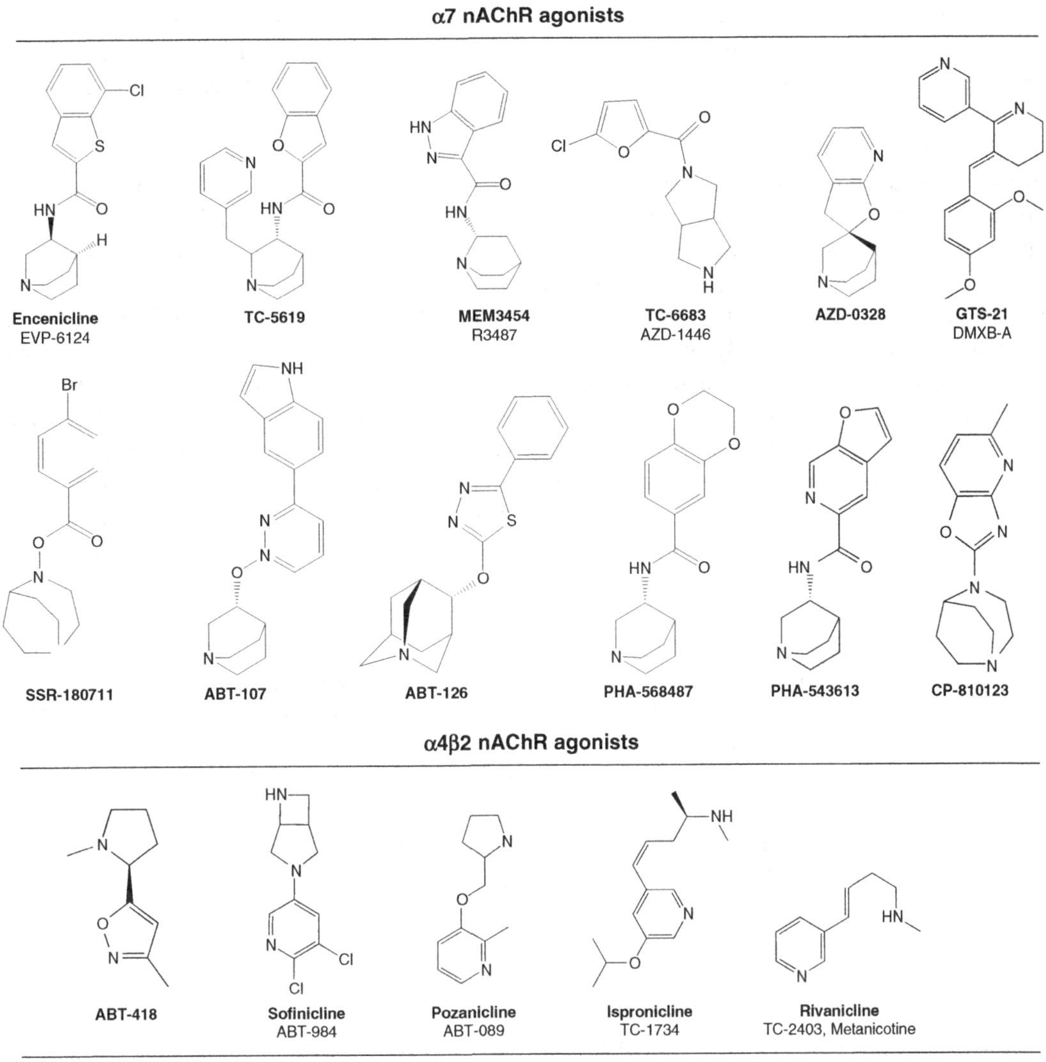

Nicotinic Agonists and Antagonists, Fig. 7 Compounds discussed under "Cognitive Deficits"

Some more recently disclosed compounds are still in clinical development. The α7 partial agonist encinicline (EVP-6124) acts as a co-agonist at low concentrations to enhance cholinergic signaling. In a recent Phase 2b clinical trial in patients with mild to moderate Alzheimer's disease, encinicline met the primary and secondary endpoints, significantly improving cognition. Importantly, cognition scores correlated with encinicline plasma concentrations and encinicline is currently undergoing Phase 3 trials for AD. Two other α7 partial agonists, AZD-0328 and SSR-180711, are of special interest, as both compounds improved memory and cognition in animal models after extremely low doses that increased the expression of α7 nAChRs in rat brain via nAChR upregulation. Finally, there are several reports of selective α7 nAChR allosteric modulators that are active in preclinical models (e.g., NS1738, PNU-120596), but these molecules have not been clinically evaluated yet.

A few α4β2 nAChR agonists have been investigated in clinical trials, but are no longer in development as cognitive enhancers in AD patients. The selective α4β2 nAChR partial agonist ispronicline (TC-1734, AZD3480) significantly improved several cognitive measures for attention and episodic memory and induced electroencephalography patterns associated with increased attention and vigilance, in healthy volunteers. Ispronicline also improved measures of attention and memory following 16 weeks' treatment in elderly subjects with age-associated memory impairment. However, ispronicline failed to meet its primary endpoint of change from baseline in ADAS-cog score in a double blind, 12-week study in AD patients with mild to moderate AD. Another selective $\alpha 4\beta 2$ nAChR agonist, ABT-418, that had positive effects on attention and in a delayed matching-to-sample task performance in aged monkeys was tested in AD patients as a transdermal application. Although ABT-418 dose-dependently improved verbal learning, memory, and reaction time, comparable to those seen in acute trials with ACh-esterase inhibitors, its development was halted due to lack of efficacy in a larger, placebo-controlled trial, possibly because of a high placebo response across studies. Finally, although preclinical data on the selective $\alpha 4\beta 2$ full agonist rivanicline (RJR-2403, TC-2403, metanicotine) indicated that it could be developed for AD, it was instead investigated as a treatment for ulcerative colitis but discontinued in Phase 2.

Schizophrenia. Currently used antipsychotics treat the positive symptoms and to some degree the negative symptoms of schizophrenia, but there is an urgent medical need to treat the cognitive impairment that is one of the core pathologies of this disorder. Interest in specifically targeting $\alpha 7$ nAChRs to address the cognitive deficits in schizophrenia was stimulated by evidence linking this receptor to deficits in auditory gating, a form of sensory filtering whereby the amplitude of an auditory evoked potential is reduced if immediately preceded by a similar auditory tone ("prepulse"). An impaired filtering process is a common endophenotype associated with schizophrenia that leads to excessive sensory input, resulting in poor performance in measures of attention. The potential association between single nucleotide polymorphisms (SNPs) in the CHRNA7 gene and the reduced expression of $\alpha 7$ nAChRs in brain regions of schizophrenic patients with gating deficits, further implicates $\alpha 7$ nAChR dysfunction as a contributing factor to this disease (reviewed by Hajós and Rogers 2010).

Results of early studies had suggested the potential value of $\alpha 7$ nAChR agonists as treatments for cognitive symptoms in schizophrenia. The mixed $\alpha 7$ nAChR partial agonist and $\alpha 4\beta 2$ nAChR/5-HT$_3$ antagonist, GTS-21 (DMXB-A; see also under AD), normalized auditory gating and improved performance and attention in a small double-blind study in schizophrenic patients, with larger effect sizes than previously seen for nicotine or atypical antipsychotics. Another mixed $\alpha 7$ agonist/5-HT$_3$ antagonist, MEM3454 (R-3487; see also under AD), was reported to improve the quality of episodic secondary memory in healthy volunteers. However, despite positive effects on sensory gating and cognitive improvement in patients, schizophrenia is no longer listed as an indication for MEM3454. The development of the $\alpha 7$ agonists PHA-543613, PHA-568487, and CP-810123 was discontinued in Phase 1 due to cardiovascular safety concerns in humans, while the $\alpha 7$ partial agonist TC-5619 was terminated due to lack of efficacy in Phase 2b. The $\alpha 7$ partial agonist encinicline (EVP-6124) (see also under AD) improved cognition in patients with schizophrenia in a Phase 2 trial and is currently undergoing Phase 3 clinical trials to assess its safety and efficacy in patients on stable atypical antipsychotics therapy.

The $\alpha 7$ agonist ABT-126 demonstrated improvements in cognition scores versus placebo in a Phase 2 study in patients with schizophrenia, and ongoing trials will examine three doses of ABT-126 in this patient population.

Evidence for a link between schizophrenia and $\alpha 4\beta 2$ nAChRs is less strong than for $\alpha 7$ nAChRs, but $\alpha 4\beta 2$ nAChRs have been proposed to be involved in cognitive deficits by decreasing the expression of the GABA-synthesizing enzyme glutamic acid decarboxylase (GAD$_{67}$) in schizophrenic patients. Since $\alpha 4\beta 2$ agonists (nicotine, varenicline, and A-85380) increased GAD$_{67}$ expression, while the $\alpha 7$ full agonist

PNU-282987 had no effect, it was suggested that α4β2 agonists could improve cognitive deficits in schizophrenics by normalizing impaired GABA function. This is consistent with the observation that selective α4β2 nAChR ligands can enhance cognitive performance in animal models. However, while the α4β2 nAChR partial agonist ispronicline (TC-1734, AZD3480), which enhances ACh release and improves memory in rodents, significantly enhanced attention and episodic memory in healthy volunteers, it failed to meet the primary endpoints in a 12-week Phase 2 trial for cognitive dysfunction and was discontinued for that indication.

Attention-Deficit/Hyperactivity Disorder (ADHD). Stimulant medications that act on dopaminergic and noradrenergic transmission were introduced more than 60 years ago to treat the motor overactivity, impulsivity, and inattentiveness associated with ADHD and remain a main treatment option. Though highly effective against these core symptoms, stimulants do not fully address the cognitive impairments, and there is growing interest in developing non-stimulant medications that will also treat cognitive aspects, which are increasingly recognized as key factors in long-term outcomes, including academic and occupational success.

The search for nAChR ligands as potential ADHD treatments was initially driven by the observation that ADHD patients use tobacco products at a much higher rate than the general population, suggesting that nicotine may alleviate symptoms of ADHD (reviewed by Wilens and Decker 2007). Several small clinical studies have indeed demonstrated the beneficial effects of transdermal nicotine on disease symptom rating scales, as well as on measures of cognitive function in both adults and adolescents. However, the pharmacokinetic profile and poor tolerability of nicotine in nonsmokers preclude the broad use of nicotine to treat ADHD. The nicotinic compounds that are being evaluated for clinical efficacy and safety in ADHD patient populations are mainly high-affinity, selective α4β2* nAChR partial or full agonists. Early placebo-controlled trials with ABT-418, a full α4β2 agonist, and pozanicline (ABT-089), a low efficacy partial α4β2 agonist, significantly improved ADHD symptom and ADHD Rating Scale scores in adults. Additional studies confirmed that pozanicline has benefits as an ADHD treatment in adults and that the α4β2 nAChR full agonist sofinicline (ABT-894) has comparable efficacy as atomoxetine in a Phase 2 study in adult ADHD patients, while the α4β2 partial agonist ispronicline (TC-1734, AZD3480) also demonstrated improvements on the Conners' Adult ADHD Rating Scale total score in a Phase 2 trial. In contrast, the development of the α4β2 partial agonist TC-6683 (AZD1446) was halted after a clinical study did not find a significant difference versus placebo in improving ADHD symptoms in adults. The positive results with pozanicline were from trials in adults, but two large multicenter placebo-controlled studies in 6–12-year-old children found that pozanicline did neither differentiate from placebo in ADHD Rating Scale scores nor in most secondary measures. These results suggest a discrepancy in the response to nicotinic agonist treatment of ADHD in children and adults, possibly due to differences in the clinical profile of adult and child ADHD and/or the developmental regulation of α4β2 expression or function. Studies are planned to examine the effects of sofinicline, which was effective in adults, in a Phase 2 dose-finding study in children with ADHD.

Only one α7 ligand has been investigated for ADHD, the α7 nAChR agonist TC-5619 (see also under AD and Schizophrenia), but since Phase 2 proof-of-concept trials showed that 5 or 25 mg QDTC-5619 for 5 weeks did not improve the Conners' Adult ADHD Rating Scale score, it is not further pursued for ADHD.

Neurodegenerative Diseases

Results from several in vitro and in vivo studies show that nicotine can provide neuroprotection against neuronal cell death, suggesting that nAChR ligands potentially represent a promising novel class of agents to treat neurodegenerative diseases, such as Parkinson's disease (PD) and Alzheimer's disease (AD). Activation of nAChRs in vitro has been shown to protect cells, primary neurons, and brain slice preparations

α4α6β2* nAChR agonists

TC-8831 **Varenicline** **A-85380** **Altinicline SIB-1508Y** **Sazetidine-A**

α7 nAChR agonists and α7 nAChR PAMs

PNU-120596 **S 249795** **Cotinine**

Nicotinic Agonists and Antagonists, Fig. 8 Compounds discussed under "Neurodegenerative Diseases"

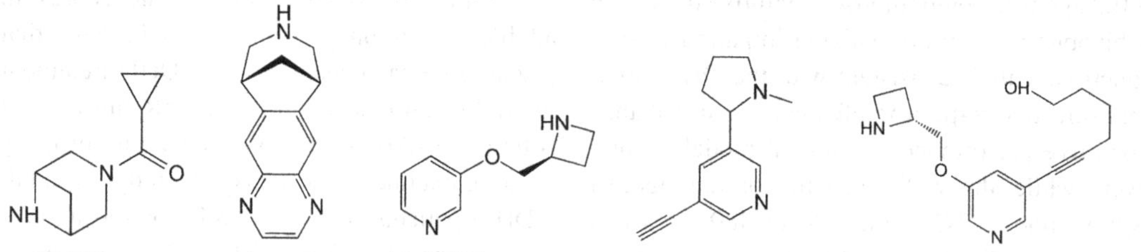

against a variety of insults including ethanol toxicity, glutamate-induced excitotoxicity, and oxygen and glucose deprivation. Since nicotine itself is less suitable as a drug, extensive research efforts are focused on identifying the nAChR subtypes that mediate nicotine's neuroprotective effects to enable the design of selective nAChR ligands (Fig. 8) that could halt or slow down the progression of a neurodegenerative disease (reviewed by Posadas et al. 2013; Hurst et al. 2013; Quik et al. 2013).

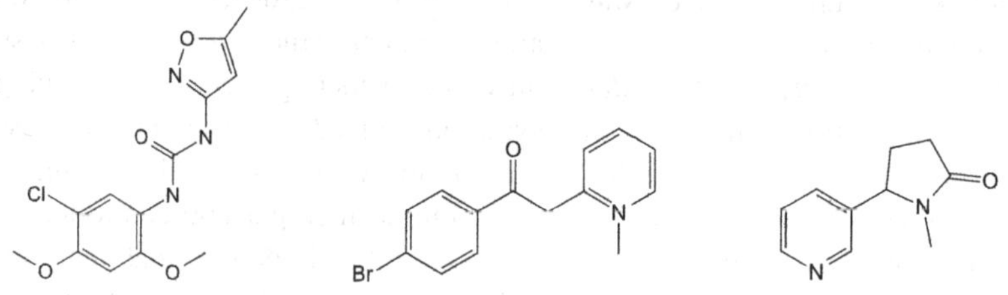

Parkinson's Disease. Compelling epidemiologic evidence exists for an inverse relationship between tobacco use and incidence of PD, with smokers of the longest duration and highest daily consumption at the lowest risk of developing the disease. Though many interpretations for this relationship have been made, it is unlikely to result from genetic factors for either PD or nicotine dependence, but seems due to nicotine's protective effect against the loss of nigrostriatal dopamine function, a hallmark feature of PD. Nicotine has also been shown in animal studies to reduce abnormal involuntary movements, associated with long-term L-dopa therapy. These beneficial effects are likely mediated either via nicotine-induced nAChR activation, which will stimulate dopamine release from the remaining dopaminergic terminals more effectively than endogenous ACh, or via nicotine-induced nAChR upregulation, which would restore the nAChR deficit in the nigrostriatal system to some extent. The stimulation of striatal dopamine release is controlled by α6β2* nAChRs that are mainly restricted to striatal dopaminergic nerve terminals and by α4β2* nAChRs located on both dopaminergic and non-dopaminergic neurons. Hence, depending on the severity of dopaminergic lesions in PD patients – and in animal PD and dyskinesia models – some or all α6β2* nAChRs may be lost, while the decline in α4β2* nAChRs will be much smaller, since most are located on GABAergic or glutamatergic neurons that are spared. This has consequences for designing nAChR ligands with optimal functional activity for neuroprotective or antidyskinetic activity. While α6β2* nAChR agonists will activate the remaining functional α6β2* nAChRs in PD with mild to moderate nigrostriatal damage, α4β2*

agonists may reduce dyskinesias in the later stages of PD with severe nigrostriatal damage. Upregulation of nAChRs, a second mechanism that could underlie the neuroprotective and antidyskinetic effects of nicotine, is less dependent on the presence of functional nAChRs, as the number of functional α6β2* or α4β2* nAChRs can be increased by nicotine and nicotinic agonists or antagonists (see "nAChR Upregulation" in "Current Concepts").

The effects of nicotine from smoking or NRT on the motor symptoms of PD have been evaluated in small clinical trials, but evidence for improvement of patients' status was at best modest and in most cases absent. When considering the disappointing clinical results, it should be kept in mind that many years of smoking seems a requirement for a reduction of PD risk, so that it is conceivable that trials with nicotine and nAChR ligands lasting only for weeks may be too short to result in beneficial effects. Promising Phase 1/2 results from a trial with nicotine bitartrate that was given for 3.5 months in an oral capsule ("NP002") and reduced dyskinesias due to L-dopa therapy in PD patients seem to support the need for long-lasting treatments. Alternatively, the mechanism by which smoking reduces the risk of developing PD may not be appropriate to alleviate symptoms once they have developed.

Several nAChR ligands are being explored in animal PD and dyskinesia models, in which dopaminergic lesions of different severity are induced with 6-hydroxydopamine or MPTP. Most nicotinic compounds studied so far are mixed α6β2*/α4β2* agonists that reduce L-dopa-induced dyskinesias in monkeys with severe nigrostriatal damage (e.g., varenicline, TC-8831) and in rodents with partial dopaminergic lesions (e.g., varenicline, TC-8831, sazetidine, A-85380). In contrast, only a few nAChR ligands have been tested in the clinic. A recent trial found that 5 weeks treatment with 10 mg of the α4β2 partial agonist altinicline (SIB-1508Y), which improved cognitive and motor function in parkinsonian monkeys, had no anti-parkinsonian or cognitive-enhancing effects in man. More data on the functional activities and effects of nicotinic compounds that target different nAChR subtypes are needed to determine the optimal functionality of a compound and which nAChR subtype to target.

Alzheimer's disease (AD) is associated with cognitive impairment, memory loss, dementia, and a host of mood and behavioral changes. While there is no single cause for all the symptoms associated with this disease, three of the four currently approved AD drugs treat the symptoms by inhibiting the enzyme ACh-esterase, thereby partially offsetting the declining cholinergic tone that is characteristic of this disease. As discussed in the AD section under "Cognitive Deficits," drugs that stimulate nAChRs may improve some of the cognitive impairment associated with this disease by substituting for the declining levels of ACh. There is preclinical evidence that agonists targeting nAChRs, in particular the α7 subtype, may also have a neuroprotective role. For example, nicotine can protect against Aβ-induced toxicity in vitro through a pathway involving α7 nAChRs and phosphatidylinositol 3-kinase Akt signaling. In addition, activation of α7 nAChRs has been suggested to decrease levels of tau protein hyperphosphorylation, a key pathological feature of AD, by inhibiting the activity of kinase GSK3beta. Likewise, PAMs of the α7 nAChR may enhance the neuroprotective effect of endogenous choline and ACh. For example, low levels of PNU-120596 complement the beneficial effect of an ACh-esterase inhibitor in aged rodent and nonhuman primates and protect against cerebral ischemia in rats. Interestingly, the main metabolite of nicotine, cotinine, has been shown to be neuroprotective and to prevent memory loss in monkeys, decreasing Aβ levels in AD mice, possibly by acting as an α7 PAM.

Contrasting their potential beneficial protective role, α7 nAChRs may also contribute to Aβ-mediated toxicity in AD by facilitating the internalization and intracellular accumulation of exogenous Aβ via a direct binding interaction between Aβ and α7 nAChRs. Consistent with these observations, neurons with significant accumulation of Aβ in AD brains tend to express high levels of α7 nAChRs, while disrupting the Aβ-α7 nAChR interaction with the partial α7 agonist

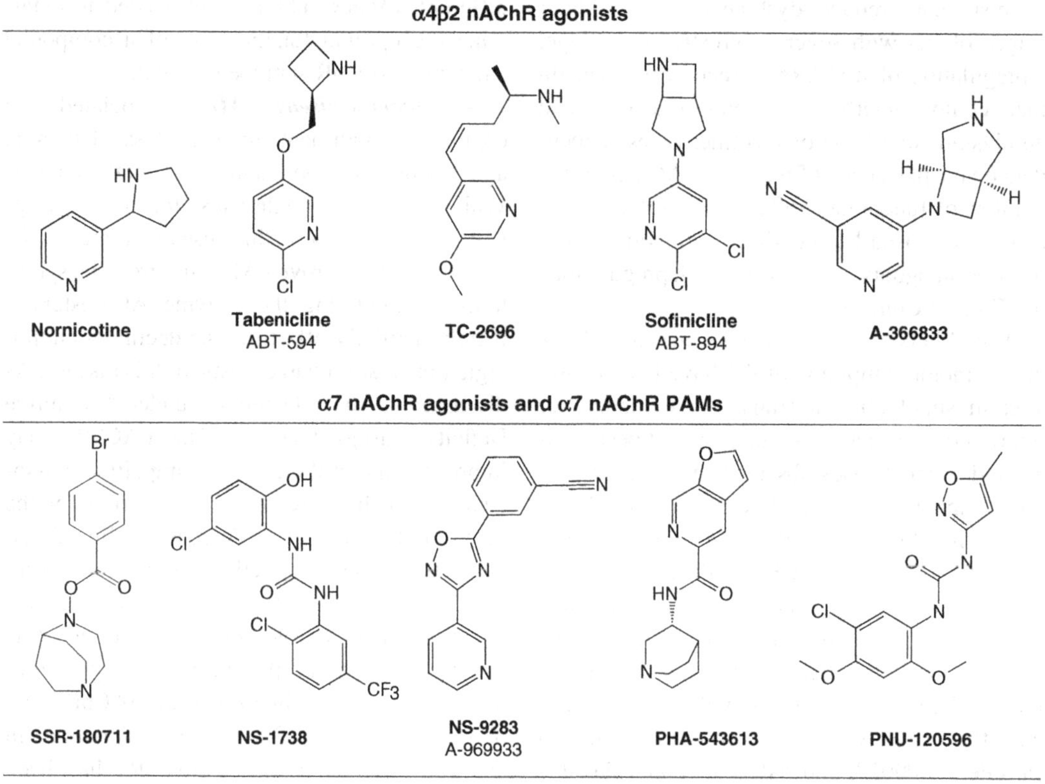

Nicotinic Agonists and Antagonists, Fig. 9 Compounds discussed under "Pain and Inflammation"

S24795 normalizes α7 nAChR function, reducing Aβ$_{42}$-mediated synaptic dysfunction and AD-like pathologies (Wang et al. 2010). In other studies, it was observed that Aβ toxicity might be caused indirectly by the release of glutamate from astrocytes and activation of the NMDA receptors. The tantalizing issue of these findings is that the Aβ-induced release of glutamate is mediated by α7 nAChRs expressed in glial cells, as shown by the blockade of Aβ toxicity with the competitive α7 inhibitor α-bungarotoxin and by the lack of effect of Aβ in α7 knockout mice (Talantova et al. 2013). Thus, a complex relationship exists between nAChRs and AD, and the α7 subtype in particular may have both neuroprotective and neurotoxic roles.

Pain and Inflammation

Preclinical pharmacology studies have shown that nicotine and several other nicotinic natural products have antinociceptive properties that are mediated via nAChRs. The best-known example is the frog alkaloid epibatidine, an extremely potent analgesic that has triggered extensive research on nicotinic compounds as potential novel pain treatments (reviewed in Hurst et al. 2013; Umana et al. 2013). Epibatidine cannot be used clinically because of safety concerns, and therefore synthetic nAChR agonists with increased subtype specificity are being pursued as analgesics with an improved therapeutic window (Fig. 9). Owing to its very high affinity to α4β2 nAChRs, this subtype is thought to play a pivotal role in epibatidine's analgesic activity. The involvement of α4β2 nAChRs in nociception is also consistent with their presence in regions that modulate pain perception and that their activation increases inhibitory GABAergic tone in the spinal cord, a known analgesic mechanism. Although the role of other nAChR subtypes has not yet been investigated in the same detail and clinical studies with highly selective ligands are lacking, other subtypes are possibly involved in the pain response.

Given the purported role of α4β2 nAChRs in mediating nociception, research has until recently mainly focused on selective α4β2 agonists for the treatment of pain. Tebanicline (ABT-594) is structurally related to epibatidine with comparable high analgesic potency and was a promising α4β2 nAChR agonist in development for diabetic neuropathic pain. However, because of insufficient efficacy with limited tolerability due to gastrointestinal side effects, tebanicline is no longer pursued. The development of two other selective α4β2 nAChR agonists, TC-2696 and TC-6499, both of which had shown analgesic efficacy in preclinical pain models, was also halted because of the lack of efficacy and an insufficient therapeutic index, respectively. Interestingly, since TC-6499 is also an α3β4 nAChR agonist that can interact with receptors in the GI tract, it will be investigated as a treatment for patients with GI disorders, such as irritable bowel syndrome. A second-generation α4β2 full agonist, sofinicline (ABT-894) that is in development for ADHD (see "Cognitive Deficits"), was investigated as a treatment for neuropathic pain, but discontinued because of insufficient analgesic efficacy. It is conceivable that analgesic doses of α4β2 agonists are associated with a small therapeutic window, although a recent selective α4β2 full agonist, A-366833, was efficacious in animal neuropathic and inflammatory pain models at doses that did not cause emesis in ferrets, but clinical data are not yet available.

In an effort to develop treatments with a greater therapeutic window, coadministration of a nAChR agonist with a PAM is being explored. Such combinations may enhance analgesia and allow lower doses of the agonist, thereby reducing the agonist-induced GI side effects. This approach is exemplified by the combination of the α4β2 agonist tebanicline (ABT-594) with an α4β2 PAM (NS-9283), which increased the analgesic activity of tebanicline in an animal model of neuropathic pain, without altering its adverse effects (Lee et al. 2011).

Given the disappointing results of clinical studies with α4β2-specific agonists, it is likely that activity at other nAChR subtypes contributes to the analgesic activity, which was assessed in detailed preclinical investigations that compared in vitro affinities, functional activities, and agonist efficacies of several nAChR ligands with their in vivo analgesic activities in rodents. Based on the effects of varenicline, ispronicline, ABT-418, tebanicline, TC-2696, and SSR-180711A, it was proposed that α4β2 nAChR activation is necessary for analgesic activity, but that α3* nAChR activation is an additional requirement, while α7 nAChRs do not play an important role in pain control (Gao et al. 2010). The analgesic activity of the nicotine metabolite nornicotine, which has been proposed as a pain treatment in combination with opioids, may thus be mediated by α4β2 nAChRs. Another study also found that analgesic efficacy of a series of α4β2* agonists in a mouse formalin model correlates with α4β2* binding affinities and concluded that desensitization of α4β2* nAChRs, and especially of α4α5β2* nAChRs, could play an important role in the analgesic effect (Zhang et al. 2012).

Interest in α7 nAChRs as a target for treating pain is mainly based on the potential benefits of α7 PAMs as analgesics per se or as potentiators of the antinociceptive properties of α7 agonists. The type-II α7 PAM, PNU-120596, has analgesic effects by itself and also potentiates the effect of the α7-selective agonist PHA-543613, while the type-I PAM NS1738 is inactive. Coadministration of type-II α7 PAMs could thus provide similar benefits as discussed above for combining α4β2 PAMs and agonists.

Finally, ongoing research is also focused on therapeutic approaches to pain related to inflammation and immunological responses. Smoking tobacco can affect the immune system and numerous data support an association between the body's immunological response and nAChRs. For instance, α7 nAChRs are present on the macrophage surface, and the α7-selective agonist PNU-282987 mimics the effect of vagus nerve activation on the parasympathetic anti-inflammatory response. Although the pharmacology that underlies cholinergic regulation of the immune response is not yet fully characterized, at least two mechanisms are considered, i.e., increased calcium influx through activated nAChRs and activation of the Janus kinase 2 pathway through stimulation of α4β2 nAChRs. Given the important role of nAChRs in inflammation, studies are examining the potential therapeutic application of nAChR ligands as anti-inflammatory agents (Bencherif et al. 2011). While initial reports pointed to a role of α4β2 and α7 receptors, more recent studies suggest the contribution of other subunits, such as α5 and α9α10. The α9 and α10 subunits have thus become of interest as novel targets for nAChR-based pain medications, especially since two selective antagonists of α9α10 nAChRs, the α-conotoxins RgIA and Vc1.1, are potent analgesics. Vc1.1 has been tested in Phase 1 and 2 clinical trials for the treatment of neuropathic pain but was found to be two orders of magnitude less potent at human than at rat α9α10 nAChRs and its development was discontinued. Taken together, these data highlight the need for further investigation of this novel therapeutic strategy to treat inflammation but also call for caution when considering the possible side effects of novel molecules targeting nAChRs.

In summary, the evidence for analgesic efficacy of selective nAChR ligands in patients is still limited and one reason for disappointing clinical results with several candidate compounds is that details of the cellular mechanisms that underlie nAChR-mediated analgesia are not completely known. Better insight in which nAChR subtypes play key roles in the nociceptive circuitry will help the design of better drug candidates.

Summary

Our increased understanding of nAChR pharmacology and how agonists, partial agonists, and antagonists modulate the function of these receptors has led to the discovery of a large number of selective ligands that are being used either as tools for exploring the role of nAChR subtypes or are in development as potential new pharmacotherapies. Several compounds represent promising new treatments for disorders in which nAChRs are thought to participate. However, the fact that to date only two nAChR ligands (mecamylamine and varenicline) have been approved for clinical use illustrates the tremendous challenge of developing selective nicotinic compounds as novel medications without major side effects. The high degree of homology between the different receptor subtypes calls for a better knowledge of their function and distribution throughout the body and for subtype-selective compounds. In addition, high selectivity is essential for an acceptable side-effect profile to avoid interactions with peripheral muscle and ganglionic nAChR subtypes associated with cardiovascular, gastrointestinal, and respiratory adverse events. Clearly, progress in the development of novel drugs that target nAChR subtypes is dependent on the design and discovery of highly selective ligands with sufficient potency and adequate pharmacokinetic properties to have efficacy at doses at which selectivity is maintained.

Cross-References

▶ Acamprosate
▶ Acetylcholine
▶ Adverse Effect
▶ Alcohol Abuse and Dependence
▶ Allosteric Modulators
▶ Alzheimer's Disease
▶ Amyloid-Beta
▶ Antidepressants
▶ Attention-Deficit/Hyperactivity Disorder
▶ Bupropion
▶ Delayed Nonmatch-to-Sample Task
▶ Depression
▶ Disulfiram
▶ Dopamine
▶ Efficacy
▶ Electroencephalography
▶ Endophenotype
▶ Episodic Memory
▶ Excitotoxicity

- ► Memory
- ► Naltrexone
- ► Neurofibrillary Tangles
- ► Neuroprotectants: Novel Approaches for Dementias
- ► Nicotine
- ► Nicotine Dependence and Its Treatment
- ► Nicotinic Receptor
- ► Parkinson's Disease
- ► Partial Agonist
- ► Pharmacokinetics
- ► Plaques
- ► Rodent Tests of Cognition
- ► Schizophrenia

References

Albuquerque EX, Pereira EF, Alkondon M, Rogers SW (2009) Mammalian nicotinic acetylcholine receptors: from structure to function. Physiol Rev 89:73–120

Bencherif M, Lippiello PM, Lucas R, Marrero MB (2011) Alpha7 nicotinic receptors as novel therapeutic targets for inflammation-based diseases. Cell Mol Life Sci 68:931–949

Changeux JP (2010) Nicotine addiction and nicotinic receptors: lessons from genetically modified mice. Nat Rev Neurosci 11:389–401

D'Souza MS, Markou A (2011) Neuronal mechanisms underlying development of nicotine dependence: implications for novel smoking-cessation treatments. Addict Sci Clin Pract 6:4–16

Gao B, Hierl M, Clarkin K, Juan T, Nguyen H, Valk M, Deng H, Guo W, Lehto SG, Matson D, McDermott JS, Knop J, Gaida K, Cao L, Waldon D, Albrecht BK, Boezio AA, Copeland KW, Harmange JC, Springer SK, Malmberg AB, McDonough SI (2010) Pharmacological effects of nonselective and subtype-selective nicotinic acetylcholine receptor agonists in animal models of persistent pain. Pain 149:33–49

Govind AP, Walsh H, Green WN (2012) Nicotine-induced upregulation of native neuronal nicotinic receptors is caused by multiple mechanisms. J Neurosci 32:2227–2238

Hajós M, Rogers BN (2010) Targeting alpha7 nicotinic acetylcholine receptors in the treatment of schizophrenia. Curr Pharm Des 16:538–554

Hurst R, Rollema H, Bertrand D (2013) Nicotinic acetylcholine receptors: from basic science to therapeutics. Pharmacol Therapeutics 137:22–54

Lee CH, Zhu C, Malysz J, Campbell T, Shaughnessy T, Honore P, Polakowski J, Gopalakrishnan M (2011) α4β2 neuronal nicotinic receptor positive allosteric modulation: an approach for improving the therapeutic index of α4β2 nAChR agonists in pain. Biochem Pharmacol 82:959–966

Mineur YS, Picciotto MR (2010) Nicotine receptors and depression: revisiting and revising the cholinergic hypothesis. Trends Pharmacol Sci 31:580–586

Philip NS, Carpenter LL, Tyrka AR, Price LH (2010) Nicotinic acetylcholine receptors and depression: a review of the preclinical and clinical literature. Psychopharmacology (Berl) 212:1–12

Posadas I, López-Hernández B, Ceña V (2013) Nicotinic receptors in neurodegeneration. Curr Neuropharmacol 11:298–314

Quik M, O'Leary K, Tanner CM (2013) Nicotine and Parkinson's disease: implications for therapy. Mov Disord 23:1641–1652

Rollema H, Bertrand D, Hurst RS (2010) Nicotinic agonists and antagonists. In: Encyclopedia of psychopharmacology. Springer, Berlin/Heidelberg, pp 887–899

Talantova M, Sanz-Blasco S, Zhang X, Xia P, Akhtar MW, Okamoto S, Dziewczapolski G, Nakamure T, Cao G, Pratt AE, Kang YJ, Tu S, Molokanova E, McKercher SR, Hires SA, Sason H, Stouffer DG, Buczynski MW, Solomon JP, Michael S, Powers ET, Kelly JW, Roberts A, Tong G, Fang-Newmeyer T, Parker J, Holland EA, Zhang D, Nakanishi N, Chen HS, Wolosker H, Wang Y, Parsons LH, Ambasudhan R, Masliah E, Heinemann SF, Piña-Crespo JC, Lipton SA (2013) Aβ induces astrocytic glutamate release, extrasynaptic NMDA receptor activation, and synaptic loss. PNAS USA 110:27. doi:10.1073/pnas.130683211

Taly A, Corringer PJ, Guedin D, Lestage P, Changeux JP (2009) Nicotinic receptors: allosteric transitions and therapeutic targets in the nervous system. Nat Rev Drug Discov 8:733–750

Umana IC, Daniele CA, McGehee DS (2013) Neuronal nicotinic receptors as analgesic targets: It's a winding road. Biochem Pharmacol 86:1208–1214

Wallace TL, Porter RH (2011) Targeting the nicotinic alpha7 acetylcholine receptor to enhance cognition in disease. Biochem Pharmacol 82:891–903

Wang HY, Bakshi K, Shen C, Frankfurt M, Trocmé-Thibierge C, Morain P (2010) S 24795 limits beta-amyloid-alpha7 nicotinic receptor interaction and reduces Alzheimer's disease-like pathologies. Biol Psychiatry 67:522–530

Weber ML, Hofland CM, Shaffer CL, Flik G, Cremers T, Hurst RS, Rollema H (2013) Therapeutic doses of antidepressants are projected not to inhibit human α4β2 nicotinic acetylcholine receptors. Neuropharmacology 72:88–95

Wilens TE, Decker MW (2007) Neuronal nicotinic receptor agonists for the treatment of attention-deficit/hyperactivity disorder: focus on cognition. Biochem Pharmacol 74:1212–1223

Zhang J, Xiao YD, Jordan KG, Hammond PS, Van Dyke KM, Mazurov AA, Speake JD, Lippiello PM, James JW, Letchworth SR, Bencherif M, Hauser TA (2012) Analgesic effects mediated by neuronal nicotinic acetylcholine receptor agonists: correlation with desensitization of α4β2* receptors. Eur J Pharm Sci 47(5):813–823

Nicotinic Receptor

Synonyms

nAChR

Definition

A subtype of acetylcholine receptors that respond to the alkaloid nicotine. They consist of ligand-gated ion channels. The two main groups are the muscle-type nicotinic receptors that are located in smooth and skeletal muscle cell membranes and the neuronal nicotinic receptors located in the plasma membranes on neurons of the CNS.

Cross-References

- ▶ Nicotine
- ▶ Nicotinic Agonists and Antagonists

Nimetazepam

Definition

Nimetazepam is a benzodiazepine that has anxiolytic, sedative, and anticonvulsant properties. It has a moderately long duration of action due to an ▶ elimination half-life of 14–30 h and conversion to the active benzodiazepine metabolite ▶ nitrazepam. Like most similar compounds, nimetazepam is subject to tolerance, dependence, and abuse.

Cross-References

- ▶ Anxiolytics
- ▶ Benzodiazepines

Nitrazepam

Synonyms

1,3-Dihydro-7-nitro-5-phenyl-2H-1,4-benzodiazepin-2-one; Mogadon

Definition

Nitrazepam is an ▶ anxiolytic drug of the benzodiazepine class. It is a nitrobenzodiazepine and like other benzodiazepines, in addition to its anxiolytic properties, it is a hypnotic drug with sedative, amnestic, anticonvulsant, and muscle-relaxant effects. It is long-acting, lipophilic, and is metabolized in the liver by oxidation. It acts as a full agonist at benzodiazepine receptors in the brain, enhancing binding of ligands for ▶ $GABA_A$ receptors. It has abuse potential and is associated with a typical benzodiazepine ▶ withdrawal syndrome.

Cross-References

- ▶ Abuse Liability Evaluation
- ▶ Benzodiazepines
- ▶ Declarative and Nondeclarative Memory
- ▶ Driving and Flying Under the Influence of Drugs
- ▶ Insomnias
- ▶ Sedative, Hypnotic, and Anxiolytic Dependence
- ▶ Social Anxiety Disorder
- ▶ Withdrawal Syndromes

Nitrites

Synonyms

Volatile nitrites

Definition

Nitrites are volatile compounds that are subject to abuse by inhalation. The first of these to see widespread use was amyl nitrite, an antianginal medication sold in ampules. The ampules were "popped" open and the contents inhaled, hence they were often referred to by illicit users as "poppers." Because they were available without a prescription in many areas in drugstores, supplies were easy to obtain. In many areas, restrictions were placed on amyl nitrite ampule sales that limited their availability. Almost immediately, commercial products containing other volatile nitrites such as butyl nitrite and cyclohexyl nitrite were sold in sexual paraphernalia and video stores as room "odorizers." Nitrite vapors have a distinct smell not unlike the odor of one of the first products of this type named *locker room*. Other commercial nitrites adopted names associated with their use in sexual activity, such as thrust, ram, and hardware. "Odorizers" and "aromas" are still widely available on the Internet and claim to contain nitrites, but their actual contents are not clear.

Cross-References

▶ Inhalant Abuse

Nitrous Oxide

Definition

Nitrous oxide is a widely abused inhalant that is used in medicine and dentistry for anesthesia. It is a gas at room temperatures and is sold under compression in cylinders. Soon after the discovery of nitrous oxide, its euphorigenic effects were noted and it became known as "laughing gas" and was used in minstrel shows where local dignitaries would agree to become intoxicated and behave abnormally for the delight of the audience. Another use of nitrous oxide is as a propellant in certain aerosols such as those for whipping crème. Nitrous oxide canisters for whipped crème dispensers became known as whippets and are subject to abuse. Nitrous oxide is usually directly inhaled from balloons or plastic bags that have been filled with the gas. The effects are almost instantaneous with the result that users can fall over and harm themselves or others. Continuous breathing of nitrous oxide can also lead to anoxia.

Cross-References

▶ Inhalant Abuse

N-Methyl-D-Aspartate Receptor

Synonyms

NMDA receptors

Definition

The *N*-methyl-D-aspartate ▶ (NMDA) receptor is one of several subtypes of glutamate receptor. It is a voltage-sensitive ionotropic receptor (ligand-gated ion channel) that facilitates excitatory transmission of electrical signals between neurons by depolarizing the postsynaptic neuronal membrane. It is composed of four subunit proteins (two NR1 plus two NR2 subunits) that assemble together to generate a channel permeable to sodium, potassium, and calcium ions. Although it is named after a selective agonist (NMDA), endogenous ligands for the receptor include glutamate and aspartate. Efficient channel opening also requires binding by the co-agonist glycine and a positive change in transmembrane potential. Activation of ▶ NMDA receptors results in the opening of an ion channel that is nonselective to cations, allowing the influx of Na^+ and Ca^{2+} and the efflux of K^+. In addition to facilitating excitatory neurotransmission, NMDA receptors are thought to play a key role in ▶ synaptic plasticity, thereby having an important role in learning and memory. The receptors are also targets for some neuroactive steroids.

Cross-References

▶ Excitatory Amino Acids and Their Antagonists
▶ Glutamate

Nociception

Definition

The neural process of signaling tissue damage or chemical irritation that is perceived as pain or itch. Nociceptors (receptors responding to noxious stimuli) initiate the process by responding to mechanical, thermal, or chemical stimuli. Nociceptors then send afferent impulses to the spinal cord and brain, leading to the perception of pain.

Cross-References

▶ Analgesics
▶ Antinociception Test Methods
▶ Opioids

Nomifensine

Synonyms

Merital®

Definition

Nomifensine is a drug that blocks dopamine and norepinephrine transporters and was once marketed as an antidepressant. It has pharmacological properties similar to amphetamine and cocaine but has fewer effects on serotonin systems than amphetamine or cocaine.

Cross-References

▶ Antidepressants
▶ Motor Activity and Stereotypy

Non-benzodiazepine Hypnotics

Jaime M. Monti and Daniel Monti
Department of Pharmacology and Therapeutics,
School of Medicine, Clinics Hospital,
Montevideo, Uruguay

Synonyms

Non-benzodiazepine agonists; Sedative-hypnotics; "Z-drugs"

Definition

The non-benzodiazepine hypnotics that work by selectively affecting γ-aminobutyric acid-A (GABA$_A$) receptors include a structurally dissimilar group of substances, such as the cyclopyrrolone agents zopiclone and eszopiclone, the imidazopyridine derivative zolpidem, and the pyrazolopyrimidine compound zaleplon.

Preclinical Pharmacology

Zolpidem was synthesized by Synthélabo Recherche in the early 1980s, and its therapeutic potential for the treatment of sleep disorders was recognized soon after.

Preclinical studies have shown zolpidem to exhibit sedative, anticonvulsant, and myorelaxant activities. However, zolpidem is more potent in suppressing locomotor activity (sedative effect) than pentylenetetrazol convulsions (anticonvulsant activity) and rotarod performance (muscle relaxation) in rats. In contrast to zolpidem, the benzodiazepine hypnotics produce a sedative effect with doses that are similar or greater than those producing anticonvulsant or myorelaxant effects.

In relation to the effects of zolpidem on sleep, a spectral analysis of sleep electroencephalograms in curarized rats has revealed that its power density in non-rapid-eye-movement (NREM) sleep ("▶ Non-rapid Eye-Movement Sleep") is predominantly increased in the low frequency band (1.0–4.0 Hz). Moreover, in freely moving animals, zolpidem has been found to augment the duration of NREM sleep and to reduce wakefulness ("▶ Wakefulness"). The effect endured during subchronic administration with no rebound occurring after abrupt withdrawal. On the other hand, variable results have been reported in regard to rapid-eye-movement (REM) sleep ("▶ REM Sleep"). In this respect, an increase and a reduction of REM sleep has been described in rats recorded during the light period (Monti et al. 2009).

Zopiclone was synthesized by Rhône-Poulenc Recherches in the early 1970s. Eszopiclone, the dextrorotatory enantiomer of racemic zopiclone, has a single chiral center with an S(+)-configuration.

Preclinical studies have shown zopiclone to exhibit sedative-hypnotic, anticonvulsant, myorelaxant, antiaggressive, and anticonflict activities. With regard to the sedative-hypnotic activity, zopiclone decreases locomotor activity, reduces waking, and increases slow-wave sleep in the rat. Spectral analysis of the electrocorticogram after zopiclone administration has shown an increase of power density in the 2.0–4.0 Hz (delta) and the 12.0–16.0 Hz (beta) bands in the rat. Eszopiclone shares the sedative-hypnotic properties of zopiclone whereas the (R)-enantiomer has no hypnotic activity (Monti and Pandi-Perumal 2007).

Zaleplon was synthesized by American Cyanamid Laboratories in the 1980s and has been evaluated for its potential sedative, muscle relaxing, and anticonvulsant activity in mice and rats. Zaleplon has been shown to reduce locomotor activity and to produce motor deficits in the rotarod and grid tests. In addition, the pyrazolopyrimidine derivative blocked electroshock-, pentylenetetrazole-, and -isoniazid-induced convulsions. In a drug discrimination procedure using rats trained to discriminate the benzodiazepine agonist chlordiazepoxide, zaleplon produced partial substitution for chlordiazepoxide at doses that significantly reduced response rates.

Both sedative and discriminative stimulus effects of zaleplon were antagonized by the benzodiazepine antagonist flumazenil (Sanger 2004).

Zaleplon increased slow-wave sleep and the relative electroencephalographic power density in the delta frequency band of rats prepared for chronic sleep recordings. REM sleep values showed no significant changes.

Mechanism of Action

GABA is the most important inhibitory neurotransmitter in the mammalian brain and localizes to approximately 30 % of central nervous system synapses.

The $GABA_A$ receptor is the site of action of the cyclopyrrolone, imidazopyridine, and pyrazolopyrimidine derivatives. These different classes of hypnotic drugs modulate GABAergic function through different GABA receptor subtypes, defined by the subunits that participate in the receptor assembly. Most $GABA_A$ receptors consist of α, β, and γ subunits which contain multiple isoforms or variants: $\alpha_1-\alpha_6$, $\beta_1-\beta_3$, and $\gamma_1-\gamma_3$. Zolpidem and zaleplon preferentially bind α_1-containing subtypes (Ator and McCann 2005). Similar to the benzodiazepine hypnotics, zopiclone and eszopiclone bind at all $GABA_A$ subtypes. Notwithstanding this, the cyclopyrrolones might have more selectivity for certain subunits of the $GABA_A$ receptor (Sanger 2004).

$GABA_A$ receptors have been analyzed by gene knockout strategies. The sedative-hypnotic activity of benzodiazepines (including flunitrazepam and diazepam) has been shown to be dependent on the integrity of the α_1 subunit. On the other hand, the anxiolytic, anticonvulsant, myorelaxant, ataxic, and withdrawal effects depend upon their predominant affinity for the α_2- and α_3-containing receptors. The finding that zolpidem and zaleplon have more selectivity for the α_1 subtype could tentatively explain the difference in effects on sleep architecture and the lower incidence of adverse events such as rebound insomnia, tolerance, dependence, and abuse (Lader 2006).

Pharmacokinetics

The pharmacokinetics of zolpidem have been investigated in healthy adults and in elderly patients. Zolpidem is rapidly absorbed and extensively distributed to body tissues, including the brain. It has a bioavailability of approximately 70 % and is extensively bound to plasma protein. Peak plasma concentrations are attained 30–60 min after a single therapeutic dose of 10 mg, and the terminal phase elimination half-life amounts to 2.6 h in healthy adults. Zolpidem is metabolized in the liver by a number of cytochrome P450 isoenzymes, but predominantly CYP 3A4, to inactive metabolites. The major routes of metabolism are oxidation and hydroxylation. After oral administration, zolpidem metabolites are largely excreted in the urine.

The metabolic clearance of zolpidem is reduced in elderly patients, resulting in increases in maximum plasma concentration, area under the concentration curve, and terminal-phase elimination half-life, the latter amounting to 2.9 h (Monti and Monti 2006).

Zopiclone 7.5 mg administered orally at night time is rapidly absorbed. It has a bioavailability of 75 % and a time of occurrence of maximum plasma concentration of 1.6 h. The compound undergoes oxidation to the N-oxide metabolite, which is pharmacologically less active, and demethylation to the inactive N-desmethylzopiclone. The elimination half-life for zopiclone and its N-oxide metabolite ranges from 3.5 to 6.0 h (Monti and Monti 2006).

Eszopiclone is rapidly absorbed and extensively distributed to body tissues. It is weakly bound to plasma protein. Peak plasma concentrations are attained 1.0–1.6 h after a single therapeutic dose of 3 mg, and the terminal elimination half-life amounts to 6.0 h. Eszopiclone is also extensively transformed in the liver to the N-oxide (less active) and the N-desmethyl (inactive) derivative. After oral administration, eszopiclone is predominantly excreted in the urine, primarily as metabolites. Less than 10 % of the compound is excreted in the urine as parent drug.

The metabolic clearance of zopiclone and eszopiclone is reduced in elderly patients, resulting in increases in maximum plasma concentration and terminal elimination half-life. The latter amounts to approximately 9.0 h in elderly patients who are being treated with eszopiclone (Monti and Pandi-Perumal 2007).

Zaleplon is rapidly and almost completely absorbed following oral administration of a 10 mg dose. The compound undergoes significant first-pass hepatic metabolism after absorption. As a result, its bioavailability amounts to only 30 %. It attains maximum plasma concentration in approximately 1 h and has a terminal elimination half-life of 1.0 h. Zaleplon is primarily metabolized by aldehyde oxidase, and all of its metabolites are pharmacologically inactive.

Clinical Uses

Various populations have been included in the studies that assessed the efficacy and safety of the non-benzodiazepine hypnotics: subjects with transient insomnia, non-elderly and elderly patients with chronic primary insomnia, and patients with secondary or comorbid insomnia ("► Comorbid Insomnia").

Zolpidem has been shown to be effective in improving sleep induction (reduction of sleep latency) at the recommended dose of 10 and 5 mg in non-elderly and elderly patients, respectively. In addition, subjective assessments and polysomnographic measures have shown that zolpidem improves sleep maintenance (reduction of wake time after sleep onset and the number of awakenings and increase of total sleep time). However, the improvement of sleep maintenance was restricted in several studies to the first part of the night. This has led to the development of a modified-release formulation that provides extended plasma concentrations beyond four hours administration. The available evidence tends to indicate that modified-release zolpidem is effective for the treatment of chronic primary insomnia characterized by difficulties with sleep onset and sleep maintenance at the recommended dose of 12.5 and 6.25 mg in adult and elderly patients, respectively (Monti et al. 2009). Recently, zolpidem sublingual tablets have been made available for the specific indication of middle-of-the-night awakening and difficulty regaining sleep. Sublingual zolpidem is administered at doses of 1.75 mg for women and 3.5 mg for men, taken once per night (Greeblatt and Roth 2012).

Zopiclone has been studied in adult insomniac patients in clinical trials and sleep laboratory studies using objective measurements of sleep variables. In all these studies zopiclone 7.5 mg reduced the time to onset of sleep and the number of nocturnal awakenings and increased total sleep time. The hypnotic drug improved also the quality of sleep. In controlled studies, elderly insomniac patients slept significantly better during treatment with zopiclone 3.75–7.5 mg than with placebo.

The subjective perception of improved sleep following eszopiclone 2 or 3 mg treatment has been demonstrated in studies of up to 6 months' duration. In these studies the drug significantly reduced sleep onset latency, the number of awakenings, and the wake time after sleep onset, whereas total sleep time and quality of sleep were increased in non-elderly and elderly patients. Sleep laboratory studies of the effects of eszopiclone have confirmed the drug's clinical efficacy in subjects with chronic primary insomnia. Eszopiclone, unlike benzodiazepine hypnotics, does not significantly alter values corresponding to slow-wave sleep and REM sleep (Monti and Pandi-Perumal 2007).

Zaleplon 10 mg was superior to placebo in decreasing latency to persistent sleep in normal adults experiencing transient insomnia.

Zaleplon 10–20 mg and 5–10 mg significantly reduced time to sleep onset in adult and elderly outpatients with chronic insomnia, respectively. Generally a significant difference from placebo on total sleep duration was not demonstrated.

It should be mentioned that the pyrazolopyrimidine derivative indiplon was initially proposed for the treatment of patients with chronic insomnia. However, the lack of data to support its use in the elderly and in patients with comorbid insomnia, as well as the absence of studies comparing side effects of indiplon with other approved non-benzodiazepine hypnotics, precluded its approval by regulatory authorities (Marrs 2008; Lemon et al. 2009).

Summary

Sleep-related complaints are common in the general population and include primary and comorbid insomnia. In patients with primary insomnia, nonpharmacological strategies and sleep-promoting medications are indicated. In patients with comorbid insomnia, the underlying disorder needs to be treated appropriately. Notwithstanding this, nonpharmacological approaches and hypnotic medication may be beneficial in comorbid insomnia. Currently used hypnotics include the cyclopyrrolones zopiclone and eszopiclone, the imidazopyridine zolpidem, and the pyrazolopyrimidine zaleplon. All these compounds reduce sleep onset latency. In addition, zopiclone, eszopiclone, and zolpidem decrease the number of nocturnal awakenings and reduce the time spent awake. The increase of total sleep time is related to greater amounts of N2 sleep. Sleep quality is also improved.

Cross-References

▶ Electroencephalography
▶ Insomnias
▶ Pharmacokinetics

References

Ator NA, McCann UD (2005) New insights into the $GABA_A$ receptor. CNS Spectr 10:20
Greeblatt DJ, Roth T (2012) Zolpidem for insomnia. Expert Opin Pharmacother 13:879–893
Lader M (2006) Rebound and withdrawal with benzodiazepine and non-benzodiazepine hypnotic medication. In: Pandi Perumal SR, Monti JM (eds) Clinical pharmacology of sleep. Birkhäuser, Basel, pp 225–234
Lemon MD, Strain JD, Hegg AM, Farver DK (2009) Indiplon in the management of insomnia. Drug Des Devel Ther 3:131–142
Marrs JC (2008) Indiplon: a non-benzodiazepine sedative-hypnotic for the treatment of insomnia. Ann Pharmacother 42:1070–1079

Monti JM, Monti D (2006) Overview of currently available benzodiazepine and nonbenzodiazepine hypnotics. In: Pandi Perumal SR, Monti JM (eds) Clinical pharmacology of sleep. Birkhäuser, Basel, pp 207–223

Monti JM, Pandi-Perumal SR (2007) Eszopiclone: its use in the treatment of insomnia. Neuropsychiatr Dis Treat 3:441–453

Monti JM, Spence DW, Pandi-Perumal SR, Langer SZ, Hardeland R (2009) Pharmacotherapy of insomnia: focus on zolpidem extended-release. Clin Med Ther 1:123–140

Sanger DJ (2004) The pharmacology and mechanisms of action of new generation, non-benzodiazepine hypnotic agents. CNS Drugs 18(Suppl 1):9–15

Nonorganic Insomnia

Synonyms

ICD-10

Definition

Nonorganic insomnia (ICD-10) is a disorder of unsatisfactory quantity and/or quality of sleep, which persists for a considerable period of time. It includes difficulty falling asleep, difficulty staying asleep, or early final wakening.

Cross-References

▶ Insomnias
▶ Sleep

Non-Rapid Eye Movement Sleep

Synonyms

NREM sleep

Definition

Sleep in humans can be broadly divided into non-rapid eye movement (NREM) and rapid eye movement (REM) sleep. NREM sleep is further divided into three stages: These range from the lightest sleep in stage N1 to the deepest sleep in stage N3 (referred to as slow wave sleep or SWS). REM sleep alternates with NREM sleep about every 90 min. Overnight sleep recording (e.g., 6–8 h) usually has four to six cycles of REM and NREM sleep. Sleep displays an ultradian rhythm (rhythm with ~90 min periodicity), in which alternations of NREM and REM sleep successively occur. This cyclic sequence of NREM and REM sleep is a highly characteristic feature of human sleep. The number and depth of sleep cycles which occur during the night, along with the body's reactions to the changes in these cycles, are clinically useful for identifying the nature of a patient's sleep problems. Disorders of arousal during NREM sleep include the following: sleepwalking, sleep terrors, and confusional arousals.

Cross-References

▶ Electroencephalography

Nonselective Blockade

Definition

The blockade of several or all types of receptor roughly equally.

Nonsteroidal Anti-inflammatory Drugs

Synonyms

NSAIDS

Definition

Nonsteroidal anti-inflammatory drugs are a group of compounds that have anti-inflammatory, analgesic, and antipyretic actions. The prototype in this group is aspirin.

Cross-References

▶ Analgesics

Nootropics

Definition

Non-stimulant drugs that may improve cognitive function in the elderly.

Cross-References

- ▶ Classification of Psychoactive Drugs
- ▶ Cognitive Enhancers: Role of the Glutamate System
- ▶ Pharmaceutical Cognitive Enhancers: Neuroscience and Society

Noradrenaline

Synonyms

Norepinephrine

Definition

A neurotransmitter and member of the catecholamine family.

Cross-References

▶ Norepinephrine

Norepinephrine

Synonyms

Noradrenaline

Definition

Norepinephrine is a catecholamine neurotransmitter that is synthesized from tyrosine via dopamine. In the brain, it arises almost exclusively from a single circumscribed region in the brain stem, referred to as the locus coeruleus. It is involved in arousal and sleep, emotion-related processes, the processing of sensory and spatial information, and other higher cognitive functions dependent on the prefrontal cortex. Outside the brain, norepinephrine is produced in postganglionic sympathetic neurons and is released into the blood from the adrenal glands and contributed to the "fight-or-flight reaction"; its main physiologic actions include constriction of small blood vessels (leading to increased blood pressure), increased heart rate, relaxation of smooth muscle, and release of glucose from energy stores.

Cross-References

▶ Beta-Adrenoceptor Antagonists

Norepinephrine Transporter

Synonyms

NET

Definition

This is the plasma membrane-bound protein that is responsible for removing norepinephrine from the synaptic cleft, transporting it back into the noradrenergic neurons from which it originated, thereby terminating its signaling activity on postsynaptic neurons. This is a target for classes of medications (including psychostimulants) used in the treatment of depression and attention-deficit/hyperactivity disorder.

Nortriptyline

Definition

Nortriptyline is a ▶ tricyclic antidepressant with a tertiary amine chemical structure. One of the earlier tricyclics to be developed, it acts by inhibiting the reuptake of norepinephrine. It is a metabolite of amitriptyline. While its primary use is in the treatment of depression, it is also used in lower doses to treat migraine headache and as an aid for smoking cessation, for which it is considered as a second-line treatment. Side effects are typical for ▶ tricyclic antidepressants and include sedation, cardiovascular, and anticholinergic effects.

Cross-References

▶ Amitriptyline
▶ Antidepressants

Novel Object Recognition Test

Synonyms

Novelty preference test

Definition

A recognition test that is often used for psychopharmacological studies in rodents. In a typical test, animals are first allowed to explore two objects within an environment for a specified period of time. After a delay, a second session is carried out in which one of the original objects has been replaced by a new one. During the second session, the animals are reintroduced into the environment and allowed to explore the objects that are present. Animals that spend more time exploring the novel object, relative to the old object, are said to recognize the old one.

Cross-References

▶ Rodent Tests of Cognition

NR2B

Definition

One of four possible NR2 subunit proteins (A–D) that, along with NR1 and possibly NR3 subunits, make up the ▶ NMDA receptor tetramer. NR2 subunits confer distinct functional and pharmacological properties to ▶ NMDA receptors, including their sensitivity to Mg^{2+} block and the kinetics of opening and closing of the ion channel.

Nuclear Magnetic Resonance

Synonyms

Magnetic resonance

Definition

Nuclear magnetic resonance (NMR) refers to a physical resonance phenomenon that is observed when atomic nuclei in a static magnetic field absorb energy from a radio frequency field (or oscillating magnetic field) applied at the resonant frequency of the nuclei. The phenomenon of NMR has been used in a variety of scientific techniques, including magnetic resonance imaging (MRI).

Cross-References

▶ Magnetic Resonance Imaging (Functional)

Nucleus Accumbens

Synonyms

NAcc

Definition

The *nucleus accumbens* is a small brain region located in the ventral striatum. It is a major projection area (terminal field) for dopaminergic neurons located in the ▶ ventral tegmental area of the midbrain. It is thought to play an important role in many behavioral and psychological processes, including reward-related behavior, incentive motivation, memory processing and habit formation, and in dopamine-mediated ▶ psychotic symptoms in ▶ schizophrenia. It has been implicated in the reinforcing effects produced by various natural stimuli, such as food, water, and sexual opportunity, as well as by ▶ drugs of abuse. In addition, it receives projections from the ▶ prefrontal cortex that may mediate the impulsivity associated with compulsive drug taking.

Cross-References

▶ Impulse Control Disorders

Number Needed to Harm

Synonyms

NNH

Definition

Number needed to harm (NNH) is a clinically useful measure for the difference between treatment groups regarding a negative outcome, such as all-cause discontinuation, relapse, or number of patients with significant weight gain. It denotes the number of patients that need to be exposed to a treatment until one additional negative event of interest occurs in excess of the rate in the comparator group. The NNH is computed as 1/(percent rate of the active treatment group - percent rate of the comparator group). An NNH of below 10 is generally considered clinically meaningful.

Cross-References

▶ Number Needed to Treat
▶ Randomized Controlled Trials

Number Needed to Treat

Synonyms

NNT

Definition

The number of patients that need to be treated in order for one to obtain a benefit that is attributable to active treatment and not placebo. It is a measure used in evidence-based medicine and epidemiology measure to assess the efficacy of interventions. The lower the number needed to treat, the more effective the intervention. Maximum efficacy is achieved when NNT = 1.

Cross-References

▶ Randomized Controlled Trials

O

ob/ob Mouse

Synonyms

Lep^{ob}/Lep^{ob} mouse

Definition

A mouse model of obesity that is deficient in the hormone leptin due to a mutation in the ▶ leptin (*ob*) gene. The *ob/ob* mouse is extremely obese and has many of the metabolic defects including ▶ hyperphagia, hypometabolism, hyperinsulinemia, and infertility.

Cross-References

▶ Hyperphagia
▶ Leptin

Obsessions

Synonyms

Obsessive ruminations; Repetitive thoughts

Definition

Obsessions are persistent ideas, thoughts, impulses, or images that are experienced as intrusive and inappropriate and that cause marked anxiety or distress. They are not simply excessive worries about real-life problems. The person usually attempts to ignore or suppress them with some other thought or action. The person recognizes that the obsessional thoughts, impulses, or images are a product of his or her own mind. At some point during the course of the disorder, the person has recognized that the obsessions are excessive or unreasonable (**Note**: This does not necessarily apply to children (▶ Obsessive-Compulsive Disorders in Childhood). The obsessions cause marked distress, are time consuming (>1 h/day), or significantly interfere with the person's normal routine, occupational or academic functioning, or usual social activities or relationships.

Cross-References

▶ Obsessive–Compulsive Disorders

Obsessive–Compulsive Disorders

Naomi A. Fineberg[1] and Ashwini Padhi[2]
[1]Department of Psychiatry, NHS Foundation Trust, Queen Elizabeth II Hospital, Welwyn Garden City, Hertfordshire, UK
[2]Community Mental Health Team, Birmingham and Solihull Mental Health Foundation Trust, Solihull, West Midlands, UK

Definition

Obsessive–compulsive disorder (OCD) is a common, debilitating illness that was considered untreatable until the 1960s. Since then, a relatively narrow range of pharmacotherapies has been found to be effective: clomipramine in the 1960s, selective serotonin reuptake inhibitors SSRIs (▶ SSRIs and Related Compounds) in the 1980s, and adjunctive antipsychotic drugs in the 1990s (Fineberg et al. 2012). The current standard of care is to offer either cognitive behavior therapy (CBT) or medication first line (NICE 2006). The form of CBT that has been most effective in OCD is *exposure and response prevention* (Drummond and Fineberg 2007). Whereas recognizing the importance of all aspects of management, we focus here on pharmacological approaches to treatment.

Role of Pharmacotherapy

OCD and the Serotonin Hypothesis
Reports that clomipramine, the 3-chloro analog of the tricyclic imipramine, but not other tricyclic antidepressants (including desipramine, imipramine, nortriptyline, and amitriptyline), relieved obsessive–compulsive symptoms directed early OCD pharmacotherapy research. Clomipramine differs from other tricyclic antidepressants by its marked potency for blocking serotonin reuptake, generating a hypothesis that serotonin might be involved in OCD pathophysiology. Although clomipramine is a powerful serotonin reuptake inhibitor (SRI), its active metabolite has noradrenergic properties. However, additional studies favorably compared the more serotonin-selective SSRIs (SSRI and related compounds) with the noradrenergic tricyclic desipramine as well as monoamine oxidase inhibitors (MAOIs) such as phenelzine. Despite this apparently selective pharmacological response, we understand little about the role of serotonin in the etiology of OCD or the mechanisms by which SSRIs (SSRI and related compounds) exert anti-obsessional effects (Fineberg and Craig 2008). Indeed, although SRIs are usually essential for OCD pharmacotherapy because clinical response is not always satisfactory, we devote the latter portion of this entry to pharmacological approaches to SRI nonresponders.

Evaluating Treatment Response and Remission (Fineberg et al. 2007) in OCD
The pivotal rating instruments for measuring OCD severity and treatment response or remission in OCD include the Yale–Brown Obsessive–Compulsive Scale (Goodman et al. 1989a, b) (Y-BOCS) and the Clinical Global Impression Scales (CGI) (Guy and editor. ECDEU Assessment Manual for Psychopharmacology 1976). Treatment response and remission rates in acute-phase OCD trials (lasting around 12 weeks) of SSRIs (SSRI and related compounds) rarely exceeded 60 % and 45 %, respectively, emphasizing the incomplete effect of these treatments over the short term in OCD.

Clomipramine in Acute-Phase Treatment Trials
Several positive double-blind placebo-controlled trials conclusively established clomipramine as an effective treatment for OCD. Efficacy was demonstrated using as few as 14 patients, emphasizing the potency of the effect. The introduction of large-scale multicenter-controlled studies enabled drug interventions to be assessed in reasonably heterogeneous groups of patients with OCD (Fineberg and Gale 2005). Two such trials, involving 238 and 263 subjects, specifically excluded comorbid depression to prevent it from confounding the outcome. Significant advantages for the clomipramine ($\leq$300 mg) group emerged by the first or second week and

benefits accrued up to the 10-week endpoint. Under clomipramine, Y-BOCS scores decreased by roughly 40 % (compared with 5 % under placebo) and correlated with similar improvements in social and emotional well-being. It is important to investigate OCD *with* comorbid depression as up to 2 out of 3 individuals with OCD are depressed when they seek treatment. Depression in OCD is profoundly impairing (Heyman et al. 2006). Of the few studies that looked at this comorbidity, most used clomipramine, which was found to significantly reduce obsessional symptoms, though the effect on depression was often not reported.

SSRIs in Acute-Phase Treatment Trials

Fluvoxamine (including slow-release formulation), fluoxetine, sertraline, paroxetine, citalopram, and escitalopram have demonstrated efficacy against placebo in large-scale studies. Small-scale SSRI (SSRI and related compounds) studies, including only around 20 patients, were also positive, again suggesting a potential effect. Fluvoxamine was effective both in the presence and absence of depression. Moreover, in depressed cases of OCD, superiority for fluvoxamine over desipramine was reported for both Y-BOCS and Hamilton Depression Scale scores, highlighting the importance of treating comorbid OCD with an SSRI (SSRI and related compounds). Venlafaxine, despite acting as an SRI at lower-dose levels, has not been found effective versus placebo.

SRIs in OCD: Which Dose?

Clomipramine has not been subjected to multiple-dose comparator trials, though fixed doses as low as 75 mg were found to be effective. Higher doses are associated with greater frequency of unwanted effects (described here). Two multicenter trials found that clomipramine dosed up to 300 mg was clinically effective. In the UK, the maximal SPC dose is 250 mg. ECG and plasma level monitoring is recommended if doses above this level are prescribed. Combined plasma levels of clomipramine plus desmethylclomipramine 12 h after the dose should be kept below 500 ng/mL to minimize the risk of seizures and cardiac conduction delay.

Apart from fluvoxamine, the SSRIs (SSRI and related compounds) have been subjected to placebo-controlled fixed-dose analysis in OCD trials. Fluoxetine −60 mg, paroxetine −60 mg, and escitalopram −20 mg appeared the most effective when compared with the lower fixed doses of active drug in acute-phase studies, but the superiority for 60 mg citalopram appeared only on secondary analyses, at the same time the sertraline study was unable to distinguish a dose–response relationship, possibly owing to design problems related to that study. An extended placebo-controlled study of escitalopram demonstrated enduring superiority for the 20 mg over the 10 mg dose lasting for 16 weeks. (Doses of escitalopram and citalopram exceeding maximal SPC limits should be used with caution, owing to observed changes in cardiac conduction at higher dose levels.) In a 12-month fixed-dose placebo-controlled fluoxetine trial in SSRI responders, only the 60 mg dose protected against relapse. Thus, higher SRI doses appear the most effective for OCD. Though fast upward titration may produce earlier responses, long-term benefits of this approach are not established and may lead to a greater burden of side effects. Expert consensus currently favors moderated doses in the first instance, with upward titrations to maximum should symptoms persist (NICE 2006). The American Psychiatric Association Practice Guideline (Koran et al. 2007) provides an empirically derived table of starting, usual, maximum, and "occasionally prescribed" SRI doses for obsessive–compulsive disorder (Table 1).

Patients should be advised at the outset of treatment that the response to SSRI (SSRI and related compounds) or clomipramine is slow and gradual, and the anti-obsessional effect takes several weeks to develop. A trial of at least 12 weeks at the maximum tolerated dose with careful assessment is advisable before judging its effectiveness.

Adverse Effects: SSRIs Versus Clomipramine

SSRIs (SSRI and related compounds) appear safe and well tolerated to the maximal SPC dose limits, according to the placebo-referenced OCD trials, which reported adverse-event-related

Obsessive–Compulsive Disorders, Table 1 SRIs in OCD, usually prescribed dosages (From Koran 2007)

	Starting dose (mg/day)[a]	Usual maintenance dose (mg/day)	Occasional maximum dose prescribed (mg/day)[b]
SRIs[c]			
Citalopram[d]	20	40–60	80–120
Escitalopram[d]	10	20	40–60
Fluoxetine	20	40–60	80–120
Paroxetine	20	40–60	80–100
Sertraline	50	100–200	200–400
Fluvoxamine	50	100–300	300–450
Clomipramine[d]	25	100–250	–

[a] In the elderly and in some individuals, there may be a need to initiate the medication at half this dose to minimize side effects
[b] In cases of treatment resistance with no or mild side effects and in rapid metabolizers
[c] SRIs appear equally efficacious in treating OCD. SSRIs are safer and better tolerated than clomipramine and therefore should usually be chosen first line (NICE 2006)
[d] Doses exceeding maximal SPC limits have been found to be associated with cardiac conduction abnormalities

withdrawal rates of around 5–15 %. SSRIs (SSRI and related compounds) may be associated with initially increased nausea, nervousness, insomnia, somnolence, dizziness, and diarrhea. Sexual dysfunction, including reduced libido and delayed orgasm, affects up to 30 % individuals. Three controlled studies compared the clinical effectiveness of different SSRIs (SSRI and related compounds), and the results were not strong enough to support the superior efficacy of any compound.

Pharmacokinetic variation may be relevant for deciding which SSRI (SSRI and related compounds) needs to be prescribed. Fluoxetine, paroxetine, and to a lesser extent sertraline inhibit the P450 isoenzyme CYP 2D6, which metabolizes tricyclic antidepressants, antipsychotics, antiarrhythmics, and beta-blockers. Fluvoxamine inhibits both CYP 1A2 and CYP 3A4, which metabolize warfarin, tricyclics, benzodiazepines, and some antiarrhythmics. Citalopram and escitalopram are relatively free from hepatic interactions and may therefore have an advantage in the elderly and physically unwell. However, doses of citalopram and escitalopram exceeding the SPC maximal limit should be used with caution as cardiac conduction changes have been observed at higher dose levels. Fluoxetine has a long half-life and fewer discontinuation effects, which can be advantageous if treatment adherence is poor. It has also been extensively used in pregnancy and generally shown to be safe.

In contrast, clomipramine is commonly associated with typical "tricyclic" effects such as dry mouth and constipation, produced by anticholinergic blockade; sedation; weight gain, resulting from antihistaminic (H_1) binding (► Histaminic Agonists and Antagonists); and orthostatic hypotension, caused by alpha-adrenergic blockade. Nausea, tremor, impotence, and anorgasmia also occur with clomipramine (as with other SRIs) and are probably mediated by serotonin. Sexual performance is impaired in up to 80 % of patients. Compared with the SSRIs (SSRI and related compounds), clomipramine is also associated with a greater risk of potentially dangerous side effects including seizures (up to 2 %) and prolongation of the cardiac QT interval. These risks increase at dosages exceeding 250 mg/day. Intentional overdose with clomipramine can thus be lethal and needs to be borne in mind when prescribing for OCD, in view of the increased suicide risk associated with the illness. Comorbidity between bipolar disorder and OCD has been reported in up to 30 % cases. SSRIs (SSRI and related compounds) are less likely than clomipramine to precipitate mania, but mood stabilizing medication is still advised to mitigate the risk (Koran 2007).

Which SRI First Line?

Clinical effectiveness depends on a balance between efficacy, safety, and tolerability. Several studies, including placebo-controlled and head-to-head comparator studies and a meta-analytic review (NICE 2006), have demonstrated this; whereas the SRIs appear equally efficacious in treating OCD, SSRIs (SSRI and related compounds) are safer and better tolerated than clomipramine and therefore should usually be chosen first line.

Long-Term Effectiveness

OCD is a chronic illness, and so treatment needs to remain effective over the longer term. Studies have shown that placebo-referenced gains accrue for at least 6 months and, according to open-label follow-up data, for at least 2 years. The SRI response is characteristically partial, particularly in the first few weeks. An uncontrolled study followed up fluvoxamine-treated patients with severe OCD for 6–8 years. By the end of the study, responder rates of 60 % were reported and 27 % of patients had entered remission. However, the majority of patients had required further treatment (either medication or CBT) in the intervening years, suggesting that long-term treatments may be required to maintain efficacy. Results from an extended double-blind placebo-controlled escitalopram study showed that the "responder" rate increased to 70 % after 24 weeks of continuous SSRI, and 40 % of these individuals achieved remission, suggesting a more favorable outcome with sustained treatment. Consistent findings were reported from an open-label trial where 320/468 (78 %) cases achieved clinical response status and 45 % cases achieved remission after 16 weeks' continuous escitalopram.

Thus, first-line treatments can lead to remission or clinically meaningful improvement in around 45 or 75 % of cases, respectively, in clinical trial populations. Benefits increase gradually with ongoing treatment, and treatment continues to be effective over the longer term. It is important to give some time for the treatment effect to develop and also not to discontinue, reduce dose levels, or change the drug prematurely. Indeed, studies looking at the effects of discontinuing SRIs under double-blind placebo-controlled conditions showed a rapid and incremental worsening of symptoms in most people who switched to placebo. Therefore, treatment probably needs to be continued for patients to remain well.

Relapse Prevention

The question as to whether continued SRI treatment protects against relapse may be investigated using a relapse prevention study design. Table 2 systematically records the peer-reviewed relapse prevention studies in OCD. Studies of fluoxetine and sertraline found that ongoing SSRI was associated with continued improvement in Y-BOCS scores and quality of life measures up to

Obsessive–Compulsive Disorders, Table 2 Double-blind studies of relapse prevention in adults with OCD

Drug	Duration of treatment prior to randomization (weeks)	Number in randomization phase	Duration of follow-up after randomization (weeks)	Relapse rates
Fluoxetine	20	71	52	PBO = pooled FLUOX, PBO > 60 mg FLUOX
Sertraline	52	223	28	PBO = SERT (acute worsening of OCD symptoms: PBO > SERT) (dropout due to relapse: PBO > SERT)
Paroxetine	12	105	24	PBO > PAR
Escitalopram	16	320	24	PBO > ESC

12 months of open-label treatment. In the fluoxetine study, only patients remaining on the highest dose (60 mg) showed significantly lower relapse rates, implying an ongoing advantage for staying at the higher dose levels. Large-scale studies of paroxetine and escitalopram clearly demonstrated a significantly better relapse outcome for those remaining on the active drug over the 6-months double-blind discontinuation phase: 59 % and 52 % relapsed on placebo compared with 38 % on paroxetine (20–60 mg) and 24 % on escitalopram (10–20 mg), respectively. The risk of relapsing on placebo was more than double than that on SSRI.

Taken together, these data emphasize the importance of maintaining SSRI (SSRI and related compounds) at an effective dose level (NICE 2006) and argue against discontinuation even after 1 year. Protection is not complete, however, with roughly one quarter of cases becoming unwell despite adherence to treatment, highlighting the need for better long-term therapies.

Treatment of SRI-Resistant OCD

It is advisable to delay changing medication until an adequate trial of 12 weeks at the maximally tolerated dose has been attempted. There is a shortage of evidence to support drug switching when compared with extending treatment with the same drug for a longer time. In one review, 11–33 % of patients not responding to the first SRI were reported to show clinically meaningful response to a second drug, with decreasing likelihoods for subsequent changes (Fineberg et al. 2006). According to meta-analyses, factors linked with SRI resistance include early onset, longer duration, more severe illness, and poor response to previous therapy. Few studies have investigated pharmacological treatments for SRI-resistant OCD using adequately controlled trial conditions.

Increase SRI Bioavailability

One strategy has been to increase the dose of SSRI (SSRI and related compounds) beyond SPC limits (Table 1). Citalopram (160 mg) and sertraline (400 mg) were helpful in small, open-label case studies of resistant OCD. Another study randomly assigned patients to sertraline (250–400 mg) and reported an improved Y-BOCS response when compared with those who remained within SPC dose limits, and high-dose sertraline was well tolerated. The decision to prescribe outside SPC limits should be discussed with the patient and take account of the potential risks and benefits. Risks are greater for increasing doses of clomipramine owing to its inherent toxicity and for increasing doses of citalopram and escitalopram owing to associated changes in cardiac conduction. In the case of clomipramine, ECG and plasma monitoring is advisable above maximal SPC doses of 250 mg/day, and other strategies are usually preferable.

Altering the mode of drug delivery may be another way to gain control of severe intractable OCD. A double-blind study showed intravenous clomipramine to be effective. Six out of 29 patients with refractory OCD were classed as responders after 14 daily infusions when compared with none in the placebo group. The superiority of intravenous citalopram observed during an open-label trial requires substantiation under double-blind conditions.

Change to SNRI

There are few rational alternatives to SRIs for first-line treatment. Two small open-label studies suggested that patients not responding to one or more SRIs might benefit from a change to venlafaxine. However, a double-blind prospective study was negative.

Add Antipsychotic Drugs to SRI

There have been no positive trials using antipsychotics as monotherapy, though evidence supports their use as adjunctive treatment with SRIs. Placebo-controlled studies of adjunctive antipsychotic drugs have been limited by methodological shortfalls including small size, focus on acute treatment, and lack of consensus on definitions of treatment resistance or response. Haloperidol, risperidone, olanzapine, quetiapine, and aripiprazole have been found beneficial, and between 18 % and 69 % of individuals responded. So far, there have been no long-term studies or fixed-dose trials in OCD, though treatment is usually started at low doses and increased cautiously subject to tolerability (e.g., 0.25–0.5 mg haloperidol, titrated slowly to 2–4 mg). One study reported a high level of relapse following open-label discontinuation, hinting that the antipsychotic may need to be continued to remain effective. This finding needs further confirmation. A meta-analysis suggested that patients with comorbid tic disorders were particularly responsive to adjunctive antipsychotic drugs (Bloch et al. 2006). Efficacy in OCD supports a theoretical link between OCD and Tourette's syndrome, for which antipsychotic drugs constitute the first-line treatments. Second-generation antipsychotics (▶ Second- and Third-Generation Antipsychotics) affect a broader range of neurotransmitters and may be preferred because of their more benign side-effect profile. Interestingly, emergent obsessions have been reported during the treatment of schizophrenia with second-generation antipsychotics (▶ Second- and Third-Generation Antipsychotics). It is unclear why dopamine antagonists might work when added to SRIs but not when used alone or in OCD associated with schizophrenia.

Novel Pharmacotherapies

Novel treatments have been tested in resistant OCD, but few show promise (reviewed in Fineberg and Craig 2008; Fineberg et al. 2012). There is no convincing evidence that augmentation with lithium, buspirone, pindolol, clonazepam, desipramine, or St. John's wort is beneficial. Monotherapy with oxytocin and naloxone has also failed to produce benefit. A small study suggested oral morphine augmentation was shown to be effective when compared with placebo. A number of lines of research, from neuroimaging to genetics, are converging to suggest that abnormal glutamatergic transmission may be important in OCD. These findings have led to a number of trials involving drugs that modulate glutamate such as riluzole, memantine, topiramate, *lamotrigine*, *N-acetylcysteine*, and D-cycloserine. To date, these are mainly at the stage of case reports and open-label series. Other compounds with initial positive results warranting further study include inositol and D-amphetamine.

Conclusion

SRIs have a rapid onset of effect and a broad spectrum of actions. SSRIs (SSRI and related compounds) offer advantages over clomipramine in terms of safety and tolerability and usually constitute first-line treatments. Higher doses usually offer greater benefits and gains continue to accrue for weeks and months. Ongoing treatment protects against relapse for most patients. For SRI-resistant cases, the strongest evidence supports adjunctive antipsychotic drugs. Their long-term effects remain unclear. Increasing the dose of SSRI (SSRI and related compounds) or switching SRIs are rational alternatives.

Cross-References

▶ Antidepressants
▶ Antipsychotic Drugs
▶ Bipolar Disorder
▶ Cognitive Behavioral Therapy
▶ Histaminic Agonists and Antagonists
▶ Lithium
▶ Monoamine Oxidase Inhibitors
▶ QT Interval
▶ Schizophrenia
▶ Second- and Third-Generation Antipsychotics
▶ SNRI Antidepressants
▶ SSRIs and Related Compounds

References

Bloch MH, Landeros-Weisenberger A, Kelmendi B, Coric V, Bracken MB, Leckman JF (2006) A systematic review: antipsychotic augmentation with treatment refractory obsessive-compulsive disorder. Mol Psychiatry 11(7):622–632

Drummond LM, Fineberg NA (2007) Obsessive-compulsive disorders. In: Stein G (ed) College seminars in adult psychiatry. Gaskell, London, pp 270–286 (Chap 15)

Fineberg NA, Craig K (2008) Pharmacotherapy for obsessive-compulsive disorder. In: Hollander E, Stein D, Rothbaum BO (eds) American psychiatric publishing textbook of anxiety disorders. American Psychiatric Publishing, Arlington (in press July 2008)

Fineberg NA, Gale TM (2005) Evidence-based pharmacotherapy of obsessive-compulsive disorder. Int J Neuropsychopharmacol 107(1):107–129

Fineberg NA, Nigam N, Sivakumaran T (2006) Pharmacological strategies for treatment-resistant obsessive compulsive disorder. Psychiatr Ann 36(7):464–474

Fineberg NA, Pampaloni I, Pallanti S, Ipser J, Stein D (2007) Sustained response versus relapse: the pharmacotherapeutic goal for obsessive compulsive disorder. Int Clin Psychopharmacol 22(6):313–322

Fineberg NA, Brown A, Reghunandanan S, Pampaloni I (2012) Evidence-based pharmacotherapy of obsessive-compulsive disorder. Int J Neuropsychopharmacol 9:1–19

Goodman WK, Price LH, Rasmussen SA, Mazure C, Fleischmann RL, Hill CL, Henninger GR, Charney DS (1989a) The Yale-Brown obsessive compulsive scale: part I. Development, use, and reliability. Arch Gen Psychiatry 46:1006–1012

Goodman WK, Price LH, Rasmussen SA, Mazure C, Delgado P, Heninger GR, Charney DS (1989b) The Yale-Brown obsessive compulsive scale: part II. Validity. Arch Gen Psychiatry 46:1012–1016

Guy W (ed) (1976) ECDEU assessment manual for psychopharmacology. U.S. Department of Health, Education, and Welfare, Rockville

Heyman I, Mataix-Cols D, Fineberg NA (2006) Obsessive-compulsive disorder. Brit Medical J 333:424–429

Koran ML (2007) focus.psychiatryonline.org 5(3):299–231

Koran ML, Hannah GL, Hollander E, Nestadt G, Simpson BH, American Psychiatric Association (2007) Practice guideline for the treatment of patients with obsessive-compulsive disorder. American Psychiatric Association, Arlington. Available online at http://www.psych.org/psych_pract/treatg/pg/prac_guide.cfm

NICE (2006) Core interventions in the treatment of obsessive-compulsive disorder and body dysmorphic disorder. Clinical guideline no. 31. Developed with the National Collaborating Center for Mental Health, National Institute for Health and Clinical Excellence, London. British Psychological Society and Royal College of Psychiatrists, Stanley L Hunt, Northamptonshire, UK. Available at www.nice.org.uk

Further Reading

Fineberg NA, Marazitti D, Stein D (eds) (2001) Obsessive compulsive disorder: a practical guide. Martin Dunitz, London, UK

Stein DS, Fineberg NA (2007) Obsessive-compulsive disorder. Oxford Psychiatry Library, Oxford University Press, UK

Obsessive-Compulsive Disorders in Childhood

James F. Leckman[1] and Michael Bloch[2]
[1]Sterling Hall of Medicine Child Study Centre, Yale University School of Medicine, New Haven, CT, USA
[2]Yale OCD Research Clinic, New Haven, CT, USA

Synonyms

Childhood-onset obsessive-compulsive disorder; Pediatric-onset obsessive-compulsive disorder

Definition

Obsessive-compulsive disorder (OCD) was classified as an anxiety disorder (Anxiety Disorders) in DSM-IV, but is now included in a separate grouping of obsessive-compulsive and related disorders in DSM-5. OCD is typically characterized by both obsessions and compulsions. Obsessions are repetitive, unwanted, intrusive thoughts, images, or impulses. Compulsions are repetitive physical or mental acts an individual feels driven to perform in a characteristic, stereotyped way, usually to relieve the anxiety or discomfort associated with an obsession, although some individuals may have compulsions in the absence of obsessions. Typical symptoms of OCD include contamination obsessions and cleaning compulsions (i.e., fear of germs and compulsive hand washing), forbidden thoughts (i.e., obsessions about harm coming to self or others, sexual obsessions, or religious obsessions), hoarding, and symmetry (i.e., the need to have things symmetrical, ordered, and arranged) (Bloch et al. 2008). These obsessions and/or compulsions must be severe enough to cause significant distress or impairment or be time-consuming (take up more than 1 h a day). Typically, patients with OCD have insight that their obsessions and compulsions are excessive and unreasonable, but feel driven to perform them by intolerable anxiety and discomfort caused by the obsessions.

OCD is generally considered to have a bimodal distribution in terms of age of onset, with a presence of 1–3 % throughout the life span. The first incidence peak occurs around puberty and the second occurs in early adulthood. Across the life cycle, symptoms of OCD are similarly expressed; however, there are several important differences between pediatric- and adult-onset OCD. Pediatric-onset cases of OCD have a male predominance (unlike adult-onset OCD cases, which are female predominant), have a stronger family history of OCD, and have higher rates of comorbid attention-deficit hyperactivity disorder and tic disorders. They are also more likely to experience a remission of their OC symptoms compared with adult-onset OCD. The specific content of obsessions and compulsions also differs across age ranges. Children with OCD have much higher rates of aggression obsessions (such as fear of a catastrophic event or fear of harm coming to self or others) and tend to have poor insight (being unable to recognize their obsessions and compulsions as excessive and unreasonable). Adolescents with OCD have a higher proportion of obsessions with sexual and religious themes, whereas cleaning and contamination OCD symptoms are quite prevalent across all age ranges. Patients with OCD and comorbid tics have a significantly higher rate of intrusive, violent, or aggressive thoughts and images; sexual and religious preoccupations; concerns with symmetry and exactness; hoarding and counting rituals; and touching and tapping compulsions. Also common in patients with OCD and comorbid tics are compulsions designed to eliminate a perceptually tinged mental feeling of unease, sometimes characterized as "just right" perceptions.

The age threshold that differentiates pediatric-onset OCD from adult-onset OCD is not clearly defined in the scientific literature, as investigators may differentiate the two groups with puberty or at ages 16 or 18. The pharmacological and behavioral treatments for OCD are remarkably similar across the life span. The combination of pharmacotherapy with selective serotonin reuptake inhibitors (SSRIs) ("▶ SSRIs and Related Compounds") and cognitive-behavioral therapy (CBT) (POTS 2004) is most common. There are slight differences in pharmacological management and behavioral techniques to treat pediatric-onset OCD that appear more important when the child is younger.

Role of Pharmacology

SSRIs are the pharmacological treatment of choice for obsessive-compulsive disorder. The number needed to treat (NNT) (▶ Number Needed to Treat) with SSRIs to induce a clinical treatment response, as defined by a 25–35 % reduction in scores on the Children's Yale-Brown Obsessive-Compulsive Scale (CY-BOCS), a standard assessment instrument, has ranged from 4 to 6 in double-blind, placebo-controlled studies (Compton et al. 2007). Based on cumulative studies, SSRI pharmacotherapy will lead to an average of a 6.5–9 point decline in CY-BOCS ratings and an average of 30–38 % reduction in symptom severity. In OCD, improvement with SSRI pharmacotherapy commonly continues to occur up to 2–3 months after the initiation of treatment. In adulthood OCD, a recent meta-analysis has demonstrated that higher doses of SSRI pharmacotherapy are more effective than lower-dose treatments (Bloch et al. 2010). No fixed-dose SSRI trials exist in pediatric-onset OCD to demonstrate if this phenomenon is similar in children. OCD patients with comorbid tics appear less likely to respond to SSRI pharmacotherapy than those without comorbid tics (March et al. 2007). Table 1 depicts the typical starting and target doses of SSRI pharmacotherapy in children and adults with OCD.

Compared to other medications used to treat child psychiatric conditions, SSRIs are generally well tolerated in children. Children, however, appear slightly more sensitive to initial treatment and dose increases than adults. Gastrointestinal disturbance, headache, fatigue, and insomnia are common side effects of SSRI medications. These side effects are ephemeral and commonly occur at the initiation of treatment or at an increase in dose. Often, with continued treatment with the same

Obsessive-Compulsive Disorders in Childhood, Table 1 Recommended dosing for serotonin reuptake inhibitors in OCD

Medication	Starting dose (in mg)		Target dose (in mg)	
	Children	Adults	Children	Adults
SSRIs				
Fluoxetine	5–10	20	10–80	40–80
Fluvoxamine	12.5–25	50	50–300	100–300
Sertraline	12.5–25	50	50–200	150–250
Paroxetine	5–10	10–20	10–60	20–60
Citalopram	5–10	20	10–60	20–60
Escitalopram	2.5–5	10	5–30	20–40
TCAs				
Clomipramine	12.5–25	25	50–200	100–250

dose, these side effects can disappear. Sexual side effects, including decreased sexual desire, pleasure, or performance, are all common dose-dependent side effects of SSRIs. Sexual side effects are a common, often unmentioned, cause of medication noncompliance in adolescents and adults. It is, therefore, important to outline these possible sexual side effects to adults and adolescents and inform them that they are not permanent. The sexual side effects disappear with SSRI discontinuation and often simply with a dose reduction.

Behavioral activation and suicidal ideation are less common but more severe side effects of SSRI pharmacotherapy. The US Food and Drug Administration (FDA) issued a "black-box warning" stating that these medications were associated with an increased risk of suicidality when utilized in pediatric populations. This warning was based on a meta-analysis of 24 placebo-controlled trials in children with depression, anxiety disorders, and ADHD (Bridge et al. 2007). The meta-analysis found a 2 % increase in the risk of suicidal ideation with treatment of SSRIs, which is equivalent to a relative risk of 2. There were no completed suicides that occurred in any of these trials. Subsequent meta-analyses have suggested that this increased risk of suicidal ideation appears restricted to only pediatric patients with depression and that the risk/benefit profile of SSRIs appears better in pediatric anxiety disorders (Bridge et al. 2007). Based on this warning, the FDA currently advises clinicians to see a child weekly for 4 weeks after starting SSRIs and monthly thereafter.

Behavioral activation is a syndrome associated with SSRI use that is characterized by hyperactivity, giddiness, insomnia, and agitation. These symptoms can sometimes involve mania, irritability, and aggression. Behavioral activation appears to be more associated with tricyclic antidepressant use and pharmacological treatment at a younger age (Martin et al. 2004).

Clomipramine, a tricyclic antidepressant with primarily serotonergic properties, was the first antidepressant the FDA approved for the treatment of pediatric-onset OCD. A meta-analysis of pharmacotherapy trials for pediatric-onset OCD found, using meta-regression techniques, that clomipramine was superior to the SSRIs in the treatment of pediatric-onset OCD, and the different SSRIs currently used to treat OCD were roughly equivalent (Geller et al. 2003). These meta-regression results, however, should be interpreted with caution, considering that no studies have directly compared the efficacy of clomipramine to other SSRIs in children with OCD. Studies comparing these agents in adults have found no significant differences. Clomipramine, due to its worse side-effect profile compared to SSRIs, is no longer used as an initial pharmacological treatment for OCD. Clomipramine has higher rates of somnolence, gastrointestinal upset, and weight gain than SSRIs. Clomipramine also causes anticholinergic effects, reduced seizure threshold, and cardiac arrhythmogenic side effects. Due to the potential arrhythmogenic properties of clomipramine, a baseline electrocardiogram (ECG) and a detailed screening for a personal and family history should be undertaken in all patients before the initiation of pharmacotherapy. Guidelines for unacceptable ECG indices for use of clomipramine have been issued by the FDA and are as follows: (1) PR interval >200 ms, (2) QRS interval >30 % increased over baseline or >120 ms, (3) blood pressure greater than 140/90, (4) heart rate >130 bpm, or (5) QTc >450 ms. Despite these concerns, clomipramine remains a valuable treatment option for children who do not respond to initial pharmacotherapy with one or more SSRIs. Clomipramine is now utilized mainly as monotherapy for OCD patients who have not responded to two SSRI trials of adequate dose and duration or as an augmentation agent to SSRIs in low doses. When using clomipramine as an augmentation agent, clinicians need to be vigilant for the onset of the serotonin syndrome.

Approximately half of the children placed on SSRI pharmacotherapy will fail to respond despite optimum pharmacologic treatment. If CBT has not been offered and conducted, it should be done so at this point. In adults with OCD, as many as 25 % of initial nonresponders may respond if treated with another SSRI or clomipramine. Other options for augmentation include antipsychotic augmentation or augmentation with glutamate-modulating agents.

No controlled trials of antipsychotic augmentation exist for children with OCD. A meta-analysis of placebo-controlled trials of adults with treatment-refractory OCD has demonstrated that these agents are effective (Bloch et al. 2006). The NNT to induce a clinically significant treatment response (35 % reduction in Yale-Brown Obsessive-Compulsive Scale [Y-BOCS] ratings) in adults with OCD was 4.5 (95 % CI: 3.2–7.1). Neuroleptic augmentation appears particularly effective in OCD patients suffering from comorbid tics (*NNT* = 2.3). This result is not surprising, since antipsychotics are the most effective medications in treating tic disorders. The doses of neuroleptic medications used to augment SSRIs are traditionally much lower than those used to treat psychosis and aggression in children. Lower doses are preferred due to the increased sensitivity of children with TS and OCD to these medications and the CYP2D6-based pharmacokinetic interactions these medications have with many SSRIs. Children with OCD appear particularly prone to the metabolic side effects associated with antipsychotics. The use of antipsychotic augmentation is, therefore, advisable only for children with comorbid tics or severe functional impairment despite adequate treatment with SSRIs and behavioral therapy.

Riluzole is a glutamate-modulating agent that is approved by the FDA for neuroprotection in the treatment of amyotrophic lateral sclerosis, also known as Lou Gehrig's disease. Riluzole is postulated to affect glutamate neurotransmission by reducing presynaptic glutamate release and by increasing glial cell reuptake of glutamate. Uncontrolled studies of both adults and children have provided encouraging data of efficacy. A case series of six children with treatment-refractory OCD found four treatment responders (Grant et al. 2007). Case series of treatment-refractory OCD adults demonstrated a 57 % response rate and treatment gains that were maintained for at least 2 years of follow-up (Pittenger et al. 2006). Riluzole's potential efficacy is yet to be demonstrated in any blinded or placebo-controlled trials. Although quite well tolerated by most patients, riluzole's widespread use is limited by its potential hepatotoxicity. Riluzole use requires monitoring of liver function tests every 3 weeks for the first 3 months of treatment.

The Role of Nonpharmacological Therapies
Cognitive behavioral therapy is a first-line treatment for pediatric-onset OCD. The NNT for CBT in pediatric-onset OCD is 2.6 (95 % CI: 1.7–4.2) to induce a treatment response (POTS 2004). CBT for children with OCD is based on exposure and response prevention. In CBT for OCD, the therapist must first examine the child and take a detailed history of the child's specific OCD symptoms. Symptom checklists, such as those that accompany the Y-BOCS, CY-BOCS, and Dimensional Y-BOCS, are useful in these efforts. Then, the therapist, working with the child and his/her parents, develops a hierarchy of exposures, both direct and imaginary, to trigger the

anxiety associated with the child's obsessions (e.g., touching a toilet seat for a child with contamination concerns or imagining a parent in a car accident for a child with fears of harm coming to others). The therapist then enters into a contract with the child and parent to work on completing these exposures over a set time frame, usually a period of a couple of months. Next, the therapist engages in the specific exposures with the child (and sometimes the parents) over the next few treatment sessions, asking the child to agree not to engage in his/her compulsions in response to the anxiety produced. During exposures, the child reports his/her level of anxiety using the Subjective Units of Distress Scale (SUDS), usually with the aid of a graphic "anxiety thermometer." The child and parent are given similar exposure assignments to complete after each session. With increased duration of exposure and subsequent exposure, the child's anxiety lessens (without performing the compulsions), helping him/her to overcome the OCD. Many CBT treatment manuals for children with OCD are readily available and additionally include techniques such as relaxation therapy, to help children deal with stress and anxiety. The manuals may also include exercises to illustrate other tenets of CBT. For example, they may help the child to recognize cognitive distortions common in OCD, such as overestimation of risk, all-or-nothing thinking, overcoming the need for certainty, and recognizing excessive feelings of responsibility or guilt. CBT is as effective as any pharmacological intervention for OCD and should be utilized before pursuing less evidence-based pharmacological augmentation strategies.

Conclusion

SSRIs and CBT are effective treatments for pediatric-onset OCD. The combination treatment of both SSRIs and CBT is more effective than either treatment alone. Although many children with OCD respond to submaximal doses of SSRI pharmacotherapy, it is important to pursue an SSRI trial of adequate dose (maximum tolerated dose) and duration (2–3 months) before progressing to augmentation therapies. Antipsychotic augmentation appears to be a particularly effective strategy for OCD patients with comorbid tics. Riluzole augmentation is an emerging treatment for OCD that may be effective for some patients. While there are several effective treatment options for OCD, there is still a great need for treatments that work better and faster.

Cross-References

▶ Adolescence and Responses to Drugs
▶ Antidepressants
▶ Antipsychotic Drugs
▶ Attention-Deficit/Hyperactivity Disorder
▶ SSRIs and Related Compounds

References

Bloch MH, Landeros-Weisenberger A et al (2006) A systematic review: antipsychotic augmentation with treatment refractory obsessive-compulsive disorder. Mol Psychiatry 11(7):622–632

Bloch MH, Landeros-Weisenberger A et al (2008) Meta-analysis of the symptom structure of obsessive-compulsive disorder. Am J Psychiatry 165(12): 1532–1542

Bloch MH, McGuire J et al (2010) Meta-analysis of the dose–response relationship of SSRI in obsessive-compulsive disorder. Mol Psychiatry 15(8):850–855. doi:10.1038/mp.2009.50

Bridge JA, Iyengar S et al (2007) Clinical response and risk for reported suicidal ideation and suicide attempts in pediatric antidepressant treatment: a meta-analysis of randomized controlled trials. JAMA 297(15): 1683–1696

Compton SN, Kratochvil CJ et al (2007) Pharmacotherapy for anxiety disorders in children and adolescents: an evidence-based medicine review. Pediatr Ann 36(9): 586–590, 592, 594–598

Geller DA, Biederman J et al (2003) Which SSRI? A meta-analysis of pharmacotherapy trials in pediatric obsessive-compulsive disorder. Am J Psychiatry 160(11):1919–1928

Grant P, Lougee L et al (2007) An open-label trial of riluzole, a glutamate antagonist, in children with treatment-resistant obsessive-compulsive disorder. J Child Adolesc Psychopharmacol 17(6):761–767

March JS, Franklin ME et al (2007) Tics moderate treatment outcome with sertraline but not cognitive-behavior therapy in pediatric obsessive-compulsive disorder. Biol Psychiatry 61(3):344–347

Martin A, Young C et al (2004) Age effects on antidepressant-induced manic conversion. Arch Pediatr Adolesc Med 158(8):773–780

Pittenger C, Krystal JH et al (2006) Glutamate-modulating drugs as novel pharmacotherapeutic agents in the treatment of obsessive-compulsive disorder. NeuroRx 3(1):69–81

POTS (2004) Cognitive-behavior therapy, sertraline, and their combination for children and adolescents with obsessive-compulsive disorder: the Pediatric OCD Treatment Study (POTS) randomized controlled trial. JAMA 292:1969–1976

Occasion Setting with Drugs

Matthew I. Palmatier[1] and Rick A. Bevins[2]
[1]Department of Psychology, East Tennessee State University, Johnson City, TN, USA
[2]Department of Psychology, University of Nebraska-Lincoln, Lincoln, NE, USA

Synonyms

Drug as cues, Drug discriminative stimulus, Drug facilitator, Drug modulator, Drug occasion setter

Definition

Occasion setting is a form of hierarchical learning involving associative and nonassociative processes. Occasion setters are discrete stimuli or contexts that disambiguate Pavlovian and/or operant relations. In cases where stimuli or responses are "sometimes" associated with biologically relevant events, occasion setters can provide discriminant information that resolves the ambiguity of the antecedent. Drug states are internal contexts that can predict positive relationships between stimuli/responses and outcomes (feature-positive occasion setters) or negative relationships between stimuli/responses and outcomes (feature-negative occasion setters). Traditionally, occasion setters are considered nonassociative because their ability to influence behavior does not depend on a direct association with the antecedent or biologically relevant events and can "transfer" to novel responses and situations.

Impact of Psychoactive Drugs

Introduction

In a standard Pavlovian (classical) conditioning experiment, the investigator establishes a relation between two stimuli. In a typical study involving rodents, one stimulus might be a tone or a light [termed conditioned stimulus (CS)] that is repeatedly paired with a more motivationally relevant stimulus such as food, water, and drug [termed unconditioned stimulus (US)]. Sometime the CS-US pairing only occurs in a particular situation or when some other stimulus is present. The stimulus that disambiguates when the CS will be reinforced is often referred to as a feature-positive occasion setter. Under these conditions, the CS evokes a conditioned response (CR) only when the feature is present. Alternatively, the feature might indicate when the CS will *not* be reinforced. In this case, the stimulus is referred to as a feature-negative occasion setter, and conditioned responding during the CS is withheld or inhibited. Many drugs have interoceptive (subjective) properties that function as stimuli capable of guiding learned behaviors. These interoceptive properties can be conceptualized as polymodal contextual (situational) cues in which the stimulus elements reflect the complex neurobiological action of that drug. As such, a drug state can function as a feature-positive or feature-negative occasion setter disambiguating when a CS-US relation will or will not be in force, respectively. Drugs such as bupropion, caffeine, chlordiazepoxide (CDP), cocaine, D-amphetamine, ethanol, flumazenil, indorenate, nicotine, methamphetamine, morphine, pentobarbital, rimonabant, sertraline, and Δ-9 tetrahydrocannabinol (THC) appear to function as feature-negative and/or feature-positive occasion setter.

Example Tasks

The occasion-setting function of drug states has been described in a variety of paradigms using a range of CS types, as well as appetitive and aversive USs. For sake of brevity, we will focus the present discussion on two variants: discriminated conditioned taste aversion (DTA; Jarbe and

Lamb 1999) and discriminated goal tracking (DGT; Palmatier and Bevins 2008). In DTA studies, the drug state sets the occasion on which a taste CS will or will not be followed by a noxious or aversive event, often lithium chloride-induced illness. Subjects are provided repeated access to a novel taste CS. During half of these conditioning trials, the taste CS is accessed in the drug state; during the remaining trials the taste CS is accessed after a placebo injection (nondrug state). Following access to the CS, the illness-inducing US or placebo (no US) is administered. The CR is typically defined as avoidance of the taste CS. In DTA studies, drug states are most often trained as positive features – the taste CS is followed by illness when it is accessed in the drug state, and the same tastant is followed by a placebo injection when it is accessed in the nondrug state. Subjects learn to avoid the taste CS when it is presented in the drug state, but readily consume the CS in the nondrug state. Because the feature-positive drug state occasions taste-illness pairings in this task, it is sometimes referred to as a "danger cue." In this paradigm, drug states can also serve as negative features which set the occasion on which the taste will not be followed by illness and are sometimes referred to as "safety cues" (see Skinner et al. (1995) and section "Hallmarks of Occasion Setters" for more information).

In DGT studies, the drug feature sets the occasion on which a CS will or will not be followed by a rewarding US, typically food or sucrose solution. In this task, a brief auditory or visual stimulus (CS) is presented in both the drug state and the nondrug state (placebo). If the drug state is a positive feature, then the CS presentations that occur in the drug state are followed by access to the US; the same CS is presented in the nondrug state, but the US is withheld (Palmatier and Bevins 2008). Conversely, when the drug state serves as a negative feature, the US follows the CS after placebo injection and the US is withheld after drug injections (Bevins et al. 2006). In these studies, the CR is goal tracking which reflects anticipatory approach to a location where the appetitive US has occurred previously (e.g., head-entries into the receptacle where sucrose is delivered).

The concept that the internal milieu can provide a context with discriminant properties is not limited to drug states. Davidson and his colleagues (1998) have argued that internal contexts of hunger and satiety might function as occasion setters. In one notable study, hunger and satiety contexts were used to set the occasion for when a rat would receive a shock (US) in a particular chamber (CS). Rats were able to use an internal state to determine when a shock would occur, and the occasion-setting function of the hunger/satiety state transferred to other chambers that were associated with shock, but not to chamber CSs that were never associated with shock. Based on these and other findings, Davidson (1998) has argued that the occasion-setting function of the internal milieu could explain why more appetitive behaviors are exerted when food stimuli are presented in "hunger" states relative to "sated" states.

Hallmarks of Occasion Setters

An important aspect of occasion setting is ascertaining whether the feature's influence on behavior depends on a direct relationship with the US. Occasion-setting features typically have a perfect positive or negative correlation with the US, meaning that they may function only as simple conditioned excitors (feature-positive) or conditioned inhibitors (feature-negative) rather than occasion setters. Determining whether a feature is an occasion setter or a simple excitor/inhibitor often demands manipulating its relationship to the US and then a test to examine whether the manipulation has altered the feature's ability to increase or decrease responding evoked by the CS. If the feature functions only as a conditioned excitor or inhibitor, then this manipulation is expected to abolish its ability to modulate CS-evoked responding.

In the DGT task, drug states (amphetamine, nicotine, and CDP) trained as positive features increased responding to a CS paired with sucrose (Palmatier and Bevins 2007). In the nondrug state, the CS evoked very little conditioned responding. To investigate whether these drug states increased responding because they served as simple CSs (i.e., direct association with US), we changed the relation between the drug state and the US. Each feature underwent an extinction treatment – rats were repeatedly exposed to the drug in the absence of the US. The discrete exteroceptive CS was never presented on these extinction trials. In subsequent tests, all three drug states still increased responding evoked by the CS. This finding suggests that the drug states functioned as occasion setters (nonassociative) rather than simple conditioned excitors (associative).

A second hallmark of occasion setters is their transitive nature; a feature that sets the occasion for one CS-US association can often transfer its discriminant function to another CS. Transfer of occasion setting normally involves a novel combination of features and CSs after subjects have been trained on multiple discriminations. For example, we recently trained two drug states (nicotine and CDP) as positive features for two different CSs in the DGT task. A noise CS was followed by sucrose after nicotine injections, and a light CS was followed by sucrose after CDP injections. After placebo injections the light and noise were presented but were never followed by sucrose. Following training on both discriminations, rats were tested on different combinations of drug features and CSs. As expected, nicotine and CDP facilitated responding to the CSs which typically occurred in those drug states. However, each drug also facilitated responding to the CS which had occurred in the *other* drug state. Neither CS evoked conditioned responding when presented after injection of a novel drug (i.e., amphetamine), and neither feature facilitated responding to a novel CS (Palmatier and Bevins 2008). In DTA procedures, positive drug features ("danger cues") typically transfer their facilitative function in a less specific manner. Skinner and colleagues (1998) have demonstrated that these danger cues will inhibit consumption of the taste CS as well as novel and familiar tastants that have never participated in an occasion-setting discrimination. They have argued convincingly that positive drug features in the DTA paradigm may set the occasion for associations between consummatory responses and the illness-inducing US (Skinner et al. 1998).

Recall that in the case of feature-negative training, conditioned responding evoked by the target CS in the drug state is significantly attenuated relative to the nondrug state. To our knowledge, transfer of negative occasion setting with drug features has not been investigated. This may be because researchers in this area suspect that negative drug features are functioning as simple conditioned inhibitors (Skinner et al. 1995; Bevins et al. 2006). The theoretical and hence therapeutic implications could be significant if a drug state could acquire inhibitory properties. To gather convincing evidence for inhibitory conditioning to the interoceptive effects of a drug, one must show that a drug trained as a negative feature passes the summation and retardation tests of conditioned inhibition. In the summation test, the putative conditioned inhibitor will also inhibit conditioned responding to an excitatory CS that has been trained independent of any occasion-setting function. For the retardation test, subsequent acquisition of an excitatory conditioned response to the negative feature will be impaired (retarded) relative to controls. Morphine trained as a negative feature in the DTA task ("safety signal") passes the summation test. That is, avoidance of a separately trained saccharin CS was diminished across repeated extinction sessions in the presence of morphine that was previously trained as a negative feature indicating that vinegar solution would not be paired with illness (Skinner et al. 1995). A retardation test has not been conducted using the DTA procedures, nor has either the retardation or summation test been used within the DGT task. Thus, the field awaits a definitive demonstration as to whether a feature-negative drug state inhibits conditioned responding through direct inhibitory properties.

Significance and Application

Training the interoceptive effects of a drug as an occasion setter requires a discrimination to at least be made between the drug and no drug state. Of course, as indicated earlier the discrimination is more specific than a mere presence versus absence of an altered internal state. Given this pharmacological specificity in control of conditioned responding, these Pavlovian drug discrimination tasks have been used as a complement to the operant drug discrimination task that use drug discriminative stimuli to help elucidate the neuropharmacological processes underlying the subjective effects of a drug (Kreiss and Lucki 1994; Reichel et al. 2007).

Although clear advances have been made in understanding drug states as occasion setters, the role of such factors in addiction and other health-related behaviors and their therapeutic implications have yet to be fully realized. Regardless, we expect their importance to be high given the demonstrations that learning can alter the functional impact of the drug state. A salient example described earlier is the transfer (substitution) of occasion-setting function to a pharmacologically unrelated drug only when the drugs share a common learning history – i.e., trained as feature-positive occasion setters with different target CSs. Another example is the alteration of the psychomotor effects of stimulant drugs depending on whether there has been a training history with a positive or negative drug feature (Reichel et al. 2007). Briefly, rats were trained with methamphetamine as the occasion setter. For one set of rats, methamphetamine served a feature-positive occasion setter indicating when a brief light was paired with sucrose. The other set of rats had methamphetamine indicate when the light was not reinforced (i.e., feature-negative occasion setter). Following acquisition of the discrimination were intermixed substitution tests in which cocaine and bupropion were tested. Substitution tests indicated that bupropion and cocaine had methamphetamine-like stimulus effect. More interestingly from the current discussion's perspective is that the typical stimulant effects of these two drugs were significantly blunted if rats had methamphetamine trained as a negative feature. Much more research is needed to know the generality, extent, and mechanism (s) of this effect. Regardless, a well-documented behavioral effect of cocaine and bupropion was affected by training methamphetamine as a negative feature.

The significance of occasion setting by drug states is critical to the study of drug addiction. The widespread acceptance of associative learning processes in addictions is reflected by their inclusion as a critical component in biobehavioral models of drug dependence (e.g., Koob and Le Moal 2008). However, the role ascribed to associative learning in most models is best described as simple and elemental. One cannot assume that for drug-dependent individuals all "drug cues" are followed by rewarding drug effects each time they are experienced. For example, a smoker may be exposed to stimuli (i.e., cigarette packaging) that have a history of pairing with the reinforcing effects of nicotine. However, in many circumstances (i.e., in public buildings, offices, hospitals, filling stations) the effects of nicotine may be unattainable. In these cases the contextual stimuli probably modulate the meaning of the CS (cigarette packaging). Given the frequently observed comorbidity between drug dependence disorders (e.g., alcoholism and smoking), it is critical that we expand the role of associative processes in addictions and further examine the role of occasion setting by drug states and other contexts. These principles undoubtedly extend to other forms of addiction (e.g., gambling), other impulsive and compulsive behaviors (e.g., eating; see Davidson 1998), as well as pain management and cancer treatment.

Cross-References

▶ Addictive Disorder: Animal Models
▶ Blocking, Overshadowing, and Related Concepts
▶ Classical (Pavlovian) Conditioning
▶ Conditioned Drug Effects
▶ Conditioned Place Preference and Aversion
▶ Conditioned Taste Aversions
▶ Drug Discrimination
▶ Instrumental Conditioning

References

Bevins RA, Wilkinson JL, Palmatier MI, Siebert HL, Wiltgen SM (2006) Characterization of nicotine's ability to serve as a negative feature in a Pavlovian appetitive conditioning task in rats. Psychopharmacology (Berl) 184:470–481

Davidson TL (1998) Hunger cues as modulatory stimuli. In: Schmajuk NA, Holland PC (eds) Occasion setting: associative learning and cognition in animals. American Psychological Association, Washington, DC, pp 223–248

Jarbe TU, Lamb RJ (1999) Discriminated taste aversion and context: a progress report. Pharmacol Biochem Behav 64:403–407

Koob GF, Le Moal M (2008) Addiction and the brain antireward system. Annu Rev Psychol 59:29–53

Kreiss DS, Lucki I (1994) Discriminative stimulus properties of the serotonin uptake inhibitor sertraline. Exp Clin Psychopharmacol 2:25–36

Palmatier MI, Bevins RA (2007) Facilitation by drug states does not depend on acquired excitatory strength. Behav Brain Res 176:292–301

Palmatier MI, Bevins RA (2008) Occasion setting by drug states: functional equivalence following similar training history. Behav Brain Res 195:260–270

Reichel CM, Wilkinson JL, Bevins RA (2007) Methamphetamine functions as a positive and negative drug feature in a Pavlovian appetitive discrimination task with rats. Behav Pharmacol 18:755–765

Skinner DM, Martin GM, Howe RD, Pridgar A, van der Kooy D (1995) Drug discrimination learning using a taste aversion paradigm: an assessment of the role of safety cues. Learn Motiv 26:343–369

Skinner DM, Goddard MJ, Holland PC (1998) What can nontraditional features tell us about conditioning and occasion setting? In: Schmajuk NA, Holland PC (eds) Occasion setting: associative learning and cognition in animals. American Psychological Association, Washington, DC, pp 113–144

Octreotide

Synonyms

SMS201-995; Sandostatin

Definition

Octreotide is an octapeptide, the first somatotropin release-inhibiting factor (SRIF) receptor agonist in the clinic. Octreotide is used to treat acromegaly, some gastrointestinal disorders, and primarily a number of tumors in the gastroenteropancreatic tract. Octreotide has high affinity for sst_2 and sst_5 receptors.

Cross-References

▶ Somatostatin

Off-Label Use of Drugs

Definition

A label is an official document that reflects information about a drug which has been approved by governmental regulatory agencies, such as the Food and Drug Administration (FDA) in the USA and the European Medicines Agency (EMEA) in the EU. The label contains the same information as the package insert which accompanies the drug when it is dispensed. The label includes the drug's approved indications, that is, the conditions, diseases, and disorders that can be treated with the drug. Off-label use refers to the quite common practice of using a drug for the treatment of a disease or condition that is not listed on its label, or using it in a way that is not described in the label (e.g., at higher doses). Synonyms for off-label are extra-label use, nonapproved use, or unapproved use. Off-label prescription raises legal and ethical issues. Some countries ban or restrict off-label use of drugs. Insurance companies may reject the reimbursement claims for off-label prescriptions. In case of unexpected side effects/adverse events of the drug treatment (i.e., adverse events which are not described in the label), the manufacturer bears no liability, as it does with prescriptions based on the label. Regulations may restrict assertions that a drug's manufacturer can make about the drug if the assertions are not supported by the label.

Olanzapine

Definition

Antipsychotic drug of the second generation, atypical category with combined dopamine D2/5-HT$_2$ receptor-blocking properties.

Cross-References

▶ Second- and Third-Generation Antipsychotics

Ontogeny

Synonyms

Morphogenesis; Ontogenesis

Definition

The origin and the developmental history of an organism from the embryo to its mature form (adult). The term is distinct from "phylogeny" which denotes the evolutionary history of a species.

Open Label

Definition

Open label is a method of research in which the identity of the treatment is known by both the researchers administering the treatment and the subjects involved in the study.

Cross-References

▶ Impulse Control Disorders

Open-Field Test

Catherine Belzung
UFR Sciences et Techniques, Université François Rabelais, Tours, France

Definition

Test consisting of scoring the behavior of a rodent that has been forced into a novel arena. Enables detection of effects of drugs on anxiety behavior, activity, and memory.

Principles and Role in Psychopharmacology

History
In 1934, Hall (1934) designed the open-field test to assess "emotionality" in rats. The procedure consists in introducing an animal into a new arena from which escape is prevented by walls (Walsh and Cummins 1976). Hall's apparatus consisted of a brightly lit circular environment (thereafter termed as "open field") of about 1.2 m diameter closed by a wall of 0.45 m high. Rats were placed individually in the open field and their behavior was recorded for 2 min, during daily repeated trials. In some cases, rats were tested after 24 or 48 h of food deprivation. Hall observed that rats exhibited increased locomotion when food was deprived, and also, surprisingly, some rats did not eat, in spite of the high alimentary motivation related to the food deprivation: these animals were termed "emotional." When compared with nonemotional rats, they displayed increased thigmotaxis (i.e., they walked in the outer ring of the apparatus, always maintaining a tactile contact with the walls via their vibrissae); these emotional rats also exhibited higher levels of defecation and micturition (Fig. 1).

Procedure
The procedure is generally based on the forced confrontation of the rodent with the open field,

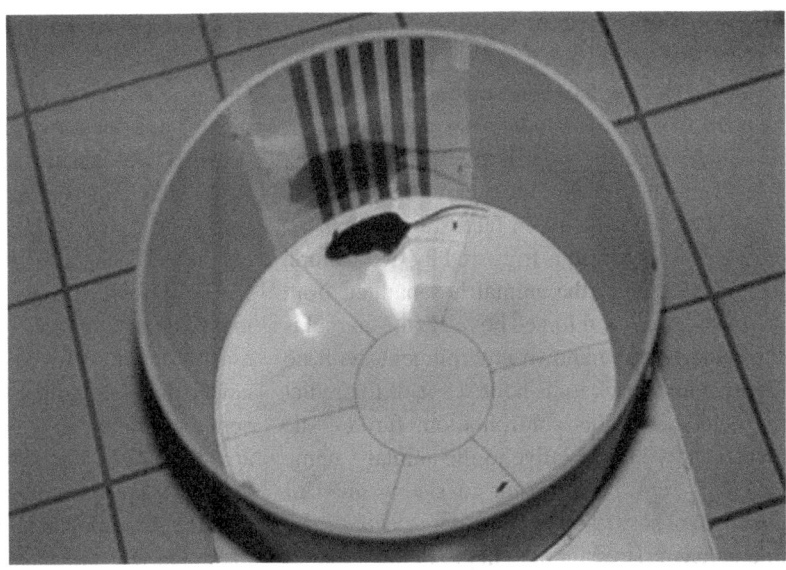

Open-Field Test, Fig. 1 A mouse in a circular open field – dimensions of the arena: 40 cm in diameter and 30 cm high

but in some cases animals have also been allowed free access to the arena from their familiar nest.

In this free confrontation situation, rodents will generally show avoidance for the brightly lit environment: before entering the arena, the animal will show risk assessment postures directed toward the open field. This is the first argument indicating that the open field may be a stressful situation. In case of forced confrontation, the animal is placed in the apparatus and the following behavioral items are recorded, usually during 5 min: locomotion, frequency of rearing or leaning (sometimes termed vertical activity), and grooming. Increase of time spent in the central part, higher ratio of central/total locomotion, or decrease of the latency to enter the central part are indications of reduced anxiety-like behavior. Therefore, experimenters are not measuring the effects of the factors they are interested in on exploration, as is sometimes claimed, but the effects on the reaction of the subjects to a stressful event. Additionally, other parameters may be measured, including physiological, neural, or endocrinological ones. For example, using telemetry, it has been shown that the forced confrontation to the open field induces increased heart rate, further indicating that the animal is in a stressful situation. Immunohistological experiments, allowing measuring of the activation of various brain areas via c-fos, have shown that exposure to the arena induces activation of several structures belonging to the limbic system, such as the hippocampus, the amygdala, or the bed nucleus of stria terminalis, further suggesting that the rodent is stressed. This is accompanied by an activation of some hypothalamic nuclei, including the paraventricular nucleus. Finally, open-field exposure also induces a rise of plasma corticosterone, which is the main stress hormone. Taken together, this indicates that forced exposure to the open field induces a complex response, characterized by a specific behavior associated with physiological, endocrine, and neural modifications, all suggesting that the animal has been stressed.

Utilization

The open-field test is now one of the most popular procedure in behavioral neuroscience (see Belzung 1999; Prut and Belzung 2003) and is used to assess the effects of different factors on anxiety-like behavior, including the action of anxiolytic drugs, brain lesions, or genetic invalidation. In fact, it has become a convenient procedure to measure not only anxiety-like behaviors but also activity, enabling us to detect sedation (decrease of activity) or hyperactivity. It can also be used to assess learning and memory. Several versions of the apparatus have been designed, differing in shape of the environment (circular, square, or rectangular), lighting, presence of objects within the arena, and so on. Utilization of this apparatus has been extended to a great number of species including not only farm animals such as calves, pigs, lambs, and rabbits but also other species including invertebrates such as honeybees and lobsters. In fact, stressful experience in the open field is triggered by two factors: individual testing (the animal is separated from its social group) and forced novelty (the animal is subjected to an unknown environment, with no ability to turn back to its home cage or to predict the outcome of this confrontation; further, the arena is very large relative to the animal's home cage or natural environment consisting in small galleries). These two factors may respectively be interpreted as stressful only in gregarious species and/or in species that show fear of open spaces into which they are forced. This is precisely the case with rodents that live in groups and in small underground tunnels. This is of course not the case in species such as lambs or cows that live in large fields.

Interpretation Bias

Behavior of rodents in the open field depends on several sensorial modalities, with the main involvement of tactile factors. Indeed, mice without vibrissae no longer display thigmotaxis, as they lose tactile contact with the walls. One must thus emphasize the possibility of misinterpretation of data related to effects of some treatments on the sensorial characteristics of the animals. As already indicated, exploration can be increased by some factors including food or water deprivation: it is therefore very important to verify that a given treatment does not act on such variables, before concluding that there may be anti-stress effects. Finally, open-field behavior also depends on lighting conditions and upon the light-dark cycle (because exploration and food patrolling is increased during the animal's high-activity period, i.e., at the beginning of the dark phase) so that it may be relevant to ensure that a treatment does not modify internal clock-related behaviors and test the treatment under different lighting conditions.

Assessment of Drug Effects on Anxiety Behavior

Using this device, the effects of different treatments have been investigated, mainly in the field of behavioral genetics (18 % of the studies investigating the effects of targeted mutation of a given gene on anxiety-like behavior have been conducted using this device, a percent rising to 26 % for mutants of the serotonergic system) and in psychopharmacology. Concerning this last aspect, one may notice that many different drugs have been investigated in this situation, including compounds not only with effective or potential anxiolytic-like effects (benzodiazepines, serotonin ligands, neuropeptides) but also with psychostimulant (amphetamine, cocaine), sedative (neuroleptic), preictal (the state characterizing the prostration state induced by some epileptogenic drugs) activity, or amnesic effects. An increased locomotion or time spent in the central part of the device without any modification of total locomotion has been interpreted as an anxiolytic-like effect, while the contrary, which is a decrease of these variables, is associated with anxiogenic-like effects. Increased total locomotion (i.e., an increase of central and peripheral locomotion, in the same proportions) can be interpreted as hyperactivity, while decreased rearing and locomotion are related to sedation or to postictal prostration. Concerning anxiety-like activity, the effects of three categories of drugs have been investigated, including compounds acting on the $GABA_A$ pentamer (not only benzodiazepine receptor ligands but also $GABA_A$ receptor agonists, barbiturate, and neurosteroid ligands), serotonergic-acting drugs such as ligands of the different 5-HT receptors or inhibitors of the serotonin transporter, and the effects of neuropeptidergic ligands (CRF, corticotropin-releasing factor; CCK, cholecystokinin; NK, neurokinin; neuropeptide Y). Interestingly, the open field is able to detect the anxiolytic-like effects produced by compounds effective in typical tests for anxiety, such as classical benzodiazepines and $5-HT_{1A}$ receptor agonists, while it lacks sensibility for the anxiolytic action of atypical benzodiazepines such as alprazolam or for the anxiolytic action of chronic antidepressants such as selective serotonin reuptake inhibitors that display efficacy in some anxiety disorders such as panic, obsessive-compulsive disorder, social phobias, and post-traumatic stress disorder. It is to be emphasized here that benzodiazepines and 5-HT_{1A} receptor agonists do not increase exploration as it is sometimes claimed; they act by reducing the stress-induced inhibition of exploration. Further, this device detects the effects of these compounds on anxiety-like behavior and not on anxiety. Indeed, anxiety-like behavior is only one component of the anxiety-like response, which also includes expressive reactions (e.g., the vocal expression such as ultrasonic distress calls), physiological alterations (modifications in body temperature, heart rate, arterial pressure), as well as cognitive/subjective aspects.

Assessment of Drug's Action on Locomotion (Sedation and Hyperactivity)

Concerning the ability to detect sedation, for example, after administration with high doses of benzodiazepines, ethanol, or neuroleptics, an important point to consider is that the free exploration situation is much more sensitive to the sedative effects of a treatment when compared with the forced situation. Indeed, in order to detect sedative effects, twofold higher doses should be used in the forced situation. This can be explained by the fact that the forced confrontation imposes the experimenter to take the animal out of its home cage, which may induce awaking of an animal that would have been sleeping in its home cage. One may also mention the fact that the effects of a sedative treatment on rearing appear at lower doses than the ones on locomotion, indicating that this parameter is more sensitive. Sedative-like effects have to be distinguished from the preictal prostration that can be observed after administration of some proconvulsive treatments. The phenomenological features of these two behaviors may seem identical, as animals exhibit a reduced exploration in both cases. Therefore, when reduced exploration is seen, the experimenter has always to be very prudent in the interpretation of data.

Further, reduced exploration can also be observed if the rodent is freezing; but in this case, the experimenter will observe a very different posture, the rodent showing a tonic immobility, while when sedated it displays a nontonic immobility.

Hyperactivity can also be detected using this device. Hyperactivity never corresponds to increased exploration. Indeed, after treatment with psychostimulants, the animal will show elevated locomotion, sometimes close to stereotypy (as it may repeat the same locomotion pattern several times) with reduced exploratory items (no sniffing, few rearing).

Assessment of Drug's Effects on Learning and Memory

The open field has also been used to detect the effects of some treatments on learning and memory. Three different processes can be assessed in this case: habituation, object recognition, and spatial memory. To test habituation, one can compare the behavior of the rodent during the first 5 min with the ones of the last 5 min of a 15-min session, enabling to detect the effect of a treatment on short-term habituation. Alternatively, one can confront a rodent to several 5-min sessions, separated by intertrial intervals of some hours or several days. In this case, one can distinguish the effects of a given treatment on encoding from its effects on consolidation or restitution, depending upon the injection schedule. When testing object recognition using an open field, the subject is usually first introduced in the arena during a 5-min session. After an intersession delay, the animal is further subjected to the open field, which now contains two identical objects. During the third session, the rodent is again introduced in the device, which contains one object identical to the one it has been confronted before and another one that is different. Higher exploration of the new object indicates that the animal has been able to remember the object it had seen before and is therefore considered an index of recognition memory. Pharmacological treatments can be tested either to detect promnesic/amnesic effects or to counteract the stress-induced decline in learning and memory (e.g., antidepressants are able to restore a stress-induced decline in object recognition). Finally, this situation can also be used to detect the effects of drugs on spatial memory. In this case, the rodent is confronted to the arena containing three identical objects, placed in a precise spatial configuration; (e.g., a line) and after a delay, it will again be introduced in the device, with the same objects, placed, for example, in a triangular configuration. In this case, exploration of all objects that had been declining after habituation will start again. This process can be altered with pharmacological treatments such as, for example, anticholinergic drugs.

Conclusion

In conclusion, the open field is a very popular and useful device, not only enabling detection of effects of drugs on anxiety behavior, but also action of pharmacological compounds on sedation, hyperactivity, learning, and memory.

Cross-References

▶ Anxiety: Animal Models
▶ Benzodiazepines
▶ Habituation
▶ Phenotyping of Behavioral Characteristics
▶ Translational Research in Drug Discovery

References

Belzung C (1999) Measuring exploratory behavior. In: Crusio WE, Gerlai RT (eds) Handbook of molecular genetics techniques for brain and behavior research (techniques in the behavioral and neural sciences). Elsevier Science, Amsterdam
Hall CS (1934) Emotional behavior in the rat. I. Defecation and urination as measures of individual differences in emotionality. J Comp Psychol 18:385–403
Prut L, Belzung C (2003) The open field as a paradigm to measure the effects of drugs on anxiety-like behaviors: a review. Eur J Pharmacol 463:3–33
Walsh RN, Cummins RA (1976) The open field test: a critical review. Psychol Bull 83:481–504

Operant

Definition

A class of behavioral actions all of which produce a common effect on the environment. A typical example is pressing a lever. The different forms of behavior that result in lever depression are members of the operant class.

Operant Behavior in Animals

Marc N. Branch
Department of Psychology, University of Florida, Gainesville, FL, USA

Synonyms

Instrumental behavior; Instrumental performance; Purposive behavior

Definition

Operant behavior in animals encompasses activities that are influenced by their consequences. More precisely, the future likelihood of such behavior is a function of the history of consequences for that behavior. Particular classes of activities that are influenced by their consequences are called operants. For instance, any form of behavior that successfully results in the depression of a lever by a rat would be considered as part of the operant "lever pressing." Effective consequences are divided into two categories: those that make the behavior more likely in the future and those that make it less likely. The former are called reinforcers or reinforcing stimuli (akin to the everyday word "rewards") and the latter are called punishers or punishing stimuli. Reinforcers and punishers are further subdivided into two subcategories: positive and negative. Positive reinforcers increase the subsequent probability of behavior if they are presented as a consequence, for example, presenting food to a hungry rat after it presses a lever, whereas negative reinforcers make behavior more likely when they are removed as a consequence, for example, when pressing a lever turns off an irritating loud noise. Positive and negative punishers therefore exert their effects upon presentation and removal, respectively. Behavior that is increased by experience with consequences is said to be reinforced and the process is called reinforcement. Conversely, decreasing behavior by means of consequences is called punishment. The effective consequences are functionally defined. For example, reinforcers are defined by three features. One, an operant has a consequence. Two, the operant subsequently becomes more probable. Three, the increase is due to the consequential relationship, not to the mere presentation of the consequence (e.g., consider trying to reinforce crying in a baby by pinching it when it cries). Punishment is analogously defined, the only difference being that the operant becomes less likely. Operant behavior in animals is used to model willful, voluntary behavior in humans. When consequences occur regularly (see next) either naturally (as when a door moves when pushed) or as a result of deliberate planning (as when an experimenter arranges that a lever press results in food presentation) and the behavior is affected by that relationship, it is often said that a contingency of reinforcement is evident.

Impact of Psychoactive Drugs

Principles and Use in Psychopharmacology

The basic techniques for establishing and studying operant behavior in animals were developed in large part by B. F. Skinner and his colleagues in the 1940s through the 1960s (see Skinner 1938). They have become a signature part of the armamentarium of psychopharmacology. Operant techniques play a prominent role in drug discrimination, drug self-administration, and behavioral economics. They are also used to study drug effects on sensory processes and on cognitive processes such as choice, timing, learning, and memory, as well as emotional responses such as anxiety and stress (Mazur 2006). Most of these applications of the techniques involve two important processes that characterize operant behavior. Those are (1) stimulus control, or discrimination (Nevin 1973), and (2) *intermittent reinforcement* via schedules of reinforcement (Ferster and Skinner 1957; Zeiler 1977). Before discussing these two processes, however, it is necessary to describe the conditions required to establish operant behavior.

There are two main methods for establishing operant behavior. The first is called free-operant acquisition, and the second is called shaping. In both, animals are generally studied, one at a time, in relatively barren environments. Usually, the animal is placed in a comparatively small space (sometimes called a Skinner box) outfitted with a device that the animal can move, such as a lever for a rodent or nonhuman primate or a lighted disk that can be depressed by a rodent or avian. Also available is an apparatus that permits delivery of a consequence, for example, a bit of food. In free-operant acquisition of lever pressing, for instance, the animal is initially habituated to the enclosure and it learns where the food is delivered. The experimenter then arranges the apparatus so that when the lever is initially pressed, for whatever reason, the food is immediately delivered. Under appropriately controlled conditions, the lever press is soon followed by many others and the behavior has been established for further study. The second method, shaping, can be used to establish behavior such as lever pressing but also can be used to engender behavior that is extremely unlikely to occur if one simply waits. Shaping is also called the method of successive approximations, which provides a good description of how the procedure works. One reinforces successive approximations to the final activity. For example, suppose one wanted to train a rat to press a panel on the ceiling of its Skinner box. The first approximation might be providing a reinforcer when the rat walks to a place under the panel. Once that behavior has been established, reinforcement might then be made to depend on lifting a forepaw off the floor. The next approximation might be two paws off the floor, followed by requiring the paws to be progressively, at each step in the process, farther off the floor until the paws touch the panel.

Once operant behavior has been established, it can be subjected to *extinction*, wherein the activity no longer results in the previously arranged consequence. Of course, the usual result of extinction is that the behavior becomes less probable over time.

Most often, operant behavior is studied under *free-operant* conditions. That is, the experiment is arranged such that after each instance of an operant, the animal is in position immediately to engage in the activity again, and the activity itself is generally of low effort and takes little time to execute. There are circumstances, however, in which the opportunity to perform the operant is limited, for example, by the use of a retractable lever. Such arrangements are usually used to study phenomena that are best examined with procedures in which independent trials can be arranged and are called discrete-operant preparations. They are often used to study processes such as remembering and choice.

There are two characteristics of conditioned operant behavior that play a very large role in how such behavior is used in psychopharmacology. The first is that operant behavior can be brought under stimulus control, or in everyday parlance, it can be used to study discrimination among stimuli. The second is that operant behavior, once established, can be maintained with intermittent consequences. These two arrangements are often combined to produce relatively complex behavior to use when studying psychopharmacological agents.

To study discrimination using operant behavior, one needs to establish at least two different programs of reinforcing (or punishing) consequences, each associated with a distinctive circumstance. As a simple example, consider a monkey which has been trained to press a lever that results in intravenous injection of cocaine. To establish stimulus control, one could arrange that a tone be on when pressing the lever is effective and that when the tone is off, pressing the lever results in presentation of nothing. If the monkey comes to press the lever when the tone is on and not when the tone is off, it indicates that the monkey can tell the difference between tone and no tone, that is, it can discriminate between the two stimulus circumstances. One could, by reducing the loudness of the tone, systematically determine how loud the tone needs to be for the monkey to be able to discriminate it. One could also test tones of other frequencies in order to test the degree to which the learning of the initial discrimination extends to other tones. All these, and many other processes, can be arranged so that the effects of psychoactive drugs on the behavior can be assessed.

Stimulus control can also be used to permit study of multiple activities. For example, one could in one stimulus circumstance (e.g., color of a light) study behavior that results in positive reinforcement and, when a different stimulus is present, study behavior that results in negative reinforcement. Simply by changing the stimulus, either kind of behavior can be "on call" for examination, and it can be accomplished in individual animals.

The "experiment" about discrimination summarized earlier would likely not be conducted exactly as described, but instead would probably involve intermittent reinforcement. That is, instead of reinforcing every lever press when the tone is on, only some presses would be reinforced. Two related experimental advantages accrue to the use of intermittent reinforcement. One, relatively large samples of behavior, often on the order of hundreds or even thousands of instances of an activity, can be generated in a single experimental observation period or session. Two, because of the large number of instances engendered, it is possible to use frequency or rate of behavior as a measure, and that measure can vary over a very wide range of values, from near zero to hundreds per minute, yielding a potentially sensitive index.

Intermittent consequences can be arranged in myriad ways, with the various arrangements called *schedules of reinforcement*, which essentially are rules that describe which instances of behavior (usually called responses) will result in reinforcement. Schedules of reinforcement, interestingly, can have substantial influences on how a psychopharmacological agent acts. For example, in an influential experiment, Dews (1955) discovered that the schedule can influence whether a drug enhances or decreases a particular activity. In his study, pigeons pecked a lighted disk to earn access to food. In one condition, every 50th peck resulted in food presentation; in another, the first peck, after 15 min had elapsed, was followed by the opportunity to eat. Sodium pentobarbital was administered to birds under either condition. The surprising result was that a dose of the drug that greatly decreased pecking under the latter circumstance resulted in increases under the former. That is, depending on the reinforcement schedule, the drug acted as either a stimulant or depressant, for the same activity, key pecking, and for the same reason, to get food. Many other examples of interactions between reinforcement schedules and drug effects exist (e.g., Kelleher and Morse 1968). A final advantage of using intermittent reinforcement is that accumulated research indicates that under a wide variety of particular reinforcement schedules for free-operant behavior, reproducible temporal patterns and rates of behavior can be reliably established in individual animals (Ferster and Skinner 1957).

A final characteristic of operant behavior, especially as employed in the laboratory, is concerned with conditions that alter the effectiveness of consequences. Such factors are called motivational operations or establishing operations (Michael 1993). For example, to make food more effective as reinforcement, one often makes the animal hungry by not letting it eat for some time prior to the experimental test. Negative reinforcers, such as loud noises or irritating electric shock, are often made effective by their mere presentation, but how well they function depends also on their intensity and duration, both of which may be considered establishing operations.

Cross-References

▶ Abuse Liability Evaluation
▶ Active Avoidance
▶ Antinociception Test Methods
▶ Anxiety: Animal Models
▶ Behavioral Economics
▶ Behavioral Tolerance
▶ Breakpoint
▶ CANTAB
▶ Conditioned Reinforcers
▶ Contingency Management in Drug Dependence
▶ Decision Making
▶ Discriminative Stimulus
▶ Drug Discrimination
▶ Extinction
▶ Fixed Ratio
▶ Go/No-Go Task
▶ Instrumental Conditioning
▶ Intracranial Self-Stimulation
▶ Learned Helplessness
▶ Passive Avoidance
▶ Primate Models of Cognition
▶ Punishment Procedures
▶ Rate-Dependency Theory
▶ Schedule-Induced Polydipsia
▶ Self-Administration of Drugs
▶ Short-Term and Working Memory in Animals
▶ Timing Behavior

References

Dews PB (1955) Differential sensitivity to pentobarbital of pecking performance in pigeons depending on the schedule of reward. J Pharmacol Exp Ther 11:393–401

Ferster CB, Skinner BF (1957) Schedules of reinforcement. Appleton, New York

Kelleher RT, Morse WH (1968) Determinants of the specificity of behavioral effects of drugs. Ergebnisse der Physiologie, Biologischen Chemie, und Experimentellen. Pharmakologie 60:1–56

Mazur JE (2006) Learning and behavior, vol 6. Pearson Prentice-Hall, Upper Saddle River

Michael J (1993) Establishing operations. Behav Anal 16:191–206

Nevin JA (1973) Stimulus control. In: Nevin JA, Reynolds GS (eds) The study of behavior: learning, motivation, emotion and instinct. Scott-Foresman, Glenview, pp 115–152

Skinner BF (1938) The behavior of organisms: an experimental analysis. Appleton-Century, New York

Zeiler MD (1977) Schedules of reinforcement: the controlling variables. In: Honig WK, Staddon JER (eds) Handbook of operant behavior. Prentice-Hall, Englewood Cliffs, pp 201–232

Operant Chamber

Definition

Experimental apparatus to record the behavior of animals. The quintessential operant chamber is the "Skinner box" designed by B. F. Skinner (1904–1990) to study instrumental behavior in animals, which typically has one or two response levers (for a rat) or pecking keys (for a pigeon), arranged side by side on one of the walls, and a method for delivering reinforcers (such as a dispenser of food or liquid). Modern chambers, which are controlled by computer, may have alternative operandi, such as nose-poke holes or touch screens, and a variety of optional stimuli, including visual, auditory, and olfactory.

Operant Conditioning

Definition

Operant conditioning is the use of environmental consequences of reinforcement and punishment to modify the occurrence and form of behavior. The strength and frequency of a response can be controlled by the schedule of reinforcement and the contingent presentation of reinforcing or punishing stimuli. Affective properties such as pain and pleasure are often attributed to such stimuli but do not enter into their definitions, which are based upon observed changes in behavior such as the acquisition of new patterns of simple or complex behavior or the suppression of existing behaviors.

Cross-References

▶ Instrumental Conditioning
▶ Operant Behavior in Animals

Operant Response

Definition

An operant response is a behavior that acts on the environment and is modifiable by its consequences. When behavior is modified by its consequences, the probability of that behavior occurring again may either increase (in the case of reinforcement) or decrease (in the case of punishment).

Opiate

Definition

This term is mostly historical ("opioid" is now more commonly used). Strictly, it refers to drugs extracted directly from opium. This includes morphine (the major alkaloid in opium), ▶ codeine, and thebaine. Drugs derived by chemical modification of morphine and codeine are also sometimes referred to as opiates although the term "opioid" is, arguably, more appropriate. These include heroin, ▶ oxymorphone, and oxycodeine.

Cross-References

▶ Opioids

Opioid

Definition

Any naturally occurring or synthetic drug or endogenous substance that acts on classical opioid receptors (mu-, delta-, or kappa-receptors).

Opioid Analgesics

Synonyms

Narcotic analgesics; Narcotics

Definition

This is a group of natural and synthetic substances that have morphine-like pharmacological properties that are brought about through their actions on opioid receptors. Morphine is derived from the opium poppy (*Papaver somniferum*). Opiates are a group of naturally occurring compounds including morphine and related opium alkaloids; the name originally indicated their derivation from opium and they were also known as analgesics, because of their ability to relieve pain. The term "opioid" was later applied to synthetic substances with morphine-like effects but which were not from opium and had distinct chemical structures. The label opioid analgesic is now used to include all of these substances, regardless of whether they are of natural, synthetic, or semi-synthetic origin.

Cross-References

▶ Mu-opioid Agonists
▶ Opioids
▶ Tolerance

Opioid Antagonist

Definition

Opioid antagonist is a class of medication that binds to the opioid receptors in the brain, effectively blocking the effects of opioids (e.g., heroin, morphine, etc.). Examples of this class of medication include naloxone and naltrexone.

Cross-References

▶ Alcohol Abuse and Dependence
▶ Impulse Control Disorders
▶ Opioids

Opioid Use Disorder and Its Treatment

Annemarie Unger[1] and Gabriele Fischer[2]
[1]Department of Psychiatry and Psychotherapy, Medical University of Vienna, Vienna, Austria
[2]Department of Public Health, Medical University of Vienna, Vienna, Austria

Definition

Opioid Use Disorder
An opioid use disorder is a medical condition characterized by an individual's preoccupation with and strong desire to take opioids coupled with persistent drug-seeking behavior. It involves a sense of compulsion to consume the psychoactive substance, an inability to stop using it or to control the mode of drug-taking behavior. Obtaining the drug takes on high priority, and the condition is frequently accompanied by psychiatric comorbidity and multiple substance abuse, resulting in numerous social, psychological, and biological problems. In moderate and severe cases these include high-risk behavior such as prostitution and drug-related crime and an increased risk of obtaining infectious diseases such as hepatitis and HIV through shared use of needles and sexual activity, leading to an increased mortality rate; accidental deaths by fatal overdose and suicides also contribute to these. The Diagnostic and Statistical Manual for Mental Disorders, 5th Edition (DSM-5) defines opioid use disorder as a problematic pattern of opioid use leading to clinically significant impairment or distress, as manifested by at least

two of the following, occurring within a 12-month period:

1. Taking larger amounts of opioids or taking opioids over a longer period than was intended.
2. Experiencing a persistent desire for the opioid or engaging in unsuccessful efforts to cut down or control opioid use.
3. Spending a great deal of time in activities necessary to obtain, use, or recover from the effects of the opioid.
4. Craving, or a strong desire or urge to use opioids.
5. Using opioids in a fashion that results in a failure to fulfil major role obligations at work, school, or home.
6. Continuing to use opioids despite experiencing persistent or recurrent social or interpersonal problems caused or exacerbated by the effects of opioids.
7. Giving up or reducing important social, occupational, or recreational activities because of opioid use.
8. Continuing to use opioids in situations in which it is physically hazardous.
9. Continuing to use opioids despite knowledge of having persistent or recurrent physical or psychological problems that are likely to have been caused or exacerbated by the substance.
10. Tolerance, as defined by either a need for markedly increased amounts of opioids to achieve intoxications or desired effect, or a markedly diminished effect with continued use of the same amount of an opioid.
11. Withdrawal, as manifested by either the characteristic opioid withdrawal syndrome, or taking opioids to relieve or avoid withdrawal symptoms.

Tolerance and withdrawal criteria are not considered to be met for individuals taking opioids solely under appropriate medical supervision.

Opioid use disorder can be classified by severity as mild, moderate, or severe. The "mild" classification requires the presence of only 2–3 of the above symptoms. "Moderate" requires 4–5 symptoms whereas "severe" requires 6 or more.

Role of Pharmacotherapy

Opioids: Psychopharmacology and Receptors

Definition of Opioids

Opioid (meaning "similar to opium") is a collective term for a group of heterogeneous natural and synthetic substances displaying morphine-like properties that unfold their actions on opioid receptors located in the brain, the spinal marrow, and peripheral organs such as the intestinal tract.

There are different subdivisions and groups of opioids:

Exogenous opioids may be divided into three groups:

1. *Natural opioids* referred to as "opiates" are substances that naturally occur in the opium poppy; examples are morphine, codeine, and thebaine. In terms of their chemical property, they are alkaloids and can be extracted from the milk of the opium poppy (*Papaver somniferum*). Morphine is considered the prototype opioid.
2. *Semisynthetic opioids* such as hydromorphone or heroin are derived from natural opioids through a chemical process.
3. Numerous *fully synthetic opioids* have been produced pharmaceutically to exploit their analgesic effects for medical purposes, at the same time trying to minimize side effects. Fentanyl, tramadol, and methadone are examples of these.

Endogenous opioids are opioid peptides (endorphins, endomorphins, and enkephalins) that are naturally present in the human body and play a role in stress reactions and pain suppression.

Opioids can be consumed intravenously, orally, sublingually, nasally, and by smoking.

Opioid agonist drugs have their main indication in the treatment of pain; other less prominent indications include suppression of coughing and sleep induction, hence the name "narcotic" drug. In the treatment of severe cases of diarrhea, the constipation induced by opioids can be a therapeutic measure. In addition, opioid maintenance therapy is considered the standard therapy for opioid dependence.

Opioid Receptors

At present, three different G-protein-coupled receptors at which opioids unfold their action are known. They are referred to as the delta (δ), kappa (κ), and mu (μ) receptors. A fourth receptor, the nociceptin/orphanin FQ (N/OFQ) receptor, has been postulated and is subject to investigation. Most opioid drugs produce their effects by acting as agonists at the μ receptor, although some also act at the κ receptor, whereas the therapeutic potential of the δ receptor is still unclear. Located in the brain, the spinal marrow, and in peripheral organs such as the intestinal tract, receptors are activated by opioid agonists that initiate central and peripheral effects through the production of G-proteins. The fact that different receptors mediate different profiles of responses may be explained by their differing distributions throughout the central nervous system (Brunton et al. 2007).

Central effects can include analgesia, euphoria, sedation, muscular rigidity, anxiolytic effects, cramps, hypothermia, miosis, respiratory depression, antitussive effects, antiemetic effects, lowering of blood pressure, and bradycardia. Peripheral effects can be obstipation, delayed emptying of the stomach, disturbance of the bile flow, urinary retention, and the inhibition of labor.

Development of Opioid Use Disorder

Biological Factors

Despite their original intended purpose as therapeutic agents, their highly addictive nature fostered by initial euphoria and strong physical withdrawal make opioids a candidate for abuse. An estimated total of eight million people worldwide abuse opioids, and the use continues to increase. The illicit abuse of opioids and opioid use disorder are major public health problems worldwide.

Opioid-induced euphoria, described by opioid users, is mainly attributable to the effects of the neurotransmitter dopamine, which is released from dopaminergic neurons that exert their mechanism of action in the nucleus accumbens located in the forebrain. Even a single exposure of morphine produces an effect on dopamine neuron activities and a desensitization of opiate receptors. Repeated exposure to morphine induces substantial adaptive changes in cellular and synaptic functions in the mesocorticolimbic system. These mechanisms are likely to be part of the neurobiological basis for the development of dependence.

Another biological determinant critical to addictive potential is the speed at which a drug crosses the blood–brain barrier. Heroin, for instance, penetrates the brain more rapidly than morphine because it is more lipid-soluble, accounting for subjective reports that following intravenous administration, the euphoric experience is particularly intense.

Opioids are among the most potent drugs with regard to physical dependence and withdrawal. When administration of the drug is discontinued or an antagonist is given, a withdrawal or abstinence syndrome is manifest, characterized by diarrhea, vomiting, fever, chills, cold sweats, muscle and bone aches, muscle cramps and spasms, restless legs, agitation, gooseflesh, insomnia, nausea, watery eyes, runny nose, and nightmares.

Psychosocial Factors

There is a strong overlap between comorbid mental disorders such as borderline and antisocial personality disorder, attention deficit disorder (ADHD), post-traumatic stress disorder (PTSD), affective disorders, and the development of opioid dependence. Predictors for the development of opioid dependence include a history of sexual or physical abuse, cohabitation with an opioid-dependent partner, and an upbringing in an environment permissive to substance abuse.

With regard to gender, men dominate in the prevalence of opioid use disorder, though the gender gap is narrowing and women show a younger age of initial use and a faster progression to addiction. Further gender-specific differences are that men with an opioid use disorder are more likely to stay employed than women, while women tend to receive more social welfare and engage in prostitution. However, men are more prone to engage in general drug-related crime and drug dealing. Though women have a faster progression to treatment, they have a lower retention rate due to gender-specific barriers such as higher stigmatization, intimate partner violence, fear of losing custody of their children, and a higher rate of comorbid affective disorders. Treatment settings designed for a predominantly male population cannot adequately accommodate for gender-specific social factors affecting women.

Consequences of Opioid Use Disorder

Somatic comorbidities generally result from poor standards of living and high-risk sex behavior and needle sharing; poor dental status and high rates of infectious disease such as HIV and hepatitis, endocarditis, and abscesses are the consequence. Studies have shown that women have a higher risk for HIV infection than men because of higher impulsivity and exposure to health-compromising situations including homelessness, condom nonuse, and exchange of sex for money and/or drugs (Gowing et al. 2008). Furthermore, suicide, accidental death, and overdose contribute to a mortality rate 13–17 times higher than the general population, with slightly higher rates for men.

Treatment of Opioid Use Disorder

Though opioid use disorder represents a chronic reoccurring illness, individually tailored comprehensive treatment can allow patients to lead normal lives. There are three main branches of treatment: withdrawal/detoxification, opiate antagonist maintenance, and opioid agonistic maintenance treatment. Currently, the most effective treatment for severe opioid use disorder is opioid agonist maintenance therapy (Amato et al. 2008).

Withdrawal/Detoxification: Drug Free

Medically assisted detoxification is successful only when it is seen as the first step in a series of behavioral interventions and counseling. Withdrawal alone will result in the same outcome with regard to relapse as having no treatment at all. Detoxification can be handled in different ways, for instance in an outpatient setting, where gradual buprenorphine taper is an option, whereas inpatient detoxification programs proceed faster and usually involve administration of ancillary medication such as clonidine, or its further development lofexidine. There is no conclusive evidence to show that either outpatient or inpatient treatment is more effective (Day et al. 2005).

Ultrarapid detoxification is a treatment in which naltrexone is administered to patients under deep sedation or anesthesia; because of a high risk of cardiopulmonary complications, this practice is not recommended.

Opiate Antagonist

Naltrexone is a long-acting pure opiate receptor antagonist that is used in rehabilitation programs to ensure abstinence from opioids and should be prescribed only to individuals who have already been detoxified to prevent relapse. Patients should be opioid free for at least 3 days before receiving naltrexone, as it can lead to unintentional rapid detoxification, precipitating massive withdrawal symptoms. The opioid antagonist naloxone is being increasingly used in a different manner, as a safe and effective treatment in the emergency room for overdoses with opioids. Possibilities for its administration to opioid users in other environments are being explored.

Opioid Maintenance Therapy

Agonist maintenance therapy is considered as first-line treatment in moderate and severe opioid use disorder and leads to statistically significant reductions in illicit opioid use, risk behavior related to drug use such as injecting and sharing injecting equipment, and an improvement in social and psychological quality of life (Amato et al. 2008; Gowing et al. 2008). The most widely used replacement, methadone, shows a statistically superior effect in reducing heroin use and retaining patients in treatment compared to medication-free programs (Mattick et al. 2002).

Proper medication-assisted treatment of opioid use disorder should ensure that patients do not experience any withdrawal symptoms or feel drugged or high during a time period of 24 h; the medication used should be present in the blood in levels sufficient to maintain normalcy over a 24 h period.

With regard to dosage regimen, drug-to-drug interactions must be taken into consideration. Opioid drug-to-drug interactions can, for example, occur with CNS depressant drugs such as benzodiazepines or alcohol and are additionally frequently evoked by a combination with substances that inhibit or induce the activity of the cytochrome P450 enzyme system. Metabolism, poor absorption, changes in urinary pH, concomitant medications or drug abuse, diet, physical condition, pregnancy, and vitamins may influence medication levels and the rate of elimination (Payte et al. 2003).

Special target populations may require additional planning in therapeutic dosing regimens. In the case of pregnant women, changes in metabolism due to enzyme induction usually require a gradual dose increase or split dosing around the third trimester. The benefits of methadone maintenance during pregnancy compared to detoxification have been well established. Detoxification is usually associated with relapse and marked fluctuations of serum methadone levels, both of which are unfavorable to fetal outcome.

Psychosocial Treatments in Addition to Pharmacological Therapy
In the treatment of opioid use disorder, ideally, behavioral interventions and a comprehensive all-round approach with a multidisciplinary team including social workers should accompany psychopharmacological treatment. Psychosocial interventions can include different psychotherapeutic methods such as cognitive behavioral therapy (CBT), interpersonal therapy, subliminal stimulation, supportive–expressive therapy, and contingency management approaches such as voucher incentives. However, a review of the literature shows that adding psychosocial interventions such as counseling, social work, and psychotherapy to maintenance therapy on a random basis has limited beneficial effect on retention rate, opiate use, or psychiatric symptoms. The only overall benefit of psychosocial support pertains to an improvement of the number of participants still on abstinence at follow-up (Amato et al. 2008).

An important aspect when evaluating the effectiveness of treatment, however, is that a chronic relapsing disease like opioid use disorder should have outcome criteria that take its nature into account. For instance, effect measurement should be done during treatment and not after discharge, especially when acute relapse episodes remain untreated.

Substances Available for Maintenance Treatment

Examples of medications commonly used for maintenance treatment are methadone, buprenorphine, buprenorphine–naloxone, and the comparatively less frequently used L-alpha-acetyl-methadol (LAAM) and slow-release morphines (SROMs).

Methadone
Methadone is a full μ receptor agonist and is the oldest, most common, and widely used substance for substitution treatment. It is a racemate available in tablet form, as an injection, or as oral solution.

Methadone has been used internationally since 1965 and is the best-studied substance available for opioid maintenance in terms of clinical effectiveness (Clark et al. 2002). It has been comprehensively shown that methadone is an effective treatment in reducing illicit opioid consumption, reducing high-risk behavior such as needle sharing and increasing rates of treatment retention (Mattick et al. 2002). It has been demonstrated that higher dosages of methadone maintenance medication lead to improved outcome in terms of treatment retention and decreases in illicit opioid use.

For methadone maintenance treatment (MMT), the medication is administered orally; concentrations in serum reach a peak level 3–5 h after administration; it has a half-life of 24–36 h (Brunton et al. 2007). Side effects can be increased sweating, mood swings, depression, lack of energy, weight gain, edema, loss of libido, and prolonged QTc time in the ECG.

Combined use with other substances that can influence the QTc time in the ECG such as antipsychotics, antiarrhythmics, antibiotics, tricyclic antidepressants, and antifungal medication should be avoided whenever possible. Cardiac safety recommendations include obtainment of cardiac histories, a pretreatment ECG, and follow-up ECGs within 30 days and annually. Other interactions involve diuretics, antibiotics, antihypertensives, HIV medicines and other antiviral medicines, other narcotic medications, and drugs against epilepsy; interactions may lead to increase in serum concentration and overdose.

Buprenorphine

Buprenorphine is a partial μ-opioid-receptor agonist and a κ-receptor antagonist that was first marketed in the 1980s as an analgesic. For opioid maintenance therapy, comprehensive treatment experience has been available in Europe and Australia since the mid-1990s, and in the USA, the US Food and Drug Administration approved buprenorphine and a buprenorphine–naloxone combination product in 2002. Because of its partial antagonism, buprenorphine counteracts the effects of concomitant opioids taken by the patient and, unlike other opioids, does not produce tolerance. It has less severe withdrawal effects than methadone, making discontinuation easier (Gowing et al. 2006).

In a systematic review, buprenorphine was found to be inferior to methadone in terms of suppressing heroin consumption and retention rates in treatment when both were administered at adequate dosages (Mattick et al. 2008). However, buprenorphine is safer than methadone, especially for patients with coexistent benzodiazepine dependence, as side effects such as respiratory depression develop only to an uncritical degree even after extreme overdosing, and it has lower dependence liability. The treatment of buprenorphine overdose with naloxone needs special medical observation because of the tight receptor binding of buprenorphine.

Buprenorphine is usually administered as a sublingual tablet. Resorption takes place through the oral mucosa. Peak serum levels are reached after 1–3 h, after which it remains effective in the body for up to 48 h. Because of this long duration of action, dosing intervals of 2 days are optional. Side effects can include nausea, vomiting, drowsiness, dizziness, headache, itch, dry mouth, meiosis, orthostatical hypotension, difficulty with ejaculation, decreased libido, urinary retention, and constipation.

When patients are induced on buprenorphine, caution should be taken because, due to its partial agonist/antagonist activity, it may precipitate significant withdrawal symptoms if the first dose is administered too soon after the intake of other opioids.

Buprenorphine–Naloxone Combination Tablet

The buprenorphine–naloxone combination tablet is a further development of buprenorphine invented with the intention of reducing intravenous misuse of the medication. Naloxone has low oral bioavailability and so does not influence the mechanisms of buprenorphine action when taken orally. However, when buprenorphine–naloxone combinations are dissolved and injected intravenously, opioid agonist actions are blocked by naloxone and can, depending on which drugs have been misused previously, precipitate unpleasant and dysphoric symptoms of opioid withdrawal. Buprenorphine–naloxone combination tablets contain buprenorphine with naloxone at a ratio of 4:1. The buprenorphine–naloxone combination tablet should not be administered to pregnant and nursing women, as naloxone exposure may alter fetal and maternal hormonal levels.

Levo-Alpha-Acetylmethadol (LAAM)

LAAM is a derivative of and has a similar mode of action as methadone. It was first approved in the USA in 1993 and in several European countries in 1997. The most significant difference between the two substances is that LAAM has a longer duration of drug effect, lasting up to 72 h; therefore, it can be administered every 2–3 days. The plasma half-life is only 14–37 h, but the substance is metabolized into two active metabolites, nor-LAAM and dinor-LAAM, which have a half-life of 24–38 h and 66–89 h, respectively.

All three substances exert their mechanisms of action on the μ-receptor; however, nor-LAAM is five times as potent as the others. It may be more effective at reducing heroin Demand than methadone, but it is associated with adverse effects such as cardiac arrhythmias and QTc prolongation, some of which may be life-threatening. Following ten cases of death caused by LAAM, it was withdrawn from the market in Europe on the recommendation of the European Agency for the Evaluation of Medicinal Products (EMEA). In the USA, the US FDA has recommended that LAAM should not be used as first-line therapy. LAAM is not approved for use in Australia and Canada.

Slow-Release Oral Morphine: SROM
Morphine on its own is not recommendable for maintenance therapy as it has low oral bioavailability, a half-life of only 2–3 h, and a very short time of drug effect. SROMs are a group of pure opiate agonists that are products of morphine (i.e., morphine sulfate and morphine hydrochloride) with retarded release characteristics. SROMs are available as tablets (i.e., morphine hydrochloride) or capsules containing microgranular compounds (i.e., morphine sulfate) and may be more tolerable than methadone in terms of side effects or as an option for patients with an inadequate suppression of withdrawal symptoms under methadone. SROMs hold the advantage of having a longer half-life, and time of drug effect lasts up to 24 h. They can therefore normally be administered once daily, except to "fast metabolizers" who need at least twice daily administration. They hold the disadvantage that urine toxicology will not detect a concomitant consumption of heroin because morphine is a metabolite of heroin.

Intravenous misuse of SROMs can lead to an increased risk of pulmonary embolism, attributable to the wax particles that are briefly liquefied in the cooking process and then solidified again in the body. Patients maintained on SROM must be warned about misusing the substance and examined for signs of needle puncture. The use of SROMs for opioid maintenance treatment is registered only in some European countries, but international interest is rising.

Conclusion

Internationally, opioid maintenance therapy has been proved doubtlessly and conclusively as the most effective treatment for opioid use disorder. It reduces illicit opioid use, lowers mortality rate and criminal behavior, and leads to improved health and social conditions of chronic opioid-dependent individuals by increasing rates of retention in treatment. The nature of opioid use as a psychiatric disorder needs to be considered carefully, and psychosocial interventions may accompany medical treatment if indicated.

Demand for the treatment of opioid dependence is on the rise in many parts of the world, and from a public health standpoint, the availability of treatment needs to be improved further. Treatment coverage worldwide changed for the better in the light of the 2002 approval of office-based treatment in the USA, where, in contrast to Europe and Australia, both continents with a long-lasting tradition in office-based opioid maintenance treatment, only specialized addiction clinics could provide treatment. This development in the USA has gone relatively well, consistent with experience of Europe and Australia, further underscoring the individual and societal benefit of increasing treatment availability for opioid use disorders.

Cross-References

▶ Buprenorphine
▶ Buprenorphine-Naloxone
▶ Detoxification
▶ L-Alpha-Acetyl-Methadol
▶ Methadone
▶ Morphine
▶ Naltrexone
▶ Opiate
▶ Opioid
▶ Slow-Release Morphines

References

Amato L, Minozzi S, Davoli M, Vecchi S, Ferri M, Mayet S (2008) Psychosocial combined with agonist maintenance treatments versus agonist maintenance treatments alone for treatment of opioid dependence. Cochrane Database Syst Rev, (4). Art. No.: CD004147. doi:10.1002/14651858.CD004147.pub3

American Psychiatric Association (2000) Diagnostic and statistical manual of mental disorders, 4th edn, text revision. American Psychiatric Association, Washington, DC

Brunton L, Goodman L, Gilman A, Blumenthal D, Parker KL, Buxton I (2007) Goodman and Gilman's manual of pharmacology and therapeutics. McGraw-Hill Professional, New York

Clark NC, Lintzeris N, Gijsbers A, Whelan G, Dunlop A, Ritter A, Ling WW (2002) LAAM maintenance vs. methadone maintenance for heroin dependence. Cochrane Database Syst Rev (2). Art. No.: CD002210. doi:10.1002/14651858.CD002210

Day E, Ison J, Strang J (2005) Inpatient versus other settings for detoxification for opioid dependence. Cochrane Database Syst Rev (2). Art. No.: CD004580. doi:10.1002/14651858.CD004580.pub2

Gowing L, Ali R, White JM (2006) Buprenorphine for the management of opioid withdrawal. Cochrane Database Syst Rev (2). Art. No.:CD002025. doi:10.1002/14651858.CD002025.pub3

Gowing L, Farrell M, Bornemann R, Sullivan LE, Ali R (2008) Substitution treatment of injecting opioid users for prevention of HIV infection. Cochrane Database Syst Rev (2). Art. No.:CD004145. doi:10.1002/14651858.CD004145.pub3

Mattick RP, Breen C, Kimber J, Davoli M, Breen R (2002) Methadone maintenance therapy versus no opioid replacement therapy for opioid dependence. Cochrane Database Syst Rev (4). Art. No.:CD002209. doi:10.1002/14651858.CD002209

Mattick RP, Kimber J, Breen C (2008) Buprenorphine maintenance versus placebo or methadone maintenance for opioid dependence. Cochrane Database Syst Rev (2). doi:10.1002/14651858.CD002207.pub3

Payte JT, Zweben JE, Martin J (2003) Opioid maintenance treatment. In: Graham AW, Schultz TK, Mayo-Smith MF, Ries RK, Wilford BB (eds) Principles of addiction medicine. American Society of Addiction Medicine, Chevy Chase, pp 751–766

Opioid Maintenance Treatment

Definition

Opioid maintenance is an evidence-based treatment in which patients with opioid dependence are provided with opioid agonist replacement medication. Proper opioid maintenance should ensure that patients do not experience any withdrawal symptoms and do not feel drugged or high during a time period of 24 h.

Opioids

MacDonald J. Christie[1] and Michael M. Morgan[2]
[1]Discipline of Pharmacology, The University of Sydney, NSW, Australia
[2]Department of Psychology, Washington State University Vancouver, Vancouver, WA, USA

Synonyms

Opiates

Definition

The term opiate refers to drugs extracted from opium, i.e., morphine and codeine, whereas an opioid is any drug or endogenous agent that acts as an agonist or antagonist on one of the three major (classical) opioid receptors.

Pharmacological Properties

Opium that contains the principal opiate, morphine, has been used for its powerful pain relieving (analgesics), somnorific, and euphoric properties since antiquity (Brownstein 1993). Since the discovery of opioid receptors in 1972 and the first endogenous opioid peptides in 1975, much has been learned about the anatomy and function of opioid actions. However, much remains to be discovered about the function of endogenous opioids and opioid drugs in specific populations of neurons and neural systems. In addition to therapeutically valuable actions on the sensory and affective components of pain, opioids also produce profound effects on neural systems involved in respiration, reward and learning, and many other behavioral and physiological processes. Current knowledge of these systems is summarized in the following sections.

Endogenous Opioid Peptides

The general distribution of opioid peptides was reported soon after the discovery of enkephalins in 1975 (Kachaturian et al. 1993). Genes encoding four endogenous opioids and opioid-related peptide precursors have since been identified in the mammalian genome, and the products of all but one act as agonists on the classical opioid receptors: mu (μ- or MOR, aka OP_3, MOP), delta (δ- or DOR, aka OP_1, DOP), and kappa (κ- or KOR, aka OP_2, KOP) receptors. As with other peptide hormones and neurotransmitters, the final active peptides are cleaved at dibasic amino acid sites from large polypeptide precursors by processing enzymes (Lewis and Stern 1983). The precursors such as pro-opiomelanocortin (POMC), preproenkephalin, and prodynorphin express the sequences for β-endorphin, enkephalins, and dynorphins, respectively. POMC contains a single copy of β-endorphin (which can be lyzed to shorter, potentially active endorphin fragments) along with other biologically important hormones/neurotransmitters including the major stress hormone, adrenocorticotropic hormone (ACTH) (Lewis and Stern 1983). Preproenkephalin encodes multiple copies of two enkephalin pentapeptides, leucine enkephalin and methionine enkephalin (longer active peptides can be found in some tissues) (Lewis and Stern 1983). Prodynorphin contains the sequences of α-neoendorphin, dynorphin-A, and dynorphin-B, as well as "big dynorphin," which is the uncleaved sequence of dynorphin A and dynorphin-B (the dynorphin-A and dynorphin-B peptides are separated by one pair of basic amino acids in prodynorphin) (Lewis and Stern 1983).

All the endogenous opioid peptide families described above contain a canonical N-terminal amino acid sequence, N-Tyr-gly-gly-phe, followed by one to 26 other amino acids. The extended amino sequences, which vary in length in different cells, confer differential selectivity among the three major opioid receptors as well as potentially modifying susceptibility to degradation by peptidase enzymes such as "enkephalinase" (EC 3.4.24.11). It should be noted that deletion (des-tyr) or chemical modification of the N-terminal tyr (N-acetylation occurs for a substantial proportion of the β-endorphin from the pituitary) renders all opioid peptides inactive at opioid receptors. These peptides subserve endocrine (e.g., β-endorphin from the pituitary; proenkephalin from the adrenal medulla) and paracrine (opioid peptides are expressed in the immune system) functions in addition to their well-known effects on the nervous system.

The most recently discovered opioid-related peptide family, orphanin-FQ or nociceptin (usually denoted N/OFQ), is distantly related to prodynorphin but contains an N-terminal Phe rather than Tyr. It is the endogenous ligand of the orphan opioid receptor opioid receptor-like-1 (ORL1, aka NOP, NOR). It has high affinity and selectivity for ORL1 but interacts very weakly with the KOR (see below). The N/OFQ-ORL1 system is not widely considered to represent an "opioid system" because classical opioid drugs and endogenous opioids do not interact significantly with ORL1. It is therefore not considered in detail here.

Two additional endogenous opioids (endomorphin 1 and endomorphin 2) have been purified from mammalian tissue. Although these tetrapeptides are very selective for MOR and can be visualized in neurons immunohistochemically (which is not proof of presence), no genomic sequences encoding endomorphin 1 and endomorphin 2 have been identified; so their physiological relevance will remain uncertain until a biological synthetic mechanism is identified.

Opioid Receptors

MOR, DOR, and KOR opioid receptor types were first identified using pharmacological approaches. σ (sigma)-Opioid receptors also were proposed, but structural and pharmacological studies indicate that σ-receptors are not part of the opioid family. Three independent genes encoding MOR, DOR, and KOR have been identified, firmly establishing that these are the three principal opioid receptors. Soon after the isolation of these genes, a fourth opioid receptor-like (ORL1) sequence was identified by homology screening of cDNA libraries for sequences resembling the other receptors. The biochemistry and regulation of opioid receptors are extensively reviewed elsewhere (e.g., Waldhoer et al. 2004). Briefly, the amino acid sequences of all opioid receptors (and ORL1) are about 60 % identical with each other. They belong to a small subfamily of G protein-coupled receptors (▶ GPCR) that includes the somatostatin receptors. Multiple RNA splice variants have been identified, in particular MOR1A, B, and C (the biological relevance of other putative splice variants is uncertain). The B and C variants differ in the amino acid composition at the C-terminus and affect receptor regulatory events such as receptor distribution, internalization, and recycling rates but have little or no influence on drug selectivity. Opioid receptor subtypes, such as μ_1 and μ_2 receptors, also have been proposed, but they remain tentative. Splice variants of DOR or KOR may explain differential pharmacology of proposed receptor subtypes (e.g., putative $\delta 1$ and $\delta 2$ receptors), but these have not yet been established. Formation of hetero-oligomers between different opioid receptor types could also explain experimental results claiming existence of opioid receptor subtypes. As for other GPCRs, oligomer formation of opioid receptors appears to be obligatory for membrane expression and function (Milligan and Bouvier 2005). While homo-oligomers are probably the most commonly formed and appear to explain most opioid pharmacology in vivo, there is growing biochemical and pharmacological evidence that different subtypes of opioid receptors can form hetero-oligomers in isolated experimental systems and perhaps in vivo.

Interaction of Opioids with Opioid Receptors
Some authors have attempted to ascribe three distinct signaling systems to the three different opioid peptide families matching them, respectively, with the three receptor types. This is incorrect because each opioid peptide family can interact with more than one receptor type, as summarized in Fig. 1. Among the three peptide groups and receptors, the dynorphin–KOR pair is the best candidate to be defined as a distinct signaling system because dynorphin is the only endogenous opioid that interacts significantly with KOR. However, dynorphins are also potent MOR agonists and can potentially be metabolized to shorter and less selective dynorphin fragments (including leucine enkephalin), which interact potently with both MOR and DOR. It should also be noted that early studies suggesting that enkephalins are the endogenous ligands for DOR were premature because they were subsequently shown to be nearly equi-effective agonists at both MOR and DOR.

A large number of small, organic molecule agonists for opioid receptors have been developed since the first isolation of morphine from opium in 1806 and synthesis of heroin in 1898 (Brownstein 1993). Nearly all of the small-molecule agonists in clinical use are selective for MOR although experimental opioid agonists selective for DOR and KOR have been developed. A more limited range of small-molecule antagonists exists, although some of these commercially available antagonists display high receptor-type selectivity. Most commercially available small-molecule opioids have the advantage that they readily penetrate the central nervous system following systemic injection. However, some have been specifically developed to avoid crossing the blood-brain barrier (e.g., methylnaloxone). The optimal selection of an opioid for a given experimental purpose depends on a number of considerations, so recommendation of particular drugs for an opioid receptor type is beyond the current scope. The major pharmacological societies provide and regularly update comprehensive guides to the most appropriate selective agonists and antagonists for each receptor. These include The International Union of Basic and Clinical Pharmacology (IUPHAR) (http://www.iuphar-db.org/index.jsp) and the British Pharmacological Society (http://www3.interscience.wiley.com/journal/122206250/issue).

Opioid Receptor Signaling and Regulation
As shown in Fig. 1, all opioid receptors when activated by an agonist transduce intracellular

Opioids,

Fig. 1 Selectivity profiles of the major endogenous opioids for the different opioid receptors are shown with thickness of the arrows indicating relative selectivities or potencies for MOR (*blue*), DOR (*green*), KOR (*yellow*), and ORL1 (*orange*). All endogenous opioids except nociceptin/OFQ (which is not generally considered an opioid) can act as agonists at more than one opioid receptor type. All opioid receptors couple to Gi proteins to modulate the major signaling mechanisms shown

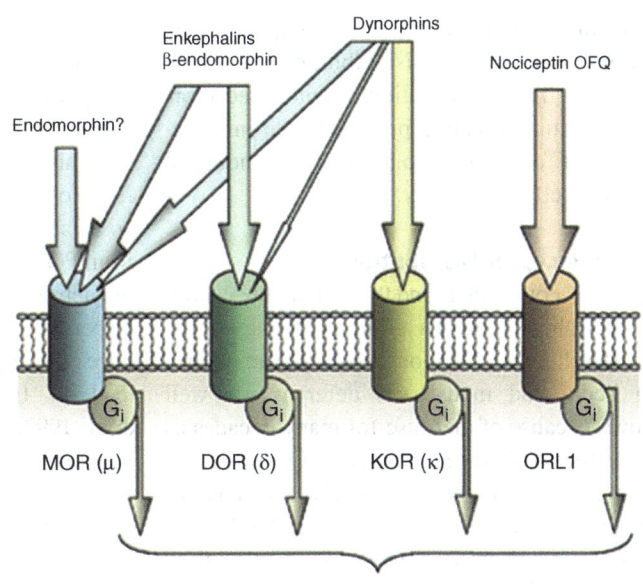

signals via activation of inhibitory G proteins. The major consequence of opioid receptor activation in neurons is inhibition in both cell bodies and nerve terminals. Downstream signaling includes modulation of many biochemical and gene regulatory cascades – a full description of which is beyond the current scope (see, however, Williams et al. 2001; Waldhoer et al. 2004). Briefly, while subtle variations occur among the specific G proteins activated by different receptor types (and perhaps hetero-oligomers), all opioid receptors activate Gi proteins, which leads to the release of GTP-bound active Giα subunits and Gβγ subunits from the receptor. The major immediate consequence (within ~50 ms of receptor activation) is inhibition of neuronal excitability via Gβγ subunit inhibition of voltage-gated calcium channels (VGCCs; particularly $Ca_V2.2$–2.3) and activation of G protein-coupled inwardly rectifying potassium channels (GIRKs) in the local membrane. Inhibition of VGCCs in nerve terminals can contribute to inhibition of neurotransmitter release probability upon invasion of action potentials. Other ionic channels and biochemical effects (e.g., inhibition of cAMP formation) also contribute to presynaptic inhibition. Free Gβγ subunits may also activate (more slowly) components of protein kinase cascades. Giα subunits inhibit most isoforms of adenylate cyclase that are expressed in many types of neurons. Other cascades are modulated by the opioid receptor trafficking and internalization that occurs after activation in an agonist-dependent manner (Waldhoer et al. 2004). It should also be noted that many of these mechanisms can contribute to the cellular and synaptic plasticity of signaling that are associated with tolerance, physical dependence, and addiction following chronic opioid treatment (Williams et al. 2001).

The complexity produced by multiple opioid receptor types and signaling mechanisms requires that opioid actions and adaptations in different neural systems be determined on a case-by-case basis. While the direct effects of opioids on cell bodies and synapses are almost invariably inhibitory, activation of neural systems can be the net outcome when the dominant opioid effect is localized to inhibitory interneurons and synapses. For example, MOR activation in the ventral tegmental area enhances dopamine release because MOR are located on inhibitory interneurons that synapse on dopaminergic neurons. This type of disinhibition is common to opioids. Thus, localization of opioid receptors to a particular structure provides little information about function without an understanding of the local cellular circuitry.

Opioid Receptor Distribution

Opioid receptors are found throughout the central and peripheral nervous systems. Some of the peripheral tissue locations such as the guinea pig ileum and mouse vas deferens are well known because of their use for many decades as assays for opioid receptor activity. Opioid receptor expression in the gut largely accounts for constipation, a major side effect of MOR agonists. Opioid receptors are also found on primary afferent nociceptors and immune cells (Stein et al. 2003).

MOR, DOR, and KOR can be found from the cerebral cortex to the spinal cord. The distribution of opioid receptors revealed by in situ hybridization, ligand binding, and immunohistochemistry (Mansour et al. 1995) is so extensive that describing the many brain structures with receptors is more tedious than useful. This point is highlighted by a list of the structures in which opioid receptors have been reported (Table 1). High levels of MOR and KOR are found from the cerebral cortex to the dorsal horn of the spinal cord. DOR distribution also is extensive but more limited than MOR and KOR. However, recent studies showing that DOR are mobilized by environmental stimuli (Cahill et al. 2007) indicate that DOR have a much broader distribution than previously thought. For example, chronic administration of morphine or prolonged stress stimulates the movement of DOR in the periaqueductal gray (PAG) from intracellular stores to the plasma membrane. Although previous studies did not report DOR in the PAG, functional membrane receptors can be found under these conditions. Thus, the intensity and distribution of DOR labeling are more extensive than that is revealed in Table 1.

Endogenous Opioid Distribution

One would expect the distribution of endogenous opioids to match closely the distribution of opioid receptors. There is much overlap, but a clear one-to-one relationship between opioid terminals and receptors is not always evident. Of course, endogenously released opioids may also act by volume transmission and therefore have a greater spatial range of influence than a single point-to-point synapse (see Williams et al. 2001). Opioids derived from POMC are synthesized in just a few sites, but the terminal fields of these neurons are widely dispersed (Table 2). By contrast, neurons that produce opioids from proenkephalin and prodynorphin tend to be local interneurons (Khachaturian et al. 1993). These neurons are found throughout the central nervous system (Table 3).

Behavioral Effects of Opioids

Given the wide distribution of opioid peptides and receptors, it is not surprising that opioids influence many behaviors. These include the inhibition of pain (analgesia), reward and addiction, mood, eating and drinking, sexual activity, sedation, thermoregulation, cardiovascular function, respiration, gastrointestinal transit, and nausea (Bodnar 2008). Most of what is known about these behavioral effects is derived from studies in which opioid drugs are administered. Administration of opioid receptor antagonists has almost no effect on ongoing behavior indicating that tonic release of endogenous opioids has a negligible impact on behavior in most circumstances, but more subtle effects have been determined from gene knockout studies (Kieffer and Gaveriaux-Ruff 2002). For example, administration of the opioid receptor antagonist naloxone does not enhance pain sensations except when used to reverse environmentally induced antinociception or to precipitate withdrawal from chronic opioid administration. A full description of the range of behaviors linked to opioids would require an entire book (Bodnar 2008). This section provides a brief description of a few key opioid receptor-mediated effects and identifies the brain structures contributing to these effects. Given that the primary medical use of opiates is the inhibition of pain, the role of opioids in pain inhibition will be highlighted. Opiates also are widely abused so the role of opioids in reward and addiction will be described briefly.

Opioids, Table 1 Location and density of opioid receptors in the rat CNS

Structure	MOR	DOR	KOR
Amygdala	High	High	High
Anterior olfactory nucleus	High	Medium	Low
Arcuate nucleus	–	–	Medium
Bed nucleus stria terminalis	Medium	Medium	High
Caudate-putamen	High	High	High
Cortex	High	High	Medium
Dorsal tegmental nucleus	Medium	–	Low
Globus pallidus	Low	Low	Low
Hippocampus	High	–	–
Hypothalamus	Low	Low	High
Inferior colliculus	High	Low	Medium
Interpeduncular nucleus	High	High	High
Islands of Calleja	Low	High	High
Locus coeruleus	High	–	Medium
Lateral reticular nucleus	Low	–	Low
Septum	High	Medium	Medium
Mammillary nucleus	High	–	Medium
Nucleus accumbens	High	High	High
Nucleus diagonal band	Medium	Medium	High
Nucleus reticularis gigantocellularis	Low	–	Low
Nucleus tractus solitarius	High	Low	High
Olfactory bulb	High	High	Medium
Olfactory tubercle	Low	High	High
Parabrachial nucleus	High	–	Medium
Paraventricular hypothalamus	–	–	High
Periaqueductal gray	Medium	–	Medium
Pituitary gland	–	–	High
Pons	–	Medium	–
Preoptic area	Low	Low	High
Presubiculum	High	High	Low
Raphe dorsalis	Medium	–	Medium
Raphe magnus	Medium	–	Low
Raphe medius	Medium		Low
Sensory nucleus trigeminal	Low	–	–
Spinal trigeminal nucleus	High	–	Medium
Spinal dorsal horn	High	Low	Medium
Spinal ventral horn	Low	–	Low
Substantia nigra	High	Low	Low
Supraoptic nucleus	–	–	Medium
Superior colliculus	High	Low	Medium
Thalamus	High	Medium	High
Zona incerta	–	–	Medium

Analgesia

Administration of agonists for all three of the primary opioid receptors (MOR, DOR, and KOR) has been shown to produce analgesia, and this has been confirmed by extensive gene knockout studies (Kieffer and Gaveriaux-Ruff 2002). However, the best and most commonly used opiates for the treatment of pain act almost

Opioids, Table 2 Synthetic pathways for POMC

Cell body location	Terminal locations
Pituitary gland	Pituitary gland
Arcuate nucleus	Periventricular gray
	Hypothalamic nuclei
	Medial preoptic area
	Medial septum
	Bed nucleus of stria terminalis
	Amygdala
	Periaqueductal gray
	Raphe nuclei
	Parabrachial nuclei
	Nucleus reticularis gigantocellularis
	Nucleus tractus solitarius
	Dorsal motor nucleus of the vagus nerve
Nucleus tractus solitarius	Medullary regions
	Parabrachial nuclei
	Spinal cord

Opioids, Table 3 Location of neurons containing proenkephalin and prodynorphin

Structure	Proenkephalin	Prodynorphin
Amygdala	Interneurons	Interneurons
Bed nucleus of stria terminalis	Interneurons	
Cerebral cortex	Interneurons	Interneurons
Dorsal tegmental nuclei	Interneurons	
Globus pallidus	Interneurons	Terminals
Hippocampus	Interneurons	Interneurons
Hypothalamus	Interneurons	Interneurons
Inferior colliculus	Interneurons	
Interpeduncular nucleus	Interneurons	
Lateral geniculate nucleus	Interneurons	
Lateral reticular nucleus	Interneurons	Interneurons
N. reticularis gigantocellularis	Interneurons	
Nucleus tractus solitarius	Interneurons	Interneurons
Parabrachial nuclei	Interneurons	Interneurons
Periaqueductal gray	Interneurons	Interneurons
Paraventricular nucleus		Interneurons
Periventricular thalamus	Interneurons	
Preoptic area	Interneurons	
Raphe nuclei	Interneurons	Terminals
Septum	Interneurons	
Spinal dorsal horn	Interneurons	Interneurons
Striatum	Interneurons	Interneurons
Substantia nigra	Interneurons	Terminals
Superior colliculus	Interneurons	
Supraoptic nucleus		Interneurons
Trigeminal nucleus	Interneurons	Interneurons
Vestibular nuclei	Interneurons	

exclusively via the MOR receptor. Morphine, heroin, and other derivatives such as meperidine (Demerol), fentanyl, and codeine are common MOR receptor agonists with potent analgesic effects.

MOR agonists (and to some extent DOR and KOR agonists) produce analgesia via both inhibiting ascending pain transmission and activating descending pain modulatory pathways. MOR agonists produce profound analgesia when applied directly to the spinal cord (used clinically in humans) because MOR is expressed on both the nerve terminals of nociceptive afferent nerves and primary transmission neurons in the dorsal horn of the spinal cord. The principal descending pathway through which MOR agonists produce analgesia runs from the PAG to rostral ventromedial medulla (RVM) to dorsal horn of the spinal cord (Fields 2004). Direct application of a small concentration of morphine into any one of these sites in the rat inhibits nociception throughout the body. Opioids in these regions inhibit tonically active GABAergic neurons by activation of GIRK channels and direct inhibition of GABAergic synapses (Williams et al. 2001). The natural function of this system is to produce analgesia in response to fear or stress. Fear appears to activate the nociceptive modulatory system in the PAG via a projection from the amygdala. Although the source of PAG opioids for fear-induced antinociception is not known, both intrinsic enkephalin-containing neurons and β-endorphin terminals arising from the arcuate nucleus could contribute. With chronic treatment, repeated morphine inhibition of GABAergic neurons in the ventrolateral PAG, but not the RVM, produces tolerance to the analgesic effects. Although tolerance is associated with a number of changes in MOR signaling, the specific mechanisms underlying tolerance to opioids are only partially known. In humans and other animals, the analgesic effects of morphine tend to be greater in males compared to females. Variations in the number of MOR, particularly in the PAG and the magnitude of morphine analgesia across the estrus cycle, suggest that estradiol influences the function of MOR.

Direct administration of morphine to many other brain regions also has been shown to inhibit pain in laboratory animals. For example, activation of MOR in the amygdala, nucleus submedius, or nucleus cuneiformis produces antinociception by activating the PAG. In addition to these central effects, MOR, DOR, and KOR receptor agonists attenuate pain by inhibition of primary afferents at the site of injury (Stein et al. 2003).

Reward and Addiction

Opioids, like other drugs of abuse (e.g., cocaine, amphetamine, ethanol), increase the release of dopamine in the nucleus accumbens (Le Moal and Koob 2007) and have been subject to animal models of addiction (▶ Addictive Disorder: Animal Models). The most commonly abused opiates, morphine and heroin, increase dopamine release in the nucleus accumbens by disinhibiting dopaminergic neurons in the ventral tegmental area. MOR and DOR agonists are thought to inhibit tonically active GABAergic neurons in the ventral tegmental area which activates dopamine neurons projecting to the nucleus accumbens. Direct opioid inhibition of nucleus accumbens neurons also contributes to the rewarding effects of opioids. Other brain structures that contribute to the reinforcing effects of opioids include the amygdala and bed nucleus of the stria terminalis.

Although the reinforcing effects of opioids are probably linked to many behaviors that are necessary for survival (e.g., food, sex, social interactions), this system also leads to opioid abuse and addiction. Addiction typically starts as a result of the euphoric effects of opioids but is maintained by dysregulation of endogenous opioid signaling (Le Moal and Koob 2007). Removal of opioid administration results in intense craving and withdrawal, which leads to further abuse. The rewarding effects of opioids are specific to MOR and DOR. KOR agonists inhibit dopamine release in the nucleus accumbens and produce aversion.

Cross-References

▶ Addictive Disorder: Animal Models
▶ Analgesics
▶ Blood–Brain Barrier
▶ Opioid Use Disorder and Its Treatment
▶ Somatostatin
▶ Synaptic Plasticity

References

Bodnar RJ (2008) Endogenous opiates and behavior: 2007. Peptides 29:2292–2375

Brownstein MJ (1993) A brief history of opiates, opioid peptides, and opioid receptors. Proc Natl Acad Sci U S A 90:5391–5393

Cahill CM, Holdridge SV, Morinville A (2007) Trafficking of delta-opioid receptors and other G-protein-coupled receptors: implications for pain and analgesia. Trends Pharmacol Sci 28:23–31

Fields H (2004) State-dependent opioid control of pain. Nat Rev Neurosci 5:565–575

Khachaturian H, Schaefer MKH, Lewis ME (1993) Anatomy and function of the endogenous opioid systems. In: Herz A (ed) Opioids 1. Springer, Berlin, pp 471–497

Kieffer BL, Gaveriaux-Ruff C (2002) Exploring the opioid system by gene knockout. Prog Neurobiol 66:285–306

Le Moal M, Koob GF (2007) Drug addiction: pathways to the disease and pathophysiological perspectives. Eur Neuropsychopharmacol 17:377–393

Lewis RV, Stern AS (1983) Biosynthesis of the enkephalins and enkephalin-containing polypeptides. Annu Rev Pharmacol Toxicol 23:353–372

Mansour A, Fox CA, Akil H, Watson SJ (1995) Opioid-receptor mRNA expression in the rat CNS: anatomical and functional implications. Trends Neurosci 18:22–29

Milligan G, Bouvier M (2005) Methods to monitor the quaternary structure of G protein-coupled receptors. FEBS J. 272:2914–2925

Stein C, Schafer M, Machelska H (2003) Attacking pain at its source: new perspectives on opioids. Nat Med 9:1003–1008

Waldhoer M, Bartlett SE, Whistler JL (2004) Opioid receptors. Annu Rev Biochem 73:953–990

Williams JT, Christie MJ, Manzoni O (2001) Cellular and synaptic adaptations mediating opioid dependence. Physiol Rev 81:299–343

Opportunity Cost

Definition

The value that would have accrued to an agent had it pursued an alternative course of action during the time spent in a chosen activity.

Cross-References

▶ Intracranial Self-Stimulation

Optogenetics

Michelle M. Sidor[1] and Melissa R. Warden[2]
[1]Department of Psychiatry, University of Pittsburgh School of Medicine, Pittsburgh, PA, USA
[2]Department of Neurobiology and Behavior, Cornell University, Ithaca, NY, USA

Synonyms

Optical neural control; Photostimulation

Definition

A neuromodulatory set of tools and techniques in neuroscience that involve the introduction of light-sensitive proteins, called opsins, into living cells and into the intact brain of freely moving animals to permit real-time control (both electrical and biochemical) of genetically defined populations of cells and neural circuits with both high spatial and temporal precision.

Principles and Role in Psychopharmacology

Neuropsychiatric Illness as a Disease of Altered Brain Circuitry

A paradigm shift is occurring in how scientists and clinicians fundamentally think of neuropsychiatric diseases. Although psychopathology has long been thought to represent underlying neural chemical imbalances, it is becoming increasingly recognized that such an approach may not fully capture the complexity of neuropsychiatric illness. Indeed, pharmaceutical research and development based on this chemical imbalance approach has produced drugs that have proven frustrating in their lack of efficacy in a large portion of the patient population. Newer approaches that involve fundamental changes to how we view neuropsychiatric illness are required before great strides can be made.

The success of non-pharmacological approaches to the treatment of neurological and psychiatric illnesses suggests that altered brain circuitry may be the pathological driving force. For instance, electricity in the form of electroconvulsive shock therapy has been historically used to "jolt" the brain back into function and is the most effective treatment for depressions that are nonresponsive to drug therapy. More recently, deep brain stimulation (DBS) treatment has shown much promise for treating brain disorders. DBS involves the delivery of precise high-frequency electrical stimulation into affected brain areas for alleviation of patient symptoms. DBS of brain regions such as cortical area 25 and of the medial forebrain bundle has been successful at treating treatment-resistant depressions. Stimulation of the ventral striatum has been successful for the treatment of obsessive-compulsive disorder, and DBS of the nucleus accumbens has proven useful in the treatment of drug addiction. As such, there is a growing awareness and agreement that certain brain disorders may be better conceptualized as diseases of chronic circuit-wide disruption. The exact mechanisms underlying the effectiveness of DBS, however, are currently unknown. Transformative leaps in our understanding of brain function have often been the direct result of the development of new tools for interrogating neural circuits. A new set of transformative tools has opened the doors to the study of how disrupted neural circuits ultimately lead to disease.

Optogenetics as a Tool to Investigate Neural Circuit Function

The development of optogenetics has revolutionized the fields of behavioral and systems neuroscience with recent years witnessing an explosion of new scientific knowledge resulting directly from the use of this new technology. Optogenetics, in which genetically encoded, light-sensitive channels, pumps, and receptors are used to control the function of specific neurons, has enabled the dissection of the neural circuits mediating cognition and behavior with unprecedented temporal and spatial precision (Boyden et al. 2005). The brain is composed of tens of thousands of different kinds of neurons, each composed of a different combination of molecules that confers a distinct identity and function. These different cell types are tightly intermingled in a complex and heterogeneous neural thicket, which makes discerning their individual contributions to brain function a challenge. Traditional causal approaches to the study of neural circuits include lesion, electrical, and pharmacological approaches, which have been highly successful at advancing our understanding of brain function. For instance, lesions and electrical stimulation have enabled us to pinpoint precise brain regions mediating specific cognitive or behavioral functions, but have not enabled a more precise dissection of the underlying neural circuitry, as these approaches are usually not capable of targeting distinct cell types. Likewise, pharmacological approaches have been invaluable for probing the function of specific neurons defined by the receptors they express, but this approach lacks the temporal specificity required to obtain a full understanding of the millisecond-precision neural dynamics used by the brain.

The discovery and application of single-component microbial opsins was pivotal to the widespread adoption of optogenetics as an essential tool for probing neural function. As the expression of these opsins can be constrained to particular cell types, either defined by genetic identity or by topology, these tools have the potential to target neural circuits with unprecedented precision. Indeed, a central hope in the field is that these tools will eventually find clinical application without the side effects that often plague current therapeutic approaches as a result of this increased specificity. In addition, these tools are capable of influencing neural processing on a millisecond time scale; they speak the fast electrical language of the brain.

Previous attempts to optically control neural activity had made some headway, but most of these technologies required the application of exogenous cofactors (chemical chromophores) that made routine use in mammalian systems difficult. The first successful use of optogenetics to control neural activity employed the single-component microbial opsin channelrhodopsin-2 (ChR2) (Boyden et al. 2005). This opsin, derived from the green alga *Chlamydomonas reinhardtii*, is a membrane-bound cation channel that can be expressed in genetically engineered neurons. Upon exposure to a short pulse of blue light, this channel opens, leading to the depolarization of the neuron and the generation of a single action potential. Initial experiments demonstrated the functional use of this channel to control neural activity in cultured hippocampal neurons; subsequent experiments have extended use to freely behaving vertebrate and invertebrate systems.

In the years since this first application, many new opsins – both naturally occurring and engineered – have been used to control neural activity in different and complimentary ways (Mattis et al. 2012). The inhibitory opsin shown to be useful for in vivo applications was a halorhodopsin (NpHR) from the halobacteria *Natronomonas pharaonis*. This opsin, a chloride pump with peak photosensitivity at ~590 nm, hyperpolarizes and therefore silences neurons when they are exposed to yellow light. Other tools for inhibition of neural activity include the outward proton pumps Arch (from *Halorubrum sodomense*) and Mac (from *Leptosphaeria maculans*), now both optimized for use in mammalian systems (eArch3.0 and eMac3.0).

Using Optogenetics in Preclinical Animal Models to Dissect the Neural Circuitry of Complex Behavioral States

Optogenetic tools have been widely used in both vertebrate and non-vertebrate systems. For the study of complex behavioral states relevant to human disease, rodents (mice and rats) have been the model organisms of choice. Indeed, optogenetics has been widely used in awake, freely moving rodents to study normal brain function and to understand how disrupted neural circuit activity drives altered behavioral states relevant to neurological conditions and psychiatric diseases (Tye and Deisseroth 2012; Nieh et al. 2013). The design and implementation of neural-optical interfaces optimized for the delivery of light to the intact brain has been instrumental in the ability to modulate circuit activity in awake, freely moving animals (Aravanis et al. 2007). Time-locked behavioral responses to neural circuit manipulation has allowed neuroscientists, for the first time, to ask causal questions linking altered circuit activity to a given behavioral state. For instance, studies examining normal brain function have investigated the neural circuitry underlying learning and memory, sleep-wake and arousal, locomotion, feeding behavior, motivation, stress, and social behavior, to name a few. These studies have provided great insight into the functional consequences of altered neural circuit activity in various neurological conditions, including Parkinson's disease and epilepsy. Additionally, optogenetic applications in rodents have been used to study core behavioral features that are observed across human neuropsychiatric diseases, such as depression, anxiety, fear, reward (natural and drug), and addiction.

Combining Optogenetic Technology with Pharmacological Approaches

Opsin-based technology has excelled at manipulating neural activity through direct changes to ion conductance or through altering synaptic activity. The advantage of manipulating the electrical activity of the brain using optogenetics is the millisecond temporal specificity obtained, which permits time-locked behavioral responses to modulation of neural activity. However, neurological and psychiatric disorders can be conceptualized as diseases of chronic circuit dysfunction that develop over extended time scales. Long-term disruption of brain activity likely induces neural adaptive molecular changes that extend beyond immediate changes to membrane voltage to fundamentally alter the way circuits behave and function. These adaptive molecular changes can include altered intracellular signaling, protein-protein interactions, receptor expression, and neurotransmitter secretion, which ultimately involve changes at the gene transcription level. Indeed, there is a small but growing literature employing chronic optogenetic stimulation paradigms to induce sustained physiological and behavioral changes that persist well after cessation of optic stimulation (Sidor and McClung 2014). These will be essential in systematically investigating how chronic circuit disruption alters biochemical processes. In turn, delineating the biochemical changes that result from prolonged circuit disruption will be essential in understanding the molecular basis of brain disease. Technology that integrates optogenetics with molecular- and genetic-based approaches for circuit modulation will be crucial to investigating the neural adaptive changes that support abnormal physiology and behavior. Exciting new technologies and tools are emerging (discussed below) that combine the temporal and spatial precision of optogenetics with biochemical tools for user-defined manipulation of intracellular signaling pathways, receptor function, and gene expression (Stuber and Mason 2013).

OptoXRs are synthetic genetically encoded receptors that permit bidirectional control of intracellular G-protein-coupled signaling in vivo (Fig.1a). Briefly, these are a group of opsin/receptor chimeras between the light-sensitive protein, rhodopsin, and the β2-adrenergic receptor (for Gs control) or α1-adrenergic receptor (for Gq control). These receptors are engineered for insensitivity to endogenous chemical ligands in place of sensitivity to light. Transduction of OptoXRs in the nucleus accumbens of mice demonstrated their utility for altering biochemical cellular events at the precisely required time scales needed to support an operant conditioning behavioral task.

Optopharmacology involves the use of tools that fuse light-sensitive proteins to effector molecules for the control of channel and receptor function. These include caged ligands and chemical photoswitches (Fig. 1b). Briefly, caged ligands are released from a photosensitive protecting group in response to light where the uncaged ligand can now act on native channels and receptors. Chemical photoswitches undergo a conformational change upon illumination with light that permits binding of the ligand to its native receptor. The temporal specificity imparted through the use of light-sensitive ligands circumvents the disadvantages plagued by traditional pharmacological approaches. However, this approach still lacks the spatial precision for cell-specific targeting.

Optogenetic pharmacology represents an improvement on existing technologies that involves the use of designer ion channels and receptors that are made sensitive to synthetic light-activated ligands (photochromic ligands; Fig. 1b). Because these exogenous receptors are genetically engineered, any receptor subtype can be inserted into a neuronal population of interest. This is important since traditional optopharmacological approaches lacked the ability to specifically target native neurotransmitter receptor subtypes. Furthermore, the use of genetic technology permits insertion of designer channels and receptors into specified cell types. Optogenetic pharmacology, therefore, solves both the receptor subtype specificity issue and cell-targeting problems associated with conventional optopharmacology (Kramer et al. 2013).

The *L*ight-*I*nducible *T*ranscriptional *E*ffectors (LITEs) system uses a two-component modular design that enables the optical control of gene transcription in select cell types with the temporal precision of optogenetics (Fig. 1c; Konermann et al. 2013). Here, the naturally light-sensitive protein, cryptochrome-2 (CRY2) protein, is fused to a transcription-activator-like effectors (TALE) DNA-binding domain (1st component) that can be custom engineered to bind any DNA sequence of interest. Illumination with blue light induces a conformational change in CRY2 that recruits a customizable effector (2nd component) to the gene target for activation or repression of transcription. Additional light-switchable transgene expression systems are currently in development (Wang et al. 2012), meaning there is great potential for using these tools to investigate how altered gene function in specific populations of neurons affects normal and diseased brain states.

Summary

Optogenetics for Targeted Drug Discovery

The future of optogenetics in the context of drug discovery starts with the exploration and identification of distinct gene regulation patterns that occur in response to cell-type-specific optogenetic manipulation of neural circuits that drive pathological behavioral states. Combining optogenetics with deep-sequencing approaches, such as high-throughput RNA sequencing technology (RNA-seq), will prove instrumental in profiling the cell-type-specific transcriptional changes that occur in response to neural circuit modulation (Stuber and Mason 2013). RNA-seq is a whole transcriptome analysis technique that can sequence the entire transcriptome from single cells of a specific cell type or from whole brain regions for the non-biased identification of candidate genes (transcriptome is the functional "readout" of the genome that includes all classes of coding and noncoding RNA molecules found in a single cell or in a collection of cells).

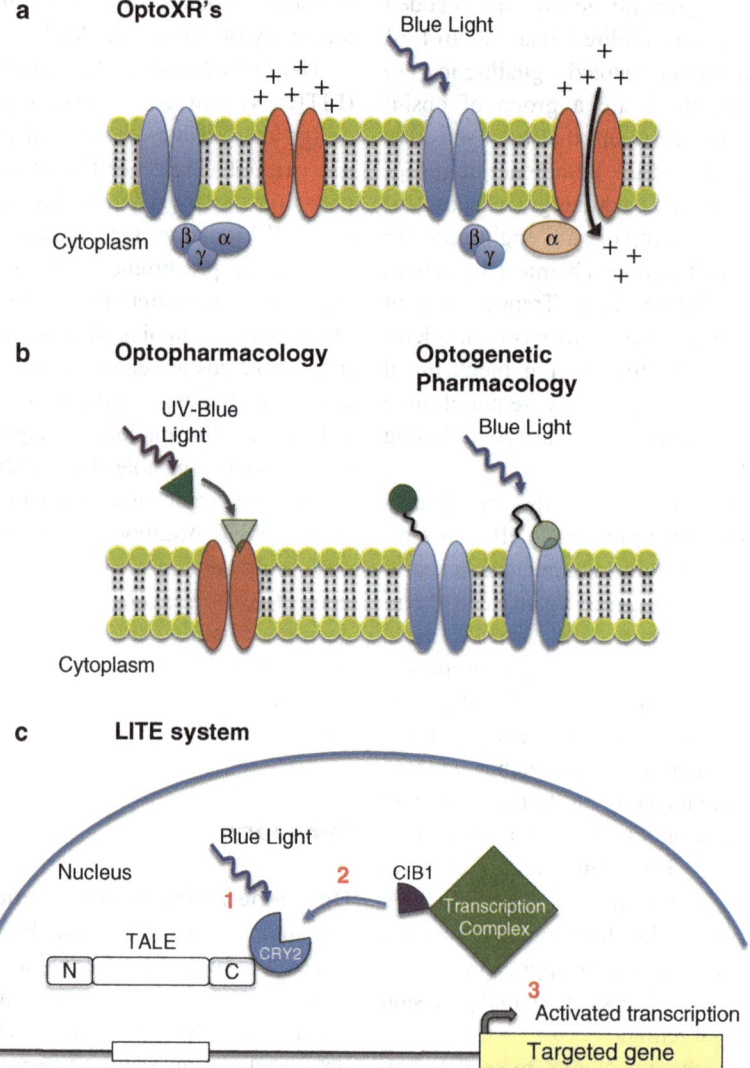

Optogenetics, Fig. 1 (a). OptoXRs are a component of the optogenetic toolbox that permit bidirectional and temporally precise control of intracellular G-protein-coupled signaling in genetically defined cell populations. These are a group of opsin/receptor chimeras between the light-sensitive protein, rhodopsin, and the β2-adrenergic receptor (for Gs control) or α1-adrenergic receptor (for Gq control). (b). An example of a photoactivatable ligand (*left*) that permits binding of the ligand to its native receptor upon illumination with UV light. Such optopharmacology tools represent an improvement on traditional agonist/antagonist pharmacological approaches by increasing the temporal precision of receptor control. However, the ability to target specific receptor cell types is not possible. Compare this with optogenetic pharmacology (*right*) that uses genetically engineered, or designer, receptors and ion channels, which are light sensitive, to control specific receptors and cell types with the temporal and spatial precision imparted by optogenetics. (c). Newer technologies, such as the *Light-Inducible Transcriptional Effectors* (LITEs) system, can turn gene transcription on and off in specific cell types through use of the natural light-sensitive protein, cytochrome 2 (*CRY2*). In this system, CRY2 is bound to a transcription-activator-like effector (*TALE*) DNA-binding domain that is engineered to bind specific DNA sequences. *Blue light* induces a conformation change in CRY2 (step 1) that enables the protein to fuse to a transcription complex (step 2) which then activates transcription (step 3) (Figure C adapted from: Konermann et al. (2013))

Once candidate genes are identified, their role in mediating a particular behavioral state will need to be systematically analyzed and validated. Through a reverse engineering approach, it may be possible to use the findings from these optogenetic-transcriptome approaches to design drugs that target specific populations of neurons based on their unique gene regulation profile. Ultimately, the ability to deliver such drugs to the brain in a cell- or circuit-specific manner will depend on emerging technologies for targeted small molecule delivery. The continued development and refinement of transformative tools will depend on a multidisciplinary approach that combines efforts across the fields of neuroscience, pharmacology, engineering, and related disciplines to inform the development of improved and novel pharmacotherapies for neurological and psychiatric diseases.

Cross-References

▶ Animal Models for Psychiatric States
▶ Anxiety: Animal Models
▶ Depression: Animal Models
▶ Electrochemical Techniques
▶ Genetically Modified Animals

References

Aravanis AM, Wang LP, Zhang F, Meltzer LA, Mogri MZ, Schneider MB, Deisseroth K (2007) An optical neural interface: in vivo control of rodent motor cortex with integrated fiberoptic and optogenetic technology. J Neural Eng 4:S143–S156. doi:10.1088/1741-2560/4/3/s02

Boyden ES, Zhang F, Bamberg E, Nagel G, Deisseroth K (2005) Millisecond-timescale, genetically targeted optical control of neural activity. Nat Neurosci 8:1263–1268. doi:10.1038/nn1525

Konermann S, Brigham MD, Trevino AE, Hsu PD, Heidenreich M, Cong L, Platt RJ, Scott DA, Church GM, Zhang F (2013) Optical control of mammalian endogenous transcription and epigenetic states. Nature 500:472–476. doi:10.1038/nature12466

Kramer RH, Mourot A, Adesnik H (2013) Optogenetic pharmacology for control of native neuronal signaling proteins. Nat Neurosci 16:816–823. doi:10.1038/nn.3424

Mattis J, Tye KM, Ferenczi EA, Ramakrishnan C, O'shea DJ, Prakash R, Gunaydin LA, Hyun M, Fenno LE, Gradinaru V, Yizhar O, Deisseroth K (2012) Principles for applying optogenetic tools derived from direct comparative analysis of microbial opsins. Nat Methods 9:159–172. doi:10.1038/nmeth.1808

Nieh EH, Kim SY, Namburi P, Tye KM (2013) Optogenetic dissection of neural circuits underlying emotional valence and motivated behaviors. Brain Res 1511:73–92. doi:10.1016/j.brainres.2012.11.001

Sidor MM, Mcclung CA (2014) Timing matters: using optogenetics to chronically manipulate neural circuitry and rhythms. Front Behav Neurosci 8:41. doi:10.3389/fnbeh.2014.00041

Stuber GD, Mason AO (2013) Integrating optogenetic and pharmacological approaches to study neural circuit function: current applications and future directions. Pharmacol Rev 65:156–170. doi:10.1124/pr.111.005611

Tye KM, Deisseroth K (2012) Optogenetic investigation of neural circuits underlying brain disease in animal models. Nat Rev Neurosci 13:251–266. doi:10.1038/nrn3171

Wang X, Chen X, Yang Y (2012) Spatiotemporal control of gene expression by a light-switchable transgene system. Nat Methods 9:266–269. doi:10.1038/nmeth.1892

Orexigenic

Definition

Systems or endogenous factors that provoke and/or sustain eating events.

Cross-References

▶ Appetite Stimulants

Organic Brain Syndromes

Definition

Organic brain syndrome is a largely obsolete term that refers to mental dysfunction resulting from a physical, as opposed to primary psychiatric, disorder. Examples of diseases causing organic brain syndrome include physical trauma, stroke, infections of the central nervous system, dementia, tumor, kidney failure, liver failure, endocrine diseases, and vitamin deficiencies.

Organizational Effects of Hormones

Definition

Relatively permanent effects of hormones on structure and function of the body. Often there is a critical period of development during which these organizational effects can take place. The most important critical periods are during fetal development and puberty.

Cross-References

▶ Sex Differences in Drug Effects

Orphenadrine

Synonyms

Banflex; Biorphen; Brocasipal; Disipal; Flexon; Lysantin; Mephenamin; N,N dimethyl-2 [α-(o-tolyl)benzyloxy]ethylamine HCl; Orphenadrine citrate; Norflex; Norgesic; Orfenace; Orfen; Orpadrex; Parekin

Definition

The main indications of orphenadrine citrate are painful musculoskeletal conditions and cramps. It is an anticholinergic but possibly also an analgesic. It may be used in Parkinson's disease as orphenadrine HCl, to treat side effects and drug-induced extrapyramidal symptoms.

Cross-References

▶ Anti-Parkinson Drugs

Orthosteric Site

Definition

A site of a receptor in which the endogenous ligand binds to produce its receptor effect.

Osmotic Minipump

Definition

Device made to slowly deliver pharmacological compounds (e.g., subcutaneously, lower back, over 2–4 weeks) and to avoid repetitive injection schedules. It is a miniature infusion pump for the continuous dosing of laboratory animals as small as mice and young rats. This minipump provides researchers with a convenient and reliable method for controlled agent delivery in vivo.

Overshadowing

Synonyms

Cue competition

Definition

The reduction in associative learning that is produced by the presentation of a more salient competing conditioned stimulus. This is an effect seen in classical (Pavlovian) conditioning and a constraint on the general importance of temporal coincidence as a determinant of new learning.

Cross-References

▶ Classical (Pavlovian) Conditioning

Oxazepam

Definition

Oxazepam is an active metabolite of diazepam formed, during the breakdown of ▶ diazepam and similar drugs. As a member of the benzodiazepine class of drugs, it has typical actions and side effects for that class. Like ▶ lorazepam, it is metabolized by glucuronidation and has an intermediate time course of action. In addition to its anxiolytic effects, it is sometimes used in the treatment of alcohol withdrawal.

Cross-References

▶ Abuse Liability Evaluation
▶ Alcohol Abuse and Dependence
▶ Benzodiazepines
▶ Declarative and Nondeclarative Memory
▶ Driving and Flying Under the Influence of Drugs
▶ Insomnias
▶ Sedative, Hypnotic, and Anxiolytic Dependence
▶ Social Anxiety Disorder
▶ Withdrawal Syndromes

Oxazolam

Definition

Oxazolam is a prodrug (precursor) for the benzodiazepine desmethyldiazepam (nordazepam) and is itself a metabolic product of other benzodiazepines. It has anxiolytic, sedative, and anticonvulsant properties.

Cross-References

▶ Anxiolytics
▶ Benzodiazepines

Oxcarbazepine

Definition

Oxcarbazepine is used primarily as an antiepileptic drug to control seizures in patients. However, it has gained utility in various mood disorders including anxiety, depression, and bipolar disorder, as well as in the management of neuropathic pain and migraine. Oxcarbazepine is structurally similar to an older antiepileptic ▶ carbamazepine and like carbamazepine works by reducing excessive or inappropriate excitability of nerve cells that normally occurs in conditions such as epilepsy and neuropathic pain.

Cross-References

▶ Anticonvulsants
▶ Bipolar Disorder
▶ Mood Stabilizers

Oxprenolol

Synonyms

(RS)-1-[2-(allyloxy)phenoxy]-3-(isopropylamino)-propan-2-ol

Definition

Oxprenolol is a nonselective β-adrenoceptor antagonist but with some ▶ intrinsic sympathomimetic activity. It is lipophilic so, unlike many of its class, it penetrates readily to the brain and exerts central effects. It is mainly used for the treatment of angina pectoris, abnormal heart rhythms and high blood pressure, and occasionally in the treatment of somatic anxiety. Typical side effects include headaches.

Cross-References

▶ Anxiolytics

Oxybate

Synonyms

Gamma-hydroxybutyrate; GHB; Sodium 4-hydroxybutyrate

Definition

Although it is a central nervous system depressant, oxybate is thought to be useful in the treatment of excessive daytime sleepiness and cataplexy in patients with narcolepsy and may also be used for pain and fibromyalgia. It is often illegally sold and abused, especially by young adults in social settings such as nightclubs.

Cross-References

▶ Abuse Liability Evaluation
▶ Hypersomnolence: Central Disorders

Oxycodone

Definition

Oxycodone belongs to the opioid (narcotic) analgesic medication class. It is used to relieve moderate to severe pain. Oxycodone is also available in combination with acetaminophen, aspirin, and ibuprofen. The use of oxycodone can lead to dependence.

Cross-References

▶ Addiction
▶ Analgesics
▶ Dependence
▶ Opioids
▶ Pain and Psychopharmacology
▶ Tolerance

Oxymorphone

Synonyms

14-Hydroxy-dihydromorphinone

Definition

Oxymorphone is a semisynthetic opioid narcotic. It is used in the treatment of chronic pain and is available in an extended-release formulation allowing 24 h management of pain in patients suffering from chronic long-term pain. Like morphine, it has the potential to lead to abuse and long-term dependence on the drug and can lead to physical dependence.

Cross-References

▶ Analgesics
▶ Opioid Use Disorder and Its Treatment
▶ Opioids
▶ Physical Dependence

Oxytocin

David Feifel and Kai MacDonald
Department of Psychiatry, University of California, San Diego, CA, USA

Synonyms

Pitocin®; Syntocinon®

Definition

Oxytocin (OT) is a nine-amino-acid neuropeptide with both peripheral and central actions. Oxytocin and a structurally very similar neuropeptide, vasopressin, compose the neurohypophyseal hormones which are highly conserved evolutionarily and the only two hormones secreted from the posterior pituitary gland. In addition to its peripheral, hormonal role, OT acts as a neurotransmitter in the brain.

Pharmacological Actions

Synthesis, Release, and Neuroanatomy

While a small amount of OT is synthesized and released in the periphery – for example, in the thymus gland, gastrointestinal tract, testes, and heart – it is primarily synthesized in two specific types of neurons within the paraventricular (PVN) and supraoptic (SON) nuclei of the hypothalamus. One type of relatively large neurosecretory cell, magnocellular neurons, synthesizes the bulk of the central OT, whereas a smaller cell type, parvocellular neurons, produces a smaller amount of OT. Most of the OT from magnocellular neurons is transported axonally to the posterior pituitary, where it is stored until secreted into the peripheral circulation. A smaller portion of magnocellular OT is released directly from collateral axons or dendrites into the extracellular fluid and diffuses passively, stimulating OT receptors (OTRs) in brain regions at some distance: a mode of signaling called "volume transmission." Undoubtedly, the concurrent release of magnocellular OT into both the peripheral circulation and key brain areas serves to coordinate complementary physiological and behavioral effects involved with birth and nursing, though most central release takes place independent of peripheral release and vice versa (MacDonald and MacDonald 2010).

In contrast to the dual peripheral and central targets of magnocellular OT, all of the OT produced in the parvocellular neurons of the PVN is directed via axonal release to key areas within the brain such as the amygdala, ventromedial hypothalamus, septum, nucleus accumbens, ventral tegmental area, and brain stem.

Regulation of OT release is complex and occurs via a balance of feedback and feed-forward mechanisms. Regarding the latter, release of OT primes OT neurons in the network to facilitate synchronized "bursting" of the entire network, making sustained peripheral and central levels possible despite OT itself having a relatively short half-life (3–6 min in the peripheral circulation and approximately 30 min in the CNS).

OT Receptors

OT is known to have only one receptor, OTR, a 389 amino-acid member of the rhodopsin-type (class I) G protein-coupled receptor (GPCR) family, and is coupled to phospholipase C through Gaq11 (Benarroch 2013). OTR is distributed in periphery (in breast, heart, kidney, GI tract, and reproductive system) as well as the brain. In addition to OTR, OT has a functionally meaningful affinity for the three vasopressin receptor subtypes (V1a-R, V1b-R, and V2-R), though it binds to them with much lower affinity than OTR. For example, several recent animal studies using strains of OT knockout mice demonstrated that OT reduction in autism-like effects and analgesia was due to its effects on the AVPV1a receptor (Stoop 2012).

Hormone Modulation

Due to its pivotal role in sex-specific reproductive events, the OT system is regulated by steroid hormones in the periphery and brain. Estrogen, in particular, stimulates OTR expression in the uterus, myoepithelial cells of the breast, and other organs which contain OTR, and the rise in estrogen late in pregnancy increases the sensitivity of these tissues to OT. Additionally, there is evidence that levels of both estrogen and testosterone regulate the expression of OTR in the brain.

Peripheral Actions and Putative Clinical Uses

Reproduction Related

OT was first discovered as a result of its peripheral actions, most notably its role in parturition and lactation, produced by stimulating contraction of hormonally primed uterine tissue and mammary myoepithelium during pregnancy and the postpartum period, respectively. Interestingly, these two peripheral actions are the basis of the established clinical applications for synthetic OT to date: induction of labor, management of postpartum hemorrhage, and stimulation of initial milk letdown.

HPA Axis-Mediated Stress Response

OT appears to have a dampening effect on the hormonal stress responses mediated by the hypothalamus-pituitary axis (HPA) (Lee et al. 2009). OT blood levels are increased in response to many stressful situations, and in rats, centrally delivered OT decreases release of the stress hormone corticosterone from adrenal glands by inhibiting the pituitary release of its main stimulatory signal, adrenocorticotropic hormone (ACTH). Central administration of an OT antagonist has the opposite effects. There is also evidence (albeit mixed) that intranasally delivered OT reduces stress hormones in humans.

Other Peripheral Effects

Other peripheral actions of oxytocin have been identified, which include anti-inflammation properties – likely mediated by lowering levels of certain cytokines (TNF-alpha and interleukins) – facilitation of wound healing, regulation of GI motility, decrease in blood pressure, reduction in blood glucose, and promotion of cardiac healing.

Central Actions and Putative Clinical Uses

General Issues

OT's significant central effects (described below) raise the potential of developing OT or OT analogs as treatment for brain disorders (MacDonald and Feifel 2013). Though OT does not efficiently cross the blood-brain barrier, growing evidence suggests that intranasal (IN) delivery of peptides such as OT facilitates their penetration into the brain. As such translational and clinical studies of OT in humans have almost exclusively utilized IN administration of OT. In parallel with these trials, additional effort is being directed to develop OTR analogs that have greater CNS penetration. To date only one OT analog, carbetocin, has been developed for human use. Structurally similar to OT – but chemically modified to resist enzymatic degradation – carbetocin has a much longer half-life (85–100 mins) than OT (whose half-life in plasma is 3–6 min). However, carbetocin's ability to evoke OTR-mediated central effects has not been studied in humans.

Modulation of Dopamine and Other Neurotransmitters

Dopamine (DA), a neurotransmitter heavily implicated in reward and reinforcement, is one of the primary central neurotransmitter systems that regulates and is regulated by OT. Moreover, neuropeptides like OT are often co-released with more classical neurotransmitters. The nature of OT's modulation of dopamine function is not fully clear, with some findings indicating OT inhibits dopamine transmission and others that it facilitates it. For example, peripherally administered OT inhibits the behavioral effects of indirect dopamine agonists such as amphetamine and cocaine. However, OT administered into the ventral tegmental area (VTA) – the main source of mesolimbic dopamine neurons and a brain site expressing OTR – stimulates dopamine release in the nucleus accumbens, the main target for dopamine-containing VTA neurons. Recently, OTR was found to form heterodimers with dopamine-2 receptors (D2R) in the striatum which resulted in greater binding of D2R ligands. The clinical implications of OT-ergic regulation of DA have been most thoroughly explored in schizophrenia, a disorder whose current psychopharmacologic treatments depend on regulating DA for their therapeutic effects on positive psychotic symptoms. To date, several separate positive clinical trials using chronic IN OT in schizophrenia have generated interest in the OT system as a novel treatment target for this illness (MacDonald and Feifel 2012). Other dopamine-related conditions like addiction are also being examined as OT sensitive

Besides DA, a variety of other central neurotransmitter systems of clinical importance are also involved with OT. These include the monoamines, serotonin, and noradrenaline, as well as GABA and glutamate.

Social Salience, Reward, and Memory

Animal and human studies reveal that OT plays a role in a broad range of social behaviors (Ross and Young 2009). For example, OT promotes several aspects of sexual interaction (e.g., erection, orgasm) and subsequent bonding between male and female animals. In several species of mammals, OT-DA interactions induce maternal behavior in females and promote the bonding/maternal behavior of females with her offspring. An outgrowth of this literature is an intense interest in the role of OT in maternal (postpartum) depression (Feldman et al. 2013).

A fundamental component of intact social functioning is social memory. A single peripheral or central administration of OT in male rodents acutely facilitates social memory of conspecifics (a member of one's species), whereas OT receptor knockout mice demonstrate impairments in recognition of conspecifics.

Human studies have also generally provided support for the hypothesis that OT plays a role in enhancing social reward, the salience of social stimuli, and social memory. A single dose of IN OT has been shown to transiently enhance many aspects of social behavior and cognition including increasing trust, gaze toward the eye region of faces, accurate recognition of facial emotions, and facial memory. Many of these OT effects on social salience and reward are thought to be mediated through OT's activity in the amygdala, but other areas have also been implicated such as the medial preoptic area (mPOA) of the hypothalamus and septum.

Although OT effects on social cognition have been generally characterized as prosocial, recent studies are revealing a more nuanced, complex suite of effects that likely depend on context in which it is given, individual factors (e.g., gender, individual differences in social cognitive capacity, early life stress), as well as the dose and duration of treatment (Guastella and McLeod 2012). Specifically, for example, there is some evidence that in individuals with early maltreatment, neural and behavioral responses to oxytocin may be anxiogenic or aversive. Moreover, some of oxytocin's "prosocial" effects may not extend beyond an individual's social group and may bias individuals toward negative responses to "outgroup" members. The clinical ramifications of these early findings, however, await larger studies.

Given the significant role of social deficits in illnesses like autism and schizophrenia, OT has been proposed as a treatment in both of these illnesses. In regard to autism, several promising single-dose trials and a few chronic trials with mixed results have been completed and more are under way. In terms of enhancement of social cognition in schizophrenia, OT has demonstrated promising effects in several studies.

Fear and Anxiety

Aside from its regulation of the HPA stress response, OT has been demonstrated to directly regulate behavioral expression of fear and anxiety in a large number of animal and human experiments. In rodents, intracerebral OT reduces anxiety-like behaviors in animal behavioral tests of anxiety like the elevated plus maze, and postcoital anxiolysis has been shown to be OT mediated in both male and female rats. Based on these and other similar findings, OT has been advanced as a potential treatment for social anxiety and possibly other anxiety and stress-related disorders. In human studies, positive trials in some models of anxiety (e.g., panic induced by CO_2 inhalation) are balanced with mixed results of others (e.g., startle tasks). To date, clinical trials of OT in patients with anxiety disorders are generally lacking, save several older, negative trials of chronic OT in obsessive-compulsive disorder. In these very small, early trials (total $n = 19$), high doses of IN OT had no effect on OCD symptoms (MacDonald and Feifel 2013). Importantly, however, imaging studies have revealed that IN OT alters activity in several of the neural substrates of fear and anxiety in a positive direction, including attenuating the reaction of the amygdala to highly fearful visual stimuli.

Learning and Memory

While OT seems to improve memory of species-relevant social stimuli (e.g., odors of a conspecific in rodents and faces in humans), it generally seems to have an amnestic effect on nonsocial memory (Lee et al. 2009). For example, in animals, OT impairs conditioned avoidance learning and spatial memory. In human trials, as well, OT has demonstrated an amnestic effect on verbal memory. Contrawise, a recent study found that 3 weeks of daily IN OT improved verbal memory in schizophrenia patients, suggesting chronic OT may have mnemonic effects even for nonsocial stimuli. Another possibility – intimated above in the discussion regarding OT's "prosocial effects" – is that certain psychiatric disease states may have different responses to IN OT, based on preexisting differences in aspects of the OT system (i.e., genetic variations in receptor subtype or receptor distribution).

Food Intake and Weight

Animal and human studies indicate that OT reduces food intake and weight, an effect thought to be produced by regulation of other satiety-related neurotransmitters and peptides (e.g., DA, cholecystokinin, leptin). OT is, therefore, a putative treatment for obesity. Furthermore, this effect combined with OT's recently revealed glucose-lowering action suggests that OT may have therapeutic effects for diabetes.

Motivation and Mood

OT's reciprocal regulation of brain DA and serotonin suggests a possible role in mood disorders. Several studies have shown OT produces antidepressant-like effects in animal paradigms that test motivation to escape aversive situations. As a result, speculation exists that OT may ameliorate depression and negative symptoms of schizophrenia, and several clinical trials examining chronic IN OT have reported benefits on negative symptoms in schizophrenia patients (MacDonald and Feifel 2013). Clinical trials resulting of chronic OT in patient with depression have not yet been reported.

Summary

A critical, multifaceted role in a variety of centrally and peripherally-mediated functions. These include physiological, cognitive, and emotional processes, as well as processes related to reproduction. Furthermore, together with its structurally similar partner hypophyseal peptide, vasopressin, it seems to powerfully regulate brain processes highly relevant for normal social and emotional function. There is increasing recognition, however, that the precise nature of this regulatory role is nuanced by a variety of influences. Despite this, there is much enthusiasm that the human OT system can be targeted to develop novel therapeutics for, among other things, psychiatric disorders that have heretofore lacked effective treatments (MacDonald and Feifel 2013).

Cross-References

▶ Amygdala
▶ Arginine-Vasopressin
▶ Hypothalamic-Pituitary-Adrenal Axis
▶ Hypothalamus
▶ Neuropeptides
▶ Social Behavior
▶ Social Impairment

References

Benarroch EE (2013) Oxytocin and vasopressin: social neuropeptides with complex neuromodulatory functions. [Review]. Neurology 80(16):1521–1528

Feldman R, Hurley RA, Taber KH (2013) Oxytocin and behavior: evidence for effects in the brain. J Neuropsychiatry Clin Neurosci 25(2):96–102. doi:10.1176/appi.neuropsych

Guastella AJ, MacLeod C (2012) A critical review of the influence of oxytocin nasal spray on social cognition in humans: evidence and future directions. Horm Behav 61(3):410–418

Lee HJ, Macbeth AH, Pagani JH, Young WS 3rd (2009) Oxytocin: the great facilitator of life. Prog Neurobiol 88(2):127–151

Macdonald K, Feifel D (2012) Oxytocin in schizophrenia: a review of evidence for its therapeutic effects. Acta Neuropsychiatr 24(3):130–146

MacDonald K, Feifel D (2013) Helping oxytocin deliver: considerations in the development of oxytocin-based therapeutics for brain disorders. Frontiers in neuroscience 7:35

Macdonald K, Macdonald TM (2010) The peptide that binds: a systematic review of oxytocin and its prosocial effects in humans. Harv Rev Psychiatry 18(1):1–21

Ross HE, Young LJ (2009) Oxytocin and the neural mechanisms regulating social cognition and affiliative behavior. Front Neuroendocrinol 30(4):534–547

Stoop R (2012) Neuromodulation by oxytocin and vasopressin. Neuron 76(1):142–159. doi:10.1016/j.neuron.2012.09.025

P

P300

Synonyms

P3

Definition

The P300 refers to a positive EEG voltage deflection seen around 300 ms after the presentation of a novel stimulus. It can be recorded using stimuli of any sensory modality but most commonly in the auditory domain. The simplest paradigm used to elicit the P300 involves presenting repeated simple auditory tones of a given frequency every second or so. The P300 is generated when a tone of a different frequency is presented randomly at a rate of around 1 in 10 of the more common tones. It is observed by comparing the event-related potentials, time locked to the presentation of the tones, generated by the "frequent" versus the "rare" frequencies. It is seen most prominently over the parietal lobe. Unlike mismatch negativity, subjects need to be consciously attending to the stimuli. The presence, magnitude, topography, and time of this signal are often used as metrics of cognitive function in decision-making processes.

Cross-References

▶ Event-Related Potential

Pain and Psychopharmacology

Stephen Tyrer
Campus for Ageing and Vitality, Wolfson Research Centre, Institute of Neuroscience, Newcastle University, Newcastle upon Tyne, Tyne and Wear, UK

Synonyms

Algesia; Dolor

Definition

An unpleasant sensory and emotional experience associated with actual and potential tissue damage or described in terms of such damage.

Classification

Painful conditions can give rise to emotional sequelae, particularly anxiety and depression. Psychological factors also affect the severity of painful feelings and often prolong pain. In DSM-IV, people with painful symptoms of this nature were diagnosed as having a somatoform disorder that was termed pain disorder associated with psychological factors, with or without an associated medical condition.

I.P. Stolerman, L.H. Price (eds.), *Encyclopedia of Psychopharmacology*,
DOI 10.1007/978-3-642-36172-2

In DSM-5, somatoform disorders are called somatic symptom and related disorders and pain disorder has been deleted. In this new classification, people with chronic pain can be diagnosed as having somatic symptom disorder (SSD) with predominant pain, with psychological factors that affect other medical conditions, or, on rare occasions, with an adjustment disorder. An important feature in DSM-5 is that an SSD diagnosis does not require that the somatic symptoms are medically unexplained. If the sufferer is gravely affected by the symptoms, the diagnosis should be made whether a known medical condition is present or not. The criteria for an SSD diagnosis are three: that there are one or more somatic symptoms that are distressing and/or result in significant disruption in daily life; that there are excessive thoughts, feelings, and/or behaviors related to these somatic symptoms (or associated health concerns because of these symptoms); and that the state of being symptomatic is more than 6 months.

Role of Pharmacotherapy

Pharmacological treatments in SSD should be directed primarily at treating the supposed condition which has given rise to the painful symptoms. In every case an estimate should be made of the degree and extent of tissue damage that has given rise to the painful state. By definition every person with an SSD diagnosis involving pain will have chronic pain, i.e., pain that has lasted for at least 3 months or beyond the time of healing of an acute condition. The clinician needs to determine how far the continuing pathological factors contribute to the painful complaint and to what extent psychopathological, personality, and behavioral factors are important. Chronic pain is associated with depression, lassitude, and lack of motivation, and these require treatment separately.

Determination of Origin of Pain

It is vital to identify the cause of pain as different treatments apply dependent on the putative etiology. Pain can be either peripheral or central in origin. Pain arising from damage to tissues of mesodermal origin, i.e., muscle, joints, bones, connective tissue, etc., is termed nociceptive pain, whereas pain originating from nerves and the brain is described as neuropathic pain.

Nociceptive Pain

Nociceptive pain arising from a peripheral origin is termed somatic pain, but when the source is from the abdominal organs, it is described as visceral pain (see below). Somatic nociceptive pain is typically well localized and constant, often with an aching or throbbing quality.

Acetaminophen and COX-2 Inhibitors

The basic guidelines for the treatment of nociceptive pain have not fundamentally changed since 1996 when The World Health Organization (WHO) described its guideline for the use of drugs in the management of pain. The WHO developed a three-step "ladder" for pain relief in adults: first, non-opioids (aspirin or acetaminophen (paracetamol); then, as necessary, mild opioids (codeine); and then strong opioids such as morphine. The guideline has been modified over time. When regular acetaminophen alone is insufficient, a standard oral nonsteroidal anti-inflammatory drug (NSAID) or a cyclo-oxygenase-2 (COX-2) inhibitor is usually recommended as the second stage of the ladder with co-prescription of a protein pump inhibitor drug, e.g., omeprazole, recommended for people over 45.

Acetaminophen is a very safe drug when administered at doses up to or less than 4G per day. It has few interactions with other drugs. It inhibits COX-2 by more than 80 %, i.e., to a degree comparable to NSAIDs and selective COX-2 inhibitors.

Nonsteroidal Anti-inflammatory Drugs (NSAIDs)

NSAIDs have been found to be effective in reducing pain in patients with osteoarthritis, rheumatoid arthritis, and back pain (Roelofs et al. 2008).

Their well-recognized side effects of gastric bleeding and diarrhea limit their use, especially in elderly patients, where gastric symptoms are frequently silent and perforation of the stomach can occur unexpectedly. Both NSAIDs and COX-2 inhibitors inhibit prostacyclin production leading to increased platelet aggregation, hypertension, and arteriosclerosis. COX-2 inhibitors have fewer gastrointestinal side effects than do traditional NSAIDs, but they are associated with increased cardiovascular risk. The choice of an NSAID or a COX-2 inhibitor in treatment depends upon the status of the gastrointestinal system and the cardiovascular state of the patient. The problem is that in any chronic painful condition, NSAIDs are likely to be prescribed for a long period of time, thereby increasing the risk. Drug holidays (omitting the dose for a few days) have been recommended, but their value is unproved.

Opioids

Although opioids are frequently prescribed, the value of these agents in chronic pain is underwhelming (Turk et al. 2011). A meta-analysis of 41 randomized controlled trials evaluating the effectiveness of opioids for the treatment of various forms of chronic noncancer pain, including osteoarthritis, low back pain, and rheumatoid arthritis, found that opioids led to a small improvement in pain severity and functional improvement compared with placebo but less improvement in function compared with other analgesic drugs (Furlan et al. 2006).

It is only within the last decade that opioid and opiate drugs (the distinction is simply based on the fact that opiate drugs are constituents of the opium poppy, *Papaver somniferum*) have become popular in the treatment of chronic pain. Earlier, there was grave concern among clinicians about using drugs that were known to cause addiction and dependence.

In the USA, the misuse of prescription opioids is the fastest growing form of drug misuse and is the leading cause of accidental overdose and mortality. There is also increasing concern about diversion and criminal trafficking by patients and physicians. A number of factors are thought to indicate a risk of addiction to prescribed opioids including current or past history of substance misuse, including alcohol; presence of a family member with history of substance misuse; poor social support; and comorbid psychiatric disorders. These require evaluation before starting drug treatment.

Not all opioid drugs have similar dangers of dependence. Tramadol, that acts through serotonin and norepinephrine reuptake inhibition as well as a μ-opioid agonist, has less addiction potential. Tramadol reduces pain substantially in osteoarthritis and fibromyalgia (Bennett et al. 2003). Methadone, a long-acting opioid, is used in patients who have problems with dependence, because of its long half-life. There are frequent side effects associated with opioids, particularly nausea, constipation, and drowsiness.

Opioid drugs should be used carefully in chronic painful states. When first prescribed, patients should be informed about the likely effects, both advantageous and otherwise. They should only be prescribed for a definite period of time, 1 month at the most, before assessment of the patient again. If they are not successful in adequate doses, alternative agents should normally be considered.

Antidepressant Drugs and Other Agents

The value of antidepressant drugs in nociceptive pain is much less than in neuropathic pain conditions (see below). There is evidence of a mild degree of benefit in low back pain, fibromyalgia, and headaches.

Anticonvulsant drugs, muscle relaxants, and some topical agents have also been used in the treatment of nociceptive pain, but the evidence for their value is meager.

Neuropathic Pain

Neuropathic pain is the result of an injury or malfunction in the peripheral or central nervous system. The pain is produced by damage to the nerve itself or by pathological changes within the nerve. The adjectives that patients use when describing neuropathic pain include words such as burning, electric, tingling, and shooting. Allodynia (painful response to normally non-noxious stimuli) and hyperesthesia (reduced pain thresholds and enhanced pain response to unpleasant thermal and mechanical stimulation) are the hallmarks of this condition, but hyperpathia (innocuous stimuli resulting in severe explosive pain) is also found. These pains are associated with reduced sensation in the affected part associated with numbness, pricking, and tingling. Common examples include painful diabetic neuropathy, postherpetic neuralgia, and trigeminal neuralgia.

Complex regional pain syndrome arises because of dysregulation of the autonomic nervous system and has similarities to neuropathic pain states. The treatment for this condition is similar to the treatment of neuropathic pain although the free radical scavengers, dimethylsulfoxide cream, and N-acetylcysteine have documented success in the treatment of this severely painful condition.

Antidepressants

Tricyclic antidepressants are first-line treatment for neuropathic pain (Dworkin et al. 2010). Oral amitriptyline, starting at a dosage of 10 mg per day, up to a maximum of 75 mg per day, is the drug of choice. The dose of amitriptyline recommended is much lower than is used in the treatment of depression. Although cardiovascular events, including postural hypotension and arrhythmias, as well as falls in elderly adults, are reported with tricyclic drugs, in practice the dosage used is small and major side effects are not usually a problem. A novel tricyclic antidepressant, cyclobenzaprine, marketed as a skeletal muscle relaxant in the USA, has been found to be effective in reducing pain in fibromyalgia.

The specific serotonin reuptake inhibitors (SSRIs) have little to no value in the treatment of neuropathic pain. The selective serotonin and norepinephrine reuptake inhibitors (SNRIs) are effective. Venlafaxine has an NNT of 3 in neuropathic pain syndromes, of similar value to tricyclic antidepressants. Duloxetine, another SNRI, has been shown to be effective (Goldstein et al. 2005) in patients with diabetic neuropathy and is endorsed for this indication by the UK National Institute of Health and Clinical Excellence (NICE).

Anticonvulsants

Three anticonvulsant drugs are used in people with SSD who have chronic neuropathic pain – gabapentin, pregabalin, and carbamazepine. Gabapentin and pregabalin act as neuromodulators by selectively binding to the $\alpha_2\delta$ subunit protein of calcium channels reducing calcium currents in the brain and the superficial dorsal horn of the spinal cord. Both drugs have been found to be effective neuropathic pain (Moore et al. 2009, 2011). Gabapentin has been used for over a decade in this condition, but pregabalin, which acts in a similar manner and has equivalent efficacy, is preferred by some because of simpler dosing and lower cost. The most common side effects of both these agents include daytime drowsiness, dizziness, fatigue, and weight gain. Carbamazepine was one of the first anticonvulsant drugs to be tested in neuropathic pain, but the evidence for its efficacy is mixed.

Opioids may be prescribed if these agents are ineffective although their efficacy is less than in nociceptive pain. Tramadol is preferred as the first option because of the lesser danger of dependence. If stronger opioids are considered, most patients should be referred to a specialized pain management unit.

Capsaicin

Capsaicin is an alkaloid derived from chilli peppers; repeated application depletes substance P from primary afferent neurons. It is of value in both neuropathic pain and musculoskeletal disorders including osteoarthritis (Derry et al. 2013).

Drug Combinations

Combinations of the drugs described above have been shown to have greater benefit in pain reduction than one agent alone. Greater pain reduction in the treatment of neuropathic pain has been shown for the combination of gabapentin and morphine and for nortriptyline and gabapentin greater than either drug alone (Gilron et al. 2009).

Visceral Pain

Unpleasant sensations arising from the viscera are generally only felt acutely or perceived as painful when stimuli exceed those in the physiological range. Adequate stimuli that induce visceral pain are distension, ischemia, and inflammation. When stimulation is prolonged, the nociceptors become more sensitive to afferent stimulation, leading to what is known as visceral hyperalgesia (VH) (Tyrer and Wigham 2007). Irritable bowel syndrome, noncardiac chest pain, and chronic pancreatitis are all conditions in which VH occurs. VH is due to sensitization of the afferent nerves (peripheral sensitization), sensitization of the spinal dorsal horn neurons (central sensitization), and altered descending excitatory or inhibitory influences from the brain. The NMDA antagonist ketamine is of value in this condition as it reduces the effect of glutamate, which is increased in central sensitization because of phosphorylation of N-methyl-d-aspartate (NMDA) receptors. The somatostatin analogue octreotide, gabapentin, and pregabalin have also been shown to reduce central sensitization. Antispasmodics and antidepressants have some value.

Cross-References

▶ Anticonvulsants
▶ Antidepressants
▶ Nonsteroidal Anti-inflammatory Drugs
▶ Opioids

References

Bennett RM, Kamin M, Karim R, Rosenthal N (2003) Tramadol and acetaminophen combination tablets in the treatment of fibromyalgia pain: a double-blind, randomized, placebo-controlled study. Am J Med 114:537–545

Derry S, Rice AS, Cole P, Tan T, Moore AR (2013) Topical capsaicin (high concentration) for chronic neuropathic pain in adults. Cochrane Database Syst Rev 2, CD007393

Dworkin RH, O'Connor AB, Audette J, Baron R, Gourlay GK, Haanpaa ML et al (2010) Recommendations for the pharmacological management of neuropathic pain: an overview and literature update. Mayo Clin Proc 85(3):S3–S14

Furlan AD, Sandoval JA, Mailis-Gagnon A, Tunks E (2006) Opioids for chronic noncancer pain: a meta-analysis of effectiveness and side effects. Can Med Assoc J 174:1589–1594

Gilron I, Bailey JM, Tu D, Holden RR, Jackson AC, Houlden RL (2009) Nortriptyline and gabapentin, alone and in combination for neuropathic pain: a double-blind, randomised controlled crossover trial. Lancet 374:1252–1261

Goldstein DJ, Lu Y, Detke MJ, Lee TC, Iyengar S (2005) Duloxetine vs. placebo in patients with painful diabetic neuropathy. Pain 116:109–118

Moore RA, Straube S, Wiffen PJ et al (2009) Pregabalin for acute and chronic pain in adults. Cochrane Database Syst Rev 3:CD007076

Moore RA, Wiffen PJ, Derry S et al (2011) Gabapentin for chronic neuropathic pain and fibromyalgia in adults. Cochrane Database Syst Rev 3:CD007938

Roelofs PD, Deyo RA, Koes BW, Scholten RJ, van Tulder MW (2008) Non-steroidal anti-inflammatory drugs for low back pain. Cochrane Database Syst Rev 2:CD000396

Turk DC, Wilson HD, Cahana A (2011) Treatment of chronic non-cancer pain. Lancet 377:2226–2235

Tyrer S, Wigham A (2007) Psychiatric aspects of visceral pain. In: Schmidt RF, Willis WD (eds) Encyclopedia of pain. Springer, Heidelberg, pp 3605–3608

Pair-Feeding

Definition

A procedure used in animal studies to control for decreases in food intake associated with chronic drug exposure that involves providing a control group with only as much food daily as is consumed by the drug-treated experimental group.

Palatability

Definition

A subjective measure of the acceptability and pleasantness of food. It implies a property of "being acceptable to the taste or mouth," with a sufficiently agreeable flavor for eating and pleasing to taste. It embraces the hedonic evaluation of a food resulting from its sensory properties, modified in relation to energy needs, past experience and learning, and the quantity of food consumed within a meal. Palatability may be inferred from the intensity and persistence of ingestion (palatability response). High-palatability foods tend to be energy dense, with high fat and carbohydrate content. It is an important determinant of caloric intake and can overwhelm processes of satiation; the greater the palatability of the food, the more likely it is that overconsumption will occur.

Cross-References

▶ Appetite
▶ Hunger
▶ Hyperphagia

Paliperidone

Synonyms

9-Hydroxyrisperidone

Definition

Paliperidone is a second-generation ▶ antipsychotic medication indicated for the acute and maintenance treatment of schizophrenia and as monotherapy or an adjunct to mood stabilizers and/or antidepressants for the treatment of psychotic or manic symptoms of ▶ schizoaffective disorder. It acts primarily as a dopamine D2 and serotonin 5-HT_{2A} antagonist. It is the 9-hydroxymetabolite of ▶ risperidone with which it shares many pharmacological properties. An extended release formulation of paliperidone allows for once-daily dosing, results in lower peak plasma levels and hence less side effects, and thereby limits the need for dose titration that is otherwise required at the start of treatment with risperidone. Paliperidone palmitate has been developed as long-acting injectable depot medication, requiring only once-monthly treatment.

Cross-References

▶ Antipsychotic Drugs
▶ Antipsychotic Medication: Future Prospects
▶ Risperidone
▶ Schizophrenia
▶ Second-Generation Antipsychotics
▶ Second- and Third-Generation Antipsychotics

Panic Attack

Definition

Panic attacks are sudden, discrete periods of intense anxiety, mounting physiological arousal, fear, and discomfort that are associated with a variety of somatic, cognitive, and behavioral symptoms. The onset is typically abrupt and may have no obvious triggers.

Cross-References

▶ Agoraphobia
▶ Panic Disorder

Panic Disorder

Delphine Capdevielle[1] and Jean-Philippe Boulenger[2]
[1]French National Institute of Health and Medical Research, INSERM U 888, La Colombière Hospital, Montpellier, France
[2]Department of Adult Psychiatry, Hôpital la Colombiere CHU Montpellier, University Montpellier 1. Inserm U888, Montpellier, France

Definition

The aim of this entry is to describe the pharmacological treatment of one of the most common anxiety disorders: panic disorder. We have focused on antidepressants (ADs), especially on serotonin selective reuptake inhibitors (SSRIs), which have been found to be effective in treating panic disorder and are the first-line treatment with regard to their efficacy and safety. The other ADs are listed along with the other pharmaceutical classes with their indications, side effects, and implementation issues.

Role of Pharmacotherapy

Introduction
Panic disorder is one of the most common anxiety disorders in the general population, with a lifetime prevalence of 4.7 % (Kessler et al. 2005). It is often disabling, especially when complicated by agoraphobia, and is associated with functional morbidity and reduced quality of life. The disorder is also costly for individuals and society, with increased use of healthcare and absenteeism.

Role of Pharmacotherapy
Medications have been known to be useful in the treatment of panic disorder for over 30 years, with the first description of efficacy of the tricyclic antidepressant imipramine for blocking panic attacks by Donald Klein in 1964. Many studies have recorded the efficacy of most ADs in panic disorder. Benzodiazepines are the other effective medication currently available. The aim of pharmacotherapy should be to eliminate panic attacks, if possible, because partial response often results in continued avoidance of frightening situations and impairment in social functioning. However, the goal of treatment should also be to reduce or eliminate associated anticipatory anxiety, phobic avoidance, and other symptoms due to comorbid conditions, such as major depression, and to improve global functioning. Medications from several classes have been shown to be effective.

Treatments

Serotonin Selective Reuptake Inhibitors (SSRIs)

Efficacy The efficacy of SSRIs in the treatment of depression is now very well documented, and they have also been found to be effective in treating panic disorder. Their efficacy and safety render them the first-line treatment for this condition. Each of the six SSRIs (▶ fluoxetine, ▶ paroxetine, ▶ sertraline, ▶ fluvoxamine, ▶ citalopram, and ▶ escitalopram) has demonstrated its effectiveness in randomized clinical trials.

Meta-analyses and reviews focusing on several of these agents have reported medium to profound effects compared with placebo and have confirmed that this efficacy may be maintained for up to 1 year (Otto et al. 2001). Therapeutic response in panic disorder seems to be a class effect, which is common to all the SSRIs, with no evidence of differential long-term efficacy within the class.

Side Effects The main side effects of SSRIs are headache, irritability, nausea and other gastrointestinal complaints, insomnia, sexual dysfunctions, increased anxiety, jitteriness, drowsiness, and tremor. Unlike tricyclic ADs, they can be prescribed to patients with prostatic hypertrophy or narrow-angle glaucoma, since they lack clinically significant anticholinergic effects.

Discontinuation reactions have been associated with all the major classes of ADs. They are reported to occur particularly with compounds having short elimination half-lives (for SSRIs, e.g., paroxetine) and can mimic the reappearance of the underlying disorder. These discontinuation symptoms may vary in nature and severity. They begin shortly after stopping the drug. Discontinuation reactions last between 1 day and 3 weeks, with a rapid reversal on restarting the original drug. Tapering of antidepressant use is the most common preventive strategy.

Implementation Issues Patients suffering from panic disorder tend to be more sensitive to SSRI side effects than depressed patients; hence, the treatment must be started at lower doses than those used in depression (half dose or less). After a few days, based on efficacy and tolerance, doses may be progressively increased to those used in depressive disorders. The onset of action is usually 2–8 weeks. There is no clear evidence as to what should be the optimal length of treatment, but generally, treatment should be continued until 6 months after the remission of symptoms (McIntosh et al. 2004).

Tricyclic Antidepressants

Efficacy Imipramine and clomipramine have been the most extensively studied of the tricyclic ADs, and both have demonstrated efficacy in treating panic disorder. Other tricyclic ADs that have shown some evidence of efficacy include desipramine, doxepin, amitriptyline, and nortriptyline (for review, see Cox et al. 1992). One meta-analysis concluded that there were no significant differences between SSRIs and tricyclic ADs in terms of efficacy or tolerability in short-term trials for panic disorder treatment (Otto et al. 2001). Another meta-analysis comparing paroxetine and imipramine concluded that paroxetine is better tolerated than imipramine and should be the first-line treatment (Wagstaff et al. 2002).

Side Effects All tricyclic ADs have common side effects such as anticholinergic effects (dry mouth, constipation, difficulty urinating, increased heart rate, and blurred vision), increased sweating, sleep disturbance, orthostatic hypotension, fatigue, cognitive disturbance, weight gain, and sexual dysfunction. Prostatic hypertrophy or narrow-angle glaucoma are contraindications for these drugs. Cardiac function should be carefully monitored.

Implementation Issues For the same reasons as with SSRIs, it is recommended that tricyclics be started at doses lower than those for patients with depression. For example, imipramine should be started at only 10 mg/day and gradually increased up to a minimal dose of 75 mg/day.

The duration of treatment should be the same as with SSRIs, and it is recommended that the dose be tapered before stopping.

Benzodiazepines

Efficacy Alprazolam and clonazepam (high-potency benzodiazepines) have been studied more extensively than low-potency benzodiazepines and have shown to be more effective for the specific treatment of panic disorder. The primary advantage of using benzodiazepines is rapid relief from anxiety and panic attacks. As already mentioned, ADs have a delayed therapeutic onset (Moroz 2004).

Side Effects The disadvantages of benzodiazepines include sedation, ataxia, slurred speech, cognitive clouding, interaction with alcohol, physiological dependence, deliberate misuse, and the potential for a withdrawal syndrome. The prescription of benzodiazepines in elderly patients must be carefully weighed against the risk of oversedation, falls, and cognitive impairment due to the decreased elimination of these compounds.

Implementation Issues Because of its short duration of action, alprazolam generally must be administered in three to five daily doses. Clonazepam, which has a longer duration of action than alprazolam, can generally be administered twice a day. Clonazepam is reported to have less abuse potential than alprazolam and is found to be easier to taper during discontinuation owing to its longer half-life. Very few studies have empirically evaluated dose requirements. Thus, it is recommended to slowly increase the doses according to the efficacy/tolerability ratio. There are very few data indicating the optimum length of treatment with benzodiazepines, but short-term treatment is recommended to minimize the risk of tolerance and dependence.

Benzodiazepines can be helpful when AD treatment is initiated and when a rapid onset of therapeutic effect is desired. They can also help to improve the short-term tolerability of SSRIs by blocking the jitteriness and exacerbation of panic sometimes observed when initiating treatment with an AD. Benzodiazepines can be used to "top up" the patient's treatment on an as-needed basis for sudden and unexpected decompensation or short-term psychosocial stressors. However, benzodiazepines are associated with a worse outcome in the long term than AD treatment (McIntosh et al. 2004).

Other Antidepressants

Venlafaxine Venlafaxine is a member of the dual serotonin-norepinephrine reuptake inhibitor (SNRI) class of ADs and has received approval for panic disorder in the USA. A number of open-label and double-blind studies have demonstrated the effectiveness of venlafaxine immediate-release (IR) and extended-release (ER) formulations for the treatment of panic disorder. One study published in 2007 compared venlafaxine ER with paroxetine in the treatment of panic disorder with or without agoraphobia, demonstrating that both venlafaxine and paroxetine are more effective than placebo and are well tolerated. Both fixed doses of venlafaxine ER (75 and 150 mg/day) demonstrated efficacy and tolerability that were comparable with each other and with paroxetine at 40 mg/day. The most common side effects reported for venlafaxine were sweating, dry mouth, dizziness, anorexia, tremor, constipation, diarrhea, and somnolence (Pollack et al. 2007). Discontinuation reactions have been described for venlafaxine. These reactions are very similar to those with SSRIs. The same strategy of decreasing the doses very slowly should be used before stopping the treatment.

Nefazodone and Trazodone A retrospective analysis and two open-label trials reported that nefazodone may be effective in the treatment of panic disorder. Furthermore, nefazodone is reported to have a better side-effect profile than other ADs, with no weight gain or sexual dysfunction. However, the potential for marked hepatotoxicity has limited the widespread use of nefazodone.

Two studies with trazodone found disparate results. One single-blind study reported that patients improved significantly compared with placebo, but a double-blind study comparing imipramine, alprazolam, and trazodone showed trazodone to be less effective than imipramine or alprazolam.

Mirtazapine A double-blind randomized controlled study with only 27 panic patients showed no statistical difference between mirtazapine and fluoxetine. Four open-label trials have also supported the effectiveness of mirtazapine in panic disorder. The average dose is 30 mg/day and the most frequently reported side effects are weight gain, initial drowsiness, and constipation (Ribeiro et al. 2001).

Reboxetine Reboxetine is a selective norepinephrine reuptake inhibitor reported to be more effective than placebo in a double-blind randomized controlled trial. However, in a single-blind study, reboxetine appeared to be less effective than paroxétine in the treatment of panic disorder. The adverse events reported most frequently with reboxetine are insomnia, constipation, and dry mouth (Bertani et al. 2004).

Monoamine Oxidase Inhibitors (MAOIs)

Efficacy There are no studies proving that the classical irreversible MAO inhibitors are effective in treating panic disorder specifically; however, six earlier pre-DSM III studies showed efficacy in treating the phobic anxiety of individuals with panic-like symptoms. Two studies with reversible inhibitors of MAO-A (RIMAs), one a double-blind comparison of brofaromine with clomipramine, and the other, an open study of brofaromine, showed antipanic and antiphobic efficacy. No medication in the RIMA class is currently approved for use in the USA, but at least one drug, moclobemide, is widely used in Europe and Canada despite the lack of clear evidence of its efficacy (for review, see Bandelow et al. 2008).

Side Effects The disadvantages of the MAO inhibitors make them second- or third-line treatments for panic disorder; these include orthostatic hypotension, weight gain, sexual dysfunction, and dietary restrictions (low-tyramine diet), with the potential for a tyramine-induced hypertensive crisis. They can also interact with other serotonergic drugs to precipitate serotonin syndrome, which can result in severe hyperpyrexia. They should be prescribed by physicians with experience in monitoring MAOI treatment. The RIMAs appear safer, with lessened potential for side effects, and do not require adherence to a tyramine-free diet.

Other Agents

Anticonvulsants Anticonvulsants, such as valproate, carbamazepine, and levetiracetam, have demonstrated preliminary evidence of efficacy in the treatment of panic disorder but with side effects such as gastrointestinal dysfunction, weight gain, dizziness, nausea, sedation, and alopecia (for review see Bandelow et al. 2008).

Levetiracetam is an anticonvulsant, currently approved by the US Food and Drug Administration for the adjunctive treatment of partial-onset seizures in patients with epilepsy. In an open-label, fixed-flexible dose study, 18 patients were treated with levetiracetam for 12 weeks. Of the 13 patients completing the study, 11 were rated "very much" or "much" improved on the CGI-I. For most patients, clinical benefits were apparent after only 1–2 weeks of treatment. The tolerance was good, with minimum side effects.

Antipsychotic Medications There is no evidence that conventional antipsychotic medications are effective in the treatment of uncomplicated panic disorder. Second-generation antipsychotics such as aripiprazole and olanzapine appear to be effective as an augmentation strategy in the treatment of SSRI-resistant panic disorder but in open-label studies or case reports only.

Recommendations and Conclusion

Unless otherwise indicated, an SSRI should be offered as a first-line treatment. When a medication is started, the efficacy and side effects should be reviewed within 2 weeks and again at 4, 6, and 12 weeks. At the end of 12 weeks, an assessment of the effectiveness of the treatment should be made and a decision taken to continue or consider an alternative intervention. If there is no improvement after a 12-week course of a first SSRI, another SSRI or an SNRI may be considered. SNRIs tend to be used as a second-line therapy after SSRIs fail to improve panic or in patients who cannot tolerate them. The tricyclics have disadvantages that make them second- or third-line treatment now. If there is no improvement after another 12-week course, an AD from an alternative class should be offered or a strategy adopted for augmenting the SSRI with a benzodiazepine, a different AD, or an atypical neuroleptic. There is no evidence that will allow the clinician to predict which of the three broad intervention groups (pharmacological, psychological, or self-help) will be effective for an individual patient, based on duration of illness, severity of illness, age, sex, gender, or ethnicity. In the same way, there is no evidence that will allow the clinician to predict which AD will be effective. Treatment choice depends on the patients' characteristics (such as previous response or contraindications), the evidence base supporting its use, and patient and physician preference. If the patient shows improvement on AD treatment, medication should be continued for at least 6 months after the optimal dose is reached, after which the dose can be tapered. There is no clear published evidence as to what is the optimal length of treatment with medication (NICE 2004).

The response rate to pharmacotherapy approaches 70 %, but many studies clearly show that discontinuation of medication results in relapse, with rates of 25–50 % recorded within 6 months. Optimal strategies for patients not responding to the first SSRI are not well documented with placebo-controlled studies, and the characteristics of treatment resistance in this condition still need to be clearly defined.

Cross-References

▶ Agoraphobia
▶ Antidepressants
▶ Anxiety: Animal Models
▶ Benzodiazepines

References

Bandelow B, Zohar J, Hollander E, Kasper S, Möller HJ, WFSBP Task Force on Treatment Guidelines for Anxiety, Obsessive-Compulsive Post Traumatic Stress Disorders (2008) World Federation of Societies of Biological Psychiatry (WFSBP) Guidelines for the pharmacological treatment of anxiety, obsessive-compulsive and post-traumatic stress disorders-first revision. World J Biol Psychiatry 9(4): 248–312

Bertani A, Perna G, Migliarese G, Di Pasquale D, Cucchi M, Caldirola D, Bellodi L (2004) Comparison of the treatment with paroxetine and reboxetine in panic disorder: a randomized, single-blind study. Pharmacopsychiatry 37(5):206–210

National Institute for Health and Clinical Excellence: Guidance (2004) Clinical guidelines for the management of anxiety: management of anxiety (panic disorder, with or without agoraphobia, and generalised anxiety disorder) in adults in primary, secondary and community care [Internet]. School of Health and Related Research (ScHARR), University of Sheffield. National Collaborating Centre for Primary Care (UK), London

Cox BJ, Endler NS, Lee PS, Swinson RP (1992) A meta-analysis of treatments for panic disorder with agoraphobia: imipramine, alprazolam, and in vivo exposure. J Behav Ther Exp Psychiatry 23(3):175–182

Kessler RC, Berglund P, Demler O, Jin R, Merikangas KR, Walters EE (2005) Prevalence, severity and comorbidity of 12 months DSM-IV disorders in the national comorbidity survey replication. Arch Gen Psychiatry 62(6):617–627

McIntosh A, Cohen A, Turnbull N, Esmonde L, Dennis P, Eatock J, Feetam C, Hague J, Hughes I, Kelly J, Kosky N, Lear G, Owens L, Ratcliffe J, Salkovskis P (2004) Clinical guidelines and evidence review for panic disorder and generalised anxiety disorder sheffield. University of Sheffield/National Collaborating Centre for Primary Care, London

Moroz G (2004) High-potency benzodiazepines: recent clinical results. J Clin Psychiatry 65(Suppl 5):13–18

Otto MW, Tuby KS, Gould RA, McLean RY, Pollack MH (2001) An effect-size analysis of the relative efficacy and tolerability of serotonin selective reuptake inhibitors for panic disorder. Am J Psychiatry 158(12):1989–1992

Pollack MH, Lepola U, Koponen H, Simon NM, Worthington JJ, Emilien G, Tzanis E, Salinas E, Whitaker T, Gao B (2007) A double blind study of the efficacy of venlafaxine extended-release, paroxetine and placebo in the treatment of panic disorder. Depress Anxiety 24(1):1–14

Ribeiro L, Busnello JV, Kauer-Sant'Anna M, Madruga M, Quevedo J, Busnello EA, Kapczinski F (2001) Mirtazapine versus fluoxetine in the treatment of panic disorder. Braz J Med Biol Res 34(10):1303–1307

Wagstaff AJ, Cheer SM, Matheson AJ, Ormond D, Goa KL (2002) Paroxetine – an update of its use in psychiatric disorders in adults. Drugs 62(4):655–703

PANSS

Synonyms

Positive and Negative Syndrome Scale (for schizophrenia)

Definition

PANSS is a scale for the measurement of positive, negative, and general schizophrenic symptoms.

Papaveretum

Definition

Papaveretum is a preparation containing three of the principal opium alkaloids: ▶ morphine, papaverine, and ▶ codeine. One of the first available analgesic formulas, papaveretum is now relatively uncommon due to the more widespread availability of single and synthetic opioids. Papaveretum continues to be used primarily as a preoperative sedative but also for moderate to severe pain. In comparison with morphine, papaveretum has fewer gastrointestinal side effects and little abuse potential as IV administration at low doses produces severe headaches in most individuals. An additional component, noscapine, was removed from the formula in 1993 due to potential genotoxic effects.

Cross-References

▶ Analgesics
▶ Opioids

Paradoxical Effects

Definition

Paradoxical effects refer to results from various measures which may lead to apparently contradictory conclusions. For example, drugs of abuse can condition both approach and avoidance behaviors. In such cases, the same drug (at the same dose and in the same animal) appears to have both rewarding and aversive effects.

Paralogism

Definition

Distorted argumentation.

Paramagnetic

Synonyms

Paramagnetism

Definition

Paramagnetic refers to having the property of being attracted to magnetic fields. It is a form of magnetism that occurs only in the presence of an externally applied magnetic field.

Cross-References

▶ Magnetic Resonance Imaging (Functional)

Parasomnias

Pandi-Perumal Seithikurippu Ratnas[1], Ahmed S. BaHammam[2] and Colin M. Shapiro[3]
[1]Center for Healthful Behavior Change (CHBC), Division of Health and Behavior, Department of Population Health, New York University Medical Center, Clinical & Translational Research Institute, New York, NY, USA
[2]University Sleep Disorders Center, College of Medicine, National Plan for Science and Technology, King Saud University, Riyadh, Saudi Arabia
[3]Department of Psychiatry, Toronto Western Hospital, University Health Network, University of Toronto, Sleep and Alertness Clinic Toronto, Toronto, ON, Canada

Synonyms

Abnormal swallowing syndrome; Arousal disorders; Behavioral modifications; Catathrenia; Confusional arousals; Exploding head syndrome; Forensic; Hypnic jerks; Jactatio capitis nocturna; Night terrors; Nightmares; Nocturnal bruxism; NREM; Pavor nocturnus; Rapid eye movement disorder; REM sleep; Rhythmic movement disorders; Sexsomnia; Sleep drunkenness; Sleep enuresis; Sleep hygiene; Sleep paralysis; Sleep terrors; Sleep-related dissociative disorder; Sleep-related eating disorder; Sleepwalking; Somnambulism; Somniloquy

Definition

The class of sleep disorders known as **parasomnias** (L. *para* = next to; *somnus* = sleep) includes some of the most unusual, challenging, fascinating, and potentially instructive of all behavioral disorders. These are clinical disorders characterized by acute, abnormal behavioral or physiological events, associated with sleep. Parasomnias may be unpleasant or undesirable events, which accompany sleep-specific sleep stages or sleep/wake transitions (American Academy of Sleep Medicine 2014; American Psychiatric Association 2000; Thorpy and Plazzi 2010). Parasomnias are associated with central nervous system (CNS) activation, increased skeletal muscle activity, and autonomic nervous system (ANS) changes. The immediate consequences of parasomnias include the possibility of sleep disruption and physical harm to affected individuals. Parasomnias can also lead to poor health and psychosocial consequences. Because parasomnias tend to run in families, it has long been suspected that genetic factors are involved.

Parasomnias, Table 1 Classification of parasomnias (Modification of American Academy of Sleep Medicine 2014)

Non-REM parasomnias (or disorders of arousals)
1. Confusional arousals (sleep drunkenness)
2. Sleepwalking (somnambulism)
3. Sleep terrors (*pavor nocturnus*, incubus attacks)
4. Sleep-related eating disorder
REM parasomnias
1. Recurrent isolated sleep paralysis
2. Nightmares disorder (terrifying dreams/anxiety dreams)
3. REM sleep behavior disorder
Other parasomnias
1. Exploding head syndrome
2. Hypnogely
3. Sleep-related hallucinations
4. Sleep enuresis
5. Sexsomnia
6. Parasomnia due to a medical disorder
7. Parasomnia due to a medication or substance
8. Parasomnia, unspecified (e.g., sleep driving and sleep texting)
Isolated symptoms and normal variants
1. Sleep talking

Clinical Diagnosis and Evaluation

A clinical diagnosis of parasomnias involves a thorough clinical history from the patient and the family. When performed, the **polysomnography** (PSG) assessment may include a more detailed montage to rule out other potential confounding factors (such as nocturnal seizures), and features of **rapid eye movement (REM)** behavior disorder should be noted.

We have had the experience of seeing patients (both adults and children) who have been reassured at another clinic that the problem was minor only to discover that the cause of the sleep complaint was epilepsy and that with appropriate treatment there was a dramatic change in the patients' well-being. In children this has changed academic performance considerably. In other areas of medicine, such an omission by specialists would be considered negligence at best and bordering on malpractice. Furthermore, in the last couple of years, it has become increasingly clear that parasomnia leads to excessive daytime sleepiness with all the attendant implications, i.e., parasomnia is not trivial.

Parasomnias can be broadly and conveniently classified based on the state from which they arise – events that occur during REM sleep, non-rapid eye movement (NREM) sleep, or both sleep states (Winkelman 2005; Zadra and Pilon 2011) (See Tables 1 and 3).

Non-REM (NREM) Parasomnias (or) Disorders of Arousals

Disorders of arousal during **NREM** sleep include the following: sleepwalking, sleep terrors, and confusional arousals. Thought to occur because of the instability of slow-wave sleep (SWS), arousal disorders are not presumed to be an all-or-none phenomenon, but rather a continuum of behaviors involving reestablishment of alertness, orientation, judgment, and self-control and/or a rapid alternation between sleep and waking states (Mahowald and Schenck 2005a).

Sleepwalking

Sleepwalking (somnambulism), which is initiated during SWS and generally occurs within the first one-third of the sleep period, consists of a series of complex, elaborate motor behaviors and results in walking during a state of altered consciousness. Sleepwalking occurs in 10–20 % of children and 1–4 % of adults and is most prevalent in children between 5 and 10 years of age. Sleepwalking events range from limited purposeless actions to elaborate behaviors. These events can include the unlocking of doors, dressing, and other complex behaviors. Typical episodes can last from 15 s to 30 min, with recall often being sketchy to absent the next day. During the somnambulistic state, there is an absence of dreaming. The affected individual's eyes are open, and behavior may be clumsy but maybe well coordinated. It may be bizarre or seemingly purposeful. When awakened, the sleepwalker typically responds with simple phrases and noninsightful mentation ("had to talk to John Doe"). Frequent sleepwalking in adults may be associated with violent or dangerous activity and warrants treatment. Sleepwalking patients are generally neurologically normal. The presence of other contributing sleep disorders should be investigated and ruled out prior to a final diagnosis.

This implies that a thorough rather than a cursory sleep assessment is merited if there is some concern. This has to be balanced with resource implications.

Sleep Terrors (Night Terrors, Pavor Nocturnus)

Sleep terrors are also characterized by arousals from SWS occurring in 5 % of children and 1–2 % of adults. A night terror starts with an incomplete arousal from SWS and is often associated with ANS activation. There may be reports of frightening imagery. A typical event might terrify the patient, who often emits an abrupt piercing scream or crying, apparent panic, and agitation. This is accompanied by autonomic and behavioral manifestations of intense fear. The events are often associated with accelerated heart rate and rapid breathing, agitation, sweating, hyperpnea, and tachycardia. In children they are usually resistant to reason or consolation. Indeed, the latter may even give rise to worse behavior or violent manifestations. Episodes typically occur in the first third of the night, with the patient having very little or no recall of the event the next morning. Witnesses tend to be more distressed by the events than the patient. Children who present with sleep terrors usually grow out of

Parasomnias, Table 2 List of safety suggestions for violent or disruptive parasomnias

Install alarm systems to alert when someone has left the room or house
Sleep in a separate bed from bed partner
Place mattress on floor
Remove obstructions from the room
Remove coat hooks from the door
Cover windows and glass doors with drapes
Lock doors with a double-cylinder lock; light outside hallways
Place gates at top of staircases and in doorways

Parasomnias, Table 3 General characteristics of NREM disorders of arousal and REM sleep behavior disorder (Adapted from Matwiyoff and Lee-Chiong 2010)

Characteristics	NREM disorders of arousal	REM sleep behavior disorder
Age of onset	Children, adolescence and young adults	Late middle age (mean age 59.3 year)
Clinical course	Usually benign and may decrease with age	May be harbinger of Parkinson's disease or other neurodegenerative diseases
Episode recall	Usually none	Often awakens completely awake
Complex motor activity	Yes	Yes
Sleep stage association	NREM first half of night	REM second half of night
Gender association	Male = female	Predominantly male (90 %)

them by the age of 14. Events may be precipitated by such stressors as alcohol use, psychological strain, sleep deprivation, and shift work. Adults presenting with sleep terrors should be assessed for comorbid psychiatric disorders.

Confusional Arousals

Similar to sleepwalking and sleep terrors, confusional arousals are brief and incomplete arousals that typically begin during SWS. The aroused individual typically looks confused.

A confusional arousal or behavior differs from sleepwalking in that the affected individual does not leave the bed. Inasmuch as the episodes generally follow arousals from SWS, they most commonly occur during the first one-third of the night. Common examples include sitting up in bed and making simple vocalizations or picking at bedclothes. The term sleep drunkenness has also been used to describe such confusions as this is notable by disorientation, impaired cognition, and behavioral disturbance. Confusingly sleep disturbances has also been used to describe disorientation after awakening especially after a short nap when the confusion may be the result of awakening from slow wave sleep. Approximately 10–20 % of children and 2–5 % of adults report a history of confusional arousals. Anxiety, sleep deprivation, fever, and endocrine factors (e.g., pregnancy) can increase the frequency of episodes. Other precipitating factors include the consumption of alcohol and/or other hypnotics, antihistamines, lithium, and potentially other medications that tend to elevate the arousal threshold. In adults, primary sleep disorders (e.g., apnea or periodic leg movement syndrome) can also exacerbate such conditions.

Sleep-Related Eating Disorder (SRED)

Sleep-related eating disorder involves partial arousals from sleep where patients engage in involuntary eating or drinking. Such episodes appeared to be triggered by a learned behavior and not real hunger or thirst. Patients have limited or no memory of the events, and foods consumed can consist of peculiar combinations or even inedible/toxic substances such as coffee grounds, cake mixes, frozen or uncooked products, eggshells or cleaning materials. Patients may have unexplained weight gain and morning anorexia. SRED has a higher prevalence in patients with a history of eating disorders or sleepwalking. Certain medications have also been reported to induce SRED (e.g., zolpidem, anticholinergics, and lithium). Patients should be evaluated for comorbid sleep disorders such as periodic limb movement disorder (PLMS) and obstructive sleep apnea (OSA). Topiramate and dopaminergic medications have been found to be effective for some patients.

Treatment of Arousal Disorders

Once an arousal disorder is assessed and properly diagnosed, a treatment plan can be developed. The structure and sequencing of the plan should take into account the nature and severity of the symptoms. Management of arousal disorders, especially in children, should primarily involve education. Although stress may be a factor in disorders of arousal, it should not be the clinician's first assumption.

It may be that as we learn more about the cognitive impact of parasomnias that a more proactive approach to management will be taken. This applies particularly in the case of children as the stakes in terms of future development are higher, and the sleep disruption they have often impacts other members of the household. There are potential treatments for children that are safe and acceptable.

Proper clinical management involves the recommendation of good sleep hygiene including refraining from alcohol and drugs. Further, the patient should keep a consistent sleep/wake schedule and take steps to reduce bedroom light and noise. Sleep deprivation should be avoided. The environment should be kept safe (e.g., mattress on the floor, locks on doors, use of alarms) when necessary. Scheduled awakenings (on a regular basis awakening the patient 15 min before each usual arousal) have been shown to be effective. Table 2 lists common safety recommendations. Stress management skills, psychotherapy, and hypnosis have also been found to be useful. Pharmacological treatment becomes necessary when arousal events are frequent, put the family or patient at a risk of being harmed, or disturb the family life. Benzodiazepines (BDZs, used continuously or as needed) and tricyclic antidepressants (TCAs) have been successfully used in treatment, though controlled trials are lacking.

Parasomnias Usually Associated with REM Sleep

REM Sleep Behavior Disorder

REM sleep behavior disorder (RBD) is characterized by vigorous motor activity, which typically occurs during REM sleep and thus in the absence of muscle atonia. RBD often consists of injurious dream-enactment motor activity associated with vivid dreaming. REM sleep behavior disorder is symptomatically complex, often involving behaviors that are dangerous to the affected patient, or to his/her bed partner, and vivid dream imagery that is almost always unpleasant. REM behavior disorder is more frequent in males and usually occurs in the middle-aged or elderly. Although the exact prevalence of RBD in the general population is unknown, it is estimated to be 0.5 % (Ohayon et al. 1997). Patients exhibiting such complaints should seek a thorough evaluation by a sleep specialist. RBD can be idiopathic or related to underlying neurological conditions (e.g., synucleinopathies). It is thus advisable that patients undergo a thorough physical examination for the purpose of identifying any comorbidity that might interrupt REM sleep (Ferini-Strambi et al. 2005). Acute RBD can be induced by medications (monoamine oxidase (MAO) inhibitors, serotonin reuptake inhibitors, and tricyclic antidepressants), alcohol, and BDZ withdrawal.

As was suggested for disorders of arousal, recommendations for therapy should emphasize the safety of the patient and his/her bed partner as paramount considerations. The patient and his/her bed partner should sleep in separate beds until events are controlled. Patients usually respond well to clonazepam (0.25–2.0 mg) or temazepam (15–45 mg). Melatonin, levodopa, and pramipexole have also been reported to be useful in some cases. Clinical subtypes of RBD include (a) subclinical RBD (increased EMG activity in REM sleep without clinical RBD behaviors), (b) parasomnia overlap disorder (patients demonstrate features of both NREM parasomnias and RBD and are usually younger than patients with idiopathic RBD), and (c) status dissociatus (an extreme form of parasomnia overlap disorder where features of NREM and REM sleep, and occasionally wakefulness, coexist).

Sleep Paralysis

Sleep paralysis occurs when there is a persistence of REM muscle atonia into wakefulness. Although eye movements and respiratory activity remain intact, somatic movements are not possible. Episodes of sleep paralysis may often accompany a residual dream-related fear or with anxiety when a patient realizes that movement is not possible. Although the episodes may last one to several minutes, they often can be terminated by an external stimulus (e.g., touch). Patients with infrequent episodes should refrain from sleep deprivation and alcohol, but further treatment (beyond reassurance) is generally not required. For recurrent episodes, evaluation for depression or other psychiatric conditions with the potential for producing sleep disturbance is warranted. The REM suppressing effect of fluoxetine particularly appears to be helpful.

Nightmares

Nightmares are frightening dreams that can awaken the sleeper from REM sleep. Although they may occur at any part of the night, they predominantly occur during the final one-third of the night where REM sleep is more prominent. Recall of a nightmare is generally vivid and detail oriented. Nightmares are often associated with predominant emotions such as fear or anxiety; however, feelings of anger, sadness, and embarrassment may also occur. Although more commonly seen in children, adults also report having nightmares, with stress and antidepressant or antihypertensive medications often being implicated as causal factors. Nightmare disorder, the experience of recurrent nightmares, can be idiopathic or related to an underlying condition such as psychological trauma or

Parasomnias, Table 4 Comparison of sleep disorders in children (Adopted from Driver and Shapiro 1993)

	Dreams	Nightmares	Sleepwalking	Sleep terrors
Sleep stage	Light non-REM and stage R sleep	Stage R sleep	Stage N3 non-REM sleep	Stage N3 non-REM sleep
Time after went to sleep (h)	3–6	3–6	1–2	1–2
Sounds	None	Occasional unintelligible sounds	Occasional meaningless speech	Scream ± continuous loud meaningless speech
Motor movement	Little or none	Little until point of waking	Usually purposeful and unpredictable; child rarely stays in bed or room	Purposeless movement; child usually stays in bed
Response to parent	Awake easily to stimuli	Awakes easily to stimuli; reorients in several minutes	Little to none	Little to none
Memory of event	Can describe immediately	Can describe immediately; often able to remember event following day	None	None

psychopathology (e.g., posttraumatic stress disorder, affective disorders). Frequent nightmares can lead to disturbed sleep onset latency, awakenings, restless sleep, insomnia, and a lower overall quality of life. Imagery rehearsal therapy has shown promising results in the treatment of nightmares. Patients are taught to "change the dream any way (he/she) wishes" and imagine the new dream for 5–20 min daily (Krakow et al. 2001). Though more commonly used to treat hypertension, prazosin has also been found effective as a routine treatment (Lancee et al. 2008; Table 4).

Other Parasomnias

A host of other parasomnias, many of which are related to medical and/or psychiatric dysfunctions, have been described in the literature. Some require specific medical intervention, whereas others only require reassurance.

Sexsomnia

Sexsomnia (somnambulistic sexual behavior) is characterized by abnormal sexual behaviors during sleep (SBS), with patients having little to no memory of the event (Shapiro et al. 2003). Sleep-related abnormal sexual behaviors are primarily classified as confusional arousals in that they typically occur without any behaviors outside of the bed and have also been less commonly associated with sleepwalking. The characteristic features of sexsomnia include sexual arousal accompanied by autonomic activation (e.g., nocturnal penile tumescence, vaginal lubrication, nocturnal emission, and dream orgasm; Andersen et al. 2007). Patients who have a history of sleepwalking and comorbid sleep disorders should be evaluated (Guilleminault et al. 2002; Zaharna et al. 2008). Patients are advised to obtain adequate sleep on a regular basis. Patients should also be advised that although sexsomnia occurs with a regular bed partner (e.g., spouse or lover), it can occur with anyone. It is therefore important to alert to the potentially catastrophic impact if a child/stepchild enters the bed of a person who has the behavior. In the state of a lack of awareness, the person with sexsomnia will not appreciate that this is not the regular partner. The ensuing ramifications are obvious.

Sleep-Related Dissociative Disorders

Sleep-related dissociative disorders are commonly divided into three categories: dissociative identity disorder, dissociative fugue, and dissociative disorder not otherwise specified. During these states (which can vary in duration from minutes to many hours), patients commonly have disturbances in consciousness, memory identity, and environmental awareness. Nocturnal reenactments range in severity and include episodes such as binging on high-caloric sweets, acting out past physical and sexual abuse, driving an automobile, or eating uncooked foods. Patients generally lack memory of the event.

Patients often have histories of psychopathology or report past experiences of sexual, physical, and/or emotional abuse. A full psychological evaluation is recommended, and no single treatment approach is accepted as the most effective in this population. A combined pharmacotherapy and psychotherapeutic approach is likely to be most effective. Treatment approaches most likely to succeed should focus on any underlying psychiatric disorders. Pharmacotherapy should similarly address the potential presence of depression and/or anxiety. Psychotherapy should be directed at tension management and reduction of stimuli that tend to provoke dissociative experience.

Some of these conditions, e.g., sleep eating or sexsomnia, can be viewed as variants of sleepwalking and respond to similar treatments.

Sleep Enuresis

Sleep enuresis is characterized by recurrent involuntary voiding that occurs during sleep. Although bed-wetting is normal in young children, it becomes pathological when it occurs at least twice weekly during sleep in patients who are 5 years of age or older. Patients with primary enuresis, those who have never been consistently dry during sleep, may have a neurological impairment or a deficient release of vasopressin. Secondary enuresis refers to those individuals who have had periods of remaining dry during sleep for at least 6 months duration. Enuresis has been linked to a number of factors including urinary tract infection, diabetes mellitus, epilepsy and other neurological disorders, and psychological stress. Treatment recommendations include reassurance and, especially in children, positive reinforcement for achieving goals. Patients should refrain from taking liquids late in the evening. Behavioral conditioning treatments such as bell and pad alarms have been found useful. Desmopressin has been used in children older than 6 years to reduce urine production. The recommended dosage is one 0.2 mg tablet at bedtime for 1 week. The dosage can be increased as per the direction of physician. Tricyclic antidepressants (desipramine or imipramine) have also been found to be beneficial for short-term management.

Catathrenia (Sleep-Related Groaning) and Hypnogely

Catathrenia (Greek: *kata* = low; *threnia* = lament) is a rare form of parasomnia (Siddiqui et al. 2008). Often most disturbing to bed partners or family members, a typical episode is characterized by a prolonged groan, which is accompanied with an expiration of oronasal flow. In the new ICSD-3 edition, catathrenia is included in the sleep-related breathing disorders (SRBD) section because it appears to be associated with prolonged expiration. Although catathrenia episodes predominantly occur during REM sleep in the later part of the night, they can in some cases occur during NREM sleep (Siddiqui et al. 2008). Evaluation should include a PSG to rule out any comorbid sleep disorders (e.g., OSA). Catathrenia is a rare condition that is more prevalent among males and not known to be associated with any apparent medical and/or psychiatric illnesses. Catathrenia is asymptomatic, and physical examination and sleep architecture are generally normal. There is too little evidence to indicate if it responds to pharmacotherapy. Reassurance is traditionally the treatment of choice. Most of the above comments apply to hypnogely (a pathological sleep laughing), which is related to dreaming and REM sleep (Trajanovic et al. 2013).

Exploding Head Syndrome

Patients suffering from what has been termed as exploding head syndrome report being awakened by the sensation of a loud, explosion-like bursting sound in the head. The syndrome appears to be more common among women than in men. Subjects may describe having a nonobjective auditory experience, with intensities ranging from a painless loud bang to more subtle loud sounds. Typically, these experiences occur just as the individual is falling asleep. Although the experience is usually painless, patients occasionally note a small jab of pain with the sound. Although the frequency of incidence varies, the subjective explosions are often exacerbated by fatigue, stress, and/or sleep deprivation and tend to diminish over time. Treatment recommendations emphasize reassuring the patient that these events are benign in nature. Occasionally clonazepam is used and in some individual needs to be in an increasing dose in time to maintain efficacy.

Sleeptalking (Somniloquy)

Sleeptalking refers to the utterance of speech or sounds during sleep, with varying degrees of comprehensibility, without simultaneous subjective detailed awareness of the event. Occurring during stage 2, SWS, or REM sleep, sleeptalking is the most common during the first half of the night. Episodes are exacerbated by stress, new medications, acute medical illness, or a comorbid sleep disorder. Simple sleeptalking generally does not require intervention unless another sleep disorder is supposed. If treatment is required, patients are counseled to follow proper sleep hygiene and reduce exacerbating factors such as alcohol and sedatives.

Forensic Implications

Violent behaviors may occur during the sleep period and can occur without the conscious awareness by affected individuals. These sleep related behaviors, which have been reported in 2 % of the population, can occasionally have significant forensic implications (Mahowald and Schenck 2005b). Prominent legal cases involving murder, assault, or apparent suicide have occasionally been linked to disorders of arousal, RBD, psychogenic dissociative states, or sleep-related seizures. As an appreciation of such linkages is being increasingly acknowledged, sleep medicine clinicians are becoming more frequently involved in these cases as expert witnesses. It is thus in their professional interest to become informed of the medical and legal implications, where assumption of intentionality in aberrant behavior is involved. Further study is necessary to understand the true prevalence of these disorders, how to best diagnose and treat them, and how to protect others around the patient.

Summary

Parasomnias are often considered as the most fascinating category of sleep disorders. The associated phenomena, often distressing to those who experience them, have been well described behaviorally but remain poorly understood in terms of their mechanisms. Parasomnias are common nocturnal events that can occur at any stage of sleep. Often a full or extended montage EEG setup accompanied by video-audio PSG is necessary for proper diagnosis of this disorder. Genetic epidemiological studies reveal that the parasomnias often run in families. While genetic susceptibility of some parasomnias has been confirmed, not all variants of the disorder demonstrate a clear linkage. Further research into the interplay between gene-environment interactions and other potential environmental causes needs to be undertaken to understand the exact nature of such disorders (Hublin and Kaprio 2003).

The clinician and patient should work together to tailor the treatment to each patient and to reduce the risk of personal injury, to increase the safety of family members and society at large, and to increase the patient's quality of life. In addition to pharmacotherapy, incorporating sleep hygiene instructions along with behavioral modification techniques often found to be successful. Given the implications of sleep disorders, in general, and parasomnias, in particular, appropriate knowledge transfer for societal and legal framework needs to be developed. Finally, as there are medicolegal issues associated with violent parasomnias, public awareness as well as policy initiatives from the government is warranted.

Cross-References

▶ Benzodiazepines
▶ Delirium
▶ Hypnotics
▶ Insomnias
▶ Modafinil
▶ Sedative, Hypnotic, and Anxiolytic Dependence

References

American Academy of Sleep Medicine (2014) ICSD-3 – international classification of sleep disorders, diagnostic and coding manual, 3rd edn. American Academy of Sleep Medicine, Westchester

American Psychiatric Association (2000) Diagnostic and statistical manual of mental disorders. Text revision, 4th edn. APA Press, Washington, DC

Andersen ML, Poyares D, Alves RSC, Skomro R, Tufik S (2007) Sexsomnia: abnormal sexual behavior during sleep. Brain Res Rev 56(2):271–282

Driver HS, Shapiro CM (1993) ABC of sleep disorders – parasomnias. BMJ 306:921–924

Ferini-Strambi L, Fantini ML, Zucconi M (2005) REM sleep behaviour disorder. Neurol Sci 26(3):186–192

Guilleminault C, Moscovitch A, Yuen K, Poyares D (2002) Atypical sexual behavior during sleep. Psychosom Med 64:328–336

Hublin C, Kaprio J (2003) Genetic aspects and genetic epidemiology of parasomnia. Sleep Med Rev 7(5):413–421

Krakow B, Hollifield M, Johnston L et al (2001) Imagery rehearsal therapy for chronic nightmares in sexual assault survivors with posttraumatic stress disorder: a randomized controlled trial. JAMA 286(5):537–545

Lancee J, Spoormaker VI, Krakow B, van den Bout J (2008) A systematic review of cognitive-behavioral treatment for nightmares: toward a well-established treatment. J Clin Sleep Med 4:475–480

Mahowald MW, Schenck CH (2005a) Non-rapid eye movement sleep parasomnias. Neurol Clin 23:1077–1106

Mahowald MW, Schenck CH (2005b) Violent parasomnias: forensic medicine issues. In: Kryger MH, Roth T, Dement C (eds) Principles and practice of sleep medicine, 4th edn. Elsevier/Saunders, Philadelphia, pp 960–968

Matwiyoff G, Lee-Chiong T (2010) Parasomnias: an overview. Indian J Med Res 131:333–337

Ohayon MM, Caulet M, Priest RG (1997) Violent behavior during sleep. J Clin Psychiatry 58:369–376

Shapiro CM, Trajanovic NN, Fedoroff JP (2003) Sexsomnia – a new parasomnia? Can J Psychiatry 48:311–317

Siddiqui F, Walters AS, Chokroverty S (2008) Catathrenia: a rare parasomnia which may mimic central sleep apnea on polysomnogram. Sleep Med 9:460–461

Thorpy MJ, Plazzi G (eds) (2010) The parasomnias and other sleep-related movement disorders. Cambridge University Press, Cambridge, UK, pp 1–341

Trajanovic NN, Shapiro CM, Milovanovic S (2013) Sleep-laughing- hypnogely. Can J Neurol Sci 40(4):536–539

Winkelman JW (2005) Parasomnias. In: Buysse DJ (ed) Sleep disorders and psychiatry, vol 24, Reviews of psychiatry. American Psychiatric Publishing, Washington, DC, pp 163–183

Zadra A, Pilon M (2011) Non-rapid eye movement (NREM) parasomnias. In: Montagna P, Chokroverty S (eds) Handbook of clinical neurology, vol 99. Elsevier, Amsterdam, pp 851–868

Zaharna M, Budur K, Noffsinger S (2008) Sexual behavior during sleep: convenient alibi or parasomnia. Curr Psychiatry 7(7):21–30

Pargyline

Synonyms

Methylbenzylpropynylamine; N-Methyl-N-2-propynylbenzylamine

Definition

Pargyline is an irreversible inhibitor of monoamine oxidase used for the treatment of depression with selectivity for MAO-B. It is less effective than ▶ tricyclic antidepressants and its interactions with dietary amines lead to serious toxicity. For example, foods such as cheeses that contain tyramine must not be consumed. It can also interact dangerously with other drugs including ▶ tricyclic antidepressants. Its use is now very limited, mainly to cases of depression that do not respond to newer antidepressants such as the inhibitors of the reuptake of serotonin and norepinephrine. Other side effects include hypotension, sedation, and weight gain.

Cross-References

▶ Antidepressants
▶ Monoamine Oxidase Inhibitors

Parkinson's Disease

Synonyms

PD

Definition

A neurodegenerative disease of adulthood and aging, characterized by neuropathology, including Lewy body formation and cell death, in the dopaminergic neurons of the *substantia nigra* in the ventral mesencephalon; dopamine denervation in the striatal forebrain targets of these neurons (in particular, the caudate nucleus and putamen); the resulting syndrome is characterized by motor symptoms of ▶ akinesia, difficulty in initiating movements, rigidity, resting tremor, and cognitive impairments in a significant subset of patients. Parkinson's disease is most usually modeled in experimental animals by neurotoxin-induced lesions of the midbrain dopamine neurons.

Paroxetine

Synonyms

Aropax; Paxil; Seroxat

Definition

Paroxetine is an antidepressant that inhibits potently and specifically serotonin reuptake (SSRI). It was introduced in 1992. It is mainly used in the treatment of major depression, ▶ social anxiety disorder or social phobia, ▶ generalized anxiety disorder, ▶ post-traumatic stress disorder, ▶ panic disorder, ▶ obsessive-compulsive disorder, and sometimes of migraine. It is claimed to be as effective as older ▶ tricyclic antidepressants but with a more favorable side-effect profile. Its major side effects are nausea, sleep disturbances, decreased ▶ libido, and possible weight gain. The most serious concern with paroxetine treatment is the emerging risk for suicidal ideation and behavior in some adolescents and adults that led to the recommendation not to prescribe it for children. Debate continues about the seriousness of symptoms upon discontinuation of paroxetine.

Cross-References

- ▶ Antidepressants
- ▶ SSRIs and Related Compounds
- ▶ Tricyclic Antidepressants

Partial Agonist

Definition

Partial agonists bind to and activate a receptor, but are not able to elicit the maximum possible response that is produced by full agonists. The maximum response produced by a partial agonist is called its intrinsic activity and may be expressed on a percentage scale where a full agonist produced a 100 % response. A key property of partial agonists is that they display both agonistic and antagonistic effects. In the presence of a *full agonist*, a *partial agonist* will act as an antagonist, competing with the *full agonist* and thereby reducing its ability to produce its maximum effect. The balance of activity between agonist and antagonist effects varies from one substance to another, according to their intrinsic activities, and is also influenced by the test system used to measure their effects. Weak partial agonists are those compounds, possessing low intrinsic activity, that are able to produce only a small percentage of the total response produced by an agonist and which act predominantly as antagonists. Strong partial agonists may come close to mimicking the maximum effects of a full agonist and may display only weak antagonistic ability. When the test system under study has a large receptor reserve, weak partial agonists show greater agonist activity then when the system has a small receptor reserve.

Cross-References

▶ Agonist
▶ Antagonist
▶ Inverse Agonists

Passive Avoidance

Sven Ove Ögren[1] and Oliver Stiedl[2]
[1]Department of Neuroscience, Karolinska Institutet, Stockholm, Sweden
[2]Department of Functional Genomics and Department of Molecular and Cellular Neurobiology, Center for Neurogenomics and Cognitive Research, VU University Amsterdam, Amsterdam, The Netherlands

Principles and Roles in Psychopharmacology

Theoretical Background

Analyses of learning experiments demonstrated very early that conditioning promotes the association of certain relationships between particular external events, e.g., stimuli and/or responses. Psychological learning theory distinguishes between classical conditioning and instrumental aversive (or appetitive) conditioning based on the different ways (contingencies) by which the aversive event (the unconditioned stimulus; US) is related to the *conditioned stimulus (CS)* and the *unconditioned response (UCR)*. Instrumental (aversive) conditioning refers to contingencies in which the behavior of the subject determines whether or not the *unconditioned stimulus (US)* will occur (Ögren 1985). In active avoidance paradigms, the animal has to perform a discrete response of a low probability, e.g., running from one side of a two-compartment box to the other when a discrete stimulus, e.g., the conditioned stimulus (CS), is presented in order to escape or avoid the US. In the *passive avoidance* task, the animal learns to suppress a motor response to avoid exposure to the test area (context) associated with or predictive of the aversive event, such as a dark compartment of the passive avoidance system that is normally preferred over the brightly illuminated compartment. A conflict situation is created in which the behavioral responses are analyzed. Analyses of avoidance learning indicate that it involves different processes. Initially, presentation of the US results in a learned emotional state (e.g., conditioned *fear*) involving Pavlovian classical conditioning followed by the acquisition of the discrete, adaptive behavioral responses, the escape or avoidance response. The adaptive response requires for its association the temporal contiguity between the *US* and the sensory stimulus or context used as the *CS*. More recent theories on aversive learning have focused on the cognitive processes by which the animal acquires information about its experimental context. The cognitive interpretation of passive avoidance suggests that this paradigm is based on place learning involving the hippocampus (see below).

Passive Avoidance Tasks

There exist several versions of the passive avoidance task such as step-down or step-through avoidance. Passive avoidance is on associative learning task (Ögren 1985; Ögren et al. 2008) similar to contextual fear conditioning (reviewed by LeDoux 2000). Unlike active avoidance tests, passive avoidance tests usually do not use an explicit CS. Instead, the training context serves as the CS. Changes in the training context or internal stimuli, e.g., time of testing, will affect *retention* performance.

Step-Through Passive Avoidance (Inhibitory Avoidance)

The step-through task is a one-trial emotional memory task combining fear conditioning with an instrumental response, e.g., the active choice of an animal to avoid entering the dark compartment associated with an aversive event (Ögren 1985). The passive avoidance task differs from the typical fear conditioning experiment in which training and testing occurs in a one-compartment box. In fear conditioning, an animal placed in the compartment is exposed to one or multiple foot shocks (shock intensities 0.5–0.8 mA). Memory *retention* is measured as the degree of active suppression of motor behavior quantified on the basis of freezing.

The typical step-through passive avoidance test is conducted in a two-compartment box, one bright compartment and one dark compartment, connected by a sliding door. During training the subject is placed in the bright compartment and will after a defined time interval gain access to the dark compartment. When entering the dark compartment (training latency), the door will be closed and the subject will be subjected to a brief aversive stimulus (unconditioned stimulus, US; foot shock intensities in mice around 0.2–0.7 mA) that will lead to the formation of an association of the dark compartment with the US. For the *retention* test (usually 24 h after training), the animal is returned to the bright compartment with the sliding door open. The animal has now the option to avoid or enter the dark compartment by discriminating between the bright (safe) compartment and the dark (unsafe) compartment. A typical example of the experimental sequence of passive avoidance learning is presented in Fig. 1.

The rapid acquisition of not making a response indicates that the test involves learned inhibition rather than loss of an innate response tendency. Usually one trial, i.e., one exposure to the inescapable shock, is sufficient to suppress the innate preference of rodents for the dark compartment of the apparatus. The acquisition of passive avoidance is either measured as a significant increase of the step-through latency compared to the training latency or associated with reduced time spent in the dark compartment by the subjects. Thus, the passive avoidance procedure has the advantage of simplicity in that both the safe and the noxious compartments are clearly defined as well as the instrumental (adaptive) response that is to refrain from entering the dark compartment.

Step-Down Passive Avoidance
In this version of passive avoidance, the animal is placed on an elevated platform from which it can step down onto the floor (shock grid) below. When stepping down, the animal will be subjected to one or several aversive stimuli (US; foot shock) that will lead to the formation of an association between the US and the training context (CS) resulting in avoidance or delay to step down when returned to the platform in the *retention* test. Unlike the step-through passive avoidance procedure, the *retention* test is performed in the original (punished) context similar to fear conditioning. In this task the compartment size is critical, i.e., the space available for exploration in the one-compartment box. It is recommended to offer sufficient space for exploration instead of having systems with minimal platform (compartment) size from which animals instantly

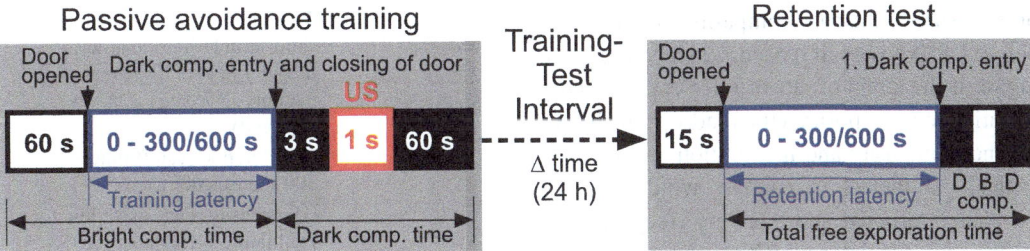

Passive Avoidance, Fig. 1 Experimental sequences for passive avoidance experiments with its training and retention test including information on the duration of subintervals as used in previous studies. The *second vertical arrows* in training and retention test indicate when the instrumental response occurs, i.e., the active choice of the animal to move from the bright into the dark compartment. Training and retention latency refer to the time interval between placement of the animal in the bright compartment and its entry of the dark compartment. The retention or memory test is usually performed 24 h after training. *US* unconditioned stimulus (foot shock), *comp.* compartment (Modified from Ögren et al. 2008)

step down (transfer) when moving. Step-down passive avoidance tasks are more sensitive to changes in locomotor activity than step-through passive avoidance tasks particularly if the platform is small. Drug-induced or lesion-induced alterations in locomotor activity may confound the actual avoidance response. Since the majority of passive avoidance investigations are based on the use of step-through avoidance paradigms, the following sections will focus on this task.

Procedural and Experimental Parameters

Quite a range of different procedures and experimental designs are described in the literature. Procedural differences have considerable consequences for passive avoidance performance and variability of results. Therefore, important procedural aspects are summarized in Table 1. Different passive avoidance systems exist that are customized or available from different vendors. It is important that they fulfill certain hardware and software requirements for flexible use.

Several preexperimental procedures have considerable effects on the results. Handling by the experimenter has been shown to reduce variations between animals in the passive avoidance task probably by lowering the adverse effects of acute stress (Madjid et al. 2006). Another procedural modification, which lowers performance variation, is to allow animals to explore both compartments for a defined time (e.g., 1 min) 1 h before the actual passive avoidance training. However, this procedure may result in *latent inhibition*, e.g., reduced transfer latency, if the preexposure to the dark compartment is too long before US exposure, if it occurs 24 h before training, or if the strain of rats or mice is particularly sensitive to preexposure (Baarendse et al. 2008).

Shock intensity and its duration are the most critical parameters in passive avoidance. The avoidance response shows a very steep increase in the transfer latency with increasing shock intensity in C57BL/6J mice (Fig. 2). Even small variations in intensity can markedly affect passive avoidance performance. It is recommendable to determine shock thresholds before the actual experiments, since there exists considerable variation in different species and strains.

Passive Avoidance, Table 1 Important experimental features in passive avoidance experiments

Apparatus	Compartment sizes (larger compartments allow for movements irrespective of lack of transfer)
	Door size and visibility (larger doors promote faster transfer during training)
	Light intensity differences between compartments (should be profound, e.g., 10–100-fold difference)
	Door mechanism (a guillotine door falling on mice when not completely in the dark compartment may act as US)
	Shocker with sufficient sensitivity (ideally at least 0.05 mA increments) and range (e.g., 0–1 mA)
Preexperimental procedures	Housing (single or group) and transfer (short versus long) to the experimental room (arousal affecting the actual training and testing)
	Preexposure(s) before training, timing of preexposure (to reduce data variability), risk of latent inhibition
	Handling (to reduce variability in training latencies)
	Timing of drug treatment (pre- versus posttraining, pretest, state-dependent learning)
	Repetitive testing of mice or rats in other behavior tests before PA testing is very likely to reduce the PA transfer latencies similarly as observed for reduced freezing[a]
Experimental procedures	Timing of retention testing (from short-term [1 h] to long-term memory [24–48 h]; remote memory [>7 days] is hardly explored)
	Test duration (cutoff time, e.g., 300 s versus 600 s)
	US intensity range used: low for facilitation versus high for impairment
	Current: constant versus scrambled shock (generally scrambled shocks are more effective)
	Extinction tests based on repetitive testing (with/without forced

(continued)

		exposure to the dark compartment)[a]
		Waiting/delay time before access to the dark compartment is provided (e.g., 60 s during training, 15 s during testing)[a]
		Delay between dark compartment entry and US exposure to avoid escaping to the bright compartment[a]
		Group housing may attenuate PA performance compared to individual housing since social support can counteract adverse experience[a]
Experimental parameters		Transfer latency detection (full transition when the animal is in the dark compartment with all 4 ft)
		Transitions between compartments (detection based on center of gravity may record transitions in mice during stretch-attend postures while exploring the door and is not ideal for automated analysis)
		Total time spent per compartment
		US response assessment (important to avoid misinterpretations based on altered nociception)
		Activity measurements (some hypoactive and neophobic mouse strains such as A/J are unsuited for PA experiments despite implying to be good learners; therefore, it is sometimes recommendable to include non-shocked controls)
		Time of testing (fluctuations in passive avoidance response throughout the circadian cycle have been reported in rats)
Statistical considerations		Group sizes (normally $n = 8$/group is minimum, ideally $n = 12$–16/group)
		Parametric (ANOVA) and nonparametric data analyses (Kruskal-Wallis and Mann-Whitney U test; Cox regression) depending on the normal distribution of individual data [group variance] or skewed data (because of maximum latency cutoff)

[a]These parameters have not been systematically investigated

It is possible to study both enhancement and impairment of passive avoidance memory by using proper shock intensities. To identify facilitatory effects on drugs on passive avoidance *retention*, a mildly aversive electrical current is used, whereas a strong current is employed to explicitly study impairment of passive avoidance *retention*. This procedure provides a sufficiently dynamic range for impairment and blockade of impairment when combining agonist and antagonist treatment (Fig. 3). An alternative approach is to increase the latency cutoff, e.g., from commonly 300–600 s, to have a sufficient dynamic range, or to combine it with preexposure ("▶ Latent Inhibition") to lower retention latencies.

The literature on passive avoidance indicates that there is large variation in retention latencies indicating considerable interindividual variability. Unless pretraining procedures are used to reduce variation (see above), a large group size (generally $n = 12$–16/group) is required to achieve sufficient statistical power. Under such conditions the statistical analysis may have to be performed with nonparametric analysis of variance (ANOVA) using the Kruskal-Wallis test followed by the Mann-Whitney U test for pairwise comparisons, particularly if the data is not normally distributed or responses exceed maximum transfer latencies (cutoff latencies). Recently, an improved statistical method has been described based on the cumulative incidence of escape (Jahn-Eimermacher et al. 2011).

The way the animals are handled just after training can also affect the *retention* latency. In mice, it is important that the animals remain in the dark compartment of the passive avoidance apparatus for at least 30 s before being picked up and placed in their holding or home cage. By this procedure, the animals have sufficient time to process and associate the training context with the US. In fear conditioning, exposure to the shock instantly after placement in the conditioning context leads to the "immediate shock deficit" in rats and mice, i.e., the absence of contextual conditioning. Thus, contextual information requires sufficient processing time to form an aversive association indicating the importance

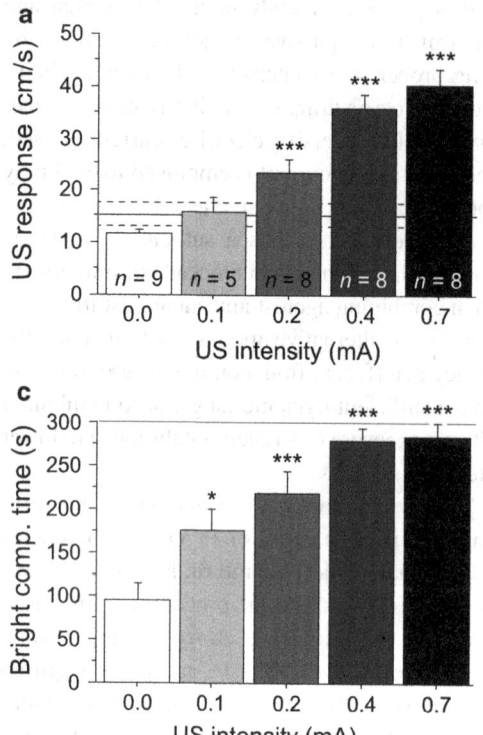
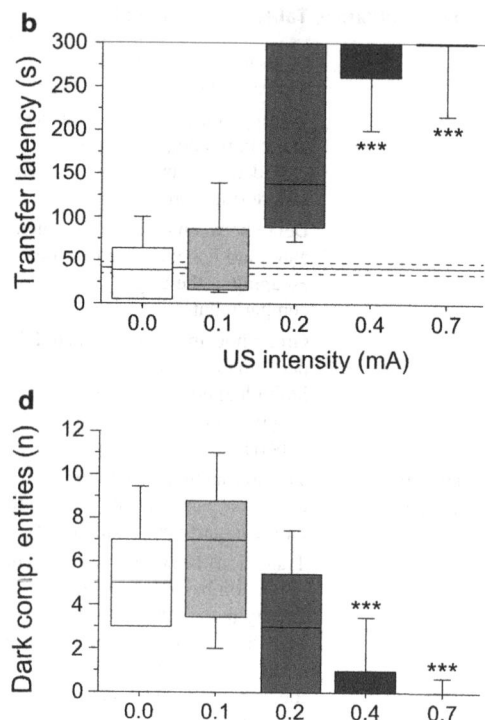

Passive Avoidance, Fig. 2 Passive avoidance performance in male C57BL/6J mice as a function of US intensity using multiple measures. US responses (activity) were measured during training and compared to baseline activity (**a**). In the retention test, transfer latencies (**b**), the total time spent in the bright compartment (**c**), and number of entries into the dark compartment (**d**) were analyzed as a function of US intensity (US duration, 2 s). The basal exploratory activity in the initial 180-s period of exploration in the light compartment is indicated by the *black horizontal line* (*dotted lines*, ± SEM) in panel (**a**). The *dashed horizontal line* (*dotted lines*, ± SEM) in panel (**b**) indicates the mean training latency. *Gray horizontal lines* at 300 s in panels (**b**) and (**c**) denote the cutoff time in the retention test. Normally distributed data (**a, c**) are shown as means + SEM, whereas non-normally distributed data are shown as box plots (**b, d**). *comp.* compartment, $*p < 0.05$ and $***p < 0.001$ versus 0.0 mA control group (based on Mann-Whitney U test for panel **b** and **d**)

of the temporal relation of CS and US for associative learning. The 30-s delay after US exposure also prevents the animals from forming an aversive association of the US with the handling procedure (removal from the compartment after the US exposure).

Neural Systems

The nature of contextual stimuli, which are critical for associative learning of passive avoidance, is not well characterized. Passive avoidance, similar to contextual fear conditioning, depends on multisensory associations rather than unisensory associations. The relative roles of sensory cues representing the CS are not defined at present. Sensory afferents, required for the detection of the different stimuli provided in the test environment, range from tactile to visual cues, to nociceptive pain receptors, and possibly also to olfactory cues. Besides processing in the thalamus (except for olfactory information), higher brain centers are involved in *encoding*, *consolidation*, and *extinction* mechanisms. The amygdala, the hippocampus, and the various cortical areas are part of the neural network that subserves passive avoidance

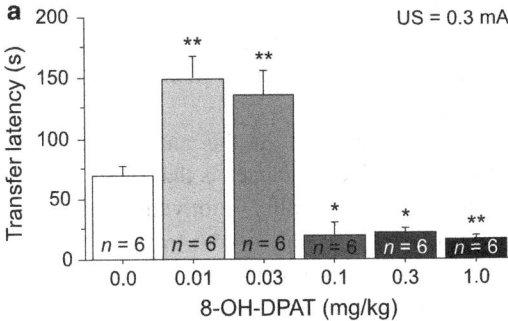
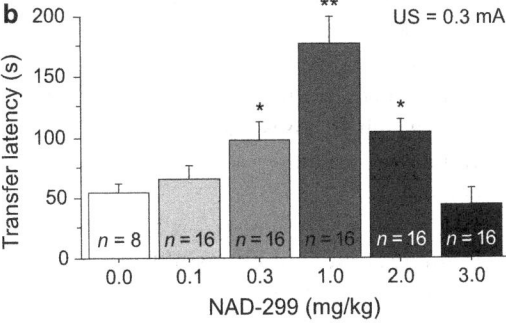
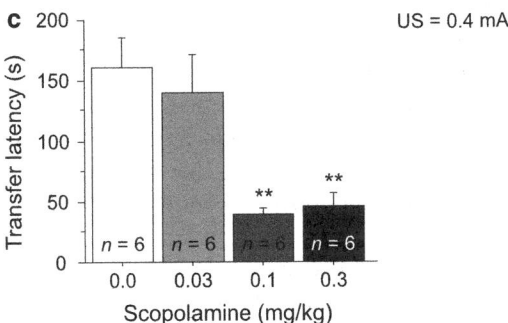

Passive Avoidance, Fig. 3 Passive avoidance performance based on transfer latencies in the retention test in male NMRI mice as a function of drug treatment using different shock (US) intensities during training as indicated. The full 5-HT$_{1A}$ agonist 8-OH-DPAT (**a**) facilitates at the low dose range (0.01–0.03 mg/kg) and impairs at high dose range ($\geq$0.1 mg/kg). The selective 5-HT$_{1A}$ antagonist NAD-299 (**b**) facilitates passive avoidance retention latencies in the dose range of 0.3–2 mg/kg. The muscarinic acetylcholine receptor antagonist scopolamine (**c**) impairs passive avoidance retention at 0.1 and 0.3 mg/kg. US duration was 2 s. Retention latency cutoff was 300 s. All drugs were injected either subcutaneously 15 min (**a**, **b**) or 40 min before training (**c**). Error bars indicate SEM; *$p < 0.05$ and **$p < 0.01$ versus 0.0 mg/kg control group (Modified from Madjid et al. 2006)

learning (Burwell et al. 2004; McGaugh 2004; Baarendse et al. 2008; Ögren et al. 2008), but the specific circuitry components are not known.

A number of studies have shown that passive avoidance depends on hippocampal function and its NMDA receptors. Infusion of the NMDA receptor antagonists AP5 into the dorsal hippocampus of mice profoundly impairs passive avoidance *retention* (Baarendse et al. 2008). Similarly, neurotoxic lesions of the corticohippocampal circuitry (perirhinal, postrhinal, and entorhinal cortices) cause profound deficits in passive avoidance learning in rats (Burwell et al. 2004). These results indicate that the passive avoidance task requires processing of spatial information about the test environment. However, the passive avoidance task differs from spatial tests such as the Morris water maze with regard to both sensitivity to parahippocampal lesions and modulation of cholinergic mechanisms, suggesting that these two tasks differ in their dependence on processing of polymodal sensory information (Burwell et al. 2004; Ögren et al. 2008). Finally, some mouse strains may acquire passive avoidance, although suboptimally, by using a hippocampus-independent strategy (see Baarendse et al. 2008).

The amygdala, a heavily innervated assembly of many different subnuclei, is essential for aversive learning (LeDoux 2000; McGaugh 2004). Among these subnuclei, the basolateral nucleus plays a major role in the convergence of CS and US for associative learning. The basolateral nucleus of the amygdala is connected to the central nucleus of the amygdala. Outputs from the central nucleus of the amygdala are essentially responsible for the expression of fear responses. This network triggers behavioral adjustments indicative of learning and memory through motor control and concomitant autonomic and endocrine adjustments via efferent pathways. The efferent pathways are specific for certain response domains subserving both learned and innate responses. The roles of these brain areas which include hypothalamic and various brain stem areas will not be discussed here.

Application of Passive Avoidance in Neuropsychopharmacology

Passive avoidance is one of the most frequently used animal tests to study learning and memory mechanisms and to identify compounds modifying cognitive processes. This task is often a first-line test in most pharmacological companies. The task has also a prominent role in neuroscience research focusing on the role of neurotransmitters and molecular signals in learning and memory processes. The major multiple neurochemical systems and some of its molecular components have been characterized to modulate this type of learning (Ögren et al. 2008). Research in the 1970–1980s showed a significant role of cholinergic transmission in storage and retrieval of information. The nonselective muscarinic receptor antagonist scopolamine, when injected before passive avoidance training, causes a dose-dependent impairment of passive avoidance *retention* (Bartus et al. 1982). Cholinomimetic drugs, such as the acetylcholine esterase inhibitor physostigmine, partially antagonize the deficit caused by scopolamine. These findings formed part of the cholinergic hypothesis of geriatric memory dysfunction (Bartus et al. 1982) that provided the rationale for introducing cholinesterase inhibitors as antidementia drugs. A number of studies have also demonstrated a significant role for brain serotonin in passive avoidance learning. Both increases and decreases in brain 5-HT transmission have resulted in passive avoidance deficits (Ögren et al. 2008) reflecting the involvement of multiple 5-HT receptors in this task (Misane and Ögren 2000). Through the use of 5-HT receptor subtype-specific ligands, such as selective $5-HT_{1A}$ receptor agonists and antagonists, the role of 5-HT transmission in passive avoidance has been more specifically investigated (see Fig. 3). With passive avoidance as the major behavioral test, it has been possible to distinguish the modulatory action of pre- and postsynaptic $5-HT_{1A}$ receptors in cognitive function (Ögren et al. 2008). Based on these findings, the $5-HT_{1A}$ receptors have emerged as an important target for drugs acting in psychopathologies characterized by mood disorders and disturbances in emotional memory (Ögren et al. 2008). In contrast, $5-HT_7$ receptor activation appears to facilitate passive avoidance retention involving a broader corticolimbic network (Eriksson et al. 2012).

A major problem in the analysis of drug effects in passive avoidance is that drug-induced changes in emotionality, motivation, or brain chemistry may be part of the training context. State-dependent learning refers to the situations in which *retention* is poor for rodents when the drug state during learning and *retention* differs. This means that information acquired under the drug state can only be retrieved when the animal is in the same drug state. Analyses of state-dependent learning can be an important tool by which to assess nonspecific effects on performance from associative learning mechanisms.

Advantages and Limitations of Passive Avoidance Experiments

The advantage of the passive avoidance procedure is that it is a single-trial task and it produces predictive results in both mice and rats. Unlike multi-trial tasks, single-trial tasks allow for a precise timing of experimental manipulations, i.e., drug injection either before or after training or before the *retention* test. Therefore, it is possible to dissect the possible contributions of drug interventions on *encoding*, *consolidation*, and *retention* (memory retrieval and expression). Since *retention* tests can be performed at variable time intervals after training, short-term memory (STM) and long-term memory (LTM) can be assessed separately. Posttraining administration provides the opportunity for studying effects on memory consolidation and retention-related administration on reconsolidation mechanisms. However, since subsequent tests are confounded by prior experiences, an animal may only be tested once.

There are also drawbacks with this one-trial task including large intersubject variability (discussed above) and variable shock sensitivity. The memory strength ("▶ Retention") is influenced by many factors including the mental state at the time of learning, emotional reactivity, and responses to stress hormones (McGaugh 2004).

Experimental manipulations that alter locomotion, pain perception, or anxiety states can confound measures of memory. Information from these behavioral domains is important to rule out alternative explanations rather than altered cognitive function(s).

Attenuated pain perception (analgesia) by pharmacological or genetic manipulations may render a low electric current ineffective as US (negative reinforcer), thereby preventing passive avoidance learning. Knowledge about the US responsiveness (Baarendse et al. 2008) is crucial to rule out passive avoidance impairments because of lack of US perception. Since US exposure occurs in the dark compartment, quantification is hardly possible unless the passive avoidance system can detect the activity of animals in the dark compartment. Photo beams that track through the black walls or infrared cameras for video tracking can be used. At least the US reactions (e.g., vocalization, jumping) of the animals should be assessed to assure a proper US response.

There is often a general assumption that enhanced anxiety-like behavior is associated with increased fear (avoidance) responses. Although this cannot be ruled out, there is no clear evidence for this assumption (Madjid et al. 2006). On the contrary, studies using different mouse strains indicate that DBA/2J mice, which are more anxious than C57BL/6J mice, show lower passive avoidance transfer latencies than C57BL/6J mice (Baarendse et al. 2008). In conclusion, despite some limitations, passive avoidance paradigms are very useful in analyzing the neurochemical and molecular basis of cognition.

Perspective

Passive avoidance tests have been used in research for more than six decades. The success of this behavior test is due to the fact that it exploits adaptive behavioral mechanisms critical for survival that have been shaped by evolution. Moreover, the passive avoidance task is very flexible and allows for investigation of a large number of cognitive domains in neuropsychopharmacology and related disciplines. Yet, the full potential of this task in studying memory mechanisms is still not always well perceived in the research community. The advent of advanced passive avoidance systems will allow a much more penetrating analysis of domains of cognition with relevance for psychopathology such as *latent inhibition* and *extinction* of passive avoidance responses. In addition, the integration of additional measures such as neural responses and autonomic adjustments in automated passive avoidance systems without human intervention will open up new avenues for an integrative analysis of avoidance behavior and risk assessment as core endophenotypes of many human psychiatric disorders. A new home cage-based approach exploiting deliberate choice between a safe compartment and an unsafe compartment is currently developed and validated. This approach without human interference allows long-term monitoring (hours to days) of risk assessment and avoidance, clinically relevant endophenotypes. Computerized and refined passive avoidance tasks will certainly have an important place in future neuroscience research and drug development.

Summary

This chapter provides an overview on passive avoidance learning in rats and mice with examples how to use this instrumental learning task for experimental modulation based on studies aiming at the analysis of mechanisms underlying impairment and facilitation of fear memory through characterizing the involvement of brain neurotransmitter systems and downstream signaling. This task offers a wide range of parameters that can be varied, many of which have not yet been systematically investigated. Advantages and disadvantages of the passive avoidance procedure are discussed with the focus on basic motivational and reinforcement variables. In conclusion, passive avoidance approaches offer important novel possibilities for research with translational aspects of significance ranging from basic neuroscience to clinical neuropsychopharmacology.

Cross-References

► Amygdala
► Anxiety
► Conditioned Fear
► Conditioned Stimulus
► Consolidation and Reconsolidation
► Fear Conditioning
► Hippocampus
► Latent Inhibition
► Pavlovian Fear Conditioning
► Punishment Procedures

References

Baarendse PJJ, van Grootheest G, Jansen RF, Pieneman AW, Ögren SO, Verhage M, Stiedl O (2008) Differential involvement of the dorsal hippocampus in passive avoidance in C57BL/6J and DBA/2J mice. Hippocampus 18:11–19

Bartus RT, Dean RL 3rd, Beer B, Lippa AS (1982) The cholinergic hypothesis of geriatric memory dysfunction. Science 217:408–414

Burwell RD, Saddoris MP, Bucci DJ, Wiig KA (2004) Corticohippocampal contributions to spatial and contextual learning. J Neurosci 24:3826–3236

Eriksson TE, Holst S, Stan TL, Hager T, Sjögren B, Ögren SO, Svenningsson P, Stiedl O (2012) 5-HT_{1A} and 5-HT_7 receptor crosstalk in the regulation of emotional memory: implications for effects of selective serotonin reuptake inhibitors. Neuropharmacology 63:1150–1160

Jahn-Eimermacher A, Lasarzik I, Raber J (2011) Statistical analysis of latency outcomes in behavioral experiments. Behav Brain Res 221:271–275

LeDoux JE (2000) Emotion circuits in the brain. Annu Rev Neurosci 23:155–184

Madjid O, Elvander Tottie E, Lüttgen M, Meister B, Sandin J, Kuzmin A, Stiedl O, Ögren SO (2006) 5-HT_{1A} receptor blockade facilitates aversive learning in mice: Interactions with cholinergic and glutamatergic mechanisms. J Pharmacol Exp Ther 316:581–591

McGaugh JL (2004) The amygdala modulates the consolidation of memories of emotionally arousing experiences. Annu Rev Neurosci 27:1–28

Misane I, Ögren SO (2000) Multiple 5-HT receptors in passive avoidance: comparative studies of p-chloroamphetamine and 8-OH-DPAT. Neuropsychopharmacology 22:168–190

Ögren SO (1985) Evidence for a role of brain serotonergic neurotransmission in avoidance learning. Acta Physiol Scand 125(Suppl 544):1–71

Ögren SO, Eriksson TM, Elvander-Tottie E, D'Addario C, Ekström JC, Svenningsson P, Meister B, Kehr J, Stiedl O (2008) The role of 5-HT_{1A} receptors in learning and memory. Behav Brain Res 195:54–77

Passive Immunization

Definition

Administration of purified drug-specific antibodies produced in some other species (e.g., rabbits, mice) or in vitro. Primary advantages of this approach are the ability to achieve necessary serum concentrations of antibody virtually immediately and precise control of the antibody dose to allow study of dose-response relationships and the effects of high antibody concentrations that can not be achieved via vaccination. Main disadvantages of passive immunization are that it requires relatively frequent injections to maintain antibody levels and is relatively expensive compared to vaccination.

Patch Clamp

Definition

Patch-clamp recording denotes a variety of electrophysiological recording modalities that exploit the property of a glass recording electrode to form an electrically tight seal (patch) against the cell membrane.

The patch-clamp technique can be used to study the opening and closing of single or multiple ion channels in mainly excitable cells. Measurements require a tight seal (gigaohm seal) between a glass pipette, filled with an intracellular solution and a Ag/AgCl wire, and the membrane of a target cell. Various approaches can be used to yield differential data. For example, whole-cell patch clamping is used to measure the activity of ion channels across the entire surface of a cell whereas cell-attached, inside-out, or outside-out patches can be used to monitor either single channel activity or massed activity of the section of membrane attached to the electrode. Inside-out patches are particularly useful for studying how cytosolic events affect ion channel activity.

Cross-References

► Intracellular Recording

Pathological Gambling

Christine A. Franco and Marc N. Potenza
Department of Psychiatry, Yale University
School of Medicine, New Haven, CT, USA

Synonyms

Compulsive gambling; Gambling addiction; Impulsive-compulsive gambling; Problem gambling

Definition

Pathological gambling (PG) is classified as an impulse control disorder (ICD) characterized by persistent and recurrent maladaptive gambling behavior and resulting in impaired social and/or occupational functioning (DSM-IV-TR; American Psychiatric Association 2000). An essential feature of ICDs is the diminished ability to resist drives or urges to perform behaviors that may be harmful; consistently, individuals with PG frequently score high on measures of impulsivity. While classified as an ICD, PG shares many similarities with substance dependence. Of the ten diagnostic criteria that characterize PG, four resemble symptoms commonly observed in substance dependence including an intense preoccupation with gambling, repeated or unsuccessful attempts to stop or cut down, and aspects of tolerance and withdrawal (American Psychiatric Association 2000). Other similarities between PG and substance use disorders include clinical characteristics (e.g., higher prevalence rates in men compared to women) and courses (e.g., earlier age of onset in men, higher rates in adolescence, and lower rates in elderly adults) and an observed telescoping phenomenon in women (i.e., shorter duration between onset of gambling and the report of a gambling problem in women compared to men). Additionally, like substance dependence, individuals with pathological gambling experience cycles of abstinence and relapse. In this sense, some have likened gambling to a nonchemical or "behavioral" addiction.

Role of Pharmacotherapy

Both psychotherapeutic and pharmacological treatments have been investigated in the treatment of PG. Effective psychotherapies have utilized behavioral (e.g., aversion therapy, in vivo and imaginal exposure, skills training) and/or cognitive (e.g., cognitive restructuring, problem solving) and motivational (e.g., motivational enhancement) techniques to address maladaptive gambling behaviors and thought patterns (Brewer et al. 2008). As with studies of psychotherapies, relatively few systematic, controlled, and adequately powered trials have investigated pharmacotherapies for PG.

Axis I and Axis II disorders often co-occur with PG (Ibanez et al. 2001); therefore, it is clinically relevant to consider treatment approaches based on types of co-occurring disorders (e.g., alcohol or drug use disorders, bipolar disorders, anxiety disorders, or major depression). Given the proposed mechanisms of action of specific psychotherapies and pharmacotherapies, it is also important to consider specific neurobiological aspects of PG (e.g., opioid system, serotonin system) when considering treatments. This chapter will focus on findings from placebo-controlled studies and cite references that review these and other studies. For additional information about specific studies, readers are encouraged to see the original sources cited in the review articles.

Opioid System

Endogenous opioids and opioid receptors in the brain may mediate reward, urges, and hedonia. Controlled studies of opiate antagonists such as naltrexone and nalmefene have yielded particularly promising results. Opiate antagonists bind to opioid receptors, influencing the release of gamma-aminobutyric acid (GABA) and subsequent dopaminergic neurotransmission from the ventral tegmental area (VTA) to the nucleus accumbens (NA) and the ventral pallidum. In this way, opioid antagonists are hypothesized to modulate the tension and/or pleasure that individuals with addictive and impulsive behaviors report feeling and are believed to help suppress urges that may lead to impulsive behaviors.

Naltrexone. Naltrexone, a long-acting opioid receptor antagonist, has been investigated in the treatment of disorders where urges are indicated as a predominant symptom including bulimia nervosa, ICDs, disorders with associated self-injurious behaviors, and various substance use disorders including alcohol and opioid dependence. Findings from such studies remain mixed, as naltrexone has not demonstrated consistent efficacy in reducing urges across all disorders. However, the success of naltrexone in treating addictive disorders (including FDA approval for the treatments of opioid and alcohol dependence) has led to the investigations of PG (Brewer et al. 2008).

In a 12-week double-blind, placebo-controlled trial, 83 individuals with PG underwent a weeklong, single-blind placebo lead-in and were randomly assigned to receive a flexible dose of either naltrexone (mean end-of-study dose of 188 mg/day) or placebo (Brewer et al. 2008). Thirty-eight individuals were removed from the analyses for multiple factors (e.g., placebo responders, intolerable side effects, inability to keep study schedule) leaving 45 participants. Individuals who received naltrexone showed significant improvement on all three measures of gambling compared to those who received the placebo. By study end, 75 % of individuals in the active naltrexone group, compared to 24 % of those in the placebo group, reported significant improvement in gambling symptoms, although there was also a significant improvement from baseline within the placebo group. The findings from the study are limited by the frequency of placebo response, dropout related to elevations in liver enzymes, and the exclusion of individuals with co-occurring Axis I disorders.

A second, 18-week, double-blind, placebo-controlled trial of naltrexone was conducted to replicate and extend earlier findings (Grant and Potenza 2008). Following a weeklong placebo lead-in, 77 individuals with PG were randomly assigned to one of three doses of naltrexone (50, 100, or 150 mg/day) or a placebo. Based upon naltrexone's proposed mechanism of action and previous findings suggesting it reduces self-reported urges to gamble, only participants who reported significant gambling urges were included in the study. While there were no significant differences in response between doses of naltrexone, naltrexone treatment compared to placebo was associated with reduced self-reported gambling urges, longer abstinence, and improved psychosocial functioning. Placebo response was also observed, although response in this group was substantially lower than in the naltrexone group. The findings were limited by the study's short-term duration. That is, while the present study employed longer treatment duration of 18 weeks, it did not assess treatment effects past this point. Despite this limitation, the findings suggest naltrexone may be useful in treating acute PG, particularly among individuals who report significant urges to gamble. In addition, the findings suggest that a high dose of naltrexone targeted in earlier studies might not be essential to observe clinical response.

Thus far, naltrexone is the only medication having shown efficacy in two or more double-blind, controlled studies. The studies support the safety and tolerability of naltrexone in the short-term treatment of PG. However, clinicians should remain aware of its dose-dependent hepatotoxicity (Grant and Potenza 2008). Continued research via large-scale, multicenter, controlled clinical trials would help to better clarify the utility of naltrexone in the treatment of PG.

Nalmefene. Nalmefene is a long-acting opioid receptor antagonist that shares proposed mechanisms of action with naltrexone and, like naltrexone, has been investigated in the treatment of alcohol dependence. Unlike naltrexone, nalmefene has not shown dose-dependent hepatotoxicity. To date, one multicenter double-blind, placebo-controlled trial has examined the efficacy and tolerability of nalmefene in the treatment of PG (Brewer et al. 2008). Two hundred and seven PG subjects were randomly assigned to one of three doses of nalmefene (25, 50, or 100 mg/day) or placebo and followed for 16 weeks. Individuals who received nalmefene demonstrated improvement across a range of gambling measures, and this difference was greater than changes observed in individuals receiving placebo. The findings were limited by high frequencies of subject dropout and placebo response. Despite these limitations, the findings suggest nalmefene may be an effective pharmacologic option for the short-term treatment of PG. While its lack of hepatotoxicity makes it an appealing alternative to naltrexone, clinicians should remain aware of the reported adverse effects of nalmefene (e.g., nausea, insomnia, dizziness), particularly at higher doses, and its limited availability due to limited indications for which it is approved.

As individual differences have been associated with opioid antagonist-treatment response in alcoholism, outcome data from nalmefene and naltrexone trials were analyzed (Grant and Potenza 2008). A positive family history of alcoholism and strong gambling urges were positively associated with response to opioid antagonists. Younger age was positively associated with placebo response. Taken together, these studies support a role for opioid antagonists in treating PG. However, more controlled studies are needed, particularly ones of longer duration and in conjunction with behavioral therapies.

Serotonin System
Serotonin systems have been implicated in various impulse control and compulsive disorders (e.g., PG, kleptomania, compulsive shopping, compulsive skin picking) with studies suggesting deficits in serotonergic functioning being associated with increased impulsivity (Brewer and Potenza 2008; Williams and Potenza 2008). Thus, serotonergic medications have been examined in treating PG.

Clomipramine. Clomipramine is a serotonin reuptake inhibitor (SRI). In an early case study with a 31-year-old female gambler, treatment with clomipramine was associated with a reduction in gambling symptoms and gambling behavior (Kim and Grant 2001). To date, no further clinical investigations of clomipramine in the treatment of PG have been published. As such, precise conclusions about its efficacy in treating PG cannot be drawn.

Fluvoxamine. The efficacy and tolerability of fluvoxamine, a selective SRI (SSRI), was initially examined in the treatment of 16 individuals with PG (Brewer et al. 2008). Participants entered an 8-week placebo lead-in phase, followed by an 8-week single-blind trial of fluvoxamine. Of the original 16 participants, 10 completed the medication trial with 7 of these considered treatment responders. Treatment with fluvoxamine (mean end-of-study active dose 220 mg/day) was associated with reductions in gambling-related thoughts and behaviors. In two of the three nonresponders (both with histories of cycling mood disorders), treatment with fluvoxamine appeared to exacerbate gambling behavior. The findings are limited by a small sample size, an open-label design, and thus the possible placebo effects.

A subsequent study involving 15 patients with PG further investigated the efficacy and tolerability of fluvoxamine (Brewer et al. 2008; Kim and Grant 2001). Individuals entered a 1-week placebo lead-in followed by a 16-week crossover trial of fluvoxamine and placebo with ten participants completing the trial. Treatment with fluvoxamine (mean end-of-study active dose of 195 mg/day) yielded greater reductions in gambling-related thoughts and behaviors as compared to placebo. A significant drug-by-phase interaction was observed in which differences between fluvoxamine and placebo were observed in the second but not in the first phase of the crossover trial, consistent with a more robust placebo effect in the first phase. The study was limited by a small sample size, short treatment duration, and the placebo effects.

In a pilot study investigating the efficacy of fluvoxamine treatment, 32 individuals with PG were enrolled in a randomized, double-blind, placebo-controlled study of fluvoxamine for a duration of 6 months (Grant and Potenza 2007). There were no observed differences in treatment outcomes between the fluvoxamine-treated and placebo-treated groups. Similar to other trials of fluvoxamine, the findings were limited by placebo response and a small sample size, and the latter factor may have precluded the identification of between-group differences.

A comparison study investigated the effectiveness of fluvoxamine to topiramate (Iancu et al. 2008). Topiramate is a medication hypothesized to mediate its therapeutic actions via influencing the mesolimbic dopamine system via GABAergic mechanisms and has shown promise in treating disorders characterized by poor impulse control. Thirty-one men with PG were randomly assigned to receive 200 mg/day of either topiramate or fluvoxamine for 12 weeks. Results demonstrated treatment completers in both groups responded positively to treatment, with 9/12 individuals in the topiramate group and 6/8 individuals in the fluvoxamine group reporting full remission of symptoms at the end of treatment. Greater dropout was observed in the fluvoxamine group relative to the topiramate group. Additionally, improvement in scores on the Clinical Global Impression (CGI) scale reached significance for individuals who received topiramate but not for those who received fluvoxamine. These findings were limited by a lack of a placebo-controlled group, raters unblinded to treatment conditions, and a small sample consisting solely of men. Nonetheless, the findings suggest that topiramate is well tolerated and may prove useful in the treatment of PG.

Paroxetine. Paroxetine, another SSRI, has been investigated as a possible treatment for PG. In a double-blind, placebo-controlled study examining the efficacy and tolerability of paroxetine treatment in PG (Brewer et al. 2008), subjects were entered into a weeklong placebo lead-in and then assigned to receive paroxetine or placebo for 8 weeks. Results revealed a greater reduction in gambling symptoms in the paroxetine group as compared to the placebo group; however, the results are limited by a small sample size with unequal numbers of men and women and a relatively short treatment course of 8 weeks. Still, the findings provided preliminary evidence to suggest paroxetine may be effective in the treatment of PG.

A larger 16-week, double-blind, placebo-controlled study was conducted across several sites in two countries to further examine the efficacy of paroxetine (Brewer et al. 2008). Seventy-six individuals with PG were entered into a weeklong placebo lead-in followed by randomization to paroxetine or placebo treatment. Although a numerically larger percentage of paroxetine-treated patients were clinical responders as compared to placebo-treated subjects (59 % vs. 48 %, respectively, at 16 weeks), the difference did not reach statistical significance. While the findings did not replicate those of the earlier study, perhaps related to the substantial placebo response, the results suggest paroxetine is well tolerated.

As with opioid antagonists, individual differences may be important to consider for SRIs. Specific subgroups of individuals (e.g., those with depression or anxiety disorders) might respond preferentially to SRIs, and this possibility warrants further investigation. A subsequent analysis of individuals involved in the earlier-mentioned paroxetine trial investigated the relationship between measures of impulsivity and compulsivity with respect to treatment outcome. Although both impulsivity and compulsivity measures were associated with gambling severity at treatment onset, changes in gambling symptomatology correlated with changes in impulsivity during the course of treatment. No relationship was observed with respect to paroxetine treatment. These findings suggest that treatments that target impulsivity may be worth pursuing further in the treatment of PG.

Sertraline. To date, only one controlled trial of sertraline, another SSRI, has been conducted (Brewer et al. 2008). Sixty individuals with PG were treated in a double-blind, flexible-dose, placebo-controlled study for 6 months. Improvement was observed in both the sertraline and the placebo groups, with no statistically significant differences in outcome measures between the two groups at trial end. These findings were limited by a high placebo response and a relatively small sample.

Escitalopram. Escitalopram, another SSRI, is FDA approved for the treatment of major depression and anxiety disorders. As PG and anxiety disorders often co-occur, escitalopram has been investigated in the treatment of individuals with PG and co-occurring anxiety disorders. Thirteen individuals were enrolled in an open-label 12-week trial of escitalopram (Brewer et al. 2008; Grant and Potenza 2007, 2008).

Approximately 62 % of individuals who completed the study experienced a 30 % or greater reduction in both gambling and anxiety symptoms and were considered treatment responders. These individuals were then enrolled in an 8-week double-blind discontinuation. While no statistically significant worsening of symptoms was reported for those who had received escitalopram, the individual randomized to placebo reported a return of gambling and anxiety symptoms within 4 weeks.

While a preliminary pilot, these data demonstrate an association between open-label escitalopram and improvements on reported gambling and anxiety symptoms and quality of life and suggest both gambling and anxiety may be addressed simultaneously during a course of pharmacologic treatment. The study is limited by a small sample size, lack of a placebo-controlled group in the initial phase, lack of blinding in the initial phase, and a small number of individuals participating in the double-blind discontinuation phase.

In a recent open-label trial, 19 individuals with PG were enrolled in a 10-week course of escitalopram following a 2-week observation period (Black et al. 2007). Improvement in gambling symptom severity, urges, and behaviors and decreases in money wagered and time spent gambling were observed. The drug was well tolerated with no dropouts attributed to side effects. The limitations of the study include a small sample size, lack of a placebo group, lack of blinding, and short treatment trial duration.

Nefazodone. Nefazodone, primarily a serotonin receptor antagonist, exhibits characteristics of a mixed noradrenergic/serotonergic reuptake inhibitor (NSRI). Fourteen PG subjects were enrolled in an 8-week open-trial flexible-dose investigation of orally administered nefazodone (Grant and Potenza 2007). Of the 12 completers, 75 % were considered responders with significant reductions in episodes of gambling and time and money spent gambling per week. Limitations include lack of both a placebo-controlled group and double blinding.

Overall, findings from studies investigating medications that target serotonin systems are mixed, generating a complex picture. While case studies and open-trial investigations have demonstrated support for the utility of SSRIs and serotonin receptor antagonists, larger controlled trials have been less consistent. The findings suggest that SSRIs and serotonin antagonists may be effective for specific groups of PG subjects. More extensive and targeted research with careful characterization of subject groups (e.g., defined by co-occurring disorders) or specific individual differences (e.g., readily measurable clinical information in self-report, behavioral, and genetic domains) is necessary to better clarify the role of serotonergic drugs in the short- and long-term treatment of individuals with PG.

Mood Stabilizers

Bipolar disorder and PG share characteristics including impulsive behavior, risk taking, fluctuations in mood, and poor judgment (Iancu et al. 2008). Additionally, PG often co-occurs with bipolar disorders. Should the impulsivity in mania and related disorders on a bipolar spectrum possess similar underlying neurobiological mechanisms as the urges in PG, then drugs effective in the treatment of bipolar disorder may be helpful for individuals with PG.

Lithium. A medication typically used to treat manic states and mood lability associated with bipolar disorder, lithium has been investigated in the treatment of PG. An early set of case studies investigated the effectiveness of lithium carbonate to treat three individuals with compulsive gambling and found lithium to be well tolerated (Brewer et al. 2008). Findings suggest lithium treatment may have blunted the reported thrill individuals experienced when gambling and decreased the frequency of gambling. The results are considerably limited by the nature of the case studies, providing only preliminary support for the continued investigation of lithium to treat PG.

To date, only one double-blind, placebo-controlled trial of lithium has been conducted (Brewer et al. 2008). Forty individuals with PG and bipolar spectrum disorders (predominantly bipolar II) were randomly assigned to receive sustained-release lithium or placebo for

10 weeks. Although no significant between-treatment differences were observed between groups on several self-reported gambling behaviors (number of gambling episodes per week, time spent gambling per episode, or money lost gambling), greater improvement was observed for individuals receiving lithium on measures of gambling-related thoughts and urges compared to individuals receiving placebo. Decreases in measures of mania over time paralleled decreases in measures of gambling. The findings were limited by the relatively short treatment duration and a small sample size. Nonetheless, the study supports the tolerability and efficacy of lithium in the short-term treatment of PG and co-occurring bipolar spectrum disorders. It also highlights the need for researchers to identify clinically relevant subgroups of individuals with PG.

Valproate. Valproate is an anticonvulsant and mood stabilizer that has been used in the treatments of bipolar disorder and epilepsy. Based upon the success of valproate and other mood stabilizers to treat disorders with impulsive features (e.g., aggressive behavior, kleptomania, trichotillomania, and borderline personality disorder), it has been explored in the treatment of PG.

In a single-blind trial, 42 individuals with PG were randomly assigned to receive lithium or valproate for 14 weeks (Brewer et al. 2008). Both treatment groups demonstrated improvement on gambling-related measures and no significant between-group differences by the end of treatment. While limited by a lack of a placebo-controlled group, a small sample size, and single-blind design, the findings suggest valproate may work as well as lithium in diminishing gambling-related symptoms in PG.

Olanzapine. Traditionally used as an antipsychotic, olanzapine demonstrates mood-stabilizing properties, possibly mediated through dopamine and serotonin receptor antagonism, and thus has been investigated as a possible treatment for PG. Support for its proposed use in PG stems from its efficacy in the treatment of other ICDs such as trichotillomania and skin picking. Two controlled trials of olanzapine to treat PG have been conducted.

In a 12-week, double-blind, placebo-controlled trial, 42 individuals with PG were randomly assigned to receive olanzapine or placebo (Grant and Potenza 2008). Treatment with olanzapine was similarly efficacious as placebo. Olanzapine treatment was associated with discontinuation. Limitations included a relatively small sample and short-term duration. A 7-week, double-blind, placebo-controlled trial of olanzapine was conducted with a group of 21 individuals with video-poker-based PG (Grant and Potenza 2008). As with the other investigations, no significant between-group differences in outcome were observed. The study was limited by a small sample, consistently, solely of video-poker gamblers and the exclusion of individuals with co-occurring psychiatric disorders. Findings suggest that olanzapine may not be any more effective than placebo in the treatment of PG.

Carbamazepine. Carbamazepine is a mood stabilizer that is used to treat bipolar disorder and has shown some efficacy in treating ICDs. Support for the investigation of carbamazepine to treat PG comes from two sources: the proposed relationship between PG and bipolar disorder and the implication that carbamazepine may dampen the impulsivity proposed to underlie PG (Black et al. 2008). A 10-week open-label study examined extended-release carbamazepine in a sample of eight individuals with PG and found the majority to be treatment responders with a reduction in gambling severity; almost half reported abstaining from gambling during the final month of treatment. The findings suggest the possible efficacy of extended-release carbamazepine in the treatment of PG, but should be interpreted with caution based upon the small sample size, the lack of blinding and a placebo-controlled group, and considerable treatment dropout.

Overall, studies investigating the effectiveness of mood stabilizers in the treatment of PG suggest some medications may reduce preoccupations with gambling, frequency of gambling, and amount of money wagered on gambling; however, not all mood stabilizers have demonstrated consistent efficacy. Therefore, it is important for research to evaluate further the efficacies and tolerabilities of specific mood stabilizers in the treatment of well-defined groups of PG subjects such as those with co-occurring bipolar disorders.

Glutamatergic Agents

N-Acetyl cysteine. Studies have suggested that extracellular glutamate within the nucleus accumbens may mediate reward-seeking behaviors by decreasing cravings. *N*-Acetyl cysteine (NAC), a glutamate-modulating agent, is proposed to influence extracellular levels of glutamate leading to a stimulation of the inhibitory glutamate receptors. NAC has been found to reduce reward seeking in studies of cocaine-dependent rats and craving in a study of individuals with cocaine addiction.

To date, only one study has examined the potential of NAC in the treatment of PG (Grant and Potenza 2008). Twenty-seven individuals with pathological gambling were entered into an 8-week open-label trial of NAC with treatment responders completing a 6-week randomized, double-blind discontinuation phase. Gambling symptoms improved in the majority of treatment completers during the treatment phase, with a trend toward a significant difference between individuals in the active versus the placebo group at the end of the discontinuation phase (response percentages of 83.3 and 28.6 % for the active drug and placebo groups, respectively). These findings suggest that there may be an effect that is attributable to the administration of the active drug, although the lack of a placebo-controlled group in the initial phase limits the interpretation of the results. Additionally, the study was limited by a small sample size. The results preliminarily suggest modulation of the glutamate system may play an important role in the reward-seeking behavior observed in PG and that NAC may be effective in the treatment of PG.

Other Agents

Bupropion. As a reuptake inhibitor of dopamine and noradrenaline, bupropion is approved for the treatment of major depression and nicotine dependence and, in the latter condition, reduces both smoking-related urges and withdrawal. Based on data that bupropion may be helpful in the treatment of attention deficit hyperactivity disorder (ADHD), proposed similarities between ADHD and pathological gambling, and bupropion's proposed mechanism of action, bupropion was investigated to test the hypothesis that it would target impulsivity and attentional deficits in individuals with PG (Iancu et al. 2008).

In an initial open-label trial, ten individuals with PG were prescribed a flexible dose of bupropion for 8 weeks (Brewer et al. 2008). The drug was well tolerated and significant improvements in gambling symptoms were observed. Study imitations include the open-label, unblinded design, small sample, and short treatment duration. A subsequent study examined the efficacy of bupropion, comparing it to that of naltrexone, in the treatment of PG (Brewer et al. 2008). Thirty-six men with PG were randomly assigned to receive either sustained-release bupropion or naltrexone for 12 weeks. The majority of individuals in both treatment groups responded well to treatment with comparable percentages of responders. The findings suggest that bupropion may be as effective as naltrexone in reducing gambling severity, the frequency of gambling, and the amount of money spent on gambling. Study limitations include the lack of a placebo-controlled group, short duration of treatment, and small sample.

A later placebo-controlled trial was conducted in which 39 individuals with PG were randomly assigned to a 12-week course of bupropion or placebo (Iancu et al. 2008). Individuals receiving bupropion and placebo showed comparable improvement on short-term measures of gambling severity and placebo response was frequently observed. Study limitations include significant treatment dropout. Together, findings suggest that future treatment trials target specific subgroups of individuals with PG, such as those with co-occurring depression or nicotine dependence, disorders representing current indications for bupropion.

Summary

Findings from pharmacologic studies emphasize the potential for pharmacotherapies, with arguably the strongest data supporting the use of opioid antagonists, particularly among individuals with strong gambling urges and familial propensities for addiction. Other medications (e.g., glutamatergic agents such as NAC) also appear promising, whereas other drugs (e.g., SRIs) have shown mixed results and might be particularly helpful for specific subgroups. The interpretation of open-label studies should be circumspect, given frequent placebo responses in controlled trials and apparent failures to replicate initially promising open-label findings in randomized placebo-controlled trials. Pharmacotherapy trials thus far have been relatively short and highlight the need for large-scale investigations with longer treatment courses in order to better understand the potential long-term benefits and risks of these medications. Medications that have been tested can be conceptualized and grouped based upon their proposed mechanisms of action (e.g., opioid antagonists, SRIs, mood stabilizers) and individual differences (e.g., with respect to co-occurring disorders) and may be particularly relevant in selecting the most appropriate drug treatments for specific patients. Investigating potential benefits and risks of specific pharmacotherapies in conjunction with specific empirically supported psychotherapies represents an important next step.

Cross-References

▶ Alcohol
▶ Alcohol Abuse and Dependence
▶ Alcohol and Drug Dependence Treatment: Pharmacogenetics
▶ Bipolar Disorder
▶ Cocaine
▶ Cocaine Dependence
▶ Depression
▶ Impulse Control Disorders
▶ Impulsivity
▶ Inter-temporal Choice
▶ Mood Stabilizers
▶ Opioids
▶ Substance Use Disorders: A Theoretical Framework

References

American Psychiatric Association (2000) Diagnostic and statistical manual of mental disorders, 4th edn, text revision. American Psychiatric Association, Washington, DC

Black DW, Shaw MC, Forbush KT, Allen J (2007) An open-label trial of escitalopram in the treatment of pathological gambling. Clin Neuropharmacol 30(4):206–212

Black DW, Shaw MC, Allen J (2008) Extended release carbamazepine in the treatment of pathological gambling: an open-label study. Progr Neuropsychopharmacol Biol Psychiatry 32:1191–1194

Brewer JA, Potenza MN (2008) The neurobiology and genetics of impulse control disorders: relationships to drug addictions. Biochem Pharmacol 75:63–75

Brewer JA, Grant JE, Potenza MN (2008) The treatment of pathologic gambling. Addict Disord Their Treat 7(1):1–13

Grant JE, Potenza MN (2007) Treatments for pathological gambling and other impulse control disorders. In: Nathan PE, Gorman JM (eds) A guide to treatments that work, 3rd edn. Oxford University Press, New York, pp 561–577

Grant JE, Potenza MN (2008) Pharmacologic treatment of impulse control disorders. Psychopharm Rev 43(9):67–74

Iancu I, Lowengrub K, Dembinsky Y, Kotler M, Dannon PN (2008) Pathological gambling: an update on neuropathophysiology and pharmacotherapy. CNS Drugs 22(2):123–138

Ibanez A, Blanco C, Donahue E, Lesieur HR, de Castro IP, Fernandez-Piqueras J et al (2001) Psychiatric comorbidity in pathological gamblers seeking treatment. Am J Psychiatry 158:1733–1735

Kim SW, Grant JE (2001) The psychopharmacology of pathological gambling. Semin Clin Neuropsychiatry 6(3):184–194

Williams WA, Potenza MN (2008) The neurobiology of impulse control disorders. Rev Brasileiria de Psiquiatria 30(S1):24–30

Patient-Controlled Analgesia

Definition

A method of pharmacological pain relief in which a patient initiates and controls delivery of the analgesic. Typically, the patient can press a button on the PCA machine, and this results in an intrathecal or intravenous infusion of a prescribed amount of the analgesic through an implanted catheter. There is a certain amount of time after each infusion during which no further infusions are possible.

Pavlovian Conditioning

Synonyms

Classical conditioning; Respondent conditioning

Definition

In its simplest form, Pavlovian conditioning typically involves the presentation of two stimuli: one is termed a ▶ conditioned stimulus (CS) and the other is termed an ▶ unconditioned stimulus (US). The stimuli are often presented relatively close in time and independent of the subject's behavior. Conditioning is said to occur when one of the stimuli either comes to control a response it did not previously control or modify an ongoing behavior. In laboratory situations, the CS tends to be a relatively neutral stimulus such as an auditory, olfactory, tactile, or visual cue. The US tends to be a stimulus with motivational properties (food, water, sex, illness, drug, etc.). This setup tends to be for methodological convenience (i.e., easy to measure conditioning) rather than ecological relevance as stimuli with motivational effects readily function as CSs and occasion setters. Although most Pavlovian conditioning research involves repeated presentation of the CS and US, one trial conditioning is well documented.

Cross-References

- ▶ Active Avoidance
- ▶ Alcohol Preference Tests
- ▶ Blocking, Overshadowing, and Related Concepts
- ▶ Classical (Pavlovian) Conditioning
- ▶ Classical Conditioning and Psychoactive Drugs
- ▶ Conditioned Drug Effects
- ▶ Conditioned Place Preference and Aversion
- ▶ Conditioned Taste Aversions
- ▶ Conditioned Taste Preferences
- ▶ Latent Inhibition
- ▶ Occasion Setting with Drugs
- ▶ Passive Avoidance
- ▶ Pavlovian Fear Conditioning

Pavlovian Fear Conditioning

Stephan G. Anagnostaras, Jennifer R. Sage and Stephanie A. Carmack
Department of Psychology and Program in Neurosciences, University of California, San Diego, La Jolla, CA, USA

Synonyms

Classical fear conditioning; Conditioned emotional response; Conditioned freezing

Definition

Pavlovian fear conditioning is a behavioral paradigm in which an initially neutral cue (the "▶ Conditioned Stimulus", CS), usually a tone, is paired with an aversive stimulus (the "▶ Unconditioned Stimulus", US), usually a footshock, that elicits a fear unconditional response (UR). As a result of this pairing, subjects form an associative memory between the CS and US. Following training, when presented with the CS alone, the subject will exhibit a fear-conditioned response (CR), which is a defensive behavior related to, but not identical with, the UR. Aside from fear of the tone ("▶ Cued Fear"), subjects also come to fear the environmental context associated with shock administration ("▶ Contextual Fear"). In the standard paradigm, rodents are trained in fear-conditioning chambers and receive 1–10 tone-shock pairings over the course of 5–10 min. The rodents are returned to the conditioning chambers days or weeks later for a brief contextual fear test, which consists of placing the animals in the chambers for 5–10 min and observing fear behaviors. On a third day, cued fear is assessed in a novel context by presenting the animals with the training tone 1–3 times after a 2-min baseline period (Anagnostaras et al. 1999, 2000; Fanselow 1984; Maren 2008). This overall experimental scheme is often modified depending on the needs of the experimenters. Fear is an inferred psychological state, which serves to organize and coordinate various species-specific defensive responses necessary for survival. Many defensive responses can be studied during the contextual and cued fear tests, including increases in heart rate, blood pressure, respiration, analgesia, ulcers, glucocorticoid activation, muscle tone, skin conductance, and potentiated startle (Davis 1992). However, the most commonly assessed fear response in mice and rats is freezing behavior, because of convenience and reliability. Freezing is defined as the absence of all movements other than that required for respiration and can be scored by observers using time sampling, a stopwatch, or a variety of automated methods (Anagnostaras et al. 2000; Fanselow 1984).

Impact of Psychoactive Drugs

Neurobiology

As the neurobiology and behavioral psychology of Pavlovian fear conditioning are well understood, Pavlovian fear conditioning has become a valuable rodent model of both memory and anxiety in humans. Pavlovian fear conditioning depends on the convergence of CS and US information in the basolateral/lateral complex of the amygdala. The CS-US association may occur in the amygdala via a synaptic process like long-term potentiation (Rogan et al. 1997). Production of the fear response, following presentation of the CS or US, depends on the central nucleus of the amygdala, which coordinates the output of defensive responses through downstream connections with response-specific brain centers (Davis 1992; Maren 2003). For example, freezing and conditioned analgesia are generated by the periaqueductal gray matter, while potentiated startle is mediated by the nucleus reticularis pontis caudalis. Thus, discrete cued fear is a reductionist model of associative memory that can be studied in the basolateral/lateral complex of the amygdala. Aside from the amygdala, contextual fear is dependent on the hippocampus; and this rodent paradigm has become a leading model of human declarative memory (Anagnostaras et al. 2001; Maren 2008). As with human amnesia, lesions of the hippocampus made shortly after conditioning (recent memory) produce a severe and selective deficit of contextual fear, whereas those made 1 month or more after training (remote memory) have little or no effect on contextual fear, a phenomenon known as temporally graded retrograde amnesia. This occurs because over time, contextual memories come to depend on cortical areas. Therefore, contextual fear conditioning can be used as a model not only for acquisition of declarative memories but also for consolidation of those memories from a hippocampus-dependent to cortical-dependent state. In contrast, the amygdala remains important for both contextual conditioning and tone-fear conditioning for the life of the animal (Gale et al. 2004). Therefore, it is believed that the role of the hippocampus in fear conditioning is specifically to acquire a spatial or configural representation of context, which is then conferred to the amygdala for association with the shock (Anagnostaras et al. 2001).

Role in Pharmacological Screening

As a Model of Higher-Level Cognition

Fear conditioning is an efficient model of higher-level cognition for many reasons (Anagnostaras et al. 2000). First and foremost, conditioned freezing is an extremely reliable behavioral measure, and the equipment used is fairly standardized, compact, and readily available. Second, the neurobiology is well understood and internally controlled; it can dissociate hippocampus-dependent and hippocampus-independent memory within the same subject. For example, a deficit observed in contextual and tone fear can be interpreted as affecting fear in general, whereas a deficit selective to contextual fear can be interpreted as likely to affect declarative memory. Third, it is rapidly acquired (1–10 tone-shock pairings in a 5–10-min period are commonly used) and lasts the lifetime of the animal. Because training is punctate and brief, it is highly suitable for studies that require the examination of memory a few minutes or hours after acquisition, or those requiring administration of an agent during training. Most other hippocampus-dependent memory paradigms, in contrast, require many hours and days of training. Fourth, it is very amenable to high-throughput screening, because of reliability, quick acquisition, and compact equipment that can be concentrated in a small lab space. Fifth, the procedures for mice and rats are quite similar and fairly interchangeable. This means fear-conditioning researchers can take advantage of the large behavioral and pharmacological database for rats, as well as the more novel database for genetically modified mice. Finally, fear conditioning has already been used extensively in rodents to model cognitive function, and a large experiential database exists on the various caveats and procedures one can use.

Some uses of fear conditioning in pharmacological screening include toxicological assessment for amnesia or specific assessment of drugs to enhance memory. For example, researchers may administer a given treatment drug during fear-conditioning training to ensure that overall cognitive functioning remains intact. Fear conditioning can be similarly useful in screening for agents that may enhance memory acquisition or expression. This can be done in normal mice or rats if screening for general cognitive enhancers or in genetic models of memory defects, such as Alzheimer's disease or mental retardation. Fear conditioning has already proven valuable in this sense, and its high efficiency and comparable performance in mice and rats have lent to this value (Maren 2008).

A variety of ancillary measures can be taken during the fear-conditioning session which can be useful in addressing confounds or other concerns the experimenter may have. For example, locomotor activity is frequently assessed prior to the tone-shock pairing on the training day, as an indicator of overall activity. If a drug or mutation produces a large drop or increase in activity during this time period, it is likely it would do so on the open-field paradigm as well. Second, activity during the shock can be used as a gross measure of pain. If shock reactivity is abnormal, this could suggest that the animals are unable to feel pain. All of these measures may be taken using automated equipment (Anagnostaras et al. 2000). Finally, freezing during the training period (known as post-shock freezing) can be used as a measure of initial acquisition. This could be particularly important in situations where animals show deficits in contextual and tone fear during testing, in order to show that the animals are actually capable of exhibiting the freezing response. Thus, fear conditioning offers many opportunities for control over various confounds that may present themselves.

As a Model of Anxiety Disorders, Especially Phobia

Pavlovian fear conditioning is also a prominent model of pathological fear, such as in anxiety disorders. Fear conditioning is a model for both etiology and treatment of pathological fear, especially phobia. Most often phobia is treated through exposure (or extinction) therapy, whereby the subject is repeatedly exposed to the cues associated with fear. In animals, this is most often modeled through repeated presentations of the tone without any reinforcement. Extinction is effective but weaker than initial conditioning and is thus subject to loss due to changes in conditions of testing (such as a context shift, known as renewal) or simply passage of time (known as spontaneous recovery). Thus, considerable effort has been made to develop extinction-enhancing drugs, which would be given during extinction training to make it more stable and robust. Among agents that have been discovered this way are D-cycloserine and cannabinoid agonists (Quirk and Mueller 2008). Fear conditioning can also be used to screen for drugs which reduce fear specifically, though this is less common than using unlearned fear tasks, such as the elevated plus maze. This is likely to change as fear conditioning continues to be adopted in pharmacological work.

Conclusion

Pavlovian fear conditioning, especially conditioned freezing, has become a valuable tool for modeling general cognitive function, especially declarative memory or pathological fear. It is also a model of learning and memory, in general. The simple and reliable nature of the paradigm lends itself to high-throughput applications, which can be followed up in secondary evaluation by additional memory or anxiety tests. Thus, Pavlovian fear conditioning is a tool of growing utility in pharmacological assessment.

Cross-References

- ▶ Amygdala
- ▶ Anxiety: Animal Models
- ▶ Consolidation and Reconsolidation
- ▶ Fear Conditioning
- ▶ Hippocampus
- ▶ Learning and Memory: Molecular Mechanisms
- ▶ Long-Term Potentiation and Memory
- ▶ Reconsolidation

References

Anagnostaras SG, Maren S, Fanselow MS (1999) Temporally graded retrograde amnesia of contextual fear after Hippocampal damage in rats: within-subjects examination. J Neurosci 19(3):1106–1114

Anagnostaras SG, Josselyn SA, Frankland PW, Silva AJ (2000) Computer-assisted behavioral assessment of Pavlovian fear conditioning in mice. Learn Mem 7(1):58–72

Anagnostaras SG, Gale GD, Fanselow MS (2001) Hippocampus and contextual fear conditioning: recent controversies and advances. Hippocampus 11(1):8–17

Davis M (1992) The role of the amygdala in fear-potentiated startle: implications for animal models of anxiety. Trends Pharmacol Sci 13(1):35–41

Fanselow MS (1984) Opiate modulation of the active and inactive components of the postshock reaction: parallels between naloxone pretreatment and shock intensity. Behav Neurosci 98(2):269–277

Gale GD, Anagnostaras SG, Godsil BP, Mitchell S, Nozawa T, Sage JR et al (2004) Role of the basolateral amygdala in the storage of fear memories across the adult lifetime of rats. J Neurosci 24(15):3810–3815

Maren S (2003) The amygdala, synaptic plasticity, and fear memory. Ann NY Acad Sci 985:106–113

Maren S (2008) Pavlovian fear conditioning as a behavioral assay for hippocampus and amygdala function: cautions and caveats. Eur J Neurosci 28(8):1661–1666

Quirk GJ, Mueller D (2008) Neural mechanisms of extinction learning and retrieval. Neuropsychopharmacology 33(1):56–72

Rogan MT, Staubli UV, LeDoux JE (1997) Fear conditioning induces associative long-term potentiation in the amygdala. Nature 390(6660):604–607

PDE3 Inhibitors

Definition

PDE3 inhibitors are drugs that inhibit phosphodiesterase3, which hydrolyzes cAMP. PDE3 is found in the heart and smooth muscles but also throughout the brain. There are two PDE3 inhibitors on the market: milrinone (Primacor) for the treatment of congestive heart failure and cilostazol (Pletal) for the treatment of intermittent claudication.

Cross-References

- ▶ Congestive Heart Failure
- ▶ Intermittent Claudication
- ▶ PDE4 Inhibitors
- ▶ PDE5 Inhibitors
- ▶ Phosphodiesterase Inhibitors

PDE4 Inhibitors

Definition

PDE4 inhibitors are drugs that inhibit phosphodiesterase4, which hydrolyzes cAMP. PDE4 is found in a wide variety of tissues including lungs and smooth muscles. It is also present in brain structures such as the hippocampus, amygdala, and cortical areas. PDE4 inhibitors have clinical potential for the treatment of inflammatory disorders including asthma and chronic obstructive pulmonary disease (COPD). In addition, they could offer possible treatments for depression or Alzheimer's disease. Until now the developed PDE4 inhibitors have difficulty in reaching the market due to the emetic side effects including nausea and vomiting. One of the most explored PDE4 inhibitors is rolipram. The only PDE4 inhibitor approved for the market is roflumilast (Daxas, Daliresp) for the treatment of COPD.

Cross-References

- ▶ Chronic Obstructive Pulmonary Disease
- ▶ PDE3 Inhibitors
- ▶ PDE5 Inhibitors
- ▶ Phosphodiesterase Inhibitors
- ▶ Rolipram

PDE5 Inhibitors

Definition

PDE5 inhibitors are drugs that inhibit phosphodiesterase5, which hydrolyzes cGMP. PDE5 is particularly present in smooth muscles of lungs and the corpus cavernosum. The PDE5 inhibitor sildenafil is on the market for the treatment of erectile dysfunction under the name of Viagra. In addition, sildenafil has been approved for the treatment of arterial pulmonary hypertension under the name of Revatio. Two other PDE5 inhibitors approved for the treatment of erectile dysfunction are vardenafil (Levitra) and tadalafil (Cialis). Tadalafil has also been approved for the treatment of arterial pulmonary hypertension under the name of Adcirca.

Cross-References

- ▶ Erectile Dysfunction
- ▶ PDE3 Inhibitors
- ▶ PDE4 Inhibitors
- ▶ Phosphodiesterase Inhibitors
- ▶ Pulmonary Hypertension
- ▶ Sildenafil

Peak-Interval Procedure

Definition

Peak-interval (PI) procedure is a ▶ reproduction protocol that comprises two types of trials randomly intermixed: In fixed-interval trials, subject's responses are reinforced at the to-be-timed duration. In peak trials, subjects are required to respond at the appropriate time, but no reinforcement is available. The typical result is that the distribution of responses in peak trials is centered around the criterion duration, with a standard deviation proportional to this criterion. In a variant of this procedure, the PI procedure with gaps and/or distracters, subjects are also presented with trials in which the to-be-timed stimulus is interrupted by a brief gap (retention interval), or with trials in which novel (unfamiliar) distracter events are presented during the uninterrupted to-be-timed stimulus, in order to evaluate their short-term memory for time.

Cross-References

- ▶ Timing Accuracy

Pediatric Schizophrenia

Christoph U. Correll
Department of Psychiatry, The Zucker Hillside Hospital, Glen Oaks, NY, USA

Synonyms

Childhood-onset schizophrenia; Early-onset schizophrenia; EOS; Schizophrenia with onset during childhood and adolescence; Very early-onset schizophrenia (VEOS)

Definition

The onset of schizophrenia before age 18 years, i.e., during childhood and adolescence. Early-onset schizophrenia refers to cases with onset between age 13 and 17 years, whereas very early-onset schizophrenia or childhood-onset schizophrenia refers to onset before age 13 years.

Role of Pharmacotherapy

Nosology, Epidemiology, and Phenomenology

The diagnosis of pediatric schizophrenia is made using unmodified criteria for adulthood-onset schizophrenia (i.e., onset at age 18 or older). The subtypes of schizophrenia are also identical in both groups. While the prevalence of childhood-onset schizophrenia (onset of psychotic symptoms before 13 years of age) is very low (approximately 1/100 cases of schizophrenia), approximately 12–33 % of individuals with schizophrenia have their illness onset between age 13 and 17 (Kumra et al. 2008). Current phenomenological, cognitive, genetic, and neuroimaging data strongly support continuity between pediatric- and adult-onset schizophrenia, suggesting similar neurobiological correlates and clinical deficits (Kyriakopoulos and Frangou 2007). Although the (yet unknown) etiology and pathophysiology of pediatric and adulthood schizophrenia are believed to be very similar, patients with pediatric-onset schizophrenia seem to have a worse illness course that generally is characterized by greater chronicity and functional impairment compared to the adulthood-onset counterpart (Kumra et al. 2008). Whether this is related to direct biological effects or to the fact that the psychotic illness occurs at a time of critical developmental tasks, which disrupts the achievement of educational and social milestones, is unclear. However, due to the difference in outcomes, schizophrenia with onset before 18 years has been used as a distinct phenotype for genetic research in order to achieve greater homogeneity. Of note, however, patients with adulthood-onset schizophrenia also commonly experience varying degrees of developmental delays, psychosocial and educational problems, and functional decline during childhood or adolescence. Moreover, patients with pediatric- and adulthood-onset schizophrenia frequently report "prodromal" psychotic symptoms and signs in childhood or adolescence. The schizophrenia prodrome often consists of depressive and negative symptoms as well as attenuated psychotic symptoms (i.e., subthreshold forms of unusual ideas, suspiciousness, grandiosity, abnormal perceptions, and disorganized thought, speech, or behavior) (Correll et al. 2010).

The Evidence Base

To date, 14 randomized controlled trials (RCTs) ($n = 1,155$) have been completed in patients with pediatric schizophrenia (Kumra et al. 2008; Sikich et al. 2008). Six trials had a placebo comparator and evaluated the efficacy and safety of haloperidol ($N = 1$, $n = 12$), haloperidol and loxapine ($N = 1$, $n = 75$), aripiprazole ($N = 1$, $n = 301$), quetiapine ($N = 1$, $n = 220$), risperidone ($N = 1$, $n = 160$), and olanzapine ($N = 1$, $n = 107$), and one trial ($n = 279$) used a very low dose of risperidone (0.15–0.6 mg/day) as a pseudo-placebo comparator (N denotes the number of trials with each drug, n denotes the number of patients).

In addition to the placebo-controlled, three-arm study comparing haloperidol and loxapine ($N = 1$, $n = 75$), another seven trials ($n = 275$) compared antipsychotics head-to-head in youth with schizophrenia. These included a comparison of thiothixene and thioridazine ($N = 1$, $n = 21$); haloperidol, olanzapine, and risperidone ($N = 1$, $n = 50$, 52 % schizophrenia spectrum psychosis, 48 % affective spectrum psychosis); molindone, olanzapine, and risperidone ($N = 1$, $n = 119$); haloperidol and clozapine ($N = 1$, $n = 21$); clozapine and olanzapine ($N = 2$, $n = 64$); and olanzapine and quetiapine ($N = 1$, $n = 50$, 64 % schizophrenia spectrum psychosis, 36 % affective spectrum psychosis).

Efficacy

While the two underpowered studies from 1976 and 1984 involving first-generation antipsychotics (FGAs) did not significantly separate from placebo, there was a trend for greater improvement on the Clinical Global Impressions Severity (CGI-S) Scale in favor of haloperidol and loxapine in one study and a significant baseline to endpoint change in the Brief Psychiatric Rating Scale (BPRS) for haloperidol, but not for placebo (Kumra et al. 2008). By contrast, all of the second-generation antipsychotic (SGA) trials completed since 2005 showed significantly greater improvements on the primary outcome measure, the Positive and Negative Syndrome Scale (PANSS) for all doses that were studied. Overall, the numbers needed to treat (NNT) for study defined response for aripiprazole, olanzapine, quetiapine, and risperidone range from 4 to 10. Based on the results from these placebo-controlled trials in pediatric schizophrenia, risperidone, quetiapine, olanzapine, and aripiprazole were approved by the Food and Drug Administration (FDA) in the USA for use in adolescents aged 13–17 years, and aripiprazole was approved for use in adolescents aged 15–17 years in Europe by the European Medicines Agency (EMEA). Moreover, after an official FDA hearing in June 2009, olanzapine and quetiapine are expected to receive FDA approval in the USA for use in adolescents aged 13–17 years with schizophrenia. Despite inadequate trial data for first-generation antipsychotics, haloperidol and thioridazine were grandfathered in, being indicated for, adolescents with schizophrenia in the USA, and the dosing and use of several first-generation antipsychotics, mainly haloperidol, for adolescents with schizophrenia are mentioned in regulatory documents in some European countries.

Across the seven studies, comparing two FGAs, one FGA with one or two SGAs, or two SGAs with each other, the only significant group differences were found in favor of clozapine compared to haloperidol, regular dose olanzapine (up to 20 mg/day), and "high"-dose olanzapine (10–30 mg/day) (Kumra et al. 2008; Sikich 2008). Since in all active controlled studies the numbers of patients in individual study arms were very small, ranging from 8 to 41, a type-2 error cannot be excluded; yet the results of relatively similar efficacy result parallel data in adult schizophrenia.

Tolerability and Side Effects

Children and adolescents seem to be more sensitive to most antipsychotic adverse effects, including sedation, extrapyramidal motor side effects (except for akathisia), withdrawal dyskinesia, prolactin abnormalities, weight gain, and metabolic abnormalities (Correll 2008a). On the other hand, adverse effects that require a longer time to develop (e.g., diabetes mellitus) and that are related to greater medication dose and lifetime exposure (e.g., "▶ Tardive Dyskinesia") are less prevalent in pediatric samples. However, there is concern that these later-onset adverse events are not seen because of short follow-up periods and that they may emerge in vulnerable patients prematurely in adulthood the earlier antipsychotics are started in childhood.

Extrapyramidal Side Effects (EPS)
In general, children and adolescents are more sensitive than adults to Parkinsonian side effects associated with FGAs and SGAs (Correll 2008a). An RCT of 40 youths with psychotic disorders comparing haloperidol (mean dose: 5 mg/d), risperidone (mean dose: 4 mg/d), and olanzapine (mean dose: 12 mg/d) found substantial EPS not only with haloperidol (67 %) but also with olanzapine (56 %) and risperidone (53 %), although haloperidol-treated patients reported more severe EPS (Sikich et al. 2004). In the Treatment of Early-Onset Schizophrenia Spectrum (TEOSS) study, patients randomized to molindone (mean dose: 59.9 mg/day) required more frequent coadministration of an anticholinergic (45 %) than patients randomized to risperidone (mean dose: 2.9 mg/day, 34 %) or olanzapine (mean dose: 11.8 mg/day, 14 %), even though patients on molindone were given prophylactic, blinded benztropine 0.5 mg bid (Sikich et al. 2008). Clozapine and quetiapine appear to be associated with relatively low EPS rates in pediatric patients. For aripiprazole and ziprasidone, EPS rates appear to increase with increasing dose.

Akathisia

Incidence rates of akathisia from placebo-controlled RCTs in pediatric schizophrenia have been reported for aripiprazole (5 % for placebo, 5 % in the 10 mg/day group, and 11.8 % in the 30 mg/day group); risperidone (6 % on placebo, 7 % in the 1–3 mg/day group, and 10 % in the 4–6 mg/day group), corresponding to NNH of 15 to no risk for aripiprazole 30 mg/day and 10 mg/day, respectively; and 25–100 for risperidone 4–6 mg/day and 1–3 mg/day, respectively (Correll 2008b). The relatively high akathisia rates for placebo, especially in the pediatric schizophrenia trials, suggest the potential presence of a relevant carryover effect from prior antipsychotic treatment or the possibility of withdrawal phenomena after a brief washout from antipsychotics and/or medications that can mitigate akathisia. In the TEOSS study, molindone, but not olanzapine or risperidone, was associated with a significantly greater rate of self-reported akathisia compared to risperidone and olanzapine (Sikich et al. 2008).

Withdrawal Dyskinesia

During FGA treatment, youths are at risk of developing withdrawal dyskinesia, yet, unlike in adults, dyskinetic movements are frequently reversible. Withdrawal dyskinesia rates appear to be lower with SGAs compared to FGAs, although a switch from an antipsychotic with strong D2 affinity (risperidone or aripiprazole) to one with less potent affinity (quetiapine or clozapine) may predispose to withdrawal dyskinesia (Correll 2008a).

Tardive Dyskinesia (TD)

Long-term TD data in patients with pediatric schizophrenia are lacking. A meta-analysis of 10 studies lasting at least 11 months reported on TD rates in 783 patients aged 4–18 (weighted mean: 10) years. Most patients were prepubertal (80 %), male (82 %), and white (79 %) and only 3 % had a schizophrenia spectrum disorder (Correll and Kane 2007). Across these studies, only three cases of TD were reported, resulting in an annualized incidence rate of 0.4 %, which was approximately half of the rate found in a prior meta-analysis of TD rates in adults. However, it is unclear how much these data can be extrapolated to pediatric patients with schizophrenia, as antipsychotic doses were low and lifetime exposure was relatively short.

Weight Gain

Although pediatric data are still limited, youth with severe psychiatric disorders seem to be at increased risk for being overweight or obese, especially when exposed to antipsychotics for longer periods of time. Age-inappropriate weight gain is of particular concern in pediatric patients, due to its association with glucose and lipid abnormalities and cardiovascular morbidity/mortality. Reasons for weight gain are complex, including psychiatric illness, unhealthy lifestyle, and treatment effects. A review of pediatric data suggests that the weight gain potential of FGAs and SGAs follows roughly the same ranking order as found in adults but that the magnitude is greater (Correll 2008a). Exceptions may be a greater relative weight gain propensity of risperidone and a greater likelihood of aripiprazole and ziprasidone to not be weight neutral in subgroups of pediatric patients (Correll et al. 2009). For example, in an 8-week study, Sikich et al. (2004) found a higher weight gain in young patients aged 5–17 years with psychotic disorders taking olanzapine for 8 weeks (7.1 ± 4.1 kg) than in those taking either risperidone (4.9 ± 3.6 kg) or haloperidol (3.5 ± 3.7 kg); all weight gain was severe and disproportionate to that expected from normal growth. Results from four 6-week studies in adolescents with schizophrenia suggest that the olanzapine group had the greatest risk for significant weight, risperidone and quetiapine were associated with intermediate risk, and aripiprazole showed the lowest risk. The respective numbers needed to harm (NNH) for ≥ 7 % weight gain was 4 for olanzapine, 7–8 for quetiapine, 8 for risperidone, and 25–34 for aripiprazole (Correll 2008b). However, the interpretation of weight gain results across various studies and agents is complicated by the effects of baseline weight, developmental stage and growth, past antipsychotic exposure, treatment duration and setting, comedications, etc., that varied across trials.

Metabolic Adverse Effects

Whereas in adults, the link between antipsychotics and adverse metabolic effects, such as dyslipidemia, hyperglycemia, diabetes, and metabolic syndrome, has been established; the few pediatric studies which reported data have produced mostly negative results. Interpretation of these findings is limited by the small sample size, varying treatment histories, and inclusion of non-fasting blood assessments. Case reports of new-onset diabetes in antipsychotic-treated youths and the known link between weight gain and metabolic abnormalities suggest that youths are at least as liable to develop metabolic abnormalities as adults. However, in pediatric RCTs, so far, only olanzapine has been associated with significant increases in glucose, insulin, and lipids (Correll 2008a, b). Nevertheless, the lack of significant metabolic abnormalities in the other short-term RCTs despite mostly significant weight gain needs to be interpreted with caution, as the negative findings could be due to the short-term trial duration, lack of strict fasting assessments, and order effects in patients with more extensive past antipsychotic exposure. A recent cohort study in 272 antipsychotic-naive youth (30.1 % with schizophrenia spectrum disorders) confirmed that during the first 12 weeks of treatment, olanzapine has the greatest adverse effect on body composition, which was associated with significant worsening of fasting glucose, insulin, insulin resistance, and all lipid parameters, except for HDL-cholesterol (Correll et al. 2009). By contrast, despite similar adverse effects on body composition, the metabolic effects differed across quetiapine, risperidone, and aripiprazole. At least during the first 3 months of treatment, quetiapine was associated with a significant increase in most lipid parameters, whereas with risperidone lead only to a significant increase in triglycerides, and changes with aripiprazole remained nonsignificant. This suggests that in addition to indirect, weight-related metabolic changes, direct, weight-independent effects on glucose and lipid metabolism exist, at least for some antipsychotics.

Prolactin-Related Side Effects

FGAs and SGAs can elevate prolactin levels, and these elevations appear to be accentuated in children and adolescents. Similar to adults, albeit at higher levels during adolescence, the relative potency of antipsychotic drugs in increasing prolactin is roughly paliperidone≥risperidone>haloperidol>olanzapine≥ziprasidone>quetiapine≥clozapine>aripiprazole. To date, adequate long-term data are lacking to determine if hyperprolactinemia at levels found during antipsychotic therapy alters bone density, sexual maturation, or the risk for benign prolactinomas (Correll 2008a). Since aripiprazole is a partial D2 dopamine agonist, prolactin levels can decrease below baseline. To date, no adverse effects of low prolactin have been described in youth. Complicating the interpretation of the relevance of prolactin elevations in youth is the fact that sexual and reproductive system side effects related to prolactin levels are rarely directly inquired about, and youth might either not express these symptoms due to sexual immaturity or because they do not know what their normal levels of functioning ought to be.

Summary and Conclusion

Although still understudied, schizophrenia with onset in childhood and, especially, with onset in adolescence seems to be biologically and phenomenologically continuous with adulthood-onset schizophrenia, albeit being more often associated with poorer illness course and outcomes. Moreover, children and adolescents appear to be more sensitive to antipsychotic adverse effects than adults, at least compared to more chronically ill samples. As in adults, antipsychotics are more effective than placebo, with meaningful clinical effects. Moreover, also like in adults, differences in efficacy between antipsychotics seem to be much smaller and less predictable than differences in side effects and, thus, in effectiveness, which takes the short- and long-term side-effect burden and treatment discontinuation rates into account. Based on this risk-benefit evaluation, it appears that second-generation antipsychotics might be preferable to first-generation antipsychotics to reduce the risk for EPS, TD, secondary negative symptoms, early treatment discontinuations, and, possibly, relapse rates. However, since a number of second-generation antipsychotics are associated with significantly greater risks for age-inappropriate weight gain and metabolic abnormalities than mid- and high-potency first-generation antipsychotics, the neuromotor side-effect advantages and related benefits are likely offset by the risk of longer-term health problems for those higher metabolic risk second-generation antipsychotics. Therefore, it appears that second-generation antipsychotics with the least risk for developmentally inappropriate weight gain and related or, even, direct metabolic abnormalities are to be considered first-line treatment options. In case these fail, higher cardiometabolic risk antipsychotics should be tried. Given the significant efficacy advantage of clozapine over first- and second-generation antipsychotics in pediatric-onset schizophrenia similar to adulthood schizophrenia, clozapine should be considered for severely ill and treatment-resistant youth with schizophrenia to improve outcomes and functioning, balancing its problematic side-effect profile against its superior efficacy.

Cross-References

▶ First-Generation Antipsychotics
▶ Schizophrenia Prodrome
▶ Second-Generation Antipsychotics

References

Correll CU (2008a) Antipsychotic use in children and adolescents: minimizing adverse effects to maximize outcomes. J Am Acad Child Adolesc Psychiatry 47:9–20

Correll CU (2008b) Assessing and maximizing the safety and tolerability of antipsychotics used in the treatment of children and adolescents. J Clin Psychiatry 69(Suppl 4):26–36

Correll CU, Kane JM (2007) One-year incidence rates of tardive dyskinesia in children and adolescents treated with second-generation antipsychotics: a systematic review. J Child Adolesc Psychopharmacol 17(5):647–656

Correll CU, Manu P, Olshanskiy V, Napolitano B, Kane JM, Molthotra AK (2009) Cardiometabolic risk of second-generation antipsychotic medications during first-time use in children and adolescents. JAMA 302(16):1765–1773

Correll CU, Hauser M, Auther AM, Cornblatt BA (2010) Research in people with psychosis risk syndrome: a review of the current evidence and future directions. J Child Psychol Psychiatry 51:390–431

Kumra S, Oberstar JV, Sikich L, Findling RL, McClellan JM, Vinogradov S, Charles Schulz S (2008) Efficacy and tolerability of second-generation antipsychotics in children and adolescents with schizophrenia. Schizophr Bull 34(1):60–71

Kyriakopoulos M, Frangou S (2007) Pathophysiology of early onset schizophrenia. Int Rev Psychiatry 19(4):315–324

Sikich L (2008) Efficacy of atypical antipsychotics in early-onset schizophrenia and other psychotic disorders. J Clin Psychiatry 69(Suppl 4):21–25

Sikich L, Hamer RM, Bashford RA, Sheitman BB, Lieberman JA (2004) A pilot study of risperidone, olanzapine, and haloperidol in psychotic youth: a double-blind, randomized, 8-week trial. Neuropsychopharmacology 29(1):133–145

Sikich L, Frazier JA, McClellan J, Findling RL, Vitiello B, Ritz L, Ambler D, Puglia M, Maloney AE, Michael E, De Jong S, Slifka K, Noyes N, Hlastala S, Pierson L, McNamara NK, Delporto-Bedoya D, Anderson R, Hamer RM, Lieberman JA (2008) Double-blind comparison of first- and second-generation antipsychotics in early-onset schizophrenia and schizo-affective disorder: findings from the treatment of early-onset schizophrenia spectrum disorders (TEOSS) study (2008). Am J Psychiatry 11:1420–1431. Epub 2008 Sep 15. Am J Psychiatry 165(11):1495 Erratum

Pemoline

Synonyms

Magnesium pemoline

Definition

Pemoline is an oxazolamine, a psychostimulant that was used for the treatment of ▶ attention deficit hyperactivity disorder. Magnesium pemoline, a hydrated mixture of pemoline and magnesium hydroxide, is thought to be better absorbed and longer-acting than pemoline. It was less effective than methylphenidate and was withdrawn from the US market due to liver toxicity. In earlier studies, it was found to produce small enhancements in selective aspects of learning and memory in human and animal subjects.

Cross-References

▶ Attention-Deficit/Hyperactivity Disorder
▶ Methylphenidate and Related Compounds

Pentazocine

Definition

Pentazocine is a synthetic opioid with partial agonist-antagonist properties. It is used to treat mild to moderately severe pain. It can partially block effects of full opioid agonists such as ▶ morphine and heroin. It exists as one of two enantiomers: (−)-pentazocine is a κ-opioid receptor agonist, while (+)-pentazocine is not but displays tenfold greater affinity for sigma (non-opioid) receptors. This action on sigma receptors may account for notable side effects that include hallucinations and other psychotomimetic effects. Apart from its analgesic effects, pentazocine has minimal clinical use. It is subject to abuse and dependence, although the severity is typically less than that for full opioid agonists. Some tablets are formulated to contain both pentazocine and naloxone to reduce abuse; the naloxone blocks opioid effects if the tablets are dissolved and injected but allows analgesia to develop if they are taken orally.

Cross-References

▶ Morphine
▶ Naloxone
▶ Partial Agonist

Pentobarbital

Synonyms

Pentobarbitone; Sodium 5-ethyl-5-(1-methylbutyl) barbiturate

Definition

First synthesized in 1928, pentobarbital is a short-acting barbiturate used formerly in humans as a sedative/hypnotic and anxiolytic agent. Now it is used more commonly as an anesthetic in veterinary practice and as an anticonvulsant. Like other barbiturates, it acts at the ▶ $GABA_A$ receptor to enhance inhibitory neurotransmission, but it also acts as an antagonist at glutamate receptors of the AMPA subtype, thereby reducing excitatory neurotransmission. It is now a regulated drug of abuse, partly because of its association with suicide and as an agent of euthanasia.

Cross-References

▶ Abuse Liability Evaluation
▶ Barbiturates
▶ Driving and Flying Under the Influence of Drugs
▶ Insomnias
▶ Sedative, Hypnotic, and Anxiolytic Dependence

Pergolide Mesylate

Synonyms

8-β[(Methylthio)methyl]-6-propylergoline monomethanesulfonate; Celance; LY127809; Nopar; Parkotil; Permax; Pharken

Definition

Pergolide is a centrally acting and long-acting D1/D2 (and 5-HT) receptor agonist. It is believed to stimulate dopamine receptors in the nigrostriatal system. The compound also inhibits the secretion of prolactin and thus is used to treat hyperprolactinemia and restless leg syndrome. Pergolide is generally used as an adjunctive treatment with levodopa/carbidopa for signs and symptoms of Parkinson's disease. However the compound was withdrawn from the US market due to stimulation of 5-HT$_{2B}$ receptors and the resulting valvulopathies. In Europe, cabergoline and pergolide are contraindicated in patients with evidence of heart valve problems.

Cross-References

▶ Anti-Parkinson Drugs

Pericyazine

Synonyms

Periciazine; Propericiazine

Definition

Pericyazine is a first-generation ▶ antipsychotic medication that acts as a dopamine D2 antagonist. Pericyazine is indicated for the treatment of schizophrenia and other psychoses and for short-term adjunctive treatment of severe anxiety, psychomotor agitation, and excited or violent states. It is a piperazine-phenothiazine derivative, and like other similar agents, it produces drowsiness and sedation. Hypotension is common when treatment is initiated. It is approved for use in Europe but not in the USA.

Cross-References

▶ Antipsychotic Drugs
▶ First-Generation Antipsychotics
▶ Impulse Control Disorders
▶ Schizophrenia

Perinatal Exposure to Drugs

Linda P. Spear
Department of Psychology, Binghamton University, Binghamton, NY, USA

Definition

Perinatal exposure to drugs involves exposure of the developing organism to substances of abuse or to psychotherapeutic compounds prior to birth and perhaps shortly thereafter. Such drug exposure typically occurs indirectly through maternal drug use, with the drug gaining access to the fetus via crossing the placenta to distribute in amniotic fluid and/or the fetal circulation and to the infant via partitioning of the drug into milk during lactation.

Current Concepts and State of Knowledge

The prenatal and early postnatal period is often a sensitive period during which exposure to a variety of drugs may induce long-term alterations in brain function and behavior that are not evident with comparable exposures later in life. In recognition of this possibility, part of the testing process required for potential new therapeutic compounds in the United States involves assessment of each substance's developmental toxicity in preclinical studies (i.e., studies in laboratory animals) before the drug can be approved for initial (Phase 1) clinical trials (e.g., see Hodgson 2004). Initially spearheaded by studies of the fetal alcohol syndrome beginning in the mid-1970s, research examining the impact of maternal drug use during pregnancy and lactation has spread to encompass all major drugs of abuse. Through such studies, along with research examining the potential developmental neurotoxicity of environmental contaminants such as lead, consensus has been reached on a number of points. Some of these generalities will be briefly summarized here before highlighting consequences of perinatal exposure to several drugs of abuse.

Despite often necessary differences in assessment measures, generally comparable functional effects are often seen following perinatal drug/chemical exposure in humans and laboratory animals. During the first wave of research in developmental toxicology, there was appropriate concern as to whether findings obtained in laboratory animals would be comparable to those seen in human clinical studies. Careful examinations of across-species comparability of findings in animal models and clinical work, however, yielded notable across-species commonalities in the developmental toxicology of chemicals, ranging from alcohol, anticonvulsants, and methadone to environmental neurotoxicants such as lead and methylmercury, even though different measures were often necessary to assess particular functions (e.g., sensory or motor function, cognition, motivation, social behavior, etc.) in clinical studies versus research with laboratory animals (Stanton and Spear 1990). Additional evidence of general across-species comparability has continued to emerge with the escalation of research in this field over subsequent decades (e.g., Slikker et al. 2005; Slotkin 2008), despite challenges in attributing clinical findings to the focal drug per se versus other factors (stressful living environment, malnutrition, use of other drugs, etc.) that often co-occur with maternal drug use (see Fried 2002).

These findings lend credence to the use of animal models in studies of perinatal drug exposures. Such studies are important to confirm and extend clinical findings, anticipate other potential consequences of early drug exposure, determine the neural mechanisms underlying behavioral effects, and suggest potential therapeutic approaches. Yet, applicability of animal models is by no means assured. Pharmacokinetic differences are common across species, raising the importance of dose, route of administration, and pattern of administration. Given the greater metabolic rate characteristic of small species such as rodents, higher doses may be necessary when using rodent models to produce drug levels in blood equivalent to those seen upon drug use in humans. Yet, too high a dose level may produce drug burdens that are out of the range associated with human exposures, decreasing the potential relevance of findings obtained with the animal model. Other important considerations when developing animal models include controlling of possible drug-induced decreases in food intake and/or residual effects of drug exposure during pregnancy on subsequent maternal behavior – issues typically addressed via the use of pair-feeding and fostering procedures, respectively (e.g., see Spear and File 1996, for discussion and references for these and other control issues). Finally, although multiple offspring are commonly generated per litter when using rodent models of developmental toxicology, the evidence is clear that the offspring from the same litter are not independent samples, and thus the assignment of more than one offspring/litter to a given experimental group markedly inflates the possibility of false positives.

Behavioral, hormonal, and neural alterations are typically induced at exposure levels well below those that induce major malformations or maternal toxicity (Adams et al. 2000). It is not surprising that at exposure levels high enough to induce maternal toxicity, chemical and other physiological consequences of that toxicity often exert multiple adverse effects on fetal development, including alterations in brain function and behavior. Yet, neurobehavioral effects are also evident in offspring at drug exposure levels lower than those necessary to induce maternal toxicity or signs of gross malformation in the offspring. The brain is unusually susceptible to disruption by drugs during its development, with drugs potentially affecting a variety of developmental processes, including neuronal migration and differentiation and glia proliferation, along with normal ontogenetic changes in cell adhesion, neural communication, energy utilization, and apoptosis (e.g., Goodlett and Horn 2001). Drug effects on ongoing developmental processes may extend well beyond the direct neural actions of that drug; for instance, perinatal exposure to nicotine not only alters cholinergic nicotine receptor function throughout life but also influences the trajectory of development for a wide variety of neurotransmitter systems other than the cholinergic system (Slotkin 2008).

Observed consequences of perinatal exposures are not only drug- and dose-dependent but also critically dependent on the timing of the exposure and when and how functional consequences are assessed. Different portions of the brain develop at different times and hence differ in the timing of their vulnerability to drug-induced disruption. Thus, consequences of developmental exposure to drugs are to some extent dependent on the timing of that exposure. Even as little as a 1 day difference in the timing of drug exposure can influence behavioral outcome, although generally speaking, the greatest vulnerability for structural damage occurs early in organogenesis, whereas functional impairments are most likely evident following drug exposure during mid- to late organogenesis (see Spear and File 1996). When varying developmental stage during which drug exposure is given, it is important to note that altricial animals (such as laboratory rats and mice) are born at a less mature stage than humans. Hence, the prenatal period in rodents is approximately equivalent to the first and second trimesters of human pregnancy, with the first 10 days or so of postnatal life in rat approximating the third human trimester (see Spear and File 1996, for discussion and citations).

The timing during which functional consequences are assessed is also critical. Functional deficits often emerge well after termination of the drug exposure and long after the drug and its metabolites have cleared the brain and body. Thus, developmental drug exposures are thought to alter the trajectory of normal brain development, changing future ontogenetic patterns in the emergence of brain neurocircuitry and function (see Slotkin 2008). Under some instances, deficits may be seen early in life and may recover thereafter, perhaps manifesting at some later developmental point (e.g., adolescence), during aging, or under stressful or challenging circumstances (Adams et al. 2000). In other cases, functional alterations may not be readily apparent until neural development is sufficiently mature to reveal underlying deficiencies. For instance, cognitive effects of prenatal exposure to marijuana are minimal during the infant and toddler stage but do become progressively more evident as normally delayed executive function capacities emerge, with youth exposed prenatally to marijuana showing disruptions in certain aspects of executive function (e.g., attentional behavior; hypothesis testing) (Fried 2002).

The developmental toxicology of perinatal drug exposure is critically moderated by genetic background and the environment. The process of neural development is orchestrated by genes, with marked changes in the profile of genes expressed as development proceeds across different brain regions. Variants across individuals in the form of expression of particular genes (gene polymorphisms) can increase vulnerability to adverse effects of perinatal drug exposures. Not all women who drink during pregnancy, for example, bear infants that meet the diagnostic criteria (i.e., growth deficiency, CNS disorders, and facial dysmorphic features) of fetal alcohol syndrome (FAS – e.g., see Mattson et al. 2001). Although timing and amount of alcohol use likely contribute to this variability, the presence or absence of genetic variants that increase the vulnerability of the developing brain to developmental perturbations also like contributes as well. And at the frontier of current research is the increasing recognition that perinatal drug exposure itself, like other environmental influences, may in turn alter developmental patterns of gene expression in specific brain regions through epigenetic regulation, exerting long-term effects on brain function through such "▶ Developmental Programming." Thus, exposure to drugs during the perinatal period provides an important component of the early environment of the developing organism that, along with other early experiences, may influence and be influenced by developmental patterns of gene expression (see Goodlett and Horn 2001). Consequences of perinatal drug exposure, like other adverse early experiences, may be moderated or exacerbated to some extent by environmental experiences as well. For instance, toddlers exposed to drugs prenatally have been found to be much more likely to exhibit secure attachment relationships if their mother had been abstinent since birth than if their mother continued to abuse drugs (Fried 2002).

Perinatal exposure to drugs of abuse often exerts lasting and drug-specific neurobehavioral effects. With some drugs of abuse, consequences for offspring may be immediate and obvious and may resolve to some extent with time. The neonatal abstinence syndrome is a good example. That is, fetuses of drug-dependent mothers may themselves become addicted and may undergo withdrawal following removal of their drug source (their mother) at birth. Such signs of withdrawal are most evident in neonates exposed prenatally to opiates such as heroin or methadone and are characterized by transient hyperreactivity and irritability as well as feeding and sleep disturbances (see Bandstra et al. 2010). In contrast to these transitory effects, other consequences of perinatal exposure to drugs of abuse may be evident early in life and persist thereafter. Other consequences may not emerge until later, becoming manifest only when portions of the nervous system critical for expression of that function are sufficiently developed. To illustrate these effects of perinatal exposure to drugs of abuse, the focus will be on four drugs whose developmental effects have been particularly well studied: alcohol, nicotine, cocaine, and cannabinoids (including those in marijuana).

Cognitive effects. Cognitive deficits are a common sequela of perinatal drug exposure. Exposure to cocaine during development has been reported to impair performance of cognitive tasks in laboratory animals and to lower IQ, impair language development, and disrupt cognitive function in children (e.g., Harvey 2004). Similarly, exposure to nicotine during development has been observed to induce deficits in cognitive function in both humans and laboratory animals (Slikker et al. 2005). Prenatal alcohol exposure has likewise been shown to disrupt learning and memory in animal studies, with FAS offspring also having difficulties learning both verbal and nonverbal problems and showing deficits in certain aspects of memory performance (Mattson et al. 2001).

In studies of children exposed prenatally to marijuana, however, little sign of cognitive alterations was evident until the children approached school age. At this time, as nonexposed children began to exhibit increasing evidence of ability in various executive function domains, deficits in these functions began to emerge in children exposed prenatally to marijuana. Impaired executive functions persisted with age and were particularly notable on measures of attentional capacity and abstract/visual reasoning. These alterations were not associated with alterations in global IQ scores (Fried 2002). Disruptions in attentional processes have also been reported following prenatal cocaine exposure in developmental studies of both laboratory animals and humans (Harvey 2004). Deficits in executive function also have been reported following heavy prenatal exposure to alcohol (Mattson et al. 2001).

Thus, although alterations in cognitive functions are a common outcome of perinatal drug exposure, the nature of those alterations seemingly differs with the drug of abuse. Indeed, from studies comparing cognitive function in offspring of mothers smoking cigarette versus marijuana during pregnancy, Fried (2002) concluded that in utero exposure to marijuana disrupts "top-down" executive functioning, whereas prenatal exposure to cigarette smoking was associated with alterations in certain basic perceptual abilities and fundamental cognitive skills – more consistent with dysfunction in "bottom-up" cognitive skills.

Increased later drug use/abuse. Human adolescents whose mothers used alcohol during pregnancy have been reported to exhibit early alcohol consumption and a greater propensity to abuse alcohol (Mattson et al. 2001). Indeed, there is some evidence that prenatal exposure to alcohol may exert a greater effect on later alcohol use/abuse than family history of alcohol abuse per se. Studies with laboratory animals have likewise generally observed that fetal or early postnatal exposure to ethanol enhances later intake of alcohol (Spear and Molina 2005). In laboratory animals, prenatal exposure to nicotine, as well as to cocaine, also has been reported to elevate levels of self-administration of the drug later in life beyond those seen in nonexposed animals (see Harvey 2004; Slotkin 2008). Similarly, adolescents whose mothers smoked during pregnancy, regardless of whether their mothers continued to smoke during their childhood, were at increased risk for beginning to smoke and were more likely to relapse when they attempted to stop smoking (Slotkin 2008). Thus, there is considerable converging evidence across a number of drugs that perinatal exposure to a drug of abuse increases later risk for drug use and abuse.

Other behavioral alterations. Children exposed to alcohol prenatally (whether or not they meet the diagnostic criteria of FAS) not only are at risk for alcohol/drug abuse problems but also are more likely than nonexposed children to exhibit other problem behaviors as well, including impulsivity, delinquency, and hyperactivity, along with poor social and communication skills. Offspring exposed prenatally to marijuana smoke likewise have been reported to exhibit increased levels of delinquency, conduct disorder, hyperactivity, and impulsivity (Fried 2002). Hyperactivity is also commonly reported following prenatal nicotine exposure in laboratory animals, as well as in children whose mothers smoked during pregnancy (Slikker et al. 2005). Such behavioral disorders are multifactorially determined and hence could reflect alterations in a diversity of neural systems precipitated by perinatal exposure to these different drugs.

Neural alterations. Cognitive and behavioral alterations following perinatal drug exposures ultimately reflect drug-induced alterations in brain function. These effects are not global but reflect alterations in specific brain regions that are typically drug-specific and are related to the timing of the exposure. For instance, fetal alcohol exposure was found in human imaging studies to exert particularly pronounced alterations in the corpus callosum, along with reductions in size of the caudate nucleus, cerebellum, and hippocampus, findings reminiscent of those seen following prenatal ethanol exposures in laboratory animals (Mattson et al. 2001). The more invasive studies possible when working with animal models have revealed a rich variety of influences of perinatal drug exposure on the developing brain, including alterations in patterns of cell replication/differentiation, neural communication, signal transduction, cell death, gene regulation, and so on (Goodlett and Horn 2001; Slikker et al. 2005). And, although observed neural alterations are sometimes highly selective to particular components of the neural systems normally targeted by action of that drug in adulthood, other neural systems are often impacted as well. For instance, in animal studies, prenatal cocaine exposure was found to uncouple D1 dopamine receptors, while leaving D2 receptor functioning largely unaffected. Long-lasting alterations in cortical cytoarchitecture were also evident in the cocaine-exposed offspring and were suggested to reflect in part altered D1-receptor functioning and the resultant disruption in balance of D1 and D2 regulatory actions on neuronal development (Harvey 2004).

Conclusions

Perinatal exposure to drugs often exerts a long-term impact on brain function and behavior via alterations in the trajectory of neural development and its functional consequences. These effects are rarely global, but instead typically are drug-specific, dependent on the timing of exposure, and moderated by genetic background and early life experiences. Expression of drug-induced alterations also varies with assessment age and functions assessed. Among the consequences of perinatal exposure to drugs of abuse are increases in the propensity for later use/abuse of that drug, potentially contributing to a perpetuating and transgenerational cycle of drug use and abuse.

Cross-References

▶ Alcohol
▶ Alcohol Abuse and Dependence
▶ Cannabinoids
▶ Cocaine
▶ Cocaine Dependence
▶ Environmental Enrichment and Drug Action

- ► Epigenetics
- ► Fetal Alcohol Syndrome
- ► Impulsivity
- ► Neurotoxicity
- ► Neurotoxins
- ► Nicotine
- ► Nicotine Dependence and Its Treatment

References

Adams J Jr, Barone S, LaMantia A, Philen R, Rice DC, Spear L, Susser E (2000) Workshop to identify critical windows of exposure for children's health: neurobehavioral work group summary. Environ Health Perspect 108:535–544

Bandstra ES, Moorow CE, Mansoor E, Accornero VH (2010) Prenatal drug exposure: infant and toddler outcomes. J Addict Dis 29:245–258

Fried PA (2002) Conceptual issues in behavioral teratology and their application in determining long-term sequelae of prenatal marihuana exposure. J Child Psychol Psychiatry 43:81–102

Goodlett CR, Horn KH (2001) Mechanisms of alcohol-induced damage to the developing nervous system. Alcohol Res Health 25:175–184

Harvey JA (2004) Cocaine effects on the developing brain: current status. Neurosci Biobehav Rev 27:751–764

Hodgson E (2004) A textbook of modern toxicology. Wiley, Hoboken

Mattson SN, Schoenfeld AM, Riley EP (2001) Teratogenic effects of alcohol on brain and behavior. Alcohol Res Health 25:185–191

Slikker W Jr, Xu ZA, Levin ED, Slotkin TA (2005) Mode of action: disruption of brain cell replication, second messenger, and neurotransmitter systems during development leading to cognitive dysfunction – developmental neurotoxicity of nicotine. Crit Rev Toxicol 35:703–711

Slotkin TA (2008) If nicotine is a developmental neurotoxicant in animal studies, dare we recommend nicotine replacement therapy in pregnant women and adolescents? Neurotoxicol Teratol 30:1–19

Spear LP, File SE (1996) Methodological considerations in neurobehavioral teratology. Pharmacol Biochem Behav 55:455–457

Spear NE, Molina JC (2005) Fetal or infantile exposure to ethanol promotes ethanol ingestion in adolescence and adulthood: a theoretical review. Alcohol Clin Exp Res 29:909–929

Stanton ME, Spear LP (1990) Workshop on the qualitative and quantitative comparability of human and animal developmental neurotoxicology, Work Group I report: comparability of measures of developmental neurotoxicity in humans and laboratory animals. Neurotoxicol Teratol 12:261–267

Perinatal Use of Antidepressants

Katherine A. Blackwell[1], Ariadna Forray[1] and Kimberly A. Yonkers[2]
[1]Department of Psychiatry, Yale School of Medicine, New Haven, CT, USA
[2]Departments of Psychiatry, Obstetrics, Gynecology and Reproductive Sciences and the School of Epidemiology and Public Health, Yale School of Medicine, New Haven, CT, USA

Synonyms

Perinatal psychopharmacology

Definition

Depressive disorders are common in pregnancy (Yonkers et al. 2009). Treatment of pregnant women with depression requires careful consideration of the benefits of antidepressant agents and the obstetrical and perinatal risks to both mothers and their infants.

Current Concepts and State of Knowledge

Mood disorders frequently affect women of reproductive age and, therefore, commonly occur during pregnancy (Yonkers et al. 2009). While psychotherapy is effective for mild-to-moderate depression, women with severe depression or an incomplete response may require pharmacotherapy (Yonkers et al. 2009). Antidepressants cross the placenta, so maternal use results in some degree of fetal exposure. Consequently, the decision to initiate or modify antidepressant treatment requires a careful assessment of risks of untreated depression and those related to antidepressant medications.

Depression poses risks to both mother and infant. Compared to women without a mood disorder, depressed women are more likely to have poor prenatal and self-care, smoke, use alcohol or illicit drugs, have poor social and interpersonal functioning, and have concurrent medical conditions (Yonkers et al. 2009). On the other hand, treatment with any medication during pregnancy may be associated with a variety of adverse outcomes, including spontaneous abortion, fetal malformation or demise, preterm birth, impaired fetal growth, and neonatal complications.

Herein, we summarize work on the perinatal effects of serotonin reuptake inhibitors (SSRIs) and, to a lesser extent, tricyclic antidepressants (TCAs) and serotonin norepinephrine reuptake inhibitors (SNRIs). However, limitations of the available literature need to be recognized. There are no randomized controlled trials, due to ethical concerns regarding research in pregnancy. Much work is based on studies using large population-based retrospective cohorts, the majority of which have limited ability to control for potential confounding, including effects of maternal psychiatric illness and medical comorbidities, medication adherence, prenatal care, prior obstetrical and perinatal history, health habits, and tobacco, alcohol, and other substance use. Therefore, studies not addressing these factors require cautious interpretation.

Spontaneous Abortion

Spontaneous abortion is fetal loss early during pregnancy, usually prior to 20 weeks gestation, although the specific definition varies by jurisdiction. Several meta-analyses show a modest increase in risk of spontaneous abortion for women who are undergoing antidepressant treatment at conception (Yonkers et al. 2014). A cohort study of nearly 70,000 women showed that those who were undergoing antidepressant treatment at conception were 68 % more likely to have a miscarriage (Nakhai-Pour et al. 2010). Women with a depressive disorder diagnosis had slightly higher odds (19 %) of spontaneous abortion compared with healthy controls. The literature on possible associations between antidepressant use and spontaneous abortion has been limited by a number of confounding and measurement factors. As many miscarriages are clinically silent, most studies focus on late events and rely on self-report. Participant-reported outcomes may be problematic because some women report a spontaneous rather than a voluntary termination of pregnancy due to the stigma associated with the latter.

Fetal, Neonatal, and Postneonatal Death

In pregnancies after 20 weeks gestation, fetal loss prior to birth is typically called stillbirth. Regional differences in risk for fetal death and ways of operationalizing and reporting fetal death make estimating the rates difficult. In high-income countries, the rate of third-trimester fetal death is approximately 5 per 1,000; in the USA, 6 per 1,000 will experience either early or late stillbirth, with higher rates among black women as compared to non-Hispanic white women. Although data that evaluate the risk of fetal death with concurrent antidepressant treatment are not available in the USA, large registry studies from other countries do not find associations between antidepressant use in pregnancy and fetal death (Yonkers et al. 2014).

There is a paucity of studies evaluating the risk of neonatal and postneonatal demise associated with in utero antidepressant exposure (Yonkers et al. 2014). A recent report using data from the Nordic population-based registry, including over 1.6 million births (Stephansson et al. 2013), found that use of an SSRI in pregnancy and risk for either neonatal or postneonatal death was nonsignificant after adjustment for potential confounding factors. This is in contrast with an earlier study that found the rate of neonatal death was increased in infants exposed to SSRIs in utero, but did not control for potential confounding.

Congenital Malformations

While medications are estimated to cause less than 1 % of congenital malformations, they are potentially modifiable risk factors. A 2013 meta-analysis studied the effects of any antidepressant exposure on congenital malformations (Grigoriadis et al. 2013). In this analysis there was no statistically significant effect of antidepressant exposure on major malformations, but there was a 36 % increase in risk for cardiovascular malformations. An overview of this meta-analysis and specific studies is published elsewhere (Yonkers et al. 2014).

Paroxetine has been more intensively studied than other antidepressants, after the US Food and Drug Administration increased the pregnancy risk category of this drug due to concerns about an elevated risk of congenital malformations. A meta-analysis conducted by its manufacturer subsequently found that paroxetine exposure was associated with a 24% increase in total malformations and 46 % increase in cardiovascular malformations (Wurst et al. 2010). The 2013 meta-analysis also analyzed risk of congenital malformations with paroxetine exposure, finding that cardiovascular malformations were increased 43 % but that there was no statistically significant association with aggregated major malformations (Grigoriadis et al. 2013).

Fewer studies address other classes of antidepressants. There are conflicting reports on the risk of congenital malformations after TCA, SNRI, and bupropion treatment; however, those studies identifying an association with congenital malformations have generally found small effect sizes. On the whole, the literature on the effects of antidepressants on risk for congenital malformations has many conflicting results, which likely reflects variability in controlling for confounding, cohort differences, limited power to detect differences in rare events, and reporting of multiple comparisons. Nonetheless, available meta-analyses suggest that antidepressants, with the possible exception of paroxetine, are unlikely to be major teratogens.

Gestational Age and Preterm Birth

This outcome has been extensively investigated in women who are pregnant and undergoing antidepressant treatment (Yonkers et al. 2014). A comprehensive analysis of a large registry that includes baseline assessment of multiple confounding factors provides valuable information on the risk associated with antidepressants (Reis and Kallen 2010). Preterm birth, typically defined as birth prior to 37 weeks gestation, was associated with use of antidepressant agents in pregnancy, with TCAs associated with 2.36-fold increased odds, SNRIs with 1.98-fold increased odds, and SSRIs with 1.46-fold increased odds relative to the odds of preterm birth with no exposure. The study did not include information about maternal diagnosis, illness severity, or other health behaviors which may distort the apparent relationship between antidepressant use and preterm birth. However, if the association is causal, the overall risk is modest and the duration of pregnancy is typically shortened by only 3–5 days.

Fetal Growth

Fetal growth abnormalities that may be associated with gestational use of antidepressants have received substantial attention. The results are mixed, and in contrast to preterm birth, an association is not strongly supported (Yonkers et al. 2014). Specifically, TCA use in pregnancy does not appear to be associated with small-for-gestational age infants. Most of the larger and better-designed studies, which controlled for confounding due to substance misuse and psychiatric disorder, do not find an elevated risk of fetal growth abnormalities in pregnant women treated with SSRIs. In their meta-analysis that included studies published before 2010, Ross and colleagues found that a causal association between antidepressant use and fetal growth would lead to a decrease in only 74 g at birth. When they controlled for psychiatric illness, there was no statistical difference between antidepressant exposure groups (Ross et al. 2013). Given these findings, fetal growth effects should not be a major concern in mothers who take antidepressants in pregnancy.

Neonatal Behavioral Syndrome

Antidepressant use in pregnancy has been associated with adverse neonatal events, including irritability, jitteriness, trouble feeding, tremor, agitation, hypertonia, hyperreflexia, respiratory distress, seizures, vomiting, and excessive crying. Symptoms are time limited, resolving within two weeks of delivery with supportive care. Special care nursery admission may be necessary in cases of severe symptoms such as apnea, seizures, and temperature instability. These signs are generally thought to be due to a combination of SSRI discontinuation and direct toxic effects.

A review by Moses-Kolko et al. (2005) found that among cohort studies that defined a neonatal behavioral syndrome, SSRI exposure was associated with an overall three times greater risk of the defined syndrome, compared with control infants. More recent studies have confirmed these findings and found that up to 30 % of infants exposed to SSRIs in pregnancy develop neonatal behavioral syndrome. The literature examining a neonatal behavioral syndrome in association with exposure to SSRIs has several limitations, including a lack of a standard definition of the syndrome, surveillance bias, and recall bias. Despite these limitations, the evidence suggests there is a time-limited neonatal behavioral syndrome associated with SSRI exposure that can be readily managed with supportive care.

Persistent Pulmonary Hypertension

Persistent pulmonary hypertension of the newborn (PPHN) is a rare but serious condition characterized by the failure of vascular resistance to decrease after birth. This results in ineffective oxygenation and can lead to respiratory failure.

There is conflicting evidence on the risk of PPHN in infants exposed to SSRIs during pregnancy. Of the seven studies on SSRI exposure and PPHN so far, 83 cases of PPHN were identified in an estimated 36,014 exposures to SSRIs. In a review of the first six available studies, Occhiogrosso et al. concluded that the evidence linking SSRIs and PPHN is not strong (Occhiogrosso et al. 2012). Only two studies have examined PPHN and exposure to SNRIs, TCAs, and other antidepressants, and they report conflicting results; therefore, no conclusions can be made at this time.

When evaluating the literature on SSRI exposure and PPHN, it is important to keep in mind that known risk factors for PPHN, including obesity, cesarean section, and smoking, are more common in depressed women. Furthermore, both untreated depression and exposure to SSRIs in pregnancy have been linked to preterm birth, which increases the risk of PPHN fourfold. In summary, the available literature shows either a small association or no association between PPHN and maternal SSRI use in late pregnancy.

Neurodevelopmental Effects

Neurodevelopment occurs throughout gestation, and serotonin plays a vital role in the development of the fetal central nervous system. As such, it is conceivable that prenatal exposure to medications that can modify synaptic serotonin concentrations may alter fetal and infant neurodevelopment. The few studies that have investigated the neurodevelopment of infants exposed in utero to maternal antidepressants have shown inconsistent results. For a review of these results, refer to Yonkers et al. 2014. While fetal neurodevelopment might be susceptible to antidepressant exposure, the effects, if any, are limited and may be compensated for over time. The influence of maternal mood and depression severity on behavioral and neurocognitive outcomes has to be taken into account, as these are known to strongly influence child behavior and development.

Summary

While there is a substantial literature on the obstetrical and perinatal outcomes of antidepressant use in pregnancy, overall the results are mixed. This likely reflects the methodological limitations of the study designs, particularly inadequate controlling for confounding. However, as a whole, several outcomes have relatively consistent results, including a probable effect on a neonatal behavioral syndrome, possible increases in spontaneous abortions and preterm birth, and a possible association of paroxetine and malformations. In most cases, the effect sizes are small and severe outcomes rare. Medical decision making related to these risks, as well as the risks of declining treatment, needs to be addressed with patients in a nuanced way, so that women are able to make informed decisions regarding pharmacological and other treatments tailored to the specific patient.

Cross-References

- ▶ Adverse Effect
- ▶ Antidepressants
- ▶ Antidepressants: Recent Developments
- ▶ Bupropion
- ▶ Depressive Disorders: Major, Minor, and Mixed
- ▶ FDA Pregnancy Category
- ▶ Generalized Anxiety Disorder
- ▶ Obsessive–Compulsive Disorders
- ▶ Panic Disorder
- ▶ Perinatal Exposure to Drugs
- ▶ Selective Serotonin Reuptake Inhibitors
- ▶ Sex Differences in Drug Effects
- ▶ SNRI Antidepressants
- ▶ Social Anxiety Disorder
- ▶ Teratogenic
- ▶ Tricyclic Antidepressants

References

Grigoriadis S, VonderPorten EH, Mamisashvili L, Roerecke M, Rehm J, Dennis CL, Koren G, Steiner M, Mousmanis P, Cheung A, Ross LE (2013) Antidepressant exposure during pregnancy and congenital malformations: is there an association? A systematic review and meta-analysis of the best evidence. J Clin Psychiatry 74(4):e293–e308

Moses-Kolko E, BogenD PJ, Bregar A, Uhl K, Levin B, Wisner KL (2005) Neonatal signs after late in utero exposure to serotonin reuptake inhibitors: literature review and implications for clinical applications. JAMA 293(19):2372–2383

Nakhai-Pour HR, Broy P, Berard A (2010) Use of antidepressants during pregnancy and the risk of spontaneous abortion. Can Med Assoc J 182(10):1031–1037

Occhiogrosso M, Omran SS, Altemus M (2012) Persistent pulmonary hypertension of the newborn and selective serotonin reuptake inhibitors: lessons from clinical and translational studies. Am J Psychiatry 169(2):134–140

Reis M, Kallen B (2010) Delivery outcome after maternal use of antidepressant drugs in pregnancy: an update using Swedish data. Psychol Med 40(10):1723–1733

Ross L, Grigoriadis S, Mamisashvili L, Vonderporten EH, Roerecke M, Rehm J, Dennis CL, Koren G, Steiner M, Mousmanis P, Cheung A (2013) Selected pregnancy and delivery outcomes after exposure to antidepressant medication: a systematic review and meta-analysis. JAMA Psychiatry 70(4):436–443

Stephansson O, Kieler H, Haglund B, Artama M, Engeland A, Furu K, Gissler M, Nørgaard M, Nielsen RB, Zoega H, Valdimarsdóttir U (2013) Selective serotonin reuptake inhibitors during pregnancy and risk of stillbirth and infant mortality. JAMA 309(1):48–54

Wurst KE, Poole C, Ephross SA, Olshan AF (2010) First trimester paroxetine use and the prevalence of congenital, specifically cardiac, defects: a meta-analysis of epidemiological studies. Birth Defects Res A Clin Mol Teratol 88(3):159–170

Yonkers KA, Blackwell KA, Glover J, Forray A (2014) Antidepressant use in pregnant and postpartum women. Ann Rev Clin Psychol 10:369–392

Yonkers KA, Wisner KL, Stewart DE, Oberlander TF, Dell DL, Stotland N, Ramin S, Chaudron L, Lockwood C (2009) The management of depression during pregnancy: a report from the American Psychiatric Association and the American College of Obstetricians and Gynecologists. Gen Hosp Psychiatry 31(5):403–413

Peripheral Markers

Definition

Accessible markers mirroring physiopathologic processes of a disease. Usually, this term refers to markers of neuropsychiatric diseases, considering the difficulties involved in accessing directly into the CNS for studying the pathologic processes. Due to their availability, CSF and blood represent two major sources for identifying suitable peripheral markers of neuropsychiatric dysfunction.

Perospirone

Definition

Perospirone is a benzisothiazole derivative that belongs to the class of second-generation

▶ (atypical) antipsychotic drugs and is indicated for the treatment of schizophrenia. It has high-affinity antagonist activity at 5-HT$_{2A}$ and D$_2$ receptors. It also displays partial 5-HT$_{1A}$ agonist properties and some affinity for D$_1$, α$_1$-adrenergic, and H$_1$ receptors, but has no appreciable affinity for muscarinic receptors. It has at least four active metabolites, but all of them have lower affinity for D$_2$ and 5-HT$_{2A}$ receptors than the parent drug. It can induce ▶ extrapyramidal motor side effects and insomnia, but it displays low toxicity.

Cross-References

- ▶ Extrapyramidal Motor Side Effects
- ▶ Schizophrenia
- ▶ Second-Generation Antipsychotics

Perphenazine

Definition

Perphenazine is a first-generation ▶ antipsychotic medication that acts as a dopamine D2 receptor antagonist. It is a piperazine-phenothiazine derivative, also available as depot medication. Perphenazine is indicated for the treatment of schizophrenia and other psychoses, mania in bipolar disorder, and the short-term adjunctive treatment of severe anxiety, psychomotor agitation, and excited or violent states. It can also be used as an antiemetic drug. Extrapyramidal motor symptoms occur, especially dystonia, particularly at high doses.

Cross-References

- ▶ Antipsychotic Drugs
- ▶ Bipolar Disorder
- ▶ Extrapyramidal Motor Side Effects
- ▶ First-Generation Antipsychotics
- ▶ Schizophrenia

Perseveration

Definition

The antithesis of flexibility, perseveration refers to responses or behaviors that persist even when they are no longer beneficial and may even have a cost. The term is usually used to refer to the continuous repetition of thoughts, statements and behavioral responses. Whenever a task involves an uncued switch, shift, or reversal, the subject will make errors until the new contingencies are learned. If the previously rewarded response persists despite a lack of reinforcement, it is perseverative. There are examples in the literature of the classification of errors made after a rule change as "perseverative" when a respondent persistently returns to making a previously rewarded response even after there is evidence that a new response has been learned (errors up until this point being classified as "learning errors"). In this, more specific, use of the term, a statistical criterion is applied to make the classification.

Cross-References

- ▶ Five-Choice Serial Reaction Time Task

Personality: Neurobehavioral Foundation and Pharmacological Protocols

Richard A. Depue[1] and Tara L. White[2]
[1]Department of Human Development, Laboratory of Neurobiology of Personality, College of Human Ecology, Cornell University, Ithaca, NY, USA
[2]Department of Community Health, Laboratory of Affective Neuroscience, Center for Alcohol and Addiction Studies, Brown University, Providence, RI, USA

Definition

Personality represents the structuring of human behavior into major traits that reflect the activity of (1) motivational-emotional systems and (2) their underlying neurobiological networks. There is consensus on the existence of four major traits: extraversion, neuroticism, social closeness or agreeableness, and constraint or conscientiousness. We describe the psychological features of these four traits, and how pharmacological protocols have aided in defining their underlying neurobiology. Discussion of the relation of dopamine (DA), serotonin (5-HT), corticotropin-releasing hormone (CRH), and mu-opiate functioning within specific brain regions to the four major traits is provided.

Current Concepts and State of Knowledge

The Nature of Personality. In the early 1970s, a significant development occurred in reconceptualizing the nature of personality. Employing an evolutionary biology perspective, Gray (1973) proposed a new structure of human-motivated behavior defined as systems that evolved to adapt to stimuli critical for survival and species preservation. Such behavioral systems are fundamentally emotional systems that incorporate both a motivational state and emotional experience that is concordant with, and engages us with or disengages us from, rewarding or aversive critical stimuli. Furthermore, Gray hypothesized that each of these motivational-emotional systems is associated with its own network of brain regions and with specific neurotransmitters and neuropeptides that modulate the functioning of those networks. Within this framework, the mean functional properties of the relevant neurotransmitters-neuropeptides, influenced by genetics and experience, may underlie stable individual differences in the behavioral reactivity of such systems. This theoretical framework led Gray to suggest further that stable neurobiological individual differences in motivational-emotional systems form the foundation of higher-order traits of personality. These notions raised the possibility that one means of testing the neurobiological tenets of this theory is the use of pharmacological protocols, where pharmacological agents that act as agonists or antagonists of neurotransmitters are used to assess whether drug-induced modulations of specific neurotransmitters are associated with particular personality traits.

The Structure of Personality. The structure of personality is hierarchical in nature, where behavioral features are assessed by inventory items which, by use of factor analysis, are clustered into lower-order primary or facet scales, which in turn are further clustered into higher-order traits. The higher-order traits represent theoretical constructs that attempt to account for the coalescence of the lower-order traits. From Gray's theoretical framework, the higher-order traits reflect the activity of a specific motivational-emotional system, and the lower-order traits represent different behavioral patterns that intercorrelate because they are influenced similarly by the motivational-emotional system represented by the higher-order trait. Although numerous classificatory systems of personality exist, all tend to agree on four higher-order traits: extraversion, neuroticism, social closeness or agreeableness, and constraint or conscientiousness. The first three traits are thought to reflect the activity of specific motivational-emotional systems, whereas the fourth may reflect not a specific emotional system, but rather a general property of the central nervous system that affects the threshold for eliciting all types of emotional behavior.

Extraversion. Depue and Collins (1999) provided a comprehensive analysis of the trait of extraversion as reflecting the activity of one of Gray's motivational-emotional systems - behavioral approach based on incentive motivation. Both unconditioned and conditioned rewarding stimuli (positive incentive stimuli) activate a motivational system of incentive motivation, which energizes approach to a rewarding goal. Extraversion, then, would represent individual differences in the threshold for positive incentives to (1) facilitate an emotional state that supports approach behavior, including a positive affective state of desire, strong, peppy, enthusiastic, excited, and self-confidence; and (2) elicit approach, forward locomotion, and engagement with rewarding stimuli.

Conceptualizing extraversion as based on incentive motivation, which is a behavioral system found at all phylogenetic levels of animals, allows for the drawing of an analogy between human personality patterns and the animal neurobehavioral research literature. Animal research demonstrates that behavioral facilitation reflecting the activation of incentive motivation is associated with the functional properties of the ventral tegmental area (VTA) dopamine (DA) projection system. Thus, just as extraversion emerges as a higher-order construct that incorporates a modulatory mechanism that operates across lower-order traits, the VTA DA projection system might also be considered a higher-order modulator of a neurobiological network that integrates behavioral functions associated with extraversion. Indeed, DA agonists or antagonists in the VTA or nucleus accumbens (Nacc), which is a major terminal area of VTA DA projections, in animals facilitate or markedly impair, respectively, incentive-elicited locomotor activity to novelty and food; exploratory, aggressive, social, and sexual behavior; and the acquisition and maintenance of approach behavior.

In humans, DA agonists, such as amphetamine, cocaine, and methylphenidate, also elicit facilitated motor activity, as well as both positive emotional feelings such as elation, and a sense of reward, and motivational feelings of desire, wanting, craving, potency, and self-efficacy (Canli 2006; Depue 2006; Depue and Collins 1999). Moreover, neuroimaging studies have found that, during acute cocaine administration, the intensity of a subject's subjective euphoria increased in a dose-dependent manner in proportion to cocaine binding to the DA uptake transporter (and hence DA levels) in the striatum.

A number of pharmacological protocols have been used to explore the relation of extraversion to DA functioning in humans. A widely used protocol involves use of a DA agonist-induced challenge of (1) variables that are known to be modulated by DA activation, such as hormones like growth hormone and prolactin, eye-blink rates, rate of switching between percepts (e.g., ascending vs. descending views of a staircase), motor velocity, positive affective reactivity, and visuospatial working memory; or (2) fMRI-imaged, DA-relevant brain regions during the performance of psychological tasks (Canli 2006; Depue 2006; Depue and Collins 1999). Such studies have consistently shown a strong, positive correlation between degree of DA-induced modulation of the variable in question and questionnaire-assessed extraversion, often accounting for over 40 % of the observed variance.

Another important protocol is the use of a pharmacological agent that binds to specific DA receptors, where the level of competitive binding with natural DA, measured by PET scanning, is inversely related to the natural level of DA functioning. This protocol has been used to assess D2-like DA autoreceptor density. Since D2 autoreceptors are localized to the soma and dendrites of VTA DA neurons and provide one of the most potent inhibitory modulations of DA cell firing, reduced autoreceptor availability in the VTA region is related to increased DA cell firing and enhanced postsynaptic activation. Extraversion-like traits have been found to be inversely related (approximately −0.68) to D2 autoreceptor availability in the VTA region (e.g., Zald et al. 2008).

Finally, use of a pharmacological challenge protocol has recently been used in much more detailed ways to assess the nature of the relation of DA to psychobiological processes associated with extraversion. A major effect of DA release in the Nacc is to enhance the binding of (1) corticolimbic inputs to Nacc dendrites that carry information about the current context and (2) the magnitude of context-induced reward (Depue and Collins 1999; Depue and Morrone-Strupinsky 2005; Kauer and Malenka 2007). That is, DA plays an important role in binding contextual ensembles to the occurrence of reward, an important predictive function of the brain. As animal research has shown, enhanced state or trait DA functioning increases the binding of contextual cues to the experience of reward. Similarly, we have found that increasing levels of extraversion are strongly associated with the ability to bind together the experience of DA agonist (methylphenidate)-induced reward and contextual cues, as evidenced in DA modulated processes of motor velocity, positive affective reactivity, and working memory (Depue 2006).

These findings have relevance for the maintenance of drugs of abuse that activate DA release into the Nacc, which most do (Kauer and Malenka 2007). Indeed, DA-induced synaptic changes in the Nacc are likely to facilitate the formation of powerful and persistent links between the rewarding aspects of the drug experience and the multiple cues associated with that experience - that is, to facilitate long-lasting memories of the drug experience. In light of our findings discussed above, it may be that extraversion represents one modifier of the strength of drug (reward)-context conditioning, and hence of relapse rates.

Neuroticism. Adaptation to aversive environmental conditions is crucial for species survival, and at least two distinct behavioral systems have evolved to promote such adaptation. One system is fear (often labeled *Harm Avoidance* in personality literature), which is a behavioral system that evolved as a means of escaping very specific and explicit aversive stimuli that are inherently dangerous to survival, such as tactile pain, predators, snakes, spiders, heights, and sudden sounds. There are, however, many aversive circumstances in which *specific* aversive cues do not exist, but rather the stimulus conditions are associated with an elevated *potential* risk of danger, such as darkness, open spaces, strangers, unfamiliarity, and predator odors. Conceptually, these latter stimuli are characterized in common by their unpredictability and uncontrollability - or, more simply, uncertainty.

To adapt to these latter stimulus conditions, a second behavioral system evolved, *anxiety*, and it is this system that is thought to underlie the trait of Neuroticism. Anxiety is characterized by negative emotion or affect (anxiety, depression, hostility, suspiciousness, distress) that serves the purpose of informing the individual that, though no explicit, specific aversive stimuli are present, conditions are potentially threatening (White and Depue 1999). This negative affective state continues or reverberates until the uncertainty is resolved. It is the prolonged negative subjective state of anxiety that distinguishes its subjective state from the rapid, brief state of panic associated with the presence of a specific fear stimulus. The trait literature supports the *independence* of anxiety and fear, which as personality traits are completely uncorrelated, and are subject to distinct sources of genetic variation (White and Depue 1999).

The psychometric independence of fear and anxiety is mirrored in their dissociable neuroanatomy (Depue and Lenzenweger 2005; Davis and Shi 1999) and neurochemistry (White and Depue 1999; White et al. 2006). Whereas fear is dependent on output from the central nucleus of the amygdala to various brainstem regions, anxiety is subserved by outputs to similar brainstem regions from the bed nucleus of the stria terminalis (BNST), a structure that receives visual information via glutamatergic efferents from the perirhinal + basolateral amygdala (e.g., light–dark conditions) and parahippocampal + entorhinal + hippocampal (contextual and unfamiliarity stimuli) regions, as opposed to specific objects or sounds converging on the basolateral amygdala (Davis and Shi 1999; White and Depue 1999). The dissociation between fear and anxiety has also been found repeatedly with monoamine agonist challenge, such that trait fear but not trait anxiety predicts mood, cortisol, and physiologic responses to general monoamine challenge and to alpha-1 noradrenergic challenge, suggesting a role for ascending catecholamine systems in trait fear but not trait anxiety in humans (White and Depue 1999; White et al. 2006).

Animal research has demonstrated that naturally occurring chronic stress activates the central CRH system, which, as shown in Fig. 1, is composed of CRH neurons located in many different subcortical brain regions that modulate emotion, memory, and central nervous system arousal (Bale 2005; Depue and Lenzenweger 2005). Importantly, CRH neurons in the central amygdala induce prolonged elevated levels of CRH in BNST, which accounts for the potentially long endurance of anxiety as opposed to fear responses (Bale 2005). For instance, marked anxiety effects lasting longer than 24 h can be produced

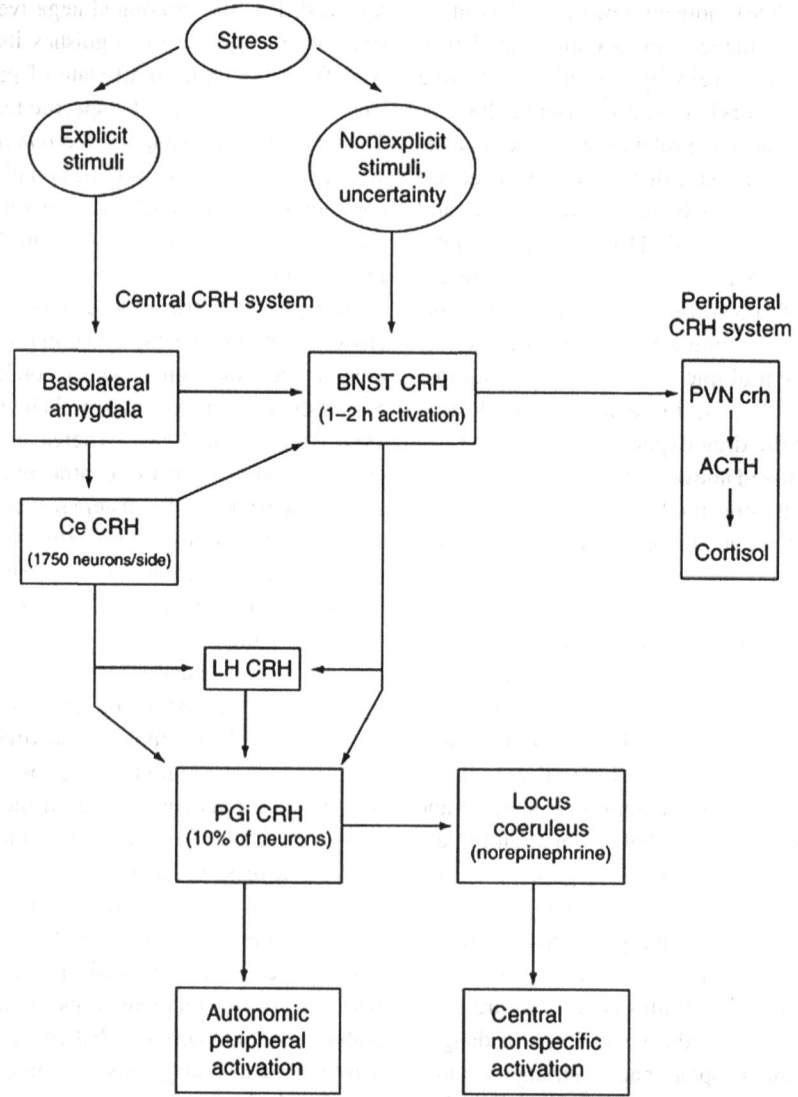

Personality: Neurobehavioral Foundation and Pharmacological Protocols, Fig. 1 CRH stress-response systems. Components of the central and peripheral corticotropin-releasing hormone (CRH) systems, which together coordinate emotional responses to stressful stimuli. Explicit stressful stimuli (e.g., predator cues, heights, tactile pain) are processed by the basolateral amygdala, whereas nonexplicit stressful stimuli (e.g., open spaces, context, darkness, strangers) are processed by the BNST. The basolateral amygdala activates Ce CRH neurons (about 1,750 neurons per hemisphere) which, together with nonexplicit stressful cues, release CRH in the BNST to produce prolonged neural activation in the BNST. In both cases, the Ce and/or BNST activate CRH neurons in the LH, which integrates and activates ANS activity in response to the stressor. In turn, Ce, BNST, and LH activation of the PGi (a major integrative nucleus in the rostroventrolateral medulla) CRH neurons (about 10 % of PGi neurons) facilitate ANS activity via afferents to the spinal cord. The PGi CRH neurons also activate LC norepinephrine neurons, which project broadly to the CNS and elicit activation of all CNS regions. The peripheral CRH system is activated by the Ce and BNST CRH output neurons to the PVN. *Ce* central amygdala nucleus; *BNST* bed nucleus of the stria terminalis; *LH* lateral hypothalamus; *PGi* medullary paragiganticocellularis nucleus; *PVN* paraventricular nucleus of the hypothalamus; *ACTH* corticotropic hormone from the anterior pituitary; *CNS* central nervous system; *ANS* autonomic nervous system experimentally after three doses of CRH administered centrally over 1.5 h, but with no lasting effect on peripheral release of CRH from the hypothalamus. Anxiogenic effects and an aversion to a CRH-paired environment, both via CRH-R1 receptors, are dependent on intra-BNST administration of CRH. Furthermore, transgenic mice with elevated CRH-R1 (but not with R2) receptors in the central forebrain (but not peripherally in hypothalamus or pituitary), or conditional activation of CRH-R1 gene expression centrally, show extreme indications of anxiety. Thus, anxiety is a stress response system that relies on a network of central CRH neuron populations, in conjunction with the peripheral CRH system, that provide integrated responses (hormonal, behavioral, autonomic and central arousal) to a stressor, and in lateral BNST mediate prolonged anxiogenic effects and aversive contextual conditioning.

Pharmacological protocols used in exploring the psychobiology of extraversion have not been applied in human studies of anxiety, and so this type of research is sorely needed. One important avenue of this research would be the development of CRH-R1 antagonists that act centrally as anxiolytic agents. That such an approach may be significant is supported by a recent study of such an antagonist (antalarmin) in monkeys: Oral antalarmin impaired development and expression of contextual fear conditioning, and decreased stress-induced increases in anxiety behaviors, ACTH, cortisol, norepinephrine, epinephrine (i.e., HPA axis), CSF CRH concentration, and locus coeruleus neuronal firing (Habib et al. 2000).

One additional area of research is worth noting. A form of chronic stress reactivity is jointly correlated with the trait of neuroticism and a polymorphism in the promoter region of the gene that codes for the serotonin (5-HT) uptake transporter, creating two common alleles - long (l) and short (s). The s-allele is associated with (1) twice the risk of anxiety, depression, and suicide attempts in a context of childhood maltreatment and stressful life events; and (2) persistent enhanced amygdala activation to emotional stimuli - and thus persistent vigilance for threat, increased aversive conditioning to context, and negative thoughts and emotional memories. One important hypothesis is that, due to the significant effects of 5-HT on early exuberant development of connectivity within forebrain regions, that the amygdala is less well modulated by emotion regulating regions such as the rostral anterior cingulate cortex. Indeed, these two regions show significantly less coherence in activity in s- versus l-allele individuals, accounting for ~30 % of the variance in neuroticism scores (Pezawas et al. 2005). There has been little use of human pharmacological protocols to explore the mechanisms of how 5-HT modulates stress reactivity, vigilance, and memory formation in these polymorphic conditions, again suggesting an area ripe for pharmacological exploration.

Social Closeness/Agreeableness. Social Closeness reflects the capacity of an individual to experience reward elicited by affiliative stimuli (e.g., soft tactile stimulation associated with grooming, caressing, and sexual intercourse). This capacity is reflected in the degree to which people value close relationships and spend time with others, as well as the desire to be comforted by others at times of stress. There has been very little human neurobiology research on the capacity to become attached to another person. While vole research has focused on the roles of oxytocin and vasopressin, genetic knockout studies have not demonstrated a *necessary* role for these neuropeptides in social bonding, suggesting that they may be affecting other variables that influence bonding, such as facilitation of sensory receptivity and subsequent acquisition of memories (Veenema and Neumann 2008). Human studies, on the other hand, are few, but some have used a promising pharmacological protocol: intranasally administered oxytocin is used to modulate short-term neural states and is then correlated with affective processes.

Noting that these neuropeptides do not mediate a sense of reward elicited by affiliative stimuli, we detailed a comprehensive role for

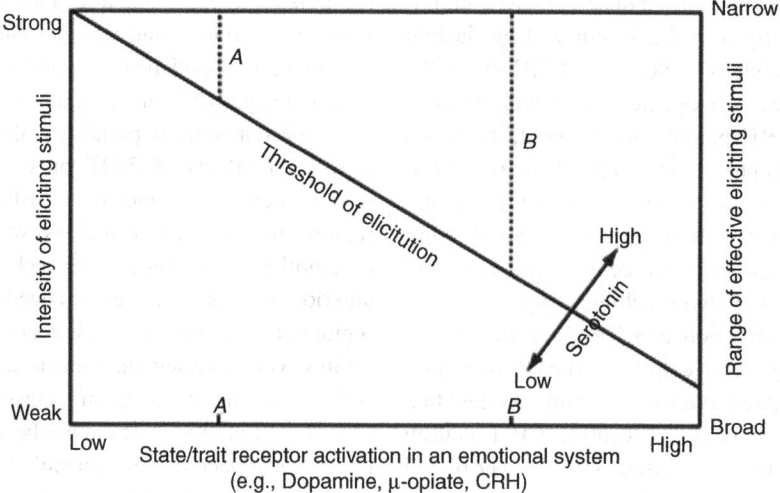

Personality: Neurobehavioral Foundation and Pharmacological Protocols, Fig. 2 Minimum threshold model. A minimum threshold for elicitation of a behavioral process (e.g., incentive motivation-positive affect, affiliative reward-affection, anxiety-negative affect) is illustrated as a trade-off function between eliciting stimulus magnitude (*left vertical axis*) and postsynaptic receptor activation in a neurobiological system (e.g., dopamine, mu-opiate, CRH) underlying an emotional trait (*horizontal axis*). Range of effective (eliciting) stimuli is illustrated on the right vertical axis as a function of level of receptor activation. Two hypothetical individuals with low and high *trait* postsynaptic receptor activation (demarcated on the horizontal axis as *A* and *B*, respectively) are shown to have narrow (*A*) and broad (*B*) ranges of effective stimuli, respectively, which influences the frequency of activation of the processes associated with a personality trait. Threshold effects due to serotonin modulation are illustrated as well

mu-opiates in affiliative reward (Depue and Morrone-Strupinsky 2005). Recent primate research supports the *necessary* aspect of mu-opiates for social bonding (Barr et al. 2008). Practically no human pharmacological work has been done on mu-opiate mediation of affiliative reward, but the recent animal studies above are concordant with findings of regulation of human affective responses by limbic-opioid neurotransmission. They are also consistent with our results of enhanced human affiliative responses in individuals scoring high versus low in trait levels of social closeness, and with the elimination of those differences by use of the mu-opiate antagonist, naltrexone. Clearly, additional pharmacological work along these lines is strongly encouraged.

Constraint/Conscientiousness. Constraint is a poorly conceptualized personality trait, but is clearly related to a generalized behavioral impulsivity. Recent models of this behavioral profile have begun to clarify the nature of this trait (Depue and Collins 1999; Depue and Morrone-Strupinsky 2005). Elicitation of behavior can be modeled neurobiologically by use of a minimum threshold construct, which represents a central nervous system weighting of the external and internal factors that contribute to the probability of response expression (Fig. 2). We and others have proposed that constraint is the personality trait that reflects the greatest CNS weight on the construct of a minimum emotional response threshold. As such, constraint exerts a general influence over the elicitation of *any* emotional behavior. In this model, other higher-order personality traits would thus reflect the influence of neurobiological variables that strongly contribute to the threshold for responding, such as DA in the facilitation of incentive motivated behavior, mu-opiates in the experience of affiliative reward, and CRH in the potentiation of anxiety.

Functional levels of neurotransmitters that provide a strong, relatively generalized *tonic inhibitory* influence on behavioral responding would be good candidates as significant modulators of a response elicitation threshold, and hence may account for a large proportion of the variance in the trait of constraint. We and numerous others have suggested that 5-HT, acting at multiple receptor sites in most brain regions, is such a modulator (see review by Carver and Miller 2006). 5-HT modulates a diverse set of functions including emotion, motivation, motor, affiliation, cognition, food intake, sleep, sexual activity, and sensory reactivity, and reduced 5-HT functioning is associated with many disorders of impulse control (Depue and Lenzenweger 2005). Thus, 5-HT plays a substantial modulatory role that affects many forms of motivated behavior. Therefore, constraint might be viewed as reflecting a modulatory influence of 5-HT over the threshold of elicitation of emotional behavior (Fig. 2).

The variety of pharmacological protocols described above have been used extensively in animal research with 5-HT, and most show that reduced 5-HT functioning is indeed related strongly to a reduced threshold of emotional reactivity. However, such studies have been few in the human personality area, despite the fact that we found strong evidence for a role of 5-HT in constraint using a protocol of 5-HT-activation of prolactin secretion. Therefore, there is a strong need for additional pharmacological research on constraint to detail the manner in which 5-HT modulates emotional and cognitive processes associated with personality traits.

Conclusion

The use of pharmacological protocols to explore the neurobiological basis of higher-order personality traits is an underutilized research strategy. Clearly, such an approach has been quite informative with respect to the neurobiological nature of extraversion and neuroticism. However, pharmacological research on the nature of social closeness and constraint lags far behind. In particular, recent work suggests that a particularly powerful research strategy is the use of pharmacological protocols in defining the relation between behavioral processes and neurobiology in personality groups defined on the basis of genetic polymorphisms associated with DA, CRH, 5-HT, and mu-opiate systems.

Cross-References

▶ Aminergic Hypotheses for Depression
▶ Anxiety: Animal Models
▶ Arginine-Vasopressin
▶ Benzodiazepines
▶ Central Catecholamine Systems
▶ Conditioned Drug Effects
▶ Conditioned Place Preference and Aversion
▶ Emotion and Mood
▶ Impulse Control Disorders
▶ Impulsivity
▶ Neuroendocrine Markers for Drug Action
▶ Opioids
▶ Phenotyping of Behavioral Characteristics
▶ Receptors: Functional Assays
▶ Social Stress
▶ Trait Independence
▶ Tryptophan Depletion

References

Bale T (2005) Sensitivity to stress: dysregulation of CRF pathways and disease development. Horm Behav 48:1–10

Barr CS, Schwandt ML, Lindell SG, Higley JD, Maestripieri D, Goldman D, Suomi SJ, Heilig M (2008) Variation at the mu-opioid receptor gene (OPRM1) influences attachment behavior in infant primates. Proc Natl Acad Sci 105:5277–5281

Canli T (ed) (2006) Psychobiology of personality and individual differences. Guilford Press, New York

Carver CS, Miller CJ (2006) Relations of serotonin function to personality: current views and a key methodological issue. Psychiatry Res 144:1–15

Davis M, Shi C (1999) The extended amygdala: are the central nucleus of the amygdala and the bed nucleus of the stria terminalis differentially involved in fear versus anxiety. In: McGinty JF (ed) Advancing from the ventral striatum to the extended Amygdala. Ann NY Acad Sci 877:281–291

Depue R (2006) Dopamine in agentic and opiates in affiliative forms of extraversion. In: Canli T (ed) Biology of personality and individual differences. Guilford Press, New York

Depue R, Collins P (1999) Neurobiology of the structure of personality: dopamine, facilitation of incentive motivation, and extraversion. Behav Brain Sci (Target Article) 22:491–569

Depue R, Lenzenweger MF (2005) A neurobehavioral model of personality disorders. In: Cicchetti D (ed) Developmental psychopathology. Wiley, New York

Depue R, Morrone-Strupinsky J (2005) Neurobehavioral foundation of affiliative bonding: implications for a human trait of affiliation. Behav Brain Sci (Target Article) 28:313–395

Gray JA (1973) Causal theories of personality and how to test them. In: Royce JR (ed) Multivariate analysis and psychological theory. Academic, New York

Habib KE, Weld KP, Rice KC, Pushkasd J, Champoux M, Listwaka S, Webster EL, Atkinson AJ, Schulkin J, Carlo Contoreg C, Chrousos GP, McCann SM, Suomi SJ, Higley JD, Gold PW (2000) Oral administration of a corticotropin-releasing hormone receptor antagonist significantly attenuates behavioral, neuroendocrine, and autonomic responses to stress in primates. Proc Natl Acad Sci 97:6079–6084

Kauer JA, Malenka RC (2007) Synaptic plasticity and addiction. Nat Rev Neurosci 8:844–858

Pezawas L, Meyer-Lindenberg A, Drabant EM, Verchinski BA, Munoz KE, Kolachana BS, Egan MF, Mattay VS, Hariri AR, Weinberger DR (2005) 5-HTTLPR polymorphism impacts human cingulate-amygdala interactions: a genetic susceptibility mechanism for depression. Nat Neurosci 8:828–834

Veenema AH, Neumann ID (2008) Central vasopressin and oxytocin release: regulation of complex social behaviours, Chp. 22. In: Neumann ID, Landgraf R (eds) Progress in brain research, vol 170. Elsevier, New York

White TL, Depue R (1999) Differential association of traits of fear and anxiety with norepinephrine- and dark-induced pupil reactivity. J Pers Soc Psychol 77:863–877

White TL, Grover VK, de Wit H (2006) Cortisol effects of D-amphetamine relate to traits of fearlessness and aggression but not anxiety in healthy humans. Pharmacol Biochem Behav 85:123–131

Zald DH, Cowan RL, Patrizia Riccardi P, Baldwin RM, Ansari MS, Li R, Shelby ES, Smith CE, McHugo M, Kessler RM (2008) Midbrain dopamine receptor availability is inversely associated with novelty-seeking traits in humans. J Neurosci 28:14372–14378

Pervasive Developmental Disorder Not Otherwise Specified

Synonyms

Atypical autism; PDD NOS

Definition

Pervasive developmental disorder not otherwise specified, or PDD-NOS, was a diagnosis in DSM-IV for individuals who met criteria in two of the three symptom domains for autistic disorder or Asperger's disorder but did not meet the full criteria. Examples include having communication and social difficulties but no repetitive behaviors or having an atypical presentation, such as developing symptoms after age 3 years. Research indicated that this diagnosis was not being used consistently across geographic regions, with some areas giving it frequently while other areas were more likely to give an autistic or Asperger's disorder diagnosis. PDD-NOS was removed from DSM-5.

Cross-References

▶ Autism Spectrum Disorders and Intellectual Disability

Pervasive Developmental Disorders

Definition

Pervasive developmental disorders is an umbrella term that in DSM-IV included autistic disorder, Asperger's disorder, pervasive developmental disorder not otherwise specified, Rett's disorder, and childhood disintegrative disorder. Key characteristics of this class of disorders included developmental difficulties in the areas of social behavior, language, play skills, and repetitive or stereotypic movements. The term "pervasive developmental disorders" was removed from DSM-5 and replaced with autism spectrum disorder.

Cross-References

▶ Autism Spectrum Disorders and Intellectual Disability

Pethidine

Synonyms

Demerol; Ethyl 1-methyl-4-phenylpiperidine-4-carboxylate; Meperidine

Definition

Pethidine is a synthetic opioid analgesic drug. Like other opioids, it can be used for treating moderate to severe pain but it has few unique indications. It has a rapid onset and short duration of action. Its analgesic effects are due to its action as an agonist at μ-opioid receptors. In accordance with this mode of action, it produces a morphine-like abstinence ▶ (withdrawal) syndrome and is subject to abuse and dependence. Pethidine also inhibits the dopamine and norepinephrine transporters, leading to psychomotor stimulant effects and a euphoric mood. Its efficacy as an analgesic is less than that of morphine and the side effects may be greater, due to the actions of a more toxic metabolite (norpethidine).

Cross-References

▶ Analgesics
▶ Opioid
▶ Opioid Analgesics
▶ Opioid Use Disorder and Its Treatment
▶ Withdrawal Syndromes

P-Glycoprotein

Definition

A 170 kD transmembrane glycoprotein from the superfamily of the adenosine triphosphate (ATP)-binding cassette (ABC) transporters that functions as ATP-dependent efflux pump able to extrude many classes of lipophilic and cationic chemicals. This protein is extensively expressed in the intestinal epithelium, hepatocytes, renal proximal tubular cells, and luminal membrane of the endothelial cells of the BBB where it transports compounds out of the brain. P-glycoprotein (abbreviated as P-gp or Pgp) is also called ABCB1 (ATP-ABC subfamily B member 1) or MDR1 (for multiple drug resistance).

Pharmaceutical Cognitive Enhancers: Neuroscience and Society

Sharon Morein-Zamir[1] and Barbara Jacquelyn Sahakian[2,3]
[1]Department of Psychiatry, University of Cambridge, Cambridge, UK
[2]Department of Psychiatry, University of Cambridge, Addenbrooke's Hospital, Cambridge, UK
[3]Behavioural and Clinical Neuroscience Institute, University of Cambridge, Cambridge, UK

Synonyms

Smart drugs

Definition

Pharmacological substances that induce improvement of various cognitive processes, such as attention and memory, are known as pharmacological cognitive enhancers (PCEs). Many pharmacological interventions are aimed at improving cognition in specific neuropsychiatric disorders where cognitive impairment is a prominent symptom, such as attention-deficit hyperactivity disorder (ADHD), schizophrenia, mild cognitive impairment (MCI), and Alzheimer's disease (AD). Enhanced cognition would be expected, in turn, to lead to improved functional outcome and quality of life. There is an increasing lifestyle use of PCEs by healthy people. The ethical implications of the use cognitive enhancers by healthy people are one of the topics addressed by neuroethics.

Current Concepts and State of Knowledge

PCEs and Treatment of Neuropsychiatric Disorders

Cognitive enhancing drugs, also known as "smart drugs," are needed to treat cognitive disabilities and improve the functional outcome, wellbeing, and quality of life for patients with neuropsychiatric disorders and brain injury (Sahakian and Morein-Zamir 2007). In neurodegenerative diseases, such as AD, cognitive enhancing drugs are used to slow down or compensate for the decline in cognitive and behavioral functioning that characterizes such disorders. PCEs are important in treating cognitive problems of neuropsychiatric disorders such as AD and ADHD to reduce the social, economic, and personal burden of such diseases.

Cognitive enhancers are beneficial in both neurodegenerative disorders and neuropsychiatric disorders. Furthermore, though it is common knowledge that people with schizophrenia typically suffer from hallucinations and delusions, it is the long-term cognitive impairments that often impede everyday function and quality of life for many patients. It has been suggested that even small improvements in cognitive functions could help patients make the transition to independent living. However, as yet, PCEs are not routinely prescribed for patients with schizophrenia as an add-on to their anti-psychotic medication. It is noteworthy that not only adults suffering from neuropsychiatric disorders can benefit from cognitive enhancing drugs. For example, as in the case of ADHD, the most prevalent neuropsychiatric disorder of childhood, which is characterized by core cognitive and behavioral symptoms of impulsivity, hyperactivity, and/or inattention. Nevertheless, over 50 % of cases of childhood ADHD will still have ADHD as an adult and could benefit from medication.

Importantly, in addition to pharmacological cognitive enhancers (PCEs), there are many other methods of enhancement such as education, physical exercise, and neurocognitive activation or cognitive training that are commonly being used (Beddington et al. 2008). However, with the popularization of the notion of cognitive enhancement, pharmacological interventions are being seen as a means not only to improve existing deficits, but also to prevent decline before its onset, and even to enhance normal functioning.

PCEs and Cognition

The effects of pharmacological substances on cognition are complex, as cognition is a multifaceted, construct-encompassing attention, executive function, and spatial and verbal learning and memory. Most cognitive enhancing drugs improve only specific aspects of cognition such as forms of executive functions or memory, which are mediated by different systems in the brain. The sizes of the effects to date range from small to moderate, but as pointed out by a 2008 report by the U.K. Academy of Medical Sciences, even small percentage increments in performance can lead to significant improvements. While the extension of enhancement from the controlled laboratory environment to daily life is controversial, several factors are contributing to the advent of increasingly effective approaches. These include the development of sophisticated neuropsychological tests and the routine inclusion of multiple, converging behavioral and brain-imaging measures. With the development of human pharmacogenetics, uncovering human genetic polymorphisms relating to cognition, cognitive enhancers may be matched to those that might benefit from them the most while reducing side effects. For example, the catecholamine-O-methyltrasferase (COMT) gene has been linked to the degree of effectiveness of COMT inhibitors and to the working-memory performance, such that the Val158Met polymorphism may also modify the effect of dopaminergic drugs (e.g., the COMT enzyme inhibitor tolcapone) in the prefrontal cortex.

Despite the fact that much research has been dedicated to the development and understanding of various cognitive enhancers, there is still only limited knowledge of how specific cognitive functions are modulated by neurotransmitters. For example, while it is known that methylphenidate improves symptoms of ADHD and also improves performance on objective behavioral tasks, such as spatial working memory and stop signal (a measure of response inhibition), it remains to be determined conclusively whether dopamine, noradrenaline, or both neurotransmitters are required for these effects on cognition. However, translation and back translation (e.g., Insel et al. 2013) can provide insight into the importance of dopamine neuromodulation of working memory and of noradrenaline modulation of impulsivity. Some of the most notable PCEs being explored to assist individuals with neurological or neuropsychiatric disorders with executive function and attention difficulties include methylphenidate, atomoxetine and modafinil. Methylphenidate, also commonly known as Ritalin, increases the synaptic concentration of dopamine and noradrenaline by blocking their reuptake. Atomoxetine on the other hand, is a relatively selective noradrenaline reuptake inhibitor (SNRI). In the case of modafinil, despite considerable research, its precise mechanism of action is unclear, though it has been found to exhibit a multitude of effects including potentiation of noradrenaline and to a degree, dopamine neurotransmission; and elevation of extracellular glutamate, serotonin and histamine levels, and decreased extracellular GABA. Recent evidence suggests that some of its cognitive effects may be modulated primarily by noradrenaline transporter inhibition.

There are many additional considerations in examining the cognitive enhancing effects of various pharmacological agents, such as stimulants like amphetamine. Neurotransmitter functions, at times, follow an inverted U-shaped curve, with deviations from optimal level in either direction impairing performance. Moreover, different neurotransmitter levels can be found across brain regions, suggesting a complex interplay between baseline levels and drug administration. While some cognitive functions may improve following drug administration, others may worsen, as they depend on different optimum neurotransmitter levels. Drug-induced neurotransmitter increases may improve functioning in some groups but have no effect or even impair performance in others, already at optimum. Namely, the baseline abilities of the individual can limit the effectiveness of certain PCEs, and their effect will be more pronounced in those with an initial below-average level of performance. Stimulants also carry abuse potential. To date, modafinil has not been demonstrated to have abuse potential. In addition, as yet, there is no clear demonstration that modafinil shows an inverted U-shape curve.

An effective method of testing the effects of cognitive enhancers on cognition is by using double-blind placebo control studies where participants undergo a battery of objective cognitive tasks targeted at measuring various facets of cognition, including memory, attention, and executive functions. For instance, it has been demonstrated that methylphenidate improves spatial working memory performance in young volunteers and in children and adult patients with ADHD, whereby patients make less task-related errors when on methylphenidate. The neural substrates mediating SWM task performance have been examined using imaging techniques such as positron emission topography (PET) and indicate that the dorsolateral- and mid-ventrolateral prefrontal cortex are particularly recruited. Studies using PET and contrasting [(11)C] raclopride binding on methylphenidate versus placebo have further indicated that methylphenidate influences dopaminergic function, particularly in the striatum. Methylphenidate has been found to improve both performance and efficiency in the spatial working memory neural network involving the dorsolateral prefrontal cortex and posterior parietal cortex in healthy volunteers. Similar studies using the double-blind, randomized, placebo-controlled methodology have reported that additional drugs such as modafinil and atomoxetine can improve performance in some tasks of executive functioning. Thus, modafinil has been found to improve spatial planning and response inhibition in ADHD patients, as measured by a variant of the Tower of London task and the stop-signal task, respectively. It has been further demonstrated that modafinil produces improvements in performance in a group of healthy volunteers on tests of spatial planning, response inhibition, visual recognition, and short-term memory. Likewise, administration of an acute dose of atomoxetine has been found not only to improve response inhibition in ADHD patients, but also in healthy adults. Using functional magnetic resonance imaging (fMRI), the brain mechanisms by which atomoxetine exerts its effects in healthy volunteers has been examined in a double-blind placebo-controlled study. Atomoxetine led to increased activation in the right inferior frontal gyrus when participants attempted to inhibit their responses in the stop-signal task. Inhibitory motor control has been shown previously to depend, at least in part, on the function of this brain region.

PCEs and Neuroethics

There is a clear trend in many Western countries toward increasing prescriptions of methylphenidate. With the advent of psychiatric medications with greater tolerability and fewer side effects, these trends are set to continue. It is not only those who suffer from neuropsychiatric disorders and brain injury who are appearing to use pharmacological cognitive enhancers (Sahakian and Morein-Zamir 2011). The use of stimulants, including methylphenidate and amphetamines by healthy students and professionals has been rising as well. The most commonly reported motives for use are to aid concentration, help with work and study, and increase alertness. Modafinil is also used to counteract the effects of jetlag and recent evidence also shows that it increases task-related motivation (Muller et al. 2013). Although drugs such as modafinil are prescribed off-label in North America, they can be freely obtained without a prescription via the internet from multiple websites in various countries (Sahakian and Morein-Zamir 2011). Current trends of growing use are likely set to increase as presently there are also novel cognitive enhancers under development, many of which are aimed at improving memory and learning. Given the aging population in the United Kingdom and elsewhere, and the fact that the lifespan of individuals is being extended, it is highly likely that cognitive enhancing drugs that can improve memory in healthy elderly will also prove to be in demand. The popular media has reported extensively both on studies finding improved performance in healthy individuals and on the rising use of PCEs in healthy individuals (see also Maher 2008).

The study of cognitive enhancing drugs, and their influence both on patients with neuropsychiatric and neurological disorders and healthy individuals, raises numerous neuroethical issues (Illes and Sahakian 2010). In the area of neuroethics, there are several important considerations in regard to the use of cognitive enhancing drugs, particularly in children and adolescents where the brain is still in development (Marcus 2002). For example, certain drugs such as stimulants have abuse potential and it is presently unclear what side effects may emerge with long-term use in healthy people. There are also many other issues, including possible direct and indirect coercion, effects on personal freedom, the perception of authenticity, and greater societal disparity, which will impact on the individual and on society. These considerations are exacerbated by the dearth of information regarding long-term efficacy and safety of PCEs in healthy individuals. The harms-benefit ratio for PCE use must be clarified, as the awareness of adverse effects must be weighed against any potential benefits.

As the use of pharmacological cognitive enhancement appears set to increase, the profile of cognitive effects of each drug on specific populations should be mapped, along with its potential for harms. This should facilitate ethical and regulatory discussion about each pharmacological substance. At present, regulatory bodies (e.g., Food and Drug Administration in the Unites States, Medicines and Healthcare products Regulatory Agency in the United Kingdom and the European Medicines Agency) are concerned with treatments for disorders or diseases, but this may change in future with the increasing use of cognitive enhancing drugs by healthy people

(Greely et al. 2008). As pharmacogenomics and individualized medicine continues to develop, cognitive enhancing drugs may prove of great benefit to society in terms of improving wellbeing and quality of life for people with neuropsychiatric disorders, brain injury, and age-related memory impairment.

Cross-References

- ▶ Atomoxetine
- ▶ Methylphenidate and Related Compounds
- ▶ Modafinil
- ▶ Neuroethics
- ▶ Schizophrenia
- ▶ Working Memory

References

Academy of Medical Sciences (2008) Brain science, addiction and drugs. Working group report chaired by Professor Sir Gabriel Horn, Foresight Brain Science, Addiction and Drugs Project. Office of Science and Technology, London

Beddington J, Cooper CL, Field J, Goswami U, Huppert FA, Jenkins R et al (2008) The mental wealth of nations. Nature 455(7216):1057–1060

Greely H, Sahakian B, Harris J, Kessler RC, Gazzaniga M, Campbell P et al (2008) Towards responsible use of cognitive enhancing drugs in the healthy. Nature 456:702–705

Illes J, Sahakian BJ (2010) The oxford handbook of neuroethics. Oxford University Press, Oxford

Insel TR, Voon V, Nye JS, Brown VJ, Altevogt BM, Bullmore ET et al (2013) Innovative solutions to novel drug development in mental health. Neurosci Biobehav Rev 37(10 Pt 1):2438–44

Maher B (2008) Poll results: look who's doping. Nature 452(7188):674–675

Marcus SJ (2002) Neuroethics: mapping the field. Dana Press, San Francisco

Muller U, Rowe JB, Rittman T, Lewis C, Robbins TW, Sahakian BJ (2013) Effects of modafinil on non-verbal cognition, task enjoyment and creative thinking in healthy volunteers. Neuropharmacology 64:490–495

Sahakian B, Morein-Zamir S (2007) Professor's little helper. Nature 450(7173):1157–1159

Sahakian BJ, Morein-Zamir S (2011) Neuroethical issues in cognitive enhancement. J Psychopharmacol 25(2):197–204

Pharmacodynamic Tolerance

S. Stevens Negus, Dana E. Selley and Laura J. Sim-Selley
Department of Pharmacology and Toxicology, Virginia Commonwealth University, Richmond, VA, USA

Definition

Tolerance is a drug-induced reduction in subsequent drug effect. Pharmacodynamic tolerance refers to instances of tolerance that involve either (a) adaptive changes in receptor binding or (b) recruitment of processes that limit or oppose the effects of the drug on receptor-mediated signaling pathways.

Impact of Psychoactive Drugs

Overview

Tolerance refers to a phenomenon in which the potency and/or maximal effectiveness of a drug to produce some effect is reduced after a regimen of prior exposure to that drug. Tolerance may be expressed as a reduced effect produced by a given drug dose, a requirement for higher drug doses to sustain a given effect, or a rightward and/or downward shift in a drug dose-effect curve. Tolerance may result from a variety of different mechanisms. Pharmacodynamic tolerance refers to instances of tolerance associated with pharmacodynamic mechanisms that are described below. It can be distinguished from pharmacokinetic tolerance (tolerance related to processes of pharmacokinetics: drug absorption, distribution, metabolism, and excretion) and behavioral tolerance (tolerance related to whole-organism learning processes).

Pharmacodynamics is the study of drug action on a biological system, and pharmacodynamic mechanisms can be divided into two dissociable but complementary components of drug action: receptor binding and pharmacodynamic efficacy.

Receptor binding quantifies the direct physical interaction between a drug and one or more target receptor(s). The degree of receptor binding produced by a given concentration of a drug is determined by the density of a given receptor type in a biological system (often expressed as Bmax, receptors per unit mass of tissue) and by the affinity of the drug for that receptor type (often expressed as Kd, the concentration of drug required to bind 50 % of the receptor population). Pharmacodynamic efficacy describes the degree to which a drug activates, inactivates, or otherwise modulates signaling pathways coupled to a receptor. These signaling pathways originate with changes in receptor conformation/function and are transduced into downstream changes in intra- and intercellular biochemistry, physiology, and behavior. Agonists stimulate these pathways, inverse agonists reduce constitutive activity in these pathways, and antagonists have no effect on their own but block effects of agonists and inverse agonists. Pharmacodynamic tolerance can be mediated by drug-induced changes in either receptor binding or receptor-mediated signaling. Moreover, these mechanisms are not mutually exclusive to each other or to nonpharmacodynamic mechanisms of tolerance. Rather, tolerance to a given effect of a given drug may involve multiple mechanisms, and mechanisms that underlie tolerance to one effect of a drug may differ from the mechanisms underlying tolerance to another effect of that same drug or to other drugs.

Tolerance and Receptor Binding

One mechanism that may contribute to tolerance is a drug-induced change in the density of receptors to which a drug binds. Depending on the drug, either decreases in receptor density (receptor downregulation) or increases in receptor density (receptor upregulation) may contribute to tolerance. Thus, receptor downregulation can reduce the effects of an agonist by reducing the number of receptors available for agonist-induced stimulation. Conversely, receptor upregulation can reduce the effects of an inverse agonist or antagonist by increasing the number of constitutively active receptors or the number of receptors available for stimulation by endogenous neurotransmitter agonists. Examples of tolerance-associated changes in receptor density include the downregulation of cannabinoid CB1 receptors produced by chronic exposure to cannabinoid agonists such as Δ^9-tetrahydrocannabinol (Martin et al. 2004) and upregulation of dopamine D2 receptors produced by chronic exposure to antipsychotic dopamine D2 receptor antagonists such as haloperidol (Sibley and Neve 1997). The precise mechanisms that underlie drug-induced changes in receptor density vary as a function of such variables as the drug, treatment regimen, and receptor type, and in many cases, these mechanisms remain unknown. However, the density of available receptors is determined by rates of gene expression, protein synthesis, receptor trafficking to and from the membrane, and degradation. Drug exposure may modulate each of these processes. One illustrative mechanism will be described below in the context of receptor desensitization.

It is also theoretically possible for tolerance to occur as a result of reduced affinity of a drug for its receptor; however, tolerance-associated reductions in affinity have usually not been observed. Chronic drug exposure may promote changes in the relative abundance of different receptor subtypes for which a drug has differing affinities (e.g., Wafford 2005), but this mechanism can best be conceptualized as a change in the relative densities of different receptor types rather than as a change in the affinity of the drug for a single receptor type.

Tolerance and Receptor-Mediated Signaling Pathways

Although changes in receptor binding may contribute to some forms of pharmacodynamic tolerance, mechanisms involving changes in receptor-mediated signaling appear to be more commonly involved. The wide variety of signaling pathways associated with different receptor types creates a multitude of potential mechanisms whereby prior drug exposure can modulate subsequent drug effects. In general, these mechanisms have the effect of limiting or opposing the effects of the drug, and these mechanisms can act at the level of the receptor itself, the intracellular signaling pathways coupled to the receptor, or the intercellular neural circuits containing cells that express the receptor. Examples will be provided from the literature on tolerance to morphine and other opioids acting at the mu-opioid receptor (Christie 2008).

Drug binding sites on receptor proteins are coupled to proximal transduction mechanisms that consist of either separate domains of the receptor itself (e.g., ion channels in the case of ligand-gated ion channels) or separate proteins located immediately adjacent to the receptor (e.g., G-proteins associated with G-protein-coupled receptors (GPCRs)). For example, mu-opioid receptors are GPCRs, and receptor-mediated signaling begins with agonist-induced interactions between the receptor and an adjacent G-protein to activate the G-protein and initiate multiple cascades of downstream neurochemical processes that affect ion channels, intracellular enzyme activity, and gene transcription. The drug binding site and proximal transduction mechanism can be uncoupled in a process also referred to as receptor desensitization, and this process is thought to be especially important for drugs that act as direct agonists at a receptor. With mu-opioid receptors, agonist binding promotes the phosphorylation of the receptor by G-protein-coupled receptor kinases (GRKs) that recognize and phosphorylate amino acids on the intracellular C-terminus of the agonist-bound conformation of the receptor. Receptor phosphorylation itself may reduce the efficiency of G-protein activation by the receptor. In addition, the phosphorylated region of the receptor may also become bound by another protein called β-arrestin, which is thought to further obstruct the receptor's ability to interact with and activate the adjacent G-protein. As a result, the receptor is effectively uncoupled from the G-protein and other downstream signaling processes (Gainetdinov et al. 2004; Martini and Whistler 2007).

Another role of β-arrestin in the regulation of GPCRs like the mu-opioid receptor is to facilitate internalization by endocytosis. Specifically, the β-arrestin-bound receptor migrates to clathrin-coated pits in the cell membrane, and these pits are then invaginated and pinched off to form intracellular vesicles. Internalized receptors can then be destroyed (via degradation by proteases in lysosomes, a process that may contribute to receptor downregulation) or recycled to the membrane (via dephosphorylation of the receptor and subsequent fusion of the vesicle with the extracellular membrane). Through this latter mechanism, β-arrestins may contribute not only to the negative regulation of receptor signaling but also to the resensitization of the receptor. Thus, it has been suggested that short-term desensitization, internalization, and recycling of receptors could prevent the development of longer-term tolerance mediated by the activation of compensatory adaptations in signaling pathways downstream of the receptor (Martini and Whistler 2007, see below). Furthermore, the interaction of internalized receptors with β-arrestins can result in signaling through alternative (non-G-protein mediated) pathways. Thus, roles of GRKs and β-arrestins in mediating acute drug action and tolerance are complex (Gainetdinov et al. 2004; Schmid and Bohn 2009). For example, the genetic knockout of the β-arrestin-2 subtype of β-arrestin increases sensitivity to some of morphine's effects (e.g., supraspinal antinociception), while decreasing sensitivity to other effects (e.g., constipation). Moreover, tolerance to the antinociceptive effects of morphine is attenuated in β-arrestin-2 knockout mice, whereas antinociceptive tolerance to more potent or efficacious opioids, such as etorphine or methadone, is unaffected by the loss of β-arrestin-2. Interestingly, antagonist-precipitated withdrawal after chronic morphine was also unaffected in β-arrestin-2 knockout mice. Overall, then, the roles of GRK/β-arrestin regulatory mechanisms in drug tolerance are likely to be multifaceted and could in some cases mediate while in other cases oppose drug tolerance.

Reductions in drug effects can also be produced by a compensatory recruitment of intracellular opponent processes downstream from the receptor. For example, in many cell types that express mu-opioid receptors, agonist binding initiates a series of biochemical events that includes (1) activation of Gi/Go proteins, which (2) inhibit adenylyl cyclase, an enzyme that converts adenosine triphosphate (ATP) into cyclic adenosine monophosphate (cAMP), resulting in (3) reduced levels of cAMP and (4) reduced activity of protein kinase A (PKA), an enzyme that is activated by cAMP. This in turn results in (5) decreased phosphorylation of other downstream proteins, including membrane-bound ion channels that can influence short-term patterns of cell activity and transcription factors that can influence gene transcription and longer-term cell structure and function. Tolerance to the effects of a mu-agonist can result from the regulation of any step in this pathway. For example, chronic exposure to mu-opioid receptor agonists may promote a compensatory upregulation in adenylyl cyclase, the enzyme normally inhibited by mu-receptors (Nestler and Aghajanian 1997). This upregulation in adenylyl cyclase may be sufficient to offset the inhibitory effect mediated by mu-opioid receptors and restore enzyme activity to basal, predrug levels despite continued drug presence.

In multicellular biological systems, opponent processes can be recruited not only in the cell that contains the receptor at which the drug acts but also in other cells involved in circuits that mediate drug effects on physiology and behavior. Thus, to the degree that a drug produces a net change in activity of a downstream cell, opponent processes may be recruited to oppose the drug effect and shift cell activity back toward basal levels. As one example from the mammalian central nervous system, mu-opioid receptors are located on inhibitory interneurons that regulate the activity of a population of dopaminergic neurons collectively referred to as the mesolimbic dopamine system. The agonist-induced activation of these mu-receptors inhibits the interneurons and thereby disinhibits the dopamine neurons to increase dopamine release. This increased mesolimbic dopamine release is a shared property of many drugs of abuse, and it is thought to play a key role in the neurobiology of addiction. However, sustained increases in dopamine (produced by mu-agonists or by other drugs of abuse) may then act on postsynaptic D1 dopamine receptors to promote synthesis in those neurons of the endogenous opioid peptide dynorphin, which can be released by recurrent collaterals that feedback onto the dopamine neurons. Dynorphin can then act at kappa-opioid receptors on the dopamine neurons to inhibit further neural activation and dopamine release. Thus, mu-agonists may activate dopamine neurons via one process, but this activated dopaminergic activity may then trigger a dynorphin/kappa-opioid receptor-mediated opponent process that opposes further activation of dopaminergic neurons (Chartoff et al. 2006; Shippenberg et al. 2007).

It should also be noted that the recruitment of opponent processes may contribute not only to tolerance but also to the expression of withdrawal syndromes and addictive disorders. Thus, when drug is present, the drug and its opponent processes oppose each others' effects in a dynamic balance that shifts the level of cell or circuit activity toward basal, predrug levels and results in the phenomenon of tolerance. However, when the drug is removed, the opponent processes are released from drug opposition, and the disinhibited opponent processes may then produce effects different from and often opposite to those originally produced by the drug. For example, withdrawal from a mu-agonist may release both upregulated adenylyl cyclase activity within mu-receptor-containing cells and upregulated dynorphin release from downstream cells, and these disinhibited opponent processes may then contribute to physiological and behavioral signs of opioid withdrawal that contribute to opioid addiction (Christie 2008; Nestler and Aghajanian 1997; Shippenberg et al. 2007).

Cross-References

▶ Addictive Disorder: Animal Models
▶ Behavioral Tolerance
▶ Opioids
▶ Pharmacokinetics
▶ Receptors: Binding Assays
▶ Withdrawal Syndromes

References

Chartoff EH, Mague SD, Barhight MF, Smith AM, Carlezon Jr WA. Behavioral and molecular effects of dopamine D1 receptor stimulation during naloxone-precipitated morphine withdrawal. J Neurosci. 2006;26:6450–7.

Christie MJ. Cellular neuroadaptations to chronic opioids: tolerance, withdrawal and addiction. Br J Pharmacol. 2008;154:384–96.

Gainetdinov RR, Premont RT, Bohn LM, Lefkowitz RJ, Caron MG. Desensitization of G protein-coupled receptors and neuronal functions. Annu Rev Neurosci. 2004;27:107–44.

Martin BR, Sim-Selley LJ, Selley DE. Signaling pathways involved in the development of cannabinoid tolerance. Trends Pharmacol Sci. 2004;25:325–30.

Martini L, Whistler JL. The role of mu opioid receptor desensitization and endocytosis in morphine tolerance and dependence. Curr Opin Neurobiol. 2007;17:556–64.

Nestler EJ, Aghajanian GK. Molecular and cellular basis of addiction. Science. 1997;278:58–63.

Schmid CL, Bohn LM. Physiological and pharmacological implications of beta-arrestin regulation. Pharmacol Ther. 2009;121(3):285–93.

Shippenberg TS, Zapata A, Chefer VI. Dynorphin and the pathophysiology of drug addiction. Pharmacol Ther. 2007;116:306–21.

Sibley D, Neve K. Regulation of dopamine receptor function and expression. In: Neve K, Neve RL, editors. The dopamine receptors. Totawa: Humana Press; 1997. p. 383–424.

Wafford KA. GABAA receptor subtypes: any clues to the mechanism of benzodiazepine dependence? Curr Opin Pharmacol. 2005;5:47–52.

Pharmacodynamics

Definition

The processes through which drugs bring about their actions on living organisms. It is distinguished from pharmacokinetics, which relates to the absorption, distribution, and metabolism of drugs.

Pharmacogenetics

Sophie Tambour
Department of Behavioral Neuroscience,
University of Liège, Liège, Belgium

Synonyms

Pharmacogenomics

Definition

Pharmacogenetics (PGx) refers to the influence of genetic variations on drug responses. More specifically, it is the study of how polymorphic genes that encode the function of transporters, metabolizing enzymes, receptors, and other drug targets in humans and other animals are related with variations in responses to drugs, including toxic and therapeutic effects. PGx, which may be considered to be a subfield of ecogenetics, tends to be used interchangeably with pharmacogenomics, with the same imprecision that conflates "genetics" with "genomics."

Current Concepts and State of Knowledge

One thread of the origin of PGx dates back to the 1950s (see Box 1) when pharmacologists began to incorporate genetics into their studies of adverse drug reactions. Examples of classical PGx studies include the observations that at doses tolerated well by others, the muscle relaxant succinylcholine, the antimalarial drug primaquine, and the antituberculosis drug isoniazid induced life-threatening effects, such as respiratory arrest, hemolytic anemia, and neuropathies, in some patients. By studying families and ethnic populations as well as through careful phenotyping, pharmacologists demonstrated that these unusual drug responses were inherited and due to certain enzyme variations controlled by single genes. The understanding of the molecular genetic basis for the phenotypes came later, with the development of DNA technology (▶ single-nucleotide polymorphism [SNP] testing and, more recently, ▶ haplotype testing) and in vitro molecular tests allowing the exact identification of the gene sequences responsible for variations in drug responses. The recent flourishing of genomics with development of new technologies such as microarrays has changed the focus of PGx from study of the effect of single-gene sequence polymorphisms in two ways. First, the search has broadened to multiple genes (i.e., from monogenic to polygenic traits), and, second, gene

Pharmacogenetics, Fig. 1 Schematic representation of how gene variants affect pharmacokinetic and pharmacodynamic factors resulting in potential modifications in the pharmacological effect of drugs (From http://www.pharmgkb.org)

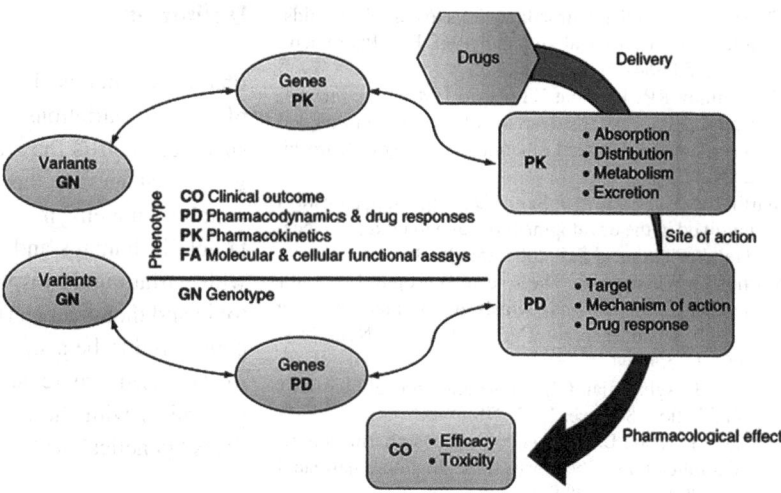

expression is now known to play an important role in affecting variations in drug responses. Indeed, modern PGx pharmacogenomics) often uses a reverse approach, studying whether sequence variations and/or gene expression is in any way related to variations in drug response phenotypes (see "Methods").

The second thread antecedent to PGx can be traced to the field of behavioral genetics. Studies of behavioral responses to a variety of drugs, often in mice and rats, employed genetically varying populations and consistently found that different genotypes had characteristic, and widely varied, responses to many drugs. Starting in the late 1940s, this field also pioneered the use of selective breeding to create rat and mouse lines with extreme drug responses. The earliest systematic summary of this work was a monograph by Broadhurst (1978). Behavioral genetic studies now also incorporate genetically modified animals, and the studies of gene-targeted mutant lines and selected lines have contributed much to our understanding in drug action, particularly in the addiction field.

Defining Pharmacogenetics

Pharmacology often divides drug interactions with the body into their pharmacokinetic and pharmacodynamic aspects. Pharmacokinetics, sometimes described as what the body does to a drug, incorporates drug absorption, distribution, metabolism, and excretion. Pharmacodynamics, described as what a drug does to the body, involves receptor binding, post-receptor effects, and chemical interactions. A drug's pharmacokinetics and pharmacodynamics are both genetically and environmentally influenced. Genes contain information that determine the structure of proteins, and any variations in the DNA sequence (mutation) may alter the expression or the function of proteins. DNA mutations that occur at a frequency of 1 % or greater are termed polymorphisms. Polymorphisms in genes coding for a protein that carries a drug to its target cells or tissues may cripple the enzyme that activates a drug or aids its removal from the body and thus may induce pharmacokinetic or pharmacodynamic variations leading to individual differences in the response to the drug (see Fig. 1).

The majority of pharmacogenetic studies have focused on drug-metabolizing enzymes. For example, cytochrome P450 (CYP450) is a large superfamily of metabolizing enzymes. Within the CYP2 family, polymorphic CYP2D6 was one of the first and most important drug-metabolizing enzymes to be characterized at the DNA level. By using response to a "marker" drug (i.e., dextromethorphan), four phenotypes may be described: "poor metabolizers," "intermediate metabolizers," "extensive metabolizers," and "ultrarapid metabolizers." Ultrarapid metabolizers have multiple copies of the *CYP2D6* gene expressed and greater-than-normal CYP2D6 activity. Therefore, ultrarapid metabolizers may not achieve therapeutic levels with usual doses and may require several doses to show a response. On the other hand, poor metabolizers are at increased risk of toxicity from CYP2D6 substrate drugs (e.g., codeine). In addition to CYP2D6, polymorphisms have now been identified in more than 20 drug-metabolizing enzymes in humans. Some of these polymorphisms show different distributions in racial groups and phenotypically relevant consequences from a clinical point of view (see Table 1). More recently, there has been increased recognition in the contribution of genetic variation in proteins involved in drug responses. For example, a polymorphism in the serotonin transporter protein that affects serotonin availability in the brain has been reported to be associated with a predisposition to depressive illness as well as with therapeutic response to antidepressive or antipsychotic pharmacotherapy.

> **Box 1. Dates**
> 1949: Initiation of first rat line selected for high alcohol consumption by J Mardones at the University of Chile (Mardones and Segovia-Riquelme 1983)
> 1950: Extensive characterization of mouse and rat genetic differences in response to psychoactive drugs
> 1957: Delineation of the field of PGx by Motulsky (1957)
> 1959: Introduction of the term "PGx" by Vogel (1959)
> 1962: Definitive establishment of the field by Kalow's monograph (Kalow 1962)
> 1968: Demonstration by Vessel of the general importance of polygenic inheritance in metabolism of many drugs (Vessel and Shapiro 1968)
> 1990s: Introduction of the term "pharmacogenomics" with emergence of the Human Genome Project

Methods

PGx may use a phenotype-driven or a genotype-driven approach for understanding the genetic contributions to variations in drug responses in

Pharmacogenetics, Table 1 Representative examples of the relation between genetic polymorphisms of drug-metabolizing enzymes with drug responses adapted from Weber (2008)

Gene	Polymorphism	Drug response
CYP2D6(L)	2–13-fold multiplication of gene	Failure to respond to nortriptyline; toxicity to codeine
CYP2C19	SNP truncates gene	Failure to respond to proton inhibitors and proguanil
CYP2C9	Two major SNPs (Arg-144-Cys) (Ile-359-Leu) impair metabolic efficiency	Toxicity to warfarin and phenytoin
VKROC1	Promoter variants impair or enhance vitamin K reductase efficiency	Toxicity to, or lack of efficacy of, warfarin
ALDH2	SNP (Glu-487-Lys). This variant is highly prevalent in East Asian gene pools. Homozygotes are essentially 100 % protected from developing alcoholism	Impaired metabolism of ethanol, leading to high levels of acetaldehyde and many highly aversive effects. The drug disulfiram (Antabuse®) mimics the slow-metabolizing variant
TPMT	SNP (Pro-238-Ala)	Drug-induced bone narrow toxicity
NAT2	Multiple SNP	INH-induced peripheral neuropathy; susceptibility to bladder cancer from aromatic amine carcinogens
DPYD and DPD	Splice site mutation IVS14 + 1G > A	Severe 5-fluorouracil toxicity
TAA	Mutated promoter (TA7)TAA	Severe irinotecan toxicity
FMO3	SNP (Pro-153-Leu)	Psychosocial disorders resulting from inability to metabolize

(continued)

Pharmacogenetics, Table 1 (continued)

Gene	Polymorphism	Drug response
		trimethylamine in foods
MDR1	MDR1 (C3435T)	Enhances therapeutic effect of antiretroviral drugs (efavirenz, nelfinavir) in HIV-infected patients
MRP2	Absence of functional MRP2	Dubin-Johnson syndrome

humans and nonhumans. In the phenotype-driven approach (i.e., from phenotype to genotype), once a genetic contribution to the phenotype of interest has been confirmed (with family analysis, animal model analysis, and/or linkage studies), genetic markers reliably associated with the phenotype are identified by genome-wide and/or candidate gene approaches. Such studies must first identify the rough genomic locations of causative genetic variation. These locations are called quantitative trait loci (QTL), and QTL analysis is the first step for identifying the subset of chromosomal areas containing genes responsible for the genetic variation in a specific trait and to locate these areas on a genomic map. QTL analyses have been widely used in animal models with inbred and recombinant strains of rats and mice, as well as in lines selectively bred for extremes in response to various drugs. When chromosomal regions associated with the magnitude of the phenotype have been identified, the QTL initially span genomic regions containing dozens or even hundreds of genes. Fine-mapping can be carried out to refine the search, narrow the confidence intervals, and, finally, identify potential susceptibility genes. Eventually, candidate genes remaining in the QTL interval are sequenced to identify the specific sequence variants associated with the phenotype. Alternatively, or additionally, in a genotype-driven approach (i.e., from genotype to phenotype), pharmacogeneticists characterize the functional significance of polymorphisms with high population frequency for a gene for which previous knowledge suggests that it could influence drug responses. In this approach, the development of genetically modified animals such as transgenic and knockout mice has been particularly useful for investigating the direct effects of gene mutation on responses to drugs. In both approaches, analyses of gene expression are now commonly evaluated as well to identify candidate genes within each QTL.

Methods used for genotyping involve a series of molecular biological, physical, and chemical procedures used to distinguish among the alleles of an SNP or the measurement of the allele-specific products. PGx principally uses methods applied in genetics and pharmacology but also borrows the knowledge and progress available in other fields such as bioinformatics. Because final proof of a quantitative trait gene (QTG) nearly always involves behavioral assessment of genetically modified animals, the tools of behavioral analysis are also employed.

Clinical Applications

The clinical consequences of interindividual differences in drug responses are manifested in a variety of ways such as acute or delayed toxicity, unusual response, resistance to a drug, or unwanted drug interactions (when combined with the use of other agents). Adverse side effects of drugs represent a significant public health and economic problem, affecting more than 2 million people annually in the United States, including 100,000 deaths and costing more than $100 billion. The use of PGx in a clinical setting could reduce these costs and help to determine the most appropriate treatment for individual patients according to their unique genetic makeup (personalized medicine). Indeed, identification of relevant polymorphisms known to affect drug effects could help treatment providers determine which available drugs would ensure maximum efficacy with minimal adverse effects.

However, high expectations surrounding "personalized medicine" remain unfulfilled, and only a few products have reached the market and clinical practice. One example is the DNA chip AmpliChip® (developed by Roche Molecular System Inc.) used for testing variations in expression of the CYP2D6 and CYP2C19 genes, which play a major role in metabolism. Another is the

HER2 test which is used prior to prescribing Herceptin® to breast cancer patients. HER2 is the receptor for the epidermal growth factor which can stimulate cells to divide and grow. HER2 testing measures the number of copies of HER2 gene (*neu*) in each cell and/or the level of HER2 protein in the tumor sample. Only people showing HER2 gene amplification and overexpression are prescribed Herceptin®, which attaches to the HER2 protein and stops human epidermal growth factor from reaching the breast cancer cells. A third is a test for variants of the enzyme thiopurine methyltransferase (TPMT) before prescribing thiopurine (i.e., antileukemic drugs such as 6-mercaptopurine and 6-thioguanine) for the treatment of acute lymphocytic leukemia. Genotype-based testing allows the identification of individuals showing a reduced TPMT enzyme activity. Because these people may show a toxic accumulation of thiopurine in the body responsible for bone-marrow suppression, they are not prescribed thiopurine drugs. Pharmacogenetic testing is currently primarily used in a limited number of teaching hospital and specialist academic centers.

Some genetic and nongenetic factors may explain why, despite the well-documented functional relevance of many drug-metabolizing enzymes, the translation of PGx into clinical practice has yet to reap its benefits. First, drug responses are likely due to many genes, each bearing many polymorphisms. Identification of all these genes and polymorphisms is difficult and time-consuming and would have to be performed before the selection of the right drug and the right dose for the individual patient. Second, genotyping is not sensitive to drug drug interactions or to environmental factors, which can affect treatment response. For example, dietary intake may influence drug metabolism, as in the case of monoamine oxidase inhibitor (MAOI) drugs for the treatment of depression or Parkinson's disease. Patients using these drugs need to avoid food containing the amino acid tyramine (i.e., such as aged cheeses), because accumulation of this amino acid leads to a hypertensive crisis with headache and high blood pressure. This example shows how pharmacogenetic information alone will not predict the variation in efficacy or safety of a drug.

Besides enabling clinicians to select the most effective drugs for treatment of individuals, PGx could also be used for salvaging drugs that have been withdrawn from the market because of adverse drug reactions – the so-called drug rescue. Retrospective genotyping of clinical trial subjects could help to distinguish patients who were susceptible to an adverse response from those patients who showed benefit from the drug. Finally, PGx could also have an impact on the process of drug research and development, by helping to develop specific drug targeting proteins with variant structures resulting from the presence of genetic polymorphisms. Such research could assist pharmaceutical companies to develop more effective drugs with fewer side effects.

Currently, there are few clinical practice regulations relating to PGx or pharmacogenomics. Such regulations, including prescription guidelines, testing, and usage labels, will be needed if the techniques are to be widely used. However, before regulation can be introduced to guide prescribing practices, the field needs more robust prospective data from well-designed studies.

Conclusion

PGx provides an experimental framework to understand variation in human drug responses as a function of inherited material. Recently, the field has experienced a period of rapid growth with the initiative of the Human Genome Project. With the improvement and the wide availability of sequencing technology, the use of PGx in clinical practice is expected to increase.

Cross-References

▶ Delayed Onset of Drug Effects
▶ Ecogenetics
▶ Gene Expression
▶ Genetically Modified Animals
▶ Pharmacokinetics
▶ Proteomics

References

Broadhurst PL (1978) Drugs and the inheritance of behavior: a survey of comparative psychopharmacogenetics. Plenum Press, New York

Kalow W (1962) Pharmacogenetics: heredity and the response to drugs. W.B. Saunders, Philadelphia

Mardones J, Segovia-Riquelme N (1983) Thirty-two years of selection of rats by ethanol preference: UChA and UChB strains. Neurobehav Toxicol Teratol 5:171–178

Motulsky AG (1957) Drug reactions enzymes, and biochemical genetics. J Am Med Assoc 165:835–837

Vessel ES, Shapiro JG (1968) Genetic control of drugs levels in man: antipyrine. Science 161:72–73

Vogel F (1959) Moderne problem der humangenetik. Ergeb Inn Med U Kinderheilk 12:52–125

Weber WW (2008) Pharmacogenetics. Oxford University Press, New York

Pharmacogenomics

Synonyms

Pharmacogenetics

Definition

Pharmacogenomics can be defined as the study of variations of DNA and RNA characteristics as related to drug response. However, there is no consensus about the semantic differences between pharmacogenetics and pharmacogenomics, and these two terms tend to be used interchangeably. The history of usage generally parallels the development of the new field of genomics from the old field of genetics. The older term "pharmacogenetics" is now generally thought of as the limited study of single or few genes and their effects on interindividual differences in drug responses, while the newer term "pharmacogenomics" generally refers to the application of genomic technologies to the study of drug discovery, pharmacological function, disposition, and therapeutic response.

Pharmacokinetics

Kim Wolff
Institute of Pharmaceutical Science, King's College London, London, UK

Synonyms

Kinetics; PK

Definition

Pharmacokinetics (in Greek "pharmacon" meaning drug and "kinetikos" meaning putting in motion and the study of time dependency) has been variously defined as the study of the relationship between administered doses of a drug and the observed blood (plasma or serum) or tissue concentrations. It is a branch of pharmacology that explores what the body does to a drug and hence concerns itself with the quantitation of drug absorption, distribution, metabolism, and excretion (ADME).

Impact of Psychoactive Drugs

Despite the enormity of the pharmaceutical industry and the vast array of psychoactive substances used in an illicit manner, the mode of action of drugs in the body can in part be understood by reference to basic pharmacokinetic principals, assuming a linear relationship between blood and tissue (brain) exposure. Understanding the pharmacokinetic principles of a drug often explains the manner of its use and aids the clinician in a number of ways: in anticipating the optimal dosage regime, in predicting what may happen if the dosage regime is not followed, in responding to over dosing, and in monitoring the consequences of harmful or dependent use. The use of two or more drugs concurrently is commonplace both in the general patient population and in substance misuse. Knowledge of the pharmacokinetic properties of a drug can help identify significant drug interactions when either the pharmacokinetics or the pharmacodynamics of one drug is altered by the consumption of another.

Pharmacokinetic parameters describe how the body affects a specific drug after administration and explains how different variables such as the site of administration and the dose and dosage form in which the drug is administered can alter this response. Pharmacokinetic analysis is performed by noncompartmental (model independent) or compartmental methods. Noncompartmental methods estimate the exposure to a drug by evaluating the area under the curve (AUC) of a concentration–time graph. Compartmental methods estimate the concentration–time graph using kinetic modeling. Compartment-free methods are often more versatile in that they do not assume any specific compartmental model and produce more accurate results.

Noncompartmental Analysis

Noncompartmental pharmacokinetic analysis is highly dependent on the estimation of total drug exposure, most often estimated by AUC methods, using the trapezoidal rule, which is the most common area estimation method (Fig. 1).

The AUC is very useful for calculating the total body clearance (CL) and the apparent volume of distribution (see later). In the trapezoidal rule, the area estimation is highly dependent on the blood/plasma sampling schedule. That is, the more frequent the sampling, the closer the trapezoids to the actual shape of the concentration–time curve (Fig. 1). The collection of blood samples from substance misusers can be problematic because of poor venous access and the difficulty of retaining subjects for sampling over prolonged periods of time. However, a monitoring window that is too short may lead to an underestimation of pharmacokinetic parameters as has been reported to be the case for delta9-tetrahydrocannabinol (Wolff & Johnston, 2013). The gold standard for such studies is the use of

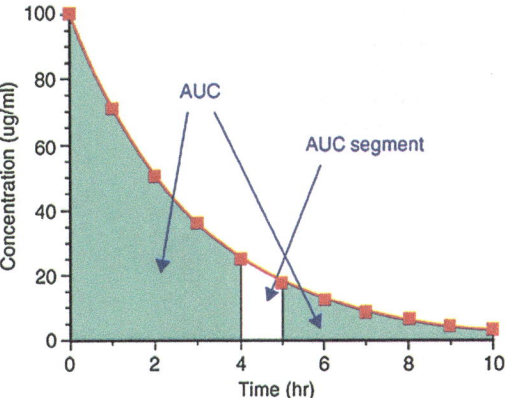

Pharmacokinetics, Fig. 1 Calculation of the area under the curve (AUC) after intravenous drug administration, using the trapezoidal rule showing a linear plot of concentration of drug in plasma (Cp) versus time showing AUC and AUC segment (http://www.boomer.org/c/p4/c02/c0208.html)

"closed" research/clinical units where subjects are "locked in" to a facility for the duration of the study period. However, such facilities are extremely expensive to resource and maintain. Intravenous administration of a developmental drug can provide valuable information on fundamental pharmacokinetic parameters but may often be biased toward an apparent rapid peak plasma concentration and short lasting exposure when compared to oral administration.

Compartmental Analysis

Compartmental pharmacokinetic analysis uses kinetic models to describe and predict the concentration–time curve. The advantage of compartmental to some noncompartmental analysis is the ability to predict the concentration at any time. The disadvantage is the difficulty in developing and validating the proper model. Compartment-free modeling based on curve stripping does not suffer this limitation. The simplest PK compartmental model is the one-compartmental PK model with IV bolus administration and first-order elimination kinetics (Neligan 2009), where a constant fraction of the drug in the body is eliminated per unit time as shown in Fig. 2.

Pharmacokinetics, Fig. 2 First-order elimination for a drug (A) administered by the intravenous route showing elimination over time

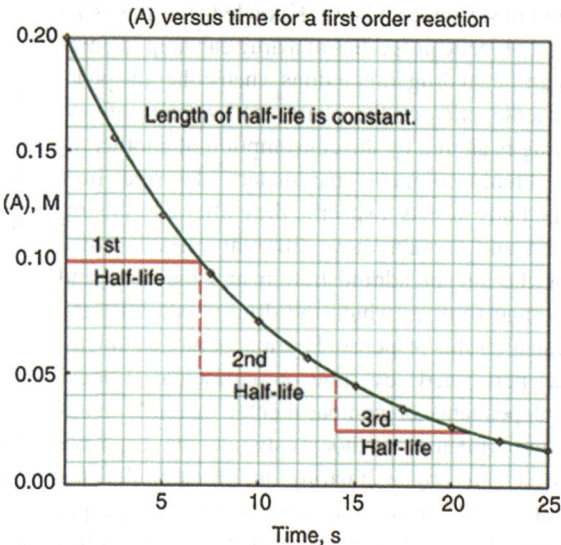

Pharmacokinetics is predominantly studied in a laboratory setting using advanced chromatographic technology (gas G, or liquid L chromatography C) coupled to mass spectrometry (MS). This is because of the complex nature of the matrix (blood (or brain tissue extracts in animals)) to be analyzed and the need for high sensitivity to detect low drug concentrations (10^{-6}–10^{-9} g) and the long time-point data. Blank or $t = 0$ samples taken before administration are important in determining a baseline and ensuring data integrity with such complex sample matrices. There is currently considerable interest in the use of very high sensitivity LC-mass spectrometry MS-MS. Using LC-MS-MS for microdosing studies, analysis can be performed with no interference from background ions because of the selectivity resulting from two stages of mass isolation. LC-MS-MS is seen as a promising alternative to animal experimentation.

Population pharmacokinetics is the study of sources and correlates of variability in drug concentrations among individuals who are the target patient population receiving clinically relevant doses of a drug of interest. This methodology seeks to identify the measurable pathophysiologic factors that cause changes in the dose–concentration relationship and the extent of these changes. The industry standard software for population pharmacokinetics analysis is NONMEM. There are also research-based companies which provide modeling and simulation software to the pharmaceutical industry for use during drug development such as Simcyp (www.simcyp.com).

A basic tenet of clinical pharmacokinetics is that the magnitudes of both desired response and toxicity are functions of the drug concentration at the site(s) of action (Rowland and Tozer 1995). Using this definition, "therapeutic failure" results when either the concentration of the drug at the site of action is too low, giving ineffective therapy, or is too high, producing unacceptable toxicity. For example, in drug treatment services, relapse and a return to illicit drug use has been frequently observed when the maintenance dose of methadone is too low to stop opioid withdrawal symptoms or craving for heroin. The concentration range in between these limits, the range associated with "therapeutic success," is often regarded as the "therapeutic window or range" (Fig. 3). However, these definitions are sometimes difficult to interpret for drugs used illicitly that are not controlled by pharmaceutical regulations (cocaine or cannabis or MDMA) or are difficult to apply to drugs consumed without restriction (alcohol and nicotine). Individual substance misusers often use illicit drugs for hedonistic purposes and experiment with the quantity

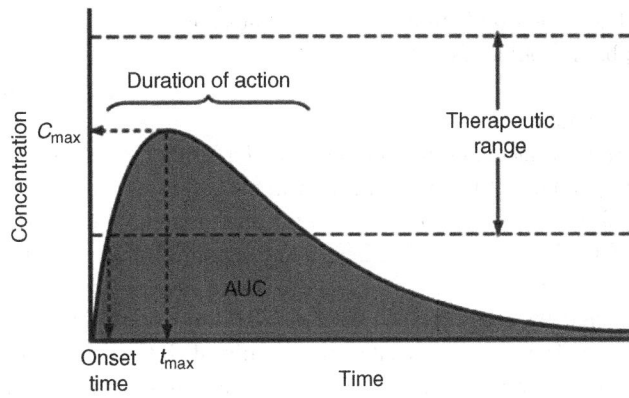

Pharmacokinetics, Fig. 3 Pharmacokinetic parameters describing a typical plasma concentration–time profile after an oral administration. C_{max} maximum concentration; t_{max} time to C_{max}; C_{max} area under the curve

and frequency of dose and with the route of administration until they achieve the desired effect (e.g., euphoria). This approach would not be acceptable in clinical settings.

In practice, the measurement of a drug in the body is usually determined in the blood or urine and in some cases in oral fluid/saliva or hair, because measurement of the drug concentration at the site of action is not easily achievable.

For some drugs with variable pharmacokinetics such as *warfarin* (*Coumadin*), biological monitoring is required to ensure that the blood concentration of the drug is strictly maintained within the therapeutic window (normal reference range, Fig. 3). Warfarin used for preventing thrombosis and embolism (abnormal formation and migration of blood clots) interacts with many common medications and some foods. Its activity has to be monitored by frequent blood testing for the international normalized ratio (INR) to ensure an adequate (appropriate prothrombin time), yet safe dose is taken. Other compounds, for example, *clozapine*, have a risk of serious side effects. Clozapine is used principally in treating schizophrenia and also for reducing the risk of suicide in patients with chronic risk for suicidal behavior. Plasma concentration of clozapine and norclozapine needs to be measured regularly in order to assess adherence to the dosing regime, prevention of toxicity, and dose optimization.

The discipline of pharmacokinetics is concerned with the quantitation of the mechanisms of absorption, the process by which a drug enters the blood stream; distribution of an administered drug, the rate at which a drug action begins; the duration of the effect; the metabolism of the drug in the body; and the effects and routes of excretion of the drug and its metabolites. These processes together are commonly referred to as *ADME* (Jacobs and Fehr 1987). More recently, biopharmaceutics has emerged as a new body of science that links traditional pharmacokinetics with pharmaceutics, and the acronym *LADME* (http://en.wikipedia.org/wiki/Pharmacokinetics) has been used to describe the processes studied by biopharmacists, essentially adding the term "liberation", the release of a drug from its formulation to the well-recognized ADME acronym.

Understanding Drug Effects

In order for a drug to exert its pharmacological effect, it must first gain entry into the body and be absorbed into the blood stream and transported to the site of action (usually in the brain for drugs that act on the Central Nervous System (CNS). The intensity of effect of a drug is governed by two major factors:

- The concentration of drug at the site of action in the body
- The sensitivity of the target cells (i.e., the magnitude of their response to a given concentration of drug)

The Concentration of Drug at the Site of Action

The concentration of drug at the site of action in the body at any given time after administration is determined by both the size of the dose and the pharmacokinetics of the substance.

Liberation

Biopharmaceutics has brought to the fore the importance of the physicochemical properties of a drug, the dosage form (or design), and the route of administration, all of which are key parameters when considering drug "liberation." Once administered, the drug must be liberated from its dosage form. Tablets and capsules may require disintegration (forming smaller particles) before dissolution and entry into the systemic circulation can occur. Similarly, the delivery of drug particulates into the lung from passive inhaler products is only achieved after "liberation" of the drug from the formulation during inhalation activation. The science of drug formulation design has rapidly grown over the last decade and is a thriving discipline.

Drug Dissolution

Drug dissolution usually occurs in the stomach and is dependent upon gastric activity. Many factors influence the dissolution of tablets and capsules, including the particle size, the chemical formulation, the inclusion of inert fillers, and the outer coating of the tablet. It is not unusual, therefore, for proprietary or generic preparations of a drug to have different dissolution characteristics and produce a range of plasma concentrations after oral administration. Dissolution characteristics of a dosage form are thus important considerations when interpreting pharmacokinetic data.

Absorption and Routes of Administration

As explained above, the extent and rate of absorption of a substance depends on the chemical properties of the drug itself and the way in which it is administered. The rate of absorption depends on the concentration of the drug, its degree of lipid solubility, the surface area of absorption, and the diffusion distance (i.e., the number of membranes it must cross before it reaches the bloodstream). Peak plasma concentration time (usually denoted as Tmax), is the most widely used general index of absorption rate; the slower the absorption, the longer the period of time to reach peak plasma concentration.

There are four principle methods of drug administration: oral/ingestion, across mucous membranes, inhalation, and by parenteral process (injection).

Oral (by Swallowing): Drugs taken orally are generally absorbed primarily in the small intestine rather than in the stomach. The benzodiazepines comprise a large family of lipophilic drugs. Diazepam, for instance, is rapidly absorbed, with peak concentration occurring within an hour for adults and, as quickly as, 15–30 min in children (Jenkins 2008). This route is characteristically slow and may be delayed by the presence of food in the digestive tract. Delays or drug loss during absorption may therefore contribute to the variability in drug response. However, the absorption of polar (water liking, lipophobic) drugs is unlikely to be complete within the 2–4 h transit period through the small intestine. This is the case with ranitidine, a H2 antagonist, where only about 60 % of the drug is absorbed following oral administration, the rest being recovered unchanged in the feces (Rowland and Tozer 1995). The oral route may be ineffective for certain drugs because the acidity of the stomach renders them ineffective. Heroin provides a good example of the latter phenomenon. Drug users looking for an immediate or intense effect or "rush" avoid the oral route. Nonlipophilic (nonionized) compounds such as alcohol readily cross the cell membranes by passive diffusion and are easily and rapidly absorbed from the gut, particularly on an "empty" stomach.

Transmucosal Absorption: Mucous membranes (that form the moist surfaces that line the mouth, nose, eye sockets, throat, rectum, and vagina) are thinner, have a greater blood supply, and are more permeable (lack keratin) compared with the epidermis (skin). For these reasons, absorption across mucous membranes is rapid, and drugs (especially lipophilic compounds) are effectively absorbed into the bloodstream. Nicotine in the form of chewing tobacco and buprenorphine placed sublingually (under the tongue) are successfully absorbed through the mucous membranes of the mouth. Drugs can be administered rectally including aminophylline, a drug used in the treatment of bronchial asthma, and tribromoethanol, an anesthetic. Drugs can also be absorbed by insufflation (sniffing, snorting), which allows cocaine, ketamine, amyl nitrite, and tobacco stuff to be absorbed across the mucous membranes of the nose and sinus cavities. Nasal administration (a preferred route for users of illicit substances) enables rapid absorption into the cerebrospinal fluid (CSF) and, hence, the cerebral circulation. After nasal administration, the concentration of some drugs in the CSF may be higher than in plasma (Calvey 2007).

Inhalation: In the case of inhalation, a drug is absorbed into the bloodstream across the alveolar membranes of the lung. This occurs in gas form (e.g., the vapors of solvents), in fine liquid drops (heroin or cocaine), or in fine particles of matter suspended in a gas (e.g., aerosols or the smoke from tobacco or cannabis). For many drugs, inhalation is the most rapid method of absorption into the general circulation. Glue "sniffing" is actually an incorrect description; technically, glue vapors are inhaled and absorbed via the linings of the lung.

Parenteral (*Injection*): The term "parenteral" denotes those pathways through which drugs are injected directly into the body. There are three principle parenteral routes: subcutaneous, intramuscular, and intravenous. Subcutaneous injection ("skin popping" to street users) involves injecting the drug under the skin into the fatty layer of tissue. The rate of absorption from the site of injection is slower than the intravenous injection but faster than the oral route. The most common example of the administration of a drug by this route is found in those with Type I diabetes mellitus, who typically inject insulin subcutaneously. Intramuscular injection involves deeper penetration of the drug into the body tissue. The drug is injected either in solution or in suspension directly into the muscle mass, where it is slowly absorbed into the bloodstream. Intravenous injection (popularly known as "mainlining") involves the direct injection of the drug into the veins. It is one of the fastest ways of getting a drug into the bloodstream and also allows relatively large amounts of a drug to be administered at one time. Intravenous drug administration poses the highest risk of toxicity and overdose (Wolff 2002).

Kinetics of Absorption The oral absorption of drugs often approximates first-order kinetics, especially when given in solution. Under these circumstances, absorption is characterized by an absorption rate constant, K_a, and a corresponding half-life. First-order kinetics depends on the concentration of only one reactant and a constant fraction of the drug in the body that is eliminated per unit time. The rate of absorption is proportional to the amount of drug in the body.

Sometimes a drug is absorbed essentially at a constant rate, called zero-order absorption. Zero-order kinetics is described when a constant amount of drug is absorbed (or eliminated) per unit time, but the rate is independent of the concentration of the drug (Fig. 4). Zero-order kinetics explains the way in which alcohol is handled in the body and several other drugs at high dosage concentrations, such as phenytoin and salicylate (Neligan 2009).

The pharmacokinetic parameter linked to the passage of a drug into the systemic circulation is known as bioavailability. It is a measurement of the amount of active drug that reaches the general circulation and is available at the site of action

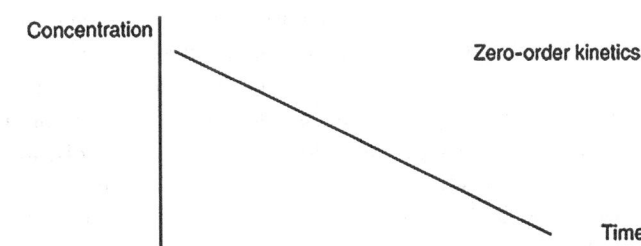

Pharmacokinetics, Fig. 4 Schematic representation of the absorption of a drug by zero-order kinetics

(although this parameter is difficult to establish for CNS active drugs, due to the need to cross the blood–brain barrier). It is expressed as the letter F.

It is assumed that a drug given by the intravenous route will have an absolute bioavailability of 100 % or 1 ($F = 1$), while drugs given by other routes usually have an absolute bioavailability of less than one. F is often calculated as the proportion of drug that reaches the systemic circulation after oral compared to IV administration. It is the fraction of the drug absorbed through non-intravenous administration compared with the corresponding intravenous administration of the same drug. The formula for calculating F for a drug administered by the oral route (p.o.) is given below:

$$F = \frac{[\text{AUC}]_{\text{po}} * \text{dose}_{\text{IV}}}{[\text{AUC}]_{\text{IV}} * \text{dose}_{\text{po}}}$$

F is measured by comparing the AUC for oral and IV doses from zero to the time point for which elimination is complete.

The calculation of F *ideally requires* an intravenous reference, that is, a route of administration that guarantees the entire administered drug reaches the systemic circulation. Such studies come at considerable cost, not least because of the necessity to conduct preclinical toxicity tests to ensure adequate safety.

Bioavailability is usually calculated by determining the maximum (peak) plasma drug concentration (C_{max}), the time that it takes to reach the peak (T_{max}), and the AUC. Plasma drug concentration increases according to the extent of absorption; the T_{max} is reached, when drug elimination rate equals absorption rate. Bioavailability determinations based on the C_{max} however, can be erroneous because drug elimination begins as soon as the drug enters the bloodstream.

For drugs excreted primarily unchanged in the urine (amphetamine), bioavailability can be estimated by measuring the total amount of drug excreted after a single dose. Ideally, urine is collected over a period of 7–10 elimination half-lives for complete urinary recovery of the absorbed drug.

The Distribution of Drugs in the Body

Once a drug is absorbed into the systemic circulation, distribution occurs throughout the body. The distribution for most drugs in the body is not even; some drugs bind to the plasma proteins, while others are sequestered into the adipose tissue, and a few have great affinity for bone tissue. Drugs must be highly fat soluble in order to enter the brain, such as Δ^9-tetrahydrocannabinol (THC), the main psychoactive constituent of cannabis. Similarly, fat-soluble drugs (like methadone) can cross the placenta to affect the fetus; these same drugs are also found in the milk of lactating women.

Within the blood, drugs can be bound to proteins as well as blood cells. The extent to which a drug is bound will affect the concentration of "free" drug in the blood (fu) and thus the amount available to reach the drug target, since only unbound drug can pass through the cell membranes. The degree of binding is often expressed as the bound-to-total drug concentration ratio and is described between 0 and 1.0. The nearer the value to 1.0, the greater the degree of binding. Binding is a function of the affinity of the protein for the drug (Rowland and Tozer 1995) and characterized by an association constant, Ka.

Unevenness of distribution is a barrier to the accurate interpretation of the concentration of a drug in the body and complicates efforts to correlate blood concentrations with behavior. High blood drug concentrations are usually related to greater behavioral effects than low concentrations. However, some drugs (the barbiturates and cannabis) have passed their peak plasma concentration before peak behavioral effects are seen. In the case of alcohol, some of the above differences do not apply, since alcohol crosses all barriers to distribute uniformly in total body water. Therefore, in the non-tolerant individual, blood alcohol concentrations are highly correlated with behavioral effects.

In pharmacokinetic terms, the volume of distribution (V_D) also known as the apparent volume of distribution is the parameter used as a direct measure of the extent of distribution. V_D is purely hypothetical and does not represent an actual physical volume inside the body. It is defined as the volume in which the amount of drug would need to be uniformly distributed to produce the observed blood concentration, by supposing that its concentration is homogeneous, i.e., the average tissue concentration is identical to that of the plasma. V_D is expressed as:

$$V_D = \text{dose}/C_0 (\text{initial concentration})$$

For example, after intravenous injection of 100 mg of a drug whose initial concentration, C_0, in plasma is 10 mg/L, the V_D would be 10 L. For a given drug, the knowledge of its desirable concentration in the blood and V_D allows evaluation of the dose to administer.

It is possible for V_D to be close to a recognizable volume, such as plasma volume (~0.05 L/kg), extracellular fluid (~0.2 L/kg), or total body water (~0.7 L/kg). This would happen if the drug is uniformly distributed in one of these "compartments," but this is rare. Indeed, when a drug binds preferentially to the tissues at the expense of plasma (a drug that is highly lipophilic), the plasma concentration will be extremely low (e.g., methadone, Δ^9-tetrahydrocannabinol). This will result in a large V_D, which may be larger than the actual volume of the individual itself (>1 L/kg), e.g., digoxin. A large V_D implies wide distribution, or extensive tissue binding, or both.

The V_D is thus a mathematical method for describing how well a drug is removed from the plasma and distributed to the tissues. However, V_D does not provide any specific information about where the drug is or whether it is concentrated in a particular organ. V_D may be increased by renal failure (due to fluid retention) and liver failure (due to altered body fluid and plasma protein binding). Conversely V_D may be decreased in dehydration.

The distribution phase is so-called, because distribution determines the early rapid decline in plasma concentration. However, changes in plasma concentration reflect primarily the movement within rather than loss from the body. In time, equilibrium is reached between the drug present in the tissue and that in plasma, and eventually plasma concentration reflects a proportional change in the concentrations of drug in all tissues and hence in the body. At this stage, decline in drug concentration is only due to the elimination of drug from the body (elimination phase). Two pharmacokinetic parameters describe the elimination phase: the V_D (as described above) and the biological or elimination half-life.

The elimination half-life of a drug is the time taken for the plasma concentration as well as the amount of drug in the body to fall by one half and is usually denoted by the abbreviation $t_{1/2}$. Knowledge of the $t_{1/2}$ is useful for the determination of the frequency of administration of a drug (the number of intakes per day) and to calculate the desired plasma concentration. Generally, the $t_{1/2}$ of a particular drug is independent of the dose administered (Table 1).

The Elimination of Drugs from the Body
Elimination occurs by metabolism and excretion. Some drugs are eliminated via the bile and others in the breath, but for most drugs, the primary route of excretion occurs via the kidneys.

Metabolism: Drugs are eliminated from the body in both changed and unchanged states, that is, part of the drug eliminated is chemically identical to the drug which was administered and part has been changed (metabolized). The proportion of a drug dose eliminated in a particular state is determined by the nature of the drug, the dose, the route of administration, and the physiological characteristics of the user. The excretion of unchanged psychoactive drugs via the urine or feces is generally inefficient because of their high fat solubility. As a result, fat-soluble substances are metabolized into water-soluble products that can be readily excreted by the kidneys and/or the intestines (Fig. 5).

Drugs are metabolized by specialized proteins called enzymes, which act as catalysts in the metabolic reaction. Most drug metabolism occurs in the liver, although enzymes in the kidneys, gut, lungs, and blood may also aid in the process. In the liver, there are two types of enzymes: microsomal (insoluble) drug-metabolizing enzymes and cytoplasmic (soluble)-metabolizing enzymes (which metabolize alcohol and similar drugs).

Pharmacokinetics, Table 1 Plasma elimination half-life for a selection of different compounds

Substance	Half-life	Notes
Diazepam (Valium)	20–50 h	Rapid absorption and fast onset of action with peak plasma concentration achieved 0.5–2 h after oral dosing. Active metabolite desmethyldiazepam has a half-life of 30–200 h
Carbamazepine	18–60 h	Carbamazepine exhibits auto induction: it induces the expression of the hepatic microsomal enzyme system CYP3A4, which metabolizes carbamazepine itself
Haloperidol	14–36 h	Fifty times more potent than chlorpromazine. Rapidly absorbed and has a high bioavailability
Fluoxetine	1–6 days	The active metabolite of fluoxetine is lipophilic and migrates slowly from the brain to the blood. The metabolite has a biological half-life of 4–16 days
Methadone	24–36 h	Orally effective. Exhibits auto induction: it induces the expression of the hepatic microsomal enzyme system CYP3A4, which metabolizes methadone itself
Water	7–10 days	Drinking large amounts of alcohol will reduce the biological half-life of water in the body
Alcohol	No half-life	Percentage of alcohol in your blood goes down by 0.015/h at a constant rate. Metabolism is zero-order kinetics (enzymes are saturated, and the rate of disappearance of ethanol in the body is *independent* of concentration. Thus, concentration falls off linearly, not exponentially. No half-life

The conversion of a fat-soluble drug to a substance that has sufficient water solubility for efficient excretion may involve several sequential chemical reactions, each step rendering the molecule slightly more water soluble than before (Feldman et al. 1997). Thus, many metabolites are often derived from the same parent drug (e.g., there are at least 25 known metabolites from THC, the main psychoactive ingredient of cannabis).

As the drug becomes progressively less fat soluble, it simultaneously loses the ability to cross the blood–brain barrier and to produce a pharmacological effect. However, some drug metabolites are pharmacologically active, producing the same effects as the parent drug. For example, heroin is metabolized to a number of metabolites, including morphine, N-morphine, codeine, and 6-monoacetylmorphine, all of which have pharmacological properties characteristic of the original drug.

Other drug metabolites can produce a completely different activity from the parent drug. Certain drug metabolites may be even more toxic than the parent drug: methanol (methyl alcohol) is an example of a drug metabolized to produce two very toxic metabolites, formaldehyde and formic acid. It is these metabolites which are considered responsible for disruption of the acid–base balance in the body and for damage to the optic nerve, both of which pose serious problems in methanol poisoning.

Both types of liver enzymes, microsomal and cytoplasmic, are inducible – therefore, repeated drug exposure causes the enzymes to increase in number; the result is a faster metabolic rate, that is, a more rapid conversion of the ingested drug into its metabolites (auto-induction). This increased metabolic rate (speeding up of the process) can result in increased intensity and speed of onset of drug effects if the original drug was inactive and the metabolites active. This process can result in decreased intensity and duration of effect, if the original drug was active and the metabolites were inactive. Many enzymes are genetically polymorphic, and their presence and number in the body is under genetic control. Four main phenotypes occur in the case of some enzymes

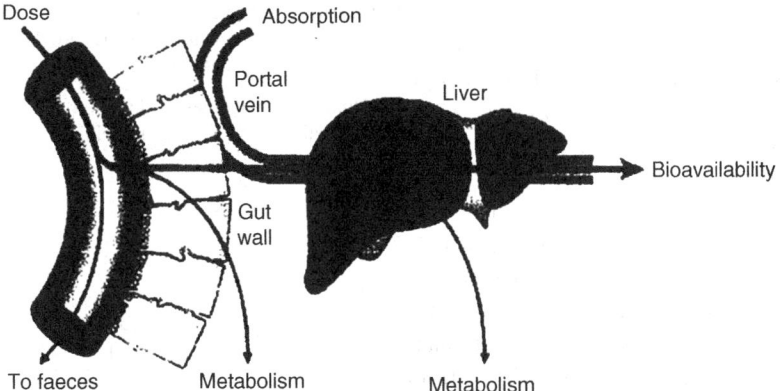

Pharmacokinetics, Fig. 5 Schematic representation of where metabolism occurs during the absorption process. The fraction of the initial dose appearing in the portal vein is the fraction absorbed, and the fraction reaching the blood circulation after the first pass through the liver defines the bioavailability of the drug administered orally (www.nature.com/.../v2/n3/images/nrd1032-i2.gif, image at www.nature.com/.../v2/n3/box/nrd1032_BX3.html)

with individuals categorized as fast, extensive (normal), intermediate, or poor metabolizers accordingly. For instance, this applies to the metabolism of MDMA (ecstasy) and nicotine metabolism as a result of the genetic polymorphism of CYP2D6 and CYP2A6, respectively.

Zero-order elimination is described when a constant amount of drug is eliminated per unit time, independent of the concentration of the compound. Zero-order reactions are typically found when the enzyme required for elimination to proceed is saturated by the drug. Zero-order kinetics explains when an individual, who drinks 20 units of beer before midnight, will fail a Breathalyzer test at 8 am the following morning. In this instance, the pathways responsible for alcohol metabolism are rapidly saturated and work to their limit. The removal of alcohol through oxidation by alcohol dehydrogenase in the liver is thus limited. Hence, the removal of a large concentration of alcohol from the blood may follow zero-order kinetics.

Excretion: The kidney acts as a pressure filter through which the blood passes. Most of the water and some of the dissolved substances contained in the blood are reabsorbed during its passage through the kidney. Substances that are fat soluble tend to diffuse back into the bloodstream, whereas residual water and unreabsorbed substances are eliminated during urination. The process of reabsorption and active excretion in the tubules of the kidneys and the time lapse between urine formation and urination make it difficult to use urine concentrations of drugs and drug metabolites as a base for accurate estimates of blood concentration of these substances. The drug concentrations determined in urine drug screening are, therefore, only a rough indication of blood concentrations (Jenkins 2008). Oral fluid/salivary drug concentrations on the other hand have a better correlation with the blood (Wolff et al 1999), although large variations in the oral fluid/blood ratios have been reported for a substance between individuals. Therefore, it is difficult to estimate drug concentrations in the blood based on drug concentrations in oral fluid (DRUID 2011).

The intestines are a site of drug excretion as well as a site of drug absorption. Some drugs and drug metabolites have chemical characteristics which cause them to be actively secreted (or "pushed out") into the bile as they pass through the liver cells, and the drug-containing bile then empties into the intestines. Thus, these drugs and metabolites may be excreted in the feces. As is the case with the kidney, however, the net excretion by this route may be greatly reduced by subsequent reabsorption into the bloodstream of the fat-soluble compounds (including psychoactive drugs) further along in the intestines. In this instance, the drugs will go through the process of excretion all over again (► Enterohepatic Cycling), and the drug effect may be prolonged.

Volatile drugs such as solvents are commonly excreted in the breath and for general anesthetics; breath may be the major route of elimination. For substances such as alcohol, however, it is a minor route. Nevertheless, it is possible through stimulation of breathing to increase the loss of drug from the body. Since the amount excreted in the breath can be reliably related to the blood level, the concentration of alcohol in the breath, as measured by the Breathalyzer test, serves to estimate the degree of intoxication. There are many routes of elimination, including sweat and saliva (and in lactating women, milk), and these all play a role in drug elimination. Though the latter are minor routes, they can be important in forensic analysis.

Just as the parameter, V_D is required to relate blood concentration to the total amount of drug in the body, so there is a need to have a parameter to relate drug concentration to the rate of elimination. Clearance, denoted by CL, is the factor. The CL of a drug is the volume of plasma from which the drug is completely removed per unit time. The elimination half-life is related to CL and V_D by the following equation:

$$t_{1/2} = \frac{\ln 2 \cdot V_D}{CL}$$

In clinical practice, this means that it takes just over 4.7 times the $t_{1/2}$ for a drug concentration in the plasma to reach a steady state after regular dosing is started and stopped or the dose is changed (Table 2). For example, methadone has a half-life of 24–36 h; this means that a change in the dose will take the best part of a week to have full effect. For this reason, drugs with a very long half-life (e.g., amiodarone, elimination $t_{1/2}$ of about 58 days) are usually started with a loading dose to achieve their desired clinical effect more quickly (Hallworth and Watson 2008).

The CL of a Drug: The amount eliminated is proportional to the concentration of the drug in the blood. The fraction of the drug in the body eliminated per unit time is determined by the elimination constant (kel). This is represented by the slope of the line of the log plasma concentration versus time:

$$CL = kel \times V_D$$

The rate of elimination equals the clearance times the concentration in the blood. It can be shown that the kel equals the log of 2 divided by the $t_{1/2} = 0.693/t_{1/2}$ and likewise, $CL = kel \times V_D$, so $CL = 0.693 V_D/t_{1/2}$, which shows $t_{1/2} = 0.693 \times V_D/CL$.

The rate of elimination (R_{oe}) is thus the clearance (CL) times the concentration in the plasma (Cp):

$$R_{oe} = CL \times Cp$$

while the fraction of the total drug removed per unit time is CL/V_D

If the volume of distribution is increased, then the kel will decrease, the $t_{1/2}$ will increase, but the CL would not change.

Physiological Tolerance or Drug Tolerance

Drug tolerance is encountered when an individual's reaction to a drug decreases so that larger doses are required to achieve the same effect. It can involve both psychological and physiological factors. The main characteristic of drug

Pharmacokinetics, Table 2 The elimination of a compound from the systemic circulation expressed as a function of its half-life

Number of half-lives elapsed	Fraction remaining	Percentage remaining
0	1/1	100
1	1/2	50
2	1/4	25
3	1/8	12.5
4	1/16	6.25
5	1/32	3.125
6	1/64	1.563
7	1/128	0.781
...	...	...
N	$1/2^n$	$100(1/2^n)$

tolerance is that the process is reversible. That is, cessation of use of the drug will diminish the degree of tolerance. The rate at which tolerance develops and is lost depends on the particular drug, the dosage, and the frequency of use. Another feature of tolerance is that differential development occurs for different drug effects.

When considered in the context of pharmacokinetics, tolerance can be conveniently divided into a number of forms that relate to the mechanisms involved (Pratt 1991):

- *Dispositional (pharmacokinetic) tolerance* results from a change in absorption, distribution, metabolism, or excretion of a drug which leads to a decreased quantity of the substance reaching the site it affects.
- *Pharmacodynamic tolerance* arises from adaptational changes occurring in the brain. These include cellular adaptive processes like *downregulation* (reduced number) and/or *desensitization* (reduced sensitivity) of drug receptors at the primary site of drug action as well as changes in neurotransmitter function in neuronal pathways downstream from the initial interaction. In opioid tolerance, several mechanisms contribute to opioid receptor desensitization, but the N-methyl-D-aspartate (NMDA) receptor cascade seems to be a common and important link.
- *Cross tolerance* results when tolerance to one drug is extended to another of a different chemical class.

Cross-References

▶ Absorption
▶ Area Under the Curve
▶ Bioavailability
▶ Clearance
▶ Distribution
▶ Drug Tolerance
▶ Elimination Half-Life
▶ Enterohepatic Cycling
▶ Excretion
▶ First-Order Elimination Kinetics
▶ Liberation
▶ Metabolism
▶ Volume of Distribution
▶ Zero-Order Elimination Kinetics

References

Calvey (2007) Drug absorption, distribution and elimination, chap 1. 27 Nov, 13:44. http://www.blackwell-publishing.com/content/BPL_Images/Content_store/Sample_chapter/9781405157278/9781405157278_4_01.pdf

DRUID, Driving under the influence of Drugs, Alcohol and Medicines (2011) Per se limits – Methods of defining cut-off values for zero tolerance, 6th framework programme, deliverable 1.4.2. Project No. TREN-05-FP6TR-S07.61320-518404-DRUID

Feldman RS, Meyer JS, Quenzer LE (eds) (1997) Principals of pharmacology, chap 1. In: Principles of neuropsychopharmacology. Sinauer Associates, Sunderland, pp 20–25. ISBN 0-87893-175-9

Hallworth M, Watson ID (eds) (2008) Therapeutic drug monitoring and laboratory medicine. ABC Venture Publications, London. ISBN 978-0-902429-42-0

Neligan P (2009) Pharmacokinetics. Basic pharmacology, vol 1.0, part 1. http://www.4um.com/tutorial/science/pharmak.htm. Accessed 10 Apr 2009

Jacobs MR, Fehr KO'B (eds) (1987) Drugs and drug abuse. A reference text, 2nd edn. Addiction Research Foundation, Toronto, pp 18–27

Jenkins AJ (2008) Pharmacokinetics: basic concepts and models, chap 1. In: Karch S (ed) Pharmacokinetics and pharmacodynamics of abused drugs. CRC Press/Taylor & Francis Group, London, pp 1–14. ISBN 0-13-978-1-4200-5458-3

Pratt JA (1991) Psychotropic drug tolerance and dependence: common underlying mechanisms, chap 1. In: Pratt J (ed) The biological bases of drug tolerance and dependence. Neuroscience perspectives (series ed Jenner P). Academic, London, pp 1–25. ISBN 0-12-564250-4

Rowland M, Tozer TN (1995) Clinical pharmacokinetics concepts and applications, 3rd edn. Williams & Wilkins, London, pp 1–17. ISBN 0-683-07404-0

Wolff K (2002) Characterisation of methadone overdose: clinical considerations and the scientific evidence. Ther Drug Monit 24:457–470

Wolff K, Farrell M, Marsden J, Monteiro RA, Ali R, Welch S, Strang J (1999) A review of biological indicators of illicit drug misuse, practical considerations and clinical usefulness. Addiction 94:1279–1298

http://en.wikipedia.org/wiki/Pharmacokinetics. Accessed 10 April 2009

PHAR 7633 chap 2. Background mathematical material. http://www.boomer.org/c/p4/c02/c0208.html. Accessed 10 Apr 2009

www.nature.com/.../v2/n3/images/nrd1032-i2.gif. Image at: www.nature.com/.../v2/n3/box/nrd1032_BX3.html. Accessed 10 Apr 2009

Pharmacological fMRI

Synonyms

phMRI

Definition

Pharmacological MRI (phMRI) refers to the use of magnetic resonance imaging for assessing the functional response to ligand-induced receptor stimulation or inhibition after drug administration. For this purpose, different functional MRI methods including CBF-, CBV-, or BOLD-fMRI can be applied to measure changes in cerebral perfusion. Image recording before, during, and after drug administration allows for drawing conclusions on the effect of pharmacological compounds on brain activity. With such an experimental design, one tests acute effects of the drug itself in the brain ("challenge phMRI"). With "modulation phMRI" on the other hand, one investigates how neurotransmitter systems are involved in neuronal systems engaged by other processes, such as cognitive challenge.

Cross-References

▶ BOLD Contrast
▶ Cerebral Perfusion
▶ Functional Magnetic Resonance Imaging

Pharmacotherapy

Definition

Use of a chemical substance (medication) in the treatment of a medical condition or illness.

Cross-References

▶ Drug Interactions

Phase I Clinical Trial

Definition

Phase I clinical trials are often the first time a potential new medication is given to humans. They are conducted following extensive animal testing for both efficacy and safety. Typically, healthy paid volunteers are administered progressively higher doses in the same session or, more commonly, over several sessions. These sessions are conducted in specialized laboratories equipped and staffed to monitor carefully the subject's physiological and behavioral response. In most phase I trials, plasma samples are taken to measure levels of the drug and its metabolites to determine rates of metabolism and elimination and develop a pharmacokinetic model of the drug's biodisposition.

Cross-References

▶ Abuse Liability Evaluation
▶ Phase II Clinical Trial
▶ Phase III Clinical Trial
▶ Randomized Controlled Trials

Phase II Clinical Trial

Synonyms

Controlled clinical trial

Definition

Phase II clinical trials are designed to assess both safety and efficacy of new medications or new uses of existing medications. They usually follow a Phase I trial but enroll patients with the disorder that the medication is intended to treat, using rigorous inclusion and exclusion criteria. Early Phase II trials may be conducted to select doses for more extensive later Phase II trials. Phase II trials include one or more groups of individuals receiving at least one dose of the proposed new treatment and at least one comparison group. The comparison may be a placebo; it may be an existing approved medication, or both. Thus, Phase II trials are often called controlled clinical trials.

Cross-References

- Abuse Liability Evaluation
- Phase I Clinical Trial
- Phase III Clinical Trial
- Randomized Controlled Trials

Phase III Clinical Trial

Synonyms

Controlled clinical trial

Definition

Phase III clinical trials are designed to assess both safety and efficacy of new medications or new uses of existing medications under conditions similar to those where it would be used clinically. The participants in Phase III trials are patients with the disorder that the medication is targeted to treat. These studies include the study medication and one or more comparison groups, which may involve a placebo, an existing approved medication, or both. Typically, Phase III trials are preceded by one or more Phase II trials; however, Phase III studies enroll a greater number of subjects. Many Phase III trials utilize several clinical sites and are referred to as multisite or multicenter trials. In one variation of a randomized controlled trial, patients are switched from one medication to another using a crossover design. In another variation, patients receive sequenced treatment utilizing a specific protocol for moving them from one treatment to another. A well-known sequenced clinical trial in psychopharmacology is known as the STAR*D trial of treatment alternatives for depression (http://www.nimh.nih.gov/health/trials/practical/stard/backgroundstudy.shtml). Phase III trials are often referred to as pivotal clinical trials, in which positive results and a reasonable balance of benefits and risk are needed to obtain regulatory approval of the new medication. After approval, a medication enters Phase IV testing which is comprised primarily of post-marketing surveillance (or pharmacovigilance) of adverse side effects.

Cross-References

- Abuse Liability Evaluation
- Phase I Clinical Trial
- Phase II Clinical Trial
- Randomized Controlled Trials

Phasic Neurotransmission

Synonyms

Phasic neuronal firing; Phasic neurotransmitter release

Definition

Phasic neurotransmission is characterized by a bursting mode of neuronal firing that quickly releases a large quantity of neurotransmitter in the synapse. This form of neural transmission is usually triggered by behaviorally relevant signals and contrasts with "tonic" neurotransmission, which instead refers to the maintenance of low steady levels of extracellular neurotransmitter.

Cross-References

- Synaptic Plasticity

Phasic Signal Transmission

Definition

Signal transmission that is triggered by neuronal activity.

Phencyclidine

Synonyms

Angel dust; PCP

Definition

Phencyclidine (PCP) was originally developed as a general anesthetic agent by Parke-Davis & Co. in the late 1950s, but during the initial clinical trials, it was discovered that approximately 30 % of the patients, as they emerged from anesthesia, developed psychotic reactions that had a close resemblance to schizophrenia. Subsequent studies found that after intravenous administration PCP induced a dysphoric, confusional state characterized by feelings of unreality, changes in body image, profound sense of aloneness or isolation, and disorganization of thoughts and amnesia; in many subjects negativism and hostility also occurred together with hallucinations and repetitive motor behavior. PCP could in healthy volunteers produce schizophrenia-like impairment of primary attention, motor function, proprioception, and symbolic and sequential thinking. Despite its dysphoric profile in patients, PCP was subject to abuse under the street name "angel dust." Its primary mechanism of actions is noncompetitive inhibition of the ion channel of the NMDA receptor, by binding inside the channel. However, PCP is also a muscarinic antagonist, a dopamine, a 5-HT and noradrenaline reuptake inhibitor, a sigma 1 and 2 ligand, and a cholinesterase inhibitor and can block voltage-gated potassium channels.

Cross-References

▶ Animal Models for Psychiatric States
▶ Excitatory Amino Acids and Their Antagonists

Phenelzine

Synonyms

Nardil

Definition

Phenelzine (2-phenylethylhydrazine) is an antidepressant that inhibits monoamine oxidase (MAOI) irreversibly and nonselectively; it is still in widespread clinical use and is claimed to be effective against major depressive disorders, particularly in treatment-resistant patients. It is also used in the treatment of dysthymia, bipolar depression, ▶ panic disorders, ▶ social anxiety disorder, bulimia, and ▶ post-traumatic stress disorder. As is the case for many other antidepressants, phenelzine requires ca. 2–6 weeks of daily treatment to become therapeutically effective.

Due to its prominent side effects, phenelzine is considered only as a treatment of last resort. Among the adverse side effects are dizziness, disturbances in vision, appetite and food, sleep, blood pressure, thermoregulation, and sexual functions. Like other MAOIs, phenelzine interacts with tyramine-containing foods (the so-called cheese effect), overindulgence of which can result in a hypertensive crisis.

Cross-References

▶ Antidepressants
▶ Monoamine Oxidase Inhibitors

Phenobarbital

Synonyms

Phenobarbitone

Definition

Phenobarbital is a barbiturate drug used widely as an anticonvulsant agent. It has a very long duration of action after a relatively slow onset of effects. It was synthesized at the Bayer pharmaceutical firm by 1904 and first marketed in 1912. It has been superseded with respect to its original use as an anxiolytic, sedative, and hypnotic agent, partly because of the dangers of overdose leading to coma and death. It shares many other properties with its shorter-acting relative pentobarbital.

Cross-References

- Abuse Liability Evaluation
- Anxiolytics
- Barbiturates
- Driving and Flying Under the Influence of Drugs
- Hypnotics
- Insomnias
- Sedative, Hypnotic, and Anxiolytic Dependence

Phenothiazines

Definition

A group of drugs that includes key members of the first generation of antipsychotic substances that brought about major changes in treatment of schizophrenia. Among the widely used drugs in this category are chlorpromazine, trifluoperazine, and fluphenazine.

Cross-References

- Antipsychotic Drugs
- First-Generation Antipsychotics

Phenotype/Genotype

Definition

Genotype is the specific genetic constitution of an organism including the gene allelic makeup. Phenotype is the physical trait or characteristic arising from the genotype.

Cross-References

- Genetically Modified Animals
- Phenotyping of Behavioral Characteristics

Phenotyping of Behavioral Characteristics

Sabine M. Hölter[1] and John Cryan[2]
[1]Helmholtz Zentrum München, German Research Center for Environmental Health (GmbH), Institute of Developmental Genetics, München, Germany
[2]Department of Anatomy & Neuroscience, University College Cork, Cork, Ireland

Synonyms

Behavioral characterization; Behavioral phenotyping

Definition

Animal behavior can be viewed as the outward manifestation of an orchestrated and complex functioning of the central nervous system (CNS) and how it interacts with the internal and external environment. Phenotyping of behavioral characteristics in its widest sense refers to behavioral paradigms in which aspects of the behavioral repertoire of a subject are analyzed, either under baseline conditions or after an experimental challenge, with the aim to assess CNS function. Most frequently, sensory, locomotor, motivational, emotional, social, cognitive, consumptive, reproductive, or reward-related behaviors are assayed. The subjects participating in a study may be either animal or human.

Principles and Role in Psychopharmacology

The goal of phenotyping of behavioral characteristics is usually either (1) the assessment of the neurobiology of certain behaviors by the use of lesions, systemic or local drug applications, or other experimental manipulations; (2) the examination of the effects of drugs in a drug discovery screening or profiling context; or (3) the analysis of CNS-specific gene function by the use of genetically altered animals, for example, knockout mice, or, in the case of human subjects, by relating human genetic haplotypes to disease-relevant behavioral characteristics assessed with neurophysiological or neuropsychological methods. All three aspects are important for the development and assessment of animal models of neurological and psychiatric disorders. These are generated by combining manipulations (e.g., stress, chemical, lesion, surgical, genetic, environmental) with a test which has a robust behavioral readout.

Most behavioral phenotyping methods carried out in animals were initially developed either for studying the neuronal circuits involved in normal and pathological behaviors or for the preclinical analysis of compounds in pharmaceutical industry – whereby rats, birds, or monkeys were primarily used as subjects (Sahgal 1993). With the completion of the genome sequences of human and several other species, the determination of gene and protein function, and understanding their role in disease mechanisms, became one of the major challenges in biomedical science. Due to its amenability to transgenic techniques, the mouse became the most used model organism in the endeavor to develop a complete functional annotation of the human genome and to employ the same information to better understand human disease and its underlying physiological and pathological basis. In this context, the experimental designs for characterizing rat behavior were transformed to successfully work in mice, and the term "behavioral phenotyping" is most frequently used in the context of behavioral characterization of mouse mutants. The development of this field owed much to the publication of reviews suggesting "test batteries" and to the publication of comprehensive guidebooks on mouse behavioral phenotyping methodology (e.g., Crawley 2000; Crusio and Gerlai 1999), which encouraged many molecular geneticists and neuroscientists to use these techniques for their research. While behavioral phenotyping is also growing among other model organisms including the fruit fly (*Drosophila melanogaster*), the worm (*C. elegans*), the sea slug (*Aplysia*), and the zebrafish (*Danio rerio*), we will focus this entry on rodents.

Examples of Frequently Used Behavioral Phenotyping Tests in Lab Rodents

Behavioral phenotyping procedures were developed to analyze animal models for human neuropsychiatric dysfunctions like posttraumatic stress disorder, other anxiety disorders, depression, schizophrenia, autism, attention-deficit hyperactivity disorder (ADHD), addiction, mental retardation, or motor, sensory, or cognitive deficits related to neurodegenerative diseases like Parkinson's disease, Alzheimer's disease, Huntington's disease, stroke, or prion diseases.

For the analysis of emotionality, most frequently tests for unconditioned anxiety-related behavior are applied, in which the animals are usually exposed to novel environments, which puts them in the conflict between the urges to explore and to avoid possible dangers. Most frequently, the open-field test and the elevated plus-maze test are used to this end, and an elevated zero maze is sometimes used as an alternative to the plus maze. Additionally, light–dark avoidance tests are used to assess anxiety, and the social interaction test is applied to assess social anxiety.

Rodent tasks relevant to human depression are primarily stress-induced reductions in avoidance or escape, termed behavioral despair. The most widely used rodent models of symptoms of depression are the forced swim test and the tail suspension test. The learned helplessness paradigm may also be used.

Prepulse inhibition is a frequently used task that assesses sensorimotor-gating phenotypes and is considered to have face, construct, and a high-predictive validity for schizophrenia and other neuropsychiatric diseases involving dysfunctions of sensorimotor integration in man.

To assess motor functions, frequently the open-field test and voluntary wheel running are used to assess locomotor activity levels. The accelerating rotarod, balance-beam tests, and the vertical pole test are often used to assess motor coordination and balance. In addition, grip strength is measured using a specialized grip strength meter, and gait abnormalities can be analyzed by manual or automated analysis of the footprint pattern. Assessment of swimming ability can reveal motor or vestibular deficits not detected by other motor coordination and balance tests.

For behavioral assessment of sensory abilities, reflex responses are used whenever possible to reduce possible confounding factors, for example, the motivation of the animal to participate in the test. Elicitation of the optokinetic nystagmus is a reflex response that is used to assess functionally intact vision. Hearing deficits can be revealed by measuring the acoustic startle reflex, either by the use of a clickbox or, more precisely, by specialized acoustic startle chambers. Pain sensitivity is frequently assessed using the hot-plate test (requiring circuitry in the brain and the spinal cord) or the tail-flick test (spinal reflex). Taste and olfactory abilities are often assessed in choice tests.

Aversively and appetitively motivated tasks are used to assess cognitive function. For the analysis of conditioned anxiety or fear learning, most frequently, procedures like active avoidance, contextual-fear and cued-fear conditioning, or fear-potentiated startle are applied, in which the animals have to learn the association between a conditioned stimulus and a mild footshock. The Morris water maze or the radial arm maze is most frequently used to assess spatial learning, reference memory, and working memory, and appropriate versions of the T-maze and the Y-maze task are also used to assess working memory. Social recognition tasks are applied to analyze social memory and olfactory function, and object recognition tasks are applied to assess object memory. Schedule-controlled operant tasks can also be used to assess different aspects of learning and motivational behavior, and one such task, the five-choice serial reaction-time task, has been developed to assess sustained and divided attention.

Repeated testing on the accelerated rotarod is applied to assess motor learning. Rodent tasks relevant to human addiction are primarily procedures that assess the rewarding properties of substances like self-administration of drugs, conditioned place preference paradigms, or generalization to drugs of abuse in drug discrimination procedures.

Applications

Behavioral phenotyping methodology has been used in the search for new and improved pharmaceuticals. It may be used not only to identify compounds with specific actions but also to exclude unwanted side effects of a new compound, for example, motor effects or abuse potential.

Since the advent of functional genomics, behavioral phenotyping is frequently applied in the analysis of genetic and molecular functions, with the aim to identify new targets for therapeutic development. Genetic studies in the mouse are important for the elucidation of molecular pathways underlying behavior. The goal of this endeavor is not only the identification of genes that control brain function and influence behavior but also the understanding of genetic factors involved in human psychiatric disorders (Tarantino and Bucan 2000). These disorders are associated with quantitative phenotypes called "intermediate traits" or "endophenotypes," some of which, in contrast to the full-complex disorder, can readily be modeled in mice. These traits are risk factors which are considered to be closer to the genetic etiology than the full syndrome. Examples of such endophenotypes are anxiety in depression, prepulse inhibition and working memory deficits in schizophrenia, and social interaction deficits in autism and schizophrenia (Gottesman and Gould 2003). Since the rather recent emergence of the endophenotype concept in neuropsychiatric research, some work has been devoted to appropriately model human endophenotypes in mice and to inspect the function of candidate genes for human endophenotypes identified (e.g., by human association or linkage studies) in adequate genetic mouse models.

Since 1997, behavioral phenotyping methodology has been used in many large-scale mouse projects (see Gondo (2008) for overview). In several large-scale, phenotype-driven, random-mutagenesis screens aiming for the discovery of novel genes in mice, behavioral phenotyping methodology was applied to detect so far unknown genes involved in defined behaviors. The development of large-scale mouse mutagenesis projects increased the demand for comprehensive robust and reliable phenotyping platforms for all body systems, including CNS function. In this context, the German National Genome Research Network established, in collaboration with the pan-European consortium EUMORPHIA, a mouse phenotyping center with open access to the scientific community, called "German Mouse Clinic" (http://www.mouseclinic.de/, Gailus-Durner et al. 2005).

To overcome the bottleneck of comprehensive phenotypic characterization of the large number of mutants generated by the large-scale mutagenesis projects, first several mouse phenotype assessment centers were built in Europe that worked together under the roof of EUMODIC (European Mouse Disease Clinic, http://www.eumodic.org/). Until 2012, these mouse clinics used the standardized tests contained in EMPReSS (the European Mouse Phenotyping Resource of Standardized Screens, empress.har.mrc.ac.uk) on 500 mutant mouse lines and made their phenome data publicly available in the EuroPhenome database (http://www.europhenome.org/). In 2011, as a next step, the International Mouse Phenotyping Consortium (IMPC) was formed (Abbott 2010; Wurst and de Angelis 2010), which includes the European mouse phenotyping centers involved in EUMODIC and additionally Asian, Australian, and North American mouse clinics. Its aim is the continuation of the production and primary phenotyping of mutant mouse lines for the remainder of the approximately 20,000 genes in the mouse genome (see www.mousephenotype.org). Members of the IMPC discuss and optimize the choice of tests, procedural details, and also the primary screen workflow of a mouse clinic in a worldwide coordinated program. This results in optimized phenotyping protocols contained in IMPReSS (International Mouse Phenotyping Resource of Standardized Screens, www.mousephenotype.org/impress), the successor of EMPReSS. Once available and quality controlled, also the IMPC phenotyping data will be made publicly available.

Pitfalls

Several factors can influence behavioral phenotyping results and therefore need to be considered in the context of data reproducibility (Wahlsten 2001). Due to developmental and degenerative processes, the age of the subject can play a role, as well as the time point of testing due to circadian rhythms of many biological processes. Therefore, experimental subjects and controls of the same age should be tested concurrently; in the case of mutants, ideally littermates are used as controls. Caution should be made in extrapolating data between species as the effects of manipulations may vary. Moreover, within a given species, there are marked differences between strains, and this should also be taken into account when phenotyping or analyzing behavioral data.

Because the previous experience of handling experimental manipulations or drug injections can influence the performance of a subject in a particular test, it might be desirable to use experimentally naive subjects for each kind of experiment. Practically, this approach increases the number of animals needed and is simply not feasible when mouse mutants are used, due to the comparatively long time needed to breed a sufficient amount of animals for concurrent testing of mutants and littermate controls in a statistically meaningful number. As a solution to this problem, frequently a series of tests is applied, in which the tests are usually ordered from those that are considered the least stressful for the subject to those considered the most stressful. Likewise, in comprehensive phenotyping platforms like mouse clinics, behavioral phenotyping is done first, before more invasive experimental manipulations are performed on the animals, and the experimental history is documented. In case of a positive phenotyping result using such "test batteries," the putative phenotype should be confirmed in a new batch of subjects.

The more comprehensive the phenotyping, the less likely are mis- or overinterpretations of individual data sets. This is particularly important when screening for phenotypes in mutants of genes with unknown functions. For example, locomotor, exploratory, or emotional phenotypes may be confounded or even caused by sensory deficits, skeletal malformations, or metabolic alterations and may thus not inevitably reflect the effects on brain function. Likewise, apparent cognitive phenotypes may not be centrally mediated but may be caused by alterations in sensory perception. Because one of the goals of behavioral phenotyping is the analysis of endophenotypes related to human neuropsychiatric disorders, it seems important to exclude alternative, not CNS-specific explanations for putative phenotypes.

Experimenter and the kind of equipment used are also known to influence phenotyping results. Therefore, careful descriptions of experimental procedures and thorough training of an experimenter are essential, as well as precise (technical) specifications of the (automated) equipment. For example, consortia like EUMORPHIA and EUMODIC (see above) that share the phenotyping effort and put particular emphasis on data reproducibility undergo considerable efforts to ensure their procedures yield comparable results and validate them in cross-laboratory comparisons (Mandillo et al. 2008). However, since also environmental factors like diet, animal husbandry, housing conditions, cage structure, and so on can have an impact, there may be practical limitations to standardization.

Advantages and Limitations of Behavioral Phenotyping Procedures

The advantages of behavioral phenotyping procedures reside in (1) their applicability to a wide range of scientific questions, from drug profiling to functional genomics; (2) their utility in the analysis of functions at the molecular level, as shown in many pharmacological studies as well as in studies with mouse mutants, for example, mutants of the dopaminergic and serotonergic neurotransmitter systems; and (3) the relative ease of collecting high-quality quantitative data. On the other hand, the more comprehensive the analysis is supposed to be, the more time and the more animals are needed. If mouse mutants are used, the time and costs required for breeding have to be taken into account, which can be particularly high in the case of conditional mutants. However, mutant mice are extremely valuable when dissection of the functional role of a molecule cannot be approached using other techniques – for example, when pharmacological compounds are unavailable or have poor selectivity for a particular receptor subtype (Cryan and Holmes 2005).

If mutants of genes of completely or mainly unknown function are analyzed, behavioral phenotyping results should be interpreted in the context of analyses of other body systems, particularly neuropathological and pathological analyses. Especially, locomotor and motor-related emotional phenotypes should be comprehensively investigated to check for possible dysmorphological, sensory, or metabolic contributions to the phenotype.

We suggest that it is prudent and most appropriate to use convergent tests that draw on different aspects of the behavioral system under investigation. In general, behavioral techniques are most effective when used in conjunction with other techniques to generate converging lines of evidence regarding the role of a molecule in a neural pathway.

Cross-References

▶ Active Avoidance
▶ Addiction
▶ Anxiety
▶ Attention
▶ Attention Deficit Hyperactivity Disorders: Animal Models
▶ Autism Spectrum Disorders and Intellectual Disability
▶ Conditioned Place Preference and Aversion
▶ Depression
▶ Drug Discrimination
▶ Elevated Plus Maze
▶ Learned Helplessness
▶ Open-Field Test
▶ Posttraumatic Stress Disorder
▶ Prepulse Inhibition
▶ Schizophrenia
▶ Self-Administration of Drugs
▶ Social Recognition Test
▶ Spatial Learning in Animals
▶ Working Memory

References

Abbott A (2010) Mouse project to find each gene's role. Nature 465:410
Crawley JN (2000) What's wrong with my mouse? Behavioral phenotyping of transgenic and knockout mice. Wiley, Hoboken
Crusio WE, Gerlai RT (eds) (1999) Handbook of molecular-genetic techniques for brain and behavior research. Elsevier Science, Amsterdam
Cryan JF, Holmes A (2005) The ascent of mouse: advances in modelling human depression and anxiety. Nat Rev Drug Discov 4(9):775–790
Gailus-Durner V, Fuchs H, Becker L, Bolle I, Brielmeier M et al (2005) Introducing the German mouse clinic: open access platform for standardized phenotyping. Nat Methods 2:403–404
Gondo Y (2008) Trends in large-scale mouse mutagenesis: from genetics to functional genomics. Nat Rev Genet 9:803–810
Gottesman II, Gould TD (2003) The endophenotype concept in psychiatry: etymology and strategic intentions. Am J Psychiatry 160:636–645
Mandillo S, Tucci V, Hölter SM, Meziane H, Banchaabouchi MA et al (2008) Reliability, robustness, and reproducibility in mouse behavioral phenotyping: a cross-laboratory study. Physiol Genomics 34:243–255
Sahgal A (ed) (1993) Behavioural neuroscience: a practical approach. IRL Press, Oxford
Tarantino LM, Bucan M (2000) Dissection of behaviour and psychiatric disorders using the mouse as a model. Hum Mol Genet 9:953–965
Wahlsten D (2001) Standardizing tests of mouse behavior: reasons, recommendations and reality. Physiol Behav 73:695–704
Wurst W, de Angelis MH (2010) Systematic phenotyping of mouse mutants. Nat Biotechnol 28:684–685

Phentermine

Definition

Phentermine is an ▶ amphetamine derivative that is used as an antiobesity agent. It was present in the fenfluramine-phentermine mixture known as fen-phen that has been withdrawn from the market because of the risk of heart valve disease. In contrast to ▶ fenfluramine, it is still available in many countries. In addition to its appetite-suppressant effects, it induces most of the sympathomimetic effects of amphetamine derivatives, with increased blood pressure, heart rate, agitation, and insomnia, and carries the risk of abuse and dependence.

Cross-References

▶ Appetite Suppressants

Phenylalanine and Tyrosine Depletion

Marco Leyton
Department of Psychiatry, McGill University, Montreal, QC, Canada

Synonyms

Acute tyrosine depletion; Acute tyrosine/phenylalanine depletion

Definition

The acute phenylalanine/tyrosine depletion (APTD) method was developed as a safe and rapid way to transiently decrease dopamine (DA) neurotransmission in humans. The method entails the administration of a protein mixture that is selectively deficient in amino acids (AA) that are used to make DA. Since ingestion of this mixture decreases the availability of the necessary raw material, the neurotransmitter's synthesis and availability for release decrease as well. A manipulation based on the same principles – acute tryptophan depletion (ATD) – is widely used to transiently decrease serotonin transmission. See entry "▶ Tryptophan Depletion" by Simon N. Young.

Current Concepts and State of Knowledge

The Catecholamine Metabolic Pathway

DA is a monoamine. As the name suggests it is made from a single amine, in this case the essential AA, phenylalanine. Essential AAs are those that we need to obtain from our diet. When we ingest protein, phenylalanine is cleaved, and the liver enzyme phenylalanine hydroxylase (PH) adds a hydroxyl group (oxygen + hydrogen, OH) to yield tyrosine. Tyrosine, in turn, can also be obtained from our diet. Irrespective of how we obtain it, tyrosine can then be carried across the blood-brain barrier by a two-tiered transport system (one active and saturable, the other diffusional) that acts on various large neutral AAs (LNAA: phenylalanine, tyrosine, tryptophan, valine, leucine, isoleucine, histidine, and methionine). Tyrosine can then enter DA neurons via a similar transport system on the cell's membrane. Inside the DA cell, tyrosine is a substrate for tyrosine hydroxylase (TH). TH adds a second OH group to yield 3,4-dihydroxy-L-phenylalanine, more commonly known as L-dopa. L-dopa is then a substrate for L-aromatic amino acid decarboxylase (AAAD or AADC) producing DA. Finally, vesicular monoamine transporters remove DA molecules

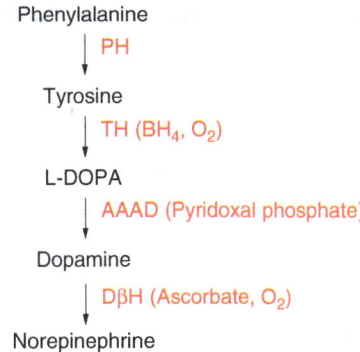

Phenylalanine and Tyrosine Depletion, Fig. 1 The catecholamine metabolic pathway. The essential amino acid, phenylalanine, is hydroxylated to tyrosine in the liver. Tyrosine from this dietary sources is then carried into the brain. Within catecholamine neurons, tyrosine is then hydroxylated by tyrosine hydroxylase into 3,4-dihydroxy-L-phenylalanine, and this constitutes the rate-limiting step in catecholamine synthesis. Decreases in tyrosine availability within a cell act like a drain, drawing in more of the amino acid. If tyrosine levels are insufficient, dopamine synthesis decreases. Tyrosine hydroxylase activity is also influenced by the availability of three other substrates, molecular oxygen, tetrahydrobiopterin, and dopamine itself. The first two increase tyrosine hydroxylase activity, while the latter is inhibitory. Similarly, the decarboxylation of L-dopa to dopamine is affected by concentrations of pyridoxal phosphate (vitamin B_6), while the hydroxylation of dopamine to norepinephrine is affected by oxygen and ascorbate acid (vitamin C). *PH* phenylalanine hydroxylase; *TH* tyrosine hydroxylase; *L-dopa* 3,4-dihydroxy-L-phenylalanine; *AAAD* aromatic amino acid decarboxylase; *DβH* dopamine beta hydroxylase; BH_4 tetrahydrobiopterin; O_2 molecular oxygen

from the cytosol and place them in storage vesicles. Within noradrenergic neurons, these vesicles contain dopamine-β-hydroxylase (DβH), an enzyme that hydroxylates the DA to norepinephrine (NE). DβH and AADC are 100–1,000 times more active than TH, and the hydroxylation of tyrosine to L-dopa is considered the rate-limiting step (Fig. 1).

Acute Phenylalanine/Tyrosine Depletion (APTD)

Under normal physiological conditions, TH is only 75 % saturated. Since TH is also the rate-limiting enzyme in DA synthesis, manipulations of tyrosine availability affect DA production.

Phenylalanine and Tyrosine Depletion, Table 1 Balanced mixtures

Amino acids	Men (g) (Leyton et al. 2000)	Men (g) (McTavish et al. 1999a)	Vervet monkeys (g) (Palmour et al. 1998)	Rats (mg) (McTavish et al. 1999b)	Rats (mg) (Jaskiw and Bongiovanni 2004)[*]
L-Alanine	5.5		0.55		
L-Arginine	4.9		0.49		
L-Cysteine	2.7		0.27		
Glycine	3.2		0.32		
L-Histidine	3.2		0.32		
L-Isoleucine	8	15	0.80	300	415.5
L-Leucine	13.5	22.5	1.35	450	623.2
L-Lysine monohydrochloride	11	17.5	1.10	350	350
L-Methionine	3	5	0.30	100	100
L-Phenylalanine	5.7	12.5	0.57	250	
L-Proline	12.2		1.22		
L-Serine	6.9		0.69		
L-Threonine	6.5	10	0.65	200	200
L-Tryptophan	2.3	2.5	0.23	50	50
L-Tyrosine	6.9	12.5	0.69	250	
L-Valine	8.9	17.5	0.89	350	350

[*]The effect of this phenylalanine/tyrosine deficient mixture is compared to saline

These effects might be particularly pronounced when the cell firing rate and precursor use is high. A method that exploits these features is called APTD.

In APTD studies, participants ingest an AA mixture that either does or does not contain phenylalanine and tyrosine (Table 1). Ingestion of the mixture induces protein synthesis in the liver. If the mixture lacks phenylalanine and tyrosine, the liver uses what it can extract from plasma. Since these plasma quantities are modest, levels decrease substantially. At the same time, plasma levels of AAs that were in the ingested mixture increase. This combination of effects markedly decreases the ratio of tyrosine to other LNAAs. Since this increases competition for access to the LNAA transporter, the amount of tyrosine that is able to enter the brain plummets (Wurtman et al. 1974; Fernstrom and Fernstrom 1995).

Studies conducted in rats, nonhuman primates, and humans indicate that the achieved tyrosine depletion significantly decreases DA synthesis and release. This includes evidence from postmortem tissue and in vivo microdialysis studies in rodents, cerebrospinal fluid (CSF) studies in monkeys, and both neuroendocrine and functional neuroimaging studies in humans. Although APTD might also affect NE metabolism, most evidence suggests an absence of significant effects on NE release except perhaps under conditions of exceptionally high NE use (Jaskiw et al. 2008). This preferential effect on DA was not predicted a priori, but the feature has been exploited to use APTD as a more selective probe than originally envisaged. Although effects of APTD on the trace amines are theoretically possible also, this has not been reported.

AA Mixture Composition: Rats, Vervet Monkeys, and Humans

Multiple APTD mixture recipes have been developed (Table 1). The two most common versions for the use in humans were developed by research groups at McGill University and

Oxford University. The McGill recipe contains 16 AAs given in the proportion that is found in human milk. The Oxford recipe has similarities, but it is without histidine and six nonessential AAs: alanine, arginine, cysteine, glycine, proline, and serine. Compared with the McGill mixture, the Oxford control mixture contains proportionally more phenylalanine and tyrosine, exactly 10.8 % for both, as compared to 5.4 % and 6.6 %, respectively, in the McGill recipe. Both APTD mixtures yield decreases in plasma phenylalanine and tyrosine levels within the range of 50–70 % and decreases in their ratio to LNAAs in the range of 80–90 %. Unequivocal differences in the elicited behavioral effects have not been identified. Direct comparisons between the two mixtures, though, have not been conducted.

The McGill group has also developed a version of the APTD mixture for use in vervet monkeys. The formula is identical to that used in humans, though adjusted accordingly for the monkey's lower body weight. At least two versions for use in rodents have also been described, and they can be administered by gavage or intraperitoneal injection. Again, direct comparisons of their effects have not been reported, but unequivocal differences are not evident in the literature.

Advantages and Disadvantages of APTD

Compared with the other methods for decreasing DA transmission, APTD has both advantages and disadvantages. On the plus side, the effects develop and dissipate rapidly, beginning as soon as 3 h after ingestion and disappearing 3–4 h later. Transient nausea is sometimes reported, and 5–10 % of subjects will regurgitate the mixture, but these sensations subside. Experience suggests that subjects who retain the mixtures for a minimum of 45–60 min achieve a decrease in plasma phenylalanine and tyrosine levels within the usual range. These mild adverse effects compare well to those associated with other methods for decreasing DA transmission. For example, administration of the competitive TH inhibitor, α-methyl-*para*-tyrosine (AMPT), typically requires 48–72 h inpatient observation and can lead to motor dyskinesias and crystalluria. The former effect likely reflects large DA depletions within the nigrostriatal pathway, and this can be either an advantage or a confound depending on the outcome of interest. DA receptor antagonists can also be used. Positron emission tomography PET imaging studies suggest that receptor blockade up to 70 % or 80 % can be achieved without eliciting extrapyramidal side effects or hyperprolactinemia, but available compounds either do not bind to all DA receptor subtypes or are nonspecific, binding also to multiple non-DA receptors.

Neurophysiological Effects of APTD

Microdialysis studies conducted in rats suggest that APTD does not affect extracellular DA levels when animals are tested at rest, but potent effects are seen during periods of increased DA cell firing (Jaskiw and Bongiovanni 2004) or release (McTavish et al. 1999b). These effects are dose dependent, and APTD can diminish stimulated DA release by up to 70 % (McTavish et al. 1999b). In humans, APTD increases circulating levels of prolactin, a neuroendocrine index of decreased DA transmission within the tuberoinfundibular pathway. In functional neuroimaging studies, APTD also decreases striatal DA release (Fig. 2), and this has been seen in the absence of an experimental challenge (Montgomery et al. 2003); whether this reflects a decrease in resting DA release or a diminished response to the mild stress of having an hour-long brain scan is difficult to disentangle. Irrespective, larger changes are reported to occur when participants are given a pharmacological challenge; that is, APTD decreases amphetamine-induced DA release, and the magnitude of this effect is twice of what is seen in subjects tested at "rest" (Leyton et al. 2004). One consequence of the APTD-induced decrease in DA transmission appears to be a disruption in the ability of different components of cortical-subcortical neurocircuits to work in coordination; that is, whereas cortical and basal ganglia structures exhibit high intercorrelations in activity levels when participants are engaging in a familiar

Acute phenylalanine/tyrosine depletion (APTD)

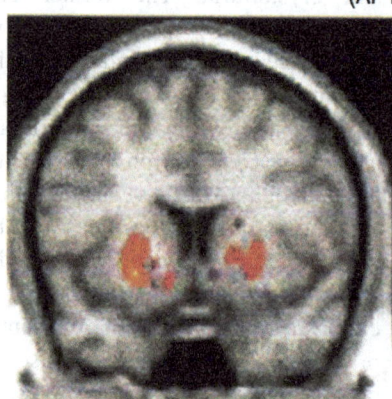
APTD vs. BAL

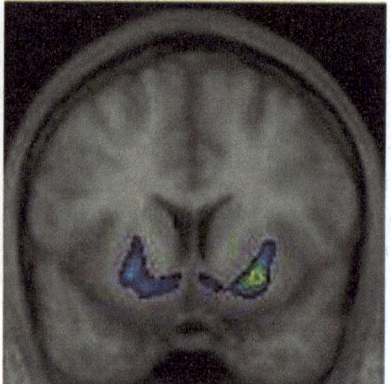
APTD + Amphetamine
vs.
BAL + Amphetamine

Phenylalanine and Tyrosine Depletion, Fig. 2 The figure depicts functional neuroimaging evidence that acute phenylalanine/tyrosine depletion (*APTD*) decreases extracellular dopamine levels in human striatum. At the left side, the colored t-map superimposed on the anatomical magnetic resonance image delineates regions where binding of the D2/D3 receptor ligand [^{11}C]raclopride was significantly higher following APTD versus ingestion of a nutritionally balanced control mixture (*BAL*), indicative of decreased extracellular dopamine levels (Montgomery et al. 2003). The image at the right shows the regions where [^{11}C]raclopride binding was higher on a test day with amphetamine plus APTD, as compared to amphetamine plus BAL, indicative of decreased drug-induced dopamine release (Leyton et al. 2004). Both effects occurred primarily within the more ventral than dorsal regions of the striatum. The right side of each image represents the right side of the brain

neurocognitive task, these correlations are reduced following APTD (Nagano-Saito et al. 2008). Based on these observations, it was proposed that normal DA tone is required to permit the efficient transfer of information throughout cortico-striatal circuitry.

Neurocognitive Effects of APTD
A number of behavioral effects have been reported. These include changes in the ability of subjects to preferentially respond to reward-related cues, to adjust responding appropriately when reward parameters change, and to sustain selective, focused interest in affectively relevant events. Some studies have also identified effects on spatial working memory, but an equal number of published studies yielded negative results; functional neuroimaging studies suggest that the variable results might reflect individual differences in the magnitude of DA depletion achieved.

Effects of APTD on Mood and Motivational States
APTD alone does not appear to lower mood in healthy individuals. In comparison, a now replicated finding is that APTD potentiates the mood-lowering effect of a psychological challenge. In people with mood disorders, a dissociation is seen also; tyrosine depletion reduces manic symptoms in bipolar patients but does not reinstate depressive symptoms in recovered patients with a history of major depression.

Effects of APTD on Substance Use
A frequent application of the APTD method has been in addiction research. APTD is reported to decrease psychostimulant effects of amphetamines, the ability of cocaine and cocaine-paired cues to elicit craving, alcohol self-administration in a free-choice task when the participants are light social drinkers, and the tendency to sustain responding on a progressive ratio breakpoint task

for successive units of an alcoholic beverage or mini-cigarette when the participants are heavy social drinkers or tobacco smokers, respectively.

Conclusions

Overall, the APTD method has proven to be an effective method to decrease DA transmission in human brain. Although, the effects might be smaller than those produced by other methods, the rapid, transient, and selective effects as well as more modest side effect profile are compelling for ethical reasons and simplify the interpretation of results.

Cross-References

▶ Addiction
▶ Amine Depletors
▶ Aminergic Hypotheses for Depression
▶ Impulsivity
▶ Mania
▶ Mood Disorders
▶ Tryptophan Depletion
▶ Working Memory

References

Fernstrom MH, Fernstrom JD (1995) Acute tyrosine depletion reduces tyrosine hydroxylation rate in rat central nervous system. Life Sci 57:97–102

Jaskiw GE, Bongiovanni R (2004) Brain tyrosine depletion attenuates haloperidol-induced striatal dopamine release in vivo and augments haloperidol-induced catalepsy in the rat. Psychopharmacology (Berl) 172:100–107

Jaskiw GE, Newbould E, Bongiovanni R (2008) Tyrosine availability modulates potassium-induced striatal catecholamine efflux in vivo. Brain Res 1209:74–84

Leyton M, Dagher A, Boileau I, Casey KF, Baker GB, Diksic M, Gunn R, Young SN, Benkelfat C (2004) Decreasing amphetamine-induced dopamine release by acute phenylalanine/tyrosine depletion: a PET/[11C]raclopride study in healthy men. Neuropsychopharmacology 29(2):427–432

Leyton M, Young SN, Pihl RO, Etezadi S, Lauze C, Blier P, Baker GB, Benkelfat C (2000) Effects on mood of acute phenylalanine/tyrosine depletion in healthy women. Neuropsychopharmacology 22:52–63

McTavish SFB, McPherson MH, Sharp T, Cowen PJ (1999a) Attenuation of some subjective effects of amphetamine following tyrosine depletion. J Psychopharmacol 13:144–147

McTavish SF, Cowen PJ, Sharp T (1999b) Effect of a tyrosine-free amino acid mixture on regional brain catecholamine synthesis and release. Psychopharmacology (Berl) 141:182–188

Montgomery AJ, McTavish SF, Cowen PJ, Grasby PM (2003) Reduction of brain dopamine concentration with dietary tyrosine plus phenylalanine depletion: an [11C]raclopride PET study. Am J Psychiatry 160:1887–1889

Nagano-Saito A, Leyton M, Monchi O, Goldberg Y, He Y, Dagher A (2008) Dopamine facilitates fronto-striatal functional connectivity during a set-shifting task. J Neurosci 28(14):3697–3706

Palmour RM, Ervin FR, Baker GB, Young SN (1998) Effects of acute tryptophan depletion and acute tyrosine/phenylalanine depletion on CSF amine metabolite levels and voluntary alcohol consumption in vervet monkeys. Psychopharmacology (Berl) 136:1–7

Wurtman RJ, Larin F, Mostafapour S, Fernstrom JD (1974) Brain catechol synthesis: control by brain tyrosine concentration. Science 185(146):183–184

Phenylketonuria

Synonyms

PKU

Definition

Phenylketonuria is a genetic disorder (autosomal recessive) in which the body lacks the enzyme required to break down the amino acid phenylalanine. Failure to break down phenylalanine leads to its, and related compounds, buildup, causing damage to the central nervous system. A strict diet begun at birth can help mitigate the negative consequences, namely, mental retardation, hyperactivity, and motor control problems. Screening newborn infants for phenylketonuria is a standard medical practice.

Modified from the NIH-NLM MedlinePlus Encyclopedia (online)

Phenylpropanolamine

Synonyms

Norephedrine

Definition

Phenylpropanolamine is an amphetamine derivative that has been used as a stimulant, decongestant, and anorectic agent. Its psychomotor stimulant effects are much weaker than those of ▶ amphetamine in humans. It has been widely used in cough and cold preparations. It has been withdrawn from several countries because of the risk of stroke. It has been subject to abuse in mixtures with other mild stimulants such as ▶ caffeine and ephedrine, with which it interacts to produce more powerful psychomotor activation.

Cross-References

- ▶ Appetite Suppressants
- ▶ Psychostimulants

Phenytoin

Definition

Phenytoin is an antiepileptic drug that controls seizure activity in epilepsy patients in much the same way as drugs such as ▶ carbamazepine, namely, by controlling the overactivity of nerve cells during an epileptic attack by acting at a protein called sodium channels. In addition to its antiepileptic effects, phenytoin like a number of other antiepileptic drugs has been used clinically to treat various ▶ mood disorders, although it is not registered for these uses.

Cross-References

- ▶ Anticonvulsants
- ▶ Bipolar Disorder
- ▶ Mood Stabilizers

Pheromones

Definition

Pheromones are chemicals transported outside of the body to send messages to individuals of the same species. There are many different pheromones, for example, alarm, food trail, or sex pheromones, that affect both behavior and physiology. Pheromones are detected by the olfactory epithelium and also in most mammals (but not humans) by the vomeronasal organ, a patch of G-protein-coupled receptor tissue in the nasal cavity.

Cross-References

- ▶ Autism: Animal Models

Phosphodiesterase Inhibitors

Jos Prickaerts
Department of Psychiatry and Neuropsychology,
University of Maastricht, Maastricht, MD,
The Netherlands

Synonyms

PDE inhibitors

Definition

There are 11 families of phosphodiesterases (PDEs; PDE1–PDE11), which degrade the second messengers cAMP and/or cGMP.

The activity of PDEs can be selectively inhibited with drugs. The most widely known PDE inhibitor is sildenafil, which is one of the three PDE5 inhibitors approved for the treatment of erectile dysfunction and also arterial pulmonary hypertension. In addition, two PDE3 inhibitors are approved for treating congestive heart failure or intermittent claudication, respectively. Recently, one PDE4 inhibitor has been approved for the treatment of chronic obstructive pulmonary disease. At the moment, PDE inhibitors are explored as possible therapeutic CNS drug targets for memory loss (PDE1, PDE2, PDE4, PDE5, PDE9), Alzheimer's disease (PDE3, PDE4, PDE5, PDE7, PDE9), Parkinson's disease (PDE1, PDE4, PDE7), Huntington's disease (PDE1, PDE4, PDE5, PDE10), anxiety (PDE2, PDE5), depression (PDE4), schizophrenia (PDE3, PDE10), pain (PDE4, PDE5), or stroke (PDE3, PDE5).

Pharmacological Properties

History
In 1886, the activity of PDEs was actually first described as it was noted that caffeine had bronchodilator properties. Later on, this effect was attributed to cyclic nucleotide cAMP and to caffeine-inhibited cAMP-specific PDEs. In 1970, PDEs were identified in rat and bovine tissue, and it was demonstrated that PDEs hydrolyze the phosphodiesteric bond of cAMP and cGMP (Bender and Beavo 2006). From then on, more PDEs were identified and characterized. Until now, 21 classes of genes for PDEs in humans, rats, and mice have been identified.

PDEs have been classified into 11 families (PDE1–PDE11) based on several criteria such as subcellular distributions, mechanisms of regulation, and enzymatic and kinetic properties. Most of these families have more than one gene product (e.g., PDE4A, PDE4B, PDE4C, PDE4D). In addition, each gene product may have multiple splice isoform variants (e.g., PDE4D1–PDE4D9). In total, there are more than 100 specific PDEs (Bender and Beavo 2006).

Caffeine is a nonselective PDE inhibitor, and it also inhibits cGMP-specific PDEs such as PDE5. cGMP causes vasodilatation in blood vessels by regulating their smooth muscle physiology. In addition, PDE5 also has an action on smooth muscles of contractile organs such as the penis. The most widely known PDE5 inhibitor is sildenafil. It was initially developed for the treatment of arterial hypertension and angina pectoris (Puzzo et al. 2008). In 1998, sildenafil was approved by the US Food and Drug Administration (FDA) for the treatment of erectile dysfunction and marketed under the name Viagra. Under the name of Revatio, it was also approved for the therapy of pulmonary artery hypertension in 2005.

The discovery of sildenafil started the research and development of numerous inhibitors of PDE5. At the same time, it stimulated researchers to explore other classes of PDEs for their therapeutic potential in different disorders. In addition, the previously explored PDEs, such as PDE4, were reevaluated after first being dismissed as a fruitful target due to side effects and a lack of specificity or efficacy of the developed PDE inhibitors (Esposito et al. 2009). For instance, in 1984, the PDE4 inhibitor rolipram was developed as a putative antidepressant, but it never made it to the market due to severe emetic side effects (e.g., nausea, vomiting).

Mechanisms of Action
PDEs hydrolyze the second messengers cAMP and/or cGMP, which are synthesized by adenylate and guanylate cyclase, respectively. However, the intracellular concentrations of both cyclic nucleotides are especially regulated by the PDE activity as its hydrolysis capacity far exceeds the capacity for synthesis. Besides this absolute and temporal regulation of cyclic nucleotides, PDEs contribute to their compartmentalized signaling as different PDEs are localized at some specific sites in the cell such as the plasma or nuclear membrane or cytosol. Thus, PDEs play a key role in the intracellular, signal transduction pathways in various biological systems as is illustrated in Fig. 1. cAMP and cGMP transfer an extracellular signal (e.g., neurotransmitter or

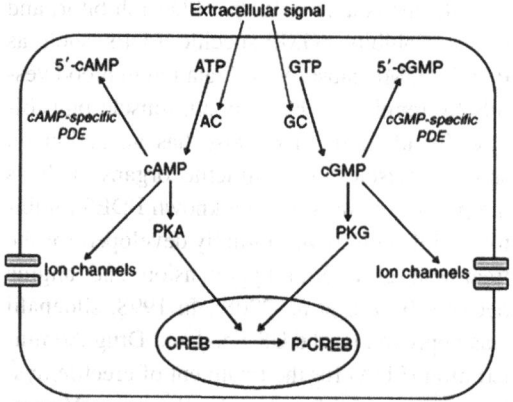

Phosphodiesterase Inhibitors, Fig. 1 Intracellular signal transduction pathways. An extracellular signal (e.g., neurotransmitter or hormone) activates adenylate cyclase (*AC*) and guanylate cyclase (*GC*), which produce their corresponding cyclic nucleotides out of ATP and GTP, respectively. cAMP activates protein kinase A (*PKA*), and cGMP activates protein kinase G (*PKG*). Both PKA and PKG can phosphorylate other enzymes or transcription factors such as CREB in the nucleus. Besides gene expression, cAMP and cGMP also regulate cAMP- and cGMP-gated ion channels, respectively, which depolarize the synaptic terminals. Eventually, these processes will result in a cellular response. Phosphodiesterases (PDEs) hydrolyze cAMP and/or cGMP leading to the formation of the inactive 5′-cAMP and 5′-cGMP, respectively. PDE inhibitors are selective for cAMP- and/or cGMP-degrading PDEs. In this way, a selective PDE inhibitor can specifically influence the cellular response of a biological system (Adapted from Puzzo et al. 2008)

hormone) to their effector proteins, protein kinase A and protein kinase G, respectively. Both kinases phosphorylate other enzymes or transcription factors, thus influencing the signal transduction. In addition, both cyclic nucleotides regulate their corresponding cyclic nucleotide-gated ion channels, which depolarize synaptic terminals and thus influence signaling pathways. For instance, cGMP regulates cGMP-gated ion channels and thus directly regulates the ion flux, which depolarizes the presynaptic terminal and influences glutamate release. Eventually, changes in signal transduction are translated into a biological system-dependent physiological and cellular response (Keravis and Lugnier 2012; Menniti et al. 2006; Puzzo et al. 2008; Reneerkens et al. 2009).

PDEs itself are regulated by intracellular cyclic nucleotide concentrations, phosphorylation (e.g., protein kinase G), interaction with regulatory proteins, subcellular compartmentalization, and binding of Ca^{2+}/calmodulin (Cheng and Grande 2007).

The specific localization of the different PDEs in the brain and the body will determine which particular physiological function may be influenced by some PDE inhibitors, but not by others. Table 1 gives an overview of the distribution of the different PDEs (for a detailed overview of the different PDE splice isoforms mRNA distribution in the brain and body of humans, see Lakics et al. 2010). Obviously, the PDE5 inhibitor sildenafil can be used for the treatment of erectile dysfunction since PDE5 is expressed in human cavernosal smooth muscle. Since PDE10A is highly expressed in the striatum where it regulates signal transduction in the cortico–striato–thalamic circuit, it is an interesting target for schizophrenia and related disorders of basal ganglia function. In contrast, PDE4 is highly expressed in the hippocampus, which is a key structure in the limbic system and is, therefore, considered as a useful target for treatment of mood disorders or cognitive deficits.

Pharmacokinetics

Only the pharmacokinetics of compounds that have been approved by the FDA and are also being evaluated for CNS applications is described.

The PDE3 inhibitor cilostazol is given orally and has a half-life of about 11–13 h. Cilostazol is metabolized and eliminated by CYP3A4 and CYP2C19, two isoenzymes of the cytochrome P450 system in the liver, after which it is predominantly excreted via the kidneys into the urine (Chapman and Goa 2003).

The PDE4 inhibitor roflumilast is metabolized in the liver by CYP3A4 and CYP1A2 to the main metabolite roflumilast N-oxide, with CYP3A4 being the major contributor. Roflumilast N-oxide is also active as PDE4 inhibitor. After oral dosing, the half-life of roflumilast and roflumilast N-oxide are about 17 and 30 h, respectively. Roflumilast is also metabolized to a minor extent to a pharmacologically inactive

Phosphodiesterase Inhibitors, Table 1 Localization of different phosphodiesterases (PDEs) in the body and brain of rodents and humans in adulthood

PDE	Localization in body	Localization in the brain
PDE1A–C	Heart, smooth muscles, lungs, pancreas, kidneys, bladder, testes	Hippocampus, cortex, olfactory bulb, striatum (highest expression levels), thalamus, amygdala, cerebellum; expression levels are in general highest for 1A and lowest for 1C
PDE2A	Heart, liver, spleen, pancreas, adrenals, skeletal muscles, bladder, platelets	Hippocampus, cortex, striatum, hypothalamus, amygdala, midbrain
PDE3	Heart, smooth muscles, lungs, liver, pancreas, kidneys, platelets	Throughout the brain low expression levels
PDE4A–D	Wide variety of tissues: e.g., smooth muscles, skeletal muscles, lungs, liver, spleen, pancreas, kidneys, bladder, testes	Hippocampus, cortex, olfactory bulb, striatum, thalamus, hypothalamus, amygdala, midbrain, cerebellum; expression levels are in general highest for 4B and lowest for 4C
PDE5A	Smooth muscles, skeletal muscles, lungs, pancreas, kidneys, bladder, platelets	Hippocampus, cortex, cerebellum
PDE6	Rod and cone cells in retina	Pineal gland
PDE7A–B	Heart, skeletal muscles, liver, pancreas, kidneys, testes	Hippocampus, cortex, olfactory bulb, striatum, thalamus, hypothalamus, midbrain; expression levels are in general highest for 7B
PDE8A–B	Heart, liver, pancreas, kidneys, adrenals, lungs, testes, thyroid	Hippocampus, cortex, olfactory bulb, striatum, thalamus, hypothalamus, midbrain; expression levels are in general highest for 8B
PDE9A	Lungs, spleen, pancreas, kidneys, bladder, prostate, various gastrointestinal tissues	Hippocampus, cortex, olfactory bulb, striatum, thalamus, hypothalamus, amygdala, midbrain, cerebellum
PDE10A	Heart, skeletal muscles, lungs, liver, pancreas, kidneys, testes, thyroid	Hippocampus, cortex, striatum (highest expression levels), midbrain, cerebellum
PDE11A	Skeletal muscles, liver, pancreas, kidneys, testes, prostate, thyroid	Throughout the brain low expression levels

Only clear expression levels are taken into consideration. Note that this table does not provide information with respect to the level of expression (protein or mRNA) of the different PDEs

metabolite by CYP3A4 and CYP1A2, with a minor contribution from CYP2C19. Roflumilast N-oxide is metabolized and eliminated by CYP3A4. The inactive metabolites are predominantly excreted via the urine and to a lesser extent via the feces (Lahu et al. 2011).

The PDE5 inhibitors sildenafil, vardenafil, and tadalafil are given orally and are rapidly absorbed in the gastrointestinal tract at the level of the small intestine. The half-life of sildenafil and vardenafil is about 3–4 h. In contrast, tadalafil has a long half-life of about 18 h. All three compounds are metabolized and eliminated in the liver by CYP3A4. For sildenafil, CYP2C9 is also partly involved. All three metabolized PDE5 inhibitors are excreted predominantly via the liver into the feces but also via the kidneys into the urine (Puzzo et al. 2008).

If a compound can be used to target CNS-related disorders, it is vital that it crosses the blood–brain barrier (BBB). Especially when the compound itself is required centrally to be effective, as otherwise alternatives have to be developed such as a central administration application for the drug. The abovementioned PDE inhibitors, which are approved by the FDA for non-CNS indications but which are currently being evaluated for CNS application, were initially considered to have insignificant brain penetration or brain penetration only dependent on BBB permeability occurring in pathological conditions such as ischemia (e.g., tadalafil). However, there is accumulating preclinical and clinical data showing that cilostazol, sildenafil, vardenafil, tadalafil, and roflumilast can enter the brain significantly to be biologically active.

Efficacy

Table 2 summarizes the PDE inhibitors currently on the market or in development. For CNS applications, also preclinical evidence is mentioned. More detailed information about the status of clinical development of a particular compound can be found at http://clinicaltrials.gov/. To check whether a drug is approved by the FDA, see http://www.accessdata.fda.gov/scripts/cder/drugsatfda/.

The PDE1 inhibitor vinpocetine (Cavinton, Intelectol, Cognitex) is not approved by FDA as a drug, but it is widely used as a supplement for vasodilation and as a nootropic for the improvement of memory. The latter effect is likely to be related to vasodilatation. However, the relevance of the possible therapeutic effect of vinpocetine can be questioned, and it has not been shown to be of real benefit in the treatment of Alzheimer's disease patients (Szatmari and Whitehouse 2003). Preclinically, vinpocetine has been reported to improve motor function and biochemical abnormalities in rat models of Parkinson's disease and Huntington's disease, respectively (Sharma et al. 2013). This suggests potential of PDE1 inhibition for the treatment of these movement disorders.

The first selective PDE2 inhibitor available was BAY 60-7550, and it has been shown to improve memory in rodents and a mouse model of Alzheimer's disease (e.g., Reneerkens et al. 2009; Sierksma et al. 2013). Recently, BAY 60-7550 and the PDE2 inhibitor ND7001 are being preclinically evaluated as possible anxiolytics (Keravis and Lugnier 2012; Xu et al. 2011). Exisulind (Aptosyn) is another developed PDE2 inhibitor, which also has the PDE5-inhibiting activity. This drug induces apoptosis in a broad range of cancer cell lines and inhibits the formation and growth of cancer in several animal models. Presently, this compound has been tested in clinical Phase-III trials for breast, lung, prostate, and colon tumors.

Cilostazol (Pletal) is a PDE3 inhibitor and has been approved by the FDA for the treatment of intermittent claudication. It is also being investigated in a Phase-IV study as a prevention of stroke recurrence and safety for bleeding complications in acute stroke. Furthermore, cilostazol has been investigated in three clinical studies to examine the effects on cognition in mild to moderate Alzheimer's disease patients who were already on a neuroprotective treatment, mostly with the acetylcholinesterase inhibitor donepezil. However, the conflicting results of those studies do not allow drawing a clear conclusion on possible cognition-enhancing effects yet. Cilostazol has also been tested on cognitive function in schizophrenic patients on antipsychotic treatment. However, results were not uniform enough to be conclusive.

Enoximone (Perfan) and milrinone (Primacor) are also PDE3 inhibitors, which have been developed for the treatment of congestive heart failure. Milrinone has been approved by the FDA for this indication, while enoximone has been tested up to Phase III of development. Their mode of action is via cAMP/PKA-mediated facilitation of intracellular Ca^{2+} mobilization. In addition, vasodilatory action plays a role in improving hemodynamic parameters in certain patients.

PDE4 inhibitors were initially considered as a possible target for the development of drugs for the treatment of depressive disorders (Esposito et al. 2009). In this respect, the PDE4 inhibitor rolipram has been widely investigated. First clinical studies showed a good antidepressant response to rolipram treatment. However, rolipram produces severe dose-limiting emetic side effects including headache, gastric hypersecretion, nausea, and vomiting in humans. This has put a serious hold on the further development of rolipram and other related PDE inhibitors. It also prevented rolipram from reaching the market. But since the approval of PDE5 inhibitors for the treatment of erectile dysfunction, PDE inhibitors in general have received renewed interest as a possible therapeutic target for the treatment of diseases. Along similar lines, a clinical Phase-II trial is ongoing to reevaluate the antidepressant properties of rolipram. Preclinically, rolipram has been tested in rodent models of Parkinson's disease and Huntington's

Phosphodiesterase Inhibitors, Table 2 Overview of the PDEs and their possible clinical applications

Type	Number of genes	Property	Substrate	Selective inhibitors	FDA-approved and possible therapeutic applications. CNS applications in bold
PDE1	3	Ca^{2+}-CaM stimulated	cAMP/cGMP	Vinpocetine, calmidazolium	**Memory improvement – vinpocetine (Cavinton, Intelectol, Cognitex)**
					Cognitive impairment in Alzheimer's disease – vinpocetine (failed)
					Parkinson's disease – vinpocetine (preclinical)
					Huntington's disease – vinpocetine (preclinical)
PDE2	1	cGMP stimulated	cAMP/cGMP	BAY 60-7550, ND7001, EHNA, exisulind	**Memory improvement/Alzheimer's disease – Bay 60-7550 (preclinical)**
					Anxiety – BAY 60-7550, ND7001 (preclinical)
					Cancer – exisulind (Aptosyn)
PDE3	2	cGMP inhibited	cAMP	Cilostazol, cilostamide, enoximone, milrinone, lixazinone, OPC-33540, SK&F 95654	**Stroke – cilostazol**
					Cognitive impairment in mild to moderate Alzheimer's disease – cilostazol (cotreatment with acetylcholinesterase inhibitor donepezil)
					Schizophrenia – cilostazol
					Intermittent claudication – cilostazol (Pletal)
					Congestive heart failure – milrinone (Primacor)
					Congestive heart failure – enoximone (Perfan)
PDE4	4	cAMP specific	cAMP	Rolipram, Ro 20-1724, cilomilast, roflumilast, ibudilast, MK0952, V11294A, L-826,141, AWD 12-281, HT0712, SCH 351591	**Cognitive impairment in mild to moderate Alzheimer's disease – MK0952**
					Parkinson's disease – rolipram (preclinical)
					Huntington's disease – rolipram (preclinical)
					Depression – rolipram
					Pain – ibudilast
					Asthma, chronic obstructive pulmonary disease (COPD) – cilomilast (Ariflo), roflumilast

(continued)

Phosphodiesterase Inhibitors, Table 2 (continued)

Type	Number of genes	Property	Substrate	Selective inhibitors	FDA-approved and possible therapeutic applications. CNS applications in bold
PDE5	1	cGMP specific	cGMP	Sildenafil, vardenafil, tadalafil, SK&F 96231, DMPPO, udenafil, avanafil, DA-8159	**Memory improvement/ Alzheimer's disease – sildenafil, vardenafil, tadalafil (preclinical)**
					Memory improvement – sildenafil (failed), vardenafil (failed), udenafil
					Huntington's disease – sildenafil, vardenafil (preclinical)
					Anxiety – sildenafil, tadalafil (preclinical)
					Pain – sildenafil (preclinical)
					Erectile dysfunction – sildenafil (Viagra), vardenafil (Levitra), tadalafil (Cialis)
					Female sexual arousal disorder (FSAD) – sildenafil
					Pulmonary hypertension – sildenafil (Revatio), tadalafil (Adcirca)
PDE6	4	Photoreceptor	cGMP	Sildenafil, vardenafil, tadalafil, DMPPO	–
PDE7	2	cAMP high affinity	cAMP	BRL 50481, IC242, BMS-586353, S14	**Alzheimer's disease – S14 (preclinical)**
					Parkinson's disease – S14, BRL 50481 (preclinical)
PDE8	2	cAMP high affinity	cAMP	PF-04957325	–
PDE9	1	cGMP high affinity	cGMP	BAY 73-6691, PF-04447947	**Cognitive impairment in mild to moderate Alzheimer's disease – PF-04447947 (failed)**
PDE10	1	cAMP inhibited	cGMP	Papaverine, PF-02545920, PQ-10, TP-10	**Schizophrenia – PF-02545920 (failed)**
					Huntington's disease – TP-10 (preclinical)
PDE11	1	Dual substrate	cAMP/ cGMP	Tadalafil	–

The properties and substrate specificity are depicted. In addition, the commonly used selective and nonselective PDE inhibitors are mentioned. FDA-approved compounds as well as compounds in clinical phases of development are given. For possible CNS applications, also preclinical evidence is given. All CNS applications are in bold
Nonselective inhibitors are the following: caffeine, theophylline, 3-isobutyl-1-methylxanthine (IBMX; all but PDE8), dipyridamole (PDE5, PDE6, PDE7, PDE9, PDE10, PDE11), zaprinast (PDE5, PDE6, PDE9, PDE10, PDE11), SCH 51866 (PDE1, PDE5, PDE7, PDE9, PDE10, PDE11), and E4021 (PDE5, PDE6, PDE10, PDE11)

disease, respectively, and was found to exert neuroprotective effects in the striatum, suggesting a beneficial role of PDE4 inhibitors in the treatment of these movement disorders (Sharma et al. 2013). However, this needs to be tested clinically.

At the moment, "second-generation" PDE4 inhibitors are being developed, which are supposed to have less-emetic side effects, and are being studied for a variety of disorders. A clinical Phase-II trial has been completed investigating whether the PDE4 inhibitor MK0952 improves cognition in patients with mild to moderate Alzheimer's disease. However, its results have never been disclosed. Additionally, the PDE4 inhibitor HT-0712 was tested on age-related memory impairment and reported to improve long-term memory. However, this has never been followed up. Furthermore, the PDE4 inhibitors cilomilast (Ariflo) and roflumilast (Daxas, Daliresp) have been clinically tested up to Phase III as anti-inflammatory drugs for the treatment of asthma and chronic obstructive pulmonary disease (COPD). Cilomilast has not been approved by the FDA, while in 2011, roflumilast was approved as an inflammatory drug for the treatment of COPD. Recently, a Phase-II study has started to investigate whether roflumilast improves memory in healthy human volunteers. Finally, ibudilast (or AV-411) is another PDE4 inhibitor in development as an anti-inflammatory drug. However, this compound is not only inhibiting PDE4 but it also suppresses drug or virus-induced inflammatory activity of glial cells, and for this reason, other clinical CNS applications are being explored in Phase-II studies, i.e., pain and drug abuse.

The PDE5 inhibitor sildenafil was the first FDA-approved treatment of erectile dysfunction. Although PDE5 inhibition causes relaxation of smooth muscles in blood vessels, it is also of particular importance for the treatment of erectile dysfunction in that it causes relaxation of smooth muscles in organs such as the penis (Puzzo et al. 2008). In women, sildenafil has an effect on the contractile state of the uterus and blood flow in the clitoris. Therefore, sildenafil has also been considered as a possible treatment for female sexual arousal disorder (FSAD) (completed Phase-II trial). Sildenafil is on the market as Viagra. Two more PDE5 inhibitors also reached the market for the treatment of erectile dysfunction as well, vardenafil (Levitra) and tadalafil (Cialis), respectively. Vardenafil is the more-potent PDE5 inhibitor when compared with sildenafil and tadalafil. The latter is considered as second-generation oral drug, and it has the longest half-life, while its effects last the longest as well.

Because of its vasodilatatory properties, sildenafil and tadalafil are also FDA approved under the names of Revatio and Adcirca, respectively, for the treatment of hypertension of the pulmonary artery. For the same application, a Phase-III trial of vardenafil was recently completed.

Sildenafil has been successfully used to treat serotonin reuptake inhibitor (SSRI) associated erectile dysfunction. In women, a Phase-IV study has been completed and was reported to show efficacy of sildenafil in women with SSRI-associated sexual dysfunction. Whether sildenafil treatment also improves mood and/or quality of life in women and men who are treated for antidepressant-induced sexual dysfunction needs further study. In this respect, a Phase-IV study has been completed that measured the impact of treatment with sildenafil on the depressive symptoms and quality of life in male patients with erectile dysfunction who have untreated depressive symptoms. However, its results have not been disclosed. Yet it will be difficult to disentangle whether a possible beneficial effect on mood is due to treatment of the sexual dysfunction or whether PDE5 inhibition directly leads to an improvement in depressive symptoms and thus an attenuation of the erectile/sexual dysfunction. Of note, no direct antidepressant potential of sildenafil has been found in preclinical rodent students (e.g., Brink et al. 2008). PDE5 inhibition is also preclinically being evaluated as a possible anxiolytic, but an anxiolytic effect appears apparent only after chronic treatment with sildenafil, whereas acute treatment is even anxiogenic (Liebenberg et al. 2012).

A Phase-I study was evaluating whether sildenafil has neuroprotective properties in the treatment for stroke. However, in 2011, this study was terminated because of the failure to recruit in the expected period. Another interesting CNS application could be neuropathic pain as sildenafil was effective in the treatment of pain in animal models (e.g., Ambriz-Tututi et al. 2005). This still awaits confirmation in clinical studies. This also holds for Huntington's disease as sildenafil as well as vardenafil had neuroprotective effects in the striatum as observed in a rat model of Huntington's disease (Sharma et al. 2013).

There is some evidence that PDE5 inhibition has cognition-enhancing and neuroprotective effects in several animal species including mouse models of Alzheimer's disease (e.g., Reneerkens et al. 2009; García-Barroso et al. 2013). However, until now, convincing evidence in humans is still lacking. Three studies have explored the effects of sildenafil or vardenafil on human cognition in healthy human volunteers, but no clear effects on cognition were found. Sildenafil was also tested on cognitive function in patients with schizophrenia who were on antipsychotic treatment. However, sildenafil was not effective. In another study, the PDE5 inhibitor udenafil was tested in patients suffering from erectile dysfunction (Shim et al. 2011). Repeated dosing of udenafil was reported to improve cognitive (executive) function. Apparently, chronic treatment and/or selection of the best target patient population needs to be considered when testing a PDE5 inhibitor.

Recently, it has been found that the PDE7 inhibitor S14 reduced cognitive impairment as well as central plaques load and tau phosphorylation in a mouse model of Alzheimer's disease (Perez-Gonzalez et al. 2013). This suggests that PDE7 inhibition might have therapeutic potential for the treatment of memory dysfunction and/or Alzheimer's disease. In addition, PDE7 inhibition might also be a target for neuroprotection in Parkinson's disease as S14, and the PDE7 inhibitor BRL 50481 prevented dopaminergic neuronal loss in a rat model of Parkinson's disease (Sharma et al. 2013).

The first selective PDE9 inhibitor available was BAY 73-6691, and it improved memory performance in rodents (e.g., Reneerkens et al. 2009). This was also found for the PDE9 inhibitor PF-04447943, which was also tested in a Phase-II study with mild to moderate Alzheimer's disease patients. However, no effect was found on cognition. It was suggested that the treatment duration may not have been long enough and/or a less cognitively impaired target population, e.g., age-associated cognitive impaired subjects, could have been a better choice for treatment.

Until very recently, the PDE10A inhibitor PF-02545920 (or MP-10) was in a second Phase-II clinical study for the treatment of schizophrenia after the first study had been terminated for reasons unknown. However, the outcome of the second study showed that chronic PF-02545920 treatment had no effect on positive and negative symptoms in schizophrenia patients. In addition, adverse events including akathisia (inner restlessness, inability to remain motionless) and dystonia (muscle contractions causing twisting and repetitive movements or abnormal postures) were observed. Of note, the action mechanism of PDE10A inhibition was attributed to be modulation/normalization of dopaminergic cortico–striato–thalamic function. In this respect, increased cAMP levels are assumed to be of more importance than cGMP, although PDE10A itself predominantly hydrolyses cGMP. Interestingly, PDE10 inhibition might still be beneficial in treating Huntington's disease as the PDE10 inhibitor TP-10 has been demonstrated to be neuroprotective in rodent models of Huntington's disease (Sharma et al. 2013).

Safety/Tolerability

Only safety and tolerability of compounds which have been approved by the FDA and are also being evaluated for CNS applications are discussed.

Possible side effects of the PDE3 inhibitor cilostazol include most commonly headache, diarrhea, abnormal stools, and since it is a quinolinone derivative also irregular heart rate and palpitations. Therefore, it is dangerous for people with severe heart failure and can only be given to people without this indication.

The most commonly reported side effects of roflumilast include diarrhea, weight loss, nausea, abdominal pain, and headache. Three suicides have been observed in a COPD patient pool versus none in the placebo pool. This was identified as a significant concern by the FDA, although none of the suicides was identified as being related to the study medication. There are no indications for cardiovascular effects associated with roflumilast use.

Sildenafil, vardenafil, as well as tadalafil have side effects such as headaches, facial flushing, nasal congestion, and dyspepsia (indigestion). However, these effects are transient. All three PDE inhibitors can act on PDE6, which is present in the retina; and high doses have been reported to cause adverse visual events, including non-arteritic anterior ischemic optic neuropathy and, thus, can cause vision problems (e.g., blurred vision). Moreover, tadalafil also potently inhibits PDE11, an enzyme with an unknown physiological function. Because of the possible vasodilatatory effects, these compounds are not suited for patients with cardiovascular indications or hypotensives.

An approach to circumvent the side effects of PDE inhibitors is to develop very selective inhibitors at the level of the splice isoform variant. At the same time, the function of interest may be more specifically targeted. For example, there are three splice isoform variants of PDE5, PDE5A1–PDE5A3 (Puzzo et al. 2008). While the first two are found nearly in all tissues, the third one is specific to smooth muscles.

The emetic side effects of the available PDE4 inhibitors, which inhibit more or less all four isoforms PDE4A, PDE4B, PDE4C, and PDE4D, prevented until now that they have reached the market (Esposito et al. 2009). Preclinical animal research already indicated that the antidepressant potential of PDE4A in the hippocampus was related to specific splice variants of PDE4A (Xu et al. 2011). Moreover, in a patient with Alzheimer's disease, it was found that the expression of most splice isoform variants of PDE4D was decreased in the hippocampus, whereas two variants were increased. These findings underscore the need to develop specific inhibitors of PDE4 splice variants as cognition enhancers or antidepressants without unwanted side effects (Reneerkens et al. 2009). An example of this is the recent development of selective inhibitors for PDE4D which are devoid of emetic effects or have at least greatly reduced emetic effects (Gurney et al. 2011).

Conclusions

Besides the already approved clinical application of erectile dysfunction, pulmonary hypertension, congestive heart failure, intermittent claudication, and chronic obstructive pulmonary disease, PDE inhibitors offer a promising drug target for a wide array of diseases including CNS-related disorders such as Alzheimer's disease, depression, schizophrenia, or stroke. Yet, the future in disease-specific PDE inhibitors lies in the development of splice isoform variant-specific inhibitors that have limited aversive side-effect profiles within the effective dose range for its clinical application.

Cross-References

▶ Chronic Obstructive Pulmonary Disease
▶ Congestive Heart Failure
▶ Erectile Dysfunction
▶ Intermittent Claudication
▶ PDE3 Inhibitors
▶ PDE4 Inhibitors
▶ PDE5 Inhibitors
▶ Phosphodiesterases
▶ Pulmonary Hypertension
▶ Rolipram
▶ Sildenafil

References

Ambriz-Tututi M, Velazquez-Zamora DA, Urquiza-Marin H, Granados-Soto V (2005) Analysis of the mechanism underlying the peripheral antinociceptive action of sildenafil in the formalin test. Eur J Pharmacol 512:121–127

Bender AT, Beavo JA (2006) Cyclic nucleotide phosphodiesterases: molecular regulation to clinical use. Pharmacol Rev 58:488–520

Brink CB, Clapton JD, Eagar BE, Harvey BH (2008) Appearance of antidepressant-like effect by sildenafil in rats after central muscarinic receptor blockade: evidence from behavioural and neuroreceptor studies. J Neural Transm 115:117–125

Chapman TM, Goa KL (2003) Cilostazol: a review of its use in intermittent claudication. Am J Cardiovasc Drugs 3:117–138

Cheng J, Grande JP (2007) Cyclic nucleotide phosphodiesterase (PDE) inhibitors: novel therapeutic agents for progressive renal disease. Exp Biol Med (Maywood) 232:38–51

Esposito K, Reierson GW, Rong Luo H, Sheng Wu G, Licinio J, Wong ML (2009) Phosphodiesterase genes and antidepressant treatment response: a review. Ann Med 41:177–185

García-Barroso C, Ricobaraza A, Pascual-Lucas M, Unceta N, Rico AJ, Goicolea MA, Sallés J, Lanciego JL, Oyarzabal J, Franco R, Cuadrado-Tejedor M, García-Osta A (2013) Tadalafil crosses the blood–brain barrier and reverses cognitive dysfunction in a mouse model of AD. Neuropharmacology 64:114–123

Gurney ME, Burgin AB, Magnusson OT, Stewart LJ (2011) Small molecule allosteric modulators of phosphodiesterase 4. Handb Exp Pharmacol 204:167–192

Keravis T, Lugnier C (2012) Cyclic nucleotide phosphodiesterase (PDE) isozymes as targets of the intracellular signaling network: benefits of PDE inhibitors in various diseases and perspectives for future therapeutic developments. Br J Pharmacol 165:1288–1305

Lahu G, Nassr N, Hünnemeyer A (2011) Pharmacokinetic evaluation of roflumilast. Expert Opin 7:1577–1591

Lakics V, Karran EH, Boess FG (2010) Quantitative comparison of phosphodiesterase mRNA distribution in human brain and peripheral tissues. Neuropharmacology 59:367–374

Liebenberg N, Harvey BH, Brand L, Wegener G, Brink CB (2012) Chronic treatment with the phosphodiesterase type 5 inhibitors sildenafil and tadalafil display anxiolytic effects in Flinders Sensitive Line rats. Metab Brain Dis 27:337–340

Menniti FS, Faraci WS, Schmidt CJ (2006) Phosphodiesterases in the CNS: targets for drug development. Nat Rev Drug Discov 5:660–670

Perez-Gonzalez R, Pascual C, Antequera D, Bolos M, Redondo M, Perez DI, Pérez-Grijalba V, Krzyzanowska A, Sarasa M, Gil C, Ferrer I, Martinez A, Carro E (2013) Phosphodiesterase 7 inhibitor reduced cognitive impairment and pathological hallmarks in a mouse model of Alzheimer's disease. Neurobiol Aging 34:2133–2145

Puzzo D, Sapienza S, Arancio O, Palmeri A (2008) Role of phosphodiesterase 5 in synaptic plasticity and memory. Neuropsychiatr Dis Treat 4:371–387

Reneerkens OA, Rutten K, Steinbusch HW, Blokland A, Prickaerts J (2009) Selective phosphodiesterase inhibitors: a promising target for cognition enhancement. Psychopharmacology (Berl) 202:419–443

Sharma S, Kumar K, Deshmukh R, Sharma PL (2013) Phosphodiesterases: regulators of cyclic nucleotide signals and novel molecular target for movement disorders. Eur J Pharmacol 714:486–497

Shim YS, Pae CU, Kim SW, Kim HW, Kim JC, Koh JS (2011) Effects of repeated dosing with Udenafil (Zydena) on cognition, somatization and erection in patients with erectile dysfunction: a pilot study. Int J Imp Res 23:109–114

Sierksma AS, Rutten K, Sydlik S, Rostamian S, Steinbusch HW, van den Hove DL, Prickaerts J (2013) Chronic phosphodiesterase type 2 inhibition improves memory in the APPswe/PS1dE9 mouse model of Alzheimer's disease. Neuropharmacology 64:124–136

Szatmari SZ, Whitehouse PJ (2003) Vinpocetine for cognitive impairment and dementia. Cochrane Database Syst Rev(1), CD003119

Xu Y, Zhang HT, O'Donnell JM (2011) Phosphodiesterases in the central nervous system: implications in mood and cognitive disorders. Handb Exp Pharmacol 204:447–485

Phosphodiesterases

Definition

Phosphodiesterases are enzymes that hydrolyze the second messenger cyclic nucleotides cAMP and cGMP. Thus, phosphodiesterases influence intracellular signal transduction pathways in various biological systems. There are 11 families of phosphodiesterases, PDE1–PDE11. This classification is based on their subcellular localizations, mechanisms of regulation, and enzymatic properties. Each family consists of multiple splice variants. In total, there are more than 100 specific PDEs. They are specifically distributed over the body and the brain.

Cross-References

▶ PDE3 Inhibitors
▶ PDE4 Inhibitors
▶ PDE5 Inhibitors
▶ Phosphodiesterase Inhibitors

Physical Dependence

Synonyms

Abstinence syndrome; Physiological dependence; Withdrawal syndrome

Definition

Physical dependence results when certain drugs are administered or self-administered regularly for extended periods of time and is revealed when drug administration is discontinued and a typical withdrawal syndrome appears. The time to onset for the withdrawal effects depends upon how rapidly the drug is eliminated from the body. For example, in ▶ alcohol physical dependence, the withdrawal signs and symptoms emerge within 12–24 h of discontinuation, whereas for ▶ benzodiazepine physical dependence, the withdrawal syndrome may not emerge for several days corresponding to the slow elimination of active metabolites from the body. The extent or magnitude of the withdrawal signs and symptoms is directly related to the amount and duration of drug exposure preceding withdrawal. Each class of drugs has a typical withdrawal syndrome with distinctive physiological and psychological features. In some laboratory studies, physical dependence can be induced using antagonist drugs, which precipitate a withdrawal syndrome. For example, naloxone or naltrexone produces precipitated withdrawal in an opioid-dependent person or laboratory animal, producing a constellation of signs and symptoms similar to those seen after spontaneous withdrawal. Despite the name "physical dependence," signs and symptoms of withdrawal syndromes are often psychological and behavioral in nature (e.g., anxiety in humans or impaired task performance in laboratory animals). Withdrawal syndromes dissipate with time, and their duration depends on several factors including the drug class and severity of dependence.

Cross-References

▶ Abuse Liability Evaluation
▶ Alcohol
▶ Benzodiazepines
▶ Cross-Dependence
▶ Nicotine
▶ Opioid Use Disorder and Its Treatment
▶ Sedative, Hypnotic, and Anxiolytic Dependence

Physostigmine

Definition

Physostigmine is an anticholinesterase that crosses the blood-brain barrier. It is reported to improve performance in short-term and working memory in animals and humans by stimulating nicotinic and muscarinic receptors and used as a cholinergic replacement therapy for conditions such as dementia. In humans however, the side effects of lethargy, nausea, and general misery are produced by physostigmine.

Pimozide

Definition

Pimozide is a first-generation antipsychotic that acts as a dopamine D2 receptor antagonist. It is a diphenylbutylpiperidine derivative with high potency, a receptor binding profile comparable to that for ▶ haloperidol but more selective for the D2 receptor, and it has a long ▶ half-life. Following reports of sudden unexplained deaths, the UK Committee on the Safety of Medicines recommended ECG before treatment. A history of arrhythmias or congenital QT prolongation is a contraindication for its use. Pimozide should not be combined with other potentially arrhythmogenic drugs. It has been regarded as a less-sedating compound, but the frequency of extrapyramidal symptoms is relatively high.

Cross-References

▶ Antipsychotic Drugs
▶ First-Generation Antipsychotics
▶ Movement Disorders Induced by Medications
▶ Schizophrenia

Pipotiazine

Definition

Pipotiazine is a first-generation antipsychotic of the phenothiazine class. It is one of the less widely used drugs of this type, but pipotiazine palmitate has been used in a depot (long-acting) formulation; a Cochrane review concluded that "it is a viable choice for both clinician and recipient of care." Side effects typical of first-generation antipsychotics have been reported, including rare instances of ▶ neuroleptic malignant syndrome.

Cross-References

▶ First-Generation Antipsychotics
▶ Neuroleptic Malignant Syndrome
▶ Phenothiazines

Piracetam

Synonyms

2-Oxopyrrolidin-1-acetamide; Aminotrophyl; Anacervix; Avigilen; Axonyl; Biocetam; Braintop; Cerbropan; Cerebroforte; Cerebrosteril; Cerebryl; Lucetam; Memoril; Nootrop; Nootropil; Nootropyl; Novocetam; Piracebral

Definition

Piracetam is a so-called nootropic with unknown mechanism of action, although it is believed to facilitate cholinergic transmission and to modulate positively glutamate (AMPA) receptors: these putative mechanisms and a wealth of preclinical data explain its use in a number of dementias, including in Parkinson's disease-associated dementias, Alzheimer's disease, Down's syndrome, vascular dementias, dyslexia, etc. The clinical data supporting Piracetam's effects are controversial. Piracetam is not registered in the USA, where it has an orphan drug status for the treatment of myoclonus.

Cross-References

▶ Dementias and Other Amnestic Disorders
▶ Dementias: Animal Models

Place Cells

Synonyms

Place fields

Definition

Place cells are neurons that exhibit a high rate of firing whenever an animal is in a specific location in an environment corresponding to the cell's "place field." Place cells were first identified in the hippocampus of rats by O'Keefe and Dostrovsky, although cells with similar properties have been found in the hippocampus of primates and humans. Furthermore, neurons that display place-selective changes in activity have been identified in a number of other brain regions, including regions of the temporal cortex, the amygdala, and the striatum. Based on this discovery, it has been proposed that a primary function of the rat hippocampus is to form a cognitive map of the rat's environment. On initial exposure to a new environment, place fields become established within minutes. The place fields of cells tend to be stable over repeated exposures to the same environment. In a different environment, however, a cell may have a completely different place field or no place field at all, a phenomenon referred to as "remapping."

Placebo Effect

Fabrizio Benedetti
Department of Neuroscience, University of Turin Medical School, and National Institute of Neuroscience, Turin, Italy

Synonyms

Placebo response

Definition

The placebo effect is the reduction of a symptom, or a change in a physiological parameter, when an inert treatment (the placebo) is administered to a subject who is told that it is an active therapy with specific properties. The placebo effect, until recently considered a nuisance in clinical research, has now become a target of scientific investigation to better understand the physiological and neurobiological mechanisms that link a complex mental activity to different functions of the body. Usually, in clinical research, the term "placebo effect" refers to any improvement in the condition of a group of subjects that has received a placebo treatment; thus, it represents a group effect. The term "placebo response" refers to the change in an individual caused by a placebo manipulation. However, these two terms are often used interchangeably. It is important to realize that there is not a single placebo effect but many, which occur through different mechanisms in different conditions, systems, and diseases.

Current Concepts and State of Knowledge

Methodological Aspects

The identification of a placebo effect is not easy from a methodological point of view, and its study is full of drawbacks and pitfalls. In fact, the apparent clinical benefits following the administration of a placebo can be due to many factors, such as spontaneous remission, regression to the mean, symptom detection ambiguity, and biases (Benedetti 2008a, 2013). All these phenomena must be ruled out through adequate control groups. For example, spontaneous remission can be discarded by means of a no-treatment group, which gives us information about the natural course of a symptom. Regression to the mean, a statistical phenomenon due to selection biases at the enrolment in a clinical trial, can be controlled by using appropriate selection criteria. Symptom detection ambiguity and biases can be avoided by using objective physiological measurements. It is also important to rule out the possible effects of co-interventions. For example, the mechanical insertion of a needle for the injection of an inert substance may per se induce analgesia, thus leading to erroneous interpretations.

When all these phenomena are ruled out and the correct methodological approach is used, substantial placebo effects can be detected, which are mediated by psychophysiological mechanisms worthy of scientific inquiry. Therefore, this psychological component represents the real placebo effect.

Mechanisms

The placebo effect is basically a context effect whereby the psychosocial context (e.g., the therapist's words, the sight of medical personnel and apparatuses, and other sensory inputs) around the medical intervention and the patient plays a crucial role. Today, we know that the context may produce a therapeutic effect through at least two mechanisms: conscious anticipatory processes and unconscious conditioning mechanisms (Benedetti 2008b; Price et al. 2008). In the first case, the expectation and anticipation of clinical benefit may affect the response to a therapy. In the second case, contextual cues (e.g., color and shape of a pill) may act as conditioned stimuli that, after repeated associations with an unconditioned stimulus (the pharmacological agent inside the pill), are capable alone of inducing a clinical improvement. In the case of pain and Parkinson's disease, it has been shown that conscious expectancies play a crucial role, even though a conditioning procedure is performed, whereas placebo responses in the immune and endocrine systems are mediated by unconscious classical conditioning.

The neural mechanisms underlying the placebo effect are only partially understood, and most of our knowledge comes from the field of pain and analgesia, although Parkinson's disease, immune and endocrine responses, and depression have more recently emerged as interesting models. In each of these conditions, different mechanisms seem to take place, so that we cannot talk of a single placebo effect but many (Benedetti 2008a, 2013).

As far as pain is concerned, there is now compelling experimental evidence that the endogenous opioid systems play an important role in some circumstances. There are several lines of evidence indicating that placebo analgesia is mediated by a descending pain-modulating circuit, which uses endogenous opioids as neuromodulators. This evidence comes from a combination of imaging and pharmacological studies. By using positron emission tomography ("▶ Positron Emission Tomography (PET) Imaging"), it was found that some cortical and subcortical regions are affected by both a placebo and the opioid agonist remifentanil, thus indicating a related mechanism in placebo-induced and opioid-induced analgesia (Petrovic et al. 2002). In particular, the administration of a placebo affects the rostral anterior cingulate cortex (rACC), the orbitofrontal cortex (OrbC), and the brain stem. Moreover, there is a significant covariation in activity between the rACC and the lower pons/medulla at the level of the rostral ventromedial medulla (RVM) and a subsignificant covariation between the rACC and the periaqueductal gray (PAG), thus suggesting that the descending rACC/PAG/RVM pain-modulating circuit is involved in placebo analgesia. In a different study with functional magnetic resonance imaging (fMRI), it was shown that placebo administration produces a decrease of activity in many regions involved in pain transmission, such as the thalamus and the insula (Wager et al. 2004; Price et al. 2008).

A clear-cut involvement of endogenous opioids is shown by several pharmacological studies in which placebo analgesia is antagonized by the opioid antagonist naloxone. In addition, it has been shown that the endogenous opioid systems have a somatotopic organization, since local placebo analgesic responses in different parts of the body can be blocked selectively by naloxone (Benedetti 2008a). By using in vivo receptor binding techniques, placebos have been found to induce the activation of *mu* opioid receptors in different brain areas, such as the dorsolateral prefrontal cortex, nucleus accumbens, insula, and rACC. It is also worth noting that opioids are activated together with dopamine, e.g., in the nucleus accumbens, thus indicating a complex interaction between different neurotransmitters (Scott et al. 2008).

The placebo-activated endogenous opioids do not act only on pain transmission but on the respiratory centers as well, since a naloxone-reversible placebo respiratory depressant effect has been described. Likewise, a reduction of beta-adrenergic sympathetic system activity, which is blocked by naloxone, has been found during placebo analgesia. These findings indicate that the placebo-activated opioid systems have a broad range of action, influencing pain, respiration, and the autonomic nervous system (Benedetti 2013). The placebo-activated endogenous opioids have also been shown to interact with endogenous substances that are involved in pain transmission. In fact, on the basis of the anti-opioid action of cholecystokinin (CCK), CCK antagonists have been shown to enhance placebo analgesia, thus suggesting that the placebo-activated opioid systems are counteracted by CCK during a placebo procedure (Benedetti 2013).

It is important to point out that some types of placebo analgesia appear to be insensitive to naloxone. For example, if a placebo is given after repeated administrations (preconditioning) of the non-opioid painkiller ketorolac, the placebo analgesic response is not blocked by naloxone, but it can be antagonized by the CB1 cannabinoid receptor antagonist, rimonabant, thus suggesting an involvement of the endocannabinoid system (Benedetti et al. 2011a).

The release of endogenous substances following a placebo procedure is a phenomenon that is not confined to the field of pain, but it is also present in other conditions, such as Parkinson's disease. As occurs with pain, in this case, patients are given an inert substance and are told that it is an anti-Parkinson drug that produces an improvement in their motor performance. A study used PET imaging in order to assess the competition between endogenous dopamine and ^{11}C-raclopride for D_2/D_3 receptors, a method that allows the identification of endogenous dopamine release (de la Fuente-Fernandez et al. 2001). This study found that the placebo-induced expectation of motor improvement activates endogenous dopamine in both the dorsal and ventral striata, a region involved in reward mechanisms, which suggests that the placebo effect may be conceptualized as a form of reward.

Placebo administration in Parkinson patients also affects the activity of neurons in the subthalamic nucleus, a brain region belonging to the basal ganglia circuitry and whose activity is increased in Parkinson's disease (Benedetti 2013). Verbal suggestions of motor improvement during a placebo procedure are capable of reducing the firing rate and abolishing the bursting activity of subthalamic nucleus neurons, and these effects are related to clinical improvement.

Depressed patients who receive a placebo treatment were shown to present metabolic changes in the regions involved in reward mechanisms, such as the ventral striatum. In addition, in patients with unipolar depression, placebo treatments were also found to be associated with metabolic increases in the prefrontal, anterior cingulate, premotor, parietal, posterior insula, and posterior cingulate cortex and metabolic decreases in the subgenual cingulate cortex, parahippocampus, and thalamus. Interestingly, these regions are also affected by antidepressants, such as the selective serotonin reuptake inhibitor, fluoxetine, a result that suggests a possible role for serotonin in placebo-induced antidepressant effects (Benedetti 2008b).

Placebos also affect anxiety, involving brain regions that also take part in placebo analgesia, e.g., the rACC and OrbC. In addition, placebo responsiveness in social anxiety has been found to be linked with genetic variants of serotonin neurotransmission, which suggests that part of the variability in placebo responsiveness may be attributable, at least in some conditions, to genetic factors (Benedetti 2008a).

Placebo responses in both the immune and endocrine systems can be evoked through classical conditioning. After repeated administrations of drugs, if the drug is replaced with a placebo, immune or hormonal responses can be evoked that are similar to those obtained by the previously administered drug, thus resembling the typical conditioned drug effects (Benedetti 2013). For example, immunosuppressive placebo responses can be induced in humans by repeated administration of cyclosporine A (unconditioned stimulus) associated to a flavored drink (conditioned stimulus), as assessed by interleukin-2 (IL-2) and interferon-γ (IFNγ) mRNA expression, in vitro release of IL-2 and IFNγ, and lymphocyte proliferation. Likewise, if a placebo is given after repeated administrations of sumatriptan, a serotonin agonist of the $5\text{-HT}_{1B/1D}$ receptors that stimulates growth hormone (GH) and inhibits cortisol secretion, a placebo GH increase and a placebo cortisol decrease can be found.

The Nocebo Effect

The nocebo effect, or response, is a placebo effect in the opposite direction, i.e., the patient or subject is instructed that the discomfort will be greater, rather than less. Mainly due to ethical constraints, the nocebo effect has not been investigated in detail, as has been done for the placebo effect. In fact, a nocebo procedure is per se stressful and anxiogenic, as it induces negative expectations of clinical worsening. For example, the administration of an inert substance along with verbal suggestions of pain increase may induce a hyperalgesic effect. In this case, anticipatory anxiety plays a fundamental role. Nocebo hyperalgesia has been found to be blocked by proglumide, a nonspecific CCK-A/CCK-B receptor antagonist. This suggests that expectation-induced hyperalgesia is mediated, at least in part, by CCK (Benedetti 2013). These effects of proglumide are not antagonized by naloxone; thus, endogenous opioids are not involved.

Dopamine also plays a role in the nocebo hyperalgesic effect. In fact, whereas placebo administration induces the activation of dopamine in the nucleus accumbens, the administration of a nocebo is associated with a deactivation of dopamine (Scott et al. 2008).

Implications

According to the classical methodology of randomized controlled trials, any drug must be compared with a placebo in order to assess its effectiveness. If the patients who take the drug show a considerable clinical improvement than the patients who take the placebo, the drug is considered to be effective. However, in light of the recent advances in placebo research, some caution is necessary in the interpretation of some clinical trials. In fact, by considering the complex cascade of biochemical events that take place after placebo administration, any drug that is tested in a clinical trial may interfere with these placebo/expectation-activated mechanisms, thus confounding the interpretation of the outcomes. As we have no a priori knowledge of which substances act on placebo-activated endogenous opioids, dopamine, or serotonin – and indeed almost all drugs, e.g., analgesics and opioids, might interfere with these neurotransmitters – one way to eliminate this possible pharmacological interference is to make the placebo-activated biochemical pathways, so to speak, "silent." This can be achieved by the hidden administration of drugs (Benedetti et al. 2011b).

In fact, it is possible to eliminate the placebo (psychological) component and analyze the pharmacodynamic effects of a treatment, free of any psychological contamination, by administering drugs covertly. In this way, the biochemical events that are triggered by a placebo procedure can be eliminated. To do this, drugs are administered through hidden infusions by machines, so that the patient is not made aware that a medical therapy is being carried out. A hidden drug infusion can be performed through a computer-controlled infusion pump that is preprogrammed to deliver the drug at the desired time. It is crucial that the patient does not know that any drug is being injected, so that he or she does not expect anything. The computer-controlled infusion pump can deliver a drug automatically, without a doctor or nurse in the room and without the patient being aware that a treatment has been started. Ethical issues need to be considered.

The analysis of different treatments, either pharmacological or not, in different conditions has shown that an open (expected) therapy, which is carried out in full view of the patient, is more effective than a hidden one (unexpected). Whereas the hidden injection represents the real pharmacodynamic effect of the drug, free of any psychological contamination, the open injection represents the sum of the pharmacodynamic effect and the psychological component of the treatment. The latter can be considered to represent the placebo component of the therapy, even though it cannot be called placebo effect, as no placebo has been given. Therefore, by using hidden administration of drugs, it is possible to study the placebo effect without the administration of any placebo.

The influence of psychological factors on drug action is also shown by the balanced-placebo design (Benedetti 2008a). In this design, four groups of patients (1) receive the drug and are told it is a drug, (2) get the drug and are told it is a placebo, (3) get a placebo and are told it is a placebo, and (4) get a placebo and are told it is a drug. This design is particularly interesting because it indicates that verbally induced expectations can modulate the therapeutic outcome, both in the placebo group and in the active treatment group. It has been used in many conditions, such as alcohol research, smoking, and amphetamine effects (Benedetti 2008a).

Cross-References

▶ Analgesics
▶ Antidepressants
▶ Anti-Parkinson Drugs
▶ Cholecystokinins
▶ Conditioned Drug Effects
▶ Endocannabinoids
▶ Expectancies and Their Influence on Drug Effects

- ▶ Magnetic Resonance Imaging (Functional)
- ▶ Opioids
- ▶ Positron Emission Tomography (PET) Imaging
- ▶ Randomized Controlled Trials

References

Benedetti F (2008a) Placebo effects: understanding the mechanisms in health and disease. Oxford University Press, New York

Benedetti F (2008b) Mechanisms of placebo and placebo-related effects across diseases and treatments. Annu Rev Pharmacol Toxicol 48:33–60

Benedetti F (2013) Placebo and the new physiology of the doctor-patient relationship. Physiol Rev 93:1207–1246

Benedetti F, Amanzio M, Rosato R, Blanchard C (2011a) Non-opioid placebo analgesia is mediated by CB1 cannabinoid receptors. Nature Med 17:1228–1230

Benedetti F, Carlino E, Pollo A (2011b) Hidden administration of drugs. Clin Pharmacol Ther 90:651–661

de la Fuente-Fernandez R, Ruth TJ, Sossi V, Schulzer M, Calne DB, Stoessl AJ (2001) Expectation and dopamine release: mechanism of the placebo effect in Parkinson's disease. Science 293:1164–1166

Petrovic P, Kalso E, Petersson KM, Ingvar M (2002) Placebo and opioid analgesia – imaging a shared neuronal network. Science 295:1737–1740

Price DD, Finniss DG, Benedetti F (2008) A comprehensive review of the placebo effect: recent advances and current thought. Annu Rev Psychol 59:565–590

Scott DJ, Stohler CS, Egnatuk CM, Wang H, Koeppe RA, Zubieta JK (2008) Placebo and nocebo effects are defined by opposite opioid and dopaminergic responses. Arch Gen Psychiatr 65:220–231

Wager TD, Rilling JK, Smith EE, Sokolik A, Casey KL, Davidson RJ, Kosslyn SM, Rose RM, Cohen JD (2004) Placebo-induced changes in fMRI in the anticipation and experience of pain. Science 303:1162–1166

Placebo-Controlled

Definition

Term used to describe a method of clinical research in which an inactive substance is given to one group of participants (i.e., the control group), while the treatment (e.g., drug, therapy, vaccine) is given to another group of participants (i.e., the treatment group).

Plaques

Definition

Pathological intrusions in the brain in which the major component is β-amyloid, that is not degraded or removed by crossing the blood-brain barrier; accumulation of β-amyloid causes oligomerization and polymerization to build up insoluble plaques mainly in extraneuronal tissue.

Platelet Activation

Definition

Platelet activation represents a central moment in the process that leads to thrombus formation. When endothelial damage occurs, platelets come into contact with exposed collagen and von Willebrand factor, becoming activated. They are also activated by thrombin or by a negatively charged surface, such as glass. Platelet activation further results in the scramblase-mediated transport of negatively charged phospholipids to the platelet surface, providing a catalytic surface for the tenase and prothrombinase complexes. Activated platelets change in shape and pseudopods form on their surface, determining a starlike appearance.

Platelet Storage Granules

Definition

Platelets contain numerous storage granules. Activated platelets excrete the contents of these granules into their canalicular systems and, then, into surrounding blood. There are two types of granules: dense granules (containing ADP or ATP, calcium, serotonin, and other ▶ monoamines) and alpha-granules (containing platelet factor 4, PDGF, fibronectin, B-thromboglobulin, vWF, fibrinogen, and coagulation factors V and XIII).

Platelets

Lucio Tremolizzo, Gessica Sala and
Carlo Ferrarese
Department of Surgery and Interdisciplinary
Medicine, Neurology Unit, S. Gerardo Hospital,
University of Milano-Bicocca, Monza, MB, Italy

Synonyms

Thrombocytes

Definition

Platelets are small cytoplasmic bodies derived from megakaryocytes. They circulate in blood and are mainly involved in hemostasis. Platelets have no nucleus and display a life span of 7–10 days. When damage to the endothelium of blood vessels occurs, platelets go through activation, change shape, release granule contents, and, finally, aggregate and adhere to the endothelial surface in order to form the blood clot.

Besides their chief role in the process of hemostasis, platelets express several signaling molecules, including various neurotransmitters. Therefore, considering the accessibility of platelets by a simple blood withdrawal, it is not surprising that several researchers have tried to take advantage of this opportunity to get a glimpse of a variety of molecules involved in neural transmission.

Current Concepts and State of Knowledge

Platelets as Peripheral Markers of Neurotransmitter Dysfunction

Human platelets have been repeatedly considered as potentially useful tools for modeling neurochemical dysfunctions in neurological and psychiatric patients. In fact, although the main function of the platelet is related to hemostasis, they are implicated with various signaling molecules that operate in the CNS as well. More specifically, platelets store, release, and uptake several neurotransmitters, such as serotonin, dopamine, glutamate, and GABA, frequently expressing cognate functional receptor molecules.

Considering that most of the neuropsychiatric disorders do not allow direct study of their pathomechanisms in the central nervous system, platelets offer the advantage of being readily accessible upon a simple blood withdrawal. Hence, in this chapter we will consider them as peripheral markers of neurotransmitter dysfunction in neuropsychiatry.

However, as a final caveat, we should remember that technical factors and reproducibility obviously play an extremely important role and influence the characterization of any suitable peripheral neurochemical marker. Platelets unfortunately display quite a strong tendency to activation, changing shape, and releasing granule contents, if careful procedures are not used when performing blood sampling and platelet separation. As an example, a generous needle gauge and avoiding tourniquet use might be critical issues for circumventing these phenomena, especially when measuring granule contents or performing electron microscopy procedures (see Fig. 1a, with respect to a less activated platelet shown in Fig. 1b).

Serotonin Uptake and Release

Human platelets are capable of producing very small amounts of serotonin (5-hydoxytryptamine, 5-HT) but represent the major storage site for this neurotransmitter outside the central nervous system. Platelets actively take up serotonin, store it in electron-dense granules, and release it, together with other platelet factors, in response to a number of signals (Paasonen 1965). Specific transporter molecules (SERT) located on platelet membranes are responsible for the highly selective uptake of serotonin, although there is some affinity for the other monoamines (e.g., norepinephrine, dopamine) (Omenn and Smith 1978). Molecular cloning has revealed that the amino acid sequence of the human platelet SERT protein is identical to the neuronal SERT, and

Platelets, Fig. 1 Bar = 500 nm (Electron microscopy photographs courtesy of Virginia Rodriguez-Menendez)

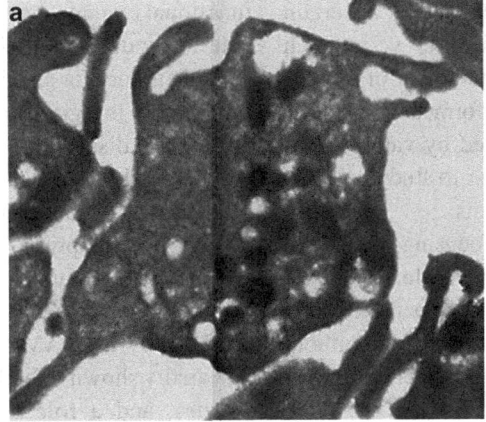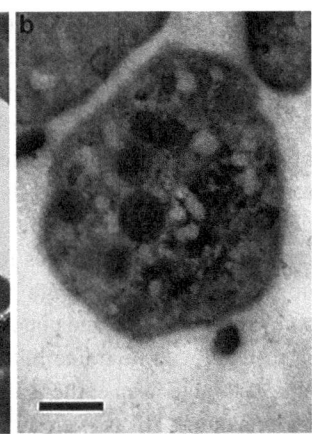

the platelet high-affinity uptake system for serotonin displays kinetic and pharmacological characteristics similar to those reported in brain synaptosomes. All this makes platelets good model systems for investigating alterations of monoamine neurotransmission occurring in neuropsychiatric diseases.

Platelet SERTs have been studied in psychiatric and neurological patients, in terms of both binding of radioligands on platelet plasma membranes and functional activity in intact platelets (Mellerup and Plenge 1986). Platelet serotonin uptake was demonstrated to be decreased in depressed patients, both adult and children, with respect to controls, before and during antidepressant treatment, although not all studies agree (Fisar and Raboch 2008). No significant correlations were found between serotonin uptake kinetics and the severity of the depressive disorder. Abnormalities of platelet serotonin uptake have been reported also in patients with Down's syndrome and Huntington's disease, and a reduced platelet content of this neurotransmitter has also been demonstrated, again, in Down's syndrome.

The use of specific serotonin receptor antagonists has shown the presence on the platelet surface of the 5-HT$_{2A}$ receptor subtype involved in the amplification of platelet aggregation by recruiting additional platelets once the process has been initiated. Using [^{3}H]-ketanserin as a radioligand, a significant decrease in the affinity of platelet 5-HT$_{2A}$ receptors was observed in patients affected by migraine, and a significant reduction in the number of binding sites was found in tension-type headache as well. Interestingly, a role for platelet-released serotonin in the pathogenesis of migraine has also been hypothesized, and differences between serotonin and other monoamines levels have been invoked to explain the differences between migraine with and without aura (Joseph et al. 1992).

Major findings related to monoamine functions in human platelets in neuropsychiatric diseases are summarized in Table 1.

Platelets, Table 1 Monoamine functions in human platelets in neuropsychiatric diseases

	Serotonin	Dopamine
Uptake	DP: ↓; DS: ↓; HD: ↓	HD: ↑; Parkinsonisms: ↓; acute SZ: ↑ in PSG; Gilles de la Tourette: ↓ in PSG; naïve PD: ↓ in PSG
Binding	M: ↓ affinity 5-HT 2A; TH: ↓ density; BPD: ↓	DP: ↑; SZ: ↑; ADHD: ↓
Level	DS, ↓	N/A

ADHD attention deficit/hyperactivity disorder, *BPD* borderline personality disorder, *DP* depression, *DS* Down's syndrome, *HD* Huntington's disease, *M* migraine, *N/A* not investigated, *PD* Parkinson's disease, *PSG* platelet storage granules, *SZ* schizophrenia, *TH* tension-type headache

Dopamine

Besides serotonin, other monoamines, such as dopamine, are actively concentrated by platelets within dense bodies. Although the significance of the uptake of dopamine in platelets is uncertain and the concentration gradient is very much lower than that of serotonin, a number of clinical investigations have been performed in samples obtained from patients affected by various neuropsychiatric disorders in order to study possible dopaminergic pathomechanisms.

The results of the studies focusing on platelet dopamine levels are quite variable. A recent study reported increased platelet dopamine levels in patients with migraine both with and without aura, although in an earlier study, interictal platelet dopamine levels were found unchanged in women affected by migraine.

Although platelets are not generally regarded as useful models for exploring dopamine transporters (DAT), given the prevalence therein of SERTs, an increase of dopamine uptake has been described in platelets from patients with Huntington's disease, while a reduction was found in Parkinson's disease.

To avoid the controversial issues deriving from assessing dopamine uptake in intact platelets (presence of a specific dopamine uptake mechanism versus dopamine entry via the serotonin uptake system), the uptake of [^{3}H]-dopamine has been measured in previously separated platelet storage granules (PSG). These granules possess the vesicular monoamine transporter 2 (VMAT2) responsible for monoamine storage into secretory vesicles (Zucker et al. 2001). An increase in dopamine accumulation has been described in platelets of acute schizophrenic patients, while a decrease was observed in Gilles de la Tourette and naïve Parkinson's disease patients. Furthermore, using high-affinity [^{3}H]-dihydrotetrabenazine binding to platelets, an elevated VMAT2 density was found in untreated major depression, schizophrenia, and former heroin addict patients treated with methadone, while a decreased density was observed in children and adolescents with attention deficit/hyperactivity disorder and in both healthy and schizophrenic habitual smokers.

The Glutamatergic System

Human platelets store, release, and take up glutamate, at the same time expressing fully functional cognate receptors. Notably, the available data indicate that platelets might represent a suitable model for specifically studying the dysfunction of the glutamatergic system in neuropsychiatric disorders, since they display similar biochemical properties with respect to the CNS (Tremolizzo et al. 2012). However, although platelets do express all the components of a functional glutamatergic system, its specific role is far from being fully understood. Even so, NMDA and AMPA receptors have been repeatedly shown in both platelets and megakaryocytes, and a role for glutamate in platelet activation and/or aggregation has already been proposed. Moreover, platelets express the three major glutamate transporters, EAAT1, EAAT2, and EAAT3, and a functional high-affinity sodium-dependent glutamate reuptake system was successfully described more than 20 years ago. Apparently this reuptake system might represent a major source of the amino acid removal from blood and also displays very similar regulatory mechanisms and susceptibility to toxins with respect to the CNS homologue. The two major vesicular glutamate transporters (VGLUT-1, VGLUT-2) are expressed as well, packing glutamate (probably in dense bodies) and releasing it following aggregation together with several other mediators (in contrast, aspirin treatment reduces glutamate release). Although the possible significance of these findings still needs to be definitively addressed in relation to platelets and hemostasis physiology, current data indicate that those functions involved in glutamate homeostasis in platelets may represent suitable windows for studying CNS alterations.

As a matter of fact, glutamate receptor sensitivity, for example, assessed by flow cytometry, has been studied in platelets obtained from patients affected by schizophrenia, psychotic depression, and major depression, suggesting the possibility of mirroring a hypofunction of glutamate receptors that may be implicated in the pathogenesis of these psychiatric disorders (see Table 1). Moreover, glutamate/aspartate uptake, assessed as [^{3}H]glutamate influx (Mangano and Schwarcz 1981), has been

Platelets, Table 2 Glutamatergic and GABAergic functions in human platelets in neuropsychiatric diseases

	Neuropsychiatric disorders	Neurodegenerative diseases	Cerebrovascular disease	Headache
Glutamate	SZ, DP: ↑ receptor sensitivity; BD, DS: ↓ uptake; BD manic: ↑ uptake	AD, ALS, PD: ↓ uptake; HD: = uptake	Acute ischemic stroke: ↓ uptake and release	MA: ↑ uptake; MoA: ↓ uptake; MA and MoA: ↑ release
GABA	Absence epilepsy: ↓ uptake; epilepsy, epileptic syndromes, and alcoholism: GABA-T modifications; BP depressive vs. manic: ↑ vs. ↓ uptake; DP + anxiety and panic disorders/SZ: ↓ PBR; DP: = PBR	AD, PD: ↓ PBR	N/A	MoA: ↑ PBR; TH: ↑ levels

AD Alzheimer's disease, *ALS* amyotrophic lateral sclerosis, *BD* bipolar disease, *DP* depression, *DS* Down's syndrome, *GABA-T* GABA transaminase, *HD* Huntington's disease, *MA* migraine with aura, *MoA* migraine without aura, *N/A* not investigated, *PBR* peripheral benzodiazepine receptors, *PD* Parkinson's disease, *SZ* schizophrenia, *TH* tension-type headache

investigated in platelets obtained from patients affected by various neuropsychiatric conditions, such as bipolar disorder, Alzheimer's disease, Parkinson's disease, amyotrophic lateral sclerosis, Huntington's disease, migraine, acute ischemic stroke, and Down's syndrome, in the attempt to mirror an excitotoxic dysfunction previously demonstrated in the CNS of these patients. Also glutamate release following aggregation, assessed by HPLC as the difference between pre- and post plasma levels, has been investigated in disease, mainly in a setting of migraine and stroke, since the release of this amino acid might mediate, at least in part, the experienced symptoms, activating both central and peripheral (i.e., endothelial, leukocytic) receptors (keeping in mind that the release function might be especially prone to the already discussed methodological issue of preparation-induced activation). Interestingly, lamotrigine, a drug able to interfere with the symptoms of migraine aura, works by blocking glutamate release, possibly, at least in part, directly from platelets.

Major findings related to the glutamatergic system in human platelets in neuropsychiatric diseases are summarized in Table 2.

GABA and Peripheral Benzodiazepine Receptors

Platelets express GABA transporters and the uptake of this amino acid has been reported disrupted in patients affected by absence epilepsy, although the functional impact of the antiepileptic drugs taken by the patients has not been completely clarified. Also GABA transaminase (GABA-T) has been studied, mainly in a setting of various epileptic syndromes, such as juvenile myoclonic, refractory localization-related, and childhood absence epilepsy, showing different modifications. GABA-T represents the main enzyme responsible for GABA catabolism, and the inhibition of this enzyme produces a considerable elevation of GABA concentrations: often the modulation of such elevation has been correlated with many pharmacological effects, mainly in the field of antiepileptic drugs. GABA-T activity in blood platelets might be altered also in some neuropsychiatric disorders, such as alcoholism, epilepsy, and Alzheimer's disease.

A related GABAergic marker, the 18 kDa translocator protein (TSPO) (the new nomenclature for the peripheral-type benzodiazepine receptor, PBR), has also been studied in human platelets, evaluating the binding of the radioactive antagonist [^{3}H]PK11195; a decrease of this parameter was shown in platelets obtained from patients affected by depression with associated adult separation anxiety disorder, panic disorder, and schizophrenia with aggressive behavior, among other psychiatric diseases. On the other hand, no changes were shown in a population of untreated depressed patients, while an upregulation was present, for example, in primary fibromyalgia patients. Interestingly, Bmax values of PBR binding have been shown to be positively correlated with scores for trait anxiety, suggesting that it might be associated with the personality trait for anxiety tolerance (Nakamura et al. 2002).

Major findings related to GABA and PBR in human platelets in neuropsychiatric diseases are summarized in Table 2.

Cross-References

▶ Excitatory Amino Acids and Their Antagonists
▶ Peripheral Markers
▶ Receptors: Binding Assays

References

Fisar Z, Raboch J (2008) Depression, antidepressants, and peripheral blood components. Neuro Endocrinol Lett 29:17–28
Joseph R, Tsering C, Grunfeld S, Welch KM (1992) Further studies on platelet-mediated neurotoxicity. Brain Res 577:268–275
Mangano RM, Schwarcz R (1981) The human platelet as a model for the glutamatergic neuron: platelet uptake of L-glutamate. J Neurochem 36:1067–1076
Mellerup ET, Plenge P (1986) Chlorimipramine-but not imipramine-rapidly reduces [3H]imipramine binding in human platelet membranes. Eur J Pharmacol 126:155–158
Nakamura K, Fukunishi I, Nakamoto Y, Iwahashi K, Yoshii M (2002) Peripheral-type benzodiazepine receptors on platelets are correlated with the degrees of anxiety in normal human subjects. Psychopharmacology (Berl) 162:301–303
Omenn GS, Smith LT (1978) A common uptake system for serotonin and dopamine in human platelets. J Clin Invest 62:235–240
Paasonen MK (1965) Release of 5-hydroxytryptamine from blood platelets. J Pharm Pharmacol 17:681–697
Tremolizzo L, Sala G, Zoia CP, Ferrarese C (2012) Assessing glutamatergic function and dysfunction in peripheral tissues. Curr Med Chem 19:1310–1315
Zucker M, Weizman A, Rehavi M (2001) Characterization of high-affinity [3H]TBZOH binding to the human platelet vesicular monoamine transporter. Life Sci 69:2311–2317

Polymorphism

Definition

In biology, polymorphism is the presence of distinct subtypes of genes and gene products within a single family. ▶ Genetic polymorphisms are the result of evolution and are hereditable. Of great interest for psychopharmacology are polymorphisms characterized by variations in the genetic sequence coding for specific receptor subtypes, which often confer higher or lower affinity for the ligand or influence functionality of the receptor (or plasma membrane transporter) itself. These differences in affinity/functionality are hypothesized to be important for the manifestation of certain psychiatric disorders, personality traits, and different responses to medications between individuals. Polymorphisms are variations in DNA that are too common to be attributable to new mutations; the frequency of a polymorphism has been defined as at least 1 % in a given population.

Cross-References

▶ Genetic Polymorphism
▶ Single-Nucleotide Polymorphism

Polypharmacy

Definition

The use of multiple drugs to treat a single or a number of conditions.

Polysomnography

Definition

Polysomnography is a diagnostic technique that outlines sleep architecture by monitoring multiple physiological variables. These may include brain electrical activity (electroencephalogram, EEG), leg movements, eye movements (electrooculogram, EOG), muscle activity (electromyogram, EMG), body temperature, heart rate, and respiration rate.

Cross-References

▶ Insomnias
▶ Parasomnias

Positive Symptoms

Definition

Positive symptoms are manifestations of psychoses such as fixed, false beliefs (delusions) or perceptions without cause that are present in mentally ill but not in healthy individuals. They contrast with negative symptoms of schizophrenia such as apathy and lack of drive.

Positron Emission Tomography (PET) Imaging

Mark Slifstein
Department of Psychiatry, Columbia University and New York State Psychiatric Institute, New York, NY, USA

Synonyms

PET; PET imaging; Radiotracer imaging

Definition

PET is a noninvasive technique that allows quantitative imaging data to be obtained from internal organs (including brain) in vivo, in humans and animal models. Molecules that interact with biologic systems of interest are tagged with positron-emitting isotopes and injected into subjects, usually in very small quantities ("▶ Tracer Dose"). Decay of the isotope results in the emission of photons that are detectable by the imaging system. These data can be used to infer the concentrations of the radiolabeled molecule ("Radiotracer") in the tissues being imaged. The technique can be used to estimate parameters related to receptor density and availability, drug occupancy, fluctuations in endogenous transmitter levels at neurotransmitter receptors/transporters, and rates of various metabolic processes.

Principles and Role in Psychopharmacology

PET imaging has been used as a noninvasive tool to probe many questions with implications for the field of psychopharmacology, including examination of receptor/transporter availability, measurement of drug receptor occupancy, detection of fluctuations in endogenous neurotransmitter levels following pharmacologic or task-based stimulation, and examination of various metabolic processes. This essay contains (1) a brief description of the physical principles underlying the imaging system, (2) a description of the pharmacokinetic methodology common to most PET studies, and (3) a survey of the currently available probes and targets studied with PET.

Physical Principles of PET

Positrons are the antiparticles of electrons – they have the same mass as electrons but positive, rather than negative, electric charge. Unstable isotopes that emit positrons when they decay can be produced in a cyclotron; many academic medical centers currently have onsite cyclotrons that can produce isotopes for PET imaging. Several isotopes can be readily incorporated into small molecules suitable as biologic tracers, including carbon (^{11}C, ½ life = 20.334 min),

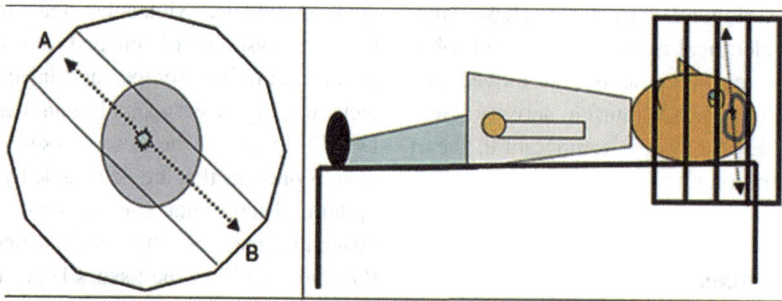

Positron Emission Tomography (PET) Imaging, Fig. 1 (*Left*) A schematic detector ring. An annihilation event occurs in an imaged object as seen along the long axis of the subject (from head to toe). If the emitted photons (paths represented by *dashed arrows*) strike detectors A and B within a coincidence timing window (several nanoseconds for most scanners), a coincidence event is recorded along the line between *A* and *B*. (*Right*): A schematic stack of rings, viewed from the side. A tomographic image is formed by combining data from all the rings

fluorine (^{18}F, ½ life = 109.8 min), and oxygen (^{15}O, ½ life = 122 s). When a positron is emitted, it briefly interacts with local atoms, losing kinetic energy during the resulting collisions. When its energy is low enough, the positron will be attracted to and captured by an electron; the pair will annihilate and emit a pair of 511 keV photons. Conservation of momentum dictates that the total momentum of the 2 photons is equal to that of the positron-electron pair at the time of their annihilation. As this momentum is small, the photons travel in nearly opposite directions (their momenta cancel each other), approximately along a line from the location of the annihilation event (Fig. 1, left panel). The imaging system contains rings of detectors composed of scintillating materials that surround the imaged object. Combined with electronic components that are attached to these detectors, the arrival of photons at the detector arrays can be counted with very high temporal resolution (on the order of nanoseconds). If a pair of photons are detected in two different detectors within a specified time interval (a "coincidence timing window," typically 4–12 ns), the scanner treats these as having originated from a single annihilation event occurring somewhere along the line between the two detectors (a "line of response"). Computer-intensive mathematical methods can then be applied to this collection of coincidence event data to infer the original distribution of the annihilation events. Typical modern systems contain from 20 to 40 of these detector rings stacked in parallel to form a cylindrical field of view (Fig. 1, right panel); the data from these can then be recombined to form a 3-dimensional image of the original distribution of the radiotracer, composed of a series of slices through the object (Fig. 2).

The preceding description is greatly simplified – there are many confounds and sources of noise and image degradation that must be accounted for. A description of the methods for correcting these confounds is beyond the scope of this essay – see Cherry et al. (2012) and Valk et al. (2003) for detailed expositions – but because PET utilizes coincidence detection, greater quantitative accuracy is attainable than with other technologies. Because of this level of quantitative accuracy, the concentration of radiotracer in tissues can be accurately inferred through detailed mathematical models of tracer pharmacokinetics. Physiologic parameters can be estimated by fitting data to these models.

Pharmacokinetic Models and Methods

General Principles

The pharmacokinetic schemes used to estimate parameters both for reversibly binding receptor radioligands and for many metabolic processes are called compartment models. Compartments can be spatially distinct, but they can also be

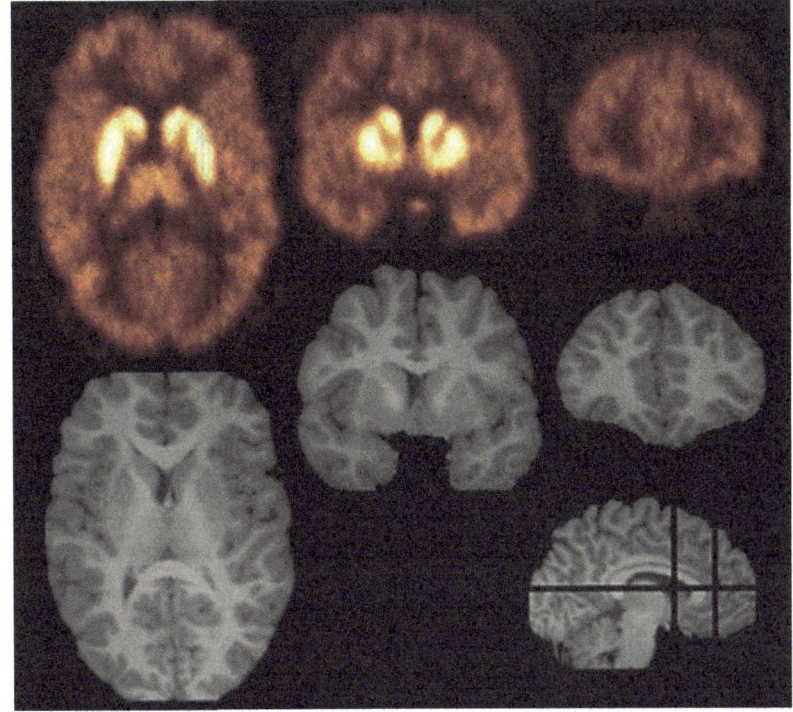

Positron Emission Tomography (PET) Imaging, Fig. 2 A PET image and co-registered high-resolution MRI from the same subject. The data, acquired using the dopamine D2 antagonist radioligand [^{11}C] FLB457, are summed over the entire 90 min of the scan. The sagittal MRI slice (*bottom right*) shows the slice levels for the transverse and coronal views

different states of the radiolabel occupying the same spatial domain, such as bound vs. unbound or parent radiotracer vs. metabolic product. Figure 3 is a schematic representation of a standard compartment model used with reversibly binding radioligands.

Corresponding to the compartment model are systems of linear first order, ordinary differential equations (ODEs). The ODEs express the temporally dynamic relationship between the input source of radioligand to the tissue (either the concentration in arterial plasma or the concentration in a tissue devoid of the biologic process of interest, but similar to the tissue of interest in other respects, a "reference tissue") and the resulting concentration in the tissue of interest (usually brain or specific brain regions for targets of interest in psychopharmacology). "First order" implies that the models operate according to the convention that movement from a source compartment to a target compartment is proportional to the concentration in the source compartment; the validity of this assumption in the case of reversibly binding receptor ligands is predicated

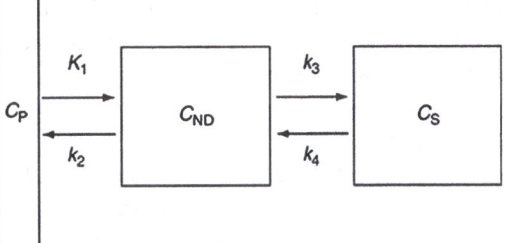

Positron Emission Tomography (PET) Imaging, Fig. 3 A 2-tissue compartment model (2TC). C_P = radioligand concentration in arterial plasma. C_{ND} = nondisplaceable compartment, the sum of free and nonspecifically bound tracer in tissue. C_S = specifically bound ligand, i.e., ligand receptor complex. K1 through k4 are rate constants governing the fractional transfer between compartments per unit time

on the use of tracer dose (free tracer concentration << equilibrium dissociation constant K_D, see below) so that the associated mass action law, which is second order when higher concentrations are used, can be treated as pseudo first order. By using some data-fitting method such as least squares minimization to regress observed

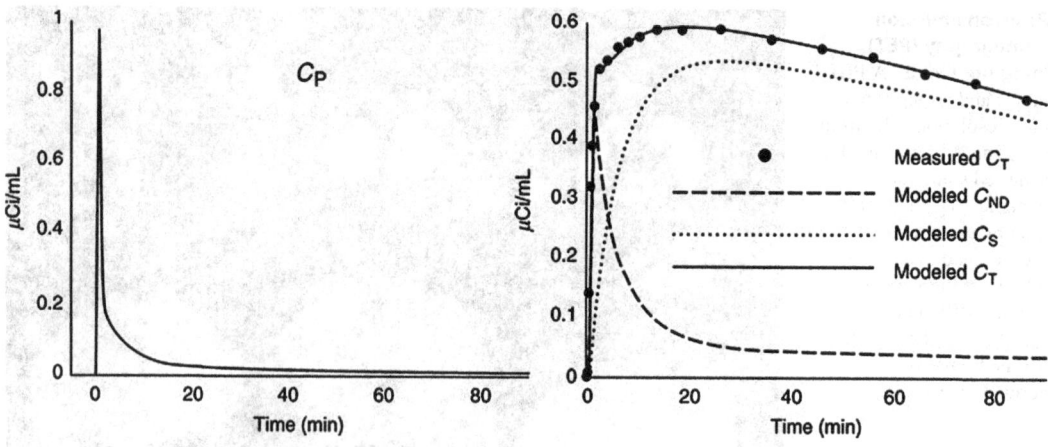

Positron Emission Tomography (PET) Imaging, Fig. 4 Arterial plasma input and resulting PET data and model fit. These data show the time course of the dopamine D1 receptor tracer [^{11}C] NNC112 in the striatum of an anesthetized baboon. Plasma input to brain (C_P) on *left*, and the compartments represented in Fig. 3 on the *right*. The discrete dots represent the measured C_T; the 3 continuous curves are the fit to the model

data onto the model, the rate constants can be estimated. In turn, various mathematical combinations of the rate constants represent physiological parameters of interest. There is an extensive literature on methods for data fitting and parameter estimation in PET. See Slifstein and Laruelle (2001), Valk et al. (2003), and Slifstein et al. (2004) for overviews.

Receptor Imaging

The form of PET imaging most frequently used in applications relevant to psychopharmacology involves the use of radioligands that bind selectively and reversibly to target neurotransmitter receptors and transporters. These are usually administered as a single bolus injection, though there are some tracers amenable to a bolus plus constant infusion administration that induces steady-state conditions in ligand concentrations. The compartment configuration shown in Fig. 3 presents a model frequently used with reversible tracers. In the figure, C_p represents the concentration of radioligand in the arterial plasma, the input to brain tissue. C_{ND} ("non-displaceable") is the sum of freely dissolved tracer in the brain and tracer that is nonspecifically bound to membranes. These quantities are combined into a single compartment based on the assumption that equilibration between free and nonspecifically bound ligand occurs on a much more rapid time scale than the specific binding process does and can therefore be treated as if constantly in equilibrium. Models in PET frequently involve simplifying assumptions of this type in order to insure that they are not over-parameterized for statistical fitting. C_S represents specifically bound ligand-receptor complex. The movement of tracer between C_p and C_{ND} is governed by a transport law, whereas exchange between the states C_{ND} and C_S is governed by a mass action law. These various states of the radioligand cannot be distinguished in the PET signal; it is comprised of the sum of all sources of radioactivity in the spatial locus being imaged and is sometimes referred to as C_T (total concentration, Fig. 4). The parameters of greatest interest in receptor imaging are the density of receptors available for binding to the radioligand (B_{max}, nM, or B_{avail}, allowing for the possibility that all of the receptors may not be accessible to the tracer) and the affinity of the radioligand for receptor (K_D^{-1} where K_D (nM) is the equilibrium dissociation constant). These quantities cannot be estimated separately from single tracer dose scans – multiple scanning sessions with increased radiotracer concentrations that bind to a significant fraction of receptors would be required for this purpose. The quantity that is readily estimated is the binding potential, a parameter that is proportional to the product of B_{avail} and affinity, or B_{avail}/K_D. There are several possible constants of proportionality, according to which of several approaches to experimental design and derivation of the binding potential estimate is used. The version appearing most frequently in the literature, however, is called BP_{ND} (Innis et al. 2007), equal to $f_{ND} B_{avail}/K_D$ where f_{ND} = free ligand/(free + nonspecifically bound ligand). The rate constants have the following physiological interpretations: $K_1 = FE$, where F is flow (actually, perfusion, mL· cm^{-3} ·min^{-1}), E is the first-pass extraction fraction (unitless); $k_2 = K_1/V_{ND}$ where V_{ND} is the non-displaceable equilibrium distribution volume, C_{ND}/C_p at equilibrium, (mL·cm^{-3}); and $k_3 = k_{on}f_{ND}B_{max}$ (min^{-1}), where k_{on} is the association rate of the receptor-ligand complex and k_4 (min^{-1}), is k_{off}, the dissociation rate of the receptor ligand complex. As K_D is equal to k_{off}/k_{on}, it is apparent that k_3/k_4 is equivalent to BP_{ND}. However, BP_{ND} is rarely estimated directly from the fitted values of k_3 and k_4. Rather, more involved methods utilizing various constraints to make the estimated BP_{ND} more statistically robust are employed. See Slifstein and Laruelle (2001), Valk et al. (2003), and Slifstein et al. (2004) for further details.

Applications of Receptor Imaging to Psychopharmacology

If an endogenous transmitter or an exogenously administered drug competes with the radioligand at the binding site on the receptor, then, due to tracer dose conditions for the radioligand, the binding potential is reduced by the factor $1/(1 + L/K_i)$ where L is the concentration of the ligand and K_i^{-1} is its affinity for the binding site. This in turn implies that the relative difference across conditions, $[BP_{ND}$(competitor on board) $- BP_{ND}$(baseline)]/BP_{ND}(baseline), is equal to $(L/K_i)/(1 + L/K_i)$, the fraction of receptors occupied by the competitor. This technique can be used to infer the occupancy of a receptor by a drug or to demonstrate that a stimulus induces transmitter release.

In a seminal study, Farde et al. (1988) used the dopamine D2 receptor radioligand [^{11}C] raclopride to measure the occupancy of D2

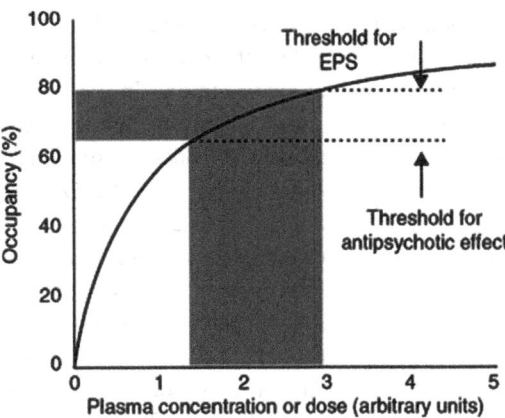

Positron Emission Tomography (PET) Imaging, Fig. 5 The concept of a therapeutic window, showing a range over which an antipsychotic drug is efficacious, but below the threshold for extrapyramidal symptoms (After Farde et al. 1988)

receptors by the antipsychotic drug haloperidol in patients with schizophrenia and presented the concept of a therapeutic window – the idea that there is a minimal receptor occupancy necessary to achieve antipsychotic efficacy, but a maximum tolerable occupancy above which extrapyramidal symptoms would appear. The in vivo competitive binding technique has subsequently been used in many published studies examining drug receptor occupancy. While a large number of these have continued to examine D2 occupancy by antipsychotics, the method has been used to look at many other receptor systems as well. The approach has been widely used by pharmaceutical companies in both drug development and in post-marketing studies characterizing efficacious occupancies (Fig. 5).

A similar imaging technique can be used to infer fluctuations in endogenous neurotransmitters as a result of either pharmacological or task-based stimuli. Again, the system that has been the most amenable to this type of study has been dopamine release at the D2 receptor. Many studies have been performed in which dopamine release and/or inhibition of reuptake has been induced by amphetamine, methylphenidate, or other compounds or by tasks hypothesized to induce dopamine release. There has been considerable evidence that the simple competitive interaction model described above is not adequate to explain the decrease in binding potential following amphetamine. In particular, the decrease lasts much longer than the apparent increased dopamine release does as measured with microdialysis in animal models. On the other hand, the binding decrease is highly correlated with amphetamine dose and with the microdialysis measurements and does not occur if dopamine stores have been pharmacologically depleted prior to the scan, suggesting that competitive binding plays at least some role in the observed effect. Also, while an effect consistent with the occupancy model has been detected with several radioligands in the benzamide ($[^{123}I]$ IBZM, $[^{11}C]$ raclopride, $[^{18}F]$ fallypride) and catecholamine ($[^{11}C]$ NPA) classes, paradoxical *increases* in radioligand binding following amphetamine have been observed using the butyrophenone radioligand N-$[^{18}F]$ methylspiperone. A number of mechanisms have been proposed to explain these anomalies, including receptor trafficking and differential responses to internalized receptors, differences in binding sites and differences in pharmacokinetic properties with concomitant differences in robustness of quantification of the various radioligands. At this time, these issues remain unresolved. Recently there has been some investigation of the use of agonist, rather than antagonist radioligands, based on the premise that they may be more sensitive to the affinity state of the receptor for endogenous neurotransmitters, and therefore endogenous dopamine might compete more successfully with these, leading to greater sensitivity to differences across conditions or populations in stimulated dopamine release. Studies using anesthetized animals have demonstrated increased sensitivity of $[^{11}C]$NPA and $[^{11}C]$PHNO, both D2 and D3 agonists, to amphetamine stimulation relative to $[^{11}C]$raclopride, and early studies with $[^{11}C]$PHNO in healthy human volunteers have also shown increased amphetamine effect in the dorsal striatum compared to previously published reports utilizing $[^{11}C]$raclopride. The competitive binding method has proved to be much more difficult to use with receptors other than the dopamine D2 receptor. Several investigators have been unable to detect the amphetamine effect on binding of dopamine D1 radioligands. Researchers have had mixed results detecting pharmacologically induced increases in serotonin levels as well, with some investigators reporting decreases in $[^{18}F]$MPPF or $[^{18}F]$MEFWAY binding to 5-HT$_{1A}$ receptors in animal models following stimulated increases in serotonin levels by either fenfluramine or SSRIs, while others have been unable to detect changes with $[^{18}F]$MPPF or $[^{11}C]$WAY100,635. However more recently, some tracers for the 5-HT$_{1B}$ receptor such as $[^{11}C]$P943 and $[^{11}C]$AZ10419369 have demonstrated sensitivity to pharmacologically induced increases in extracellular serotonin using compounds such as fenfluramine or citalopram. See Laruelle (2000) for a comprehensive examination of competitive binding techniques in PET and SPECT, especially as pertains to dopamine D2/D3 receptor imaging. Finally, it is worth noting that a small number of studies have used receptor imaging to infer some more subtle pharmacological effects, such as allosteric interaction between tracers and neurotransmitters (Frankle et al. 2009).

Metabolism Imaging

The models used for metabolism imaging can be quite varied according to the mechanism being studied. Here, two widely used metabolism tracers will be described: $[^{18}F]$DOPA and $[^{18}F]$FDG. Both are substrates for some, but not all enzymes that act on an endogenous compound, partially following the same metabolic pathway. Both reach a stage in the metabolic process in which they are assumed to be irreversibly trapped. In each case, there has been a considerable body of literature demonstrating that this assumption is oversimplified and that more accurate estimates of the measured process can be obtained using more complex models. Nonetheless, the trapping models are still widely used due to their simplicity and ease of implementation. A basic compartment model for irreversible trapping is shown in Fig. 6. Unlike the receptor models above, there is no notion of equilibrium. As long as free radioligand is in the tissue, the concentration in the trapped

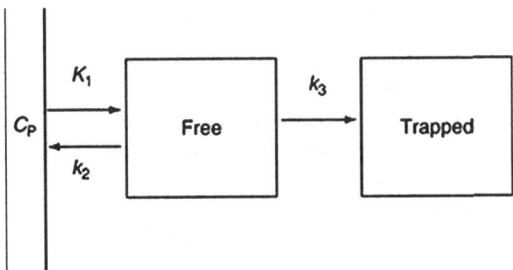

Positron Emission Tomography (PET) Imaging, Fig. 6 An irreversible trapping model

compartment will continue to increase as k_3 times the free concentration. However, one can envisage a hypothetical steady state in which influx to the tissue from plasma just balances the sum of efflux back to the plasma plus the conversion into the trapped form, so that the free concentration is constant. Under these conditions, the free concentration equals $K_1/(k_2 + k_3)$ times the plasma concentration C_p; the steady state rate of conversion into the trapped form equals $K_1 k_3/(k_2 + k_3)$ times the plasma concentration. This parameter is often referred to as K_{in}, the steady state uptake rate for an irreversible compartment model.

[^{18}F]DOPA is a substrate for amino acid decarboxylase (AADC), the enzyme that catalyzes L-dihydroxyphenylalanine (DOPA) into dopamine. [^{18}F]DOPA readily passes the blood-brain barrier and cell membranes and is metabolized by AADC into 6-fluorodopamine (6-FDA) in dopaminergic terminals. 6-FDA, like endogenous dopamine, does not cross the blood-brain barrier and so is "trapped" in the proximity of the terminals at the stages of loading into vesicles, release through exocytosis and reuptake into the terminal, the main path followed by endogenous dopamine in the striatum, where reuptake through dopamine transporters is the dominant mode of clearance from the synapse. Thus K_{in} represents a lumped marker of presynaptic dopaminergic condition. 6-FDA is, however, a substrate for monoamine oxidase (MAO) and for catechol-O-methyltransferase (COMT). Both are present in the extracellular environment, and the radiolabeled metabolic by-products from 6-FDA metabolism can readily diffuse across the blood brain barrier. Numerous studies have demonstrated that failure to account for this loss of radiolabel from the brain results in underestimation of the true uptake rate, and several approaches have been proposed for its incorporation into the modelling and design of [^{18}F] DOPA experiments. See Cumming and Gjedde (1998) for a detailed discussion of the metabolism of [^{18}F] DOPA.

^{18}F labeled 2-fluoro-2-deoxy-D-glucose (FDG) is arguably the most extensively used PET radioligand. The development of FDG for use with PET followed the groundbreaking work of Sokoloff et al. (1977) with [^{14}C]DG, a ^{14}C labeled tracer which is an analogue of glucose and partially follows its metabolic pathway. FDG is a useful probe for measuring the cerebral glucose metabolism rate (CMR_{glu}). More recently, techniques such as fMRI have become prevalent for studies measuring brain metabolic activity, owing to their better temporal and spatial resolution and the less invasive nature of the procedure. FDG has gone on to be used extensively as a radiological diagnostic tool in other fields such as oncology. But given the historic nature of role of FDG in the use of PET to study the brain, a brief description of the model is included here. FDG is a substrate for the same carrier protein that transports glucose into the brain tissue. It is also a substrate for hexokinase, the enzyme that metabolizes glucose, and is phosphorylated into FDG-6-PO4. FDG-6-PO4 does not follow further steps of glucose metabolism. FDG-6-PO4 dephosphorylates slowly, so that FDG-6-PO4 is treated as a trapped state. Here, K_{in} represents the steady state phosphorylation rate of FDG. When multiplied by a conversion factor accounting for the different phosphorylation and transport rates (1/LC, the "lumped constant," because it lumps two conversion factors together), K_{in}/LC times the plasma glucose concentration is an estimate of brain glucose metabolism. The dephosphorylation step, while small, still contributes to underestimation of K_{in} if it is not accounted for, but the irreversible model, and further simplifications of it, is still the most widely used, due to convenience of implementation.

Positron Emission Tomography (PET) Imaging, Table 1 A sample of PET probes currently used in research with human subjects

	Ligand	Target	Description
Dopamine	[^{11}C]raclopride	D2/D3 receptors	D2/D3 antagonist; useful in striatum only
	[^{18}F]fallypride	D2/D3	D2/D3 antagonist; useful in striatum, thalamus, and limbic regions
	[^{11}C]fallypride	D2/D3	Similar to the [^{18}F] version; [^{11}C] label allows multiple scans on 1 day for quantitative analysis in extra-striatal regions
	[^{11}C]FLB457	D2/D3	D2/D3 antagonist; useful in cortical regions
	[^{11}C]-(+)-PHNO	D2/D3	D2/D3 agonist; D3 preferring (strong signal in globus pallidus)
	[^{11}C]NPA	D2/D3	D2/D3 agonist
	[^{11}C]MNPA	D2/D3	D2/D3 agonist
	[^{11}C]NNC112	D1 receptors	D1 antagonist; useful in cortex and striatum
	[^{11}C]SCH23390	D1	D1 antagonist; useful in cortex and striatum
	[^{11}C]PE2I	DA transporters (DAT)	Striatal and extra-striatal DAT ligand
	[^{11}C]altropane	DAT	
	[^{18}F]CFT	DAT	
	[^{11}C]CFT	DAT	
	[^{18}F]DOPA	DA terminals	Substrate for AADC; indicator of DA synthesis and turnover; presynaptic DA function
	FMT		Substrate for AADC
	[^{11}C]DTBZ	VMAT2	
	[^{18}F]AV133	VMAT2	
Serotonin	[^{11}C]P943	5-HT$_{1B}$	Recently developed 5-HT$_{1B}$ ligand
	[^{11}C]McN5652	5-HT transporter (SERT)	The first PET tracer for SERT
	[^{11}C]DASB	SERT	Improved signal to noise ratio compared to [^{11}C]MCN; allows more accurate quantitation
	[^{11}C]AFM	SERT	Recently developed tracer; shows promise for imaging in moderate density cortical regions
	[^{11}C]HOMADAM	SERT	Recently developed tracer; shows promise for imaging in moderate density cortical regions
	[^{11}C]WAY100635	5-HT$_{1A}$ receptors	5-HT$_{1A}$ antagonist
	[^{18}F]MPPF	5-HT$_{1A}$	5-HT$_{1A}$ antagonist
	[^{18}F]FCWAY	5-HT$_{1A}$	Requires coadministration of other compounds to reduce defluorination
	[^{11}C]CUMI101	5-HT$_{1A}$	Recently developed 5-HT$_{1A}$ partial agonist
	[^{11}C]MDL100907	5-HT$_{2A}$ receptors	Selective 5-HT$_{2A}$ antagonist
	[^{18}F]Altanserin	5-HT$_{2A}$	F18 label makes altanserin conducive to bolus + infusion design; there may be a confound associated with bbb penetrant radiolabeled metabolites
	[^{11}C]P943	5-HT$_{1B}$	Recently developed 5-HT$_{1B}$ ligand
	[^{11}C]AZ10419369	5-HT$_{1B}$	Recently developed 5-HT$_{1B}$ ligand
	[^{11}C]GSK215083	5-HT$_6$	
Norepinephrine	[^{18}F]MeNER	Norepinephrine transporter (NET)	
	[^{11}C] MRB	NET	

(continued)

Positron Emission Tomography (PET) Imaging, Table 1 (continued)

	Ligand	Target	Description
Opioid	[^{11}C]carfentanil	mu opioid receptors	Potent mu agonist
	[^{11}C]diprenorphine	Opioid receptors	Non selective
Acetylcholine	2-F-18-FA-85380	Nicotinic receptors	Specific to alpha 4 beta 2* subtype; requires very long (8 h) scans
	[^{18}F]FP-TZTP	Muscarinic M2	Selective M2 agonist
Glycine Transporter	[^{11}C]GSK931145	Glycine transporter 1 (GlyT1)	Recently developed glyt 1 ligand
	[^{18}F]FCPyPB	GlyT1	Recently developed glyt 1 ligand
	[^{11}C]RO5013853	GlyT1	Recently developed glyt 1 ligand
GABA	[^{11}C]flumazenil	Benzodiazapine site on GABA$_A$ receptors	
Glutamate	[^{11}C]ABP688	mGluR5	Recently developed allosteric antagonist
Substance P	[^{18}F]SPA-RQ	NK1 receptors	
Enzymes	[^{11}C]deprenyl	MAO B	
	[^{11}C]clorgyline	MAO A	
Inflammation	[^{11}C]PK11195	TSPO	
	[^{11}C]PBR28	TSPO	
	[^{18}F]PBR06	TSPO	
	[^{18}F]PBR111	TSPO	
	[^{18}F]FEPPA	TSPO	
Cannabinoid	[^{11}C]OMAR	CB1 receptor	
	[^{18}F]MK9470	CB1	
	[^{18}F]CURB	FAAH	Binds to fatty acid amide anhydrogenase

Survey of Probes and Targets

Table 1 in this section is intended to give a general sense of the CNS targets currently accessible by PET imaging. It is extensive, but not necessarily exhaustive. Criteria for inclusion are that radioligands have been used recently in human subjects. The table demonstrates that by far the most explored system in PET is the dopaminergic, followed by the serotonergic, but many other systems have or are beginning to be imaged as well.

Summary

PET imaging is unique in its ability to provide quantitative measurements of pharmacokinetic parameters noninvasively in vivo. The studies described here have provided important data about neurochemical processes in living human brains that are not attainable through any other currently available technology. The field is limited only by the availability of probes; any system can be studied provided a suitable radiotracer can be developed. The dopaminergic and serotonergic receptor systems have proved particularly amenable to probing with PET, and this has been fortuitous for the field of psychopharmacology due to the prominent role these systems play in psychiatric and neurologic conditions. Less progress has been made in the study of other systems such as amino acid transmitters, but these are of course of great interest as well. The challenge of the future for PET is to continue to expand into these and other neurochemical systems.

Cross-References

► Antipsychotic Drugs
► Magnetic Resonance Imaging (Functional)
► Pharmacokinetics
► SPECT Imaging
► Structural and Functional Magnetic Resonance Imaging

References

Cherry S, Sorenson J et al (2012) Physics in nuclear medicine, 4th edn. Saunders, Philadelphia

Cumming P, Gjedde A (1998) Compartmental analysis of dopa decarboxylation in living brain from dynamic positron emission tomograms. Synapse 29:37–61

Farde L, Wiesel FA et al (1988) Central D2-dopamine receptor occupancy in schizophrenic patients treated with antipsychotic drugs. Arch Gen Psychiatry 45:71–76

Frankle WG, Cho R et al (2009) Tiagabine increases [11C] flumazenil binding in cortical brain regions in healthy control subjects. Neuropsychopharmacology 34:624–633

Innis RB, Cunningham VJ et al (2007) Consensus nomenclature for in vivo imaging of reversibly binding radioligands. J Cereb Blood Flow Metab 27(9):1533–1539

Laruelle M (2000) Imaging synaptic neurotransmission with in vivo binding competition techniques: a critical review. J Cereb Blood Flow Metab 20(3):423–451

Slifstein M, Laruelle M (2001) Models and methods for derivation of in vivo neuroreceptor parameters with PET and SPECT reversible radiotracers. Nucl Med Biol 28:595–608

Slifstein M, Frankle W et al (2004) Ligand tracer kinetics: theory and application. In: Otte A, Audenaert K, Peremans K, Heeringen K, Dierckx R (eds) Nuclear medicine in psychiatry. Springer, Berlin/Heidelburg

Sokoloff L, Reivich M et al (1977) Deoxyglucose-C-14 method for measurement of local cerebral glucose-utilization – theory, procedure, and normal values in conscious and anesthetized albino-rat. J Neurochem 28(5):897–916

Valk PE, Bailey DL et al (eds) (2003) Positron emission tomography: basic science and clinical practice. Springer, London

Postnatal Period

Definition

An interval of time following birth, typically subsuming the neonatal and infant stages of development.

Postpsychotic Depressive Disorder of Schizophrenia

Samuel G. Siris
Hofstra North Shore LIJ School of Medicine,
The Zucker Hillside Hospital, Glen Oaks,
NY, USA

Synonyms

Associated depression in schizophrenia; Depression superimposed on residual schizophrenia; Depression NOS (in the context of schizophrenia, when the episode of depression follows the active psychotic phase); Post-psychotic depression; Secondary depression in schizophrenia

Definition

Postpsychotic depressive disorder of schizophrenia (PPDDS) is a clinical condition which occurs when an individual with the preexisting diagnosis of schizophrenia manifests the syndrome of depression subsequent to the remission, or partial remission, of a psychotic episode of schizophrenia. PPDDS is diagnosed only during the residual phase of schizophrenia which follows a positive psychotic episode (the active phase representing the presence of symptoms meeting Criterion A of schizophrenia according to the Diagnostic and Statistical Manual of Mental Disorders, 5th Edition – DSM-5; American Psychiatric Association 2013). Negative symptoms or attenuated manifestations of active phase symptoms (e.g., odd or overvalued beliefs or unusual perceptual experiences) may, however, persist at the time of diagnosis of PPDDS. The diagnosis of PPDDS requires the presence of features sufficient to meet the DSM-5 criteria for the diagnosis of a major depressive episode and must also, specifically, include the presence of depressed mood (emotion and mood). Symptoms which are due to the direct physiological effects of a medication, a substance of abuse, or general medical condition are not counted toward the diagnosis of PPDDS.

Context of the Definition of "Depression"

"Depression" is a term which can be used in a variety of contexts. It can refer to an affect, a symptom, a syndrome, or a disease. As an affect, the term "depression" refers to the experience of blue mood which is an appropriate response to stimulus such as a sad event or sad story. This is component of a full range of appropriate affect and is not, in and of itself, pathological.

As a symptom, the term "depression" refers to a mental state involving an experience of sadness, joylessness, and/or emptiness which is exaggerated in comparison to the circumstances and which is associated with psychic pain or distress. The degree to which this type of "depression" is pathological depends on the degree of exaggeration and the amount of suffering which is involved.

The syndrome of "depression" occurs when a defined group of signs and symptoms is present simultaneously with adequate severity and duration. The DSM-5 definition of "depression" is the most common contemporary definition of "depression," and this definition is at the level of a syndrome. It includes features such as sad mood or tearfulness, sleep disturbance (may be either insomnia or hypersomnia), appetite disturbance or fluctuations in weight (can be in either direction), reduced energy level or excessive fatigue, psychomotor agitation or retardation, anhedonia, reduced interest level, impaired concentration or ability to think or make decisions, feelings of excessive or inappropriate guilt and/or worthlessness, pessimism, feelings of helplessness and/or hopelessness, recurrent thoughts of death or dying, and suicidal ideation, intent, or behavior. Interestingly, many of these features were initially incorporated into the DSM-III diagnosis of "depression" (the predecessor of DSM-IIIR and, subsequently, DSM-IV and DSM-5) on the basis of their being associated with, among other things, a favorable response to antidepressant medication (originally meaning a tricyclic or MAO inhibitor antidepressant but more recently "second-generation" antidepressants as well). Other widely recognized features of depression at the syndrome level which were not quite as predictive of response to early antidepressant medications (e.g., diurnal variation) did not become part of the DSM-III and subsequent diagnostic definitions (and it may be argued that some possible opportunities for the understanding of various biological, psychological, and/or social aspects of the syndrome of depression may have been missed as a result).

The definition of "depression" as a disease state would require an in-depth understanding of causation and pathophysiology of depression as a distinct biomedical condition – which is arguably still beyond our grasp. To wit, we as yet have no "tissue-level" diagnostic criteria (such as a blood test, electrophysiological or imaging test, biopsy, a specific predisposing gene, or even an autopsy finding) which confirms the diagnosis of "depression." Indeed, we cannot even be certain that the diagnosis of "depression" really belongs at the disease level or whether "depression" is better characterized (like fever, seizures, or congestive heart failure) as being a clinical syndrome (syndrome-level diagnosis) which can occur in a variety of different "disease" states.

The definition of PPDDS given above is at the level of a syndrome.

Epidemiology

"Depression" has long been described as a feature in many patients diagnosed as having schizophrenia (McGlashan and Carpenter 1976; schizophrenia), and many studies have been undertaken to explore and/or document the frequency of occurrence of depressive symptoms and/or depressive syndromes in people with schizophrenia (Buckley et al. 2009; Siris and Bench 2003). These studies have varied considerably in terms of the populations which were assessed, the definitions of schizophrenia and depression which were employed, the interval of observation which was involved, and the setting and treatment situation of the patients. The results for the observation of depression occurring in schizophrenia have ranged from a rate of 6 % to a rate of 75 %, but both the modal and the median frequency of occurrence was 25 %. "Depression" has been associated with higher rates of adverse outcomes in schizophrenia, such as poor social adjustment, lower levels of occupational functioning, reduced quality of life, undesirable life events, relapse into psychosis, and rehospitalization. Additionally, it is of importance that the symptom or syndrome of depression, along with the symptom of hopelessness and events involving loss, has been identified as the most frequent correlates of suicidal ideation and behavior in schizophrenia, a tragic outcome which has been estimated to occur at a frequency of 4–12 % in schizophrenia (Siris 2001; Suicide).

Differential Diagnosis

There are a variety of conditions which can present with the clinical features representative of PPDDS (Siris 2000; Siris and Bench 2003). These include medical disorders; effects of treatments used for medical disorders; effects of other substances; neuroleptic side effects including akinesia, akathisia, and neuroleptic-induced dysphoria; negative symptoms of schizophrenia; acute and chronic disappointment reactions; the prodrome of psychotic relapse; schizoaffective disorder; and the expression of an independent primary diathesis for depression in an individual who also has schizophrenia.

Blue mood (emotion and mood) and/or a syndrome which can mimic depression can be a feature of a variety of medical conditions including anemias, endocrinopathies, metabolic abnormalities, infectious diseases, cancer, cardiovascular, autoimmune, or neurological disorders. Many commonly prescribed medications can also have depression occur as a side effect. These include beta-blockers, various other antihypertensive medications, sedative hypnotics, antineoplastic agents, nonsteroidal anti-inflammatory agents, interferon, sulfonamides, and indomethacin. Other medications can be associated with depression at the time of their discontinuation (withdrawal syndromes). Examples of these include corticosteroids and psychostimulants. Various substances of use and/or abuse (e.g., alcohol, cannabis, or cocaine) can also be associated with depression-like presentations, either at the time of their acute use, chronic use, or discontinuation. Additionally, it merits mention that the withdrawal state from two commonly used legal substances, nicotine and caffeine, can also involve dysphoria and other features which can easily be misinterpreted as the syndrome of depression (withdrawal syndromes).

Antipsychotic medications, perhaps more notably first-generation antipsychotics, have been associated with side effects which can also be phenocopies of depression (Awad 1993; first-generation antipsychotics). It may be relevant to this observation that dopamine is an important neurotransmitter in the "pleasure" pathways of the brain. Thus, blocking dopamine receptors, as all antipsychotic medications do, could lead to an experience which is the opposite of pleasure. This effect has been described as either a primary impact of neuroleptic agents on mood or as a component of one of the two classical extrapyramidal neuroleptic side effects of akinesia and akathisia (medication-induced movement disorder). Akinesia is marked by a general diminution of motor behavior – causing patients to appear to be non-spontaneous, i.e., "as if their starter motor is broken." Akathisia reflects the opposite of this – patients' appearance (frequent fidgeting and/or restless movements) and subjective experience and is "as if their starter motor won't turn off." Both akinesia and akathisia can be associated with substantial dysphoria, and akathisia has additionally been associated with suicide risk (perhaps because of the predilection for motor behavior which is involved). Both akinesia and akathisia can also be present in subtle rather than blatant forms. Subtle akinesia can sometimes occur in the absence of large muscle stiffness or cogwheel rigidity. Subtle akathisia may be reflected in a generalized tendency toward behavioral excesses, such as overtalkativeness or wandering into other people's territory.

"Negative" symptoms of schizophrenia can also present as a phenocopy of depression. Loss of pleasure, loss of interest, and decreased activity and/or initiative are features of the negative symptom syndrome which have their counterparts in depression, and this is a central reason why affective features (manifest blue mood) and cognitive features (e.g., guilt, hopelessness) are so important in distinguishing depression from negative symptoms in schizophrenia. The phenotypic similarities between "negative" symptoms, "parkinsonian" symptoms, and "retarded depression" have given rise to speculation that each of these states may represent the product of the common syndrome of "akinesia" being manifest in each of these situations (Bermanzohn and Siris 1992).

In addition to the above-biological and pharmacological issues, persons with schizophrenia often have much to be disappointed about in terms of how their lives are progressing in comparison to their original hopes and expectations. Consequently, both acute and chronic disappointment reactions are common. Acute disappointment reactions are generally relatively brief and can be linked to some recent event which was damaging to the patient's wishes, prospects, self-concept, or self-esteem (taking into account, of course, that such an insult is not always obvious in the face of potential idiosyncrasies of the patient's thinking or communication). Chronic disappointment reactions (also sometimes referred to as the demoralization syndrome) are of longer, even open-ended duration, are based on a history of repeated failures or losses, and consequently can be more difficult to disentangle from other types of depression occurring in the course of schizophrenia.

Another important condition which can mimic depression in schizophrenia is the prodrome of psychotic relapse (pre-psychotic states and prodromal symptoms). When decompensating into a new psychotic episode, a patient may become dysphoric, anhedonic, restless and/or withdrawn, pessimistic, or apprehensive. Such an individual may also experience sleep or appetite disturbances, unstable energy levels, and/or difficulty concentrating. The feature which distinguishes this state from depression is the eventual emergence of frank psychotic symptomatology, but this feature may not make its appearance for a week or more. In the interim, the patient's condition may strongly mimic that of depression.

Schizoaffective disorder also enters into the differential diagnosis of PPDDS (schizoaffective disorder). To be diagnosed with schizoaffective disorder, a patient must have a period of overlap between florid psychotic symptoms and either a major depressive episode or a manic episode, as well as a period of at least two weeks of florid psychotic symptoms (including hallucinations or delusions) in the absence of prominent mood symptoms during the same episode of illness. Additionally, for the diagnosis of schizoaffective disorder, the symptoms which meet criteria for the mood episode must be present for the majority of the total duration of the active and residual portions of the illness (American Psychiatric Association 2013). Some authors also consider it to be a case of PPDDS when a patient with schizoaffective disorder (rather than the diagnosis of schizophrenia) manifests the syndrome of depression subsequent to the resolution of florid psychotic symptomatology (Siris and Bench 2003).

Finally, there is the case where PPDDS may be manifested in a patient who has coexisting independent diatheses for the psychosis of schizophrenia and for the syndrome of depression. The argument that this, logically, would be a statistical rarity is countered by the argument that the respective diatheses may well be continuous variables rather than categorical ones – and that each diathesis may promote or aggravate the expression of the other (Siris 2000).

Role of Pharmacotherapy

Initial Approaches

When a patient presents with a new episode of PPDDS, the first response should not necessarily be to change medications. Rather a careful history needs to be taken which would include assessment of any recent changes in medications – psychiatric or medical (including attention to adherence and/or abuse issues), a consideration of possible medical conditions, an exploration of potential psychosocial stressors, and an investigation of the possible use (or discontinuation) of illicit and/or recreational substances (including caffeine and nicotine). An exploration of risk factors for suicide is also indicated and protective steps to safeguard the patient be taken if necessary (Siris 2001). The initial intervention would be to raise the level of monitoring and provide psychosocial supports. If the "depression" is an acute disappointment reaction, it will resolve itself over a period of several days to several weeks. If it is a component of the prodrome of a new psychotic episode, that also will soon declare itself, and the increased monitoring will maximize the opportunity to nip this new episode in the bud with appropriate treatment, thereby limiting psychiatric and social/vocational morbidity. Medical conditions, the role of medications employed to treat medical conditions, and the possible role of substance use or abuse can also be addressed in this initial phase.

Treatment of the Persisting Syndrome of PPDDS

When it is apparent that the syndrome of PPDDS is stably present and in particular it is clear that the patient is not in the process of deteriorating into a new psychotic episode, it is appropriate to consider whether the dosage of antipsychotic medication is excessive (antipsychotic drugs). Unnecessarily high doses of antipsychotic medication may contribute to neuroleptic-induced dysphoria or the neuroleptic side effects of akinesia or akathisia. Once antipsychotic medications have been established at the lowest doses which are consistent with adequate antipsychotic activity in a given patient, treatment of akinesia can be undertaken with adjunctive antiparkinsonian medication (anti-Parkinson drugs). For example, benztropine may be tried in doses up to a full dose of 2 mg po TID, anticholinergic side effects permitting. Occasionally higher doses of antiparkinsonian medications (to 8 or even 10 mg/day) may be tried if anticholinergic side effects such as constipation, difficulty urinating, or dry mouth – crude but meaningful tests of bioavailability of the anticholinergic effect – are not present and the akinesia persists. Alternatively, a non-anticholinergic antiparkinsonian agent such as amantadine may be tried as an adjunct to the antipsychotic medication. In the case of akathisia, anticholinergic antiparkinsonian medications are unlikely to be helpful, but benzodiazepines are often useful (benzodiazepines), and beta-blockers sometimes work as well (beta-adrenoceptor antagonists).

Switching from a "typical" (first-generation) antipsychotic agent to an "atypical" (second-generation) antipsychotic agent is another adjustment of the medication regimen to consider in cases of PPDDS (Siris 2000; second- and third-generation antipsychotics). The literature is a bit inconsistent but suggests that, at least in some cases, negative symptoms are reduced with second-generation antipsychotics (SGAs) and/or extrapyramidal side effects are lessened (Möller 2008). There is also a literature suggesting, subject to a variety of methodological limitations, that rating scores for depression may be improved by the use of SGAs in schizophrenia (Möller 2008). Interestingly, SGAs have sometimes been touted as possessing augmenting antidepressant effects when used as adjunctive agents in patients with major depressions.

There is also a role for adjunctive antidepressant medications in the treatment of PPDDS, particularly when extrapyramidal side effects have been ruled out (Siris and Bench 2003; Buckley et al. 2009; antidepressants). Patients must be maintained on adequate doses of antipsychotic medications when antidepressants are used, but the antidepressant drugs can gradually be raised to full therapeutic dosages. Vigilance should be maintained, however, for the possibility that the antipsychotic and antidepressant might each interfere with the metabolism of the other, and there is also the possibility of adverse pharmacokinetic interactions (drug interactions), perhaps particularly when specific serotonin reuptake inhibitors (SSRIs) are involved (Möller 2008). Although most of the controlled studies of adjunctive antidepressant use in PPDDS come from the era of first-generation antipsychotics (FGAs) and tricyclic antidepressants, the subsequent wide use of combinations including both FGAs and SGAs and a wide variety of more novel antidepressants nevertheless suggest both safety and utility for a variety of regimens (Siris et al. 2001).

It is additionally relevant to note that all the studies of utility of these medications were done utilizing pre-DSM-5 diagnostic criteria. This is a situation which takes a certain toll with respect to our advancing "evidence-based treatment" in our psychopharmacological practices as we advance (as required by other circumstances) with the most recent diagnostic definitions. But, fortunately, variations in diagnostic boundaries due to different diagnostic schemes, have not, as yet, appeared to result in substantially different outcomes in terms of the medications and medication classes utilized (Siris, et al. 2001). So our position, in the face of this conundrum, is to remain practical and assume that classes of medications are equivalent across diagnostic schemes – while remaining vigilant to the possibility that meaningful differences might arise as we move forward.

Although the proper studies have not been done to support the use of lithium or anticonvulsants (mood stabilizers) in PPDDS, it is rational to consider a trial of these agents, perhaps particularly in patients who manifest features of schizoaffective disorder, have a history of excitement or an episodic course as a component of their illness, or have family histories of affective disorders. Similarly, although there are not specific prospective, randomized, double-blind data to support it, ECT belongs on the list of other treatments which might be useful in cases of PPDDS.

It is additionally necessary to emphasize the importance of adequate psychosocial support, increased structure, stress reduction, skill building, and raising confidence and self-esteem for patients during their treatment for PPDDS. This support is particularly crucial when the patient is suffering from a chronic disappointment reaction (demoralization syndrome) and needs to learn new strategies of thinking to foster success and happiness. But proper psychosocial supports may be pivotal as well during pharmacological interventions because, from the patients' point of view, these interventions literally change their world, and they may need help in moving ahead from long-held, but now suboptimal, old ways of adapting to their old worlds toward new ways of productively adapting to their new worlds.

Summary

A syndrome of post-psychotic depressive disorder of schizophrenia (PPDDS) is commonly noted to occur in the residual phase of schizophrenia, after the resolution of florid psychotic symptoms. PPDDS can be a source of considerable morbidity, and even mortality, and consequently merits clinical attention.

The phenomenology of PPDDS may represent any one of a number of conceptually distinct situations or states, including organic or medical factors; acute use, chronic use, or discontinuation of a variety of medications or substances; mood or parkinsonian side effects to antipsychotic medications; an expression of the "negative" symptoms of schizophrenia; an acute or chronic disappointment reaction; the prodrome of relapse into a new psychotic episode; a presentation of schizoaffective disorder; or the expression of a diathesis of affective disorder distinct from, but occurring in the context of, the schizophrenia diathesis. In each case, appropriate psychopharmacologic and psychosocial management of the condition is crucial for reducing suffering and promoting functioning.

Cross-References

▶ Anticonvulsants
▶ Antidepressants
▶ Anti-Parkinson Drugs
▶ Antipsychotic Drugs
▶ Benzodiazepines
▶ Beta-Adrenoceptor Antagonists
▶ Drug Interactions
▶ Emotion and Mood
▶ First-Generation Antipsychotics
▶ Lithium
▶ Mood Stabilizers
▶ Prepsychotic States and Prodromal Symptoms
▶ Schizoaffective Disorder
▶ Schizophrenia

- ▶ Second- and Third-Generation Antipsychotics
- ▶ Suicide
- ▶ Withdrawal Syndromes

References

American Psychiatric Association (2013) Diagnostic and statistical manual of mental disorders, 5th edn. American Psychiatric Association, Washington, DC

Awad AG (1993) Subjective response to neuroleptics in schizophrenia. Schizophr Bull 19:609–618

Bermanzohn PC, Siris SG (1992) Akinesia, a syndrome common to Parkinsonism, retarded depression, and negative symptoms. Compr Psychiatry 33:221–232

Buckley PF, Miller BJ, Lehrer DS, Castle DJ (2009) Psychiatric comorbidities and schizophrenia. Schizophr Bull 35:383–402

McGlashan TH, Carpenter WT Jr (1976) Postpsychotic depression in schizophrenia. Arch Gen Psychiatry 33:231–239

Möller H-J (2008) Drug treatment of depressive symptoms in schizophrenia. Clin Schizophr Relat Psychoses 1:328–340

Siris SG (2000) Depression in schizophrenia: perspective in the era of "atypical" antipsychotic agents. Am J Psychiatry 157:1379–1389

Siris SG (2001) Suicide and schizophrenia. J Psychopharmacol 15:127–137

Siris SG, Bench C (2003) Depression and schizophrenia. In: Hirsch SR, Weinberger D (eds) Schizophrenia. Blackwell Scientific, London, pp 142–167

Siris SG, Addington D, Azorin J-M, Falloon IRH, Gerlach J, Hirsch RD (2001) Depression in schizophrenia: recognition and management in the U.S.A. Schizophr Res 47:185–197

Postsynaptic Proteins

Definition

Proteins present at the dense postsynaptic membrane complex that is a characteristic of excitatory glutamatergic synapses in the brain, including postsynaptic density 95 (a marker of excitatory postsynaptic sites in hippocampal pyramidal cell dendritic spines) and protein phosphatases (enzymes that remove a phosphate group from their substrate by hydrolyzing phosphoric acid monoesters into a phosphate ion and a molecule with a free hydroxyl group).

Posttranslational Modification

Synonyms

Posttranslational amino acid modification; Posttranslational protein modification; PTM

Definition

Posttranslational modifications (PTMs) are chemical alterations to a primary protein structure, often crucial for conferring biological activity on a protein. PTMs are usually catalyzed by substrate-specific enzymes (which themselves are under strict control by PTMs). It comprises covalent alteration of one or more amino acids occurring in a protein after the protein has been completely translated and released from the ribosome.

Cross-References

- ▶ Electrospray Ionization
- ▶ Imaging Mass Spectrometry
- ▶ Mass Spectrometry
- ▶ Matrix-Assisted Laser Desorption Ionization
- ▶ Metabolomics
- ▶ Neuropeptidomics
- ▶ Proteomics
- ▶ Two-Dimensional Gel Electrophoresis

Posttraumatic Stress Disorder

Synonyms

PTSD

Definition

This is an emotional condition or illness resulting from quite frightening or life-threatening experiences. Those who suffer from the disorder continually reexperience the precipitating traumatic events in some way and tend to avoid all places and people that are associated with those events. They are usually quite sensitive to normal life experiences (hyperarousal). Although the symptoms of PTSD were described during the American Civil War and World War I, only since 1980 has PTSD been recognized as a formal diagnosis. Approximately 7–8 % of people in the USA develop PTSD in their lifetime, while the prevalence of the disorder among combat veterans and rape victims ranges from 10 to as high as 30 %.

Potency

Synonyms

Measure of drug activity

Definition

A measure of the concentration or dose of a drug at which it is effective. Commonly expressed as the ED_{50} or EC_{50}, the concentration of an agonist or a positive allosteric modulator that produces 50% of its maximal possible effect, or IC_{50}, the concentration of a compound that produces 50% of its maximal possible inhibition. A highly potent drug evokes a larger response at low concentrations. A compound's potency is proportional to its affinity and efficacy.

Cross-References

▶ Agonist
▶ Allosteric Modulators
▶ Antagonist
▶ ED_{50}
▶ Inverse Agonists
▶ Partial Agonist

Power Spectral Analysis

Synonyms

Fourier analysis of time series; Fourier transform; Frequency estimation

Definition

Power spectral analysis refers to a collection of methods used to decompose a time series of data into its elementary frequency (i.e., defined physically in cycles/s or Hertz) components. A power spectrum is a function that describes the frequencies of oscillation present in the original time series and the power or variance that each frequency contributes to the amplitude changes recorded in the original time series. The method allows one to discover in the power spectrum the presence of repeatable rhythms in what may appear to be a largely random process when seen in the original time series.

Cross-References

▶ Electroencephalography

Pramipexole

Synonyms

(S)-N^6-propyl-4,5,6,7-tetrahydro-1,3-benzothiazole-2,6-diamine; Mirapex; Mirapexin; Sifrol

Definition

Pramipexole functions as a partial agonist at D_{2-4} dopamine receptors with a strong preference for D_3. Belonging to the non-ergoline class, pramipexole is indicated in the treatment of ▶ Parkinson's disease (PD) and restless legs syndrome. Its ability to stimulate dopamine receptors in the striatum is believed to contribute to its therapeutic efficacy in PD, but the mechanism of action for restless legs syndrome is not clear. Side effects associated with pramipexole are insomnia, hypotension, nausea, constipation, hallucinations, and impulse control behaviors such as eating, gambling, shopping, and hypersexuality. Pramipexole has been approved by the Food and Drug Administration for use in the United States. It has been investigated for efficacy in the treatment of bipolar disorder, major depression, and fibromyalgia.

Cross-References

▶ Dopamine Agonist

Prazepam

Definition

Prazepam is a benzodiazepine with anxiolytic, anticonvulsant, sedative, and muscle-relaxant properties. Prazepam's therapeutic properties are largely attributed to its active metabolite desmethyldiazepam that displays a very long ▶ half-life. The clinical indication for prazepam is the short-term treatment of ▶ anxiety. Like most similar compounds, its long-term use is subject to tolerance, abuse, dependence, and withdrawal.

Cross-References

▶ Benzodiazepines
▶ Sedative, Hypnotic, and Anxiolytic Dependence

Prazosin

Synonyms

Hypovase®; Minipress®; Vasoflex®

Definition

Prazosin hydrochloride is an antagonist at α-1 noradrenaline receptors. Its clinical use is in the treatment of hypertension, but in behavioral pharmacology experiments it is used to block brain α-1 receptors.

Cross-References

▶ Motor Activity and Stereotypy

Prediction Error

Definition

The mismatch between the unconditioned stimulus (UCS) predicted and the UCS that in fact occurs, which generates new learning. This concept of prediction error arises directly from discrepancy theories of associative learning, for example, Rescorla-Wagner (1972). This is a terminology applied in classical (Pavlovian) conditioning and the importance of prediction error for new learning provides a constraint on the general importance of temporal coincidence as a determinant of new learning.

Cross-References

▶ Classical (Pavlovian) Conditioning

Predictive Validity

Definition

The term refers to one of the criteria used to assess the validity of animal models for psychiatric states. A similarity in response to a manipulation in the model and in the human syndrome is indicative of predictive validity. Most commonly, this refers to the actions of drugs that alleviate or worsen the condition; predictive validity is high if drugs known to alleviate the disease in humans also attenuate the measures taken in the model and if drugs that do not work in the human are also ineffective in the model. Such a model may have the ability to predict the effectiveness of a drug in humans.

Cross-References

▶ Animal Models for Psychiatric States
▶ Construct Validity
▶ Face Validity

Preference Reversal

Synonyms

Switch in preference

Definition

A preference reversal is a change in choice from one option to another.

Cross-References

▶ Behavioral Economics

Prefrontal Cortex

Synonyms

Prefrontal lobe

Definition

The prefrontal cortex is the entire part of the cerebral cortex that is located in front of the premotor areas. It plays an important role in executive function, complex cognitive behaviors, working memory, attention, expression of personality, and appropriate social behavior. It has been subdivided into the medial prefrontal cortex, involved in behavioral error monitoring; ventral/orbital prefrontal cortex, related to emotional control; and dorsolateral prefrontal cortex, related to spatial working memory and executive functions. Recent evidence suggests that additionally, the frontal pole (the anterior tip of the brain) is an important component involved in executive functions. The prefrontal cortex is needed for behavioral planning, categorizing, and sequencing complex actions. It is supplied by inputs from association cortices and receives strong dopaminergic, noradrenergic, serotonergic, and cholinergic inputs that modulate its activity. Its outputs are directed to the corpus striatum, amygdala, and other subcortical centers. The prefrontal cortical projections to the nucleus accumbens mediate the compulsivity and impulsivity associated with drug taking.

Cross-References

▶ Short-Term and Working Memory in Animals
▶ Short-Term and Working Memory in Humans

Pregabalin

Definition

An antagonist of alpha-2 delta calcium channels, with anticonvulsant and analgesic properties and efficacy in acute treatment and relapse prevention in ▶ generalized anxiety disorder (GAD). It diminishes both physical and somatic symptoms (such as tachycardia and tremor) and psychological or psychic symptoms such as worrying and irritability. Drowsiness can sometimes be troublesome, but can also be useful in GAD patients with marked sleep disturbance.

Premenstrual Dysphoric Mood Disorder

C. Neill Epperson, Liisa Hantsoo and
Deborah R. Kim
Departments of Psychiatry and Obstetrics/
Gynecology, Penn Center for Women's
Behavioral Wellness, Perelman School of
Medicine at the University of Pennsylvania,
Philadelphia, PA, USA

Synonyms

Late luteal phase dysphoric disorder; Premenstrual syndrome; Premenstrual tension

Definition

Premenstrual dysphoric disorder (PMDD) affects 2–5 % of premenopausal women and is characterized by moderate to severe mood symptoms, with or without physical symptoms. Symptoms are present during the week prior to menstrual flow and absent in the postmenstrual week. Women must experience at least one of four key mood symptoms (mood lability, irritability, anxiety/tension, or depressed mood) and have at least five symptoms in total in order to meet the diagnostic criteria included in the Diagnostic and Statistical Manual for Mental Disorders, Fifth Edition [DSM-5] (American Psychiatric Association 2013). Physical symptoms frequently include breast tenderness, bloating, fatigue, and cramping, while mood lability and irritability are the two most frequently reported mood symptoms. Women with PMDD are symptomatic during the majority of menstrual cycles and note significant psychological distress and/or impairment in important areas of daily functioning during the premenstruum. PMDD is not merely the worsening of an ongoing psychiatric disorder or a state that is secondary to a general medical condition. Importantly, 2 months of prospective daily ratings are required to confirm the diagnosis of PMDD. In early 2012, the APA DSM-5 committee agreed with recommendations from the expert panel review of the extant literature regarding PMDD and voted to move the diagnosis of PMDD from the Appendix B of the DSM-IV "Criterion Sets and Axes Provided for Further Study" to become a formal DSM-5 diagnosis under the heading of Depressive Disorders (Epperson et al. 2012).

Relationship of PMDD to Other Premenstrual Syndromes

PMDD and PMS are often used interchangeably to describe negative mood or physical symptoms during the premenstrual period; however, there are diagnostic differences that warrant consideration (Table 1). Much of the world refers to premenstrual mood and physical distress as premenstrual tension syndrome (PMTS), a diagnosis found in the World Health Organization (WHO)'s International Classification of Diseases, tenth edition (ICD-10). PMDD has several features that distinguish it from PMS (American College of Obstetrics and Gynecology [ACOG] criteria) and PMTS (ICD-10 criteria), as outlined in the following text. The timing of symptoms with respect to the menstrual cycle is consistent among the two diagnoses. However, the number, type, and severity of symptoms are highly divergent, with the diagnosis of PMDD being far more stringent, emphasizing the affective symptoms that are not merely the worsening of an ongoing psychiatric disorder (Freeman 2003; Halbriech 2007).

PMDD is also associated with significant impairment in at least one domain of daily functioning. Women typically report that impairment is greatest in the interpersonal realm, at home with family and friends.

Role of Pharmacotherapy

The symptoms of PMDD occurring during the luteal phase suggest that ovulation and its associated fluctuations in ovarian steroids are important for the manifestation of symptoms. A seminal study found that women with PMDD experienced symptom improvement with gonadotropin-releasing

Premenstrual Dysphoric Mood Disorder, Table 1 Diagnostic criteria for PMS/PMTS and PMDD

	PMS/PMTS	PMDD
Number of symptoms	1	5
Type of symptoms	Physical or psychological	At least one symptom has to be irritability, mood lability, anxiety/tension, depression
Symptom pattern	Symptoms present in the premenstruum and remit within a few days of menstrual flow	Symptoms present in the premenstruum and remit within a few days of menstrual flow
Functional impairment	Required for ACOG criteria (PMS) but not for ICD-10 (PMTS)	Required
Daily symptom ratings	Required for ACOG criteria (PMS) but not for ICD-10 (PMTS)	Required
Comorbidity	Not discussed in the ACOG or ICD-10 criteria for either PMS or PMTS, respectively	Must not be merely the exacerbation of an ongoing disorder

ACOG American College of Obstetrics and Gynecology, *ICD-10* International Statistical Classification of Diseases and Related Health Problems 10th Revision, *PMS* premenstrual syndrome, *PMTS* premenstrual tension syndrome

hormone (GnRH) agonist treatment, only to experience a return of mood symptoms when estradiol or progesterone was added back (Schmidt et al. 1998). These data support the theory that women with PMDD, but not healthy controls, are "sensitive" to ovarian hormone exposure. While the suppression of ovulation with GnRH agonist treatment is effective in the majority of women with PMDD, it is not a long-term treatment option secondary to the negative health effects of extended hypoestrogenism in young premenopausal women. Interestingly, ovulation suppression with oral contraceptives is not effective and can worsen mood symptoms in women with PMDD. An exception to this is contraceptive treatment with agents containing the progestin drospirenone, which have shown greater efficacy than placebo in controlled trials (Pearlstein et al. 2005; Yonkers et al. 2005).

The mainstay of treatment for PMDD includes daily or luteal phase administration of a selective serotonin reuptake inhibitor (SSRI). SSRIs are well tolerated, and a majority of women respond within the first few days of medication use (Nevatte et al. 2013). That PMDD is preferentially responsive to treatment with SSRIs versus antidepressants with alternative mechanisms of action is one factor that implicates serotonin in the pathogenesis of PMDD. However, the rapid onset of symptom relief with SSRI administration suggests that the pathogenesis of PMDD is distinct from other disorders (e.g., major depression, panic disorder, obsessive-compulsive disorder) for which SSRIs are effective therapies (Epperson et al. 2012). Moreover, the rapid onset of effectiveness also appears to implicate other neurotransmitter systems, such as gamma-aminobutyric acid (▶ GABA), in symptom expression and response to treatment (Epperson et al. 2002). SSRIs increase allopregnanolone, a potent $GABA_A$ receptor agonist. Table 2 provides doses of each of the SSRIs that have been shown to be effective in the treatment of PMDD.

Premenstrual Dysphoric Mood Disorder, Table 2 Selective serotonin reuptake inhibitors

Drug	Usual therapeutic dose (mg/day)	
	Daily administration	Luteal phase administration
Fluoxetine	10, 20[a]	10, 20
Sertraline	25, 50[b]	25, 50[b]
Paroxetine	10, 30	Not studied
Paroxetine CR	12.5, 25	25
Citalopram	10–30[c]	10–30

[a]One study found 60 mg/day to be more effective than placebo but not more effective than 20 mg/day
[b]Some studies used graded dosing between 50 and 100 mg/day
[c]Dosing was titrated from 10 to 30 mg/d

Daily Administration

A recent meta-analysis of 12 randomized, placebo-controlled studies confirmed that the daily administration of an SSRI is effective in the treatment of PMDD (Brown et al. 2013). Several, but not all, of these studies found clinically meaningful improvement in physical symptoms such as breast tenderness, cramping, and bloating, in addition to improvements in irritability and other behavioral symptoms. Daily administration is ideal for women with shorter cycles who have severe symptoms starting at the time of ovulation that extend into the first week of menstrual flow. Such women are symptomatic much of the month and could benefit from continued treatment. Certainly, if there is any question that a woman may have mild depression, recurrent brief depression, dysthymia, or generalized anxiety disorder, it would be preferable to use daily administration until one of these other disorders that frequently masquerade as PMDD could be ruled out. In a study of women claiming to have PMDD who presented to a premenstrual dysphoric disorder specialty program, over 30 % were diagnosed with a mood and/or anxiety disorder (Bailey and Cohen 1999). As the overlap in the type of symptoms seen in PMDD and a number of other psychiatric disorders is considerable, the importance of using daily ratings to examine the pattern of symptom onset and offset cannot be overestimated.

Luteal Phase Administration

A review of 12 randomized, placebo-controlled clinical trials provides strong evidence that intermittent or luteal phase treatment with an SSRI is effective in the treatment of PMDD (Brown et al. 2009). Luteal phase administration is best reserved for women whose daily ratings and clinical evaluation clearly confirm the diagnosis of PMDD. The benefit of luteal phase administration is considerable. Drug costs are lower, and many women prefer to use a medication on an as-needed basis. Luteal phase administration with a shorter half-life SSRI (e.g., sertraline, paroxetine CR, citalopram) is preferable for women who experience significant changes in sexual function/interest when using an SSRI. Interestingly, the tolerability of going on and off an SSRI seems to be more than adequate (Yonkers et al. 2006). In addition to luteal phase administration, some women may benefit from symptom-onset treatment which has been shown to significantly reduce negative affect in randomized controlled trials (Yonkers et al. 2006; Steinberg et al. 2012).

Conclusions

PMDD is a relatively common clinical phenomenon with the onset during the teen years and the early 1920s. That PMDD is preferentially responsive to SSRI treatment and returns upon medication discontinuation provides support for serotonin and menstrual cycle-related fluctuation in ovarian hormones in the pathogenesis of PMDD. The completion of 2 months of prospective daily ratings is not only required for the diagnosis, but it is also crucial for women who wish to use the luteal phase administration of an SSRI.

Cross-References

▶ Antidepressants
▶ SSRIs and Related Compounds
▶ Neuroactive steroids
▶ Sex hormones

References

American Psychiatric Association (2013) Diagnostic and statistical manual of mental disorders, 5th edn. American Psychiatric Publishing, Arlington

Bailey JW, Cohen LS (1999) Prevalence of mood and anxiety disorders in women who seek treatment for premenstrual syndrome. J Womens Health Gend Based Med, 8(9):1181–4

Brown EB, Wei SM, Kohn PD, Rubinow DR, Alarcón G, Schmidt PJ, Berman KF (2013) Abnormalities of dorsolateral prefrontal function in women with premenstrual dysphoric disorder: a multimodal neuroimaging study. Am J Psychiatry 170(3):305–314

Epperson CN, Haga K, Mason GF, Sellers E, Gueorguieva R, Zhang W, Weiss E, Rothman DL, Krystal JH (2002) Cortical gamma-aminobutyric acid levels across the menstrual cycle in healthy women and those with premenstrual dysphoric disorder: a proton magnetic resonance spectroscopy study. Arch Gen Psychiatry 59:851–858

Epperson CN, Steiner M, Hartlage SA, Eriksson E, Schmidt PJ, Jones I, Yonkers KA (2012) Premenstrual dysphoric disorder: evidence for a new category for DSM-5. Am J Psychiatry 169(5):465–475

Freeman EW (2003) Premenstrual syndrome and premenstrual dysphoric disorder: definitions and diagnosis. Psychoneuroendocrinology 28(suppl 3):25–37

Halbriech U (2007) The diagnosis of PMS/PMDD-the current debate. In: O'Brien PMS, Rapkin AJ, Schmidt PJ (eds) The premenstrual syndromes: PMS and PMDD. Informa, London

Nevatte T, O'Brien PM, Bäckström T, Brown C, Dennerstein L, Endicott J, Epperson CN, Eriksson E, Freeman EW, Halbreich U, Ismail K, Panay N, Pearlstein T, Rapkin A, Reid R, Rubinow D, Schmidt P, Steiner M, Studd J, Sundström-Poromaa I, Yonkers K, Consensus Group of the International Society for Premenstrual Disorders (2013) ISPMD consensus on the management of premenstrual disorders. Arch Womens Ment Health 16(4):279–91

Pearlstein TB, Bachmann GA, Zacur HA, Yonkers KA (2005) Treatment of premenstrual dysphoric disorder with a new drospirenone-containing oral contraceptive formulation. Contraception 72(6):414–421

Schmidt PJ, Nieman LK, Danaceau MA, Adams LF, Rubinow DR (1998) Differential behavioral effects of gonadal steroids in women with and in those without premenstrual syndrome, N Engl J Med. 22;338(4): 209–16

Steinberg EM, Cardoso GM, Martinez PE, Rubinow DR, Schmidt PJ (2012) Rapid response to fluoxetine in women with premenstrual dysphoric disorder. Depress Anxiety 29(6):531–540

Yonkers KA, Brown C, Pearlstein TB, Foegh M, Sampson-Landers C, Rapkin A (2005) Efficacy of a new low-dose oral contraceptive with drospirenone in premenstrual dysphoric disorder. Obstet Gynecol 106:492–501

Yonkers KA, Holthausen GA, Poschman K, Howell HB (2006) Symptom-onset treatment for women with premenstrual dysphoric disorder. J Clin Psychopharmacol 26(2):198–202

Premenstruum

Definition

The premenstruum is a term used by clinicians and researchers for the portion of the menstrual cycle that occurs immediately prior to the onset of menstrual flow. While there is no specified number of days referred to as the premenstruum, it typically includes the few days to a week prior to the onset of menses.

Premonitory Urge

Definition

By age 10 years, many patients with tic disorders will describe a feeling or urge prior to the execution of their tics. This sensation or warning may be located at the anatomical site of the tic or may be described as a more general feeling of unease or discomfort. Patients who describe the premonitory urge often report that there is momentary relief from this sensation following the performance of the tic.

Prenatal Exposure to Alcohol

Definition

Exposure to alcohol in utero by maternal abuse or use of alcohol or according to an experimental protocol in studies with animal models of psychiatric states.

Prenatal MAM Model

Definition

An animal model in which pregnant female rats are treated with the drug methylazoxymethanol (MAM). MAM blocks mitosis for a period of about 24 h and thus interferes with normal (brain) development. Depending on the timing of the MAM treatment, different brain regions will be more or less affected, leading to different behavioral alterations in adulthood. Although different protocols exist, the model most often used as a simulation model for schizophrenia involves treatment on gestational day 17.

In adulthood, these animals develop a large number of schizophrenia-like deficits.

Cross-References

▶ Schizophrenia: Animal Models
▶ Simulation Models

Prenatal Period

Definition

An interval of time from conception to birth.

Prepsychotic States and Prodromal Symptoms

Patrick D. McGorry[1] and Alison R. Yung[2]
[1]Department of Psychiatry, Orygen Youth Health Research Centre, Centre for Youth Mental Health, University of Melbourne, Melbourne, VIC, Australia
[2]Institute of Brain, Behaviour and Mental Health, University of Manchester, Manchester, UK

Synonyms

At-risk mental state; Psychosis; Schizophrenia prodrome; Ultrahigh-risk young people

Definition

The "prodromal phase" that precedes a first psychotic episode is a period characterized by increasing levels of nonspecific subthreshold symptoms associated with significant distress and growing functional impairment. This phase often continues for several years prior to the emergence of diagnostically specific psychotic symptoms, with significant social disability becoming apparent well before the first psychotic episode. Because the psychotic disorders usually manifest during adolescence, a period of major developmental change, they may have particularly devastating consequences on lifetime functioning, highlighting the need for early intervention in order to minimize ongoing disability.

Role of Pharmacotherapy

The psychotic disorders occur at a frequency of around 2 % in the general population and thus are relatively rare. However, their onset is most common during late adolescence and early adulthood, a period of life where critical developmental tasks are being accomplished in the psychological, social, educational, and vocational domains. Because serious mental illness substantially disrupts these processes and often leads to ongoing long-term disability, the early detection and treatment of people at risk of psychosis, before the onset of frank psychotic disorder, has long been a major goal in psychiatric practice.

The existence of a prodromal phase prior to a first episode of psychosis or a relapse of schizophrenia was noted over a century ago, prompting the first calls for early treatment as a means of preventing serious illness and ongoing disability. However, until relatively recently, research into the possibilities for early intervention has been limited by the lack of effective treatments, as well as the widespread perception that the ongoing disability associated with the psychotic disorders was inevitable. Over the last decade, the advent of more effective drugs, particularly the atypical antipsychotics, and the development of better psychosocial treatments have led to the realization that good long-term outcomes are possible for patients and rekindled interest in the area of early intervention. An important result of this renewed research effort is a series of careful epidemiological studies that have enabled the characterization of the "psychosis prodrome," a significant advance that has finally allowed early therapeutic intervention in the psychotic disorders to become a real possibility. These research findings have now been translated into evidence-based clinical practice in a growing number of specialized early intervention services worldwide and have made a major contribution to the improved outcomes that are now expected for young people experiencing the onset of a psychotic illness (McGorry et al. 2008).

Retrospective studies of first-episode psychosis patients, examining the course of illness from the pre-morbid period through to the emergence of frank psychosis, have shown that the first episode is almost always preceded by a prodromal period of several years that is characterized by increasing levels of psychological symptoms, significant distress, and a marked decline in social and vocational functioning compared to pre-morbid levels. In general, negative symptoms such as decreased concentration, reduced drive, and lack of energy predominate early in the prodromal phase, accompanied by nonspecific symptoms including sleep disturbance, anxiety, and irritability. Affective symptoms, primarily depression, are also common. These symptoms tend to accumulate exponentially until relatively late in the prodrome, when subthreshold positive symptoms (psychotic symptoms) emerge. Ultimately, these positive symptoms intensify and culminate in the transition to frank psychosis. Typically, increasing levels of social and vocational disability accompany the increase in symptomatology, with significant disability becoming apparent well before the first psychotic episode. The degree of disability that develops during the prodromal period appears to set a ceiling for the extent of the eventual recovery, highlighting the need for early intervention (McGorry et al. 2008).

Because these symptoms, including subthreshold psychotic-like experiences, are nonspecific and occur frequently in the general population, especially among adolescents and young adults, they cannot be considered as diagnostic of a prepsychotic state in their own right. Additional risk factors and specific criteria are necessary to exclude false-positive cases in order to avoid unnecessary treatment and the stigma often associated with the diagnosis of a mental illness. In order to increase the prognostic specificity of these prodromal symptoms, two additional risk factors have been added to these screening criteria, based on clinical experience and the available epidemiological evidence. The first of these is being aged between 15 and 30, since young people in this age range are at greatest risk of developing a psychotic disorder. The second is a need for clinical care, since young people who are not distressed by their symptoms and who have not experienced a decline in their functioning are less likely to become seriously unwell in the near future.

A careful prospective study of a young, help-seeking population identified a subset of young people who appear to be at incipient risk of frank psychosis. The specific criteria defining this ultrahigh-risk (UHR) group fall into three groups (McGorry et al. 2008): having experienced attenuated psychotic symptoms during the previous year (Yung and McGorry 1996); having brief episodes of frank psychotic symptoms that resolve spontaneously over the previous year; and (Fusar-Poli et al. 2013) having a schizotypal personality disorder, or a first-degree relative with a psychotic disorder, and recently experiencing a significant decline in functioning (Table 1). Up to 40 % of the young people who met these UHR criteria made a transition to psychosis within the following year, a rate several hundredfold greater than the expected incidence rate for first-episode psychosis in the general population (Yung and McGorry 1996). These criteria have since been validated in a series of international studies. However, it should be borne in mind that while they do identify a group of young people who are at incipient risk of psychosis, they are by no means a diagnosis per se; the majority of young people who fulfil the UHR criteria do not develop a full--threshold psychotic disorder. While a significant proportion (up to 50 %) experiences remission of their subthreshold psychotic symptoms within a year of seeking help, most of these young people continue to report clinically relevant symptoms, primarily a blend of anxiety and depression, as well as difficulties in social and occupational functioning, emphasizing their need for ongoing clinical care (Fusar-Poli et al. 2013).

This elaboration of operationalized criteria that significantly reduce the risk of inappropriate treatment has not only proven to be clinically useful but has also catalyzed renewed efforts into developing

Prepsychotic States and Prodromal Symptoms, Table 1 *Broad criteria for the ultrahigh-risk group*: must be aged between 14 and 29 years, have been referred to a specialized service for help, and meet the criteria for one or more of the following three groups

Group 1: *Attenuated positive psychotic symptoms*	Presence of at least one of the following symptoms: ideas of reference, odd beliefs or magical thinking, perceptual disturbance, paranoid ideation, odd thinking and speech, odd behavior and appearance
	Frequency of symptoms: at least several times a week
	Recency of symptoms: present within the last year
	Duration of symptoms: present for at least 1 week and no longer than 5 years
Group 2: *Brief limited intermittent psychotic symptoms*	Transient psychotic symptoms. Presence of at least one of the following: ideas of reference, magical thinking, perceptual disturbance, paranoid ideation, odd thinking or speech
	Duration of episode: less than 1 week
	Frequency of symptoms: at least several times per week
	Symptoms resolve spontaneously
	Recency of symptoms: must have occurred within the last year
Group 3: *Trait and state risk factors*	Schizotypal personality disorder in the identified individual or a first-degree relative with a psychotic disorder
	Significant decline in mental state or functioning, maintained for at least 1 month and not longer than 5 years
	This decline in functioning must have occurred within the past year

effective early intervention strategies designed to prevent, or at least delay, the onset of psychosis and other serious mental illness. Clearly, the young people who fulfil these criteria have demonstrable clinical needs, and thus effective treatment is called for, not only on human but also on medical and ethical grounds. Current early intervention strategies range from the psychologically based, including psychoeducation, supportive psychotherapy, cognitive behavioral therapy (CBT), and family work, to the biologically based, including symptomatic treatment for depression, anxiety, and any subthreshold psychotic symptoms, through to experimental neuroprotective approaches. The global aim of treatment in the prodromal phase is to provide comprehensive clinical care designed to reduce presenting symptoms and if possible, to prevent these symptoms from worsening and developing into an acute psychosis (Yung et al. 2004).

Currently Accepted Strategies for Treatment of Prodromal Symptoms

Studies have shown that around 25 % of these at-risk young people have a concomitant diagnosis of depression and that over 60 % of UHR patients will experience a depressive disorder during their lifetime. Thus, treatment with cognitive behavior therapy and/or antidepressants, most commonly the SSRIs, may be indicated. These therapies are generally well accepted and well tolerated by this patient group and lead to significant clinical improvement. Some preliminary evidence suggests that antidepressants may have a protective effect (see below), while effective treatment of depression may help to limit the development of negative symptoms and social withdrawal. Anxiety is also extremely common in this patient group, with around 25 % of these young people having a current diagnosis of an anxiety disorder, while around 30 % will experience an anxiety disorder in their lifetime. Benzodiazepines may be prescribed to relieve short-term anxiety and sleep disturbance and to reduce agitation. Because anxiety tends to increase as positive symptoms develop, effective treatment of anxiety may help to relieve the stress associated with any subthreshold psychotic symptoms that may be present and allow the patient to better cope with social and vocational difficulties as they arise, limiting the functional decline that occurs during the prodromal period (Yung et al. 2004).

Subthreshold psychotic symptoms are inevitably present in UHR patients. However, in general, antipsychotic treatment should be avoided if at all possible. Indications for antipsychotic treatment include rapid deterioration, hostility, and aggression that poses a risk to the patient or others, severe suicidality, or depression that does not respond to other treatments. Antipsychotics may also be trialled for patients who have not responded to psychosocial interventions and who are still unwell and functioning poorly. If medication is warranted, the atypical antipsychotics should be used on a trial basis for a limited time only and at the lowest dose possible, in order to minimize the risk of extrapyramidal symptoms and metabolic side effects (see below). If there is clinical benefit and resolution of symptoms after 6 weeks, the medication may be continued for a further 6 months to 2 years, with the patient's consent (Yung et al. 2004). The atypical antipsychotics are the agents of first choice for young patients since they have been shown to be associated with fewer extrapyramidal side effects than the potent typical antipsychotics and are generally better tolerated. Apart from movement disorders, the major side effects reported for the atypical agents include significant weight gain and an increased risk of diabetes and metabolic disturbance. Other common side effects include sedation, fatigue, and decreased libido and less commonly, prolactinemia and cardiac arrhythmias.

Experimental Strategies Designed to Treat Prodromal Symptoms and Prevent the Onset of Psychosis

Seven pharmacological intervention studies have been carried out in UHR groups to date and are summarized briefly below and in Table 2. The first of these was a randomized controlled trial conducted in Melbourne, Australia, comparing combined CBT and low-dose (1–2 mg/day) atypical antipsychotic medication (risperidone) (n = 31) with usual case management (n = 28) (McGorry et al. 2002). The rate of onset of psychosis was significantly lower in the treatment group than in the control group at the end of the 6-month treatment phase (9.7 % vs. 35 %, $p = 0.026$). However, this finding was no longer statistically significant by the end of the 6-month follow-up period, due to participants who were not fully adherent to risperidone during the treatment phase transitioning to full-threshold psychosis during the follow-up period. Those who were fully adherent all remained nonpsychotic over the follow-up period, even though they had ceased drug treatment, demonstrating that the onset of psychosis can at least be delayed by specific intervention. Since medication and CBT were combined in this trial, the active component of the treatment regime could not be identified. Longer-term follow-up of the study cohort over 3–4 years failed to show any persisting benefit in the experimental group, suggesting that a longer treatment time may be necessary (Phillips et al. 2007).

A second randomized double-blind placebo-controlled trial was conducted by researchers from Yale University, USA. Low-dose olanzapine (5–15 mg/day) (n = 31) was compared to placebo (n = 29) for 12 months, followed by a 12-month monitoring period (McGlashan et al. 2006). While there was a statistically and clinically significant improvement in the levels of psychotic symptoms in the olanzapine group compared to the placebo group, there was only a trend towards a reduced rate of transition to psychosis in the olanzapine-treated group that did not reach statistical significance, and the adverse effects associated with olanzapine treatment, primarily weight gain, were of some concern in this trial.

Trials of aripiprazole (Woods et al. 2007) and amisulpride (Ruhrmann et al. 2007) have also been conducted in UHR cohorts. In the aripiprazole trial, 15 UHR patients were treated with a flexible dose regime of 5–30 mg/day for 8 weeks (Woods et al. 2007). Improvements on clinical measures were evident by the first week, adverse events were minimal, and no participants transitioned to psychosis. Similar findings were seen in the amisulpride trial, a randomized controlled trial involving a cohort of 124 UHR patients considered to be in the late initial prodromal stage who received either amisulpride (50–800 mg/day) together with a needs-focused

Prepsychotic States and Prodromal Symptoms, Table 2 Clinical trials of interventions for the relief of prodromal symptoms in young people at ultrahigh risk of psychosis

Authors	Intervention	Timeframe of study	Outcome (intervention vs. control groups)
Medication plus psychosocial intervention			
(McGorry et al. 2002; Phillips et al. 2007)	RCT, N = 59 CBT + 1–2 mg/day risperidone vs. needs-based psychosocial support	6-month treatment phase, + 6-month follow-up phase Medium-term follow-up at 3–4 years post-baseline	Transition to psychosis: 9.7 % vs. 35 %; p = 0.026
(McGlashan et al. 2006)	RCT, N = 60 5–15 mg/day olanzapine vs. placebo	12-month treatment phase + 12-month follow-up phase	Transition to psychosis: 16.1 % vs. 37.9 %; p = 0.09
(Ruhrmann et al. 2007)	RCT, N = 124 50–800 mg/day amisulpride + needs-focused intervention vs. needs-focused intervention	12 weeks	Improvement in both groups, with a significantly greater improvement seen in the aripiprazole group for positive ($F(1.98) = 7.83$, $p < 0.01$), negative ($F(1.98) = 4.85$, $p < 0.05$), and general symptoms ($F(1.98) = 4.63$, $p < 0.05$)
(Woods et al. 2007)	Open-label pilot study, N = 15 5–30 mg/day aripiprazole	8 weeks	Significant reductions in positive, negative, and general symptoms ($F = 9.2$, $p < 0.001$), with 11/15 participants responding to treatment
(Cornblatt et al. 2007)	Naturalistic study, N = 48 Antidepressants or atypical antipsychotics	2 years	Transition to psychosis: 43 % of those in the antipsychotic group vs. 0 in the antidepressant group
(McGorry et al. 2013)	RCT, N = 115 CBT + ≤2 mg/day risperidone vs. CBT + placebo vs. supportive therapy + placebo	1 year	Transition to psychosis: CBT + risperidone, 10.7 %; CBT + placebo, 9.6 %; supportive therapy + placebo, 21.8 %. No statistically significant difference in transition rates, though all groups showed significant clinical improvement
Dietary supplementation			
(Amminger et al. 2010)	RCT, N = 81 1.2 g/day omega-3 fatty acids vs. placebo	12-week treatment phase + 40-week follow-up phase	Rate of conversion to psychosis: 4.9 % vs. 27.5 %; p = 0.007

intervention or the needs-focused intervention alone for 12 weeks (Ruhrmann et al. 2007). At the end of the study period, the amisulpride group showed significantly greater improvements in positive ($F(1.98) = 7.83$, $p < 0.01$), negative ($F(1.98) = 4.85$, $p < 0.05$), and general symptoms ($F(1.98) = 4.63$, $p < 0.05$), as well as in overall functioning ($F(1.98) = 5.70$, $p < 0.05$) than the control group. Adverse events were minor, with prolactinemia and a small weight gain being the most important.

Antidepressants have also been suggested to reduce the risk of psychosis in UHR patients. A naturalistic study of 48 young people with prodromal symptoms who were treated with either antidepressants (primarily SSRIs) or atypical antipsychotics has been carried out by Cornblatt and colleagues (Cornblatt et al. 2007), who found that 12 of the 28 patients (43 %) who were prescribed antipsychotics progressed to full-threshold psychosis in the following 2 years, while none of the 20 patients prescribed antidepressants developed psychosis. Similar results were reported by Fusar-Poli et al. (2007) from a file audit study. However, these results need to be interpreted with caution due to the uncontrolled nature of these studies. Firstly, there may have been differences in baseline symptoms, functioning or other variables between the treatment groups, and secondly, nonadherence to treatment was far more prominent in those patients who had been prescribed antipsychotics than in the participants prescribed antidepressants. In this regard, the initial Melbourne trial found that antidepressants, again prescribed according to clinical need, had no influence on the transition rate (McGorry et al. 2002).

The most recent study published involved a head-to-head comparison of low-dose risperidone plus cognitive behavioral therapy with cognitive therapy plus placebo, or supportive therapy plus placebo, in a cohort of 115 young people who met UHR criteria. While lower-than-expected transition rates reduced the power of the study, meaning that no statistically significant differences in the rate of transition to psychosis could be found, all three randomized groups showed significant clinical and functional improvements during the course of the study, an important outcome in itself (McGorry et al. 2013).

Finally, a recent study of omega-3 essential fatty acids (EFA) for indicated prevention of psychosis in a group of 81 UHR young people has proven particularly promising (Amminger et al. 2010). Participants were randomized to either the placebo group or the EFA group and underwent a 12-week trial of treatment with 1.2 g/day EFA (a balanced mix of 700 mg EPA + 480 mg DHA) and a 12-month follow-up period. All participants were offered the same psychosocial support package, and antidepressants and benzodiazepines were allowed for the treatment of depression and anxiety if necessary. Significantly, at the end of the 12-month follow-up period, 2 of the 41 (4.9 %) participants in the EFA group had made the transition to frank psychosis, compared to 11 of the 40 (27.5 %) participants in the placebo group. Significant reductions in the levels of positive, negative, and general symptoms were seen in the EFA group, along with an increase in overall functioning, and most interestingly, these benefits were continued after the cessation of the 12-week intervention phase. Furthermore, the number needed to treat (NNT) of 4 calculated in this study compared favorably to the NNTs of 4 and 4.5 that were reported in the trials of the antipsychotics described above (McGorry et al. 2002; McGlashan et al. 2006). No adverse effects were reported in either group, while the high level of compliance (over 80 %) and low withdrawal rate (6 % overall) indicate that this intervention was very well accepted by the participants. This extremely promising study is the first trial of a natural substance for the indicated prevention of psychosis and the alleviation of prodromal symptoms. The evident clinical benefits and the absence of side effects suggest that the omega-3 fatty acids may indeed be a viable alternative to antipsychotic medication, offering a similar degree of overall therapeutic gain without the potentially serious and often distressing side effects associated with these agents.

Conclusions

Current clinical experience indicates that young people experiencing prodromal symptoms, and in particular those at ultrahigh risk of developing a psychotic disorder, should be treated with the aim of ameliorating their symptoms and preventing further deterioration in the course of their illness. The results of the few experimental trials currently available indicate that these young people may benefit from various therapeutic intervention strategies, including cognitively oriented psychotherapy and/or specific indicated prevention with low-dose atypical antipsychotic medication or neuroprotective agents such as the omega-3 fatty acids, and that treatment should be continued for longer than 6 months, given that these patients remain symptomatic and vulnerable to the onset of psychosis well after their initial detection and treatment. However, the evidence for the use of antipsychotics in prodromal patients is still preliminary and their use cannot be endorsed in routine clinical practice. Given that current evidence suggests that the psychological therapies are equally effective as the antipsychotic agents in this patient group (Preti and Cella 2010; Van der Gaag et al. 2013), it seems prudent to recommend that on the balance of risk-benefit considerations, antipsychotic treatment should not be offered as a first-line therapy for these young people, but instead be reserved for those with more severe or persistent symptoms that do not respond to more benign treatments.

At this stage, indicated prevention for the psychotic disorders remains a major goal for researchers in this area. Apart from relieving the increasing distress and disability associated with prodromal symptoms, early intervention provides numerous other advantages to these vulnerable young people. Effective treatment allows them to remain in education, training, or employment with minimal disruption due to illness and thus they maintain better levels of social functioning and a higher quality of life than might otherwise be expected. Early engagement in therapy means that even those who do become psychotic can be treated promptly without needing emergency or inpatient care, thereby, avoiding the distress and trauma associated with psychiatric hospitalization. Finally, early intervention and effective treatment allow the best chance of a full social and functional recovery, the best possible outcome in both economic and human terms.

Cross-References

▶ Antipsychotic Drugs
▶ Antipsychotic Medication: Future Prospects
▶ Benzodiazepines
▶ Neuroprotection
▶ Schizophrenia

References

Amminger GP, Schafer MR, Papageorgiou K, Klier CM, Cotton SM, Harrigan SM et al (2010) Long-chain omega-3 fatty acids for indicated prevention of psychotic disorders: a randomized, placebo-controlled trial. Arch Gen Psychiatry 67:146–154

Cornblatt BA, Lencz T, Smith CW, Olsen R, Auther AM, Nakayama E et al (2007) Can antidepressants be used to treat the schizophrenia prodrome? Results of a prospective, naturalistic treatment study of adolescents. J Clin Psychiatry 68:546–557

Fusar-Poli P, Valmaggia L, McGuire P (2007) Can antidepressants prevent psychosis? Lancet 370:1746–1748

Fusar-Poli P, Borgwardt S, Bechdolf A, Addington J, Riecher-Rossler A, Schultze-Lutter F et al (2013) The psychosis high-risk state: a comprehensive state-of-the-art review. JAMA Psychiatry 70:107–120

McGlashan TH, Zipursky RB, Perkins D, Addington J, Miller T, Woods SW et al (2006) Randomized, double-blind trial of olanzapine versus placebo in patients prodromally symptomatic for psychosis. Am J Psychiatry 163:790–799

McGorry PD, Yung AR, Phillips LJ, Yuen HP, Francey S, Cosgrave EM et al (2002) Randomized controlled trial of interventions designed to reduce the risk of progression to first-episode psychosis in a clinical sample with subthreshold symptoms. Arch Gen Psychiatry 59:921–928

McGorry PD, Yung AR, Bechdolf A, Amminger P (2008) Back to the future: predicting and reshaping the course of psychotic disorder. Arch Gen Psychiatry 65:25–27

McGorry PD, Nelson B, Phillips LJ, Yuen HP, Francey SM, Thampi A et al (2013) Randomized controlled trial of interventions for young people at ultra-high risk of psychosis: twelve-month outcome. J Clin Psychiatry 74:349–356

Phillips LJ, McGorry PD, Yuen HP, Ward J, Donovan K, Kelly D et al (2007) Medium term follow-up of a randomized controlled trial of interventions for young people at ultra high risk of psychosis. Schizophr Res 96:25–33

Preti A, Cella M (2010) Randomized-controlled trials in people at ultra high risk of psychosis: a review of treatment effectiveness. Schizophr Res 123:30–36

Ruhrmann S, Bechdolf A, Kuhn KU, Wagner M, Schultze-Lutter F, Janssen B et al (2007) Acute effects of treatment for prodromal symptoms for people putatively in a late initial prodromal state of psychosis. Br J Psychiatry (Suppl) 51:s88–s95

Van der Gaag M, Smit F, Bechdolf A, French P, Linszen D, Yung AR et al (2013) Preventing a first episode of psychosis: meta-analysis of randomized controlled prevention trials of 12 month and longer-term follow-ups. Schizophr Res 149:56–62

Woods SW, Tully EM, Walsh BC, Hawkins KA, Callahan JL, Cohen SJ et al (2007) Aripiprazole in the treatment of the psychosis prodrome: an open-label pilot study. Br J Psychiatry Suppl 51:s96–s101

Yung AR, McGorry PD (1996) The initial prodrome in psychosis: descriptive and qualitative aspects. Aust N Z J Psychiatry 30:587–599

Yung AR, Phillips LJ, Mcgorry PD (2004) Treating schizophrenia in the prodromal phase. Taylor and Francis, London

Prepulse Inhibition

Mark A. Geyer
Department of Psychiatry, University of California San Diego, La Jolla, CA, USA

Synonyms

Sensorimotor gating; Sensory gating; Startle modulation

Definition

When mammals are exposed to a sudden stimulus, typically a loud acoustic noise, a startle response is elicited. Prepulse inhibition of the startle response, an operational measure of sensorimotor gating, is a cross-species measure of the normal decrement in startle when a barely detectable prestimulus immediately precedes (30–500 ms) a startling stimulus (see Fig. 1) (Graham 1975; Ison and Hoffman 1983). While the startling event is typically an acoustic or tactile (e.g., air-puff or shock) stimulus having a rapid onset, the prepulse or prestimulus can be in the same or different modality, including light, electrical shock, air-puff, sound, or brief gap in the background noise.

Impact of Psychoactive Drugs

Attentional and information processing dysfunctions have long been considered important in understanding schizophrenia and other psychiatric disorders. Prepulse inhibition is disrupted in certain neuropsychiatric disorders that are characterized by an inability to filter or "gate" sensory (and, theoretically, cognitive) information. Theoretically, impairments in basic information processing functions such as sensorimotor gating contribute to disordered thought and cognitive fragmentation observed in psychotic disorders such as schizophrenia (Braff and Geyer 1990). Prepulse inhibition is used commonly as an operational measure of sensorimotor gating in studies in rodents, infrahuman primates, and humans. Patients with schizophrenia exhibit reduced sensorimotor gating as indexed by prepulse inhibition when compared to healthy control subjects and several categories of nonpsychotic psychiatric disorders (Braff et al. 2001). Strikingly, similar prepulse inhibition abnormalities have also been observed in unmedicated and nonpsychotic schizotypal patients and asymptomatic first-degree relatives of schizophrenia patients, supporting a strong role for genetic influences on sensorimotor gating. The reproducibility of the finding in schizophrenia, the fact that abnormal prepulse inhibition parallels a putative central abnormality in the disease, and the fact that prepulse inhibition is a conserved phenomenon among vertebrates make prepulse inhibition a promising candidate endophenotype to use in genetic association studies and animal models of schizophrenia (Swerdlow et al. 2009). The identification of genetic contributions to startle and prepulse inhibition in humans can be readily confirmed and extended in parallel studies using genetically engineered mice or other relevant strains of rats or mice (Geyer et al. 2002).

Using startle plasticity measures such as prepulse inhibition is advantageous in neuroscientific research for a number of reasons. First, startle plasticity in rodents has proven face, predictive, and construct validity for startle plasticity in humans. Second, startle behavior remains relatively stable across repeated testing sessions in mice, rats, healthy humans, and clinically stable psychiatric patients. This stability enables one to use longitudinal designs to explore developmental and environmental perturbations on prepulse inhibition over time and across experience. Third, startle and prepulse inhibition involve fairly rapid tests that do not involve complex stimuli, increasing their ease of use and their reliability. Since startle relies on a simple reflex measure, its reliability and reproducibility are greater than more complex behavioral measures that are modulated by competing behaviors or motivations (e.g., approach/avoidance behavior) and increase the chances of translation of these effects to humans.

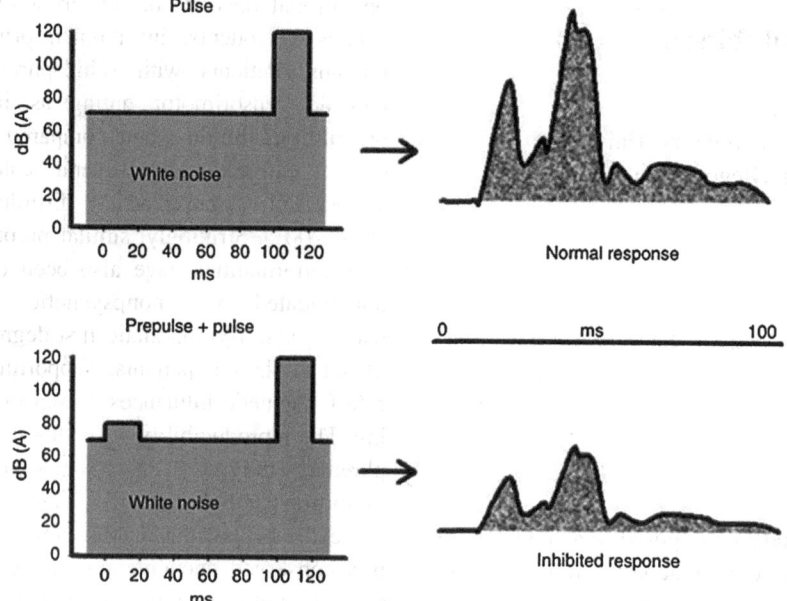

Prepulse Inhibition, Fig. 1 Diagrammatic representation of prepulse inhibition of startle. The *top panel* illustrates a pulse-alone trial in which startle is elicited by a 120 dB(A) noise burst above a 70 dB(A) background. The startling stimulus is presented for 20 msec in this example. The startle response is typically measured for 100 (rodents) or 250 (humans) msec after the onset of the pulse. The *lower panel* illustrates the prepulse-plus-pulse trial, in which startle inhibited by a weak (e.g., 80 dB(A)) prepulse given 30–1,000 msec (in this example, 100 msec onset-to-onset) before the same startle-eliciting noise used in the pulse-alone trial. In experiments examining tactile rather than acoustic startle, the pulse is an air-puff or mild electric shock to the neck (humans) or back (rodents). The *right side* of each panel illustrates the measured response, typically assessed using the electromyographic signal from orbicularis occuli muscle as a measure of the eyeblink response in humans or using an accelerometer-based signal as a measure of the whole-body flinch response in rodents. A typical test session includes presentations of several pulse-alone trials and several occasions of prepulse trials that may vary in the intensity of the prepulse stimulus or the interval between the prepulse and pulse onsets

Fourth, the neuroanatomical and neurochemical substrates mediating and modulating startle plasticity are well defined, allowing greater hypothesis generation and interpretability before and after obtaining results.

The neuroanatomical substrates that contribute to the modulation of prepulse inhibition in rats have been studied extensively, providing an excellent example of the regulation of behavior by integrated neuronal circuits (Swerdlow et al. 2001). The deficits in prepulse inhibition observed in psychiatric patient populations appear to reflect abnormal information processing and may result from pathology within forebrain cortico-striato-pallido-pontine circuitry that modulates this form of startle plasticity (Koch 1999; Swerdlow et al. 2001). Furthermore, a wide range of developmental and pharmacological manipulations have been found to alter prepulse inhibition in rats, leading to multiple rat models having utility in the identification of antipsychotic medications (Geyer et al. 2001). Prepulse inhibition has shown good predictive validity as a screen for antipsychotic drugs. In keeping with the dopamine hypotheses of psychotic disorders such as schizophrenia and mania, dopamine agonists such as apomorphine and amphetamine disrupt prepulse inhibition in rodents. These effects can be reversed by antipsychotics having selective antagonist effects at dopamine D2, but not dopamine D1 receptors. One important aspect of animal models of schizophrenia is their ability to distinguish between typical and atypical antipsychotic drugs. Prepulse inhibition deficits induced by apomorphine are reversed by both typical and atypical antipsychotics. Thus, although the ability of antipsychotics to restore prepulse inhibition in apomorphine-treated rats strongly correlates with their clinical potency, when used with the dopamine agonist apomorphine, this paradigm fails to make the important distinction between these two classes of antipsychotic drugs. In contrast, the prepulse inhibition disruptions produced by glutamate antagonists (e.g., phencyclidine, dizocilpine, and ketamine) differentiate between typical and atypical antipsychotics to some degree (Geyer et al. 2001). Specifically, typical antipsychotics such as haloperidol do not attenuate the prepulse inhibition-disruptive effects of glutamate antagonists in rats, while clozapine and some other atypical antipsychotics reduce the disruption in prepulse inhibition produced by these psychotomimetics in both rats and mice. Thus, prepulse inhibition already serves an important role in the identification of novel treatments for schizophrenia (Braff and Light 2004) and may ultimately contribute to our understanding of other psychiatric disorders such a bipolar disorder, panic disorder, and post-traumatic stress disorder.

Cross-References

▶ Habituation
▶ Schizophrenia
▶ Startle

References

Braff DL, Geyer MA (1990) Sensorimotor gating and schizophrenia: human and animal model studies. Arch Gen Psychiatry 47:181–188
Braff DL, Light GA (2004) Preattentional and attentional cognitive deficits as targets for treating schizophrenia. Psychopharmacology (Berl) 174:75–85
Braff DL, Geyer MA, Swerdlow NR (2001) Human studies of prepulse inhibition of startle: normal subjects, patient groups, and pharmacological studies. Psychopharmacology (Berl) 156:234–258
Geyer MA, Krebs-Thomson K, Braff DL, Swerdlow NR (2001) Pharmacological studies of prepulse inhibition models of sensorimotor gating deficits in schizophrenia: a decade in review. Psychopharmacology (Berl) 156:117–154
Geyer MA, McIlwain KL, Paylor R (2002) Mouse genetic models for prepulse inhibition: an early review. Mol Psychiatry 7:1039–1053
Graham FK (1975) The more or less startling effects of weak prestimuli. Psychophysiology 12:238–248
Ison JR, Hoffman HS (1983) Reflex modification in the domain of startle: II. The anomalous history of a robust and ubiquitous phenomenon. Psychol Bull 94:3–17
Koch M (1999) The neurobiology of startle. Prog Neurobiol 59:107–128
Swerdlow NR, Geyer MA, Braff DL (2001) Neural circuitry of prepulse inhibition of startle in the rat: current knowledge and future challenges. Psychopharmacology (Berl) 156:194–215
Swerdlow NR, Weber M, Qu Y, Light GA, Braff DL (2009) Realistic expectations of prepulse inhibition in translational models for schizophrenia research. Psychopharmacology (Berl) 199:331–388

Presynaptic Bouton

Definition

A widening of an axon that contains the components to form the presynaptic part of a synapse.

Primary Insomnia

Definition

Primary insomnia is a complaint of difficulty in initializing or maintaining sleep or of non-restorative sleep that lasts for at least 1 month and causes clinically significant distress or impairment in social, occupational, or other important areas of functioning. The disturbance in sleep does not occur exclusively during the course of another sleep disorder or mental disorder and is not due to the direct physiological effects of a substance or a general medical condition.

Cross-References

▶ Insomnias
▶ Sleep

Primate Models of Cognition

Angela Roberts
Department of Physiology, Development and Neuroscience, University of Cambridge, Cambridge, UK

Definition

Cognition is an array of higher-order processes that occur between sensory processing and motor output and that are inferred from animals' behavior. It includes constructs such as attention, memory, and executive control, and the behavioral tests used to study such constructs need to provide specificity, selectivity, and reproducibility. All behavioral tests have in common the need for perception, action, and attention to varying degrees, and so it is important for any drug study that the test employed should differentiate effects caused by the modulation of one or the other of these processes as well as the particular cognition under investigation. This is especially true when studying the actions of drugs given peripherally, which can have widespread influences in all regions of the brain including regions primarily involved in sensory processing or motor output.

Principles and Role in Psychopharmacology

Behavioral tests of cognition have had a long tradition in the field of primate neuropsychology. Originally, testing took place in the Wisconsin General Test Apparatus (WGTA) with the experimenter sitting in front of the monkey, behind a one-way mirror. Given that vision is an extremely important sense for monkeys, tests tended to be designed around objects or spatial locations. This contrasted with that of human neuropsychology in which the tests so often involved language and pen and paper. Consequently, the extrapolation of results from primate studies into the clinic was fraught with difficulty, as the tests used to measure a particular cognitive process differed considerably between humans and monkeys. After a seminal paper by L Weiskrantz, in 1977, the recognition of the advantages of testing humans and monkeys on the same tests led to much closer integration of human and monkey neuropsychological studies. In addition, the important advantages of automated and computer-controlled testing devices over manually operated ones (e.g., WGTA) were recognized, eliminating experimenter-subject interactions and increasing the degree of experimental control and efficiency (Bartus and Dean 2009). However, the ease of presenting tests in the WGTA or other manually operated environments should not be underestimated, not least due to the comparative ease at which monkeys can learn a range of different cognitive tests when the response and the reward are spatially contiguous. The spatial (and thus also temporal) separation of a response and the associated reward recruits additional cognitive processes, probably dependent on the frontal lobes, which may not be the focus of interest, but may need to be taken into account when interpreting the results.

Currently, there is an array of cognitive tests designed to measure specific aspects of primate cognition that are available for the psychopharmacologist. In many cases, it is possible to test monkeys on a battery of such tests allowing the effects of drugs to be compared across a range of different cognitive functions within the same animal.

Attention

Attention can be selective or divided, sustained or not. If selective, the attention may be directed at a specific spatial location or a particular sensory cue, e.g., red circle. Alternatively, attention may transcend specific sensory cues and occur instead at the level of higher-order perceptual dimensions, e.g., color or shape, in which case, it is said that an animal has developed an attentional set (see "Cognitive Flexibility"). A variety of tests have been developed to study attentional abilities in monkeys and the majority of them are dependent on an intact frontal and parietal cortex. The serial reaction time task, first developed in humans and later used to study attention in rats, investigates the ability of monkeys to locate a briefly presented target in one of a number of spatial locations (Spinelli et al. 2004; Weed et al. 1999). It tests aspects of both divided and selective attention. In a version for monkeys (Cambridge Neuropsychological Automated Test Battery, CANTAB: Lafayette Instrument Company), five circles are presented on a touch-sensitive computer screen, and a small colored stimulus is briefly presented in one of those locations (Fig. 1a). To start each trial, the monkey must perform an orienting response to ensure readiness to perform the trial, i.e., press down a lever and, after a variable delay, respond to where they saw the colored stimulus. The demands on divided attention can be increased or decreased by altering the number of spatial locations in which the target stimulus may appear. In contrast, the level of selective attention can be modulated by varying the duration of the target presentation. Other manipulations involve altering the lengths of the intertrial interval and the length of time the animal waits at the start of a trial for the onset of the target stimulus, both of which increase overall difficulty and reduce successful performance. Such manipulations are especially useful if the cognitive-enhancing effects of a drug are under investigation. Another test of attention that assesses the monkey's ability to focus attention using advanced information is a cued reaction time test. Here, four outlines of circles are presented on a touch-sensitive computer screen, and similar to that described earlier, the monkey must depress a lever to start the trial and respond to the circle that turns white as quickly as possible. In a cued version, a cue light appears above the circle that will become the next target thereby improving the speed of reaction time to that target. By comparing reaction and movement times in the cued and uncued conditions, a specific measure of selective attention can be obtained (Decamp and Schneider 2004). A variation of this task, first developed by MI Posner and colleagues, tests the abilities of monkeys to shift visuospatial attention and includes trials in which the cue is invalid and the target appears on the side opposite to that of the cue. The difference in the speed of responding to a target at expected (valid) and unexpected (invalid) locations is taken as a measure of ability to shift attention.

Memory

Recognition Memory

The classic test of recognition memory or the judgment of the prior occurrence of an object involves monkeys being presented with a novel object and then, after a variable delay (e.g., 5 s to 24 h), being presented with two objects, the previously seen object and a novel object. In the delayed "match to sample" (DMS) version, monkeys have to select the previously seen object, while in the delayed "non-match to sample" (DNMS) version, monkeys have to choose the novel object. It was first described by M Mishkin and J Delacour in 1975. This test differs from the working memory tasks described in the following text in that each pair of objects is only seen once or at least only once within a session and therefore tests the ability of monkeys to recognize objects as familiar or not. The presence of delay-dependent deficits is taken to indicate a mnemonic impairment, while poor performance at very short delays might indicate, instead, a perceptual deficit. However, to avoid artifacts attributable to the provision of extensive experience at short delays, followed by testing with less familiar long delays, it is important to intermix different delay lengths. For psychopharmacological studies, it is often beneficial to titrate the duration of the delay interval of each monkey to obtain matched levels of performance accuracy. This helps to equate levels of difficulty across monkeys and to avoid ceiling effects in the highest performing monkeys (Buccafusco 2008). More recently, D(N)MS has been run with computer graphic stimuli presented on touch screen monitors (e.g., Fig. 1b) by a number of different research laboratories including those of D Gaffan and EA Murray. In the DMS version, the number of stimuli at the time of choice can be varied, along with the degree to which the different choices differ from one another perceptually.

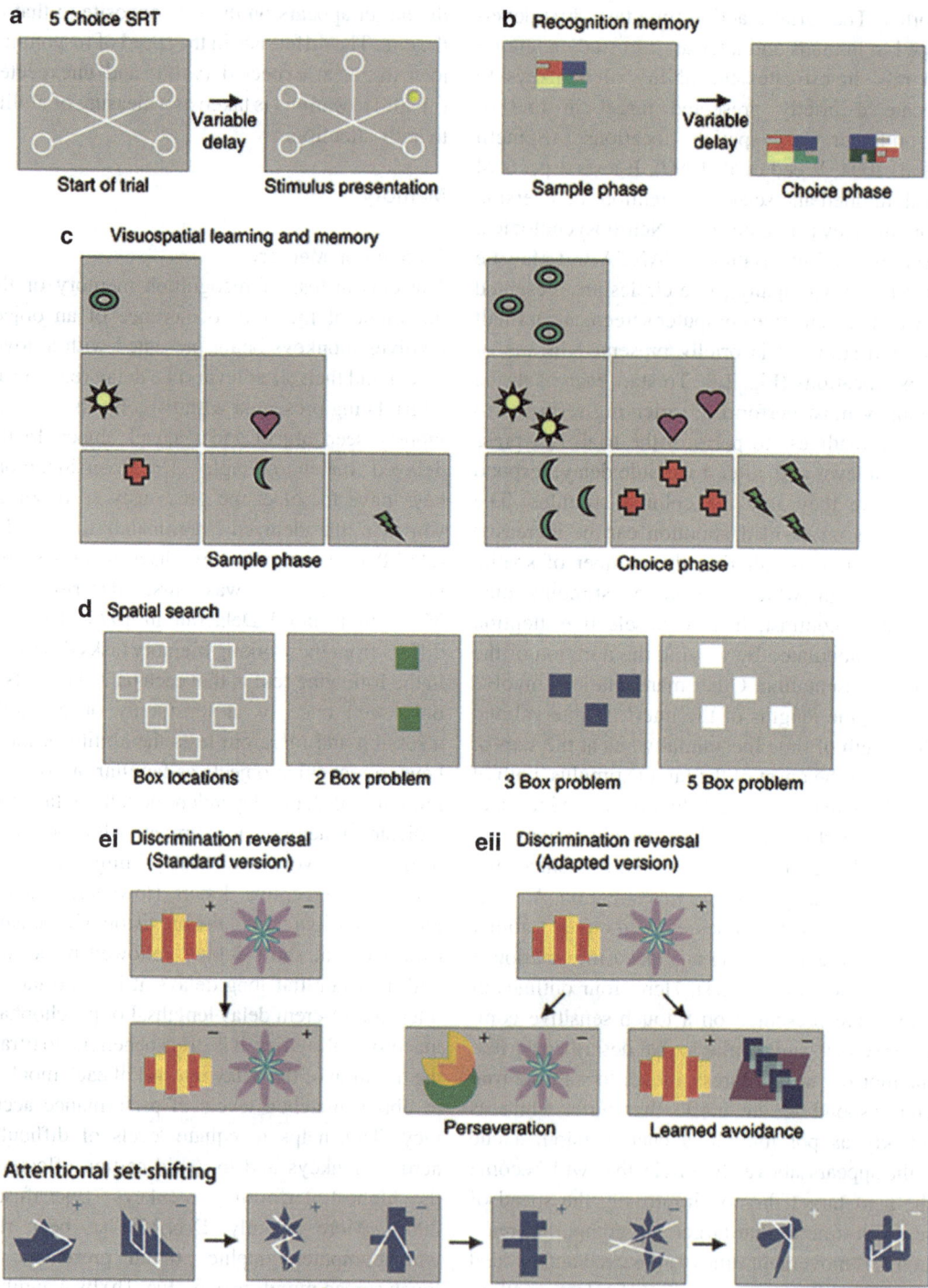

Primate Models of Cognition, Fig. 1 Examples of some of the cognitive tasks used successfully in studies of primate psychopharmacology. (**a**) The five-choice serial reaction time (*SRT*) test used in monkey CANTAB in which on each trial, after a variable delay, monkeys have to detect the brief presentation of a stimulus in one of five spatial locations. (**b**) An example of a recognition memory test depicting stimuli similar to those used in monkey CANTAB. See text for details. (**c**) A visuospatial learning and memory test also from

Drug manipulations during sample presentation, choice phase, and during the retention interval can target encoding, retrieval, and consolidation, respectively. Two distinct processes may underlie recognition memory, recollection, and familiarity judgment. Performance on this test is dependent upon an intact perirhinal cortex.

Visuospatial Memory

A variety of tests have been used to study visuospatial memory, which is the ability to integrate visual and spatial information and to recall that information subsequently. It is dependent on the hippocampus and related circuitry. One such test is a scene discrimination task in which, on each trial, an artificially constructed unique "scene" is presented consisting of randomly selected attributes including a colored background containing ellipse segments of different colors, sizes, and orientations and typographical characters (Gaffan and Parker 1996). In the foreground, one of two objects is rewarded. Monkeys learn these discriminations over a series of sessions in which each unique scene is presented once each session. Subsequently, it is possible to investigate the ability of monkeys to recall these discriminations, to relearn them as well as to acquire new discriminations, and, thus, as for recognition memory, to investigate encoding and retrieval. An alternative test, known as the visuospatial paired-associate learning (vsPAL, CANTAB) test, involves learning to associate different patterns in distinct spatial locations (Fig. 1c). On any one trial (sample phase), one or more patterns may be presented, one at a time, in distinct spatial locations, and the monkey must learn which pattern(s) is associated with which spatial location(s). The monkeys are subsequently tested on this knowledge (choice phase) by presenting the pattern in two or more spatial locations and requiring them to respond to the pattern in the correct spatial location (Taffe et al. 2004). Performance on the first attempt of the choice phase is a test of how well monkeys recall the information. However, subsequently, monkeys are presented the sample and choice phases an additional number of times to determine how rapidly they can learn this information. Because on the first attempt, monkeys have to remember multiple information across a short delay, this test loads quite heavily on working memory and is sensitive to frontal lobe dysfunction. This particular test is able to differentiate probands who will subsequently be diagnosed with Alzheimer dementia as opposed to other forms of mnemonic deficit related to normal aging and depression.

Executive Functions

Executive functions can be thought of as general-purpose control mechanisms that coordinate specific cognitive processes in order to optimize performance. They include holding information online and updating information in working memory, marshaling attentional resources, monitoring behavior and its outcomes, and inhibiting inappropriate strategies and responses.

Working Memory

Working memory is a short-term memory system that allows for the active maintenance and manipulation of information that is not present in the outside world. This concept of working memory involves temporary buffers of information and an attentionally constrained central executive. Tests of working memory measure a monkey's ability

---

Primate Models of Cognition, Fig. 1 (continued) monkey CANTAB. The stimuli used here are for illustrative purposes and are not the actual stimuli used in monkey CANTAB. Examples of one-, two-, or three-stimulus trial types are illustrated. (**d**) A spatial search task in which monkeys have to respond once and once only to each of a number of spatial locations on the screen (out of a possible eight locations) in order to get reward. An example of 2, 3, and 5 box problems are shown. The different colors help to differentiate the different box problems from one another. (**ei**) Example of a discrimination reversal task, as described in the text. Which stimulus is rewarded and which is not are shown by the "+" and "−" signs, respectively. An example of a test to separate out whether a reversal deficit is due to perseveration or learned avoidance is shown in (**eii**). (**f**) Examples of a series of discriminations involving an ID shift, probe test, and ED shift, as described in the text

to remember information for a short period of time when that information is no longer present in the environment. Whether the information is remembered or not is determined by how well their response is guided by this information after a variable delay. The information to be remembered is usually visual, i.e., remembering the spatial location of a stimulus or the form/color of a stimulus, but other sensory modalities can be used. It involves presenting a stimulus to the monkey for a brief period of time (seconds) and then after a variable delay requiring the subject to use that information to guide their subsequent responding. One of the most well-known versions of this test is the spatial working memory test, whereby a monkey either watches food reward being hidden in one of two food wells to the left and right of the animal or alternatively fixates a central cross while a spot of light is briefly flashed (e.g., 100 ms) in one of eight spatial locations in the surround. In both cases, the animal must remember the location of the stimulus for a brief delay, e.g., 1–30 s, and then make a response (arm reach/saccade) to the remembered location at the end of the delay period. In object working memory tests, an object is presented centrally in a WGTA or a visual pattern is presented on a computer screen (sample phase) for a short period of time during which the monkey may have to respond to the stimulus to demonstrate that the stimulus has been seen (Weed et al. 1999). Following a variable delay the same visual stimulus, along with another one is presented to the left and right of the center, and the animal has to choose the object/pattern that matches the sample object/pattern. The smaller the stimulus set used and, thus, how frequently the same stimulus is seen across trials increases the level of interference between trials and increases the load on working memory. By comparing responding following a delay with that at zero delay, it is possible to look at the effects of a drug on those processes specifically involved in short-term/working memory as compared to other more general perceptual and rule learning processes. In the spatial versions of the test, it is important to ensure that the monkey does not use a mediating response to bridge the delay between the stimulus disappearing and the response being made and thus reducing the load on working memory. The addition of distraction during the delay period allows for attentional mechanisms and working memory processes to be assessed within the same session. The distractor can take the form of additional, related stimuli presented during the delay that are irrelevant to the task, or alternatively, a burst of loud noise may be presented.

Monitoring Behavior

Self-ordered search tasks. Self-ordered search tasks were originally used in studies of human frontal lobe and were later adapted for studies in monkeys by M Petrides. In these original versions, monkeys were presented with variable numbers of objects (usually three), and across a series of three trials, they were presented with the same three objects. Food could be retrieved each trial as long as they selected a different object on each occasion. Thus, the monkey must monitor within working memory its earlier choices in order to avoid returning to objects that have already been selected. Spatial versions require animals to search through a series of spatial locations in order to find food reward. In those versions of self-ordered tasks in which food is obtained by selecting an object or spatial location for the first time, but not on repeated selections, the task is very similar to a foraging test. Alternative, computerized versions in which stimuli are presented on a touch-sensitive computer screen (Fig. 1d) tend not to reward monkeys until they have responded once and once only to a series of spatial locations (Collins et al. 1998). In the latter, monkeys are having to perform sequences of actions in order to gain reward, which may depend upon distinct prefrontal circuitry to those tasks in which the stimuli themselves are associated with reward. Besides monitoring of actions, the spatial search tasks can also be used to assess simple planning ability, especially if the number of boxes/locations is increased, as implementation of a strategy, such as following a clockwise or anticlockwise strategy, reduces demands upon working memory.

Cognitive Flexibility

Cognitive flexibility is the ability of animals to adapt their responding to changes in the environment. In the tasks described in the following text, previously rewarded responses or strategies have to be inhibited in favor of the development of new responses and strategies. Different aspects of cognitive flexibility are associated with different regions of prefrontal cortex, with many neuropsychiatric disorders being associated with cognitive inflexibility.

Discrimination reversal tasks. These typically involve presenting two stimuli to a monkey, usually two visual objects, and the animal learns, through trial and error, that a response to one of the objects leads to food reward while a response to the other does not. Having learned this discrimination to a particular level of performance, usually around 90 % correct over a series of trials, the reward contingencies reverse such that the previously rewarded stimulus is no longer associated with reward but the previously unrewarded stimulus is now associated with reward. How rapidly an animal learns to reverse their responding to match the change in reward contingencies is a measure of how flexible their behavior is. The stimuli can be from any sensory dimension, smell, audition, somatosensation, and vision. Visual discriminations are used most commonly in primate studies as visual cues are particularly salient for primates. The stimuli are either spatial in nature, i.e., left is rewarded, but right is not, or they involve visual features such as shapes or color. A typical discrimination reversal task is shown in Fig. 1ei where an animal may receive a series of reward contingency reversals.

A selective deficit in cognitive flexibility is shown by intact performance on the original discrimination, ruling out perceptual deficits, and an impairment on the subsequent reversal or series of reversals, a deficit seen following damage to the orbitofrontal cortex and ventromedial striatum. A more fine-grained analysis is then required to determine the nature of the underlying reversal deficit. Some useful information can be gleaned from a careful analysis of the errors that are made while performing the reversal. The pattern of errors on a reversal task is very distinctive.

First, monkeys tend to respond to the previously rewarded stimulus, almost exclusively, known as the perseverative stage. Then, they respond randomly to both stimuli (chance stage) and finally begin to respond more to the previously unrewarded, but now rewarded stimulus (learning stage). By analyzing the numbers and proportions of errors made at these three stages, some insight can be gained as to whether animals have problems (1) disengaging their attention and inhibiting their responding to the previously rewarded stimulus (perseveration) or alternatively (2) learning to respond to a stimulus that they had previously learned was not associated with reward (learned avoidance). If it is the former, animals would make more errors in the perseverative stage, but if it is the latter, then they may make more errors in the chance and/or learning stages. It has been highlighted that in the classic discrimination reversal task, there are only two stimuli to choose from, and thus, the only error is a response to the previously rewarded stimulus. A better measure of whether the deficit is truly perseverative in nature may be gained by requiring the animal to perform a three-stimulus visual discrimination reversal task as highlighted by JD Jentsch and JR Taylor. In which case, an error would include not only a response to the previously rewarded stimulus but also a response to the other, previously unrewarded stimulus. Thus, a more generalized impairment in reversal learning may manifest itself as errors made equally to both of the currently, unrewarded stimuli while a perseverative deficit would still be characterized by errors made primarily to the stimulus that had been previously rewarded. This version can rule out perseverative responding as an underlying cause of a reversal deficit.

However, if perseverative responding is seen, the underlying cause of the perseverative deficit is still unclear. It may still be a consequence of the subject avoiding the two, previously unrewarded stimuli, rather than due to a failure to inhibit responding to the previously rewarded stimulus. To differentiate these two possibilities, the following design can be used. At the reversal stage of the discrimination task, one of two different versions of the discrimination is given. One version includes the previously rewarded stimulus and a novel stimulus and the monkey has to choose the novel stimulus. The other version includes the previously unrewarded stimulus and a novel stimulus and the monkey has to select the previously unrewarded stimulus (Fig. 1eii). If the perseverative deficit seen in the original, discrimination reversal task is truly perseverative, then the monkey should be impaired on the former but not the latter. In contrast, if the deficit is due to the animal actively avoiding the previously unrewarded stimulus, then they should be impaired on the latter and not the former (Clarke et al. 2007).

Object retrieval test. Another commonly used test of cognitive flexibility that has proven useful in psychopharmacological studies is a test of object retrieval in which monkeys have to make a detour reach around the sides of a clear perspex box in order to retrieve the food reward inside. Performance on this test is dependent upon the orbitofrontal cortex and the caudate nucleus (and also large lesions of dorsolateral prefrontal cortex including Walker's areas 9, 46, and 8). It investigates not only the ability of monkeys to inhibit a prepotent response tendency to reach directly for the food reward but also their ability to switch their responding between the left and right sides of the box, as on some trials the opening is on the left and other trials, on the right (Jentsch et al. 1999). *A modified version of this test* involves training monkeys, first, on an opaque version of the box, in which they learn to retrieve reward from within the box by trial and error. There are doors on each side of the box but only one door is open on any one trial and the monkey has to push on each door in order to find the open door and retrieve the food reward. Monkeys eventually learn the strategy on each trial of pushing on each door once and once only to find the open door. Once monkeys have developed this strategy, the opaque box is replaced with a transparent box and for the first time, the food reward is visible within. However, the rule is the same as for the opaque box. Monkeys initially perform a direct reach for the food reward but soon learn to inhibit this response and adopt the same rule as before, namely, to push on each door once and once only, until they find the open door. In this version of the task, involving the transfer of a rule or strategy from one context to another, excitotoxic lesions of the ventrolateral prefrontal, and not the orbitofrontal cortex, disrupt performance (Wallis et al. 2001) and may recruit similar functions to those required in shifting attentional sets (see below). Together, these two versions of the object retrieval test tap into the functions of two distinct regions of prefrontal cortex.

Attentional set (rule)-shifting tasks. The psychological and neural mechanisms underlying flexible responding to changes in stimulus-reward contingencies are not the same as those required for the flexible use of rules to guide responding. One of the most commonly used tests to study rule switching in humans is the Wisconsin Card Sorting Test (WCST), which requires subjects to sort a pack of cards according to one particular perceptual dimension, e.g., shape, and then to switch to sorting them according to another dimension, e.g., number. A number of different versions of this test have been adapted for use in nonhuman primates, the different versions focusing on different aspects of the original task. In one such version (Roberts et al. 1988), monkeys are presented with a series of visual discriminations composed of bidimensional stimuli using abstract dimensions of "shape" and "line" (Fig. 1f). The monkey has to learn that one particular perceptual dimension is relevant to the task and that an exemplar from that dimension is associated with food reward, i.e., the blue boat, regardless of the white line superimposed over it. Animals then perform a series of such discriminations in which the same dimension remains relevant throughout, but each discrimination is composed of novel bidimensional stimuli in which one exemplar from the relevant dimension is associated with food each time (intradimensional shift, IDS). In the critical set-shifting test (or extradimensional shift, EDS), a discrimination is presented in which the previously relevant dimension is no longer relevant, and the subject has to learn, through trial and error, that an exemplar from the previously irrelevant dimension is now rewarded. This requires monkeys to shift their attentional set from one dimension to another and is similar to the shift of category in the WCST. Impairment at the EDS stage is dependent upon the ventro lateral PFC. The advantage of this test is that it separates out some of the component parts of the WCST, including the ability to develop an attentional set, to apply the rule across different discriminations, and then to switch from using one rule to using another. A distractor probe test, in which the exemplars from the irrelevant dimension of a well-learned discrimination are replaced for one session only with novel irrelevant exemplars, can be used to investigate how much an animal is distracted by the irrelevant dimension. This particular task has a major emphasis on learning. In contrast, other primate rule-switching tests are more akin to the WCST and require monkeys to learn to select, from a set of three stimuli, the stimulus that matches the sample stimulus according to a particular perceptual dimension (Mansouri et al. 2006). In these tasks, the monkeys receive extensive training on the matching to sample rules before they are able to perform the task successfully. Both these and the previously described discrimination tests require subjects to use feedback in the form of reward to guide their responding at the time of the shift. In contrast, another set of task-switching paradigms focus on the mechanisms underlying task switching per se, and in these cases, cues signal which particular rule is in operation at any one time (Stoet and Synder 2008).

Cross-References

▶ Behavioral Flexibility: Attentional Shifting, Rule Switching and Response Reversal
▶ Rodent Tests of Cognition
▶ Short-Term and Working Memory in Animals
▶ Spatial Learning in Animals

References

Bartus RT, Dean RL III (2009) Pharmaceutical treatment for cognitive deficits in Alzheimer's disease and other neurodegenerative conditions: exploring new territory using traditional tools and established maps. Psychopharmacology (Berl) 202:15–36

Buccafusco JJ (2008) Estimation of working memory in macaques for studying drugs for the treatment of cognitive disorders. J Alzheimers Dis 15:709–720

Clarke HF, Walker SC, Dalley JW, Robbins TW, Roberts AC (2007) Cognitive inflexibility after prefrontal serotonin depletion is behaviorally and neurochemically specific. Cereb Cortex 17:18–27

Collins P, Roberts AC, Dias R, Everitt BJ, Robbins TW (1998) Perseveration and strategy in a novel spatial self-ordered sequencing task for nonhuman primates: effects of excitotoxic lesions and dopamine depletions of the prefrontal cortex. J Cogn Neurosci 10:332–354

Decamp E, Schneider JS (2004) Attention and executive function deficits in chronic low-dose MPTP-treated non-human primates. Eur J Neurosci 20:1371–1378

Gaffan D, Parker A (1996) Interaction of perirhinal cortex with the fornix-fimbria: memory for objects and "object-in-place" memory. J Neurosci 16:5864–5869

Jentsch JD, Taylor JR, Redmond DE Jr, Elsworth JD, Youngren KD, Roth RH (1999) Dopamine D4 receptor antagonist reversal of subchronic phencyclidine-induced object retrieval/detour deficits in monkeys. Psychopharmacology (Berl) 142:78–84

Mansouri FA, Matsumoto K, Tanaka K (2006) Prefrontal cell activities related to monkeys' success and failure in adapting to rule changes in a Wisconsin Card Sorting Test analog. J Neurosci 26:2745–2756

Roberts AC, Robbins TW, Everitt BJ (1988) The effects of intradimensional and extradimensional shifts on visual discrimination learning in humans and non-human primates. Q J Exp Psychol B 40:321–341

Spinelli S, Pennanen L, Dettling AC, Feldon J, Higgins GA, Pryce CR (2004) Performance of the marmoset monkey on computerized tasks of attention and working memory. Brain Res Cogn Brain Res 19:123–137

Stoet G, Synder L (2008) Task switching in human and non-human primates: understanding rule encoding and control from behaviour to single neurons. In: Bunge SA, Wallis JD (eds) Neuroscience of rule guided behaviour. Oxford University Press, New York, pp 227–254

Taffe MA, Weed MR, Gutierrez T, Davis SA, Gold LH (2004) Modeling a task that is sensitive to dementia of the Alzheimer's type: individual differences in acquisition of a visuo-spatial paired-associate learning task in rhesus monkeys. Behav Brain Res 149:123–133

Wallis JD, Dias R, Robbins TW, Roberts AC (2001) Dissociable contributions of the orbitofrontal and lateral prefrontal cortex of the marmoset to performance on a detour reaching task. Eur J Neurosci 13:1797–1808

Weed MR, Taffe MA, Polis I, Roberts AC, Robbins TW, Koob GF, Bloom FE, Gold LH (1999) Performance norms for a rhesus monkey neuropsychological testing battery: acquisition and long-term performance. Brain Res Cogn Brain Res 8:185–201

Priming

Definition

The facilitation of performance following prior exposure to a stimulus.

Problem Gambling

Definition

Gambling behavior that persists despite negative consequences or a desire to stop gambling. Problem gambling has been used at times to include and at other times to exclude ▶ pathological gambling.

Cross-References

▶ Pathological Gambling

Procedural Memory

Definition

Long-term memory of skills and procedures.

Prochlorperazine

Definition

Prochlorperazine is an antipsychotic that acts as a dopamine D2 receptor antagonist. It is a piperazine-phenothiazine derivative. It has prominent antiemetic and anti-vertigo activity. It is used infrequently as an antipsychotic, but lower doses are used to treat severe nausea, vomiting, vertigo, and labyrinthine disorders.

Cross-References

▶ Antipsychotic Drugs
▶ First-Generation Antipsychotics

Procyclidine

Synonyms

1-cyclohexyl-1-phenyl-3-pyrrolidin-1-yl-propan-1-ol hydrochloride; Kemadrin

Definition

Procyclidine is an anticholinergic drug and functions as an antagonist at M_{1-4} muscarinic acetylcholine receptors. It is used for the treatment of extrapyramidal motor symptoms commonly associated with drug-induced side effects and ▶ Parkinson's disease. Common side effects associated with procyclidine are confusion, agitation, hallucinations, fever, tachycardia, dry mouth, nausea, constipation, and blurred vision. Procyclidine has been approved by the Food and Drug Administration for use in the United States.

Cross-References

▶ Antimuscarinic/Anticholinergic Agent

Progestins

Definition

Progestins are a class of hormones with progesterone being the primary progestin produced by the ovary. When progesterone is metabolized via the 5a-reductase enzyme, many of the metabolites have anxiolytic actions through their effects at the $GABA_A$ receptor.

Progressive-Ratio Schedule

Definition

A ▶ schedule of reinforcement used in operant conditioning. Laboratory animals or humans are trained to emit an ▶ operant response to obtain a ▶ reinforcer such as food or drug. Following each reinforcer delivery, the number of responses required to earn the next reinforcer is increased until the subject stops responding for a prolonged period of time. The ratio requirement for the delivery of successive ▶ reinforcers increases through a predetermined series. The progression is usually either arithmetic or exponential. The point at which responding ceases is called the "▶ breakpoint" and commonly serves as the main dependent variable in studies that utilize progressive-ratio schedules. The evaluation of the reinforcing efficacy of abused drugs is one application of these schedules.

Cross-References

▶ Drug Self-Administration

Promazine

Definition

Promazine is a first-generation antipsychotic that acts as a dopamine D2 receptor antagonist. It is an aliphatic ▶ phenothiazine with very low antipsychotic potency. Its primary use is for the treatment of agitation and restlessness in the elderly. Promazine is usually well tolerated, with a low incidence of extrapyramidal symptoms.

Cross-References

▶ Antipsychotic Drugs
▶ First-Generation Antipsychotics
▶ Schizophrenia

Promethazine

Definition

Promethazine is a first-generation antihistamine and antiemetic medication, acting as a histamine H_1 receptor antagonist. It can also have strong sedative effects and in some countries, it is prescribed for insomnia when benzodiazepines are contraindicated. It is a main ingredient of "purple drank" (a slang term for a recreational drug popular in the hip-hop community of the southern United States), being part of a prescription-strength cough syrup also containing ▶ codeine.

Cross-References

▶ Abuse Liability Evaluation
▶ Driving and Flying Under the Influence of Drugs
▶ Insomnias
▶ Sedative, Hypnotic, and Anxiolytic Dependence

Propranolol

Definition

Propranolol is an ▶ anxiolytic acting as a nonselective antagonist at β-adrenoceptors. Originally prescribed to treat hypertension and, subsequently, for the prevention and treatment of migraine, the use of propranolol has expanded due to its anxiolytic properties, so as to include treatment of specific forms of acute stress reactions such as performance anxiety. As an anxiolytic, propranolol is usually not administered chronically but only acutely, prior to specific ▶ panic- and anxiety-inducing events. Furthermore, this drug is often prescribed in treating alcohol withdrawal-induced tremors and tachycardia and also in the treatment of antipsychotic-induced movement disorders. In addition to this, recent preclinical studies suggest that it may also hold promise for the treatment of some drug dependencies. The side effects of propranolol are usually mild and transient and include light-headedness, depression, insomnia, nightmares, disorientation, nausea, decreased heart rate (bradycardia), and hypotension. Withdrawal symptoms after abrupt termination of chronic therapy are usually mild but may include chest pain, increased heart rate (tachycardia), headache, and trembling.

Prosaccade Task

Definition

A simple ▶ eye movement task in which participants make a saccade (typically from a central location) toward a sudden onset target. Important measurements include the saccade latency, peak velocity, amplitude, and duration.

Prospective Daily Ratings

Definition

Prospective daily ratings can consist of a daily diary in which symptoms experienced over the course of the day are recorded. However, most daily ratings require that the individual rate themselves with respect to the presence and/or severity of specific symptoms each day. That ratings are done each day over a period of time makes them prospective in nature. Completing daily diaries in this manner helps to decrease the bias that can occur with retrospective recall of symptoms. One of the most commonly used prospective daily ratings in PMDD research is the Daily Record of Severity of Problems.

Prospective Memory

Definition

Remembering to do something in the future.

Prostanoids

Definition

A class of endogenous chemicals that includes prostaglandins, which mediate inflammation; the thromboxanes, which mediate vasoconstriction; and the prostacyclins, which are involved in the resolution of inflammation. Cyclooxygenase catalyzes the synthesis of prostaniods.

Protein Microarray

Synonyms

Protein array; Protein-binding microarray

Definition

Protein microarrays constitute a multiplex approach to identify protein-protein interactions, to identify the substrates of protein kinases, to identify transcription factor protein activation, or to identify the targets of biologically active small molecules. The most common protein microarray is the antibody microarray, where many different antibodies are spotted onto a protein chip and are used as capture molecules to detect proteins from cell lysate solutions.

Cross-References

▶ Posttranslational Modification

Protein Synthesis and Memory

Ivan Izquierdo, Jociane C. Myskiw,
Fernando Benetti and Cristiane Furini
Memory Center, Brain Institute, Pontifical
Catholic University of Rio Grande do Sul,
Porto Alegre, RS, Brazil

Definition

The long-lasting storage of information acquired through learning processes is carried out in several brain structures involving a sequence of interconnected and highly regulated molecular steps leading to the initiation of protein synthesis. Protein synthesis is necessary for the consolidation, reconsolidation, and persistence of many forms of long-term memory (Flood et al. 1973; Igaz et al. 2002; Hernandez and Abel 2008; McGaugh 2013), including that of extinction learning (Vianna et al. 2001). The regions of the brain involved vary with the type of learning (see Izquierdo et al. 2006; McGaugh 2013).

Protein Synthesis as a Mechanism of Memory

A large amount of recent evidence indicates that memory processes rely on, or are otherwise very closely related to, the electrophysiological process known as long-term potentiation (Izquierdo et al. 2006), as had in fact been hypothesized many times over the past four decades. Protein synthesis is necessary for both the long-term potentiation and for the long-term consolidation of memory (Flood et al. 1973; Igaz et al. 2002; Abraham and Williams 2008; Hernandez and Abel 2008; Myskiw et al. 2008; Almaguer-Melian et al. 2012; Myskiw et al. 2013).

From time to time, learning theorists have speculated on the possibility that memory formation (consolidation) may not require protein synthesis; a large array of recent papers have dispelled such doubts (see Alberini 2008; Hernandez and Abel 2008). The administration of protein synthesis inhibitors at the translation level in the ribosomes, such as anisomycin, cycloheximide, puromycin, emetine, and others, into the brain areas involved or systemically at appropriate times (immediately posttraining or 3–4 h later) causes irreversible amnesia (inhibition of memory) or declarative memory. Anisomycin has been the most widely used protein synthesis inhibitor in the last 40 years. The amnesic effect of the protein synthesis inhibitors is shared by blockers of gene transcription, such as 5,6-dichloroimidazole-1-β-D-ribofuranoside (DRB) (Igaz et al. 2002; Myskiw et al. 2013). Some of the drugs used to inhibit protein synthesis cause brain lesions (cycloheximide and its derivatives), and by far the least toxic and therefore most widely used compound has been, for the last 40 years, anisomycin. By far, the brain structure best studied in terms of the involvement of protein synthesis both in long-term potentiation and in long-term memory consolidation is the hippocampus, particularly the CA1 area. Depending on the task under study, similar effects of protein synthesis inhibitors on either long-term potentiation or memory formation have been reported in a number of mammalian and nonmammalian brain structures, such as in septum, amygdala, and thalamus, as well as in the auditory, prefrontal, entorhinal, and insular cortices. It is likely that changes in the expression of proteins related to memory consolidation are specific to brain region and to behavioral tasks. In several tasks more than one brain area is involved in consolidation, and protein synthesis is usually required in most or in all of them. This appears to be the case in auditory fear conditioning and in conditioned taste aversion, in which the insular cortex and the central amygdaloid nucleus are involved (see Vianna et al. 2001 for references), and also in various types of fear conditioning, alimentary conditioning, and the extinction of at least fear-motivated behavior (Vianna et al. 2001; Izquierdo et al. 2006). Studies with anisomycin on reconsolidation show that, depending on the task, protein synthesis in the hippocampus or the amygdala is also involved. In conditioned taste aversion, protein synthesis in the central amygdala is important for consolidation but not reconsolidation of the task. The findings on the influence of anisomycin and other protein synthesis inhibitors on hippocampal long-term potentiation and long-term memory consolidation correlate very well, particularly when examined together with the many timed biochemical events and effects that have been measured in the subregion of the hippocampus known as *cornu ammonis 1* (CA1) in both physiological processes (Igaz et al. 2002, "Synaptic Plasticity").

Non-ribosomal protein synthesis relying on a molecular apparatus called the mammalian target of rapamycin (mTOR), in the dendrites, is also activated by and necessary for consolidation and reconsolidation of object recognition learning (Myskiw et al. 2008). This is also the case with the synaptic tagging of many forms of learning, including extinction learning (Myskiw et al. 2013).

Synaptic tagging is a process whereby long-term potentiation or learning takes place in a given set of hippocampal synapses that "tag" those synapses, so that they can attract and "capture" proteins synthesized in other sets of hippocampal synapses by other long-term potentiations or other learning (Almaguer-Melian et al. 2012).

The consolidation of many forms of learning relies on hippocampal long-term potentiation of hippocampal synapses (Izquierdo et al. 2006; Almaguer-Melian et al. 2012). The tagging-and-capture process underlies the persistence of some memories or long-term potentiations and their interaction (Almaguer-Melian et al. 2012; Myskiw et al. 2013).

The effects of protein synthesis inhibitors, particularly anisomycin, on memory are well established. The amount of inhibition needed in order to observe amnesia or lack of memory formation is around 85 % during 1–3 h (Flood et al. 1973). Several of the proteins synthesized in the hippocampus 3–4 h after behavioral training that play a role in memory formation have been identified. In fact, at least 30 different genes have been shown to be modulated by one-trial avoidance training; most of them were upregulated. New memories are associated with the need for gene expression and protein synthesis in areas of the brain in charge of each learning experience. The proteins synthesized by memory formation processes in different areas of the brain are in general constituents of synaptic membranes or other elements of synapses and incorporate to them changing their physiological function lastingly (Alberini 2008; Hernandez and Abel 2008).

Curiously, from time to time there appear proposals that protein synthesis may not be necessary for memory formation and that protein synthesis inhibitors cause amnesia for some other reasons. Those criticisms have all been recently dispelled by rather overwhelming evidence (see Alberini 2008; Hernandez and Abel 2008 for reviews). Short-term memory (short-term and working memory in animals or humans), lasting a few minutes or a few hours, are not dependent on gene expression or protein synthesis, but long-term memory certainly depends on both.

Summary

A large amount of evidence, mainly obtained through the use of drugs that block the ribosomal and non-ribosomal synthesis of proteins as well as gene transcription blockers, has demonstrated memory formation in different parts of the mammalian and nonmammalian brain.

Cross-References

▶ Behavioral Flexibility: Attentional Shifting, Rule Switching, and Response Reversal
▶ Classical (Pavlovian) Conditioning
▶ Conditioned Place Preference and Aversion
▶ Conditioned Taste Aversions
▶ Declarative and Nondeclarative Memory
▶ Gene Expression
▶ Gene Transcription

- ▶ Inhibition of Memory
- ▶ Instrumental Conditioning
- ▶ Long-Term Potentiation and Memory
- ▶ Learning and memory: molecular mechanisms
- ▶ Passive Avoidance
- ▶ Pavlovian Fear Conditioning
- ▶ Reconsolidation
- ▶ Short-Term and Working Memory in Animals
- ▶ Short-Term and Working Memory in Humans
- ▶ Synaptic Consolidation
- ▶ Synaptic Plasticity

References

Abraham WC, Williams JM (2008) LTP maintenance and its protein synthesis-dependence. Neurobiol Learn Mem 89:260–268

Alberini CM (2008) The role of protein synthesis during the labile phases of memory: revisiting the skepticism. Neurobiol Learn Mem 89:234–246

Almaguer-Melian W, Bergado-Rosado J, Pavón-Fuentes-N, Alberti-Amador E, Mercerón-Martínez D, Frey JU (2012) Novelty exposure overcomes foot shock-induced spatial-memory impairment by processes of synaptic-tagging in rats. Proc Natl Acad Sci U S A 109:953–958

Flood JF, Rosenzweig MR, Bennett ML, Orme A (1973) The influence of duration of protein synthesis inhibition on memory. Physiol Behav 10:555–562

Hernandez PJ, Abel T (2008) The role of protein synthesis in memory consolidation: progress amid decades of debate. Neurobiol Learn Mem 89:293–311

Igaz LM, Vianna MRM, Medina JH, Izquierdo I (2002) Two time periods of hippocampal RNA synthesis are required for memory consolidation of fear-motivated learning. J Neurosci 22:6781–6789

Izquierdo I, Bevilaqua LR, Rossato JI, Bonini JS, Medina JH, Cammarota M (2006) Different molecular cascades in different sites of the brain control memory consolidation. Trends Neurosci 29:496–505

McGaugh JL (2013) Making lasting memories: remembering the significant. Proc Natl Acad Sci U S A 110 Suppl 2:10402–10407.

Myskiw JC, Rossato J, Bevilaqua LR, Medina JH, Izquierdo I, Cammarota M (2008) On the participation of mTOR in recognition memory. Neurobiol Learn Mem 89:338–351

Myskiw JC, Benetti F, Izquierdo I (2013) Behavioral tagging of extinction learning. Proc Natl Acad Sci U S A 110:1071–1076

Vianna MRM, Szapiro G, McGaugh JL, Medina JH, Izquierdo I (2001) Retrieval of memory for fear-motivated training initiates extinction requiring protein synthesis in the rat hippocampus. Proc Natl Acad Sci U S A 98:12251–12254

Protein Tau

Definition

A normally soluble microtubule-binding protein that when hyperphosphorylated loses solubility altering the neuronal cytoskeletal through the formation abnormal intracellular filamentous inclusions, main component of ▶ neurofibrillary tangles which is one of the histopathological findings associated with ▶ Alzheimer's disease.

Proteomics

Per E. Andrén[1], Peter Verhaert[2] and Per Svenningsson[3]
[1]Biomolecular Imaging and Proteomics, National Laboratory for Mass Spectrometry Imaging, Department of Pharmaceutical Biosciences, Uppsala University, Uppsala, Sweden
[2]Department of Biotechnology, Netherlands Proteomics Centre, Delft University of Technology, Delft, Netherlands
[3]Department of Physiology and Pharmacology, Karolinska Institute, Hasselt University, Stockholm, Sweden

Synonyms

Studies on proteins expressed by the genetic material of an organism

Definition

The proteome is the genome complement of proteins, and proteomics is the study of proteomes in a cell at any given time. Neuroproteomics defines the entire protein complement of the nervous system. It represents a higher complexity than the transcriptome, and it displays a higher degree of dynamics due to posttranslational modifications (PTMs). Two main approaches of this field are profiling and functional proteomics. Profiling proteomics includes the description of the whole proteome of a cell, tissue, organ, or organism and comprises organelle mapping and differential measurement of expression levels between cells or conditions. Functional proteomics aims to characterize protein activity, by determining protein interactions and the presence of PTMs.

Current Concepts and State of Knowledge

The analysis of proteomes is significantly more challenging than that of genomes. Measuring gene expression at the protein level is potentially more informative than the corresponding measurement at the mRNA level. Though certain RNAs are known to function as effector molecules, proteins are the major actors and catalysts of biological function. Proteins contain several dimensions of information, which represent the actual rather than the potential functional state as indicated by mRNA analysis. These dimensions include the abundance, state of PTM, (sub)cellular localization, and association and interaction with each other. PTMs of a protein can determine its activity state, localization, turnover, and interactions with other proteins. More than 300 different types of PTMs are currently known. Changes in gene expression at the level of the message, mRNA expression, may not directly correlate with protein expression since mRNA is not the functional endpoint of gene expression. Recent investigations show that differences in protein concentrations are only 20–40 % assigned to variable mRNA levels, emphasizing the importance of posttranscriptional regulation. In general, protein concentrations depend on the translation rate and the degradation rate, i.e., the protein turnover. In addition to the complexity at the transcriptional level, proteome approaches have to deal with the considerable increase in isoforms due to multiple PTMs (Tyers and Mann 2003).

Proteomic Methodologies

Proteomic technologies (both expression profiling and functional approaches) have widely expanded in recent years. Because of protein diversity, a range of techniques have emerged, which depend on integration of biological, chemical, and analytical methods. The main proteomic technologies of today utilize mass spectrometry (MS) coupled with global protein separations (Aebersold and Mann 2003) and methods based on protein arrays (Phizicky et al. 2003). The global protein separation methods are conventionally divided into gel-based and gel-free, where the gel-free are all MS based. The protein arrays are not global in the sense that they typically depend on the availability of antibodies. The methodologies are complementary and are increasingly used in combination with each other. Even though their analytical windows overlap, each of them covers selected sets of proteins that are not identified by the other techniques (Choudhary and Grant 2004).

Two-dimensional gel electrophoresis. The global protein separation method two-dimensional gel electrophoresis (2-DE) makes possible the resolution of several thousand proteins in a single sample. 2-DE mainly produces data which allow the investigator to determine whether a particular protein shows an increase or decrease when comparing two different conditions. The limited dynamic range and poor reproducibility between gels have been of major concern with traditional 2-DE experiments (Westermeier et al. 2008).

In a typical gel-based proteomic experiment, the proteins in a sample are separated by 2-DE, stained, and each observed protein spot is quantified by its staining intensity. Selected spots are excised and analyzed by MS after "digestion" (see below). Pattern-matching algorithms as well as interpretation by skilled researchers are required to relate the 2-DE patterns to each other in order to detect characteristic patterns and differences among samples. The major limitation of 2-DE techniques is a relatively low throughput particularly in cases where many proteins have to be identified with subsequent MS analysis which is very time-consuming (Westermeier et al. 2008).

Detecting changes in protein expression is improved by the introduction of fluorescence difference gel electrophoresis (DIGE). 2-D-DIGE enables the pre-labeling and separation of up to

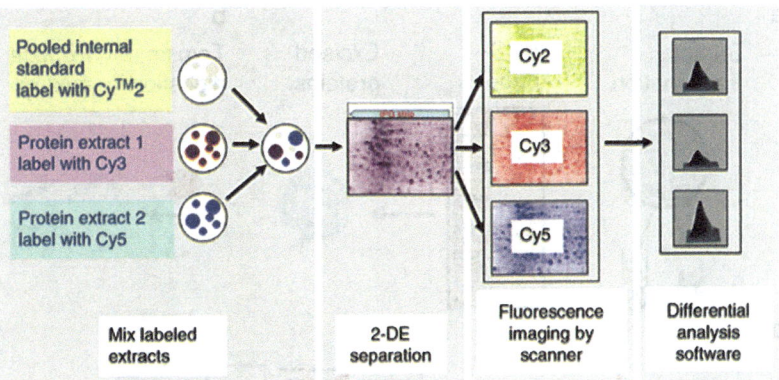

Proteomics, Fig. 1 Workflow for a standard 2-D difference gel electrophoresis (DIGE) experiment. Samples are labeled with molecular weight and charge-matched CyDye DIGE Fluors, minimal dyes. This permits multiplexing of up to two samples and a pooled internal standard on the same first- and second-dimension gel. Gels are scanned on an imager and processed using analysis software

three samples on a single 2-D gel providing quantitation of proteins. Up to three protein samples are labeled with size- and charge-matched CyDye DIGE fluorochromes and co-separated on the same 2-D electrophoresis gel. Gels using the 2-D-DIGE method usually contain three samples labeled with three distinct fluorescent dyes: Cy2, Cy3, and Cy5. The Cy2 dye is typically used to label an internal standard, which is a mix of all samples in the experiment, and the other two dyes are then employed to label two biological samples of interest. The strength of the internal standard is to help the mapping of spots/proteins between gels and thus make the different gels more comparable. The internal standard is also used in some methods for normalization within and between gels (Westermeier et al. 2008; Fig. 1).

Mass Spectrometry-Based Proteomics. The coupling of chromatographic separation methods with MS is commonly utilized for qualitative and quantitative characterization of highly complex protein mixtures. The advances in chemical-tagging and isotope-labeling techniques have improved the quantitative analysis of proteomes. High-performance liquid chromatography ("▶ HPLC") methods provide powerful tools of protein and peptide separation of protein mixtures. Primary advantages of liquid chromatographic (LC) separation are the flexibility of the methods and the possibility of linking LC directly to MS. Proteins and peptides can be separated through their physical properties, including affinity, charge (ion exchange), hydrophobicity (reversed phase), and size (size exclusion) (Aebersold and Mann 2003).

Presently two ionization techniques – electrospray ionization (ESI) and matrix-assisted laser desorption ionization (MALDI) – are playing a significant role in the field of MS-based proteome analysis. ESI ionizes the analytes out of a solution and is therefore readily coupled to chromatographic and electrophoretic separation tools. MALDI sublimates and ionizes the samples out of a dry, crystalline matrix via laser pulses. There are four basic types of mass analyzers currently used in proteomic research. These are the ion-trap, time-of-flight (TOF), quadrupole, and Fourier transform ion cyclotron (FT-MS) analyzers. These analyzers can be stand alone or, in some cases, put together in tandem to take advantage of the strengths of each (Aebersold and Mann 2003).

There are two complementary approaches for the MS analysis of proteins (Aebersold and Mann 2003). The bottom-up method is generally used for identifying proteins and determining details of their sequence and PTMs. Proteins of interest are digested with a sequence-specific enzyme such as trypsin, and the resulting tryptic peptides are analyzed by ESI or MALDI, which allow intact peptide molecular ions to be put into the

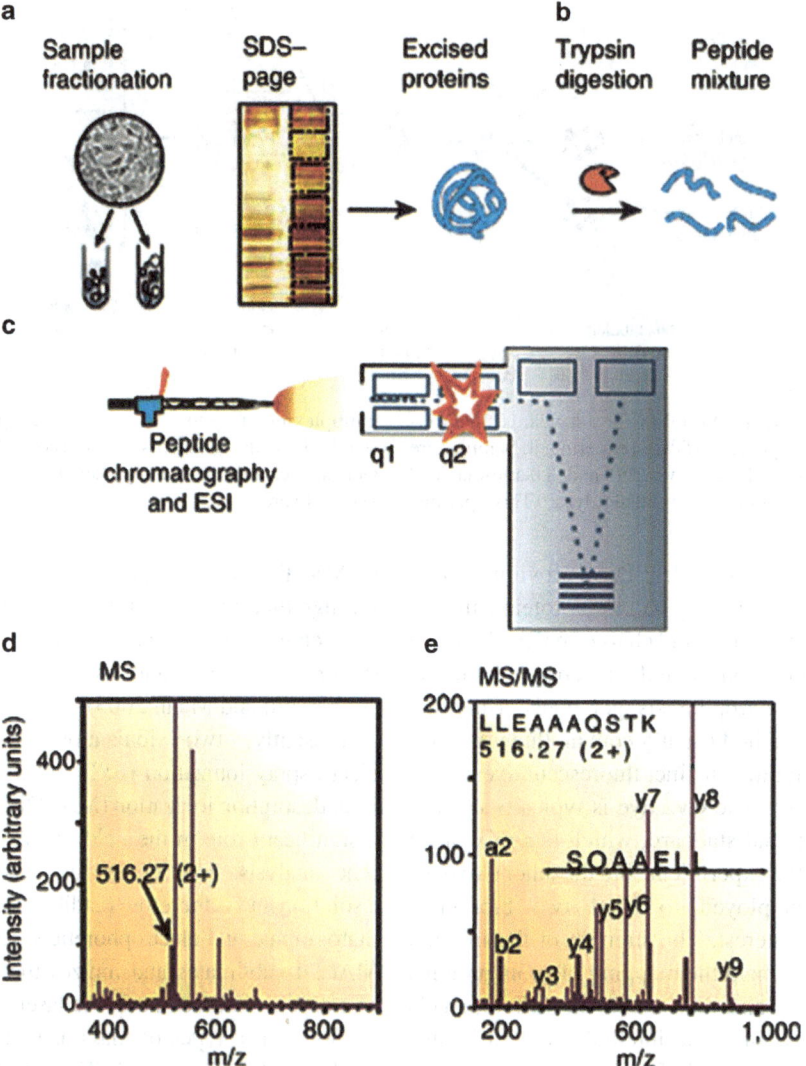

Proteomics, Fig. 2 Typical MS-based proteomic experiment. The general proteomic experiment consists of five stages: (**a**) the proteins to be analyzed are isolated from cell lysate or tissues by biochemical fractionation (such as SDS polyacrylamide gel electrophoresis (PAGE) or multidimensional LC) for reduction of the sample complexity or affinity selection (for enrichment of a sub-proteome); (**b**) the proteins are degraded enzymatically to peptides, usually by trypsin; (**c**) the peptides are separated by one or more steps of high-performance liquid chromatography (HPLC) in very fine capillaries and eluted into an electrospray ionization (ESI) ion source; (**d**) after evaporation, multiple protonated peptides enter the mass spectrometer and a mass spectrum of the peptides eluting at this time point is taken; and (**e**) the computer generates a prioritized list of these peptides for fragmentation and a series of tandem mass spectrometric (MS/MS) experiments follows. The outcome of the experiment is the identity of the peptides and therefore the proteins making up the purified protein population (Modified from Aebersold and Mann 2003)

gas phase. The MS analysis takes place in two steps. The masses of the tryptic peptides are determined; next, these peptide ions are fragmented in the gas phase. The masses of the peptide fragments yield information on their sequence and modifications. Prior to being introduced to the MS, the tryptic peptides usually are separated on reversed-phase LC directly coupled to the ESI source (Fig. 2). For quantitative analysis of proteins by MS, stable-isotope labeling of proteins can be used. These methods utilize either metabolic labeling, tagging by chemical reaction, or stable-isotope incorporation via enzyme reaction of proteins or peptides (Aebersold and Mann 2003). Labeled proteins or peptides are combined, separated, and analyzed by MS and/or tandem MS for identifying the proteins and determining their relative abundance.

In the top-down approach, intact protein ions are introduced into the gas phase and are then fragmented in the mass spectrometer. If enough numbers of informative fragment ions are observed, the analysis can provide a complete description of the primary structure of the protein and reveal all of its modifications. Until recently, it has proved difficult to produce extensive gas-phase fragmentation of intact large protein ions, but novel techniques such as electron transfer dissociation promise to drastically improve this situation.

Neuropeptidomics. The core proteomic tools, including 2-DE in combination with MS, are limited to the analysis of proteins >10 kDa. Other technologies are therefore necessary to identify small endogenous proteins and peptides such as those present in brain samples. Neuropeptidomics is the technological approach for detailed analyses of endogenous peptides from the nervous system/brain. It is a relatively new direction in proteomic research that covers the gap between proteomics and metabolomics and overlaps with both areas. Peptidomics methodologies are generally based on separating complex endogenous peptide mixtures by multistep LC approaches, usually nL/min flow capillary reversed-phase LC (nanoLC), or gel- or liquid-based isoelectric focusing combined with MS for sequence analysis. The levels of peptides in the brain reflect certain information about physiological status; this information can be revealed when MS is used to generate broad profiles of the dynamic neuropeptide patterns (Svensson et al. 2007).

MALDI Imaging Mass Spectrometry. MALDI mass spectrometric tissue imaging (MALDI-IMS) of peptides and proteins in the brain is performed in thin (10–20 μm) tissue sections in situ (Stoeckli et al. 2001). The sections are typically coated with a raster of matrix droplets before an ordered array of mass spectra is acquired from each matrix spot. This way each spectrum reflects the local molecular composition at known x, y coordinates. Image profiles of selected peptides and proteins in the section are generated by extracting their corresponding mass-to-charge (m/z) ranges from the spatially acquired MS data files. The approach yields information on the spatial localization of the peptides and proteins in the tissue analyzed, without the requirement of extensive sample manipulation. Applications of MALDI-IMS range from low-resolution peptide and protein profile images of selected areas in, e.g., mouse brain to single neural cell peptide-profiling analyses and high-resolution imaging of proteins and drugs.

Protein Microarrays. Protein and peptide microarrays involve the spotting of proteins (including antibodies or other affinity reagents directed against defined proteins) and peptides at high density on surfaces such as glass slides and can be used for both profiling and functional proteomics. Antibody microarrays hold potential for high-throughput protein profiling. A complex mixture, such as a brain cell lysate, is passed over the microarray surface to allow the antigens present to bind to their cognate antibodies or targeted reagents. The bound antigen is detected either by using lysates containing fluorescently tagged or radioactively labeled proteins or by using a secondary antibody against each antigen of interest. Functional protein arrays allow for testing of activities and interactions with lipids, nucleic acids, and small molecules as well as other proteins (Phizicky et al. 2003).

Understanding the Molecular Organization and Complexity

There are a variety of applications for proteomic technology in psychopharmacology and neuroscience. These range from defining the proteome of a particular cell type, identifying changes in brain protein expression under different experimental (including pathological) conditions, profiling protein modifications, and mapping protein-protein interactions. All of them have their strengths and limitations, and a major challenge is to determine the most appropriate proteomic technology to the system studied. Large-scale proteomic analysis can help unravel the complexities of brain function as many of the activities of the brain involve intricate signaling networks and changes in PTMs (Choudhary and Grant 2004). Clinical research may benefit from proteomics through the identification of new drug targets and new diagnostic markers.

Proteome Global Mapping. Detailed analysis of the mouse brain proteome has established 2-DE protein indices of thousands of proteins, with MS annotations of ~500. Proteins from all functional classes have been identified by such analyses. However, membrane proteins are typically underrepresented in 2-DE due to poor solubility in the initial isoelectric focusing step of the method, which requires relatively large amounts of starting material. Another approach using multidimensional LC coupled to ESI MS/MS does not have this bias against membrane proteins and identified close to 5,000 proteins from 1.8 mg of rat brain homogenate with an average of 25 % protein sequence coverage. The proteins identified included membrane proteins, such as neurotransmitter receptors and ion channels implicated in important physiological functions and disease.

Neuropeptidomics has been used to profile a large number of neuropeptides from the brain and the central nervous system. Strategies that reduce complexity and increase the dynamic range of endogenous peptide detection, particularly fractionation methods and separations based on the peptides' chemical and physical properties and bioinformatics approaches, have resulted in the discovery and chemical characterization of novel endogenous peptides. Some of these, such as peptides originated from the secretogranin-1, somatostatin, prodynorphin (endogenous opioids), and cholecystokinin precursors, appear differentially expressed in the striatum with and without 3,4-dihydroxy-L-phenylalanine ("▶ Levodopa, L-Dopa") administration in 6-hydroxydopamine (6-OHDA) animal models for Parkinson's disease (Nilsson et al. 2009).

The MALDI-IMS technique has been applied to animal models of neurodegenerative ("▶ Neurodegeneration") disorders to investigate peptide and protein expression, particularly to compare patterns in pre- and post-lesions using 6-OHDA and 1-methyl-4-phenyl-1,2,3,6-tetrahydropyridine (MPTP) (Fig. 3). In these affected brains, the IMS molecular tool identified changes in complex protein patterns and identified specific proteins involved in specific brain regions (Svensson et al. 2007). IMS has also been utilized to reveal the regional distribution of psychopharmacology agents in the brain such as clozapine ("▶ Atypical Antipsychotic Drugs"). IMS images revealing the spatial localization in rat brain tissue sections following administration of clozapine were found to be in good correlation with those using an autoradiographic approach. The results are encouraging for the potential applicability of this technique for the direct analysis of drug candidates in intact tissue slices (Cornett et al. 2007).

The utility of *functional proteomics* has been recently exploited to elucidate cellular mechanisms in the brain, of particular importance in the area of signal transduction. Reversible phosphorylation of proteins is the most widely studied type of signal transduction. Recent synapse phosphoproteomic studies analyzing phosphorylation sites of proteins identified 974 sites in mouse synaptosomes and 1,563 in postsynaptic densities isolated from mouse cortex, midbrain, cerebellum, and hippocampus. Phosphoproteomics has identified phosphorylation sites that are altered in Alzheimer's disease, particularly in the microtubule-associated protein Tau (Bayes and Grant 2009).

Characterization of protein complexes provided information about molecular organization as well as cellular pathways. A multi-bait yeast with two hybrid screen of proteins relevant to Huntington's disease found 186 protein-protein interactions, of which 165 had not been previously described and which might be relevant to the pathology of Huntington's disease (Bayes and Grant 2009).

Advantages and Limitations in Neuroproteomics

While many of the proteomic approaches have focused on determining the proteins or peptides

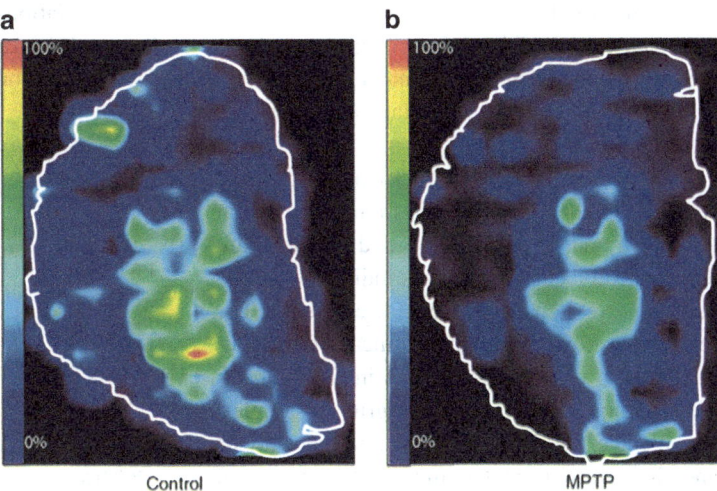

Proteomics, Fig. 3 IMS analysis of brain tissue sections after 1-methyl-4-phenyl-1,2,3,6-tetrahydropyridine (MPTP) treatment. The relative ion density of PEP-19 from one control (**a**) and one MPTP-treated animal (**b**) shows that there is a reduction of PEP-19 expression in the striatum. Typically, in this experiment, about 400 distinct mass signals were detected in the mass range of 2–30 kDa (Modified from Svensson et al. 2007)

present in the cell and their relative expression levels, the specific aim of proteomics would be to simultaneously identify, quantify, and analyze a large number of proteins within their functional context. The shift in focus from a proteomic discovery mode to a functional approach creates challenges that are at present unmet in all aspects of the proteomic experiment, including the experimental design, data analysis, storage, and publication of data (Choudhary and Grant 2004).

Sample preparation is a major source of variation in the outcome of proteomic experiments. After tissue or body fluid sampling, proteases and other protein-modifying enzymes can rapidly change composition of the proteome. As a direct consequence, analytical results will reflect a mix of in vivo proteome and ex vivo degradation products. Vital information about the pre-sampling state may be destroyed or distorted, leading to variation between samples and incorrect conclusions. Sample stabilization and standardization of sample handling are imperative to reduce or eliminate this problem. A recently introduced tissue stabilization system which utilizes a combination of heat and pressure under vacuum has been used to stop degradation in brain and loss of PTM (e.g., phosphorylations) tissue immediately after sampling (Soloview et al. 2008; Svensson et al. 2007). This methodology provides an improvement to proteomics by greatly reducing the complexity and dynamic range of the proteome in tissue samples and enables enhanced possibilities for discovery and analysis of clinically relevant protein and peptide biomarkers. Rapid removal of neuronal tissue, dissection, and freezing are obvious important procedures for the maintenance of the proteome state in the animal. Human *postmortem* studies present problematical challenges in neuroproteomics, where careful documentation of *postmortem* time interval, brain pH, and agonal state is of greatest importance (Soloview et al. 2008).

Proteomic studies by definition result in large amounts of data. The analysis as well as interpretation of the enormous volumes of proteomic data to effectively use their content remains an unsolved challenge, particularly for MS-based proteomics. The development of tools for the integration of different experimental approaches enabling analyses of such proteomic data sets using statistical principles is an important task for the future (Aebersold and Mann 2003).

MS-based peptidomics technologies in combination with sophisticated bioinformatics tools have great potential for the discovery of novel biologically relevant neuropeptides. It is likely that a considerable number of neuropeptides are still to be discovered. The human genome contains ~550 genes belonging to the G protein-coupled receptor class of proteins. For 25 % of these the natural endogenous ligands remain elusive until today, and novel neuropeptides are very plausible candidates. Improved peptidomics approaches and technologies may therefore identify novel biologically important neuropeptides (Svensson et al. 2007).

MALDI-IMS has become an important tool for assessing the spatial distribution of molecular species in brain tissue sections and for the elucidation of molecular signatures indicative of disease progression and drug treatment. The technique allows simultaneous measurements of hundreds of different molecules in tissue specimens without disrupting the integrity of samples. It can trace the distribution of pharmaceuticals and their various metabolites in the brains of dosed animals and can be successfully applied to monitor the changes in the proteome organization upon drug application. Functional information obtained in MALDI-IMS studies can be correlated with proteomic profiles and routine immunohistochemical staining, thereby providing an in-depth comprehension of molecular mechanisms underlying health and disease (Cornett et al. 2007; Stoeckli et al. 2001).

A catalog of the complete neuroproteome will propose new directions to understand brain function. Differential proteomics permits correlations to be drawn between the range of proteins produced by a cell or tissue and the initiation or progression of a disease state. It permits the discovery of new protein markers for diagnostic purposes and the study of novel molecular targets for brain drug discovery. The markers identified may have a wide range of potential applications, such as clinical diagnostic and prognostic tools. Proteomic information has superior functional value and can generate knowledge of cellular protein networks. Proteomic technologies are progressing fast and their increasing usage as a functional high-throughput approach is adding to vital biological findings in areas not accessible to genomics studies.

Cross-References

▶ Matrix-Assisted Laser Desorption Ionization
▶ Metabolomics
▶ Neuropeptidomics
▶ Posttranslational Modification
▶ Protein Microarray
▶ Two-dimensional Gel Electrophoresis

References

Aebersold R, Mann M (2003) Mass spectrometry-based proteomics. Nature 422:198–207
Bayes A, Grant SG (2009) Neuroproteomics: understanding the molecular organization and complexity of the brain. Nat Rev Neurosci 10:635–646
Choudhary J, Grant SG (2004) Proteomics in postgenomic neuroscience: the end of the beginning. Nat Neurosci 7:440–445
Cornett DS, Reyzer ML, Chaurand P, Caprioli RM (2007) MALDI imaging mass spectrometry: molecular snapshots of biochemical systems. Nat Methods 4:828–833
Nilsson A, Falth M, Zhang X, Kultima K, Skold K, Svenningsson P, Andren PE (2009) Striatal alterations of secretogranin-1, somatostatin, prodynorphin, and cholecystokinin peptides in an experimental mouse model of Parkinson disease. Mol Cell Proteomics 8:1094–1104
Phizicky E, Bastiaens PI, Zhu H, Snyder M, Fields S (2003) Protein analysis on a proteomic scale. Nature 422:208–215
Soloview M, Shaw C, Andrén P (2008) Peptidomics: methods and applications. Wiley, Hoboken
Stoeckli M, Chaurand P, Hallahan DE, Caprioli RM (2001) Imaging mass spectrometry: a new technology for the analysis of protein expression in mammalian tissues. Nat Methods 7:493–496
Svensson M, Skold K, Nilsson A, Falth M, Nydahl K, Svenningsson P, Andren PE (2007) Neuropeptidomics: MS applied to the discovery of novel peptides from the brain. Anal Chem 79(15–6):18–21
Tyers M, Mann M (2003) From genomics to proteomics. Nature 422:193–197
Westermeier R, Naven T, Höpker HR (2008) Proteomics in practice: a guide to successful experimental design, 2nd edn. Wiley-VCH, Weinheim

Pseudo-Hallucination

Definition

A subject experiencing pseudo-hallucinations retains the capacity to recognize that these experiences are transient and drug-induced, as opposed to true hallucinations in which no such discernment is possible.

Pseudopregnant

Definition

A hormonal state similar to pregnancy that is induced in mice by mating a female with a vasectomized male. In this state, the uterus is receptive to an implanted embryo.

Cross-References

▶ Genetically Modified Animals

Psychoactive Drug

Definition

A chemical substance that affects the central nervous system and modifies behavior and mental processes. When an individual is under the influence of psychoactive drugs, changes in one or more of the following may be observed; neural functionality, behavior, mood, perception, and consciousness. While some psychoactive drugs are used as therapeutic agents, recreational use and subsequent abuse are also common.

Cross-References

▶ Sex Differences in Drug Effects

Psychometric Function

Definition

The mapping between an independent variable under experimental control, such as stimulation strength or cost, and a measure of behavioral output, such as response rate or time allocation.

Psychometric Tests

Definition

Tests or apparatus that measure and quantify physical dimensions (e.g., time taken, number, rate, accuracy, speed) of information processing by sensory, cognitive, and motor systems.

Psychomotor Performance in Humans

Ian Hindmarch
University of Surrey, Guildford, UK

Synonyms

Psychomotor function; Skilled performance

Definition

Humans have an innate capacity to react to environmental stimulation which, by way of learning and experience, develops into a range of coordinated motor behaviors that are appropriate responses to the sensory information perceived by an individual. These motor behaviors range in complexity from the instantaneous reflex reactions that follow the accidental grasping of a hot poker, to simple sensorimotor behaviors such as poking a fire, climbing stairs, and steering a car, to the skilled psychomotor functions necessary to climb a rickety staircase at night or drive a car in heavy motorway traffic during a rainstorm. Such highly complex motor behaviors, i.e., psychomotor performance, result from the cognitive processing of sensory and perceptual information.

The sense organs receive and convey information to the brain but rarely, except in the case of reflex reactions and simple sensorimotor behaviors, are such data transmitted without interpretation by the cognitive system. The perception, i.e., interpretation, of environmental events and circumstances allows sensory information to be manipulated within the cognitive system and decisions made regarding the appropriate motor response(s) necessary for an immediate adaptive motor reaction. An individual will select from the array of stimuli available and choose to place more reliance on some cues and ignore others in accordance with prior learning, previous experiences, prejudices, and established habits. Such over-/underemphasis of some stimuli augments interindividual differences in overall psychomotor performance due to the individual's concurrent needs, motivations, and expectations. Perception is dependent on maintaining an appropriate level of attention and remaining vigilant. The extent to which an individual is able to divide or shift attention between two or more relevant stimuli is an important variable in determining the overall capacity for psychomotor performance.

The motor component of psychomotor performance becomes more coordinated with sensory input because of learning and practice. Basic sensorimotor skills (e.g., steering a car) are quite robust and resistant to environmental influences, but any faulty cognitive judgment or decision can cause a major impairment of psychomotor performance (e.g., car driving) if "unrealistic" demands are made on motor reactions.

It is the speed, reliability, and validity of the processing of environmental information, particularly where the circumstances are complex or rapidly changing, which determine the appropriateness and precision of psychomotor performance. Any alteration in the reliability or validity of perceptual input will increase the possibility of errors in cognitive processing which, in turn, will be reflected as an impairment of the overall psychomotor performance. However, if the cognitive system itself is directly affected by trauma, disease, or the administration of a psychoactive drug, then the processing of perceptions will be similarly influenced and the associated motor reactions will necessarily change. Such changes in the integrity of the coordination between sensory and motor functions will lead to an impairment of psychomotor performance together with a subsequent loss of competence and increased risk of accident when undertaking the activities of everyday living in the home, at work, and on the road.

It is virtually impossible, within the context of everyday scenarios, to measure psychomotor performance reliably: there are too many uncontrolled influences and variables, and inter- and intraindividual differences are too great. The measurement of psychomotor performance requires an instrument where the relevant stimulus is readily perceived and its inherent information processed in a straightforward manner, without the necessity of having to rely on learning, previous experience, or the adoption of particular viewing strategies. The cognitive processing requirements have to be easy to understand and the motor responses undemanding and uncomplicated. Overall, psychomotor performance is dependent on vigilance, attention and the speed, accuracy, and coordination of sensorimotor behavior, and, most importantly, the speed, capacity, and flexibility of the cognitive system. The basic elements (sensory, cognitive, and motor) can be measured using the standardized application of psychometric tests within a controlled environment to minimize the effects of the numerous extraneous influences both on the individual and on the test performance.

Psychopharmacologists are primarily interested in the behavioral effects of drugs but it is a basic assumption of human psychopharmacology that the effects of a particular psychoactive drug will be manifest by the changes it produces in psychomotor performance, as measured by psychometric tests. A major task of a human psychopharmacologist is not only to devise valid and reliable psychometrics by which such characteristic drug-induced changes can be measured but also to develop appropriate methodologies to control the studies by which the effects of drugs are to be assessed.

The successful assessment of the effects of drugs by way of the changes produced in psychomotor performance can only be done with studies in which the intrinsic and extrinsic sources of variability in psychomotor performance are controlled.

Interindividual Variability

A major variable in determining the effect of a drug on psychomotor performance is the psychological state of the individual at the time of testing. Most pharmacodynamic studies of drugs are performed in healthy, psychologically "normal" volunteers. This certainly does not preclude the use of psychometric tests in patient populations but study design issues will be different. As it is not possible to obtain psychologically identical individuals with comparable levels of psychomotor ability, the underlying intersubject variability must be reduced by controlling it via the use of a crossover design where each subject acts as their own control and receives each and every treatment in a random order. A parallel group design, where individuals are randomly allocated to receive only one of the treatment conditions, is much less suitable to control individual differences, even when large group numbers are used and subjects are screened and preassigned to treatments to ensure similar mean psychometric test scores across treatment groups. Pre-assignment does nothing to reduce the magnitude or speed of individual reactions to a particular drug. Individual differences can also be reduced by using a neuropsychological test to screen out potential subjects with a preexisting abnormal or reduced cognitive function (e.g., Mini-Mental State Examination, Wisconsin Card Sorting Test, Wechsler Adult Intelligence Test, and especially the subscales for block design and picture completion). Neuropsychological test results indicate an individual's degree of abnormality in comparison with normative values from large-scale databases. They are not, in themselves, reliable in the test-retest protocols necessary for studying the pharmacodynamic effects of a drug on psychomotor performance. However, a similar prescreening of experimental subjects' levels of neuroticism and extraversion/introversion (e.g., Eysenck Personality Inventory) can help reduce the influence intersubject personality differences have on psychometric test performance. The established effects of age, sex, etc., on psychomotor performance can be simply controlled by preselecting and stratifying groups of subjects.

Treatment Variables

All treatments must be identical and matching with no discernible difference in color, size, shape, or taste, and they must be administered double-blind where neither the subject nor the person collecting the psychometric test scores are aware of the order in which the treatments are given. The subjective experiences of taking a substance believed to be a "drug" – even when it is actually pharmacologically inert – change psychometric test scores in susceptible subjects. These effects are controlled by the use of a placebo effect condition, which is regarded as another drug and included in the randomization and allocation of treatments. The psychometric test scores following active drug treatments can then be compared to those obtained following treatment with the placebo condition to indicate the psychoactive effects of the drug itself. A placebo control is an essential feature of all psychometric assessment of drug effects. Changes in psychomotor performance observed with active drugs are valid only if seen to be significantly (from the application of appropriate statistical analysis, probability levels, confidence intervals, etc.) different with respect to the results obtained with a matching placebo.

Testing Variables

Not all psychometric tests have an established sensitivity for the detection of changes produced by psychoactive drugs and an individual test will not always detect the particular effects of a specific drug. As well as reinforcing the need to use a psychometric test battery assessing a range of psychomotor activities, such considerations clearly show the need to demonstrate the sensitivity of a particular test battery to drug-induced change for a specific, especially unknown, drug. This is done by the use of a verum (positive internal control), a drug with well-understood and robust effects on the psychometric tests to be used. It is usual for the positive controls to belong to the same pharmacological or therapeutic class as the drug under investigation although they are often used at supra-therapeutic doses to ensure that performance is reliably changed. The results from a verum are to validate the psychometric battery and the verum is expected to cause impairment, such results must not be used to compare side effects with drugs under consideration. Most positive internal controls (e.g., lorazepam, alprazolam, zolpidem, amitriptyline, promethazine, and mirtazapine) are used because they are known to lengthen reaction time, impair working memory and thinking, disrupt sensorimotor coordination, and cause an overall impairment of psychomotor performance. However, the use of psychostimulants (e.g., amphetamine and methylphenidate) as positive internal controls will indicate the extent to which a test battery is able to detect changes in psychomotor performance due to CNS activation. Unpredictable variation in psychometric test scores can occur with repeated testing if there are auditory/visual distractions present during testing or if noticeable changes in ambient temperature or lighting occur. Psychometric testing must be conducted in a stable environment that remains constant both during each test period and between each occasion testing is conducted. As some external events are unpredictable, e.g., power failures and thunderstorms, the sequence of administration of treatments should be balanced across the various days when testing takes place in accordance with a Latin square design to ensure that any random effects of extraneous variables on psychometric test scores will be spread equally across all treatment conditions.

Psychometric Tests

Below is a list of some of the psychometric tests that have been used in published studies of the effects of psychoactive drugs: there are many more (Baselt 2001; Hindmarch 2004). The three-part division of the tests does not imply a rigid classification, but shows that although some psychometric tests might emphasize individual sensory, cognitive, or motor functions, it is overall psychomotor performance – the integration of sensory and motor systems via the cognitive processing of information – which is measured (Table 1).

Although the effects of a particular drug on psychomotor performance can be assessed by an individual psychometric test, it is usual to employ a test battery (group of tests each reflecting a different aspect of psychomotor performance) particularly when a new substance with unknown or supposed psychoactive effects is to be investigated.

There is no consensus as to which tests are the most suitable for combination and inclusion in a psychomotor performance test battery. Some researchers place emphasis on the sensitivities and robustness of a particular test that other authors may find less reliable. The choice of tests may be influenced by the level of existing information regarding the known, or expected, effects of the drug(s) to be assessed.

It is most usual to regard impaired psychomotor performance as a necessary consequence of the administration of a sedative drug but psychostimulants, by causing over-activation of the sensory input of information, can also impair psychomotor performance.

Typically, a psychomotor test battery will involve assessments of sensorimotor coordination, vigilance and/or divided attention, complex reaction time, executive or short-term memory, and mental ability (arithmetic and logical reasoning). The exact composition of any test battery is not crucial but the proven sensitivity of the component tests to the psychoactive effects of drugs is a prerequisite for the inclusion of a particular individual test.

Psychophysiological measures, e.g., saccadic eye movements, sleep polysomnography (including the multiple sleep-wakefulness test and the maintained-wakefulness test), and continuous 24-h electroencephalography, can provide objective measures of the effects of drugs on sleep and daytime arousal. There is no doubt that lowering of arousal, as revealed by these psychophysiological assessments, will directly impair psychomotor performance. However, psychophysiological assessments are not, in themselves, tests of psychomotor performance. Similarly, subjective ratings of sleepiness, e.g., Stanford Sleepiness Scale, visual analogue rating scales, Epworth Sleepiness Scale, and especially the Leeds Sleep Evaluation Questionnaire (Parrott and Hindmarch 2008), which was specifically developed to measure the changes in the

Psychomotor Performance in Humans, Table 1 Psychometric tests used to measure the effects of drugs

Sensory ability	Attention and cognition	Motor ability
Pursuit rotor; Gibson Spiral Maze; dynamic visual acuity; angle perception; Maddox wing test; auditory discrimination; letter cancelation; simple reaction speed (visual, auditory)	Critical flicker fusion frequency; two-flash threshold; rapid visual information processing; memory tests with immediate and/or delayed recall for short-term (Sternberg test), executive, recognition (pictures, words, faces, numbers, prose, semantic categories, nonsense syllables, music), and spatial (Corsi blocks) abilities; syntactic/logical reasoning (Tower of London); learning (Rey's test and paired associates); CDR test systems; vigilance (continuous performance, Pauli test, auditory vigilance, divided attention tasks, Dinges' psychomotor vigilance test, Erikson and Erikson's response competition test); mental arithmetic; time estimation; Stroop test; verbal fluency; free word association; on-the-road car driving (brake reaction time); digit-symbol-letter manipulation (copying, substitution, matching, differentiation); mental arithmetic (serial subtraction of numbers, simple mathematics with or without auditory interference)	Peg board; card dealing; trail making; on-the-road car steering (standard deviation of lateral position); body balance; finger tapping; compensatory, continuous or adaptive tracking; wrist actigraphy

subjective ratings of sleep and early morning performance following the administration of psychoactive drugs, will not automatically indicate the psychomotor effects of a drug, but such subjective data can provide valuable information regarding the total impact of a psychoactive drug in clinical use.

Current Concepts and State of Knowledge

Psychometric tests have to be parsimonious in the demands they make on sensory, cognitive, and motor systems if they are to be reliable measures of psychomotor performance suitable for use in a wide population of individuals of different ages, personality traits, skill levels, and intellectual abilities. The increased availability of computer programs to present, control, and collect results from psychometric tests has led to the automation of many psychometric test batteries. Such accurate and consistent presentation of the stimulus component of a psychometric test and the error-free recording of results are efficient ways of ensuring the reliability of a test battery. However, the programming of computerized test batteries must take into account important intersubject differences in speed of reaction, comprehension of test instructions, etc., particularly when several tests are presented one after the other. The pace at which the individual tests are presented may give rise to pressures on performance and cause unpredictable effects which are not due to the effects of study medications. If psychometric tests are developed in a straightforward manner with realistic stimulus recognition, information processing, and motor response requirements (especially the demands made on "▶ Short-Term and Working Memory in Humans"), then the effects of learning on psychomotor performance can be minimized although pre-study practice on even the simplest of psychometric tests is always required.

Computerized tests are able to present information at a rate and magnitude which can exceed the information processing capacity of some subjects, especially the elderly and those under the influence of drugs. When faced with too much information, individuals will adopt various strategies – some of which will require extensive learning – as in the playing of a computer game. Strategic perceptual and/or complex information processing requirements and/or the need for a finely controlled or coordinated motor response will increase the effects of inter- and intraindividual variability and affect the validity and reliability of the results obtained.

In short, psychometric tests for the measurement of psychomotor performance must be straightforward in their stimulus perception, information processing, and motor response requirements. Furthermore, their sensitivity to drug activity has to be demonstrated by way of the use of verum controls. It is also necessary to evaluate a range of doses of a particular drug, not only because in clinical use patients often take "supra-dose" regimens but also because psychomotor impairment effects might be dose related (Holgate et al. 2003).

Although measuring different aspects of psychomotor performance, many individual psychometric tests utilize speed of reaction as one of the motor response measures. Research on age-related changes in mental capacity suggests that a fundamental part of the cognitive system governs the speed of processing information (Kail and Salthouse 1994). Use can be made of this concept when assessing the effects of drugs on psychomotor performance by combining the reaction time components from each of the response-based tests in a particular test battery.

Tests of psychomotor performance are primarily used to assess the effects of CNS-active drugs. A major step forward in validating this use of psychometrics has been the advent of human positron emission tomography (PET) studies of the effects of psychoactive drugs, e.g., antihistamines (Yanai and Tashiro 2007). PET studies have differentiated between antihistamines in their capacity to penetrate the brain and occupy, ranging from virtually 0 to 90 %, histamine-1 receptors ("▶ Histaminic Agonists and Antagonists").

The extent a particular antihistamine occupies H1 receptors in the brain is generally correlated with the degree of psychomotor impairment found from placebo- and verum-controlled studies of the same drug (Shamsi and Hindmarch 2000). This not only helps elucidate the mechanisms by which drugs, in this instance antihistamines, act but also validates the use of psychometric tests to measure drug-induced changes in human psychomotor performance.

Cross-References

▶ Attention
▶ Histaminic Agonists and Antagonists
▶ Pharmacodynamic Tolerance
▶ Placebo Effect
▶ Short-term and Working Memory in Humans
▶ Spatial Memory in Humans
▶ Verbal and Nonverbal Learning in Humans

References

Baselt RC (2001) Drug effects and psychomotor performance. Biomedical Publications, Foster City, p 475

Hindmarch I (2004) Psychomotor function and psychoactive drugs. Br J Clin Pharmacol 58:S720–S740

Holgate ST, Canonica GW, Simons FE et al (2003) Consensus group on new generation antihistamines (CONGA): present status and recommendations. Clin Exp Allergy 33:1305–1324

Kail R, Salthouse TA (1994) Processing speed as a mental capacity. Acta Psychol (Amst) 86(2–3):199–225

Parrott AC, Hindmarch I (2008) The leeds sleep evaluation questionnaire for psychopharmacology research. In: Pandi SR (ed) Sleep disorders: diagnosis and therapeutics. Informa Health Care, London, pp 685–689, Chap 62

Shamsi Z, Hindmarch I (2000) Sedation and antihistamines: a review of inter-drug differences using proportional impairment ratios. Hum Psychopharmacol Clin Exp 15(1):S3–S30

Yanai K, Tashiro M (2007) The physiological and pathophysiological roles of neuronal histamine: an insight from human positron emission tomography studies. Pharmacol Ther 113:1–15

Psychopharmacology

Synonyms

Neuropsychopharmacology

Definition

Psychopharmacology is a scientific subfield of pharmacology that utilizes drugs or other chemical agents to understand neural function, to prevent and treat mental illness and drug abuse, and to understand how nontherapeutic ▶ psychoactive drugs and natural substances alter human mood, memory, motor activity, endocrine, and other centrally mediated functions.

Cross-References

▶ History of Psychopharmacology
▶ Psychoactive Drug

Psychophysiological Methods

R. Hamish McAllister-Williams
Academic Psychiatry, Institute of Neuroscience, Wolfson Research Centre, Newcastle University, Newcastle upon Tyne, UK

Definition

Stern (1964), in the first edition of the journal *Psychophysiology* established by the Society for Psychophysiological Research, defined the field as where behavioral variables are manipulated, and the effects of these independent variables are observed on physiological measures as dependent variables. This definition has been expanded by Furedy (1983) in the first edition of the International Organization of Psychophysiology's *International Journal of Psychophysiology* to "the study of physiological processes in the intact organism as a whole by means of unobtrusively measured physiological processes." Clearly, the nature of what constitutes "unobtrusive" in this context is subjective, but is an acknowledgment that the process of measurement can influence the measure itself. At least in some senses, the relationship between physiology and psychology implied by the term "psychophysiology" is the converse of that described by "physiological psychology" or "biological psychology" which relate to the study of the biological and physiological underpinnings of psychological processes. Psychophysiological methods are utilized in both human and animal studies. Furedy (1983) argues that this is only the case when the interest of the experimenter is focused on psychological processes of the whole intact organism.

Principles and Role in Psychopharmacology

The Role of Psychophysiology and Breadth of Techniques

As suggested earlier, the very definition of psychophysiology is a subject of some debate. Nathan Kline (1961) took an existential approach in his paper "On the relationship between neurophysiology, psychophysiology, psychopharmacology, and other disciplines." He argued that ambiguities in bioscience arise from asking a question in one "universe of discourse" (e.g., psychology) and seeking the answer in quite a different one (e.g., physiology). This is a potential issue with respect to the term psychophysiology and, indeed, psychopharmacology. Kline developed a number of laws regarding the relationship between disciplines including the "Law of Technique." This law states that the technique used to obtain information does not necessarily determine the "universe of discourse" in which it is used. This remains relevant, especially to the use of psychophysiological methods within psychopharmacology, as this spans three or more disciplines. To illustrate this, consider an example based on one given in Furedy's (1983) discussion of the definition of psychophysiology. Anxiety, via effects on the autonomic nervous system (ANS), causes changes in heart rate and other cardiac parameters such as the electrocardiograph (ECG) T-wave amplitude. If during a stress-exercise test in a patient with suspected cardiac pathology, a decrease in T-wave amplitude is seen, a physiological explanation will relate to cardiac pathology. However, from a psychophysiological perspective, the effects of the ANS on the myocardium need to be considered. For example, if the patient has great anxiety regarding his or her heart, the test may constitute a psychological as well as a physical stressor. Further, the psychopharmacologist would also take into account the antidepressant or anxiolytic medication that the patient may be taking which may be modifying the psychophysiological response. The technique or measurement (T-wave amplitude) is common to these three "universes of discourse," but each views or uses it in different ways.

It can therefore be seen that within the research area of psychopharmacology, psychophysiology is a tool to assist in the measurement of drug-induced changes in psychological processes. When studying the effects of potentially psychoactive drugs, one of the greatest challenges is having objective "outcome measures" that demonstrate the effect of the drug. It is essential to measure the effect of a drug on mood, for example, when developing an antidepressant. Such an outcome has fundamental importance regarding the therapeutic use of the drug. However, the psychological process involved can only be measured in a subjective way in humans and only very indirectly in animals. Alternative and additional outcome measures, which can be assessed more objectively and in a range of situations, are essential to facilitate a number of investigations including the study of the pharmacokinetics of a drug, understanding dose-response relationships and the mechanism of action of the drug itself, drug safety, and investigating drug interactions.

There is almost an infinite number of psychophysiological (as defined earlier) outcome measures used in psychopharmacology, limited only by the ingenuity of the scientists involved. These include, for example, direct methods of investigating the effects of a drug on the physiological function of the central nervous system such as measuring changes in the brain electrical activity using electroencephalography (EEG) and magnetoencephalography (MEG). These techniques can be utilized in a myriad of ways. A common method of using the EEG to acquire information is to record event-related potentials (ERPs), that is, EEG activity recorded time locked to some event such as the presentation of a stimulus to a subject or a subject's response. Examples of ERPs include the "P300" (referring to a positive voltage deflection occurring around 300 ms after the presentation of rare or task relevant stimuli) and "mismatch negativity" (a negative voltage deflection occurring after presentation of a stimulus that is deviant, e.g., in terms of loudness, duration, or frequency), both of which have been widely used as outcome measures in psychopharmacological research. In recent years, there has been an explosion of novel analysis techniques of EEG data, such as independent component analysis (ICA) and exploration of the frequencies of the signals contained within the EEG, which further enhances its potential utility as an outcome measure. At their heart, EEG and MEG techniques offer the opportunity for investigating the effects of psychopharmacological agents with high temporal resolution.

Recent decades have seen an explosion of neuroimaging techniques including their application in psychopharmacology. In the simplest types of paradigms, the notion is that a psychoactive substance leads to changes in psychological processes that are reflected by changes in the cerebral blood flow which can be measured using positron emission tomography (PET) or functional magnetic resonance imaging (fMRI). Such techniques offer the opportunity of providing high-resolution spatial information regarding the site of action of the drugs. PET and other imaging methodologies can also be utilized with radio-labeled ligands to investigate a whole range of pharmacological processes including transmitter release and receptor binding. Recently, there have also been reports of drug effects on connectivity within the brain as measured using diffusion tensor imaging (DTI).

Perhaps the most direct method of exploring the effects of a psychopharmacological agent on psychological processes is the use of neuropsychological tests to examine changes in cognition. Cognitive enhancement is a particular goal of a branch of psychopharmacology, for example, in the treatment of dementias. However, in addition, cognition can be utilized as an objective outcome measure in treatment studies using reliable neuropsychological tests of proven validity. Such tests can be combined with, for example, EEG or fMRI to provide additional information regarding which temporal and spatial component of a cognitive task is being influenced by the psychopharmacological agent.

Historically, some of the most widely used psychophysiological techniques have utilized the close interaction between the central nervous system and the neuroendocrine system. Such neuroendocrinological techniques offer the advantage of the simplicity of sample collection by measuring hormonal levels most commonly in plasma. The technique can be used in a number of different ways. For example, stress in a number of forms leads to the activation of the hypothalamic–pituitary–adrenal (HPA) axis with a consequent release of corticosteroids. The effect of a psychopharmacological agent on stress can be assessed by examining changes in peripheral corticosteroids. This is a classic example of where Furedy's (1983) point about the measures being "unobtrusive" is a key issue. If the method by which samples are collected is stressful, this may lead to an alteration in the measured levels of corticosteroids. An alternative neuroendocrinological strategy is to assume that changes in transmitter systems in higher centers are reflected in similar changes in these transmitter systems that control neuroendocrine function. An example of this approach, much in evidence in the 1980s and 1990s, is the use of growth hormone and prolactin responses to serotonergic probes such as tryptophan or buspirone to assess the presumed functional status of hypothalamic 5-HT_{1A} receptors in mood disorders and following administration of antidepressants.

All of the abovementioned methods can be considered as within Furedy's (1983) definition of "psychophysiological." However, within the specialty of psychopharmacology, the term in common usage usually refers to the more peripheral physiological effects of changes in the central (psychological) function. Many, but certainly not all, of these effects relate to changes in the autonomic nervous system function as indexed by, for example, changes in pupil diameter, heart rate variability, and electrodermal responses.

Principles of Psychophysiological Methods as Applied in Psychopharmacology

Furedy (1983) argued that psychophysiology is different from physiological psychology in that the latter is related to the study of the physiological underpinnings of psychological processes. However, this is a "legitimate" area of study for the psychophysiologist, as perhaps the most important principle of psychophysiology is that the mechanism connecting the psychological process with the physiological response is known. Essentially, the issue here is the same as in any area of research when two measures are correlated; correlation does not imply causation. Heart rate may well be observed to increase when a person is anxious. However, it is necessary to be convinced that it is anxiety (a psychological process) in any particular circumstance that is leading to an increase in heart rate to make it a viable psychophysiological measure. In psychopharmacology, the situation is even more complex when using psychophysiological outcome measures because it is also important to know if the drug could be directly influencing the physiological response. A good example of this is the use of PET and fMRI imaging exploring blood flow changes in response to administration of a pharmacological agent. Before concluding that changes in the image signal relate to the effects of the drug on a particular psychological process, it is essential to know if the drug has direct effects on cerebral blood flow.

To illustrate the issues around the importance of understanding the mechanisms underlying psychophysiological effects and other principles of these methods, work in the area of arousal will be described.

A potential psychophysiological outcome measure to assess arousal is the diameter of

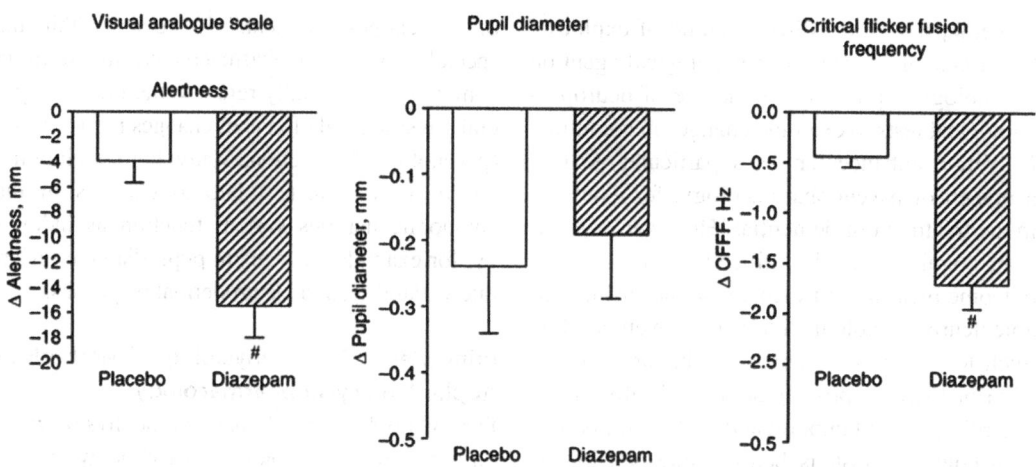

Psychophysiological Methods, Fig. 1 Level of alertness following a single oral dose of diazepam 10 mg given to 16 healthy male volunteers. Alertness assessed using a visual analogue scale, pupil diameter, and critical flicker fusion frequency. # $p < 0.05$ versus placebo. Full details of the study and the methods used can be found in Hou et al. (2007) from where the figures have been obtained

a subject's pupil. Pupil diameter has a close relationship with arousal, with decreased arousal being accompanied by constriction of the pupil (miosis). This is believed to reflect decreased sympathetic activity as levels of arousal drop. The benzodiazepines (e.g., diazepam), as anxiolytic and sedative drugs, cause a decrease in arousal. This can be assessed subjectively, for example, using visual analog scales (Fig. 1). Psychophysiology offers a more objective outcome measure. However, diazepam does not lead to any change in papillary diameter (Hou et al. 2007, Fig. 1). Does this paradox suggest that either diazepam is not sedating or pupillary diameter is a poor psychophysiological outcome measure? The former option seems an unlikely explanation. In the study by Hou et al. (2007), additional psychophysiological measures of alertness were also conducted. These included the critical flicker fusion frequency (CFFF) test. This involves determining the frequency at which a flickering light gives rise to the sensation of a steady light. This is measured by exposing the subjects both to a low-frequency flicker and increasing the frequency to the point at which the subject has the sensation that the flickering stops and a high-frequency flicker that decreases to the point at which the flicker is detected. The CFFF is the mean of the two. As in all areas of science, having well-designed methodology that is objective, valid, and replicable is essential. The CFFF is a test that has been in usage since the 1950s, and it is well accepted as a psychophysiological measure of arousal (Tomkiewicz and Cohen 1970). Hou et al. (2007) showed a significant effect of diazepam on the CFFF (Fig. 1), suggesting that diazepam does indeed have an effect on arousal, and this can be measured both subjectively and objectively. They provided data that suggest that the paradox of the lack of effect of diazepam on pupil diameter relates to a lack of effect on either the sympathetic or parasympathetic influence on the iris. This highlights the importance of being aware of both the pharmacology of the drug being studied and the mechanism of the psychophysiological tests being used.

However, it is incorrect to think that diazepam has no effect on the pupil. In addition, to simply measuring the pupil diameter, Hou et al. (2007) also measured spontaneous pupillary fluctuations, by analyzing these data in two ways: first, by performing an analysis of the frequency of fluctuations using a Fast Fourier Transformation to obtain an assessment of power and second, by a "pupillary unrest index" (PUI), which is the distance travelled by the margin of the pupil in 1 min. These two measures reflect "pupillary fatigue waves" (Ludtke et al. 1998) that are

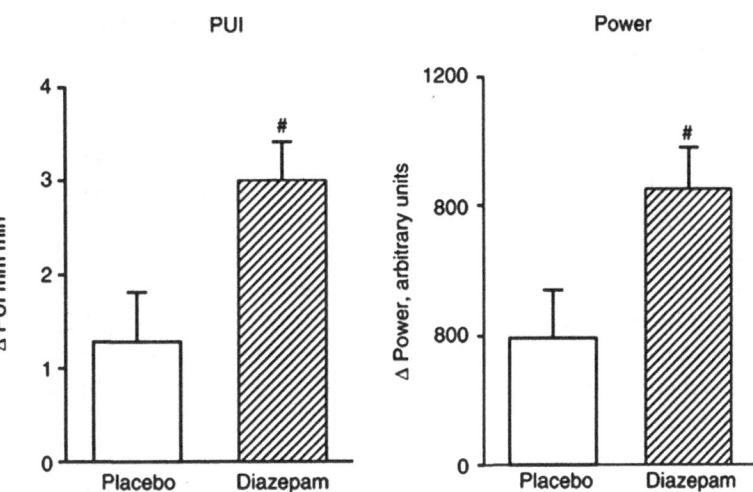

Psychophysiological Methods, Fig. 2 Pupillary spontaneous fluctuations following a single oral dose of diazepam 10 mg given to 16 healthy male volunteers. Data show the effect of diazepam on both the power of the fluctuations and the pupillary unrest index (PUI). # $p < 0.05$ versus placebo. Full details of the study and the methods used can be found in Hou et al. (2007) from where the figures have been obtained

believed to closely parallel fluctuations in the activity of the noradrenergic neurons of the locus coeruleus (Aston-Jones and Cohen 2005), influenced by arousal. While diazepam has no effect on static pupil diameter (Fig. 1), it significantly increases the pupillary fluctuation power and PUI (Hou et al. 2007, Fig. 2), in line with its proposed sedating effects and the results from the CFFF test. The study illustrates the importance of utilizing a range of methodologies relating to different mechanisms and pathways whenever possible to be able to draw legitimate conclusions regarding the effect of the drug. Further, the study illustrates the complexity and ingenuity of many modern psychophysiological tests.

Advantages and Limitations of Psychophysiological Methods

As described earlier, the most obvious advantage of psychophysiological methods is their use in providing objective outcome measures of the effects of psychopharmacological agents, and as such, they relate to the very essence of the discipline. There is an enormous range of such techniques, and it is impossible to delineate all their advantages and limitations here. The various methods span a range of ease of use, cost, and utility. Their main limitations are when there is a lack of a clear understanding of the mechanism underlying the response being measured and the direct and indirect effects that a drug may be having on it.

Cross-References

▶ Benzodiazepines
▶ Drug Interactions
▶ Electroencephalography
▶ Event-Related Potential
▶ Magnetic Resonance Imaging (Functional)
▶ Neuroendocrine Markers for Drug Action
▶ Pharmacokinetics

References

Aston-Jones G, Cohen JD (2005) An integrative theory of locus coeruleus-norepinephrine function: adaptive gain and optimal performance. Annu Rev Neurosci 28:403–450

Furedy JJ (1983) Operational, analogical and genuine definitions of psychophysiology. Int J Psychophysiol 1:13–19

Hou RH, Samuels ER, Langley RW, Szabadi E, Bradshaw CM (2007) Arousal and the pupil: why diazepam-induced sedation is not accompanied by miosis. Psychopharmacology (Berl) 195:41–59

Kline NS (1961) On the relationship between neurophysiology, psychophysiology psychopharmacology, and other disciplines. Ann N Y Acad Sci 92:1004–1016

Ludtke H, Wilhelm B, Adler M, Schaeffel F, Wilhelm H (1998) Mathematical procedures in data recording and processing of pupillary fatigue waves. Vision Res 38:2889–2896

Stern JA (1964) Toward a definition of psychophysiology. Psychophysiology 1:90–91

Tomkiewicz RL, Cohen W (1970) The effects of changes in arousal level on critical flicker fusion frequency and figural reversal tasks. Psychophysiology 6:421–428

Psychosis

Definition

The word "psychosis" is used to describe a condition that affects the mind to the extent that there has been some significant loss of contact with reality. A psychotic episode can be caused by numerous different psychiatric disorders. Somatic or physical disorders can also cause ▶ psychotic symptoms, as can prescribed and nonprescribed drugs.

Psychosocial Interventions

Definition

Psychosocial interventions are measures of a psychological or social nature that are intended to improve the course or outcome of a pathological condition. They are often, but not always, provided in addition to pharmacological or somatic treatments. Examples include psychotherapy, behavior therapy, family therapy, counseling, and education.

Psychostimulant Dependence and Its Treatment

Matthew E. Layton and John M. Roll
Program of Excellence in Addictions Research,
Washington State University-Spokane,
Spokane, WA, USA

Definitions

Psychostimulants – psychoactive drugs that dramatically increase brain neurotransmitters and produce stimulating physiological and psychological effects.

Indirect Dopamine Agonists – drugs that increase the availability of dopamine in the brain, either by blocking dopamine reuptake providing a metabolic precursor to dopamine, or by blocking dopamine breakdown.

Stimulant Use Disorders – diagnostic terminology in the Diagnostic and Statistical Manual (DSM-5) for the category of addiction to psychostimulants, including cocaine and amphetamine-like drugs, replaces the terms substance "abuse" and "dependence".

Category 1 Drug – defined by the US Food and Drug Administration (USFDA) as a controlled substance with no recognized therapeutic use.

Introduction

Overview of Psychostimulants

Psychostimulants dramatically increase dopamine in the brain. However, depending on the specific drug and how the category is further defined, they also dramatically increase norepinephrine and serotonin brain levels. This psychoactive drug category includes more than just amphetamines, cocaine, and methylphenidate. Cathinone, one of the common ingredients of "bath salts," and *Catha edulis* (khat) are also considered psychostimulants. Cathinone and related drugs mephedrone, methylenedioxypyrovalerone (MDPV), and methylone boost both dopamine and norepinephrine levels dramatically in the brain and produce many of the same physiological and psychological effects as amphetamines (UFSDA 2013). In the United States, cathinone itself is considered Category 1. While the use of both cathinone and khat is now illegal in some countries, the most well-known cathinone-like drug is the antidepressant bupropion. Bupropion increases brain dopamine levels less dramatically and will be discussed later as a putative treatment for stimulant-related disorders (Perez-Mana 2011).

While there are subtle differences in mechanism of action, psychostimulants typically block neurotransmitter reuptake transporters. For example, this is how methylphenidate works at the dopamine reuptake transporter (Stahl 2008). However, in addition to blocking reuptake, amphetamines also cause the direct release of dopamine. The degree to which each agent affects reuptake of monoamines at each respective dopamine, norepinephrine, and serotonin transporter depends on their affinity for that specific transporter protein. There is some degree of nonspecificity in how natural monoamines are taken back up into the neurons, and once inside, vesicular monoamine transporters do not discriminate among dopamine, norepinephrine, and serotonin when packaging them into vesicles prior to release into the synaptic cleft.

Other drugs could be considered psychostimulants in a broader definition if the key similarity is that they increase monoamine levels in the brain and produce a stimulating effect, both physiologically and psychologically. These would include methylenedioxymethamphetamine (MDMA). Both methamphetamine and MDMA, also known as "ecstasy" and "molly," are chemically substituted amphetamines. MDMA is thus chemically closely related to amphetamines and produces similar physiological effects, such as increased heart rate and blood pressure. However, since MDMA increases serotonin levels in the brain relatively more than amphetamines, many of the psychoactive effects differ. MDMA is considered to be an "empathogen" or "entactogen" because in addition to stimulating physiological responses similar to psychostimulants, it also increases self-reported feelings of empathy, love, and emotional closeness. For purposes of this entry, empathogens/entactogens and hallucinogenic drugs such as lysergic acid diethylamide (LSD), psilocybin, and phencyclidine (PCP) will not be considered as psychostimulants.

Stimulant Use Disorders as Psychiatric Conditions

The American Psychiatric Association (APA) *Diagnostic and Statistical Manual of Mental Disorders-5* (DSM-5) was released in May 2013 (APA 2013). The terms substance "abuse" and "dependence" are now considered obsolete by the APA. Substance-related addictions are now considered "substance use disorders" and depend on the specific substance being used. The diagnosis is then subcategorized by level of severity. For example, psychostimulant use disorders fall under the heading "stimulant-related disorders."

An individual would be diagnosed when they demonstrate "a pattern of amphetamine-type, cocaine, or other stimulant use leading to clinically significant impairment or distress, as manifested by at least two of the following during a 12 month period:

1. The stimulant is often taken in larger amounts or over a longer period than intended.
2. There is a persistent desire or unsuccessful efforts to cut down or control stimulant use.
3. A great deal of time is spent in activities necessary to obtain the stimulant, use the stimulant, or recover from its effects.
4. Craving, or a strong desire or urge to use the stimulant.
5. Recurrent stimulant use resulting in a failure to fulfill major role obligations at work, school or home.
6. Continued stimulant use despite having persistent or recurrent social or interpersonal problems caused or exacerbated by the effects of the stimulant.
7. Important social, occupational, or recreational activities are given up or reduced because of stimulant use.
8. Recurrent stimulant use in situations in which it is physically hazardous.
9. Stimulant use is continued despite knowledge of having a persistent or recurrent physical or psychological problem that is likely to have been caused or exacerbated by the stimulant.
10. Tolerance, as defined by either of the following: a. a need for markedly increasing amounts to achieve intoxication or desired effect; or b. a markedly diminished effect with continued use of the same amount of the stimulant.
11. Withdrawal, as manifested by either of the following: a. the characteristic withdrawal syndrome for the stimulant, or b. the stimulant (or closely related substance) is taken to relieve or avoid withdrawal symptoms."

If an individual demonstrates 2–3 of the 11 problematic use symptom criteria listed, the disorder is considered "mild," 4–5 symptoms "moderate," and six or more "severe." Further specifiers include "in early remission," "in sustained remission," and "in a controlled environment." To be classified as a "disorder," the impact of stimulant use must, in addition to causing clinically significant distress, also negatively affect an individual's ability to function (APA 2013).

Treatment of Stimulant Use Disorders

When the use disorder is severe, as in cocaine use disorder with characteristic tolerance and withdrawal symptoms and continued use despite many negative consequences, inpatient care may be warranted. This involves acute detoxification and treatment of the patient in a supervised setting with medical oversight. Acute agitation and psychosis due to stimulant use may be managed using a combination of benzodiazepines and antipsychotic medication. The Treatment Episode Data Set (TEDS) is published by the US Department of Health and Human Services Substance Abuse and Mental Health Services Administration (SAMHSA 2007). In 2005, cocaine and stimulants accounted for 23 % of all TEDS admissions or approximately 425,400 admissions for psychostimulant treatment in the United States alone. Access to inpatient substance use treatment varies widely and depends not only on the availability of qualified providers in appropriately licensed and credentialed facilities but also on available resources. If the stimulant use disorder is mild, it may actually be more disruptive to the individual's life to pursue inpatient treatment. Many programs are for at least 30 days, possibly taking them away from their family, work environment, and support network. While some of these may be stressors adding to the individual's tendency to use, in the case of a mild pattern of use, it is essential to include family and friends when effectively addressing substance use. If the individual does not have the means to pay for inpatient treatment, either through insurance coverage, health-care entitlement programs, or other sources, or does not want to pursue an inpatient level of care, the next step would be to pursue an intensive outpatient program.

Psychosocial Assessment and Treatment of Stimulant Use Disorders

Since the effects of problematic stimulant use cover the range of biological, psychological, and social, treatment philosophies for stimulants must also address each of these aspects of the disorder. Overall treatment goals differ, often depending on the severity of the disorder. These range from complete abstinence to harm-reduction models that tolerate some use within the recovery process. Treatment of stimulant-related disorders needs to cover this entire span. In a best-case scenario, someone with a stimulant use disorder would achieve abstinence and employ relapse reduction techniques to stay healthy.

Treatment of stimulant-related disorders ideally includes first establishing a solid clinician-patient therapeutic alliance, with fully informed consent. Openly discussing a substance use disorder with a treatment provider is difficult for people to do and often has to be done under court-ordered circumstances, frequently due to negative consequences of the individual's compulsive behavior. A comprehensive diagnostic evaluation needs to include a thorough patient and family history, a mental status examination, appropriate laboratory testing, and cross-cultural case formulation. Each diagnosis made should have evidence to support it. This includes sufficient information to educate the individual and their support system as to how best approach treating this person. The typical clinical course and consequences of not treating their substance use disorder need to be communicated to the patient. Individually tailored treatment recommendations need to be clearly articulated and coordinated among treatment providers.

Clinical and research approaches to address stimulant-related disorders can be divided into broad categories of behavioral (community reinforcement approach, contingency management), psychosocial (motivational interviewing, cognitive behavioral therapy, matrix model – actually eclectic), and abstinence-oriented treatments (detoxification, 12-step programs, and residential rehabilitation programs, or "clean houses"). Evidence supports interventions that may be brief and self-directed, often when an individual is experiencing emergency medical or legal consequences due to their use.

The Cochrane Collaboration review of psychosocial interventions for cocaine- and psychostimulant-related disorders provides a thorough evaluation of the available evidence (Knapp et al. 2008). 27 randomized, controlled trials met the criteria for inclusion in this meta-analysis; 26 of these studies addressed cocaine use. Results showed that more comprehensive behavioral treatments, in which a motivational incentive-based contingency management program is provided in addition to a community reinforcement approach, demonstrated significantly better effects on increasing abstinence and improving treatment retention rates than drug counseling alone or behavioral interventions without the added incentive program. These approaches also appear useful in reducing methamphetamine use (Roll et al. 2006). In combination with contingency management, the Community Reinforcement Approach intervention focuses on five areas: (1) providing skills training to minimize drug use, (2) improving family relations, (3) providing educational and employment counseling, (4) assisting with developing new recreational activities and social networks, and (5) monitoring disulfiram therapy for those who abuse alcohol.

Putative Pharmacotherapy for Stimulant Use Disorders

Psychoactive medication interventions for stimulant-related disorders are still at the research level. As of this time, there are no FDA-approved compounds for the treatment of stimulant-related disorders. Indirect dopamine agonists (IDAs) including modafinil, selegiline, bupropion, disulfiram (which in addition to inhibiting alcohol metabolism also decreases dopamine breakdown), dexamphetamine, methylphenidate, mazindol, L-dopa/carbidopa, and methamphetamine have all been assessed for efficacy in psychostimulant-dependent individuals. A meta-analysis of 29 randomized, parallel-group, placebo-controlled clinical trials revealed that IDAs increased psychostimulant abstinence, but did not increase retention in treatment (Perez-Mana 2011). The efficacy of IDAs was greater in comorbid heroin-dependent individuals and was positively related with treatment length.

Vocci et al. (Roll 2009) summarized findings from clinical trials in methamphetamine dependence for bupropion, methylphenidate, baclofen, topiramate, lobeline, gamma-vinyl-GABA (vigabatrin), naltrexone, and modafinil. Their conclusion from these results was that only bupropion (Newton et al. 2006) and methylphenidate warranted further study for use in methamphetamine dependence. Recent results of relatively small clinical trials indicate that other psychoactive agents may hold promise for treating stimulant-related disorders, in particular the widely available generic antidepressant mirtazapine (Graves et al. 2012).

Summary

Stimulant-related disorders continue to take a huge toll on individuals and society. In addition to the detrimental impacts on people's ability to function at work and school, in parenting, or even in basic social relationships, stimulant use disorders also contribute to crime and poverty. Evidence-based treatments do exist for stimulant use disorders, whether mild, moderate, or severe. Many obstacles remain if we are to provide adequate treatment resources for these individuals, as access to thorough, comprehensive diagnostic evaluation and treatment services at needed levels of care and in appropriate facilities remains a challenge. Substance use disorders also frequently co-occur with other mental disorders and medical disorders, including pain conditions. Providers need to tailor treatment interventions to deliver patient-centered care, and the therapeutic alliance is essential in order to produce the best outcomes for all involved.

Cross-References

► Abstinence
► Addiction
► Agonist
► Amphetamine
► Appetite Suppressants
► Benzodiazepines
► Bupropion
► Carbidopa
► Cocaine
► Cognitive Behavioral Therapy
► Contingency Management
► Detoxification
► Disulfiram
► Dopamine
► Dopamine Agonist
► Dopamine Transporter
► DSM
► Entactogen
► Hallucinogens
► Khat
► Levodopa, L-DOPA
► Metabolism
► Methamphetamine
► Methylenedioxymethamphetamine (MDMA)
► Methylphenidate and Related Compounds
► Mirtazapine
► Modafinil
► Neurotransmitter
► Neurotransmitter Transporters
► Norepinephrine
► Norepinephrine Transporter
► Pharmacotherapy
► Phencyclidine
► Physical Dependence
► Psychosis
► Psychosocial Interventions
► Psychostimulant Dependence and Its Treatment
► Psychostimulant Use Disorder
► Psychostimulants
► Psychotic Symptoms
► Receptor
► Relapse
► Retention
► Selective Serotonin Reuptake Inhibitors
► Selegiline
► Serotonin
► Serotonin Transporter
► Topiramate

References

American Psychiatric Association (APA) (2013) Diagnostic and statistical manual of mental disorders-5. Washington, DC
Graves SM, Rafeyan R, Watts J, Napier TC (2012) Mirtazapine, and mirtazapine-like compounds as possible pharmacotherapy for substance abuse disorders: evidence from the bench and the bedside. Pharmacol Ther 136(2012):343–353
Knapp WP, Farrell M, Silva De Lima M (2008) Psychosocial interventions for cocaine and psychostimulant related disorders. The Cochrane Collaboration, Wiley, Chichester
Newton TF, Roache JD, De La GR et al (2006) Bupropion reduces methamphetamine-induced subjective effects and cue-induced craving. Neuropsychopharmacology 31:1537–1544
Pérez-Mañá C (2011) Efficacy of indirect dopamine agonists for psychostimulant dependence: a systematic review and meta-analysis of randomized controlled trials. J Subst Abuse Treat 40:109–122
Roll JM (ed) (2009) Methamphetamine addiction: from basic science to treatment. The Guilford Press, New York/London
Roll JM, Petry NM, Stitzer ML et al (2006) Contingency management for the treatment of methamphetamine use disorders. Am J Psychiatry 163:1993–1999
Stahl S (2008) Essential Psychopharmacology. Cambridge University Press, Cambridge
Substance Abuse and Mental Health Services Administration, Department of Health and Human Services (2007) Treatment episode data set. Accessed 28 Oct 2013
United States of America Food and Drug Administration Database (2013) http://www.accessdata.fda.gov/cms_ia/importalert_180.html. Accessed 28 Oct 2013

Psychostimulant Use Disorder

Synonyms

Psychostimulant addiction; Psychostimulant dependence

Definition

Psychostimulant use disorder is a chronic, relapsing disorder diagnosed by the hallmark of compulsive drug taking and drug seeking to the exclusion of other important life activities. The fifth edition of the *Diagnostic and Statistical Manual of Mental Disorders* (*DSM-5*) combined two subcategories (substance abuse, substance dependence) into a single category of "substance use disorder" with graded clinical severity (mild to severe). The diagnosis of a psychostimulant use disorder is based upon the presence of 2 or more of 11 criteria (e.g., hazardous use, social/interpersonal problems related to use, neglected major life activities, craving). The feeling states of craving (an intense desire for the drug) and euphoria are primary reinforcers of the addiction. Brain areas important for decision making, learning, memory, and behavioral control are altered in individuals addicted to psychostimulants and other abused drugs.

Cross-References

- Craving
- DSM
- Substance Abuse
- Substance Dependence
- Substance Use Disorders: A Theoretical Framework

Psychostimulants

Kathryn A. Cunningham[1] and Marcy J. Bubar[2]
[1]Center for Addiction Research and Department of Pharmacology and Toxicology, University of Texas Medical Branch, Galveston, TX, USA
[2]Center for Addiction Research, University of Texas Medical Branch, Galveston, TX, USA

Synonyms

Psychomotor stimulants; Stimulants

Definition

Psychostimulants are drugs that diminish fatigue and elevate mood and alertness.

Pharmacological Properties

History

Psychostimulants are a broad family of drugs that include compounds from naturally occurring plant alkaloids (e.g., caffeine, cocaine, ephedrine, nicotine) to synthetic amphetamines such as *d*-amphetamine, methamphetamine, and 3,4-methylenedioxymethamphetamine (MDMA; ecstasy) as well as methylphenidate. Some of these psychostimulants are available over the counter (e.g., caffeine) or under prescription for a given medical indication (e.g., methylphenidate), while others are illicit and available only on the black market (e.g., MDMA). The use of psychostimulants has been recorded throughout history, dating back to Paleolithic times when Ethiopian nomads brewed *Coffea arabica* (coffee) into a stimulating beverage, North American Indians smoked wild tobacco *Nicotiana* variants, and Peruvian Indians chewed the leaves of *Erythroxylum coca* (cocaine) to enhance endurance and energy, while Chinese physicians utilized *Ephedra vulgaris* (ephedrine) to treat fever, nasal congestion, and asthma. The scientific literature has focused on several members of the broad family of psychostimulants, including caffeine, cocaine, amphetamine derivatives, and nicotine.

Caffeine has been consumed in concoctions derived from plant sources since prehistoric times. This mild stimulant and diuretic found naturally in coffee, tea, and cocoa is currently the most widely used psychoactive substance in the world. Employed to enhance alertness and reduce fatigue, caffeine is now an additive to many other food substances, including beverages which boast names such as 5-Hour Energy, Powershot, and *Full Throttle*. Caffeine is used medically to enhance the efficacy of medications used in the treatment of migraine and pain, and is occasionally used to stimulate respiration, or to overcome the sedative effects of antihistamines.

Cocaine is one of about thirty alkaloids of the coca plant and was first isolated in 1855. By the late 1800s, cocaine was widely used for its anesthetic properties and was sold over the counter until 1916. By the early twentieth century, the addictive properties of cocaine were well known, and, in 1970, cocaine was officially made illegal and classified as a Schedule II drug (limited medical uses) in the United States under the Comprehensive Drug Abuse Prevention and Control Act of 1970 (Controlled Substances Act). Cocaine is used in various forms including powder, freebase, and crack and remains a popular abused street drug in Western society. The powder form of cocaine is snorted or injected, while freebase or crack cocaine are smoked. The therapeutic uses of cocaine are limited to its application as a topical analgesic in ear, nose, and throat surgeries. Street names abound for the various abused forms of cocaine, including coke, snow, flake, blow, candy, rock, and toot.

The use of amphetamine derivatives dates back to Neolithic times in China and Africa where herbs in the genus *Ephedra* (Ma huang) and *Catha edulis* (khat or qat) were employed for their stimulant properties. Ephedrine was isolated from the *Ephedra* plant in the late 1800s and was soon utilized widely to treat asthma. In the 1920s, interest in finding a synthetic substitute for ephedrine brought about the extensive characterization of the structurally similar compound amphetamine which had first been synthesized in 1887. Thereafter, amphetamine was marketed to treat nasal congestion, narcolepsy, and attention deficit hyperactivity disorder (ADHD), and subsequently a number of amphetamine analogs were developed and characterized, including the now-popular street drugs methamphetamine and MDMA (ecstasy).

Amphetamine use was common during World War II when soldiers were issued the stimulant to decrease fatigue and increase alertness. In the 1950s, legally manufactured tablets of amphetamine and methamphetamine were utilized by college students, truck drivers, and athletes across the United States to stay awake and to increase energy. Amphetamines were widely prescribed through the 1960s to suppress appetite and enhance weight loss. While effective for this purpose, the high abuse liability of amphetamine outweighed their utility for this indication. Due to the epidemic of amphetamine use, the US Bureau of Narcotics and Dangerous Drugs [BNDD; later named the Drug Enforcement Administration (DEA)] shifted all amphetamine products to Schedule II in 1971 under the Controlled Substances Act. A prescription then became required for possession of an amphetamine. In 1988, the growth in MDMA abuse and evidence to suggest its systemic toxicity prompted the classification of this amphetamine derivative as Schedule I controlled substance (due to its high abuse potential and no currently accepted medical use). A multitude of street (slang) names for amphetamine (speed, uppers, crosses, black beauties), methamphetamine (meth, ice, crank, speed) MDMA (ecstasy, E, X, molly), and other derivatives are used.

Nicotine is the main active psychostimulant alkaloid of the tobacco plant. Introduced to Western society in the 1500s, tobacco had been used by Native Americans and Asians for centuries before. By 1964, scientists had concluded that tobacco smoking was causally linked to lung cancer and heart disease, and health warnings were deemed mandatory for tobacco products. The principal therapeutic use of nicotine and nicotine agonists (e.g., the partial agonist varenicline) is in the treatment of nicotine dependence.

Effects in Humans

The constellation of subjective effects that characterize the "intoxication" induced by psychostimulants and that capture the attention of users as pleasant includes euphoria, mood elevation, enhanced feelings of well-being, and motor and mental stimulation, although each psychostimulant provokes a unique subjective profile. For example, caffeine is generally not euphorigenic, while MDMA users also report feelings of closeness with other people, increased empathy, and perceptual alterations when under the influence of this drug. Despite subjective effects that are perceived as positive, psychostimulants also have many negative consequences. Some stimulant users report anxiety, decreased appetite and weight loss, feelings of restlessness, irritability, and sleep disturbances. Other negative consequences include cardiac arrhythmia, seizures, and myocardial infarction. Higher doses, which are often used to intensify the high, may also induce aggressiveness, disorientation, hallucinations, paranoia, muscle twitches, tremors, and vertigo. Additionally, high doses of MDMA evoke hyperthermia, which can lead to life-threatening conditions such as multiple organ failure; cocaine can also elicit hyperthermia, but this effect is less reported than that evoked by MDMA.

When taken chronically, psychostimulants (especially, cocaine and amphetamine derivatives) may lead to stereotypical and repetitive behaviors. Compulsive drug taking and drug seeking can develop, while the emergence of additional psychiatric disorders can be observed, including anxiety, depression, and a paranoid psychosis that shares some features of schizophrenia. Toxic physiological reactions are also observed. Individuals addicted to psychostimulants are remarkably willing to risk death, incarceration, medical and psychiatric complications, job loss, and family turmoil in pursuit of the drug.

Mechanisms of Action

Dopamine is the most prominent and well-studied neurotransmitter involved in the effects of psychostimulants. The dopamine mesoaccumbens pathway, originating in dopamine neurons of the ventral tegmental area and terminating in the nucleus accumbens, mediates the reinforcing effects of natural rewards (i.e., food and sex), and also serves as a common substrate for the acute rewarding effects of all drugs of abuse, regardless of their individual mechanisms of action (for review, see Kelley 2004). In addition, dopaminergic influence, in conjunction with glutamatergic neurotransmission, within nodes of the extended, limbic–corticostriatal circuitry (see ▶ Prefrontal Cortex, ▶ Amygdala, ▶ Hippocampus) is integral in coordinating reward-related associative learning and motivated behaviors that contribute to aspects of addictive behaviors such as craving, withdrawal, and relapse (Kelley 2004). While dopamine is fundamental to the normal function of this brain circuitry, regulatory oversight of the signalling involved in reward circuits is provided by multiple neurotransmitters.

Each psychostimulant presents a distinct profile of actions that mediates its psychostimulant effects. The caffeine molecule bears structural resemblance to the endogenous neurotransmitter adenosine and appears to evoke its psychostimulant effects due to its action as an adenosine receptor antagonist in brain (Ferre 2008). This action of caffeine occurs at the adenosine A_{2A} receptor which plays a critical role in the control of dopamine function in reward and arousal. Additional actions of caffeine in the body abound, while metabolites of caffeine also contribute to its biological profile. The possibility has been raised that selective A_{2A} adenosine receptor antagonists may prove useful in the pharmacotherapy of stimulant use disorders (Ferre 2008), although this hypothesis remains to be fully explored (see Psychostimulant Use Disorders, below).

Amphetamines and cocaine present a multifaceted modulatory impact over monoamine function, and the chemical structure of the amphetamine molecule is a determinant of its exact pattern of influence over neuronal function (Sulzer et al. 2005). The effects of cocaine and amphetamine derivatives in humans are thought to be mediated by their ability to enhance synaptic concentrations of monoamines via actions at neurotransmitter transporters for dopamine, serotonin (5-HT), and norepinephrine. These transporters are found in the outer membranes of monoamine neurons and act to reuptake neurotransmitters released by the nerve terminal. In fact, reuptake serves as the primary mechanism of inactivation for released monoamine neurotransmitters. Amphetamine derivatives display different profiles of affinity for and efficacy at the three monoamine transporters. In general, amphetamine derivatives increase synaptic monoamine levels through both the blockade of reuptake and the reversal of the membrane-localized transporter for these monoamines.

The reversal of the transporter results in release of monoamines from the nerve terminal, providing an additional mechanism for elevation of synaptic neurotransmitters. The parent molecule, amphetamine, has greatest affinity for the dopamine transporter, while methamphetamine and MDMA have greater affinity for the serotonin transporter compared to the dopamine or norepinephrine transporters. Amphetamines can interfere with the storage of monoamines in compartments (vesicles) inside neurons and can retard degradation of monoamines. Both of these actions further result in increased accumulation of free neurotransmitter in the nerve terminal (Sulzer et al. 2005).

Cocaine binds to the monoamine transporters and inhibits reuptake of dopamine, serotonin, and norepinephrine. Thus, by blocking reuptake, cocaine increases the length of time that the monoamines remain in the synapse available to activate their receptors. Cocaine is also reported to have affinity for the serotonin 5-HT_3 receptor, M_1 and M_2 muscarinic receptors, and σ-receptors, although their roles in mediating the stimulant effects of cocaine are not clear.

Nicotine acts as a psychostimulant at very low doses through activation of endogenous nicotine receptors, proteins that are normally activated by the endogenous neurotransmitter acetylcholine. Its actions result in increased levels of dopamine in reward circuits of the brain, providing a neuroanatomical basis for its stimulant effects.

Psychostimulant Use Disorders

The new millennium dawned on a continuing societal struggle with the consequences of use and abuse of psychostimulants. Stimulant abuse remains an alarming concern, and the issues are complex. On one hand, stimulants have diverse therapeutic utility in a large population for which they are essential (e.g., attention deficit disorders). On the other hand, the abuse of illicit stimulants such as cocaine, methamphetamine, and other substituted amphetamines does not show signs of abatement. For example, the United Nations Drug Control Program (UNDCP) estimates that in 2011, 33.8 million people used amphetamine-type substances, 19.4 million used ecstasy, and 17 million used cocaine in the preceding year, representing 0.7 %, 0.4 %, and 0.37 % of the global population, respectively (United Nations Office on Drugs and Crime 2013). The continued public health problems with cocaine, combined with the escalating abuse of methamphetamine and prescription amphetamines, are a colossal challenge which must be tackled with all available resources. The recidivism rate of stimulant users is very high, and access to effective treatment modalities is so rare, that new, easily available and effective treatments are needed. To achieve this goal, an essential understanding of the neurobiological mechanisms underlying stimulant abuse disorders and their persistence is required. The long history and cyclical nature of stimulant abuse, the emergence and subsidence of a given psychostimulant as an abused favorite, concurrent with the emergence of yet others, suggests that there is a need to take a long and realistic view. There will be a continuing need to deal with the problems of stimulant use disorders, whether with those known or those that will emerge in the future.

The fifth edition of the *Diagnostic and Statistical Manual of Mental Disorders* (*DSM-5*) combined two subcategories (▶ Substance Abuse, ▶ Substance Dependence) into a single category of "substance use disorder" with graded clinical severity (mild to severe) (Hasin et al. 2013). Diagnosis of a psychostimulant use disorder is based upon the presence of two or more of 11 criteria (e.g., hazardous use, social/interpersonal problems related to use, neglected major life activities, craving). Although patients cite many behavioral reasons why they use a psychostimulant, the feeling states of craving (an intense desire for the drug) and euphoria are primary reinforcers of the disorder. Drug-induced euphoria and craving alternate over time, exerting positive and negative reinforcement to form a cycle of abuse that becomes increasingly entrenched and uncontrollable. The severity of a stimulant use disorder can progress very rapidly in vulnerable individuals. While it may take years or decades for alcoholics to develop end-stage alcoholism, progression to severity among stimulant abusers typically occurs more rapidly. Rapid progression is particularly common with the smoked, freebase form of cocaine known as crack, probably because the intrapulmonary route is the quickest means of delivering a bolus of cocaine to the brain. Because psychostimulants engage brain reward centers, the strength and characteristics of stimulant use take on an almost primary survival drive. The inability of patients to control stimulant use illustrates the power of reward centers over behavior and partially explains the relapsing nature of cocaine and amphetamine dependence (Kalivas and Volkow 2005; Koob and Volkow 2010).

The etiology of stimulant use disorders includes genetic, psychological, pharmacological, neural, and environmental determinants. Major advances in understanding the neurobiology of substance use disorders have been generated over the last 25 years revealing the complex biological processes that trigger and sustain addictive behavior and the physiological ramifications of chronic exposure to abused drugs (Kelley 2004; Kalivas and Volkow 2005; Everitt and Wolf 2002; Koob and Volkow 2010). The escalation from mild to severe substance use disorders is linked inextricably to brain pathophysiology that develops during chronic drug exposure, composed of a myriad of neuroadaptive responses (termed neuroplasticity) that alter normal homeostasis. Some of these neuroadaptations underlie the evolution of strong mnemonic associations between environmental cues and the drug-taking experience as well as emerging psychiatric complications. Cessation of drug use during abstinence represents another phase of substance use disorders during which drug withdrawal, "craving," poor impulse control, and reactivity to drug-associated environmental cues challenge the best intentions of the addict to remain abstinent. Each of these hallmark phases (predrug exposure, early drug exposure, chronic abuse, dependence, withdrawal, abstinence) are stages during which brain circuits that normally control the intertwined processes of reward, motivation, memory, stress responsivity, and emotional processing accumulate adaptations that contribute significantly to the behavioral sequelae of these disorders. The normality of these circuits is dependent upon the function of overlapping and interacting neurotransmitter systems in the brain, and the impact of a given psychostimulant on these neurotransmitters sets in motion the turmoil of abuse. Drug-induced neuroadaptations of genes and proteins in these circuits have been described and include neurotransmitter transporters and receptors, synthetic enzymes, and numerous transcription factors. However, which specific protein changes contribute to neural plasticity associated with drug exposure, and the reversibility of such modifications, and their integration remain to be fully elucidated.

The complex neuronal interconnectivity is genetically determined, but environmentally tweaked, to provide infinitely unique individuals. This epigenetic background provides the fertile soil into which the use of a drug of abuse is introduced; if the individual is vulnerable to the rewarding and reinforcing effects of the drug, then the transition from use to abuse to dependence can be set in motion. Time-locked modifications in neurotransmitter functionality culminate in neuronal plasticity in these circuits, a devolutionary process that presumably results in the hazardous patterns of drug-taking and drug-seeking behavior which characterize a substance use disorder.

Modeling Psychostimulant Use Disorders in Animals

The development of effective approaches to treatment of stimulant use disorders relies on well-designed clinical research in drug-using subjects as well as preclinical research in which rigorous experimental control and specific behavioral and pharmacological manipulations can be undertaken to model and control the progression of drug taking and drug seeking across time. These models have greatly enhanced our understanding of the roles of neurotransmitter systems in the rewarding, reinforcing, and conditioned effects of stimulants (Cunningham and Anastasio 2014; Kalivas and Volkow 2005; Koob and Volkow 2010). These preclinical advances originally focused the field on the dopamine neurotransmitter system as a relevant target for medications development for psychostimulant abuse. However, clinical trials have not yet identified dopamine-based pharmacotherapies which lack abuse liability and exhibit efficacy to enhance abstinence, reduce craving, and prevent relapse. Alternative investigations now suggest that in addition to dopamine, other neural systems play vital roles in addictive processes, including adenosine, glutamate, γ-aminobutyric acid (► GABA), neuropeptides, and serotonin; these results have generated interesting new prospects for pharmacotherapy. These systems may function to modulate the neural output of the anatomically defined reward circuit and in the wider circuit of reciprocal interactions termed the limbic cortical–ventral striatopallidal circuit (Cunningham and Anastasio 2014; Kalivas and Volkow 2005; Koob and Volkow 2010).

The reinforcing and rewarding effects of psychostimulants have been well characterized in the intravenous drug self-administration paradigm in rodents and nonhuman primates. This model exhibits face validity in that drugs with high abuse potential in humans are readily self-administered by animals (Mello and Negus 1996). One advantage is that studies can be designed to model multiple stages of human drug dependence from acquisition through maintenance to relapse, and studies employing this model have led to a greater understanding of the neurobiological underpinnings of addictive behaviors. Drug self-administration is solidly based upon the operant conditioning principle of "reinforcement." In drug self-administration, the presentation of a stimulus (e.g., cocaine as an appetitive "reward") contingent upon a behavioral response increases the probability that behavior will reoccur. Based upon the ability of an animal to make an operant response (e.g., nose poke or lever press) which is reinforced by the presentation of a psychostimulant reward, this assay is widely used to study environmental and pharmacological variables important in the initiation, maintenance, and escalation of drug-taking behavior. For intravenous self-administration studies, the subject acquires a drug infusion by performing a discrete response. The number and pattern of responding required for each infusion is determined by the schedule of reinforcement imposed by the experimenter. Drug availability typically is signaled by an environmental stimulus, and the dependent variables are the number of infusions obtained or the rate of responding (i.e., lever presses or nose pokes) during a session (Mello and Negus 1996).

Modeling Drug Seeking and Relapse

A basic definition of relapse in the context of substance use disorders is the return to drug-seeking and drug-taking behavior following a period of abstinence. Often, the role of "craving," a strong desire or urge to use the substance, is invoked as a primary motivating force behind relapse and DSM-5 now recognizes craving as a criteria in diagnosing substance use disorders. Although biological measures of craving remain to be identified, pharmacological means to reduce craving are being sought as potential "abstinence enhancers." It is widely known that conditioned cues associated with psychostimulant use are a major factor in relapse to continued use, and it is clear that associations between environmental cues and the drug-taking experience become well consolidated into memory. In fact, the molecular pathways of drug addiction overlap prominently with those involved in learning and memory processes, as elegantly reviewed by (Kelley 2004). The animal self-administration paradigm possesses good predictive and face validity for modeling the activation of craving states by conditioned environmental stimuli (e.g., places in which a drug was consumed or drug-associated paraphernalia) in drug-dependent individuals. In these models, conditioned cues previously associated with psychostimulant delivery (e.g., lights, tones) are presented in the absence of the drug either to trigger the operant responses engendered in the drug-taking situation of self-administration or reinforce continued drug seeking by presentation of the cues upon the behavioral response. The use of nonresponse contingent ("priming") injections of drugs to induce reinstatement of drug self-administration has also been found to produce a robust degree of reinstatement and is a good model for pharmacologically induced relapse in stimulant use disorders (Mello and Negus 1996). Researchers are continually working to adapt these models to more closely align with aspects of human drug taking, drug seeking, and relapse to further the understanding of processes involved in the development of stimulant use disorders. In addition, the imperative to discover the means to reverse neuroplastic events resultant from psychostimulant exposure and to therapeutically improve function in the affected brain requires refined preclinical models.

The Future

We have entered the world of personalized medicine in which a patient's unique genetic and phenotypic profile will ultimately provide the signposts to tailor diagnosis and treatment for health disorders. The science in chronic diseases such as cancer and diabetes is rapidly advancing in these goals to provide new targeted diagnostics and medications for treatment tailored to provide the best clinical response in an individual. The vision to improve our diagnostic and prognostic capabilities with identified disease biomarkers for stimulant use disorders, to understand vulnerability factors involved in their progression and relapse, and to treat this complex disorder with maximal efficacy is now within reach of science.

Clinical research endeavors continue to refine psychotherapeutic strategies (e.g., cognitive behavioral therapy, contingency management) to suppress relapse and enhance recovery from stimulant use disorders. Medications are also useful to suppress relapse which requires reestablishment of normal neural function consequent to long-term, drug-induced neuroplasticity and diminish the power of relapse triggers. Effective medications for alcohol, opioid, and nicotine use disorders are available. For example, nicotine gum, patches, and lozenges are therapeutically useful in smoking cessation programs, and the nicotine receptor partial agonist varenicline provides an additional option. However, medication development efforts have not yet yielded effective pharmacotherapies for use disorders involving cocaine or other abused psychostimulants. A great deal of interest remains in filling this gap to maximize the probability of treatment success by minimizing lapses to drug use (Hendershot et al. 2011; Volkow and Skolnick 2012). Several potential medications with various mechanisms of action have been or are currently under study in cocaine- and/or methamphetamine-dependent subjects in outpatient clinical trials (www.clinicaltrials.gov) in which efficacy to decrease drug-positive urines and/or reduce craving is measured.

The initial focus of medications development for cocaine addiction was to utilize compounds that acted at dopamine receptors or the dopamine reuptake transporter. However, agents that act as dopamine D_1 or D_2 receptor antagonists and selective dopamine transporter inhibitors were either ineffective or resulted in unwanted side effects. Other more recent dopaminergic strategies explored in Phase I and Phase II clinical trials for cocaine use disorder include D-amphetamine or its prodrug lisdexamfetamine as agonist replacement therapies (similar to the use of methadone for opioid use disorder); the nonselective and selective dopamine β hydroxylase (DβH) inhibitors, disulfiram and nepicastat, respectively; the nonselective dopamine receptor antagonist L-tetrahydropalmatine (L-THP); and the nonselective D_3 receptor antagonist, buspirone. Novel selective D_3 receptor antagonists are currently under development and have shown efficacy in preclinical studies.

Antidepressants have also been employed in open-label and placebo-controlled clinical trials; however, both tricyclic antidepressants such as desipramine and selective serotonin reuptake inhibitors (SSRIs), such as fluoxetine, did not exhibit efficacy to suppress relapse in the populations under study. However, the SSRI citalopram in combination with behavioral therapy significantly reduced cocaine-positive urines in subjects who exhibited high baseline impulsivity scores (Green et al. 2009); these findings suggest that phenotypic vulnerability factors (such as impulsivity) are important factors in defining personalized therapeutic strategies for stimulant use disorders.

Various classes of drugs thought to serve as cognitive enhancers are currently being explored in clinical trials for treatment of cocaine and/or methamphetamine use disorders. These include cholinesterase inhibitors (e.g., rivastigmine, huperzine A, tacrine, galantamine), nicotine acetylcholine receptor ligands (e.g., varenicline), norepinephrine reuptake inhibitors (e.g., atomoxetine), the NMDA receptor co-agonist D-serine, and the cystine–glutamate exchanger N-acetylcysteine. Early preclinical and clinical research suggests that peroxisome proliferator-activated receptor gamma (PPAR-γ) agonists (e.g., glitazones) may also be useful for treatment of psychostimulant use disorder.

Numerous other promising pharmacological targets are in various stages of exploration, including serotonin 5-HT_{2C} receptor agonists (e.g., lorcaserin), mGluR5 antagonists (e.g., fenobam), adenosine A_{2A} receptor antagonists (e.g., caffeine, SYN115), alpha-2-adrenergic agonists (e.g., guanfacine), and alpha-1-adrenergic antagonists (e.g., doxazosin). In addition, biologics are being pursued as alternative strategies to conventional pharmacotherapies. For example, a clinical trial is underway with TV-1380, a recombinant human serum albumin fused with mutated butyrylcholinesterase. TV-1380 acts to enhance the normal metabolic pathway of cocaine to blunt its central actions and toxicity. TV-1380 exhibits favorable pharmacokinetics, high catalytic efficiency, and an extended plasma half-life and is currently in phase II safety and efficacy trials. Immunotherapies to produce antibodies that bind to the psychostimulant to reduce the amount of the drug that reaches the brain are also being developed. This can be accomplished via administration of antidrug vaccines or monoclonal antibodies (mAb). A phase I placebo-controlled trial of the anti-cocaine vaccine TA-CD is currently being conducted in cocaine-dependent patients, while a first-in human study to determine the safety and tolerability of the anti-methamphetamine antibody Ch-mAb7F9 was recently completed. An important aspect of such vaccines is the necessity to achieve high antibody levels as a means to slow drug entry into brain and reduce the subjective and reinforcing effects of the psychostimulant (Orson et al. 2009).

The observation that about 60 % of alcohol-, opioid-, or cocaine-dependent patients who received treatment medications relapsed into drug use within 1 year suggests that further research and advancement of these efforts are essential in order to improve treatment efficacy. The development of new pharmacological agents for the treatment of substance use disorders would greatly increase treatment and access options, particularly if medications to enhance abstinence, reduce craving, and prevent relapse could be perfected. This is an important quest and one ripe for with near-term therapeutic potential with positive outcomes.

Cross-References

▶ Abstinence
▶ Adenosine
▶ Adenosine A_{2A} Receptors
▶ Alcohol
▶ Amphetamine
▶ Amygdala
▶ Antidepressants
▶ Atomoxetine
▶ Caffeine
▶ Cocaine
▶ Cognitive Enhancers: Role of the Glutamate System
▶ Comprehensive Drug Abuse Prevention and Control Act of 1970
▶ Crack
▶ Craving
▶ Disulfiram
▶ Dopamine
▶ Dopamine Transporter
▶ Drug Self-Administration
▶ Freebase
▶ Impulsivity
▶ Methylenedioxymethamphetamine (MDMA)
▶ Methamphetamine
▶ Methylphenidate and Related Compounds
▶ Neuroplasticity

- ▶ Neurotransmitter Transporters
- ▶ Nicotine Dependence and its Treatment
- ▶ N-Methyl-D-Aspartate Receptor
- ▶ Norepinephrine
- ▶ Norepinephrine Transporter
- ▶ Nucleus Accumbens
- ▶ Opioid
- ▶ Prefrontal Cortex
- ▶ Psychostimulant Use Disorder
- ▶ Reinstatement of Drug Self-Administration
- ▶ Relapse
- ▶ Schedule I
- ▶ Schedule II
- ▶ Substance Abuse
- ▶ Substance Dependence
- ▶ Substance Use Disorders: A Theoretical Framework
- ▶ Varenicline
- ▶ Ventral Tegmental Area
- ▶ Withdrawal Syndromes

References

Cunningham KA, Anastasio NC (2014) Serotonin at the nexus of impulsivity and cue reactivity in cocaine addiction. Neuropharmacology 76:460–478

Everitt BJ, Wolf ME (2002) Psychomotor stimulant addiction: a neural systems perspective. J Neurosci 22:3312–3320

Ferre S (2008) An update on the mechanisms of the psychostimulant effects of caffeine. J Neurochem 105:1067–1079

Green CE, Moeller FG, Schmitz JM, Lucke JF, Lane SD, Swann AC, Lasky RE, Carbonari JP (2009) Evaluation of heterogeneity in pharmacotherapy trials for drug dependence: a Bayesian approach. Am J Drug Alcohol Abuse 35:95–102

Hasin DS, O'Brien CP, Auriacombe M, Borges G, Bucholz K, Budney A, Compton WM, Crowley T, Ling W, Petry NM, Schuckit M, Grant BF (2013) DSM-5 criteria for substance use disorders: recommendations and rationale. Am J Psychiatry 170:834–851

Hendershot CS, Witkiewitz K, George WH, Marlatt GA (2011) Relapse prevention for addictive behaviors. Subst Abuse Treat Prev Policy 6:17

Kalivas PW, Volkow ND (2005) The neural basis of addiction: a pathology of motivation and choice. Am J Psychiatry 162:1403–1413

Kelley AE (2004) Memory and addiction: shared neural circuitry and molecular mechanisms. Neuron 44:161–179

Koob GF, Volkow ND (2010) Neurocircuitry of addiction. Neuropsychopharmacology 35:217–238

Mello NK, Negus SS (1996) Preclinical evaluation of pharmacotherapies for treatment of cocaine and opioid abuse using drug self-administration procedures. Neuropsychopharmacology 14:375–424

Orson FM, Kinsey BM, Singh RA, Wu Y, Kosten TR (2009) Vaccines for cocaine abuse. Hum Vaccin 5:194–199

Sulzer D, Sonders MS, Poulsen NW, Galli A (2005) Mechanisms of neurotransmitter release by amphetamines: a review. Prog Neurobiol 75:406–433

United Nations Office on Drugs and Crime (2013) 2013 World drug report (United Nations publication, sales no. E.13.XI.6). http://www.unodc.org/unodc/secured/wdr/wdr2013/World_Drug_Report_2013.pdf

Volkow ND, Skolnick P (2012) New medications for substance use disorders: challenges and opportunities. Neuropsychopharmacology 37:290–292

Psychotic Symptoms

Definition

Refers to a cluster of symptoms such as paranoia, hallucinations, delusions, and incoherent speech and behavior.

Cross-References

- ▶ Antipsychotic Drugs
- ▶ Psychosis
- ▶ Schizophrenia

Pulmonary Hypertension

Definition

Pulmonary hypertension is an increase in blood pressure in the pulmonary artery, pulmonary vein, or pulmonary capillaries, leading to shortness of breath, dizziness, fainting, and other symptoms. Pulmonary hypertension can involve a markedly decreased exercise tolerance and heart failure. The PDE5 inhibitor sildenafil, which is used to treat erectile dysfunction (marketed under the trade name Viagra), is also used for the treatment of pulmonary arterial hypertension (marketed under the trade name Revatio).

Cross-References

- ▶ Congestive Heart Failure
- ▶ Erectile Dysfunction
- ▶ PDE5 Inhibitors
- ▶ Sildenafil

Punishment

Definition

A behavioral process wherein the consequences for an operant make instances of the operant less frequent in the future. Such consequences take the form of presentations of environmental stimuli, which are thereby defined operationally as aversive (punishers). Such stimuli can either be exteroceptive (e.g., pain-inducing events such as excessive heat or pressure) or interoceptive (e.g., noxious drugs).

Cross-References

- ▶ Punishment Procedures

Punishment Procedures

Jeffrey M. Witkin[1] and James E. Barrett[2]
[1]Lilly Res Labs, Eli Lilly & Co., Indianapolis, IN, USA
[2]Department of Pharmacology and Physiology, Drexel University College of Medicine, Philadelphia, PA, USA

Synonyms

Conditioned emotional response; Experimental conflict; Punished behavior; Suppressed behavior

Definition

Punishment is defined as the suppression of behavior that results from the response-dependent delivery of a stimulus (punisher). Punishment procedures in rodents and other species have been used for nearly 50 years to detect effects of classical antianxiety agents (e.g., benzodiazepine receptor agonists such as diazepam) as well as other mechanisms that are or might prove to be anxiolytic in humans. Classical anxiolytic agents including ethanol increase behavior suppressed by punishment. The punishment procedures include passive avoidance, conflict tests, and other related methods. The conflict tests have been most widely used and include the Geller-Seifter and Vogel conflict tests of which a host of variants exist. In the Geller-Seifter test, stable baselines of responding are maintained in the presence of one stimulus and suppressed by punishment in the presence of an alternate stimulus. In the Vogel conflict procedure, drinking is suppressed by drink-dependent delivery of electric shock. The two procedures differ principally in that drug effects in the Geller-Seifter conflict test are evaluated against steady-state behavioral suppression whereas under the Vogel conflict tests, drug effects are evaluated for their ability to prevent response suppression upon the introduction of shock. A related test, albeit not explicitly a punishment procedure, is the conditioned emotional response test (CER) in which a stimulus that is temporally paired with shock suppresses ongoing responding.

Impact of Psychoactive Drugs

Preclinical evaluation of the potential efficacy of putative anxiolytic agents has relied heavily upon the use of a variety of punishment procedures in experimental animals. Punishment is operationally defined as a decrease in the future probability of behavior subsequent to the response-dependent presentation of a stimulus (punisher). Punishment procedures have been studied widely and have been used for both behavioral analyses as well as for assessing drug action, yielding a wealth of information on both behavior and pharmacology. For the most part, the behavioral procedures have employed electric shock as the punishing event, mainly because it is discrete, readily controllable, and manipulable over a wide range of intensities. Other events have been employed (e.g., blasts of compressed air, electric brain stimulation), but these have typically yielded a similar behavioral and pharmacological outcome. One of the earliest punishment methods to be used was the so-called passive avoidance procedure. Under this procedure, the movement of a rodent from one portion of an apparatus to another resulted in the delivery of electric shock. Subsequent placement of the rodent in the initial portion of the apparatus resulted in a decrease in the probability of moving to the second location. It was found that drugs with antianxiety effects in humans would increase this suppressed behavior (Kelleher and Morse 1968). Despite the relative simplicity of this procedure, passive avoidance is highly influenced by variables controlling memory; hence, memory disruptors can be false positives (e.g., scopolamine) in this procedure when attempting to assess potential anxiolytic drug effects. Thus, due to a host of factors including the lack of discrete responses and the detection of false positives, other punishment methods have dominated the anxiolytic drug discovery process for decades. These latter methods utilize the discrete response of a lever press, nose poke, or lick on a drinking tube; these responses are typically controlled by schedules of reinforcement and are not as readily influenced by memory disruption as passive avoidance responses.

Another punishment procedure that utilizes the movement of rodents as the dependent variable is the *four-plate test*. Under this procedure, rodents are placed in an open arena in which four metal squares are situated. Contact with the metal results in electric shock to the paws and subsequent suppression of behavior. As in the case of the passive avoidance procedure, anxiolytic agents can increase movement within such an environment and the probability of contact with the metal plates.

A variant of the four-plate test involves active behaviors engendered by the punisher. Exposure to electric shock can result in burying behaviors if material is made available that permits burying to occur. For example, in the *shock-probe burying test* (sometimes referred to as defensive burying), rodents, generally rats, are placed in a chamber with bedding material in which an electrified probe exists. Contact with the probe as a result of exploratory behavior results in the delivery of an electric shock, which generally leads to *increases* in the probability of burying of the probe. Anxiolytic drugs decrease the probability of shock-induced burying (File et al. 2004). Although not a punishment procedure per se, another highly used method for screening potential anxiolytic agents is marble burying. Mice placed in a bedding-filled chamber with marbles on the top of the bedding will bury the marbles, and such burying behavior is reduced by antianxiety drugs (Borsini et al. 2002).

Although first- and second-generation antianxiety drugs were not developed through the use of punishment procedures, the clinical efficacy of meprobamate, barbiturates, and later benzodiazepine-based anxiolytics led to more refined methods capable of detecting the distinct behavioral effects engendered by these compounds. Geller and his colleagues were the first to develop and pharmacologically characterize a punishment procedure (*Geller-Seifter conflict*) that embodied several distinctive features (Geller et al. 1962). Under the Geller-Seifter or Geller conflict procedure, lever pressing of food-deprived rats is maintained under a variable-interval 2-min schedule of food delivery in which a lever press results in food delivery on the average of every 2 min. Interspersed throughout this initial training are 3-min periods in which a tone is presented and every lever press during the tone produces food. Subsequently, foot shock is introduced during the tone periods such that each lever response produces both food and shock. Under stable conditions, a relatively constant rate of responding occurs in the unpunished component, and lower rates of responding occur in the tone period associated with punishment. The effects of drugs are assessed under this stable baseline of unpunished and punished responding. Typically, drugs are studied in the same animals with a drug-free period of a day to several days allowing stable predrug rates of responding to become reestablished. Many variations of the Geller-Seifter conflict test have been developed and utilized over almost 50 years since its introduction into the behavioral pharmacology arena. The most common variations have involved changes in the schedule of food delivery in either the unpunished or punished components or variations in the schedule or the intensity of shock delivery. One common variation in food scheduling is one involving fixed-ratio schedules of food delivery. Under one variation, a variable-interval schedule of food delivery is used to control unpunished responding, and a fixed-ratio 10 schedule control punished responding where every tenth lever press produces food and shock. Another common variation schedules response-dependent shock delivery for only the first response in the fixed-ratio of responses required for food delivery. In all cases, if other experimental conditions are set appropriately (see below), benzodiazepine anxiolytics such as diazepam increase the rate of suppressed or punished behavior.

Great advantages of the Geller-Seifter conflict methods are the highly reproducible behavioral baselines that are maintained and the reliable sensitivity of this behavior to drug effects engendered by the judicious scheduling of reinforcing and punishing events. In addition, the control by discriminative stimuli during the unpunished and punishment periods permits a simultaneous assessment of anxiolytic-like effects and effects on behavior not controlled by punishment. In addition, there is the benefit that comes from the ability to use the experimental subjects as their own controls, which greatly minimizes variability and the need for large numbers of animals. With these advantages, however, come the burden of time that is needed to establish stable behavioral performances.

In an attempt to decrease training time, the *Vogel conflict test* was developed (Vogel et al. 1971). The procedure utilizes water drinking in water-deprived rodents as the response. There is little training required to develop drinking from a drinking tube. Drug testing can occur after only a day of exposure to this procedure. During the actual assessment of drug effects, the animals are allowed to drink for a brief period (e.g., 3 min), and then an unsignalled punishment period begins in which licking results in electric shock (to the sipper tube or to the grid floor). It is critical to note that the Vogel conflict test is distinct from the Geller-Seifter procedure in that in the Vogel procedure, drug effects are evaluated for their ability to prevent the response suppression following the introduction of shock, whereas in the Geller models, effects of drugs attenuate the suppression of behavior resulting from shock. Despite this difference, drug effects are often comparable across methods (see Table 1 and reviews in McMillan and Leander 1976; Millan 2003).

Although all punishment procedures cannot be described due to space constraints, it is important to mention the CER method. Under this method, originally developed by Estes and Skinner in 1941, steady rates of behavior are maintained under a variable-interval schedule of food delivery. A stimulus is then presented that terminates with the delivery of electric shock independently of responding. Repeated stimulus-shock pairings eventually result in a behavioral baseline characterized by steady rates of responding punctuated by the suppression of responding in the presence of the pre-shock stimulus. Although the outcome of this CER procedure is similar to that in the Geller-Seifter method (i.e., responding is suppressed during the stimulus associated with shock), a key difference is that in the CER procedure, the shock is independent of responding, whereas under the Geller-Seifter procedure, shock only occurs following a response. Though the behavior appears to be the same, the effects of drugs on these two different types of suppressed behavior can be quite different (Kelleher and Morse 1968).

Punished behavior as described earlier has been shown to be an effective method for studying the behavioral effects of drugs across many

Punishment Procedures, Table 1 Effects of some pharmacological agents in Geller-Seifter conflict and Vogel conflict procedures in rodents

Compound[a]	Mechanism(s)/class[b]	Activity[c]	Procedure[d]	Comment[e]
Diazepam	Benzodiazepine receptor agonist	++	G-S	Clinical efficacy
		++	V	
Chlordiazepoxide	Benzodiazepine receptor agonist	++	G-S	Clinical efficacy
		++	V	
Lorazepam	Benzodiazepine receptor agonist	++	G-S	Clinical efficacy
		++	V	
Bretazenil	Benzodiazepine receptor partial agonist	++	G-S	Clinical efficacy
		++	V	
Alprazolam	Benzodiazepine receptor agonist	++	G-S	Clinical efficacy
		++	V	
DOV 51892	Alpha-1-preferring GABAA agonist	++	V	Clinical efficacy – one study
TPA023	Alpha 2/3-preferring GABAA agonist	+	V	Conditioned drink suppression
NS11394	Alpha 3/5-preferring GABAA agonist	++	CER	CER data
Allopregnanolone	Neuroactive steroid GABAA agonist	++?	G-S	No comparator molecule
Pentobarbital	Barbiturate GABAA agonist	++	G-S	Clinical efficacy
		++	V	
Muscimol	Direct GABAA agonist	−	G-S	
		−	V	
Baclofen	GABA-B agonist	−	G-S	
Chlorpromazine	Dopamine antagonist/typical antipsychotic	−	G-S	
		−	V	Mouse procedure with food
Haloperidol	Dopamine antagonist/typical antipsychotic	−	G-S	Clozapine produced small increases
		−	V	
S32504	D3/D2 agonist	+	V	
Clozapine	Atypical antipsychotic	+	G-S	Clinical efficacy
		−	V	
Olanzapine	Atypical antipsychotic	+	G-S	Clinical efficacy
D-Amphetamine	Dopamine stimulant	−	G-S	
		−	V	
Yohimbine	Alpha-2 antagonist	−	G-S	
		+	V	
Amibegron	Beta 3-agonist	+	V	
Propranolol	Beta antagonist	+	G-S	Effect under limited conditions
Scopolamine	Muscarinic antagonist	−	G-S	
		−	V	
Fluoxetine	5-HT uptake inhibitor/ antidepressant	−	G-S	Clinical efficacy
		−	V	
Imipramine	5-HT/NE uptake inhibitor/ antidepressant	−	G-S	
		−	V	

(continued)

Punishment Procedures, Table 1 (continued)

Compound[a]	Mechanism(s)/class[b]	Activity[c]	Procedure[d]	Comment[e]
Valproate	Mood stabilizer/anticonvulsant	+	G-S	Conditioned emotional response method
		−	V	Mixed results in a small literature
Gabapentin	Mood stabilizer/pain	+	G-S	CER procedure
8-OH-DPAT	5-HT1A agonist	+		Literature inconsistent
		+		
Flesinoxan	5-HT1A agonist	+		
Buspirone	5-HT1A partial agonist	−	G-S	Clinical efficacy
		−	V	Literature inconsistent
WAY100635	5-HT1A antagonist	−	G-S	
		−	V	
MDL 100,907	5-HT2A antagonist	−	V	
Ritanserin	5-HT2A/2C antagonist	+	G-S	
BW 723C86	5-HT2B agonist	++	V	
SB 206553	5-HT2B/2C antagonist	+	V	
S32006	5-HT2C antagonist	+	V	Literature inconsistent
Ondansetron	5-HT3 antagonist	−	G-S	
		+	V	Effects not dose dependent
SB 204070A	5-HT4 antagonist	−	G-S	
SB-399885	5-HT6 antagonist	+	V	
SB 269970	5-HT7 antagonist	+	V	
Ethanol	Anesthetic/GABAA/NMDA	+	G-S	Clinical efficacy
		++	V	
Morphine	Mu opioid agonist	−	G-S	
Naloxone	Mu opioid antagonist	−	G-S	
Diprenorphine	Kappa opioid antagonist	+	G-S	
Ethylketocyclazocine	Kappa opioid agonist	−	G-S	
Phencyclidine	Uncompetitive NMDA antagonist	−	G-S	Literature inconsistent
Dizocilpine	Uncompetitive NMDA antagonist	+	G-S	Literature inconsistent
		++	V	
CGP37849	Competitive NMDA antagonist	++	G-S	
		++	V	
Ifenprodil	NR2b/Polyamine NMDA antagonist	−	G-S	
ACPC	Glycine B-site NMDA partial agonist	−	G-S	Pigeon
		++	V	
EGIS-10608	AMPA antagonist	+	V	Literature inconsistent
LY388284	Kainate antagonist	+	V	
EMQMCM	mGlu1 antagonist	−	G-S	
LY354740	mGlu2/3 agonist	−	G-S	
LY341495	mGlu2/3 antagonist	−	G-S	
Fenobam	mGlu5 antagonist	++?	G-S	Clinical efficacy; no comparator
		++?	V	Clinical efficacy; no comparator
MPEP	mGlu5 antagonist	++	G-S	

(continued)

Punishment Procedures, Table 1 (continued)

Compound[a]	Mechanism(s)/class[b]	Activity[c]	Procedure[d]	Comment[e]
Sch 221510	Nociceptin 1 agonist	+	V	
GR205171	NK1 antagonist	−	G-S	Safety signal withdrawal procedure
		−	V	
SR48968	NK2 antagonist	−	G-S	
		−	V	
R278995/CRA0450	CRF1 antagonist	−	V	
SSR149415	Vasopression 1b antagonist	+	V	
PD 135,158	CCK-B antagonist	−	V	Literature inconsistent
SNAP 37889	Galanin 3 antagonist	++	V	
Rimonabant	CB1 antagonist	+	V	
MCL0042	MC4 antagonist/SSRI	+	V	
Pregabalin	Ca++ channel blocker/anticonvulsant	+	G-S	Clinical efficacy

ACPC 1-aminocyclopropanecarboxylic acid, *BW 723C86* 1-[5-(2-thienylmethoxy)-1H-3-indoyl] propan-2-amine hydrochloride, *EGIS-10608* (±)1-(3-methyl-4-amino-phenyl)-4-methyl-7,8-methylenedioxy-4,5-dihydro-3-acetyl-2,3-benzodiazepine, *EMQMCM* ((3-ethyl-2-methyl-quinolin-6-yl)-(4-methoxy-cyclohexyl)-methanone methanesulfonate), *8-OH-DPAT* 8-hydroxy-2-(di-*n*-propyl-amino) tetralin, *GR205171* (S)-(2-methoxy-5-(5-trifluoromethyltetrazol-1-yl)-phenylmethylamino)-2(S)-phenylpiperidine, *LY341495* (2S)-2-amino-2-[(1S,2S)-2-carboxycycloprop-1-yl]-3-(xanth-9-yl) propanoic acid, *LY35474* (+)-2-aminobicyclo[3.1.0]hexane-2,6-dicarboxylate monohydrate, *LY388284* 3S,4aR, 6S,8aR-6-((carboxyphenyl)methyl)-1,2,3,4,4a,5,6,7,8,8a-decahydroisoquinoline-3-carboxylic acid, *MCL0042* 1-[2-(4-fluorophenyl)-2-(4-methylpiperazin-1-yl)ethyl]-4- [4-(1-naphthyl) butyl]piperazine, *MDL 100,907* R-(+)-alpha-(2,3-dimethoxyphenyl)-1-[2-(4-fluorophenyl)ethyl-4-piperidinemethanol, *MPEP* (2-methyl-6-(phenylethynyl)pyridine), *NS11394* [3_-[5-(1-hydroxy-1-methyl-ethyl)-benzoimidazol-1-yl]-biphenyl-2-carbonitrile], *PD 135,158* N-methyl-D-glucamine, R278995/CRA0450 1-[8-(2,4-dichlorophenyl)-2-methylquinolin-4-yl]-1,2,3,6-tetrahydropyridine-4-carboxamide benzenesulfonate, *S32006* N-pyridin-3-yl-1,2-dihydro-3H-benzo[e]indole-3-carboxamide, *S32504* (+)-trans-3,4,4a,5,6,10b-hexahydro-9-carbamoyl-4-propyl-2H-naphth[1,2-b]-1,4-oxazine, *SB-399885* N-[3,5-dichloro-2-(methoxy)phenyl]-4-(methoxy)-3-(1-piperazinyl)benzenesulfonamide, *Sch 221510* 8-[bis(2-methylphenyl)-methyl]-3-phenyl-8-azabicyclo[3.2.1]octan-3-ol, *SNAP 37889* (1-phenyl-3-[[3--(trifluoromethyl)pheny-1]imino]-1H-indol-2-one, *SSR149415* (2S,4R)-1-[5-chloro-1-[(2,4-dimethoxyphenyl)sulfonyl]-3-(2-methoxy-phenyl)-2-oxo-2,3-dihydro-1H-indol-3-yl] -4-hydroxy-N,N-dimethyl-2-pyrrolidine carboxamide, *TPA023*: 7-(1,1-Dimethylethyl)-6-(2-ethyl-2H-1,2,4-triazol-3-ylmethoxy)-3-(2-fluorophenyl)-1,2,4-triazolo[4,3-b]pyridazine, *WAY 100635* N-[2[4(2-methoxyphenyl)-1-piperazinyl]xethyl]-N-(2-pyridinyl) cyclohexanecarboxamide trihydrochloride

[a] Only selected compounds are shown for purposes of comparing mechanisms/drug classes using data from two major classes of punishment procedures. Only data from systemic dosing are included. Only data from acute dosing are included. Anxiolytic-like effects of some compounds in these assays have been reported in the literature but with other procedures suggested to detect anxiolytic mechanisms (e.g., endocannabinoid uptake or enzyme inhibition; mGlu8 agonists; gabapentin). It must also be noted that sometimes the findings summarized have not been replicated

[b] A comprehensive review of drug effects under the Vogel conflict procedure and associated mechanisms can be found in Millan (2003). General summaries of data in the Geller-Seifter procedures can be found in McMillan and Leander (1976) and Kelleher and Morse (1968)

[c] Data included in this table represent results of findings from a host of sources readily accessible in the archival literature; a listing is available upon request from the authors. Data from experiments using rats are used when possible to keep the compound comparisons as uncomplicated as possible. Data are classified by activity on the basis of comparison to the maximal effect observed under the procedure. Maximal effect in each experiment was estimated on the basis of effects of a positive allosteric modulator of GABAA receptors, generally, a benzodiazepine anxiolytic. ++ maximal increase in punished responding at some dose (in comparison with a benzodiazepine anxiolytic if available), + significant increase but less than maximal, − no increase in punished responding or decrease at some dose, ++? An increase in punished responding was observed but the comparative magnitude of effect relative to a standard was not tested in the same study

[d] Data from two major types of procedures are compared if available. *G-S* Geller-Seifter conflict tests of which there are variants, *V* Vogel conflict tests of which there are variants

[e] Clinical efficacy is in reference to anxiety disorders. Literature inconsistency refers to both the potential qualitative differences observed with the specific compound and/or with discrepancies across compounds within a given mechanism of action

species including mice, rats, rabbits, pigeons, fish, nonhuman primates, and humans. Various responses have been successfully employed that include lever pressing, nose poking, key pecking in the case of pigeons, and chain pulling. In addition to the sensitivity of punishment to the effects of anxiolytic drugs, it is important to note that this behavior is relatively insensitive to the effects of other drug classes, such as the antipsychotics and antidepressants, which typically only further reduce punished behavior, thereby making these methods valuable assays for pharmacological evaluation (Table 1). As such, these methods have been of exceptional value in drug-screening efforts to predict potential therapeutic activity in humans.

Although large variations in the experimental conditions being studied generally do not alter the effects of drugs on punished behavior, there are conditions that easily modify the quantitative effects as well as the qualitative changes in the behavioral effects of drugs. Quantitative modifications in the magnitude of drug effects occur with variations in the reinforcement schedule, the frequency, and the intensity of the punisher, as well as the deprivation levels that are used. It is important to acknowledge that the conditions under which punished behavior is established and maintained are critical to the sensitivity to drug effects as well as to the qualitative and quantitative nature of those effects (Barrett 1987; McMillan 1975; Witkin and Katz 1990). While recognizing that human behaviors are not typically punished by electric shock but rather suppressed by a variety of social and other forms of stimuli, it appears that, based on a wide range of studies, the procedures described in this section have been generous in yielding not only an insight into the effects of punishment on behavior but also on the sensitivity and selectivity of punished behavior to the effects of various drugs. In addition, these procedures have also been applied to the investigation of anxiogenic behavioral effects and behavioral effects of drugs.

As noted earlier, the Geller-Seifter and Vogel conflict punishment procedures demonstrate a fair degree of pharmacological specificity that is exemplified in Table 1. These methods are robust detectors of compounds that are positive allosteric modulators of $GABA_A$ receptors (benzodiazepine anxiolytics, barbiturates, neuroactive steroids). Other compounds known to have antianxiety effects in humans also increase punished responding in these preclinical models (e.g., (alcohol) ethanol). However, anxiolytic-like effects (increases in punished responding) are generally not observed with a host of other drug classes and mechanisms (antidepressants, antipsychotics, opiate analgesics, psychomotor stimulants, and muscarinic antagonist amnestics) although exceptions exist (Table 1). Some of the other methods described in this chapter have broader scope in detecting pharmacological effects. Marble burying, for example, can detect the effects of acutely administered selective serotonin reuptake inhibitors (SSRIs) as well as their subacute administration (Borsini et al. 2002). Since the SSRIs are used in the therapeutic control of anxiety symptoms and obsessive-compulsive disorder (OCD) as are some antipsychotic agents, the pharmacological specificity of the punishment methods clearly does not universally detect the breadth of antianxiety mechanisms known to be medically valuable (though some variations in these methods have increased sensitivity). Nonetheless, punished responding is still widely used to evaluate potential antianxiety compounds and mechanisms. Table 1 illustrates that novel mechanisms sometimes display positive effects in these punishment assays.

Cross-References

▶ Alcohol
▶ Animal Models for Psychiatric States
▶ Antidepressants
▶ Anxiety: Animal Models
▶ Barbiturates
▶ Benzodiazepines
▶ Dissociative Anesthetics
▶ Elevated Plus Maze
▶ Emotion and Mood
▶ Mood Stabilizers

▶ Passive Avoidance
▶ Social Anxiety Disorder
▶ SSRIs and Related Compounds

References

Barrett JE. Non-pharmacological factors determining the behavioral effects of drugs. In: Meltzer HY, editor. Psychopharmacology, the third generation of progress. New York: Raven; 1987. p. 1493–501.

Borsini F, Podhorna J, Marazziti D. Do animal models of anxiety predict anxiolytic-like effects of antidepressants? Psychopharmacology (Berl). 2002;163:121–41.

File SE, Lippa AS, Beer B, Lippa MT (2004) Animal tests of anxiety. Curr Protoc Neurosci Chapter 8, Unit 8.3.

Geller I, Kulak Jr JT, Seifter J. The effects of chlordiazepoxide and chlorpromazine on a punishment discrimination. Psychopharmacologia. 1962;3:374–85.

Kelleher RT, Morse WH. Determinants of the specificity of behavioral effects of drugs. Ergeb Physiol. 1968;60:1–56.

McMillan DE. Determinants of drug effects on punished responding. Fed Proc. 1975;34:1870–9.

McMillan DE, Leander JD. Effects of drugs on schedule-controlled behavior. In: Glick SD, Goldfard SD, editors. Behavioral pharmacology. St Louis: CV Mosby; 1976. p. 85–139.

Millan MJ. The neurobiology and control of anxious states. Prog Neurobiol. 2003;70:83–244.

Vogel JR, Beer B, Clody DE. A simple and reliable conflict procedure for testing anti-anxiety agents. Psychopharmacologia. 1971;21:1–7.

Witkin JM, Katz JL. Analysis of behavioral effects of drugs. Drug Dev Res. 1990;20:389–409.

Puromycin

Definition

An aminonucleoside isolated from *Streptomyces alboniger* that inhibits protein synthesis by causing premature chain termination during polypeptide translation.

Q

qEEG

Synonyms

Digital EEG; Quantified (wake) EEG

Definition

First an analog-amplified EEG trace is obtained in the form of equidistant samples. To minimize errors like aliasing, the signal is in practice low-pass filtered (e.g., high cutoff at 70 Hz, steepness 24 dB/oct or 48 dB/oct commensurate with Bessel characteristics). The interval is determined by the sampling rate, usually 256 samples/s or higher. From here, PC-assisted digital filtering, artifact rejection, and quantification in the frequency domain become feasible.

Cross-References

▶ Digital EEG Nomenclature
▶ Electroencephalography
▶ Spectrograms

QT Interval

Definition

The QT interval is a measure of the time between the start of the Q wave and the end of the T wave in the heart's electrical cycle. The QT interval is dependent on the heart rate (the faster the heart rate, the shorter the QT interval) and has to be adjusted to aid interpretation. The standard clinical correction is to use Bazett's formula (named after physiologist Henry Cuthbert Bazett). The corrected QT interval is denoted as QTc. Normal values for the QT interval are between 0.30 and 0.44 s (0.45 s for women). The QT interval is influenced by various antidepressants and antipsychotic drugs, and thus, caution must be exercised while prescribing these drugs particularly in vulnerable individuals with cardiac ailments.

Cross-References

▶ Antidepressants
▶ Antipsychotic Drugs

© Springer-Verlag Berlin Heidelberg 2015
I.P. Stolerman, L.H. Price (eds.), *Encyclopedia of Psychopharmacology*,
DOI 10.1007/978-3-642-36172-2

Quality of Life

Synonyms

QOL

Definition

Quality of life (QOL) is the degree of well-being, defined in terms of health, comfort and happiness, felt by an individual or group of people. The perception of QOL is made up of two components: the physical and the psychological. QOL can be measured only indirectly using different rating scales or questionnaires.

Quazepam

Definition

Quazepam is a benzodiazepine with hypnotic and anticonvulsant properties. Clinically, quazepam is particularly effective in the short-term treatment of insomnia by inducing and maintaining sleep. Unlike other benzodiazepines, quazepam selectively targets $GABA_A$ type 1 receptors, but one of its major long-acting metabolites does not share this selectivity. Due to its long-lasting action, impairment of motor function is a significant side effect. Quazepam use is subject to tolerance, abuse, dependence, and withdrawal, although possibly with reduced severity as compared with other benzodiazepines.

Cross-References

▶ Benzodiazepines
▶ $GABA_A$ Receptor
▶ Insomnias
▶ Sedative, Hypnotic, and Anxiolytic Dependence

Quetiapine

Definition

Quetiapine is a second-generation antipsychotic that acts as an antagonist at dopamine D2 and serotonin $5\text{-}HT_{2A}$ receptors, with a generally broad receptor-binding profile. Quetiapine is indicated for the acute and maintenance treatment of schizophrenia and bipolar disorder. Because of its low propensity to cause extrapyramidal symptoms, the drug can be used to treat psychosis in Parkinson's disease. It is a dibenzothiazepine derivative that dissociates quickly from the D2 receptor. Fast-dissociating D2 antagonists have been hypothesized to allow dopamine to still interact with the D2 receptor under conditions of phasic bursts of dopamine release, thereby eliciting its normal effects in the nigrostriatal and tuberoinfundibular pathways and reducing the risk of side effects. Possibly because of these properties, the risk of extrapyramidal symptoms and effects on prolactin release is lower than for first-generation antipsychotics. Quetiapine has relatively prominent histamine H1 antagonistic effects that are likely to contribute to its sedative properties. The drug also has an active metabolite, norquetiapine.

Cross-References

▶ Antipsychotic Drugs
▶ Antipsychotic Medication: Future Prospects
▶ Bipolar Disorder
▶ Parkinson's Disease
▶ Schizophrenia
▶ Second- and Third-Generation Antipsychotics

R

Racemic Mixture

Definition

A mixture of equal amounts of each enantiomer of a molecule.

Cross-References

▶ Stereoisomers

Radial Arm Maze

Synonyms

Olton maze; Radial maze

Definition

The radial arm maze, originally developed by David Olton, is an apparatus that taps into natural foraging abilities of rodents and can be used to test different types of memory, including short-term (working), long-term (reference), nonspatial, and spatial memories. The arena consists of a number of arms (typically ranging from 4 to 16) that extend from a central hub, and food is placed in the ends of some or all of the arms. The food cannot be seen from the central platform. Versions that test spatial memory require rats or mice to navigate through the environment using allocentric spatial cues placed around the maze. Thus, animals must recall which arms had been entered or remain to be entered using the location of each arm relative to the different cues around the maze. Numerous studies have shown that rats do indeed use a spatial strategy to solve this task, as elaborate control experiments have been used to ensure that the rats do not simply use their sense of smell either to sense unclaimed food objects or to sense their own tracks.

Cross-References

▶ Rodent Tests of Cognition
▶ Short-Term and Working Memory in Animals
▶ Spatial Learning in Animals

Radionuclide

Definition

A radionuclide is an unstable isotope that undergoes radioactive decay.

Cross-References

▶ Positron Emission Tomography (PET) Imaging

Radiopharmaceutical

Synonyms

Radioactive tracer; Radioligand; Radiotracer; Radiolabeled probe; Reporting ligand

Definition

A molecule containing a positron-emitting isotope, used as a tracer in PET imaging. It consists of a radionuclide attached to a compound of interest. In the case of a perfusion radiopharmaceutical, the compound of interest should have properties that allow distribution in proportion to brain blood flow, but the compound should not have specific binding to brain components. In the case of a target molecule-binding radiopharmaceutical, the compound of interest should bind with high specificity to a target molecule, such as a neurotransmitter receptor or a neurotransmitter reuptake transporter.

Cross-References

▶ Positron Emission Tomography (PET) Imaging

Randomized Controlled Trials

Andrea Cipriani[1] and John Geddes[2]
[1]Department of Psychiatry, Medical Sciences Division, University of Oxford, Oxford, UK
[2]Warneford Hospital, University of Oxford, Oxford, Oxfordshire, UK

Synonyms

Controlled clinical trials; Randomized clinical trials

Definition

Randomized controlled trials (RCTs) are studies in which people are allocated at random (*by chance alone*) to receive one of two or more treatments. People who take part in an RCT are called *participants* (or *subjects*). The Consolidated Statement of Reporting Trials (CONSORT) provides readers of RCTs with a list of criteria that will be useful to assess trial validity (for full details, visit www.consort-statement.org) (Altman 1996).

As quantitative, comparative studies, RCTs are one of the simplest and most powerful tools in clinical research. In the field of medicine, RCTs are recognized as the most rigorous method of assessing the efficacy of interventions and are designed to determine whether an association exists between treatment and outcome (Rees 1997). RCTs essentially provide evidence of causality. These clinical studies point to a link between events rather than an explanation of how or why these events may be linked.

Random allocation ensures that there are no systematic differences between intervention groups in factors, known and unknown, that may affect outcome. Other study designs, including nonrandomized controlled trials, can detect associations between an intervention and an outcome; however, they cannot rule out the possibility that a third factor, or *confounder*, linked to both intervention and outcome, might have caused the association. To be convincing, observational studies need very large samples, which can provide sufficient power to control for known variables that might influence the outcome of treatment (e.g., age, sex, social class, ethnic group, and other clinically meaningful variables, such as severity of illness or setting of health care). Even then, there is the risk of residual confounding, and it is impossible, of course, to control for unknown factors. Randomization is the only way of controlling for both known and unknown factors, and it is also very efficient. Nonetheless, RCTs are generally more difficult, costly, and time consuming than other study designs. Careful consideration needs to be given, therefore, to their use and timing.

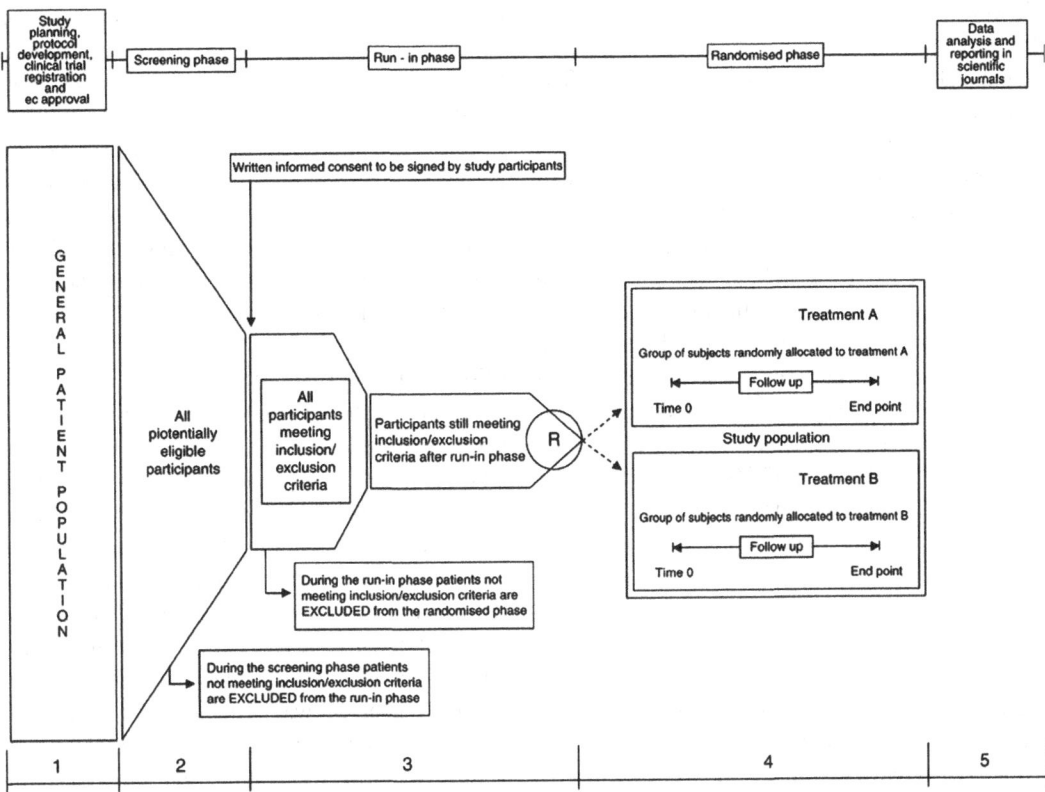

Randomized Controlled Trials, Fig. 1 The main phases of a randomized controlled trial

Current Concepts and State of Knowledge

The main phases of an RCT are shown in the Fig. 1.

Study Planning and Protocol Development

Before starting to design an RCT, the two key questions to be asked are (1) whether the tested intervention is sufficiently well developed to permit evaluation and (2) whether there is enough preliminary evidence that the intervention is likely to be beneficial (most frequently, this evidence comes from observational studies). The development of a new pharmaceutical is conventionally divided into phases, with the initial preclinical toxicity and pharmacological work leading to small preliminary trials in human volunteers (phase I) and then small (perhaps 30–50 participants), usually randomized, studies in patients, before the pivotal phase III RCTs of efficacy and safety. Noncommercial trials will try to replicate this approach. The retrieval of preliminary data will need to inform the estimation of the size of the likely treatment effect, dose, and safety. Such information is needed to estimate the sample size of the study and also to justify the expenses of a larger trial. When planning an RCT, it is now common for trialists to carry out a systematic review of the available evidence to answer these questions.

The trial protocol should be as detailed as possible, describing every step of the study (e.g., the identification of the problem, application of the results, etc.) and answering relevant questions: Are the objectives consistent with the study question? Does the study design achieve these objectives? What is the public health impact of the findings? In the protocol, researchers need to report all information about the conduct and analysis of the trial (e.g., criteria for the inclusion or exclusion of the study population, data handling, and data analysis plan). The lead author of the protocol needs to identify individuals with relevant experience to contribute to the protocol and to provide statistical input or technical information. All protocols must meet a standard that adheres to the principles of Good Clinical Practice, prior to being submitted to the Ethics Committee (EC) or Institutional Review Board (IRB) (for details, see the International Conference on Harmonization of Technical Requirements for Registration of Pharmaceuticals for Human Use – ICH, available at http://www.ich.org/cache/compo/276-254-1.html). After preparation, the first draft of the protocol should be peer reviewed, ensuring that all points have been incorporated or discussed further within the review team. In parallel with finalizing the protocol, it is necessary to prepare all other necessary trial documents, such as subject information sheets and informed consent forms (an example can be found on the Web at the following address: http://www.dh.gov.uk/en/Publichealth/Scientificdevelopmentgeneticsandbioethics/Consent/Consentgeneralinformation/DH_4015950).

Ethics Committee Approval

Once the protocol and other relevant documentation are ready, applications should be made to the EC. As for experimental studies that have the potential of risk to human subjects, all related RCTs have to be approved by an EC before patient recruitment is started. All study participants must be informed about the study characteristics; they must provide written informed consent before enrollment into the trial. In some circumstances, an RCT may be ethical but difficult to carry out (or even unfeasible). One reason might be difficulties with randomization or recruitment (Fairhurst and Dowrick 1996). For example, once an intervention becomes widespread, it can prove impossible to recruit clinicians who are willing to "experiment" with alternatives. This might be true for some widespread practices in clinical psychiatry that are actually not backed by randomized evidence. For example, it is not clear whether anticholinergic drugs work better than levodopa against extrapyramidal side effects in patients with schizophrenia who take long-term antipsychotic medications; however, in daily clinical practice, clinicians are reluctant to use specific antiparkinsonian drugs (such as levodopa) that might trigger psychotic symptoms. Strong patient preferences may also limit recruitment and bias outcomes if not accommodated within the study design (Brewin and Bradley 1989).

Clinical Trial Registration

The protocol should also be registered for publication in dedicated Web sites. Based on both ethical and scientific reasons, since 2000 major registries (such as ClinicalTrials.gov and controlled clinical trials) have been recording trial protocol information. This progress in the public registration of clinical trials worldwide is important because it allows clinicians and researchers to see in advance details about study conduct and analysis in a more transparent way. The number of registered trials increased dramatically in 2005 after the requirement for registration was introduced by several medical journals, led by the International Committee of Medical Journal Editors.

Screening Phase

All patients enrolled in the study must meet the eligibility criteria. Screening procedures are performed to determine the eligibility of participants for the study and document their baseline medical conditions. Information gathered during the screening phase is used to make decisions on participants' eligibility. Therefore, the threshold for access to assessment should be set at a low enough level to ensure people are not screened out before full information on their condition is obtained. There are usually one or two screening visits, and all screening and enrollment procedures take place within a predefined period (7–56 days, depending on the clinical status and diagnosis).

Run-In Phase

Run-in phases are pre-randomization periods used to select or exclude patients in a clinical trial. Run-in periods might be used to exclude noncompliant subjects, placebo responders, or subjects who could not tolerate or did not respond to active drug. For this reason, run-in periods have implications for interpreting the results of clinical trials and applying these results in clinical practice (Pablos-Méndez et al. 1998).

Run-in phases can use either placebo or active therapy and are usually single blind (i.e., only the study staff is aware of the nature of the medication, but the patient is not) or unblinded (both the staff and the patient are aware of the nature of the medication). A placebo run-in allows trial staff to be sure that reported side effects are not caused by treatment. By contrast, an active run-in can identify and exclude individuals who may be unable to tolerate the medication being tested in a long-term trial. For instance, the long-term treatment of bipolar disorder is supposed to be a treatment specifically intended to prevent recurrence of mood episodes. Subjects who are entering a long-term trial are asked to take the study medication (or the combined study medications) for a long period of time. In real-world clinical settings, people who are prescribed long-term treatments will usually be selected from those who have been able to tolerate the medication in the short term – perhaps during acute phase therapy. In these studies, a run-in design (i.e., a run-in phase with active long-term treatments before randomization) may maximize the study power through participants' better compliance. Individuals who lose interest early on (potential "dropouts") can then be excluded before random allocation. Similarly, subjects who become convinced about receiving the intervention treatment (potential "drop-ins") can also withdraw before randomization. This potentially lowers rates of anticipated noncompliance to allocated treatment during a trial, resulting in a smaller required sample size. Once randomized, these participants would generally be included in an intention-to-treat analysis (for details, see below).

Run-in periods can dilute or enhance the clinical applicability of the results of a clinical trial, depending on the patient group to whom the results will be applied. This approach usually enhances clinical applicability by selecting a group of study subjects who closely resemble patients undergoing active clinical management for this problem. However, compared with results that would have been observed without the run-in period, the reported results might overestimate the benefits and underestimate the risks of treatment. Because the reported results apply to subgroups of patients who cannot be defined readily based on demographic or clinical characteristics, the applicability of the results in clinical practice might be diluted. For this reason, reports of clinical trials using run-in periods should indicate how this aspect of their design might affect the application of the results to clinical practice.

Randomized Phase

RCTs have several important features:

- Randomization
- Allocation concealment
- Blinding
- Choice of interventions and outcomes

Randomization

The simplest approach to evaluating a new treatment is to compare two consecutive series of patients (historical controls are undesirable because of the consistent tendency for historically controlled trials to yield more optimistic results than randomized trials). If clinicians are allowed to choose which treatment to give to each patient, there will probably be differences in the clinical and demographic characteristics of the patients receiving the different treatments. Such systematic differences are called *bias*, and bias may lead to a distortion (overestimate or underestimate) of the difference between treatments (Schulz et al. 1995). Thus, the main reason for using randomization to allocate treatments in a controlled study is to try to prevent biases. Randomization is a process that aims at producing groups that are similar in terms of clinical, demographic, and prognostic factors. Random allocation means that each patient has a known, equal chance of being given each treatment, but the treatment to be given to each patient cannot be predicted.

The process of randomization begins with the generation of a random allocation sequence. The simplest procedure to equally allocate to two treatment groups involves using a table of computer-generated random numbers. This is called simple randomization, and it is equivalent to tossing a coin. Similar lists of random numbers can be found in most statistics textbooks. From a methodological point of view, a simple randomization may remove the risk of bias from the allocation procedure; however, it cannot guarantee that the individuals in each group have a similar distribution of clinical or demographic variables. Some chance imbalance may occur, especially in small studies. An imbalance in the distribution of baseline characteristics might make the interpretation of results complicated or even impossible. There are ways to modify simple randomization, such as *block randomization* and *stratified randomization*. *Block randomization* ensures closely similar numbers of patients in each group; by contrast, *stratified randomization* is used to keep the groups balanced for certain important clinical or prognostic patient characteristics, producing a separate randomization list for each prespecified subgroup based on such characteristics (each such subgroup is called a *stratum*). For instance, in multicenter studies, the stratifying variable may be the study center itself, because the patients within each center need to be equally randomized to one or the other of the two study treatments. Stratified randomization can be extended to two or more stratifying variables, even though in small studies it is not practical to stratify on more than one (or perhaps two variables), as the number of strata can quickly approach the number of subjects. When it is really important to achieve close similarity between treatment groups for several variables, *minimization* can be used. *Minimization* is based on a different principle from randomization, and it has the advantage, especially in small trials, that there will be only minor differences between groups in those variables used in the allocation process. When using minimization, the first participant is randomly allocated to treatment, but each subsequent participant is allocated to the treatment that would lead to a better balance between the groups in the variables of interest. The chosen treatment could simply be taken as the one with the lower allocated sample (100 % of probability to be allocated to this group); otherwise, a random element can be introduced (a high chance, say 80 %, of probability to be allocated to the treatment that minimizes the imbalance – the so-called weighted randomization). The use of a random element makes the allocation more unpredictable. Even though it may slightly worsen the overall imbalance between the groups, the balance (for the chosen variables) is usually much better than with simple randomization. Minimization is a secure allocation system when used by an independent person. In RCTs, individual patients are usually randomized to a treatment or control group, but sometimes cluster-randomized controlled trials are carried out, in which groups (or clusters) of individuals, rather than individuals themselves, are randomized. The groups may be villages, colleges, medical practices, or families. A cluster-randomized design may be used simply for convenience. One of the main consequences of a cluster design is that participants within the cluster often tend to respond in a similar manner, and thus their data can no longer be assumed to be independent of one another. In such studies, the "unit of allocation" is the cluster and not the patient. Cluster trials are increasingly common because they are often the only valid approach to evaluate certain types of interventions (such as those used in health promotion and educational or psychosocial interventions). However, cluster trials are generally more difficult to design and carry out than individually randomized studies, and some of their design features may make them particularly vulnerable to bias. A proper analysis can help overcome this risk.

Allocation Concealment

The process of randomization does not end until participants are actually assigned to their groups. Allocation concealment refers to the process used to ensure that subjects, investigators, and all other study personnel are unaware which group participants are being allocated to at the time they are enrolled in the study. Adequate methods of concealment might be centralized allocation (for instance, via telephone), computerized allocation systems (via Web-based facilities), or coded identical containers (using sealed envelopes).

The use of adequate methods of concealment is considered an indicator of a well-conducted study (see CONSORT Statement). There is evidence that where allocation concealment is inadequate, investigators can overestimate the effect of interventions by up to 40 %. Study reports should include a description of the method used to conceal allocation, with enough technical details to let readers determine the likelihood of the success of this process.

Blinding

Blinding refers to the process whereby people are kept unaware of which treatment an individual patient has been receiving. Blinding (or "masking") is different from allocation concealment. Blinding is the prevention of knowledge of treatment assignment after randomization has been done; allocation concealment refers to the prevention of knowledge of upcoming assignment from the randomization sequence before the treatment is allocated. Allocation concealment is part of the randomization process and must always be included in the design or conduct of an RCT. By contrast, blinding, in some clinical circumstances, may not be feasible, for example, in trials investigating the effect of some psychological treatments. However, blinding might be very important when the outcome for a patient has to be assessed (most of all, when using a clinician-based rating scale).

Blinding can be carried out up to four levels: (1) patient, (2) investigator (or clinical trial nurse), (3) outcome assessor, and (4) biostatistician. In medical journals, the term "double blind" is often used. This term does not have a standard definition and cannot always be relied upon to convey which groups in an RCT were truly blind (for further details, please visit www.consort-statement.org). Because of this ambiguity, descriptions of blinding in reports of RCTs ideally should be explicit, reporting precisely who was masked and how this was done.

Choice of Interventions and Outcomes

A key component of an RCT is to specify the interventions of interest and the interventions against which these will be compared (Huston and Locher 1996). These interventions are called *comparisons*. One of these treatments is a treatment commonly used in clinical practice – the so-called *standard* (or *control*) treatment; the new treatment to be tested in the study is named *experimental* (or *investigational*) treatment. The control might be an inactive control intervention (such as placebo, no treatment, standard care, or a waiting list control), or an active control intervention (such as a different variant of the same intervention, a different drug, a different kind of therapy, or a combination treatment, when the intervention of interest is combined with another intervention). When specifying drug interventions, dosing factors such as drug preparation, route of administration, and dose, duration, and frequency of administration should be considered and clearly specified in the study protocol. Regarding dose, for example, it is necessary to know whether there is a critical dose below which the intervention may not be effective. Dosing factors are important because study authors must always consider whether the characteristics of the intervention might result in substantially different effects on the participants and outcomes of interest. For instance, when comparing two drugs, it would be unacceptable to use one drug at the lower end of the therapeutic range and the second one at the higher end of the therapeutic range. This imbalance might affect the reliability of the study results, reducing the internal and external validities of the RCT. For more complex interventions (such as psychoeducational or psychotherapeutic interventions, like cognitive-behavioral or psychodynamic psychotherapy), the core features of the interventions have to be defined. In general, for psychotherapies it is useful to report exactly what is delivered, at what intensity, how often it is delivered, who delivers it, and whether people involved in delivery of the intervention need to be trained (manualized interventions).

Another key component of an RCT is the delineation of particular outcomes that are of interest. In general, an RCT should have a primary outcome (on which to base the calculation of sample size) and also include other outcomes that are likely to be meaningful to patients and clinicians. Outcomes may include survival (mortality), clinical events (recurrence, relapse, and hospitalization), patient-reported outcomes (rating of symptoms, quality of life), adverse events, and economic outcomes (costs and resource use). When planning an RCT, authors should consider how outcomes may be measured, in terms of both the type of scale to be used and the timing of measurement. Outcomes may be measured objectively (number of relapses, number of hospitalizations) or subjectively as rated by a clinician, patient, or caregiver (for instance, severity of illness scales or disability scales). In other fields of medicine, it is common to use so-called hard outcomes (i.e., objective outcomes like mortality rates, concomitant medications, return to work). By contrast, in psychiatric trials changes in rating scale scores are more commonly reported. There are four different kinds of rating scales that can be used in psychiatric trials: (1) observer-based disease-specific rating scales, where a clinician rates a patient on items about a specific disease; (2) patient-based disease-specific rating scales, which analyze the same domains from a patient's perspective; (3) observer-based non-disease-specific scales, which are rating scales filled by clinicians and designed to measure a patient's global functioning; and (4) patient-based non-disease-specific scales of global functioning. If a psychiatric treatment really worked, one might expect the effect of treatment to show up on rating scales from all four domains of measurement. However, this often does not happen because of the complexity of psychiatric illness and because of the degree of subjectivity in assessing outcomes. This is the reason why it is important to use published or validated rating instruments. Furthermore, when defining the timing of outcome measurement, authors may consider whether all time frames or only selected time points are to be considered.

Data Analysis and Reporting in Scientific Medical Journals

Analysis

Attrition, a ubiquitous problem in RCTs of psychotropic agents, can cause biased estimates of the treatment effect, reducing statistical power and restricting the generalizability of results. There are various approaches to data analyses in RCTs (Rennie 1999). One way is to include data only on participants, whose results are recorded, using as a denominator the total number of people who had data recorded for the particular outcome in question. Another way is to use the total number of randomized participants, irrespective of whether some participants withdrew from the study. This will involve imputing outcomes for these missing participants. There are several approaches to imputing outcome data, either for dichotomous or continuous outcomes. One common approach is to assume either that all the missing participants experienced the event or that all of them did not. An alternative approach is to impute data according to the event rate observed in the control group, or according to event rates among completers in the separate groups (the latter provides the same estimate of intervention effect but results in an unwarranted inflation of the precision of effect estimates). The choice among these assumptions should be based on clinical judgment and should be preplanned and clearly reported in the study protocol. Intention-to-treat analysis maintains the advantages of random allocation, which may be lost if subjects are excluded from analysis through, for example, withdrawal or failure to comply.

Journal Reporting

When data are analyzed, results of the RCT are disseminated and published in scientific journals. To be published, study reports must undergo a peer-review process. Almost all the most important scientific journals want to see the study protocol before accepting a study for publication, in order to check the consistency between the protocol and the full report of the RCT (Gilbody and Song 2000).

The first use of an RCT in psychiatry was to demonstrate that cortisone did not work for schizophrenia. Nowadays in psychiatry, new agents are usually compared with placebo. This is because there is a high placebo response rate that varies considerably depending on chronicity, severity, and subtype of the disorder. While it has been suggested that placebo run-in periods can help to control for this, evidence is inconsistent. In an illness such as major depression, up to 50 % of those who respond while taking a medication may improve due to placebo factors. This placebo response rate in depression RCTs varies, but is growing. In other disorders (such as ▶ obsessive-compulsive disorder or schizophrenia), the placebo response is much less. The use of a placebo arm in an RCT raises ethical, clinical, and methodological issues. Ethically, it may be argued that placebo is only appropriate when no established effective treatment is available. On the other hand, the dangers of marketing ineffective drugs or psychotherapies that carry risks of side effects may outweigh the modest risk associated with placebo treatment in randomized, placebo-controlled trials. A placebo group is essential to establish that a treatment is effective for any condition. However, a placebo-controlled design may also influence the outcome of the trial. For example, individuals likely to participate in a placebo-controlled trial have less severe, less chronic, and less disabling disease. Therefore, they may be the very people who are apt to respond to placebo.

One important issue for RCTs in psychiatry is the way in which the treatments are delivered. In drug trials, a dosage schedule must be specified. This can be left up to the clinician or predetermined by the investigators. Clinician titration is probably closer to clinical practice and may reduce dropouts since sensitive patients will be maintained at lower doses and dosage increased more slowly. The exact details of treatment and comparator are also important in trials of psychotherapeutic interventions. Since different types of psychotherapies may give rise to different results, it is important to define the exact form of therapy to be given, to ensure adherence of therapists to that therapy, and to maintain its quality. In reporting the trial, it is important to describe the therapy in such a way that it can be replicated if it proves to be clinically useful. RCTs have particular problems when comparing psychotherapies, of which the most serious is that interventions do not represent distinctly different treatments (because of the many factors common to all forms of psychotherapy) and therefore violate the premise of controlled trials. Often it may not be possible to sufficiently control how treatments are conducted, leading to poor external validity of psychotherapy RCTs. In psychotherapy, for example, RCTs typically involve manual-based interventions. However, the relationship of these therapies to general practice may be unclear, since in reality most clinicians do not have the level of training and supervision provided to research therapists.

Apart from the proposed treatments under investigation in an RCT, it is important to anticipate the possibility that additional, less specific treatments will be needed during a trial. In trials for mania or depression, certain "rescue" medications (such as benzodiazepines) may be used. The use of rescue medications may vary between study groups and could obscure a difference in outcome. This may therefore need to be considered as an actual treatment outcome. The use of additional treatments (and which medications to use) should also be predetermined and included in the protocol.

Many published trials in psychiatry have included several outcome measures. This makes it likely that at least one of them will show a difference by chance. It is not easy to decide on a single outcome measure. Differing definitions of what constitutes a response to treatment is a problem in psychiatric trials where clear outcomes, such as death or highly reliable physiological changes, are rarely suitable. In many trials of antidepressant treatment, it has been total score or change from baseline on a rating scale at a predefined time point. Psychiatry researchers are increasingly advocating the use of simple objective outcomes to make trials large enough (and therefore simple enough to be carried out at many sites) to show relatively small differences.

A typical psychiatric RCT has fewer than 100 subjects. Small samples are more likely to have initial group differences (despite randomization) and differential dropout rates. Statistical techniques can be used to control for differences in assignment or from dropouts, but small samples can only have adequate power to detect moderate to large treatment differences and are unlikely to detect differences either in rare events, such as a suicide attempt, or factors that have high variability.

Randomized controlled trials often recruit different subjects from those who present to clinical services. Exclusion criteria, usually set to create a more homogeneous sample, reduce the comparability of the enrolled subjects to those in a clinical service. Excluding patients with co-occurring disorders should not occur when these comorbidities are common or they affect treatment response and prognosis. For example, in patients with major depression, comorbid Axis I and Axis II conditions are present in most patients and may impact response to treatment. A further problem is that most RCTs are of short duration, but clinicians are interested in longer-term outcome. Many disorders are recurrent or chronic, but trials usually last between 6 and 12 weeks and can only address short-term response or remission, since they do not deal with the clinical outcome of recovery (a sustained remission). This would not be a significant problem if short-term response was strongly predictive of recovery. However, the relationship between initial response and recovery is not consistent.

Since all treatments potentially have adverse effects, it is important to anticipate, measure, and report them accurately. A treatment that increases the speed of recovery from depression but gives rise to enduring side effects may not be deemed useful. This is the case with ECT, which may lead to a rapid recovery for people with severe depression, but which may lead to cognitive side effects measured in a longer-term perspective. For many conditions (e.g., anorexia nervosa, somatoform disorders, or personality disorders), insufficient data are available to guide treatment. For other disorders, despite research, no consensus has been reached. RCTs are unlikely to influence practice unless they address an area of perceived uncertainty. However, RCTs are often driven by commercial interest. This leads to a focus on discrete choices, such as treatment modality or intensity, rather than practical decisions, such as when to use physical restraint or to hospitalize a patient.

To summarize, patients need individually tailored treatment. RCTs are the best way to tell clinicians what treatments are effective, but not necessarily which patients should receive them and when they should be given.

Cross-References

▶ Antidepressants
▶ Anti-Parkinson Drugs
▶ Antipsychotic Drugs
▶ Bipolar Disorder
▶ Ethical Issues in Human Psychopharmacology
▶ Mood Stabilizers
▶ Rating Scales and Diagnostic Schemata
▶ Suicide

References

Altman DG (1996) Better reporting of randomised controlled trials: the CONSORT statement. BMJ 313:370–371

Brewin CR, Bradley C (1989) Patient preferences and randomised clinical trials. BMJ 299:313–315

Fairhurst K, Dowrick C (1996) Problems with recruitment in a randomised controlled trial of counselling in general practice: causes and implications. J Health Serv Res Policy 1:77–80

Gilbody SM, Song F (2000) Publication bias and the integrity of psychiatry research. Psychol Med 30:253–258

Huston D, Locher M (1996) Redundancy, disaggregation and the integrity of medical research. Lancet 347:1024–1026

Pablos-Méndez A, Barr RG, Shea S (1998) Run-in periods in randomized trials implications for the application of results in clinical practice. JAMA 279:222–225

Rees WL (1997) The place of controlled trials in the development of psychopharmacology. Hist Psychiatry 8:1–20

Rennie D (1999) Fair conduct and fair reporting of clinical trials. JAMA 282:1766–1768

Schulz KF, Chalmers I, Haynes RJ, Altman DG (1995) Empirical evidence of bias. Dimensions of methodological quality associated with estimates of treatment effects in controlled trials. JAMA 273:408–412

Rapamycin

Synonyms

23,27-Epoxy-3H-pyrido[2,1-c][1,4]oxaazacyclohentriacontine; Antibiotic AY 22989; NSC 2260804; Rapamune; Sirolimus

Definition

A potent immunosuppressant drug that possesses both antifungal and antineoplastic properties. Sirolimus belongs to a class of macrolide antibiotics and used to prevent the rejection of organ transplants by the body. It is a type of serine/threonine kinase inhibitor. It forms a complex by interacting with the FK-binding protein 12 (FKBP12) and inhibits the mammalian target of rapamycin complex 1 (mTORC1) signaling.

Rapid Eye Movement Sleep

Synonyms

Dream sleep; Paradoxical sleep; REM sleep; Stage R sleep

Definition

Stage R is one of the most interesting of the sleep stages; it is characterized by fast, rapid eye movements, giving this stage its name. Stage R is associated with low voltage and mixed frequencies (similar to Stage 1) of EEG, along with a sawtooth wave pattern. During Stage R, the EMG activity reaches its lowest level and additionally episodic rapid eye movements (REMs) begin to occur. The frequency of REMs per hour of REM sleep is designated as REM density, which is a reflection of REM sleep intensity. REM sleep is prominent in the last third of the night and most of the dreaming occurs at REM sleep. There are both phasic (episodic) and tonic (persistent) components to Stage R. When the phasic or tonic component occurs, the EEG tracing is similar to that of Stage N1, as noted above. Additionally, a generalized atonia of muscles occurs in Stage R. An important exception to this is that the muscles of the diaphragm and extraocular muscles remain tonic.

REM sleep Onset Latency (REMOL) is the interval between first epoch of sleep and the appearance of first REM sleep episode in a recording. The REM sleep interval is the interval between the end of a REM sleep episode and the beginning of the subsequent REM sleep episode. Time awake is generally excepted from the interval. The REM ▶ parasomnias include nightmares and REM behavior disorder.

REM sleep is usually associated with dreaming (hence known as "dream sleep"). REM sleep is characterized by paralysis or nearly absent muscle tone (atonia, except for control of breathing and erectile tissue), periods of phasic eye movements, and a rise in cortical activity associated with dreaming. The brain however remains highly active, and the electrical activity recorded by ▶ electroencephalography (EEG) during REM sleep is similar to that seen during wakefulness. Irregular respiration and heart rate are also features. In adults, about 20 % of nighttime sleep is devoted to REM. In the rat, REM sleep is characterized by low-voltage fast frontal waves, a regular theta rhythm in the occipital cortex and a silent electromyogram except for occasional myoclonic twitchings.

Cross-References

▶ Electroencephalography
▶ Parasomnias
▶ REM Sleep Behavior Disorder

Rasagiline

Synonyms

(R)-N-(prop-2-ynyl)-2,3-dihydro-1H-inden-1-amine; AGN 1135; Azilect

Definition

Rasagiline is an irreversible monoamine oxidase (MAO) inhibitor selective for MAO-B. MAO-B inhibition prevents the breakdown of dopamine in the nerve terminals, resulting in increased extracellular dopamine. Rasagiline has been approved by the Food and Drug Administration for use in the United States. It is prescribed as a monotherapy for the treatment of early ▶ Parkinson's disease (PD) or an adjunct therapy with levodopa for advanced PD. Headache, motor impairments, nausea, vomiting, constipation, drowsiness, depression, and hallucinations are some common side effects of rasagiline use. Rasagiline has been investigated for the treatment of restless legs syndrome.

Cross-References

▶ Monoamine Oxidase Inhibitors

Rate-Dependency Theory

Galen R. Wenger
Department of Pharmacology and Toxicology, University of Arkansas for Medical Sciences, Little Rock, AR, USA

Definition

An early theory in behavioral pharmacology that seeks to predict effects of drugs on operant behaviour as a function of the baseline rates of that behavior in the absence of drug.

Definitions of Terms

Fixed ratio (FR) – a schedule of reinforcement under which the reinforcing stimulus is presented upon the completion of a fixed number of responses.

Fixed interval (FI) – a schedule of reinforcement under which the reinforcing stimulus is presented upon the first response to occur following the elapse of a fixed amount of time.

Interresponse time (IRT) – the time between two successive responses.

Punishment – the decrease in the probability of a response in real time as a consequence of the presentation of a punishing stimulus upon a response.

Rate of responding – the number of responses per unit of time (e.g., responses/second).

Reinforcement – the increase in the probability of a response in real time as a consequence of the presentation of a reinforcing stimulus upon a response.

Impact of Psychoactive Drugs

Evolution of the Theory

The 1950s was a time of great advances in the development of behaviorally active drugs, and a great deal of effort was spent trying to understand why drugs altered behavior in specific ways. During this period, drug action on behavior was interpreted, for the most part, in terms of the drug's capacity to reduce a primary drive (e.g., hunger, fear, or thirst) or by the drug's capacity to interfere with a learned association between a stimulus in the environment and one of the primary drives. The work of Clark L. Hull and Neal Miller figured prominently during this period (for review see McMillan and Katz 2002). Thus, when drugs with clinically relevant effects on anxiety were examined in laboratory animals and shown to produce differential effects on behavior controlled by food presentation as compared to behavior controlled by electric shock, the observations were viewed as being consistent with the reduction of primary drives and confirmed the validity of the drive reductionist concepts of drug action. It was in this historical context that Peter B. Dews initiated a series of studies that would eventually lead to the development of what is now known as rate-dependency theory.

In 1955, Dews published the first of a series of papers in which he examined the effect of drugs in food-deprived pigeons responding under different operant schedules of food presentation (Dews 1955). In his first paper, he showed that doses of pentobarbital that produced dramatic decreases in the rate of responding under a fixed-interval 900 s (FI900) schedule did not affect the rate of responding under a fixed-ratio 30 s (FR30) schedule of food presentation. Since responding under the two schedules was maintained by the same event, it was difficult to explain such profound differences based upon drive reduction theories proposed by the current theorists of the day. Alternatively, Dews suggested that rather than interacting with drive states, pentobarbital appeared to be producing the differential effect by interacting with the control of the rate of responding. In a subsequent paper, Dews showed that the effects of methamphetamine in pigeons responding under four different schedules of food presentation depended upon the length of the time between individual responses (IRT; rate of responding). When control performance was characterized by low rates of responding, low to moderate doses of methamphetamine increased rates of responding. At higher doses, methamphetamine did the same thing but it also decreased the high rates of responding (Dews 1958). Over the next several years, Dews along with colleagues, William Morse and Roger Kelleher, as well as investigators from other labs showed that the control rate of responding was a key factor in determining drug effects and that the control rate of the behavior was an important independent variable. However, it was not until 1964 that Dews formally proposed a way to express this relationship (Dews 1964). Dews argued that since one frequently had to deal with a range of doses that spanned several log units, by analogy to dose–response curves in classical pharmacology, the control rate of responding should be plotted on the abscissa using a log scale. He went on to show that if one plotted the log of the rate of responding following drug administration divided by the rate of responding under control conditions on the ordinate (Fig. 1), not only did this log–log plot aid in the uniform distribution of the data points, it yielded a linear relationship that

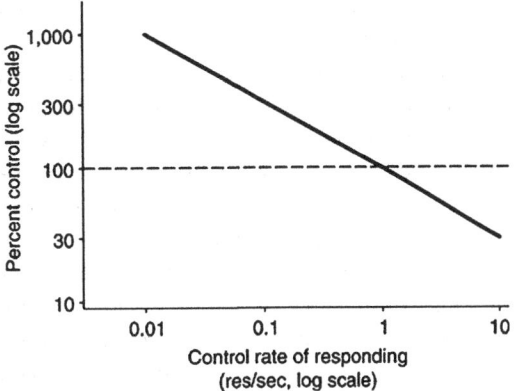

Rate-Dependency Theory, Fig. 1 Typical rate-dependency plot. *Abscissa*: control rate of responding (log scale); *Ordinate*: rate of responding following drug administration divided by the control rate and converted to a percentage (log scale)

had predictive value. As a result, one typically sees a log–log linear relationship with a negative slope. Thus, very low rates of responding tend to be increased more than moderate rates, and high rates of responding are either increased less than moderate rates or actually decreased compared to the control rate of responding. This log–log linear relationship and the theory behind it are referred to as the rate-dependency theory of drug action.

Generality of the Theory

During the 1960s and early 1970s, numerous studies showed that the rate of responding was more important in determining the effect of drugs than the specific reinforcing event used to maintain the behavior. Furthermore, in addition to the amphetamines and barbiturates, rate-dependent drug effects were shown for many drugs including benzodiazepines, meprobamate, phenothiazines, atypical antipsychotics, anticholinergics, phencyclidine, and ketamine.

Although Dews and others showed that the rate-dependency theory described the actions of many drugs on schedule-controlled operant responding, a question remained about whether the rate of the ongoing behavior could account for drug effects on other behaviors (nonoperant). Looking at a wide range of behavioral effects of amphetamine in rats, Dews and Wenger (1977) could not show any separation between operant studies and nonoperant studies (e.g., locomotor activity or rearing). For the most part, data from operant and nonoperant studies fell on the same log–log regression line. Similarly, if one examined very low rates of behavior, although the variance was very high, there was no obvious break in the log–log relationship. However, there were a couple of notable exceptions for the rate dependency of amphetamines. Control rates of behavior that were very low as a result of the presentation of an aversive event upon a response (punishment) did not appear to be increased to the same extent as equally low rates of responding in the absence of a punishment procedure. These low punished rates may, however, have their own parallel relationship between drug effect and control rate of responding. Interestingly, barbiturates, benzodiazepines, and other drugs with clinical antianxiety properties do not share this attenuated capacity of amphetamines to increase low rates under punishment situations. The barbiturates and benzodiazepines, for example, may produce an even larger increase in low rates maintained by punishment procedures compared to their effect on equally low rates maintained by procedures not involving punishment. In nonoperant experiments, there was one notable exception to the rate-dependent effects of amphetamine. The rate of licking by rats appeared to be relatively resistant to change by amphetamines. The resistance to change of licking responses has been reported for several other drug classes.

In addition to the punishment situations discussed earlier for amphetamines, there are a couple of other situations where the rate of ongoing behavior is not the sole determinant of the drug effect (for review see McKearney 1981). As with the effect of amphetamines on low rates of behavior that have been suppressed by punishment, the ability of barbiturates to increase low rates of responding that are under strong discriminative control has been shown to be less than predicted based solely upon the rate of responding. Nevertheless, the effect of the barbiturates in this situation seemed to be strongly influenced by the control rate of responding. A third situation of note is where the rate of responding directly influences the rate of reinforcement. In many situations, the rate of responding can vary widely with minimal effect on reinforcement rate. However, in situations where an increase in rate decreases reinforcement frequency, these low rates are relatively resistant to increases by drug. Finally, although under most situations the drug effects are independent of the event maintaining the behavior, there are a few known examples where similar rates of responding are altered differently depending upon the reinforcing event. For example, the effect of morphine has been shown to alter rates of responding differentially depending upon whether the reinforcing event used to maintain the behavior is food or electric shock presentation.

Challenges to the Theory

Although a vast body of literature shows that drugs obey the rate-dependency theory of drug action as proposed by Dews, there have been several challenges and alternative theories proposed. The most significant challenge was, in part, a mathematical issue versus a biological issue, but it also raised questions about the nature of drug effects. Gonzalez and Byrd (1977) stated that the log–log linear relationship proposed by Dews (1964) can be summarized by Eq. 1. By rearranging Eq. 1 to yield Eq. 2, it can be seen that when the slope of the rate-dependency plot equals -1, then the rate following drug administration becomes independent of the control rate and all rates are equal. They further stated that in order to make it easier to see when the rate after drug administration becomes independent of the control rate, the data should be plotted as shown in Fig. 2. As can be seen with increasing dose and increasing effect, the slope becomes less and less positive and potentially approaches zero. Furthermore, they argued that both the control rate and the rate after drug administration are dependent upon the maximum possible response rate. Thus, Gonzalez and Byrd stated that the effect of a drug on the rate of responding is best considered independent of the control rate, and it is necessary to know the maximum possible rate

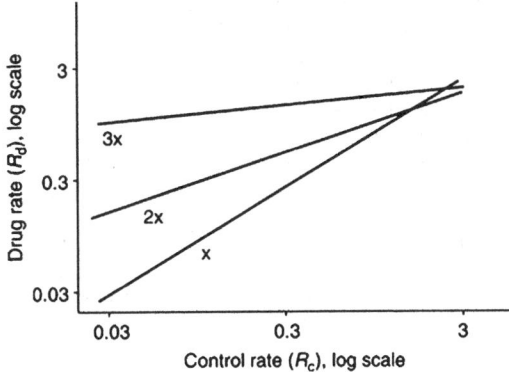

Rate-Dependency Theory, Fig. 2 Rate-dependency plot according to Gonzalez and Byrd (1977). *Abscissa*: rate of responding under control conditions (R_c); *Ordinate*: rate of responding following drug (R_d). X dose administered

of responding before the drug effect can be interpreted.

$$\log\left(\frac{R_d}{R_c}\right) = \log k + j \log R_c \quad (1)$$

Rearrangement yields the following:

$$\log R_d = \log k + (j+1)(\log R_c) \quad (2)$$

where

R_d is the rate after drug administration
R_c is the control rate
k is the Y-intercept
j is the slope of the regression line

In defense of Dews' position (Dews 1964), it must be remembered that drugs do not create behavior, they only modify existing behavior. Thus, the drug effect cannot be considered in the absence of a comparison to the control behavior. With respect to the contribution played by the maximum possible response rate, such a rate can only be determined experimentally, and it never will be possible to establish the true maximum rate of responding. Furthermore, there is no experimental evidence to support the need to know the maximum rate of responding (see Dews 1978).

The second major challenge to the theory was published in 1981 (Ksir 1981). Ksir's challenge can be summarized by saying that drugs tend to make the rate of responding a constant independent of the control rate of responding. Thus, with increasing doses of drug, the rate of responding approaches some constant value. If this were the case, such a nonspecific effect of a drug would be of little interest. Furthermore, few pharmacologists would accept such a conclusion. Thus, like the Gonzalez and Byrd (1977) challenge, the arguments may be mathematically correct, but they may not be correct from a biological standpoint.

Summary

In summary, a question that arises is how strongly can the theory be stated? Based upon the literature, Dews and Wenger (1977) stated that it is probably true that with all other variables unchanged, a change in the rate of responding may change the behavioral effect of a drug. It is probably also true that not only may differences in the rate of responding lead to differences in the effects of drugs, but in general, such differences will be determined by the control rate, and there will be a systematic relationship between the rate of responding and drug effect. Although there are some exceptions as noted earlier, the control rate of responding relates to the drug effect in such a manner that in most cases the log of the effect is a linear function of the log of the control rate of the behavior. However, probably it is a mistake to conclude that the control rate of responding is the sole determinant of the effect of a drug and that other factors are only important to the extent that they modify the control rate of responding.

The rate-dependency theory describes a relationship between the control rate of behavior and the rate of responding after drug administration. It was never intended to be an absolute relationship, but rather it was a way of showing order and predictability in drug effects. McKearney (1981) summarized the concept by saying rate dependency is more of a theoretical law than it is a theory or hypothesis. It describes a predictable relationship between the control behavior and the effects of drugs. Viewed in this manner, it is not unlike the way the temperature of a tissue bath influences the response of a tissue to a drug. The temperature of the bath does not explain the action of the drug, but to consider the action of the drug without considering the bath temperature will lead to something that makes little sense. Thus, the rate of responding should be viewed not as the sole determinant of drug action, but, under the right conditions, it can assume a prime role in the determination of the response.

Although the issues and theories of the 1950s seem of little consequence today, the lessons learned need not be forgotten. For example, as we explore the role of genetics in the control of behavior and the response to drugs in newly developed animal strains, it must be remembered that ignoring the role of the control behavior in influencing drug effects is done at one's own peril.

Cross-References

▶ Barbiturates
▶ Operant Behavior in Animals
▶ Psychostimulants

References

Dews PB (1955) Studies on behavior. I. Differential sensitivity to pentobarbital of pecking performance in pigeons depending on the schedule of reward. J Pharmacol Exp Ther 113:393–401
Dews PB (1958) Studies on behavior. IV. Stimulant actions of methamphetamine. J Pharmacol Exp Ther 122:137–147
Dews PB (1964) A behavioral effect of amobarbital. Archiv für Experimentelle Pathologie und Pharmackologie 248:296–307
Dews PB (1978) Rate-dependency hypothesis (Comment on Gonzales and Byrd, 1978). Science 198:1182–1183
Dews PB, Wenger GR (1977) Rate-dependency of the behavioral effects of amphetamine. In: Thompson T, Dews PB (eds) Advances in behavioral pharmacology, vol 1. Academic, New York, pp 167–227
Gonzalez FA, Byrd LD (1977) Mathematics underlying the rate-dependency hypothesis. Science 195:546–550
Ksir C (1981) Rate-convergent effects of drugs. In: Thompson T, Dews PB, McKim WA (eds) Advances in behavioral pharmacology, vol 3. Academic, New York, pp 39–59
McKearney JW (1981) Rate dependency: scope and limitations in the explanation and analysis of the behavioral effects of drugs. In: Thompson T, Dews PB, McKim WA (eds) Advances in behavioral pharmacology, vol 3. Academic, New York, pp 91–109
McMillan DE, Katz JL (2002) Continuing implications of the early evidence against the drive-reduction hypothesis of the behavioral effects of drugs. Psychopharmacology (Berl) 163:251–264

Rating Scales and Diagnostic Schemata

David V. Sheehan and Kathy H. Sheehan
Depression and Anxiety Disorders Research Institute, University of South Florida College of Medicine, Tampa, FL, USA

Definition

Diagnostic precision is achieved using structured diagnostic interviews. In contrast, rating scales are not intended for diagnostic purposes. Their function is to measure the severity of a condition or symptom cluster after the diagnosis is made by independent means. Typically, in psychopharmacology research, patients are screened into studies using structured diagnostic interviews. The response to treatment over time is then assessed by measuring the severity of the primary and associated symptom clusters or domains of interest from baseline through treatment follow-up (American Psychiatric Association 2000).

Principles and Role in Psychopharmacology

Structured Diagnostic Interviews
Structured diagnostic interviews were developed to improve reliability in making psychiatric diagnoses for clinical and research purposes.

Psychiatric diagnosis was very unreliable prior to the introduction of structured diagnostic interviews. Different criteria were used in different countries and within different schools of psychiatric thought. Within these groups, diagnosis varied based on the manner and on the amount of information elicited from patients. Information provided by patients was interpreted and assembled differently in reaching a diagnosis. The systems used to elicit information, to assemble this information, and to make diagnoses were too many, too varied, too inconsistent, and too divergent to permit accurate communication between large numbers of clinicians and to foster constructive and efficient international collaboration in research. The adoption of efficient and easily administered structured diagnostic interviews, which were translated into a large number of languages and used as consistent diagnostic criteria, has harnessed the energy of international research collaboration in pursuit of finding better treatments for, and understanding of, psychiatric disorders (Sheehan et al. 1998; Spitzer et al. 1992). Structured diagnostic interviews reduce variance in information gathering and reduce criterion variance in assembling this information to make a diagnosis. In this way, they improve both the reliability and quality of diagnostic decision-making.

Choice of Scales

A typical treatment-outcome study uses at least three scales: a scale to measure improvement in the symptom cluster of primary interest (Hamilton 1959, 1967; Montgomery and Asberg 1979), another scale to measure the functional impairment or disability (Endicott et al. 1993; Sheehan and Sheehan 2008), and a third scale to measure the global improvement (Guy 1976). In the past 2 years, the tracking of suicidality in psychiatric and neurological studies has also become routine and is now required by many regulatory agencies (Coric et al. 2009; Posner et al. 2007). Additional scales are frequently used to assess other secondary symptom clusters of interest, e.g., quality of sleep, cognitive impairment (Folstein et al. 1975), negative symptoms in schizophrenia (Kay 1990), anxiety symptoms in depression, or sexual symptoms. With the rise in managed care and increasing concern about cost containment, there is more inclusion of pharmaco-economic measures, especially on health-care utilization and cost data. Since major drivers of costs in the management of many psychiatric disorders are poor adherence, increasing switch rates, and more augmentation with other medications, cost drivers are also increasingly tracked in long-term studies. A wide array of scales is available to assess adverse events. Examples include assessments of extrapyramidal side effects on antipsychotic medications (Barnes 1989), assessments of sexual side effects with antidepressants (Clayton et al. 1997), and assessments of discontinuation withdrawal symptoms and of abuse liability, craving or "liking" with controlled substances (Selzer 1971).

Scale Metrics

Psychometric constructs may be measured in two major ways. Categorical classification (e.g., yes or no) qualitatively assesses the presence of symptoms, signs, or attributes and is widely used in diagnostic assessment through structured diagnostic interviews. Continuous measures (e.g., height or weight) permit quantitative assessment of the severity, intensity, or frequency of symptoms, signs, or attributes. In between these two poles lie the ordinal scales that use an ordered set of categories (e.g., none, mild, moderate, and severe). Such ordinal scales are effectively treated and analyzed like continuous measures especially if they have ten or more points. Continuous or ordinal scales that yield a total score may themselves be used for categorical classification. For example, a score of seven or less on the Hamilton Depression Rating Scale (Hamilton 1967) may be classified as a remission in depression.

A variety of scale structures or metrics are used in continuous or ordinal scales. The most common scale structure involves the use of a 5- or 10-point scale. Dichotomous categorical scales are less common and a few widely used scales have six or seven response options. Likert scaling is frequently used to measure change in symptoms in efficacy studies. Visual analog scales are popular in the assessment of pain. The DISCAN design metric is increasingly popular in scales designed to be sensitive to drug-placebo differences (Sheehan and Sheehan 2008). Although measures of frequency, e.g., of panic attacks, of seizures, or of tics, are often used as outcome measures, in general, they disappoint in their ability to discriminate between active and placebo treatments. The main reason is that the frequency of most behaviors is of skewed distribution in nature. Nonparametric tests statistically punish sensitivity, and this price makes them less able to discriminate drug from placebo and to be less sensitive to change.

Scale Testing

Scales and structured interviews in psychopharmacology must be subjected to proper reliability and validity testing. Validity tests the degree to which the scale or structured interview is consistent with a "gold standard" measure (e.g., how well a structured diagnostic interview maps the diagnostic classification or how a severity scale accurately reflects the severity of a disorder). The problem in psychiatry is that the "gold standard" is not really "gold." Three kinds of reliability testing are usually done on scales and structured interviews. These include inter-rater reliability, test-retest reliability, and internal consistency. Inter-rater reliability measures the agreement between two raters in rating the same subjects using the same scale or interview at the same time. It is most useful in assessing clinician-rated scales. Test-retest reliability measures the agreement between assessments done at two different points in time (e.g., the same scale done on consecutive days, but assessing the same time period). It has special value in assessing self-rated scales. Internal consistency measures the agreement among the items within a scale. It assesses the degree to which all the items measure a single dimension.

In spite of the limitations of these paper and pencil interview tests, they have performed well in guiding our search for and selection of effective treatments. Researchers in psychopharmacology continue to refine their precision and predictive value. Someday, they will be largely replaced by more precise laboratory tests.

References

American Psychiatric Association (2000) In: Rush AJ et al (eds) Handbook of psychiatric measures. American Psychiatric Association, Washington, DC

Barnes TRE (1989) A rating scale for drug-induced akathisia. Br J Psychiatry 154:672–676

Clayton AH, McGarvey EL, Clavet GJ (1997) The changes in sexual functioning questionnaire (CSFQ): development, reliability, and validity. Psychopharmacol Bull 33(4):731–745

Coric V, Stock EG, Pultz J, Marcus R, Sheehan DV (2009) Sheehan suicidality tracking scale (Sheehan STS): preliminary results from a multicenter clinical trial in generalized anxiety disorder. Psychiatry (Edgemont) 6(1):26–31

Endicott J, Nee J, Harrison W (1993) Quality of life enjoyment and satisfaction questionnaire: a new scale. Psychopharmacol Bull 29:321–326

Folstein MF, Folstein SE, McHugh PR (1975) "Mini-Mental State": a practical method for grading the cognitive state of patients for the clinician. J Psychiatr Res 12:189–198

Guy W (ed) (1976) ECDEU assessment manual for psychopharmacology. National Institute of Mental Health, Rockville

Hamilton M (1959) The assessment of anxiety states by rating. Br J Med Psychol 32:50–55

Hamilton M (1967) Development of a rating scale for primary depressive illness. Br J Soc Clin Psychol 6:278–296

Kay SR (1990) Positive–negative symptom assessment in schizophrenia: psychometric issues and scale comparison. Psychiatr Q 61:163–178

Montgomery SA, Asberg M (1979) A new depression scale designed to be sensitive to change. Br J Psychiatry 134:382–389

Posner K, Oquendo MA, Gould M, Stanley B, Davies M (2007) Columbia classification algorithm of suicide assessment (C-CASA): classification of suicidal events in the FDA's pediatric suicidal risk analysis of antidepressants. Am J Psychiatry 164(7):1035–1043

Selzer ML (1971) The Michigan alcoholism screening test: the quest for a new diagnostic instrument. Am J Psychiatry 127(12):1654–1658

Sheehan KH, Sheehan DV (2008) Assessing treatment effects in clinical trials with the Discan metric of the Sheehan disability scale. Int Clin Psychopharmacol 23(2):70–83

Sheehan DV, Lecubrier Y, Sheehan KH, Amorim P, Janavs J, Weiller E, Hergueta T, Baker R, Dunbar GC (1998) The mini-international neuropsychiatric interview (M.I.N.I): the development and validation of a structured diagnostic psychiatric interview for DSM-IV and ICD-10. J Clin Psychiatry 59(suppl 20):22–33

Spitzer RL, Williams JBW, Gibbon M et al (1992) The structured clinical interview for DSM-III-R (SCID), I: history, rationale, and description. Arch Gen Psychiatry 49:624–629

Reactive Oxygen Species

Synonyms

ROS

Definition

Reactive oxygen species (ROS) are, for example, hydroxyl radicals or superoxide radicals. They are extremely reactive in "oxidizing" membrane constituents, like lipids, thus causing membrane damage and cell death.

Cross-References

▶ Free Radicals

Reboxetine

Synonyms

Davedax; Edronax; Norebox; Prolift; Solvex; Vestra

Definition

Reboxetine, $(R*,R*)$-2-[(2-ethoxyphenoxy)-phenyl-methyl]morpholine mesylate, is an antidepressant that selectively inhibits the reuptake of norepinephrine (SNRI). It is claimed to have mood-lightening effects in the elderly, but it is significantly less clinically effective against the acute phase of major depressive disorders and less acceptable relative to other new antidepressants. It is also used in the treatment of panic disorder and ▶ attention deficit hyperactivity disorder.

The side effects of reboxetine derive from the anticholinergic effects of the compound, particularly antimuscarinic effects, and range from dry mouth, constipation, headache, drowsiness, dizziness, and excessive sweating to insomnia. Rare cases of seizures have been reported.

Cross-References

▶ Attention-Deficit/Hyperactivity Disorder
▶ Antidepressants
▶ Panic Disorder

Receptor

Definition

A protein embedded in the cell membrane or the cytoplasm, which is sensitive to a drug or chemical neurotransmitter. It docks (binds) to the drug or neurotransmitter agonists then generates an electrical or chemical signal in the neuron in which it is located, whereas antagonists bind but do not activate. A given substance can act on more than one type of receptor.

Receptor Trafficking

Definition

For example, ▶ AMPA receptor trafficking, the dynamic movement of receptors into and out of the synapse from intracellular compartments.

Cross-References

▶ Long-Term Depression and Memory

Receptors: Binding Assays

Josee E. Leysen[1], Xavier Langlois[2], Lieve Heylen[2] and Adriaan A. Lammertsma[1]
[1]Nuclear Medicine and PET Research, VU University Medical Centre, Amsterdam, The Netherlands
[2]Neuroscience Department, Johnson & Johnson Pharmaceutical Research and Development, Program Management Office, Beerse, Belgium

Synonyms

Labeled ligand binding assays; Ligand binding assays; Radioligand binding assays

Definition

Receptor binding refers to a technique in which a labeled compound, a ligand, which binds to a receptor, is used to detect that receptor. Usually, the ligand is labeled by means of a radioactive isotope, such as ^3H, ^{125}I, ^{35}S, etc., but a fluorescent moiety is also possible. The receptor can be localized in a tissue that is homogenized or sliced or in cells in culture that either have an endogenous expression of the receptor or have been transfected with a cloned receptor gene. Tissue preparations are incubated with a labeled ligand that has a high binding affinity for the target receptor. The labeled ligand bound to tissue is then collected and detected using various techniques, such as filtration techniques combined with radioactivity counting, scintillation proximity analysis and autoradiography for radioactive ligands, and time-resolved fluorescence resonance energy transfer (TR-FRET) or amplified luminescent proximity homogeneous assay (AlphaScreen) for ligands labeled with a fluorescent or a chemiluminescent probe, respectively. These techniques are applied in vitro or ex vivo. In vivo receptor binding can be investigated using positron emission tomography (PET) or single-photon emission computerized tomography (SPECT) imaging.

Principles and Role in Psychopharmacology

Receptor binding techniques were first introduced in the 1970s (historic reviews: Lefkowitz 2004; Snyder and Pasternak 2003). They were the first tools enabling receptors to be demonstrated in tissues using biochemical methods. In earlier years, there were no clues about the nature or structure of receptors. Extensive research throughout the 1980s led to the cloning of receptor genes and the identification of their structure. Today, four major classes of receptors that are located in cell membranes and one class of cytosolic receptors have been identified:

1. The G-protein-coupled receptors (GPCRs), also called "7 transmembrane (7 TM)" receptors, are integral membrane protein monomers. With 802 known and predicted human GPCRs, derived from the human genome, this is the largest receptor family and at least 50 of them have been identified as major drug targets (Lagerström and Schiöth 2008). Sequence comparison revealed five subfamilies, all of which have a central core domain in common, which consists of 7 TM helices connected by three intracellular and three extracellular loops. The protein has an extracellular N-terminal and an intracellular C-terminal, which can be of widely varying lengths. GPCRs have diverse natural ligands comprising of small molecules (amines, amino acids, nucleotides, nucleosides, prostaglandins, peptides, lipid-like molecules, etc.), light, Ca^{++}, odorants and pheromones, and proteins (for review of the GPCR classes, structures, features, and natural ligands, see Lagerström and Schiöth 2008). Upon activation, a GPCR associates with a G-protein that further activates effectors in the cell. Various different proteins can interact with GPCRs so that receptor complexes can exist in multiple states leading to possible complex kinetics for ligand binding (Christopoulos and Kenakin 2002).
2. Ligand-gated ion channels are multimeric, often pentameric, protein complexes that form a pore in the membrane. The members

of this family are, e.g., the nicotinic acetylcholine, $GABA_A$, $5-HT_3$, glycine, puringergic P2X, and the NMDA, AMPA, and kainate subtypes of glutamate receptors, each of itself has several subtypes.
3. Receptors that are enzymes with a ligand binding domain that couples to an intracellular membrane-anchored enzyme such as tyrosine kinase, tyrosine phosphatase, serine/threonine kinase, and guanylyl cyclase. The members of this class are, e.g., growth factor receptors, neurotrophic factor receptors, and transforming growth factor β (TGFβ) receptors.
4. Cytosolic receptors regulate transcription in the cell nucleus. The members are steroid, retinoid, and thyroid hormone receptors.

Large numbers of members of each of these receptor classes as well as (neurotransmitter) transporters have been studied with receptor binding techniques and receptor binding assays, and hundreds of different molecular targets have been described. In this chapter, in vitro and ex vivo methods used for investigating membrane-bound receptors will be discussed.

Principles

Analytical Rules

Receptor binding assays require working according to analytical rules and procedures. Water must be of freshly double-distilled or milliQ quality and all chemicals (salts, metal ions, organic solvents, etc.) of analytical grade. Labeled ligands must have a purity of >98 % and purity must be checked regularly (radioligands, in particular with high specific activity, suffer from radiolysis). Stability of compounds must be checked and purified where necessary (e.g., catecholamines and indoleamines are very unstable and solutions need to be purified just prior to use). (Micro)balances and pipettes (manual and automated) must be regularly gauged. Compounds must be solubilized and diluted according to their physicochemical properties. Lipophilic compounds are best solubilized and diluted in pure dimethyl sulfoxide (DMSO) and buffer should be used only for the last dilution step, or the organic solvent solution should be added directly into the incubation mixture. However, the organic solvent concentration in the assay mixture should not exceed 0.5 %. Dilution in the organic solvent is indicated to avoid the loss of lipophilic compound due to adsorption to glass or plastic ware. Plastic tubing must be avoided when using robotic devices. Hydrophilic compounds must be solubilized and diluted in buffer or water. Sometimes protecting agents such as antioxidants must be added. Biological materials must be kept on ice in buffer of neutral pH until the start of the incubation. Buffer pH must be adjusted to the temperature of the incubation.

Law of Mass Action

Ligand–receptor binding experiments are usually analyzed according to the simple model of law of mass action:

$$[R] + [L] \underset{k_{off}}{\overset{k_{on}}{\leftrightarrow}} [RL]$$

where [R], [L], and [RL] represent the free concentration of receptor, ligand, and ligand–receptor complex, respectively, and where it is assumed that the reaction components can freely diffuse within the medium. k_{on} and k_{off} are the rate constants for association and dissociation of the ligand–receptor complex. It is assumed that after dissociation, receptor and ligand are not altered.

Association and dissociation rates are temperature dependent. The reaction is driven by the concentration of the reacting agents. Equilibrium is reached when the rates at which ligand–receptor complexes are formed and dissociate are equal. At equilibrium, the following applies:

$$[R][L]k_{on} = [RL]k_{off}$$

The equilibrium dissociation constant, K_d, a measure for binding affinity, is defined as

$$K_d = \frac{k_{off}}{k_{on}} = \frac{[L][R]}{[RL]}$$

The constants have the following units: k_{on}: $M^{-1} \cdot s^{-1}$; k_{off}: s^{-1}; K_d: M, where M stands for

Molar or mol·L^{-1}. When receptor binding experiments are performed, [R] usually is unknown (needs to be determined), [RL] is measured in the assay, and [L] usually is assumed to be equal to the applied ligand concentration. This implies, however, that only a minor fraction (<5 %) of the total applied ligand should become bound (either by specific binding, nonspecific binding, or adsorption to tissue and assay container) so that the free concentration is not significantly altered.

Criteria Required by the Law of Mass Action: Can They Be Met?
It is important to realize that the reaction conditions never meet the required criteria. The reaction model itself is largely simplified, as receptors can occur in different states and ligands may bind to orthosteric and allosteric sites (see Christopoulos and Kenakin 2002). "Free in solution" and "freely moving in solution" do almost never apply to biological tissue, either as an homogenate or a membrane suspension, and certainly not when tissue slices, whole cells, or tissue fixed on a support are used. Membrane preparations consist of membrane vesicles of varying dimension, possibly with strips of membrane suspended in aqueous medium. Surface phenomena will inevitably take place when ligands (hydrophobic or hydrophilic) approach the membrane vesicle. Examples of surface phenomena are electrostatic interactions of ionized compounds or surface excess of lipophilic compounds. Surface phenomena, which are difficult to measure or estimate, have received virtually no attention in receptor binding research. Yet, they are a matter of fact and, for ▶ haloperidol, a 1,000× surface excess was calculated to exist in the monolayer around a membrane vesicle as compared to the concentration in aqueous medium (Leysen and Gommeren 1981).

Furthermore, reaction temperature needs to be carefully controlled. A binding reaction is temperature dependent and tissue is temperature sensitive; the transition temperature of the lipid cell membrane (usually around 15 °C) must be considered. Tissue and compounds may also behave differently according to pH, in particular when they have ionizable groups with an acid association constant, pKa, between 5 and 8. Therefore, it is advised never to consider experimentally measured values for, e.g., K_d or K_i as "absolute" but refer to them as "apparent" under the conditions of a particular assay. Those conditions should always be fully reported.

Types of Binding
Labeled ligands show different types of binding, depending on their physicochemical and pharmacological properties. There is specific binding to the target receptor, which occurs in a concentration range of 2 log units around the K_d value. This binding is saturable, as the number of receptors present is limited. Binding to the target receptor should be reversible and can be inhibited by a compound (competitor) that has the same pharmacological property, related to the target receptor, as the labeled ligand. Full inhibition of receptor binding is obtained at 100× K_i of the competitor (K_i, inhibition constant, see below), provided that the labeled ligand is used at concentrations around its K_d. A particular labeled ligand (or competitor) may bind to several different receptors within a narrow concentration range. In particular agonists or antagonists for various subtypes of dopamine, noradrenaline, serotonin, histamine, and acetylcholine GPCRs can show, what is called, a "broad receptor profile" (see Leysen 2002, 2004). Binding of a labeled ligand to several different receptors can be a problem, in particular when natural tissues are investigated (this is less of an issue when cultured cells transfected with a particular receptor gene are used and high receptor expression is obtained). If required, occluding agents can be added to prevent the labeling of nontargeted receptors. Labeled ligands also show "nonspecific" binding due to adsorption to tissue, this binding is linearly proportional with the labeled ligand concentration and non-saturable, i.e., non-displaceable. Nonspecific binding can be measured in an experiment where an appropriate competitor (with high ▶ affinity for the target receptor but preferably of

a different chemical structure than the labeled ligand) is added at a concentration 100–1,000 times its K_i and incubated together with the labeled ligand and the tissue. Resulting binding represents nonspecific binding. Labeled ligands may show an additional type of binding, apparently of high affinity and saturable, hence displaceable, but related to a particular moiety in its structure and apparently unrelated to binding to a known biological target (e.g., [^{3}H]spiperone, the 5-HT$_{2A}$ and dopamine D2 receptor ligand, labels "spirodecanone sites"); this binding can be occluded by adding a structural analogue of the labeled ligand, but which does not bind to the target receptor. When studying receptor binding in natural tissues, the various possible types of binding should be carefully investigated (Leysen 1984).

Types of Experiments

Saturation Binding Experiments or "Labeled Ligand Concentration Binding Isotherms"
Aim: To determine K_d and B_{max} (maximum number of receptor binding sites in the tissue preparation) values.

These assays are performed using a constant amount of tissue at a particular temperature (preferably 37 °C but sometimes lower temperatures are used) and pH (around 7.4) with increasing concentrations of labeled ligand. The specific binding isotherm, according to the simple law of mass action, follows the course of a hyperbola when the labeled ligand's free concentration on the abscissa and specific binding on the ordinate are plotted using linear scales. Half-maximal binding is reached at a concentration that equals the K_d value, and maximal binding is approached at four times the K_d value. Applied labeled ligand concentrations should span a range from 0.2 to 8 times K_d in a series of at least 12 points covering the rising part and plateau of the hyperbola (e.g., for a labeled ligand with $K_d = 1$ nM, appropriate test concentrations are 0.2, 0.4, 0.6, 0.8, 1.0, 1.5, 2.0, 2.5, 3.0, 4.0, 6.0, 8.0 nM). The "total binding" is measured in assays with the labeled ligand at its various concentrations, the tissue preparation, and a solvent sample; the "nonspecific binding" is measured in assays with the labeled ligand (at each of its concentrations), the tissue preparation, and a (non-labeled) competitor at a concentration of 100 times its K_i value. Specific binding is calculated by subtracting nonspecific binding from total binding at each labeled ligand concentration. To date, binding isotherms are calculated by nonlinear regression analysis and computerized curve fitting, which provide the K_d and B_{max} values. In case a "one-binding-site model" does not fit the data, a "two-binding-site model" can be attempted. Curve-fitting programs for labeled ligand binding are available commercially, e.g., GraphPad Prism. The programs usually provide background on calculations with radioactivity, e.g., on converting bound radioactivity, measured in counts, into pmoles of bound labeled ligand. In the past, when computing power was not readily available, the hyperbola was transformed into a linear equation, e.g., a Scatchard plot.

Kinetic Binding Experiments
Aim: To determine the association and dissociation rate constants.

Rate constants are relatively easy to measure for a labeled ligand.

To measure the association rate, a labeled ligand, at a given concentration (e.g., 3–4 × K_d), together with the tissue preparation, is incubated for various periods of time, and for each time, bound labeled ligand is measured. Association of ligands with nM affinity is fast and incubation times should be in the range from seconds up to 10 min. The maximal bound radioligand at equilibrium is usually reached within 2–4 min. Ligand–receptor association is a second-order reaction, involving two reaction partners: the free ligand and the free receptor. However, measured binding is the result of association and dissociation, and usually the free receptor concentration is small, and consequently the free ligand concentration drives the association. Therefore, the reaction proceeds according to the pseudo first-order kinetics. The plot of bound labeled ligand versus time is hyperbolic and can be transformed according to a pseudo first-order rate equation, in a linear plot:

$$\ln\left(\frac{B_{eq}}{B_{eq} - b_t}\right) \text{ versus time(s)}$$

B_{eq} and b_t are bound labeled ligand at equilibrium and at a given time. The pseudo first-order rate constant, k_{obs} (s^{-1}), is given by the slope of the line. For calculation of the association rate constant, k_{on} ($M^{-1} \cdot s^{-1}$), the dissociation rate constant, k_{off} (s^{-1}), is needed.

$$k_{on} = \frac{k_{obs} - k_{off}}{[L]}$$

The dissociation rate constant can be measured by incubating the tissue preparation with the labeled ligand at a given concentration (preferably not exceeding the K_d value) until equilibrium is reached and then either strongly diluting the reaction medium (more than 10 times) to favor dissociation and minimize reassociation or, more common, adding an excess of strong competitor to the reaction medium, which will induce displacement and hence dissociation of the labeled ligand. From that point onward, a series of incubation samples is taken to measure the remaining bound labeled ligand. The dissociation reaction is a first-order reaction, a semilogarithmic plot of log[bound labeled ligand] versus time is linear and the dissociation rate constant, k_{off}, is given by the slope of the line × 2.3 (transformation of $\log_{10}$ into ln). The half-life of dissociation is given by $t_{1/2} = 0.693/k_{off}$. Measurement of the dissociation rate of an unlabeled competitor requires a more elaborate procedure. The approximate half-life of dissociation of an unlabeled competitor can be assessed by an indirect method. Tissue with competitor at a given concentration (not exceeding $2 \times K_i$) is incubated until equilibrium and filtered over a glass fiber filter. The tissue (with bound competitor) adsorbed on the filter is rinsed for various periods of time, followed by short incubation of the tissue on the filter with a labeled ligand. The amount of labeled ligand that becomes bound is then an indication of the amount of competitor that has dissociated (Leysen and Gommeren 1986). Knowledge of the receptor dissociation rate of compounds has gained interest, in light of deriving information on factors that contribute to the duration of action of drugs and on the ease with which a drug bound to the receptor in vivo can be displaced, e.g., by the endogenous ligand for the receptor.

Competitive Binding Experiments
Aim: To determine the IC_{50} value (concentration producing 50 % inhibition) of a competitor Competitive binding experiments are used to compare different ligands acting at similar sites. The sigmoid curve obtained with a competitor should asymptotically approach the level of nonspecific binding, without surpassing it. Nonlinear curve-fitting programs can be used to estimate IC_{50} values. Curve fitting also yields a value for the slope (n) of the curve; $n = 1$, for competition at one binding site; $n > 1$ or $n < 1$ for multiple binding sites, binding of the competitor with different affinity to multiple states of the receptor, or involvement of cooperative binding. Curve-fitting programs allow for the analysis of data for multiple-binding-site models. IC_{50} values are dependent on applied concentration and K_d of the labeled ligand. The Cheng–Prusoff equation provides for the calculation of the equilibrium inhibition constant, K_i (equaling the equilibrium dissociation constant), which is independent of the applied labeled ligand:

$$K_i = IC_{50}/(1 + [L]/K_d)$$

The Cheng–Prusoff equation is derived from a simple model of competition at one and the same binding site; the same warnings of caution for the interpretation of K_i values as described earlier apply.

For further reading on different types of binding experiments and illustrations and details on the analyses of receptor binding data, see Keen and MacDermot (1993).

Antagonist Binding
For GPCRs and ligand-gated ion channels, the majority of studied labeled ligands and competitors

are antagonists or ▶ inverse agonists. Antagonist binding to GPCRs has appeared to be insensitive to the occurrence of the receptor in various states (e.g., G-protein coupled or uncoupled). Antagonist binding curves (labeled antagonist saturation binding and antagonist–antagonist competition) have often been found to obey apparent single-site binding kinetics.

Agonist Binding

Agonist binding to GPCRs is sensitive to the state in which the receptor occurs. Agonists have substantially higher affinity for the G-protein-coupled than for the G-protein-uncoupled receptor. As a consequence, competition between a labeled antagonist and an unlabeled agonist often yields a shallow inhibition curve, pointing to multiple binding sites or binding to the receptor in multiple states. Inhibition curves of agonists can be influenced by additives in the incubation medium, in particular by agents that affect the coupling of the receptor to the G-protein. Guanosine triphosphate (GTP), or a stable analogue thereof, uncouples the receptor–G-protein complex. The addition of GTP to the medium will shift a shallow or biphasic agonist inhibition curve to the right, make it monophasic, and decrease the slope to approach 1 (see Lefkowitz 2004). Agonist binding to GPCRs is also affected by divalent cations, in particular Mg^{++} or Mn^{++}, which may favor receptor–G-protein coupling and increase the agonist binding affinity. The effect of divalent cations on agonist binding should be experimentally investigated. A good example is the investigation of the effects of divalent cations on the affinity states of the M1 muscarinic acetylcholine receptor (Potter et al. 1988). Labeled agonist concentration binding isotherms will similarly be influenced by additives in the incubation medium and affect the measured apparent K_d and B_{max} values. When labeled agonist saturation curves are measured in a low concentration range, usually only the high-affinity binding constant of the agonist, K_H, will be detected. The B_{max} value measured with a labeled agonist may only represent the G-protein-coupled portion of the receptor and, for the same tissue sample, usually will be lower than the B_{max} value measured with a labeled antagonist.

GPCRs for which labeled agonist binding has been amply studied and for several of which agonists are used as therapeutic agents are μ − opiate, $5\text{-}HT_{1A}$, $5\text{-}HT_{1B}$, $5\text{-}HT_4$, and various peptide receptors (e.g., NK1, NK2, NK3, CRF1). Examples of GPCRs that have been studied with both labeled agonists and antagonists are dopamine D_2, $5\text{-}HT_{2A}$, $5\text{-}HT_{2C}$, $5\text{-}HT_4$, $5\text{-}HT_6$ and muscarine M1ACh receptors, and β1- and β2-adrenoceptors.

Allosteric Competitors

The natural ligand or a synthetic agonist that binds to the orthosteric site and a compound that attaches to the allosteric binding site can concomitantly occupy the same receptor without mutual inhibition of binding. Positive ▶ allosteric modulators will enhance and negative allosteric modulators will reduce the activity/affinity of an orthosteric agonist. The allosteric receptor interaction of compounds is usually studied in functional, signal transduction assays ("▶ Receptors: Functional Assays"). Labeled ligand binding to an allosteric site may be used in competition binding assays to detect competitors that bind to the same allosteric site.

Ligand-gated ion channel receptors are known to have several different allosteric binding sites. The $GABA_A$ receptor was the first receptor for which the phenomenon was discovered with drugs that were in use as therapeutic agents, namely, the ▶ benzodiazepines. GPCRs of family 3, the $GABA_B$ receptor, and the metabotropic glutamate receptor subtypes are amply studied for positive and negative allosteric modulators.

Methods for In Vitro Receptor Binding Using Tissue or Cell Homogenates or Membrane Preparations

Tissues and Preparation

Tissues used for receptor binding are, e.g., dissected brain regions or organs, cultured cells (cell lines or primary cultures) or blood cells that express a particular receptor endogenously, or cultured cells with (abundant) expression of a particular receptor following the transfection of the cells with that receptor's cloned gene from a specific species (often human). Although, when conditions are appropriate, receptor binding can be performed on intact cells, in most cases the tissue is homogenized, and either the whole homogenate or, more often, a membrane preparation is used for the assay. Used membrane preparations are the "total particulate fraction" (i.e., the total fraction of membranes spun down at high centrifugation speed following extensive homogenization of the tissue in buffer with a blender) or a membrane preparation from separated subcellular particles (e.g., heavy and light mitochondrial fractions, plasma membrane fraction) following careful homogenization of the tissue in 0.25 M sucrose followed by differential centrifugation (see Laduron et al. 1978). Membrane preparations are washed by resuspension in medium and recentrifugation. All tissue preparation steps are to be carried out at 0–4 °C. Tissue preparations can be stored below −20 °C.

Incubation in Tubes and Multiwell Plates

Incubations can be performed in test tubes (volumes 0.5–2 mL) or 96-multiwell plates (0.1–0.2 mL). Incubation mixtures are composed of a sample of cells or membrane preparation suspended in buffer, an aliquot of labeled ligand to give a desired final concentration and, depending on the type of experiment, an aliquot of a competitor to give a desired final concentration (see types of experiments). The buffer has a particular pH (usually around 7.4) and can contain certain additives such as metal ions. Depending on the type of experiment, the incubation is run at a given temperature for a given time period (see types of experiments). For competition binding experiments, the incubation time should be sufficiently long to reach binding equilibrium.

Filtration Methods

Labeled ligand, bound to the tissue preparation and free-labeled ligand, can be separated by filtration over glass fiber filters (filters are sometimes presoaked in polyethyleneimine to reduce absorption of the labeled ligand to the filters) under suction, followed by rapid rinsing with ice-cold buffer. For incubation in test tubes, a Brandel harvester (Brandel, Gaithersburg, USA) is used; for incubations in 96-multiwell plates, various harvesting devices are available (e.g., Micromate 196 and Mac II). The radioactivity of bound labeled ligand collected on the filters is counted; for β-ray emitting isotopes (e.g., ^{3}H-labeled ligands), the filters are dried, liquid scintillation fluid is added, and radioactivity is counted in a liquid scintillation counter; for γ-ray emitting isotopes (e.g., ^{125}I), radioactivity can be counted directly in a γ-counter. For competition binding assays, 96-multiwell incubation, filtration, and counting methods and devices are appropriate; for saturation binding experiments with the determination of K_d and B_{max} values and the determination of ligand association and dissociation rates, incubation in test tubes and filtration over separate filters that are counted in vials in a calibrated liquid scintillation counter are advised.

Scintillation Proximity Assay (SPA)

SPA is a homogenous assay that avoids the filtration step. For SPA, either poly-vinyl-toluene beads filled with scintillant or 96-multiwell plates with scintillant coated on the inner surface of the wells are used. The beads or plates are coated with membrane preparation. Samples of coated beads added in 96-multiwell plates are incubated with labeled ligand with or without competitor at a desired concentration in buffer. Since the tissue is not "free in solution," several hours of incubation time are required to reach binding equilibrium. After incubation, the multiwell plates are counted directly in a TopCount NXT. The scintillant will only detect radioactivity in its immediate vicinity, i.e., the radioactive ligand that is bound to the tissue coated on the beads or on the plates. The mix-and-read format technology is useful for high-throughput screening, but long incubation times have to be taken into account.

Nonradioactive Proximity Assays

The burden on the environment and the inherent high cost for the removal of radioactive waste encouraged the development of nonradioactive techniques for receptor binding. TR-FRET, making use of fluorescent probes, and AlphaScreen™, which is based on chemiluminescence, are examples that are mainly applied for high-throughput competition binding assays. The technologies are based on the excitation of the "donor" probe (e.g., attached to the ligand), which triggers an energy transfer to the acceptor probe (e.g., attached to the receptor preparation) if they are within a given proximity. The acceptor probe in turn emits light of a given wave length, which is detected. The application of these techniques is limited, since it requires the development of ligands labeled with relatively large probes, resulting in a compound with a chemical structure different from that of the original ligand. The receptor binding characteristics of such fluorescent or chemiluminescent ligands will be altered and need to be fully investigated. The development of apt fluorescent or chemiluminescent ligands is a matter of trial and error.

Labeled Ligand Autoradiography

Autoradiography is a general technique allowing the visualization of the distribution of a radioactive ligand, bound to a molecular target in a tissue section. The radioactive ligand can be injected (iv) in an animal, followed by sacrifice of the animal, tissue dissection, and sectioning. Otherwise, tissue sections can be incubated with radioactive ligand in vitro. Tissue (often brain) sections (20 μm thick) usually are cut with a cryostat and mounted on a coated microscope glass support. Incubation is performed by overlaying the tissue section with a drop of incubation medium containing the radioactive ligand. After incubation, the tissue sections are quickly rinsed and dried. To "visualize" radioactive ligand bound to the tissue, different techniques have been used. The highest resolution, but requiring the longest exposure time (weeks up to several months), is obtained by exposing the tissue sections to a photographic emulsion. Shorter exposure times (several days up to 1 week) can be achieved by using a phosphor imager. The most recently developed device, the β-imager (Biospace Lab, Paris, France), can produce an image of the bound radioactivity following overnight counting, and up to 15 microscope glass slides can be counted simultaneously. The images are analyzed and quantified by counting the number of β-particles emerging from the delineated brain areas by using the β-vision program.

For technical details on receptor autoradiography protocols, see Wharton and Polak (1993).

Apart from neuroanatomical mapping, autodiographic techniques can also be applied to measure receptor occupancy by non-labeled drugs (Schotte et al. 1993). A drug, at varying dosages, systemically is given to laboratory animals. The animals are sacrificed, tissue dissected and sectioned, and tissue sections are briefly incubated with a radioactive ligand for the target receptor. The difference in labeling in a defined area on matching tissue sections from drug-treated and nontreated animals is a measure for the occupancy of the receptor by the drug administered to the animal. This "ex vivo" incubation technique is only applicable for drugs with a sufficiently slow dissociation rate from the receptor. Alternatively, the radioactive ligand can be injected iv to the animal that is systemically treated with the investigational drug; in this way, the competition for receptor binding between the drug and the radioactive ligand will take place in vivo. Subsequently, the animal is sacrificed, tissue dissected and sliced, followed by the quantification of the image. With the introduction of the highly sensitive β-imager in the 1990s, allowing images to be obtained in a few hours, receptor occupancy assays ex vivo became a feasible and highly valuable drug screening technique (Langlois et al. 2001).

Application of Receptor Binding

Pharmacologists sometimes demean receptor binding data and may even view it scornfully because binding does not say anything about function. Yet, receptor binding has contributed substantially to pharmacology. It has demonstrated the existence of receptors at the molecular level. It has contributed to the isolation and purification of receptor proteins, leading to the cloning of the first receptor gene. It has allowed the study of the localization of receptors at anatomical and cellular levels and of receptor dynamics, such as receptor desensitization, receptor endo- and exocytosis, and transport of receptors along axons (see Lefkowitz 2004).

In terms of drug discovery and drug profiling, receptor binding has revolutionized the field. Drug screening with receptor binding started in the mid-1970s and soon led to high-throughput, automated assays and super high-throughput assays, allowing for the screening of thousands of compounds in one day at many different receptors with a minimum of personnel.

Receptor binding profiles of drugs have revealed that many of the drugs that were in clinical use, such as ▶ antidepressants and antipsychotics, bound to several different receptors within a narrow potency range, e.g., the antipsychotic clozapine was found to bind to more than 20 receptors with such potency, and that all of them could be hit at a therapeutic dose (Leysen 2002, 2004). Receptor binding is the method of choice to investigate and demonstrate the selectivity of action versus the multiplicity of action of drugs. Measurement of receptor occupancy in the brain by drugs that are administered to animals, using ex vivo autoradiography, has become a key component of CNS drug discovery programs. It allows the determination of the dose range within which central receptors become occupied and does provide information not only on the extent of receptor occupation but also on the brain penetration of the drug.

Cross-References

▶ Agonist
▶ Allosteric Modulators
▶ Antagonist
▶ Antidepressants
▶ Antipsychotic Drugs
▶ Benzodiazepines
▶ B_{max}
▶ Catecholamines
▶ Clozapine
▶ Dopamine
▶ Equilibrium Dissociation Constant
▶ GABA$_A$ Receptor
▶ Glutamate Receptors
▶ Inverse Agonists
▶ Metabotropic Glutamate Receptor
▶ Muscarinic Cholinergic Receptor Agonists and Antagonists
▶ Neurotrophic Factors
▶ Nicotinic Acetylcholine Receptor Subtypes
▶ N-Methyl-D-Aspartate Receptor
▶ Positron Emission Tomography (PET) Imaging
▶ Reversible Binding

References

Christopoulos A, Kenakin T (2002) G protein-coupled receptor allosterism and complexing. Pharmacol Rev 54:323–347
Keen M, MacDermot J (1993) Analysis of receptors by radioligand binding. In: Wharton J, Polak JM (eds) Receptor autoradiography, principles and practice. Oxford University Press, Oxford
Laduron PM, Janssen PFM, Leysen JE (1978) Spiperone: a ligand of choice for neuroleptic receptors. 3. Subcellular distribution of neuroleptic drugs and their receptors in various brain regions. Biochem Pharmacol 27:323–328
Lagerström MC, Schiöth HB (2008) Structural diversity of G protein-coupled receptors and significance for drug discovery. Nat Rev Drug Discov 7:339–357
Langlois X, te Riele P, Wintmolders C, Leysen JE, Jurzak M (2001) Use of the β-imager for rapid ex vivo autoradiography exemplified with central nervous system penetrating neurokinin 3 antagonists. J Pharmacol Exp Ther 299:712–717
Lefkowitz RJ (2004) Historical review: a brief history and personal retrospective of seven-transmembrane receptors. TRENDS Pharmacol Sci 25:413–422
Leysen J (1984) Problems in in vitro receptor-binding studies and identification and role of serotonin receptor sites. Neuropharmacology 23:247–254
Leysen JE (2002) Atypical antipsychotics. In: Di Chiari G (ed) Handbook of experimental pharmacology, vol 154/II dopamine in the CNS. Springer, Berlin, pp 474–490
Leysen JE (2004) 5-HT2 receptors. Curr Drug Targets CNS Neurol Disord 3:11–26
Leysen JE, Gommeren W (1981) Optimal conditions for [^{3}H]apomorphine binding and anomalous equilibrium binding of [^{3}H]apomorphine and [^{3}H]spiperone to rat striatal membranes: involvement of surface phenomena versus multiple binding sites. J Neurochem 36:201–219
Leysen JE, Gommeren W (1986) Drug-receptor dissociation time, new tool for drug research: receptor binding affinity and drug-receptor dissociation profiles of

serotonin-S2, dopamine-D2, histamine-H1 antagonists and opiates. Drug Devel Res 8:119–131

Potter LT, Ferrendelli CA, Hanchett HE (1988) Two affinity states of M1 muscarine receptors. Cell Mol Neurobiol 8:181–191

Schotte A, Janssen PF, Megens AA, Leysen JE (1993) Occupancy of central neurotransmitter receptors by risperidone, clozapine and haloperidol, measured ex vivo by quantitative autoradiography. Brain Res 631:191–202

Snyder SH, Pasternak GW (2003) Historical review: opioid receptors. TRENDS Pharmacol Sci 24:198–205

Wharton J, Polak JM (eds) (1993) Receptor autoradiography, principles and practice. Oxford University Press, Oxford

Receptors: Functional Assays

Hilde Lavreysen[1] and John Atack[2]
[1]Janssen Research and Development, A Division of Janssen Pharmaceutica, Beerse, Belgium
[2]Translational Drug Discovery Group, School of Life Sciences, University of Sussex, Beerse, Belgium

Synonyms

Measurement of biological effect resulting from interaction with the receptor; Measurement of receptor signaling

Definition

Psychoactive drugs generally exert their effects by binding to a specific recognition site, whether that is associated with a receptor, a transporter, or an enzyme, and thereby alter the function of that receptor, enzyme, or transporter. However, receptors are the predominant targets for psychoactive drugs (Grigoriadis et al. 2009), and these may be either G-protein-coupled receptors (GPCRs) or, to a lesser extent, ion channels (the latter being either ligand or voltage gated).

Although the actual effect produced by drugs affecting GPCRs and ion channels is very different, their mechanisms of action follow similar principles. Hence, a drug acting at a GPCR or ion channel may block the effect of an endogenous agonist (i.e., acts as an antagonist), mimic the effect of the endogenous activator (i.e., acts as an agonist), or inhibit spontaneous (nonagonist-stimulated) receptor activation (i.e., acts as an inverse agonist). In addition, a drug may bind to a recognition site that is physically distinct from the agonist-binding site and, as a result, produce allosteric changes in receptor structure that enhance or reduce the effects of the endogenous agonist; such drugs are known as positive allosteric modulators (PAMs) and negative allosteric modulators (NAMs), respectively.

Stated simply, a drug binds to a receptor and then "does something." It is the purpose of functional assays to measure and quantitate that "something," with the choice and design of assay being highly dependent upon the type of receptor (i.e., GPCR or ion channel), as well as the effect being studied (agonism, antagonism, inverse agonism, or allosteric modulation).

Principles and Role in Psychopharmacology

Introduction

Drugs may exert their effects by specifically interacting with receptors, enzymes, or transporters; by interfering with DNA, RNA, or ribosomal processes; by exerting physicochemical effects; or, in the case of monoclonal antibodies, by targeting a variety of different proteins. With regard to receptors, this classification includes GPCRs, ion channels, and nuclear receptors. Since the latter are primarily the therapeutic focus for obesity, diabetes, and cancer, the GPCR and ion-channel family represent the main receptor targets in psychopharmacology. As a therapeutic class, GPCRs are more successful than ion channels as measured in terms of FDA-approved drugs, and hence, GPCRs appear to be more "druggable" targets. However, there is little doubt that ion channels have been generally underexploited, especially as targets for CNS disorders, especially when one considers that most of the neuropsychopharmacological ion-channel drugs are $GABA_A$ modulators that were discovered prior to the revolution in molecular biology and genomics.

In the search for new psychopharmacological drugs, the identification of novel chemical structures that interact with the target of interest is arguably the most crucial step in the drug discovery process. Such compounds are usually identified in the screening of large chemical collections (ranging from hundreds of thousands to up to several million) in high-throughput screening (HTS) campaigns. Accordingly, the focus of this article will be the measurement of the functional effects of compounds at GPCRs and ion channels in such HTS formats.

Functional Assays: What to Measure?

The design of an HTS-compatible functional assay is determined not only by the functional response being measured (i.e., change in second messenger or ion concentration or a downstream consequence of such changes) but also by the pharmacological profile – agonist, antagonist, and modulator – of the desired compound. The simplest type of assay is one in which the test compound is being evaluated for its agonist response. The potency of this effect is generally characterized in terms of the concentration at which a response that is 50 % of the maximum (EC_{50}) is observed as well as the size of the maximum response. For instance, the response may either be of a magnitude comparable to the reference agonist or may produce a reduced response relative to the reference agonist (full and partial agonist, respectively; Fig. 1a). Compounds may also bind to the receptor and produce allosteric modulatory changes that might reduce or increase the EC_{50} (PAM or NAM, respectively) (Fig. 1b), as well as the maximal response (Fig. 1c). If an assay is being established in order to detect modulatory compounds, then the choice of agonist concentration should be picked according to whether a PAM or NAM is desired, in which case EC_{20}- or EC_{80}-equivalent agonist concentrations are usually employed (Fig. 1d). Although benzodiazepines represent a well-established class of ion-channel ($GABA_A$ receptor) PAMs, more recently allosteric modulators of GPCRs have attracted considerable attention as potential therapeutics (Conn et al. 2009).

As regards the shift toward the identification of allosteric modulators as drug targets, one of the attractive features of certain types of instrumentation is the ability to record data in real time rather than producing just single, end-point readouts. This permits compounds to be assessed using two-addition or sometimes even three-addition protocols. For example, in the absence of exogenous agonist, an initial addition of compound will detect direct agonist effects. If there is no direct agonist effect, then a second addition, this time of agonist, can detect a PAM (Fig. 2).

As regards antagonists, the ability of a compound to block an agonist-stimulated response is related to the functional affinity of the agonist as well as the affinity of the antagonist. For example, a low affinity antagonist will have difficulty in blocking the effects of a high affinity agonist. Although the potency of a competitive antagonist can be quantified in terms of the pA_2 derived from a Schild plot of the rightward shift of an agonist concentration-effect curve at increasing antagonist concentrations, such analyses are not suitable for an HTS screen. Accordingly, the choice of the reference agonist and the concentration and preincubation time of test compound should be given consideration. In certain expression systems, some receptors demonstrate activity in the absence of an agonist and are therefore described as possessing constitutive activity. Compounds that block this constitutive activity are described as inverse agonists.

General Methodologies

The general principle of functional assays for receptors is that they should demonstrate that a compound binds to a receptor and "does something." Whether that "something" is direct activation, inhibition, or modulation of an agonist response, the key is being able to reliably demonstrate and quantitate a response. These methods most frequently rely upon a biosensor to detect that there has been an increase or a decrease in some component of an intracellular signaling pathway.

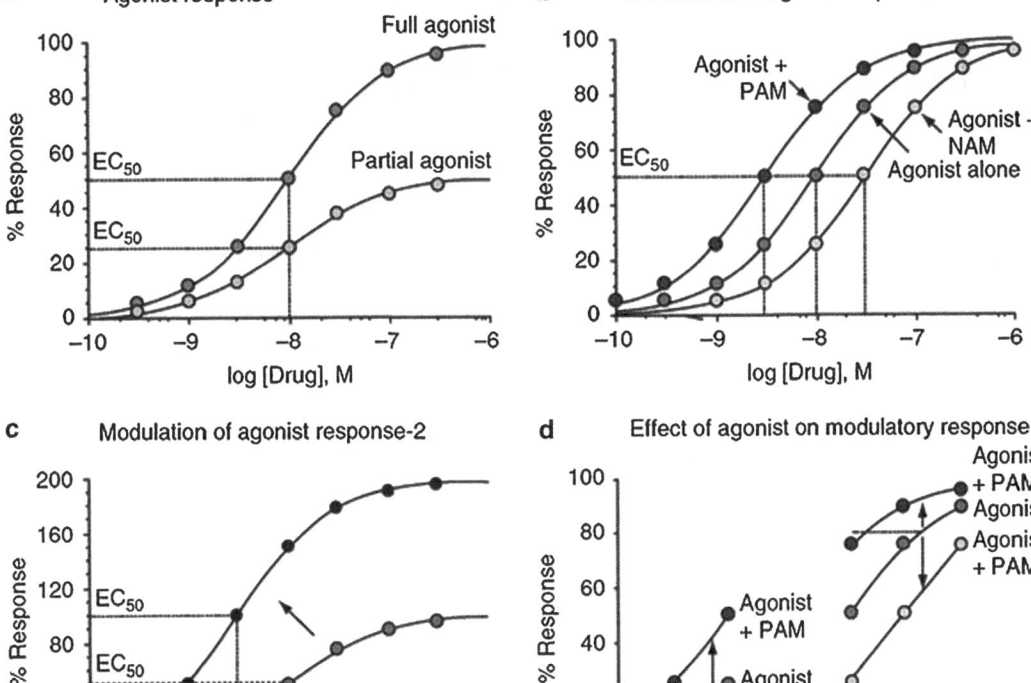

Receptors: Functional Assays, Fig. 1 Schematic representation of agonist-related responses. (**a**) Compounds may directly activate the receptor and produce a response that is comparable to or submaximal when compared with the endogenous ligand (full agonist and partial agonist, respectively). (**b**) Compound might produce a leftward or rightward shift in the agonist concentration-effect curve. Such compounds are designated positive or negative allosteric modulators (PAMs and NAMs, respectively). (**c**) In addition to shifting the EC_{50}, a compound may also have an effect on the maximal response (in this case increasing the maximum response). (**d**) An expanded view of parts of the concentration-effect curves shown in panel (**b**) illustrates how the absolute modulatory effects of PAMs and NAMs are a function of the agonist concentration. Hence, PAMs produce a proportionately greater potentiation of an EC_{20} when compared with EC_{80} response, whereas NAMs give a greater attenuation of an EC_{80} when compared with EC_{20} response

Some methods are unique to the class of receptor, for example, $[^{35}S]GTP\gamma S$ is a hydrolysis-resistant analog of GTP (guanosine triphosphate). Its binding is specific for GPCRs in comparison to, for example, the measurement of ion flux which is specific for ion channels. However, some techniques are more generally applicable to the measurement of changes in intracellular molecules. These include, but are not limited to, techniques such as fluorescence resonance energy transfer (FRET) and the related technique of homogenous time-resolved fluorescence (HTRF) and bioluminescence resonance energy transfer (BRET).

Fluorescence and Bioluminescence Resonance Energy Transfer (FRET and BRET)

FRET refers to the process whereby energy is transferred from a fluorescent donor to a suitable energy acceptor, with the absorption spectrum of the acceptor overlapping the emission spectrum of the donor. Since the efficiency of this process is inversely proportional to the

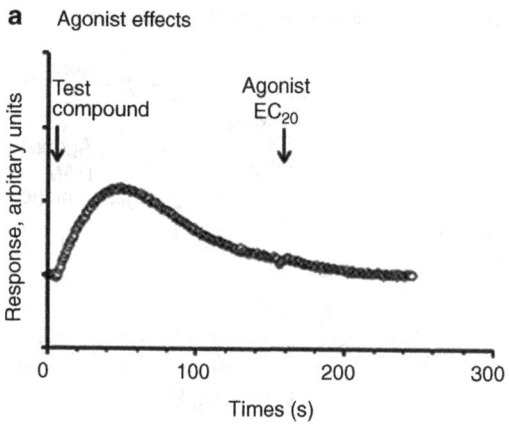
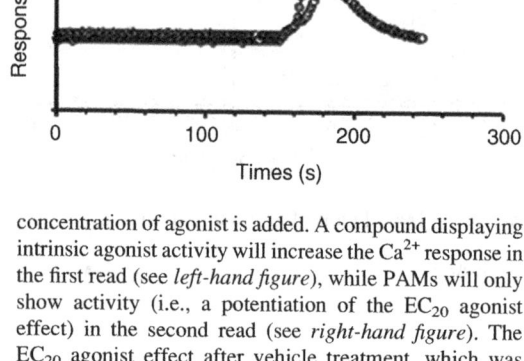

Receptors: Functional Assays, Fig. 2 Raw data traces from a real-time Ca^{2+} kinetic readout, showing the effect of a compound with (a) agonist activity or (b) PAM activity. Ca^{2+} flux assays can be designed to identify compounds exhibiting agonist or PAM activity in one and the same assay well. Test compounds are added to cells loaded with a Ca^{2+}-sensitive dye and incubated for 2.5 min. Thereafter, a submaximal (approximately EC_{20}) concentration of agonist is added. A compound displaying intrinsic agonist activity will increase the Ca^{2+} response in the first read (see *left-hand figure*), while PAMs will only show activity (i.e., a potentiation of the EC_{20} agonist effect) in the second read (see *right-hand figure*). The EC_{20} agonist effect after vehicle treatment, which was measured in a separate well of a 96-well plate, is indicated in the *right panel* in *gray*

sixth power of the distance between the two molecules, then the acceptor and donor molecules need to be in close proximity, in the region of 1–10 nm, for significant FRET to occur. The readout is a change in fluorescence emission of the acceptor, which may either increase or decrease as the proximity of the donor and acceptor also increases or decreases. The most common FRET acceptor-donor pair is the cyan fluorescent protein and yellow fluorescent protein, both of which are variants of green fluorescent protein. However, these proteins are not suitable for detecting ion-channel-mediated changes in membrane potential, and therefore, special dyes are employed for this purpose (see below).

BRET is based on the same donor-acceptor energy transfer principle as FRET, with the difference being that in BRET the donor is a luminescent molecule, generally coelenterazine, which is excited by the enzyme Renilla luciferase (Rluc) rather than a fluorescent molecule; this avoids the need for the external illumination source required for FRET. The BRET-acceptor molecule can be a fluorescent protein such as green or yellow fluorescent protein.

Ca^{2+} Sensors

Fluorescent Ca^{2+}-sensitive probes linked to a fluorescence plate reader, such as the Fluorometric Imaging Plate Reader (FLIPR™, Molecular Devices, Corp.) or Functional Drug Screening System (FDSS, Hamamatsu), have become probably the gold standard for HTS of GPCRs. This is in part related to the combined liquid-handling and imaging capabilities of machines such as the FLIPR and FDSS as well as the properties of Ca^{2+}-sensitive dyes, such as organic fluorophores, fura-2, indo-1, fluo-3, fluo-4, and Calcium Green. In addition, Ca^{2+}-sensitive proteins may be used as Ca^{2+} sensors. Fluorescent proteins are generally engineered to contain the calcium-binding protein calmodulin as a molecular Ca^{2+}-sensing switch, which may be associated with either a single or two different fluorescent proteins, with the latter producing a FRET signal due to Ca^{2+} binding to calmodulin-producing conformational changes in the sensor.

Aequorin is a Ca^{2+}-activated bioluminescent photoprotein derived from the jellyfish *Aequorea victoria*, which emits luminescence as a result of the irreversible binding of Ca^{2+} ions. It comprises two components, the 22 kDa apoaequorin protein

Receptors: Functional Assays, Table 1 Summary of methods available for assessing functional effects at G-protein-coupled receptors (GPCRs)

Assay type	Comment
[^{35}S]GTPγS binding	Generic assay for studying interaction of G-proteins but Gα$_{i/o}$-linked receptors give best signal/noise ratio
cAMP measurement – direct (radioimmunoassay)	Radiometric assays not preferred for high-throughput screening (HTS)
cAMP measurement – direct (nonradiometric)	High-affinity enzyme complementation (HitHunter™)
	Bioluminescence resonance energy transfer (BRET) (e.g., AlphaScreen)
	Fluorescence resonance energy transfer (FRET) (e.g., LANCE™, PerkinElmer; HTRF®, CisBio)
	Fluorescence polarization (FP)
cAMP measurement – reporter gene assay	Reporter gene (e.g., β-galactosidase, β-lactamase, GFP, luciferase) linked to a cAMP-response element (CRE)
Ca^{2+} – direct measurement	Organic fluorophores (fura-2, fluo-3, fluo-4, etc.)
	Fluorescent proteins incorporating Ca^{2+} sensor (e.g., calmodulin)
	Bioluminescence – emission upon irreversible binding of Ca^{2+} to aequorin
Ca^{2+} – reporter gene assay	Reporter gene (e.g., β-galactosidase, β-lactamase, GFP, luciferase) linked to a calcium-sensitive element such as activator protein 1 (AP1) or nuclear factor of activated T cells (NFAT)
β-arrestin	BRET
	FRET
	Enzyme fragment complementation (PathHunter™)
ERK phosphorylation	BRET (AlphaScreen SureFire)
Electrical impedance or refractive index	Label-free cell-based assays (refractive index: EPIC; impedance: ECIS, RT-CES, CellKey)

and a coelenterazine cofactor, and cells can be engineered to express apoaequorin, whereas coelenterazine needs to be added exogenously. When Ca^{2+} binds to the apoprotein, coelenterazine is oxidized to coelenteramide and this results in the emission of a blue light that is detected as luminescence.

Functional Assays for GPCRs

GPCRs activate intracellular signal transduction mechanism via either G-protein- or β-arrestin-mediated pathways, with examples of the various methodologies used to measure components of these pathways being presented in Table 1. With respect to the G-protein-mediated pathway, the binding of an agonist produces conformational changes in the GPCR which permit interaction between the C-terminal, intracellular receptor domain, and the heterotrimeric G-protein, which comprises an α, β, and γ subunit. This results in the exchange of GDP for GTP associated with the Gα subunit, and in this activated state, the Gα subunit may, depending upon its subtype, stimulate or inhibit the production of cAMP (Gα$_s$ and Gα$_i$, respectively) or result in the Gα$_q$-mediated, phospholipase C (PLC)-dependent cleavage of phosphatidylinositol-4,5-bisphosphate to produce inositol trisphosphate and diacylglycerol, which mobilize intracellular Ca^{2+} stores and activate protein kinase C, respectively. These alterations in intracellular second messengers can in turn trigger cascades that result in changes in the gene expression.

Following agonist binding, GPCRs become substrates for G-protein-coupled receptor kinases (GRKs), and the resulting phosphorylated GPCRs are able to bind β-arrestins which causes the membrane trafficking machinery to internalize receptors away from the cell surface in clathrin-coated pits. In addition, the formation of the GPCR/β-arrestin complex can result in activation of the extracellular, signal-regulated (ERK)/mitogen-activated protein (MAP) kinase pathway.

[^{35}S]GTPγS Binding Assay

The nucleotide exchange process at G-proteins in cell membranes, resulting from agonist binding to GPCRs, can be measured by monitoring the binding of [^{35}S]GTPγS. The advantage of this assay is that it measures an early event in the signal transduction cascade and hence is less subject to amplification or regulation by more distal cellular processes that may compromise other functional measurements, such as reporter gene assays. The disadvantage of this assay is that it is generally more feasible for receptors coupled to the abundant G_i family of G-proteins, although several approaches have been introduced to extend its utility, such as the use of immune capture of [^{35}S]GTPγS-bound G-proteins of interest at the end of the assay or the use of GPCR-G-protein fusion proteins (Milligan 2003).

cAMP Measurements

Historically, cAMP was measured based on radioisotopes using, for example, GE Healthcare SPA™ and PerkinElmer FlashPlate™ cAMP assays; however, these methods are relatively expensive for screening large numbers of compounds, in addition to having issues with safety, and are therefore generally not the preferred option for HTS.

Alternatively, reporter gene assays are popular functional cell-based assays suitable for cAMP measurements. In order to monitor $G_{\alpha s/i}$-dependent changes in cAMP, the cAMP-response element (CRE) can be used as the promoter for an easily detectable reporter gene, such as the firefly luciferase, β-galactosidase, β-lactamase, or GFP. Various detection methods are used to measure expressed reporter gene protein, including luminescence, absorbance, and fluorescence. CRE-luc assays are generally very sensitive given the low background signal and the amplification of the signal between GPCR activation and gene expression. They have the disadvantage of measuring downstream effects and hence require a few hours of incubation to accumulate the reporter gene product in the cells. As a consequence, these assays are characterized by a high level of false positives.

Other HTS technologies for direct cAMP measurements rely on homogenous competitive immunoassays. TR-FRET combines HTRF and FRET principles and is based on the competition between an europium- or terbium-labeled cAMP tracer and sample cAMP for binding sites on cAMP-specific antibodies labeled with a fluorescent dye. The formation of antibody-cAMP complex is inversely proportional to cAMP concentration in the sample. Several assay kits allowing cAMP detection based on this principle (e.g., LANCE™ and HTRF®) are commercially available.

The DiscoveRx technology platform HitHunter™ is based on high-affinity complementation of two separately inactive enzyme (β-galactosidase) fragments (enzyme donor and acceptor) to form a stable heteromeric enzyme complex capable of hydrolyzing substrates to produce a chemiluminescent or fluorescent signal. The analyte (cAMP) competes with an analyte-conjugated donor fragment for binding to an analyte-specific antibody. Upon antibody binding, the donor portion of the enzyme is incapable of association with the enzyme acceptor. Hence, the amount of cAMP-conjugated donor available for enzyme complementation is proportional to the concentration of cAMP in the sample.

Alternative assay kits include AlphaScreen™, a highly sensitive bead-based chemiluminescent assay based on the competition between endogenous cAMP and exogenously added biotinylated cAMP, and fluorescence polarization (FP), which measures the parallel and perpendicular components of fluorescence emission generated by polarized light. The magnitude of the polarization signal is used to quantitatively determine the binding extent of a fluorescent cAMP tracer complex (Heilker et al. 2009; Thomsen et al. 2005).

Ca^{2+} Mobilization

If the GPCR of interest signals via PLC, then the most broadly used cell-based technique is the measurement of transient changes in intracellular Ca^{2+} concentrations using a fluorescent plate reader such as FLIPR™ or FDSS. In FLIPR-type assays, cells expressing a receptor of interest are usually incubated with a membrane-permeable fluorescent dye that after hydrolysis is converted to a Ca^{2+} sensitive, membrane-impermeable probe. Alternatively, Ca^{2+} generation can be measured with the use of Ca^{2+}-responsive luminescent photoproteins, such as aequorin. While Ca^{2+} fluxes traditionally could only be measured for G_q-coupled GPCRs, the coexpression of promiscuous G-proteins has extended the use of this technique to all GPCRs.

Reporter gene assays (using AP1 or NFAT response elements) can be used as an alternative method to measure Ca^{2+} mobilization (Heilker et al. 2009; Thomsen et al. 2005).

ERK Phosphorylation and β-Arrestin Signaling

GPCRs can generate diverse signals that activate the ERK MAP kinase pathway. Depending on the receptor and cell type, GPCR-mediated ERK activation may involve classical second messenger-dependent routes, utilize tyrosine protein phosphorylation, or involve β-arrestins as scaffolds for the MAP kinase module. A popular assay to measure ERK phosphorylation is the AlphaScreen SureFire™ technology. Advantageously, ERK phosphorylation is independent of the type of G-protein linked to the receptor.

Signaling via the β-arrestin pathway is usually assessed by measuring β-arrestin translocation to receptors. This can be done via fluorescently tagged β-arrestins and either microscopic imaging of β-arrestin redistribution or FRET or BRET assays to detect β-arrestin-receptor interactions. An HTS-suitable enzyme fragment complementation β-arrestin assay (PathHunter™) has been developed by DiscoveRx (Heilker et al. 2009).

Label-Free Functional Assays

Label-free technologies based on either electrical impedance or refractive index are novel tools for measuring cell-based, real-time, kinetic functional responses (Minor 2008).

While these assays eliminate the need for expensive and "artificial" tags, dyes, or specialized reagents, the downside of this technology is that one measures a response "signature" and that it needs to be confirmed whether that is specific for the G-protein subtype to which the receptor is coupled. And although these systems can be configured in multiwell format, most of them lack onboard pipetting capabilities and are therefore not yet equipped to support HTS campaigns (Table 1).

Functional Assays for Ion Channels

As their name suggests, ion channels are intramembrane proteins which when activated undergo a conformational change that results in the opening of a channel through which ions cross the cell membrane. Ion channels can be classified into two types: ligand-gated and voltage-activated ion channels, which are activated by the binding of a ligand or changes in the membrane potential, respectively, and representative technologies associated with measuring ion-channel function are presented in Table 2. The relevance of ion channels as drug targets is highlighted by their pathological involvement in a variety of disorders (the so-called channelopathies), as well as the efficacy of drugs that target specific ion channels, most notably from a psychopharmacological view the benzodiazepines, which modulate $GABA_A$ receptor function, as well as antiepileptic drugs that target sodium and calcium channels (Camerino et al. 2007).

High-Throughput Electrophysiology

The flow of ions across the membrane is most precisely measured using electrophysiological methods, such as patch-clamp electrophysiology in mammalian cells or two-electrode voltage clamp in *Xenopus laevis* oocytes. While electrophysiology remains the gold standard for assessing ion-channel function, these methods are relatively low throughput and although higher-throughput platforms have been developed, such as IonWorks (Molecular Devices), QPatch (Sophion Bioscience), Patchliner (Nanion Technologies), or PatchXpress (Axon Instruments), nonelectrophysiological, high-throughput cell-based assays using fluorescent dyes or ion flux have been widely used for HTS assays (Dunlop et al. 2008a, b).

Receptors: Functional Assays, Table 2 Summary of methods available for assessing functional effects at ion-channel receptors

Assay type	Comment
Electrophysiology	"Gold standard" for ion-channel measurements. Labor intensive but recent technological advances (e.g., QPatch, IonWorks, PatchXpress) have markedly increased throughput
Optical probes – intracellular Ca^{2+}	Organic fluorophores (fura-2, fluo-3, fluo-4, etc.)
	Fluorescent proteins incorporating Ca^{2+} sensor (e.g., calmodulin)
	Bioluminescence – emission upon irreversible binding of Ca^{2+} to aequorin
Optical probes – other	Thallium-sensitive dyes as surrogate for K^+ (e.g., BTC-AM, ThalKal)
	MEQ for Cl^-
Optical probes – membrane potential	FRET based – immobile donor (coumarin-linked phospholipid) mobile, voltage-sensitive oxonol acceptor (e.g., $DiSBAC_2(3)$, $DiSBAC_4(3)$)
	Fluorescence based – voltage-sensitive oxonol or FMP dye plus quencher
Ion flux – radiometric	^{22}Na for sodium channels, ^{86}Rb for potassium channel, ^{45}Ca for calcium channels, ^{36}Cl or ^{125}I for chloride channels
Ion flux – atomic absorption spectroscopy	Lithium for Na channels, rubidium for potassium channels, chloride for chloride channels[a] (e.g., Ion Channel Reader from Aurora Biomed)

[a]Chloride ion measurement is indirect and involves precipitation with silver nitrate and measuring free-silver concentrations

Optical Methods: Membrane Potential Sensors

The changes in membrane potential that occur when ion channels are activated may be detected using dyes that are sensitive to membrane potential, such as the negatively charged oxonols, $DiSBAC_2(3)$ or $DiSBAC_4(3)$. These molecules are mobile and can move from the outer to inner face of the membrane according to the membrane potential. Hence, as the membrane potential changes, so does the partitioning of the oxonol across the plasma membrane. In the presence of a cell-impermeable fluorescence quencher, the dye will only produce a fluorescent signal when associated with the inner face of the membrane. FMP dyes operate on the same principle but have a faster response than the oxonol dyes.

Fluorescent probes may also be used to detect changes of membrane potential using FRET. One component, for example, a coumarin-labeled phospholipid, binds specifically to the exterior surface of the membrane whereas the other partner, typically $DiSBAC_2(3)$ or $DiSBAC_4(3)$, is mobile. Hence, as the membrane potential changes, so does the partitioning of the oxonol across the plasma membrane and as a result, the FRET between the membrane-bound donor and the mobile voltage-sensitive acceptor changes. Commercialization of this technology has resulted in the voltage-sensitive ion probe reader (VIPR; Aurora Biosciences Corp.) and the electronic-stimulation voltage ion probe reader (E-VIPR; Vertex Pharmaceuticals).

Optical Methods: Ion Flux Indicators

The use of detectors that are selective for particular ions provides a reliable means of detecting intracellular changes in ion concentrations in response to ion-channel activation. In this regard, methods for Ca^{2+}-permeable ion channels are the best established due to the wide range of well-characterized Ca^{2+}-sensitive organic fluorophores and Ca^{2+}-sensitive proteins. In contrast to Ca^{2+}, there are currently no generally accepted fluorescent indicators for K^+ suitable as primary functional assays for potassium channels, although thallium-sensitive indicators can be used as a surrogate since K^+-conducting ion channels are also permeable to other ions such as rubidium (for which no sensor currently exists) and thallium. As regards halide ions, the interaction of halide ions with 6-methoxy-N-ethylquinolinium iodide (MEQ) reduces the fluorescence of the latter, and this reduction in fluorescence can be used as an indicator of chloride channel-mediated Cl^- flux.

Ion Flux Indicators: Nonoptical

As an alternative to optical methods, the flux of ions through the channel may be measured either as an influx or efflux assay, depending upon the direction of ion flow. This might be using a radiometric assay in which, for example, $^{22}Na^+$ or ^{14}C-guanidinium may be used for sodium channels, $^{86}Rb^+$ for potassium channels, $^{45}Ca^{2+}$ for calcium channels, and $^{36}Cl^-$ or $^{125}I^-$ for chloride channels. However, such assays require a separation step in order to remove radioisotopes that have not been taken up into the cell and are therefore generally unsuitable for HTS. An additional factor that makes radiometric ion flux assays unattractive is that they often use high-energy radioisotopes. As an alternative, nonradioactive flux assays may be used in which highly sensitive flame atomic absorption spectrometers capable of measuring the ion content in small volumes are used to quantify ion flow (e.g., the Ion Channel Reader from Aurora Biosciences Corp.).

The disadvantages of flux methods, whether optical or nonoptical, include the possibility of large, nonspecific background fluxes or a rapid inactivation of the target ion channel. Moreover, although this latter aspect can be addressed by using mutated target proteins and/or pharmacological agents that prolong the channel open state or decrease the rate of channel inactivation, it is possible that some interactions may be missed or artifactually detected in such a modified system (Table 2).

Limitations and Other Considerations

The main caveat to most functional assays is that they are performed in artificial systems, generally using human receptors stably, or sometimes transiently, expressed in generally nonneuronal cell lines, such as Chinese hamster ovary (CHO) or human embryonic kidney (HEK) cells. Under such conditions, the functional effects of compounds can be influenced by the level of receptor expression and, in the case of GPCRs, by the efficiency of coupling to G-proteins. The level of receptor expression and G-protein coupling efficiency dictate the level of GPCR reserve (i.e., the amount of "spare" receptors, in excess of those necessary to produce a maximum response when occupied by an agonist), and this may further complicate the interpretation of functional screening data. For example, a partial agonist in one tissue may appear to act as a full agonist in another tissue with a higher receptor reserve.

Consideration should also be given to the subunit composition of the particular receptor of interest, as well as the impact that auxiliary proteins may have upon the function of the receptor. For instance, ion channels may be heteromeric assemblies, the subunit composition of which is a key determinant of receptor pharmacology. Furthermore, an emerging area of GPCR research is that of heterodimerization of subunits which can also markedly influence the pharmacology (Waldhoer et al. 2005). However, since information concerning the physiological subunit composition and the relevance of ancillary proteins may well be absent, or at best poorly understood, a minimalistic approach is often adopted in which homomeric receptors are studied.

An emerging area of GPCR research is that of collateral efficacy, also known as biased agonism, functional selectivity and stimulus trafficking, in which different agonists can produce differential effects on different intracellular signaling pathways (in other words, all agonists for a particular GPCR do not necessarily produce the same effects on signaling pathways; Galandrin et al. 2007). The implication of this is that compounds that produce a functional response using a methodology based upon a GPCR-linked signaling pathway, for example, cAMP, may not necessarily produce a similar functional response in another pathway such as the β-arrestin-mediated pathway.

In summary, it is important to emphasize that there is no one-size-fits-all functional assay for GPCRs or ion channels. Hence, the choice of assay should be based upon practical considerations such as the number of compounds requiring testing, the instrumentation available, and the properties of the cell lines being used. As a result, a degree of trial-and-error evaluation of different assay formats is to be expected. Finally, while functional data obtained from recombinant model systems are crucial early on in the search for novel pharmaceuticals, it is critical that these functional effects of compounds are also evaluated at native receptors.

Cross-References

► Affinity
► Agonist
► Allosteric Modulators
► Antagonist
► Binding
► Desensitization
► Efficacy
► Inverse Agonists
► Partial Agonist

References

Camerino DC, Tricarico DA, Desaphy J-F (2007) Ion channel pharmacology. Neurotherapeutics 4:184–198

Conn PJ, Christopoulos A, Lindsley CW (2009) Allosteric modulators of GPCRs: a novel approach for the treatment of CNS disorders. Nat Rev Drug Discov 8:41–54

Dunlop J, Bowlby M, Peri R, Tawa G, LaRocque J, Soloveva V, Morin J (2008a) Ion channel screening. Comb Chem High Throughput Screen 11:514–522

Dunlop J, Bowlby M, Peri R, Vasilyev D, Arias R (2008b) High-throughput electrophysiology: an emerging paradigm for ion-channel screening and physiology. Nat Rev Drug Discov 7:358–368

Galandrin S et al (2007) The evasive nature of drug efficacy: implications for drug discovery. Trends Pharmacol Sci 28:423–430

Grigoriadis DE, Hoare SR, Lechner SM, Slee DH, Williams JA (2009) Drugability of extracellular targets: discovery of small molecule drugs targeting allosteric, functional, and subunit-selective sites on GPCRs and ion channels. Neuropsychopharmacology 34:106–125

Heilker R, Wolff M, Tautermann CS, Bieler M (2009) G-protein-coupled receptor-focused drug discovery using a target class platform approach. Drug Discov Today 14:231–240

Milligan G (2003) Principles: extending the utility of [^{35}S] GTPγS binding assays. Trends Pharmacol Sci 24:87–90

Minor LK (2008) Label-free cell-based functional assays. Comb Chem High Throughput Screen 11:573–580

Thomsen W, Frazer J, Unett D (2005) Functional assays for screening GPCR targets. Curr Opin Biotechnol 16:655–665

Waldhoer M, Fong J, Jones RM, Lunzer MM, Sharma SK, Kostenis E, Portoghese PS, Whistler JL (2005) A heterodimer-selective agonist shows in vivo relevance of G protein-coupled receptor dimers. Proc Natl Acad Sci U S A 102:9050–9055

Recombinant Cell Line

Definition

A cell line into which recombinant DNA has been introduced, either stably or transiently.

Reconsolidation

Definition

The process by which some types of consolidated memories are rendered stable once again after being reactivated.

Recurrent Brief Depressive Disorder

Lucie Bartova and Lukas Pezawas
Department of Psychiatry and Psychotherapy, Division of Biological Psychiatry, Medical University of Vienna, Vienna, Austria

Synonyms

RBD; Recurrent brief depression

Definition

Recurrent brief depressive disorder (RBD) is a well-defined and prevalent mood disorder with an increased risk of suicidal behavior and significant clinical impairment (Angst 1994; Pezawas et al. 2003). The syndrome is defined by depressive episodes that occur at least monthly and last only a few days. RBD represents a distinct and frequent clinical diagnosis in ICD-10 (F38.1) (WHO 2010). In the recently published DSM-5 (311.1), RBD is featured in the category "other specified depressive disorder" together with short-duration depressive episodes and depressive syndromes with insufficient symptoms

(APA 2013). In both diagnostic manuals, psychopathological symptoms required for the diagnosis of RBD are the same as for major depressive disorder (MDD) (WHO 2010; APA 2013). Hence, RBD is primarily diagnosed by its duration criterion (episode duration 2–13 days) and frequency criterion (at least 1 episode/month for at least 12 consecutive months). The average duration appears to be 3–4 days.

Whereas RBD may appear alone without any associated condition, its symptomatology occurs as a part of other mood disorders in approximately two-thirds of RBD patients (Pezawas et al. 2005; Lovdahl et al. 2009). Consequently, the longitudinal course of mood disorders is characterized by a substantial diagnostic overlap between different diagnostic groups (Angst 1994). Despite the comparable symptom profile, longitudinal epidemiological studies have demonstrated that RBD is neither a prodromal nor a residual state of MDD. However, given the fact that RBD is often accompanied by recurrent brief hypomanic episodes and related clinical phenomena such as comorbidity patterns, personality profiles, and a balanced sex ratio, RBD has also been suggested to be part of the bipolar spectrum (Lovdahl et al. 2009). Accordingly, brief hypomania has been suggested to be a severity marker of RBD since more anxiety disorders, higher family load of depression, and more atypical and melancholic symptomatology were reported for RBD patients with a history of brief hypomanic phases in between the brief depressive episodes (Lovdahl et al. 2009). This assumption is further supported by neuroimaging evidence, which indicates that the presence of brief hypomania impacts on illness severity from a neurobiological perspective (Koisnes et al. 2013). The concept of RBD and its clinical significance have been further delineated by studies on the patterns and consequences of a lifetime co-occurrence of both RBD and MDD, called combined depression (CD) (Angst 1994; Pezawas et al. 2003; Pezawas et al. 2005). Epidemiological and clinical studies have demonstrated a dramatic increase in suicide attempt rates and measures of impairment in cases of CD in comparison to either RBD or MDD alone.

Based on this data, it has been suggested that it is important to ascertain the existence of previous recurrent brief depressive episodes in patients with current MDD, because such information contributes to the assessment of suicidal behavior risk as well as drug response, which appear to be worse for CD in comparison to RBD patients.

The differential diagnosis of RBD and MDD is based on their distinct courses of depressive episodes (1–3 days vs. weeks or months) (Pezawas et al. 2003, 2005). Other psychiatric disorders with short-lived syndromes have to be ruled out, such as the following: (1) panic disorder, characterized by recurrent unexpected panic attacks lasting for minutes up to a few hours; (2) premenstrual dysphoric disorder (PMDD), appearing exclusively in relation to the menstrual cycle; (3) borderline personality disorder (BPD), in which affective instability can be manifested by depressive mood fluctuations that usually last for hours but can persist for days as well (although clinical studies suggest low comorbidity of RBD with borderline personality disorder, as well as different endocrine response patterns); (4) rapid cycling (RC) (>4 episodes/year), ultra-RC (episodes in a weekly rhythm), or ultra-ultra-RC (ultradian episodes) bipolar disorders, which must show at least hypomanic (or manic) episodes according to DSM-5; (5) other affective disorders, such as minor depression and dysthymia, which can be easily distinguished from RBD by the criterion of duration of the depressive episode; (6) drug-induced depression, such as in cocaine withdrawal; and (7) various somatic diseases which might be accompanied by brief depressive episodes, such as postictal dysphoria or the so-called interictal dysphoric disorder in epilepsy, migraine plus aura, Parkinson's disease, Fabry disease, celiac disease, Prader–Willi syndrome, corpus callosum agenesis, and chronic hepatitis C.

Role of Pharmacotherapy

Two-thirds of all RBD patients seek professional help in the course of their life (Angst 1994; Kasper et al. 2000). Of this group, one quarter consult a general practitioner, one quarter see a psychologist, and the remaining half consult a psychiatrist or neurologist. Epidemiological studies have shown that nearly all such patients receive treatment with psychotherapy. However, RBD patients hardly ever receive pharmacological treatment in community samples. Almost three decades after the concept of RBD was introduced, data from clinical trials are still limited.

However, there are a few placebo-controlled studies and several case reports, a single-case analysis, and one open trial assessing the drug treatment of patients with RBD (Kasper et al. 2000; Baldwin 2003; Pezawas et al. 2003). The most rigorous of these studies, which have used a classical double-blind placebo-controlled two-tailed design, have not been able to demonstrate successful drug treatment for RBD. It has been suggested that the negative results in placebo-controlled studies have been due to a study design that is inappropriate for this clinical phenomenon. This suggestion is mainly supported by reported treatment responses in double-blind placebo-controlled one-tailed single-case analyses with nimodipine, verapamil, and carbamazepine. Furthermore, the controlled studies were carried out with highly selected RBD patient samples, i.e., patients with repeated suicide attempts or comorbid borderline personality disorder or somatic diseases. This might have led to a bias reflecting treatment resistance. These are views supported by case reports or open trials demonstrating successful treatment of RBD with mirtazapine, reboxetine, fluoxetine, paroxetine, lithium, carbamazepine, lamotrigine, tranylcypromine, and olanzapine.

The optimal treatment regime for RBD remains an open question. The aggregate of available data and clinical experience suggest that second-generation antidepressants might serve as a first-line treatment of RBD followed by mood stabilizers as a second-line alternative. However, since recent evidence has suggested RBD to be a part of the bipolar spectrum, the use of mood stabilizers could be favored in the future. With respect to treatment outcome, longer response rates (weeks to a few months) than in MDD are typical due to the monthly occurrence of short-lasting depressive symptoms. Nevertheless, significant neuropsychological dysfunction has been shown to persist even in euthymic states in some patients similar to MDD (Andersson et al. 2010).

Because available study results remain unclear and contradictory, there is a great need for additional controlled clinical trials without the methodological limitations of previous studies. Early pessimism concerning the pharmacological treatment of RBD could be due to the false-negative study results and is probably not justified.

Cross-References

▶ Aminergic Hypotheses for Depression
▶ Antidepressants
▶ Bipolar Disorder
▶ Depression
▶ Depression: Response and Remission
▶ Depressive Disorders Not Otherwise Specified
▶ Depressive Disorders: Major, Minor, and Mixed
▶ Emotion and Mood
▶ Mood Disorders
▶ Mood Stabilizers
▶ Recurrent Brief Depressive Disorder
▶ Treatment-Resistant Depression

References

Andersson S, Lovdahl H, Malt UF (2010) Neuropsychological function in unmedicated recurrent brief depression. J Affect Disord 125(1–3):155–164
Angst J (1994) Recurrent brief depression. In: Hippius H, Stefanis CN (eds) Research in mood disorders: an update. Hogrefe & Huber, Seattle, pp 17–30
APA (2013) Diagnostic and statistical manual of mental disorders, fifth edition: DSM-5. American Psychiatric Association, Arlington
Baldwin DS (2003) Recurrent brief depression – more investigations in clinical samples are now required. Psychol Med 33(3):383–386
Kasper S, Stamenkovic M, Pezawas L (2000) Recurrent brief depression: diagnosis, epidemiology and potential pharmacological options. In: Palmer KJ (ed) Managing depressive disorders. Adis International, Auckland/Philadelphia, pp 29–36

Korsnes MS, Lovdahl H, Andersson S, Bjornerud A, Due-Tonnesen P, Endestad T, Malt UF (2013) Working memory in recurrent brief depression: an fMRI pilot study. J Affect Disord 149(1–3):383–392

Lovdahl H, Andersson S, Hynnekleiv T, Malt UF (2009) The phenomenology of recurrent brief depression with and without hypomanic features. J Affect Disord 112(1–3):151–164

Pezawas L, Angst J, Gamma A, Ajdacic V, Eich D, Rossler W (2003) Recurrent brief depression – past and future. Prog Neuropsychopharmacol Biol Psychiatry 27(1):75–83

Pezawas L, Angst J, Kasper S (2005) Recurrent brief depression revisited. Int Rev Psychiatry 17(1):63–70

WHO (2010) The ICD-10 classification of mental and behavioral disorders: clinical descriptions and diagnostic guidelines, 4th edition of ICD-10. World Health Organization, Geneva

Re-encoding

Definition

The fact or process whereby a memory item is again encoded after it has been retrieved. The resulting ▶ engram may or may not be identical to the one that preceded it.

Reference Electrode

Definition

In neurophysiological studies, including those of the electroencephalogram (EEG), voltage potentials are only defined with respect to a reference: an arbitrarily chosen zero level. For each EEG recording, a reference electrode has to be selected in advance. Ideally, this electrode should be affected by ambient electrical activity in the same manner as all the other electrodes, such that this activity is subtracted out by the referencing procedure. In most studies, a reference on the head, but at some distance from the other recording electrodes, is chosen. Such a reference can be the earlobes, the nose, or the mastoids (i.e., the bone behind the ears).

Cross-References

▶ Electroencephalogram
▶ Event-Related Potential

Reinforcement

Definition

A behavioral process whereby the probability of a response is increased because it previously produced a particular environmental consequence. Such consequences take the form of presentations of environmental stimuli, which are thereby defined operationally as reinforcers. Such stimuli can either be exteroceptive (e.g., receiving a sum of money or a sweet-tasting substance) or interoceptive (e.g., psychoactive effects of an abused drug).

Reinforcement Disorders

Roy A. Wise
Behavioral Neuroscience Branch, NIDA IRP, Baltimore, MD, USA

Definition

Reinforcement refers to the process or constellation of processes that "stamp in" learned associations and response habits, processes that enhance or establish the transfer of immediate short-term memories to relatively permanent long-term memories (White 1996). The mechanisms of reinforcement remain mysteries not yet fully understood, but a variety of animal models identify reinforcement closely with the consequences of the extracellular actions of the neurotransmitter dopamine. Dopamine is released from nerve terminals in response to various reinforcers – the prototypical reinforcer is food for a hungry animal – and, once the association is formed, in response to conditioned stimuli that reliably predict those rewards. The habit-forming and habit-sustaining effects of most positive reinforcers are lost in animal models when the receptors for dopamine are blocked pharmacologically.

Current Concepts and State of Knowledge

Until recently in our evolutionary history, the ability to learn associations and response habits leading to sweet and calorie-dense substances has had obvious adaptive value. By learning to identify the predictors and repeat the acts that bring them in contact with sweet taste, infants learn to find the mother's nipple efficiently and adults learn to find and eat ripe fruits and to save them for another day. When the food supply has been limited, this has served us well. However, in recent decades, people in industrial societies are increasingly exposed to more tasty and abundant energy-rich foods than they need for simple subsistence, and the tendency to overindulge in sweet and fatty foods has resulted in an epidemic of obesity that can be seen as a disorder of reinforcement, of reinforcement gone awry. When we have easy access to tasty energy-laden foods, our reinforcement mechanisms prove to be *too* effective, encouraging maladaptive overeating.

In the last few thousand years, we have learned to store and refine alcohol and inhale the smoke of tobacco, and in the last few hundred years, we have learned to refine and synthesize several addictive substances, substances that can themselves serve as reinforcers. Easy access to these substances and to effective ways to ingest them has given rise to a more immediately knotty problem of over-reinforcement: addiction. Drugs such as heroin, cocaine, and methamphetamine can elevate dopamine levels further and more rapidly than even the most palatable of foods and can establish drug-taking habits that are arguably more compulsive than the eating habits of the clinically obese. The smoking of tobacco and the drinking of alcohol also elevate dopamine levels much more rapidly than does food, again often leading to compulsive and maladaptive habits. Here again is a disorder resulting from reinforcement that is too effective, from reinforcers that are too habit-forming. Also often viewed as disorders of over-effective reinforcement are compulsive gambling, compulsive video gaming, compulsive sexual habits, compulsive thrill seeking, and other learned compulsions. Our understanding that the mechanisms of reinforcement have gone awry derives in large part from studies of addictive drugs: substances that can act directly at the various sites of action of endogenous neurotransmitters.

Drug Addiction

Almost all addictive drugs share the ability to elevate extracellular levels of the neurotransmitter dopamine (Di Chiara and Imperato 1988; Wise and Rompré 1998). Amphetamines cause dopamine to be released from nerve terminals. Cocaine blocks the reuptake of dopamine when it has been released by neuronal activity. Nicotine stimulates dopamine neurons to fire and release dopamine. Opiates also cause dopamine neurons to fire, not by stimulating them but by inhibiting neighboring cells that usually limit dopamine cell firing. Ethyl alcohol and cannabis also – through yet other mechanisms – increase dopamine cell firing and extracellular dopamine release (Wise 2002).

Moment-to-moment changes in dopamine levels have been monitored during periods of self-administration of cocaine and heroin (Wise 2004). This is best seen in animals trained to earn intravenous injections by lever pressing and given limited daily access to the drug. Experienced rats initiate a drug self-administration session by earning one or more injections, causing rapid elevation of dopamine levels to three or four times their normal extracellular concentration. Once dopamine levels are elevated, the animals cease lever pressing until dopamine levels fall back to about twice normal. Then they make another response, again elevating dopamine levels and again waiting to respond further until the dopamine level again falls to about twice normal. This sequence is repeated until the session ends or until the animal is exhausted, with dopamine levels maintained above the trigger point of about twice normal elevations. If the animal is given an unpredictable sequence of sometimes small, sometimes medium, and sometimes large cocaine doses, the delay between successive lever presses will be short, medium, or long, respectively (Wise et al. 1995).

The regular psychomotor stimulant intake seen when animals are limited to 1–4 h of daily drug access can escalate and become irregular when animals are given prolonged access to the drug (Ahmed and Koob 1998; Tornatzky and Miczek 2000). This can be seen with many weeks of access or with six or more hours of daily access. A variety of neuroadaptations in the reward circuitry itself have been identified (Hyman et al. 2006) and hypothesized to contribute to the escalation of intake and the increasing compulsiveness of intake at a time when the subjective perception of drug reward is decreasing. The degree to which such neuroadaptations affect other compulsive behaviors remains unclear.

In the cases of cocaine and amphetamine – drugs that appear to rely entirely on dopamine as their reinforcement substrate – the pharmacological blockade of dopamine receptors eliminates the reinforcing effects of the drugs. Lesions of the terminals of the mesocorticolimbic dopamine system can have similar effects if the lesions are large enough and are correctly placed. It is not yet clear, however, what constitutes correct placement. Large lesions of the ventral striatum are effective, but they usually include damage not only to dopamine fibers terminating at nucleus accumbens, the intended lesion target, but also to dopamine fibers of passage to the olfactory tubercle and the medial prefrontal cortex, two sites in which cocaine is known to have habit-forming actions. Small lesions tend to implicate the core but not the shell of nucleus accumbens as the site of cocaine reinforcement, but while rats will learn to work for cocaine microinjections into the shell of accumbens (and even more readily learn to work for injections into the underlying olfactory tubercle), they do not similarly learn to work for injections into the core of accumbens. Thus, extracellular dopamine plays a necessary role in cocaine reinforcement, but it is not yet completely clear *where* dopamine plays this necessary role (Wise 2004).

In the cases of opiates and phencyclidine, dopamine does not appear to play an essential role. The reinforcing effects of phencyclidine appear to occur in the ventral striatum, but postsynaptic to the dopamine terminals that innervate the GABAergic ventral striatal output neurons. Phencyclidine blocks the excitatory glutamatergic input to the GABAergic output neurons, an input that parallels and is thought to interact with the adjacent dopaminergic input to these neurons. Morphine (like dopamine) inhibits some of these output neurons, and morphine appears to be reinforcing in part because of this ventral striatal action. Nicotine and dopamine may have additional inputs to reinforcement circuitry, inputs that are also "downstream" to the dopamine synapses of the mesocorticolimbic dopamine system. The effects of dopamine antagonists and dopaminergic lesions on the self-administration of other addictive drugs are more complex and less well understood.

Compulsive Eating

Foods are the prototypical reinforcers. The most powerful food reinforcers are sugars, particularly when taken in combination with lipids (which are also reinforcers in their own right); however, even untasted vitamins and minerals can serve as reinforcers. These are substances that, once encountered, are returned to because of their reinforcing effects. Sugars and lipids result in the release of dopamine from nerve terminals in many brain regions, elevating extracellular dopamine to as much as twice normal levels. If the receptors that normally respond to dopamine are pharmacologically blocked, food is no longer habit-forming or habit-sustaining. Sweet foods appear to have a double reinforcing effect; sweet taste is naturally reinforcing to many species and the post-ingestional effects of natural sugars are also reinforcing (Volkow and Wise 2005; Wise 2006).

While compulsive eating habits are usually associated with sweet and fatty foods, compulsive eating can also be entrained by normally unpalatable substances. For example, calcium-deprived institutionalized orphans used to compulsively eat plaster off the walls of their rooms. Sodium-deprived animals and humans develop compulsive cravings for salt. A variety of such "specific hungers" have been reported in animals lacking the nutrient provided by the craved substance. Most of these hungers are learned, but the mechanisms of their reinforcement are not known.

Other Learned Compulsions

Many other compulsions have been labeled or likened to addiction. Compulsive gambling, compulsive sexual behavior, and compulsive thrill seeking are examples. Several investigators are pursuing the hypothesis that the brain mechanisms subserving a variety of such behaviors are shared with the brain mechanisms of drug addiction and compulsive eating.

Reinforcement Leads to Craving

Most compulsive cravings are the result of experience. Some exceptions can be attributed to reflexes, such as the reflexive approach induced by sexual pheromones or the reflexive avoidance induced in some species by fox odor and in other species by putrid smell. Some cravings can develop through human communication; cravings for illicit drugs, for example, can be established through conversations with peers. Most cravings, however, develop as a result of the reinforcing effects of contact with the substance or event in question. The cravings associated with compulsive intake of drugs or foodstuffs soon come under the control of the subject's history of reinforcement.

Extinction

Once a compulsive habit becomes established, many repetitions of the habit in the absence of reinforcement are required before the habit is "extinguished." Indeed, there is strong evidence that a compulsive habit is never completely extinguished even after long periods of non-reinforcement. For example, presumably extinguished habits can be reestablished with less reinforcement than was required to establish them in the first place.

The rate of extinction of a response habit depends on several things. If reinforcement immediately follows every response during the learning period, extinction (never to zero but to some low probability) can be fairly rapid. If reinforcement is delayed or intermittent during training, a response habit can become very resistant to extinction. Intermittent reinforcement is termed "partial" reinforcement, and the greater resistance to extinction after partial reinforcement training is known as the "partial reinforcement extinction effect." The partial reinforcement extinction effect has important implications for the treatment of compulsive habits. Each binge after an attempt to diet, for example, results in an intermittent reinforcement that makes subsequent extinction more difficult.

Thus, it takes much less reinforcement to sustain a habit than it takes to establish one. Reinforcement can become less frequent or less intense and still maintain a compulsive response habit. Thus, addicts report that even when, because of drug tolerance, their drug of choice no longer seems to bring the subjective pleasure that it once did, they still seek the drug and have compulsive thoughts about it. Clinicians refer to the persistence of drug habits despite reduced subjective pleasure as "chasing the remembered high." Whatever form the "remembered high" takes – whether it be a conscious memory or merely a residual trace in the nervous system – the consequences of reinforcement in the past are of great importance for the cravings and behaviors of the present.

Potential Medications

The fact that some degree of reinforcement is required to maintain reinforcement-based response habits suggests that drugs that block reinforcement should cause extinction of those habits. Indeed, drugs that block the central actions of dopamine do cause extinction-like decreases in habitual responding in laboratory animals. However, consistent with the fact that it takes less reinforcement to maintain a habit than it takes to establish one, dopamine antagonists do not block an established response habit right away and must be given repeatedly to cause anything approaching complete extinction. Moreover, dopamine antagonists have not been found effective in treating reinforcement disorders in humans. For one thing, these drugs attenuate the effectiveness of many of life's pleasures and are not pleasant to take. These drugs are used effectively to treat the (seemingly unrelated) symptoms of schizophrenia, and patients do not like them and often discontinue their use when given the chance. Moreover, schizophrenic substance abusers continue to seek and use cocaine and amphetamine despite treatment with dopamine antagonists. This may be, of course, because cocaine and amphetamine are antagonists for the dysphoric effects of the dopamine blockers. It is also the case that dopamine antagonists have little if any effect on the immediate motivational impact of reward-predictive environmental cues that contribute to drug seeking and drug euphoria. In any case, dopamine antagonists have not proven useful for the treatment of reinforcement disorders in the clinic.

Two classes of drugs that are currently under study as potential medications for reinforcement disorders are opioid antagonists and cannabinoid antagonists. Exogenous opiates and cannabinoids are drugs of abuse, but endogenous opioids and cannabinoids are found in the brain and serve neurotransmitters or neuromodulators that, like other addictive drugs, have actions in the reward circuitry of the brain. Opioid antagonists are known to block the reinforcing effects of opiates and also appear to attenuate cravings for alcohol and for nicotine. Cannabinoid antagonists not only block the reinforcing effects of cannabinoids (marijuana, hashish) but also attenuate heroin, cocaine, and food cravings. The fact that each drug appears to affect natural reinforcers as well as drug reinforcers strengthens the evidence that drug and natural reinforcers share a common reward mechanism in the brain; however, this makes the search for medications that can target a troublesome reinforcer without affecting others more challenging. Over most of the course of our evolutionary history, the substances our brain finds reinforcing have been the necessities of life (Volkow and Wise 2005).

Cross-References

▶ Addictive Disorder: Animal Models
▶ Eating and Appetite
▶ Intracranial Self-Stimulation
▶ Pathological Gambling
▶ Reinstatement of Drug Self-Administration
▶ Self-Administration of Drugs

References

Ahmed SH, Koob GF (1998) Transition from moderate to excessive drug intake: change in hedonic set point. Science 282:298–300

Di Chiara G, Imperato A (1988) Drugs abused by humans preferentially increase synaptic dopamine concentrations in the mesolimbic system of freely moving rats. Proc Natl Acad Sci U S A 85:5274–5278

Hyman SE, Malenka RC, Nestler EJ (2006) Neural mechanisms of addiction: the role of reward-related learning and memory. Annu Rev Neurosci 29:565–598

Tornatzky W, Miczek K (2000) Cocaine self-administration "binges": transition from behavioral and autonomic regulation toward homeostatic dysregulation in rats. Psychopharmacology (Berl) 148:289–298

Volkow ND, Wise RA (2005) How can drug addiction help us understand obesity? Nat Neurosci 8:555–560

White NM (1996) Addictive drugs as reinforcers: multiple partial actions on memory systems. Addiction 91:921–949

Wise RA (2002) Brain reward circuitry: insights from unsensed incentives. Neuron 36:229–240

Wise RA (2004) Dopamine, learning and motivation. Nat Rev Neurosci 5:483–494

Wise RA (2006) Role of brain dopamine in food reward and reinforcement. Philos Trans R Soc Lond B Biol Sci 361:1149–1158

Wise RA, Rompré P-P (1989) Brain dopamine and reward. Ann Rev Psychol 40:191–225

Wise RA, Newton P, Leeb K, Burnette B, Pocock P, Justice JB (1995) Fluctuations in nucleus accumbens dopamine concentration during intravenous cocaine self-administration in rats. Psychopharmacology 120:10–20

Reinforcer

Definition

In ▶ instrumental conditioning, a reinforcer is an event which maintains or strengthens a behavioral response that produces it. For example, cocaine can function as a reinforcer because when an animal is allowed to self-administer it by means of pressing a lever, the animal will continue to press the lever again and again. The act of pressing the lever was reinforced by cocaine. In the context of ▶ Pavlovian conditioning, the term "reinforcer" includes events that strengthen stimulus-stimulus and stimulus-reward associations.

Reinstatement of Drug Self-Administration

Sunila G. Nair[1] and Yavin Shaham[2]
[1]Harborview Medical Center, Seattle, WA, USA
[2]Behavioral Neuroscience Branch, IRP/NIDA/NIH, Baltimore, MD, USA

Synonyms

Reinstatement of drug seeking; Reinstatement of drug-taking behavior

Definition

In the learning literature, the term "reinstatement" refers to the resumption of a learned response that occurs when a subject is exposed noncontingently to the unconditioned stimulus. In the addiction field, "reinstatement of drug self-administration" in experimental animals (typically termed "reinstatement of drug seeking") refers to the resumption of a previously drug-reinforced behavior by noncontingent exposure to drugs, different types of drug cues, or stressors after the extinction of the drug-reinforced behavior. The animal model that addiction researchers use to study the reinstatement of drug seeking is termed the "reinstatement model" (Shaham et al. 2003).

Impact of Psychoactive Drugs

Background
A major problem in the treatment of drug addiction is the high rate of relapse to drugs of abuse following a prolonged abstinence period. Results from studies in humans suggest that in drug-free individuals, relapse to drug use during periods of forced or voluntary abstinence can be triggered by acute exposure to the self-administered drug, stimuli previously associated with drug taking, or stressors. Acute exposure to the self-administered drug or related drugs, drug-associated cues, or stress also reinstates drug seeking in laboratory rats and monkeys (Shaham et al. 2003). Because of the similarities between the human condition and the laboratory animal model, many investigators currently use the reinstatement model to study mechanisms underlying relapse to drug use. Conceptual issues related to the validity of the reinstatement model as an animal model of relapse to drug use are discussed by Epstein et al. (2006).

Experimental Procedures
In the drug self-administration version of the reinstatement model, which is based on an operant-conditioning procedure, laboratory animals (rats, mice, and monkeys) are initially trained to self-administer drugs by lever pressing (or another operant response such as nose poking) for intravenous drug infusions (or for oral delivery of alcohol). Subsequently, the drug-reinforced behavior is extinguished by substituting the drug solutions with saline or by disconnecting the infusion pumps. After the extinction of the drug-reinforced behavior, the ability of acute noncontingent exposure to drugs (termed "drug priming"), drug-associated cues, or stress to reinstate lever responding is measured under extinction conditions (Shaham et al. 2003). The non-reinforced responding on the "active lever" – the lever that previously delivered the drug – is interpreted to reflect the reinstatement of drug seeking. The responding on the "inactive lever" – the lever that has not been associated with drug injections – is often interpreted to reflect nonspecific activity, but it may also reflect response generalization.

In laboratory animals, the reinstatement of drug seeking has been studied using the "between-session," "within-session," and "between-within-session" variations of the reinstatement model (Shalev et al. 2002). In the between-session procedure, training for drug self-administration, the extinction of the drug-reinforced behavior, and tests for reinstatement are conducted on different daily sessions. In the within-session procedure, all three phases are conducted within a daily session. This daily session consists of 1–2 h of drug self-administration that is followed by 3–4 h of extinction of the drug-reinforced behavior; after extinction of lever responding, a test for reinstatement is conducted. In the between-within procedure, the laboratory animals are trained daily for drug self-administration; subsequently, the extinction of the drug-reinforced lever responding and tests for reinstatement are conducted on the same day. In addition, researchers also assess the reinstatement of drug seeking in variations of the reinstatement model that are based on the runway and conditioned place preference procedures. Shalev et al. (2002) provide a detailed description of the different procedural variations of the reinstatement model and discuss the advantages and disadvantages of using them for assessing the reinstatement of drug seeking.

Neuropharmacology of Reinstatement of Drug Seeking

In this section, we summarize the extant literature on the neuropharmacology of the reinstatement of drug seeking induced by exposure to the self-administered drug or related drugs (drug priming), drug-associated cues, or stress. Due to space constraints, we often do not differentiate between results obtained from studies in which rats were trained to self-administer cocaine and other psychostimulants, heroin and morphine, or alcohol. A detailed description of the neuropharmacology of the reinstatement of drug seeking is provided by Shalev et al. (2002) and Bossert et al. (2005).

Drug-Priming-Induced Reinstatement

Results from studies on drug-priming-induced reinstatement of drug seeking indicate that this reinstatement is primarily mediated by dopaminergic and glutamatergic neurotransmission in the mesocorticolimbic system (Kalivas and McFarland 2003; Schmidt et al. 2005; Self 2004). This system consists of dopamine cell bodies in the ventral tegmental area (VTA) that project to limbic areas, including nucleus accumbens, amygdala, and bed nucleus of stria terminalis (BNST), and cortical areas, including medial prefrontal cortex (mPFC) and orbital frontal cortex (OFC).

Regarding dopamine, systemic injections of either D1-family dopamine receptor agonists or antagonists decrease cocaine-induced reinstatement; D1-family dopamine receptor antagonists also decrease heroin-priming-induced reinstatement. On the other hand, systemic injections of D2-family receptor antagonists decrease heroin- or cocaine-priming-induced reinstatement, while injections of D2-family receptor agonists reinstate drug seeking. The effect of cocaine priming on reinstatement is also attenuated by systemic injections of selective D3 receptor antagonists. Additionally, injections of D1-family or D2-family receptor antagonists into the dorsal mPFC, accumbens shell (but not core), and amygdala (both central and basolateral subregions) decrease cocaine-priming-induced reinstatement. Accumbens core injections of D1-family or D2-family receptor antagonists decrease heroin-priming-induced reinstatement. Finally, activation of VTA dopamine neurons by local injections of morphine reinstates heroin and cocaine seeking, while local inhibition of these neurons by a mixture of $GABA_A$ + $GABA_B$ receptor agonists (muscimol + baclofen) decreases cocaine-priming- and heroin-priming-induced reinstatement.

Regarding glutamate, systemic injections of metabotropic glutamate receptor 5 antagonists or group II metabotropic glutamate agonists (which decrease evoked glutamate release via a presynaptic autoreceptor mechanism) decrease cocaine-priming-induced reinstatement. Additionally, local injections of nonselective ionotropic glutamate receptor antagonists into the VTA, selective AMPA receptor antagonists into the accumbens, or reversible inactivation of the glutamate projection from the dorsal mPFC to the accumbens decrease heroin- or cocaine-priming-induced reinstatement. Finally, there is evidence that cocaine self-administration and subsequent withdrawal lead to long-lasting neuroadaptations in glutamatergic transmission in the mPFC-accumbens pathway and that the pharmacological reversal of these neuroadaptations prevents cocaine-priming-induced reinstatement.

Results from studies in which investigators used reversible inactivation manipulations (the sodium channel blocker tetrodotoxin or muscimol + baclofen) suggest a role of the ventral pallidum in cocaine- and heroin-priming-induced reinstatement. In the case of heroin-priming-induced reinstatement, there is also evidence for a role of the BNST (dorsal and ventral subregions), dorsal striatum, substantia nigra, and ventral mPFC. Additionally, there is evidence from pharmacological studies that other neurotransmitter systems, including endocannabinoids, serotonin, and GABA, contribute to drug-priming-induced reinstatement. However, the brain sites involved in these effects are unknown. Finally, there is evidence for a role of accumbens protein kinase A and calcium-calmodulin-dependent kinase II signaling pathways in cocaine-priming-induced reinstatement.

Cue-Induced Reinstatement

In humans, relapse-provoking drug-associated stimuli can be divided into two general categories: discrete drug cues (e.g., drug paraphernalia) that are associated with the acute rewarding effects of the drug and contextual drug cues (e.g., a specific environment such as a local bar) that predict drug availability. In laboratory animals, procedures to assess cue-induced reinstatement of drug seeking are classified into three types according to the type of the conditioned cue: discrete cue, discriminative cue, and contextual cue.

Discrete-cue-induced reinstatement: Results from pharmacological studies suggest a role of several neurotransmitters and receptors in discrete-cue-induced reinstatement (Feltenstein and See 2008). Systemic injections of D1-family receptor antagonists, selective D3 receptor antagonists, nicotinic cholinergic antagonists, several serotonergic agents, group II metabotropic receptor agonists, cannabinoid 1 (CB1) receptor antagonists, and mu opiate receptor antagonists decrease the discrete-cue-induced reinstatement of drug seeking. In the case of D1-family receptors, two brain sites are critical for their effects: the amygdala (both central and basolateral subregions) and the accumbens core (but not shell). The brain sites involved in the modulation of the discrete-cue-induced reinstatement by the other neurotransmitter systems are unknown. Additionally, results from studies in which reversible inactivation manipulations were used suggest a role of the dorsal mPFC and dorsal striatum in the discrete-cue-induced reinstatement. In heroin-experienced rats, there is evidence that discrete-cue-induced reinstatement also involves the ventral mPFC, the dorsal BNST, the ventral pallidum, and the substantia nigra.

Discriminative-cue-induced reinstatement: Results from pharmacological studies suggest a role of several neurotransmitter systems in discriminative-cue-induced reinstatement (Weiss 2005). Systemic injections of D1-family and D2-family receptor antagonists, selective D3 receptor agonists, group II metabotropic receptor agonists, metabotropic glutamate receptor type 1 or type 5 antagonists, CB1 receptor antagonists, sigma 1 receptor antagonists, 5-HT2B/2C receptor agonists, and mu opioid receptor antagonists decrease the discriminative-cue-induced reinstatement. The discriminative-cue-induced reinstatement is also attenuated by ventricular injections of the opioid peptide nociceptin/orphanin FQ (the endogenous ligand for the opioid-like orphan receptor). In the case of CB1 receptor antagonists, two brain sites are critical for their effects: the mPFC and the accumbens. The brain sites involved in the modulation of discriminative-cue-induced reinstatement by the other neurotransmitter systems are unknown. However, in the case of dopamine receptor antagonists, likely brain sites are the amygdala, the accumbens, and the mPFC. Thus, exposure to cocaine discriminative cues increases dopamine release in the accumbens and amygdala. Additionally, discriminative-cue-induced Fos (a neuronal activity marker) expression in the basolateral amygdala and prefrontal cortex is decreased by systemic injections of a D1-family receptor antagonist.

Context-induced reinstatement: Results from pharmacological studies suggest a role of several neurotransmitter systems in context-induced reinstatement (Crombag et al. 2008). Systemic injections of D1-family and D2-family receptor antagonists, group II metabotropic receptor agonists, CB1 receptor antagonists, and 5-HT2B/2C receptor agonists decrease context-induced reinstatement. In the case of D1-family receptors, a critical brain site is the accumbens shell. In the case of group II metabotropic receptors, the critical brain sites are the accumbens shell and VTA. The brain sites involved in the modulation of the context-induced reinstatement by the other neurotransmitter systems are unknown. Results from studies in which reversible inactivation methods were used suggest a role of the dorsal mPFC, the dorsal striatum, the basolateral amygdala, and the dorsal hippocampus in context-induced reinstatement. Results from studies on the effect of context-induced reinstatement testing on Fos expression in the brain suggest a role of the lateral hypothalamus in this reinstatement.

Stress-Induced Reinstatement

The phenomenon of stress-induced reinstatement of drug seeking was initially demonstrated in studies in which investigators used an intermittent footshock stressor. The effect of this stressor generalizes to certain other stressors, including swim stress, acute food deprivation, social defeat, and several pharmacological stressors: the stress neurohormone corticotropin-releasing factor (CRF), the anxiogenic drug yohimbine (an alpha-2 adrenoceptor antagonist), and the agonists of the kappa opioid receptor.

Evidence from studies in which pharmacological agents were used suggests a role of several neurotransmitter systems in stress-induced reinstatement. The stressor used in most of the neuropharmacological studies is intermittent footshock. Systemic injections of nonselective CRF antagonists, selective CRF1 receptor antagonists, alpha-2 adrenoceptor agonists (which decrease noradrenaline cell firing and release), hypocretin 1 receptor antagonist, nonselective dopamine receptor antagonist, selective serotonin reuptake blockers (SSRIs), and $5HT_3$ receptor antagonists decrease stress-induced reinstatement. The stress-induced reinstatement of alcohol (but not cocaine) seeking is attenuated by ventricular injections of nociceptin/orphanin FQ.

The effect of CRF antagonists on stress-induced reinstatement is independent on the activity of the hypothalamic-pituitary-adrenal axis (HPA) and the effect of these antagonists on the stress hormone corticosterone. The critical extrahypothalamic brain sites and projections for CRF's role in footshock-induced reinstatement are the BNST, VTA, and a projection from the central amygdala to the BNST. Within the VTA, footshock causes local CRF release, which leads to increased glutamate transmission; this enhanced glutamate transmission is critical for stress-induced reinstatement, presumably via an activation of the mesocorticolimbic dopamine system. Support for this notion is provided by the findings that injections of D1-family receptor antagonists into the dorsal mPFC or OFC or preferential D3 receptor antagonists into the accumbens decrease stress-induced reinstatement. The critical brain sites and projections for noradrenaline's role in footshock-induced reinstatement are the central amygdala and BNST and the noradrenergic projection from the lateral tegmental nuclei to these brain sites.

Finally, results from studies in which reversible inactivation methods were used confirm the findings on the role of the dorsal mPFC, the BNST, the central amygdala, the accumbens, and the VTA in stress-induced reinstatement and further suggest a role of the ventral pallidum in this reinstatement. Finally, as in the case of drug-priming-induced reinstatement, the glutamatergic projection from the mPFC to the accumbens plays an important role in stress-induced reinstatement.

Acknowledgment The writing of this chapter was supported by the Intramural Research Program of the NIH, NIDA.

Cross-References

▶ Conditioned Place Preference and Aversion
▶ Context-Induced Reinstatement
▶ Discrete-Cue-Induced Reinstatement

- Discriminative-Cue-Induced Reinstatement
- Drug Self-Administration
- Extinction
- Runway Procedure
- Stress

References

Bossert JM, Ghitza UE, Lu L, Epstein DH, Shaham Y (2005) Neurobiology of relapse to heroin and cocaine seeking: an update and clinical implications. Eur J Pharmacol 526:36–50

Crombag HS, Bossert JM, Koya E, Shaham Y (2008) Review. Context-induced relapse to drug seeking: a review. Philos Trans R Soc Lond B Biol Sci 363:3233–3243

Epstein DH, Preston KL, Stewart J, Shaham Y (2006) Toward a model of drug relapse: an assessment of the validity of the reinstatement procedure. Psychopharmacology (Berl) 189:1–16

Feltenstein MW, See RE (2008) The neurocircuitry of addiction: an overview. Br J Pharmacol 154:261–274

Kalivas PW, McFarland K (2003) Brain circuitry and the reinstatement of cocaine-seeking behavior. Psychopharmacology (Berl) 168:44–56

Schmidt HD, Anderson SM, Famous KR, Kumaresan V, Pierce RC (2005) Anatomy and pharmacology of cocaine priming-induced reinstatement of drug seeking. Eur J Pharmacol 526:65–76

Self DW (2004) Regulation of drug-taking and -seeking behaviors by neuroadaptations in the mesolimbic dopamine system. Neuropharmacology 47(Suppl 1):242–255

Shaham Y, Shalev U, Lu L, De Wit H, Stewart J (2003) The reinstatement model of drug relapse: history, methodology and major findings. Psychopharmacology (Berl) 168:3–20

Shalev U, Grimm JW, Shaham Y (2002) Neurobiology of relapse to heroin and cocaine seeking: a review. Pharmacol Rev 54:1–42

Weiss F (2005) Neurobiology of craving, conditioned reward and relapse. Curr Opin Pharmacol 5:9–19

Relapse

Definition

In drug dependence, to regress back to a level of drug use at or comparable to use prior to a quit attempt following a period in which reduced use or abstinence was attained.

Cross-References

- Addictive Disorder: Animal Models
- Drug Self-Administration
- Reinstatement of Drug Self-Administration

Relapse Prevention Studies

Definition

The standard approach to assessing the potential value of continued treatment in mood and anxiety disorders. Patients who have responded to open treatment are randomized to either continue with that compound or to be switched to placebo and followed up over 6–18 months, monitoring for signs of relapse, that is, the reappearance or significant worsening of symptoms. Efficacy is assessed through examining the mean time to relapse and the proportion of patients who relapse, in both groups.

Relapse Prevention Study Design

Definition

A study design that randomizes open-label responders to ongoing active or placebo treatment and follows them over several weeks under double-blind conditions to observe relapse rates.

Cross-References

- Double-Blinded Study
- Open Label

Relative Validity

Synonyms

Degraded contingency; Partial reinforcement

Definition

A relative reduction in associative learning (compared with what would ordinarily be supported) arising because the reliability with which an unconditioned stimulus (UCS) is predicted depends on the level of contingency between the UCS and the preceding conditioned stimulus (CS) at issue. To the extent that this informational relationship or contingency is partial, associative learning to the nominal CS will be reduced and alternative available stimuli, including those provided by the context, are likely to gain associative strength. Relative validity is a term applied to classical (Pavlovian) conditioning and a constraint on the general importance of temporal coincidence as a determinant of new learning.

Cross-References

▶ Classical (Pavlovian) Conditioning

Release Probability

Definition

Quantal release probability refers to the likelihood of vesicular release events of quanta of neurotransmitter occurring in a population of nerve terminals or a single nerve terminal, usually during invasion of an action potential. Drugs that have presynaptic actions on nerve terminals either increase or decrease quantal release probability.

REM Sleep Behavior Disorder

Synonyms

RBD

Definition

REM sleep behavior disorder (RBD) is characterized by vigorous motor activity, which typically occurs during REM sleep and thus in the absence of muscle atonia. RBD often consists of injurious dream-enactment motor activity associated with vivid dreaming. RBD is symptomatically complex, often involving behaviors that are dangerous to the affected patient, or to his/her bed partner, and vivid dream imagery that is almost always unpleasant. RBD is more frequent in males and usually occurs in the middle-aged or elderly. Acute RBD can be induced by medications (monoamine oxidase (MAO) inhibitors, serotonin reuptake inhibitors, and tricyclic antidepressants), alcohol, and BDZ withdrawal.

Remission (from Depression)

Definition

Remission from depression is essentially the complete elimination of all symptoms for at least 6–12 months. In clinical trials, it may be defined according to the achievement of a particular score on a depression rating scale (e.g., a score of 7 or less on the Hamilton Depression (HAM-D) Rating Scale).

Cross-References

▶ Depression

Remission (from Obsessive-Compulsive Disorder)

Definition

The concept of remission for OCD is debatable, and there is no universally accepted definition. It has been described as a brief period during which sufficient improvement has occurred so that the individual no longer suffers with OCD. Studies have chosen varying remission criteria based on the ▶ Y-BOCS. A score of 16 on Y-BOCS is generally considered too high to meaningfully represent true remission, and a score of 7 too low to be achieved by all but a very few cases. In a recent analysis, a Y-BOCS score of 7 did not discriminate between active and control treatments, whereas in two recent multicenter extended SSRI (SSRI and related compounds) studies, a score of 10 did. Therefore, for the purposes of definition, a reduction in OCD symptoms such that the individual no longer suffers with OCD and has a total score of 10 on the Y-BOCS is usually accepted as remission.

Cross-References

- ▶ Clinical Global Impressions Scales
- ▶ SSRIs and Related Compounds
- ▶ Yale-Brown Obsessive-Compulsive Scale

Reproduction Protocol

Definition

Reproduction protocol refers to a protocol in which subjects are presented with an event and required to reproduce a sequence of behavior compatible with the estimated duration of the event. Reproduction protocols avoid verbalizing the durations.

Reserpine

Definition

Reserpine is an indole alkaloid with antipsychotic and antihypertensive properties. It is currently not in clinical use for psychiatric conditions. Reserpine was isolated in 1952 from the dried root of *Rauvolfia serpentina* (Indian snakeroot), which was used in India for centuries to treat insanity and high blood pressure. Reserpine blocks the vesicular monoamine transporter, thereby acutely inhibiting the reuptake of the ▶ monoamines into the synaptic vesicles and increasing monoamine concentrations in the synaptic cleft, but eventually leading to depletion of vesicular monoamine stores in synaptic nerve endings upon chronic use, as the neurotransmitters that cannot reenter the vesicles will be degraded by monoamine oxidase. Chronic treatment can lead to ▶ depression, possibly due to depletion of these neurotransmitter stores. The drug can cause a range of additional side effects, for example, in the gastrointestinal tract (ulcerations, cramps, diarrhea) or the cardiovascular system (hypotension, bradycardia), all of which make it obsolete as an antipsychotic for today's treatment of ▶ schizophrenia or other psychoses. It was also used as a tool in experimental pharmacology, to deplete stored neurotransmitters and thus allow tests of their roles in pharmacological reactivity; this usage ended when more selective drugs became available.

Cross-References

- ▶ Antipsychotic Drugs
- ▶ Depression
- ▶ Schizophrenia

Retaining Current

Definition

In neurophysiological studies, a current (either positive or negative) that reduces or blocks the spontaneous efflux of charged transmitters, drugs, or other compounds of interest from the iontophoretic pipette into the extracellular environment.

Retardation of Acquisition Test

Synonyms

Retardation test

Definition

The retardation of acquisition test is one of the two widely accepted tests for whether a stimulus functions as a ▶ conditioned inhibitor, the other being the ▶ summation test. In the retardation test, a stimulus is first trained as a putative conditioned inhibitor. Once complete, the stimulus is then repeatedly paired with the ▶ unconditioned stimulus (US). If the stimulus functions as a conditioned inhibitor, acquisition of an excitatory ▶ conditioned response should be impaired (retarded) relative to controls. Because alternative accounts remain for delayed acquisition, the strongest case for a stimulus functioning as a conditioned inhibitor also requires the use of the summation test.

Cross-References

▶ Blocking, Overshadowing, and Related Concepts
▶ Classical (Pavlovian) Conditioning
▶ Occasion Setting with Drugs
▶ Pavlovian Fear Conditioning
▶ Summation test

Retention

Definition

Memory refers to a theoretical state in which a particular behavior is based on retrieval of previously learned information and its expression. These two processes can be distinguished in humans, whereas in animals this distinction is almost impossible. Retention performance, which depends on memory recall, is influenced by the ability of the subject to retrieve contextual stimuli. It is important to dissociate between memory and performance, since the subject may not be able properly express a learned response.

Retrieval

Definition

The fact or process whereby learned information or response changes are expressed or remembered.

Rett's Disorder

Definition

Rett's disorder is a genetic condition, occurring almost exclusively in girls, that can be diagnosed with genetic testing. It is characterized by normal development for the first 5 months of life, following which, between 5 and 48 months, there is a deceleration of head growth, loss of previously acquired purposeful hand skills (along with the development of hand-wringing), loss of social skills, loss of coordination, and loss of both receptive and expressive language skills. Rett's disorder was removed from DSM-5 because autism spectrum disorder (ASD)-like symptoms are not salient in the disorder except early in development. It is possible for a patient with Rett's disorder to also be diagnosed with an ASD with the specifier "with known genetic or medical condition" if sufficient autistic-like symptoms are present.

Cross-References

▶ Autism Spectrum Disorders and Intellectual Disability

Reversible Binding

Definition

The mode of interaction with receptors undergone by most neuroreceptor imaging tracers.

Reversible means that the radiotracer will dissociate from the receptor-ligand complex with some regularity during the course of the imaging experiment, that is, that the ratio of the "on" rate of binding to the "off" rate of dissociation is not exceedingly large.

Reward Intensity

Definition

The subjective strength of a reward, for example, the subjective attractiveness of different concentrations of sucrose to a hungry rat.

Reward-Related Incentive Learning

Synonyms

Incentive learning

Definition

Reward-related incentive learning is said to occur when a behavior is selected, initiated, and maintained by external incentive stimuli that through learned associations with desired outcomes acquire incentive salience and come to predict the availability and location of primary rewards.

Reye's Syndrome

Definition

A rare but potentially fatal syndrome targeting the liver and brain; changes in behavior and cognitive state can be seen as a result of encephalopathy and chemical alterations in the blood derived from liver damage. Other symptoms include vomiting, seizures, hyperventilating, and coma. It primarily presents in children and adolescents as the result of consuming acetylsalicylic acid during a viral infection such as chickenpox or influenza.

Modified from the NIH-NLM MedlinePlus Encyclopedia (online).

Rhythmicity

Synonyms

Frequency of oscillation

Definition

In describing rodent behaviors that are expressed with a relatively invariant approximately sine wave-like regularity, "rhythm" and "rhythmicity" are used as descriptors to avoid the ambiguity associated with the word "frequency," which can mean "oscillations in cycles/s (Hz)" in physical contexts but in statistical contexts refers to frequency of occurrence of events that have no synchronous relation to one another.

Cross-References

▶ Motor Activity and Stereotypy

Ribozyme

Synonyms

Catalytic RNA molecule

Definition

Ribozymes are RNA molecules of viral, prokaryotic, or eukaryotic origin that possess catalytic activity. The term "ribozyme" was coined by Thomas R. Cech upon the discovery of a self-splicing intron in the ciliate protozoan *Tetrahymena thermophila*. Most commonly, ribozymes catalyze sequence-specific endonucleolytic cleavage, but also other catalytic activities, such as RNA polymerization, have been described. Similar to antisense oligonucleotides and siRNA, ribozymes find frequent application in the knockdown of expression of selected genes. There are several different classes of ribozymes, with the "hammerhead" ribozyme being the most widely studied, in recent years.

Cross-References

▶ Antisense Oligonucleotides
▶ siRNA

Rigidity

Definition

Involuntary resistance to passive movement.

Rilmazafone

Definition

Rilmazafone is a benzodiazepine prodrug. Its activity is due to an active metabolite that is a benzodiazepine with sedative, hypnotic, and anxiolytic properties. It is a long-acting compound used in the short-term treatment of insomnia. Like most similar compounds, its use is subject to tolerance, abuse, dependence, and withdrawal.

Cross-References

▶ Benzodiazepines
▶ Insomnias
▶ Sedative, Hypnotic, and Anxiolytic Dependence

Rimonabant

Synonyms

SR141716

Definition

Rimonabant was the first selective CB_1 ligand introduced into clinical practice. This CB_1 receptor antagonist was found to be efficacious as a treatment for obesity and for improving dyslipidemias and metabolic disorders. It acts as a ▶ partial agonist and also has ▶ inverse agonist action. There was also some promise for treatment of nicotine dependence and cannabis addiction. However, due to some increased rates of depression and anxiety and cases of suicide related to drug use, it was withdrawn from the market. It is used in experimental psychopharmacology as a probe for the involvement on CB_1 receptors in physiological and pharmacological phenomena.

Cross-References

▶ Appetite Suppressants
▶ Inverse Agonists
▶ Partial Agonist

Risk Assessment

Synonyms

Stretched-attend posture

Definition

Risk assessment comprises a range of ▶ defensive behaviors displayed when animals are confronted with a novel and/or threatening stimulus. The biological function of these behaviors is to gather information about the potential threat by cautiously approaching it or by scanning the surrounding area. As rodents may display an increase in stretched-attend postures even when not avoiding an unprotected area, it has been advocated that risk assessment, thus representing the most enduring behavioral expression of anxiety, may even be more sensitive to anxiety-modulating drugs than avoidance behavior.

Cross-References

▶ Anxiety
▶ Defensive Behaviors

Risk Taking

Synonyms

Adventuresome; Thrill seeking

Definition

Selection of alternative associated with lower probability of occurrence but, usually, larger subjective return. Unlike novelty seeking, the alternative selected is not necessarily unfamiliar to the individual. Sensation seeking incorporates both novelty seeking and risk taking.

Risperidone

Definition

Antipsychotic drug of the second-generation, atypical category with combined dopamine D2/serotonin2 receptor-blocking properties.

Cross-References

▶ Antipsychotic Medication
▶ Second- and Third-Generation Antipsychotics

Ritual Uses of Psychoactive Drugs

Marlene Dobkin de Rios[1] and Charles S. Grob[2]
[1]Department of Psychiatry and Human Behavior, University of California, Irvine, CA, USA
[2]Department of Psychiatry, Harbor/UCLA Medical Center, Torrance, CA, USA

Definition

Substances, mainly plants that provoke an altered state of consciousness, have been used ritually for several thousand years throughout the globe in a wide variety of cultures. Their use has included religious, recreational, and healing aspects.

Current Concepts and State of Knowledge

Cultural anthropologists with interest in these substances document their important role in all continents among hunter/gatherers, incipient horticulturalists, advanced agricultural societies, and pristine state societies as well as contemporary cultures. Over time, as societies became more complex, the use of plant hallucinogens changed from open access and widespread experiences to the usurpation of such use by elite segments of society. As societies became more complex, access to drug-induced altered states of consciousness became part of sumptuary laws, as fewer individuals were permitted entry to these states. This contrasts with societies of hunters/gatherers, for example, whereas many as one third of adult men might use plant psychedelics in ritual ceremonies for spiritual purposes. With cultural change, wars, and conquest, much of our information about historical and prehistorical use of these substances was lost, only to be reformulated since the 1960s. We can suppose that the abrogation of such drug access was related to the supposed power of the

---

Marlene Dobkin de Rios: deceased.

hallucinogenic state and the power believed to be conferred upon the user to control or harm others through magical means or witchcraft. There is a cultural evolutionary movement from exoteric rituals, open and accessible to all adults, to esoteric rituals, much like the Eleusinian mysteries of ancient Greece. Unauthorized drug use under these circumstances may have become a crime against the commonwealth. Cross-cultural studies of thematic materials linked to societies where these substances have been utilized and reported upon enable us to make a reasonable connection, however (Dobkin de Rios 1984). One illustrative example has been through the careful examination of ancient Mayan art, which has revealed that the water lily, found more frequently than the rain god, contained aporphine, which after causing heavy vomiting provided the user with a languid state conducive to visual ruminations, influencing the art and architecture of this ancient civilization.

Societies for which we have excellent data on the ritual use of plant hallucinogens include the Chumash Indians of Santa Barbara, the Tsogana Tsonga of East Africa, and the Australian Aborigines (see Dobkin de Rios 1974; Dobkin de Rios and Katz 1975; Mckenna et al. 1998; Siegel 1989). The examination of these tribal societies have identified that there existed managed altered states of consciousness where plant hallucinogens were given by elders to youth as part of an intensive short-term socialization for religious and pedagogical purposes. The use of hypersuggestibility as a cultural technique to "normalize" youth in these societies can be compared to the role of pathology of drug ingestion patterns among Euro-American adolescents (Dobkin de Rios and Smith 1977; Grob and Dobkin de Rios 1991).

In tribal societies for which we have good data, these plants have been used mainly as facilitators for religious ecstasy and to permit individuals to come into firsthand contact with spirit or divinity. In hunter/gatherer societies, male hunters used hallucinogens in shamanistic religious rituals to divine the future and to locate the animals they hunted. The chief focus in these societies was on power and its exercise by religious practitioners. These substances, however, have been viewed as a two-edged sword. On the one hand, they have been utilized by different societies over time and space due to their perceived ability to access spiritual realms. The obverse is a faulty wiring hypothesis that argues that plant chemicals deceive and trick. These substances however have been seen by serious scholars as a psychotechnology that allows tribal elders to manage the altered states of consciousness of their adolescents through hypersuggestibility, utilizing the properties of the plant hallucinogens to decondition youth and to heighten religious experiences deemed important for social survival.

A major theme among traditional societies ritually utilizing plant hallucinogens is that of shape shifting. The transformation of human being into animals is found associated with plant drugs where the shamans, technicians of ecstasy, are metamorphosed into animal familiars. This action symbolized the power source of the individual who calls upon his animal familiar to do his bidding. The shaman is believed to be able to control and beckon a series of familiars for his own personal use in curing or bewitching. The concept of morphing, a phenomenon where one image remains in the mind's eye, while a second is superimposed upon it, with the first then fading away, can be cited to explain this common occurrence. Animal familiars are viewed as an empowering element. The religion of hunters and gatherers, shamanism, has a focus on personal ecstasy and direct knowledge of the preternatural. Human beings appear to be wired for the ability to experience ecstasy and the ability to have unordinary experiences to apprehend the divine, what psychiatrists call dissociative or potentially therapeutic transpersonal states. Hallucinogenic ingestion for purposes of religious ecstasy has been reported in all segments of human society.

Under the effects of the plant hallucinogens, perception of time changes. There is a circularity and reversible element of time, in which an eternal mythical present exists which is periodically reintegrated into the religious rites of tribal peoples. One of the characteristics of hallucinogenic drug use entails the perception of time which slows up almost to an imperceptible flow or else is experienced as indescribably fast.

A second theme deals with animals that appear to have played a vital role in teaching or revealing to human beings the properties of plant hallucinogens. Animals seek out psychotropic experiences, and several societies that incorporate plant hallucinogens into their rituals report learning about drug plants from deer, reindeer, or wild boars in their environment. Despite the apparent nonadaptive aspects of such animal behavior as they are likely to be placed in danger of predators, this is widespread and may point out the antiquity of hallucinogenic use in human society since hunters and gatherers are most likely to have been the ones to observe animal plant use most carefully and imitate this behavior.

The spiritual animation of hallucinogenic plants is another theme of interest. Many tribal groups believe that animated spirits of hallucinogenic plants exist that are either minuscule or gigantic in size, called micropsia or macropsia. Cultures across the world report this phenomenon.

Music is very important in ritual use of hallucinogens. Music is deemed by healers or sorcerers to evoke stereotypic visions. Music which is generally of a percussive nature may be viewed as necessary for the individual to attain certain cultural goals, such as seeing an individual deemed responsible for bewitchment, to help in healing, to foresee the future, etc. Sudden access to the unconscious by means of hallucinogenic use, despite the esthetic and expressive dimensions, is a dangerous space for human beings to enter. Psychodynamically oriented researchers stress the emotional response to such entry in terms of somatic stress registered by nausea, vomiting, diarrhea, tachycardia, and elevated blood pressure. The role of music with its implicit structure may be to provide a substitute psychic structuring during periods of ego dissolution. Music does not simply create mood within the drug setting. Given the change in ego structure and the anxiety, fear, and somatic discomfort attendant upon unexpected access to unconscious materials, the shaman guide also creates a corpus of music which some investigators have called the jungle gym in consciousness whose intrinsic structure provides the drug user with a series of paths and banisters to help him negotiate his way during the actual experience. Shamans themselves claim that the music created under their guidance provokes specific, highly valued patterned drug visions that permit their clients access to particular supernatural entities, to view the source of witchcraft, to permit contact with ancestor forces, etc. This goes along with death and resurrection themes reported in tribal societies where powerful hallucinogens are used in rituals. When used in adolescent rituals, the youth is seen to die in his social status of a child and reborn and returned to social life as a new person with a new name, responsibilities, and knowledge of the supernatural world. The psychedelic states heightened the learning of sacred knowledge and created a bonding among members of the cohort group.

Psychiatric/Psychological Implications

Cross-cultural research has shown that at the most private and personal level of being – the hallucinogenic experience – cultural membership determines the nature of the visionary experience. Among ayahuasca users in the Peruvian Amazon (various *Banisteriopsis* spp.), beliefs hold that the mother spirit of the plant is a boa constrictor, which appears in visions overseen by the rural and urban healers. When Westerners take ayahuasca, they often report idiosyncratic visions that go along with the lack of hallucinogenic traditions in their own culture. These substances, drawing upon enhanced suggestibility, have a long history in the transformation of adolescent boys and girls into fully participating members of adult society. Legal constraints of use in Euro-American society contrast with the ritualistic use of such plant drugs in traditional tribal societies of the world. With indigenous traditions, we find managed altered states of consciousness where tribal elders provide a didactic experience to prepare youth for new adult roles. The psychedelic states heighten the learning of youths and create a bonding among the cohort members so that individual psychic needs are subsumed to the needs of the social group. The youth may undergo austerities and rituals with painful activities such as genital mutilation, sleeplessness, and beatings. This results in an aboriginal boot camp where a youth would share and identify with his cohorts upon whom survival success might often depend. Docility and bounded rationality is a mechanism for social selection and is implicated in the voluntary success of altruistic behavior. This docility – receptivity to social influence – contributes greatly to fitness in the human species. Plant hallucinogens used ritually create docility states in adolescents in tribal society for the purpose of maturity preparation and is widely found in the anthropological record.

Drug use patterns have also changed over time, and some observers have documented controversies in recent years with foreigners seeking out ayahuasca experiences in the Peruvian rain forest, unfortunately with some abuses occurring on occasion with unscrupulous practitioners.

Modern Syncretic Religions and Ayahuasca

During the twentieth century, the use of ayahuasca was incorporated into a number of new syncretic churches in Brazil. By the late 1980s, the Brazilian government sanctioned the use of this powerful Amazonian plant hallucinogenic decoction discovered from indigenous plant medicinal traditions for use as a ceremonial sacrament in the structure of modern religions. In 2006, the US Supreme Court also ruled, by unanimous vote, to protect the legal rights of the Brazilian syncretic church, Uniao do Vegetal, to practice its religion and ritual use of ayahuasca within the context of religious ceremony in the USA. Similar legal advances have been made in Europe by another Brazilian syncretic church, Santo Daime, allowing formal ayahuasca ceremonies. The origin of the Brazilian ayahuasca religions lie in the first half of the twentieth century when their respective founders, Maestre Gabriel of the Uniao and Maestre Irrianeu of the Daime churches, during expedition to the remote regions of the forest, encountered indigenous Amazonian people who used ceremonial ayahuasca for healing and divination. Bringing back knowledge of these ancient plants, they created the necessary religious doctrine and rituals for modern church structures that utilized this powerful psychoactive sacrament (Dobkin de Rios and Grob 2005).

A critical series of issues examined during proceedings to determine the legality of ayahuasca have focused on its effects on the health and safety of its users. After an extensive examination of the available evidence, all court rulings have accepted that there are safe parameters of ayahuasca use within the context of the groups studied. In the 1980s, representatives of the Brazilian Confen (Narcotics Commission) tasked to investigate the use of ayahuasca by syncretic religions, conducted extensive examination of the Uniao do Vegetal in particular, and determined that the responsible administration of ayahuasca in Uniao religious ceremonies caused no apparent harm to participants and should be provided legal status. The US Federal Courts investigation of ayahuasca also emphasized the importance of data provided by the medical section of the Uniao do Vegetal as well as published reports in the medical and neuroscience literature of an international multidisciplinary research investigation of the short- and long-term effects of ayahuasca in subjects recruited in Manaus, Brazil, in the early 1990s (Grob et al. 1996).

Findings from these studies supported that the human use of ayahuasca, administered within the context of the Uniao do Vegetal, appears to be tolerated without evident negative consequence. Indeed, a number of subjects evaluated for this research investigation reported dramatic improvements in psychological and physical health, including most notably a number of cases of abolition of dangerous addictive and antisocial behaviors. Overall, mood regulation, cognitive acuity, and responsible behavior all appeared to improve. An intriguing neuroscience finding from this project was the increased density of serotonin transporters in the blood platelets of long-term ayahuasca users compared to ayahuasca naïve controls, which may provide the biological substrate for the observed positive and potentially therapeutic outcomes (Callaway and Grob 1998). A cautionary note should be provided, however, regarding adverse interactions between ayahuasca and other biologically active prescription and nonprescription medicines. One example has been the reports of serious and potentially life-threatening serotonin syndrome in individuals administered ayahuasca who were already taking selective serotonin reuptake inhibitor (SSRI) antidepressants. Another concern is with individuals combining psychostimulants with ayahuasca. Given the ubiquity of stimulants in modern society, from legally prescribed Ritalin for attention deficit disorder to illegal cocaine and methamphetamine, there is a risk if combined with ayahuasca of causing serious cardiovascular problems as well as sustained adverse psychiatric reaction.

In the early 2000s, the Brazilian judiciary requested a formal research evaluation of the health and well-being of adolescents who participated with their parents in religious ceremonies of the Uniao do Vegetal. In collaboration with Brazilian psychological and psychiatric colleagues, an investigation was conducted of ayahuasca-exposed youth contrasted with a matched control group of young people without a history of ayahuasca use. Results included no differences between the two groups of neuropsychological function with ayahuasca-exposed adolescents demonstrating fewer psychiatric symptoms and less use of alcohol and other psychoactive recreational drugs. Indeed, one Brazilian medical authority of the Uniao do Vegetal, Glacus de Souza Brito, has described the use of ayahuasca by young people within the Uniao as "prophylaxis against drug abuse" (Dobkin de Rios and Grob 2005).

In forming an understanding of the relative safety of ayahuasca within the context of the modern syncretic religion, Uniao do Vegetal, it is important to appreciate the contributions of both the psychobiological effects of the plant compound as well as the sociocultural context within which it is taken. Ayahuasca, a decoction of native plants of the Amazon forest, *Banisteriopsis caapi* and *Psychotria viridis*, contains biologically active harmala alkaloids and dimethyltryptamine (DMT). When DMT is administered orally, its potent psychedelic effects are entirely neutralized by the monoamine oxidase (MAO) enzyme system in the gut. However, when the brew is infused with monoamine oxidase-inhibiting (MAOI) harmala alkaloids, the DMT is absorbed through the gut allowing for the intense activation of the central nervous system. Such precise knowledge of the native plant pharmacopeia of the Amazon forest was passed down over millennia from the earliest inhabitants through the indigenous tribal people to the founders of the syncretic religions (Callaway et al. 1994, 1996). The Uniao do Vegetal in particular has developed a scrupulous process of evaluating prospective participants in their ayahuasca ceremonies, screening for individuals with medical and psychiatric vulnerabilities. For those who pass the screening process, careful monitoring of their reaction to the ayahuasca experience is established. The relative care the Uniao takes with novice participants and vulnerable long-term members, along with the relative safety of ayahuasca when administered under ideal conditions, has led to a very low incidence of adverse health reaction. Indeed, the Brazilian syncretic ayahuasca religions represent an illustrative modern example of the value of ritual in establishing an essential structure that will minimize the potential adverse effects and maximize the therapeutic and positive transformative values of the psychedelic experience (Dobkin de Rios and Rumrrill 2008).

Conclusion

Two major themes emerge from a study of the ritual use of hallucinogens. The first is the reluctance of Western scholars to acknowledge the important role of plant hallucinogens in human history and expressive behavior. In the hallucinogens, hallucinophobic westerners see the forbidden and irrational. Yet in tribal societies, access to supernatural power and the unitive experience (so-called oceanic experience in psychoanalytic discourse) was highly valued. Psychedelic plants were used to enhance perception and intuition and played an important role in healing (Winkelman and Roberts 2008). The second area of interest is the enormous potential of these plants to create hypersuggestible states.

These can reinforce religious and spiritual beliefs in their communities and can also be used to control and direct youth while contributing to the survivability of the social group.

Cross-References

▶ Hallucinogens

References

Callaway JC, Grob CS (1998) Ayahuasca preparations and serotonin uptake inhibitors: a potential combination for severe adverse interaction. J Psychoactive Drugs 30:367–369

Callaway JC, Airaksinen MM, McKenna DJ, Brito GS, Grob CS (1994) Platelet serotonin uptake sites increased in drinkers of ayahuasca. Psychopharmacology (Berl) 116:385–387

Callaway JC, Raymon LP, Hearn WL, McKenna DJ, Grob CS, Brito GS, Mash DC (1996) Quantitation of N, N-dimethyltryptamine and harmala alkaloids in human plasma after oral dosing with ayahuasca. J Anal Toxicol 20:492–496

Dobkin de Rios M (1974) The influence of psychotropic flora and fauna on Maya religion. Curr Anthropol 15:147–164

Dobkin de Rios M (1984) Hallucinogens: cross-cultural perspective. Waveland Press, Prospect Heights

Dobkin de Rios M, Grob CS (eds) (2005) Ayahuasca. J Psychoactive Drugs 37:119–237

Dobkin de Rios M, Katz F (1975) Some relationships between music and hallucinogenic ritual: the jungle gym in consciousness. Ethos 3:64–76

Dobkin de Rios M, Rumrrill R (2008) A hallucinogenic tea laced with controversy. Praeger Press, Westport

Dobkin de Rios M, Smith D (1977) Drug use and abuse in cross-cultural perspective. Hum Organ 36:15–21

Grob CS, Dobkin de Rios M (1991) Adolescent drug use in cross-cultural perspective. J Drug Issues 22:121–138

Grob CS, McKenna DJ, Callaway JC, Brito GS, Neves ES, Oberlaender G, Saide OL, Labigalini E, Tacla C, Miranda CT, Strassman RJ, Boone KB (1996) Human psychopharmacology of hoasca, a plant hallucinogen used in ritual context in Brazil. J Nerv Ment Dis 184:86–94

McKenna DJ, Callaway JC, Grob CS (1998) The scientific investigation of ayahuasca: a review of past and current research. Heffter Rev Psychedel Res 1:65–77

Siegel R (1989) Intoxication. Life in pursuit of artificial paradise. E. P. Dutton, New York

Winkelman M, Roberts T (eds) (2008) Psychedelic medicine. Praeger Publishing, Westport

Rivastigmine

Definition

Rivastigmine is an anti-dementia drug used for the treatment of mild to moderately severe ▶ Alzheimer's disease as well as for dementia in ▶ Parkinson's disease. Clinical trials have indicated that it produces a very modest improvement in symptoms but appears to be particularly effective in patients with either type of dementia who suffer from hallucinations. Rivastigmine is a high-potency anticholinesterase (inhibiting both ▶ acetylcholinesterase and butyrylcholinesterase), the clinical effectiveness of which is based on its ability to enhance central cholinergic function by increasing the availability of acetylcholine. However, with the progression of the disease and further loss of functional cholinergic neurons, its effectiveness is reduced. The most common side effects of rivastigmine are nausea, vomiting, and diarrhea. However, recently developed transdermal rivastigmine patches reduce the incidence of unpleasant side effects without the loss of clinical effectiveness.

RNA Editing

Definition

RNA editing is a molecular process changing the information contained in the coding region of the primary RNA transcript. Editing can occur in tRNA, rRNA, and mRNA molecules of eukaryotes and takes place in the nucleus and in mitochondria. Mechanisms of RNA editing include nucleoside modifications such as C to U and A to I deaminations but could also be non-templated nucleotide additions and deletions. RNA editing in neuronal iGluR transcripts is catalyzed by adenosine deaminase that recognizes partially double-stranded primary RNA transcripts, modifying adenosines to cause A to I, an editing process that results in this case in codon changes.

RNA Interference

Synonyms

RNAi

Definition

RNA interference (RNAi) is a surveillance system in most living eukaryotic cells that assists to regulate gene activity and to control expression of endogenous and parasitic genes, for example, viral genes or transposable elements. RNAi was discovered in 1998 by Andrew Z. Fire and Craig C. Mello in the nematode worm *Caenorhabditis elegans*. Mechanistically central to the RNAi system are small RNA molecules, such as siRNA (and also microRNAs). Long double-stranded RNAs, cleaved and processed by the RNase III family enzyme Dicer, are the typical precursors of siRNA molecules; shRNAs (short or small hairpin RNAs) introduced to cells by recombinant expression vectors are frequently used precursors in experimental gene-silencing studies (Fig. 2 in "▶ Antisense Oligonucleotides"). One of the two strands of each siRNA fragment, known as the guide strand, gets incorporated into the multiprotein RNA-induced silencing complex (RISC). This complex specifically binds to and efficiently cleaves single-stranded RNA, for example, mRNA transcripts, which match the siRNA bound by RISC (Fig. 2b in "▶ Antisense Oligonucleotides"). Thus, RNAi represents a natural mechanism to knock down the expression of selected genes, which is more potent than conventional antisense oligonucleotides or ribozyme approaches because it is assisted by an efficient cellular machinery able to execute multiple rounds of knockdown. In addition, most eukaryotic cells possess RNAi-related pathways also recruited by siRNA (and microRNA, to some extent), including activation of cellular antiviral defense mechanisms (e.g., the interferon response) or interference with the chromatin structure of the genome.

siRNA can also be used for pharmacological target validation by inhibiting target gene expression, thus representing an alternative strategy to pharmacological tools including antisense oligonucleotides, antagonists, and knockout mice.

siRNA can either be intracellularly expressed by vectors containing shRNA expression cassettes or chemically synthesized and transfected into cells or whole animals, which is associated with technical challenges similar to the delivery of standard antisense oligonucleotides.

Cross-References

▶ Antagonist
▶ Antisense Oligonucleotides
▶ Gene Expression
▶ Gene Transcription
▶ Genetically Modified Animals
▶ Ribozyme
▶ siRNA

Rodent Tests of Cognition

Trevor W. Robbins
Department of Psychology, University of Cambridge, Cambridge, UK

Synonyms

Rodent behavioral test paradigms or procedures; Rodent models of attention, memory, and learning; Rodent models of cognition

Definition

Rodent tests of cognition are behavioral means of assessing different components of cognition, including memory, learning, and attention, for evaluating the ability of drugs to remediate impaired cognitive functions or to produce cognitive enhancement.

Principles and Role in Psychopharmacology

Cognition is currently a much-employed term in psychopharmacology, referring to a collection of higher-order processes that intervene between sensory processing and motor output to produce behavior. Cognition is thus not a unitary construct and has to be carefully decomposed into its constituents, which can be modeled in terms of well-designed procedures that provide objective measures with good test-retest reliability. These constituents are derived from theories that provide operational definitions of constructs such as perception, attention, working memory, associative learning, and executive control. Further requirements are for tests that are validated in terms of their presumed psychological processes, neural basis, and sensitivity to drug effects. The ultimate requirement is to find procedures that can predict cognitive-enhancing effects in humans with neurological or neuropsychiatric disorders. A secondary consideration is for the procedures to be sensitive to detrimental effects of certain drugs, neurotoxins, or other manipulations to provide models that can be remediated by appropriate drug treatments. The final, and perhaps key, consideration is that the tests have some translational validity for humans (McArthur and Borsini 2008a, b). The last is a contentious issue, as it is of course unclear to what extent cognitive functions in rodents might map onto functionally homologous processes in humans. Nevertheless, there has been an excellent degree of translatability thus far in certain domains, which bodes well for the further development of this discipline.

Although cognition can be segregated into its different aspects, it is important in any evaluation to measure a range of functions to be best able to define the functional selectivity of any drug effect; for example, is an effect on a memory or learning test in fact dependent on a perceptual or attentional effect of the drug? Moreover, one has to be certain that basic sensory, motor, motivational, or sedative actions are not in fact responsible for any behavioral change. This can be achieved by a battery approach, in which performance on several tests with different requirements is compared, or by the more elegant and economic method of incorporating "control" tests of such factors as motivation and sensorimotor capacity within a test of a cognitive construct such as attention or working memory. Several examples of this will be provided later. The possibility of preclinical cognitive test "batteries" is currently popular in the wake of such approaches to measure cognition in clinical trials and experimental studies in humans, embodied, for example, by the MATRICS battery for schizophrenia and more generally by the CANTAB battery.

Perception

This is generally best tested by means of discrimination learning or performance, whereby responding in the presence of one stimulus is reinforced, whereas the other is not. The stimuli have to be presented randomly across two locations to avoid a confound by spatial factors. The test can be made more sensitive by varying the degree of similarity of the stimuli and also using a titration method to determine the limits of discrimination (by which the stimuli are made more similar following a correct response and less so by an incorrect one). By this means, it may be feasible to determine a psychological threshold for detection. Of course, one has to rule out other factors such as motivation that might produce possible changes (often deficits) in perceptual function as a result of a drug effect. If a deficit, for example, occurs regardless of level or type of motivation, then a perceptual impairment is more likely. Comparison of different types of discrimination performance, for example, in the visual, auditory, or olfactory modalities will also serve to test the specificity of a perceptual explanation. Similarly, learning factors are also ruled out if the deficit occurs following training to asymptotic performance. It is quite difficult however to distinguish between an effect on perception and attention, unless special manipulations are used to influence the latter. A useful quantitative technique for separating sensory/perceptual from motivational or other response biasing factors is signal detection theory, which has been applied with success to determine whether drugs affect primarily perceptual/sensory factors (d') or response bias (β)

Rodent Tests of Cognition, Fig. 1 Schematic of the rodent five-choice task (See Robbins (2002); Figure provided by courtesy of Dr. Christelle Baunez)

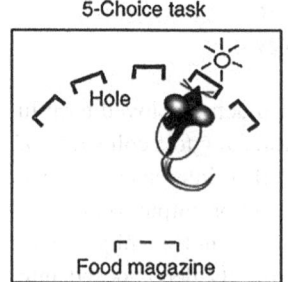
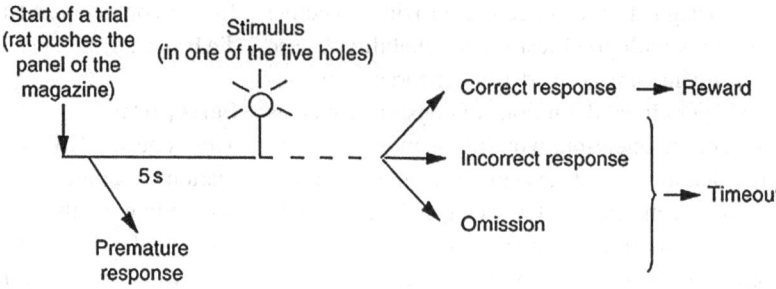

(Appel and Dykstra 1977). Finally, operant psychologists tend to avoid using such terms as "perception" in such tasks and prefer the more theoretically neutral term, "stimulus control." However, a loss of stimulus control is not necessarily due to perceptual factors.

Attention

There are many forms of human attention, including selective attention (focusing on one input or feature, while ignoring the rest), sustained attention (maintaining attention over a long period), vigilance (detecting rare inputs), and divided attention (maintaining attention to more than one input or task). Most, if not all, of these can be measured in experimental animals. Continuous performance tests measure the capacity to sustain attention and generally reveal impairments in disorders such as schizophrenia or attention-deficit hyperactivity disorder. A simple analogue of this in experimental animals is the five-choice serial reaction-time task (Robbins 2002), based on a paradigm once used to assess attention in human volunteers in a variety of experimental situations, including stress, distracting white noise, and following drug treatment. The five-choice task (Fig. 1) measures the accuracy (errors of commission) and latency of detecting visual targets, as well as errors of omission and impulsive responding (i.e., responding prior to target onset). The latency to collect food pellets provides a control measure of motivation. The difficulty of the task can be enhanced in various ways, including shortening of the duration of the visual target, varying its rate of presentation and temporal predictability, and also the occurrence of defined distractors, such as burst of white noise interpolated into the intertrial interval. This task has now been widely used in rats and, more recently, mice to measure effects of drug, regional brain lesions, and manipulations of the central neurotransmitters or genetic mutations. Its major uses have been to reveal beneficial effects on response accuracy of some putative "cognitive-enhancing" drugs such as dopamine D1 agonists and also to characterize the neuropharmacology of impulsive behavior, which has also been shown to predict escalation of cocaine self-administration. There are several variants of this standard task, the main one of which requires the rat to make an observing response into a central location to detect a peripheral target and has also been used to quantify the "attentional neglect" that can occur after unilateral manipulations of corticostriatal brain regions. A rather different form of test requires the cross-modal integration of auditory and visual stimuli.

Some tests of rodent selective attention appear to mimic strongly specific tests in humans that are sensitive to frontal lobe damage, involving the formation and shifting of attentional "sets" and the capacity to avoid a prepotent response to one aspect of a stimulus in order to respond to another (Birrell and Brown 2000). The attentional set-shifting task is based on the use of compound visual stimuli (i.e., that vary in at least two perceptual dimensions). Humans or nonhuman primates are trained to attend to one dimension on the basis of reinforcement and to ignore the other one, including tests of reversal (where the two stimuli within a dimension have their reinforcement contingencies reversed) and intra-dimensional shifting (where novel stimuli are introduced, but the same dimension is reinforced). Finally, an extra-dimensional shift is arranged in which novel stimuli are again introduced, but now, the previously irrelevant dimension is reinforced. This latter stage is analogous to the category shift on the Wisconsin Card Sort Test, which is much used to assess cognitive flexibility in human patient populations, especially those with presumed damage to the prefrontal cortex. In the rodent version (being available for mice as well as rats), the test is implemented using olfactory cues and texture in a "digging for food" test paradigm. Performance across the various stages is qualitatively comparable to that seen in primates; the extra-dimensional shift is the most sensitive stage to drug effects, performance at other stages usually being employed as internal controls. There are now various versions of these tests of "cognitive flexibility," which use similar logic for shifts between, for example, responding according to body turns or to space on a cross-maze and alternatively attending to discrete (e.g., visual) cues versus contextual cues on a maze. These tests do not use different stimuli at each test and so are also confounded by any response to interference. However, they do exemplify the general requirements of tests of executive function (see below).

Another major source of tests of attention comes from animal learning theory, in which the repeated non-reinforced presentation of stimuli retards subsequent learning about them, a process called ▶ latent inhibition. One theory of latent inhibition ascribes the loss of salience of the preexposed stimuli to a loss of attention, although it is possible that an associative account might suffice. However, a particular advantage of latent inhibition paradigms, which are very sensitive to dopamine D2 receptor antagonists (i.e., antipsychotic drugs), is that impairments are indexed subsequently by successful learning, in drug-free state, thus controlling for many other explanations of impairment.

Learning

Both Pavlovian and ▶ instrumental conditioning are now considered to have cognitive aspects, given that the basis for the former is prediction and expectancy and for the latter cognitive control over environmental contingencies (see Gluck et al. 2007). The detection of instrumental contingency, in particular, can be thought of as a higher-order cognitive process, which plays an important component of our ability to make voluntary actions that form part of goal-directed behavior. Disruptions of aspects of Pavlovian and instrumental learning almost certainly underlie all of the major forms of neuropsychiatric disorder, including drug addiction. However, these forms of conditioning will be considered elsewhere in more detail under specific entries. Learning is generally measured in simplified chambers or operant settings (especially for rats); however, maze learning is often employed when spatial cognition is the main subject of study, being especially compatible with the well-developed foraging tendencies of rodents.

Memory

This section serves to overview the many distinct forms of memory paradigms used in rodents, which are considered in greater detail under individual entries (see Gluck et al. 2007; Morris 2007). Memory can be divided in many ways; a basic distinction is between relatively transient short-term memory and long-term or more permanent memory. Memory "traces" are thus hypothesized to be "consolidated" into long-term memory.

Another distinction that has been made in human long-term memory by Tulving, "episodic" (generally autobiographical, the "what, where and when" of memory) versus semantic memory (memory for meaning), has not been exploited so far to any great extent in rodents.

Perhaps the greatest contribution made to the study of memory from rodent studies has been the posttrial or posttraining paradigm popularized by McGaugh and Roozendaal (2009). Here, what is generally a single trial or training session is *followed* immediately by a drug treatment that can either be amnestic (e.g., protein synthesis inhibitors) or promnestic (e.g., amphetamine). Retention is tested on a subsequent trial, perhaps 24 h or 3 days later. The posttraining manipulation is thus designed to influence consolidation, either beneficially or adversely. The procedure most often used is of aversive memory; the rodent is punished for stepping down from a platform or through a door by presentation of electric footshock. Memory is expressed on the retention trial by a longer response latency to step down or through the door. The great advantage of this design is that the drug cannot be said to have affected memory indirectly by its actions on perceptual, attentional, or motivational mechanisms, as it is administered at a time when these no longer impinge on learning. It is necessary however to perform controls with longer posttrial treatments to check that the drug effects are not affecting retention proactively (i.e., by being active at the time of retention and affecting memory retrieval). Studies of the consolidation of appetitive memory are also feasible but are used less often because of the unreliability of one-trial appetitive learning. The posttrial paradigm has been employed, for example, to demonstrate the contribution of noradrenergic, opioidergic, and GABAergic mechanisms to emotional memories laid down in the amygdala.

Reference memory is a form of long-term memory, which refers to rodent task requirements that stay constant from trial to trial. This definition was originally applied by Olton to rats remembering the constant location of the food-baited arms in an eight-arm radial maze. However, it can also be applied to the Morris water maze, a notable assay of hippocampal function, in which rodents are required over a number of learning trials to learn the location of a hidden platform in order to escape from a vat of water (D'Hooge and De Deyn 2001; Morris 2007). The rodent is allowed to swim the maze beginning from different vantage points, and so successful learning depends on the construction of a "cognitive map" to navigate the environment. In contrast, the term "working memory" in the Olton maze refers to the requirements of another memory test procedure in which rodents are required to visit each of the eight arms once and once only in order to retrieve a maximum of eight pellets. So, the animals have to remember only where they have recently been, and this memory is irrelevant to performance on subsequent test days. It can be argued that this form of "working memory" is not quite the same as that defined by human memory theorists such as Baddeley, where there is a coordination of different, modality-specific short-term memory buffers for use in various tasks such as planning, linguistic discourse, and logical reasoning. However, it does seem to overlap the human form of working memory in some important respects (see also Ko and Evenden (2009)).

Olton's working memory tasks are strongly reminiscent of the tests of spatial delayed response and delayed alternation that have been used to establish the role of the primate prefrontal cortex in working memory. Delayed alternation in rodents is easily implemented in a maze or operant chamber, where it is often referred to as "delayed nonmatching to position." Nonmatching is an easier task for rodents than matching because of their preexisting foraging tendency to alternate spatial choices. The operant versions of the task allow the systematic variation of delay intervals, which can extend from 0 to 60 s. A "delay-dependent" effect in such a task is generally taken as evidence of a specific memory effect, independent, for example, of attention. However, for that inference to be valid, it is necessary for performance on the task at zero seconds to be shown not to be similarly susceptible when the perceptual difficulty of the task is enhanced. An additional artifact that is difficult to surmount in the operant task is that of mediating responses, by which the rodent adopts postures or positions that minimize the memory requirement of the task. One way of overcoming this problem is to use sensitive touch screens to record responding, which can more precisely vary the spatial requirements of the memory tasks, as in the CANTAB battery for humans and nonhuman primates.

Recognition memory tasks have a superficial resemblance to those employed for working memory. A commonly used variant is that of object recognition (devised by Aggleton and others based on the paradigm of Delacour) in which a rodent explores a novel object during a sample trial and is then given a choice between this familiar object and a novel object, in terms of the amount of time it allocates to exploring both objects. Lesser exploration of one object indicates greater familiarity and hence recognition of it (in some, restricted sense). The test can also be adapted to measure "social recognition" by using experimental animals as the "object." Recognition memory is generally manifested over long delays, up to 24 h, although it can be tested at much shorter intervals also and has been shown to depend on structures such as the rodent perirhinal cortex. What distinguishes such tasks from those used to test working memory, apart from the precise test material, is that recognition memory tasks generally employ stimuli only once, so that the test is "trial unique." If the same set of objects were to be used over many trials (as occurs in the spatial delayed alternation or delayed response task), this would produce considerable proactive interference, and the test therefore becomes one of recency memory (how *recently* the stimulus has been experienced) rather than one of recognition memory. In that case, the test also becomes one more of frontal rather than temporal lobe function.

Recognition memory is a less sensitive test of memory than either cued or free recall, in which the memory has to be generated from long-term memory store. Unlike recall, recognition is not particularly sensitive to hippocampal damage and nor is it the earliest manifestation of Alzheimer's disease, where amnesia for episodic memories is more evident. Some human and primate data indicate that the hippocampus is implicated in forms of associative memory, particularly in animals involving space, for example, remembering the location of objects. Recent advances have begun to focus on these forms of associative memory, building on the classical Morris water maze. For example, recent touch-screen tests have been developed for mice and rats of their ability to remember the location of objects and also to form associative memories of tastes with specific locations (Bussey et al. 2008).

Recognition and recall correspond to what Squire has denoted for humans as "declarative memory" as distinct from "procedural memory" (memory for "how" or for "skill"). We have not discussed procedural memory in any detail in this article, but it may readily be tested in rodents in motor-learning situations such as the rotor-rod test or as memory for "habits," being part of the process of instrumental learning.

Executive Functions

In human cognitive neuropsychology, this term refers to that set of cognitive processes that serve to optimize performance. The term embraces a number of functions already described earlier, including cognitive flexibility (e.g., attentional set shifting), response inhibition, working memory, planning, decision making, and aspects of social cognition, some of which are difficult to test in rodents (but see Crawley 2007).

A recent development has been a rodent analogue of Logan's "stop-signal" task, which is much used in the testing of patients with attention-deficit hyperactivity disorder, for assessing their impulsivity. The stop-signal task requires the cancellation or termination of a motor response that has already been initiated by a "Go" cue, by a less frequently occurring "Stop" cue. It is feasible to measure a stop-signal reaction time, as well as a conventional (go) reaction time. The rodent stop-signal task (Fig. 2) so far has shown a similar pharmacology to that of the human task, both being sensitive, for example, to the beneficial effects on stopping of the selective noradrenaline uptake blocker atomoxetine (Eagle et al. 2008). Intriguingly, the classical go/no-go task, which has formal

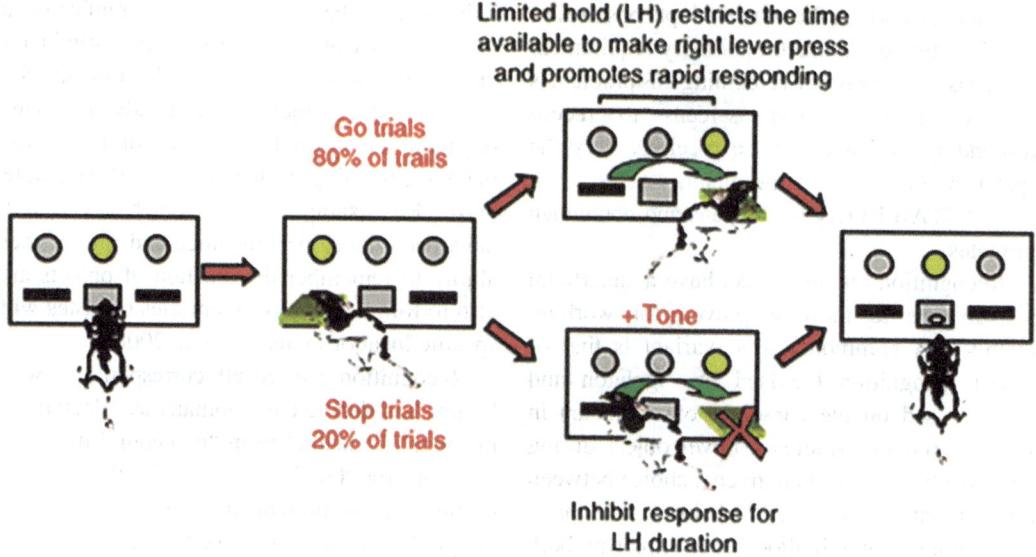

Rodent Tests of Cognition, Fig. 2 Schematic of the rodent stop-signal reaction-time task (See Eagle et al. (2008); Figure provided by courtesy of Dr. Dawn Eagle)

similarities with the stop task, appears to be differentially sensitive to certain drug effects in a manner suggesting the mechanisms of stopping a motor response to be distinct from those of selecting it (Eagle et al. 2008).

Until recently, it was considered difficult to implement a version of the human Stroop task for rodents. The human Stroop requires subjects to resist responding to the dominant aspect of a colored word stimulus (the word itself) and to report the color. A similar conflict has been achieved in rats by training them on two different conditional discriminations with different rules for responding to each discriminative stimulus and then placing the rat into conflict by presenting combinations of the discriminative stimuli that are either congruent (i.e., both stimuli indicate the same response or incongruent, both stimuli indicating opposing options) (Haddon and Killcross 2006).

Decision-making cognition has attracted considerable recent interest, in the context of both neuropsychological studies of human patients and in neuroeconomics. This trend is followed in rodent studies by the provision of tests of important component processes, including the discounting of rewards over time, as well as time perception itself. Recently, rat analogues of the Iowa Gambling Task, as used to show deficits in patients with lesions of the ventromedial prefrontal cortex, have begun to be implemented, but there are, as yet, few pharmacological studies.

Cross-References

▶ Animal Models for Psychiatric States
▶ Attention
▶ Attention Deficit Hyperactivity Disorders: Animal Models
▶ Behavioral Flexibility: Attentional Shifting, Rule Switching, and Response Reversal
▶ Blocking, Overshadowing, and Related Concepts
▶ Consolidation and Reconsolidation
▶ Cycloheximide
▶ Declarative and Nondeclarative Memory
▶ Instrumental Conditioning
▶ Latent Inhibition
▶ Long-Delay Learning
▶ Long-Term Depression and Memory
▶ Long-Term Potentiation and Memory
▶ Primate Models of Cognition
▶ Puromycin

- ▶ Rate-Dependency Theory
- ▶ Reconsolidation
- ▶ Schizophrenia: Animal Models
- ▶ Short-Term and Working Memory in Animals
- ▶ Spatial Learning in Animals
- ▶ Spatial Memory

References

Appel JB, Dykstra LA (1977) Drugs, discrimination and signal detection theory. In: Thompson T, Dews PB (eds) Advances in behavioral pharmacology, vol 1. Academic, New York, pp 139–166

Birrell JM, Brown VJ (2000) Medial frontal cortex mediates perceptual attentional set shifting in the rat. J Neurosci 20:4320–4324

Bussey TJ, Padain TL, Skillings EA, Winters BD, Morton AJ, Saksida LM (2008) The touch-screen cognitive testing method for rodents: how to get the best out of your rat. Learn Mem 15:516–523

Crawley J (2007) What's wrong with my mouse? Behavioral phenotyping of transgenic and knockout mice. Wiley, Hoboken, ISBN 2

D'Hooge R, De Deyn PP (2001) Applications of the Morris water-maze in the study of learning and memory. Brain Res Brain Res Rev 36:60–90

Eagle DM, Bari A, Robbins TW (2008) The neuropsychopharmacology of action inhibition: cross-species translation of the stop-signal and go/no-go tasks. Psychopharmacology (Berl) 199:439–456

Gluck M, Mercado E, Myers CE (2007) Learning and memory: from brain to behaviour. Worth Publishers, New York

Haddon J, Killcross S (2006) Prefrontal lesions disrupt the contextual control of response conflict. J Neurosci 26:2933–2940

Ko T, Evenden JL (2009) The effects of psychotomimetic and putative cognitive-enhancing drugs on the performance of an n-back working memory task in rats. Psychopharmacology (Berl) 202:67–78

McArthur RA, Borsini F (2008a) Animal and translational models for CNS drug discovery. Psychiatric disorders, vol 1. Academic/Elsevier, Amsterdam

McArthur RA, Borsini F (2008b) Animal and translational models for CNS drug discovery. Neurological disorders, vol 2. Academic/Elsevier, Amsterdam

McGaugh JL, Roozendaal B (2009) Drug enhancement of memory consolidation: historical perspective and neurobiological implications. Psychopharmacology (Berl) 202:3–14

Morris RGM (2007) Theories of hippocampal function. In: Andersen P, Morris RGM, Amaral D, Bliss T, O'Keefe J (eds) The hippocampus book. Oxford University Press, New York, pp 581–713

Robbins TW (2002) The five-choice serial reaction time task: behavioural pharmacology and functional neurochemistry. Psychopharmacology (Berl) 163:362–380

Rolipram

Definition

Rolipram is a PDE4 inhibitor. It is being widely researched as a possible alternative to current antidepressants. In addition, it has been suggested to improve memory and hence could be beneficial in patients with mild Alzheimer's disease either directly or via a neuroprotective mechanism. Like most PDE4 inhibitors, it also has anti-inflammatory properties. However, rolipram did not reach the market due to emetic side effects including nausea and vomiting.

Cross-References

- ▶ PDE4 Inhibitors
- ▶ Phosphodiesterase Inhibitors

Ropinirole

Synonyms

4-[2-(dipropylamino)ethyl]-1,3-dihydro-2H-indol-2-one; Adartrel; Requip; Ropark

Definition

Ropinirole is a non-ergoline D_{2-4} dopamine receptor agonist indicated in the treatment of ▶ Parkinson's disease (PD), restless legs syndrome, and antipsychotic drug-induced extrapyramidal motor symptoms. It has been approved by the Food and Drug Administration for use in the United States. Ropinirole is prescribed as a monotherapy for the treatment of early PD or an adjunct therapy with levodopa for advanced PD. Potential side effects of ropinirole are hallucinations, nausea, vomiting, constipation, hypotension, anxiety, dizziness, and impulse control behaviors such as gambling and hypersexuality.

Rotigotine

Synonyms

(S)-6-[propyl(2-thiophen-2-ylethyl)amino]-5,6,7,8-tetrahydronaphthalen-1-ol; Neupro

Definition

Rotigotine is prescribed as a transdermal patch for the treatment of ▶ Parkinson's disease and restless legs syndrome. It functions via partial/full agonism at D_{2-5} dopamine receptors at clinically relevant doses, although it has affinity for several receptor subtypes (adrenaline, serotonin, and histamine). Side effects of rotigotine include insomnia, confusion, hallucinations, nausea, vomiting, constipation, dizziness, and impulse control behaviors such as hypersexuality, gambling, and punding. Rotigotine has been approved by the Food and Drug Administration for use in the United States.

Cross-References

▶ Dopamine Agonist

RSA

Synonyms

Hippocampal EEG domain; Theta rhythm

Definition

RSA is the acronym of rhythmic slow activity, seen in various mammalian species for intra-hippocampal recordings. Upon visual inspection, one can recognize easily these quasi-sinusoidal ▶ field potentials that are in the range from 4 to 7 Hz, somewhat higher in rodents (in the rat, e.g., up to 12 Hz), named commonly theta rhythm.

Cross-References

▶ Electroencephalography

Runway Procedure

Definition

In the form of these procedures used in connection with the reinstatement of drug self-administration, the dependent measure is the run time from a *start box* to a *goal box* associated with a reward (e.g., food or drug). During daily training sessions, food-restricted rats are given food pellets when they reach the goal box, and over sessions, they decrease their run time. During the subsequent extinction sessions, food is not available in the goal box, and over daily sessions, the rats' run time progressively increases (reflecting extinction). A single exposure to food pellets during extinction leads to a faster run time on the subsequent daily session; this decrease in run time is the operational measure of food-priming-induced reinstatement. A unique feature of the runway reinstatement procedure is that the effects of pharmacological manipulations are assessed on a subsequent, drug-free day. This helps to rule out the possibility that any observed effects of pharmacological manipulations on reinstatement are due to locomotor activation or sedation during testing.

S

Saporin

Definition

A neurotoxin that is used in conjugation with different ligands to selectively deplete target neurons.

Satiety

Definition

The end state of satisfaction after consumption of a meal. The further suppression of the drive to consume post-meal (*between-meal inhibition*).

Cross-References

▶ Hunger

Saturation Binding Curve

Synonyms

Labeled ligand concentration binding isotherm

Definition

Isotherm: a line that connects points of equal temperature or points of measurements performed at equal temperature. The specific binding of the labeled ligand to tissue samples is measured at a given temperature at various concentrations of the labeled ligand. The values of the specific binding, on the ordinate, are plotted versus the labeled ligand's free concentration on the abscissa; connecting the points results in a curve with the form of a hyperbola. At a certain concentration of the labeled ligand, all available specific binding sites in the tissue sample will be occupied by the labeled ligand and specific binding will not further increase with increasing labeled ligand concentration; saturation of specific binding is reached.

Cross-References

▶ Receptors: Binding Assays

Scalar Property

Definition

Scalar property refers to the linear relationship between the estimation error and the estimated duration. In most species investigated, including humans, the estimation error (standard deviation) increases linearly with the estimated duration across a large range of intervals (see Gibbon et al. 1997). Although currently thought to be a fundamental property of ▶ interval timing, the scalar property may reflect underlying principles of neural processing of information (Buhusi and Oprisan 2013).

References

Buhusi CV, Oprisan SA (2013) Time-scale invariance as an emergent property in a perceptron with realistic, noisy neurons, Behavioral Processes, 95:60–70 (http://dx.doi.org/10.1016/j.beproc.2013.02.015)

Gibbon J, Malapani C, Dale CL, Gallistel CR (1997) Toward a neurobiology of temporal cognition: advances and challenges. Curr Opin Neurobiol 7:170–184

Schedule I

A category of controlled substances outlined in the US Comprehensive Drug Abuse Prevention and Control Act of 1970. Schedule I drugs:

- Have a high potential for abuse
- Have no currently accepted medical use in treatment in the USA
- Have a lack of accepted safety for use under medical supervision

Examples of Schedule I drugs include crack cocaine, GHB, heroin, LSD, MDMA

Cross-References

- ▶ Comprehensive Drug Abuse Prevention and Control Act of 1970
- ▶ Crack
- ▶ Cocaine
- ▶ Hallucinogens
- ▶ Methylenedioxymethamphetamine (MDMA)
- ▶ Oxybate

Schedule II

Definition

A category of controlled substances outlined in the US Comprehensive Drug Abuse Prevention and Control Act of 1970. Schedule II drugs:

- Have a high potential for abuse
- Have a currently accepted medical use in treatment in the USA or a currently accepted medical use with severe restrictions
- May lead to severe physiological or physical dependence

Examples of Schedule II drugs include amphetamine, cocaine, morphine, oxycodone, and methylphenidate (Ritalin).

Cross-References

- ▶ Amphetamine
- ▶ Cocaine
- ▶ Comprehensive Drug Abuse Prevention and Control Act of 1970
- ▶ Methylphenidate and Related Compounds
- ▶ Morphine
- ▶ Oxycodone

Schedule of Reinforcement

Definition

Describes the relationship between responding by a subject and the delivery of a reward to that subject, for example, in a fixed ratio schedule a reward is delivered after every ten responses.

Schedule-Induced Polydipsia

Synonyms

Adjunctive behavior; Adjunctive drinking

Definition

Schedule-induced polydipsia is excessive drinking induced by a particular schedule of reinforcement. Typically, the schedule involves the delivery of a food pellet at predictable intervals (e.g., fixed-interval 60 s).

Cross-References

▶ Operant Behavior in Animals

Schizoaffective Disorder

Andreas Marneros
Klinik und Poliklinik für Psychiatrie,
Psychotherapie und Psychosomatik, Martin Luther
Universität Halle-Wittenberg, Halle, Germany

Definition

There is no agreement about how schizoaffective disorders should be defined. According to the World Health Organization (WHO) and its International Classification of Mental and Behavioral Disorders (ICD-10), schizoaffective disorders are "episodic disorders in which both affective and schizophrenic symptoms are prominent but which do not justify a diagnosis of either schizophrenia or depressive or manic episodes." The diagnostic criteria are the following:

G1	The disorder meets the criteria of one of the affective disorders (F30, F31, F32) of moderate or severe degree, as specified for each category
G2	Symptoms from at least one of the groups listed below must be clearly present for most of the time during a period of at least 2 weeks:
	Thought echo, thought insertion or withdrawal, thought broadcasting
	Delusions of control, influence or passivity, clearly referred to body or limb movements or specific thoughts, actions or sensations
	Hallucinatory voices giving a running commentary on the patient's behavior or discussing the patient between themselves or

(continued)

	other types of hallucinatory voices coming from some part of the body
	Persistent delusions of other kinds that are culturally inappropriate and completely impossible, but not merely grandiose or persecutory, e.g., has visited other worlds, can control the clouds by breathing in and out, and can communicate with plants or animals without speaking
	Grossly irrelevant or incoherent speech or frequent use of neologisms
	Intermittent but frequent appearance of some forms of catatonic behavior, such as posturing, waxy flexibility, and negativism
G3	Criteria G1 and G2 above must be met within the same episode of the disorder and concurrently for at least part of the episode. Symptoms from both G1 and G2 must be prominent in the clinical picture
G4	The most commonly used exclusion clause. The disorder is not attributable to organic mental disorder or to psychoactive substance-related intoxication, dependence, or withdrawal

ICD-10 defines three different types of schizoaffective disorders: manic type (F25.0), depressive type (F25.1), and mixed type (F25.2).

The American Psychiatric Association (APA) in the fifth revision of its Diagnostic and Statistical Manual of Mental Disorders (DSM-5) defines the following diagnostic criteria for schizoaffective disorder:

A	An uninterrupted period of illness during which there is a major mood episode (major depressive or manic) concurrent with criterion A of schizophrenia
B	Delusions or hallucinations for 2 or more weeks in the absence of a major mood episode (depressive or manic) during the lifetime duration of the illness
C	Symptoms that meet criteria for a major mood episode are present for the majority of the total duration of the active and residual portions of the illness
D	The disturbance is not attributable to the effects of a substance (e.g., a drug of abuse, a medication) or another medical condition

DSM-5 specifies two subtypes: bipolar type, if a manic episode is part of the presentation (major depressive episodes may also occur), and depressive type, if only depressive episodes are part of the presentation.

While the main problem with the ICD-10 definition of schizoaffective disorders is its longitudinal aspect, the problem with DSM-5 concerns both cross-sectional and longitudinal aspects. The problem with the cross-sectional definition of DSM-5 concerns the time span indicated in criterion B (delusions or hallucinations for 2 or more weeks in the absence of a major mood episode (depressive or manic) during the lifetime duration of the illness.). Obviously, that is an attempt of DSM-5 to separate schizoaffective disorders from psychotic mood disorders. The DSM-5 definition of depressive and bipolar disorders is broad, including even those with mood-incongruent symptoms (even first-rank schizophrenic symptoms). The chronological criterion, however, is rather arbitrary (2 weeks of psychotic symptoms without mood disorders is schizoaffective; less than 2 weeks is psychotic mood disorder). Yet, the beginning of a psychotic episode is hard to assess exactly. Every clinician knows that there is usually a gap of many days, weeks, or months between the beginning of a psychotic episode and admission to hospital. Reconstruction of the psychopathological picture, retrospectively, is fraught with difficulties. Given the likelihood that the psychotic period is underestimated, many patients who are really schizoaffective could be diagnosed as schizophrenic or as having psychotic mood disorder.

Furthermore, the intensity of both concurrent syndromes (mood and schizophrenic syndromes) can vary enormously during an episode; hence, it seems arbitrary to give chronological priority to the psychotic symptoms over the mood component. It is strange that DSM-5 rejected Jasper's hierarchical diagnostic principle, which suggested a diagnostic superiority of schizophrenic symptoms over affective symptoms, but, regarding the chronological criterion of the schizoaffective definition, obviously made an exception.

Considering what is known so far about schizoaffective disorders (see overviews in Marneros and Tsuang 1986, 1990), we suggest that the definition of schizoaffective disorders should contain two components: a cross-sectional and a longitudinal aspect.

The cross-sectional definition should be the definition of an episode, while the longitudinal definition should be that of a disease or disorder. The cross-sectional definition of a schizoaffective episode should be based on the simultaneous occurrence of symptoms of a schizophrenic and a mood episode, independent of the chronological manifestation. Thus, we agree with the definition of ICD-10, which yields three types of schizoaffective episodes: schizodepressive, schizomanic, and mixed ones.

The longitudinal definition of schizoaffective disorder should consider the sequential occurrence of mood and schizophrenic episodes during course. Longitudinal research demonstrates that the course of schizoaffective disorders can be very unstable because schizoaffective episodes, pure mood episodes, and pure schizophrenic episodes can each occur at different points in the patient's longitudinal course.

What are such disorders when viewed longitudinally? Are they considered to be mood disorders because of the pure mood episodes, schizophrenic disorders because of some pure schizophrenic episodes, or schizoaffective disorders because of some schizoaffective episodes? Relevant to this question is the finding that there are no differences between patients who have only had schizoaffective episodes and those in whom schizoaffective episodes occur along with pure mood and schizophrenic episodes. There are therefore no differences between the "concurrent" and the "sequential" type of schizoaffective disorder. Patients, who change from pure mood episodes to pure schizophrenic episodes and vice versa, do not differ from patients having schizoaffective episodes. In this sense, Marneros et al. suggest a longitudinal definition of schizoaffective disorders, including a concurrent and a sequential type, the "concurrent type" being characterized by the coincidence of schizophrenic and affective episodes and the "sequential type" being characterized by the longitudinal change from schizophrenic to affective episodes and vice versa (Marneros et al. 1989).

How essential it is to have a longitudinal definition of schizoaffective disorder has been illustrated by the Halle Bipolarity Longitudinal Study

(HABILOS), in which the investigators tried to allocate disorders with manic symptomatology to "pure mood disorders" or to "schizoaffective disorders" according to DSM-IV, ICD-10, and the "empirical definition" as described above. Applying the ICD-10 definition, only 8.3 % of the 277 patients could longitudinally be allocated to schizoaffective bipolar disorder and 36.1 % to affective bipolar disorder, while the majority of patients (55.6 %) could not be allocated longitudinally due to the occurrence of different types of episodes (schizophrenic, schizoaffective, affective) at different times.

Using the empirical definition with its cross-sectional and sequential aspects, however, all patients can be allocated: 36.1 %, as in the ICD-10 categorization, could be allocated to bipolar mood disorder and 63.9 % could be allocated to schizoaffective disorder.

Recent research has confirmed earlier assumptions that schizoaffective disorders occupy a position between affective and schizophrenic disorders with regard to relevant sociodemographic and premorbid features, as well as with regard to patterns of course, outcome, treatment response, and prophylaxis (Marneros et al. 1988, 1989).

It seems certain that schizoaffective disorders are not simply a type of schizophrenic disorder, although in some cases that are schizo-dominant, the relationship to schizophrenia is quite clear. With respect to the relationship between schizoaffective and mood disorders, the similarities are more compelling than the differences (Marneros and Tsuang 1986, 1990).

Role of Pharmacotherapy

Although the clinical relevance of schizoaffective disorders is – in spite of controversies – meanwhile well-established, their treatment has received less attention in pharmacological studies, especially double-blind studies, than other psychotic or nonpsychotic major mental disorders. One of the main reasons might be the problem of their definition and, most important for the pharmaceutical industry, the clinical fact that schizoaffective disorders usually need combined treatment with more than one substance, for example, with antipsychotics, antidepressants, and mood stabilizers. Pharmacological studies dealing with schizoaffective disorders mostly investigated them as a subgroup of schizophrenia and seldom as a subgroup of mood disorders. Pharmacological studies only on schizoaffective disorders are rare. Nevertheless, it can be said that schizoaffective disorders are the domain of antipsychotics and mood stabilizers (Baethge 2003; Jäger et al. 2009; Levinson et al. 1999; McElroy et al. 1999; Mensink and Slooff 2004).

All antipsychotics seem to be effective in the treatment of schizoaffective disorders, but some atypical antipsychotics like olanzapine, quetiapine, risperidone, or ziprasidone are superior or have some advantages in comparison to typical ones. The heterogeneity of the studies and the investigated populations do not permit a science-based statement on the topic. The clinical effectiveness of mood stabilizers like lithium, carbamazepine, or valproate has been reported in some, however heterogeneous, studies. Clinical experience is compatible with such a conclusion.

Pharmacotherapy depends on the type of schizoaffective disorder. The official types of schizoaffective disorder registered in ICD-10 are manic type, depressive type, and mixed type, and those in DSM-5 are bipolar type and depressive type. Empirical work and longitudinal investigations considering the course of the disorder over many years, however, support other subtypes:

1. Schizo-dominant type
2. Affective-dominant type
3. Bipolar type
4. Unipolar type
5. Sequential type

In the schizo-dominant type, the main medication must be an antipsychotic. In the affective-dominant type, mood stabilizers and antidepressants or antipsychotics are effective. The bipolar type is treated with antipsychotics combined with mood stabilizers, whereas the unipolar type needs to be treated with antipsychotics and antidepressants. The sequential type is totally ignored. The reasons are given at the beginning of this chapter. It is characterized by the occurrence of schizophreniform or mood episodes during its course. Treatment focuses on the treatment of the particular episode. The longitudinal treatment is a prophylactic one, mainly with mood stabilizers and antipsychotics (McElroy et al. 1999; Mensink and Slooff 2004).

Clinical studies have also reported a positive effect of electroconvulsive treatment (Swoboda et al. 2001). Other treatments like augmentation with L-thyroxine found only small benefit (Bauer et al. 2002). The role of psychological treatment in schizoaffective disorders has not yet been systematically investigated.

Conclusions

Schizoaffective disorder is a very common diagnosis in clinical practice, but not sufficiently investigated, especially with regard to treatment.

Cross-References

▶ Antidepressants
▶ Antipsychotic Drugs
▶ Bipolar Disorder
▶ Lithium
▶ Mood Stabilizers
▶ Schizophrenia

References

Baethge C (2003) Long-term treatment of schizoaffective disorder: review and recommendations. Pharmacopsychiatry 36:45–56
Bauer M, Berghofer A, Bschor T, Baumgartner A, Kiesslinger U, Hellweg R, Adli M, Baethge C, Muller-Oerlinghausen B (2002) Supraphysiological doses of L-thyroxine in the maintenance treatment of prophylaxis-resistant affective disorders. Neuropsychopharmacology 27:620–628
Jäger M, Becker T, Weinmann S, Frasch K (2009) Treatment of schizoaffective disorder – a challenge for evidence-based psychiatry. Acta Psychiatr Scand 121 (1):22–32. doi: 10.1111/j.1600-0447.2009.01424.x. Epub 2009 Jun 30.
Levinson DF, Umapathy C, Musthaq M (1999) Treatment of schizoaffective disorder and schizophrenia with mood symptoms. Am J Psychiatry 156:1138–1148
Marneros A, Tsuang MT (1986) Schizoaffective psychoses. Springer, Berlin
Marneros A, Tsuang MT (1990) Affective and schizoaffective disorders. Similarities and differences. Springer, New York
Marneros A, Deister A, Rohde A, Junemann H, Fimmers R (1988) Long-term course of schizoaffective disorders. Part I: definitions, methods, frequency of episodes and cycles. Eur Arch Psychiatry Neurol Sci 237:264–275
Marneros A, Deister A, Rohde A (1989) Unipolar and bipolar schizoaffective disorders: a comparative study. I. Premorbid and sociodemographic features. Eur Arch Psychiatry Neurol Sci 239:158–163
McElroy SL, Keck PE Jr, Strakowski SM (1999) An overview of the treatment of schizoaffective disorder. J Clin Psychiatr 60(Suppl 5):16–21, discussion 22
Mensink GJ, Slooff CJ (2004) Novel antipsychotics in bipolar and schizoaffective mania. Acta Psychiatr Scand 109:405–419
Swoboda E, Conca A, Konig P, Waanders R, Hansen M (2001) Maintenance electroconvulsive therapy in affective and schizoaffective disorder. Neuropsychobiology 43:23–28

Schizo-Dominant

Definition

Type of schizoaffective disorder dominated by schizophreniform symptoms.

Schizophrenia

Alex Hofer
Biological Psychiatry Division, Department of Psychiatry and Psychotherapy, Medical University Innsbruck, Innsbruck, Austria

Synonyms

Dementia praecox

Definition

Schizophrenia is a severe and persistent debilitating psychiatric disorder consisting of disturbances in thoughts, cognition, mood, perceptions, and relationships with others. According to DSM-IV-TR (American Psychiatric Association 2000), the patient must have experienced at least two of the following symptoms: delusions, hallucinations, disorganized speech, disorganized or catatonic behavior, or negative symptoms. The recently released DSM-5 (American Psychiatric Association 2013) stipulates that at least one of these symptoms must be delusions, hallucinations, or disorganized speech. Furthermore, the DSM-IV-TR recommendation that only one symptom is required if the delusions are bizarre or if auditory hallucinations occur in which the voices comment in an ongoing manner on the person's behavior or if two or more voices are talking with each other has been eliminated in DSM-5. The patient must experience at least 1 month of symptoms during a 6-month period, and social or occupational deterioration problems occur over a significant amount of time. These problems must not be attributable to another condition for the diagnosis of schizophrenia to be made.

By contrast, *schizophreniform disorder* is a short-term type of schizophrenia with the characteristic symptoms (including prodromal, active, and residual phases) being present for at least 1 month but not for the full 6 months required for the diagnosis of schizophrenia. In contrast to schizophrenia, the onset of schizophreniform disorder can be relatively rapid and the individual's level of functioning may or may not be affected. According to the American Psychiatric Association, about two-thirds of people with schizophreniform disorder do not recover and are subsequently diagnosed with schizophrenia.

Current Concepts and State of Knowledge

Historical Aspects

The clinical picture of schizophrenia was described for the first time by the German physician Emil Kraepelin who used the term "dementia praecox" to define a disorder with early beginning, a uniformly deteriorating course, and a poor prognosis (Kraepelin 1893). Subsequently, Eugen Bleuler coined the term "schizophrenia" and distinguished between fundamental and accessory symptoms of the disease (Bleuler 1908, see Table 1). He believed that the fundamental symptoms were present in all patients and were unique to the disorder while the accessory symptoms could also occur in other disorders. Bleuler realized that the condition was not a single disease and referred to a whole "group of schizophrenias." Some decades later, Kurt Schneider established the differentiation between first- and second-rank symptoms (Schneider 1959, see Table 2), representing a preliminary stage of contemporary operationalized classification systems.

In 1980, Timothy Crow postulated two dimensions of pathology underlying schizophrenia. According to his concept, the type I syndrome is mainly characterized by positive symptoms, potentially antipsychotic-responsive and reversible, while the type II syndrome is mainly

Schizophrenia, Table 1 Fundamental and accessory symptoms of schizophrenia (Bleuler 1908)

Fundamental symptoms	Accessory symptoms
Loosening of associations	Hallucinations
Disturbances of affectivity	Delusions
Ambivalence	Catatonic symptoms
Autism	Speech abnormalities (e.g., mutism, neologisms)

Schizophrenia, Table 2 First- and second-rank symptoms (Schneider 1959)

First-rank symptoms	Second-rank symptoms
Voices heard arguing	Other forms of hallucinations
Voices heard commenting on one's actions	Sudden delusional ideas
Audible thoughts	Perplexity
Thought insertion	Depressive or euphoric mood changes
Thought withdrawal	Emotional blunting
Thought diffusion	
Delusional perception	

characterized by negative and cognitive symptoms, sometimes progressive and relatively irreversible (Crow 1980, see Table 3).

The concepts described above have used a categorical approach, thereby indicating homogeneous, mutually exclusive subtypes of the disease. Essentially, schizophrenic disorders are heterogeneous and consequently, Peter Liddle has introduced a dimensional approach comprising three neuroanatomically classifiable syndrome clusters: psychomotor poverty, disorganization, and reality distortion (Liddle 1987, see Table 4).

Psychopathology

Disorders of Thinking and Speech

Basically, thought processes may be disordered in form and content. Common formal thought disorders observed in schizophrenia patients include incoherent thinking. The patient's thoughts are illogical and confused due to a defect in processing and organizing language ("schizophasia"). During *thought blocking*, thinking is decelerated and stagnant and the patient's language is diminished accordingly. Thought content may be diminished as well. On the other hand, the term *flight of ideas* describes excursive and uncontrollable thinking, during which associations become loose and mental activity is generally accelerated. Furthermore, *improper responding to questions*, perseverations, paralogism, neologisms, and concretism are commonly observed in schizophrenia patients.

Essentially, content-disordered thought processes are equivalent to delusions. They are characterized by abnormal, apparently unreasonable interpretations of one's own experiences and perceptions to which the person concerned adheres despite refutation by others ("▶ Delusional Disorder"). According to Karl Jaspers, these interpretations fulfill three main criteria: certainty, incorrigibility, and impossibility (Jaspers 1913). The most frequently observed delusions in schizophrenia patients are delusions of reference, of persecution and of guilt, as well as megalomania, nihilism, and delusions with religious content.

Many patients experience prepsychotic states and prodromal symptoms before the first episode of schizophrenia is apparent. For example, this stage of the illness includes attenuated psychotic symptoms (APS) or brief limited intermittent psychotic symptoms (BLIPS) ("▶ Pre-psychotic States and Prodromal Symptoms").

Disorders of Affect and Mood

Schizophrenia is characterized by abnormalities of affect, emotional response, and mood. In this

Schizophrenia, Table 3 Two-syndrome concept of schizophrenia (Crow 1980)

	Type I	Type II
Characteristic symptoms	Hallucinations, delusions, thought disorders (positive symptoms)	Affective flattening, poverty of speech, loss of drive (negative symptoms)
Type of illness in which most commonly seen	Acute schizophrenia	Chronic schizophrenia, the "defect" state
Response to antipsychotics	Good	Poor
Outcome	Reversible	? Irreversible
Intellectual impairment	Absent	Sometimes present
Postulated pathological process	Increased dopamine receptors	Cell loss and structural changes in the brain

Schizophrenia, Table 4 Dimensional approach (Liddle 1987)

Neuroanatomical dysfunction	Left dorsolateral prefrontal cortex	Medial temporal lobe	Right ventrolateral prefrontal cortex
Syndrome	Psychomotor poverty	Reality distortion	Disorganization
Symptoms	Poverty of speech	Delusions	Formal thought disorder
	Blunted affect	Hallucinations	Distractibility
	Slowness		Incongruous affect

Schizophrenia, Table 5 Symptoms of affective flattening (Andreasen 1987)

Unchanging facial expression
Decreased spontaneous movements
Paucity of expressive gestures
Poor eye contact
Affective nonresponsivity
Inappropriate affect
Lack of vocal inflections

context, the profound disturbance of emotional rapport perceived in an intuitive way by an experienced psychiatrist interacting with a schizophrenia patient was named "praecox feeling." Due to fluctuations in attention and misinterpretation of stimuli schizophrenia patients might be confused, depressed, and anxious. Sometimes, even situations of minimal novelty cause anxiety (novophobia), e.g., unfamiliar people, items, and conversations. Common objects might take on undue significance and therefore scare patients. A further fundamental symptom of the schizophrenia-related affective disorder is parathymia, the inappropriateness of facial expression, gesture, and speech which is contrary to the patient's real experience.

Affective flattening becomes manifest as a diminution of emotional response and indifference to events or topics that normally evoke such a response (Andreasen 1987, see Table 5).

Anhedonia refers to a pervasive and refractory reduction in the capacity to experience pleasure which becomes apparent in reduced leisure activities or sexual interest.

Ambivalence is characterized by simultaneous conflicting feelings toward a person or thing, e.g., love and hate or happiness and fear. These contradictory impulses are usually unconscious and uninterpretable for other people.

Depressive symptoms can be found in up to 50 % of schizophrenia patients. On the other hand, some patients develop manic conditions characterized by hyperactivity, a reduced ability to think critically, and an overestimation of their own capabilities.

Hallucinations

Hallucinations are conscious perceptions affecting the different senses in the absence of external stimuli. Auditory hallucinations are perceived as noise, words, sentences, whisper, or voices. They are apparent in about 70 % of schizophrenia patients and mainly become manifest in thoughts becoming aloud as well as commenting, dialogic, or imperative voices, respectively. Visual hallucinations occur relatively infrequently. Tactile hallucinations create the sensation of tactile sensory input and are perceived as touch, burning, or electrifying sensations. Coenesthesia is characterized by abnormal bodily sensations, e.g., parts of the body that are perceived as changing their shape or size. Olfactory and gustatory hallucinations are quite uncommon in schizophrenia patients but can be associated with delusions of poisoning.

Disorders of the Ego

Patients with schizophrenia experience themselves in a disordered manner and often believe that they are affected by external forces. Accordingly, the term "disorders of the ego" comprises various symptoms. For example, patients experiencing thought diffusion are convinced that other people know their thoughts. Other symptoms belonging to this category are thought withdrawal, thought insertion, depersonalization, and derealization.

Disorders of Psychomotor Functioning

Many patients with schizophrenia experience psychomotor disturbances even at an early stage of the illness. Accordingly, changes in facial expressions, gestures, posture, voice, and speech are often observed. Everyday activities have to be consciously considered by the patient (loss of automatisms), which is why patients appear mannered.

Catatonic phenomena comprise disorders of movement, speech, and autonomic function. These motor disturbances consist of hyper- or hypokinesias (excitement and inhibition, respectively) and parakinesias (abnormal postures, mannerisms, grimacing, stereotypies). Catatonic speech disorders include perseveration, echolalia, and mutism. Characteristic autonomic signs are dilatation of pupils, seborrhea, sweating, and alterations in muscle tone (rigidity or hypotonia, respectively). Catatonic stupor and excited states represent the extreme ends of the spectrum.

Cognitive Symptoms

Cognitive impairment is a cardinal feature of schizophrenia which is found in 60–80 % of patients. Generally, it is assumed that schizophrenia patients show deficits across a large number of neurocognitive domains including attention, executive functioning, memory, and fine motor skills. Performance has been reported to be two standard deviations below the mean of healthy control subjects. Cognitive variables have been related to the heterogeneity of functional outcomes, with difficulties in profiting from rehabilitation programs and with quality of life. Evidence suggests that cognitive deficits may be of equal or greater importance in predicting functional outcome as positive or negative symptoms.

Such cognitive disturbances are present both in children who have a schizophrenic parent ("high-risk children") and in first-degree relatives of patients, who do not suffer from a schizophrenic disorder. Furthermore, these deficits are apparent long before the onset of psychotic symptoms and they endure after a psychotic episode when patients are in remission. Therefore, cognitive deficits represent a possible trait marker of schizophrenia.

Next to neurocognitive impairment, schizophrenia has consistently been associated with deficits in the recognition, discrimination, and experience of emotional stimuli. Impairments in affect perception have been demonstrated in chronic and first-episode patients and their unaffected siblings but not in high-risk individuals with initial prodromes, and it has been suggested that the initial psychotic episode represents a critical point for the emergence of emotion perception deficits in schizophrenia spectrum illnesses. Several studies have shown that these deficits in emotion recognition are associated with illness-related measures such as duration of illness, symptomatology, symptom severity, and cognitive disturbances but are not influenced by age or antipsychotic treatment.

Several studies have shown significant and stable "theory of mind" (ToM) impairments in patients with schizophrenia. ToM refers to the ability to perceive other people's opinions, beliefs, and intents and to establish a connection between these mental states and a person's behavior. These deficits are probably independent of neurocognitive dysfunctions and they might contribute to social and behavioral deviations in schizophrenia patients.

Subtypes

Depending on the combination of symptoms, DSM-IV-TR and ICD-10 (World Health Organisation 1992) have defined different subtypes of schizophrenia:

Paranoid schizophrenia is characterized by relatively stable, often paranoid delusions, usually accompanied by (auditory) hallucinations and perceptual disturbances. Disturbances of affect, volition and speech and catatonic symptoms are not prominent.

Hebephrenic schizophrenia is dominated by affective changes with shallow and inappropriate mood. Thought is disorganized and speech rambling and incoherent, whereas delusions and hallucinations are fragmentary if present at all.

Catatonic schizophrenia is mainly determined by psychomotor disturbances with alterations between extremes such as hyperkinesias and stupor.

Undifferentiated schizophrenia meets the diagnostic criteria for schizophrenia but does not conform to any of the above subtypes or exhibits the features of more than one of them without a clear predominance of a particular set of diagnostic characteristics.

The diagnosis of post-schizophrenic depression ("▶ Postpsychotic Depressive Disorder of Schizophrenia") should be made if the patient has had a schizophrenic illness meeting the general criteria for schizophrenia within the past 12 months and some schizophrenic symptoms are still present. Depressive symptoms must be prominent and distressing, fulfilling at least the criteria for a depressive episode, and have been present for at least 2 weeks.

The term "residual schizophrenia" describes a chronic stage in the development of a schizophrenic disorder in which there has been a clear progression from an early stage (comprising one or more episodes with psychotic symptoms meeting the general criteria for schizophrenia described above) to a later stage characterized by long-term, though not necessarily irreversible, negative symptoms, i.e., psychomotor slowing, underactivity, blunting of affect, passivity and lack of initiative, poverty of quantity or content of speech, poor nonverbal communication by facial expression, eye contact, voice modulation, and posture, and poor self-care and social performance.

Simple schizophrenia represents an uncommon disorder characterized by a slowly progressive development of negative symptoms without any history of hallucinations, delusions, or other manifestations of an earlier psychotic episode and with significant changes in personal behavior, manifest as a marked loss of interest, idleness, and social withdrawal.

Due to their limited diagnostic stability, low reliability, and poor validity, these subtypes have been eliminated in DSM-5. Instead, a dimensional approach has been included for rating severity of the core symptoms of schizophrenia.

Epidemiology, Risk Factors, and Course of Symptoms

The incidence of schizophrenia in industrialized countries ranges from 10 to 70 new cases/100,000, whereas prevalence rates range from 1.6 to 12.1/1,000. The lifetime risk is 0.5–1 %. Both "social drift" and environmental risk factors (e.g., drug abuse, migration) might account for an increased prevalence in the lower socioeconomic classes.

Epidemiological studies have shown gender differences in the age of onset of schizophrenia. The peak age of onset is 10–28 years for men and 26–32 years for women. However, there is a second peak of onset in women after the menopause, resulting in an equal lifetime incidence

Schizophrenia, Table 6 Factors affecting prognosis

Favorable prognosis	Unfavorable prognosis
Acute onset of illness	Insidious onset of illness
Positive symptoms	Negative symptoms
Lack of family history of schizophrenia	Family history of schizophrenia
Inconspicuous premorbid personality	Poor premorbid personality
Average IQ	Low IQ
High social class	Low social class
Social integration	Social isolation
Lack of comorbidities	Comorbid drug abuse
Female sex	Male sex
Being married	Single marital status

for both genders. Affected men make contact with psychiatric services at an average of 5 years earlier than women.

Both twin and adoption studies indicate genetic effects in the liability to schizophrenia. Although the mode of transmission has not been clarified, several studies corroborate the so-called multifactorial polygenic model that involves an interaction between many contributing genes and environmental factors (obstetric complications, prenatal infection, neurodevelopmental abnormality, substance misuse, social and psychological factors).

Approximately 20 % of patients with schizophrenia improve without relapse and 40 % achieve a clinically stable residual state and are socially integrated, whereas approximately one-third of patients suffer from continuous psychotic symptoms and/or increasing social disability. Factors affecting prognosis are listed in Table 6.

Treatment Considerations

The discovery of chlorpromazine in the early 1950s introduced an era of effective pharmacological treatment for schizophrenia. Today, antipsychotics of different chemical structures, ranging from tricyclic phenothiazines to thioxanthenes, butyrophenones, dibenzazepines, substituted benzamides, benzisoxazole derivatives, and chinolones are considered to form the backbone of treatment ("▶ Antipsychotic Drugs," "▶ First-Generation Antipsychotics,"

"▶ Second and Third Generation Antipsychotics"). They shorten the length of an acute episode of the illness and reduce the risk of relapse. Clearly, modern concepts of schizophrenia management include psychosocial and rehabilitative measures, and pharmacotherapy should always be embedded in integrative treatment procedures.

Essentially, the prognosis of schizophrenia depends on when pharmacological treatment is started and on the number of psychotic episodes. Two-thirds of patients experiencing a first episode of the illness experience symptomatic remission within half a year if antipsychotic treatment is started immediately. In addition, consistent prophylaxis with antipsychotic medication results in a reduction of the 1-year relapse rate from 75 % to 20 %. Since less than 50 % of patients with schizophrenia in long-term treatment take their medication according to the physician's recommendations, interventions to enhance compliance should be implemented at the beginning of treatment (Fleischhacker et al. 2007).

Parenteral administration of rapid-acting antipsychotics should be confined to emergency situations, where acute agitation and lack of insight lead to a high risk of patients harming themselves or others. In contrast, the administration of long-acting injectable antipsychotics is an important treatment option for long-term management. The advantages of these injectable drugs include dose reduction due to avoidance of the first-pass effect, the fact that patients do not need to take medication every day, and the facilitation of management due to certainty concerning compliance. The disadvantages of this type of treatment include the fact that some patients refuse intramuscular injections or develop irritations at the injection site. Another problem is that the dose of a long-acting injectable antipsychotic cannot be rapidly reduced once administered.

Despite continuous new findings in the field of psychopharmacology, there are no reliable predictors for individual drug response/tolerability. Therefore, the choice of medication depends on an individual risk-benefit analysis. In patients who do not adequately respond to two antipsychotic drugs of different chemical classes, switching to clozapine is indicated. This compound has been shown to have unique efficacy in so far as it reduces symptoms in 30–60 % of treatment refractory schizophrenia patients (Fleischhacker 1999) ("▶ Antipsychotic Drugs"). However, although pharmacological agents have substantially advanced the treatment of schizophrenia and have had a significant impact on the lives of patients in terms of relapse prevention, quality of life, and resocialization, there are still considerable unmet needs in the management of these patients.

Cross-References

▶ Antipsychotic Drugs
▶ Delusional Disorder
▶ First-Generation Antipsychotics
▶ Hallucinations
▶ Postpsychotic Depressive Disorder of Schizophrenia
▶ Prepsychotic States and Prodromal symptoms
▶ Second- and Third-Generation Antipsychotics

References

American Psychiatric Association (2000) Diagnostic and statistical manual of mental disorders (DSM-IV-TR), 4th edn. American Psychiatric Association, Washington, DC
American Psychiatric Association (2013) Diagnostic and statistical manual of mental disorders, 5th edn. American Psychiatric Association, Arlington
Andreasen NC (1987) The diagnosis of schizophrenia. Schizophr Bull 13:9–22
Bleuler E (1908) Die Prognose der Dementia praecox (Schizophreniegruppe). Allgemeine Zeitschrift für Psychiatrie und psychisch-gerichtliche Medizin 65:436–464
Crow TJ (1980) Molecular pathology of schizophrenia: more than one disease process? BMJ 280:66–68
Fleischhacker WW (1999) Clozapine: a comparison with other novel antipsychotics. J Clin Psychiatry 60(12):30–34
Fleischhacker WW, Hofer A, Hummer M (2007) Managing schizophrenia: the compliance challenge, 2nd edn. Current Medicine Group, London

Jaspers K (1913) Allgemeine psychopathologie. Ein Leitfaden für Studierende, Ärzte und Psychologen. Springer, Berlin

Kraepelin E (1893) Psychiatrie. Ein Lehrbuch für Studierende und Ärzte, 4th edn. Abel, Leipzig

Liddle PF (1987) The symptoms of chronic schizophrenia. A re-examination of the positive–negative dichotomy. Br J Psychiatry 151:145–151

Schneider K (1959) Clinical psychopathology. Grune and Stratton, New York

World Health Organisation (1992) ICD-10 classifications of mental and behavioural disorder: clinical descriptions and diagnostic guidelines. World Health Organisation, Geneva

Schizophrenia Prodrome

Definition

Subthreshold psychotic symptoms that emerge prior to meeting full criteria for schizophrenia (or before relapse after remission of symptomatology).

Cross-References

▶ Prepsychotic States and Prodromal Symptoms
▶ Schizophrenia

Schizophrenia: Animal Models

Bart Ellenbroek
School of Psychology, Victoria University of Wellington, Wellington, New Zealand

Definition

Animal models for schizophrenia refer to animal preparations which attempt to mimic aspects of the human disorder, schizophrenia. In general, these animal models are subdivided into screening models, in which the therapeutic effects are modelled, and simulation models, in which the emphasis is on the symptoms of schizophrenia.

Current Concepts and State of Knowledge

Screening Models

In screening models, sometimes referred to as animal models for antipsychotic action, the focus is on the pharmacology of schizophrenia (Arnt 2000). The principal aim is to predict the therapeutic effects of novel drugs, based on the efficacy of antipsychotic drugs that have already been in clinical use. Such clinically successful drugs are often referred to as "gold standard." The development of a screening model therefore starts with the identification of an appropriate gold standard. In the early years, this gold standard was usually chlorpromazine or haloperidol. In more recent years, screening models often use two gold standards, a so-called classical (most often haloperidol) and an atypical antipsychotic drug (most often clozapine). The rationale behind this approach is that many clinical studies have shown differences between these two (classes of) antipsychotic drugs. Most notably, while classical antipsychotic drugs induce severe neurological (the so-called extrapyramidal) side effects, atypical antipsychotic drugs do not, or to a much lesser extent. Having two gold standards might help identify which aspects of the screening models are more related to the therapeutic efficacy of antipsychotic drugs (i.e., those aspects in which both gold standards show similar effects) and those that are more related to the side effects (i.e., those aspects that are induced by the classical but not by the atypical gold standard).

After having selected the appropriate gold standard (or standards), the drug is given to an animal, usually a rat or a mouse. In many cases this is a normal, naive, and untreated animal, but in some cases the animal is trained beforehand (such as in the conditioned avoidance response) or treated with a specific drug (such as a dopamine agonist or a glutamate antagonist) to induce a specific behavioral response. Table 1 lists the most often used screening models for schizophrenia.

The effects of the gold standard are then evaluated in this behavioral setup, and if a clear

Schizophrenia: Animal Models, Table 1 A summary of the most important screening models for schizophrenia and their validation on the basis of the criteria discussed in the text

Screening test	No false-positives	No false-negatives	Potency correlation	Anticholinergic	Chronic treatment
Spontaneous behavior					
Paw test	+	+	+	+	+
Learned behavior					
Conditioned avoidance response	−	+	+	−	±
Intracranial self-stimulation	−	+	±	−	+
Drug-induced behavior					
DA agonist-induced hyperactivity	−	±	+	±	Ø
DA agonist-induced stereotypy	−	−	−	−	−
DA agonist-induced prepulse inhibition deficit	±	+	+	−	+
NMDA antagonist-induced prepulse inhibition deficit	−	±	−	Ø	+

+ indicates that there is convincing evidence that the criterion is fulfilled; − indicates that there is convincing evidence that the criterion is not fulfilled; ± indicates that contradictory evidence exists in the literature regarding this criterion; Ø indicates that there is not enough literature to assess this criterion

reproducible response is obtained, the model can go into the next and most important phase of model development: the validation phase. In this most crucial phase of the model development, the validity of the model is established on the basis of clinical evidence. ▶ Antipsychotic drugs have been in clinical practice since the 1950s, and therefore, a vast knowledge about their clinical effects now exists. From this knowledge base, several criteria can be formulated that can be used to validate a screening model (Ellenbroek 1993). The two most obvious criteria are the absence of false-negatives and false-positives. False-negatives are drugs that in the model do not resemble the gold standard but in clinical practice are effective antipsychotic drugs. False-positives, on the other hand, are drugs that act like the gold standard in the model, but in clinical practice are not effective. Especially this criterion is difficult to test, since there are thousands of compounds available most of which have not been extensively tested in patients. Moreover, it was common practice that many pharmaceutical companies did not publish negative results from phase II or phase III studies, and in several instances, contradictory data have been published. The evaluation of this criterion is therefore usually based on a selection based on similarity in chemical structure or drugs with a known influence on motor activity (such as diazepam or morphine), since such drugs can interfere with behavioral performance in general. Another general criterion is that there should be a correlation between the potency of the antipsychotic drugs in the clinic and in the model. Finally, several other criteria can be specifically deduced from the clinical literature on antipsychotic drugs. One of the most interesting criteria is the lack of effect of anticholinergic drugs. The rationale behind this criterion is that while anticholinergic drugs do not influence the therapeutic effect of antipsychotic drugs, they do diminish the extrapyramidal side effects. Thus, anticholinergic drugs can be useful to distinguish between parameters modelling the therapeutic and the unwanted side effects. The same holds true for chronic treatment with antipsychotic drugs. It is well known that the therapeutic effect usually increases with prolonged treatment, whereas the extrapyramidal side effects usually wane with repeated administration. Table 1 tries to summarize the vast literature on the validation of the screening models for schizophrenia (Ellenbroek 1993).

Once the validation phase has (satisfactorily) finished, the model goes into the final stage in which novel compounds can be evaluated and compared to the gold standard(s). Once a promising new drug candidate has been identified (that also shows a good pharmacokinetic and safety profile), it can then be tested in clinical trials.

Although most of the currently available treatments for schizophrenia have been developed using screening models, they have received much less attention in recent years. The principal reason for this is that such screening models, however predictive they may be, will always produce drugs similar to the gold standard (often referred to as a "me too" approach). Indeed, this probably explains why there are so few differences between the currently available drugs. However, in spite of this disadvantage, screening models are still being used because they are usually quick and easy but also because they can help us to understand the mechanism of action of antipsychotic drugs and especially the differences between classical and atypical antipsychotic drugs.

Simulation Models

Since screening models will unlikely lead to a major breakthrough in the treatment of schizophrenia, attention has in recent years shifted more and more to simulation models. This shift was aided by the increasing clinical knowledge of the disease. In simulation models, one tries to mimic as many aspects of a disease as possible. In its simplest form, this requires two steps: (1) the induction of the disease and (2) the measurement of the symptoms. Unfortunately, both steps have proven to be very difficult in the case of schizophrenia. With respect to the symptoms, most of the cardinal symptoms (especially the so-called positive symptoms such as hallucinations and delusions) in humans can only be assessed in a psychiatric interview. Since such symptoms cannot be modelled in animals, researchers have resorted to negative symptoms (such as anhedonia and social withdrawal) and symptoms within the cognitive domain (such as attention and (working) memory). In addition, much attention has recently been paid to the so-called endophenotypes and potential biomarkers for schizophrenia. It is beyond the scope of this entry to discuss all the aspects of endophenotypes, but in general, they are physiological or psychological phenomena that have a genetic basis and are associated with the specific disease. The idea behind such endophenotypes is that they may be more related to specific brain structures and neurotransmitter functions than the symptoms of the disease. Moreover, they are usually easier to model in animals. In the case of schizophrenia, several endophenotypes have been proposed, such as smooth pursuit eye movement, P_{50} sensory gating, and ▶ prepulse inhibition. Especially the latter two can also quite easily be assessed in animals. Biomarkers are also increasingly considered to be of great value for diseases, although the traditional peripheral (for instance, blood or urine) analyses may not be very relevant for central nervous system disorder such as schizophrenia. However, with the advent of modern functional imaging, new possibilities have occurred. For instance, using PET and SPECT scan analyses, it has been shown that psychotic patients have an increased basal as well as amphetamine-enhanced dopamine release in the basal ganglia. Moreover, this increase in dopamine reactivity is site specific (occurring only in the anterodorsal part of the caudate nucleus) and can be observed in drug-naive, first-episode patients. Interestingly, it has recently been shown to even occur in ultrahigh-risk patients before the outbreak of florid symptoms. Using microdialysis techniques such changes in dopamine release can also be measured in freely moving animals. Thus, although the cardinal symptoms of schizophrenia seem to be beyond modelling in animals, other symptoms, biomarkers, and endophenotypes can be assessed in humans and animals often with more or less identical techniques and offer the possibility to assess vital aspects of schizophrenia in rats and mice. Especially in the field of electrophysiological biomarkers and endophenotypes, great improvements have been made. It is becoming increasingly clear that many patients with schizophrenia suffer from a reduced communication between brain regions (often referred to as dysconnection) rather than a dysfunction of one specific brain region. This is now increasingly being studied in animal models as well.

However, even though we can measure social behavior, prepulse inhibition, dopamine release, and other phenomena, the crucial step in animal modelling is to induce a disturbance in these phenomena that resemble the disturbance in the disease. There are several different approaches to achieve this, and those approaches will be discussed in the following sections.

Spontaneous Models

One way of identifying a potentially interesting model is to compare different strains of rats or mice, to see whether certain strains have deficits in specific signs or symptoms of a disease. In the case of schizophrenia for instance, Brown Norway and Brattleboro rats show deficits in prepulse inhibition compared to Sprague-Dawley and Lewis rats, respectively. Interestingly, these animals also show other abnormalities related to schizophrenia and have thus been suggested to represent "spontaneous models" for schizophrenia.

An alternative to investigating different strains of animals is to investigate individual animals within one strain to identify extremes within a normal population. Once such individuals have been identified, the next step is usually to try to breed these extremes selectively in order to establish a colony of high and low responders to a specific situation. For instance, there have been several approaches to selectively breed animals with a poor performance in the prepulse inhibition. Many years ago, we started to select Wistar rats on the basis of their gnawing response to apomorphine. After only a few generations, there was an almost complete separation in susceptible (APO-SUS) and unsusceptible (APO-UNSUS) animals. Further experiments showed that the APO-SUS animals shared a large number of symptoms with schizophrenic patients (such as enhanced sensitivity to stress, enhanced dopamine release, reduced prepulse inhibition).

Thus, the spontaneous approach has led to several potentially interesting models. However, very few have been investigated in detail (for instance, with respect to their validity), and the lack of a clear link (even if it is only theoretical) to the etiology of schizophrenia has limited the widespread use of them. In this respect, it is important to emphasize that none of the symptoms of schizophrenia are pathognomic and that deficits in prepulse inhibition, for instance, occur in many diseases. Thus, an animal with a deficit in prepulse inhibition does not immediately constitute an animal model for schizophrenia. On the other hand, if a specific strain of rats is identified (or selectively bred) that encompasses a number of schizophrenic symptoms, it would allow for a forward genetic analysis, identifying the genes that are involved in these phenomena, and by analogy perhaps in schizophrenia. Given that schizophrenia is certainly not a "one-gene" disorder, such a phenotypical approach might bear more relevance than the more popular "single-gene" genetic approaches described below.

Pharmacological Models

It has long been known that certain drugs can induce symptoms resembling those seen in psychiatric or neurological disorders. Thus, antipsychotic drugs can induce a parkinsonian-like rigidity and akinesia, and reserpine can induce depressive-like symptoms. Likewise, dopamine agonists such as amphetamine and NMDA antagonists such as phencyclidine can induce schizophrenia-like symptoms. Whereas amphetamine induces predominantly (positive) psychotic symptoms (though it can induce negative or cognitive deficits too), phencyclidine seems to be able to induce the full spectrum of schizophrenia. Such pharmacological tools have very often been used in animal modelling as well, and both drugs induce a large number of schizophrenia-like abnormalities in animals. Hence, amphetamine increases dopamine release and reduces prepulse inhibition, P_{50} gating, and social behavior (especially in monkeys). Phencyclidine induces virtually the same symptoms as amphetamine, but several pharmacological studies have shown that this model is mediated via different mechanisms. For instance, all known antipsychotic drugs can reverse the amphetamine-induced deficits in prepulse inhibition, whereas most of these drugs do not reverse the phencyclidine-induced prepulse inhibition (Geyer et al. 2001). A more fundamental problem with these pharmacological models is that they induce only very short-lived symptoms, and the validity for the schizophrenic process (which is a chronic process probably already starting before birth, see below) is questionable. Indeed, models such as the amphetamine- or phencyclidine-induced deficit in prepulse inhibition are now often regarded more as screening models (similar to amphetamine-induced hyperactivity) than as simulation models (see also Table 1). In order to overcome this limitation, more chronic models have been developed, in which either amphetamine or phencyclidine is given repeatedly (typically 5–7 days). The resulting symptoms are then either studied immediately after the last injection or after an incubation period of, say, 1 week. These models are thought to have more relevance for the chronic nature of the disorder. However, the models have not been extensively studied and vary greatly in dose, duration, and period of treatment.

Etiological Models
The most interesting class of simulation models are those in which one tries to mimic the etiology of schizophrenia. Unfortunately, schizophrenia is a highly complex disorder with a multifactorial etiology, in which genetic and environmental factors interact. Moreover, the nature of most of the genetic and environmental factors is still largely unknown. Fortunately, clinical genetic research is increasing rapidly and there is now more, and more evidence for certain candidate genes associated with schizophrenia. It would go beyond the limits of this entry to discuss the evidence in favor of each of these genes, but the most often associated ones are DISC-1 (disrupted in schizophrenia-1), dystrobrevin, neuregulin-1, and regulator of G protein signalling-4 (Harrison and Weinberger 2005). Although such association studies have led to the development of specific transgenic mice (and increasingly rats), it should be noted that in different studies, different polymorphisms have been linked to schizophrenia and the functional consequence of these polymorphisms is in most cases still unknown, leading to intriguing results. For instance, whereas the heterozygous neuregulin knockout mouse shows a reduction in prepulse inhibition, patients with the schizophrenia associated neuregulin haplotype seem to have increased mRNA and protein levels of neuregulin. One specific limitation of these genetic animal models is that it is clear that schizophrenia is not due to a single genetic deficit. Indeed, a recent simulation study suggested that, depending on the penetrance and odds ratio of the individual genes, as many as 100 genes may contribute to the etiology of schizophrenia (Paek and Kang 2012). In this respect, selective breeding for a "schizophrenia-like" phenotype (as discussed above) may represent a more naturalistic approach to modelling the genetic factors in schizophrenia. Another very interesting model is the 22q11 mutant mouse model. This model is based on the finding that patients with a microdeletion of the 22q11 region have a much higher risk of developing schizophrenia (Karayiorgou et al. 2010).

The search for specific environmental factors associated with schizophrenia is even more complicated than the search for genetic factors, since they often predate the onset of the disease by many years (if not decades) and are often only indirectly measurable, as an association between a specific event and the occurrence of the disease. For instance, several studies have reported that exposure to the 1958 influenza pandemic increased the risk of the schizophrenia in the offspring. However, whether this was due to the influenza virus, the medication, the stress of the situation, or other factors is impossible to deduce. Nonetheless, there is quite a large group of environmental factors that can increase the risk of developing schizophrenia. These include winter birth, migration, prenatal malnutrition, pre- and postnatal stressors, perinatal infection, and obstetric complications (Tandon et al. 2008).

Schizophrenia: Animal Models, Table 2 An overview of the most widely used etiologically based simulation models for schizophrenia

Etiologically based simulation models	DA hyperactivity	PPI deficit	P_{50} gating deficit	Memory impairment	Social withdrawal	Anhedonia
Genetic models						
DISC-L100P	Ø	+	Ø	+	−	−
DISC-Q31L	Ø	+	Ø	+	+	+
DISC−/−	+	+	Ø	Ø	+	Ø
Sdy mutant rat (DTNBP1−/−)	−	−	Ø	Ø	+	Ø
III-neuregulin$^{\pm}$	Ø	+	Ø	+	Ø	Ø
TM-neuregulin$^{\pm}$	Ø	+	Ø	−	Ø	Ø
CRD-neuregulin$^{\pm}$	Ø	+	Ø	+	Ø	Ø
22q11.2 deletion	Ø	+	Ø	+	Ø	Ø
Environmental models						
Prenatal stress	+	+	+	+	+	Ø
Prenatal polyI:C	+	+	Ø	+	Ø	Ø
Cesarean section	+	+	Ø	Ø	Ø	Ø
Maternal deprivation	+	+	+	Ø	±	Ø
Isolation rearing	+	+	+	±	±	±
Early ventral hippocampal lesion	+	+	Ø	+	+	+
Prenatal MAM treatment	+	±	Ø	+	+	Ø

+ the sign or symptom is observed in the animal model; − the sign or symptom is not observed in the animal model; ± there is contradictory evidence whether the sign or symptom is observed in the animal model; Ø the sign or symptom has not been investigated in the animal model
PPI prepulse inhibition, *DISC* disruption in schizophrenic, *DTNBP1* dystrobrevin-binding protein 1, *MAM* methylazoxymethanol

Some of these factors have also been modelled in animals, such as prenatal stress, prenatal protein restriction, Cesarean section, maternal deprivation, and isolation rearing (Carpenter and Koenig 2008). Another environmental factor which has received substantial attention in clinical and preclinical research is adolescent drugs use, especially cannabis. Although clinical studies can never unequivocally prove a causal link between adolescent cannabis use and the development of schizophrenia, rodent models show that adolescent THC treatment can induce schizophrenia-like symptoms. Few of these models have been investigated in great detail, yet some of these manipulations seem to induce quite a variety of the schizophrenia-like signs and symptoms (see Table 2).

In addition to these etiologically based simulation models, there are a few additional environmental models where one can argue whether they are really etiologically based, especially the early ventral hippocampal lesion model and the prenatal MAM model. Both models certainly aim to induce a neurodevelopmental disorder, but there is no evidence that patients with schizophrenia have an early hippocampal lesion nor that neurogenesis was blocked for some time during gestation. Nonetheless, both treatments induce a variety of schizophrenia-like signs and symptoms.

The inspection of Table 2 shows several interesting points. Firstly, in contrast to clinical studies, much less data are available for the genetic models than for the environmental models. Secondly, of the signs and symptoms, most interest has focused on prepulse inhibition, presumably because of the ease of measurement and the similarity to the clinical situation. In contrast, P_{50} gating and anhedonia have received much less attention.

Validation of Simulation Models

As with screening models, simulation models also have to be validated. Although this is often done using the same criteria used for screening models, it is important to note that there is much less clinical evidence that these criteria can be applied. For instance, most of the signs and symptoms in Table 2, including memory, social withdrawal, and anhedonia, are related to the cognitive and negative symptoms of schizophrenia, i.e., domains that generally do not respond to antipsychotic drugs. Even with respect to prepulse inhibition, there is insufficient evidence from longitudinal studies whether (some or all) antipsychotic drugs improve this information-processing deficit. Thus, with a lack of clinical data on these signs and symptoms, the validity of simulation models is difficult to assess.

Gene: Environment Models

The models described earlier and in Table 2 are based on either genetic or environmental factors. However, there is a vast amount of evidence that schizophrenia results from an interaction between genetic and nongenetic factors (Van Os et al. 2008). Although so far, this interaction has received less attention in animal modelling, the tide is slowly turning. Some of the environmental manipulations have been performed in different strains of rats, such as early hippocampal lesions, isolation rearing, and maternal deprivation, but extensive gene-environment interactions still need to be performed. Nonetheless, some interesting findings have already been reported. Thus, compared to Sprague-Dawley or Wistar rats, Lewis rats seem to be much less sensitive to early hippocampal lesioning, isolation rearing, and maternal deprivation (Ellenbroek and Cools 2000; Lipska and Weinberger 1995; Varty and Geyer 1998). This finding suggests that the genetic makeup of the Lewis rats may somehow be protective to these different manipulations. It would be of great interest to try to identify the underlying protective mechanism. More recently, researchers have started to combine the genetic and the environmental models described in Table 2 (Hida et al. 2013). For instance, DISC-1 mutant mice have been subjected to a prenatal polyI:C challenge or to a later life stressor. Although it is too early to draw any firm conclusions, these and other studies have clearly shown that the combined genetic and environmental challenge leads to a more severe phenotype.

In addition to these gene-environment interaction studies, some authors have combined different environmental challenges, such as early postnatal lesions with adult cannabinoid treatment or perinatal NMDA treatment with isolation rearing.

Summary

Schizophrenia is a severe psychiatric disorder with an unknown etiology and complex symptomatology. This has severely hindered the development of appropriate animal models. In the past, most therapeutic agents were developed using the so-called screening models, in which novel compounds were compared to known therapeutic agents. Although these models are still in use nowadays, it is highly unlikely that they will lead to a major breakthrough in therapy. With the identification of specific signs and symptoms, including biomarkers and endophenotypes that could also be assessed in animals, the field of animal modelling obtained the boost it needed. Moreover, genetic and epidemiological studies have started to identify specific genes and environmental factors that contribute to the disorder. This has led to a wealth of novel simulation models in the last decade, most of which are still in a very early exploratory phase and have not yet been properly validated. The most important challenge for the coming years will be to merge the two types of etiologically based simulation models in order to investigate the gene-environmental interaction, which is crucial for the occurrence of schizophrenia. Given that schizophrenia is unlikely due to a single gene-environment interaction, the challenge will be to identify the most important gene(s) and environmental challenges and combine these (for instance, using the phenotypical approach mentioned above).

Cross-References

▶ Animal Models for Psychiatric States
▶ Antipsychotic Drugs
▶ Antipsychotic Medication: Future Prospects
▶ Genetically Modified Animals
▶ Latent Inhibition
▶ Prepulse Inhibition
▶ Rodent Tests of Cognition
▶ Schizophrenia

References

Arnt J (2000) Screening models for antipsychotic drugs. In: Ellenbroek BA, Cools AR (eds) Atypical antipsychotics. Birkhauser, Basel, pp 99–119

Carpenter WT, Koenig JI (2008) The evolution of drug development in schizophrenia: past issues and future opportunities. Neuropsychopharmacology 33:2061–2079

Ellenbroek BA (1993) Treatment of schizophrenia: a clinical and preclinical evaluation of neuroleptic drugs. Pharmacol Ther 57:1–78

Ellenbroek BA, Cools AR (2000) The long-term effects of maternal deprivation depend on the genetic background. Neuropsychopharmacology 23:99–106

Geyer MA, Krebs-Thomson K, Braff DL, Swerdlow NR (2001) Pharmacological studies of prepulse inhibition models of sensorimotor gating deficits in schizophrenic patients: a decade in review. Psychopharmacology (Berl) 156:117–154

Harrison PJ, Weinberger DR (2005) Schizophrenia genes, gene expression, and neuropathology: on the matter of their convergence. Mol Psychiatry 10:40–68

Hida H, Mouri A, Noda Y (2013) Behavioral phenotypes in schizophrenic animal models with multiple combinations of genetic and environmental factors. J Pharmacol Sci 122:185–191

Karayiorgou M, Simon TJ, Gogos JA (2010) 22q11.2 microdeletions: linking DNA structural variation to brain dysfunction and schizophrenia. Nature reviews Neuroscience 11:402–416

Lipska BK, Weinberger DR (1995) Genetic variation in vulnerability to the behavioral effects of neonatal hippocampal damage in rats. Proc Natl Acad Sci U S A 92:8906–8910

Paek MJ, Kang UG (2012) How many genes are involved in schizophrenia? A simple simulation. Prog Neuropsychopharmacol Biol Psychiatry 38:302–309

Tandon R, Keshavan MS, Nasrallah HA (2008) Schizophrenia, "Just the Facts" what we know in 2008. 2. Epidemiology and etiology. Schizophr Res 102:1–18

Van Os J, Rutten BPF, Poulton R (2008) Gene-environment interactions in schizophrenia: review of epidemiological findings and future directions. Schizophr Bull 34:1066–1082

Varty GB, Geyer MA (1998) Effects of isolation rearing on startle reactivity, habituation, and prepulse inhibition in male Lewis, Sprague-Dawley, and Fischer F344 rats. Behav Neurosci 112:1450–1457

Schizophrenia: Management of Treatment Resistance

Kazuyuki Nakagome
National Center of Neurology and Psychiatry, Kodaira, Tokyo, Japan

Synonyms

Drug resistance schizophrenia; Treatment refractory schizophrenia; Treatment unresponsive and intolerant schizophrenia

Definition

Treatment-resistant patients are defined as patients who do not respond well enough to standard treatment. However, defining treatment-resistant schizophrenia is a complicated undertaking, especially because refractoriness to treatment in schizophrenia is best viewed as a continuum rather than a discrete category. Moreover, in clinical settings, there has been a trend to broaden the treatment goal to include the concept of recovery, which encompasses not only symptom alleviation but also improvement in functional outcome and subjective well-being. The appropriate definition of treatment resistance depends on the circumstances in which the definition is to be applied. For example, a narrow definition primarily focusing on suboptimal response of positive symptoms and tolerability to medication treatment is suitable for research purposes relating to an antipsychotic drug for which the indication will be treatment-resistant schizophrenia, whereas a broader definition that incorporates assessment of psychosocial functioning, cognitive deficits, affective symptoms, and behavior problems may be appropriate for clinical practice.

Schizophrenia: Management of Treatment Resistance, Table 1 Kane et al. (1988) criteria for treatment-resistant schizophrenia

Historical criteria
1. At least three periods of treatment in the preceding 5 years with neuroleptic agents (from at least two different chemical classes at dosages $\geq$ 1,000 mg/day of chlorpromazine) for a period of 6 weeks, each without significant symptomatic relief
2. No period of good functioning within preceding 5 years
Severity criteria
1. Total Brief Psychiatric Rating Scale (BPRS) score of at least 45 (18-item version) plus a minimum Clinical Global Impressions (CGI) scale rating of 4 (moderately ill)
2. Item scores of at least 4 (moderate) on two of the following four BPRS items: conceptual disorganization, suspiciousness, hallucinatory behavior, and unusual thought content

The original concept of the narrow definition of treatment resistance was proposed by Kane et al. (1988) in their seminal study that demonstrated the efficacy of clozapine compared with chlorpromazine against treatment-resistant schizophrenia (see Table 1). Since then, the core principles in defining treatment resistance have not been changed, despite reductions in the threshold of dosage level, number of trials, and period of assessment, presumably in accordance with the movement toward earlier institution of clozapine.

Although we acknowledge the various unmet needs from the patient's viewpoint and thus feel inclined to shift to a broader definition, cognitive deficits are known to be relatively unresponsive to medication treatment, and as a good consensus has not yet been reached about which assessment tool is optimal for psychosocial functioning and other issues, it may be premature to incorporate them in the definition of treatment-resistant schizophrenia.

Role of Pharmacotherapy

Before moving to the topic of methods to treat treatment-resistant schizophrenia, we need to explore confounding factors relevant in forming a clinical picture as observed in a treatment-resistant patient. The patient factors include illicit substance misuse, physical comorbidity, and poor quality of the social environment, and the treatment factors include noncompliance, drug-drug interactions, delay in initiating treatment, drug bioavailability problems, and poor therapeutic alliance between physician and patient, all of which should be addressed before undergoing various interventions.

Clozapine

Strong evidence suggests that clozapine is more efficacious than other antipsychotic drugs in treating patients with treatment-resistant schizophrenia. In their 6-week double-blind, randomized study of 268 patients with treatment-resistant schizophrenia, Kane et al. (1988) found that 30 % of the patients improved following treatment with clozapine, compared with 4 % of those who received chlorpromazine. However, its potential for agranulocytosis and other serious side effects has generally limited its use to patients with treatment-resistant schizophrenia. Clozapine has shown benefits over other antipsychotic drugs not only for positive symptoms but also for suicidality, violent behaviors, or a comorbid substance misuse. Clozapine also was found to be associated with a remarkably low incidence of tardive dyskinesia and plasma prolactin elevation, which may be due to weak dopamine type 2 (D2) receptor blockade. In contrast to clozapine's superiority in treatment-unresponsive and treatment-intolerant schizophrenia, no such evidence of its greater efficacy is found in first episode (Lieberman et al. 2003) or among other patient populations. It raises a question of exactly when in the course of the illness will clozapine's benefits for treatment-resistant schizophrenia begin to appear.

Several studies suggested that clozapine serum concentrations can be useful to help guide dosing (Perry et al. 1991). Dosage levels of 300–600 mg/d are generally needed to achieve the plasma concentrations for good response ($> = 350$ ng/ml), although care is needed since nicotine lowers the concentrations (Chung and Remington 2005). The dosage levels should be increased gradually – not exceeding 600 mg/d – to avoid serious side effects because at least the risk of seizures is dose dependent.

Although many researchers strongly recommend clozapine as the agent of choice for treatment-resistant schizophrenia, and progression to clozapine use in the treatment course is explicitly encouraged, reluctance to use clozapine in the clinical field is apparent. For example, phase 2 results of Clinical Antipsychotic Trials of Intervention Effectiveness (CATIE) showed that many participants chose the tolerability pathway (n = 444) over the efficacy pathway (n = 99) (Swartz et al. 2008), presumably to avoid being assigned to clozapine treatment, which was an option in the efficacy pathway but not in the tolerability pathway. The provider restrictions, high costs, and complexities of clozapine treatment are necessary for treatment safety and efficiency but may have had the unintended consequence of reducing training opportunities for many residents and leading to underuse of clozapine for patients who might otherwise gain benefit from the drug.

Augmentation and Adjunctive Strategies

Besides clozapine, options are limited for the many patients with treatment-resistant schizophrenia, and none has been supported by systematic evidence. Various augmentation strategies that have limited or no evidence supporting their efficacy are often used. Overall effectiveness in a certain patient group does not always translate into effectiveness in each individual patient. No best drug or best dose of any drug exists for all patients. Predicting which antipsychotic medication might be optimal for a given patient is impossible. Decisions about antipsychotic therapy consequently entail a trial-and-error process with careful monitoring of clinical response and side effects and an ongoing risk-benefit assessment. Therefore, clinicians may consider a time-limited trial of a drug to determine if it may offer any benefit exceeding risk to an individual patient.

It is recommended that patients with treatment-resistant schizophrenia be given a trial of clozapine monotherapy for up to 6 months insofar as no serious side effects occur. If optimal response is not attained after an adequate trial of clozapine, adjunctive agents such as mood stabilizers (e.g., lithium, valproate, lamotrigine), benzodiazepines (e.g., clonazepam), propranolol, antidepressants, and antipsychotic drugs may be tried depending on the residual symptom profile. It should be noted that these augmentation strategies for clozapine are not supported by evidence. For example, the Texas Medication Algorithm Project (TMAP) recommends clozapine augmentation with second-generation antipsychotics (SGA) or first-generation antipsychotics (FGA) for patients whose symptoms do not respond to clozapine alone, although a review of the TMAP documentation suggests that the evidence favoring either risperidone or lamotrigine is weak. In one study, the placebo group actually showed greater improvement than the risperidone group on PANSS positive syndrome subscale scores (Anil Yaciolu et al. 2005). Royal Australian and New Zealand College of Psychiatrists Clinical Practice (RANZCP) guidelines, by contrast, recommend the use of the most effective prior drug and an appropriate adjunctive therapy, such as lithium. Such adjunctive strategies should be considered on an individual basis, with goals of treatment carefully defined and subsequently monitored so that ineffective polypharmacy is not sustained.

Novel Pharmacological Approaches

Considering the significance of expanding the concept of treatment resistance, various aspects of the illness should be involved such as disorganization, negative pathology, and cognitive deficits, which are thought to be relatively independent from each other. Although scarcely any evidence favors antipsychotic combination therapy, which also increases the side-effect burden, monotherapy of antipsychotic drugs fails to address all these pathologies. These unmet treatment needs are clinical targets for drug discovery involving novel therapeutic strategies including combination therapy. Considering the unique properties of clozapine's mechanism of action, clozapine's major active metabolite N-desmethylclozapine (NDMC), which has glycine reuptake inhibition property and cholinergic muscarinic 1 receptor agonistic function, was a candidate of an adjunct to existing antipsychotic medications for patients with treatment-resistant positive symptoms (Natesan et al. 2007); however, results from a 6-week, multicenter, double-blind, placebo-controlled, phase 2 study showed that it was no more effective than placebo in patients with schizophrenia in an acute stage (Bishara and Taylor 2008). In regard to persistent negative symptoms, the effectiveness of adjunctive antidepressants has been reported in several studies, although findings remain inconsistent. Augmentation of antipsychotic medications with mirtazapine, paroxetine, fluvoxamine, or the selective monoamine oxidase type B inhibitor selegiline has shown benefit in respective controlled studies that have isolated the effect on negative symptoms from impact on secondary factors, including positive symptoms, depression, and extrapyramidal symptoms (Webber and Marder 2008). The Measurement and Treatment Research to Improve Cognition in Schizophrenia (MATRICS) project identified nine promising molecular targets for cognition-enhancing agents with a potential for pronounced efficacy in cognitive deficits in schizophrenia, which has been unfulfilled by atypical antipsychotic drugs (Webber and Marder 2008):

1. Alpha-7 nicotinic receptor agonists
   - Partial α-7 nicotinic cholinergic agonist (3-[(2,4-dimethoxy)benzylidene]-anabaseine, DMXB-A)
   - Acetylcholinesterase inhibitor (galantamine)
2. D1 receptor agonists
   - Full D1 agonist (dihydrexidine, DAR-0100)
3. AMPA glutamatergic receptor agonists
   - Positive allosteric modulators of AMPA receptors, ampakines (CX-516)
4. Alpha-2 adrenergic receptor agonists
   - Alpha-2 receptor stimulators, antihypertensive drugs (clonidine, guanfacine)
5. NMDA glutamatergic receptor agonists
   - D-cycloserine, glycine, and D-serine
6. Metabotropic glutamate receptor agonists
   - mGlu2/3 agonist (LY2140023)
7. Glycine reuptake inhibitors
   - Sarcosine
8. M1 muscarinic receptor agonists
   - NDMC (ACP-104)
9. GABA-A R subtype selective agonists
   - Alpha-2 subunit specific stimulator, positive allosteric modulators

Summary

Clozapine shows superiority in both treatment-unresponsive and treatment-intolerant schizophrenia. However, clozapine has been underused for various reasons, including occurrence of agranulocytosis, restriction of the providers, high costs, and complexities of clozapine treatment. Besides clozapine, limited treatment options with scarce evidence are often used in trial-and-error process, including augmentation or adjunctive strategies with various agents. Recently, development of novel drugs that target unmet treatment needs has been in progress with the hope that they may show efficacy against treatment-resistant schizophrenia of broader definition.

Refining the definition of treatment resistance holds advantages and disadvantages. Consensually developed definitions may yield good criteria to guide the use of specialized treatment agents such as clozapine. Broader definitions may provide more chances for patients to receive trials of such treatments. But what if those trials end up in failure? Because we have few pharmacological options beyond the step of using clozapine, definition may lead to additional serious stigma.

The definitions of treatment resistance mentioned in the current text actually represent "drug resistance," to the extent that psychosocial approaches are ignored or inadequately implemented. Now that we have more resources in terms of psychosocial treatments, we surely need to take these approaches into consideration in future efforts to define treatment resistance. We should keep in mind that our ultimate goal is not merely to broaden the criteria; rather, it is to ensure that no patient remains in a state of treatment nonresponse.

Cross-References

▶ Clozapine
▶ Schizophrenia

References

Anil Yaciolu AE, Kivircik Akdede BB, Turgut TI, Tümüklü M, Yazici MK, Alptekin K, Ertuğrul A, Jayathilake K, Göğüş A, Tunca Z, Meltzer HY (2005) A double-blind controlled study of adjunctive treatment with risperidone in schizophrenic patients partially responsive to clozapine: efficacy and safety. J Clin Psychiatry 66:63–72

Bishara D, Taylor D (2008) Upcoming agents for the treatment of schizophrenia. Mechanism of action, efficacy and tolerability. Drugs 68:2269–2292

Chung C, Remington G (2005) Predictors and markers of clozapine response. Psychopharmacology (Berl) 179:317–335

Kane J, Honigfeld G, Singer J, Meltzer H (1988) Clozapine for the treatment-resistant schizophrenic. A double-blind comparison with chlorpromazine. Arch Gen Psychiatry 45:789–796

Lieberman JA, Phillips M, Gu H, Stroup S, Zhang P, Kong L, Ji Z, Koch G, Hamer RM (2003) Atypical and conventional antipsychotic drugs in treatment-naive first-episode schizophrenia: a 52-week randomized trial of clozapine vs chlorpromazine. Neuropsychopharmacology 28:995–1003

Natesan S, Reckless GE, Barlow KB, Nobrega JN, Kapur S (2007) Evaluation of N-desmethylclozapine as a potential antipsychotic-preclinical studies. Neuropsychopharmacology 32:1540–1549

Perry PJ, Miller DD, Arndt SV, Cadoret RJ (1991) Clozapine and norclozapine plasma concentrations and clinical response of treatment-refractory schizophrenic patients. Am J Psychiatry 148:231–235

Swartz MS, Stroup TS, McEvoy JP, Davis SM, Rosenheck RA, Keefe RS, Hsiao JK, Lieberman JA (2008) What CATIE found: results from the schizophrenia trial. Psychiatr Serv 59:500–506

Webber MA, Marder SR (2008) Better pharmacotherapy for schizophrenia: what does the future hold? Curr Psychiatry Rep 10:352–358

Schizophrenia: Remission

Robin Emsley
Department of Psychiatry, Faculty of Medicine and Health Sciences, University of Stellenbosch, Tygerberg, Cape Town, South Africa

Synonyms

Remission, as an outcome state for schizophrenia, does not have any synonyms. It is somewhat beyond "response" but somewhat short of "recovery."

Definition

In general medicine, the term "remission" is used to describe a state or period during which the symptoms of an incurable disease are diminished or abated. A disease is regarded as incurable if there is always a risk of relapse, no matter how long the duration of remission. Schizophrenia is a chronic and disabling illness and is ranked among the world's top ten causes of disability-adjusted life-years (Murray and Lopez 1996). Schizophrenia imposes a disproportionate emotional and financial burden on affected individuals, their families and carers, health-care systems, and society. A recent systematic review of relapse after treatment discontinuation in first-episode schizophrenia confirms the incurable nature of schizophrenia, with reported relapse rates of 77 % after 1 year and over 90 % by 2 years (Zipursky et al. 2013). Traditionally, treatment goals were modest, and remission was not generally emphasized as a treatment target. Clinicians often settled for outcomes such as symptom or behavioral control or "stability." However, in recent years, outcome expectations have heightened for a number of reasons. Firstly, the increasing involvement of families, caretakers, and advocacy bodies has resulted in a greater alignment of patient and family needs with physician treatment goals. Secondly, new pharmacological and psychosocial research developments have raised outcome expectations. Thirdly, and most importantly, the importance of early intervention is recognized. It is now generally accepted that there is a neuroprogressive component to the illness, and the key to achieving optimal outcomes is early, effective intervention and prevention of accruing morbidity.

In keeping with these trends, there has been renewed interest in defining more appropriate outcome measures in schizophrenia, and a need was identified to establish reliable, valid, and easy to apply measures that would be realistically attainable for most patients. Several long-term outcome studies included remission as an outcome measure, with remission being defined in various ways. Definitions usually incorporated a symptom improvement component reflecting absent or low levels of symptoms and a requirement that the symptom improvement be maintained over a period of time. A significant advance was made in 2005, when the "Remission in Schizophrenia Working Group" (RSWG) published consensus-derived, operationally defined criteria for remission in schizophrenia (Andreasen et al. 2005). Remission was defined using an absolute threshold of severity for the core symptoms of the illness including positive, negative, and disorganized symptoms. These core symptoms are assessed by means of psychopathology rating scales such as the Positive and Negative Syndrome Scale (PANSS) and the Brief Psychiatric Rating Scale (BPRS). Remission is defined as, at most, a mild level of symptom intensity, not negatively influencing the individual's behavior, and enduring for at least 6 months. The RSWG criteria were immediately popular and have been applied in both clinical and research settings. They have generated much interest and debate and have stimulated research investigating treatment outcome in diverse patient samples. The criteria define a clinical state based on symptom reduction that is associated with other outcome benefits including better social and occupational functioning, quality of life, and cognitive performance. The RSWG remission criteria appear to be a realistic treatment target, with studies reporting rates of RSWG remission between 17 % and 78 % (weighted mean = 35.6 %) in first-episode schizophrenia and between 16 % and 62 % (weighted mean = 37 %) in multi-episode samples. Studies comparing second-generation antipsychotics (SGAs) to haloperidol reported higher remission rates for SGAs, and patients treated with long-acting risperidone showed high rates of sustained remission over time (AlAqeel and Margolese 2012).

Role of Pharmacotherapy

Antipsychotics and Remission

Antipsychotic medication is the mainstay of treatment for schizophrenia with many studies over the past 50 years and more reporting efficacy for these agents in the treatment of schizophrenia. Acute treatment is associated with variable but substantial symptom reduction in most patients. Maintenance treatment with antipsychotic medication not only significantly decreases the risk of relapse but it also promotes sustained remission (Leucht et al. 2012). Increasingly, maintenance studies, while retaining relapse as the primary outcome measure, are including RSWG remission as a secondary outcome measure. Several studies suggest that patients are able to retain their remission status over time once remission is attained, as long as antipsychotic medication is adhered to. However, other studies suggest that remission status fluctuates considerably over time, with many patients moving in and out of remission. In addition to medication adherence, factors such as substance abuse and psychosocial adversity are likely to play a role in precipitating episodes of symptom exacerbation. The acute treatment response is most favorable in the early phase of illness, particularly the first episode. Both first-generation antipsychotics (FGAs) and SGAs are effective in reducing symptoms. This is particularly the case for positive and disorganized symptoms as well as for the symptom domains of excitement/hostility and depression/anxiety. Negative symptoms respond less well to treatment, and cognitive symptoms are generally nonresponsive. Psychosocial interventions such as social skills training, cognitive behavioral therapy, cognitive remediation, and social cognition training can further improve outcome. The unfavorable long-term outcome in most patients with schizophrenia stands in stark contrast to the established efficacy of antipsychotic medication in acute treatment and particularly in the early stages of illness. The reason for this discrepancy is that the initial treatment response is not maintained, with most patients following a course characterized by multiple relapses, persistence of positive and negative symptoms, and enduring functional and cognitive deficits. The most important factor here is nonadherence to treatment. Treatment discontinuation is the strongest predictor of relapse by far (Robinson et al. 1999), and there is now increasing evidence to suggest that relapse events are the critical factor in disease progression. After each relapse, some patients develop persistence of both positive and negative symptoms and become refractory to treatment (Emsley et al. 2013). Maintenance treatment with antipsychotic medication not only significantly decreases the risk of relapse but also promotes sustained remission. Treatments associated with improved adherence are likely to increase the chances of sustained remission. While better-tolerated medications are more likely to be adhered to, the adherence advantages of oral SGAs over FGAs are only modest. Several psychosocial interventions improve adherence and should, together with antipsychotic medication, form part of standard care for schizophrenia. Also, more use should be made of long-acting injectable antipsychotics as a way of ensuring adherence.

Predictors of Remission

In clinical practice, it would be very useful to know which factors accurately predict remission. This would greatly assist clinicians in deciding whether to persist with a particular treatment or whether to switch to alternative options. Perhaps the most valuable predictor of later treatment response and remission status is the early treatment response (e.g., at 2 weeks). Antipsychotics act rapidly, and the greatest improvements in symptoms occur during the first few weeks of treatment. Other factors reported to predict remission include younger age, being employed, shorter duration of illness, shorter length of current episode, and lower levels of psychopathology at baseline. Researchers are also investigating biological concomitants of remission. For example, structural brain abnormalities have been described which may represent a neural marker for non-remission status.

The Association Between Remission and Other Aspects of Outcome

While the RSWG criteria define remission only according to levels of core symptoms (positive, negative, and disorganized), several studies have indicated a significant association between symptomatic remission and other favorable outcomes, including social and occupational functioning and quality of life. The RSWG remission criteria have been criticized for not including measures of functionality. However, while there is a need to develop assessment instruments to more broadly define outcome, objective measures of "functional remission" are more difficult to define and are likely to be substantially influenced by environmental factors.

Remission and Recovery

It has been argued that remission is not in itself a sufficient outcome goal, but should be regarded as a necessary step toward recovery. Recovery has been described as the new vision for mental health services. Recovery is a more complex concept than remission and means different things to different people. The RSWG regard recovery as a more demanding and longer-term phenomenon than remission. The group did not propose operational criteria for recovery as they felt that more research is required regarding aspects such as relationships between psychosocial functioning and cognitive ability on the one hand and core symptoms used to define remission on the other hand. Outcome studies assessing remission and recovery rates invariably report much lower rates of recovery than of remission.

Conclusions

The utilization of remission as an outcome measure in schizophrenia represents an important step forward in research and in clinical practice. The widespread adoption of the RSWG remission criteria is at least in part due to the fact that they are easy to apply and relatively unambiguous. The next step will be for a further consensus-driven process to develop a broader definition of remission or recovery, including functional outcomes and quality of life measures. However, there is a risk that broadening the criteria will make it more difficult to define remission and thereby less attractive to researchers and clinicians alike. While a huge amount of research has been published on the RSWG criteria, much remains to be done. The RSWG remission criteria have proven to be very useful and define an outcome that is highly desirable and is realistic within the context of currently available treatments.

Cross-References

- ▶ Antipsychotic Drugs
- ▶ Antipsychotic Medication
- ▶ Atypical Antipsychotic Drugs
- ▶ Psychosis
- ▶ Schizophrenia
- ▶ Typical Antipsychotics

References

AlAqeel B, Margolese HC (2012) Remission in schizophrenia: critical and systematic review. Harv Rev Psychiatry 20(6):281–297

Andreasen NC, Carpenter WT Jr, Kane JM, Lasser RA, Marder SR, Weinberger DR (2005) Remission in schizophrenia: proposed criteria and rationale for consensus. Am J Psychiatry 162(3):441–449

Emsley R, Chiliza B, Asmal L (2013) The evidence for illness progression after relapse in schizophrenia. Schizophr Res. 148(1–3):117–21

Leucht S, Tardy M, Komossa K et al (2012) Maintenance treatment with antipsychotic drugs for schizophrenia. Cochrane Database Syst Rev 5:CD008016

Murray CJL, Lopez AD (1996) The global burden of disease and injury series, vol 1. Harvard University Press, Cambridge, MA

Robinson D, Woerner MG, Alvir JM et al (1999) Predictors of relapse following response from a first episode of schizophrenia or schizoaffective disorder. Arch Gen Psychiatry 56:241–247

Zipursky RB, Menezes NM, Streiner DL (2014) Risk of symptom recurrence with medication discontinuation in first-episode psychosis: a systematic review. Schizophrenia Res. 152(2–3):408–14

Schizophrenia: Violent Behavior

Jan Volavka[1] and Leslie Citrome[2]
[1]Department of Psychiatry, School of Medicine, New York University, New York, NY, USA
[2]Department of Psychiatry and Behavioral Sciences, New York Medical College, Valhalla, NY, USA

Synonyms

Aggression; Aggressive behavior; Violence

Definitions

Aggression or violence is overt behavior that is intended to injure or damage another individual or object. The overt behavior may involve verbal aggression, physical aggression, or both.

Hostility denotes a variety of unfriendly attitudes ranging from uncooperativeness to aggression. It can be measured by rating scales and is frequently assessed in psychopharmacological studies. It has been used as a proxy measure of violence.

Role of Pharmacotherapy

Heterogeneous mechanisms contribute to the elevated risk for aggressive behavior in schizophrenia (Volavka and Citrome 2011). Only 20 % of assaults among schizophrenia *inpatients* are clearly attributable to psychotic symptoms; other assaults are related to confusion, impaired impulse control, or personality factors consistent with comorbid psychopathy or antisocial personality disorder (Volavka et al. 2012). The etiological heterogeneity is even more complex among schizophrenia *outpatients* who frequently abuse street drugs and alcohol. Furthermore, many schizophrenia patients are not adherent to treatment.

These multiple sources of heterogeneity have not always been accounted for in treatment trials or in routine clinical treatment planning. It is clear that not all of the factors causing aggressive behavior in schizophrenia are amenable to pharmacotherapy alone; concurrent psychosocial interventions are important (Volavka et al. 2012). Nevertheless, pharmacotherapy has been the mainstay of the treatment of aggression in schizophrenia.

Pharmacotherapy of Acute Agitation and Violent Behavior in Schizophrenia

Patients presenting with an acute behavioral emergency require a rapid resolution of any symptoms that place the patient and others at risk for harm. In addition to behavioral and psychological interventions, medication treatment should be offered early so that dangerous behavior does not escalate further (Citrome 2008). Available rapid-acting medication interventions include parenteral formulations of antipsychotics and benzodiazepines (Citrome 2007), at times given simultaneously (Battaglia et al. 1997).

Downsides to the use of benzodiazepines include the potential for respiratory depression, dependence and cognitive impairment, and possible behavioral disinhibition or "paradoxical reactions." However, benzodiazepines can treat underlying alcohol or sedative withdrawal. Benzodiazepines are not appropriate for prolonged use because of tolerance, withdrawal, and no/little effect on core symptoms of psychosis. Typical antipsychotics include low-potency agents (e.g., chlorpromazine) where cautions include hypotension, anticholinergic effects, and decrease in seizure threshold. High-potency agents (e.g., haloperidol) are associated with acute dystonia and akathisia. Several atypical antipsychotics are available as a rapid-acting intramuscular formulation (ziprasidone, olanzapine, and aripiprazole) and have received regulatory approval for the indication of agitation associated with schizophrenia. Akathisia and dystonia can be avoided by using these agents rather than haloperidol. Recently approved in the USA and EU is inhaled loxapine for the treatment of agitation; this intervention is associated with a rapid onset of efficacy but requires observation for potential bronchospasm (Citrome 2013). Asenapine, an atypical antipsychotic that is absorbed sublingually, has been demonstrated to be also effective in reducing agitation.

Long-Term Pharmacotherapy of Violent Behavior in Schizophrenia

Antipsychotics

Atypical antipsychotics are currently the principal treatment of aggressive behavior in schizophrenia. The results of randomized trials of six atypical antipsychotics for this indication are summarized in Table 1. A measure of aggression or hostility was used for each of the comparisons. Each of the comparisons is supported by published evidence; references for individual comparisons may be found elsewhere (Volavka et al. 2011, 2012).

Clozapine

Clozapine was found to have superior effects in reducing hostility or overt aggression in most, but not all, randomized studies. Similar superiority over other antipsychotics was found in most uncontrolled studies and case series. The antiaggressive effect of clozapine is specific in the sense that it is (statistically) independent of the changes of positive symptoms of psychosis, and it is not mediated by sedation (Topiwala and Fazel 2011). Full antiaggressive effect of clozapine is not reached until the dose escalation is completed. The dose escalation may take about

Schizophrenia: Violent Behavior, Table 1 Antiaggressive effects of atypical antipsychotics in randomized trials

Comparator	Aripiprazole	Clozapine	Olanzapine	Quetiapine	Risperidone	Ziprasidone
Placebo	Ari > Pl			Que > Pl	Ris > Pl	
Typical antipsychotic	Ari = Hal	Clo > Hal	Olz > Hal	Que < Per	Ris = Per	Zip = Hal #
			Olz > Hal			Zip = Per
			Olz = Hal			
			Olz = Per			
Atypical antipsychotic		Clo > Ris	Olz < Clo	Ris = Olz = Que = Zip	Ris < Clo	Ris = Olz = Que = Zip
		Clo = Olz	Olz = Clo = Ris	Que < Olz	Clo = Ris = Olz	
		Clo = Ris = Olz	Ris = Olz = Que = Zip		Ris = Olz = Que = Zip	
			Olz > Que			

<, > statistically significant difference was detected
= statistically significant difference was not detected
effect greater than haloperidol only in the first week of treatment
Ari aripiprazole, *Clo* clozapine, *Hal* haloperidol, *Olz* olanzapine, *Per* Perphenazine, *Que* quetiapine, *Ris* risperidone, *Zip* ziprasidone

three weeks; faster rate of dose increase may not be possible due to cardiovascular side effects. The adequate dose for antiaggressive effect is about 400 mg/day. The antiaggressive effects of clozapine may be particularly expressed in schizophrenia patients who are treatment resistant (Frogley et al. 2012).

Other Antipsychotics

Olanzapine's antiaggressive effects are weaker than those of clozapine, but at least as good or better than those of other antipsychotics. The superiority of olanzapine over typical antipsychotics in antiaggressive effects was shown in some, but not all, randomized clinical trials. Comparisons of olanzapine with other atypical antipsychotics showed no difference or superiority of olanzapine in antiaggressive effects. In open studies, olanzapine's antiaggressive effects appeared similar or superior to risperidone.

Antiaggressive effects of aripiprazole, quetiapine, and risperidone were superior to placebo in randomized trials. Aripiprazole, quetiapine, risperidone, and ziprasidone did not show consistent differences from each other or from typical antipsychotics in their antiaggressive effects. Typical antipsychotics such as haloperidol carry a substantial risk of extrapyramidal symptoms which may elicit irritability and aggression.

In summary, the available evidence indicates that clozapine should be considered first-line treatment for persistent aggression in schizophrenia. If clozapine cannot be used, olanzapine may be a second choice. However, in individual clinical decisions, the effectiveness of clozapine and olanzapine must be weighed against their adverse effects such as agranulocytosis (for clozapine), weight gain, and effects on lipid and glucose metabolism (for clozapine and olanzapine).

Anticonvulsants and Lithium

Valproate is frequently used as an adjunctive treatment (in combination with an antipsychotic) for aggressive schizophrenia patients. However, this treatment is off-label and lacks robust evidence of effectiveness. Lamotrigine and carbamazepine were reported to show antiaggressive effects as adjunctive treatments, but these effects were difficult to replicate. Lithium does reduce aggressive behavior in mania, but controlled studies of this effect in schizophrenia showed no benefit of this adjunctive treatment. Thus, adjunctive treatment of aggressive behavior in schizophrenia using anticonvulsants or lithium is not supported by adequate empirical evidence. Perhaps it may be effective in individual cases, but potential adverse effects must be considered in a clinical decision to use such treatment. The use of anticonvulsants or lithium in the treatment of schizophrenia or aggression is off-label.

Adrenergic Beta-Blockers

Propranolol was demonstrated to have antiaggressive properties in several studies. Adjunctive pindolol was demonstrated to significantly reduce the number and severity of aggressive incidents in schizophrenia patients.

The dose escalation of beta-blockers is a slow process because these compounds significantly reduce blood pressure and pulse rate. They are therefore not suitable for rapid control of aggressive behavior, and antipsychotics are considered a better choice in most cases. However, antipsychotics are not always effective, and they have adverse effects of their own. Therapeutic potential of beta-blockers for the treatment of aggression in schizophrenia remains largely unexplored. It merits further consideration and study.

Serotonin-Specific Reuptake Inhibitors (SSRI)

SSRI reduce aggression in patients with personality disorders and in mental retardation. Citalopram reduced the frequency of aggressive incidents in schizophrenia patients in a single promising study that awaits replication.

Substance Abuse, Treatment Adherence, and Violent Behavior in Schizophrenia

Substance abuse is linked to nonadherence, and both of these problems elevate the risk of violent behavior in schizophrenia (Volavka and Citrome 2011). Substance abuse is associated with increased extrapyramidal symptoms in schizophrenia; this could lead to nonadherence and then to violence. Substance abuse is strongly associated with violence and violent crime in the mentally ill as well as in general population. Clozapine reduces craving (particularly for cocaine) and reduces comorbid substance use in patients with schizophrenia. Psychosocial interventions for comorbid substance use and treatment nonadherence are required in patients with schizophrenia.

Summary

Aggressive behavior in schizophrenia is not amenable to pharmacotherapy alone; concurrent psychosocial interventions are important. Nevertheless, pharmacotherapy has been the mainstay of the treatment of aggression in schizophrenia. Patients presenting with an acute behavioral emergency require a rapid resolution of any symptoms that place the patient and others at risk for harm. Ziprasidone, olanzapine, and aripiprazole are available as a rapid-acting intramuscular formulation and have received regulatory approval for the indication of agitation associated with schizophrenia. Akathisia and dystonia can be avoided by using these agents rather than haloperidol.

Atypical antipsychotics are currently the principal long-term treatment of aggressive behavior in schizophrenia. Clozapine is the treatment of choice for aggressive behavior in schizophrenia. Olanzapine's antiaggressive effects are weaker than those of clozapine, but at least as good or better than those of other antipsychotics. Antiaggressive effects of other atypical antipsychotics are not significantly different from each other. Anticonvulsants or lithium have not been shown to be effective in the treatment of schizophrenia or aggression. Adrenergic beta-blockers were demonstrated to have antiaggressive properties, but are rarely used.

Cross-References

▶ Aggression
▶ Aggressive Behavior: Clinical Aspects

References

Battaglia J, Moss S, Rush J, Kang J, Mendoza R, Leedom L, Dubin W, McGlynn C, Goodman L (1997) Haloperidol, lorazepam, or both for psychotic agitation? A multicenter, prospective, double-blind, emergency department study. Am J Emerg Med 15:335–340

Citrome L (2007) Comparison of intramuscular ziprasidone, olanzapine, or aripiprazole for agitation: a quantitative review of efficacy and safety. J Clin Psychiatry 68:1876–1885

Citrome L (2008) Pharmacologic treatment of agitation. In: Glick RL, Berlin JS, Fishkind A, Zeller S (eds) Emergency psychiatry: principles and practice. Lippincott Williams & Wilkins/Wolters Kluwer Health, Baltimore

Citrome L (2013) Addressing the need for rapid treatment of agitation in schizophrenia and bipolar disorder: focus on inhaled loxapine as an alternative to injectable agents. Ther Clin Risk Manag 9:235–245

Frogley C, Taylor D, Dickens G, Picchioni M (2012) A systematic review of the evidence of clozapine's anti-aggressive effects. Int J Neuropsychopharmacol 15:1351–1371. doi:10.1017/S146114571100201X

Topiwala A, Fazel S (2011) The pharmacological management of violence in schizophrenia: a structured review. Expert Rev Neurother 11:53–63

Volavka J, Citrome L (2011) Pathways to aggression in schizophrenia affect results of treatment. Schizophr Bull 37:921–929

Volavka J, Czobor P, Derks EM, Bitter I, Libiger J, Kahn RS, Fleischhacker WW (2011) Efficacy of antipsychotic drugs against hostility in the European First-Episode Schizophrenia Trial (EUFEST). J Clin Psychiatry 72:955–961

Volavka J, Swanson JW, Citrome LL (2012) Understanding and managing violence in Schizophrenia. In: Lieberman JA, Murray RM (eds) Comprehensive care of Schizophrenia: a textbook of clinical management. Oxford University Press, New York, pp 262–290

Schizotypal Personality Disorder

Nobumi Miyake and Seiya Miyamoto
Schizophrenia Treatment Center, Department of Neuropsychiatry, St. Marianna University School of Medicine, Kawasaki, Kanagawa, Japan

Synonyms

Borderline schizophrenia; Latent schizophrenia; Schizotaxia; Schizotype; Schizotypy; SPD; STPD

Definition

In DSM-5 (American Psychiatric Association 2013), schizotypal personality disorder (SPD) is characterized by pervasive social and interpersonal deficits and by subtle, psychotic-like symptoms (Table 1).

Individuals with SPD often have ideas of reference. They frequently misinterpret situations as being strange or having unusual experience for them. Suspiciousness/paranormal and superstitious beliefs are common for these people. They therefore feel extreme discomfort with maintaining close relationships with other people. As a result, they usually have no or few close friends or confidants other than a first-degree relative (American Psychiatric Association 2013).

SPD may be first apparent in childhood and adolescence with solitariness, poor peer relationships, social anxiety, underachievement in school, hypersensitivity, peculiar thoughts and language, and bizarre fantasies. These children may appear "odd" or "eccentric" and attract teasing. Patients with SPD usually exhibit a low level of occupational functioning and at best work at occupations considerably below their levels of education. SPD has also a relatively stable course, with only a small proportion of individuals going on to develop schizophrenia or another psychotic disorder (American Psychiatric Association 2013). There is actually high consistency in genetic and neurobiological data that SPD can be reasonably regarded as a schizophrenia spectrum disorder (Herpertz et al. 2007).

Role of Pharmacotherapy

Clinical Features, Etiology, and Pathogenesis

In community studies of SPD, reported rates range from 0.6 % in Norwegian samples to 4.6 % in a US community sample, which implies an appreciable social cost and public health impact. The prevalence of SPD in clinical populations seems to be infrequent (0–1.9 %), with a higher estimated prevalence in the general population (3.9 %) found in the National Epidemiologic Survey on Alcohol and Related Conditions. SPD may be slightly more common in males (American Psychiatric Association 2013). With regard to psychiatric comorbidities, patients tend to suffer from anxiety and depressive disorders, and some of them also fulfill the criteria of schizoid, borderline, paranoid, and avoidant personality disorders (Herpertz et al. 2007).

Schizotypal Personality Disorder, Table 1 Classification and criteria of schizotypal personality disorder according to DSM-V (American Psychiatric Association 2013)

Diagnostic criteria 301.22 (F21)
A. A pervasive pattern of social and interpersonal deficits marked by acute discomfort with, and reduced capacity for, close relationships as well as by cognitive or perceptual distortions and eccentricities of behavior, beginning in early adulthood and present in a variety of contexts, as indicated by five (or more) of the following
1. Ideas of reference (excluding delusions of reference)
2. Odd beliefs or magical thinking that influences behavior and is inconsistent with subcultural norms (e.g., superstitiousness, belief in clairvoyance, telepathy, or "sixth sense"; in children and adolescents, bizarre fantasies or preoccupations)
3. Unusual perceptual experiences, including bodily illusions
4. Odd thinking and speech (e.g., vague, circumstantial, metaphorical, overelaborate, or stereotyped)
5. Suspiciousness or paranoid ideation
6. Inappropriate or constricted affect
7. Behavior or appearance that is odd, eccentric, or peculiar
8. Lack of close friends or confidants other than first-degree relatives
9. Excessive social anxiety that does not diminish with familiarity and tends to be associated with paranoid fears rather than negative judgments about self
B. Does not occur exclusively during the course of schizophrenia, a bipolar disorder or depressive disorder with psychotic features, another psychotic disorder, or autism spectrum disorder

The exact pathogenetic mechanisms underlying SPD are unknown. However, there are a number of biological data suggesting that SPD is a schizophrenia spectrum disorder. For example, structural brain MRI and diffusion tensor imaging studies in SPD have shown temporal lobe abnormalities similar to those observed in schizophrenia, while frontal lobe regions appear to show more sparing (Hazlett et al. 2012). Cognitive function is also impaired in the areas of working memory, verbal learning, and attention in SPD, and they may be particularly susceptible to cognitive tasks with high context dependence, as in schizophrenia (Siever et al. 2002). Furthermore, patients with SPD have been reported to show higher risk for schizophrenia-related disorders in first-degree relatives, deficits in prepulse inhibition and P50 suppression, and antisaccade paradigms (Herpertz et al. 2007). Evidence to date supports further investigation of genetic associations between symptoms of SPD and schizophrenia and suggests that social-interpersonal symptoms may be particularly promising in genetic analyses of schizophrenia (Tarbox and Pogue-Geile 2011).

Treatment Conditions

The conceptualization of SPD within the schizophrenia spectrum supports treatment with antipsychotic medications. Antipsychotics appear to be useful in the treatment of SPD, particularly in terms of psychotic-like symptoms (Ripoll et al. 2011). Individuals with SPD often seek treatment for the associated symptoms such as depression or anxiety. Over half may have a history of at least one major depressive episode. From 30 % to 50 % of individuals diagnosed with this disorder have a concurrent diagnosis of major depressive disorder when admitted to a clinical setting (American Psychiatric Association 2013). Antidepressants and/or anxiolytic medications may be prescribed to treatment for these symptoms.

Treatment with Antipsychotics

Overall, clinical trials for SPD have been complicated by comorbidity, especially with other personality disorders (Ripoll et al. 2011). There have been few clinical studies of the pharmacological treatment of SPD. Five studies of treatment with low-dose first-generation antipsychotics such as haloperidol and thiothixene have been reported, but only three of the studies were placebo controlled (Hori 1998; Koenigsberg et al. 2003). Most patients in these studies had comorbid diagnoses of SPD and borderline personality disorder (BPD), making it difficult to determine which disorder the medication might be treating. In general, the patients in these trials appeared to show modest improvement, with the greatest effects on psychotic-like symptoms and anxiety (Hori 1998; Koenigsberg et al. 2003). However, dropout rates were high because of considerable sensitivity to side effects of antipsychotics in this population. In addition, the sample size of these studies was small, which limited the power and generalizability of the findings.

The effects of second-generation antipsychotics (SGA) on SPD have been investigated in one well-controlled study of risperidone (Koenigsberg et al. 2003) and one open-label study of olanzapine (Keshavan et al. 2004). On the whole, there is some evidence that SGA may be effective in reducing symptom severity, targeting psychotic-like symptoms and general functioning in SPD (Herpertz et al. 2007). However, the sample size of these studies was also small, and one study was an open-label design. Further large, controlled, prospective studies are required to evaluate the effects of antipsychotics on symptoms and functional outcome in patients with SPD.

Treatment with Antidepressants and Other Medications

Several open-label studies have suggested a role for antidepressants in treating self-injury, psychotic-like, and depressive symptomatology in SPD (Ripoll et al. 2011). However, there have been no randomized, placebo-controlled trials of antidepressants, mood stabilizers, or anxiolytic medications in SPD. Thus, there is no reliable evidence for the efficacy of these medications in the treatment of SPD (Herpertz et al. 2007).

Both pergolide, a dopaminergic agonist at D_1 and D_2 receptor, and guanfacine, a noradrenergic α_{2A} agonist, may be effective for improving cognitive performance in patients with SPD (Ripoll et al. 2011). Given that cognitive impairments may be responsible for the overall clinical functioning in SPD like schizophrenia, research on pharmacotherapy for this target symptom, particularly for social cognition, is warranted in the future.

Cross-References

- ▶ Antidepressants
- ▶ First-Generation Antipsychotics
- ▶ Haloperidol
- ▶ Olanzapine
- ▶ Pergolide Mesylate
- ▶ Prepulse Inhibition
- ▶ Risperidone
- ▶ Schizophrenia
- ▶ Second-Generation Antipsychotics
- ▶ Thiothixene

References

American Psychiatric Association (2013) Diagnostic and statistical manual of mental disorders, 5th edn. American Psychiatric Association, Arlington, pp 655–659

Hazlett EA, Goldstein KE, Kolaitis JC (2012) A review of structural MRI and diffusion tensor imaging in schizotypal personality disorder. Curr Psychiatry Rep 14(1):70–78

Herpertz SC, Zanarini M, Schulz CS et al (2007) World Federation of Societies of Biological Psychiatry (WFSBP) guidelines for biological treatment of personality disorders. World J Biol Psychiatry 8(4):212–244

Hori A (1998) Pharmacotherapy for personality disorders. Psychiatry Clin Neurosci 52(1):13–19

Keshavan M, Shad M, Soloff P et al (2004) Efficacy and tolerability of olanzapine in the treatment of schizotypal personality disorder. Schizophr Res 71(1):97–101

Koenigsberg HW, Reynolds D, Goodman M et al (2003) Risperidone in the treatment of schizotypal personality disorder. J Clin Psychiatry 64(6):628–634

Ripoll LH, Triebwasser J, Siever LJ (2011) Evidence-based pharmacotherapy for personality disorders. Int J Neuropsychopharmacol 14(9):1257–1288

Siever LJ, Koenigsberg HW, Harvey P et al (2002) Cognitive and brain function in schizotypal personality disorder. Schizophr Res 54(1–2):157–167

Tarbox SI, Pogue-Geile MF (2011) A multivariate perspective on schizotypy and familial association with schizophrenia: a review. Clin Psychol Rev 31(7):1169–1182

Scopolamine

Synonyms

Hyoscine; Levo-duboisine

Definition

A tropane alkaloid drug that acts as an antagonist at the muscarinic subtype of acetylcholine receptor. It is named after the plant genus *Scopolia*. It acts as a nonselective competitive antagonist. In addition to its use in experiments as a disruptor of cognitive functioning, its main medical uses are in the treatment of nausea and motion sickness and intestinal cramping, as an antiemetic, and for ophthalmic purposes to induce pupil dilation and paralysis of the eye-focusing muscles.

Cross-References

- ▶ Animal Models for Psychiatric States
- ▶ Muscarinic Cholinergic Receptor Agonists and Antagonists

Screening Models

Synonyms

Animal model with predictive validity

Definition

An animal model focused on drug development. The principal aim is to predict the therapeutic

Cross-References

- ► Abuse Liability Evaluation
- ► Anxiety: Animal Models
- ► Attention Deficit Hyperactivity Disorders: Animal Models
- ► Autism: Animal Models
- ► Dementias: Animal Models
- ► Depression: Animal Models
- ► Eating Disorders: Animal Models
- ► Predictive Validity
- ► Schizophrenia: Animal Models

SDLP

Definition

Standard deviation of lateral position.

Secobarbital

Synonyms

Secobarbitone

Definition

Secobarbital is an intermediate- to short-acting (half-life 15–40 h) barbiturate drug that is used primarily as a hypnotic agent to treat insomnia. It is also used as a premedication before surgery and for the emergency management of seizures. Secobarbital is an agonist at ► $GABA_A$ receptors thereby increasing inhibitory ► GABAergic transmission in the CNS. It binds to a distinct site upon the ► $GABA_A$ receptor ion channel complex increasing the duration of time for which the Cl^- channel is open. Secobarbital, like other barbiturate drugs, has a narrow therapeutic window, that is, a small difference between the minimum effective dose for sedation and maximum tolerated dose producing coma and death. Long-term use of this drug is contraindicated because of its ability to produce tolerance and dependence, and abrupt withdrawal from chronic treatment precipitates a life-threatening ► withdrawal syndrome. It is used illicitly for the purposes of getting high, to reduce anxiety, to counteract the effects of stimulant drugs, for example, cocaine and ecstasy, and in suicide.

Cross-References

- ► Barbiturates
- ► Hypnotics
- ► Withdrawal Syndromes

Second- and Third-Generation Antipsychotics

Gary Remington
Department of Psychiatry, Centre for Addiction and Mental Health, University of Toronto, Toronto, ON, Canada

Synonyms

Atypical antipsychotics; New-generation antipsychotics; Novel antipsychotics

Definition

These antipsychotics, with clozapine as the prototype, are distinguishable from their first-generation counterparts by a lower liability for extrapyramidal symptoms (EPS) and claims of improved efficacy that extend to other domains beyond psychosis (e.g., cognition, negative symptoms). These drugs have been designated "atypical" when compared to the older "typical" or conventional antipsychotics like chlorpromazine, haloperidol, perphenazine, and fluphenazine.

The distinction between second- and third-generation antipsychotics has been made based on mechanistic differences. Specifically, aripiprazole is the first approved antipsychotic that is a partial dopamine agonist and, as such, has been designated a third-generation antipsychotic. This presumes that all other atypical compounds to this point, including clozapine, share some common attributes pharmacologically that account for their unique clinical profile. However, beyond dopamine D_2 antagonism, which characterizes all antipsychotics including aripiprazole, this is not the case. For example, the serotonin 5-HT_2/dopamine D_2 model does not adequately represent all of these compounds (see "Atypicality: Mechanisms of Action", below). Thus, as a partial dopamine agonist, aripiprazole is unique amongst the atypical antipsychotics, but the distinction between second- and third-generation agents may misrepresent the homogeneity of all other atypicals.

Pharmacological Properties

First-Generation Antipsychotics and Limitations

Antipsychotics were introduced for clinical use in the 1950s. At the time, descriptive terminology included "major tranquilizers" (because of their sedating/calming effect) and "neuroleptics," literally meaning "to take the neuron" and reflective of their liability for motor side effects (i.e., EPS). They rapidly became the treatment of choice for psychotic conditions such as schizophrenia although it was initially unclear as to what aspect of their diverse pharmacology accounted for this effect. Within the next years, it was established that dopamine D_2 binding seemed critical to the antipsychotic effect, and this generated an additional subgroup of these drugs characterized by high affinity for the dopamine D_2 receptor, drawing a distinction between low-potency (e.g., chlorpromazine) and high-potency (e.g., haloperidol) neuroleptics. A liability for EPS, both acute (e.g., parkinsonism) and chronic (e.g., tardive dyskinesia or TD) characterized all these drugs and represented a substantial burden to their use in the clinical setting.

Clozapine

Clozapine, developed in the 1960s, was notable for its antipsychotic efficacy in the face of minimal EPS at therapeutic doses. It was introduced for clinical use in the early 1970s but soon withdrawn in a number of countries because of a cluster of deaths, subsequently linked to its associated risk of agranulocytosis. Work in the late 1980s, though, underscored this drug's superiority in treatment-resistant schizophrenia and suggested it might also have broader efficacy (i.e., negative and positive symptom improvement). By the 1990s it was reintroduced in different countries with the associated requirement of regular hematologic monitoring.

Atypicality

Atypicality has been defined as lack of EPS at therapeutic doses (Meltzer et al. 2003). At the time of clozapine's development, it was unique amongst existing antipsychotics in this regard, establishing it as the prototype of atypicality. Over time, a new class of antipsychotics (second generation) was developed to mirror clozapine's clinical benefits while circumventing its adverse side effects, in particular risk of agranulocytosis. In this process, it was assumed that these drugs would also mirror clozapine in other regard (e.g., greater efficacy in treatment-resistant schizophrenia, improvement in negative and positive symptoms). With a new class of antipsychotic available, the previous medications came to be designated as "first generation" (vs. second generation), "typical" (vs. atypical), or "conventional" (vs. novel).

Atypicality: Mechanisms of Action

Several aspects of clozapine's diverse pharmacology were highlighted as possibly contributing to its unique clinical profile. These included its low binding affinity for the dopamine D_2 receptor in addition to comparatively high binding at the serotonin 5-HT_2 receptor. An elegant body of preclinical work led Meltzer et al. to postulate that the profile of greater serotonin 5-HT$_2$ versus dopamine D$_2$ binding accounted for clozapine's atypical features (Meltzer et al. 2003), a model that was rapidly embraced in drug development and one that describes most second-generation antipsychotics including asenapine, iloperidone, lurasidone, olanzapine, paliperidone, quetiapine, risperidone, sertindole, ziprasidone, and zotepine. Critical to this model is the ratio of serotonin 5-HT$_2$ and dopamine D$_2$ binding; earlier antipsychotics demonstrated combined serotonin 5-HT$_2$/dopamine D$_2$ antagonism but did not meet this identified ratio and were not seen as unique clinically.

Clozapine's low affinity for the dopamine D$_2$ receptor also called into question the long-standing argument that dopamine D$_2$ antagonism is absolutely critical for antipsychotic efficacy. This encouraged efforts to look at non-dopaminergic strategies and, following the importance this model ascribed to serotonin, led to work investigating the notion that serotonin 5-HT$_2$ antagonism alone might be sufficient (e.g., ritanserin, MDL-100907, fanaserin). While this was not substantiated, there continues to be interest in the pursuit of antipsychotic development that is not hinged on dopamine blockade.

Although the serotonin 5-HT$_2$/dopamine D$_2$ model was widely embraced, there was reason to pursue other explanatory models. In countries outside of North America, amisulpride and sulpiride (substituted benzamides) are accessible clinically and seen as atypical; however, their pharmacological profile precludes an explanation based on serotonin 5-HT$_2$/dopamine D$_2$ antagonism. At least two other models have since been posited to account for the atypical features of newer antipsychotics. The "low affinity-fast dissociation" model (Kapur and Seeman 2001) holds that clinical benefits such as decreased EPS, lack of prolactin elevation, and possible improvement in cognitive and negative symptoms may occur as a result of transient binding at the dopamine D$_2$ receptor, which mitigates against impairment of phasic dopamine release. Thus, while all drugs depress tonic dopamine release, those with more rapid dissociation (i.e., fast K$_{off}$ drugs) are less likely to alter phasic dopamine release, essential for dopamine to exert its physiologic effects in the course of daily activities. There is, for example, abundant evidence linking dopamine to reward and goal-directed behaviors, as well as cognition (Berridge 2007). This line of thinking suggests that there may be clinical gains through what these drugs *do not do* (i.e., sustained dopamine blockade) rather than by unique gains attributable to some other aspect of their rich receptor-binding profiles (e.g., concomitant serotonin 5-HT$_2$ binding).

A second hypothesis ascribes the clinical gains associated with atypicality to limbic selectivity (Bischoff 1992) and support for this comes from several lines of investigation (Remington and Kapur 2005). Looking at early gene expression (c-fos and c-jun) as a marker of synaptic activity, it has been noted that as a class the newer antipsychotics are associated with regional differences (e.g., increased c-fos expression in limbic vs. striatal regions). Similarly, more recent opportunities to examine D$_2$ binding extrastriatally have provided evidence that atypical antipsychotics demonstrate preferential binding for extrastriatal structures, for example, the temporal cortex. This could account for the diminished risk of EPS and hyperprolactinemia, as well as a more direct effect on other brain regions. Of note, while the atypicals as a class are associated with decreased EPS and a diminished risk of hyperprolactinemia, this is variable; risperidone, paliperidone, and the substituted benzamides carry a greater liability. It has been suggested that the limbic selectivity model is, at least in part, a variation of the low affinity-fast dissociation hypothesis in that the decreased binding in striatal regions reflects competitive displacement by endogenous dopamine at dopamine D$_2$ receptors in dopamine-rich regions.

The distinction between aripiprazole and other atypical antipsychotics is, in contrast, clearer and draws upon a shift in thinking regarding dopamine's role in schizophrenia. For decades, the most widely held biochemical model for schizophrenia posited a disorder of hyperdopaminergic activity. However, by the 1980s a clearer distinction was drawn between positive and negative symptoms, and subsequently our conceptualization of schizophrenia has broadened even further to include other domains (e.g., cognition). In the context of these changes, dopamine's role has been reframed to accommodate its differential involvement in this expanded model, suggesting more complex feedback loops involving various dopamine receptors and other neurochemical systems. For example, while limbic structures, dopamine D_2/D_3 receptors, and hyperdopaminergic activity have been implicated in the positive symptoms, cognitive and negative symptoms have been linked to prefrontal regions, a role for dopamine D_1 receptors, and hypodopaminergia.

In this context, drugs that affect dopamine differentially may have distinct clinical advantages and aripiprazole's proposed dual action on dopamine fits with this line of thinking. A partial dopamine agonist acting on postsynaptic D_2 receptors as well as presynaptic dopamine autoreceptors, its effects are linked to whether dopamine activity is high (i.e., mesolimbic) or low (i.e., mesocortical). Thus, through its antagonist properties it diminishes positive symptoms (mesolimbic), while its agonist profile results in improved negative and cognitive symptoms (directly at the mesocortical level and indirectly at the nigrostriatal level in the form of attenuated EPS).

The popularity of each of these models speaks to the importance of dopamine and serotonin in explaining the unique clinical benefits of the newer antipsychotics, but these drugs are characterized by diverse pharmacological profiles that impact a variety of receptors and systems (e.g., acetylcholine, norepinephrine, histamine, glutamate) (Remington and Kapur 2005). Although the currently available newer antipsychotics can each be characterized by one of the aforementioned models, the potential role(s) of these other systems remains a subject of investigation, both in terms of therapeutic efficacy and side effects. The involvement of other receptors and systems has taken on added interest with the growing body of evidence conceptualizing schizophrenia as a disorder of multiple symptom domains (e.g., positive, negative, cognitive, affective), paralleled by increased awareness regarding the clinical and functional importance of these other nonpsychotic features. The potential benefits of an effective antipsychotic without direct dopamine blockade remains alluring, ensuring the search for new antipsychotics focused on other systems thought to play an important role in schizophrenia.

Atypicality: Second-/Third-Generation Antipsychotics

While Meltzer et al. defined atypicality as absence of notable EPS at therapeutic doses (Meltzer et al. 2003), evidence clearly indicates that the newer antipsychotics are not the same in this regard. For example, risperidone has been identified as having a greater liability for EPS than many other atypical agents, a risk that is dose dependent – this also distinguishes it from other atypicals such as clozapine and quetiapine. It must also be noted that the benefits of the atypical antipsychotics, particularly agents like risperidone, in terms of EPS have been challenged on methodological grounds; often comparator trials with first-generation antipsychotics have used haloperidol, which has a notable propensity for EPS, and at doses in excess of what are currently recommended. It remains, though, that as a class the atypicals have been associated with diminished risk of EPS, and it is this defining feature that best distinguishes atypical antipsychotics from their conventional counterparts.

Although classifying medications based on presence or absence of a particular side effect may seem misguided, it has occurred again more recently with the growing recognition that the newer antipsychotics, in particular agents like clozapine and olanzapine, demonstrate a substantially greater risk for weight gain and associated metabolic disturbances. At least some regulatory bodies have chosen to identify this as a class effect characterizing all atypicals although evidence clearly indicates that, as with EPS, there are marked differences between these drugs on this dimension. For example, at the other end of the continuum from olanzapine and clozapine are drugs like ziprasidone and aripiprazole that are considered to be "weight neutral" (Newcomer 2007).

Other definitions, for example, based on mechanism(s) of action, are even less homogeneous. While the serotonin 5-HT$_2$/dopamine D$_2$ hypothesis dominated initial efforts to replicate clozapine's clinical advantages, agents now classified as atypical cannot be collectively grouped within this model. Asenapine, clozapine, lurasidone, olanzapine, iloperidone, paliperidone, quetiapine, risperidone, sertindole, ziprasidone, and zotepine can be categorized as serotonin 5-HT$_2$/dopamine D$_2$ antagonists, but this is not the case for amisulpride and sulpiride, where their benefits might best be explained by the low affinity-fast dissociation model. In fact, it has been argued that this same model could as readily explain the benefits of drugs that meet criteria for serotonin 5-HT$_2$/dopamine D$_2$ antagonists although this is the subject of ongoing debate. As a partial dopamine agonist, aripiprazole currently stands alone pharmacologically, which has led some to suggest that it represents a third-generation antipsychotic. This is, of course, a misnomer if we are to use serotonin 5-HT$_2$/dopamine D$_2$ antagonism as the mechanism of action that categorizes all other newer antipsychotics (given the substituted benzamides). The low affinity-fast dissociation explanation might better capture other antipsychotics, but this debate itself speaks to the pitfall of classifying such diverse pharmacological agents on models that are hypothetical and singularly focused. It is very likely that the clinical gains we are seeking to achieve can be attained through different pathways and that benefits may be accrued through a combination of pharmacological manipulations.

Attempting to collectively group these agents based on clinical similarities faces similar challenges (Tandon et al. 2008), made more difficult by the rapid expansion in purported benefits that occurred in parallel to the development of the different atypicals.

The fact that the atypical antipsychotics differ on various measures of efficacy and side effects, in combination with evidence that clinical differences between them and first-generation antipsychotics may not be as prominent as once thought, has challenged the utility of terminology like "second generation" (Leucht et al. 2009). As an alternative, it has been argued that this "all or none" approach to categorization should give way to a strategy that allows individual drugs to be defined across a number of relevant measures (i.e., different clinical measures and side effects) (Waddington and O'Callaghan 1997).

Formulations

All atypical antipsychotics are available in oral preparations that call for once daily administration. There are examples where it is recommended that the dose be twice daily, based on shorter elimination half-life (e.g., quetiapine, ziprasidone), although extended-release oral formulations have been developed to address this (e.g., quetiapine). There is some question as to the value of establishing daily dosing based on peripheral pharmacokinetics, as in vivo neuroimaging techniques such as positron emission tomography (PET) have demonstrated that central kinetics do not necessarily mirror what is observed peripherally.

With certain atypical antipsychotics, variations in formulations have been marketed to ease administration and adherence. These include rapidly dissolving oral formulations (e.g., risperidone, olanzapine), short-acting intramuscular for acute treatment (e.g., risperidone, olanzapine, ziprasidone), and longer-acting depot formulations for maintenance treatment (e.g., risperidone, administered every 2 weeks). At least some of the first-generation antipsychotics had also developed shorter- and longer-acting injectable formulations, but the rapidly dissolving oral formulation is unique to the newer medications.

Pharmacokinetics

(see "▶ Antipsychotic Drugs")

Efficacy and Effectiveness

Central to antipsychotic development was the goal of more tolerable agents from the standpoint of EPS, but limitations in the efficacy of first-generation antipsychotics also represented an impetus. As many as 25–30% of individuals treated with these drugs were designated treatment resistant or refractory, with an additional subgroup deemed only partially responsive. Furthermore, these drugs, while at least reasonably effective in treating positive symptoms (e.g., delusions, hallucinations), were not particularly useful in controlling negative symptoms (e.g., amotivation, alogia).

With clozapine there was evidence that it was superior to first-generation antipsychotics on both these domains, in addition to its improved EPS profile. Thus, implicit in the development of other antipsychotics that were intended to mirror clozapine was the assumption of similar clinical benefits. As the list of new agents expanded so too did the purported benefits though, with suggestions that these drugs demonstrated clinical superiority on other symptoms (e.g., cognitive, affective) and outcome measures (e.g., quality of life, adherence).

Initial efficacy trials generally demonstrated superiority of the different atypical agents versus a first-generation antipsychotic on total clinical (e.g., Brief Psychiatric Rating Scale (BPRS), Positive and Negative Syndrome Scale (PANSS)) and positive symptom scores and frequently negative symptom scores. However, many of these efficacy studies were subsequently challenged methodologically; not only were the trials relatively short (i.e., 6–8 weeks), they often employed haloperidol as the comparator, a typical antipsychotic with notable propensity for EPS, at doses in excess of what are currently recommended.

Similarly, questions arose over claims of superior efficacy in other symptom domains. These followed two lines of thinking: (a) the magnitude of change and extent of difference between the atypical agents and first-generation antipsychotics and (b) factors underlying such changes. With regard to the former, claims regarding the extent of improvement on neuropsychological measures were tempered over time, with current thinking that at best these changes are modest and not of a magnitude that would translate to notable differences in functional outcome. For negative symptoms as well, subsequent evidence fostered a rethinking of how much different atypical antipsychotics were. In both cases, initial claims implicated unique aspects of these new drugs and their pharmacology (e.g., greater serotonin 5-HT$_2$ antagonism). However, an alternative explanation suggested that methodological issues, discussed earlier, favoring the atypicals could account for much of these differences. Further, the possibility was raised that gains could be explained, at least in part, not by what these drugs did through different receptors and systems compared to first-generation antipsychotics, but by what they did not do (i.e., invoke high and sustained levels of dopamine D$_2$ binding). Evidence demonstrated that this was associated with various adverse effects beyond motor and endocrine problems (EPS and hyperprolactinemia), including dysphoria, amotivation, and cognitive impairment.

More recently, the research focus shifted from an assessment of atypical antipsychotics in the context of smaller and circumscribed efficacy trials to larger effectiveness studies thought to be more representative of "real-world" clinical practice. Once again, results have suggested that differences between atypicals and first-generation antipsychotics are notably smaller than once thought. They have also provided further evidence that clozapine is clinically superior, even amongst other atypicals, in treating treatment-resistant schizophrenia. Of note, there is some suggestion that amongst other atypical antipsychotics there may be clinical differences, a position that has been supported through other larger-scale approaches such as meta-analytic studies. As of yet, such differences are not well established, with methodological issues (e.g., dosing) making conclusions of this sort difficult (Leucht et al. 2009; Tandon et al. 2008).

Treatment Algorithms

Current guidelines advocate atypical antipsychotics as first-line treatment for schizophrenia and related psychotic conditions. There is a lack of evidence that these drugs are more effective in this population, although decreased liability for EPS, including TD, and hyperprolactinemia, in combination with improved tolerability represent advantages from the standpoint of side effects. However, the increased risk of weight gain and metabolic disturbances that can be observed with at least some of the atypicals, in conjunction with their higher cost and lack of clear clinical superiority, have more recently challenged this position.

Clozapine is generally advocated as treatment of choice following suboptimal response to two antipsychotic trials. This is in line with evidence that it is superior to all other antipsychotics in the treatment-resistant population and that treatment resistance can be seen quite soon after the illness onset. There is no consistent evidence that combinations of antipsychotics, including those that include clozapine, offer superior clinical efficacy in individuals who have not responded to monotherapy (i.e., one antipsychotic).

Paralleling the increased number of atypical antipsychotics has been a rise in off-label use of these medications and efforts to expand their indications. Their efficacy in a variety of psychiatric disorders is being examined (e.g., pervasive developmental disorder; generalized anxiety disorder; psychotic depression), and as a result their approved indications have grown (e.g., autism; bipolar disorder, manic and mixed states), depending on agent and regulatory body.

Side Effects

As a class, the atypical antipsychotics continue to be seen as better from the standpoint of EPS, although such a distinction is less clear depending on a variety of factors that include specific agent, comparator first-generation antipsychotic, and dose. Current evidence, albeit limited, does favor the atypicals as well in terms of TD risk (Correll and Schenk 2008), a significant clinical advantage should this finding hold true. There is also an indication that as a class the atypical antipsychotics might be better tolerated. While there was an expectation that such benefits would translate to improved adherence, this has not been confirmed, a reminder that nonadherence is complex and multifactorial.

Atypical antipsychotics, to varying degrees, also share side effects that were identified with their first-generation counterparts (e.g., sedation, postural hypotension). In general, the atypicals have a lower risk of hyperprolactinemia, although this is variable; risperidone, paliperidone, and the substituted benzamides carry a greater liability. There was anticipation that neuroleptic malignant syndrome (NMS), a potentially fatal adverse event that can occur with all first-generation antipsychotics, would be diminished or even absent with the atypical agents; however, all of the newer drugs to date, including clozapine, appear to carry this risk and it is not yet clear as to whether the overall liability has been reduced.

Although the atypical antipsychotics appear superior from the standpoint of movement disorders, concern has been raised about their increased liability for weight gain and associated metabolic disturbances, including dyslipidemia and type 2 diabetes. This, in turn, translates to a greater risk of metabolic syndrome, cardiovascular risk, and associated mortality. There is added concern because schizophrenia has already been associated with a shortened life span (independent of suicide) which may be related to various factors (e.g., economic, lifestyle, access to medical care). Moreover, there is some evidence to suggest that schizophrenia itself may be a risk factor for diabetes. As with EPS, the atypicals vary in risk of weight gain, with olanzapine and clozapine identified as carrying the greatest risk, while at the other end of the spectrum, aripiprazole and ziprasidone have been identified as "weight neutral" (Newcomer 2007).

In the geriatric population, atypical antipsychotics have been associated with increased risk of death (in the range of 1.6-fold) from varied causes (e.g., cardiovascular, infectious). In fact, though, both typical and atypical antipsychotics have been linked to increased mortality rates in schizophrenia (Weinmann et al. 2009). Data regarding safety and efficacy in the pediatric population are limited.

Conclusions

With clozapine as the prototype, a number of new antipsychotics have been developed. Collectively, they have been termed "atypical" and further distinguished as "second-" or "third-generation" antipsychotics. Atypical differentiates these drugs from their conventional counterparts based on clinical profile, specifically diminished risk of EPS, although this definition subsequently broadened to include a variety of other measures. The distinction between second- and third-generation antipsychotics is mechanistic, with aripiprazole (third generation) the only atypical antipsychotic to date that is a partial dopamine agonist.

The difference between these newer drugs and the so-called typical antipsychotics may not be as notable as once thought, with the exception of clozapine's clinical superiority to all other antipsychotics in treatment-resistant schizophrenia. Further, "atypical antipsychotics" implies a homogeneity amongst these newer drugs and separation from first-generation antipsychotics that may be both misleading and confusing. There are differences between atypicals on a variety of clinical/biological measures, and even the distinction between second- and third-generation antipsychotics implies two distinct mechanisms of action. However, this belies the complexity of pharmacological features that may underlie clinical differences.

This said, the atypical antipsychotics represent a clinical advance and the first notable shift in schizophrenia's pharmacotherapy since chlorpromazine. Moreover, they have stimulated numerous lines of research that have substantially impacted our understanding and conceptualization of schizophrenia. The clear limitations of these medications as well in terms of clinical response and side effects, though, underscore the need for better drugs. The widespread claims of atypical antipsychotics were in keeping with a "magic bullet" approach, where a single drug could be effective across the multiple symptom domains now defining schizophrenia. However, it may well be that a multidimensional approach, premised on the notion that each domain reflects different pathophysiologic mechanisms, proves a more useful strategy.

Cross-References

▶ Atypical Antipsychotic Drugs
▶ Effectiveness Studies
▶ Extrapyramidal Motor Side Effects
▶ Functional Outcome
▶ Hyperprolactinemia
▶ Tardive Dyskinesia
▶ Treatment-Resistant Schizophrenia

References

Berridge KC (2007) The debate over dopamine's role in reward: the case for incentive salience. Psychopharmacology (Berl) 191:391–431

Bischoff S (1992) Limbic selective neuroleptics. Clin Neuropharmacol 15(Suppl 1 Pt A):265A–266A

Correll CU, Schenk EM (2008) Tardive dyskinesia and new antipsychotics. Curr Opin Psychiatry 21:151–156

Kapur S, Seeman P (2001) Does fast dissociation from the dopamine D_2 receptor explain the action of atypical antipsychotics?: A new hypothesis. Am J Psychiatry 158:360–369

Leucht S, Kissling W, Davis JM (2009) Second-generation antipsychotics for schizophrenia: can we resolve the conflict? Psychol Med 39:1–12

Meltzer HY, Li Z, Kaneda Y, Ichikawa J (2003) Serotonin receptors: their key role in drugs to treat schizophrenia. Prog Neuropsychopharmacol Biol Psychiatry 27:1159–1172

Newcomer JW (2007) Metabolic considerations in the use of antipsychotic medications: a review of recent evidence. J Clin Psychiatry 68(Suppl 1):20–27

Remington G, Kapur S (2005) The pharmacology of typical and atypical antipsychotics. In: Factor SA, Lang A, Weiner WJ (eds) Drug induced movement disorders. Blackwell Futura, Malden, pp 55–71

Tandon R, Belmaker RH, Gattaz WF, Lopez-Ibor JJ Jr, Okasha A, Singh B, Stein DJ, Olie JP, Fleischhacker WW, Moeller HJ (2008) World Psychiatric Association Pharmacopsychiatry Section statement on comparative effectiveness of antipsychotics in the treatment of schizophrenia. Schizophr Res 100:20–38

Waddington JL, O'Callaghan E (1997) What makes an antipsychotic "atypical"? CNS Drugs 7:341–346

Weinmann S, Read J, Aderhold V (2009) Influence of antipsychotics on mortality in schizophrenia: a systematic review. Schizophr Res 113:1–11

Second Messenger

Synonyms

Secondary messenger molecule

Definition

Intracellular substance like Ca^{2+}, cAMP, inositol phosphates, or nitric oxide, the concentration of which is controlled by the activation of membrane receptors. Second messenger production relays downstream intracellular receptor events like protein phosphorylation or neurotransmitter release.

Cross-References

- Agonist
- Antagonist
- G-Protein-Coupled Receptors

Secondary Amine Tricyclic Antidepressants

Synonyms

TCAs

Definition

Secondary amine TCAs result from the metabolism of tertiary amine TCAs, during which there is loss of one methyl group on the nitrogen side chain. They include ▶ nortriptyline, protriptyline, and desipramine. Secondary amine TCAs are better tolerated than tertiary amine TCAs due to decreased histaminic, cholinergic, and alpha-1 adrenergic receptor blockade.

Second-Generation Anticonvulsants

Synonyms

Newer anticonvulsants

Definition

This category includes all compounds introduced after 1990.

Cross-References

- Gabapentin
- Oxcarbazepine
- Pregabalin
- Tiagabine
- Topiramate
- Zonisamide

Second-Generation Antipsychotics

Synonyms

Atypical antipsychotics; Novel antipsychotics

Definition

Clozapine was the first medication to have both good antipsychotic efficacy and a minimal risk for extrapyramidal motor side effects. It is therefore considered to have started a second generation of antipsychotics. Other drugs in this category include amisulpride, asenapine, blonanserin, iloperidone, lurasidone, olanzapine, paliperidone, perospirone, quetiapine, risperidone, sertindole, ziprasidone, and zotepine. Current evidence suggests that the differences in efficacy among these antipsychotics are generally small but the differences in adverse effects are larger. Thus, second-generation antipsychotics are not a homogeneous class.

Cross-References

- ► Amisulpride
- ► Blonanserin
- ► Clozapine
- ► Extrapyramidal Motor Side Effects
- ► Olanzapine
- ► Paliperidone
- ► Perospirone
- ► Quetiapine
- ► Risperidone
- ► Second- and Third-Generation Antipsychotics
- ► Sertindole
- ► Ziprasidone
- ► Zotepine

Second-Order Schedules

Definition

In a second-order schedule of reinforcement, a contingency arrangement under which a series of responses under a schedule of conditioned reinforcement is treated as a unit response under a second schedule that is simultaneously in effect. For example, under a second-order fixed-ratio 10 schedule of food delivery with fixed-ratio 5 units, every fifth response produces a conditioned reinforcer (e.g., a visual stimulus), and the completion of the tenth fixed-ratio 5 units produces the conditioned reinforcer accompanied by food delivery.

Cross-References

- ► Operant Behavior in Animals
- ► Schedule of Reinforcement

Sedative, Hypnotic, and Anxiolytic Dependence

Dan J. Stein
Department of Psychiatry, University of Cape Town, Cape Town, South Africa

Synonyms

Anxiolytic dependence; Benzodiazepine dependence; Hypnotic dependence; Sedative dependence

Definition

Benzodiazepines are widely prescribed for sedative, hypnotic, and anxiolytic purposes. In this entry, I focus on benzodiazepine dependence, although similar considerations may pertain to certain nonbenzodiazepine hypnotics (e.g., zolpidem, zopiclone), and even to alcohol dependence (although I will not discuss this here). Criteria for substance dependence are provided in the most recent editions of the *Diagnostic and Statistical Manual of Mental Disorders* (fifth edition) (American Psychiatric Association 2013) and the *International Classification of Diseases* (tenth edition) (World Health Organization 1992). These criteria include tolerance (as defined by a need for either markedly increased amounts of the substance to achieve the same clinical effect or markedly diminished effect with continued use of the same amount of the substance), withdrawal (as defined either by the characteristic withdrawal syndrome for the substance or by the same or similar substance taken to avoid withdrawal symptoms), escalation of dosage, and various features indicative of impairment (e.g., unsuccessful attempts to cut down or control substance use, time spent in activities necessary to obtain the substance, important activities are given up or reduced, continued use despite knowledge of having a problem).

Background

Benzodiazepines were introduced in the 1950s and, over subsequent decades, became widely prescribed for various psychiatric symptoms, including agitation, insomnia, and anxiety. There has long been a controversy about whether such widespread use of benzodiazepines is appropriate (Ashton 1989; Lader 1983). On the one hand, these agents appear to be effective for the short-term treatment of several anxiety disorders and insomnia and are far safer than the older barbiturates. On the other hand, the benzodiazepines are not effective for all anxiety disorders, have important adverse effects, and may be difficult to withdraw. Although the nonbenzodiazepine hypnotics or "z-drugs" (e.g., zolpidem, zopiclone) may have a more advantageous adverse event profile, similar concerns again arise (Lader 1997). The animal behavioral literature is consistent with the relevant clinical observations (Lader 1994).

Soon after the introduction of the selective serotonin reuptake inhibitors for depression, many of these agents received extensive study for the treatment of several anxiety and related disorders (generalized anxiety disorder, obsessive-compulsive disorder, panic disorder, social anxiety disorder). These agents are effective for a broader range of anxiety disorders than are the benzodiazepines, and their robust antidepressant effects are valuable, given the high prevalence of comorbid depression in anxiety disorders. Further, they are relatively safe agents, and withdrawal is not typically problematic. Thus, they are widely advocated as the first line of pharmacotherapy for most anxiety disorders, for both acute symptoms and maintenance management (Bandelow et al. 2012).

Although there is clear support for the SSRIs in the treatment of anxiety disorders, there remains ongoing debate about the specific role of the benzodiazepines in these conditions. A number of studies indicate that when prescribed together with the SSRIs, the benzodiazepines may facilitate early efficacy and tolerability. However, by 6 or 8 weeks of treatment, no advantage is seen in patients initially co-prescribed benzodiazepines. Similarly, many clinicians would recommend benzodiazepines only in the short term for patients with psychotic disorders, depression with anxiety, and other psychiatric disorders where acute sedation, hypnosis, or anxiolysis is useful. A number of treatment guidelines and consensus statements have emphasized that benzodiazepines should generally be reserved for short-term use in minimal dosage (Ashton 2005; Lader and Russell 1993; Baldwin et al. 2013). They may also be useful as SSRI augmenting agents in treatment-refractory patients (Pollack et al. 2014).

The issue of whether medium- to long-term prescription of benzodiazepines can be appropriate under any circumstances remains contentious. There are clinicians who argue that there is a role for long-term benzodiazepine prescription; not all patients respond to other agents such as the SSRIs. There is some evidence of the value of long-term treatment in disorders such as panic disorder, and certain features of dependence (e.g., dose escalation, tolerance) are rarely present. Although the benzodiazepines may not be highly robust antidepressants, they may have some antidepressant effects. However, some authors emphasize the significant adverse events associated with benzodiazepines (with no change over time in cognitive impairment or vulnerability to accidents), argue that long-term prescription inevitably leads to benzodiazepine dependence (with inefficacy and impairment), and conclude that there is rarely a place for such treatments.

Given the availability of the SSRIs for anxiety disorders and of a range of treatment options for insomnia, it is reasonable to be cautious about medium- to long-term prescription of benzodiazepines. Certainly, both clinicians and patients should be aware of the associated adverse effects and the potential difficulties of withdrawal. Particular caution is required in certain populations (e.g., the elderly and those with a prior history of substance dependence). Although dose escalation is not very common and although there are perhaps patients in whom long-term use of benzodiazepines is useful, given data that benzodiazepine discontinuation is associated with psychiatric and cognitive improvements, it may be useful to consider benzodiazepine withdrawal at regular intervals in those who are on longer-term treatment (Ashton 2005; Lader and Russell 1993).

Role of Pharmacotherapy

The pharmacological mechanisms underlying benzodiazepine tolerance and withdrawal have received considerable study and may include both desensitization of inhibitory GABA receptors and sensitization of excitatory glutamatergic receptors. Benzodiazepine withdrawal symptoms include symptoms of anxiety states (e.g., anxiety, insomnia, dysphoria) as well as a range of other characteristic symptoms (e.g., perceptual distortions, depersonalization, confusion). Withdrawal symptoms may be somewhat different in younger patients, and relevant scales have been developed specifically for this population.

As above, some authors have emphasized that few patients on long-term dependence show tolerance, that not all meet criteria for dependence, and that, in some cases, long-term treatment is needed for chronic symptoms. Nevertheless, given data that reduction of benzodiazepines is accompanied by decreased psychiatric and cognitive symptoms, it also seems reasonable to consider benzodiazepine discontinuation in those who are taking these agents in the long term. Certainly, tapering of benzodiazepines is indicated when there are concerns about cognitive impairment or other adverse effects of long-term benzodiazepines, when there is loss of efficacy, or where there is recreational abuse and dependence. Several strategies are useful for encouraging such benzodiazepine withdrawal.

First, gradual dose tapering is important. Rapid tapering can precipitate benzodiazepine withdrawal symptoms, including the more severe ones (e.g., convulsions, psychotic symptoms, delirium). A reasonable approach may be to decrease the dose by around one-tenth every 1–2 weeks. There has been surprisingly little work comparing different withdrawal regimens (Lader et al. 2009). Thus, a reasonable approach is to individualize the withdrawal regimen; while some patients may be able to discontinue benzodiazepines over 6–8 weeks, others may require many months.

Second, switching to a benzodiazepine with a longer half-life prior to tapering may be useful. This may be particularly necessary in patients using high-potency benzodiazepines (e.g., lorazepam, alprazolam), where trials indicate a higher dropout rate during attempts at discontinuation (Voshaar et al. 2006). Diazepam, for example, has a slow elimination, allowing a gradual fall in blood concentration. Furthermore, it is available in low-dosage forms, including liquid formulations, which allows gradual tapering. Equivalent doses of diazepam can be determined for each of the benzodiazepines.

Third, psychological support can be useful. A range of simple interventions, including clinical audit, information sheets, letter from the general practitioner (GP), reminder cards, or brief consultations, have been shown to be effective in controlled studies (Lader et al. 2009). Similarly, elements of cognitive-behavioral therapy (CBT), such as psychoeducation and providing support, appear important. Furthermore, CBT may address comorbid disorders and may be particularly useful in preventing relapse. Patients with high-dose benzodiazepine abuse may require more intensive management, including hospitalization.

Fourth, treatment with concurrent psychotropics may be considered. Anticonvulsants, antidepressants, beta blockers, buspirone, flumazenil, and gabapentin have all been investigated. A Cochrane review suggested that only carbamazepine was useful in assisting in benzodiazepine discontinuation (Denis et al. 2006). However, the available data are insufficient to suggest routine administration of this agent in benzodiazepine withdrawal. Antidepressants may be useful for preexisting or emergent comorbid depression.

In general, a stepped approach, beginning with simple interventions and then proceeding to more intensive ones, is appropriate when attempting benzodiazepine discontinuation (Lader et al. 2009). There is a growing database of information about predictors of response to attempts to discontinue benzodiazepines, which may be useful in guiding future interventions (Mol et al. 2007). In the interim, however, it is reassuring to note the relative success of benzodiazepine discontinuation and the positive effects that this has for many patients.

Cross-References

▶ Benzodiazepines
▶ Selective Serotonin Reuptake Inhibitors

References

American Psychiatric Association (2013) Diagnostic and statistical manual of mental disorders, 5th edn. American Psychiatric Association, Washington, DC
Ashton H (1989) Risks of dependence on benzodiazepine drugs – a major problem of long-term treatment. Br Med J 298:103–104
Ashton H (2005) The diagnosis and management of benzodiazepine dependence. Curr Opin Psychiatry 18:249–255
Baldwin DS, Aitchison K, Bateson A, Curran HV, Davies S, Leonard B, Nutt DJ, Stephens DN, Wilson S (2013) Benzodiazepines: risks and benefits. A reconsideration. J Psychopharmacol 27(11):967–971
Bandelow B, Sher L, Bunevicius R, Hollander E, Kasper S, Zohar J, Möller HJ, WFSBP Task Force on Mental Disorders in Primary Care, WFSBP Task Force on Anxiety Disorders, OCD and PTSD (2012) Guidelines for the pharmacological treatment of anxiety disorders, obsessive-compulsive disorder and posttraumatic stress disorder in primary care. Int J Psychiatry Clin Pract 16(2):77–84
Denis C, Fatseas M, Lavie E, Auriacombe M (2006) Pharmacological interventions for benzodiazepine monodependence management in outpatient settings. Cochrane Database Syst Rev 3, CD005194
Lader M (1983) Dependence on benzodiazepines. J Clin Psychiatry 44:121–127
Lader M (1994) Biological processes in benzodiazepine dependence. Addiction 89:1413–1418
Lader M (1997) Zopiclone: is there any dependence and abuse potential? J Neurol 244:S18–S22
Lader M, Russell J (1993) Guidelines for the prevention and treatment of benzodiazepine dependence – summary of a report from the mental-health-foundation. Addiction 88:1707–1708
Lader M, Tylee A, Donoghue J (2009) Withdrawing benzodiazepines in primary care. CNS Drugs 23:19–34
Mol AJJ, Voshaar RCO, Gorgels WJMJ, Breteler MHM, Van Balkom AJLM, van de Lisdonk EH et al (2007) The role of craving in relapse after discontinuation of long-term benzodiazepine use. J Clin Psychiatry 68:1894–1900
Pollack MH, Van Ameringen M, Simon NM, Worthington JW, Hoge EA, Keshaviah A, Stein MB (2014) A double-blind randomized controlled trial of augmentation and switch strategies for refractory social anxiety disorder. Am J Psychiatry 171(1):44–53
Voshaar RCO, Couvee JE, Van Balkom AJLM, Mulder PGH, Zitman FG (2006) Strategies for discontinuing long-term benzodiazepine use – meta-analysis. Br J Psychiatry 189:213–220
World Health Organization (1992) ICD-10 classification of mental and behavioural disorders: diagnostic criteria for research. World Health Organization, Geneve

Selective Association

Definition

Selective association refers to the fact that not all classically conditioned associations can be acquired. Specifically, animals appear better able to associate specific stimuli, for example, tastes and toxicosis, than others, for example, audiovisual cues and toxicosis. These selective associations are thought to be shaped by evolution and to prepare animals for the acquisition of stimulus associations important for survival.

Selective Attention

Definition

This is the cognitive process by which we can select certain things in the environment to concentrate on, for example, a specific object, specific spatial location, a category of objects, etc., and ignore others.

Cross-References

▶ Attention

Selective Breeding

Definition

Selective breeding occurs when individual animals with specific behavioral (or anatomical or physiologic) traits are bred with animals with similar characteristics. The traits selected are extremes (such as high responders or low responders in a specified task) and are determined by some criterion of deviation from the group mean (e.g., behaviors that are two standard deviations above or below the mean).

Selective Serotonin Reuptake Inhibitors

Synonyms

SSRIs

Definition

Selective serotonin reuptake inhibitors (SSRIs) are a group of antidepressants that selectively inhibit the reuptake of serotonin into the presynaptic cell. This leads to an increased concentration of serotonin available to bind to postsynaptic receptors. Before the introduction of the SSRIs, the pharmacologic treatment of depression was mainly based on tricyclic antidepressants that inhibit the reuptake of serotonin and noradrenaline into the presynaptic cell.

Cross-References

▶ SSRIs and Related Compounds

Selegiline

Synonyms

Eldepryl; Emsam patch; L-deprenyl; Zelapar

Definition

Selegiline or L-deprenyl is a phenylethylamine, specifically (R)-N-methyl-N-(1-phenylpropan-2-yl)prop-2-yn-1-amine. It is an antidepressant that selectively inhibits the enzyme monoamine oxidase (MAO), preferentially the type B isoenzyme in the recommended low doses. Unlike other MAOIs, selegiline is less likely to interact adversely with the consumption of tyramine-containing food, the so-called cheese reaction. In 2006, the FDA approved this compound for the treatment of major depression in the form of a transdermal patch (Emsam patch). It is most often used in the treatment of ▶ Parkinson's disease in interaction with L-DOPA.

The side effects of selegiline are similar to those of amphetamine-like stimulants and may comprise cardiovascular, respiratory, motor, and perceptual problems.

Cross-References

▶ Antidepressants
▶ Monoamine Oxidase Inhibitors

Self-Administration of Drugs

David C. S. Roberts
Department of Physiology and Pharmacology,
Wake Forest University Health Sciences,
Winston-Salem, NC, USA

Synonyms

Drug seeking; Drug taking; Self-injection

Definition

"Self-administration" is an experimental technique which gives subjects, to some extent, control over the ingestion of a drug. In essence, the method is an operant conditioning procedure in which drug delivery is made contingent on a particular response emitted by the subject. A variety of species have been tested in self-administration procedures including human; however, the vast majority of published self-administration studies have examined oral or intravenous drug intake in rats or nonhuman primates. Self-administration techniques are used to study the reinforcing effects and underlying neurobiology of different classes of reinforcing drugs, to evaluate abuse liability of specific agents and to model the addiction process.

Impact of Psychoactive Drugs

History

There was a time when human intelligence was considered an essential requirement for the voluntary ingestion of drugs of abuse, such as heroin or cocaine. Given this bias, it was considered unlikely that any animal would self-administer psychoactive drugs.

In one of the earliest reports to challenge the prevailing dogma that drug addiction was exclusively a human condition, Spragg (1940) reported that chimpanzees would "work intentionally" to receive a morphine injection. The chimps received daily injections of morphine until there was clear evidence that physical dependence had developed (i.e., the appearance of withdrawal signs when the drug was withheld). Spragg then showed that chimps would engage in a sequence of responses that resulted in a morphine injection. If the animal were drug deprived, it would select the correct "key" (a colored wooden stick), turn, walk to a particular box, open it with the key, remove a syringe containing morphine, and hand it to the investigator who would deliver an intramuscular injection of morphine. When not drug deprived, the chimp would select a different key which provided access to food inside a different box.

Spragg wrote

> ...since morphine addiction seems to depend essentially upon forming an association between the administration of the drug and the alleviation of withdrawal symptoms, and since this sequence involves a time lag of 10-15min or more, the value of using subjects high enough in the phyletic scale to be able to make a delayed association of this nature is obvious. By this token, animals such as the rat, for example, could probably never become addicted to morphine, simply because they are not capable of forming associations of this order.... (p. 126)

The quote illustrates two important presumptions of the early literature. The first is the notion that drug taking is phylogenetically biased toward primates, and the second is that physical dependence is a necessary precondition of drug self-administration in animals. These assumptions went unchallenged for two decades.

Weeks (1962) of the Upjohn Company published a seminal paper which marked the beginning of modern self-administration studies. By combining operant conditioning techniques, a pump mechanism that delivered precise amounts of drug, and a chronically indwelling jugular catheter that allowed animals to move freely within a cage, Weeks demonstrated that rats learned to respond on a lever which resulted in the delivery of an intravenous injection of morphine. Clearly rats were capable of making the correct associations and were able to self-administer morphine. Weeks' initial demonstration used animals that were first made physically dependent. Later, however, it was shown that physical dependence was not a necessary precondition for drug self-administration. Readers are referred to the 1978 NIDA Monograph No. 20, Self-Administration of Abused Substances: Methods for Study (http://archives.drugabuse.gov/pdf/monographs/20.pdf) for a collection of reviews which summarize the foundational work in the field.

Patterns of Intake

The pattern of drug self-administration depends on the pharmacological class of the drug as well

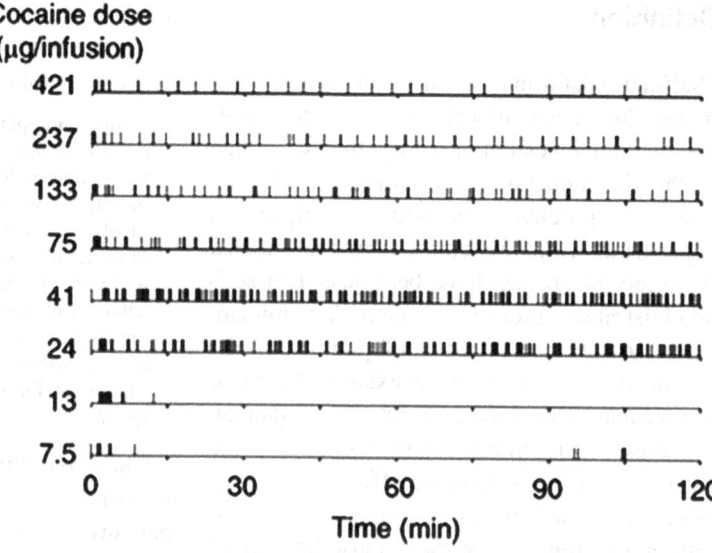

Self-Administration of Drugs, Fig. 1 Examples of a rat self-administering various doses of cocaine on an FR1 schedule. Each line is an event record of a daily 2-h session with each injection indicated by an *upward* tick. The unit dose for each daily session is indicated to the *left*. The figure illustrates that (above some threshold) lower doses are self-administered more frequently

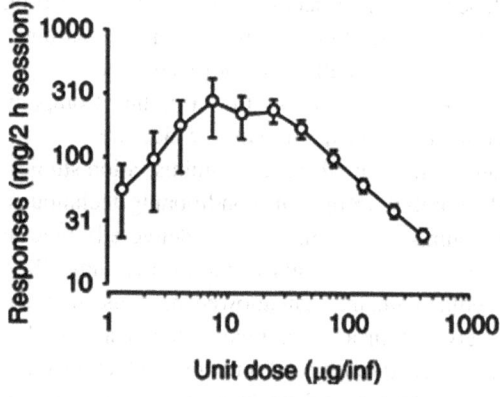
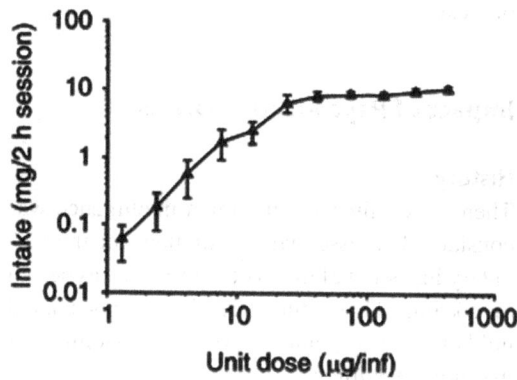

Self-Administration of Drugs, Fig. 2 The effect of manipulating dose on rate of responding (*left*) and total intake (*right*) for a group of rats. Each rat in the group was implanted with an IV catheter and trained to self-administer cocaine. Each animal was then tested using a descending series of doses as illustrated in Fig. 1. The graph on the *left* illustrates that above a mean threshold dose (apex), the rate of self-administration deceases as the unit injection dose is increased. The graph on the *right* illustrates that in spite of the decrease in the number of injections self-administered, the amount consumed within a session increased gradually with dose (Oleson and Roberts 2009)

as experimental factors such as dose, price, and availability. Figure 1 shows the pattern of cocaine self-administration in a rat using the most basic schedule, a fixed ratio 1 (FR1). Note the brief burst of responding in the first few minutes of the session; for the remainder of the session, the interval between each injection is relatively consistent. The dose per injection is referred to as the "unit dose." Figure 1 illustrates that as the unit dose is reduced, the number of self-administered injections increases within the 2-h session until; at some point, the dose no longer supports sustained responding (bottom two event records). This dose-response relationship for a group of animals is shown in Fig. 2 (*left*). The gradual ascending limb apparent to the left of the apex is an artifact of averaging across animals in the group. Data from each individual animal shows an abrupt change from a non-reinforcing to a threshold reinforcing dose (see Fig. 1). To the right of the apex is a descending limb with a very small variance surrounding the mean rate at each dose. Figure 2 (*right*) shows the same data plotted according to total drug intake during the session. This graph illustrates that above some threshold dose, animals tend to self-administer similar amounts or show a slight increase in intake at each suprathreshold dose.

Appetitive Versus Consummatory Responses

It is important to distinguish between two related concepts in drug self-administration: appetitive versus consummatory responses (see Roberts et al. 2013). A consummatory response is determined by the route of administration. For example, the consummatory response for alcohol might involve opening a bottle and drinking from it. For smoking tobacco or crack cocaine, it is the act of preparing and lighting a cigarette or pipe and inhaling the smoke. For intravenous drug use, it is filling the syringe and injecting the drug into a vein. Such consummatory behaviors can become extremely ritualistic and powerful habits.

By contrast, appetitive responses serve to make the drug available. In humans, appetitive responses might involve working for money to buy the alcohol, tobacco product, or drug; in a laboratory situation, it could be pressing on a lever and completing a response requirement. The difference between appetitive and consummatory responses is relatively straightforward in humans; however, the distinction becomes blurred in animals receiving drug via intravenous catheters. The appetitive response is obvious – animals can be shown to work quite hard to receive an injection. But since the drug is infused automatically via a catheter, the consummatory act is harder to define. It might be argued the consummatory act is the last response on the lever prior to the injection. With that being the case, the single operant response on an FR1 schedule is both appetitive and consummatory.

Schedules will be reviewed below which attempt to address separately the appetitive and motivational factors which influence drug seeking. For the moment it is assumed that responding on an FR1 is largely consummatory and that the rate and pattern of responding reflects the preferred level of consumption.

Factors Affecting Drug Consumption

The pattern shown in Fig. 1 has been taken as evidence that animals regulate (or titrate) their intake around some preferred blood or brain level. The early burst is a "loading phase," during which blood levels of cocaine increase presumably to a preferred plateau. The subsequent regular pattern reflects a maintenance phase with each post-infusion pause determined by the half-life of the unit injection dose. Longer-acting drugs, such as amphetamine, produce longer post-infusion pauses.

The distinct patterns of psychostimulant and opiate use shown by humans can be seen in the patterns of drug self-administration shown by nonhuman primates and rodents. For example, human cocaine users typically develop a "binge/abstinence" pattern of intake. Cocaine binges are characterized by sustained drug use, with high doses of drug being repeatedly ingested for variable periods (4–24 h). A binge, which typically disrupts sleeping and eating patterns, is followed by an abstinence period of unpredictable length (perhaps a day or two or longer). By contrast, human use of nicotine and opiates tends to show a more consistent daily pattern and rhythm without necessarily causing a disruption of sleep and food consumption. These characteristic patterns of drug intake are reflected in the self-administration patterns shown by nonhuman primates and rodents. Subjects given 24 h/day access to cocaine or other stimulant drugs on an FR1 schedule tend to self-administer many injections per hour for many consecutive hours or days. Animals maintain high blood levels of drug and generally do not eat or sleep during this period. It is unclear what ultimately terminates these extended drug-taking bouts; however, sleep deprivation and exhaustion no doubt contribute. Subjects inevitably relapse to drug taking after some recuperative drug-abstinence period. Unlimited access to stimulant drugs, therefore, results in a series of binge/abstinence cycles much like that seen in human addicts. Toxic effects can accumulate with repeated cycles resulting in an increased likelihood of convulsions. The probability of lethal overdose is extremely high for virtually all psychostimulants tested including cocaine, amphetamine, and methamphetamine. This is true for every species tested. By contrast, unlimited access to opiates results in a very different outcome. Heroin and morphine intake is relatively circadian rather than binge-like. Drug intake increases gradually over days as tolerance and physical dependence develop. The danger of lethal overdose is far less for opiates than for stimulants. Instead, a severe withdrawal reaction during experimenter-imposed opiate deprivation becomes a concern.

Any self-administration experiment using extended access to stimulant drugs must necessarily limit intake in some fashion to avoid toxicity. The majority of cocaine and amphetamine studies constrain intake by restricting the daily session to 1 or 2 h a day. This produces a very consistent pattern of intake from day to day. Longer sessions (6 h) have been shown to produce an escalation of intake across days (Ahmed and Koob 1998), modeling an important aspect of the addiction process. Another method of constraining daily drug intake is to restrict the number of injections offered each hour. For example, four discrete trials per hour effectively restrict hourly intake (and the danger of drug overdose or toxicity), while permitting round-the-clock access to drug. When the number of trials and the dose is chosen so as to cap cocaine intake at 3 mg/kg/h, self-administration generally becomes restricted to the active part of the light-dark cycle and will remain circadian over many weeks (Roberts et al. 2002). Binge-abstinence cycles (without lethality) are engendered by higher hourly access (7.5 mg/kg/h).

Factors Influencing Appetitive Responding

A variety of schedules of reinforcement have been used in order to directly address appetitive responding. It is important to recognize that different drug classes can increase or decrease rates of operant responding independently of their reinforcing effects; therefore, once the first injection is delivered, the response rates are "contaminated" by having the drug on board. Second-order schedules have provided a useful method to circumvent this problem. A second-order schedule, for example, [FR5 (FR10:S)], might have an initial component wherein every tenth response results in the delivery of a stimulus light; completion of five of these components results in the delivery of the stimulus light and a drug injection. Relatively high rates of responding over prolonged periods can be maintained prior to the delivery of the first drug injection. Second-order schedules have demonstrated that stimuli associated with drug injections can become potent conditioned reinforcers, and brief presentations of these stimuli can effectively maintain drug-seeking behavior. Second-order schedules have been particularly useful in evaluating pharmacological treatments for drug addictions in nonhuman primates and for evaluating the neural mechanism in rats (Schindler et al. 2002). There has been a recent trend for an appetitive and consummatory phase to be formally incorporated into an experimental design. A second-order schedule might be used in the initial phase of the session. Completion of the second-order requirement then results in access to drugs, perhaps on an FR1. The two phases are designed to assess "drug seeking" versus "drug taking."

Choice procedures offer another way of assessing reinforcing efficacy of a drug. Typically, a subject is offered a choice between a particular dose of drug and a nondrug reinforcer such as food. Such methods, together with an assessment using drug discrimination techniques, have been used to evaluate abuse liability. Drugs that are abused by humans are almost invariably self-administered by rats and nonhuman primates, indicating self-administration procedures provide a good predictor of abuse liability for new therapeutic agents being considered for the clinical market (Bergman and Paronis 2006).

The progressive ratio (PR) schedule has been used extensively to assess the reinforcing strength of a variety of drugs. The distinguishing feature of the PR schedule is that the response requirements increase following the delivery of each reinforcement. In a typical self-administration study in rats, the first injection

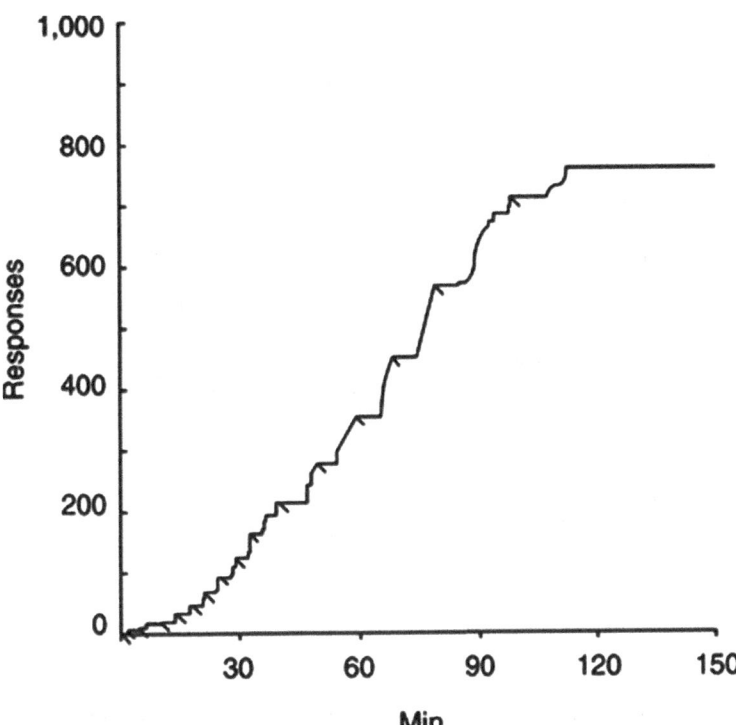

Self-Administration of Drugs, Fig. 3 A cumulative record illustrating the response pattern of a rat self-administering cocaine on a progressive ratio schedule. *Upward* movements of the record represent responses; *downward* angle ticks indicate a drug injection. The first response resulted in an injection. The response requirements for subsequent injections escalated through an exponential series. In this case, the animal self-administered 16 injections. The final ratio was 145

"costs" only one response, but with each injection, the response requirement escalates through an exponential series (1, 2, 4, 6, 9, 12, 15, 20, 32, 40, 50, 62, 77, 95, 118, 145, 178, 219, 268, 328, 226 402, 492, 603). Studies in nonhuman primates have used the same or similar series except that the initial ratio often begins at 50. Figure 3 shows a cumulative record from a session in which a rat was self-administering cocaine on a PR schedule. Note that at some point in the session, responding ceases. The last completed ratio (termed a breakpoint) is considered an index of the reinforcing value.

Figure 4 shows the dose-response curves for several different stimulant drugs. Higher unit doses are associated with higher breakpoints. The large range of breakpoints across various drugs suggests a wide spectrum of reinforcing efficacies.

Whether responding is measured while drug is on board is a critical theoretical issue. In some cases, it is imperative that drug seeking be measured in a drug-free state. However, it has become clear that blood levels of drug can greatly affect responding and in fact may be one of the most important factors which drives further drug taking once it has started. In this context, it is important to emphasize that the breakpoint on a PR schedule measures the motivation to continue a binge and not necessarily the motivation to start a binge.

In the past decade, there has been a pronounced increase in the number of studies examining reinstatement of drug self-administration. In these "relapse" experiments, animals are tested either in a drug-free state or under the influence of an experimenter delivered drug. Typically rats are trained to self-administer (e.g., cocaine or heroin), and in many cases, responding is then extinguished. The effect of a priming drug injection or conditioned stimuli is then assessed on lever responding. The term relapse is perhaps a misnomer, since animals do not self-administer during these test sessions; however, the procedure has provided a robust model to examine important influences on drug seeking. It has been shown that conditioned stimuli, stress, or a priming injection of drug can reinstate responding (Shaham et al. 2003).

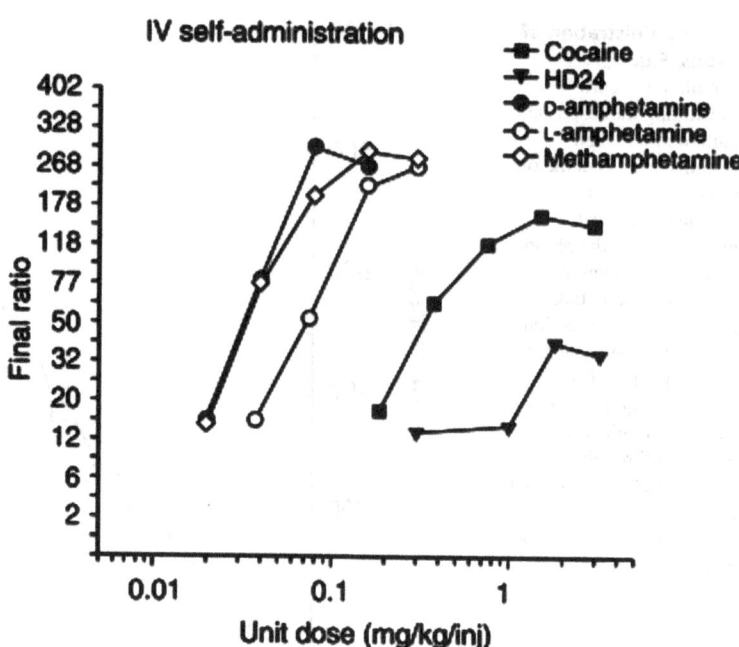

Self-Administration of Drugs, Fig. 4 Graph shows the dose-response relationship in rats for a variety of drugs self-administered on a PR schedule. Groups of rats were first trained to self-administer cocaine. Various doses of test drug were then evaluated. The dependent measure was the final ratio (or breakpoint). HD24 is an experimental tropane compound synthesized by Dr. H. M. L. Davies. Note that the selected drugs have a range of reinforcing efficacies and potencies

Another important trend in the self-administration literature has been the application of behavioral economic principles. The theoretical constructs of supply, demand, consumption, and price have been usefully applied to the analysis of drug intake. Price can be manipulated in two ways: (1) by changing the response requirement for a fixed unit dose of drug and (2) by changing the unit injection dose while the response ratio remains fixed (e.g., FR1) as might occur in an assessment of threshold (illustrated in Fig. 1). Either way the theoretical issues are the same. A behavioral economic analysis of the relationship between consumption and price offers a relatively new way of assessing reinforcing strength (Hursh 2000).

As stated previously, the majority of self-administration studies have used the intravenous route of administration. Studies of alcohol intake have quite naturally used the oral route. Rodents are generally reluctant to drink alcohol without some encouragement. Schedule-induced polydipsia has been used to induce higher levels of alcohol intake. Similarly, intake can be increased by sweetening the alcohol solutions. After a reasonable intake level has been established, then the sweetener is faded out over days. The neurobiology of both appetitive and consummatory aspects of alcohol intake is currently being assessed in rodent studies through analysis of different strains and knockout mice and through selective breeding programs for alcohol preferring animals.

Intragastric, inhalation, intraventricular, and intracerebral routes have all been used in self-administration experiments. The pharmacokinetics of each route is different; the rate of drug delivery to the brain necessarily affects the reinforcing efficacy.

Self-administration by the intracerebral route is a special case which offers a unique perspective on brain circuitry involved in drug reinforcement. Since the first report by Olds et al. (1964), there have been about 50 papers describing self-administration of drugs into various brain regions. Interestingly, drugs that are not normally abused have been reported to support operant responding when infused into particular regions of the brain. Because drugs are introduced directly into circumscribed brain regions, they can have a purely local effect which is quite different from a systemic injection. For example, Liu and Ikemoto (2007) have shown that muscimol (a $GABA_A$ agonist which would produce local inhibition) is self-administered into the median raphe nucleus. By contrast, picrotoxin (a $GABA_A$ antagonist which would produce neuronal disinhibition) is self-administered into the ventral tegmental area. Thus, the brain region and presumably the local circuit determine whether a drug has a reinforcing effect. Note that neither muscimol nor picrotoxin is self-administered when offered systemically.

Cross-References

▶ Abuse Liability Evaluation
▶ Behavioral Economics
▶ Cocaine
▶ Cocaine Dependence
▶ Inhalant Abuse
▶ Nicotine
▶ Opioids
▶ Psychostimulant Dependence and Its Treatment
▶ Reinstatement of Drug Self-Administration
▶ Withdrawal Syndromes

References

Ahmed SH, Koob GF (1998) Transition from moderate to excessive drug intake: change in hedonic set point. Science 282:298–300

Bergman J, Paronis CA (2006) Measuring the reinforcing strength of abuse drugs. Mol Interv 6:273–283

Hursh SR (2000) Behavioral economic concepts and methods for studying health behavior. In: Bickel WK, Vuchinich RE (eds) Reframing health behavior change with behavioral economics. Lawrence Erlbaum, Mahwah, pp 27–60

Liu ZH, Ikemoto S (2007) The midbrain raphe nuclei mediate primary reinforcement via GABA (A) receptors. Eur J Neurosci 25:735–743

Olds J, Yuwiler A, Olds ME, Yun C (1964) Neurohumors in hypothalamic substrates of reward. Am J Physiol 207:242–254

Oleson EB, Roberts DCS (2009) Behavioral economic assessment of price and cocaine consumption following self-administration histories that produce escalation of either final ratios or intake. Neuropsychopharmacology 34:796–804

Roberts DCS, Brebner K, Vincler M, Lynch WJ (2002) Patterns of cocaine self-administration in rats produced by various access conditions under a discrete trials procedure. Drug Alcohol Depend 67:291–299

Roberts DCS, Gabriele A, Zimmer BA (2013) Conflation of cocaine seeking and cocaine taking responses in IV self-administration experiments in rats: methodological and interpretational considerations. Neurosci Biobehav Rev 37:2026–2036

Schindler CW, Panlilio LV, Goldberg SR (2002) Second-order schedules of drug self-administration in animals. Psychopharmacology (Berl) 163:327–344

Shaham Y, Shalev U, Lu L, De Wit H, Stewart J (2003) The reinstatement model of drug relapse: history, methodology and major findings. Psychopharmacology (Berl) 168:3–20

Spragg SDS (1940) Morphine addiction in chimpanzees. Comp Psychol Monogr 15:1–132

Weeks JR (1962) Experimental morphine addiction: method for automatic intravenous injections in unrestrained rats. Science 138:143–144

Self-Monitoring

Definition

Self-monitoring is the ability to recognize that actions such as inner speech are initiated by the self.

Semantic Memory

Definition

Our memory for language, concepts, and facts about the world.

Sensitization

Synonyms

Reverse tolerance

Definition

Sensitization, in relation to ▶ Pavlovian conditioning, is an incremental change in the response to a stimulus in the absence of any associated contingency. In psychopharmacology, sensitization is more frequently used as the name for the progressive enhancement of an unconditioned response to a drug when it is administered on two or more occasions. It regularly occurs with ▶ psychostimulant and opioid drugs and is believed to involve enhanced function of dopaminergic neurons projecting from the ▶ ventral tegmental area to the ▶ nucleus accumbens.

Sensitization to Drugs

Terry E. Robinson
Department of Psychology, University of Michigan, Ann Arbor, MI, USA

Synonyms

Behavioral augmentation; Behavioral facilitation; Behavioral sensitization; Reverse tolerance

Definition

The word "sensitization" is used to refer to a number of different but related effects. For example, in immunology, sensitization refers to the hypersensitivity to an antigen (often an allergen) that can develop upon repeated exposure to the antigen. In the study of learning, sensitization refers to a form of nonassociative learning whereby exposure to a stimulus (an unconditional stimulus, US) increases subsequent responsiveness to the same or other stimuli, even though they were not explicitly paired. Similarly, in pharmacology, the word sensitization has come to refer to an increase in a drug effect upon successive exposures to a drug or hypersensitivity to a drug in animals that were exposed to the drug in the past (Fig. 1). For example, one unconditional effect of drugs such as amphetamine or cocaine is to produce *psychomotor activation*, often measured as an increase in forward locomotion. Under some circumstances, the repeated administration of psychostimulant drugs results in a progressive increase in this drug effect, whereby successive injections of the same dose produce greater and greater psychomotor activation. Furthermore, exposure to one drug (e.g., amphetamine) can also render animals hypersensitive to the locomotor-activating effects of other drugs (e.g., cocaine or morphine). When exposure to one drug (or another stimulus, such as stress) renders an animal hypersensitive to another drug or stimulus, this is called "cross-sensitization" (Kalivas and Barnes 1988; Robinson and Becker 1986; Stewart and Badiani 1993).

Impact of Psychoactive Drugs

Behavioral Sensitization

By itself, with no adjective, the word sensitization provides no information about the nature of the drug effect that is changed or about the underlying mechanisms. Indeed, it is not very useful to refer to "drug sensitization" or to just say an animal is "sensitized," because sensitization (or tolerance) does not develop to the drug itself. Specific drug *effects* show sensitization or tolerance. Drugs produce many different effects, and with repeated administration some effects may show sensitization while at the same time other effects show tolerance and yet others do not change. In fact, the same drug effect may show tolerance or sensitization depending on the conditions under which the drug is administered (e.g., if it is given continuously or intermittently – intermittent injections favor sensitization). Thus, an animal that is "sensitized" may also be "tolerant" – depending on which drug effect is under consideration.

It is important to ask, therefore, which effects of drugs tend to show sensitization and under what conditions is sensitization induced and expressed. Most studies on sensitization involve some behavioral measure, and if a behavioral effect of a drug increases with repeated drug administration, this may be called *behavioral sensitization* (Fig. 1). In other studies, a neurobiological effect of a drug may be measured, and if increased by prior drug exposure, this may be referred to as *neural sensitization*, although,

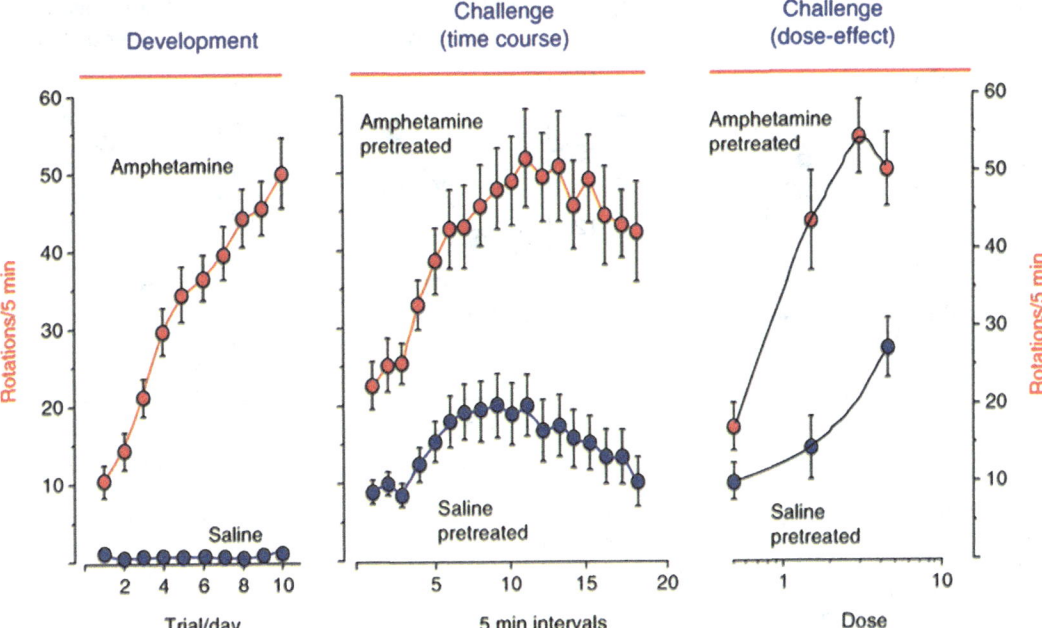

Sensitization to Drugs, Fig. 1 An illustration of the different ways of viewing psychomotor sensitization. The specific behavior quantified is amphetamine-induced rotational behavior in rats with a unilateral 6-OHDA lesion. The dose-effect function for this behavior is linear over a much wider range of doses than for locomotor activity (see Fig. 2). *Left panel: a within-subjects* measure of sensitization, in which rats were given an injection of 3 mg/kg of D-amphetamine (or saline) once every 3–4 days for a total of ten injections. Amphetamine produced more and more rotational behavior ("psychomotor activation") with successive injections, whereas the response to saline did not change. The increase in behavioral effect is called behavioral or psychomotor sensitization. *Middle panel: a between-subjects* measure of sensitization, in which rats that were previously treated with saline or amphetamine both received a challenge injection of amphetamine (1.5 mg/kg). The behavioral response in rats that previously received ten injections of amphetamine (*left panel*) is much greater than in those that previously received saline, and the magnitude of the group difference indicates the degree of sensitization. *Right panel: dose-effect analysis*, in which rats previously treated as in the *left panel* received a challenge injection with different doses of amphetamine. In this case, the degree of sensitization is indicated by the magnitude of the shift to the left in the dose-effect function (Data are from Anagnostaras and Robinson 1996)

again, these terms provide no information about the exact effect that is changed. Therefore, unless one wants to just refer to the general phenomenon, it is best to use terms that convey information about the specific drug effect under study, because most of them are dissociable.

In many studies of behavioral sensitization, the drug effect measured is a psychomotor effect (*psychomotor sensitization*; Figs. 1 and 2). But the psychomotor-activating effects of drugs represent a very complex set of different behaviors, and they can compete with one another, making the accurate assessment of psychomotor effects a complicated endeavor (Flagel and Robinson 2007; Robinson and Becker 1986). The most frequent psychomotor effect studied involves some measure of locomotor activity (e.g., beam breaks, crossovers, distance traveled, etc.). To the extent that locomotor activity increases with repeated drug administration, this can be called *locomotor* sensitization. But drugs can produce many other psychomotor effects, including complex patterns of repetitive motor actions (stereotyped behaviors), including rearing, head movements, limb movements, sniffing, oral movements (licking and biting), rotational behavior, etc. Which psychomotor effect dominates behavior depends on many factors, including the drug, dose, the test environment, the time after drug administration, and many others.

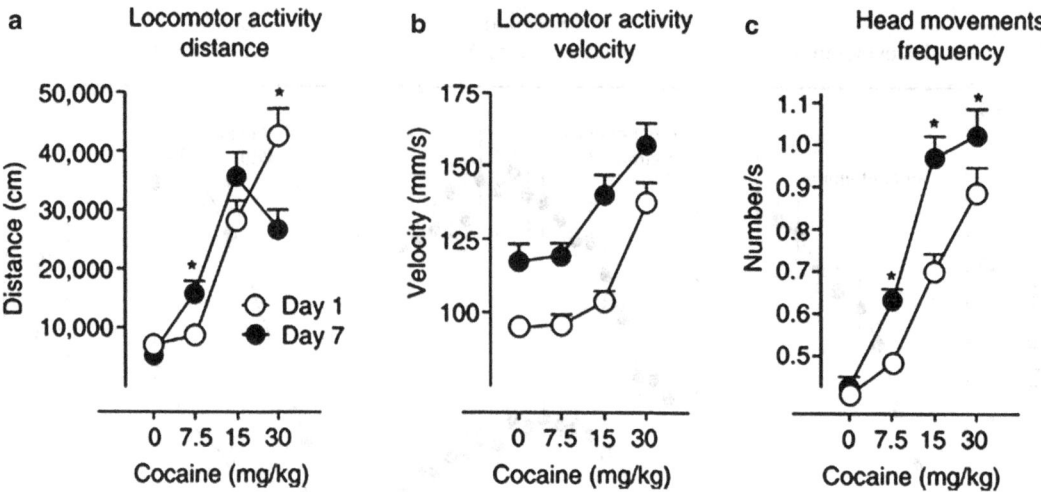

Sensitization to Drugs, Fig. 2 Different measures of psychomotor activation in a within-subjects study of behavioral sensitization. On the first day of treatment, rats were given three injections of cocaine in ascending doses (0, 7.5, 15, and 30 mg/kg), with 45 min between each treatment. For the next 6 days, they received one injection of 15 mg/kg each day. Then, on the seventh day, they again received multiple doses, as on day 1. The graph shows three different behavioral measures on day 1 and day 7 of testing, obtained from video. *Panel A* shows a measure of locomotion (distance traveled in cm). Between day 1 and 7, the animals showed an increase in locomotor activity when given 7.5 mg/kg, no change when given 15 mg/kg, and a decrease in locomotor activity when given 30 mg/kg. This highlights the complexity of the dose-effect function when using locomotor activity as a measure of psychomotor activation (compare with Fig. 1). For example, under the conditions of this study, if only a dose of 15 mg/kg were used, one would conclude that the animals did not sensitize. However, other measures of psychomotor activation reveal robust sensitization. *Panel B* shows the average velocity of each bout of locomotion, and there is a large effect of repeated drug treatment on this measure (the effect at a dose of zero also reveals a conditioned effect not evident in the distance traveled). *Panel C* shows the average frequency of head movements, and again this measure reveals robust psychomotor sensitization, at all doses. These data illustrate the importance of using multiple measures of psychomotor activity in studies of sensitization, especially if negative findings are obtained (Data are from Flagel and Robinson 2007)

For example, in animals previously exposed to amphetamine, a subsequent injection may initially increase locomotor activity, which then decreases and is replaced by complex patterns of stereotyped behaviors performed "in place," and then this is followed by the reemergence of locomotor hyperactivity ("post-stereotypy hyperactivity"). What is critical to realize is that each of these different psychomotor effects can be dissociated, and they do not all change uniformly as a function of repeated drug treatment (see Fig. 2). Repeated drug administration does not have the same effect on each of these behaviors, presumably in part because they are mediated by different neural systems that are changing in different ways and in part because of the competitive relationship between different behaviors.

For this reason, negative findings can be very difficult to interpret in studies of sensitization. If a given manipulation is reported to not produce or to prevent the expression of "sensitization," one must immediately ask – by what measure? It could be that the manipulation influenced the induction or expression of locomotor sensitization, but had no effect on some other measure of psychomotor sensitization or on other psychological processes that undergo sensitization (see below). Or, it could be that the measure of locomotion was not the most appropriate measure under the conditions of the study. For example, repeated exposure to cocaine may have no influence on one measure of locomotion, such as distance traveled, while at the same time dramatically increasing the velocity of each individual bout of locomotion and the frequency of stereotyped head movements (Flagel and Robinson 2007; Fig. 2). Thus, the apparent absence of locomotor sensitization may not allow one to conclude no effect on "sensitization" but only on one behavioral measure. Unfortunately, in many studies, only one measure is provided, making it nearly impossible to interpret negative results.

It is also important to remember that many other behavioral effects of drugs can sensitize, besides just psychomotor effects. Other behaviors that have been reported to sensitize include acoustic startle, drinking behavior, lick rate, discriminative effects, and the disrupting effect of amphetamine on latent inhibition and selective attention, to name a few (Badiani and Robinson 2009; Kalivas and Barnes 1988; Robinson and Becker 1986; Stewart and Badiani 1993). In humans, the psychotomimetic effects of psychostimulant drugs also sensitize (Featherstone et al. 2007; Robinson and Becker 1986). In addition, there is considerable evidence that repeated exposure to a variety of drugs of abuse increases some aspect of their rewarding or incentive motivational effects (Vezina and Leyton 2009). Repeated exposure to a number of potentially addictive drugs facilitates the later acquisition of drug self-administration behavior and a conditioned place preference, facilitates the learning of S-R habits, increases motivation for drug based on performance on a progressive ratio schedule and running in an alley, and increases the incentive salience attributed to stimuli associated with drug and nondrug rewards. This latter form of sensitization – *incentive sensitization* – may be especially important in the development of addiction because it may result in people being maladaptively attracted to drugs and cues associated with drugs, thus instigating and maintaining drug-seeking behavior even when there is a desire to remain abstinent (Robinson and Berridge 2008).

Neural Sensitization

Presumably the reason that so many different behaviors and psychological functions can be sensitized by repeated exposure to drugs is that drugs change many different neural systems, neural systems that mediate the behaviors and psychological functions that sensitize (Thomas et al. 2008). Many studies of neural sensitization have focused on mesotelencephalic dopamine systems, and a number of sensitization-related changes in dopamine systems have been described, including an increase in stimulated dopamine release and striatal D2 receptors (Robinson and Becker 1986; Robinson and Berridge 2008). However, sensitization-related changes have been described in nearly every neurotransmitter system within the relevant mesencephalic-striatal-amygdala-cortical circuitry, including glutamate, GABA, serotonin, acetylcholine, norepinephrine, etc. systems, and in a host of intracellular signaling cascades. Indeed, sensitization has been associated with changes in patterns of synaptic connectivity in these circuits, suggesting a level of reorganization that may fundamentally and persistently change their operation. Despite considerable research on neural sensitization, cause-effect relations are not well understood. What changes in what neural systems are causally related to what changes in behavior and psychological function remains a topic of active investigation.

The Induction Versus the Expression of Sensitization

Most major drugs of abuse are capable of inducing sensitization, including psychostimulants (amphetamine, cocaine, MDMA, cathinone, fencamfamine, methylphenidate, phenylethylamine, etc.), opiates, phencyclidine, alcohol, and nicotine. Direct agonists that act on the D2 receptor also produce robust psychomotor sensitization, although it is not clear the mechanism is the same as with drugs of abuse, which influence dopaminergic activity indirectly. In addition, repeated intermittent exposure to stress can produce cross-sensitization to drugs, and vice versa (Kalivas and Barnes 1988; Stewart and Badiani 1993).

For some drugs, it is thought that an action at the level of dopamine cell bodies in the midbrain is necessary to induce sensitization. However, once induced, sensitization may be expressed by drug actions in structures that receive dopaminergic inputs (Vezina and Leyton 2009). Whatever the case, one remarkable feature of sensitization is that, once induced, it can persist for very long periods of time in the absence of any further exposure to the drug – for weeks, months, or years. The persistence of sensitization depends on many factors, including the drug, the dose, the number of exposures, and the pattern of exposure. Furthermore, sensitization can be induced when a drug is administered by an experimenter or when it is self-administered, including when it is self-administered under conditions that promote the development of symptoms of addiction (i.e., under extended access conditions).

Modulation of Induction and Expression

Although the repeated, intermittent administration of a number of drugs of abuse may produce sensitization, it is important to emphasize that there are a host of factors that modulate both the induction and expression of sensitization. Whether a given dose of a drug *induces* sensitization, or how robust the effect is, is dependent on many factors besides dose and the treatment regimen, including the strain of the animal, its sex (females generally sensitize more), hormonal status, past experience with stress, age, the rate of drug delivery (faster rates produce greater sensitization), and the context in which the drug is administered, among others (Badiani and Robinson 2009; Kalivas and Barnes 1988). Indeed, there are large individual differences in susceptibility to sensitization. After neural sensitization is induced, there are also a number of factors that determine if it is *expressed* in behavior at any particular place or time. For example, when an animal that has developed psychomotor sensitization is reexposed to a drug in a context where it has never before experienced the drug, sensitization may not be expressed in behavior (this is so-called context-specific sensitization; Fig. 3). Exactly, how contextual factors modulate the expression of sensitization is not well understood. The available data suggest that contexts *not* associated with drugs may act to actively inhibit expression, perhaps through a kind of inhibitory occasion-setting-type mechanism. This is because specific extinction procedures and ECS can "release" this inhibition so that sensitization is now expressed in a nondrug context (Stewart and Badiani 1993; Vezina and Leyton 2009; Fig. 3). Of course, drug-associated contexts can also elicit drug-like conditioned responses that may add to sensitization effects, but this kind of conditioned response appears to be a different process than contextual modulation of sensitization. Indeed, it may be misleading to speak of "context-specific sensitization" versus "context-independent sensitization" as if they represent two different forms of sensitization. It appears there is one form of sensitization, a nonassociative increase in responsiveness to various stimuli because of changes in the relevant neural systems. However, the *expression* of sensitization can be powerfully modulated by associative learning, and if it is, sensitization may be expressed only under specific conditions. How contextual factors (and other stimuli) modulate the expression of sensitization is an important (but little investigated) topic, because of the potential importance of such factors in relapse. Addicts are much more prone to relapse in contexts associated with drugs than in other contexts. This may be related not only to the ability of drug-associated contexts to evoke conditioned responses, which is well documented, but also because context can gate the expression of sensitization. Thus, conditions that promote the expression of neural sensitization in behavior may also promote relapse, whereas conditions that inhibit the expression of sensitization may inhibit the propensity to relapse. For this reason, a better understanding of factors that modulate the induction and expression of sensitization (especially incentive sensitization) may prove useful in developing treatments for addiction (Robinson and Berridge 2008; Stewart and Badiani 1993; Vezina and Leyton 2009).

Sensitization in Humans

By necessity, most research on sensitization has involved preclinical studies in nonhuman animals.

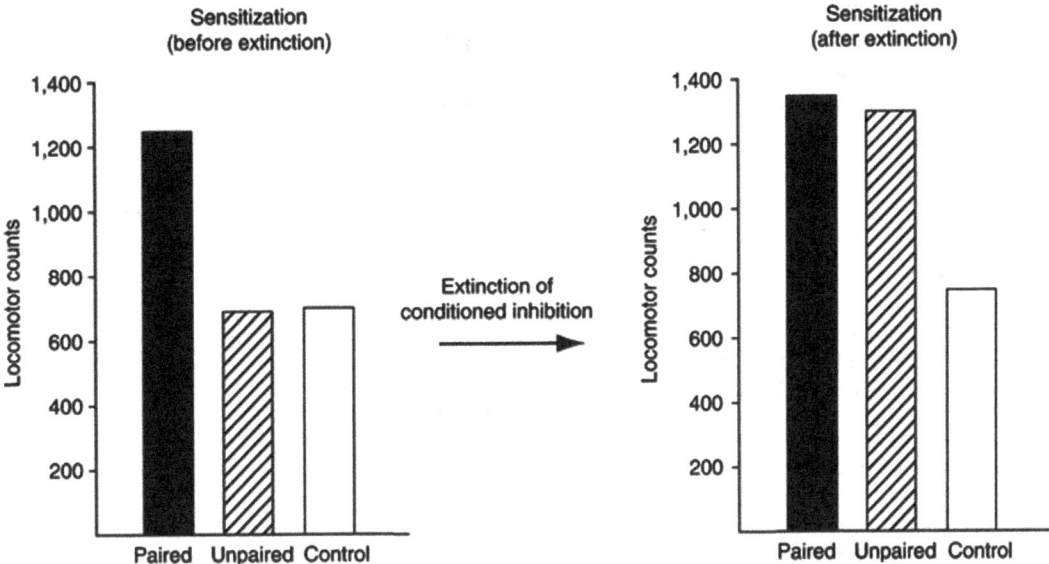

Sensitization to Drugs, Fig. 3 Context specificity of sensitization (Adapted from Vezina and Leyton 2009). Paired rats received amphetamine in the test environment on one day and saline in the home cage (or somewhere else) the following day. Unpaired animals were subjected to the opposite configuration, while control animals received saline in both environments. After repeated treatments, all animals were then administered a challenge injection of the drug in the test environment to assess the expression of sensitization. The results on the left illustrate that sensitization was only expressed in paired animals. Rats that received the same number of drug injections but explicitly unpaired with the test environment showed levels of responding similar to animals receiving the drug for the first time (controls). Evidence that this is due to conditioned inhibition of the expression of sensitization in unpaired rats comes from the finding that after these rats were subjected to a procedure known to extinguish conditioned inhibition (but spare conditioned excitation), they now expressed sensitization (*right panel*). This suggests that the failure of unpaired animals to express sensitization in the first test was due to some form of conditioned inhibition (see Vezina and Leyton 2009 for further discussion)

However, it is worth noting that similar effects have been described in humans (Leyton 2007). As mentioned above, it has long been recognized that the psychotomimetic effects of psychostimulant drugs sensitize, as does their ability to produce complex stereotyped behavioral patterns (in humans this is called "punding"). Indeed, sensitization of psychotomimetic effects has been interpreted as involving sensitization of incentive salience, whereby otherwise innocuous stimuli in the environment acquire pathological importance (Featherstone et al. 2007). In addition, there are now a number of reports that repeated exposure to amphetamine induces both behavioral sensitization (e.g., increased eyeblink responses, vigor, and energy ratings) and neural sensitization (e.g., an increase in evoked dopamine "release" as indicated by raclopride displacement). Finally, as in nonhuman animals, the expression of sensitization in humans appears to be highly modulated by the context under which drugs are experienced (Leyton 2007; Robinson and Berridge 2008; Vezina and Leyton 2009).

In conclusion, the sensitization produced by repeated exposure to many drugs of abuse turns out to be a very complex form of neurobehavioral plasticity that can alter many different neural systems and thus be manifest as a change in many different behaviors and psychological processes. Some of these changes may be related to why, in addicts, drugs and drug-associated stimuli come to acquire such inordinate control over behavior (Robinson and Berridge 2008).

Cross-References

▶ Psychostimulants
▶ Substance Use Disorders: A Theoretical Framework
▶ Tolerance

References

Anagnostaras S, Robinson TE (1996) Sensitization to the psychomotor stimulant effects of amphetamine: modulation by associative learning. Behav Neurosci 110:1397–1414

Badiani A, Robinson TE (2009) Drug addiction: behavioral pharmacology of drug addiction in rats. In: Squire LR (ed) Encyclopedia of neuroscience, vol 3. Academic Press, Oxford, pp 683–690

Featherstone RE, Kapur S, Fletcher PJ (2007) The amphetamine-sensitized state as a model of schizophrenia. Prog Neuropsychopharmacol Biol Psychiatry 31:1556–1571

Flagel SB, Robinson TE (2007) Quantifying the psychomotor activating effects of cocaine in the rat. Behav Pharmacol 18:297–302

Kalivas PW, Barnes CD (eds) (1988) Sensitization in the nervous system. The Telford Press, Caldwell

Leyton M (2007) Conditioned and sensitized responses to stimulant drugs in humans. Prog Neuropsychopharmacol Biol Psychiatry 31:1601–1613

Robinson TE, Becker JB (1986) Enduring changes in brain and behavior produced by chronic amphetamine administration: a review and evaluation of animal models of amphetamine psychosis. Brain Res Rev 11:157–198

Robinson TE, Berridge RC (2008) The incentive-sensitization theory of addiction: some current issues. Phil Trans R Soc B Biol Sci 363:3137–3146

Stewart J, Badiani A (1993) Tolerance and sensitization to the behavioral effects of drugs. Behav Pharmacol 4:289–312

Thomas MJ, Kalivas PW, Shaham Y (2008) Neuroplasticity in the mesolimbic dopamine system and cocaine addiction. Br J Pharmacol 154:327–342

Vezina P, Leyton M (2009) Conditioned cues and the expression of stimulant sensitization in animals and humans. Neuropharmacology 56(suppl 1):160–168

Sensorimotor Behavior

Definition

Motor responses that immediately follow sensory stimulation.

Sentient

Definition

The ability of an organism to perceive or feel things. Sentient animals have higher-order nervous systems and behaviors.

Serotonin

Synonyms

5-HT; 5-Hydroxytryptamine

Definition

A monoamine (indoleamine) neurotransmitter found in the brain, platelets, and the gastrointestinal tract. Serotonin is synthesized from the amino acid tryptophan and is involved in a wide range of behavioral, physiological, and cognitive functions, including mood, emotion, appetite, sleep, memory, temperature regulation, platelet aggregation, vascular tone, intestinal secretion, peristalsis, and bone metabolism. Serotonin is well known for its role in psychiatric and behavioral disorders such as depression, anxiety, obsessive compulsive disorders, eating disorders, and addiction. This neurotransmitter may also be involved in physiological disorders such as irritable bowel syndrome and cardiac arrhythmias. The actions of serotonin (5-HT) are mediated through at least 16 receptor subtypes grouped into 7 families ($5\text{-}HT_1$-$5\text{-}HT_7$) according to their structural and functional characteristics and include 13 distinct G-protein-coupled receptors, coupled to various effector systems, and 3 ligand-gated ion channels (the $5\text{-}HT_3$ receptors). The drugs targeting the 5-HT system that are in the widest clinical are the selective serotonin reuptake inhibitors that are used for the treatment of depression and other disorders.

Cross-References

- Antidepressants
- Anxiety
- Depression
- Fluoxetine
- Monoamines
- Neurotransmitter
- Obsessive–Compulsive Disorders
- SSRIs and Related Compounds
- Substance Use Disorders: A Theoretical Framework

Serotonin Syndrome

Maryana Duchcherer and Nicholas Mitchell
Department of Psychiatry, University of Alberta, Edmonton, AB, Canada

Synonyms

Serotonin toxicity

Definition

A neurobehavioral syndrome consisting of motor symptoms and autonomic and mental status changes secondary to an excess of serotonergic activity in the central nervous system.

Current Concepts and State of Knowledge

Serotonin syndrome (SS) is a neurobehavioral syndrome consisting of motor symptoms and autonomic and mental status changes secondary to an excess of serotonergic activity in the central nervous system. In its most severe form, it can have lethal complications. One of the sentinel cases bringing SS to widespread attention involved an 18-year-old woman in New York, whose death was thought to be due to coadministration of an opioid analgesic, meperidine, with a monoamine oxidase inhibitor (MAOI), phenelzine (Boyer and Shannon 2005).

Increased central serotonergic transmission can occur as a result of both pharmacological and non-pharmacological triggers. Prescription medications, drugs of abuse, and natural products with serotonergic activity are identified as the most common causes of SS (Table 1). Non-pharmacological contributors include medical conditions, such as carcinoid tumors, and therapeutic procedures, such as electroconvulsive therapy (ECT), which directly or indirectly increase activity at serotonergic synapses. Many reported cases of SS appear to result from the interaction of multiple factors, with iatrogenic causes being the most common.

The severity of SS symptoms appears to mirror the concentration-dependent activity of serotonin in the synaptic cleft. In animal studies, peak levels of serotonin are directly linked to SS symptoms (Birmes et al. 2003). In humans, the combination of selective serotonin reuptake inhibitors (SSRIs) and MAOIs synergistically increases activity at serotonergic neurons and is most commonly implicated in SS. A number of lethal outcomes from SS have been reported when MAOIs are combined with SSRIs or amphetamines (Gillman 2006).

Epidemiology

The first recorded observation of SS was in 1960 from patients treated with MAOIs who received supplemental L-tryptophan, a serotonin precursor (Sternbach 1991). Tricyclic antidepressants have also been associated with SS. From their initial development, the SSRIs have become the most commonly prescribed class of antidepressants and have been implicated in cases of SS across all age ranges.

Estimating the frequency of SS in the public is challenging as up to 85 % of physicians are unaware of how to make this clinical diagnosis (Mackay et al. 1999). As such, rates of SS in the literature vary widely (Arora and Kannikeswaran 2010). The American Association of Poison Control Centers Toxic Exposure Surveillance System published a report of clinical cases from outpatient and inpatient settings in 2003. A total of

Serotonin Syndrome, Table 1 Agents implicated in serotonin syndrome

Class	Agent
Antidepressants	SSRIs: paroxetine, fluoxetine, fluvoxamine, citalopram, escitalopram, sertraline
	SNRIs: venlafaxine, desvenlafaxine, duloxetine, sibutramine, milnacipran
	MAOIs: phenelzine, tranylcypromine, moclobemide, toloxatone, selegiline
	TCAs: imipramine, doxepin, maprotiline, nortriptyline, amitriptyline, clomipramine, trimipramine, protriptyline
	Trazodone, nefazodone, vilazodone
	Buspirone
	Bupropion
Opioid analgesics	Tramadol, methadone, fentanyl, pentazocine, meperidine, dextropropoxyphene
Phenethylamines	Amphetamines
	Methylphenidate
	Methamphetamine
	MDMA (ecstasy)
	Cocaine
Triptans	Sumatriptan
	Rizatriptan
	Eletriptan
	Zolmitriptan
	Naratriptan
	Almotriptan
	Frovatriptan
Indoles	Psilocybin
	LSD
Mood stabilizers	Lithium
	Carbamazepine
	Valproic acid
Dopaminergic medications	L-Dopa
Antiemetic	Metoclopramide
	Ondansetron
	Granisetron
Amino acids (precursor of serotonin)	L-Tryptophan
	5-Hydroxytryptophan
Over the counter	Dextromethorphan (cough syrup ingredient)
	St John's wort (*Hypericum perforatum*)
	Yohimbe
	Syrian rue (*Peganum harmala*)

Serotonin Syndrome, Table 2 Symptoms of serotonin syndrome

Autonomic	Neuromuscular	Mental status changes
Tachycardia	Hyperreflexia	Anxiety
Tachypnea	Rigidity	Restlessness
Diaphoresis	Tremor	Excitement
Fever	Akathisia	Confusion
Hypertension	Myoclonus	Agitated delirium
Vomiting	Bilateral Babinski sign	
Diarrhea	Nystagmus	

26,733 exposures to SSRIs were described, including 7,349 cases of toxic side effects and 93 lethal outcomes. Other reports indicate that 14–16 % of individuals exposed to SSRIs in overdose will develop symptoms of SS. A separate estimate from primary care under the British National Health Service published in 1999 placed the incidence of SS at 0.5–0.9 cases per 1,000 patient months of treatment with SSRIs.

Presentation and Diagnosis

Serotonergic neurons are involved in the regulation of a variety of biological functions including mood, the sleep-wake cycle, thermoregulation, and appetite. Therefore, any alteration in serotonin homeostasis impacts a number of physiological processes and behavior, making the diagnosis of SS a clinical one based on a number of symptoms.

Serotonin syndrome has been characterized by a triad of clinical features (Table 2): autonomic dysregulation, neuromuscular symptoms, and an altered mental state. The autonomic changes include tachycardia (rapid heart rate), tachypnea (rapid breathing), diaphoresis (sweating), fever, hypertension (elevated blood pressure), vomiting, and diarrhea. Neuromuscular symptoms include hyperreflexia, rigidity, tremor, myoclonus (involuntary twitching of muscle groups), and bilateral upward plantar reflexes. Altered mental states range from mild anxiety and restlessness to confusion and agitated delirium.

The onset of SS is usually observed within 24 h after initiation or dose adjustment of a serotonergic agent, with up to 60 % of cases developing in the first 6 h. The addition of a second or subsequent agent will increase the risk of SS on a patient stable on a serotonergic drug. In some cases, an individual can present with chronic mild symptoms, while others have a rapid progression in severity. The addition of a second or subsequent agent will increase the risk of SS on a patient stable on a serotonergic drug.

The symptoms of SS occur along a spectrum of severity. In early stages, alertness is maintained with mild anxiety and neuromuscular signs which are more evident in the lower extremities. With further progression, autonomic dysregulation becomes apparent. Exaggerated deep tendon reflexes and myoclonus appear, and mental status is characterized by hypervigilance, easily evoked startle reflex, slightly agitated behavior, and pressured speech. Severe and life-threatening complications occur when mental state deteriorates to agitated delirium with rigid muscle tone. Severe shivering and shaking can produce elevated temperature. Severe autonomic symptoms can include impaired respiratory function and shock. Permanent neurological injury or death occurs rarely and largely due to secondary complications of SS (such as anoxic brain injury).

Various criteria have been developed for the diagnosis SS with no single consensus definition in the literature (Frank 2008). Historically, criteria have been derived from published case reports. Two diagnostic criteria sets appear most frequently in the SS research. The Sternbach criteria (Table 3) are based on retrospective analysis of 38 case reports of SS and represent all three clusters of SS symptoms. Research suggests that these criteria are 75 % sensitive and 96 % specific for the diagnosis of SS, but they tend to represent the severe symptoms of SS and may underdiagnose mild to moderate cases. The Hunter Serotonin Toxicity Criteria (Table 4) are based on a review of 2,222 case reports and are the most widely used and clinically accepted diagnostic schema for SS. This criteria set has demonstrated 84 % sensitivity and 96 % specificity for SS and reflects neuromuscular and autonomic symptoms of SS as opposed to mental status changes.

The symptoms of SS are common to a number of conditions including other toxidromes. Malignant hyperthermia evolves rapidly after the administration of an anesthetic agent and presents with tachycardia, elevated muscle tone, hyporeflexia, and hyperthermia. Neuroleptic malignant syndrome is a cluster of similar alterations in mental state, increased muscle tone, hyperthermia, and autonomic instability that usually occurs hours to days after the initiation of a dopamine antagonist. Anticholinergic delirium is also associated with an altered mental state and agitation following administration of an agent with anticholinergic activity, but lacks a number of other symptoms of SS.

Serotonin Syndrome, Table 3 Sternbach criteria for the diagnosis of serotonin syndrome

1. Recent addition or increase in a known serotonergic agent
2. Absence of other possible etiologies (infection, substance abuse, withdrawal, etc.)
3. No recent addition or increase of a neuroleptic agent
4. At least 3 of the following symptoms
Mental status changes (confusion, hypomania)
Agitation
Myoclonus
Hyperreflexia
Diaphoresis
Shivering
Tremor
Diarrhea
Ataxia
Fever

Serotonin Syndrome, Table 4 Hunter criteria for the diagnosis of serotonin syndrome

In the presence of a serotonergic agent, the occurrence of one of the following symptoms
Spontaneous clonus
Inducible clonus and agitation or diaphoresis
Ocular clonus and agitation or diaphoresis
Ocular clonus or inducible clonus
Tremor and hyperreflexia
Hypertonia and temperature >38 °C
Ocular clonus or inducible clonus

Laboratory findings in SS are nonspecific, reflecting the sequelae of increased muscle tone and autonomic dysregulation, including severe metabolic acidosis, rhabdomyolysis (a condition of muscle breakdown that can result in kidney failure), liver failure, renal failure, and disseminated intravascular coagulation resulting from uncontrolled hyperthermia. As such, the diagnosis of SS remains a clinical one.

Neurobiology and Pathophysiology

Serotonin (5-hydroxytryptamine), a neurotransmitter derived from the amino acid tryptophan, is abundant in the central nervous system and gastrointestinal tract of mammals. Serotonin is primarily released from neurons in the raphe nuclei in the brain stem; however, serotonergic neurons have projections throughout the brain. In the gut, serotonin is produced by the enterochromaffin cells. Serotonin receptors are classified into seven families designated $5HT_1$ to $5HT_7$. The majority of serotonin receptor families represent G-protein-coupled receptors, with the activation of the receptor influencing an intracellular second messenger system. The $5HT_3$ receptor family is a ligand-gated sodium-potassium channel. Serotonergic neurotransmission can result in either excitation or inhibition of postsynaptic neurons. Serotonin is removed from the synapse by the serotonin transporter protein on the presynaptic neuron. Ultimately, serotonin is metabolized by the enzyme monoamine oxidase to 5-hydroxyindole acetic acid.

The activity of central serotonergic neurons affects a number of behavioral and physiological processes including the sleep-wake cycle, pain, learning, appetite, thermoregulation, and mood and anxiety. Alterations in serotonin are also implicated in a number of medical and psychiatric disorders such as migraine, epilepsy, insomnia, and depression and anxiety. As such, pharmacological agents used to treat a number of these disorders have serotonergic activity.

The symptoms of serotonin syndrome appear to be mediated through increased serotonergic activity in the synapse (Iqbal et al. 2012). This can be accomplished through (1) direct serotonin agonism (i.e., triptans), (2) increasing serotonin release from the presynaptic neuron (i.e., amphetamine), (3) decreasing serotonin reuptake (i.e., SSRIs), or (4) decreasing serotonin breakdown (i.e., MAOIs).

The receptor family responsible for the symptoms of serotonin syndrome is not clear. The $5HT_{2A}$ receptor has been implicated, with clinical evidence derived from case reports of therapeutic benefit following the administration of the antagonist cyproheptadine. Activation of postsynaptic $5HT_{2A}$ receptors is associated with increased muscle tone and hyperthermia. Similarly, activation of postsynaptic $5HT_{1A}$ receptors is associated with hyperactivity, anxiety, and hyperreflexia. Evidence from case studies of drug interactions leading to SS has also implicated the glutamatergic and GABAergic neurotransmitter systems in the onset of SS.

Treatment

Given a limited understanding of the pathophysiology of SS, there is no single direct antidote (Iqbal et al. 2012). The first line of treatment involves removing the offending serotonergic agent. In mild to moderate cases, no further treatment is needed and symptoms often resolve in 24–72 h. The duration of SS will therefore depend on the pharmacokinetics of the serotonergic agent, including its duration of action and half-life of elimination. Autonomic dysregulation, changes in mental state, or increased muscle tone can all indicate a need for hospitalization. Supportive care should be offered during the washout period after discontinuation of the agent, with benzodiazepines or neuroleptics most used for sedation and behavioral control. Cyproheptadine has been used in an attempt to reduce serotonergic activity at the $5HT_{2A}$ receptor, though there is a lack of definitive evidence of its clinical effectiveness. In severe cases, individuals may need to be admitted to an intensive care unit for monitoring and management of autonomic instability. Other supportive strategies for severe SS include neuromuscular paralysis and intubation in order to treat hyperthermia, and active cooling. The prognosis of SS is good if secondary complications are avoided.

Summary

While rates of serotonergic antidepressant prescription have climbed in North America through the past three decades, SS remains a rare but clinically significant and potentially life-threatening complication of these agents. Other potential causes include prescription medications with serotonergic activity, natural products, and some drugs of abuse. The mechanism of SS is related to overactivity of serotonergic neurons, and the treatment involves removing the causative agent. In moderate to severe cases, supportive therapy, including hospital or intensive care unit admission, may be warranted.

Rates of SS are difficult to estimate as the severity of symptoms varies widely from mild agitation to autonomic instability, confusion, and severe muscle rigidity. Diagnostic criteria for SS have been developed but are based on case reports, which may predispose these criteria to under-identifying cases at the milder end of the spectrum. A high degree of suspicion on the part of the clinician is essential in recognizing and avoiding this condition.

Cross-References

▶ 5-HT$_{2A}$ Receptor
▶ Anticholinergic Side Effects
▶ Antidepressants
▶ Delirium
▶ Monoamine Oxidase Inhibitors
▶ Neuroleptic Malignant Syndrome
▶ Selective Serotonin Reuptake Inhibitors
▶ Serotonin
▶ SSRIs and Related Compounds
▶ Tricyclic Antidepressants

References

Arora B, Kannikeswaran N (2010) The serotonin syndrome – the need for physician's awareness. Int J Emerg Med 3(4):373–377

Birmes P, Coppin D, Schmitt L, Lauque D (2003) Serotonin syndrome: a brief review. CMAJ 168(11): 1439–1442

Boyer EW, Shannon M (2005) The serotonin syndrome. N Engl J Med 352(11):1112–1120

Frank C (2008) Recognition and treatment of serotonin syndrome. Can Fam Phys 54(7):988–992

Gillman PK (2006) A review of serotonin toxicity data: implication for the mechanisms of antidepressant drug action. Biol Psychiatry 59(11):1046–1051

Iqbal MM, Basil MJ, Kaplan J, Iqbal MT (2012) Overview of serotonin syndrome. Ann Clin Psychiatry 24(4):310–318

Mackay FJ, Dunn NR, Mann RD (1999) Antidepressants and the serotonin syndrome in general practice. Br J Gen Pract 49(448):871–874

Sporer KA (1995) The serotonin syndrome. Implicated drugs, pathophysiology and management. Drug Saf 13(2):94–104

Sternbach H (1991) The serotonin syndrome. Am J Psychiatry 148(6):705–713

Serotonin Transporter

Definition

A presynaptic membrane protein that transports the neurotransmitter serotonin from the synapse back into the presynaptic neuron, thereby "recycling" serotonin.

Sertindole

Definition

Sertindole is a second-generation antipsychotic that acts as a dopamine D2 and serotonin 5-HT$_{2A}$ antagonist. A phenylindole derivative, sertindole has been reintroduced as a second-line drug treatment after it had been voluntary withdrawn from the market because of concerns over cardiac arrhythmias and sudden deaths. However, the risk for cardiovascular mortality was found not to differ from that associated with other atypical antipsychotics, while it may be useful for patients in whom other antipsychotics have failed. Sertindole also lacks the sedative properties inherent to many other antipsychotic drugs. However, ECG monitoring is required before and during the treatment.

Cross-References

- Antipsychotic Drugs
- Antipsychotic Medication: Future Prospects
- Schizophrenia
- Second-Generation Antipsychotics
- Second- and Third-Generation Antipsychotics

Sertraline

Definition

Sertraline is a naphthalenamine derivative and acts as a ▶ selective serotonin reuptake inhibitor (SSRI) that is used as an antidepressant. Its effects are mostly typical for the class of SSRIs, sharing their therapeutic effects and side effects, although some studies have found a particularly favorable balance between benefits, acceptability, and cost. A recent systematic review and meta-analysis found evidence favoring sertraline over several other antidepressants for the acute-phase treatment of major depression. Among side effects, there was a higher rate of diarrhea with sertraline than some other SSRIs. It is absorbed slowly after oral administration and there are several metabolites including N-desmethyl-sertraline, which has weak pharmacological activity. The ▶ elimination half-life is in the range of 22–36 h. Sertraline has little inhibitory effect on the ▶ cytochrome P450 enzymes, with the result that interactions with other drugs are less likely than with some other SSRIs.

Cross-References

- Antidepressants

Setiptiline Maleate

Definition

Setiptiline maleate is a tetracyclic antidepressant that is used for the treatment of depression and other depressive states. It has antihistamine and hypnotic-sedative effects, but almost no anticholinergic effects. It is a weak inhibitor of ▶ norepinephrine reuptake in vitro and strongly stimulates the release of central norepinephrine by blocking presynaptic α_2-adrenoceptors, similar to ▶ mianserin. It also acts as a 5-HT$_{2A}$, 5-HT$_{2C}$, and 5-HT$_3$ receptor antagonist. Unlike many conventional antidepressants, it has no efficacy as a serotonin reuptake inhibitor. It can induce drowsiness and thirst, but it displays low toxicity.

Cross-References

- Depression
- Mianserin

Sex Differences in Drug Effects

Sakire Pogun[1,2] and Gorkem Yararbas[2]
[1]Center for Brain Research, Ege University, Bornova, Izmir, Turkey
[2]Institute on Drug Abuse, Toxicology and Pharmaceutical Science, Ege University, Bornova, Izmir, Turkey

Synonyms

Gender differences; Sexual dimorphism

Definition

In the 2001 National Academy of Sciences report (Exploring the Biological Contributions to Human Health: Does Sex Matter?), *sex* is defined as "the classification of living things, generally as male or female according to their reproductive organs and functions assigned by their chromosomal complement, and *gender* as a person's self-representation as male or female, or how that person is responded to by social institutions on the basis of the individual's gender presentation."

(Wizemann and Pardue 2001). The term "sex differences" is used to explain biological differences between males and females, while "gender differences" imply differential effects of social and cultural milieu on males and females; in other words, gender can be considered as a psychosocial construct. In nonhuman animal research, "sex differences" is the preferred term. Because of the complex interplay between biological and environmental factors, it is hard to define a clear-cut distinction between sex- and gender-based differences between males and females. Furthermore, recent findings from epigenetic research show that the biological factors are influenced by the environment. In either case, there is a variation in the characteristic associated with either the male or female sex. Morphological characteristics or features that are different in males and females of the same species are referred to as "sexual dimorphism" (two forms).

Sex differences are not limited to reproductive function and endocrine systems, but are also observed in the brain and behavior. As a result of these distinctive features between males and females, their biological responses to drugs are not the same.

Current Concepts and State of Knowledge

Until recently, women were excluded from many experiments with drugs, and, therefore, information about appropriate treatment of diseases and reactions to drug therapy was based on studies conducted only on men. Without systematic research, it was assumed that sex differences in responses to drugs might be attributed to differences in height, weight, and hormones. The exclusion of women from drug trials was, in part, the result of the thalidomide tragedy of the 1950s. Subsequently, in 1977 the FDA recommended that women of childbearing potential be excluded from drug research. It was assumed that studies conducted in adult males could be extrapolated to other patient populations, including women, children, and the elderly. For example, a 1989 study, excluding women, suggested the use of aspirin as the first-line treatment of cardiovascular disease, but in 1992 two studies showed that the benefits of aspirin were greater in men than in women. In 1993, the FDA withdrew the rule against the inclusion of women of childbearing potential in clinical trials and went forward to ensure that women were included in all future drug research. In more recent studies, drug researchers show increased awareness of gender-specific responses to drugs and include more women in clinical trials (Wetherington 2007).

The chemical structure and dose of a drug are important factors in producing the expected effects, but the actual amount of drug which binds to its target sites, in other words its bioavailability, is critical. The primary target site for psychoactive drugs is the brain. Dynamic factors (e.g., absorption, distribution, metabolism, and excretion) that determine the bioavailability of a drug (▶ Pharmacokinetics) are not the same in males and females. Gastric acid secretion and gastrointestinal blood flow are lower and gastric emptying time is longer in women; on average, women weigh less and have a higher percentage of body fat; activity of enzymes that metabolize drugs is sexually dimorphic; women have a lower glomerular filtration rate and a lower creatinine clearance compared with men.

The composition of the body, more specifically the ratio of fat and water, is an important factor in determining the plasma levels of a drug. As mentioned before, women have relatively more adipose tissue and the water content is less than in males. Therefore, as the proportion of water is smaller in females than males, the same dose of a water soluble drug would be expected to result in higher blood levels in females than males; subsequently, higher drug levels could result in greater effects in females, even if the dose was corrected for body weight. On the other hand, because there is more adipose tissue in a women's body (approximately 25 %) and many psychoactive drugs, e.g., most of the antipsychotic drugs, are lipophilic, some psychoactive drugs may accumulate in the adipose tissue and may show extended and greater bioavailability from the adipose tissue in women.

The endocrine milieu of women is much more variable than that of men; the gonadal (gonad) hormones fluctuate during the menstrual cycle, pregnancy, postpartum period, lactation, menopause, and while using oral contraceptives. Ovarian hormones have substantial modulatory roles on central neurotransmission. In case of a challenge such as an epileptic seizure, the permeability of the blood brain barrier is slightly more increased in females compared to males. This difference may impact the rate of drug entry to the brain and result in a stronger effect. Drug metabolism is another factor where sex differences are observed. The liver is the major site of drug inactivation and sexual dimorphisms in liver function may partly account for different metabolism of drugs in women and men. Sex-specific differences in enzyme activity can lead to sex differences in drug response. For example, women wake from general anesthesia several minutes before a man, given an equivalent dose. Gastric enzymes which metabolize alcohol before it reaches the bloodstream are weaker in women than men. Thus, women will have higher blood concentrations than men after ingesting similar amount of alcohol. Women are also more susceptible to alcohol-related hepatic damage than men.

Several cytochrome P450 isoenzymes are responsible for the metabolism of many drugs. For example, the activity of certain cytochrome enzymes which break down nicotine to its inactive metabolites is significantly higher in women than men, and higher metabolism is related to nicotine dependence.

Although the activational effects of sex hormones cannot fully account for all sex differences observed in the adult brain, and though many sex differences persist in the absence of these hormones, they have a substantial impact on the brain, behavior, and response to drugs.

The interplay between genetic and environmental factors may underlie sex differences observed in some neuropsychiatric disorders including schizophrenia. Recent studies show that sex hormones may directly control epigenetic mechanisms, and consequently, considering sex differences in brain epigenetics will be critically important in developing drugs that target epigenetic mechanisms (Menger et al. 2010). Responses to environmental cues are sexually dimorphic and result in sex-specific epigenetic programming. Although currently drugs that modify gene expression through epigenetic mechanisms such as histone acetylation are mainly used for treating leukemia and some solid cancers, animal studies suggest that similar drugs can ameliorate cognitive, behavioral, and hormonal symptoms of depression. Van den Hove et al. (2012) showed that prenatal stress, a rat model for affective disorders, increases anxiety and depression in adult male, but not female offspring. However, although the behavioral effects of prenatal stress were greater in male than female rats, alteration of gene expression was significantly higher in females than males suggesting the possible protective effects of altered molecular mechanisms against prenatal stress in female rats. In this regard, sex hormones and sex-specific epigenetics are apparently among the most important factors to be considered in developing therapeutic agents acting through epigenetic mechanisms.

Ovarian hormones affect the pharmacokinetics of a variety of drugs. Changes in the menstrual cycle are related to differential absorption and bioavailability of certain drugs. Gastric emptying is slower during the luteal phase of the menstrual cycle compared to the follicular phase. Gastrointestinal transit time is prolonged during the luteal phase compared to the follicular phase, allowing for greater drug absorption. Sodium retention, water content, and urinary volume may also vary with the menstrual cycle in women and influence the distribution of drugs.

Overall, rates of side effects to most drugs are reported to be higher in women than in men, and sex differences in drug metabolism may be involved in these adverse reactions. Additionally, increasing age, polypharmacy, and liver and renal disease may have sexually dimorphic impacts on side effects.

Dopaminergic System

Central dopaminergic systems are the target of many psychoactive drugs, including abused substances.

Males and females differ in the functioning of the dopaminergic system, and, therefore, sex differences are observed in response to drugs which act primarily through this system. The expression of the catechol-O-methyltransferase (COMT) gene, which encodes the enzyme that metabolizes dopamine, is higher in men than women. COMT, possibly together with other autosomal genes, has sexually dimorphic effects on normal as well as disturbed brain function.

Dopamine neurotransmission mediates the reinforcing effects (reward system) of drugs of abuse, and altered dopaminergic neurotransmission may underlie neuropsychiatric disorders such as schizophrenia, Parkinson's disease, and attention-deficit disorder, where sex differences have been reported in incidence, prevalence, clinical course, and treatment outcome. Addiction to some abused psychostimulants such as cocaine and nicotine is more severe in women, the age of onset of schizophrenia is later in women than men, and the prevalence of Parkinson's disease is lower in women than men. Thus, sex differences in the baseline activity of the dopaminergic system may underlie these observed differences as well as responses to dopaminergic drugs.

Dopamine release and uptake rates, which regulate extracellular dopamine levels, are higher in females than males in some brain regions such as the striatum. Women have higher levels of dopamine transporters in the striatum, lower striatal dopamine D2 receptor affinity, and higher dopamine 2 (D2) receptor levels in the frontal cortex compared to men; this may result in elevated levels of dopamine in these brain regions in women. In fact, postmortem and neuroimaging studies in human subjects suggest that dopamine release in women is higher relative to men. In human subjects, d-amphetamine-induced dopamine release in striatal and extrastriatal regions is higher in women suggesting sex differences in cognitive function and sensation-seeking behavior. The decline observed in DA receptor levels with age is also slower in women than men. In rats, indirect dopamine agonists like cocaine, which block the dopamine transporter, induce greater behavioral and neuroendocrine effects in females than males. Female rats acquire self-administration of cocaine faster than males (Lynch 2008). Responses to direct-acting dopamine D2 ligands display sex differences with females being more sensitive than males.

Ovarian hormones underlie some, but not all sex differences observed in drug abuse, suggesting qualitative and quantitative sexual dimorphisms in neural systems. Estrogen enhances the response to the psychomotor stimulants amphetamine or cocaine; this effect is observed both during estrogen replacement following ovariectomy and during the estrus phase of the estrus cycle in rats (Becker et al. 2001).

In studies on rodents, estrogen is reported to increase dopamine synthesis, and baseline dopamine release in the striatum, and to increase neuronal firing in substantia nigra (Walker et al. 2006). Estrogen inhibits gamma-aminobutyrate (GABA) neurons in the striatum and accumbens, which subsequently increase dopaminergic activity. Another established effect of estrogens is the downregulation of dopamine receptors. Dopaminergic transmission varies with estrous cycle phase. For example, amphetamine-stimulated dopamine release is greatest during the estrus phase of the cycle, a time when the behavioral response is also greatest. Estrogen plays an important role in modulating sex differences in neurochemical responses to psychomotor stimulants.

Stress, negative mood, and exposure to drug-related cues predict relapse in cocaine dependence. The effect of stress is more significantly associated with relapse in women and drug cues in men. Estrogen and progesterone have opposite effects on behavioral responses to cocaine; while the former increases the effects, the latter decreases them, and the effects are more pronounced in females than males. Stress and cocaine enhance the stress response, mainly through the corticotrophin releasing factor (CRF) and central noradrenergic pathways, both of which activate the mesolimbic dopaminergic systems involved in the rewarding effects of cocaine. Progesterone also plays a key role in stress regulation and D2 receptor function, which results in an increase of DA release. On the other hand, in male rats, the density of dopamine D1 receptors is higher than in female rats in the nucleus accumbens, a brain region implicated in reward.

Sex differences observed in particular drug-induced behavioral responses cannot necessarily be generalized to other behaviors. For example, while the locomotor response to amphetamine and methamphetamine in rats shows sex differences with females being more sensitive than males, no sex differences are reported for conditioned place preference induced by these drugs (Schindler et al. 2002). Sex differences are observed in responses to nicotine both in locomotor activity and in conditioned place preference, albeit in different directions: The effect of nicotine on locomotion is more pronounced in females, but unlike males, females do not show nicotine-induced place preference.

Sensitization of behaviors such as rotation, stereotyped grooming, head bobs, and forelimb movements observed following chronic exposure to psychostimulants is greater in female rats than in males.

Methamphetamine, another abused drug like cocaine, increases dopamine levels in some brain regions but through different mechanisms: While cocaine blocks reuptake, methamphetamine induces release from nerve terminals. The sex differences observed in the effects of methamphetamine are similar to cocaine: Females are more sensitive to the locomotor activating effects, but the effect gradually declines with repeated treatment in rats.

Drugs of Abuse

Until recently, preclinical (▶ Addictive Disorder: Animal Models) and clinical research on drug abuse has been conducted mainly on male subjects. Drug abuse was generally accepted as a male problem and males were thought to be more vulnerable to drug abuse (▶ Abuse Liability Evaluation). However, this view did not take into consideration the opportunity for use. There is growing evidence that points to sex differences in the effects of abused drugs. Biological and environmental predictors of drug use such as depression, conduct disorder, physical and sexual abuse, prenatal drug exposure, and family dysfunction affect males and females differently. Therefore, treatment strategies that are effective in one sex may not be equally successful in the other. Sex differences in sensitivity to abused drugs and in drug self-administration are also observed in laboratory animals and during different stages of the addiction process (i.e., acquisition, maintenance, and relapse). In both humans and laboratory animals, females, compared to males, are more susceptible to key phases of the addiction process that mark transitions in drug use such as initiation, drug bingeing, and relapse. Females appear to be more vulnerable than males to the reinforcing effects of psychostimulants, opiates, and nicotine as revealed by their behavioral, neurological, and pharmacological responses to these drugs. Women take less time to progress to dependence (tobacco, "▶ Caffeine," alcohol, cannabinoids, opiates, sedatives, cocaine, inhalants, amphetamine, "▶ Hallucinogens," and "▶ Phencyclidine") than men. Furthermore, in many cases, women appear to be more sensitive to the adverse health effects of drugs than men, pointing to the importance of developing gender-specific prevention and treatment programs in drug-related health problems. Methamphetamine (amphetamine derivative) abuse may be an exception. It is reported that women begin to use methamphetamine at their earlier ages and appear to be more dependent but also to respond better to treatment than men do. In women, use of methamphetamine is associated with depression, suggesting a type of self-medication.

Epidemiological studies indicate that during adulthood, men outnumber women in using illicit substances (except prescription medications), alcohol, and tobacco and that men are more likely to have a drug abuse/dependence disorder or an alcohol use disorder than women. However, this gender difference in prevalence is not observed among adolescents; in fact, some studies point to a slightly higher prevalence of smoking among girls than boys. Nonetheless, alcohol abuse and dependence is higher in males than females at all age groups.

Women are more sensitive to the physiological effects of alcohol than men, have higher blood levels following similar doses of alcohol, and report feeling more intoxicated. Negative mood-induced craving for alcohol is also greater in women than men. Even though higher blood levels in women may underlie these differences, sex differences in region-specific cerebral perfusion changes following alcohol intake have also been reported.

Following different routes of cocaine intake, women detect the subjective effects of cocaine later than men and report less euphoria and dysphoria, but feel more "nervous" compared to men. Women show greater reactivity than men to cocaine-related cues, and the craving during abstinence is significantly higher in women.

Women have greater vulnerability for smoking-related diseases (specifically myocardial infarction and lung cancer) than men, but are less successful in quitting smoking. Men benefit from nicotine replacement therapy more than women. Studies on rodents point to similar sex differences, suggesting the involvement of underlying sexual dimorphisms in biology. Females may take a shorter time to become dependent than males, they make fewer quit attempts, and they can stay abstinent for shorter periods than males; the rate of relapse is higher in females than males (Pogun and Yararbas 2009). Sexually dimorphic pharmacokinetics that causes variance in blood/brain levels of nicotine or the effects of gonadal hormones may underlie some of the sex differences observed in nicotine/tobacco addiction. Research findings (Cosgrove et al. 2012) show that the beta2-subunit of the nicotinic acetylcholine receptors, implicated in the addictive properties of nicotine, is regulated differently in the brains of male and female smokers; while these receptors are upregulated in the brains of male smokers, no effect of smoking on receptor density is observed in females. Nicotine replacement, nicotinic or non-nicotinic drugs, and behavioral therapies are currently used smoking cessation treatments. In line with the information above, treatments not involving nicotine, such as non-nicotinic medications or behavioral approaches, are reported to be more useful in women smokers. Recent research by the same group also demonstrates a sex-by-smoking interaction at the $GABA_A$-benzodiazepine receptors (Esterlitz et al. 2013). Imaging studies show that women have higher $GABA_A$-benzodiazepine receptor availability than men. Progesterone is likely to contribute to benzodiazepine use and abuse in women because it enhances the behavioral effects of benzodiazepines. Following one week of smoking abstinence, a negative correlation between $GABA_A$-benzodiazepine receptors and craving was observed in female smokers but not male smokers suggesting a sexually dimorphic role for these receptors in craving. Subsequently, $GABA_A$-benzodiazepine receptors may be a useful target to consider in drug development, specifically for smoking cessation in women smokers.

There are established sex-dependent differences in opiate reward. Female rats are more sensitive to the rewarding effects of morphine, and work more than males to self-administer morphine, and acquire conditioned place preference at lower doses than males. Additionally, there are also sex-related differences in the effect of morphine on locomotor activity, cardiovascular system, temperature, stimulus discrimination, physical dependence, and analgesia.

The cytoarchitecture of male and female brains is not the same; for example, the dendritic arborization in the anterior cingulate cortex, implicated in craving, is greater in male rats than in females. Psychoactive drugs modify the action of various neurotransmitter systems at the cellular and molecular levels and impact synaptic structure and function. There are clearly established sex differences in the action of neurotransmitters, with specific characteristic. The organizational and activational effects of gonadal hormones underlie some of the sex differences in the brain and behavior. Since sex steroids, and especially estrogens, modify the binding of ligands to their receptor sites, sex differences in the action of pharmaceuticals deserve attention.

Overall, results from a large number of studies indicate that females are more vulnerable to drug abuse (critical phases) than males. On the other hand, males show more withdrawal symptoms than females. Anker and Carroll (2011) suggest that males are more sensitive o the aversive effects of drugs and females are more sensitive to the rewarding effects. This is an important issue and should be considered in sex-specific medication development and treatment.

Antipsychotics

The response and end-organ sensitivity to antipsychotic treatment, similar to other drugs, are influenced by genetics, age, height, weight, lean-fat ratio, diet, exercise, concurrent disease, smoking and alcohol, and the administration of concomitant drugs. There are sex differences in all of these factors, and taken together, these factors are estimated to contribute to a tenfold variability in responses to medication.

Sex differences in response to antipsychotic agents have been reported in both animals and humans. In drug-naive subjects, response to antipsychotic drugs is reported to be superior in women. In chronically ill patients with schizophrenia or related psychoses, men require twice as high a dose as women for effective maintenance. Recommendations for prescribing antipsychotic medication to women include using lower doses for women than men, longer intervals for women when depot doses are used, and careful modulation of drug dose in aging women. In general, schizophrenic women have higher degree of symptom improvement but also of extrapyramidal symptoms (Seeman 2004).

Until recently, women have been underrepresented in clinical trials of second-generation antipsychotics, and therefore, there are limited data on possible sex differences in drug efficacy and side effects. A main reason for this underrepresentation was the fear of potential teratogenicity. Currently, sex-specific analysis of efficacy and safety data is a requirement, and therefore, an increasing number of studies are now available concerning sex differences in pharmacokinetics or pharmacodynamics of second-generation antipsychotics.

There are reported sex differences in pharmacokinetics for second-generation antipsychotics clozapine, olanzapine, and sertindole, and CYP1A2 activity has been implicated in these differences. Women have higher plasma levels of all these second-generation antipsychotics than men. On the other hand, no sex differences were observed for quetiapine. However, as the studies involve relatively small number of patients, larger samples are required before generalizations can be made on sex differences in responses to novel antipsychotics.

Recent reports suggest that women on olanzapine have a significantly better treatment response than men, and premenopausal women have a better treatment response than postmenopausal women. Overall, schizophrenic women under 40 years of age require lower antipsychotic doses than men regarding both acute response and maintenance (Aichhorn et al. 2006).

Because of important side effects of antipsychotic medication, regular monitoring of weight gain and body mass index, plasma glucose level, lipid profiles, signs of prolactin elevation and hyperprolactinemia, or sexual dysfunction is recommended. Cardiac side effects and extrapyramidal symptoms should be considered as well. However, despite established sex differences in response to antipsychotic medication, these effects do not receive much emphasis in prescribing these drugs to men or women. Women, perhaps because of the influence of sex hormones, have a lower risk of sudden cardiac death, but a higher risk of acquired long QT syndrome from antiarrhythmic drugs. Estrogens may facilitate bradycardia-induced prolongation of the QT interval. Potent antipsychotic drugs were likely to block cardiac voltage-gated potassium channels, prolong the QT interval, and result in ventricular arrhythmias. Subsequently, women may be at a higher risk of cardiac side effects of antipsychotic medication. Older antipsychotics like butyrophenone and phenothiazine derivatives and some of the second-generation antipsychotics elevate prolactin levels (hyperprolactinemia) and result in sexual dysfunction, which in turn results in noncompliance with treatment, particularly in men. In the long run, hyperprolactinemia also causes galactorrhea, amenorrhea, breast engorgement, and osteoporosis. In women, the elevation of prolactin concentrations is generally noted to be higher than in men. On the other hand, aripiprazole which has a different profile from second-generation antipsychotics by being a partial D2 receptor agonist is reported to lower prolactin levels.

Antidepressants

Although the prevalence of major depression is twice as much in women than men, sex differences observed in antidepressant medication are not overwhelming. Women experience more vegetative and atypical symptoms, anxiety, and anger than men and report higher severity of depression on self-report measures; however, no significant sex differences are observed in the course of the illness and treatment response. The efficacy of bupropion and selective serotonin reuptake inhibitors (SSRI) appear to be equally effective in treating depression, anxious/somatic symptoms, and insomnia; there is a slight sex difference as greater improvement was seen in women during SSRI treatment regarding anxious/somatic symptoms of depression. Additionally, the efficacy of antidepressants is more pronounced in younger patients. On the other hand, a recent study suggests that women with generalized anxiety disorder, particularly those with a later age of onset, may have a poorer response to the SSRI fluoxetine compared to men, but benefit more from sertraline. In a number of earlier studies, men are reported to show greater therapeutic response than women to the tricyclic antidepressant imipramine. Similar to depression, anxiety affects more women than men. Preclinical studies in rodents also point to the importance of sex differences in responses to anxiolytic drugs. While diazepam and chlordiazepoxide had similar anxiolytic-like effects in both male and female rats, buspirone, propranolol, fluoxetine, and paroxetine showed activity only in male rats (Basso et al. 2011). MAO inhibitors in women and tricyclic antidepressants in men may be better therapeutic choices in the treatment of depression. However, there are yet other studies that have not detected significant sex differences in the effects of antidepressants. Methodological differences may underlie the apparent discrepancy between results from different laboratories. Considering the enhancing effect of estrogens on the efficacy of antidepressant treatment, menopausal status should not be overlooked. Elucidating the sex-specific effects of antidepressants regarding efficacy, metabolism, and side effects will aid in developing sex-specific treatment strategies and advance our tools for coping with depression.

Analgesics

Pain management by pharmacotherapy is another important medical problem where sex differences are noted. Furthermore, there may be gender differences in pain tolerance. Women report suffering from migraine and arthritis, more intensely than men do. Men have a better response to nonsteroidal anti-inflammatory analgesics (NSAIDS) than women, and women benefit more from narcotics than men do. Therefore, over-the-counter analgesics do not help women as much as men, and lower doses of narcotics, which are effective on women, are not as effective on men.

Males and females respond differently to drugs acting at opioid receptors; these quantitative and qualitative differences are not restricted to the analgesic and antinociceptive properties of opioids, but are also present in opioid-induced side effects (i.e., effects on respiration, locomotor activity, learning/memory, addiction, and cardiovascular system). However, the direction and magnitude of sex differences regarding the potency of opioids depend on the interacting variables which are specific to the drug (i.e., dose, pharmacokinetics and pharmacodynamics, route and time of administration) or to the subject (i.e., species, type of pain, genetics, age, gonadal/hormonal status, and psychological factors). The differential organizational and/or activational effects of gonadal steroid hormones in males versus females have received emphasis in explaining sex differences in opioid antinociception. There are sex differences in response to drugs acting on kappa-opioid receptors: Women respond more robustly than men to kappa-opioid agonists and antagonists with analgesic and hyperalgesic properties, respectively. When drugs acting at different receptors are compared, women have better pain scores with kappa-opioid agonists (butorphanol) than mu-opioid agonists (morphine). However, rodent data from laboratory pain models contrast with human data.

Conclusion

Recent studies on both animals and humans clearly indicate that many normal physiological and pathological functions are influenced by sex-based differences in biology, either directly or indirectly. Males and females are not only different regarding reproductive function, but sex differences exist in the brain and behavior, including emotion, memory, vision, hearing, processing faces, pain perception, navigation, neurotransmitter levels, stress hormone action on the brain, and disease states. Sexual dimorphisms are observed in many neurotransmitter systems, including monoamines, serotonin, GABA (gamma-aminobutyric acid), acetylcholine, vasopressin, and opioids. Furthermore, dynamic factors that influence the bioavailability of drugs, body composition, gastrointestinal, renal and liver function, hormonal status, and activity of enzymes involved in drug metabolism are not the same in males and females. Subsequently, it is clear that there are sex and gender differences in response to neuropsychopharmacological treatments. Sex is an important variable that should be considered when designing and analyzing studies in all areas of biomedical and health-related research, as well as in clinical applications.

Cross-References

- Abuse Liability Evaluation
- Addictive Disorder: Animal Models
- Alcohol Abuse and Dependence
- Analgesics
- Antidepressants
- Antipsychotic Drugs
- Blood–Brain Barrier
- Caffeine
- Cannabinoids and Endocannabinoids
- Cocaine
- Conditioned Place Preference and Aversion
- Epigenetics
- Hallucinogens
- Inhalant Abuse
- Insomnias
- Nicotine
- Opioids
- Pharmacokinetics
- Schizophrenia
- Second- and Third-Generation Antipsychotics
- Sedative, Hypnotic, and Anxiolytic Dependence
- Sensitization to Drugs
- Sex Hormones
- SSRIs and Related Compounds

References

Anker JJ, Carroll ME (2011) Females are more vulnerable to drug abuse than males: evidence from preclinical studies and the role of ovarian hormones. Curr Top Behav Neurosci 8:73–96.

Aichhorn W, Whitworth AB et al (2006) Second-generation antipsychotics: is there evidence for sex differences in pharmacokinetic and adverse effect profiles? Drug Saf 29(7):587–598

Basso AM, Gallagher KB et al (2011) Vogel conflict test: sex differences and pharmacological validation of the model. Behav Brain Res 218(1):174–183.

Becker JB, Molenda H et al (2001) Gender differences in the behavioral responses to cocaine and amphetamine. Implications for mechanisms mediating gender differences in drug abuse. Ann N Y Acad Sci 937:172–187

Cosgrove KP, Esterlis I et al (2012) Sex differences in availability of beta2*-nicotinic acetylcholine receptors in recently abstinent tobacco smokers. Arch Gen Psychiatry 69(4):418–427.

Esterlis I, McKee SA et al (2013) Sex-specific differences in GABA(A)-benzodiazepine receptor availability: relationship with sensitivity to pain and tobacco smoking craving. Addict Biol 18(2):370–378.

Lynch WJ (2008) Acquisition and maintenance of cocaine self-administration in adolescent rats: effects of sex and gonadal hormones. Psychopharmacology (Berl) 197(2):237–246

Menger Y, Bettscheider M, Murgatroyd C, Spengler D (2010) Sex differences in brain epigenetics. Epigenomics 2:807–821

Pogun S, Yararbas G (2009) Sex differences in nicotine action. In: Henningfield JE, London ED, Pogun S (eds) Nicotine psychopharmacology. Handbook of experimental pharmacology, vol 192. Springer, Berlin, pp 261–291

Schindler CW, Bross JG et al (2002) Gender differences in the behavioral effects of methamphetamine. Eur J Pharmacol 442(3):231–235

Seeman MV (2004) Gender differences in the prescribing of antipsychotic drugs. Am J Psychiatry 161(8):1324–1333

Van den Hove DL, Kenis G, Brass A, Opstelten R, Rutten BP, Bruschettini M, Blanco CE, Lesch KP, Steinbusch HW, Prickaerts J (2012) Vulnerability versus resilience to prenatal stress in male and female rats; Implications from gene expression profiles in the hippocampus and frontal cortex. Eur Neuropsychopharmacol. doi:10.1016/j.euroneuro.2012.09.011

Walker QD, Ray R et al (2006) Sex differences in neurochemical effects of dopaminergic drugs in rat striatum. Neuropsychopharmacology 31(6):1193–1202

Wetherington CL (2007) Sex-gender differences in drug abuse: a shift in the burden of proof? Exp Clin Psychopharmacol 15(5):411–417

Wizemann T, Pardue M-L (eds) (2001) Exploring the biological contributions to human health: does sex matter? National Academy of Sciences, Institute of Medicine Report, Washington, DC. http://lab.nap.edu/openbook/0309072816/html/1.html

Sex Hormones

Jill B. Becker
Psychology Department and Molecular & Behavioral Neuroscience Institute, University of Michigan, Ann Arbor, MI, USA

Synonyms

Gonadal hormones

Definition

The brains of mammals are sexually dimorphic due to the actions of testosterone and estradiol during sensitive periods of development. Additionally, males and females produce different sex hormones, so the circulating hormonal milieu is also sexually dimorphic. The gonadal hormones act in the brain at both intracellular and membrane receptors to induce changes in brain and behavior. One of the things that the ovarian hormones, estradiol and progesterone, do is modulate the response to cocaine and amphetamine by acting at membrane receptors in the striatum and nucleus accumbens. These effects of the ovarian hormones result in sex differences in the behavioral responses to these drugs. While androgens are reinforcing in their own right, they do not seem to modulate the behavioral response to cocaine or amphetamine.

Pharmacological Properties

Background

The brains of mammals undergo sexual differentiation during sensitive periods of development. In rodents, this occurs during the perinatal period and again during the peripubertal period. In humans, this occurs during the second trimester and, then again, during the peripubertal period. Exposure to testosterone (which is converted to estradiol in some areas of the brain) during early brain development influences neuronal survival, differentiation, and connectivity. During the peripubertal period, less is known about the extent of hormonal influences, but neuronal pruning and dendritic pruning, as well as neuronal reorganization, are thought to occur when the brain is exposed to ovarian or testicular hormones at the onset of puberty. As a result of these periods of hormone exposure, the brains of adult males and females are sexually dimorphic (Becker et al. 2005, 2008).

In concert with the sexual dimorphisms caused by the processes involved in sexual differentiation, there may also be sex differences in brain function produced in response to exposure to gonadal hormones in the adult. This second type of sex difference can be either due to the effects of gonadal hormones acting on sexually dimorphic brain regions or as a result of the different effects induced by ovarian versus testicular hormones, which in some neural regions may decrease differences between the sexes.

Gonadal hormones are by definition sexually dimorphic, since males have testes that produce androgens and females have ovaries that produce estrogens and progestins. They are also dimorphic in their patterns of secretion.

Testicular hormones are secreted in a tonic pattern, and feedback of testosterone to the brain maintains the rate of pituitary hormone secretion that sustains a constant rate of spermatogenesis in the testes. Ovarian hormones are released in a cyclic manner. First, estradiol is released gradually during the development of the follicle that nourishes the development of the egg (follicular phase). Then, a rapid increase in estradiol induces the hormonal trigger to induce ovulation (periovulatory phase). Finally, in some species, estradiol and progesterone are released to prepare the womb for implantation of a fertilized egg (luteal phase) (Becker et al. 2005, 2008).

Unlike many pharmacological agents, in general, gonadal hormones do not activate or inhibit specific behaviors; instead, they prepare an individual to be able to engage in a particular behavior should the appropriate conditions arise (e.g., sexual behavior) or modulate the behavior that is exhibited (e.g., the behavioral response to amphetamine in females). One exception to this rule is the anxiolytic effects of progesterone and its 5α-reductase-derived metabolites, which are produced in the brain and are known as neurosteroids.

Mechanisms of Action

Gonadal hormones are steroidal in chemical structure, so they are lipophilic and act at both intracellular and membrane-associated receptors. The primary intracellular receptors have been well characterized. When bound to the ligand, the receptor-ligand complex dimerizes and then binds to the DNA to initiate transcription. Activation of this receptor mechanism requires hours to days for a response to be observed. There are two forms of the estradiol receptor (ER), known as ERα and ERβ. One of the products of ER activation is the progesterone receptor (PR) which has two isoforms, PR-A and PR-B (Mani 2008). Only one form of the intracellular androgen receptor (AR) has been identified to date. Males and females tend to express intracellular ER, PR, and AR in the same brain regions, but there are sex differences in the number of receptors expressed, with females expressing more ER and PR than males and males expressing more AR than females.

Membrane receptors for the gonadal hormones have been identified only recently. In the case of progesterone, a novel receptor with seven transmembrane domains that activates a G-protein-mediated intracellular signaling pathway has been identified in vertebrates with two isoforms. The nuclear PRs are also thought to be found at the cell membrane after palmitylation, which anchors the nuclear receptor to the membrane. The nuclear PRs at the membrane also activate G-protein-mediated intracellular signaling pathways, but the exact nature of the interaction is not yet known (Mani 2008). For estradiol, two novel membrane receptors have been identified, GPR30 and Gq-mER; both activate G-protein-mediated, intracellular signaling pathways. As was seen with progesterone, nuclear ERs are also found at the cell membrane to activate G-protein-mediated intracellular signaling pathways. In the case of the ERs, palmitylation may also be involved, and the ERs are thought to associate with caveolin-1 and/or caveolin-3 as well as to maintain their position in association with the membrane (Boulware et al. 2007; Micevych et al. 2009). It is not yet clear how the nuclear ERs (or PRs) are positioned relative to the membrane so that the ligand can be bound and intracellular pathways activated. In some brain regions, it is thought that ERs associate with metabotropic glutamate receptors to gain access to intracellular signaling pathways (Boulware et al. 2005). The rapid actions of androgens have been well documented in the brain and other tissues. The nuclear AR is implicated in many of these effects, and it is likely that there are also novel membrane ARs yet to be discovered.

Gonadal Hormone Modulation of Drug Effects in the Brain

Stimulant Drugs

In both rats and humans, the behavioral effects of drugs of abuse, and the psychomotor stimulants in particular, are both sexually dimorphic and modulated by the gonadal steroid hormones.

With repeated intermittent exposure to cocaine or amphetamine, the behavioral responses exhibited may be enhanced with each drug exposure. This is referred to as behavioral sensitization, is different in males and females, and is differentially affected by gonadal steroid hormones. For example, intact females exhibit more robust sensitization than do intact males. Following ovariectomy (OVX) of female rats, the expression of sensitization to amphetamine is attenuated or suppressed altogether relative to intact female rats. Estradiol treatments in OVX rats enhance sensitization of locomotor activity induced by amphetamine or cocaine (Becker and Hu 2008; Carroll et al. 2004).

These effects of gonadal hormones are relevant to addiction, where sex differences have been reported during all phases of the addiction process. Women tend to become addicted to drugs more rapidly than men and to escalate drug use at a greater rate. In animal models, when a low dose of a drug is used, female rats acquire cocaine self-administration at a faster rate than do males. The role of gonadal hormones can be seen when estradiol is given just before a self-administration session, as estradiol treatment enhances acquisition of cocaine self-administration in OVX female rats and treatment with the estradiol antagonist, tamoxifen, inhibits acquisition in intact females. In male rats, estradiol does not enhance acquisition or cocaine intake, so the brains of males and females are sexually dimorphic in this regard (Becker and Hu 2008; Carroll et al. 2004; Walker et al. 2006).

During maintenance conditions, when given a choice between two doses of cocaine, female rats in estrus preferred higher doses of cocaine when compared with females in other phases of the estrous cycle or male rats. When the role of estradiol in "binge" cocaine intake and subsequent motivational changes is examined, estradiol treatment increases the initial binge length and total levels of cocaine self-administration. Another way to assess motivation is to use a schedule of reinforcement in which the number of responses required to obtain a cocaine infusion progressively increases with each dose received. Under this "progressive ratio schedule," intact female rats reach much higher final ratios than do males, indicating that females are more motivated to obtain cocaine. Females also work harder for access to cocaine during the phase of the estrous cycle when estradiol is elevated, indicating that ovarian hormones modulate the motivation to obtain cocaine. In fact, estradiol treatment enhances responding for cocaine on a progressive ratio schedule. These results show that estradiol influences acquisition of cocaine self-administration and that there are motivational effects of estradiol on cocaine intake (Becker and Hu 2008).

In contrast to estradiol, progesterone treatment given concurrently with estradiol counteracts the effect of estradiol on acquisition of cocaine self-administration behavior. We have recently confirmed this finding and understand that progesterone not only affects cocaine self-administration but also enhances cocaine intake in estradiol-primed OVX rats. Taken together, ovarian hormones contribute to sex differences in cocaine self-administration, and estradiol in particular is a key factor influencing the reinforcing effects of cocaine in female rats. So, over the course of the estrous cycle, there are peaks and valleys during which females are more or less susceptible to the reinforcing properties of cocaine and other drugs of abuse.

As discussed above, our understanding of the mechanisms mediating the effects of estradiol and progesterone in the areas of the brain that mediate the response to drugs of abuse is evolving rapidly. For the most part, we believe that the effect of drugs of abuse on the ascending mesotelencephalic dopamine (DA) system is necessary for the addictive and motivational properties of these drugs. The acute administration of estradiol to OVX rats induces a rapid increase in amphetamine- or cocaine-induced DA release in nucleus accumbens or striatum. Estradiol also induces an increase in striatal DA turnover and downregulates D2-class DA receptors. These effects have been shown to be due to the direct effect of estradiol on the striatum and nucleus accumbens, presumably at membrane receptors for estradiol. Finally, estradiol induces an increase in PRs in the striatum, and progesterone also acts directly on the estradiol-primed striatum to modulate DA release (Becker and Hu 2008; Walker et al. 2006).

In terms of androgens, castration (CAST) of males has been reported to enhance sensitization of amphetamine- or cocaine-induced psychomotor behavior, although this result has not been found consistently. It has been hypothesized that if CAST enhances the induction and/or expression of behavioral sensitization, that testosterone treatment should reverse this effect. This is not the case, however, as testosterone treatment has not been found to affect behavioral sensitization in CAST males. Furthermore, there is no effect of CAST on acquisition of cocaine self-administration behavior, and a dose of estradiol that enhances self-administration in female rats has no effect on cocaine self-administration behavior in males. So, it does not seem that androgens play a role in modulating the motivational effects of drugs of abuse, even though hamsters, rats, and humans self-administer testosterone and other anabolic steroids all by themselves (Becker and Hu 2008).

Other Psychoactive Drugs
Gonadal hormones can also affect the response to drugs other than the stimulants, but the mechanisms mediating the effects are not as well worked out as for the psychomotor stimulants. Nevertheless, menstrual cycle effects and sex differences in the effectiveness of antidepressants and analgesics suggest that the gonadal hormones can modulate a wide variety of drug effects. In addition, the effects of gonadal hormones on the pharmacokinetics and pharmacodynamics of a drug's action must always be considered (Becker and Meisel 2007).

Conclusion
In summary, the gonadal hormones act in the brain at both intracellular and membrane receptors to induce changes in brain and behavior. The ovarian hormones, estradiol and progesterone, modulate the response to cocaine and amphetamine by acting at membrane receptors in the striatum and nucleus accumbens. These effects of the ovarian hormones result in sex differences in the behavioral responses to these drugs.

Cross-References
▶ Activational Effects of Hormones
▶ Addictive Disorder: Animal Models
▶ Adolescence and Responses to Drugs
▶ Analgesics
▶ Antidepressants
▶ Behavioral Tolerance
▶ Cocaine
▶ Cocaine Dependence
▶ Drug Abuse
▶ Drug Self-Administration
▶ Estrogens
▶ Motor Activity and Stereotypy
▶ Self-Administration of Drugs
▶ Sensitization to Drugs
▶ Sex Differences in Drug Effects
▶ Stress: Influence on Relapse to Substance Use

References

Becker JB, Hu M (2008) Sex differences in drug abuse. Front Neuroendocrinol 29:36–47

Becker JB, Meisel RL (2007) Neurochemistry and molecular biology of reward. In: Lajtha A, Blaustein J (eds) Handbook of neurochemistry and molecular neurobiology, vol 21, 3rd edn. Springer Science, New York, pp 739–774

Becker JB, Arnold A, Berkeley KJ, Blaustein JD, Eckel LA, Hampson E, Herman JP, Marts S, Sadee W, Steiner M, Taylor J, Young E (2005) Strategies and methods for research on sex differences in brain and behavior. Endocrinology 146:1650–1673

Becker JB, Berkley K, Geary N, Hampson E, Herman JP, Young EA (eds) (2008) Sex differences in the brain: from genes to behavior. Oxford University Press, Oxford

Boulware MI, Weick JP, Becklund BR, Kuo SP, Groth RD, Mermelstein PG (2005) Estradiol activates group I and II metabotropic glutamate receptor signaling, leading to opposing influences on cAMP response element-binding protein. J Neurosci 25:5066–5078

Boulware MI, Kordasiewicz H, Mermelstein PG (2007) Caveolin proteins are essential for distinct effects of membrane estrogen receptors in neurons. J Neurosci 27:9941–9950

Carroll ME, Lynch WJ, Roth ME, Morgan AD, Cosgrove KP (2004) Sex and estrogen influence drug abuse. Trends Pharmacol Sci 25:273–279

Mani S (2008) Progestin receptor subtypes in the brain: the known and the unknown. Endocrinology 149(6):2750–2756

Micevych P, Kuo J, Christensen A (2009) Physiology of membrane estrogen receptor signaling in reproduction. J Neuroendocrinol 21:249–56. PMID: 19207814

Walker QD, Ray R, Kuhn CM (2006) Sex differences in neurochemical effects of dopaminergic drugs in rat striatum. Neuropsychopharmacology 31:1193–1202

Sexual Behavior

Anders Ågmo
Department of Psychology, University of Tromsø, Tromsø, Norway

Synonyms

Mating behavior; Reproductive behavior

Definition

Anaphrodisiac, any compound reducing the intensity of sexual behavior.

Aphrodisiac, any compound enhancing the intensity of sexual behavior or the pleasure derived from such behavior. The term is sometimes used to designate anything that increases sexual urges, like in "visual aphrodisiacs," referring to sexually exciting pictures or movies.

Lordosis, a concave dorsiflexion of rodents associated with lifted rump and the tail moved to one side. This posture exposes the vaginal orifice and is the basic female copulatory posture in many mammals. The ease by which lordosis is activated is frequently called receptivity.

Paracopulatory behavior, stereotyped motor patterns, in rodents mainly ear wiggling and hop-darting, typical of many female mammals and supposed to indicate a high propensity to engage in copulatory behavior. Also known as *proceptive behavior*.

Copulatory behavior, the display of reflex-like, stereotyped motor patterns making possible contact between male and female genitalia. In male mammals, the basic motor pattern is mount with pelvic thrusting. The mount may or may not lead to vaginal penetration and ejaculation. In female mammals, lordosis is the basic pattern. The term copulatory behavior is mainly used for nonhuman animals.

Sexual behavior, any action leading to sexual reward. Sexual reward is a state of positive affect activated by the physical stimulation of the genitalia or mental representations of such stimulation.

Sexual motivation, a concept referring to a set of central nervous processes determining the likelihood of display of sexual behaviors and their intensity if displayed.

Impact of Psychoactive Drugs

A Long Tradition

The human seems to have been interested in manipulating sexual urges through pharmacological means since the dawn of history. There is anecdotal mention of plants and herbs purportedly stimulating or reducing the interest in sexual activities in many ancient sources. The search for and use of aphrodisiacs have attracted much attention, but there has also been efforts made to find ways to diminish sexual desire. In the classical Chinese empires, this was achieved by a simple procedure, modifying the availability of androgens through castration. Eunuchs were favored as high state officials during many centuries. Likewise, some men of faith, like the Egyptian theologian Origen, castrated themselves in order to be relieved from sexual temptations. During the Dark Ages, castration was no longer an approved practice, but monks are said to have prepared and consumed extracts of a plant, the common rue (*Ruta graveloens*), supposed to inhibit sexual desire. Despite the existence of occasional efforts to reduce sexual behavior, it can safely be stated that the main interest has been directed toward ways to

Sexual approach behaviors	Copulation	Aftereffects of copulation
Activated by distant sexual incentives (unlearned or conditioned) variable activities in skeletal muscles, may or may not be accompanied by autonomous reflexes	Activated by proximal (mainly tactile) stimuli stereotyped activity in skeletal muscles + autonomous reflexes (mainly vascular)	Positive affect (reward) + temporary reduction of sexual motivation

Sexual Behavior, Fig. 1 The elements of sexual behavior. Distant sexual incentive stimuli, emitted by a potential mate, activate approach behaviors. The specific motor patterns involved are determined by the context. If successful, the approach behaviors will lead to the establishment of physical contact with the potential mate. Eventually a sequence of stereotyped copulatory motor patterns (such as mounting and lordosis in rodents) may be activated. These motor patterns will bring the male genitals in contact with the female genitals and normally continue until the deposit of sperm in the female reproductive tract has been accomplished. Sperm deposit (ejaculation in the male) is followed by a state of positive affect. Every part of this sequence may be modified by drugs. The reactivity to sexual incentives may be enhanced or reduced, having as consequence increased or decreased likelihood for the display of sexual approach behaviors. Even the incentive properties of an individual may be altered through pharmacological means. However, very little is known about this. The stereotyped copulatory motor patterns themselves are rarely affected by drug treatments, but their frequency or number may be affected. Finally, the hedonic consequences of sexual activity may be enhanced or reduced by drugs. There are few experimental studies of drug-induced modifications of postcopulatory affective states. However, the opioid antagonist naloxone has been shown to block the postcopulatory hedonic state in male and female rats (See Ågmo (2007) for an extensive discussion of the sequence of sexual behaviors)

enhance it. Unfortunately, none of the many suggested aphrodisiacs have been shown to have any reliable effect, although most of them have never been subjected to clinical or experimental tests. This is also the case for the vast majority of anaphrodisiacs. However, it is conceivable that plants causing a strong sensation of ill-being, as *Ruta graveolens*, temporarily reduce or eliminate interest in sexual activity. Nevertheless, it is only with the advent of modern pharmacology that scientifically sound studies of the effects of xenobiotics on sexual functions have been performed (see Riley et al. 1993; Crenshaw and Goldberg 1996, for classical reviews of the pharmacology of human sexual behavior).

Essential Notions About Sexual Behavior

Any discussion of drug actions on sexual behavior becomes difficult without having some basic notions about this behavior. A schematic representation of the basic elements of sexual behavior is found in Fig. 1. Prominent among these is copulation. In all nonhuman animals, copulation is highly stereotyped. Some would even maintain that the copulatory acts basically are a series of somatic reflexes. However, copulation requires at least two individuals in physical proximity. In fact, behaviors leading to approach to a potential mate always precede the execution of copulatory behaviors. Approach behaviors are extremely variable and determined by the prevailing context. The intensity of approach to a potential sexual partner is an exquisite indicator of the intensity of the urge to engage in sexual activities or sexual motivation. In nonhuman animals, the display of copulatory reflexes is as dependent on sexual motivation as sexual approach and paracopulatory behaviors are. However, the human may engage in copulatory behaviors because she/he is forced to do so (the term for most such activities is rape) or as part of a business arrangement (often called prostitution), even in the absence of sexual motivation. It is not entirely clear whether nonhuman animals also may engage in sexual behaviors without an active sexual motivation. In the case of rodents, all available data suggest that sexual motivation is required for the display of sexual behaviors. With these rare exceptions, the intensity of sexual motivation determines the likelihood of display of sexual approach behaviors and their intensity as well as the likelihood of copulatory acts and their intensity. It can be maintained that all drug actions on sexual behaviors are, in fact, actions

Sexual Behavior, Table 1 Effects on sexual behaviors of manipulations of the activity in major neurotransmitter systems

Transmitter	Drug action	Effect in males	Effect in females
Acetylcholine	Muscarinic agonists	Facilitation	Facilitation[a]
	Muscarinic antagonists	Inhibition	Inhibition
Dopamine	Nonselective or selective agonists	Stimulation or no effect	Inhibition or no effect
	Nonselective or selective antagonists	Inhibition	Facilitate lordosis
GABA	Agonists	Inhibition	Inhibition
	Antagonists	Facilitation	Inhibition or facilitation[a,b]
Noradrenaline	α_2-antagonists	Facilitation	Facilitation
Endorphins/enkephalins	Agonists	Inhibition	Inhibition or facilitation[b]
	Antagonists	No reliable effect	Inhibition or facilitation[b]
Oxytocin	Agonists	Facilitation	Facilitation
	Antagonists	Inhibition	
Serotonin	Precursor 5-hydroxytryptophan (5-HTP)	Inhibition	Inhibition
	Synthesis inhibition (PCPA)	Stimulation	Stimulation
	Neurotoxin (5,7-dihydroxytryptamine)	Facilitation	Facilitation
	5-HT$_{1A}$ agonists	Facilitation	Inhibition
	5-HT$_{2A/C}$ agonists	Inhibition	Facilitation

Only the most consistently reported effects are mentioned. Facilitation includes enhanced receptivity and/or increased frequency of paracopulatory behaviors in females and reduced latencies to mount, intromit, or ejaculate in males. Inhibition includes reduced receptivity and/or reduced frequency of paracopulatory behaviors in females and enhanced latencies to mount, intromit, or ejaculate in males. Many other effects on subtle details of sexual behaviors have been described in addition to those mentioned, but they are ignored here (see Argiolas 1999; Paredes and Ågmo 2004; Hull and Dominguez 2006, for reviews). Several additional transmitters/modulators have been shown to modify sexual behaviors, but our knowledge about these is at a very preliminary stage, making it too adventurous to summarize their actions
[a]Facilitation observed in partly receptive rats
[b]Depending on site of intracerebral infusion

on motivation. It must be noted that different neurobiological processes underlie different aspects of sexual behavior. The neural systems crucial for sexual approach may be different from those controlling seminal emission and ejaculation, for example (Ågmo 2007). This means that a drug may affect the latter without modifying the former. No effort will be made to present a comprehensive list of drugs that may affect sexual behaviors. The following brief review is limited to compounds that have attracted more than occasional attention, and for which reasonably consistent and sound experimental data are available. A short summary of the effects of manipulations of the most studied neurotransmitters is found in Table 1.

Drugs Enhancing Sexual Motivation

Any increase in sexual motivation leads to an increase in one or more of the sexual behaviors. However, not all increases in the intensity of sexual behaviors should be attributed to an increment of sexual motivation. In males, including men, penile erection is a requisite for sexual behavior in the form of penile-vaginal intercourse and for some other forms of penetrative sex. Thus, the frequency of copulation is by necessity low in males suffering from erectile dysfunction. Such males will display more sexual behavior when using sildenafil, tadalafil, or some other of the phosphodiesterase-5 inhibitors than when treated with placebo. This is also the case for men using intra-penile treatments or centrally acting, erection-enhancing drugs such as apomorphine. Nevertheless, drugs facilitating erection, regardless of their site of action, are usually not considered as stimulating sexual motivation. This also holds for treatments promoting vaginal lubrication, although such treatments well may enhance the frequency of intercourse. A drug considered as stimulating sexual motivation should not render sexual activity possible. Instead, it should enhance the urge to engage in sexual activity and/or improve the performance of copulatory acts.

Acetylcholine. It has been reported that muscarinic agonists facilitate ejaculatory mechanisms in male rats. The number of intromissions necessary to trigger ejaculation has repeatedly been found to be reduced after treatment with oxotremorine. Muscarinic agonists have also been found to facilitate lordosis in females injected with a low dose of estradiol, by itself producing a low level of sexual behavior.

Dopamine. Agonists reliably facilitate copulatory behavior in males with a low baseline level of sexual activity, but their effect in males with a normal activity level is highly variable from one study to another, and frequently it is none. A few studies have evaluated the intensity of sexual approach behaviors in male and female rats after treatment with drugs facilitating dopaminergic neurotransmission, such as amphetamine or apomorphine, usually without finding any consistent effect. To the contrary, dopamine antagonists have been found to facilitate lordosis in female rats. Strangely enough, apomorphine administered directly into the ventromedial nucleus of the hypothalamus has been found to facilitate lordosis in estrogen-primed, ovariectomized females.

Noradrenaline. Adrenergic α_2-antagonists such as yohimbine and atipamezole have been reported to facilitate copulation and sexual approach in male rats, but these studies need to be extended before a firm conclusion can be proposed. There are also some data showing facilitated sexual behavior in both male and female stump-tailed macaques after treatment with atipamezole. Treatment with the α_2-agonist dexmedetomidine reduces sexual approach and prolongs ejaculation latencies in male rats. Noradrenergic transmission in the ventromedial nucleus of the hypothalamus is altered by ovarian steroids and is, consequently, important for the display of lordosis in female rats (Etgen et al. 2001). It appears that the role of noradrenaline in the control of both male and female sexual behaviors has been underestimated.

Peptides. Among the peptide neurotransmitters, oxytocin stands out as producing the most reliable stimulatory effects on male and female sexual behavior. In the case of the female, it is known that estrogens enhance the expression of the oxytocin receptor gene as well as the oxytocin gene at brain sites relevant for female sexual behavior. This has led to the hypothesis that oxytocin is involved in the physiological control of female sexual behaviors. There are also data from male rats suggesting that oxytocin may enhance sexual behavior. It must be observed, though, that male mice lacking the oxytocin gene show sexual behavior indistinguishable from that of mice with a functional oxytocin gene. Thus, even though oxytocin agonists stimulate sexual behavior, the peptide is not necessary for the display of this behavior. Several other peptides, such as orexin, leptin, and galanin-like peptide, have been reported to facilitate different aspects of copulation in rats after intracerebroventricular or intracerebral administration, but the number of studies is still too low for any firm conclusion. A melanocortin-4 agonist, bremelanotide, has been reported to stimulate some aspects of copulatory behavior in female rats. The compound has also been tested as a treatment for sexual dysfunctions in clinical trials in women and a Phase III trial started at the end of 2014. It should also be mentioned that opioid peptides have been found to regulate the expression of lordosis in rats. Proopiomelanocortin neurons located in the arcuate nucleus project to the medial preoptic nucleus, in which they release β-endorphin when activated by estrogens. This release leads initially to inhibition of lordosis. When this inhibition is removed, lordosis is displayed. Metabotropic glutamate receptors, $GABA_B$ receptors, and NPY are involved in this pathway (Sinchak et al. 2013).

Serotonin. Drugs reducing serotonergic neurotransmission generally stimulate copulatory behavior in both males and females. In male rats, a particularly consistent effect has been reported after treatment with 5-HT$_{1A}$ agonists. The numbers of preejaculatory intromissions and of the time needed to achieve ejaculation are much reduced after such diverse drugs as buspirone, lisuride, and 8-OH-DPAT. These effects are usually explained as a consequence of actions at serotonergic autoreceptors within the raphe nucleus, leading to reduced serotonin release in the forebrain. Interestingly, in female rats 5-HT$_{1A}$ agonists have actions opposite to those found in males, inhibiting all aspects of sexual behavior. Despite this fact, the 5-HT$_{1A}$ agonist/5-HT$_{2A}$ antagonist flibanserin has been reported to enhance female rat paracopulatory behaviors after subchronic treatment. Flibanserin has also been subjected to clinical trials for the treatment of hypoactive sexual desire disorder in women. These trials provided unclear results, and no further testing of this compound is under way. The possible effects of other receptor subtype-specific 5-HT agonists are also unclear. One possible exception is the 5-HT$_{2A/2C}$ agonist (+/−)-1-(2,5-dimethoxy-4-iodophenyl)-2-aminopropane (DOI). This compound has repeatedly been found to facilitate both lordosis and paracopulatory behaviors in female rats.

Drugs Reducing Sexual Motivation

In contrast to the mostly failed efforts to enhance sexual motivation with drugs, many compounds reliably inhibit it. In fact, almost any drug will inhibit sexual motivation if the dose is large enough. Sedation, gross motor disturbances, stereotypies, or other drug actions will eventually interfere with sexual behaviors. Therefore, the present discussion is limited to drugs that appear to reduce sexual functions without having dramatic effects on other behaviors.

Acetylcholine. Muscarinic antagonists have generally an inhibitory action on male and female copulatory behavior. There are also some data showing that scopolamine reduces sexual approach behaviors in female rats and paracopulatory behaviors in monkeys.

Dopamine. Antagonists reduce copulatory behavior in males. Whether the inhibitory effects in males are specific to sexual behavior or consequences of a generally reduced arousal is a matter of debate. Agonists are known to inhibit lordosis in females when given in high doses. The role of dopamine-dependent activation of competing behaviors, in any case, should not be ignored. In moderate doses, dopamine agonists have no effect in fully receptive females.

GABA. Systemic treatment with GABA agonists is always inhibitory in both males and females, but the doses needed are almost always of such a magnitude that motor functions are compromised, and effects on sexual behavior may be secondary to motor effects. The intracerebral administration of GABAergic compounds has variable effects, again depending on the site of infusion. It is most likely so that GABA agonists will reduce nervous activity at the site of infusion, while the antagonists will have an excitatory effect. Indeed, the effects of locally infused GABA agonists are similar to those of a lesion, and GABA antagonists have effects similar to those of electrical stimulation.

Opioids. Chronic use of opiates is known to reduce sexual behaviors in human males, and acute and chronic treatment reduces it in male rats. Since opiates inhibit gonadotropin release, it is possible that the deleterious effects of chronic opiate use are caused by reduced testosterone availability. In male rats, acute treatment with an opiate agonist that does not penetrate the blood–brain barrier blocks sexual behavior as efficiently as morphine, suggesting that peripheral actions may be of importance. However, the intracerebral administration of large doses of opioid receptor agonists also reduces male sexual behavior. In females, studies employing the intracerebral infusion of compounds acting at opioid receptors have provided evidence for both facilitation and inhibition of lordosis. Facilitation was observed after infusion into the ventromedial nucleus of the hypothalamus, while inhibition was observed when drugs were infused into the medial preoptic area. Both effects may be mediated by the δ-opioid receptor.

Peptides. Oxytocin receptor antagonists or antisense oligonucleotides directed against the oxytocin receptor consistently reduce lordosis in female rats. Receptor antagonists are also known to reduce male rat sexual behaviors. However, oxytocin or oxytocin receptor gene knockout does not seem to reduce female or male sexual behaviors. Neuropeptide Y has also been reported to inhibit sexual behavior in male and female rats.

Serotonin. Compounds facilitating serotonergic neurotransmission inhibit copulatory behavior and sexual approach behaviors in rodents. These effects do not seem to be secondary to drug actions on other behaviors. Efficient drugs range from the serotonin precursor 5-hydroxytryptophan (5-HTP) through the selective serotonin reuptake inhibitors (SSRIs) to several receptor agonists. Lordosis is consistently inhibited by 5-HT_{1A} agonists. It should be observed that the effects of 5-HT_{1A} agonists are completely different in males and females. 5-$HT1_A$ antagonists usually fail to facilitate lordosis, suggesting that there is no tonic inhibitory activity at 5-HT_{1A} receptors. The inhibitory effects of serotonin agonists are not limited to rodents. In the clinical use of the SSRI antidepressants, one of the main adverse effects is reduced sexual function, both in men and women. The specific serotonin receptor involved in this effect is unknown. However, in male rats a 5-HT_{1B} antagonist blocks the inhibitory effect of 5-HTP, suggesting that the 5-HT_{1B} receptor is crucial for the inhibitory actions. Other serotonin receptors may also inhibit sexual behavior. For example, a selective 5-HT_{2C} antagonist has been found to reduce sexual motivation in female rats tested in a procedure similar to the natural mating situation. There is also some evidence suggesting that DOI inhibits copulatory behavior in male rats through actions on the 5-HT_{2A} receptor. It should be noted that DOI facilitates female copulatory behavior. Here we might have another example of the opposite actions of serotonergic drugs in males and females. Data from genetically modified mice with a severe depletion of central serotonergic neurons and from mice lacking tryptophan hydroxylase further complicate the potential role of serotonin in sexual behaviors. These mice prefer to interact with members of their own sex, whereas the wild-type controls show a strong preference for the opposite sex. Heterosexual copulatory behavior is enhanced in males lacking serotonin, whereas it remains unaltered in females (Zhang et al. 2013). These observations would suggest that serotonin is necessary for being attracted to the opposite sex, a most intriguing notion.

Conclusion

It is certainly too simplistic to imagine that a complex behavior like sex is controlled by a single neurotransmitter. This proposal holds both for sexual approach behaviors and for the copulatory reflexes. As well it is probably too simplistic to imagine that a particular aspect of sexual behavior, for example, lordosis, is controlled by a single transmitter. Likewise, although the medial preoptic area is required for the performance of all elements of male sexual behavior and the ventromedial nucleus of the hypothalamus is necessary for all aspects of female sexual behavior, it is too simplistic to believe that drug actions within these structures can explain the effects observed after systemic drug treatments. Manipulations of one transmitter system will certainly modify the activity of a host of other transmitter systems, and alterations within a particular structure will alter the activity of other structures. Finally, receptor agonists may have actions that the endogenous ligand never has because of a mismatch between receptor localization and sites of endogenous transmitter release. All these facts make it extremely risky to try to elucidate the physiological control of the sexual behaviors through pharmacological means. However, if we are interested in finding compounds that can enhance or reduce some particular aspect of the sexual behaviors, then there is no need to consider whether the drug's action is part of the physiological control of sexual behaviors or not. A pure description of drug effects is sufficient and interesting by itself.

Cross-References

▶ Muscarinic Cholinergic Receptor Agonists and Antagonists
▶ Sex Differences in Drug Effects

- Sex Hormones
- Sexual Disorders
- SSRIs and Related Compounds

References

Ågmo A (2007) Functional and dysfunctional sexual behavior. A synthesis of neuroscience and comparative psychology. Academic, San Diego

Argiolas A (1999) Neuropeptides and sexual behavior. Neurosci Biobehav Rev 23:1127–1142

Crenshaw TL, Goldberg JP (1996) Sexual pharmacology: drugs that affect sexual functioning. Norton, New York

Etgen AM, Ansonoff MA, Quesada A (2001) Mechanisms of ovarian steroid regulation of norepinephrine receptor-mediated signal transduction in the hypothalamus. Implications for female reproductive physiology. Horm Behav 40:169–177

Hull EM, Dominguez JM (2006) Getting his act together: roles of glutamate, nitric oxide, and dopamine in the medial preoptic area. Brain Res 1126:66–75

Paredes RG, Ågmo A (2004) Has dopamine a physiological role in the control of sexual behavior? A critical review of the evidence. Prog Neurobiol 73:179–226

Riley AJ, Peet M, Wilson C (eds) (1993) Sexual pharmacology. Clarendon, Oxford

Sinchak K, Dewing P, Ponce L, Gomez L, Christensen A, Berger M, Micevych P (2013) Modulation of the arcuate nucleus-medial preoptic nucleus lordosis regulating circuit: a role for $GABA_B$ receptors. Horm Behav 64:136–143

Zhang S, Liu Y, Rao Y (2013) Serotonin signaling in the brain of adult female mice is required for sexual preference. Proc Natl Acad Sci U S A 110:9968–9973

Sexual Disorders

Vera Astreika[1] and Robert Taylor Segraves[2]
[1]Cleveland, OH, USA
[2]Department of Psychiatry, Case Western Reserve University School of Medicine, Cleveland, OH, USA

Definition

Sexual disorders are common in the general population and have a high prevalence in certain psychiatric populations. The Diagnostic and Statistical Manual of Mental Disorders, fifth edition (DSM-5), divided sexual disorders into three groups: (1) sexual dysfunctions, defined as psychophysiological impairment of sexual desire and/or of sexual response cycle; (2) paraphilias, defined as recurrent, intense sexual urges or behaviors that cause marked distress and involve unusual objects or activities; and (3) gender dysphoria, defined as a strong desire and/or conviction that one is of the opposite sex. This review will focus on the pharmacological treatment of sexual dysfunctions and paraphilias.

Sexual Dysfunctions

Sexual dysfunction can be understood by having knowledge of the stages of the normal sexual responsiveness, which vary with age and physical status. Masters and Johnson described a five-stage model of response: excitement (arousal), plateau (maximum arousal prior to orgasm), orgasm (involving muscular contractions at 0.8 s intervals), resolution (return to baseline), and refractory period (a recovery stage during which another orgasm is not possible; in men this stage increases with age, whereas some women do not have a refractory period). The triphasic model of sexual response was introduced by Helen Kaplan and includes desire, excitement (arousal, a vascular phenomenon, mediated by innervation of the parasympathetic nervous system, specifically the second, third, and fourth sacral segments of the spinal cord), and orgasm (a muscular reaction, mediated by innervation of the sympathetic nervous system, whose reflex center is in the lumbar cord). Changes in sexual response are associated with aging. In females, having decreased levels of estrogen leads to less vaginal lubrication and narrowing of the vagina. Males are slower to achieve an erection and need more direct stimulation of the penis to achieve an erection. The models introduced by Masters and Johnson as well as Kaplan have been criticized as being male centered and linear. Many women report responsive desire, e.g., desire which accompanied rather than precedes arousal. Sexual dysfunction can be primary (lifelong) or secondary (acquired), and it may be generalized (occurring in all circumstances) or situational (limited to certain types of stimulation, situations, or partners). Medications, drugs of abuse, diseases, injuries, and psychological conditions can affect the sexual response in any of its component phases and can lead to different dysfunctional syndromes. The emphasis on biological treatment in this chapter is not intended to minimize the importance of psychological interventions. In DSM-5, sexual dysfunctions are categorized as follows:

1. Male hypoactive sexual desire disorder denotes persistently deficient sexual fantasies and an infrequent desire for sexual activity.
2. Female sexual/interest arousal disorder denotes lack of sexual interest, lack of receptivity to sexual cues, and decreased subjective arousal and pleasure.
3. Male erectile disorder consists of the inability to attain or maintain a satisfactory erection until completion of sexual activity.
4. Female orgasmic disorder is a condition in which there is a persistent delay in, or absence of, orgasm.
5. Premature ejaculation is comprised of persistent ejaculation with minimal stimulation before or after penetration and before the person wishes it. It is the most common male sexual disorder.
6. Delayed ejaculation is a condition in which the man either cannot ejaculate or it takes an inordinate amount of time to ejaculate in spite of intense stimulation.
7. Pelvic-genital pain/penetration disorder consists of intense fear about vaginal penetration or pain experienced during attempted vaginal penetration.

Paraphilias

DSM-5 delineates eight categories of specific paraphilias, all involving recurrent, intense, sexually arousing fantasies, sexual urges, or behaviors occurring over a period of at least 6 months, in addition to "other specified" and "unspecified" paraphilias:

1. Exhibitionistic disorder involves sexual arousal from exposure of one's genitals to an unsuspecting stranger.
2. Fetishistic disorder denotes the use of nonliving objects or a focus on a highly specific nongenital body part (e.g., foot) for sexual arousal.
3. Frotteuristic disorder consists of touching and rubbing against a nonconsenting person.
4. Pedophilia describes sexual activity with prepubescent children.
5. Sexual masochism reflects the act of being humiliated, beaten bound, or otherwise made to suffer.
6. Sexual sadism denotes acts in which the psychological or physical suffering of the victim is exciting to the person.
7. Transvestitic disorder fetishism consists of cross-dressing to become sexually aroused.
8. Voyeuristic disorder involves the act of observing an unsuspecting person who is naked, in the process of disrobing, or engaging in sexual activity.

Role of Pharmacotherapy

Treatment of Sexual Dysfunction in Women

When education, lifestyle, communication, and behavioral changes do not achieve the desired level of success, pharmacological therapy can be utilized to treat sexual dysfunction in women.

Estrogen

Estrogen replacement therapy (ERT) may positively affect sexual function in a number of ways. Estrogens rapidly restore the superficial cell layer of the vaginal epithelium, reestablish elasticity, restore the balance in vaginal pH, improve mood, and increase blood flow to enhance lubrication. Short-term studies of estrogen replacement therapy have confirmed a benefit in some postmenopausal women with sexual dysfunction. However, not all studies have demonstrated positive results, possibly because women most likely to respond are those with symptoms of hypoestrogenism. Any short-term positive effect of oral estrogen may diminish in long-term use because of increasing sex hormone-binding globulin (SHBG) levels, which lead to reduced estrogen and androgen bioavailability and consequent decreased desire and activity. The increase in SHBG appears to be less significant in women who use non-oral delivery system for ERT. There continues to be controversy about possible long-term risks of hormone replacement therapy in postmenopausal women. Vaginal estrogen is highly effective for treating genitourinary atrophy symptoms, in particular, the vaginal dryness and dyspareunia, and involves minimal systemic absorption. Estrogen for vaginal use is available in cream, in tablet, and as an estradiol ring. Water-soluble lubricants are also helpful for continued sexual activity.

Progestin

Progestin agents downregulate the estrogen receptor, a desired result in the endometrium, but potentially undesirable in the brain, heart, bone, and genitalia. Progestins generally have an overall negative effect in the central nervous system (CNS) with respect to depression and mood and have been shown to decrease sexual desire and diminish vaginal blood flow. Available options include micronized progesterone (MP) and 19-nortestosterone derivatives, norethindrone acetate (NA), and norgestimate (NGM). When estrogen is given with progestin, the effect on SHBG depends upon the type of progestin used. 19-nortestosterone-derived progestins decrease the SHBG levels. Newer studies in progress with more modern combinations of progestin with estrogen and androgens will provide better insight into the progestational effects on sexuality.

Androgens

Androgens play an important role in physiologic aspects of the female sexual response. However, the role of androgens in the treatment of female sexual dysfunction, especially the treatment of complaints of low sexual desire, is controversial. Many placebo-controlled, double-blind, multicenter studies have reported improvements in libido, sexual arousal, and the frequency of sexual fantasies with testosterone therapy in a variety of forms. Potential side effects of androgens include a decline in serum high-density lipoprotein (HDL) cholesterol with oral preparations and mild cosmetic side effect such as hirsutism and acne. The long-term safety of androgen therapy in women remains unclear. Testosterone preparations include creams, gels, and tablets that can be taken orally or used sublingually. They are not approved in the USA, but one product approved in Europe for postmenopausal women is a transdermal testosterone patch. Some clinicians utilize testosterone products approved for male hypogonadism "off label" to treat female complaints of low sexual desire. Assay of serum testosterone levels prior to treatment is of limited utility as most studies have not found a correlation between serum testosterone levels and sexual responsiveness in the general population. Studies on the use of oral dehydroepiandrosterone (DHEA) have not shown a benefit on female sexual desire. However, DHEA administered intravaginally has been shown to decrease pain during sexual activity in postmenopausal women. No guidelines for androgen therapy for female sexual dysfunction are available. Women most likely to benefit from androgen therapy are probably those who have undergone bilateral oophorectomy with hysterectomy. Women with hepatic disease, a history of breast cancer, uncontrolled hyperlipidemia, acne, or hirsutism should not be treated. Liver function tests, a complete blood count, and lipid profile should be obtained prior to initiating therapy. These tests should be monitored every 6 months during therapy. Women should be current on cervical and breast screening.

Herbal Therapies

Various herbal therapies have been suggested to have benefit in the treatment of human sexual problems. The beneficial effects of these products have not been shown in clinical populations with diagnosed disorders. ArginMax is a nutritional supplement containing ginseng, ginkgo, L-arginine, and other compounds. It appears to enhance female responsiveness. Zestra, a topical botanical product, has been shown to have efficacy in nonclinical populations.

Future Therapies

Tibolone is a synthetic hormone, currently available in Europe and Australia. Tibolone is used in women to relieve symptoms of menopause and to prevent osteoporosis. It has not been approved by the US FDA. Its metabolites have estrogenic, androgenic, and progestational effects. Oral tibolone is found to increase vaginal lubrication, arousability, and sexual desire.

Sildenafil, a phosphodiesterase-5 inhibitor, in preliminary findings demonstrated positive effects in the areas of sexual arousal and orgasm in appropriately selected women, but several large studies have yielded inconclusive results. It has been reported to reverse anorgasmia associated with SSRIs.

Apomorphine SL, a non-ergoline dopamine agonist, has been reported to improve sexual desire and function in premenopausal women with hypoactive sexual desire. Apomorphine SL is not FDA approved for this indication. Its side effects include hypotension, dizziness, bradycardia, headache, sedation, and nausea.

Other compounds being studied include flibanserin, melanocyte-stimulating hormone analogs, alternative forms of testosterone therapy, and a combination of testosterone with a phosphodiesterase inhibitor.

Treatment of Sexual Dysfunction in Men

The three major forms of male sexual dysfunction are erectile dysfunction, ejaculatory dysfunction, and decreased libido.

Erectile Dysfunction
For first-line therapy, the phosphodiesterase-5 (PDE-5) inhibitors are recommended because of their efficacy, ease of use, and favorable side effect profile. The four PDE-5 inhibitors, sildenafil, vardenafil, tadalafil, and avanafil appear to be equally effective, but tadalafil has a longer duration of action. The rationale for the use of PDE-5 inhibitors is based upon the role of nitric oxide-induced vasodilatation. Nitric oxide release triggers the production of cyclic guanosine monophosphate (cGMP), which leads to decreased intracellular calcium, smooth muscle relaxation, and penile erection. All available PDE-5 inhibitors work by inhibiting the degradation of cGMP. They are highly specific, vary somewhat in selectivity for other phosphodiesterase enzyme types, and differ in duration of action. Common side effects include dyspepsia, flushing, rhinitis, and headache. The use of PDE-5 inhibitors with nitrates is contraindicated by the risk of severe hypotension. PDE-5 inhibitors should be used cautiously with alpha-blockers (because of the risk of hypotension) and in men with aortic stenosis, recent myocardial infarction, unstable angina, heart failure, arrhythmias, degenerative retinal disease, or poorly controlled hypertension. The rare possibility of priapism exists, and patients should be instructed to seek emergent medical attention if erection persists for more than 4 h without sexual stimulation. Cases of sudden loss of vision or hearing have been associated with PDE-5 inhibitors.

Intrapenile injection therapy with alprostadil (prostaglandin E), papaverine, and phentolamine has been used to induce erection. Since the advent of PDE-5 inhibitors, these approaches are used much less frequently than in the past. In the early days of penile self-injection therapy, papaverine proved to be more reliable than phentolamine in producing an erection. Papaverine is a parenteral vasodilator with marginal efficacy in peripheral vascular disease, while phentolamine is an alpha-adrenergic blocker. Alprostadil and papaverine are still used in monotherapy, and all three drugs can be given together. Intrapenile injection is recommended 10–20 min before intercourse and may require additional penile stimulation. Limitations of these agents include penile pain, lack of efficacy, and need for self-injection. Priapism occurs in 6 % of men who use intrapenile alprostadil and about 11 % of those who use intrapenile papaverine.

Intraurethral alprostadil provides a less invasive alternative to intrapenile injection. After insertion of the alprostadil into the urethra 5–10 min before intercourse, the penis should be massaged for up to 1 min to ensure equal distribution in the corpora cavernosae. The drug can be used twice daily. It is not recommended with pregnant partners. Systemic effects are uncommon and complications such as priapism and penile fibrosis are less common than after alprostadil given by penile injection.

Off-label use of some other agents may be moderately effective in treating erectile dysfunction in men who do not respond adequately to PDE-5 inhibitors. Efficacy has been reported for cabergoline, a dopamine D2 receptor agonist used to treat hyperprolactinemia and Parkinson's disease; valvular heart disease has been observed at higher doses of this drug. Melanocortin receptor agonists, which act on the CNS rather than on the vascular system, are being developed as a possible new therapy for erectile dysfunction. PT-141, an intranasal preparation, appears to be effective as monotherapy or in combination with PDE-5 inhibitors. However, significant side effects, including flushing and nausea, may limit its clinical utility. This agent is not commercially available.

Premature Ejaculation
Locally applied anesthetic creams, such as prilocaine/lidocaine mixtures, have been found to increase ejaculatory latency by approximately 7–10 min. Their major side effect is penile hypoanesthesia. The man also must use a condom or wash off the cream before vaginal penetration to minimize vaginal absorption.

Although no medications are FDA approved for this indication, case reports have described the use of monoamine oxidase inhibitors, tricyclic antidepressants, and antipsychotics for this off-label use. Clomipramine, a tricyclic antidepressant with strong serotonergic activity, has been shown in double-blind trials to be effective in treating premature ejaculation on an as-needed basis. Clomipramine usually is taken 4–6 h before coitus in doses of 25–50 mg. Common side effects are dry mouth, nausea, and fatigue. The selective serotonin reuptake inhibitors (SSRIs) paroxetine, sertraline, and fluoxetine also delay ejaculation. Most trials have found that the dose needed to delay ejaculation is similar to the dose necessary to treat depressive disorders. SSRIs require chronic dosing to be effective. Among the SSRIs, paroxetine appears to have the greatest effect on ejaculatory latency. Paroxetine 20 mg daily can be started as initial treatment. Dapoxetine, a short-acting selective serotonin reuptake inhibitor, has been approved in many European countries for the treatment of premature ejaculation. Low doses (0.5–1 mg) of the benzodiazepine lorazepam, taken 30 min before coitus, also can be effective in some men. Lorazepam has the common side effect of sedation.

Decreased Libido
For men with sexual dysfunction and low serum testosterone levels, testosterone replacement therapy should be the initial treatment. If sexual dysfunction is associated with testosterone deficiency, testosterone therapy to restore normal-range testosterone levels can be achieved with transdermal gels, patches, or injections. Current recommendations are to attempt to restore serum testosterone to mid-normal range. A variety of therapeutic approaches are possible. These include 75–100 mg of testosterone enanthate or cypionate administered weekly, one to two 5 mg patches nightly, 5–10 mg of testosterone gel applied daily, or a 30 mg bioadhesive buccal testosterone tablet applied every 12 h. Potential side effects of testosterone replacement therapy include prostatic hypertrophy, increased erythropoiesis, worsened sleep apnea, gynecomastia, and fluid retention possibly worsening hypertension and cardiac failure. Follow-up care should include routine prostate-specific antigen (PSA) and hematocrit and serum testosterone levels.

Pharmacological Treatment of Paraphilias
Pharmacological interventions for paraphilias are symptom focused and directed toward ameliorating or managing comorbid conditions. The number of medications used to treat paraphilias has been steadily increasing. Pharmacological interventions fall into three primary categories: antiandrogens (testosterone-lowering agents), antidepressants, and neuroleptics and other agents.

Antiandrogens or Testosterone-Lowering Agents
Use of these drugs in treating paraphilias usually must be long-term. Relapse is common upon cessation of the medication. Testosterone-lowering agents are the gold standard for treating any paraphilic disorder in which sex-drive reduction is a desirable component.

Medroxyprogesterone acetate (MPA) is the most commonly used hormonal agent for the reduction of sex drive. It reduces levels of testosterone by inducing hepatic testosterone reductase. The goal of this strategy is to reduce baseline testosterone to 50 % of initial value. Common dosages are 50–300 mg orally or 300–400 mg weekly via intramuscular injections. Depot preparations of MPA are also available.

Cyproterone acetate (CPA) blocks androgen receptors, directly decreasing the biological effect of testosterone. CPA can be given orally 100 mg daily or 200 mg every other week via intramuscular injections. CPA is not available in the USA. Side effects of treatment with MPA and CPA include reduced sexual drive and erectile ability, fatigue, depression, liver dysfunction, gynecomastia, weight gain, hyperglycemia due to an exaggerated insulin response to a glucose load, headaches, and increased risk of deep vein thrombosis.

Leuprolide and triptorelin are long-acting luteinizing hormone-releasing hormone (LHRH) agonists. LHRH agonists suppress testosterone by decreasing the number of pituitary LHRH receptors and testicular receptors, thereby desensitizing the testes to luteinizing hormone (LH). As a result of their depleting effects, levels of circulating testosterone and dihydrotestosterone decrease to prepubertal levels. LHRH agonists more completely suppress androgen than MPA or CPA. Leuprolide and triptorelin are given intramuscularly in doses of 3.75 or 7.5 mg monthly. They can cause a surge in gonadotropin secretion in the first 2–4 weeks of treatment. Because of this effect, patients receiving treatment with LHRH agonists should temporarily be prescribed an androgen receptor blocker, such as flutamide. The typical flutamide oral dosage is 250 mg three times a day. It can be discontinued after 2–4 weeks. Side effects of LHRH agonists are related to hypoestrogenic states and consist of erectile failure, hot flashes, and decreased bone mineral density.

Antidepressants
SSRIs have been proposed for alleviating paraphilic symptoms because of their adverse side effects of diminished sexual desire and arousability. A study comparing the effectiveness of fluvoxamine, fluoxetine, and sertraline in paraphilics found all three effective in reducing the severity of fantasies and no significant differences in overall efficacy. It is unclear whether the SSRIs are selectively useful in individuals with a clear obsessive-compulsive disorder component, comorbid anxiety, or depressive disorder underlying the paraphilia or, rather, whether they have a more generalized usefulness for the paraphilias. The SSRIs and other antidepressants are highly variable in the degree to which they cause the sexual side effects that might be beneficial in managing paraphilias.

Neuroleptics and Other Agents
Neuroleptic agents have been reported to diminish paraphilic behavior and fantasies. Lithium and anticonvulsants have been reported to be useful in treating sexual impulsivity. However, the successful use of neuroleptics and mood stabilizers in treating paraphilias may reflect their efficacy for the comorbid mania or other psychotic states that are often associated with paraphilia.

Acknowledgements Dr. Vera Astreika was involved in the first edition of this manuscript.

Cross-References

▶ Antipsychotic Drugs
▶ Benzodiazepines
▶ Fluoxetine
▶ Lorazepam
▶ Monoamine Oxidase Inhibitors
▶ Mood Stabilizers
▶ Paroxetine
▶ PDE3 Inhibitors
▶ Sertraline
▶ Tricyclic Antidepressants

References

Alan Altman MD (2008) Treatment of sexual dysfunction in women. www.UpToDate
Balon R (2008) Sexual dysfunction. The brain-body connection. Adv Psychosom Med 29:33–49

Balon R, Segraves R (2008) Sexual dysfunction and paraphilias. In: Ebert M, Loosen P, Nurcombe B, Leckman J (eds) Current diagnosis and treatment psychiatry. McGraw Hill, New York

Diagnostic and statistical manual of mental disorders, 5th edn (DSM-5)

Fabian M, Saleh MD, Fred S, Berlin MD, Martin Malin H, Kate Thomas J (2007) APRN, Paraphilias and paraphilia-like disorders. In: Glen O, Gabbard MD (eds) Treatment of psychiatric disorders

Kaplan and Saddock's (2007) Synopsis of psychiatry, 10th edn

Levine SB et al (2010) Handbook of clinical sexuality for mental health professionals

Osborne CS, Wise TN (2008) Paraphilias. In: Balon R, Segraves R (eds) Handbook of sexual dysfunction. Taylor & Francis, Boca Raton

Richard F, Spark MD (2008) FACE. Treatment of male sexual dysfunction. www.UpToDate

Sharp Microelectrode Recording

Definition

Sharp microelectrode recording denotes the implementation of ▶ intracellular recording techniques using fine microelectrodes that access the interior of the cell by penetrating through the membrane.

Cross-References

▶ Intracellular Recording

Short-Term and Working Memory in Animals

Yogita Chudasama
Department of Psychology, McGill University, Montreal, QC, Canada

Definition

Short-term memory refers to a storage mechanism that retains information for a limited period of time. In animals, working memory (WM) is conceptualized as a form of short-term memory that encompasses both storage and processing functions. In WM, information is not only held transiently in mind but also actively manipulated; items may be monitored, compared, and/or associated to other items and incoming information as they are recalled. Information in WM is active on the timescale of several seconds and is sensitive to attentional distraction. The main role of WM is to keep track of relevant information, continuously update it, and use it to shape behavior in the present. In animals, the prefrontal cortex plays a central role in WM, and disruption to its intrinsic circuitry or its connectivity with other brain regions leads to several cognitive deficits. In animals, a wide range of psychoactive agents affect WM.

Impact of Psychoactive Drugs

Recently, much attention has been focused on the psychopharmacology of WM. Two main developments have converged to create this interest. First, there have been significant advances in our understanding of the anatomy and organization of the neurotransmitter pathways in the mammalian brain. These pathways include the major monoaminergic (dopamine, serotonin, and norepinephrine) and cholinergic projections, which innervate diverse forebrain regions that have discrete psychological functions. This has refined our understanding of the neurochemical basis of pathophysiological conditions such as Parkinson's disease, Alzheimer's disease, and schizophrenia, which are characterized by memory loss and cognitive disorganization. Second, research has linked poor psychosocial functioning to persistent cognitive deficits, including WM, which fail to improve with current pharmacotherapies. Consequently, there has been an important need to identify pharmacological compounds intended to improve cognitive deficits in psychiatric illness and dementia, and WM impairments have become important targets for treatment.

In animals, WM can be assessed using the spatial delayed response task, which evaluates the animal's capacity to act on the basis of stored information rather than information available in the environment. Typically, a monkey is shown the location of a food morsel. The food morsel is then hidden from view. After a delay of several seconds, the monkey is required to choose one of the two locations. To be rewarded, the animal must remember where the food was located prior to the delay period. In a variation of the task, spatial delayed alternation, the monkey is required to alternate between left and right food wells in successive trials that are separated by delay periods. In this latter version, the monkey is not permitted to observe the experimenter baiting the food well. Therefore, to be correct on any given trial, the monkey must remember which response was made last. In rats, spatial delayed alternation is a commonly used rodent test of cognition which exploits the ease with which rats can keep track of several different locations and therefore use foraging tasks such as remember different locations and hence have used spatial stimuli as in the T-maze and radial arm maze (Dudchenko 2004). In a similar vein, self-ordered WM tasks require monkeys to actively search through a number of boxes, producing a self-ordered sequence of responses, thereby effectively keeping track of previously visited locations (see short-term and working memory in humans). The delayed (non)match to sample is another test of WM, but it is used more commonly in monkeys than rats (see primate models of cognition). Lesions of the dorsolateral prefrontal cortex in monkeys and medial prefrontal cortex in rats profoundly impair WM performance in these tasks (Dalley et al. 2004; Goldman-Rakic 1987). This finding has proven to be extremely useful in the study of the psychopharmacology of WM; the prefrontal cortex is highly sensitive to its neurochemical environment, and small changes in the degree of prefrontal monoamine and acetylcholine neurotransmission contributes significantly to prefrontal WM function.

Dopaminergic Drugs

The work of Brozoski et al. (1979) was the first to demonstrate that catecholamines played a critical role in the modulation of spatial WM function in the prefrontal cortex. Large catecholamine depletions of the prefrontal cortex using 6-hydroxydopamine (6-OHDA) in monkeys impair performance in spatial WM tasks to the same degree as large prefrontal lesions and can be pharmacologically reversed with dopamine (DA) agonists such as levodopa (L-DOPA) and apomorphine. Spatial WM deficits can be mimicked in rats and monkeys by acute, local infusions of DA D_1 receptor antagonists into the prefrontal cortex, and similar cognitive impairments can be observed following systemic administration of haloperidol.

In addition to the detrimental effects of reducing DA transmission in the prefrontal cortex, increased DA turnover in the prefrontal cortex impairs spatial WM as well. This is because WM relies on an optimal level of mesocortical DA, which functions according to an inverted "U-shaped" relationship between DA and cognitive performance (Arnsten 1998). This means that "too much or too little DA" is associated with impaired WM performance. For example, in animals, mild stressors such as conditioned fear or a weak footshock preferentially increase DA turnover in the prefrontal cortex compared with other DA terminal fields. Mild stress exposure to monkeys or rats produces WM deficits that can be blocked with low pretreatment doses of DA D_1 receptor antagonists such as Sch 23390, typical antipsychotics such as haloperidol, or atypical ones such as clozapine. This stress-induced WM deficit can be mimicked in rats by infusing DA D_1 agonists (e.g., SKF 81297) directly into the prefrontal cortex which can be blocked with a pretreatment dose of a DA D_1 receptor antagonist. Relative to the pretreatment effects of DA antagonists, however, the β-adrenergic antagonist propanolol, or the SSRI fluoxetine, is not effective in reversing stress-induced WM deficits, illustrating the importance of dopaminergic mechanisms of WM in the prefrontal cortex.

Research has now established the importance of the DA D_1 receptor family in the regulation of prefrontal WM function. Selective antagonists that work on the D_2 receptor in the prefrontal cortex are mostly ineffective. Overall selective D_1/D_5, but not D_2/D_3 receptor antagonist infusion in the prefrontal cortex impairs WM in rats and monkeys. However, although D_2/D_3 antagonists have little effect on WM when infused in the prefrontal cortex, systemic treatment of these drugs improves WM in monkeys and humans (see short-term and working memory in humans). More research is needed to establish where in the brain the actions of these drugs are taking place.

Disruptions in delayed response tasks are not necessarily indicative of a specific impairment in WM. WM relies not just on holding information online but also overlapping mechanisms of attention which enable the active maintenance of information within a WM system. Recent research indicates that different levels of prefrontal DA activity may be required for different cognitive processes so that DA D_1 receptor stimulation sufficient to improve one cognitive function may serve to impair another. Thus, although treatments with psychostimulants such as amphetamine facilitate mechanisms of arousal (or attention), they disrupt WM and enhance long-term memory. This has led to the idea that arousing conditions in animals such as white noise or highly emotional contexts can impair short-term recall (i.e., WM) but benefit long-term retention and suggest the existence of independent memory systems in animals, including rats.

Noradrenergic Drugs

While antipsychotic medications exert their benefits primarily by modulating the DA neurotransmitter, they also affect norepinephrine (NE) receptors. Recent studies in animals have shown that NE has just as powerful an influence on the prefrontal cortex as DA. In the prefrontal cortex, NE has opposing actions at α_2- and α_1-adrenergic receptors (AR) with respect to WM function. Thus, while NE enhances WM via α_2-AR stimulation, it impairs WM via α_1-AR stimulation. For example, α_2-AR agonists such as clonidine, guanfacine, or medetomidine administered systemically improve WM in rats and monkeys (including aged monkeys), which can be blocked with α_2-AR antagonists such as yohimbine. Yohimbine alone is sufficient to impair WM performance. The improved WM effect of α_2-AR agonists is pronounced in monkeys with large catecholamine depletions and under conditions of high interference or distraction, conditions that require prefrontal function. Performance of tasks that do not depend on the prefrontal cortex is not improved by treatment with α_2-AR agonists (Arnsten 1998).

NE has a lower affinity for α_1-AR than α_2-AR, which means a high level of NE release is necessary to engage α_1-AR in the prefrontal cortex. In addition to increasing DA, exposure to stress preferentially increases NE in the prefrontal cortex. Hence, α_1-AR antagonists infused into the prefrontal cortex have little effect on WM under nonstressed conditions. By contrast, rats and monkeys exposed to uncontrollable stress, which causes NE activation at α_1-AR, markedly impair WM function. Infusing phenylephrine, a highly selective α_1-AR agonist, directly into the prefrontal cortex, mimics the detrimental effect of stress on WM. This impairment can be reversed by coinfusion of the α_1-AR antagonists, urapidil. Together, these findings suggest not only that the therapeutic effects of antipsychotic medication may be due in part to α_1-AR blockade but also that α_2-AR agonists such as clonidine may serve to protect prefrontal function and therefore WM from the detrimental effects of stress.

Cholinergic Drugs

Atypical antipsychotics such as clozapine and risperidone, in addition to increasing DA release in the prefrontal cortex, also increase acetylcholine (ACh) release in the prefrontal cortex. It is thought, therefore, that ACh release in the prefrontal cortex may also contribute to the ability of atypical antipsychotics to improve cognitive symptoms of WM. In rats, for example, local prefrontal infusions of the cholinergic muscarinic receptor antagonist scopolamine produce spatial WM deficits in delayed match to sample tasks. These deficits can be ameliorated with cholinergic treatments such as physostigmine, an acetylcholinesterase inhibitor. However, it is now apparent that cortically projecting cholinergic neurons play a crucial role in attentional processing, and failures in attention may contribute to an overall WM deficit (McGaughy et al. 2000). Thus, delay-independent deficits in delayed match to sample tasks following prefrontal cholinergic manipulations are thought to reflect impairments in attention, rather than mnemonic processing per se. Nevertheless, more research is needed to fully appreciate the role of prefrontal ACh in cognitive function, especially WM. Recent observations suggest that the function of prefrontal ACh is to distribute attentional capacity in tasks that require effortful processing such as holding a stimulus online in WM.

Unlike the muscarinic ACh receptor, nicotinic cholinergic receptors have received much less attention. Recently, however, the effects of nicotine as a cognitive enhancer have made it particularly attractive for clinical use. Both acute and chronic nicotine treatments significantly improve WM performance in rats using the radial arm maze. Chronic treatment with nicotinic agonists has been shown to improve memory performance in other tasks such as passive avoidance and in other species including monkeys, mice, and zebra fish (Levin et al. 2006). In addition to enhancing WM performance, nicotine also improves attention. Chronic nicotine infusion diminishes the impairing effects of antipsychotics such as haloperidol, risperidone, and clozapine on attention, whereas mecamylamine, a nicotinic antagonist used sometimes as an anti-addictive drug, decreases attentional performance and WM. The cognitive impairing effects of mecamylamine have reduced its clinical utility for smoking cessation.

The temporal lobe structures such as the hippocampus and perirhinal cortex also receive a dense cholinergic input from the medial septal (MS) nucleus and the vertical limb of the diagonal band of Broca (VDB). Although the hippocampus and related structures are known to be critical for certain memory functions (see short-term memory and long-term memory), robust WM impairments following manipulations to the cholinergic septo-hippocampal pathways have been difficult to demonstrate.

Serotonergic Drugs

One characteristic feature of atypical antipsychotic drugs is their high affinity for various serotonin 5-HT receptors, especially 5-HT$_{1A}$ and 5-HT$_{2A}$, which are densely localized within the limbic cortex of humans and animals. These 5-HT receptors received much attention because they are involved in the action of hallucinogens such as lysergic acid diethylamide (LSD) and psilocybin, but there is growing evidence from animal experiments indicating a role for serotonin in modulating cognitive functions. While there is considerable evidence connecting the serotonergic system to cognitive functions of behavioral inhibition, its modulatory role in WM has not been well characterized. Much of the earlier work on the role of 5-HT in WM used nonselective global strategies in which the entire 5-HT system was manipulated by increasing or reducing central 5-HT transmission. Thus, in rats, global 5-HT depletion using 5,7-dihydroxytryptamine (5,7-DHT) or p-chlorophenylalanine (PCPA) had little effect on delayed nonmatch to position tests of WM. By contrast, 5-HT ligands with varying degrees of receptor specificity have been shown to disrupt WM performance and in some cases improve it. Importantly, although the prefrontal cortex is substantially innervated by serotonergic fibers from the raphe nuclei in monkeys and rats, 5-HT depletions in the prefrontal cortex do not impair WM performance. In addition, although the 5-HT$_{2A}$ receptor is abundant on the dendrites of prefrontal pyramidal cells implicating a role for it in WM function, there is no direct evidence linking 5-HT$_{2A}$ receptor activation (or deactivation) with changes in spatial WM. Furthermore, the physiological evidence for the apparent deleterious effects of prefrontal 5-HT$_{2A}$ blockade on WM stands in contrast to the proposed benefit of 5-HT$_{2A}$ properties of atypical antipsychotics such as clozapine, olanzapine, and risperidone; these drugs share potent 5-HT$_{2A}$ and relatively weaker DA D$_2$ receptor antagonism. While it is thought that the clinical efficacy of atypical antipsychotics might be due to increased prefrontal DA release, further experiments are needed to evaluate the impact of pharmacological treatments targeted to alter both DA and 5HT signaling on WM performance (Meltzer et al. 2003).

Other Neurotransmitters and Psychoactive Drugs

GABA receptors are the major targets for benzodiazepines and related anxiolytics. Alcohol and benzodiazepine such as diazepam both potentiate actions at $GABA_A$ receptors, and both impair WM performance in rats. In monkeys, WM performance depends on appropriate GABA transmission in the dorsolateral prefrontal cortex.

Dissociative anesthetics such as phencyclidine (PCP) and ketamine work primarily as NMDA receptor antagonists and produce WM deficits in rats performing spatial delayed alternation or delayed match to position tasks. Low doses of AMPA/kainate antagonists produce delay-dependent WM impairments. In addition, potential anxiolytics such as LY-354740, which act as $mGLU_{2/3}$ receptor agonists, autoregulate glutamate release in the prefrontal cortex and reverse WM deficits caused by NMDA antagonist treatment.

Conclusion

The prefrontal cortex is critical for guiding behavior using WM. In WM tasks, animals can code and recall constantly changing locations or stimuli as in delayed response and delayed alternation tasks. Animals can also track in WM their recent choices from a set of stimuli and update relevant information in order to respond on a moment-to-moment basis. As in humans, a disruption to WM in animals prevents the execution of effective organized behavior, a behavioral deficit that is evident in patients with dementia and psychiatric illnesses. While improvements in WM predict favorable behavioral outcomes, the neurochemical basis of WM is not yet fully understood. Animal studies have demonstrated convincingly that catecholamines (DA and NE) are involved in WM function, but there is also evidence that pro-cholinergic drugs can ameliorate WM deficits. Although there is little evidence for a direct role for serotonin in WM, serotonin receptors are used as targets for antipsychotic drug development. That existing antipsychotic drugs possess many pharmacological profiles, it is likely that any therapeutic action on WM will involve interacting receptor mechanisms.

Cross-References

▶ 6-Hydroxydopamine
▶ Acetylcholine
▶ Amphetamine
▶ Anxiolytics
▶ Attention
▶ Atypical Antipsychotic Drugs
▶ Behavioral Inhibition
▶ Benzodiazepines
▶ Hallucinogens
▶ Haloperidol
▶ Long-Term Potentiation and Memory
▶ Nicotine
▶ *N*-Methyl-D-Aspartate Receptor
▶ Norepinephrine
▶ Primate Models of Cognition
▶ Psychostimulants
▶ Radial Arm Maze
▶ Rodent Tests of Cognition
▶ Scopolamine
▶ Short-Term and Working Memory in Humans
▶ Spatial Delayed Alternation
▶ SSRIs and Related Compounds

References

Arnsten AFT (1998) Catecholamine modulation of prefrontal cortical cognitive function. Trends Cogn Neurosci 2:436–447

Brozoski TJ, Brown RM, Rosvold HE, Goldman PS (1979) Cognitive deficit caused by regional depletion of dopamine in prefrontal cortex of rhesus monkey. Science 205:929–932

Dalley JW, Cardinal RN, Robbins TW (2004) Prefrontal executive and cognitive functions in rodents: neural and neurochemical substrates. Neurosci Biobehav Rev 28:771–784

Dudchenko PA (2004) An overview of the tasks used to test working memory in rodents. Neurosci Biobehav Rev 28:699–709

Goldman-Rakic P (1987) Circuitry of primate prefrontal cortex and regulation of behavior by representational memory. In: Plum F, Mountcastle V (eds) Handbook of physiology: the nervous system. The American Psychological Society of Bethesda, Bethesda, pp 317–373

Levin ED, McClernon FJ, Rezvani AH (2006) Nicotinic effects on cognitive function: behavioral characterization, pharmacological specification, and anatomic localization. Psychopharmacology (Berl) 184:523–539

McGaughy J, Everitt BJ, Robbins TW (2000) The role of cortical cholinergic afferent projections in cognition: impact of new selective immunotoxins. Behav Brain Res 115:251–263

Meltzer HY, Li Z, Kaneda Y, Ichikawa J (2003) Serotonin receptors: their key role in drugs to treat schizophrenia. Prog Neuropsychopharmacol Biol Psychiatry 27:1159–1172

Short-Term and Working Memory in Humans

Roshan Cools
Department of Psychiatry, Donders Institute for Brain, Cognition and Behaviour, Centre for Cognitive Neuroimaging, Radboud University Medical Center, Nijmegen, The Netherlands

Definition

The term short-term memory is considered to refer to just one component of working memory and is generally used only to describe a range of tasks that focus explicitly on short-term storage, e.g., digit span tests. Working memory refers to a broader set of processes that facilitate the ability to encode, maintain, store, hold on-line, and retrieve representations of information for further processing or recall (Baddeley 2007). Working memory processes are active temporarily for a short period of time, on the order of seconds. These processes include short-term storage of information, but also executive processes that operate on the stored material. Working memory is mediated by a widely distributed neural system in the human brain, including (not exclusively) the prefrontal cortex (PFC), the striatum, the posterior parietal cortex, and the inferotemporal cortex.

Impact of Psychoactive Drugs

Evidence indicates that the major ascending neuromodulators, in particular dopamine, but also noradrenaline, acetylcholine, and serotonin, can influence working memory performance in humans. However, the effects of drugs that affect these neuromodulators depend on (1) the particular task demands under study, i.e., the stage (encoding, delay, probe) and type (e.g., spatial vs. nonspatial) of working memory, (2) the neural region that is targeted by the drug (e.g., the striatum or the prefrontal cortex), (3) the baseline neurochemical state of the system, and (4) the receptor specificity of the drug. Accordingly, drugs that are known as cognitive enhancers can induce impairments on some tasks and in some individuals. Similarly drugs that are known as cognitive inhibitors can induce improvements on some tasks and in some individuals.

A variety of paradigms are used to study working memory in humans. Examples of tasks in behavioral psychopharmacological studies include the self-ordered spatial working memory task from the Cambridge Neuropsychological Test Automated Battery (CANTAB), which requires self-ordered searching through a set of boxes to locate hidden tokens, tests of digit span, and *the delayed match-to-sample* (DMTS) *test*. Although the *Wisconsin Card Sorting Test* (WCST) and related tests were not designed to test working memory (behavioral flexibility), adequate performance also critically relies on intact working memory, in particular its executive components. Neuroimaging studies which employ positron emission tomography (PET imaging) or functional magnetic resonance imaging ("▶ Magnetic Resonance Imaging" (functional)) have often employed blocked versions of *the n-back task*, which do not directly allow the disentangling of its different components (e.g., maintenance during delay vs. flexible updating during encoding). Other neuroimaging studies have used variants of the DMTS paradigm or adaptations, which do allow such decomposition.

Dopaminergic Drugs

Best studied are the effects of dopaminergic drugs on working memory. Acute administration of single low doses of dopamine receptor agonists, such as bromocriptine or pergolide, affects working memory performance on DMTS tests as well as tests measuring the executive components of working memory (n-back, WCST). Low doses seem more beneficial than high doses, which might induce sedative side effects. Beneficial effects of bromocriptine and other dopamine-enhancing drugs like levodopa have been observed in patients with Parkinson's disease and traumatic brain injury. These effects are most pronounced on tests measuring the executive, flexible updating components of working memory (e.g., n-back, self-ordered searching, WCST, DMTS with stimulus-reordering requirements), but not on tests measuring only the maintenance component of working memory (e.g., DMTS). Nonspecific catecholamine enhancers like amphetamine and methylphenidate, which affect both dopamine and noradrenaline ("▶ Methylphenidate and Related Compounds"), can also improve working memory performance on the more executive working memory tasks (i.e., n-back, self-ordered searching, WCST) in healthy volunteers, as can the COMT inhibitor tolcapone. Improvements after methylphenidate, amphetamine, and tolcapone in young healthy volunteers and after levodopa in Parkinson's disease are accompanied by and correlated with reductions in activity in frontoparietal brain regions, possibly reflecting enhanced processing efficiency and higher signal-to-noise ratio. Conversely agents that block dopamine D2 receptors, like sulpiride, can impair working memory. However, paradoxical improvements have also been observed following sulpiride administration in some individuals, perhaps reflecting predominant action at self-regulating autoreceptors. Effects of acute tyrosine depletion are less consistent, questioning the reliability of tyrosine depletion as a modulator of dopamine function.

Some data have suggested that spatial variants of the DMTS and self-ordered search tests are more sensitive to manipulations of dopamine than are tests of nonspatial working memory. However, a number of prominent caveats have been highlighted (Mehta and Riedel 2006). For example, the spatial variants of the DMTS test often loaded on movement preparation to a greater extent than did the nonspatial variants, thus confounding modality dependency with dependency on movement preparation.

The direction of the effects of dopaminergic drugs depends on baseline working memory capacity, as measured, for example, with the *listening span test*: Dopamine D2 receptor agonists like bromocriptine and cabergoline, as well as D2 receptor antagonists like haloperidol, have opposite effects in subjects with high and low working memory capacity. Similar dependency on baseline working memory capacity has been observed for the nonspecific catecholamine enhancers methylphenidate and amphetamine. For example, low-span subjects with low working memory capacity benefit from bromocriptine in terms of set-shifting performance on the WCST, while high-span subjects with high working memory capacity are impaired by bromocriptine. The direction of drug effects, however, cannot be predicted from working memory capacity alone; rather the type of working memory test must also be taken into account. Thus, while benefiting in terms of flexible set-shifting on the WCST, low-span subjects are actually impaired by bromocriptine on the DMTS test, which loads more highly on the robust maintenance rather than on the flexible updating of current representations.

The dependency of these dopaminergic drug effects on baseline working memory capacity has been hypothesized to reflect dependency on baseline levels of dopamine. Evidence for this hypothesis comes from neurochemical PET studies showing that working memory capacity, as measured with the listening span, correlates positively with baseline dopamine synthesis capacity in the striatum. Higher working memory capacity is associated with higher dopamine synthesis capacity (Cools and D'Esposito 2009). The implication of these data is that subjects with low baseline levels of dopamine benefit from dopamine-enhancing drugs, at least in terms of the flexible updating component of working memory, while subjects with already optimized levels of dopamine are impaired by the same dopamine-enhancing drugs. This is consistent with the existence of an "inverted U"-shaped relationship between dopamine and cognitive performance ("▶ Short-Term and Working Memory in Animals").

Individual variability in baseline dopamine levels might reflect individual variation in genetic predisposition. Opposite effects of amphetamine, which increases both dopamine and noradrenaline transmission, as well as tolcapone, are seen on PFC activity during n-back performance as a function of genetic variation in COMT (catechol-O-methyltransferase). Amphetamine and tolcapone increase the efficiency of PFC processing during n-back performance in subjects homozygous for the Val allele of the COMT gene polymorphism, who have hypothetically reduced PFC dopamine function. By contrast, amphetamine and tolcapone reduce efficiency of PFC processing in subjects homozygous for the met allele, who have hypothetically enhanced PFC dopamine function (e.g., Mattay et al. 2003).

Theoretical models and preliminary results suggest that the effects of dopaminergic drugs on working memory will depend also on the neural locus of drug action. For example, bromocriptine, which stimulates dopamine receptors, modulated the flexible updating of information during encoding of a DMTS test, and these effects are accompanied by modulation of neural activity in the striatum. Conversely, bromocriptine modulated activity in the PFC during (the resistance to distraction in) the delay period of a DMTS test (Cools and D'Esposito 2009). Further study is necessary to identify the neural loci of dopaminergic drug action during component processes of working memory.

Research with experimental animals and computational modeling work has suggested a particularly important role for D1 receptor stimulation in the on-line maintenance of information in working memory ("▶ Short-Term and Working Memory in Animals"). However, agonists used in human research so far (i.e., bromocriptine, pergolide) have affinity for both D1 and D2 receptors. The lack of selective D1 receptor agents available for human research has limited the ability to investigate the supposed D1 receptor selectivity of dopaminergic drug effects on working memory. Future study should address this issue, for example, by making use of receptor-selective genetic polymorphisms or neurochemical PET imaging with receptor-selective ligands.

Similar lack of methods available for human research prevents the study, in human volunteers, of another current hypothesis, which states that effects of dopamine depend on the mode of transmission affected. Specifically, phasic and tonic modes of dopamine transmission might influence different (updating vs. maintenance) component processes of working memory.

Noradrenergic Drugs

Although much work with experimental animals has focused on a role of noradrenaline (in particular the alpha-adrenergic system) in (the maintenance components of) working memory ("▶ Short-Term and Working Memory in Animals"), work with humans has concentrated more often on long-term and emotional memory. Nevertheless, empirical studies using selective noradrenergic manipulations in human volunteers have revealed a role for, in particular, the beta-adrenergic system in modulating (executive components of) working memory, perhaps via their well-accepted modulation of (selective and sustained) attentional processes ("▶ Attention"). Specifically working memory performance is consistently impaired by the beta-blocker propranolol, which penetrates the blood–brain barrier easily, but generally not by oxprenolol and atenolol, which have more limited central activity. The α_2 adrenoceptor agonist clonidine also generally impairs working memory, possibly via a presynaptic mechanism of action. However, dose-dependent effects of clonidine as well as effects of other α_2 adrenoceptor agonists (like guanfacine), α_2 adrenoceptor antagonists (like yohimbine), and acute noradrenaline potentiation with reboxetine are less consistent (Barch 2003; Chamberlain et al. 2006; Ellis and Nathan 2001).

Cholinergic Drugs

Cholinergic drugs are known to affect working memory performance. Specifically, working memory performance benefits from agents that enhance cholinergic function, such as physostigmine, while being impaired by agents that block cholinergic function, such as scopolamine (Barch 2003; Ellis and Nathan 2001). The effects might be receptor-specific, with muscarinic and nicotinic receptors mediating distinct (spatial and nonspatial) types of working memory. Cholinergic effects are particularly pronounced in the presence of distraction.

Consistent with the well-known link between acetylcholine and stimulus detection (or vigilance), cholinergic effects in working memory are hypothesized to reflect changes in the selectivity of attentional processing during encoding, thus simplifying demands during maintenance and retrieval. In keeping with this hypothesis are findings that cholinergic antagonists, like scopolamine which is sometimes referred to as a "memory inhibitor," impair the encoding but not the maintenance or the retrieval of new information. Furthermore, the anticholinesterase inhibitor physostigmine, sometimes referred to as a "memory enhancer," improved accuracy on a task that maximized demands for attention while leaving unaffected accuracy on a matched working memory task. Neuroimaging results from studies with physostigmine have substantiated hypotheses from work with experimental animals that acetylcholine promotes stimulus-driven shifts in attention by enhancing encoding- and attention-related activity in the posterior sensory cortex (Bentley et al. 2004).

Serotonergic Drugs

Although manipulations of the indolamine serotonin (5-HT, 5-hydroxytryptamine) have been shown to alter performance on tasks involving long-term (verbal) memory consolidation and affective components of impulse control processing, it remains unclear whether serotonin is involved in working memory. Effects on memory of the most common procedure to manipulate central serotonin in humans, the dietary acute tryptophan depletion (ATD) procedure, are generally restricted to delayed recall, not extending to immediate recall. ATD also does not affect spatial working memory, consistent with a lack of effect of PFC 5-HT loss on spatial working memory performance in monkeys ("▶ Short-Term and Working Memory in Animals"). Furthermore, ATD administered an hour after word learning left delayed recall unaffected. These data suggested that ATD affects memory consolidation rather than the encoding or retrieval of information. Few studies have examined effects of selective serotonin receptor agents on working memory. One study revealed that fenfluramine impaired performance on a spatial working memory task in a delay-dependent manner. Effects of ketanserin, a $5HT_{2A}$ receptor blocker, on spatial working memory tasks are inconsistent. It is not unlikely that any effects of $5HT_2$ receptor stimulation or blockade reflect indirect effects on dopamine transmission. Further work is required to investigate interactions between 5-HT and dopamine as well as acetylcholine and the possible receptor selectivity of any effects (there are at least 17 serotonin receptors which may serve different functions).

Other Psychoactive Drugs

The benzodiazepine sleeping pills and tranquillizers (e.g., triazolam, diazepam [trade name Valium], lorazepam) are some of the best-known drugs to cause memory impairment and facilitate the transmission of the major inhibitory neurotransmitter GABA (γ-aminobutyric acid). However, like the so-called memory-enhancing (anti-dementia) drugs, which enhance cholinergic transmission, these benzodiazepines are thought to act primarily on processes other than those affecting qualitative aspects of working memory. These processes include episodic memory (long-term memory humans) and/or processing speed (Curran and Weingartner 2002).

The NMDA receptor antagonist ketamine, which affects the transmission of the major excitatory neurotransmitter glutamate, produces impairments on manipulation in working memory as well as other so-called frontal tests, known to rely on the executive components of working memory (Curran and Weingartner 2002).

The wake-promoting agent modafinil, which affects the catecholamine, glutamatergic, GABAergic, orexin, and histamine systems in the brain, while having little abuse potential, has been shown to improve working memory in a number of studies in healthy volunteers (with or without sleep deprivation), with some evidence for dependency on baseline performance. Improved working memory following modafinil has also been observed in patients with ADHD, schizophrenia, and depression and might reflect primary modulation of the catecholamines (Minzenberg and Carter 2008).

Notes of Caution

Psychoactive drugs modulate the effects of neurotransmitter systems that are diffuse and widespread, innervating large undifferentiated parts of cortex (and subcortical structures). Receptors of the different neurotransmitters are often co-localized on the same cells. Therefore, despite clear evidence for distinct functions, the different systems must interact to determine optimal working memory. Further, drugs with preferential effects on one system (e.g., serotonin, glutamate) might exert secondary effects on other systems (e.g., dopamine). In addition, most drugs are not selective for one particular system. For example, propranolol also exhibits affinity for serotonin receptors.

Conclusion

Working memory is disturbed in a range of neurological and psychiatric disorders, including schizophrenia, Parkinson's disease, attention deficit hyperactivity disorder, and traumatic brain injury. The treatment of these cognitive deficits will benefit from a better understanding of the psychopharmacology of human working memory. Consistent with the observation that all these disorders implicate abnormal dopamine neurotransmission is the conclusion that human working memory is most sensitive to drugs that affect dopamine neurotransmission. Particularly pronounced are effects of dopaminergic drugs on tests measuring executive and flexible updating components of working memory. This might reflect the fact that compounds used for human research all act at D2 receptors.

The direction and extent of these drug effects vary greatly between different tasks and between different individuals. Factors that contribute to this variability include task demands and baseline dopamine levels in underlying brain regions. Further work is required to establish the regional selectivity of working memory processes and dopaminergic drug effects. Drugs affecting acetylcholine also modulate performance on working memory tests, probably via altering attentional processes. Working memory is less consistently sensitive to drugs affecting serotonin neurotransmission. Further research is necessary to elucidate the role of noradrenaline, glutamate, and GABA in human working memory.

Cross-References

▶ Attention
▶ Behavioral Flexibility: Attentional Shifting, Rule Switching, and Response Reversal
▶ Cognitive Enhancers: Role of the Glutamate System
▶ Methylphenidate and Related Compounds
▶ Modafinil
▶ Short-Term and Working Memory in Animals
▶ Tryptophan Depletion

References

Baddeley AD (2007) Working memory, thought, and action. Oxford University Press, Oxford

Barch D (2003) Pharmacological manipulation of human working memory. Psychopharmacology (Berl) 174:126–135

Bentley P, Husain M, Dolan RJ (2004) Effects of cholinergic enhancement on visual stimulation, spatial attention, and spatial working memory. Neuron 41:969–982

Chamberlain SR, Muller U, Robbins TW, Sahakian BJ (2006) Neuropharmacological modulation of cognition. Curr Opin Neurol 19:607–612

Cools R, D'Esposito M (2009) Dopaminergic modulation of flexible cognitive control in humans. In:

Björklund A, Dunnett S, Iversen L, Iversen S (eds) Dopamine handbook. Oxford University Press, Oxford

Curran H, Weingartner H (2002) Psychopharmacology of human memory. In: Baddeley AD, Kopelman MD, Wilson BA (eds) The handbook of memory disorders. Wiley, Chichester

Ellis KA, Nathan PJ (2001) The pharmacology of human working memory. Int J Neuropsychopharmacol 4:299–313

Mattay V, Goldberg T, Fera F, Hariri A, Tessitore A, Egan M, Kolachana B, Callicot J, Weinberger D (2003) Catechol O-methyltransferase Val^{158}-met genotype and individual variation in the brain response to amphetamine. Proc Natl Acad Sci U S A 100:6186–6191

Mehta MA, Riedel WJ (2006) Dopaminergic enhancement of cognitive function. Curr Pharm Des 12:2487–2500

Minzenberg MJ, Carter CS (2008) Modafinil: a review of neurochemical actions and effects on cognition. Neuropsychopharmacology 33:1477–1502

Sibutramine

Definition

Sibutramine is an amphetamine derivative that has been used as an appetite suppressant. It has been associated with an increased risk of nonfatal heart attacks and strokes and has been withdrawn from the market in many countries. It was indicated as an antiobesity agent in adjunctive therapy within a weight management program for those that did not respond to diet alone. The beneficial weight loss is thought to be outweighed by the risks. It acts as a serotonin and norepinephrine reuptake blocker.

Cross-References

▶ Appetite Suppressants

σ-Receptor

Synonyms

Sigma-receptor

Definition

A receptor on which several opioid drugs act in addition to their actions at opioid receptors. It was originally thought to be a subtype of opioid receptor but was subsequently shown not to be a member of the opioid receptor family. Although one type of σ-receptor (type 1) has been isolated, it is not a G-protein-coupled receptor (▶ GPCR) and its biological function is still being resolved.

Signal Detection Theory

Definition

A quantitative method of analyzing behavioral data that discriminates between independent indices of overall sensitivity and discriminability of stimuli (e.g., of discriminative stimuli or memory traces) and response bias, which might reflect motivational factors or other confounding influences on responses (e.g., spatial bias). Useful, for example, in separating pain measures into their sensory and emotional concomitants.

Signaling Cascades

Synonyms

Signal transduction pathways

Definition

The molecular events that link extracellular stimuli to intracellular responses by means of a multistep sequence of events, generally integrating and amplifying.

Sign-Tracking

Shelly B. Flagel
Department of Psychiatry, Molecular and Behavioral Neuroscience Institute, University of Michigan, Ann Arbor, MI, USA

Synonyms

Autoshaping; Pavlovian conditional approach

Definition

Sign-tracking refers to behavior that is directed towards a stimulus as a result of a learned association between that stimulus and a reward. The sign-tracking response develops even though reward delivery is not contingent upon a response. For example, following classic Pavlovian conditioning procedures, wherein a *conditional stimulus (CS)* is repeatedly paired with presentation of food reward (*unconditional stimulus, US*), it is often observed that the *conditional response (CR)* that develops consists of orientation to, and eventually approach and engagement of, the CS (i.e., the cue or "sign" that signals impending reward delivery). The term sign-tracking was initially coined by Hearst and Jenkins (1974) to describe behaviors observed in pigeons during "*autoshaping*" experiments, but has since been applied to a number of species, including humans.

Current Concepts and State of Knowledge

When a discrete cue or *conditional stimulus* (CS; like a localized light source or a lever) comes to predict the delivery of a food reward (*unconditional stimulus*, US), some animals respond to presentation of the cue (CS) with approach and engagement of it. This behavior was first systematically observed in pigeons, such that repeated response-independent presentations of a key light followed by food delivery resulted in pecking behavior directed at the key light (see Fig. 1a). This phenomenon was initially referred to as "*autoshaping*," but shortly thereafter it was recognized that this was a misnomer and that this term is best used to describe the procedure rather than the behavioral response (Hearst and Jenkins 1974). Hence, Hearst and Jenkins (1974) used the term "sign-tracking" to more accurately describe behavior that is directed towards a stimulus as a result of a learned association between that stimulus and a reward. (Note that Hearst and Jenkins (1974) also used "sign-tracking" to refer to behavior directed away from a cue signaling reward unavailability; but this behavior is not considered to be representative of "sign-tracking" in the current state of the field.)

A sign-tracking response emerges from classical Pavlovian conditioning procedures wherein the presentation of a cue (CS) is associated with delivery of a reward (US), and with repeated pairings comes to elicit a conditional response (CR). In the case of sign-tracking, the conditional response is directed towards the cue (CS). In Pavlov's original studies, a metronome served as the CS, food reward as the US, and salivation was measured as the response (CR) to the CS. Indeed, Pavlov's original studies suggested that the CS evoked a simple reflexive response similar to the *unconditional response* (UR) produced by the reward itself (Pavlov 1928). However, the dogs in Pavlov's studies were typically restrained, and when they were freed from their restraints, it became readily apparent that what was learned was not a simple reflexive response, but a complex emotional and motivational response. A visiting researcher in Pavlov's laboratory, Howard Liddell, described the behavior of one of the dogs upon being freed from its harness, "The dog at once ran to the machine (CS), wagged its tail at it, tried to jump up on it, barked, and so on..." (Liddell, unpublished, see Hearst and Jenkins 1974). Many studies have since shown that, in addition to simple conditional responses, Pavlovian CSs can evoke more complex responses that can be

Sign-Tracking, Fig. 1 Representative drawings of (**a**) a pigeon and (**b**) a rat exhibiting sign-tracking behavior. (**a**) In the classic pigeon experiments a key light was used as the conditional stimulus (*CS*) and food delivery occurred below the key light. Repeated pairings of the key light with food delivery resulted in pecking behavior directed at the key light (i.e., sign-tracking). (**b**) When rats are presented with a retractable lever as the conditional stimulus (*CS*) and its presentation is immediately followed by food delivery into an adjacent receptacle, some rats develop approach and consummatory behavior directed at the lever (*CS*). (Drawings by Katie Long)

manifest in a number of ways, one of which is sign-tracking.

Factors that Influence the Sign-Tracking Response

The topography of the conditional response that emerges from Pavlovian conditioning procedures depends on the species and the nature of the CS and the US. There is often striking similarity between the behavioral patterns involved in consuming the reward and those directed towards the CS. When pigeons are exposed to a key light (CS) that has been paired with presentation of water (US), they exhibit a drinking-specific motor pattern, complete with gullet movement, directed at the key light. In contrast, when a key light is associated with food reward, pigeons will distinctly peck at the key light, as if they were consuming grain (Fig. 1a; see also Hearst and Jenkins 1974). Similarly, for rats, if presentation of a lever (CS) is immediately followed by the response-independent delivery of a food pellet, the sign-tracking response often consists of grasping and gnawing the lever (CS) as if it were the food itself (Fig. 1b). Remarkably, when a CS is paired with the opportunity to copulate with a female (US), male Japanese quail (*Coturnix japonica*) will, under some conditions, exhibit a sign-tracking response that consists of approach and copulation with the inanimate object (CS) (for review, see Koksal et al. 2004).

The fact that the sign-tracking response often resembles behavior elicited by the reward itself (e.g., consummatory behavior) led to the notion that the CS simply acts as a surrogate for the US (i.e., "stimulus-substitution"). However, there are many examples suggesting that sign-tracking behavior is not due to simple "stimulus-substitution" (see Hearst and Jenkins 1974). For instance, the form of the CR to a CS that predicts food is different if the CS is a lever, a live rat, or a block of wood. Further, Wasserman (1973; for review, see Hearst and Jenkins 1974) conducted a series of experiments with young chicks using overhead heat in a cold chamber as the reward (US), and presentations of a key light as the CS, and showed that early presentations of the CS elicited a key pecking response. Thus, the topography of the sign-tracking CR in the young chicks was quite different from the vocalizations and wing extension (i.e., the UR) that occurred in response to the heat lamp itself. There are additional examples wherein the sign-tracking response is not obviously reflective of the reward (US) being signaled, but in these cases, the US itself is not localizable, nor is it manipulable. Peterson (1975; for review, see Hearst and Jenkins 1974) compared sign-tracking behavior of rats that received brain stimulation (at the lateral hypothalamus) to those that received food reward as the US. For both conditions, the CS was the insertion of an illuminated retractable lever. For the food-US subjects, the sign-tracking response consisted of licking and gnawing of the lever (CS), whereas the brain-stimulation subjects engaged in more "exploratory" behavior directed towards the CS, with fewer contacts. Similarly, when intravenous drug (e.g., cocaine, opiates) delivery is the US, rats will rarely contact the lever (CS), but some exhibit a sign-tracking response that consists of orientation and approach behavior (see Robinson et al. 2013).

Maladaptive Behavior
When a reward signal (CS) and the reward itself (US) are spatially separated (as in the examples above), sign-tracking will typically steer the animal away from the site of the reward. Hence, sign-tracking is often described as "unintelligent" or "maladaptive" behavior. A classic example of this, known as "the long box experiment," was originally described by Hearst and Jenkins (1974), who paired illumination of a key light at one end of a long box with subsequent delivery of food reward at the other end. No response was required to receive the food, but pigeons nonetheless began to approach and repeatedly peck the key light, even though doing so prevented them from retrieving the food at the other end of the box, which was available for only a limited amount of time. Thus, these hungry pigeons compulsively pecked at the key light, despite the loss of reward. This study is in agreement with a number of others that have shown that sign-tracking is not simply due to "accidental reinforcement" of the response. For example, when an omission schedule is employed wherein contact with the CS will now prevent the delivery of reward, responses directed towards the CS persist, although often at a lower rate or with different topography than that exhibited during the initial conditioning phase (Hearst and Jenkins 1974; see also (Flagel et al. 2009) for discussion). These studies were taken as evidence that sign-tracking behavior is not controlled by response reinforcement.

The fact that animals will continue to sign-track even when it results in loss of reward is an enigma from an evolutionary standpoint. However, in many situations, it is an adaptive response. In the natural environment of many organisms, including humans, the "sign" and the "reward" are often part of the same unified object (Hearst and Jenkins 1974). In pigeons, for example, the sign indicating that grain is available some distance away will normally be signaled by the sight of the grain itself. The same goes for humans and food reward, where it is often the sight and smell of the food itself that results in consummatory responses. Thus, in these situations, it is advantageous to "sign-track," as it results in procurement of valuable resources. Nevertheless, there are also plenty of examples outside of the laboratory whereby reward-related cues can engender maladaptive and presumably compulsive sign-tracking behavior, in part due to the abundance of cues in the modern environment. For example, Tomie (1996) has argued that when cues are embedded in the device that delivers reward, such as specialized glassware used to consume alcohol, these cues can promote especially compulsive patterns of behavior. Indeed, sign-tracking has been associated with a number of addiction-related behaviors (e.g., impulsivity); and it has been suggested that sign-tracking behavior may confer vulnerability to addiction (Tomie 1996; Flagel et al. 2009; Robinson et al. 2013).

Individual Differences in the Propensity to Attribute Incentive Salience to Reward Signals: Sign-Tracking Versus Goal-Tracking
In attempt to explain the occurrence of sign-tracking behavior, Bindra (1974) was among the first to suggest that it is the transfer of incentive motivational properties from the reward (US) to

the reward signal (CS) that elicits this CR. However, ample evidence supporting this notion did not emerge until recent years, when researchers began to focus their attention on individual differences in the topography of CRs. As stated above, there are a number of factors that can influence the nature of the CR; but even when trained under identical conditions, there can be large individual differences in the resulting behavior. The first detailed observation of these individual differences came from Zener (1937), who systematically described distinct CRs that emerged when unrestrained dogs were exposed to procedures similar to Pavlov's. Zener (1937) reported that some dogs responded to the CS with "an initial glance at the bell" followed by a "constant fixation...to the food pan," whereas others exhibited "...approach toward the conditioned stimulus...followed by a backing up later to a position to eat"; and others vacillated, looking back and forth between the bell and the food pan. Individual variation in the topography of the CR did not receive much further attention until 1977 when Boakes described similar individual differences in rats. Using a standard autoshaping procedure consisting of presentation of an illuminated lever immediately followed by the response-independent delivery of a food pellet, he reported, similar to Zener, that some rats exhibited sign-tracking behavior with approach and contact of the lever (CS), whereas others, upon lever (CS) presentation, immediately went to the place where food would be delivered. For the sake of symmetry, Boakes referred to CS-evoked approach to the location of food delivery as "*goal-tracking*" (Boakes 1977). This natural variation in the tendency to exhibit a sign- versus a goal-tracking response has been exploited in recent years in order to parse the psychological and neurobiological processes underlying these CRs.

Using procedures similar to Boakes (1977), Robinson and Flagel have shown that a reward signal (CS) is equally predictive for both rats that sign-track and those that goal-track, because it supports learning of a Pavlovian CR in both (Flagel et al. 2009; Robinson et al. 2013). However, only for sign-trackers does the reward signal (CS) acquire incentive motivational properties (i.e., *incentive salience*), because only in sign-trackers does the CS itself become attractive and elicit approach towards it. Moreover, the CS attains conditioned reinforcing properties to a greater extent in sign-trackers than goal-trackers. That is, sign-trackers will work for presentation of the reward signal (CS) alone, in the absence of reward. Thus, this model provides the means to dissociate the predictive from the incentive properties of reward signals and to uncover the factors that contribute to the propensity to sign-track (Saunders and Robinson 2013).

Neurobiology of Sign-Tracking Behavior

There have been a number of studies identifying the brain mechanisms critical for *Pavlovian conditional approach* behavior; and most have implicated the mesocorticolimbic circuitry, demonstrating a role for both dopaminergic and glutamatergic neurotransmitter systems in mediating this response (e.g., for review, see Cardinal et al. 2002). However, the methods used in these studies limit the ability to distinguish between sign- versus goal-tracking CRs, making it unclear whether the identified neural systems are involved in one or both Pavlovian CRs. Importantly, there is indeed large individual variation in the extent to which reward signals engage neural systems. In fact, only when a reward signal is attributed with incentive motivational value (i.e., only for sign-trackers) is the cortico-striatal-thalamic circuitry, or the so-called motive circuit, activated. This animal model has also been used to demonstrate that the sign-tracking response, but not the goal-tracking response, is dopamine dependent (for review, see Saunders and Robinson 2013). For example, dopamine antagonists, including atypical and typical antipsychotics (olanzapine and haloperidol, respectively), block learning of a sign-tracking CR, but have no effect on learning the CS-US association that underlies a goal-tracking CR (for review, see Saunders and Robinson 2013). In contrast, potentiation of dopamine release, via administration of pscyhostimulant drugs, increases sign-tracking, but not goal-tracking behavior (for review, see Saunders and Robinson 2013). Taken together, these data support the notion that dopamine is necessary for the attribution of incentive, but not predictive value to reward signals. Moreover, the neural circuitry underlying the sign-tracking response has been implicated in a number of psychopathologies, including schizophrenia and addiction.

Clinical Implications of Sign-Tracking Behavior

Although many people experiment with potentially addictive drugs in their lifetime, only a relatively small subset of the population becomes addicted. It has been proposed that individual variation in the propensity to attribute incentive salience to reward signals confers addiction liability. Indeed, preclinical studies have shown that rats that sign-track to cues associated with food reward also sign-track to drug-associated cues (Robinson et al. 2013). Further, rats that attribute enhanced incentive motivational value to food cues are more susceptible to control by drug-associated cues, as evidenced by the ability of drug cues to maintain drug self-administration and reinstate drug-seeking behavior to a greater extent in sign-trackers than goal-trackers (Robinson et al. 2013; Saunders and Robinson 2013). Likewise, there is considerable individual variation in the ability of drug cues to bias attention, elicit craving, and instigate relapse in humans. Emerging evidence suggests that some humans may be more "cue reactive" than others. For example, individuals with the highest craving in response to food cues, when hungry, were the same individuals that showed the highest craving in response to smoking cues during abstinence (for review, see Saunders and Robinson 2013). These findings are reminiscent of the sign-tracking rats that attribute excessive incentive motivational value to both food- and drug-paired cues and lend credence to the notion that variation in this trait may underlie susceptibility to addiction.

Summary

Sign-tracking is one type of conditional response that emerges following Pavlovian conditioning, wherein a discrete cue comes to predict the noncontingent delivery of a reward. Sign-tracking is characterized by approach and consummatory behavior directed towards the reward-associated cue, but the topography of the response is dependent on a number of factors, including the nature of the conditional and unconditional stimuli and the species being examined. In recent years, an animal model exploiting individual variation in the propensity to attribute incentive motivational value to reward cues has provided the means to parse the psychological and neurobiological mechanisms mediating the sign-tracking response. It is now well accepted that sign-tracking reflects the attribution of incentive salience to reward cues and that this response is dopamine dependent; but much remains to be determined regarding the neurobiology underlying this phenomenon. The propensity to sign-track is thought to confer vulnerability to addiction, as those individuals who attribute maladaptive levels of motivational value to reward cues will have the most difficulty abstaining from drug use and be the most likely to relapse. As preclinical studies have been focusing on individual variation in order to better understand the mechanisms underlying aberrant reward-cue processing, in moving forward, it is critical that similar methodology is applied to clinical studies in order to improve treatment strategies for humans suffering from psychopathology such as addiction and other disorders of impulse control.

Glossary

Autoshaping refers to an experimental procedure consisting of Pavlovian conditioning in which presentation of a cue (e.g., light) precedes the presentation of reward (e.g., food) and results in approach and consummatory behavior directed at the cue, even though no response is required for reward delivery (i.e., the animal "shapes" itself).

Conditional response (CR) is the learned response to a previously neutral stimulus that has attained predictive (and sometimes incentive motivational) properties as a result of Pavlovian conditioning.

Conditional stimulus (CS) a neutral stimulus (e.g., light) that comes to signal the occurrence of a second stimulus (e.g., food reward) and attains predictive (and sometimes incentive motivational) properties as a result of Pavlovian conditioning.

Goal-tracking describes a conditional response that is directed towards the location of reward delivery (i.e., the goal) rather than the signal for reward. Thus, if a lever (CS) precedes delivery of food reward, goal-tracking rats will approach the location of food delivery upon lever (CS) presentation.

Incentive salience describes the motivational value that can be attributed to reward-associated cues. A cue attributed with incentive salience becomes a "motivational magnet" and attains the ability to control behavior, sometimes to an inordinate degree.

Pavlovian conditional approach describes conditional responses that develop as a result of Pavlovian conditioning. Pavlovian conditional approach responses are characterized by cue-elicited approach behavior directed towards either the CS itself or the location of the US.

Stimulus-substitution is a term used to describe the Pavlovian process by which a conditional stimulus (cue) elicits a conditional response similar to the response elicited by the unconditional stimulus (reward). In this case, the conditional stimulus (cue) is thought to act as a substitute for the reward itself.

Unconditional stimulus (US) is usually a biologically significant stimulus (e.g., food, pain) that can elicit a specific response without any training. During Pavlovian conditioning procedures, the US follows presentation of the conditional stimulus (CS or cue).

Unconditional response (UR) is a response elicited by the unconditional stimulus. For example, food (US) elicits salivation (UR).

Cross-References

▶ Addictive Disorder: Animal Models
▶ Attentional Bias to Drug Cues
▶ Classical (Pavlovian) Conditioning
▶ Impulse Control Disorders
▶ Impulsivity

References

Bindra D (1974) A motivational view of learning, performance, and behavior modification. Psychol Rev 81:199–213

Boakes R (1977) Performance on learning to associate a stimulus with positive reinforcement. In: Davis H, Hurwitz H (eds) Operant-pavlovian interactions. Erlbaum, Hillsdale

Cardinal RN, Parkinson JA, Hall J, Everitt BJ (2002) Emotion and motivation: the role of the amygdala, ventral striatum, and prefrontal cortex. Neurosci Biobehav Rev 26:321–352

Flagel SB, Akil H, Robinson TE (2009) Individual differences in the attribution of incentive salience to reward-related cues: implications for addiction. Neuropharmacology 56(Suppl 1):139–148

Hearst E, Jenkins H (1974) Sign-tracking: the stimulus-reinforcer relation and directed action. Monograph of the Psychonomic Society, Austin

Koksal F, Domjan M, Kurt A, Sertel O, Orung S, Bowers R, Kumru G (2004) An animal model of fetishism. Behav Res Ther 42:1421–1434

Pavlov I (1928) Lectures on conditioned reflexes. Twenty-five years of objective study of the higher nervous activity (behaviour) of animals. International Publishers, New York

Peterson GB (1975) Response selection properties of food and brain-stimulation reinforcers in rats. Physiol Behav 14:681–688

Robinson TE, Yager LM, Cogan ES, Saunders BT (2013) On the motivational properties of reward cues: individual differences. Neuropharmacology 76:450–459

Saunders BT, Robinson TE (2013) Individual variation in resisting temptation: implications for addiction. Neurosci Biobehav Rev 37:1955–1975

Tomie A (1996) Locating reward cue at response manipulandum (CAM) induces symptoms of drug abuse. Neurosci Biobehav Rev 20:505–535

Wasserman EA (1973) Pavlovian conditioning with heat reinforcement produces stimulus-directed pecking in chicks. Science 181:875–877

Zener K (1937) The significance of behavior accompanying conditioned salivary secretion for theories of the conditioned response. Am J Psychol 50:384–403

Sildenafil

Definition

Sildenafil is a PDE5 inhibitor that is sold under the name Viagra for the treatment of erectile dysfunction. Its primary competitors on the market are vardenafil (Levitra) and tadalafil (Cialis). It is also on the market under the name Revatio to treat arterial pulmonary hypertension.

Cross-References

▶ Erectile Dysfunction
▶ PDE5 Inhibitors
▶ Phosphodiesterase Inhibitors
▶ Pulmonary Hypertension

Simulation Models

Synonyms

Animal model with construct validity

Definition

An animal model intended to model a specific human disorder as closely as possible. It typically involves mimicking the etiology or the pathology of the disease and subsequently assessing whether the resulting changes in brain and/or behavior are similar to those found in humans with the disease.

Cross-References

▶ Schizophrenia: Animal Models

Single-Barrel Micropipette

Definition

A single micropipette, filled with saline solution, usually employed for the recording of neuronal activity.

Single-Nucleotide Polymorphism

Synonyms

Polymorphism; SNP

Definition

DNA sequence variations that occur when a single nucleotide (adenine, thymine, cytosine, or guanine) in the genome sequence is altered, usually present in at least 1 % of the population. For example, two sequenced DNA fragments from different individuals, AAGCCTA to AAGCTTA, contain a difference in a single nucleotide.

Cross-References

▶ Gene Expression
▶ Gene Transcription
▶ Haplotype
▶ Pharmacogenetics
▶ Pharmacogenomics

siRNA

Synonyms

Short interfering RNA; Silencing RNA; Small interfering RNA

Definition

siRNA stands for small interfering RNA, also known as silencing or short interfering RNA. In contrast to classical antisense oligonucleotides, siRNA is a class of double-stranded RNA molecules, 19–25 nucleotides in length, with 2-nucleotide overhangs at either $3'$ end and a phosphate group at each $5'$ end. siRNA can efficiently knock down expression of target genes via the RNA interference (RNAi) pathway, which was first described in the nematode worm *Caenorhabditis elegans* in 1998 by Andrew Z. Fire and Craig C. Mello. Any gene of interest with known sequence can be targeted by siRNA that matches the mRNA sequence. siRNA may also recruit RNAi-related pathways, including activation of cellular antiviral mechanisms (e.g., the interferon response) or interference with the chromatin structure of the genome.

Cross-References

▶ Antisense Oligonucleotides
▶ RNA Interference

Sleep

Definition

Sleep is a recurring state of inactivity accompanied by loss of awareness and a decreased reaction to the environment. It is accompanied by characteristic, complex patterns of physiological and hormonal activity and behavior and is found in all mammalian species. Its function is unclear but it may involve restoration of brain and bodily functioning, or memory processing.

Cross-References

▶ Insomnias

Sleep Apnea

Definition

Sleep disorder involving temporary cessation of breathing during sleep.

Sleep Bruxism

Definition

A stereotyped movement disorder characterized by grinding or clenching of the teeth during sleep. The disorder has also been identified as *nocturnal bruxism*, *nocturnal teeth grinding*, and *nocturnal teeth clenching*. Among the symptoms of sleep bruxism are abnormal wear on the teeth, the grinding sounds associated with bruxism, and jaw muscle discomfort.

Cross-References

▶ Parasomnias

Slow-Release Morphines

Synonyms

SROMs

Definition

Slow-release oral morphines (SROMs) are a group of pure opiate agonists that are derived from ▶ morphine but formulated so as to be slowly absorbed through the gastrointestinal system. They are registered in some European countries for the indication of opioid maintenance treatment. SROMs may be suitable for patients who do not tolerate other agents, such as ▶ methadone, because of side effects or because of an inadequate suppression of withdrawal symptoms.

Smooth Pursuit Task

Definition

A simple eye movement task in which participants follow a target with their eyes as it moves smoothly backward and forward on the horizontal plane. Important measurements include velocity gain (the ratio of eye velocity to target velocity) and the number of saccades that occur during pursuit.

SNRI Antidepressants

Huaiyu Yang[1] and George I. Papakostas[1,2]
[1]Department of Psychiatry, Massachusetts General Hospital, Harvard Medical School, Boston, MA, USA
[2]Depression Clinical and Research Program (DCRP), Harvard Medical School, Massachusetts General Hospital, Boston, MA, USA

Synonyms

Serotonin-norepinephrine reuptake inhibitors

Definition

The serotonin-norepinephrine reuptake inhibitors (SNRIs) are a class of antidepressants that simultaneously inhibit the reuptake of both serotonin and norepinephrine. SNRIs are primarily used to treat major depressive disorder (MDD).

Following the development of the tricyclic antidepressants (TCAs), the monoamine oxidase inhibitors (MAOIs), the selective serotonin reuptake inhibitors (SSRIs), and the norepinephrine-dopamine reuptake inhibitors (NDRIs), SNRIs are one of the most recent classes of antidepressants approved for the treatment of MDD. Many TCAs, such as clomipramine, imipramine, and amitriptyline, also inhibit the reuptake of serotonin and norepinephrine. However, the SNRIs differ from the TCAs in two major ways: (1) they do not possess a tricyclic-like structure, and (2) they do not have a clinically relevant affinity for any monoaminergic, histaminergic, or cholinergic receptors or for the sodium channel. As a result, they are safer (i.e., have a lower risk of arrhythmia and seizures) and better tolerated (i.e., have a lower risk of sedation, somnolence, and weight gain).

Pharmacological Properties

Four SNRIs have been developed and marketed to date: venlafaxine, desvenlafaxine, duloxetine, and milnacipran. First introduced in the early 1990s, venlafaxine is the prototypical agent of this class. Milnacipran was introduced as an antidepressant in the late 1990s in Europe and received approval in 2009 for the treatment of fibromyalgia in the United States. Duloxetine was first marketed in 2004 in the United States. Desvenlafaxine, the newest SNRI, is an active metabolite of venlafaxine.

The efficacy of all four SNRIs in the treatment of MDD in adults has been well established in multiple randomized, double-blind, placebo-controlled trials (Montgomery and Andersen 2006; Papakostas and Fava 2007; Perahia et al. 2006; Rickels et al. 2004). In addition to their use as antidepressants, some SNRIs have been approved by the US Food and Drug Administration (FDA) for the treatment of other Axis I disorders in adults, including generalized anxiety disorder (venlafaxine, duloxetine), panic disorder (venlafaxine), social anxiety disorder (venlafaxine), and fibromyalgia (duloxetine, milnacipran).

It has been proposed that SNRIs may be more effective in the treatment of MDD than the SSRIs, perhaps due to their simultaneous action on both serotonin and norepinephrine transporters (Papakostas et al. 2007). However, whether this difference is clinically relevant (Papakostas et al. 2007) and whether it applies to all SSRIs remain unclear (Montgomery and Andersen 2006; Lam et al. 2008).

The efficacy and safety of the SNRIs have not been established for children, adolescents, or the elderly. The US FDA placed a "black box warning" (signifying the possibility of severe adverse events) on the labeling for all four SNRIs based on pooled analysis of data for nine antidepressants (SSRIs and others) describing an increased risk of suicidal ideation and/or suicide gestures in a pediatric population. Therefore, close monitoring is required when prescribing SNRIs or other antidepressants for MDD, especially in vulnerable populations.

Pharmacokinetics

All four SNRIs are well absorbed after oral ingestion and show modest to excellent bioavailability; they are primarily metabolized in the liver and eliminated by the kidney.

The bioavailability of venlafaxine extended-release capsule formulation is about 45 %, while for the immediate-release tablet it is 100 % (relative to an oral solution). Approximately 27 % is protein bound. Venlafaxine is metabolized extensively by the hepatic cytochrome P450 (CYP) isozyme 2D6; O-desmethylvenlafaxine is its active metabolite. Approximately 87 % of a dose of venlafaxine is excreted by the kidneys, 82 % as metabolites, and 5 % remaining unchanged. The elimination half-lives for venlafaxine and O-desmethylvenlafaxine are about 5 h and 11 h, respectively, but they are prolonged in patients with hepatic and/or renal impairment.

After oral administration, desvenlafaxine has a bioavailability of approximately 80 %, and 30 % is protein bound. It is metabolized primarily by the liver via conjugation and to a lesser extent via the hepatic cytochrome P450 3A4. Forty-five percent of desvenlafaxine is excreted via the kidneys in an unchanged form. The elimination half-life of desvenlafaxine is 10–11 h, but it is prolonged in patients with impaired liver function.

Duloxetine is well absorbed after oral administration; its bioavailability ranges between 50 % and 80 % (decreased among smokers). It is more than 90 % protein bound. Duloxetine is metabolized by the hepatic cytochromes P450 2D6 and 1A2, and it is ultimately eliminated in the urine (70 %) and feces (20 %). Its half-life is approximately 12 h.

Milnacipran is well absorbed after oral use. Its bioavailability is about 85 %. Protein binding is low (approximately 13 %), and it is nonsaturable. The drug is metabolized by the liver to an unknown extent, primarily via glucuronidation, and it does not appear to be metabolized via the hepatic cytochrome P450 system, suggesting that it might have some pharmacokinetic advantages (including a low drug-drug interaction potential). Milnacipran has no active metabolites. The excretion of milnacipran is mainly (90 %) through the kidneys, and its half-life is 8 h (Puozzo et al. 2002).

Cross-References

▶ Antidepressants
▶ Desvenlafaxine
▶ Duloxetine
▶ Generalized Anxiety Disorder
▶ Milnacipran
▶ Monoamine Oxidase Inhibitors
▶ Panic Disorder
▶ Selective Serotonin Reuptake Inhibitors
▶ Serotonin Syndrome
▶ Social Anxiety Disorder
▶ Tricyclic Antidepressants
▶ Venlafaxine

References

Lam RW, Andersen HF, Wade AG (2008) Escitalopram and duloxetine in the treatment of major depressive disorder: a pooled analysis of two trials. Int Clin Psychopharmacol 23(4):181–187

Montgomery SA, Andersen HF (2006) Escitalopram versus venlafaxine XR in the treatment of depression. Int Clin Psychopharmacol 21(5):297–309

Papakostas GI, Fava M (2007) A meta-analysis of clinical trials comparing milnacipran, a serotonin-norepinephrine reuptake inhibitor, with a selective serotonin reuptake inhibitor for the treatment of major depressive disorder. Eur Neuropsychopharmacol 17(1):32–36

Papakostas GI, Thase ME, Fava M, Nelson JC, Shelton RC (2007) Are antidepressant drugs that combine serotonergic and noradrenergic mechanisms of action more effective than the selective serotonin reuptake inhibitors in treating major depressive disorder? A meta-analysis of studies of newer agents. Biol Psychiatry 62(11):1217–1227

Perahia DG, Gilaberte I, Wang F, Wiltse CG, Huckins SA, Clemens JW, Montgomery SA, Montejo AL, Detke MJ (2006) Duloxetine in the prevention of relapse of

major depressive disorder: double-blind placebo-controlled study. Br J Psychiatry 188:346–353
Puozzo C, Panconi E, Deprez D (2002) Pharmacology and pharmacokinetics of milnacipran. Int Clin Psychopharmacol 17(Suppl 1):S25–S35
Rickels K, Mangano R, Khan A (2004) A double-blind, placebo-controlled study of a flexible dose of venlafaxine ER in adult outpatients with generalized social anxiety disorder. J Clin Psychopharmacol 24(5):488–496

Social Anxiety Disorder

Siegfried Kasper
Department of Psychiatry and Psychotherapy, Medical University of Vienna, Vienna, Austria

Synonyms

Avoidant personality disorder; Social phobia

Definition

Social anxiety disorder (SAD) is recognized as a highly prevalent (up to 13.3 % lifetime prevalence rate) as well as a chronic disorder with onset during the teenage years (Kasper 1998). Although the disorder is associated with significant disability, which includes educational and occupational impairments that have a negative impact on the quality of life, it is highly underdiagnosed and undertreated. SAD is recognized as a serious medical condition that is accompanied by significant disability since it starts in early life and therefore impairs the otherwise occurring socialization including the development of coping skills. This view has been strengthened by data showing that SAD cannot simply be equated with shyness or with avoidant personality traits and by evidence that this disorder is a discrete entity associated with distinct psychobiological dysfunctions (Lanzenberger et al. 2007) and responsive to specific pharmacotherapeutic interventions (Kasper et al. 2003).

A significant change for the diagnosis of social anxiety disorder in the newly established DSM-5 is that the "generalized" specifier has been deleted and replaced with a "performance-only" specifier. This means in practical terms that most of the patients fall in the old DSM-IV TR generalized group, e.g., the fears include most social situations; however, individuals who fear only performance situations (e.g., speaking or performing in front of an audience) are now considered as a distinctive subset of social anxiety disorder, with a specific etiology, age at onset, physiological response, as well as treatment response. Interestingly, beta-blockers have been demonstrated to be useful in performance-related social anxiety, but not in generalized SAD. Smaller changes of DSM-5 include that adults no longer have to recognize that their anxiety or fear is excessive or unreasonable and that the symptoms have to be present for at least six months.

Role of Pharmacotherapy

Most pharmacotherapeutic studies, specifically the systematic ones with selective serotonin reuptake inhibitors (SSRIs) and serotonin-norepinephrine reuptake inhibitors (SNRIs), have been carried out in patients suffering from generalized SAD. The most commonly used primary efficacy scale in medication studies of SAD is the Liebowitz Social Anxiety Scale (LSAS) (Liebowitz 1987), which has facilitated comparisons across studies.

Early work demonstrated that irreversible monoamine oxidase inhibitors (MAOIs; e.g., phenelzine) are effective in the treatment of SAD, but these agents are limited by their side-effect profile, the need for dietary precautions, and drug interactions (Versiani 2000). Studies with the reversible MAOI moclobemide indicated effectiveness in earlier but not later development program studies and have therefore not been granted indication status by health regulatory authorities (Brunello et al. 2000). More recent work has established the efficacy of several SSRIs (Montgomery et al. 2004), and these

Social Anxiety Disorder, Table 1 Social anxiety disorder

	SSRIs	TCAs	Benzodiazepines	Others
Acute efficacy	Escitalopram, fluoxetine, fluvoxamine, paroxetine, and sertraline	–	Alprazolam, Bromazepam and clonazepam	CBT, phenelzine, moclobemide, brofaromine, venlafaxine, gabapentin, pregabalin, olanzapine
Long-term efficacy	Escitalopram, fluvoxamine, paroxetine, and sertraline	–	–	CBT, and phenelzine, moclobemide, and venlafaxine
Relapse prevention	Escitalopram, paroxetine, and sertraline	–	Clonazepam	CBT
Enhanced efficacy of psychological treatment	Sertraline	–	–	–
After nonresponse	–	–	–	–

Placebo-controlled studies (*SSRIs* selective serotonin reuptake inhibitors, *TCAs* tricyclic antidepressants, *CBT* cognitive behavioral therapy) (Baldwin et al. 2005)

agents have been recommended as first-line pharmacotherapy agents.

The available placebo-controlled studies have been carried out for acute as well as long-term efficacy, including relapse prevention (see Table 1). Interestingly, the question of whether psychological treatment is able to enhance pharmacotherapy has only been addressed in one study, and there are no placebo-controlled studies to guide clinicians as to which treatment to administer after treatment nonresponse. The main group of medications which has been studied is SSRIs, and interestingly, there are no studies with tricyclic antidepressants (TCA), reflecting the fact that pharmacotherapy for SAD was developed after the introduction of SSRIs for the indication of depression. A few studies with benzodiazepines and SNRIs as well as pregabalin have been conducted. Pregabalin is not a benzodiazepine; it differs from other medications used to treat SAD in that it binds with high specificity to the α_2-δ subunit of P/Q-type voltage-gated Ca^{2+} channels, thereby reducing the influx of Ca^{2+} at nerve endings and attenuating the release of excitatory neurotransmitters such as norepinephrine, glutamate, aspartate, substance P, and calcitonin gene-related peptide.

In addition to pharmacotherapeutic approaches, cognitive behavioral psychotherapy has been studied in a placebo-controlled design indicating superior efficacy to placebo; however, no medication control group has been studied in these treatment trials.

Tretament of Comorbid Conditions

There is a high rate of comorbidity (up to 60 %) between depression and anxiety disorders as well as between anxiety disorders themselves, and antidepressants have been used and granted indication by health authorities for both depression and anxiety disorders. It is apparent that both SSRIs and SNRIs work in depression as well as in different forms of anxiety disorders, including SAD, and for SAD, additionally benzodiazepines, (▶ Clonazepam) as well as to a lesser extent pregabalin, have been studied (Zohar et al. 2002).

An algorithm for the pharmacotherapy of SAD in primary practice has been developed by Stein et al. (2001) and divided into different steps as outlined in Fig. 1. Firstly, this algorithm indicates the necessity to establish a diagnosis and to clarify if certain complications may have an impact on the choice of medication to be used, like comorbid severe depression or suicidality which requires specific therapeutic regimens. If this can be ruled out, SSRIs could be initiated, and if patients are responsive, a maintenance therapy can be continued with the same dosage of medication with which remission has been achieved. Interestingly, the same dosages are used for SAD as for depression. Whereas in depression, a clear

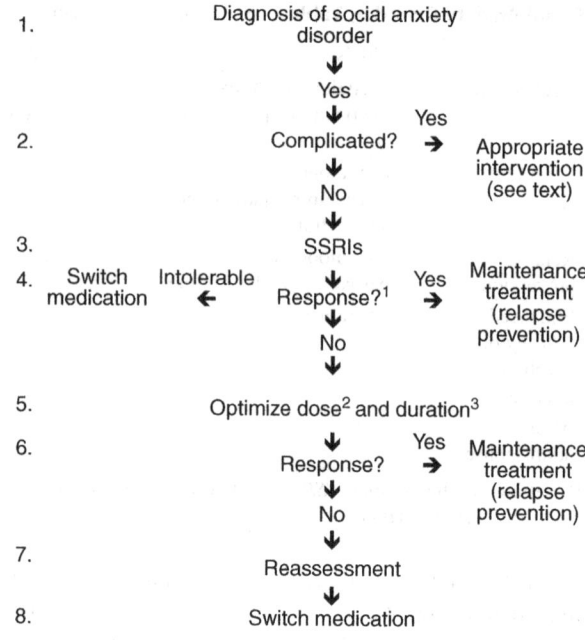

Social Anxiety Disorder, Fig. 1 Algorithm for the pharmacotherapy of social anxiety disorder in primary practice (After Stein et al. 2001)

response might be observed as early as after 2–3 weeks' time, this usually takes at least 6–8 weeks in SAD, since patients need to experience the social situation in which the symptoms appeared before. If intolerable side effects are emerging, the switch to another SSRI or SNRI can be considered, although there are no placebo-controlled studies to justify this approach. Thereafter, an optimization of the dosage (most likely an increase) as well as the duration of treatment (over 8 weeks) needs to be taken into account before a treatment response can be expected. If this regime does not lead to remission, a reassessment as well as a switch to another type of medication, like pregabalin, can be considered. However, it needs to be noted that pregabalin does not have the indication for SAD and that this approach is also not substantiated by placebo-controlled switch studies.

Conclusion

SAD is not a mild form of mental disorder but a clinically meaningful medical illness with an established pathophysiology which deserves early diagnosis and distinct treatment. Since SAD starts early in life and may result in a disabling disorder, appropriate pharmacotherapeutic and psychotherapeutic interventions need to be initiated early in the course of the illness. Our continued progress in understanding and managing SAD will result in progress in delineating the subtypes and treatment responses to this condition.

Declarations of Interest

Dr. Kasper has received grant/research support from Eli Lilly, Lundbeck, Bristol-Myers Squibb, GlaxoSmithKline, Organon, Sepracor, and Servier; has served as a consultant or on advisory boards for AOP Pharma, AstraZeneca, Bristol-Myers Squibb, GlaxoSmithKline, Eli Lilly, Lundbeck, Pfizer, Organon, Schwabe, Sepracor, Servier, Janssen, and Novartis; and has served on speakers' bureaus for AstraZeneca, Eli Lily, Lundbeck, Schwabe, Sepracor, Servier, Pierre Fabre, and Janssen.

References

Baldwin DS, Anderson IM, Nutt DJ, Bandelow B, Bond A, Davidson J, Den Boer JA, Fineberg NA, Knapp M, Scott J, Wittchen H-U (2005) Evidence-based guidelines for the pharmacological treatment of anxiety disorders: recommendations from the British Association for Psychopharmacology. J Psychopharmacol 19:567–596

Brunello N, den Boer JA, Judd LL, Kasper S, Kelsey JE, Lader M, Lecrubier Y, Lepine JP, Lydiard RB, Mendlewicz J, Montgomery SA, Racagni G, Stein MB, Wittchen HU (2000) Social phobia: diagnosis and epidemiology, neurobiology and pharmacology, comorbidity and treatment. J Affect Disord 60(1):61–74

Kasper S (1998) Social phobia: the nature of the disorder. J Affect Disord 50:s3–s9

Kasper S, den Boer JA, Sitsen JMA (eds) (2003) Handbook of depression and anxiety, 2nd edn, revised and expanded. Marcel Dekker, New York/Basel

Lanzenberger RR, Mitterhauser M, Spindelegger C, Wadsak W, Klein N, Mien LK, Holik A, Attarbaschi T, Mossaheb N, Sacher J, Geiss-Granadia-T, Kletter K, Kasper S, Tauscher J (2007) Reduced serotonin-1A receptor binding in social anxiety disorder. Biol Psychiatry 61:1081–1089

Liebowitz MR (1987) Social phobia. Mod Probl Pharmacopsychiatry 22:141–173

Montgomery SA, Lecrubier Y, Baldwin DS, Kasper S, Lader M, Nil R, Stein D, van Ree JM (2004) ECNP consensus meeting, March 2003. Guidelines for the investigation of efficacy in social anxiety disorder. Eur Neuropsychopharmacol 14:425–433

Stein DJ, Kasper S, Matsunaga H, Osser DN, Stein MB, van Ameringen M, Versiani M, Westenberg H, Zhang M (2001) Pharmacotherapy of social anxiety disorder: an algorithm for primary care – 2001. Prim Care Psychiatr 7:107–110

Versiani M (2000) A review of 19 double-blind placebo-controlled studies in social anxiety disorder (social phobia). World J Biol Psychiatry 1:27–33

Zohar J, Kasper S, Stein DJ (2002) Pharmacotherapy of social anxiety disorder. In: Kasper S, Zohar J, Stein DJ (eds) Decision making in psychopharmacology. Martin Dunitz, London, pp 47–56

Social Behavior

Frank Sams-Dodd
Willingsford Limited, Southampton, UK

Definition

Social behavior is behavior between members of the same species. However, depending upon the context, it may more specifically signify nonaggressive, nonterritorial, and nonsexual behavioral interactions between members of the same species.

Impact of Psychoactive Drugs

Social Behavior

Social behavior is in its broadest sense behavior taking place between members of the same species. It can therefore include territorial, parental, and sexual behavior as well. However, individuals living in groups have a special need to communicate because they have to share resources, coordinate activities such as hunting or foraging, defend the territory against intruders, and protect the group against predators. Social species such as humans, monkeys, and rats have therefore developed communication systems and internal organizational structures that allow them to live in proximity without having to engage in frequent fights over resources. Fighting may cause injury and can jeopardize the welfare of the individual and of the group, and it is therefore something that must be avoided if possible.

Drug effects in animals are typically studied to predict how the drugs may affect human behavior, and the relevance of studying drug effects on social behavior in animals such as rats, mice, and monkeys may seem rather distant to the complexity of human social behavior and structure. However, humans are also a social species that has a highly hierarchical organization and are very territorial, and while the exact details will differ between species, the more general responses can be comparable. Human behavior is often viewed to be fundamentally different from animals, but a very obvious example of human territorial behavior is how countries and borders are defined and the vigor with which a territory is defended by humans. Another more mundane example is how it is felt if someone sits in "our chair" in a classroom. The social hierarchy is similarly very obvious in many work situations and when observing an interaction between two individuals with different status in an organization: from body posture and movements, it can easily be understood, which of the individuals has the higher status. The military is an example of an organization where this hierarchy has been practiced to its fullest extent.

In this entry, social behavior will be used to indicate nonaggressive, nonterritorial, and nonsexual social behaviors between members of a species that live in social structures. It is therefore the normal behavior expressed when interacting with other people in a daily life situation. This type of social behavior is disturbed in many diseases such as depression, anxiety, social phobias, schizophrenia, and autism; it may be affected by drug side effects, e.g., corticosterone, and it may be changed as a result of another condition, e.g., chronic pain.

In relation to psychopharmacology, social behavior represents a very interesting group of behaviors. They are important aspect of human behavior with a clear biological function, they are disrupted in many diseases, and they are complex behaviors that must be normalized by a treatment, and because it is a set of complex active behaviors, it is relatively easy to dissociate the nonspecific, e.g., sedation from specific drug effects.

Social species usually have a rich repertoire of communication signals, but typically for other species than humans, only a fraction of these has been observed and recorded. For example, rats and mice rely to a great extent on olfactory signals and ultrasound in their communications and presumably also use very discrete facial and body postures that is not recognized, and the level of information that is extracted from an encounter between two individuals is therefore very limited. This therefore does not mean that their communication systems and social organization are simple and primitive, but rather that it was not possible to record and interpret the true level of complexity. For example, most studies on social behavior between pairs of rats record time spent investigating, following, and grooming the other individual as an index of the level of social behavior. These measures are very crude and probably disregard 95 % of the true communication that exist between the rats. However, for the purpose of studying drugs that interfere with the social structure, this may be sufficient – it should just be realized that it does not represent the true level of communication.

A number of factors need to be considered, when studying social behavior. One factor is, as mentioned above, that each species will use different routes of communication, e.g., visual, auditory, olfactory, and postures and movements. Humans have, compared to most animals, a rather inferior olfactory sense and they will rarely be able to monitor communication using this channel. Similarly, it will for humans be difficult to register minute changes in postures and movements in a different species, and for auditory signals, humans cannot perceive signals above 15 kHz unless special equipment is used. A second factor is strain differences in the chosen species. There are considerable differences between different lines of rats and mice in terms of their level of aggression and anxiety, and this will affect the study. A third factor is the influence of other behavioral categories on the observed social behavior. Sex differences and estrus cycle will affect the level of sexual behaviors in an experimental situation, and repeated exposure to a particular arena may prompt the appearance of strong territorial behaviors, thus increasing the level of aggressive behaviors. A fourth factor is that diseases, just as for humans, will affect the behavior of an animal and particularly for transgenic animals or aged animals, this can be a concern. It has, for example, been reported that some of the transgenic lines of mice carrying an Alzheimer's disease mutation have inferior vision and reduced level of hearing. Also, in case a drug causes pain or induces, e.g., nausea, this will affect behavior as well.

A final factor is that experimental conditions and handling will affect the outcome of an experiment. A stressed animal or an animal placed in an aversive experimental situation will necessarily exhibit a different behavior compared to a nonstressed or a nonanxious animal. These factors need to be controlled and what may constitute an aversive stimulus for that particular species needs to be considered. However, it is also possible to use aversive stimuli to control the types of behavior that the animals express and that are studied. For example, the light level in an arena will have considerable impact upon the behavior of a rat. High-light conditions will produce an anxiogenic state in the rat, which will reduce the level of active behaviors, whereas a low-light condition will increase the level of natural behaviors, e.g., explorative and social behaviors (Whishaw and Kolb 2005). The olfactory level in the arena and the number of repeated visits to the arena will affect the speed with which the animal begins to perceive the arena as a part of its territory and thereby the level of aggressive behaviors it expresses in order to defend its territory, if confronted with a conspecific. The level of aggressive behavior is also affected by the housing condition of the animals, e.g., in rats single housing will increase the aggression level. Light/dark cycle will affect the sensitivity of the animals to drugs. If animals are tested during their normally active period, i.e., during the night for a rat, it will be more sensitive to drug effects compared to testing conducted in the daytime, i.e., the rats inactive period. Finally, handling can have unexpectedly strong effects on behavior, e.g., handling a rat by the tail can stress it so much that it becomes impossible to study social interactions, and the presence of a computer in the experimental room can affect behavior because the rats may be able to hear it.

Standard Test Paradigms

Social behavior can in principle be studied in any situation that involves two or more members of the same species. Complex situations involving many members living in a stable structure are likely to provide more detailed information on the effects of drugs on normal behavior or on behavioral abnormalities in animals carrying a disease, but it can be very difficult to standardize a test of this type to allow comparison of different drugs, and the amount of time required to study drug effects may become so demanding that it in practice becomes impossible to compare drug series. For these reasons a number of simplified, standardized behavioral tests have been developed that allow the experimental observation of some aspects of social behavior thought to be relevant to human behaviors.

One of the most routinely used tests for social behavior is the social interaction test where two rats or mice are placed simultaneously into an arena and where their level of social behavior is recorded for 7–10 min (File and Seth 2003). The animals may be familiar with the arena and to each other. The test is usually performed in a high-light and a low-light configuration. For the high-light configuration, the arena is brightly illuminated and this will create an aversive environment since rats and mice in general avoid open, brightly lit areas. Their natural tendency will be to avoid the center of the arena and to stay in the periphery of the arena, where they usually remain inactive. Using this design, it has been shown by many researchers that anxiolytic drugs, such as the benzodiazepines, diazepam and chlordiazepoxide, will increase the level of social interaction, indicating a reduced level of anxiety in the animals, whereas drugs that lack anxiolytic effects are inactive. However, it should be noted that the avoidance of open spaces is a biologically appropriate response for a rat and a mouse, and the test therefore studies drug effects on natural behaviors and not on pathological anxiety, which is the type sought to treat human patients.

In the low-light configuration, the arena is only illuminated by dim red light to simulate a nighttime situation, which is non-aversive to rats and mice, and they will typically explore the entire arena and engage in frequent social interactions, e.g., rats will typically investigate the novel area together. Under these conditions it is possible to evaluate manipulations that disrupt social behavior, e.g., to model a disease condition or to determine if a drug causes side effects that affect normal behavior. An example of a disease model is phencyclidine-induced social isolation (Ellenbroek and Cools 2000). Phencyclidine (PCP) is an NMDA-antagonist that often is abused by humans and it has frequently been observed that PCP will induce a type of symptomatology, which resembles the negative symptoms of schizophrenia, i.e., the inability to engage in normal social relations. PCP will potently disrupt social behavior in rats, and this disruption can be reversed by antipsychotic drugs that in patients have an effect on the negative symptoms, e.g., clozapine and risperidone, whereas drugs that only affect the positive symptoms, e.g., haloperidol, or that are not effective in the treatment of schizophrenia, e.g., citalopram and diazepam, are ineffective.

A different test paradigm is the social recognition test, which typically is performed in rats, where the level of social behavior or social interest is used as a measure of recognition. In this test, an adult rat is allowed, e.g., 5 min, to investigate a juvenile rat. The juvenile is removed and after a certain time period, e.g., 30 min, 2 h, or 24 h, either the same juvenile or a different juvenile is presented to the adult rat. If the adult rat recognizes the juvenile, it will spend less time investigating it compared to the situation where it is a novel juvenile. Drug studies have shown that nicotine, which is known to improve memory in humans, can prolong the time period in which the adult rat recognizes the juvenile, whereas scopolamine, which interferes with memory, reduces the time period. The social recognition test is typically used in two configurations with either a long delay or a short delay until the reintroduction of the juvenile. The purpose of using a long delay is to identify drugs that can enhance memory, e.g., for symptomatic treatment of memory deficits in Alzheimer's disease, by demonstrating that a drug extends the time period in which the adult animal recognizes the juvenile. The purpose of using a short delay at which normal animals will recognize the juvenile is to determine whether a drug inhibits normal memory processes, e.g., caused by side effects, or to determine if animals display memory deficits, e.g., in connection with disease models, and whether drugs are able to reverse these deficits. Examples of the latter are scopolamine-induced memory deficits and cognitive deficits in disease models of e.g., Alzheimer's disease, schizophrenia, and stroke. Finally, it is possible to administer the investigative drug at different time points in the testing sequence to determine which aspects of the memory process it affects, e.g., administering it after the first presentation of the juvenile to determine if it affects the consolidation of memory.

Cross-References

▶ Animal Models for Psychiatric States
▶ Antidepressants
▶ Antipsychotic Drugs
▶ Anxiety: Animal Models
▶ Autism Spectrum Disorders and Intellectual Disability
▶ Benzodiazepines
▶ Dementias and Other Amnestic Disorders
▶ Dementias: Animal Models
▶ Depression: Animal Models
▶ Ethopharmacology
▶ Nicotine
▶ Schizophrenia
▶ Schizophrenia: Animal Models

References

Ellenbroek BA, Cools AR (2000) Atypical antipsychotics. BirkHauser Verlag, Basal
File SE, Seth P (2003) A review of 25 years of the social interaction test. Eur J Pharmacol 463:35–53
Whishaw IQ, Kolb B (eds) (2005) The behavior of the laboratory rat: a handbook with tests (hardcover). Oxford University Press, New York

Social Defeat

Definition

Social defeat is a procedure in which a smaller "intruder" rodent (rat or mouse) is introduced into the cage of a larger and aggressive "resident" rodent. The resident rat usually attacks the intruder rodent until it manifests a submissive posture. Additionally, the mere presence of the resident rodent (a threat condition) is sufficient to provoke physiological and behavioral stress responses in the intruder rodent that are similar to those obtained after the physical confrontation (Miczek et al. 2008).

Cross-References

► Aggression

References

Miczek KA, Covington HE, Yap JJ (2008) Social stress, therapeutics and drug abuse: preclinical models of escalated and depressed intake. Pharmacol Ther 120:102–28

Social Impairment

Definition

Individuals with autism spectrum disorder (ASD) have core social impairments that are persistent and severe. These impairments can be seen through deficits in multiple behavioral and relational areas. The core social impairment is demonstrated through nonverbal behaviors, such as deficits in eye contact, facial expression, and gestures that are used to help regulate social interaction. Failure to develop age-appropriate friendships may be present, along with an ability to get along with adults but not same-aged peers. Spontaneous seeking to share enjoyment or interests may not be demonstrated and social reciprocity is typically not present. Individuals may lack an awareness of others and their feelings.

Cross-References

► Autism Spectrum Disorders and Intellectual Disability

Social Interaction Test

Definition

Two animals, typically rats or mice, are placed into an arena, and their interactions, for example, investigation, following, and grooming, are recorded for a period of time, usually 5–10 min. Social behaviors such as following, adjacent lying, and anogenital sniffing are recorded by an observer or via automated image analysis. Many drugs modulate behavior in the social interaction test: benzodiazepines, MDMA, and oxytocin tend to increase social interaction, while amphetamines, cannabinoids, NMDA antagonists, and withdrawal from various ► drugs of abuse tend to decrease social interaction. In the high-light version of the test, the arena is brightly illuminated, and this creates an aversive situation that results in low levels of social interaction. In this configuration, it is possible to identify drugs or manipulations that reduce the inferred level of anxiety in the animals, that is, they result in an increased level of social interaction. In the low-light version, the arena is only illuminated with low, typically red light in order to minimize aversive cues. In this configuration, the level of interaction will be maximal, and it is possible to test drugs or manipulations that reduce the normal level of social interactions. In addition to light level, which is the strongest experimental factor, the animals' familiarity with the arena can be varied, for example, by having been introduced to the arena prior to the testing session and whether the animals had encountered each other prior to the testing session. Familiarity with the arena will reduce the level of aversive cues but will also increase the level of territorial behavior, resulting in more fighting between the animals. Familiarity between the animals being tested can reduce the level of aversive cues during the testing situation and the level of fighting, because a hierarchy does not have to be established, but it may also increase variability in the data, because the animals will have a preestablished rank that not will be present if they are unfamiliar to each other.

Cross-References

► Amphetamine
► Anxiety: Animal Models
► Benzodiazepines

- Cannabinoids
- Methylenedioxymethamphetamine (MDMA)
- Oxytocin
- Social Behavior
- Withdrawal Syndromes

Social Recognition Test

Definition

The test uses social memory, that is, remembering having seen another individual before, as a measure of cognitive function. A juvenile is introduced to an adult animal, and typically rats are used, for 5–7 min, and the level of investigative behavior performed by the adult animal toward the juvenile is recorded. After a delay of typically 30 min, 2 h, or 24 h, either the same juvenile or a novel juvenile is introduced to the same adult animal, and the level of investigative behavior is recorded. If the adult animal recognizes the juvenile, the level of social investigation will be lower compared to the situation, where the adult male does not recognize the juvenile or it is a novel juvenile. In a variant of this test, an object is used instead of a juvenile animal. In this design social cues are avoided and it may therefore function as a nonsocial control test.

Social Stress

Klaus A. Miczek
Tufts University, Medford, MA, USA

Synonyms

Social challenge

Definition

In most general terms, social stress refers to the immediate response to a change in the social life of an individual, either in response to separation or confrontation. If the social stress persists, pathological consequences become evident in suppressed reproductive, immune, and metabolic functions and in social intercourse; many cardinal symptoms of psychosis, depression, and drug abuse, such as anhedonia-like responses, dysregulated circadian rhythms, weight, and movements, emerge in vulnerable individuals (Anisman and Matheson 2005; McEwen and Chattarji 2007).

Impact of Psychoactive Drugs

Types of Social Stress

In the definition of social stress and the response to stress, pressures, loads, and strains, as characterized in the physical, engineering, and military sciences, serve as metaphors for physiological, cellular, and molecular events. In biological systems, the concepts of homeostasis and allostasis are important in the definition of stress – the former defining a set point surrounded by minimal and maximal limits and the latter referring to constantly changing boundaries (McEwen and Chattarji 2007). Like many other stressors, maternal separation stress and stress arising from social conflict in adulthood incorporate the rapid sympathetic and slower hypothalamic-pituitary-adrenal (HPA) responses. In addition to these changes in the autonomic nervous system, many physiological and neurobiological features are unique to the specific social stress.

Social Isolation, Crowding, and Instability

Most primate species and other socially cohesive species, such as rats, require social contact, and deprivation of this contact constitutes a type of social stress that eventually results in behavioral, physiological, and neurochemical pathologies (*social isolation*). In contrast, in animal species that disperse after puberty, single housing mimics the life of males who mark, patrol, and defend their territory and exclude other males. The opposite type of social stress is represented by *crowding*, which in most cases, leads to lethargy and eventual increased mortality.

High population density profoundly alters the toxicity to drug action as demonstrated earlier by increased amphetamine toxicity in crowded mice (Miczek et al. 2008). In species with well-defined social hierarchies, the establishment of social stratification by *instability* can be stressful. Experimentally, the constant rearrangement of social groups of laboratory animals has been used to induce constant social instability, although with questionable validity.

Maternal Separation Distress

Two types of *maternal social separation* stress have been identified with behavioral and neurobiological consequences that continue into adulthood and that are particularly relevant to drug abuse and affective disorders. Experimental studies in rodents have identified a critical developmental period in rodents, starting about 3–4 days after birth, during which individuals are hyporesponsive to stress in terms of glucocorticoid activation. Brief separations of pups from the dam (ca. 15 min/day) during this period render the individual more resistant to stress in adulthood, possibly related to anxiety-like responses. In contrast, prolonged separations from the dam (ca. 180 min/day) enhance the hypothalamic-pituitary response to stress in adulthood, possibly relevant to anhedonia, a core symptom of affective disorders.

The amount of play and its intensity during adolescence determine how individuals cope with stress later on as adults. Some adolescents appear to be particularly vulnerable to social defeat stress, and these episodes of stress can impair neural maturation. Of particular concern is the impact of social stress during adolescence on the later acquisition of compulsive drug taking.

Social Defeat Stress

Social stress in adulthood is experienced most often in aggressive confrontations, typically during the formation of dominance hierarchies and during the breeding season. The *socially defeated* animal rapidly learns to display the species-typical defensive and submissive postures, acts, and signals. The physiological and behavioral consequences manifest themselves by impaired social interactions, particularly reproductive activities, reduced exploration, disrupted foraging, less feeding, and drinking (Miczek et al. 2008). Even a single episode of social defeat stress impairs circadian activity, nociception, and motor activity and activates cellular activity in mesencephalic and corticolimbic nuclei. Repeated episodes of social defeat stress amplify these neurobiological and behavioral consequences over a long term. These enduring neuroadaptive changes lead to sensitized stimulant responses, similar to those after intermittent stimulant drug treatment. Repeated social defeat stress engenders also deficits in memory that relies on intact hippocampal activity.

Social Subordination Stress

When social stress is continuous rather than episodic such as in *subordinate* animals that are continuously exposed to the presence of higher ranking individuals, its behavioral repertoire and physiological functions are severely compromised, in extreme cases, leading to a morbid course (Fuchs et al. 2004). Subordinate animals are characterized by scars and wounds, inactivity, immunosuppression, elevated ACTH and glucocorticoid activity, lower androgens, eventual testicular regression, and adrenal hypertrophy. The persistent behavioral, immunological, and endocrine characteristics of a subordinate animal lead to lowered success in transmitting genes into the next generation and ensuring the survival of its offspring.

Neurobiology and Psychopharmacology of Maternal Separation Stress

GABAergic, glutamatergic, monoaminergic, and opioid peptidergic mechanisms are activated during the separation stress of an infant from the dam (Miczek et al. 2008). In most myomorph rodents, maternal separation stress is evident behaviorally by the emission of loud and frequent ultrasonic vocalizations during the first 2 weeks after birth. Similar distress vocalizations are emitted in avian and primate species. These vocalizations are sensitive to the suppressive effects of agonists at the 5-HT_1 receptor family, positive modulators of the

GABA$_A$ receptors, and mGluR$_{2/3}$ receptor agonists. Drugs which are clinically effective as anxiolytics (e.g., benzodiazepines, buspirone) and antidepressants (e.g., SSRIs, SNRIs) effectively reduce the ultrasonic distress vocalizations by rodent pups that are separated from the dam and litter mates. These vocalizations may be expressions of affective distress and serve primarily the immediate purpose to prompt retrieval by the dam and, during prolonged separations, thermoregulatory functions may become relevant. While the GABAergic, glutamatergic, monoaminergic, and opioid receptor systems continue to undergo significant maturation during the first weeks of life of rodents, the pharmacological manipulations of these targets may result in effects on affective vocal responses that differ from those in the mature adult.

Repeated maternal separation stress in the first 2 weeks after birth results in profound and long-lasting changes in the expression of the message for GABA, glutamate, and neuropeptides such as CRF and opioid peptides and their receptors. A key variable is the duration of the maternal separation; for example, prolonged daily maternal separations resulted in increased mRNA for CRF in hippocampal cells relative to the lower mRNA levels of CRF after short separations (180 vs. 15 min/day). Two types of long-term consequences are significant from a psychopharmacological perspective. First, rodent pups that are repeatedly separated from the dam and littermates can show depressive-like behavioral and neural phenotypes, and antidepressant treatment can prevent some of these changes. Secondly, of particular significance are the large and persistent increases in alcohol and stimulant self-administration and in sensitized responses to stimulant challenge in adults that were exposed during the stress hyporesponsive period to repeated long episodes of uncontrollable maternal separation stress, although these increases are limited to specific experimental conditions.

In addition to the hyporesponsiveness to psychomotor stimulants during adolescence, social defeat stress during this developmental period renders adolescents far less sensitive to the psychomotor stimulant effects of amphetamine or cocaine relative to adults in several rodent models. Repeated exposure to social defeat stress or to cocaine activates adolescent neural circuits in a remarkably similar manner suggesting shared mechanisms of emerging resilience and vulnerability.

Neurobiology and Psychopharmacology of Social Defeat and Subordination Stress

In adulthood, even a single episode of social defeat stress may increase the release of DA in mesolimbic, but not striatal structures as assessed by in vivo microdialysis and fast-scan voltammetry, complementing earlier evidence for noradrenergic activation in both the perpetrator and victim during social confrontations (Miczek et al. 2008). Stimulant-evoked DA rises in the prefrontal cortex, and nucleus accumbens are very large after the intruder rat has experienced several episodes of social defeat providing evidence for neural sensitization. This stress-sensitized release of cortical and accumbal DA needs to be integrated with the often reiterated hypothesis of DA's special role in reward processes.

Increased serotonergic impulse flow characterizes socially stressed tree shrews and mice, as is evident in the activity of raphé cells as well as in the terminals in the hippocampus. Chronic social subordination stress results also in reduced affinity and message for receptors of the 5-HT$_1$ family in several animal species. Persistent changes in cellular activity after repeated social defeat stress are evident not only in the raphé cells but also in the VTA (ventral tegmental area), prefrontal cortex, and amygdala.

The study of brief social defeat stress and of continuous subordination stress represents a promising source of information on account of anxiety- and depressive-like physiological and behavioral profiles in several animal models (Martinez et al. 2002). Social defeat can engender features indicative of profound emotion, ranging from anhedonia to intense fear. The loss of social initiatives and the lack of experiencing pleasure ("anhedonia"), most often assessed by the preference for sweet tastes, in animals that were exposed to uncontrollable, unpredictable, and chronic social defeat stress, can be attenuated by tricyclic antidepressants and SSRIs. In parallel, antidepressants can reverse the social subordination and stress-induced suppression of neurogenesis in the hippocampus. Endogenous brain-derived nerve growth factor (BDNF) is profoundly suppressed by continuous social subordination stress, and increased BDNF activity is detected in the VTA in animals exposed to intermittent episodes of brief social defeat stress suggesting an important role of this factor in the contrasting neuroadaptations after brief versus continuous social stress (Berton et al. 2006).

In anticipation of an imminent episode of social defeat stress, animals become hyperthermic and start to emit distress calls; these physiological and behavioral indices are interpreted to reflect intense emotional states. In fact, clinically effective anxiolytic drugs which act as positive modulators at $GABA_A$ receptors or at $5-HT_{1A}$ receptors attenuate these anxiety-like responses. These anxiolytic-like effects of benzodiazepines, buspirone, and also alcohol of the anticipatory hyperthermia and distress vocalizations contrast with the absence of ameliorative effects on defensive responses in reaction to social stress.

In laboratory rodent models, similar to other mild stressors such as tail pinch or foot-shock pulses, intermittent defeat stress induces a persistent, sensitized, psychomotor stimulant response upon challenge with a moderate dose of cocaine, amphetamine, morphine, or ethanol (Anisman and Matheson 2005). As little as one defeat experience is sufficient to induce a sensitized locomotor activation to a challenge with D-amphetamine, and this effect grows in magnitude and extends in time after repeated episodes of social defeat stress. Blockade of glutamate receptors in the VTA prior to each stress episode can prevent the sensitizing effect of stress, and this protective effect critically depends on BDNF and phosphorylation of ERK and CREB. Behavioral and neural sensitization has been hypothesized to contribute to the neuroadaptations that escalate drug seeking and taking.

The long-postulated facilitative effects of social stress on drug abuse are more directly examined under experimental conditions that quantify the rate, amount, and persistence of drug intake as a result of exposure to discrete episodes of social stress (Miczek et al. 2008). Similar to other types of intermittent stress, brief episodes of social defeat stress in laboratory rats and nonhuman primates can facilitate the acquisition of stimulant and possibly alcohol self-administration, increase the rate of regular daily drug intake, and especially increase the amount and duration of cocaine self-administration under unlimited access conditions ("binging"). Mechanistic studies have begun to delineate the glutamatergic, GABAergic, and peptidergic modulation of DA in the VTA-accumbens-prefrontal cortex-amygdala neural circuit as being critical for social defeat stress to escalate psychomotor stimulant intake.

In contrast to the intense and prolonged cocaine taking after episodic social defeat stress, continuously threatened subordinate rats responded less to a stimulant challenge in terms of locomotor hyperactivity and accumbal DA release, and they ceased self-administering intravenous cocaine sooner when given unlimited access relative to controls. There is an indication that BDNF in the VTA, prefrontal cortex, and amygdala is suppressed as a result of chronic subordination stress. The divergent adaptations in BDNF cells after episodic relative to continuous social stress appear, characteristic of escalated versus suppressed cocaine taking that follow different types of stress. To which extent, the controllability of stress determines the BDNF activation in the prefrontal cortex remains to be determined (Anisman and Matheson 2005).

In sum, considerable evidence points to different kinds of social stress as risk factors for initiating, escalating, and resuming drug abuse in vulnerable individuals (Miczek et al. 2008). The hyperdefensive behavior, weight dysregulation, signs of anhedonia, altered sleep, and activity patterns and the profile of HPA activity in subordinate animals bear similarities to cardinal signs of depressed patients. Other signs in the physiological and behavioral repertoire of socially stressed animals may reflect anxiety-like responses. Understanding the intracellular cascade of events in mesocorticolimbic circuits for the transition from episodic to continuous, inescapable social stress in infancy and adulthood promises to provide insight into basic reward processes that are relevant for addictive and affective disorders.

Cross-References

▶ Aggressive Behavior: Clinical Aspects
▶ Antidepressants
▶ Antipsychotic Drugs
▶ Anxiolytics
▶ Depression
▶ Distress Vocalization
▶ Drug Self-Administration
▶ Reinforcement Disorders
▶ Selective Serotonin Reuptake Inhibitors
▶ Stress: Influence on Relapse to Substance Use

References

Anisman H, Matheson K. Stress, depression, and anhedonia: caveats concerning animal models. Neurosci Biobehav Rev. 2005;29:525–46.

Berton O, McClung CA, Dileone RJ, Krishnan V, Renthal W, Russo SJ, et al. Essential role of BDNF in the mesolimbic dopamine pathway in social defeat stress. Science. 2006;311:864–8.

Fuchs E, Czeh B, Fluegge G. Examining novel concepts of the pathophysiology of depression in the chronic psychosocial stress paradigm in tree shrews. Behav Pharmacol. 2004;15:315–25.

Martinez M, Calvo-Torrent A, Herbert J. Mapping brain response to social stress in rodents with c-fos expression: a review. Stress. 2002;5:3–13.

McEwen BS, Chattarji S. Neuroendocrinology of stress. In: Blaustein JD, editor. Behavioral neurochemistry, neuroendocrinology and molecular neurobiology. Heidelberg: Springer; 2007. p. 571–93.

Miczek KA, Covington HE, Yap JJ. Social stress, therapeutics and drug abuse: preclinical models of escalated and depressed intake. Pharmacol Ther. 2008;120:102–28.

Sodium Oxybate

Definition

The sodium salt of gamma-hydroxybutyrate (GHB).

Somatostatin

Daniel Hoyer[1], Jacques Epelbaum[2] and Cécile Viollet[2]
[1]Department of Pharmacology and Therapeutics, School of Biomedical Sciences, Faculty of Medicine, Dentistry and Health Sciences, The University of Melbourne, Parkville, VIC, Australia
[2]Faculté de Médecine, Centre de Psychiatrie et Neuroscience, UMR 894, Université Paris Descartes, Paris, France

Synonyms

Somatotropin release-inhibiting factor; SRIF

Definition

Somatostatin is a cyclic peptide widely expressed throughout the central nervous system (CNS), in endocrine tissues, and in the gastrointestinal tract (GIT). The 14 amino acid peptide, somatostatin (SRIF-14), was first identified and isolated from ovine hypothalamic extracts by Guillemin and collaborators (Brazeau et al. 1973) during their search of GHRH hormone/growth hormone (GH)-releasing hormone, instead they identified SRIF as a potent inhibitor of growth hormone release from the pituitary. Subsequently, a longer N-terminally extended form (SRIF-28) was identified in the periphery, both deriving from the same prepropeptide.

Pharmacological Properties

Introduction

About 20 years later, a rat neuropeptide with strong homology to SRIF was identified, cloned, and named cortistatin (CST) due to its supposed brain selective expression; however, CST is also present in the periphery. In rodents, CST is a tetradecapeptide (CST-14), sharing up to 11 amino acids with SRIF, whereas the human homolog is a heptadecapeptide (CST-17). Similar to SRIF, a longer isoform of CST has also been identified, i.e., CST-29. Both peptides (SRIF and CST) are produced as prepropeptides, which are then processed to the final forms by peptide cleavage; it is strongly suggested that SRIF and CST are the products of gene duplication (De Lecea 2008; Hoyer et al. 1995). Finally, neuronostatin, a somatostatin gene-derived peptide isolated from gut, has recently been reported to be expressed in various brain regions and involved in regulating depressive behavior and nociception (Yang et al. 2011a, b, 2012), but activation of SRIF receptors does not seem to mediate its reported neuroendocrine effects. Although there is evidence for coexpression with SRIF, neuronostatin probably acts via the melanocortin system and orphan receptors, possibly GPR 107. SRIF exerts a wide range of biological actions, including inhibition of secretion of growth hormone, insulin, glucagon, and gastrin as well as other hormones secreted by the pituitary and the GIT. SRIF also acts as a neuromodulator in the CNS and, in addition, has marked antiproliferative effects on a wide range of cancer cells (Viollet et al. 2008). Although various physiological parameters, including transitions between sleep phases, memory and memory consolidation, control of emotion, locomotor activity, and sensory processes (Martel et al. 2012), respond in an apparently peptide-specific manner to SRIF and CST, both peptides in the short and long forms have nanomolar affinity for each known SRIF receptor subtype (sst_1–sst_5), The existence of specific CST receptors has not been demonstrated so far in spite of major efforts (Siehler et al. 2008). Clinically, SRIF receptor modulation is targeted primarily in the endocrine and gastrointestinal sphere, especially in a number of gastroenteropancreatic (GEP) cancers, although preclinical evidence points at a number of other diseases/conditions, such as inflammation, pain, migraine, epilepsy, additional cancers, and neuropsychiatric disorders such as depression and Alzheimer's disease (Viollet et al. 2008).

SRIF and CST Receptors

SRIF and CST act via a family of G-protein-coupled receptors (GPCRs); it was suggested initially that there are two receptors, based on pharmacology and distribution studies; receptors were distinguished by their high affinity or low affinity for octreotide (SMS-201-995); however, in the early 1990s, at least five receptors (sst_1–sst_5) have been cloned and characterized from various species. Sequence homology is 39–57 % among the five subtypes, each being highly conserved across species. Based on structural and operational features, they are divided into two groups: SRIF-1 (sst_2, sst_3, and sst_5 receptors) and SRIF-2 (sst_1 and sst_4 receptors), distinguished by high affinity for small analogs such as octreotide, seglitide (MK-678), and lanreotide (BIM-23014) for the SRIF-1 group. They have nanomolar affinities for SRIF and CST and will be generically referred to as SRIF receptors here. In humans, SRIF receptors are encoded by five nonallelic genes, located on chromosomes 14, 17, 22, 20, and 16, respectively. Genes coding for sst_1, sst_3, sst_4, and sst_5 are intronless, whereas in rodents the sst_2 gene contains three introns, which result in the generation of two receptor splice variants, sst_{2A} and sst_{2B} (Hoyer et al. 1995). Studies utilizing subtype selective SRIF analogs in both in vivo and in vitro experiments demonstrate that sst_2 receptors are the major player in the SRIF receptor family with broad inhibitory effects on the endocrine secretion, e.g., growth hormone, insulin, glucagon, gastrin, cholecystokinin, vasoactive intestinal peptide, and secretin, as well as the exocrine secretion, e.g., gastric acid, intestinal fluid, and pancreatic enzymes. sst_2 receptors also seem to play a major role in various forms of GEP cancers, in epilepsy, and pain. The sst_1 receptor may function as an autoreceptor in the basal ganglia, the hypothalamus, and the eye and possibly in the hippocampus. It is also involved in regulating insulin secretion (Thermos et al. 2006). sst_3 receptors are enigmatically localized to neuronal cilia, and sst_3 antagonists have marked behavioral effects (Viollet et al. 2008; Einstein et al. 2010). sst_4 receptors are highly expressed in the cortex and the hippocampus and in the lung, where their role still remains to be defined; in the mouse they modulate epileptic activity whereas in the rat it seems that this effect is largely assigned to sst_2 receptors (Moneta et al. 2002; Aourz et al. 2011). Hippocampal sst_4 also appears to be involved in the selection of memory strategies, switching from the use of hippocampus-based multiple associations to the use of simple dorsal striatum-based behavioral responses (Gastambide et al. 2009). sst_5 receptors mediate the inhibition of insulin release from the pancreatic b-cells in addition to regulating growth hormone release (Viollet et al. 2008). Currently, octreotide and lanreotide are in clinical use for the treatment of acromegaly, diarrhea, and various GEP tumors, whereas [^{111}In] penteotride is used for whole body tumor imaging. Other SRIF receptor agonists are in development/registered for additional indications, such as Cushing's disease (SOM230, Pasireotide). CST and SRIF appear to produce different effects both in the brain and in the periphery, especially in the immune system. Due to these differences, the existence of CST-specific receptors has been suggested: MgrX2 and GHS-R1a are two unrelated orphan receptors reported to have affinity for CST; however, these reports have not received positive confirmation. On the other hand, there is convincing evidence for CST and SRIF to act with similar high affinity and efficacy at the known sst_1–sst_5 receptors (De Lecea 2008; Siehler et al. 2008). Since the existence of specific CST receptors is still an open question, we will generically refer to SRIF receptors here.

SRIF and SRIF Receptors in the Brain

Two types of somatostatinergic neurons exist in the CNS: long-projecting neurons and short GABAergic interneurons. SRIF neurons are primarily located in the hypothalamus, hippocampus, striatum, cerebral cortex, amygdala, preoptic area, olfactory bulb, and brainstem. Concerning receptor mapping, data from receptor autoradiography, RT-PCR, in situ hybridization, and immunocytochemistry studies and receptor knockout/LacZ knockin rodent models allow to state the following: all five SRIF receptors are present in the brain, sst_2 being the most prominent whereas sst_5 shows the lowest expression levels. sst_2 is often coexpressed with other given subtypes in a tissue-specific way, and the functional meaning of such differential expression repertoires is currently under study. In rodents and humans, the olfactory bulb expresses sst_2, sst_3, and sst_4, whereas the anterior olfactory nucleus does express primarily sst_2, the neocortex, sst_2, and sst_4; in the hippocampus, DG and CA3 region express sst_2 whereas the CA1 region displays both sst_2 and sst_4; in the amygdala and locus coeruleus, sst_2 expression is largely predominant, whereas the hypothalamus expresses sst_1 and sst_2 receptors. sst_5 receptors are very scarce in the brain. The sst_3 receptor is peculiar, as its mRNA is highly detected in several brain regions; however, the sst_3 protein does not distribute in a somatodendritic fashion, as the other receptors, but is restricted to somatic neuronal cilia, whose function remains elusive (see Viollet et al. 2008). Somatostatin and its receptors (sst_1, sst_2, and sst_3) are also transiently expressed concomitantly during early sensory and brain development, suggesting that they may participate in ontogenetic processes.

The Somatostatin System in Health and Disease

Besides its neuroendocrine functions, somatostatin and its receptors have neuromodulatory roles, influencing motor activity, sleep, sensory processes, and cognitive functions. SRIF and its receptors are altered in affective and neurological disorders, e.g., depression, migraine, pain, epilepsy, and Alzheimer's disease. SRIF acts as a neuromodulator, e.g., of gamma-aminobutyric acid ("▶ GABA"), glutamate, and acetylcholine, and has both pro- and anticonvulsant activity that are potentially relevant to the neurobiology of affective disorders. A consistent alteration observed in depression is the state-dependent decrease of CSF SRIF (Roca et al. 1999). However, such a decrease is seen also in other patient populations where cognition is or may be impaired, such as schizophrenia, Alzheimer's dementia, drug refractory epilepsy, and multiple sclerosis. The SRIF decrease in affective illness and active multiple sclerosis has been considered state dependent, in that levels normalize with recovery from the acute episode. On the other hand, increased SRIF levels have been reported in psychiatric illness where cognition is accelerated, as in mania and ruminative or obsessive-compulsive disorder (Viollet et al. 2008; Martel et al. 2012; Lin and Sibille 2013). Finally, acromegalic patients treated with somatostatin agonists are largely devoid of migraine attacks.

Which Role for the Various SRIF Receptors?

sst_2 receptors: Somatostatin induces an inwardly rectifying K^+ current in almost all projection neurons of the lateral amygdala: these effects are blocked by an sst_2 antagonist or mimicked by a sst_2 agonist. A role for sst_1, sst_3, or sst_4 receptors was ruled out since selective agonists were ineffective. Preclinical data suggest a role of somatostatin neurons in the central nucleus of the amygdala in rats in fear, since octreotide applied locally on caudal pontine reticular nucleus neurons blocks fear potentiation and induces an anxiolytic effect (Yeung and Treit 2012). These cellular effects support anxiolytic as well as anticonvulsant and antiepileptogenic actions of somatostatin or analogs in the amygdala (Fendt et al. 1996). Somatostatin also plays a critical role in the acquisition of contextual fear memory but does not tone fear learning, which supports a role for SRIF in hippocampal synaptic plasticity in processing contextual information. These data are in line with an anxiety phenotype presented by the sst_2 KO mouse, which is also impaired with respect to memory. SRIF modulates glutamate release in the striatum, an effect that is abolished in sst_2 receptor KOs. The sst_2 receptor modulates neprilysin, which affects amyloid $A\beta_{42}$ metabolism. SRIF KO mice have higher concentrations of the $A\beta_{42}$ peptide, accumulation of which leads to late-onset sporadic Alzheimer's disease. sst_2 and sst_4 receptors also seem to interact positively on the phosphorylation of Tau proteins at the Ser 262 site, a site modified in Alzheimer's disease. On the other hand, 3 and 6 h after middle cerebral artery occlusion (MCAO) in the rat, sst_2 receptors were internalized excessively in cerebrocortical neurons adjacent to the infarct, and sst_2-deficient mice exhibit a 40 % reduction of infarct size after permanent distal MCAO and a 63 % reduction after transient proximal MCAO as compared to wild-type animals. Thus, the activation of sst_2 receptors by an endogenous ligand after focal ischemia may contribute to increased sst_2 gene expression and postischemic neurodegeneration (see Viollet et al. 2008).

sst_1 receptors: sst_1 receptors are present in the brain, retina, neuroendocrine cells, endothelial cells, and various human tumors. They are involved in the intrahypothalamic regulation of growth hormone (GH) secretion and modulate somatostatin release in basal ganglia. We have proposed an autoreceptor role for sst_1 receptors located on somatostatin neurons in the hypothalamus, basal ganglia, retina, and possibly hippocampus. Thus, sst_1-selective analogs may play a role in various diseases, such as retinal and endocrine dysfunctions, cancer, and neuropsychiatric disorders. We have reported NVP-SRA880, an sst_1 antagonist to promote social interactions, reduce aggressive behavior, and stimulate learning. SRA880 reduced contextual/fear conditioning without affecting cue-elicited freezing in rats (Thermos et al. 2006). Further, sst_1-selective analogs have been shown to mimic the inhibitory effect of SRIF on GH secreting pituitary tumors; in medullary thyroid carcinoma, calcitonin secretion/gene expression is inhibited by sst_1-selective agonists. Finally, sst_1-selective agonists inhibit endothelial activities, suggesting utility for sst_1-selective agonists in angiogenesis.

sst_3 receptors: A potential role for sst_3 has been proposed in hippocampal dependent memory. sst_3 null mutant mice showed a severe impairment in their ability to recall previously learned spatial information. NVP-ACQ090, an orally active somatostatin sst_3 receptor antagonist, has marked sociotropic effects in intruder mice, shows sociotropic effects in aggressive resident mice, and increases the social exploration of an intruder toward a nonaggressive resident rat. ACQ090 has anxiolytic-like activity in stress-induced hyperthermia and pronounced antidepressant-like effects in both the modified rat forced swim test and in olfactory bulbectomized rats. Thus, sst_3 receptor blockade has profound central effects in animal models for neurological and psychiatric disorders.

sst_4 receptors: There is evidence in the mouse that the main role in mediating the antiepileptic effects of SRIF is by the sst_4 in conjunction with sst_2 receptor, whereas in the rat only sst_2 is involved. Unprovoked seizures are observed in the sst_4 KO. The sst_4 receptor couples to the K^+ M-current (IM, Kv7), which is an important regulator of cortical excitability; mutations in these channels cause a seizure disorder in humans. SRIF augments IM in hippocampal CA1 pyramidal neurons. When seizures were induced by a systemic injection of kainate, only sst_4 knockouts showed an increase in seizure sensitivity. sst_2 and sst_4 appear to mediate the majority of SRIF inhibition of epileptiform activity in CA1. sst_4 receptors could therefore be an important novel target for developing new antiepileptic and antiepileptogenic drugs, but it remains to be seen whether this applies to humans. sst_4 and sst_2 receptors can interact in a cooperative or in a competitive manner. Thus, when applied in the hippocampus, L-803087, a selective sst_4 receptor agonist, doubled seizure activity and facilitated AMPA-mediated synaptic responses in wild-type mice on average, and this effect was blocked by octreotide. However, the sst_4 agonist was no longer active in sst_2 KO mice.

sst_5 receptors: Little is known about sst_5 receptors in the brain, which seem to express very low sst_5 levels, although the LacZ expression of the sst_5 KO looks very high in the hippocampus. Brain penetrant selective ligands are essentially missing or have not produced any remarkable central effect. Peripherally, sst_5 receptors modulate insulin release, and various other endocrine effects appear to be mediated by hypothalamic sst_5 receptors.

Outlook

The involvement of SRIF and its receptors in neurological and psychiatric disorders is still largely circumstantial, primarily since selective and brain penetrating ligands have not reached the clinic yet. However, preclinical evidence and changes in SRIF and receptor levels in various diseases strongly suggest that modulating SRIF receptor activity will have clinical relevance in neuropsychiatric diseases.

Cross-References

▶ Somatostatin Receptors

References

Aourz N, De Bundel D, Stragier B, Clinckers R, Portelli J, Michotte Y, Smolders I (2011) Rat hippocampal somatostatin sst3 and sst4 receptors mediate anticonvulsive effects in vivo: indications of functional interactions with sst2 receptors. Neuropharmacology 61:1327–1333

Brazeau P, Vale W, Burgus R, Ling N, Butcher M, Rivier J, Guillemin R (1973) Hypothalamic polypeptide that inhibits secretion of immunoreactive pituitary growth-hormone. Science 179:77–79

De Lecea L (2008) Cortistatin – functions in the central nervous system. Mol Cell Endocrinol 286:88–95

Einstein EB, Patterson CA, Hon BJ, Regan KA, Reddi J, Melnikoff DE, Mateer MJ, Schulz S, Johnson BN, Tallent MK (2010) Somatostatin signaling in neuronal cilia is critical for object recognition memory. J Neurosci 30:4306–4314

Fendt M, Koch M, Schnitzler HU (1996) Somatostatin in the pontine reticular formation modulates fear potentiation of the acoustic startle response: an anatomical, electrophysiological, and behavioral study. J Neurosci 16:3097–3103

Gastambide F, Viollet C, Lepousez G, Epelbaum J, Guillou JL (2009) Hippocampal SSTR4 somatostatin receptors control the selection of memory strategies. Psychopharmacology (Berl) 202:153–163

Hoyer D, Bell GI, Berelowitz M, Epelbaum J, Feniuk W, Humphrey PPA, Ocarroll AM, Patel YC, Schonbrunn A, Taylor JE, Reisine T (1995) Classification and nomenclature of somatostatin receptors. Trends Pharmacol Sci 16:86–88

Lin LC, Sibille E (2013) Reduced brain somatostatin in mood disorders: a common pathophysiological substrate and drug target? Front Pharmacol 4:110

Martel G, Dutar P, Epelbaum J, Viollet C (2012) Somatostatinergic systems: an update on brain functions in normal and pathological aging. Front Endocrinol 3:154

Moneta D, Richichi C, Aliprandi M, Dournaud P, Dutar P, Billard JM, Carlo AS, Viollet C, Hannon JP, Fehlmann D, Nunn C, Hoyer D, Epelbaum J, Vezzani A (2002) Somatostatin receptor subtypes 2 and 4 affect seizure susceptibility and hippocampal excitatory neurotransmission in mice. Eur J Neurosci 16:843–849

Roca CA, Su T-P, Elpern S, Mcfarland H, Rubinow DR (1999) Cerebrospinal fluid somatostatin, mood, and cognition in multiple sclerosis. Biol Psychiatry 46:551–556

Siehler S, Nunn C, Hannon J, Feuerbach D, Hoyer D (2008) Pharmacological profile of somatostatin and cortistatin receptors. Mol Cell Endocrinol 286:26–34

Thermos K, Bagnoli P, Epelbaum J, Hoyer D (2006) The somatostatin sst(1) receptor: an autoreceptor for somatostatin in brain and retina? Pharmacol Ther 110:455–464

Viollet C, Lepousez G, Loudes C, Videau C, Simon A, Epelbaum J (2008) Somatostatinergic systems in brain: networks and functions. Mol Cell Endocrinol 286:75–87

Yang AM, Ge WW, Lu SS, Yang SB, Su SF, Mi ZY, Chen Q (2011a) Central administration of neuronostatin induces antinociception in mice. Peptides 32:1893–1901

Yang AM, Ji YK, Su SF, Yang SB, Lu SS, Mi ZY, Yang QZ, Chen Q (2011b) Intracerebroventricular administration of neuronostatin induces depression-like effect in forced swim test of mice. Peptides 32:1948–1952

Yang SB, Yang AM, Su SF, Wang HH, Wang NB, Chen Q (2012) Neuronostatin induces hyperalgesia in formalin test in mice. Neuroscience Letters 506:126–130

Yeung M, Treit D (2012) The anxiolytic effects of somatostatin following intra-septal and intra-amygdalar microinfusions are reversed by the selective sst2 antagonist PRL2903. Pharmacol Biochem Behav 101:88–92. doi:10.1016/j.pbb.2011.12.012, Epub 2011 Dec 22

Somatostatin Receptors

Synonyms

sst_1; sst_2; sst_3; sst_4; sst_5

Definition

Five SRIF receptors (sst_1–sst_5) have been cloned from various species. Sequence homology is 39–57 % among the five subtypes; each is highly conserved across species. They have nanomolar affinity for SRIF and CST. Human SRIF receptors are encoded by five nonallelic genes, located on chromosomes 14, 17, 22, 20, and 16, respectively. Genes coding for sst_1, sst_3, sst_4, and sst_5 are intronless, whereas in rodents the gene for sst_2 contains three introns, which result in the generation of two receptor splice variants, sst_{2A} and sst_{2B}. SRIF receptors are GPCRs; they couple to G_o/G_i proteins and modulate cyclic AMP production and other transduction pathways.

Source Localization Techniques

Synonyms

Source analysis

Definition

Source localization techniques refer to a set of mathematical techniques for identifying the locus or loci of neural activity that give rise to a particular ERP (event-related potential) voltage distribution measured on the surface of the scalp. By making assumptions about the physical properties of the skull, brain, and underlying active neural populations, these methods attempt to constrain the number of possible source configurations that can explain the observed scalp distribution. Preexisting knowledge about the brain and further assumptions, for example, about the number of active sources, are needed to arrive at a plausible solution. An important shortcoming of source localization techniques is that there is no means of quantifying the likelihood that a solution is correct.

Cross-References

▶ Event-Related Potential
▶ Inverse Problem

Spatial Delayed Alternation

Definition

Classical paradigm for measuring short-term (often called "working") memory in rodents generally employing a T-maze and basing the measure of memory on the well-known tendency of rats to spontaneously alternate their spatial choices. The memory load can be increased by lengthening the delay between the first and second trials. Analogous but not equivalent to spatial delayed response task in nonhuman primates.

Spatial Learning in Animals

Stan B. Floresco
Department of Psychology, University of British Columbia, Vancouver, Canada

Synonyms

Place learning; Spatial navigation

Definition

Spatial learning refers to the process through which animals encode information about their environment to facilitate navigation through space and recall the location of motivationally relevant stimuli. This form of learning is critically dependent on the integrity of the hippocampus, although surrounding regions of the temporal cortex and certain forebrain structures also play a role in these processes. It is generally believed that spatial learning entails encoding of the locations of cues relative to the position of other cues in a particular environment that leads to the formation of a cognitive map of an individual's surroundings. Thus, during spatial learning, animals use allocentric spatial cues to navigate in space, keeping track of their position relative to other distal stimuli. Support for this notion comes from the finding of place cells in the hippocampus, where different groups of neurons in this region display consistent changes in firing when an animal is oriented in a particular location in space. Moreover, changing the arrangement of distal spatial cues in an arena reorganizes the patterns of activity in these place cells to match the new orientation of stimuli in the environment. In rodents, one of the most common means for assessing spatial learning is with different types of maze tasks, where rats or mice use allocentric cues to guide an escape response (e.g., finding a hidden platform in a water maze task) or to locate food (e.g., on a radial arm maze). The majority of studies that have investigated the psychopharmacology of spatial learning have used these types of tasks. An important caveat is that drugs that alter performance of these tasks do not appear to exert a selective influence on spatial learning and memory, as they typically can affect other forms of learning independent of the hippocampus. As such, preclinical tests of spatial learning in rodents, either in intact animals or those subjected to treatments that impair spatial learning, are often used as an initial assay to investigate the effects of potential cognition-enhancing drugs.

Impact of Psychoactive Drugs

Excitatory Amino Acids

Drugs that disrupt glutamatergic transmission, particularly those which block the n-methyl-D-aspartate (NMDA) receptor subtype, severely interfere with spatial learning. The systemic administration of either competitive (e.g., CPP) or noncompetitive (MK-801, phencyclidine) NMDA antagonists impairs learning of spatially cued escape response in a water maze task (reviewed by McNamara and Skelton 1993). Similarly, administration of these compounds also impairs search behavior guided by spatial memory on appetitively motivated tasks such as the radial arm maze. The effects of these treatments seem largely attributable to the blockade of NMDA receptors in the hippocampus, as local administration of these compounds also impairs spatial learning. However, the administration of NMDA receptor antagonists after training has a substantially blunted effect, suggesting that hippocampal NMDA receptors are required for the acquisition and perhaps the initial consolidation of hippocampus-dependent memory, but not for its maintenance. These effects do not appear attributable to disruptions in motor or motivational processes because they do not interfere with the ability of animals to learn the approach at a single proximal cue, when, for example, rats must swim to a visible platform in a water maze. In contrast to the abovementioned findings, blockade of non-NMDA (e.g., AMPA) receptors impairs both the acquisition and retrieval of spatial information. In this regard, drugs that potentiate currents mediated by AMPA-type glutamate receptors (ampakines) have shown efficacy in improving spatial and other forms of learning in animal models (Arai and Kessler 2007). The effects of drugs acting on metabotropic glutamate receptors have received less attention, although antagonists at mGlur1 and five sites have been reported to impair the acquisition of a water maze task.

The ability of NMDA antagonists to impair spatial learning has been proposed to be related to disruptions in certain forms of synaptic plasticity in the hippocampus, given that compounds that impair spatial learning also impede the induction of long-term potentiation (LTP) in this region. Notably, blockade of NMDA receptors does not affect maintenance of LTP. Similarly, systemic blockade of these receptors with competitive antagonists induces a selective disruption of the formation of place cell activity in the hippocampus. Place fields that formed de novo in the presence of NMDA antagonists are unstable, but these treatments do not affect the maintenance of previously established place fields, although they do disrupt the expansion of place fields that occurs with repeated exposure to an environment. These findings are in keeping with the dissociable effects of NMDA antagonism on learning versus retrieval of spatial memories (Nakazawa et al. 2004).

Acetylcholine

The hippocampus receives dense cholinergic enervation from the septum that mediates theta frequency oscillatory activity, thought to be important for the exploration of novel environments. The blockade of muscarinic cholinergic receptors with nonselective antagonists such as scopolamine or atropine severely disrupts both spatial learning and memory in a manner independent of the effects of these compounds on sensorimotor and procedural learning (McNamara and Skelton 1993). The use of these compounds in combination with spatial learning tasks has become a popular preclinical animal model for cognitive dysfunction for which to test the effects of potential cognitive enhancers. Perturbations in spatial learning induced by these drugs appear to be due to the blockade of postsynaptic M_1 receptors, as similar effects have been observed following peripheral or central administration of M_1 antagonists such as pirenzepine. Conversely, in some instances, enhancing cholinergic transmission can induce a beneficial effect on spatial learning. For example, blockade of M_2 autoreceptors with drugs such as methoctramine or AFDX 116 increases acetylcholine release and improves learning on tests of spatial abilities using different maze paradigms. Likewise, cholinesterase inhibitors ("▶ physostigmine," tacrine) also can improve spatial learning, particularly in aged animals. Drugs of this nature have been used to offset aging-related cognitive decline occurring in human patients. Nicotinic cholinergic receptors have also been implicated in spatial learning. Nicotine or receptor subtype-selective agonists improve spatial abilities in preclinical animal models. Current drug discovery research aimed at developing novel pro-cognitive compounds has focused on the alpha-7 subunit of the nicotinic receptor as a potential target. Genetically modified mice lacking this subunit display impairments in spatial and other forms of learning, whereas compounds, which stimulate receptors that include this subunit, can enhance spatial learning and memory. The development of these compounds may prove beneficial for treating cognitive deficits observed in a number of disorders, including Alzheimer's and schizophrenia.

Although it is well established that the blockade of cholinergic transmission can impair spatial learning, it remains unclear where these compounds may be exerting their effects. The local administration of muscarinic antagonists into the hippocampus interferes with place cell firing. However, the deafferentation of cholinergic inputs to the hippocampus via 192 IgG-saporin lesions of the septum does not reliably impair spatial learning using water or radial maze tasks (Baxter 2001). This would suggest that impairments in spatial learning induced by the systemic administration of cholinergic antagonists occur via the blockade of extra-hippocampal cholinergic receptors, possibly in different regions of the cerebral cortex. Notably, cholinergic transmission has been heavily implicated in attentional processing. It has been proposed that acetylcholine serves to enhance the influence of feedforward afferent input to the cortex while decreasing background activity by suppressing excitatory feedback connections within cortical circuits. By enhancing the response to sensory input, high levels of acetylcholine enhance attention to sensory stimuli in the environment and enhance the encoding of memory for specific stimuli (Hasselmo 2006). Thus, even though manipulations of cholinergic transmissions can influence acquisition and performance of tasks requiring the use of spatial cues to guide behavior, the effects of these compounds may not be due to a specific effect on the encoding or retrieval of spatial memories per se.

Monoamines

The systemic administration of dopamine antagonists acting at D_1 (SCH 23390) or D_2 ("▶ haloperidol," "▶ sulpiride") receptors has been reported to impair spatial learning assessed with water maze tasks. In some instances, these effects were accompanied by motoric deficits as well, whereas other reports have shown that lower doses of these drugs that do not induce motor impairments are effective at impairing spatial learning. The hippocampus receives dopaminergic innervation, and studies combining behavioral and electrophysiological methodologies point to a role for D1 receptors in facilitating hippocampal synaptic plasticity, which may enhance the formation of novel spatial memories. Dopamine has been shown to modulate acetylcholine release, which may contribute to some of the effects of pharmacological manipulation of dopamine transmission on spatial learning and memory. Furthermore, dopamine receptor activity in forebrain regions (ventral striatum, "▶ prefrontal cortex") also appear to be required for some forms of spatial learning and memory, as the local infusion of dopamine antagonists into these regions can also impede acquisition and recall.

Noradrenergic inputs to the hippocampus stem exclusively from the locus coeruleus, the majority of which terminates in the dentate gyrus. Pharmacological manipulations of noradrenergic receptors, particularly the beta subtype with agonists such as isoproterenol, can modulate the formation of long-term potentiation in the hippocampus. Furthermore, increases or decreases in noradrenergic tone via blockade (atipamezole) or stimulation (dexmedetomidine) of alpha-2 adrenergic autoreceptors lead to instability of place fields in the hippocampus. Interestingly, however, the depletion of central noradrenaline with selective neurotoxins does not impair the learning of spatial tasks. Nevertheless, drugs that enhance central noradrenergic activity can improve performance on spatial memory tasks. Much of this research has focused on the contribution of noradrenergic transmission in the memory modulating effects of stress (Roozendaal et al. 2006). For example, the ability of glucocorticoid stress hormones to disrupt spatial and other forms of memory is blocked by the coadministration of beta-1 receptor adrenergic antagonists such as atenolol either systemically or directly into the hippocampus. Conversely, the posttraining activation of adrenergic receptors with adrenaline improves subsequent performance of spatial memory tasks, and these effects are blocked by the coadministration of beta-receptor antagonists. These findings suggest that under some conditions, acute increases of noradrenergic activity facilitate the consolidation of new spatial memories. A particularly interesting aspect of these findings is that posttraining infusions of these compounds into the basolateral amygdala (a region that normally does not contribute to spatial learning) also enhance the consolidation of spatial memories (as well as other forms of learning; Ferry et al. 1999). Enhancements in learning induced by the activation of amygdala adrenoreceptors are thought to be part of a neural mechanism through which stressful or emotionally charged events augment the encoding of new memories.

The hippocampus receives considerable serotonergic innervation and contains a high density of 5-HT receptors, yet global depletions of brain serotonin typically do not impair spatial learning. Nevertheless, the pharmacological manipulation of different 5-HT receptors has been shown to affect spatial learning processes. In general, the blockade of different 5-HT receptors tends to improve spatial learning, whereas the administration of relatively selective 5-HT agonists induces detrimental effects. For example, systemic or intra-hippocampal administration agonists active at either 5-HT_{1A} (8-OH-DPAT) or 1B (anpirtoline) receptors generally lead to impairments in spatial learning in the water maze (Ogren et al. 2008). Furthermore, the administration of the 5-HT_{1A} antagonist WAY 100635 ameliorates impairments in spatial memory induced by muscarinic antagonism with scopolamine, indicated that 5-HT_{1A} receptor agents may modify spatial memory through interactions with the cholinergic system. Similar improvements in spatial learning have been observed following intra-hippocampal blockade of $5\text{-HT}_{2A/2C}$ receptors with ritanserin or systemic blockade of 5-HT_6 receptors (SB-271046). Thus, it appears that the blockade of multiple 5-HT receptors may actually exert a beneficial effect on spatial learning. Accordingly, identifying novel 5-HT antagonists may prove to be a fruitful strategy in the development of cognitive enhancing drugs.

Cross-References

▶ Beta-Adrenoceptor Antagonists
▶ Excitatory Amino Acids and Their Antagonists
▶ Long-Term Potentiation and Memory
▶ Nicotinic Agonists and Antagonists

References

Arai AC, Kessler M (2007) Pharmacology of ampakine modulators: from AMPA receptors to synapses and behavior. Curr Drug Targets 8:583–602

Baxter MG (2001) Effects of selective immunotoxic lesions on learning and memory. Methods Mol Biol 166:249–665

Ferry B, Roozendaal B, McGaugh JL (1999) Role of norepinephrine in mediating stress hormone regulation of long-term memory storage: a critical involvement of the amygdala. Biol Psychiatry 46:1140–1152

Hasselmo ME (2006) The role of acetylcholine in learning and memory. Curr Opin Neurobiol 16:710–715

McNamara RK, Skelton RW (1993) The neuropharmacological and neurochemical basis of place learning in the Morris water maze. Brain Res Rev 18:33–49

Nakazawa K, McHugh TJ, Wilson MA, Tonegawa S (2004) NMDA receptors, place cells and hippocampal spatial memory. Nat Rev Neurosci 5:361–372

Ogren SO, Eriksson TM, Elvander-Tottie E, D'Addario C, Ekström JC, Svenningsson P, Meister B, Kehr J, Stiedl O (2008) The role of 5-HT(1A) receptors in learning and memory. Behav Brain Res 195:54–77

Roozendaal B, Okuda S, de Quervain DJ, McGaugh JL (2006) Glucocorticoids interact with emotion-induced noradrenergic activation in influencing different memory functions. Neuroscience 138:901–910

Spatial Memory

Definition

Spatial memory refers to memory for the location of objects and self-orientation within the environment.

Cross-References

▶ Spatial Memory in Humans

Spatial Memory in Humans

Mitul A. Mehta
Centre for Neuroimaging Sciences, Institute of Psychiatry at King's College London, London, UK

Synonyms

Spatial memory is the term often used to describe spatial navigation and encompasses egocentric and allocentric memories

Definition

Spatial memory describes information storage and retrieval required for identification and navigation of proximal or distal space. This is distinct from spatial working memory which refers to active representations stored and manipulated over seconds. Two main frames of reference have been described: egocentric, which is related directly to the observer, and allocentric, which is dependent on the relational position of objects in space. Routes depend more on egocentric frames of reference, whereas maps are more flexible to landmark changes and thus depend more on allocentric frames of reference. Although often discussed separately, an emerging view is that both egocentric and allocentric spatial memories are coded, but these may interact and depend on interacting brain regions. Spatial memories can be representations of salient cues for navigation or be more detailed representations, including topography, allowing reexperiencing of the environment (Moscovitch et al. 2005). A useful analogy has been proposed such that schematic representations of topography correspond to semantic memory and detailed representations correspond to episodic, autobiographical memory. Classic tasks of spatial memory in experimental animals include the Morris water maze, Olton maze, and radial arm maze (although these latter tasks can incorporate a large working memory component). In humans, both static (route reporting and drawing) and real-world tasks can be used, and while being informative, they are severely limited in interpretability and control of motivation and strategy. Emerging use of virtual reality provides a mechanism to integrate realistic representations of space and environment with controlled laboratory assessments, including functional brain imaging, which requires the participant to remain still during the course of an experiment.

Functional Neuroanatomy

The human visual system comprises two processing streams, a dorsal "where" system important for spatial navigation and a ventral "what" system for object recognition. Goodale and Milner (1992) emphasized the transformations that may be required for information in each stream, relabeling the dorsal stream as a "how" system. This system was proposed to be involved in action preparation via transformation of visual information using an egocentric frame of reference, whereas the ventral system processed transformation for long-lasting representations using multiple frames of reference. Functional brain imaging, including studies using virtual reality, largely supports this general distinction. For example, Aguirre and D'Esposito (1997) initially trained subjects on learning the location of 16 objects in a virtual town. Functional MRI (fMRI) scanning was performed during two conditions in which they judged (1) if the appearance of a location within the town matched the name of a location, the appearance condition, and (2) the direction of travel from a displayed location in the town to a given new location, the position condition. When contrasted against each other, the appearance condition showed greater activation in the fusiform and parahippocampal gyrus as well as occipital regions, whereas the position condition showed greater activation in more dorsal areas (i.e., superior and inferior parietal lobes). This study was performed in only four volunteers, but in general there is an agreement that the hippocampus and medial temporal lobe structures are required for the acquisition of spatial memories in humans (Parslow et al. 2005), and while fewer studies have examined retention and retrieval of spatial memories, the results of the fMRI study fit with studies showing that detailed episodic spatial memories cannot survive damage to the hippocampus/medial temporal lobe (Moscovitch et al. 2005) and retrieval of spatial memories of taxi drivers in London activates the hippocampus/parahippocampal gyrus (Maguire et al. 1997). An alternate view is that the hippocampus may not have a selective role in allocentric spatial memory. For example, in a group of six patients with hippocampal damage performing an image-location memory task, impairments were dependent on the number of items rather than the degree of rotation between study and test phases (Shrager et al. 2007), although the use of smooth rotation might have favored egocentric mental rotation strategies (Burgess 2008).

Overall, these studies align with the presence of hippocampal place cells in the rodent, monkey, and human brain. In humans, these cells were first described in seven patients with intractable epilepsy undergoing presurgical invasive monitoring, who played the role of a taxi driver in a spatial navigation game (Ekstrom et al. 2003). Cells that responded to specific locations were primarily located in the hippocampus, whereas cells that responded to views of landmarks were primarily located in the parahippocampal region. Indeed, forming memories of places, or landmarks or associating objects with particular locations, appears to require the parahippocampal cortex. Other extrahippocampal regions also important for spatial memory task performance include the posterior parietal cortex and striatum. The former is necessary in the representation of spatial information in terms of egocentric coordinates, allowing reaching and movement plans to be formulated as well as egocentric imagery. The striatum is generally thought to have a role in the learning of the procedural aspects of tasks and may have dissociable roles in egocentric and allocentric spatial memory consolidation (De Leonibus et al. 2005). Using a task of object exploration with animals entering the same location (egocentric reference frame) or different locations (allocentric reference frame), focal injection of an NMDA receptor antagonist into the dorsal striatum impaired consolidation of the egocentric procedure, whereas intra-accumbens administration impaired both procedures.

Impact of Psychoactive Drugs

In experimental animals, the neurobiology and psychopharmacology of spatial memory have been extensively studied, dominated by the use of the Morris water maze (McNamara and Skelton 1993). The neurotransmitter and molecular mechanisms of medial temporal lobe and cortical structures involved in spatial memory implicate specific drug systems in the accurate acquisition and expression of these memories. Focusing on the hippocampus as the most studied of these regions implicates glutamatergic neurotransmission through AMPA and NMDA receptors, as well as modulatory influences of the cholinergic and serotonergic inputs. Muscarinic and nicotinic cholinergic receptor blockade as well as serotonin receptor antagonism can all modulate hippocampal glutamatergic and GABAergic neurotransmission as well as the theta rhythm, which is hypothesized to allow rapid transitions between encoding and retrieval.

Glutamate Receptors

The NMDA receptor is a major target for glutamate in the brain, and its role in long-term potentiation is thought to be related to mnemonic encoding. It is no surprise that NMDA blockade using compounds such as AP5, ketamine, or MK-801 either intrahippocampal or intracerebral in rats impairs spatial learning using the Morris water maze, without affecting perceptual or motivational processes. For example, a degree of selectivity is suggested by the lack of impairment on a nonspatial discrimination task following AP5 infusion (McNamara and Skelton 1993). Metabotropic glutamate receptor agents can also modulate spatial learning. For example, an mGluR5 allosteric modulator enhances spatial learning in the rat, probably by enhancing LTP and LTD, and an mGluR2 receptor antagonist can impair spatial learning when infused into the nucleus accumbens. Although they are likely to play a role, the function of AMPA and kainate receptors in specific modulation of glutamatergic activity during spatial memory tasks is not well understood, similarly for the influence of NMDA receptor subunits. Aging, at least in mice, is known to be associated with changes in NMDA subunit expression which may contribute to memory decline with aging (Zhao et al. 2009).

Cholinergic Receptors

Studies employing lesions of the cholinergic cell groups in the basal forebrain nucleus basalis magnocellularis or medial septum, pharmacological blockade of acetylcholine receptors, age-related decreases in acetylcholine activity,

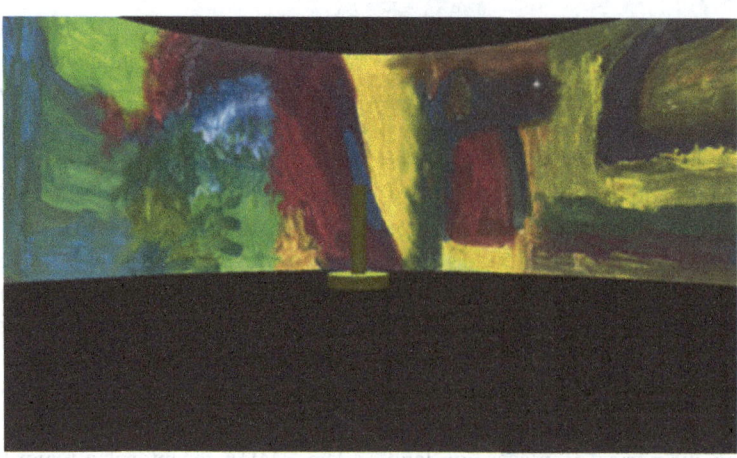

Spatial Memory in Humans, Fig. 1 A screenshot of the arena task used by Parslow et al. (2005) reproduced with kind permission from Professor Robin Morris. The goal of the task is to move toward the pole, at first with the pole visible and subsequently after removal of the pole. Changes in the starting position and background environment allow the control of the frames of reference likely to be used in spatial memory recall

and microdialysis to index acetylcholine release provide compelling converging evidence for the important role of the acetylcholine system in spatial memory. However, it is now generally accepted that cholinergic modulation can also influence visual attention, altering cue processing (e.g., Muir et al. 1993), and thus, it is important that these additional effects are taken into account. In humans, fMRI and positron emission tomography (PET) have confirmed the effects of cholinergic enhancement with physostigmine on stimulus-driven shifts in attention, which can reduce demands on working memory processes in the prefrontal cortex.

Within the hippocampus, microdialysis in the rat clearly shows a positive relationship between acetylcholine and spatial learning and memory performance, and in humans it is predicted that patients with cholinergic dysfunction (e.g., Alzheimer's disease) or volunteers administered scopolamine will be impaired on tasks of spatial memory via alteration of hippocampal activity based on the proposal that acetylcholine activity dynamics modulates hippocampal-based encoding of new information (Hasselmo 2006). Using a virtual reality-based spatial memory task based on the Morris water maze, deficits in hippocampal activation following scopolamine (0.4 mg s.c.) have been demonstrated in healthy volunteers (Antonova et al. 2011). The task requires the movement (using a trackball) to a platform in a virtual reality arena surrounded by abstract images (see Fig. 1). The memory for place is assessed by removing the platform and asking the volunteer to move to the platform location from the same starting point with different environmental cues (viewpoint-dependent or egocentric condition) or from different starting points with the same environment cues (viewpoint-independent or allocentric condition). Using a different task of spatial navigation (Gron et al. 2006), the acetylcholinesterase inhibitor and nicotinic cholinergic receptor agonist, galantamine (4 mg b.i.d. for 7 days), has been shown to modulate neural activity in a group of ten patients with mild cognitive impairment (these patients have an increased risk for developing clinically defined Alzheimer's disease). Using one-sample t-tests, galantamine was shown to decrease activity in the supplementary motor area and increase activity in a network of regions known to be associated with this task in young healthy volunteers: occipitotemporal junction, fusiform gyrus, posterior cingulate, parahippocampal gyrus, and hippocampus. While the task depended in part on spatial memory, the use of egocentric or allocentric frames of reference was not controlled, and the role of motivational and decision-making processes in the completion of such tasks is poorly understood.

Other Neurotransmitters

γ-Aminobutyric acid (GABA) receptors control the flow of excitation through the brain. Within the hippocampus, GABA receptors are categorized as $GABA_A$ and $GABA_B$, which have differential distribution and control of neuronal activity. Expression of the α_5 subunit of the $GABA_A$ receptor is enriched in pyramidal cells of the hippocampus; thus, the knockout mouse can be considered a model of hippocampal $GABA_A$ functionality. Performance is improved in the α_5 subunit knockout mouse after administration of a full inverse agonist of the α_5 subunit on the delayed match-to-place Morris water maze task, in which the platform is moved between trials. Dopamine systems have previously been associated with the learning aspects of spatial memory tasks, with degeneration of the nigrostriatal dopamine system associated with learning deficits in the Morris water maze. Long-term effects on the consolidation of aspects of spatial memory, including object displacement tasks impaired after intra-accumbens injection of dopamine D1 and D2 antagonists, as well as improvements with levodopa, highlight a role for the striatal dopamine system. However, in terms of normal aging, analysis of impaired performance on the Morris water maze with age, in relation to hippocampal receptor density, showed no associations with dopamine D1 receptors. $Serotonin_{1A}$ receptor binding sites were, however, increased in poor performers in keeping with the proposed role of $5-HT_{1A}$ receptors in encoding and consolidation of spatial memory rather than retrieval. Interestingly, $5-HT_{1B}$ knockout mice appear protected against age-related hippocampal memory decline. However, a selective role in spatial memory has been questioned by 5-HT release being associated more closely with motor behavior and feeding behavior during spatial memory tasks rather than task performance. To date, there are no studies in humans that test other agents against spatial memory tasks.

Conclusions

Spatial memory describes a collection of functions associated with spatial information processing including navigation and localizing of objects. Egocentric and allocentric frames of reference have been proposed for spatial processing but are likely to interact in many situations. Nonetheless, separable functional neuroanatomy appears plausible for egocentric (parietal lobes) and allocentric (hippocampal region/striatum) memory tasks, with the latter emerging recently in functional imaging investigations using virtual reality. In keeping with the research in experimental animals, the hippocampal and proximal medial temporal lobe structures have a central role in spatial memory processing in humans. Predictions from hippocampal neuropharmacology of important neurotransmitters in spatial memory processing imply a degree of opportunity for enhancement in human populations although this remains largely unexplored. A positive association between cholinergic activity and spatial memory in healthy volunteers (using scopolamine) and patients with mild cognitive impairment (using galantamine) is a strong indicator for the scope of behavioral modulation possible with the correct target compounds. Further work with GABAergic agents, particularly α_5 inverse agonists, and dopaminergic and serotonergic compounds is required to validate predictions from experimental animals.

Cross-References

▶ Short-Term and Working Memory in Humans

References

Aguirre GK, D'Esposito M (1997) Environmental knowledge is subserved by separable dorsal/ventral neural areas. J Neurosci 17(7):2512–2518

Antonova E, Parslow D, Brammer M, Simmons A, Williams S, Dawson G, Morris R (2011) Scopolamine disrupts hippocampal activity during allocentric spatial memory in humans: an fMRI study using a virtual reality analogue of the Morris Water Maze. J Psychopharmacol 25(9):1256–1265

Burgess N (2008) Spatial cognition and the brain. Ann N Y Acad Sci 1124:77–97

De Leonibus E, Oliverio A, Mele A (2005) A study on the role of the dorsal striatum and the nucleus accumbens in allocentric and egocentric spatial memory consolidation. Learn Mem 12(5):491–503

Ekstrom AD, Kahana MJ, Caplan JB, Fields TA, Isham EA, Newman EL, Fried I (2003) Cellular networks underlying human spatial navigation. Nature 425(6954):184–188

Goodale MA, Milner AD (1992) Separate visual pathways for perception and action. Trends Neurosci 15(1):20–25

Gron G, Brandenburg I, Wunderlich AP, Riepe MW (2006) Inhibition of hippocampal function in mild cognitive impairment: targeting the cholinergic hypothesis. Neurobiol Aging 27(1):78–87

Hasselmo ME (2006) The role of acetylcholine in learning and memory. Curr Opin Neurobiol 16(6):710–715

Maguire EA, Frackowiak RS, Frith CD (1997) Recalling routes around London: activation of the right hippocampus in taxi drivers. J Neurosci 17(18):7103–7110

McNamara RK, Skelton RW (1993) The neuropharmacological and neurochemical basis of place learning in the Morris water maze. Brain Res Brain Res Rev 18(1):33–49

Moscovitch M, Rosenbaum RS, Gilboa A, Addis DR, Westmacott R, Grady C, McAndrews MP, Levine B, Black S, Winocur G, Nadel L (2005) Functional neuroanatomy of remote episodic, semantic and spatial memory: a unified account based on multiple trace theory. J Anat 207(1):35–66

Muir JL, Page KJ, Sirinathsinghji DJ, Robbins TW, Everitt BJ (1993) Excitotoxic lesions of basal forebrain cholinergic neurons: effects on learning, memory and attention. Behav Brain Res 57(2):123–131

Parslow DM, Morris RG, Fleminger S, Rahman Q, Abrahams S, Recce M (2005) Allocentric spatial memory in humans with hippocampal lesions. Acta Psychol (Amst) 118(1–2):123–147

Shrager Y, Bayley PJ, Bontempi B, Hopkins RO, Squire LR (2007) Spatial memory and the human hippocampus. Proc Natl Acad Sci U S A 104(8):2961–2966

Zhao X, Rosenke R, Kronemann D, Brim B, Das SR, Dunah AW, Magnusson KR (2009) The effects of aging on N-methyl-D-aspartate receptor subunits in the synaptic membrane and relationships to long-term spatial memory. Neuroscience 162(4):933–945

Spatiotemporal Coincidence

Definition

Signals that arrive at the same time at the same place. A specific example is high-frequency stimulation to induce long-term potentiation by depolarizing postsynaptic synapses at the time when the presynaptic axon is depolarized.

SPECT Imaging

Ronald L. Cowan[1] and Robert Kessler[2]
[1]Departments of Psychiatry and Radiology and Radiological Sciences, Psychiatric Neuroimaging Program, Vanderbilt Addiction Center, Vanderbilt University School of Medicine, Nashville, TN, USA
[2]Neurochemical Brain Imaging and PET Neurotracer Development, University of Alabama School of Medicine, Birmingham, AL, USA

Synonyms

Single-photon emission computed tomography; Single-photon emission tomography (SPET); SPET

Definition

SPECT is a nuclear imaging method that uses a radioactive tracer (radiopharmaceutical) to measure blood flow or to label brain molecules of interest. The radioactive tracers used in SPECT produce a single photon that is detected by a camera sensitive to photon emissions (gamma camera). SPECT can be used in animal and human research and in human clinical diagnosis to noninvasively assay cerebral blood flow (as an indirect marker of neuronal activity) and cell molecular components of interest (targets), such as neurotransmitter receptors, neurotransmitter reuptake transporters, and other proteins of interest. SPECT methods and uses overlap considerably with those of positron emission tomography (PET), another nuclear imaging method widely used in psychopharmacology.

Principles and Role in Psychopharmacology

Basic Method
The fundamental principle of SPECT imaging consists of detecting, localizing, and quantifying

gamma ray emissions from radiolabeled compounds. Gamma ray emissions are produced when an unstable isotope decays, emitting a photon. The radiolabeled compound is injected intravenously into an animal or human where it circulates throughout the body and to the brain. The photons emitted by the radiolabeled compound are then detected with the gamma camera system. Figure 1 shows a schematic of a SPECT imaging method, and Fig. 2 is an example of a human SPECT imaging system with two gamma cameras that can be positioned to image an area of interest.

SPECT tracers in psychopharmacology are primarily of two types: (1) those that cross the blood–brain barrier and distribute nonspecifically in the brain (perfusion radiopharmaceuticals) and (2) those that cross the blood–brain barrier and distribute according to specific molecular targets (target molecule-binding radiopharmaceuticals).

SPECT perfusion radiopharmaceuticals (used to measure blood flow) are compounds that cross the blood–brain barrier and then distribute into the extracellular space in proportion to the blood flow to a region (Masdeu and Arbizu 2008). A successful SPECT perfusion tracer must have the correct lipophilic properties so that it crosses the blood–brain barrier easily, but it must also have no specific binding to the brain components. Once the photon counts are detected by the gamma camera system, the relative cerebral blood flow to an area can be determined. Since local neuronal activity is strongly coupled to cerebral blood flow, SPECT tracers can therefore be used as indirect indicators of changes in local neuronal activity.

Unlike SPECT perfusion radiopharmaceuticals, target molecule-binding radiopharmaceuticals must have high affinity and specificity for the target molecule. Usually, the affinity for the specific target molecule is at least tenfold greater than the concentration of the target molecule to be imaged. This requires that target molecule-binding radiopharmaceuticals have affinities for their target molecules in the low nanomolar to picomolar range. In addition, these radiotracers must have sufficient lipophilicity to allow penetration of the

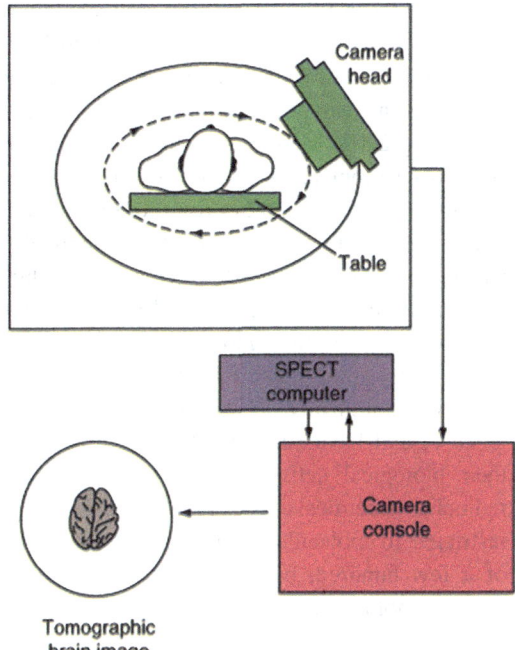

SPECT Imaging, Fig. 1 SPECT imaging system (Mettler and Guiberteau 2006, p. 26)

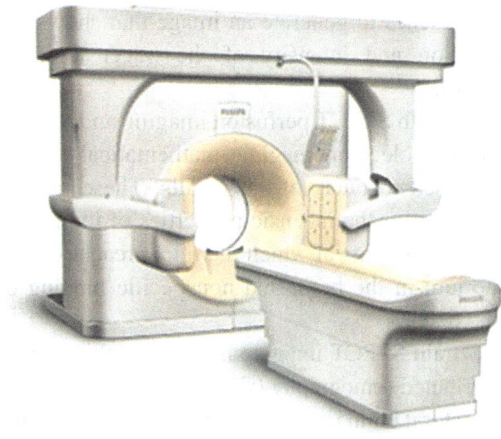

SPECT Imaging, Fig. 2 Human SPECT Scanner. A two-camera human SPECT scanner (http://www.impactscan.org/images/Philips_Precedence_SPECT-CT.jpg. Picture provided courtesy of Philips Healthcare)

blood–brain barrier, but not so high a lipophilicity as to produce high nonspecific binding.

Radiopharmaceutical Synthesis

SPECT radiopharmaceutical synthesis begins with a radionuclide. The two most widely used radionuclides in SPECT imaging are technetium (Tc) as [^{99m}Tc] and iodine (I) as [^{123}I]. SPECT radionuclides are made primarily by irradiating atoms with charged particles (Leslie and Greenberg 2003; Mettler and Guiberteau 2006). A pharmacological molecule of interest, for example, a neurotransmitter receptor or reuptake transporter binding ligand, is then chemically attached to the radionuclide to create the radiopharmaceutical. Because ^{123}I, ^{99m}Tc, and other single-photon-emitting radionuclides are relatively heavy atoms, it is difficult to attach them to small biologically active molecules, such as glucose, and have such small molecules retain their biological activity. As a result, SPECT radiopharmaceuticals for brain imaging are restricted to molecules with molecular weights of a few hundred, but cannot be too large as they will not pass the blood–brain barrier.

Image Construction

Image construction for SPECT requires the use of a specialized detection system called a gamma camera that uses hardware and software approaches to generate an image that shows the location and intensity of the tracer (Accorsi 2008).

In both SPECT perfusion imaging and SPECT target molecular imaging, mathematical models are used to interpret the results. These models account for the intrinsic properties of the tracer, such as the rate at which the tracer reaches equilibrium in the brain and nonspecific binding of the tracer.

Brain SPECT imaging is often combined with computed tomography (CT) to provide a detailed structural brain image for overlaying SPECT data.

Resolution

The resolution of SPECT imaging is determined largely by the design of the hardware and software for the gamma camera system. Modern SPECT scanners in general clinical use have resolutions of around 1 cm, but special small-animal research SPECT scans, using modified hardware and software for image construction, can achieve resolutions of less than 1 mm (Masdeau and Arbizu 2008; Spanoudaki and Ziegler 2008).

Time Course and Availability

SPECT radionuclides have relatively long half-lives; for example, the half-life for two widely used radionuclides is about 13 h for ^{123}I compounds and about 6 h for ^{99m}Tc compounds. Because of this stability, SPECT ligands can be injected up to several hours before the actual scanning procedure. Some commonly used SPECT radiopharmaceuticals and their indications are shown in Table 1. It is important to note that the availability of these ligands for human research or clinical applications is determined by the rules of each country, so that not all ligands are available in all regions.

Uses

SPECT has broad applicability in basic and clinical research, diagnosis, and treatment. Based on the principle that increased regional cerebral neuronal activity is associated with increased blood flow, SPECT perfusion studies can provide indirect information related to local neuronal activity and metabolism. SPECT perfusion studies are used to measure baseline blood flow, blood flow changes associated with task-related activity, or blood flow changes associated with drug effects (Malizia 2006).

SPECT imaging with target molecule-binding radiopharmaceuticals can be used to measure the baseline levels of target molecules and can measure the influence of experimental manipulations, genetics, and disease states on target molecules. Two useful examples of SPECT methods as applied to human clinical and research conditions include the study of Parkinson's disease and substance abuse disorders.

In Parkinson's disease, a neurodegenerative illness characterized by loss of brainstem dopaminergic innervation of the striatum, SPECT imaging has been used to demonstrate reductions in the levels of the presynaptic dopamine transporter (DAT) and increased levels of postsynaptic dopamine receptors (D2 subtype). SPECT may also have utility in monitoring Parkinson's disease progression and assessing the success of treatment interventions. In clinical conditions suggestive of early Parkinson's disease, SPECT assays of dopamine status may aid in diagnosis by

SPECT Imaging, Table 1 Sample brain SPECT compounds and applications used in psychopharmacology

Radionuclide	Half-life	Compound	Acronym	Measurement
^{133}Xe	5.24 days	^{133}Xenon	^{133}Xe	rCBF
^{99m}Tc	6.02 h	^{99m}Tc-hexamethylpropyleneamineoxime	^{99m}Tc-HMPAO	rCBF
^{99m}Tc	6.02 h	^{99m}Tc-diethylenetriaminepentaacetic acid	^{99m}Tc-DTPA	CSF, brain death
^{67}Ga	78.3 h	^{67}Ga-ethylenediaminetetraacetic acid	^{67}Ga-EDTA	Blood–brain barrier permeability
^{123}I	13.2 h	^{123}Iomazenil	^{123}I-IMZ	Central-type benzodiazepine receptor binding
^{123}I	13.2 h	^{123}I-2-((2-((dimethylamino)methyl)phenyl)thio)-5-iodophenylamine	^{123}I-ADAM	Serotonin transporter imaging
^{123}I	13.2 h	^{123}I-β-carbomethoxy-3-β-(4-iodophenyl)-tropane	^{123}I-CIT	Dopamine and serotonin transporters
^{123}I	13.2 h	^{123}I-iodobenzamide	^{123}I-IBZM	Dopamine D_2 receptor ligand
^{99m}Tc	6.02 h	^{99m}Tc-TRODAT	^{99m}Tc-TRODAT	Dopamine transporter sites
^{111}In	2.83 days	[^{111}In-DOTA^{0},D-Phe^{1},Tyr^{3}] octreotide	^{111}In-DOTA-TOC	Somatostatin receptor imaging
^{123}I	13.2 h	[^{123}I]5-iodo-3-[2(S)-2-azetidinylmethoxy]pyridine	^{123}I-5IA	Nicotinic acetylcholine receptors
^{123}I	13.2 h	6-iodo-2-(4′-dimethylamino-)phenylimidazo[1,2-a]pyridine	^{123}I-IMPY	β-amyloid plaque imaging

rCBF relative cerebral blood flow, *CSF* cerebrospinal fluid (Modified and adapted from Accorsi 2008. Used with permission)

determining whether early alterations in the DAT or D2 receptors are consistent with this illness (Booij and Knol 2007). Figure 3 illustrates the difference between a healthy control subject and patients with Parkinson's disease at different levels of severity. As disease severity worsens, there is near complete absence of detectable DAT binding as measured using the DAT ligand [^{99m}Tc] TRODAT-1 (Huang et al. 2004).

As reviewed by Volkow and others (2003; Felicio et al. 2009), SPECT (and PET) methods are used widely to study the effects of substance abuse and dependence and have contributed greatly to our understanding of drug effects in humans. For example, SPECT has been used to detect drug-induced changes in the serotonin transporter, serotonin receptors, and cerebral blood flow following recreational exposure to 3,4-methylenedioxymethamphetamine (MDMA; Ecstasy). For multiple drugs of abuse, SPECT imaging has been successful in documenting the effects of drug exposure on receptor and transporter systems (for multiple neurotransmitters, but especially the dopamine system) and on cerebral blood flow. SPECT imaging has also proven critical in documenting the time course of neurotransmitter and blood flow changes associated with acute and chronic drug exposure and following drug withdrawal.

Strengths and Limitations

Strengths of SPECT are considerable. Some SPECT radiopharmaceuticals are less costly and more easily synthesized and provide a lower radiation dosage than some PET tracers. This enhances the ease of use and acceptability of

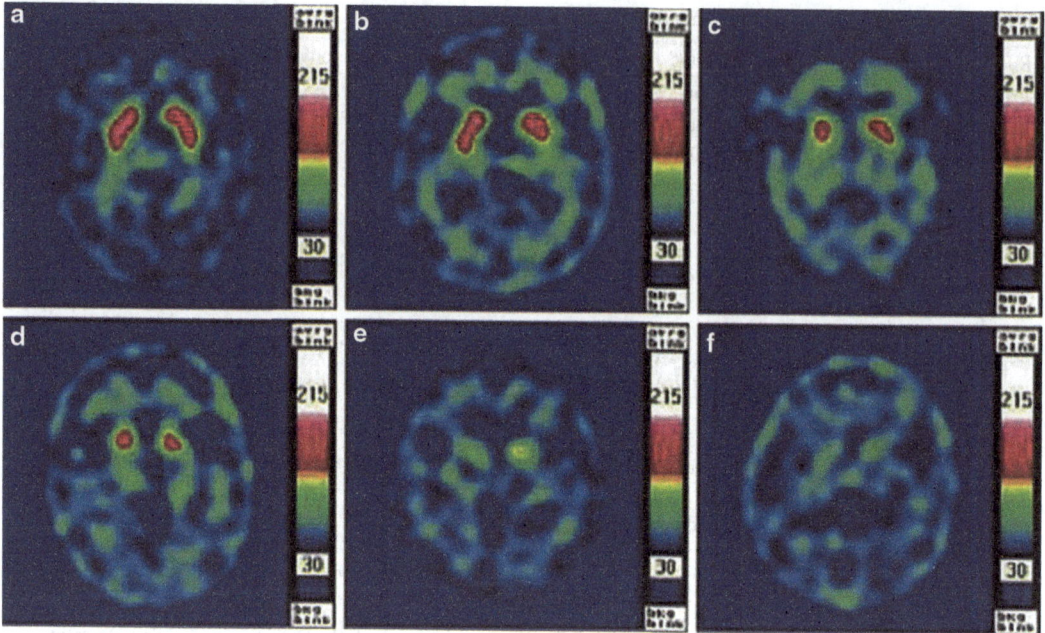

SPECT Imaging, Fig. 3 SPECT imaging in Parkinson's disease. These images show the utility of SPECT, using [^{99m}Tc]TRODAT-1 to measure dopamine transporter (DAT) loss in Parkinson's disease. The highest signal intensities are shown in *red*, corresponding to high levels of the DAT. (**a**) Healthy control subject. The bilaterally symmetric *bright red* regions are the striatum. (**b–f**) Represent increasingly severe stages of Parkinson's disease, with progressive loss of striatal DAT binding (From Huang et al. 2004. Used with kind permission of Springer Science+Business Media)

SPECT for human psychopharmacological research. Limitations of SPECT include the relatively small number of available radiopharmaceuticals and the resolution of this technique as commonly available for basic and clinical research. Unlike PET, where 18-fluorine fluorodeoxyglucose [18FDG] is commonly used to examine cerebral glucose metabolism, there is no comparable SPECT tracer for measuring cerebral glucose metabolism.

Emerging Technology

Emerging technologies include the use of modified gamma camera systems to provide very high spatial resolutions (in the submillimeter range) in animal nuclear imaging research studies. Dual tracer or multi-isotope imaging, in which different tracers are used to image two components simultaneously (e.g., perfusion and receptor binding), holds promise (Accorsi 2008).

Future Directions

With the development of high-resolution gamma cameras and additional radiopharmaceuticals, SPECT imaging can be expected to play an increasingly significant role in psychopharmacological research. In addition to the potential for high-resolution pharmacological mapping of neurotransmitter systems in animal models, human applications may increasingly expand to include clinical diagnostic applications such as detecting early types of degenerative illnesses (such as Alzheimer's disease or Parkinson's disease) or neurodevelopmental illnesses (such as autism or schizophrenia).

Cross-References

▶ Functional Magnetic Resonance Imaging
▶ Positron Emission Tomography (PET) Imaging

References

Accorsi R (2008) Brain single-photon emission CT physics principles. AJNR Am J Neuroradiol 29(7):1247–1256

Booij J, Knol RJ (2007) SPECT imaging of the dopaminergic system in (premotor) Parkinson's disease. Parkinsonism Relat Disord 13(suppl 3):S425–S428

Felicio AC, Shih MC, Godeiro-Junior C, Andrade LA, Bressan RA, Ferraz HB (2009) Molecular imaging studies in Parkinson disease: reducing diagnostic uncertainty. Neurologist 15(1):6–16

Huang WS, Lee MS, Lin JC, Chen CY, Yang YW, Lin SZ, Wey SP (2004) Usefulness of brain ^{99m}Tc-TRODAT-1 SPET for the evaluation of Parkinson's disease. Eur J Nucl Med Mol Imaging 31(2):155–161

Leslie WD, Greenberg ID (2003) Nuclear medicine. Landes Bioscience, Georgetown

Malizia AL (2006) The role of emission tomography in pharmacokinetic and pharmacodynamic studies in clinical psychopharmacology. J Psychopharmacol 20(Suppl 4):100–107

Masdeu JC, Arbizu J (2008) Brain single photon emission computed tomography: technological aspects and clinical applications. Semin Neurol., Sep;28(4):423–34. Review

Mettler FA, Guiberteau MJ (2006) Essentials of nuclear medicine imaging, 5th edn. Saunders Elsevier, Philadelphia

Spanoudaki VC, Ziegler SI (2008) PET & SPECT instrumentation. In: Semmler W, Schwaiger M (eds) Handbook of experimental pharmacology, molecular imaging I. Springer, Berlin/Heidelberg

Volkow ND, Fowler JS, Wang GJ (2003) Positron emission tomography and single-photon emission computed tomography in substance abuse research. Semin Nucl Med 33(2):114–128

Spectrograms

Synonyms

Deconvolution; Fourier spectrum; Frequency spectrum; Power spectrum

Definition

Representation of an EEG trace signal by a "virtual" decomposition (by a mathematical operation called ▶ Fast Fourier Transformation or FFT) of all possible contributing quasi-sinusoidal waves of different carrier frequencies. The result is a descriptor in the frequency domain and allows the distinction, for example, of low, middle fast, or high frequencies; these are easily expressed in a magnitude (equal to square root of power) that is a convenient metric of EEG for research purposes.

Cross-References

▶ Digital EEG Nomenclature
▶ Electroencephalography

Spectrographic

Definition

A spectrogram is an image that depicts the spectral density of a sound varying with time. ▶ Spectrograms are used to analyze *spectographically* the vocalizations of animals yielding quantitative statistics for comparison between a variety of experimental and natural conditions.

Spinal Cord Primary Sensory Neurons

Definition

These are neurons with cell bodies located in sensory ganglia with processes terminating in the periphery (skin, viscera, other) and in the spinal cord in the CNS. They convey the first (primary) sensory information to the spinal cord. They vary in cell body size, axon diameter, conduction velocity, and their content of neurotransmitters and neuroactive peptides. These differences and the distribution of central and peripheral ends account for sensory modality content. Small-diameter primary sensory neurons might, in addition, liberate neuropeptides antidromically with proinflammatory consequences.

Splice Variant

Definition

Different forms of an expressed protein derived from splicing of different regions of its primary RNA transcript.

Cross-References

▶ ISOFORM

SRI-Resistance

Definition

It has been suggested that failure to improve Y-BOCS (Yale-Brown Obsessive Compulsive) scores by 25 % from baseline after treatment with at least two serotonin reuptake inhibitors (SRIs) given at maximally tolerated SPC (Summary of Product Characteristics) doses for at least 12 weeks constitutes clinically meaningful SRI resistance.

Cross-References

▶ Obsessive–Compulsive Disorders

SSRIs and Related Compounds

Sidney H. Kennedy[1] and Sakina J. Rizvi[2]
[1]Department of Psychiatry, University Health Network, University of Toronto, Toronto, ON, Canada
[2]Department of Psychiatry, Departments of Pharmaceutical Sciences and Neuroscience, University Health Network, University of Toronto, Toronto, ON, Canada

Synonyms

Selective serotonin reuptake inhibitors; Serotonin antidepressants

Definition

Selective serotonin reuptake inhibitors are a class of drugs primarily developed for the treatment of major depressive disorder (MDD), although most agents now have additional indications, particularly for the treatment of various anxiety disorders. These drugs share a common mechanism of action in that they inhibit the serotonin transporter in its reuptake of synaptic serotonin.

Pharmacological Properties

History

First-generation tricyclic antidepressants (TCAs), particularly clomipramine, are potent inhibitors of the serotonin transporter, but these agents have a wide range of effects on various neurotransmitter receptors, resulting in a substantial side effect burden. Although zimelidine was the first SSRI to receive approval as an antidepressant in Europe, plans to launch it in the United States were abandoned when several cases of Guillain-Barre syndrome were reported, so fluoxetine became the first SSRI to be licensed in the United States in 1987 and paved the way for a new era of antidepressant drugs and a major shift in public awareness of the burden and costs associated with major depressive disorder (MDD) (Kramer 1999). Over the next decade, a whole class of SSRI antidepressants emerged as first-line agents for the treatment of MDD, accounting for 60 % of prescribed antidepressants in the US Medicaid program (Chen et al. 2008).

Mechanism of Action

As their name implies, the SSRIs selectively block serotonin (5HT) reuptake. This occurs through inhibitory actions on the Na^+/K^+ adenosine triphosphatase-dependent carrier on presynaptic neurons. Among the six available SSRIs, citalopram, escitalopram, and paroxetine are the most potent blockers of 5HT reuptake. Some SSRIs have additional antagonist effects on neurotransmitter receptors (Table 1). For example, paroxetine and citalopram have moderate

SSRIs and Related Compounds, Table 1 Mechanism of action and indications for six available SSRIs

Drug	Mechanism of action	Indications
Citalopram	**5HT blockade**; mild NE blockade; post receptor blockade: ACH; H1, DA, and alpha-1 (weak blockade)	MDD; panic disorder
Escitalopram	**5HT blockade**; mild NE blockade	MDD; GAD; panic disorder
Fluoxetine	**5HT blockade**; mild NE blockade; mild post DA receptor blockade	MDD; OCD; bulimia nervosa; panic disorder
Fluvoxamine	**5HT blockade**; mild NE blockade	MDD; OCD
Paroxetine	**5HT blockade**; postsynaptic receptor blockade: ACH	MDD; panic disorder; SAD; GAD; OCD; PTSD; premenstrual dysphoric disorder
Sertraline	**5HT blockade**; mild NE blockade; post receptor blockade: alpha-1; presynaptic receptor blockade: DA	MDD; OCD; panic disorder; PTSD; SAD; premenstrual dysphoric disorder

NE norepinephrine, *ACH* acetylcholine, *H1* histamine 1, *DA* dopamine, *MDD* major depressive disorder, *GAD* generalized anxiety disorder, *OCD* obsessive-compulsive disorder, *PTSD* post-traumatic stress disorder, *SAD* social anxiety disorder

anticholinergic effects and sertraline blocks presynaptic dopamine receptors. Escitalopram is a stereoisomer of citalopram and has been shown to exert actions at both the primary binding sites for the serotonin transporter and also on secondary allosteric binding sites, a property not shared by other SSRIs (Sanchez 2006).

The SSRIs, like their predecessors the TCAs, inhibit neurotransmitter reuptake almost immediately, but often take 2–3 weeks to exert clinically meaningful benefit. In addition, SSRIs, like other antidepressants, stimulate neurogenesis, particularly in the CA_3 layer of the hippocampus after 2–3 weeks of exposure. Animal models of depression using stress paradigms show suppression of neurogenesis which is reversed by antidepressants (Schmidt and Duman 2007).

Contribution of Neuroimaging

There is evidence from positron emission tomography (PET), using a ligand for the serotonin transporter, that 80 % or greater occupancy of the transporter occurs with citalopram, paroxetine, and sertraline at therapeutic doses (Meyer 2007). These findings are based on a small sample of depressed patients, and the SSRIs were not examined across a wide range of doses.

The effects of SSRIs on brain circuitry have been evaluated using functional magnetic resonance imaging. Emotion provocation tasks done in the scanner have shown that amygdala activity in response to negative stimuli is higher in MDD patients compared to healthy controls, an effect that reverses with successful antidepressant treatment. SSRIs also increase the connectivity between the amygdala to the prefrontal cortex and striatum (Harmer 2008). Further research demonstrates that early changes in amygdala and subgenual cingulate activity in response to negative emotional stimuli with SSRI treatment are predictive of symptom resolution.

Pharmacokinetics

The SSRIs are generally well absorbed and not affected by food administration, with the exception of sertraline, where food can increase levels of the drug in plasma. They are metabolized by hepatic microsomal enzymes that are part of the cytochrome P_{450} system, particularly the CYP_{2D6} isoenzyme, although the $2C_9$, $2C_{19}$, and $3A_4$ isoenzymes are also substrates for several SSRIs (Table 2) (Kennedy and Gorwood. 2014).

It is important to note that certain SSRIs prevent their own clearance through the inhibition of their metabolizing enzyme, resulting in elevated plasma levels and increased side effects. Both fluoxetine and paroxetine are strong inhibitors of the CYP_{2D6} enzyme, and fluvoxamine is a potent inhibitor of CYP_{1A2}, CYP_{3A4}, and CYP_{2C19}. Therefore, caution should be exercised when combining these drugs with other medications that are metabolized through any of these enzymes, as in the following two examples. Codeine requires CYP_{2D6} for its conversion to

SSRIs and Related Compounds, Table 2 Pharmacokinetic profile of SSRIs

Drug	% bioavailability	Metabolism	Active metabolites	Half-life (hours)
Citalopram	80	CYP3A4	Desmethylcitalopram	35
		CYP2C19	Di-desmethylcitalopram	
		CYP2D6		
		Weakly inhibits CYP2D6		
Escitalopram	80	CYP3A4	Di-desmethylescitalopram	27–32
		CYP219		
		CYP2D6		
		Weakly inhibits CYP2D6		
Fluoxetine	72	CYP2D6	Norfluoxetine	24–72
		CYP2C19		
		Inhibits CYP2D6, CYP2C9, CYP2C19		
Fluvoxamine	53	CYP2D6	None	15.6
		CYP2C9		
		Inhibits CYP2C19, CYP2D6, CYP1A2, CYP3A4		
Paroxetine	50	CYP2D6	None	21–24
		Inhibits CYP2D6		
Sertraline	44	CYP3A4	Norsertraline	26
		Weakly inhibits CYP2D6, CYP2C9		

morphine, which produces analgesic effects, and inhibiting this pathway prevents these effects. In the case of desipramine, a secondary tricyclic antidepressant, combining it with fluoxetine will result in increased plasma levels of desipramine due to the inhibition of CYP_{2D6} by fluoxetine. This can cause cardiotoxicity and, in rare instances, death.

The half-life of SSRIs ranges from about 15 h (fluvoxamine) to over 60 h (fluoxetine), with other agents in the 30 ± 6 h range. This means that in all cases, SSRIs can be prescribed once daily and, in the case of fluoxetine, even less frequently. This also means the drug is capable of causing drug-drug interactions 2–3 weeks after the last dose.

Genetic Polymorphisms

There is emerging evidence to support a relationship between genetic polymorphisms of the serotonin transporter (5HTT) and response to SSRIs. The 5HTT gene-linked polymorphic region (5HTT-LPR) significantly influences transcription of 5HTT, resulting in differential expression of the serotonin transporter. There have been several reports linking the 5HTT-LPR "l" allele to superior or faster response to SSRI therapy and the 5HTT-LPR "s" allele to greater SSRI-related side effects. However, these findings have not been consistent. A subsequent meta-analysis of studies evaluating serotonin transporter polymorphisms conducted in the last decade reported that the 5HTT-LPR "l" allele was associated with response and remission to SSRIs in Caucasians, but not Asians (Porcelli et al. 2012). A parallel meta-analysis of additional polymorphisms demonstrated that a BDNF polymorphism was associated with SSRI response, particularly in Asian samples (Niitsu et al. 2013).

Efficacy and Effectiveness

The SSRIs have primarily been used to treat patients with MDD, but many also are indicated for the treatment of anxiety disorders (Table 1) (Lam et al. 2009). While it was initially believed

SSRIs and Related Compounds, Table 3 Common side effect profile of SSRIs across average dose range

Drug	Average dose range	10–30 %[a]	30 %[a]
Citalopram	20–60 mg	CNS overarousal	Nausea
Escitalopram	10–20 mg	Headaches and nausea	None
Fluoxetine	20–60 mg	CNS overarousal	Nausea and headaches
Fluvoxamine	100–300 mg	CNS overarousal, dizziness, and constipation	Drowsiness, headaches, nausea, nervousness
Paroxetine	20–60 mg	CNS overarousal, constipation, and dizziness	Drowsiness, nausea, sexual dysfunction
Sertraline	50–200 mg	CNS overarousal, dizziness, sexual dysfunction	Headaches and nausea

CNS central nervous system
[a]Drug-placebo differences modified from product monographs

that there was no difference in therapeutic efficacy among SSRIs, subsequent meta-analyses suggest this is not the case. Since most trials were designed with sufficient sample sizes (power) only to detect differences between the novel agent and placebo, the inclusion of an active comparator agent to ensure assay sensitivity did not provide sufficient power to compare efficacy between the two active drugs. However, when similar trial design and outcome measures are applied across a series of trials, it is justifiable to pool the data from all subjects and conduct a meta-analysis. Using this methodology to detect differences between active agents, Cipriani and colleagues (2009) demonstrated a combined advantage for escitalopram and sertraline in efficacy and acceptability across 117 randomized control trials evaluating 12 new generation antidepressants.

Tolerability

Even though the tolerability and safety profile of SSRIs is superior to the previous generation of TCAs, lack of compliance due to treatment-emergent side effects remains a significant issue. There are both early transient adverse effects and persistent effects, which frequently result in drug discontinuation. Acute effects most often involve the central nervous system (CNS) and the gastrointestinal (GI) system, while later sustained effects are more likely to influence metabolism and sexual dysfunction (Table 3) (Kennedy and Gorwood 2014).

Central Nervous System

Headache, sleep disturbance, sedation, and paresthesia are commonly reported in the acute phase of treatment with any SSRI. Although some aspects of sleep are improved with SSRIs, there are reports that SSRIs disrupt sleep continuity and may exacerbate bruxism and restless leg syndrome. Additionally, while SSRIs generally improve cognitive dysfunction associated with depression, there is limited evidence, particularly with paroxetine, of drug-related cognitive side effects, likely due to its additional anticholinergic effects.

Gastrointestinal and Metabolic Effects

Nausea is a common gastrointestinal side effect during the first 2 weeks of treatment with an SSRI and is most pronounced with fluvoxamine and sertraline. This is generally a transient effect.

Weight gain is a factor that can severely decrease drug compliance. Although initial treatment with an SSRI can result in weight loss, mainly due to nausea and an enhanced feeling of early satiety, long-term use of SSRIs (particularly paroxetine) has been associated with weight gain (Fava et al. 2000; Kennedy and Gorwood. 2014).

There is also evidence that long-term SSRI treatment may be associated with a decrease in bone mineral density. This appears to reflect inhibitory actions of serotonin on osteogenesis. Whether this might contribute to the development of clinical osteopenia or osteoporosis is still unclear.

Sexual Dysfunction

The consensus from a series of well-designed comparative studies is that up to 60 % of patients receiving SSRIs report some form of treatment-emergent sexual dysfunction, with paroxetine and sertraline exerting the greatest burden of sexual side effects (Kennedy and Rizvi 2009). All of the SSRIs have been associated with delayed or absent orgasm/ejaculation and, in some instances, a reduction in libido and arousal. These effects appear to be related to the stimulation of serotonin, particularly its agonist effects on $5HT_2$ receptors. However, other mechanisms that block cholinergic receptors and inhibit nitric oxide synthase are also likely to be involved as well. This may explain why patients receiving paroxetine compared with other SSRIs had a significantly greater incidence of erectile dysfunction or reduced vaginal lubrication.

Safety

Suicidality

Depressed patients are at greatest risk for suicide attempts during the month before and the first month after starting medication, and the risk progressively declines with treatment. Although most population studies have shown a reduction in rates of completed suicide associated with increased use of modern antidepressants, the issue of suicidality with SSRIs generated considerable media attention, particularly in relation to antidepressant use in children and adolescents. The "black-box" warning for SSRIs in the United States, Canada, and elsewhere in 2003–2004 likely contributed to a decrease in antidepressant prescriptions for children and adolescents. Since then, rates of completed suicide have increased for adolescents (Henry et al. 2012). In adults, meta-analyses of randomized controlled trials (RCTs), or analyses of research databases, found no support for increased suicides with antidepressant use (Möller 2006).

Serotonin Syndrome

Excessive serotonin release may result in a clinically significant "serotonin syndrome," characterized by diarrhea, delirium, tremor, muscle rigidity, and hypothermia. This may occur when patients are taking two SSRIs or taking an SSRI with another serotonin-enhancing agent. Therefore, caution must be applied when combining treatments and during dose increases.

Emerging Drugs

Drugs that affect serotonin continue to be prominent in development pipelines. Both vilazodone (approved in 2011) and vortioxetine (approved in 2013) have combined serotonin transporter inhibition and additional serotonin receptor modulation properties (Stahl et al. 2013).

Vilazodone combines 5HT1A agonism and serotonin transporter inhibition, but with greater affinity for 5HT1A than drugs such as buspirone. Potential benefits of vilazodone include a low rate of drug-induced sexual dysfunction (Clayton et al. 2012), although direct comparisons with other antidepressants are awaited (Singh and Schwartz, 2012).

In addition to antidepressant efficacy, positive effects of vortioxetine on cognition have been reported (McIntyre et al. 2014). A subsequent comparator study of vortioxetine 10–20 mg against agomelatine in patients with inadequate response to an SSRI or SNRI demonstrated superiority of vortioxetine after 8 weeks (Montgomery et al. 2014).

Conclusions

Since the launch of fluoxetine, SSRI use has increased to the extent that this is now the most widely prescribed antidepressant class, reflecting a favorable balance between efficacy and tolerability. The pharmacokinetic profile of SSRIs allows for once daily dosing and long-term use. Future research should be directed toward recognizing biomarker profiles of patients who are most likely to respond to specific antidepressant agents or classes, including SSRIs.

Cross-References

▶ Aminergic Hypotheses for Depression
▶ Antidepressants

- ▶ Depression: Animal Models
- ▶ Depressive Disorders in Children
- ▶ History of Psychopharmacology
- ▶ Randomized Controlled Trials
- ▶ SNRI Antidepressants
- ▶ Tryptophan Depletion

References

Chen Y, Kelton CM, Jing Y, Guo JJ, Li X, Patel NC (2008) Utilization, price, and spending trends for antidepressants in the US Medicaid Program. Res Social Adm Pharm 4:244–257

Cipriani A, Furukawa TA, Salanti G, Geddes JR, Higgins JPT, Churchill R, Watanabe N, Nakagawa A, Omori IM, McGuire H, Tansella M, Barbui C (2009) Comparative efficacy and acceptability of 12 new-generation antidepressants: a multiple-treatments meta-analysis. Lancet 373:746–758

Clayton AH, Kennedy SH, Edwards JB, Gallipoli S, Reed CR (2012) The effect of vilazodone on sexual function during the treatment of major depressive disorder. J Sex Med 10:2465–2476 doi:10.1111/jsm.12004 (published online)

Fava M, Judge R, Hoog SL, Nilsson ME, Koke SC (2000) Fluoxetine versus sertraline and paroxetine in major depressive disorder: changes in weight with long-term treatment. J Clin Psychiatry 61:863–867

Harmer CJ (2008) Serotonin and emotional processing: does it help explain antidepressant drug action? Neuropharmacology 55:1023–1028

Henry A, Kisicki MD, Varley C (2012) Efficacy and safety of antidepressant drug treatment in children and adolescents. Mol Psychiatry 17:1186–1193

Kennedy SH, Gorwood P (2014) Pocket guide: successful management of major depressive disorder. Evolving Medicine, London

Kennedy SH, Rizvi S (2009) Sexual dysfunction, depression and the impact of antidepressants. J Clin Psychopharm 29:157–164

Kramer PD (1999) Listening to Prozac. Viking, New York

Lam RW, Kennedy SH, Grigoriadis S, McIntyre RS, Milev R, Ramasubbu R, Parikh SV, Patten SB, Ravindran AV; Canadian Network for Mood and Anxiety Treatments (CANMAT) (2009) Canadian Network for Mood and Anxiety Treatments (CANMAT) clinical guidelines for the management of major depressive disorder in adults. III. Pharmacotherapy. J Affect Disord 117 (Suppl 1):S26–43.

McIntyre RS, Lophaven S, Olsen CK (2014) A randomized, double-blind, placebo-controlled study of vortioxetine on cognitive function in depressed adults. Int J Neuropsychopharmacol (Epub ahead of print].

Meyer JH (2007) Imagining the serotonin transporter during major depressive disorder and antidepressant treatment. J Psychiatry Neurosci 32:86–102

Montgomery SA, Nielsen RZ, Poulsen LH, Häggström L (2014) A randomised, double-blind study in adults with major depressive disorder with an inadequate response to a single course of selective serotonin reuptake inhibitor or serotonin-noradrenaline reuptake inhibitor treatment switched to vortioxetine or agomelatine. Hum Psychopharmacol. doi: 10.1002/hup.2424.

Möller HJ (2006) Is there evidence for negative effects of antidepressants on suicidality in depressive patients? A systematic review. Eur Arch Psychiatry Clin Neurosci 256:476–496

Niitsu T, Fabbri C, Bentini F, Serretti A (2013) Pharmacogenetics in major depression: a comprehensive meta-analysis. Prog Neuropsychopharmacol Biol Psychiatry 45:183–194

Porcelli S, Fabbri C, Serretti A (2012) Meta-analysis of serotonin transporter gene promoter polymorphism (5-HTTLPR) association with antidepressant efficacy. Eur Neuropsychopharmacol 22:239–258

Sánchez C (2006) The pharmacology of citalopram enantiomers: the antagonism by R-citalopram on the effect of S-citalopram. Basic Clin Pharmacol Toxicol 99:91–95

Schmidt HD, Duman RS (2007) The role of neurotrophic factors in adult hippocampal neurogenesis, antidepressant treatments and animal models of depressive-like behavior. Behav Pharmacol 18:391–418

Singh M, Schwartz TL (2012) Clinical utility of vilazodone for the treatment of adults with major depressive disorder and theoretical implications for future clinical use. Neuropsychiatr Dis Treat 8:123–130

Stahl SM, Lee-Zimmerman C, Cartwright S, Morrissette DA (2013) Serotonergic drugs for depression and beyond. Curr Drug Targets 14:578–585

Startle

Synonyms

Startle response

Definition

The startle response refers to a group of reflexive motor and physiological responses elicited by an intense and sudden stimulus, usually in the auditory or tactile modalities.

State Dependence of Memory

Francis C. Colpaert
SCEA-CBC, Puylaurens, France

Synonyms

Dissociated learning; State-dependent learning; State-dependent retrieval

Definition

Traditionally, state dependence is said to occur when an organism remembers better when it is in a state similar to the one in which it learned what is to be remembered than when it is in a different state. While this observation remains essential, progress indicates that the phenomenon of state dependence is not necessarily related or limited to the ability to learn per se in a particular state. Dependence on state occurs with memory processes other than learning as the latter is conventionally understood and may determine what is remembered and when; beyond conscious cognition, it also concerns such functions as emotion, mood, and motor behavior. Thus, the term "state dependence of memory" refers to an attribute of the ability to remember and act upon past experience, which is broader than that implied by "state-dependent learning."

Impact of Psychoactive Drugs

A Word of History
Awareness of state dependence arose in seventeenth-century popular and medical culture and concerned states of the internal, physiological, and mental milieu (Siegel 1985). Both that century's most acknowledged accounts and the first experimental investigation of state dependence (Girden and Culler 1937) involved states

---

Francis C. Colpaert: deceased.

induced by drugs. Influentially capturing the perception of state dependence at that time, Overton (1983) attributed state dependence to high, "toxic" doses of drugs; considered state dependence to be genuine only if bidirectional (i.e., if occurring upon both drug-to-placebo and placebo-to-drug changes of state); struggled with the issue as to whether state dependence requires that drug-to-placebo and placebo-to-drug state changes should produce symmetrical failures to remember; and argued on theoretical grounds that state dependence and drug discrimination are fundamentally similar. As will be apparent in the following lines, the state dependence concept has much evolved; how and when state dependence occurs has since been studied extensively in humans and animals in well-controlled experimental investigations in laboratory settings.

A Case of State Dependence
Studies of state dependence involve learning. The learning assays used may include single trials or multiple-trial learning opportunities, single or multiple-session classical conditioning, or acquisition of an appetitively or aversively motivated instrumental response. As in many memory researches, the key dependent variables assess the ability to remember, and recall is often measured operationally by the latency with which a remembered response occurs. The special issue of interest here about the functioning of memory is how memory may vary depending on the similarity or dissimilarity between the state that prevailed at the time of learning and the state that prevails at the time of recall. States thus constitute the independent variable. Failures of "transfer" of the engram, or memory trace, are said to occur when recall is hampered by a change of state.

Figure 1 offers an example of state dependence; here, rats are trained in 15 min sessions every day to press a lever for food reward; every tenth lever press yields access to food (fixed ratio, 10, or FR10 schedule). The rat reaches the acquisition criterion when, on a given training session, it completes 10 lever presses within 120 s after the session begins. Two days later, in a test

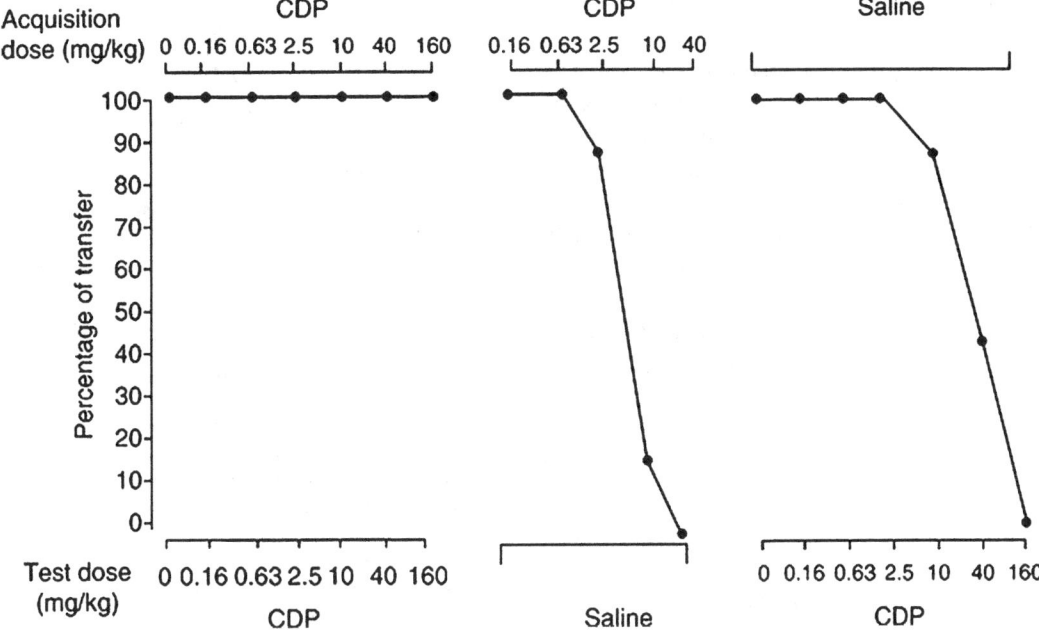

State Dependence of Memory, Fig. 1 Results of transfer tests in groups of rats that acquired an operant response in one pharmacological condition (i.e., saline or one of several doses of chlordiazepoxide (CDP)) and were tested in either the same or another condition. Upper and lower abscissa: CDP dose at which the animals were trained and tested, respectively; 0 refers to saline. Ordinate: percentage of rats ($n = 7$/group) that in test sessions completed the first FR10 schedule within 120 s (Redrawn from Colpaert 1990)

session, recall is measured by determining whether the animal again completes the first FR10 schedule within 120 s. Different groups of rats are trained with (i.e., before the training sessions receive an injection and thus are "under the influence of") either saline or one of different "training" doses of the benzodiazepine, chlordiazepoxide (CDP). Later, the rats are tested with (i.e., "under the influence of") either saline or one of different CDP "test" doses.

As shown in the left panel of Fig. 1, recall was perfect by rats that were both trained and tested with saline (points at "0") and by rats that acquired the response while treated with one CDP dose (0.16–40 mg/kg for different rats) and then tested with the same dose. However, as shown in the middle panel, rats trained with 0.16–40 mg/kg doses of CDP and then tested with saline failed to recall in a manner that depended on the training dose, complete failure occurring with 40 mg/kg. Conversely (right panel), rats trained with saline and then tested with 0.16–160 mg/kg doses also failed to recall, now in a manner that varied with the CDP test dose. Thus, state dependence can occur with both drug-to-saline and saline-to-drug changes of state, albeit not necessarily at the same dose; here, drug-to-saline transfer failure occurred at doses that were about threefold lower than those at which saline-to-drug state changes caused transfer to fail (ED_{50}, 9.8 and 29 mg/kg, respectively).

State changes can produce powerful memory failure; under the conditions described above, extreme food deprivation to the point of starvation fails to overcome the inability to remember. Remarkably, studies involving agents such as benzodiazepines and opiates, to which tolerance is considered to readily develop, have so far failed to reveal any evidence that tolerance develops to those agents' ability to induce state dependence; training with drug followed by numerous further drug ("over-") training sessions does not prevent failures of recall when the subjects are later tested with saline.

The state-dependent memory failure is, however, surmountable. Rats trained with saline or CDP 40 mg/kg fail to remember when tested for the first time with CDP 40 mg/kg or saline, respectively; but if later provided the opportunity to learn the same response in the alternative state (i.e., after administration of saline or CDP 40 mg/kg, respectively), rats will eventually learn to recall the response in either state. It is uncertain, though, whether such recall in either state implements only one as opposed to one of two separate engrams (Colpaert 1990).

State Specificity

When a response is acquired with saline or with any given agent, the extent to which transfer occurs to other agents may vary widely, but it can also be the case that transfer is confined to a narrow class of pharmacologically similar agents. For example, rats trained with morphine demonstrate transfer when tested with another μ-opiate receptor agonist, but not with lower-efficacy opiates or any other drug. Recall may be further limited to a particular dose; rats trained with 5 mg/kg morphine remember well when tested with 5 mg/kg, but not with lower or higher doses of morphine (Fig. 2). Thus, retrieval can be confined to an exquisitely exclusive, molecularly defined magnitude of activation of a single, particular neurotransmitter receptor (sub)type; these findings highlight memory's state specificity in addition to its state dependence (Bruins Slot and Colpaert 1999).

Specificity can also extend to conditions other than the drug state. For example, a given agent may render state dependent a response that is acquired in a particular set of conditions of arousal and drive, for instance, but not the same response in another set of conditions. Equally, with a given agent, a normal-to-drug change of state may impair the retrieval of some but not all items of memory that were acquired in one or another normal set of conditions.

Both these two instances of specificity underlie the ability of states to operate in an often incisive, discrete manner. Partly because of this, state dependence is in many instances overlooked in accounts of CNS function and of CNS drug actions.

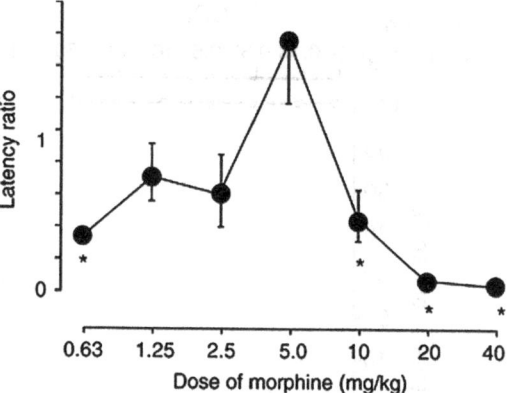

State Dependence of Memory, Fig. 2 Results of transfer tests in groups of rats ($n = 7$/group) that acquired an operant response after 5 mg/kg morphine injections and were tested for retrieval with either the same or a lower or higher morphine dose. Retrieval here is measured by the (log-transformed) ratio of the response latency in the last training session to that in the test session (ordinate); data are mean + SEM (Redrawn from Bruins Slot and Colpaert 1999)

State Change: A Mechanism

States and state changes regulate, shape, and direct memory throughout ontogeny; they govern mnemonic normalcy as well as psychiatric and neurological pathogenesis and may constitute the essential neurobiological, or "system," mechanism of different CNS agents.

State Changes in Cognition

Compounds such as ketamine and – most prominently – the muscarinic acetylcholine receptor antagonist, scopolamine, offer drug models of memory disorders, mimicking in unaffected humans the deficiencies that are found in Alzheimer's disease. Surprisingly, scopolamine actually may allow fully adequate learning, retrieval encoding, re-encoding, and retention of complex tasks provided that all these memory processes take place in the scopolamine state. However, each of these processes is profoundly impaired when subjects switch from the normal to the drug state (or from the drug to the normal state) as they move from one to the other process. As a result, the state change that acute scopolamine administration produces in subjects having learned the task in the normal state simulates the regressive, temporally graded retrograde amnesia that is characteristic of Alzheimer's disease. That is, as a function of the scopolamine dose, what can be remembered is limited to what was learned at an increasingly remote point of time in the past, while what was learned more recently has become inaccessible (Figs. 3, 4 and 5). Importantly, the state dependence analysis emphasizes the role of state instability, in addition to but more so than that of putative cholinergic hypofunction, in the pathophysiology of such disorders (Colpaert et al. 2001).

Some drug states that are induced after learning or retrieval that occurred in the normal state (i.e., at the time of encoding or re-encoding, respectively) impair subsequent retrieval in the normal state. Drugs may thus preempt or erase the normal-state memory of traumatic, pathogenic life events.

State Changes in Anxiety

Anxiety has long been considered as an acquired drive, i.e., a motivation that is learned in the course of ontogeny. Benzodiazepines exert anxiolytic effects only in as much as they also impair the retrieval of, for example, a food-reinforced operant response. In each situation, the benzodiazepines institute a state of memory in which material previously acquired in the normal state cannot be retrieved. Thus, it has been proposed that benzodiazepines impair the retrieval of what in a normal state has previously been acquired what is referred to as anxiety, rather than selectively reducing, or "lysing," anxiety (Colpaert 1990).

State Changes in Pain and Opiate Actions

Pain can be of either physiological (e.g., nociceptive) or pathological (e.g., neuropathic) origin and relates to both its somatosensory and affective/motivational dimensions; the affective/motivational dimension of pathological pain is particularly susceptible to state dependence.

Mammalian newborns learn to synchronize their breathing, which at the prenatal stage is erratic, chaotic, and utterly ineffectual, with the acidity (pH) of arterial blood. This synchrony governs mature breathing, and it has been suggested that rather than merely depress respiration, opiates induce a state in which this learned synchronization cannot be retrieved (Bruins Slot and Colpaert 1999).

State Changes in Affective Disorders

Mood states fluctuate and the memory that prevails at any time is governed by mood congruency, or sameness of mood state. Depressed subjects better memorize new negative-affect than new positive-affect items and remember negative-affect items more effectively in a subsequent depressed episode than in a more elated mood (Blaney 1986; Bower 1981; Colpaert et al. 2000; Eich 2007). While allowing for possible predisposing biological variables, the state dependence of memory uncovers a remarkable pathophysiological mechanism of mood disorders as well as an equally remarkable neurobiological mechanism of antidepressant drug action both of which emphasize the ability to acquire and remember, and disremember, mood.

State Changes in Neurological Disorders

Importantly, states and state changes not only govern cognition but also pervade and affect motor behavior while spanning large swaths of ontogeny. The peculiar features of horizontal locomotion in Parkinson's disease (difficulty with both initiating and arresting locomotion, continuous postural disequilibrium during movement) closely resemble those of 10-month old infants who learn to walk. Under L-DOPA treatment and fluctuating between ON and OFF states, the parkinsonian patient when in an ON state may walk adequately while not remembering what precisely the difficulty it is that he or she has in the OFF state; when in OFF, the patient fails to remember just how he or she succeeds to walk when in ON. Gilles de la Tourette patients similarly alternate episodes of "remembering" with those of "forgetting" their condition, making no distinction between the memory, the knowledge, the impulse, and the act. The patient may or may not be aware of his current state, but any awareness fails to affect the state's implications.

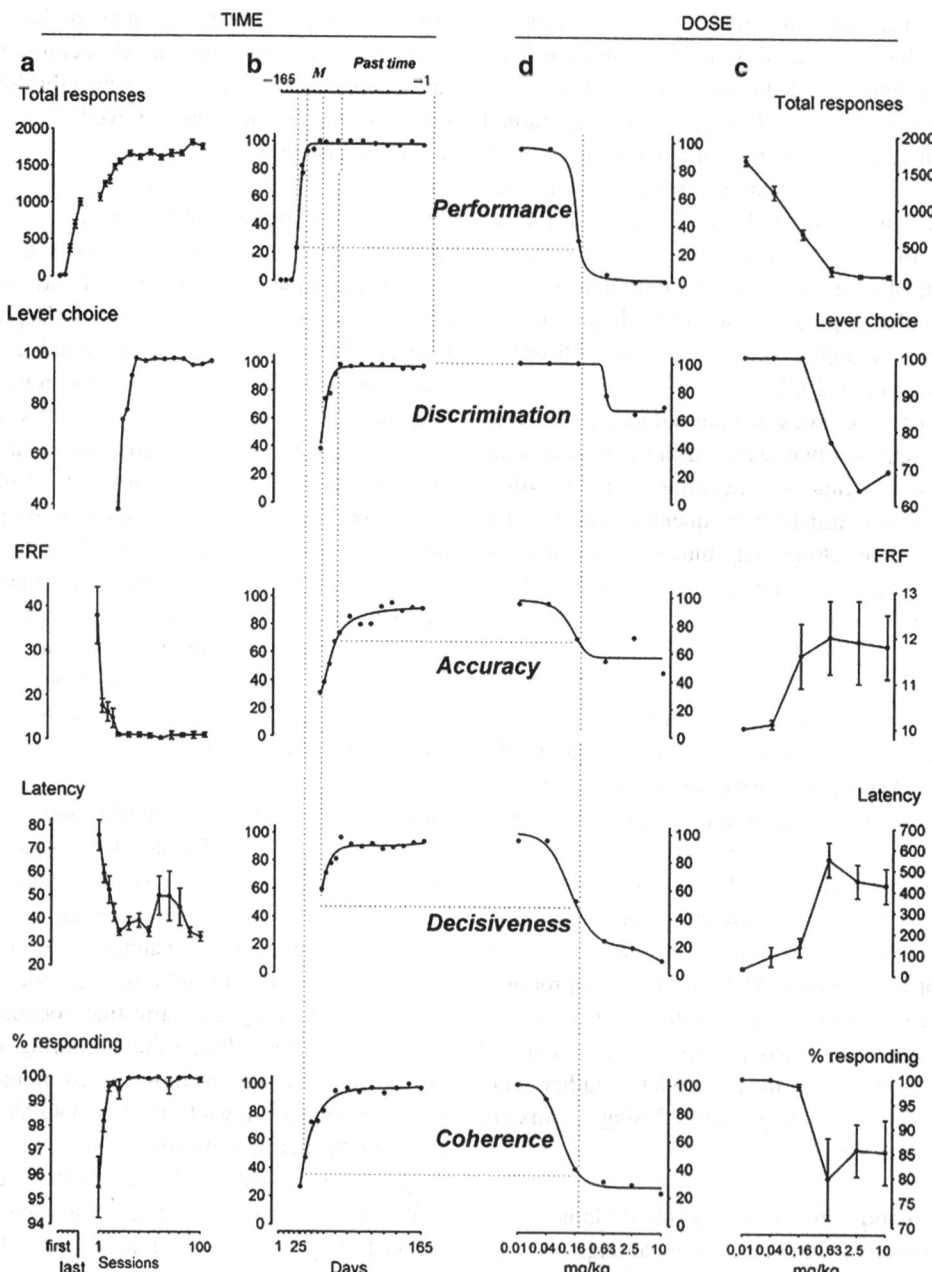

State Dependence of Memory, Fig. 3 Studies examining the effects of saline-to-scopolamine changes of mnemonic state on the recall of five aspects of learned behavior. (**a**) Rats were trained for in all 165 days in a complex task in which the different independent measures (e.g., total responding) reflect different aspects (e.g., performance) of what was learned. (**b**) The acquisition data in panel A are here plotted as the % of animals satisfying the learning criterion that was applied to each of the five measures. These % data were fit by best-fitting functions to generate the five learning curves shown in panel B. (**c**) Over the 8-month period that followed training, rats were tested for retrieval after injection of different doses of scopolamine; data from these tests are shown here and expressed as in panel A. (**d**) The scopolamine test data in panel C are here plotted as the % of animals satisfying the criteria used to generate panel B and expressed as in the same manner. These % data were fit with the dose-response functions shown. At 0.01 mg/kg scopolamine, the outcome for each of the parameters was

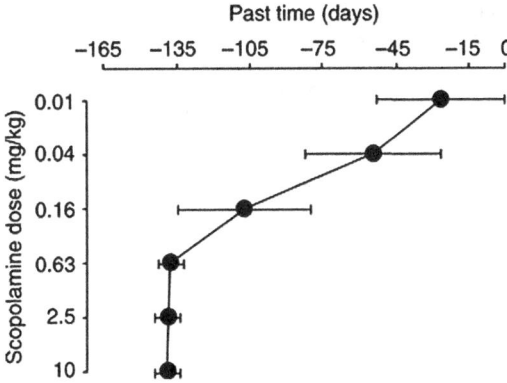

State Dependence of Memory, Fig. 4 Regressive, temporally graded retrograde amnesia induced by a normal-to-scopolamine state changes. Rats were in the normal state trained in a complex task in which learning was assessed from five parameters that characterize the newly acquired behavior (Fig. 3). The acute injection of different doses (ordinate) of scopolamine caused the parameters to assume values that resembled those observed at one or another point in the course of training. The abscissa indicates the remoteness in time (relative to day 0, the last day of training) of such corresponding points (mean ± SEM of the five parameters; day − 165 corresponds to the beginning of training). Increasing doses resulted in a behavior similar to that observed at increasingly remote stages of the acquisition that preceded the drug test (Redrawn from Colpaert et al. 2001)

Dependence

Whether occurring in a normal or pathogenic manner in the course of ontogeny, or drug-induced, states create a dependence: one of memory and of the capabilities that it enables to the very point of survival. As such, state dependence is akin to drug dependence in as much the state is required to maintain the homeostasis of memory. When induced by a drug, state dependence constitutes an instance and learned form of drug dependence.

State-dependent memory may or may not encompass the subject's seeking to modify that state; such action appears to depend on the state's perceived affective valence and on the opportunities that are available to acquire behaviors that are instrumental in overcoming or modifying those consequences. Alzheimer patients may be little able to gauge their cognitive deficit and, available treatments being only marginally effective, to learn how to instrumentally overcome the deficit. But, state-dependent, compulsive seeking to restore the original state does appear to underlie opiate addiction, in effect constituting its very "system" mechanism. This implies that the signal transductions that are involved in pain processing and opiate actions thereon are currently thought to proceed in a bidirectional manner; any input to such systems induces not a single but two dual effects that are opposite in sign. Thus, morphine causes not only analgesia as a "first-order" effect but also a "second-order" hyperalgesia or frank pain that outlasts the analgesia for some time. With chronic opiate exposure, the second-order pain grows and masks the first-order analgesia, resulting in analgesic tolerance. In a parallel manner, opiates also induce two distinct memory states that coincide temporally with their first- and second-order effects on pain processing. The first-order opiate memory state is one in which analgesia is encoded, but the second-order state is one in which the individual experiences a possibly excruciating, opiate-induced pain and is enabled to learn and remember that the next opiate administration powerfully relieves that pain, however temporarily. Thus emerges the view that opiate addiction represents the self-medication of compelling opiate pain, with both that pain and the pain-relieving instrumental behavior being encoded in the second-order (withdrawal) state of memory (Colpaert et al. 2006).

State Dependence of Memory, Fig. 3 (continued) similar (i.e., about 100 %) to that at the end of acquisition. Higher doses decreased each parameter value, such that the outcomes resembled those observed earlier in training. Projections are plotted (dashed lines) to identify the point in "time since the end of acquisition" ("past time") with which the test data corresponded. These retrograde projections yielded outcomes that differed among the five parameters, the mean (M along past time) being 136 days for the 0.16 mg/kg dose. In this manner, the mean was found for all doses, generating the function relating scopolamine dose to past time shown in Fig. 4 (Redrawn from Colpaert et al. 2001)

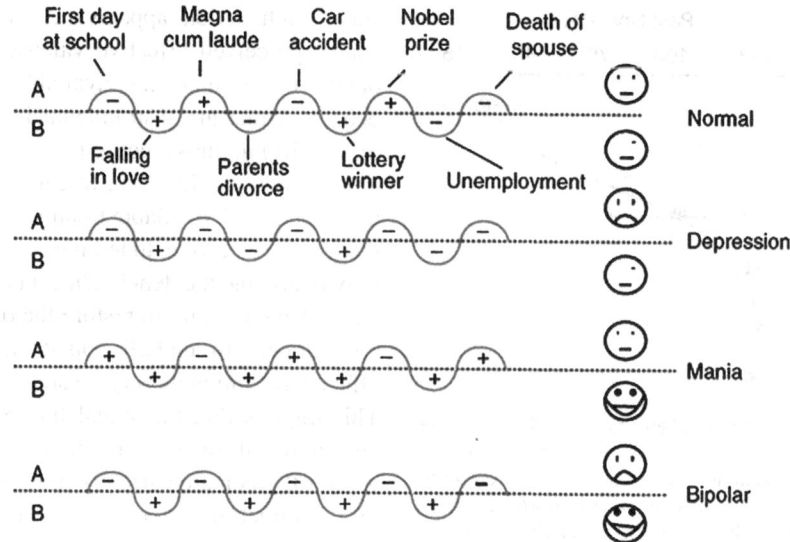

State Dependence of Memory, Fig. 5 The shaping of mood states in ontogenesis. Though multiple states can exist, it is sufficient to assume here that memory operates in one of just two cyclically recurrent states (*A* or *B*) that are induced by some biological rhythm and are mutually exclusive. Throughout ontogeny (*cp*), the individual experiences and memorizes life events that have an affective valence that is positive (+) or negative (−). There is no a priori relationship between a state on the one hand and the (valence of) the events that are encountered in a state on the other. In this manner is most likely established in both states *A* and *B*, a cumulative memory that retains both positive and negative items (*panel I*); the overall affective valence of memory is rather neutral in either state and does not greatly differ, or swing, between *A* and *B*. This outcome being most probable, *panel I* represents what because of this greatest likelihood is perceived as a normal individual. However, chance may have it that the events experienced in *A* or *B* happen to predominantly be negative or positive (*panels II* and *III*, respectively); thus it is established in one state a memory that retains items of polarized valence and that differs from the alternative, normal-affect state. These outcomes are less likely and result in recurring episodes of monopolar, depressed or manic, mood. An even smaller chance may result in large-amplitude bipolar mood swings (*panel IV*) (Colpaert et al. 2000)

States

Clearly, an organism's memory can operate in any one of a decidedly large number of different states in which engrams can be specifically stored. Such state dependence of memory challenges the common view that an individual possesses a single, unified memory that integrates all past experience and can be acted upon at any time. Rather, at any point of time, memory appears to be locked into, limited to, but also enabled with the unique capabilities of one particular state. As time moves from one to another point, so memory comprises another, equally unique set of retrievable capabilities based on current state.

The state dependence of memory invites the question as to whether forgetting occurs at all; it is unclear whether such failure to remember results from a time-dependent dissipation of the engram or from the non-restoration of the state that prevailed in ontogeny at the time that the memory was established. State dependence definitely does not require forgetting to account for profound retrieval deficits, and state changes constitute a mechanism whereby an engram can be erased; when in an initial state the subject has learned and remembers and when then an alternative state is instituted shortly after retrieval (i.e., at the time of re-encoding), it may be the case that the engram in question is no longer available in the initial state. Further, such a state change may establish, in the alternative state, a memory that is then false (false memory).

A weakness of state dependence research resides with a certain vagueness as to what defines a state. Awareness of state dependence historically arose from observations of humans who were "under the influence" of exogenous agents, and the use of drugs in studying state dependence has allowed that in this case states can at least operationally be defined in a highly precise, accurate, and reproducible manner. However, states also arise endogenously and current neurobiology and medicine invoke the existence of a varied array of states that operate at levels of integration ranging from the whole organism to cell assemblies, single cells, or molecular signal transduction (Lydic and Baghdoyan 1999). Fascinating examples include hormone-controlled phases, pacemaker assemblies, and circadian and seasonal rhythms, as well as the rise and fall of entire neurotransmitter and other signaling systems in the course of ontogeny. Such states are usually investigated with regard to the particular stimulus-response relationships that they demonstrate, but a wide realm of knowledge remains to be gained in terms of whether and how those states regulate memory.

Learning and memory are also investigated from such perspectives as context-dependence and coincidence detection; how these perspectives shed light on and interlock with state dependence remains to be explored.

Acknowledgment I thank Drs. V. Curran, I. Kiss, A. Newman-Tancredi, and A. Young for comments and discussions. This essay is dedicated to the memory of Dr. Liesbeth Bruins Slot.

Cross-References

▶ Addictive Disorder: Animal Models
▶ Analgesics
▶ Animal Models for Psychiatric States
▶ Antidepressants
▶ Behavioral Tolerance
▶ Benzodiazepines
▶ Bipolar Disorder
▶ Declarative and Nondeclarative Memory
▶ Dementias and Other Amnestic Disorders
▶ Dissociative Anesthetics
▶ Drug Discrimination
▶ Dysthymic Disorder (Persistent Depressive Disorder)
▶ Emotion and Mood
▶ Generalized Anxiety Disorder
▶ Inhibition of Memory
▶ Mood Stabilizers
▶ Opioid Use Disorder and Its Treatment
▶ Opioids
▶ State-Dependent Learning and Memory

References

Blaney PH (1986) Affect and memory: a review. Psychol Bull 99:229–246

Bower GH (1981) Mood and memory. Am Psychol 36:129–148

Bruins Slot LA, Colpaert FC (1999) Opiate states of memory: receptor mechanisms. J Neurosci 19:10520–10529

Colpaert FC (1990) A mnesic trace locked into the benzodiazepine state of memory. Psychopharmacology (Berl) 102:28–36

Colpaert FC, Bruins Slot LA, Koek W, Dupuis DS (2000) Memory of an operant response and of depressed mood retained in activation states of 5-HT$_{1A}$ receptors: evidence from rodent models. Behav Brain Res 117:41–51

Colpaert FC, Koek W, Bruins Slot LA (2001) Evidence that mnesic states govern normal and disordered memory. Behav Pharmacol 12:575–589

Colpaert FC, Deseure K, Stinus L, Adriaensen H (2006) High-efficacy 5-HT$_{1A}$ receptor activation counteracts opioid hyperallodynia and affective conditioning. J Pharmacol Exp Ther 316:892–899

Eich E (2007) Mood and memory at 26. In: Lesgold A (ed) Memory and mind. CRC Press, London, pp 1–7

Girden E, Culler EA (1937) Conditioned responses in curarized striate muscle in dogs. J Comp Psychol 23:261–274

Lydic R, Baghdoyan HA (1999) Handbook of behavioral state control. CRC Press, Boca Raton, p 700

Overton D (1983) State dependent learning and drug discrimination. In: Iversen LL, Iversen SD, Snyder SH (eds) Handbook of psychopharmacology, vol 18. Plenum, New York, pp 59–127

Siegel S (1985) Psychopharmacology and the mystery of the moonstone. Am Psychol 40:580–581

State-Dependent Learning and Memory

Alice M. Young
Department of Psychological Sciences, Texas Tech University, Office of the Vice President for Research, Lubbock, TX, USA

Synonyms

Drug dissociative learning and memory; State-dependent learning; State-dependent memory; State-dependent retrieval

Definition

In psychopharmacology, **state-dependent learning and memory** are phenomena in which skills or relationships learned under the influence of a psychoactive drug, or memories formed under its influence, are more likely to be executed, recalled, or retrieved under the influence of the same or similar drugs than under the influence of no or dissimilar drugs. The concept can be expanded to include both interoceptive "states" produced by psychoactive drugs and those produced by physiological processes linked to emotions, drives, affects, and biological rhythms.

Principles and Role in Psychopharmacology

The following quotation in the 1835 text *Human Physiology* by John Elliotson (page 646) captures the concept of state-dependent learning and memory:

> "Dr. Abel informed me," says Mr. Combe, "of an Irish porter to a warehouse, who forgot, when sober, what he had done when drunk: but, being drunk, again recollected the transactions of his former state of intoxication. On one occasion, being drunk, he had lost a parcel of some value, and in his sober moments could give no account of it. Next time he was intoxicated, he recollected that he had left the parcel at a certain house, and there being no address on it, it had remained there safely, and was got on his calling for it." This man must have had two souls, one for his sober state, and one for him when drunk.

Dr. Abel's Irish porter, along with the Millionaire encountered by the Tramp in Charlie Chaplin's *City Lights* and the protagonist in the first English language detective story, *The Moonstone* by Wilkie Collins, demonstrates the key features of state-dependent learning and memory (Overton 1984, 1991). First, an individual learns a new skill, has a notable experience, or encounters unique information while intoxicated with a psychoactive drug. Later, when sober, the individual cannot execute the skill and has no recollection of the experience or information and, indeed, may behave like a novice or display complete amnesia for events. When the state of intoxication is reintroduced, however, the individual can immediately execute a skilled performance or recall the episode.

State-dependent learning can be defined as a phenomenon in which skills or relationships learned in one context are more likely to be executed in the same or similar contexts than in different contexts. The relevant contexts are created by interoceptive "states" produced by psychoactive drugs or by physiological processes linked to affect, biological rhythms, drives, or emotions. The first phenomenon, in which the learner requires a matching drug state to demonstrate learning, is often described as state-dependent learning (Overton 1984), whereas the latter, in which the learner requires a matching affect or mood, has been termed "mood-dependent memory" (Eich 2008).

State dependence can also be described from the perspective of memory. State-dependent memory (also termed state-dependent remembering or state-dependent retention) is a phenomenon in which memories are more likely to be recalled or retrieved in the drug or mood states in which they were formed than in dissimilar states. Put another way, memories may appear to be inhibited, repressed, or even forgotten until the original state is restored.

Perhaps because different scientific communities have studied state-dependent learning and

State-Dependent Learning and Memory, Table 1 2 × 2 Design used in many studies of state dependence. To emphasize the key changes during the "test phase," the state present during training is indicated by red stipples and a change in state by grey shading

Experimental group	Training phase Subjects receive	Test phase Subjects receive	Description of state change or of performance or memory transfer
D – D	Drug	Drug	Drug to drug
D – N	Drug	Vehicle	Drug to vehicle
N – D	Vehicle	Drug	Vehicle to drug
N – N	Vehicle	Vehicle	Vehicle to vehicle

mood-dependent memory, at least two descriptive languages have developed. The impact of changes in drug states on learning or remembering is often examined using tools of behavioral pharmacology and concepts of classical and instrumental conditioning. In contrast, the impact of changes in mood (e.g., affect, emotions, biological rhythms), is often examined with tools of verbal learning and memory and cognitive psychology.

Key Techniques

Experiments to explore state dependence often use an experimental design that incorporates four groups of subjects and two experimental phases (Table 1). Table 1 illustrates an experiment that uses administration of a drug to induce a state, but investigators have used this basic design to study the impact of manipulations that alter mood. In the first, or training, phase, different subjects receive either a drug or its vehicle and a new learning task. In the second, or test, phase, subgroups receive either drug or vehicle and tests of their efficiency and/or accuracy in performing the task and/or recalling its elements. State dependence is inferred if groups trained and tested in a similar state (i.e., D-D and N-N) display better recall or stronger performance than do groups trained and tested in different states (i.e., D-N and N-D).

Because this 2 × 2 design contains groups trained and tested in all possible combinations of "drug" and "no-drug" states, it can, all else being equal, distinguish between effects of a "training state" and those of a "test state," and it has been modified to achieve diverse experimental aims. Chief among these has been identification of drug classes that can support state dependence, with robust state dependence reported for ethanol and constituents of marihuana (the most frequently studied drugs in humans), antimuscarinics, benzodiazepines, N-methyl-D-aspartate receptor (NMDAR) antagonists, nicotine, and opioid analgesics, among others. Studies in human participants often convert the original between-subjects design to a within-subjects design, in which an individual subject learns a task or cues in the presence of one state and is tested both states.

Caution is needed in interpreting experiments that use the 2 × 2 design and its variations, because of interpretative problems that arise when drugs and other manipulations have multiple impacts on behavior. Among these are state-induced learning or performance deficits, motoric changes, and changes in learning or memory processes – all of which may occur alongside state-dependent effects and require careful control experiments or design changes to rule out competing interpretations. For example, design changes such as lengthening the training phase and incorporating a learning criterion (see below) appear to reduce interpretative difficulties that arise if learning does not occur during training.

Table 2 illustrates several of the important features of state dependence, using data from what is generally regarded as its first published demonstration (Girden and Culler 1937). Girden and Culler, interested in whether classical conditioning requires execution of the unconditional

State-Dependent Learning and Memory, Table 2 Data from the first published demonstration of drug-induced state dependence (Girden and Culler 1937). Page numbers indicate where an experiment is described in the text. As in Table 1, the state present during training is indicated by red stipples and a change in state by grey shading. Dogs were given intravenous injections of a "saturated" solution of curare, and a hind leg semitendinosus muscle was exposed to pairings of a bell and shock. The solution of curare contained impurities, and the injection volume was chosen to produce loss of function as indicated by respiratory paralysis, loss of gag, and corneal-lid reflexes and flaccidity and loss of responsiveness of striate muscles (page 263); dogs were maintained on respiratory support. The UR and CR decreased markedly in magnitude after curare and comprised a twitch or flick of the tip rather than the body of the muscle

Subject	Phase 1[a]	Phase 2[c]	Phase 3	Phase 4
Pages 266–267	Curare Training: CS and US	No curare CS alone	Curare Continued training: CS and US	No curare CS alone
11	10/10 10/10 4/5	0/4	0/15 6/10 10/10 10/10 4/5	0/4
12	5/10 10/10 3/5	3(?)/3	2/25 5/10 4/10 5/10 10/10 3/5	2/8
14	7/10 1/10 8/11	0/3	CR at first CS, then 24/24	
15	3/10 7/10 6/10	0/3	4/4 with CS only, then 3/3 8/10 7/10 3/5	

	Phase 1	Phase 2	Phase 3	
Page 268	No curare[b] Training: CS and US	Curare CS alone	No curare CS alone	
No ID	"Conditioned"	0/5	5/5	

	Phase 1	Phase 2	Phase 3	Phase 4
Pages 265, 268–269	No curare Training: CS and US	Curare CS alone	No curare CS alone	Curare CS alone
20: right	8/10 8/10 8/10	No CR	CR on several trials	No CR
		Curare CS and US	No Curare CS alone	Curare CS alone
20: left		4/5 4/5 8/8	No CR	CR

[a]Numbers indicate the number of trials on which CR occurred/number of trials in a training or test bout. A (?) indicates that the authors reported "some contraction" rather than a CR. Data are presented for the final 3 bouts of initial conditioning trials and for all re-conditioning or test trials described in the text. Measurement of UR during training and presentation of US controls during tests demonstrated that the US itself (shock) elicited the expected UR under all conditions. Dogs were maintained on life-support during experiments and recovered from curarization between phases of the experiment

[b]The text states "[t]he semitendinosus muscle, having first been conditioned in the normal state..."

[c]In the final experiment in dog 20, training of the left leg was begun after tests of the right leg

response (UR) or the conditional response (CR), explored whether pairings of a bell and shock could establish a conditional reflex in an isolated muscle in dogs after an intravenous injection of curare. Curare did not change the speed of conditioning, but certainly changed the magnitude of the UR and form of the CR. Of importance here, in the first subjects conditioned under curare (see subjects 11 and 12), the "CR in the semitendinosus muscle vanished ... as the animal was restored to normal" (page 266). When curare was reintroduced, CRs rapidly reappeared, demonstrating that the conditional reflex had not been lost or forgotten, but rather seemed to require, i.e., be dependent on, the presence of curare. Moreover, in subject 15, when curare was reintroduced, repeated presentations of the CS alone elicited "... a strong and clear-cut CR" (page 267). Subject "no ID" showed that state dependence could be conditioned also to the absence of a drug. Finally, a single individual could have different memories that depended, separately, on either the presence or absence of a drug. As shown by subject 20, training in the "no-drug" state yielded a CR in the right hind limb dependent on the absence of curare, whereas contemporaneous training of the left hind limb in the "drug state" yielded a CR that required (i.e., was dependent on) the presence of curare.

These early experiments demonstrate also several other characteristics seen in subsequent studies of state dependence. These include considerable inter- and intra-subject variability; the phenomenon of asymmetric transfer, in which

State-Dependent Learning and Memory, Table 3 Variation of state during training and/or test phases allows quantitative characterization of transfer in studies of drug-induced state dependence (Colpaert this volume, Koek 2011). Each training + test combination is studied in a unique group of subjects, and the test phase comprises a single session. As in Table 1, the state present during training is indicated by red stipples and a change in state by grey shading

Training phase	Test phase	Experimental goals
Saline	Saline	**Test phase**: assess saline to saline transfer
	Range of doses of drug X	**Test phase**: assess saline to drug transfer and drug's potency as test state and/or to disrupt behavior
Drug X, at different doses in different groups of subjects	Saline	**Training phase**: assess drug's potency as a training state **Test Phase**: Assess drug to saline transfer.
	Drug X, at training dose	**Test phase**: assess drug-to-drug transfer
	Dose X, at range of other doses	**Test phase**: assess drug's potency as test state and/or to disrupt behavior
	Range of doses of drug X (same pharmacological class) and drug Y(different pharmacological class)	**Test phase**: assess pharmacological selectivity of drug to state transfer

transfer from a "drug state" to a "no-drug state" differs from transfer from a "no-drug state" to a "drug state"; and use of only two states, the presence and absence of one high dose of one drug. These characteristics and the resulting interpretative challenges, coupled with confusing failures to demonstrate robust state-dependent effects, led some investigators to conclude that state dependence was restricted to high doses of drugs, displayed considerable intra- and inter-experiment variability, and/or was unpredictable (Eich 2008; Overton 1984). Scientific interest in state dependence has been rekindled by investigations that have sought to identify its potential constraints and boundary conditions. Two prominent examples illustrate the advances in understanding that have emerged.

First, in studies of drug state-dependent learning, Colpaert and colleagues developed procedures that allow quantitative descriptions of the magnitude of state-dependent effects (Colpaert this volume, Koek 2011). In these procedures, the training phase extends over multiple sessions and includes a learning criterion and assessment of how well individual subjects learn the task. The number of experimental groups expands dramatically to allow variation of drug dose and/or type (Table 3). Experiments using such designs have shown that psychoactive drugs can produce state-dependent effects at doses similar to those at which the drugs produce other psychological effects. For example, chlordiazepoxide, the best-studied drug to date, can produce state dependence at doses similar to those required to disrupt operant responding or to exert anti-punishment or anxiolytic-like effects. Morphine can do so at doses similar to those required for antinociceptive effects and the 5-HT1A agonist 8-OH-DPAT, at doses similar to those required for antidepressant-like immobility effects. These findings suggest that certain therapeutic effects of psychoactive drugs might arise because their introduction or removal creates state changes that hamper memory retrieval or that therapeutic effects achieved while a drug is taken might be lost due to state changes after drug treatment ends.

Variation of doses in either the training or test phase (Table 3) allows comparison of a drug's potency to produce state dependence as a training state versus a test state. Several drugs, including NMDAR antagonists and morphine, appear more potent as training states than as test states; i.e., state-dependent effects occur at somewhat lower doses in tests of drug to saline transfer than in tests of saline to drug transfer. Such asymmetry does not occur with all drugs, however. As noted above, the range of doses observed to affect transfer as either training states or test states overlaps substantially with the range of doses required for other behavioral effects of the drugs, suggesting that state dependence may co-occur with these other drug effects, including therapeutic effects.

A second line of research has found evidence that state dependence is affected not only by pharmacological factors, but also by behavioral and psychological factors. For example, in studies of drug-induced state dependent learning in rats, ethanol failed to produce state dependence when behaviors were reinforced by food pellets but did so when the reinforcer was sweetened condensed milk. This shows that state dependence is influenced by behavioral factors such as the nature of the reinforcer. Numerous studies of drug- and mood-dependent memory in humans (Eich 2008) also found that state dependence is influenced by behavioral and psychological factors. For example, state dependence is more likely when materials are self-generated as opposed to when they are externally presented, and when other prominent retrieval cues are unavailable as opposed to when such cues are present. Additional work is required to delineate how such behavioral, psychological, and pharmacological factors interact to influence the likelihood of state-dependent retrieval.

Limitations and Implications of State-Dependent Learning and Memory

In many psychopharmacology experiments, drug treatments vary across experimental conditions. If these treatments allow an individual's pharmacological state at the time of learning to vary from the pharmacological state at the time(s) of later assessment, analyses and interpretations must consider the possibility that changes in performance or remembering might arise from state dependence. Experimental controls for state dependence may be particularly important if a key assessment of learning or memory will occur simultaneously with the sudden introduction or removal of a drug.

The usefulness of controls for state dependence can be illustrated by analyses of drug-induced blockade of psychomotor sensitization. When stimulant drugs such as d-amphetamine and cocaine are administered repeatedly at intermittent intervals, the magnitude and duration of their psychomotor stimulant effects can increase markedly. Numerous experiments have identified the doses, intermittent treatment intervals, and treatment durations that can combine to produce sensitization. Of importance to this discussion, noncompetitive NMDAR antagonists can block development and/or expression of sensitization, an outcome that has been interpreted as evidence that neuronal NMDARs and related long-term potentiation (LTP) processes are involved in psychomotor sensitization.

However, the designs required to study drug-induced blockade of psychomotor sensitization may allow unexpected state dependence (Tzschentke and Schmidt 2000). The hypothesis that NMDAR antagonists decrease sensitized locomotor behaviors by a process of state-dependent learning is based on several features of sensitization experiments, and the following discussion omits some important controls in order to emphasize those features. In experiments designed to explore blockade of the *development* of sensitization, one group of animals is treated repeatedly with a stimulant and its vehicle, and a second group is treated with the stimulant combined with a challenge drug. After a treatment duration known to produce sensitization to the stimulant alone, all animals are tested with the stimulant. When blockade of sensitization is observed, animals treated initially with the stimulant alone show a sensitized locomotor response, whereas animals treated with the stimulant and the challenge drug do not. A state dependence interpretation asks whether this absence of sensitization arises from the change in state in animals experienced with the drug combination, rather than from blockade of sensitization processes. Note that the usual experimental design could support state dependence, in that the repeatedly administered stimulant or stimulant + challenge combination may serve both to elicit a progressive increase in locomotor behavior and also as a state. At the test, the stimulant alone would match the training state for one group but create a state change for the second. Evaluation of this hypothesis requires tests in both of the drug states that are present during training.

A state dependence analysis can also be applied to studies of blockade of the *expression* of sensitization. In one design, all animals are treated repeatedly with stimulant alone for a treatment duration known to produce sensitization. Then, one group of animals is tested with the stimulant drug alone; a second group, with the stimulant plus a challenge drug; and others, with the vehicle alone or the challenge drug alone. When blockade of the expression of sensitization is observed, animals tested with the stimulant alone show sensitized locomotor responses, whereas animals tested with the combination do not. This design also could support state dependence, as the repeatedly administered stimulant may serve both to elicit a progressive increase in locomotor behavior and as a state, while the drug given at the test matches the training state for one group but creates a state change for others.

Table 4 illustrates three experiments that have examined the hypothesis that state dependence may account for at least some cases of blockade of psychomotor sensitization. The combined results of studies by Carlezon et al. (1995) and Wise et al. (1996) suggest that co-treatment with the NMDAR antagonist dizocilpine may not block sensitization but rather render it state dependent so that it disappears when the antagonist is omitted in tests of the stimulant alone. The experiment by Stephens et al. 2000 demonstrated that state dependence of psychomotor sensitization is not limited to agents that act through NMDAR systems. While further research is needed to understand the generalizability of these results, taken together they make a good case that an imposed manipulation that disrupts behavior or remembering may do so either by changing a targeted neurochemical circuit or by changing a learner's state – and experimental designs that control for state dependence are a necessary tactic to disambiguate these possibilities.

Critics of the concept of state dependence have noted that states are imperfectly defined and characterized; one point of contention is whether a state must be "noticeable" to a learner in order to influence learning or remembering. Additionally, the literature contains lively controversies about whether state dependence is a unique process or an instance of stimulus or contextual control that develops rapidly, without explicit exposure to different stimuli and associated differential reinforcement. Whereas many researchers have concluded that drug-induced state dependence arises from stimulus control by drugs or other interoceptive states and is thus a case of drug- or state-discrimination learning (Overton 1984), Colpaert and colleagues argued that state-dependent learning and drug discrimination are distinct processes (Colpaert this volume, Koek 2011). In a related controversy, it remains an open question whether state dependence is a special case of context-dependent learning and memory or whether it uses unique processes or circuits. Context dependence occurs when retrieval of a memory is most successful in the presence of the same contextual cues present when learning occurred and the memory was acquired (Bouton and Moody 2004). From this perspective, state dependence may be a special case of context dependence that relies on interoceptive rather than exteroceptive cues, such that retrieval is most successful in the presence of interoceptive cues produced by the physiological state that obtained when learning occurred and memories were acquired. Future studies that manipulate the behavioral conditions present when physiological state is altered by drug administration or other operations are needed to determine whether factors known to influence drug discrimination or context dependence can

State-Dependent Learning and Memory, Table 4 Tests of pharmacological "blockade" of psychomotor sensitization under experimental designs that incorporate both a sensitizing regimen (left column) and tests of state dependence. The combination allows examination of the possibility that state dependence may underlie the decreased locomotor behavior observed when a new drug is introduced after treatment with a sensitizing drug regimen. As in Table 1, the state present during training is indicated by red stipples and a change in state by grey shading

Stimulant and challenge treatments	Did treatment produce locomotor sensitization?	Was locomotor sensitization observed during tests of _____ ?		
		Stimulant + vehicle	Vehicle + challenge	Stimulant + challenge
Bromocriptine + vehicle	Yes	Yes	No[a]	No
Vehicle + vehicle	No	No	No	
Vehicle + dizocilpine (MK-801)	No	No	No	
Bromocriptine + dizocilpine (MK-801)	Yes	No	No[a]	
		Source: Carlezon et al. (1995), Wise et al. (1996)		Source: Carlezon et al. (1995)

Stimulant and challenge treatments	Did treatment produce locomotor sensitization?	Was locomotor sensitization observed during tests of _____ ?	
		Stimulant + vehicle	Vehicle + challenge
d-Amphetamine + vehicle	Yes	Yes	No[b]
Vehicle + vehicle	No	No	No
Vehicle + chlordiazepoxide	No	No	No
d-Amphetamine + chlordiazepoxide	Yes	No	No
		Source: Stephens et al. (2000)	

[a] In Wise et al. (1996), activity counts on the test day were highest in animals that had received bromocriptine treatment, but, unlike patterns when a stimulant was given on the test day, counts decreased as the session progressed

[b] Activity counts were highest in animals that had received d-amphetamine + vehicle treatment but were much lower than counts at the end of treatment or following tests of d-amphetamine alone

influence the state dependence of learning and remembering.

Finally, state dependence may influence therapeutic or other effects of drugs. If the need to use a newly acquired skill or memory is coincident with the sudden removal or introduction of a psychoactive drug, state dependence may lead to an unexpected behavioral or memory failure. Such combinations of new learning and a change in drug state may occur in therapies that combine psychological and drug treatments, in abusive drug use, or in situations when an individual tries a new "endurance" or "cognitive enhancing" drug as he/she races to learn new material. Improved understanding of state dependence will facilitate understanding of these and other outcomes that emerge from interactions of drugs and learning and memory processes.

Cross-References

▶ Amphetamine
▶ Benzodiazepines

- ▶ Bromocriptine
- ▶ Chlordiazepoxide
- ▶ Drug Discrimination
- ▶ Glutamate Receptors
- ▶ Morphine
- ▶ Muscarinic Cholinergic Receptor Agonists and Antagonists
- ▶ Nicotine
- ▶ Opioid Analgesics
- ▶ Sensitization to Drugs
- ▶ Serotonin

References

Bouton ME, Moody EW (2004) Memory processes in classical conditioning. Neurosci Biobehav Rev 28(7):663–674

Carlezon WA Jr, Mendrek A, Wise RA (1995) MK-801 disrupts the expression but not the development of bromocriptine sensitization: a state-dependency interpretation. Synapse 20(1):1–9

Eich E (2008) Mood and memory at 26: revisiting the idea of mood mediation in drug-dependent and place-dependent memory. In: Gluck MA, Anderson JR, Kosslyn SM (eds) Memory and mind: a festschrift for Gordon H. Bower. Lawrence Erlbaum Associates, New York

Girden E, Culler E (1937) Conditioned responses in curarized striate muscle in dogs. J Comp Psychol 23(2):261–274

Koek W (2011) Drug-induced state-dependent learning: review of an operant procedure in rats. Behav Pharmacol 22(5–6):430–440

Overton DA (1984) State dependent learning and drug discriminations. In: Iversen LL, Iversen SD, Snyder SH (eds) Handbook of psychopharmacology, vol 18. Plenum Publishing Corporation, New York

Overton DA (1991) Historical context of state dependent learning and discriminative drug effects. Behav Pharmacol 2(4 And 5):253–264

Stephens DN, Elliman TD, Dunworth SJ (2000) State-dependent behavioural sensitization: evidence from a chlordiazepoxide state. Behav Pharmacol 11(2):161–167

Tzschentke TM, Schmidt WJ (2000) Blockade of behavioral sensitization by MK-801: fact or artifact? A review of preclinical data. Psychopharmacology (Berl) 151(2–3):142–151

Wise RA, Mendrek A, Carlezon WA Jr (1996) MK-801 (dizocilpine): synergist and conditioned stimulus in bromocriptine-induced psychomotor sensitization. Synapse 22(4):362–368

Stationary Phase

Definition

The packing material in a chromatographic column. In RP-HPLC, the most common stationary phases comprise beads of silica (glass-like material), covered by a layer of organic molecules (the bonded phase), comprising alkyl chains, chemically bonded to the silica. These are normally of 8 (C8: octylsilane) or 18 (C18: octadecylsilane – ODS) carbon molecules, tightly packed into the pores in the silica bead. The silica beads are normally 5 or 3 μm diameter.

Cross-References

▶ High-Pressure Liquid Chromatography

Stereoisomers

Synonyms

Enantiomers; Enantiomorphs; Optical isomers

Definition

These usually refer to pairs of drugs, which have identical physicochemical properties but rotate polarized light in different directions (left or right, designated as levo or dextro, "L-" or "D-" and "−" or "+," respectively). They are important as they often have different biological activities. Thus, an atom found on the drug molecule has bonds that do not permit the attached functional groups to assume different positions relative to one another. For example, morphine derived from opium has one important carbon atom with functional groups in positions that rotate polarized light to the left (termed levorotatory). The configuration of this carbon atom also allows the morphine molecule to fit precisely into the pockets and contours of the opioid receptor and to activate it. In other words, the receptor has a structure that matches the levorotatory form. One can also construct morphine synthetically with the functional groups on the same carbon atom in different relative positions. The arrangement of the bonds is such that they precisely mirror those in the levorotatory form and the molecules of this morphine rotate polarized light to the right (show dextrorotatory activity). This synthesized substance is also morphine in a chemical sense, because it has all of the physicochemical properties of morphine made from opium. However, the dextrorotatory form of morphine evokes only the effects of normal morphine that have low chemical structure requirements (e.g., membrane stabilization, convulsions, taste aversion produced by systemic application of opioid); it does not activate the typical pharmacological effects attributed to activity on the opioid receptor. The use of opioid stereoisomers has been recommended in order to exclude so-called "nonspecific" pharmacological effects that are not due to interaction with the opioid receptor.

Cross-References

- ▶ Conditioned Taste Aversions
- ▶ Receptors: Binding Assays

Stereotypical and Repetitive Behavior

Definition

Stereotypical and repetitive behavior is commonly seen in individuals with autism spectrum disorder (ASD) and intellectual developmental disorder (IDD). In lower functioning individuals with ASD or IDD, this usually consists of self-stimulatory, nonfunctional, motor behaviors. However, in mild IDD or higher functioning individuals with ASD, this can consist of verbal and motor rituals, obsessive questioning, rigidly held routines, preoccupation with details, and desire for sameness and completeness.

Cross-References

- ▶ Antidepressants
- ▶ Antipsychotic Drugs
- ▶ Autism Spectrum Disorders and Intellectual Disability
- ▶ Psychostimulants

Stimulation Paradigm

Definition

In fMRI experiments, stimulation paradigms are applied to evoke brain activity and, accordingly, increased or decreased cerebral perfusion. The variety of stimulation paradigms ranges from no stimulation (resting-state fMRI) over sensory (thermal, visual, or auditory), motor, language, or memory task to pharmacological stimuli. Aside from resting-state fMRI performed while no stimulation is applied, fMRI images are recorded during two states: the resting state (no stimulation) and the activated state during which the specific stimulation is applied. The applied stimulation paradigm is often incorporated as prior knowledge in the subsequent MR image analysis.

Cross-References

- ▶ Cerebral Perfusion
- ▶ Functional Magnetic Resonance Imaging
- ▶ MR Image Analysis
- ▶ Pharmacological fMRI

Stimulation Strength

Definition

The intensity dimension(s) of an intracerebral stimulus. In the case of rewarding electrical or optical brain stimulation, these include the amplitude (current or optical power), pulse duration, pulse frequency, and train duration.

Stimulus Control

Definition

A behavioral process ensuring that different behavior occurs in the presence of different stimuli. It is usually established by arranging different contingencies of reinforcement in the presence of different stimuli.

Stimulus Generalization

Definition

Stimulus generalization occurs when a behavioral response conditioned to one stimulus is elicited by another stimulus that was not used during conditioning. For example, animals are conditioned to use a particular stimulus as a discriminative cue and are then tested with a novel stimulus that may resemble the training stimulus in one or more respects. If the animal responds to a test stimulus as it does to the training stimulus, then stimulus generalization is said to have occurred. Generalization may be either complete or partial and can reflect both qualitative and quantitative differences between stimuli (e.g., color and brightness of visual stimuli).

Cross-References

▶ Drug Discrimination

Stop-Signal Task

Synonyms

Response inhibition task; SSRT

Definition

A neurocognitive task designed to provide a sensitive measure of the time required to inhibit or suppress inappropriate motor responses. The stop-signal paradigm was originally developed by Gordon Logan in the 1980s based on a cognitive task first used by Lappin and Erikson in 1966. Though manual responses are by far the most commonly used in humans, versions using left and right saccadic eye movements have also been developed in humans and monkeys. Versions of the task have also been developed in rats where differential presentations of reinforcers are used to train appropriate responding to "go" and "stop" signals.

In the version of the task for humans, participants are instructed to respond as fast as possible to a simple stimulus on a computer screen, typically one of two stimuli each associated with a response (e.g., press a left button for "X" and a right button for "O"). On a minority of trials, a stop signal is presented after the stimulus but before the response, and participants are instructed to try and stop, or countermand, their response. Using this task, impairments in response inhibition have consistently been observed in neuropsychiatric disorders such as attention deficit hyperactivity disorder.

The stop-signal task not only provides measures of reaction times and accuracy but importantly also the latency to inhibit a prepotent response: stop-signal reaction time (SSRT). The stopping process is not directly observable and has to be estimated from a "race model" that assumes stochastic finishing times for the go and stop processes. The model derives SSRT from the distribution of "go" reaction times and the observed probability of responding on "stop" trials for a given stop-signal delay. The estimated SSRT provides a measure of the duration of the inhibitory process, which is initiated with the presentation of stop signal. Contemporary versions often employ a staircase that adjusts the stop-signal delay online allowing a slightly modified computation of SSRT that assumes 50 % successful inhibition rates. In addition to providing an SSRT estimate, the stop-signal task is different from go/no-go tasks as the stop signal is distinct from and presented after go signals and requires the canceling of an already planned prepotent response.

A network in the brain mediating the suppression of a motor response has been investigated, involving numerous cortical and subcortical areas including the prefrontal cortex and the basal ganglia.

Cross-References

- ▶ Attention Deficit Hyperactivity Disorders: Animal Models
- ▶ Impulse Control Disorders
- ▶ Impulsivity
- ▶ Rodent Tests of Cognition
- ▶ Translational Research in Drug Discovery

Stress

Definition

Stress as a psychological construct is complex, and despite many years of research, it has yet to be defined operationally in a satisfactory way. In the context of animal models of psychiatric disorders, stress can be defined broadly as forced exposure to events or conditions that are normally avoided. In humans, the definition is extended to incorporate cognitive and emotional responses – for example, "stress is a condition in which the environmental demands exceed the coping abilities of the individual." In laboratory animals, the precipitating events or conditions can be divided into two categories. The first category includes environmental events such as restraint, footshock, tail pinch, and defeat as well as pharmacological events such as administration of a normally avoided drug (e.g., yohimbine). The second category includes food deprivation, social isolation, and maternal deprivation; each of these entails the removal of an environmental condition that is important for maintaining the animal's normal physiological and psychological steady-state conditions, a state that the subject will attempt to ameliorate by seeking food, conspecific partners, or the dam.

Cross-References

- ▶ Reinstatement of Drug Self-Administration
- ▶ Social Stress
- ▶ Stress: Influence on Relapse to Substance Use

Stress Response

Definition

The spectrum of physiological and behavioral adaptations coordinated by stress system mediators that defend homeostasis and/or promote allostasis.

Stress: Influence on Relapse to Substance Use

John R. Mantsch[1] and Yavin Shaham[2]
[1]Department of Biomedical Sciences, Marquette University, Milwaukee, WI, USA
[2]Behavioral Neuroscience Branch, IRP/NIDA/NIH, Baltimore, MD, USA

Definition

In humans, stress often refers to a condition in which environmental demands exceed an individual's coping abilities. In animal models of psychiatric disorders, stress can be defined broadly as a forced exposure to aversive events or conditions that are normally avoided. The term stress includes three related elements: stressors, stress responses, and genetic and environmental factors that modulate the effect of stressors on the organism. Stressors are events, physical or psychological, that profoundly interfere with the organism's normal steady state. These disruptions generate a stress response manifested at the physiological (e.g., activation of the sympathetic nervous system), psychological (e.g., anxiety, depression), and behavioral (e.g., performance deficits) levels. Factors that modulate the stress response include, among others, genetic predisposition to stress reactivity and predictability and controllability of the stressors. These factors influence the relationship between the stressors and the stress response, leading to large individual differences in response to a given stressor.

Current Concepts and State of Knowledge

Background

Anecdotal reports that drug use is more likely to occur in individuals exposed to environmental stressors are supported by results from epidemiological studies indicating a high rate of comorbidity between stress-related psychiatric disorders such as post-traumatic stress disorder (PTSD), anxiety, and depression and drug addiction. There is also evidence from laboratory studies in humans that exposure to stressors increases cigarette smoking and subjective measures of cocaine, opiate, and alcohol craving. The influence of stress on abused drugs appears to occur through all phases of the addiction process: stress facilitates the initiation of illicit drug use, increases the frequency and amount of ongoing drug use, and precipitates relapse to drug use during periods of abstinence. Stress can also emerge as a consequence of drug addiction, because illicit drug use is associated with problems with the law, the workplace, and the family. The stress caused by the lifestyle of illicit drug users in turn promotes more severe forms of drug addiction. Finally, chronic use of addictive drugs can sensitize physiological and psychological responses to stress.

However, methodological and ethical considerations limit the scope of research that can be conducted in humans addicted to drugs of abuse. Consequently, it has been difficult to demonstrate a cause-effect relationship between stress and drug addiction in humans. Cause-effect relationships between stress and drug-taking behavior can be established in laboratory models of drug addiction. These models can be used to identify behavioral and neuronal mechanisms that mediate the effect of stress on drug-taking behavior. Below, we summarize results obtained from preclinical studies in laboratory rats on the mechanisms underlying the effect of environmental stressors on drug-taking behavior, as assessed using the intravenous drug self-administration procedure and the reinstatement procedure. The drug self-administration procedure is the gold standard animal model used to assess the rewarding effects of drugs and their abuse liability. The reinstatement procedure is a widely used animal model of relapse to drug-taking behavior during periods of abstinence.

Before reviewing the empirical data on biological mechanisms of the effects of stress on drug-taking behavior, we would like to point out that we will not cover in this short chapter several research topics that are relevant to the relationship between stress and drug addiction. These include (1) the extant and conflicting literature on the effects of stress on oral opiate and alcohol self-administration; (2) the conflicting literature on the effects of adverse early life experience (maternal deprivation, social isolation) on drug self-administration in adulthood; (3) the literature on the role of stress systems in the mediation of physiological and psychological withdrawal symptoms of abused drugs; and (4) the literature on the role of the stress hormone corticosterone in the mediation of psychostimulant self-administration in the absence of stress.

Mechanisms of Stress Effects on Drug Self-administration

Investigators have used the drug self-administration procedure to assess the effects of stressors on the acquisition and maintenance of intravenous drug self-administration. When evaluating the findings on the effects of stressors on drug self-administration, it is important to remember that these effects are highly dependent on several variables, including stressor controllability, duration of stress exposure, the time interval between stress exposure and drug availability, the reinforcement schedule, and the dose of the self-administered drug. Additionally, there are large individual variations in the effects of stressors on drug self-administration. In studies on the mechanisms of stressor-induced increases in drug self-administration in rats, investigators have used several stressors including intermittent unpredictable footshock, food deprivation, and social defeat.

The stress hormone corticosterone (or cortisol in humans) is released from the adrenals when stressors activate the hypothalamic-pituitary-adrenal (HPA) axis. There is evidence that stressor-induced corticosterone secretion is required for the ability of intermittent footshock and food deprivation stressors to facilitate the initiation of cocaine self-administration. Additionally, corticosterone may contribute to footshock stress-induced escalation of cocaine self-administration once the self-administration behavior has been established. The mechanisms through which corticosterone promotes the self-administration of psychostimulant drugs may involve modulation of dopaminergic neurotransmission in both the ventral tegmental area (VTA), the cell body region of the mesolimbic dopamine system, and one of its projection areas, the nucleus accumbens. The mesolimbic dopamine system is known to mediate the rewarding effects of psychostimulant drugs.

In most previous studies on the effects of stress on intravenous self-administration of psychostimulant drugs, investigators limited access to the drugs for a few hours per day; under these conditions, rats typically maintain stable drug intake over time. Human addicts, however, self-administer psychostimulant drugs such as cocaine in a binge-like pattern that is characterized by extended periods of intense and escalated drug intake over many hours and even days. This binge-like pattern of drug self-administration is also observed in rats that are given unlimited access to cocaine. Thus, after about 20–24 h of continuous cocaine self-administration, rats shift from regulated drug intake, characterized by stable inter-infusion intervals, to dysregulated "out-of-control" drug intake, characterized by unstable inter-infusion intervals and higher hourly drug intake. The emergence of this binge-like pattern of cocaine intake is facilitated by prior exposure to social defeat stress. There is evidence that this effect of social defeat stress on binge-like cocaine self-administration is due to sensitization of dopamine neurons in the VTA. This sensitization is likely mediated by stressor-induced activation of glutamatergic projections to the VTA. As the neuronal and hormonal responses to stressors are typically stressor specific, a question for future research is whether the neurobiological mechanism identified for the effect of social defeat stress on binge-like cocaine self-administration generalizes to other stressors.

Stress and Reinstatement of Drug Seeking

The phenomenon of stress-induced reinstatement of drug seeking has been demonstrated using several stressors, including intermittent footshock and acute food deprivation. Stress-induced reinstatement is critically dependent on the stress neuropeptide corticotrophin-releasing factor (CRF). This neuropeptide acts in the paraventricular nucleus of the hypothalamus to activate the HPA-endocrine stress axis. Additionally, many physiological and behavioral responses to stress are mediated by CRF's effects on extrahypothalamic sites within the central nervous system (CNS).

CRF receptor antagonists decrease both footshock- and food deprivation-induced reinstatement of drug seeking, while ventricular injections of CRF reinstate drug seeking. Results from studies using pharmacological and endocrine methods indicate that the reinstating effects of stressors are independent of their ability to activate the HPA axis. The critical extrahypothalamic brain sites and projections for CRF's role in footshock-induced reinstatement include the bed nucleus of the stria terminalis (BNST), VTA, and a projection from the central amygdala to the BNST. Within the VTA, footshock causes local CRF release, which leads to increased glutamate transmission. This enhanced glutamatergic neurotransmission is critical for stressor-induced reinstatement, presumably due to activation of the mesocorticolimbic dopamine system. This possibility is supported by the findings that injections of D1 family dopamine receptor antagonists into the dorsal, medial prefrontal cortex (mPFC) or the orbitofrontal cortex, or of a preferential D3 dopamine receptor antagonist into the nucleus accumbens decrease stressor-induced reinstatement.

Footshock stress-induced reinstatement is also dependent on noradrenaline transmission: systemic injections of alpha-2 adrenoceptor agonists (which decrease central noradrenaline release) decrease footshock-induced reinstatement, while systemic injections of the alpha-2 adrenoceptor antagonist yohimbine (which increases central noradrenaline release) cause reinstatement of drug seeking. The critical brain sites and projections for noradrenaline's role in footshock-induced reinstatement include the central amygdala and BNST, and the noradrenergic projection from the lateral tegmental nuclei to these brain sites.

Results from studies in which discrete brain areas were reversibly inactivated confirm the previous findings discussed above on the role of the dorsal mPFC, BNST, central amygdala, accumbens, and VTA in stress-induced reinstatement and further suggest that the ventral pallidum plays a role in this reinstatement. Results from these studies also indicate that the glutamatergic projection from the mPFC to the accumbens plays an important role in stress-induced reinstatement. This glutamatergic projection is also involved in reinstatement of drug seeking induced by acute reexposure to the self-administered drug.

Recent evidence suggests that the magnitude of footshock- and CRF-induced reinstatement of cocaine seeking depends on the history of prior drug exposure: these effects were significantly stronger in rats that were previously given extended access to cocaine for 6 h/day during training than in rats given drug access for 2 h/day. These and related findings indicate that prior exposure to high doses of cocaine over many days sensitizes brain CRF stress systems, leading to increased vulnerability to stress-induced reinstatement of cocaine seeking. These findings parallel results from human studies demonstrating more pronounced stress-induced craving, anxiety, and physiological responses in abstinent, previously high-frequency drug users than in individuals with a history of less frequent drug use.

Finally, there is evidence from studies in which pharmacological agents were used for a role for several other neurotransmitter systems in stress-induced reinstatement. These include the neurotransmitter serotonin and the neuropeptides dynorphin, hypocretin (orexin), and nociceptin/orphanin FQ.

Implications for Treatment

The recognition that stress is a key contributor to drug-taking behavior highlights the need for the development and implementation of therapeutic strategies aimed at minimizing the contribution of stress to drug addiction. This is especially important in subpopulations of addicts whose drug use is stress driven. Identifying which addicts will benefit most from such approaches poses a challenge to drug addiction treatment providers and will likely require the establishment of new assessment tools for identifying the role of stress in an individual's drug use. The development and approval of new drugs that block CRF receptors and other receptors implicated in stress-induced drug seeking will hopefully provide important tools for the management of drug addiction.

Acknowledgment The writing of this chapter was supported in part by the Intramural Research Program of the NIH, NIDA.

Cross-References

▶ Conditioned Place Preference and Aversion
▶ Drug Discrimination
▶ Extinction
▶ Reinstatement of Drug Self-Administration
▶ Stress

References

Koob GF (2008) A role for brain stress systems in addiction. Neuron 59:11–34
Lu L, Shepard JD, Scott Hall F, Shaham Y (2003) Effect of environmental stressors on opiate and

psychostimulant reinforcement, reinstatement and discrimination in rats: a review. Neurosci Biobehav Rev 27:457–491
Mantsch JR, Katz ES (2007) Elevation of glucocorticoids is necessary but not sufficient for the escalation of cocaine self-administration by chronic electric footshock stress in rats. Neuropsychopharmacology 32:367–376
Mantsch JR, Baker DA, Francis DM, Katz ES, Hoks MA, Serge JP (2008) Stressor- and corticotropin releasing factor-induced reinstatement and active stress-related behavioral responses are augmented following long-access cocaine self-administration by rats. Psychopharmacology (Berl) 195:591–603
Marinelli M, Piazza PV (2002) Interaction between glucocorticoid hormones, stress and psychostimulant drugs. Eur J Neurosci 16:387–394
McFarland K, Davidge SB, Lapish CC, Kalivas PW (2004) Limbic and motor circuitry underlying footshock-induced reinstatement of cocaine-seeking behavior. J Neurosci 24:1551–1560
Miczek KA, Yap JJ, Covington HE III (2008) Social stress, therapeutics and drug abuse: preclinical models of escalated and depressed intake. Pharmacol Ther 120:102–128
Shaham Y, Erb S, Stewart J (2000) Stress-induced relapse to heroin and cocaine seeking in rats: a review. Brain Res Brain Res Rev 33:13–33
Sinha R (2008) Chronic stress, drug use, and vulnerability to addiction. Ann N Y Acad Sci 1141:105–130
Wang B, Shaham Y, Zitzman D, Azari S, Wise RA, You ZB (2005) Cocaine experience establishes control of midbrain glutamate and dopamine by corticotropin-releasing factor: a role in stress-induced relapse to drug seeking. J Neurosci 25:5389–5396

Stress-Induced Antinociception

Definition

Decreased pain transmission in response to stressful stimuli. Stress-induced antinociception results in attenuated perception of pain. Both opioid and non-opioid endogenous mechanisms have been implicated in stress-induced antinociception.

Cross-References

▶ Antinociception Test Methods
▶ Opioids

Stressors

Definition

Any perceived threat to the individuals' integrity.

Cross-References

▶ Stress: Influence on Relapse to Substance Use

Striatum

Synonyms

Neostriatum

Definition

The striatum is a subcortical brain structure. The corpus striatum, which includes the putamen rostrally and the caudate nucleus caudally, is a component of the ventral cerebral hemisphere, receiving strong projections from the cerebral cortex and projecting back to it via the thalamus. In addition, the striatum receives a robust dopaminergic innervation from the *substantia nigra* and the ventral tegmental area. It is fundamental for the selection of motor programs in response to external signals, which is triggered by dopaminergic signaling. Its ventral component, the ventral striatum or nucleus accumbens, is a key element in the response to salient stimuli predicting reward, hence inducing intensely motivated states.

Cross-References

▶ Psychostimulants

Stroke

Synonyms

Cerebrovascular accident

Definition

Stroke results from the occlusion or bursting of a cerebral blood vessel so that the cerebral tissue is starved of both oxygen and nutrients and this tissue then dies. The large majority of strokes (around 85 %) are caused by an occlusion of a major cerebral artery either by a thrombus or embolism. The other strokes are the result of a hemorrhage where a blood vessel bursts in the brain or on the surface of the brain. Stroke is the third major cause of death in major industrialized countries, and its incidence is predicted to increase over the next decade.

Structural and Functional Magnetic Resonance Imaging

Christof Baltes[1], Thomas Mueggler[2] and Markus Rudin[3,4]
[1]Varian Medical Systems Imaging Laboratory GmbH, Dattwil, Switzerland
[2]Hoffmann-La Roche Ltd, DTA Neuroscience, Basel, Switzerland
[3]Molecular Imaging and Functional Pharmacology, Institute for Biomedical Engineering, University and ETH Zurich, Zurich, Switzerland
[4]Institute for Pharmacology and Toxicology, University of Zurich, HCI D426, Zurich, Switzerland

Definition

Structural and functional magnetic resonance imaging (MRI) are medical imaging approaches widely used in radiology and neuroradiology. While structural MRI allows acquiring three-dimensional images of neuroanatomy with high spatial resolution and excellent soft tissue contrast, functional MRI (fMRI) is applied to assess brain activity. FMRI provides information on the current status and on neuropharmacological modulation of neuronal activity across the entire brain in a spatially and temporally resolved manner. Structural and functional MRI methods are readily translatable to clinical systems as they are inherently noninvasive and can therefore be applied in human subjects without exposure to radiation such as in nuclear imaging techniques. However, methodological constraints and limitations require careful interpretation of fMRI data. Today, structural and functional MRI have emerged as powerful tools for neuropharmacological research and hold great potential for clinical applications.

Principles and Role in Psychopharmacology

Basic Principles of MRI

Magnetic resonance imaging (MRI) is a biomedical imaging technique applied in both research laboratories and in clinical radiology to visualize structure and function of the body. MR images represent a weighted distribution of hydrogen atoms (protons) in living tissue, the major contribution being due to water constituting approximately 60–80 % of tissue mass (Vlaardingerbroek and den Boer 1999). Hydrogen atoms possess an intrinsic magnetic moment. From a technical point of view, three key components are required to generate the MR signals allowing the formation of an image: the static magnetic field aligning the proton magnetic moments in the direction of the field, the radio-frequency (RF) coil(s) for excitation and reception of the MR signal, and the magnetic gradient field coils for spatial encoding of the MR signal. MRI systems measure the magnetic properties of the protons, which are influenced by electrical and magnetic characteristics of their environment. The local environment varies between tissues and structures the protons are embedded in and

Structural and Functional Magnetic Resonance Imaging, Table 1 Intrinsic MR contrast parameter and information derived

Contrast-generating process	MRI parameter	Information derived
Longitudinal relaxation: return to magnetic equilibrium state	$R_1(=1/T_1)$	Basic structural information: e.g., gray vs. white matter differentiation
		Enhancement following contrast agent administration: e.g., blood–brain barrier integrity or retrograde axonal tracing
Transverse relaxation: loss of phase coherence due to stochastic processes	$R_2(=1/T_2)$	Basic structural information: sensitive to tissue water content
		Edema formation, inflammation
Magnetic susceptibility: loss of phase coherence due to magnetic field differences	$R_2^* (=1/T_2^*)$	Basic structural information: gray vs. white matter contrast, hemorrhages
		Cerebral blood flow and volume (using intravascular contrast agent)
		BOLD contrast
Incoherent motion–diffusion: loss of phase coherence due to molecular diffusion	ADC, FA	ADC: cellularity (intra- vs. extracellular volume fraction)
		FA: restricted anisotropic diffusion
Incoherent motion–perfusion: loss of phase coherence due to flow in capillaries	f	Local tissue perfusion
Water exchange: water exchange between two states of different water mobility	MTR	Macromolecule content: e.g., degree and integrity of myelination
Local magnetic susceptibility	χ	Paramagnetic and diamagnetic inclusions

R_i relaxation rate (s^{-1})
T_i relaxation time (s)
ADC apparent diffusion coefficient (mm^2/s)
FA fractional anisotropy
f tissue perfusion (in ml/s/g tissue or ml/min/100 g tissue)
MTR magnetization transfer ratio: ratio between the equilibrium magnetization and the steady-state magnetization under saturation conditions
BOLD blood oxygen-level dependent (contrast)

are furthermore influenced by physiological processes such as diffusion, perfusion, or blood flow. Consequently, image contrast is governed by a number of MR parameters (Table 1) such as intrinsic relaxation rates (R_1, R_2, R_2^*), incoherent motion of water such as diffusion or perfusion, coherent blood flow in major vessels, and water exchange processes between cellular/interstitial fluid and water bound to macromolecules.

MRI acquisition parameters can be adapted to emphasize the specific contrast optimal for the structure or the process of interest. In addition tissue relaxation rates can be altered using exogenous contrast agents based on paramagnetic compounds such as gadolinium chelates or iron oxide nanoparticles. The strong effect of contrast agents on relaxation rates is due to unpaired electron(s) contained in their electron shell, the magnetic moment of which is approximately 650 times higher than that of protons.

Structural MRI

Three-dimensional imaging, excellent spatial resolution, and superior soft tissue contrast render MRI the method of choice for structural neuroimaging. A weak point of the method is its inherent low sensitivity, which has a negative impact on spatial resolution. Correspondingly, identifying small brain structures or assessing subtle/minor pathological alterations, putting high demands on spatial resolution, is challenging. Substantial efforts have been made in recent

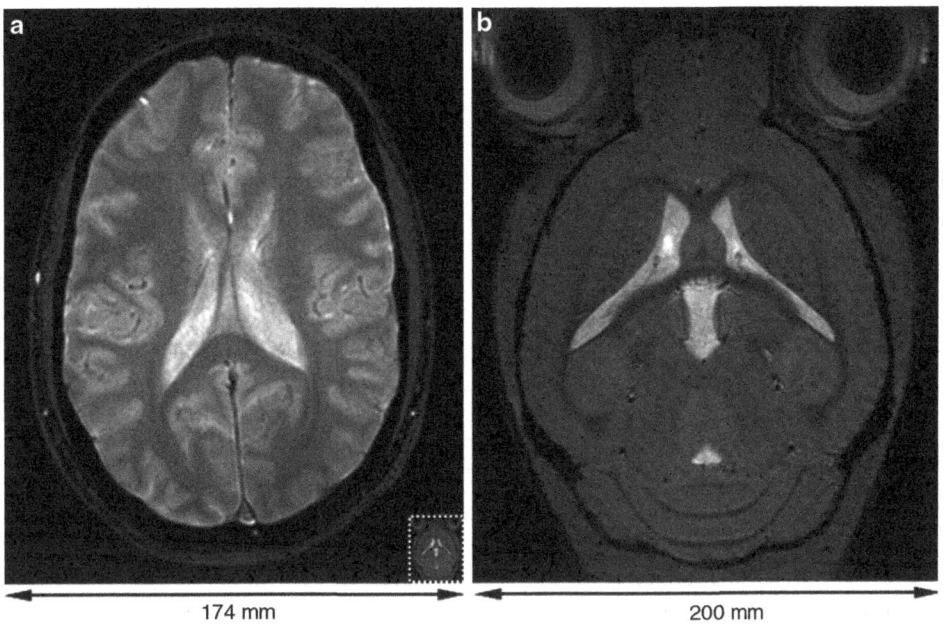

Structural and Functional Magnetic Resonance Imaging, Fig. 1 The visualization of neuroanatomical structures in human* (**a**) and in mouse brains (**b**) puts different demands on spatial resolution. While an in-plane resolution of 0.9×0.9 mm^2 in human subjects is sufficient for gray versus white matter discrimination, a resolution of 52×52 μm^2 is required in mouse brains to depict cortical structures such as the hippocampus. For comparison of the dimensions, the mouse brain image (**b**) is depicted in the human brain image (**a**) using the same scale (*Courtesy of R. Luechinger, PhD, University and ETH Zurich, Switzerland)

years to increase MR sensitivity by either moving to higher static magnetic field strength or by refining RF detection devices such as the cryogenic detection technology. State-of-the-art MR systems for routine clinical neuroimaging (Konarski et al. 2007) operate at 3 T allowing for spatial resolutions on the order of 1 mm^3 (Fig. 1a), while clinical research systems up to 9.4 T providing significantly higher spatial resolution are currently being evaluated (Budde et al. 2014). Currently an 11.7 T human MRI system is under construction and foreseen to become operational in 2014. In contrast, the high spatial resolution required in rodent brain imaging led to the development of MR systems operating at up to 17.6 T. For example, the combination of high magnetic field strength (9.4 T) with cryogenic RF detection devices enabled the routine recording of high-resolution ($52 \times 52 \times 170$ μm^3) mouse brain images (Fig. 1b) in a measurement time of 10 min (Baltes et al. 2009; Ratering et al. 2008).

The quality of images is characterized by its spatial resolution and the contrast-to-noise ratio (CNR). A high CNR is essential to discriminate individual (brain) structures, for qualitative diagnostic purposes or for quantitative analysis. In structural MRI usually morphometric comparisons between groups are performed using hypothesis-based selection of regions of interest (ROIs), which is a lengthy process and prone to evaluation errors. More recently, voxel-based morphometry (VBM) has been developed as a fully automated comparison of whole brains on a voxel-by-voxel basis. After spatial normalization of the brains to a stereotactic standard space, brain regions are compared with respect to differences in residual tissue concentrations rather than differences in shape. Clinical application of structural readouts using MRI covers a broad range of disorders from neurodegenerative (i.e., stroke and dementia) to neuropsychiatric diseases, i.e., schizophrenia, PTSD, or mood disorders such as bipolar affective disorders (as reviewed by Malhi and Lagopoulos 2008). For example VBM studies in schizophrenia generally confirmed and extended ROI-based studies showing less gray matter concentration in multiple cortical and subcortical regions (Pearlson and Calhoun 2007). In patients suffering, e.g., from bipolar disorders or other psychiatric diseases, structural MRI has been applied to assess neuroanatomical abnormalities between healthy subjects and patients (American Psychiatric Association 2013) While overall brain volumes appeared to be normal, regional differences have been observed in the prefrontal cortex and subcortical and medial temporal structures involved in the behavioral network which is known to be affected in bipolar disorders (Strakowski et al. 2005). Similarly in preclinical MRI morphometric readouts hold promise to phenotype rodent models of CNS diseases (Nieman et al. 2007; Pitiot et al. 2007) with applications in mouse models of neurological disorders such as Alzheimer's disease (AD) (Lau et al. 2008; Redwine et al. 2003) or Huntington disease (Lerch et al. 2008), in a rat model of schizophrenia (Schubert et al. 2009) and in a mouse model of autism spectrum disorder (Ellegood et al. 2011). Its value for detection of subtle or diffuse morphometric abnormalities, i.e., in neuropsychiatric and neurodevelopmental models which are predictive for disease progression or can serve as surrogate marker for end-stage disease status, has to be further validated in carefully planned and analyzed longitudinal studies.

Specific structural information may be derived from MR angiography and diffusion tensor imaging. MR angiography methods discriminate signals from flowing blood from that of stationary tissue. So-called time-of-flight methods are based on rapid acquisition methods that do not allow relaxation of the signal from stationary tissue. As a result the corresponding signal contributions are largely suppressed. In contrast, all water protons in the blood compartment that are rapidly replaced will yield the full signal intensity: As a result, an MR angiogram is obtained that depicts the major (arterial) vasculature in some detail (Baltes et al. 2009). The sensitivity of the method may be enhanced by use of contrast agents (Klohs et al. 2012).

An attractive method for exploring structural connectivity is diffusion tensor imaging (DTI) and related techniques. A diffusing water molecule in a biological tissue inevitably encounters lipid membranes; hence, diffusion is hindered. For neurons this implies that the mean diffusion path along the direction of an axon is significantly larger than perpendicular to it; hence, diffusion is largely anisotropic. This anisotropy is captured in a diffusion tensor, which is defined by the values of its principal components (the corresponding diffusion coefficients D_1, D_2, D_3) and three rotation angles describing its orientation in space. By carrying out at least six independent measurements, the diffusion tensor can be determined for every voxel. An impressive number of mathematical algorithms have been developed that enable tracing of such fiber tracts over large brain areas. A basic problem with conventional DTI is that the spatial resolution is such that a voxel comprises a large number of fibers. Correspondingly, DTI cannot resolve fibers crossing, bending, or twisting within an individual voxel. Advanced DTI methods such as high angular resolution diffusion imaging (HARDI) or Q-ball imaging (Van et al. 2010) overcome this limitation by more fully characterizing the angular dependence of intravoxel diffusion. Q-ball imaging can resolve multiple fiber orientations within a voxel and does not require assumptions on the nature of diffusion process. MRI-based fiber tracking yielding information of structural connectivity has raised considerable interest in the context of the *Human Connectome Project* (http://www.humanconnectomeproject.org/), a multicenter program that aims at constructing a map of the "complete structural and functional neural connections in vivo within and across individuals, thereby providing unique opportunity to understand details of neural connectivity in a comprehensive manner."

While structural MRI provides information on neuroanatomical alterations preceding or accompanying neuropsychiatric diseases, functional MRI allows assessing changes in neuronal activation patterns between healthy subjects and

Structural and Functional Magnetic Resonance Imaging, Fig. 2 Schematic representation of the relationship between neuronal activity and the hemodynamic response function. Functional MRI allows to assess various processes involved such as changes in cerebral blood flow (CBF), cerebral blood volume (CBV), and changes in the blood oxygenation level (BOLD)

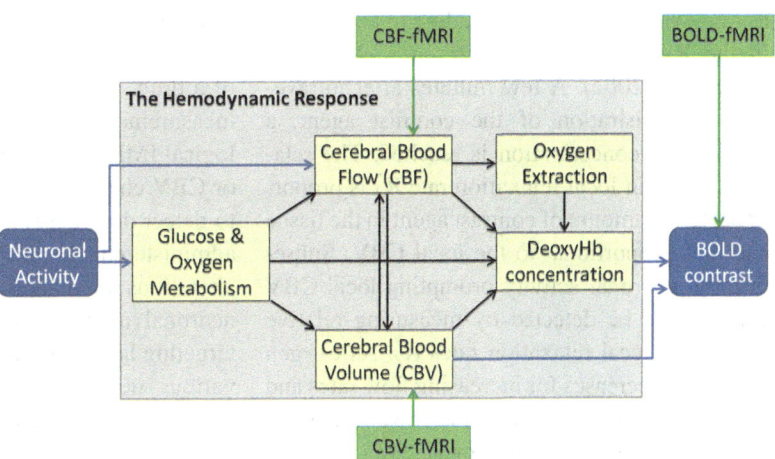

patients, which might be more closely related to the disease progression or effects of drug administration.

Functional MRI

Underlying Biological Processes

Functional MRI (fMRI) is widely applied in clinical and preclinical studies to assess brain function, keeping in mind that the MR method is sensitive to hemodynamic changes prompted by neuronal activity rather than the neuronal activation itself. Local neuronal activity leads to an increased consumption of oxygen and nutrients triggering an increase in local perfusion, i.e., regional cerebral blood flow (CBF) and cerebral blood volume (CBV) (Fig. 2). As the efficiency of oxygen extraction decreases with increasing flow rates, the venous blood contains more oxygenated hemoglobin in the activated state compared to the resting state. The higher concentration of oxyhemoglobin, and correspondingly the lower concentration of deoxyhemoglobin, which is paramagnetic and therefore an endogenous contrast agent, during activation leads to a decrease in R_2^* relaxation rates, thus an increase of the MR signal (Norris 2006; Ogawa et al. 1990). This mechanism, called blood oxygenation level-dependent (BOLD) contrast, has found widespread use in the neuroscience community to study brain function under physiological and pathological conditions. It is important to note that an intact neurovascular coupling is essential for the reliability of functional MRI signals.

To further elucidate the complex mechanism of neurovascular coupling, additional fMRI methods have been developed, directly assessing the vascular response to neuronal activation by measuring CBF and CBV changes. In recent reports MR arterial spin labeling (ASL) techniques have been described to measure regional CBF. For this purpose the arterial blood flowing into the brain is magnetically labeled at the level of the common carotid artery (Williams et al. 1992). In this way arterial blood can be used as endogenous contrast agent. Regional CBF values can be estimated from the difference of the MR signal intensities before and during labeling of the inflowing arterial spins taking the finite lifetime of the magnetically labeled state into consideration. ASL is independent of contrast agent administration and allows therefore continuous CBF monitoring. As CBF changes are supposed to be proportional to neuronal activity (Stephan et al. 2007), ASL presents itself an attractive method for providing a more direct readout of neuronal activity than BOLD and CBV measurements, which relate to activity in a nonlinear manner. Furthermore, CBF measurements are less susceptible to magnetic field variations than BOLD-fMRI.

For the assessment of local CBV values, MRI methods have been developed using exogenous contrast agents with a long plasma half-life such as iron oxide nanoparticles leading to an increase in the R_2 and R_2^* relaxation rates (Mueggler et al. 2001, 2002). A few minutes after intravenous administration of the contrast agent, a steady-state concentration is reached. The relative change in local relaxation rate R_2^* is proportional to the amount of contrast agent in the tissue and thus proportional to the local CBV. Subsequently, neuronal activity prompting local CBV changes can be detected by measuring relative changes in local relaxation rates R_2^*. As oxygen extraction decreases for increasing flow rates and the BOLD contrast decreases at lower magnetic field strength, CBV measurements are more sensitive than fMRI based on the BOLD contrast.

Stimulation Paradigms

A key component of fMRI experiments is the stimulation paradigm applied to evoke brain activity. A variety of different stimuli ranging from no stimulation (resting state fMRI) over thermal, sensory, mechanical, visual, or auditory to pharmacological stimuli have been used during fMRI studies. The rationale behind resting state fMRI, which analyzes spontaneous activity/hemodynamic changes in the brain, is to investigate potential differences of *intrinsic* brain network activity between a healthy control group and patients suffering, e.g., from schizophrenia (Pearlson and Calhoun 2007) as these experiments do not rely on the ability of the patient to perform certain tasks. Aside from resting state fMRI, fMRI studies always rely on at least two measurements comparing a state A (e.g., the baseline state) with a state B (e.g., during stimulus induced activation). When designing fMRI studies, it is important to consider the dynamics of the hemodynamic response to neuronal activity, which determines the time resolution that can be achieved in the experiment.

In pharmacological fMRI (phMRI) the functional response to ligand-induced receptor stimulation or inhibition after drug administration is assessed throughout the brain using the above described fMRI methods. Consequently, phMRI like all fMRI methods relies on intact neurovascular coupling and assesses the functional response induced by drug administration. In humans phMRI involves BOLD signal acquisition before, during, and after the administration of a drug, while in animals also CBF and CBV measurements are widely established. Pharmacological fMRI studies in the rat measuring BOLD or CBV changes confirmed sufficient sensitivity to detect dose-dependent effects of systemically administered receptor ligands. Assessment of alterations in magnitude and spatial extent of neuronal activity induced by pharmacological targeting has been successfully demonstrated for various neurotransmitter systems such as the dopaminergic (Chen et al. 1997), opioid (Xu et al. 2000), GABA-ergic (Reese et al. 2000), glutamatergic (Houston et al. 2001), or cannabinoid system (Shah et al. 2004). Figure 3 shows an example of phMRI in the mouse. Acute administration of the $GABA_A$ receptor antagonist bicuculline led to region-specific CBV changes (Mueggler et al. 2001). The impact and potential of phMRI in the area of psychopharmacology will be discussed in the following as exemplified for mood and anxiety disorders. As therapeutics targeting serotonin (5-hydroxytryptamine; 5-HT) receptors and transporters have been shown beneficial and 5-HT neurotransmission is closely linked to the pathophysiology of these two neuropsychiatric disorders (Hoyer et al. 2002), investigations using neuroimaging techniques within the field of depression and anxiety disorders have been focused measuring neurotransmission and receptor occupancy and their response to pharmacological intervention. Based on the availability of radiolabeled 5-HT receptor ligands, distinct serotonin receptor populations have been imaged using positron emission tomography (PET) and single positron emission tomography (SPECT). On the other hand the nature of serotonergic neuro-circuitries can be investigated using pharmacological fMRI by modifying endogenous neurotransmitter levels or manipulating their receptor activity by specific ligands. Most phMRI studies investigated the $5-HT_{2C}$ receptor system as one of the main targets for novel anxiolytic drugs (Wood 2003). Among the different ligands used, meta-chlorophenylpiperazine (m-CPP) a mixed $5-HT_{1B/2C}$ receptor agonist has been advanced as a useful pharmacological

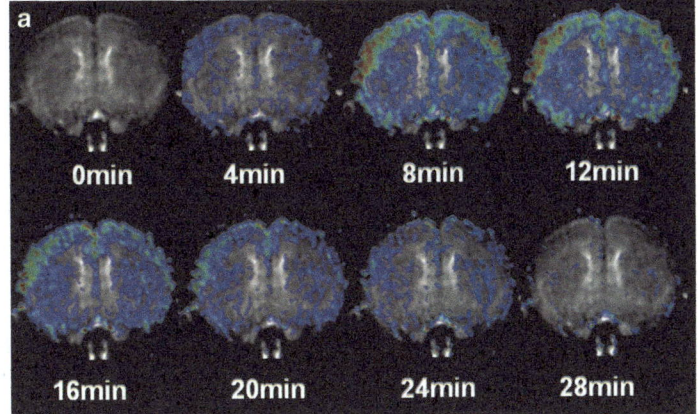
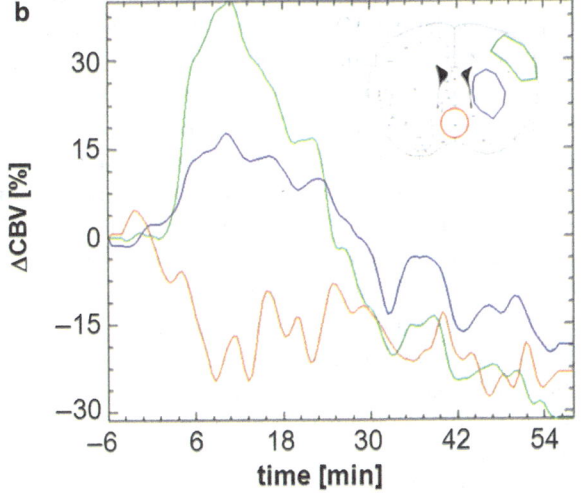

Structural and Functional Magnetic Resonance Imaging, Fig. 3 Example of pharmacological fMRI in the mouse. CBV changes in the mouse brain during infusion of the $GABA_A$ receptor antagonist bicuculline. (**a**) Percentage ΔCBV activity maps of brain section +0.74 mm relative to the bregma showing highest activity in cortical areas. (**b**) Temporal profile for three different ROIs (see brain atlas inlet: *green* cortex, *blue* striatum, *orange* control ROI) highlighting the region specificity of the induced CBV changes

substance for fMRI studies of regional activation in rat (Hackler et al. 2007; Stark et al. 2006) and human (Anderson et al. 2002). Furthermore the acute and chronic effect of SSRIs on region-specific neuronal activation measured by BOLD-fMRI in humans has been examined using citalopram (McKie et al. 2005) or in the rat using fluoxetine (Schwarz et al. 2007). The literature on phMRI investigating specifically the $5\text{-}HT_{1A}$ receptor is rather sparse, but represents an attractive field of research given that $5\text{-}HT_{1A}$ autoreceptor desensitization is one of the suggested mechanisms of action of chronic antidepressants (i.e., SSRIs) and might be responsible for the delayed onset of antidepressants under clinical conditions as a challenge for the serotonergic neurotransmitter system. Using CBV-fMRI, Scanley and colleagues (Scanley et al. 2001) mapped decreased activity across several brain areas of the rat with the strongest effects in the hippocampus and septum after acute administration of the $5\text{-}HT_{1A}$ receptor agonist 8-OH-DPAT. Subsequently, these paradigms have been applied to mice including heterozygous $5\text{-}HT_{1A}$ receptor knockout mice (Mueggler et al. 2011), and correlation analysis was applied to identify the functional network associated with $5HT_{1A}$ receptor activation (Razoux et al. 2013). Circuits identified comprised the brain structures known to be involved in stress-related disorders (e.g., prefrontal cortex, amygdala, and hippocampus). Also the dorsomedial thalamus, a structure associated with fear processing, was identified as a hub in the $5\text{-}HT_{1A}$-receptor functional network. Administration of a $5\text{-}HT_{1A}$ receptor antagonist or use of heterozygous 5-HT$_{1A}$-R knockout mice not only reduced the amplitude of the CBV response in individual brain structures but also reduced functional connectivity elicited by 8-OH-DPAT.

Data Analysis and Interpretation

The analysis of fMRI data can be especially challenging due to the complex relationship among the various physiological processes involved (Fig. 2) and, furthermore, due to the small signal changes detected in fMRI, e.g., BOLD signal changes in humans are in the order of 1–5 % (Jezzard and Buxton 2006). For this reason the fMRI analysis is often simplified by fitting a general linear model (GLM) based on a priori information of the stimulation paradigm convolved with an assumed hemodynamic response function (Friston et al. 1995a, b). More sophisticated approaches incorporate the various processes associated with the hemodynamic response to form a biophysically plausible framework such as the Balloon model (Buxton et al. 1998; Stephan et al. 2007). However, pharmacological MRI aims at resolving neuronal networks and connectivities throughout the brain. As the functional relationship and the causal dependencies between different brain areas are a priori unknown, hypothesis-driven methods such as GLM might lead to false results or to a loss in sensitivity in activated areas. In this case, data-driven or exploratory approaches are assumed to be superior: these include independent component analysis (ICA) (Calhoun and Adali 2006) or methods detecting temporal or spatial correlations. For example, in wavelet-based cluster analysis (WCA), activated pixels are grouped according to their own activity pattern avoiding assumptions derived from the experimental design (Whitcher et al. 2005).

As phMRI methods monitor neuronal activity via neurometabolic or neurovascular coupling processes, careful interpretation of phMRI data is indicated taking various processes influencing the phMRI signal into account. Direct systemic effects of the drug administered or influences of the pathology on the vascular tone might lead to changes of the functional response measured by phMRI, which do not reflect changes in neuronal activity. Furthermore, one needs to keep in mind that phMRI detects brain regions showing hemodynamic changes which usually exceed local regions of neuronal activity, thus overestimating the region involved, or neurons in projection areas, downstream of the local receptor activation and hence distant to the primary target region of the drug are modulated, too. In small rodents phMRI is commonly carried out in anesthetized animals; thus it has to be considered that the anesthetic might potentially interfere with the ligand–receptor interaction or transmitter level of interest.

Remaining topics with respect to fMRI data analysis and interpretation are the understanding of the functional signal changes acquired and their link to neuronal activity, even in healthy subjects (Logothetis 2008) as without better knowledge of these fundamental processes under physiological conditions, the interpretation of fMRI changes under pathological conditions must remain incomplete. Furthermore, fMRI data analysis and interpretation is carried out in groups of healthy or diseased subjects. As described above, normalization of all brains of a group to a standard space reduces within-group difference, thus enhancing between-group difference and accordingly statistical significance of the differences (Konarski et al. 2007). Biological variability and the low level of the signal changes hamper subject-specific interpretation or diagnosis of fMRI examinations.

Resting State fMRI

Conventional fMRI studies assess changes in the brain state elicited by a stimulus. This may involve potential confounds. fMRI is an indirect measure of brain function, which critically depends on the integrity of neurovascular coupling. While under normal conditions, this criterion is typically fulfilled (Schulz et al. 2012), this may not be the case under pathological conditions. Also, nonspecific hemodynamic responses to the stimulus (e.g., in response to arousal, stress, fear) might compromise the specificity of the fMRI signal. This might be in part accounted for by the proper choice of the reference state. An alternative strategy is to use *resting state* fMRI (rs-fMRI), which, in absence of a stimulus, measures the spontaneous fluctuations of the brain signal and does, apart from true noise components, contain physiological information (respiration, heart rate) as well as information on changes in blood oxygenation due to local brain activity. Rs-fMRI yields information of functional connectivity within brain networks. The underlying hypothesis is that a strong functional interaction between two brain regions is reflected by a high degree of synchronization between the corresponding fMRI signals. Hence rs-fMRI analysis searches for temporal correlation among fMRI time series.

Today, rs-fMRI is widely applied in clinical neuroscience covering topics such as aging and neurodegeneration, schizophrenia, depression, and anxiety disorders.

Advantages and Limitations with Respect to Alternative Neuroimaging Modalities

The neuroimaging technique to be applied depends on the biomedical question to be addressed. Assessing structural information commonly requires high spatial resolution and adequate contrast for the structure of interest, while image acquisition time, typically in the order of several minutes, is not an issue as anatomical structures can be assumed to be static during the data collection period. As alternative imaging modality, computerized tomography (CT) can be applied measuring X-ray attenuation by tissue. Both structural MRI and CT provide three-dimensional information with excellent spatial resolution and the accessibility to deep brain structures allowing to quantitatively document volumetric and morphological changes (Konarski et al. 2007). However, the advantage of structural MRI is its absolute noninvasiveness, thus enabling repeated measurements in longitudinal studies to monitor disease progression or therapeutic effects and superior soft tissue contrast.

In contrast, imaging of dynamic processes such as brain perfusion requires not only sufficient sensitivity to detect but also sufficient temporal resolution to resolve the physiological changes under investigation. The time available for imaging is determined by the physiological process of interest, which has an impact on the fMRI method that can be applied and the spatial resolution that can be achieved. Alternative imaging modalities are SPECT or PET. Both modalities are using exogenous contrast agents labeled with radionuclides such as, e.g., fluorine-18 for PET and iodine-123 for SPECT (Rudin 2005). Although PET is relatively expensive compared to SPECT, it is superior in terms of sensitivity and spatial resolution and provides inherently quantitative data. PET provides information on physiological (perfusion), biochemical processes such as, e.g., neuronal glucose metabolism, and information on receptor distribution and occupancy. In comparison to SPECT and PET, fMRI is less sensitive but provides information on functional consequences downstream of the primary target engagement of the drug in a spatially highly resolved manner. Due to their noninvasiveness, fMRI and nuclear imaging methods developed in small rodents are readily translatable to clinical applications.

The dimensions of small rodents further allow the use of fluorescent molecular tomography (FMT) as alternative neuroimaging method. In this method, a compound labeled with a fluorescent dye is administered. After specific binding of this reporter to the target and after clearance of the unbound fraction, the fluorophore is excited using laser light in the near infrared range (Rudin 2005). FMT is superior to MRI in terms of sensitivity, and lower detection limits are in the nanomolar range but provides relatively poor spatial resolution. However, in human subjects deep brain areas are not accessible due to the small penetration depth of light in the near infrared range.

Future Directions

Noninvasive neuroimaging has been rapidly developing in the past decade, as various imaging modalities provide an overwhelming amount of information on functional neuroanatomy, neuronal activity, and neuronal networks. However, technical limitations such as low sensitivity in fMRI or poor localization in EEG prevented the assessment of an integrated view of the brain function. These issues have prompted the development of data fusion methods aiming to combine complementary information from different imaging modalities. For example, EEG data have been constrained using fMRI activation maps (Pearlson and Calhoun 2007). Although these methods hold great potential to improve data interpretation, one has to avoid unrealistic assumptions. Alternative approaches under development are trying to combine the strengths of two modalities such as the high sensitivity of molecular imaging with the high spatial resolution and localization of CT and MRI. Substantial efforts have been made to bring forward such hybrid imaging systems combining, e.g., CT-PET or MRI-PET for clinical applications (Schlemmer et al. 2008). In small animal imaging, CT-FMT (Hyde et al. 2009) or MRI-PET solutions are also pursued.

Cross-References

▶ BOLD Contrast
▶ Cerebral Perfusion
▶ Functional Magnetic Resonance Imaging
▶ MR Image Analysis
▶ Neuropsychiatric Disorders
▶ Pharmacological fMRI
▶ Stimulation Paradigm
▶ Translational Neuroimaging

References

American Psychiatric Association (2013) Diagnostic and statistical manual of mental disorders, 5th edn. American Psychiatric Association, Arlington
Anderson IM, Clark L, Elliott R, Kulkarni B, Williams SR, Deakin JF (2002) 5-HT(2C) receptor activation by m-chlorophenylpiperazine detected in humans with fMRI. Neuroreport 13:1547–1551
Baltes C, Radzwill N, Bosshard S, Marek D, Rudin M (2009) Micro MR imaging of the mouse brain using a novel 400 MHz cryogenic quadrature RF probe. NMR Biomed 22(8):834–842
Budde J, Shajan G, Zaitsev M, Scheffler K, Pohmann R (2014) Functional MRI in human subjects with gradient-echo and spin-echo EPI at 9.4 T. Magn Reson Med 71(1):209–18
Buxton RB, Wong EC, Frank LR (1998) Dynamics of blood flow and oxygenation changes during brain activation: the balloon model. Magn Reson Med 39:855–864
Calhoun VD, Adali T (2006) Unmixing fMRI with independent component analysis. IEEE Eng Med Biol Mag 25:79–90
Chen YC, Galpern WR, Brownell AL, Matthews RT, Bogdanov M, Isacson O, Keltner JR, Beal MF, Rosen BR, Jenkins BG (1997) Detection of dopaminergic neurotransmitter activity using pharmacologic MRI: correlation with PET, microdialysis, and behavioral data. Magn Reson Med 38:389–398
Ellegood J, Lerch JP, Henkelman RM (2011) Brain abnormalities in a Neuroligin3 R451C knockin mouse model associated with autism. Autism Res 4(5):368–376
Friston KJ, Frith CD, Turner R, Frackowiak RS (1995a) Characterizing evoked hemodynamics with fMRI. Neuroimage 2:157–165
Friston KJ, Holmes AP, Poline JB, Grasby PJ, Williams SC, Frackowiak RS, Turner R (1995b) Analysis of fMRI time-series revisited. Neuroimage 2:45–53
Hackler EA, Turner GH, Gresch PJ, Sengupta S, Deutch AY, Avison MJ, Gore JC, Sanders-Bush E (2007) 5-Hydroxytryptamine2C receptor contribution to m-chlorophenylpiperazine and N-methyl-beta-carboline-3-carboxamide-induced anxiety-like behavior and limbic brain activation. J Pharmacol Exp Ther 320:1023–1029
Houston GC, Papadakis NG, Carpenter TA, Hall LD, Mukherjee B, James MF, Huang CL (2001) Mapping of brain activation in response to pharmacological agents using fMRI in the rat. Magn Reson Imaging 19:905–919
Hoyer D, Hannon JP, Martin GR (2002) Molecular, pharmacological and functional diversity of 5-HT receptors. Pharmacol Biochem Behav 71:533–554
Hyde D, de Kleine R, MacLaurin SA, Miller E, Brooks DH, Krucker T, Ntziachristos V (2009) Hybrid FMT-CT imaging of amyloid-beta plaques in a murine Alzheimer's disease model. Neuroimage 44:1304–1311
Jezzard P, Buxton RB (2006) The clinical potential of functional magnetic resonance imaging. J Magn Reson Imaging 23:787–793
Klohs J, Baltes C, Princz-Kranz F, Ratering D, Nitsch RM, Knuesel I, Rudin M (2012) Quantitative assessment of cerebral amyloid angiopathy related changes in vessel density with contrast-enhanced magnetic resonance microangiography. J Neurosci 32:1705–1713
Konarski JZ, McIntyre RS, Soczynska JK, Kennedy SH (2007) Neuroimaging approaches in mood disorders: technique and clinical implications. Ann Clin Psychiatry 19:265–277

Lau JC, Lerch JP, Sled JG, Henkelman RM, Evans AC, Bedell BJ (2008) Longitudinal neuroanatomical changes determined by deformation-based morphometry in a mouse model of Alzheimer's disease. Neuroimage 42:19–27

Lerch JP, Carroll JB, Dorr A, Spring S, Evans AC, Hayden MR, Sled JG, Henkelman RM (2008) Cortical thickness measured from MRI in the YAC128 mouse model of Huntington's disease. Neuroimage 41:243–251

Logothetis NK (2008) What we can do and what we cannot do with fMRI. Nature 453:869–878

Malhi GS, Lagopoulos J (2008) Making sense of neuroimaging in psychiatry. Acta Psychiatr Scand 117:100–117

McKie S, Del-Ben C, Elliott R, Williams S, del Vai N, Anderson I, Deakin JF (2005) Neuronal effects of acute citalopram detected by pharmacoMRI. Psychopharmacology (Berl) 180:680–686

Mueggler T, Baumann D, Rausch M, Rudin M (2001) Bicuculline-induced brain activation in mice detected by functional magnetic resonance imaging. Magn Reson Med 46:292–298

Mueggler T, Sturchler-Pierrat C, Baumann D, Rausch M, Staufenbiel M, Rudin M (2002) Compromised hemodynamic response in amyloid precursor protein transgenic mice. J Neurosci 22:7218–7224

Mueggler T, Razoux F, Russig H, Buehler A, Franklin TB, Baltes C, Mansuy IM, Rudin M (2011) Mapping of CBV changes in 5-HT(1A) terminal fields by functional MRI in the mouse brain. Eur Neuropsychopharmacol 21(4):344–53

Nieman BJ, Lerch JP, Bock NA, Chen XJ, Sled JG, Henkelman RM (2007) Mouse behavioral mutants have neuroimaging abnormalities. Hum Brain Mapp 28:567–575

Norris DG (2006) Principles of magnetic resonance assessment of brain function. J Magn Reson Imaging 23:794–807

Ogawa S, Lee TM, Nayak AS, Glynn P (1990) Oxygenation-sensitive contrast in magnetic resonance image of rodent brain at high magnetic fields. Magn Reson Med 14:68–78

Pearlson GD, Calhoun V (2007) Structural and functional magnetic resonance imaging in psychiatric disorders. Can J Psychiatry 52:158–166

Pitiot A, Pausova Z, Prior M, Perrin J, Loyse N, Paus T (2007) Magnetic resonance imaging as a tool for in vivo and ex vivo anatomical phenotyping in experimental genetic models. Hum Brain Mapp 28:555–566

Ratering D, Baltes C, Nordmeyer-Massner J, Marek D, Rudin M (2008) Performance of a 200-MHz cryogenic RF probe designed for MRI and MRS of the murine brain. Magn Reson Med 59:1440–1447

Razoux F, Baltes C, Mueggler T, Seuwen A, Russig H, Mansuy I, Rudin M (2013) Functional MRI to assess alterations of functional networks in response to pharmacological or genetic manipulations of the serotonergic system in mice. Neuroimage 74:326–36

Redwine JM, Kosofsky B, Jacobs RE, Games D, Reilly JF, Morrison JH, Young WG, Bloom FE (2003) Dentate gyrus volume is reduced before onset of plaque formation in PDAPP mice: a magnetic resonance microscopy and stereologic analysis. Proc Natl Acad Sci USA 100:1381–1386

Reese T, Bjelke B, Porszasz R, Baumann D, Bochelen D, Sauter A, Rudin M (2000) Regional brain activation by bicuculline visualized by functional magnetic resonance imaging. Time-resolved assessment of bicuculline-induced changes in local cerebral blood volume using an intravascular contrast agent. NMR Biomed 13:43–49

Rudin M (2005) Molecular imaging – basic principles and applications in biomedical research. Imperial College Press, London

Scanley BE, Kennan RP, Gore JC (2001) Changes in rat cerebral blood volume due to modulation of the 5-HT(1A) receptor measured with susceptibility enhanced contrast MRI. Brain Res 913:149–155

Schlemmer HP, Pichler BJ, Schmand M, Burbar Z, Michel C, Ladebeck R, Jattke K, Townsend D, Nahmias C, Jacob PK, Heiss WD, Claussen CD (2008) Simultaneous MR/PET imaging of the human brain: feasibility study. Radiology 248:1028–1035

Schubert MI, Porkess MV, Dashdorj N, Fone KC, Auer DP (2009) Effects of social isolation rearing on the limbic brain: a combined behavioral and magnetic resonance imaging volumetry study in rats. Neuroscience 159:21–30

Schwarz AJ, Gozzi A, Reese T, Bifone A (2007) In vivo mapping of functional connectivity in neurotransmitter systems using pharmacological MRI. Neuroimage 34:1627–1636

Shah YB, Prior MJ, Dixon AL, Morris PG, Marsden CA (2004) Detection of cannabinoid agonist evoked increase in BOLD contrast in rats using functional magnetic resonance imaging. Neuropharmacology 46:379–387

Stark JA, Davies KE, Williams SR, Luckman SM (2006) Functional magnetic resonance imaging and c-Fos mapping in rats following an anorectic dose of m-chlorophenylpiperazine. Neuroimage 31:1228–1237

Stephan KE, Harrison LM, Kiebel SJ, David O, Penny WD, Friston KJ (2007) Dynamic causal models of neural system dynamics: current state and future extensions. J Biosci 32:129–144

Strakowski SM, Delbello MP, Adler CM (2005) The functional neuroanatomy of bipolar disorder: a review of neuroimaging findings. Mol Psychiatry 10:105–116

Van AT, Granziera C, Bammer R (2010) An introduction to model-independent diffusion magnetic resonance imaging. Top Magn Reson Imaging 21:339–345

Vlaardingerbroek MT, den Boer JA (1999) Magnetic resonance imaging. Springer, Heidelberg

Whitcher B, Schwarz AJ, Barjat H, Smart SC, Grundy RI, James MF (2005) Wavelet-based cluster analysis: data-driven grouping of voxel time courses with application to perfusion-weighted and pharmacological MRI of the rat brain. Neuroimage 24:281–295

Williams DS, Detre JA, Leigh JS, Koretsky AP (1992) Magnetic resonance imaging of perfusion using spin inversion of arterial water. Proc Natl Acad Sci USA 89:212–216

Wood MD (2003) Therapeutic potential of 5-HT2C receptor antagonists in the treatment of anxiety disorders. Curr Drug Targets CNS Neurol Disord 2:383–387

Xu H, Li SJ, Bodurka J, Zhao X, Xi ZX, Stein EA (2000) Heroin-induced neuronal activation in rat brain assessed by functional MRI. Neuroreport 11:1085–1092

Subjective Value

Synonyms

Reinforcer efficacy

Definition

Within the theoretical framework of behavioral economics it is the worth of a commodity for an individual. Subjective value is a weighted function of the benefits associated with the commodity (magnitude, quality, ability to fulfill individual's needs) and the costs of obtaining the commodity (price).

Cross-References

▶ Behavioral Economics

Substance Abuse

Definition

The use of a drug or other substance e.g. alcohol, tobacco, chemical, in a manner or amount that can be harmful or problematic to an individual.

Cross-References

▶ Agoraphobia
▶ Substance Use Disorders: A Theoretical Framework

Substance Dependence

Definition

An unhealthy or addictive pattern of repeated substance use most typically indicated by tolerance to the effects of the substance and symptoms of withdrawal when the substance is terminated for a period of time.

Cross-References

▶ Substance Use Disorders: A Theoretical Framework

Substance P

Synonyms

Neurokinin 1; SP

Definition

Substance P (SP) is an 11 amino acid neuropeptide, the first to be discovered from the tachykinin family.

It is generated by alternative splicing of the preprotachykinin gene RNA and mostly expressed in smooth muscles and the central nervous system. SP has been mainly associated in the regulation of pain, emotional processes, and emesis by activation of NK1 receptors.

Substance Use Disorders: A Theoretical Framework

Stephen T. Higgins[1] and Warren K. Bickel[2]
[1]Department of Psychiatry and Psychology, Vermont Center on Behavior and Health, University of Vermont, Burlington, VT, USA
[2]Addiction Recovery Research Center, Virginia Tech Carillion Research Institute, Roanoke, VT, USA

Synonyms

Drug abuse; Drug addiction; Drug dependence

Definition

A scientific theory summarizes a hypothesis or group of hypotheses that has been strongly supported by empirical observation, in this case those having to do with substance use disorders (see Hassin et al 2013). Within the DSM, perhaps the most widely used system for diagnosing substance use disorders, an individual must exhibit at least two of the following 11 symptoms to meet the criteria for a substance use disorder: (1) failure to fulfill major role obligations due to substance use; (2) use in situations where it is physically dangerous; (3) continued use despite persistent or recurrent social or interpersonal problems; (4) tolerance; (5) withdrawal; (6) substance taken in larger amounts or over a longer period than initially intended; (7) persistent desire or unsuccessful efforts to cut down or control use; (8) a great deal of time spent in activities necessary to obtain substances, using them, or recovering from their use; (9) important social, occupational, or recreational activities given up or reduced because of substance use; (10) continued substance use despite knowledge of a persistent or recurrent physical or psychological problems that are caused or exacerbated by the substance; and (11) craving or strong desire or urges to use a specific substance. Meeting two to three criteria is considered a disorder of moderate severity and four or more is deemed severe.

Current Concepts and State of Knowledge

Introduction

Substance use and associated problems are as old as the history of mankind. Yet efforts to programmatically research this long-standing aspect of human behavior are relatively new, dating from approximately the first half of the twentieth century. Prior to that time problems related to substance use were largely considered as manifestations of moral or characterological flaws of individuals or groups and hence more aligned with philosophy or theology than biomedical research. As with virtually all complex human behavior, the scientific evidence is clear that substance use disorders are multiply determined, involving genetic, learning and conditioning, micro- and macroeconomic, and sociological/cultural influences. While no widely accepted formal theoretical model exists that integrates these multiple individual- and population-level influences, pivotal research developments on substance use disorders cumulatively have come to provide a conceptual framework for much of the ongoing research in this area of inquiry (Bickel et al. 2013; Higgins and Heil 2004). Below we outline a chronology of these developments, acknowledging in advance that the outline is not comprehensive and surely omits developments that others may feel should have been included.

Drug Use Has Orderly Biological Consequences That Perpetuate Use

A pivotal development that facilitated entry of the study of substance use disorders into the realm of scientific inquiry was the recognition that repeated and heavy use of drugs like alcohol and opioids results in the development of tolerance and a dependence/withdrawal syndrome. That is, overtime regular substance users must escalate the dose of drug consumed in order to continue experiencing the same desired effects (i.e., pharmacodynamic tolerance). This can be understood as a physiological adaptation that occurs to the repeated pharmacological perturbations that regular drug use represents, that is, as an effort to regain homeostasis. However, this new homeostasis is dependent on the presence of the drug (i.e., physical dependence). An abrupt discontinuation of drug use will disrupt homeostasis and can result in a highly unpleasant syndrome of signs and symptoms that with certain drugs (e.g., alcohol, barbiturates) may be life threatening (i.e., drug withdrawal syndrome). A critical feature of this syndrome is that renewed drug use during a period of withdrawal brings relatively immediate relief from the aversive signs and symptoms (negative reinforcement).

This set of observations allowed substance use disorders to be conceptualized in terms of a syndrome with associated sequela (e.g., liver disease)that was on a par with other medical diseases (Jellinek 1960). These observations and concepts remain an important aspect of contemporary understanding of substance use disorders and provide the rationale for such clinical interventions as opioid detoxification and maintenance interventions for opioid dependence, nicotine replacement therapy for smoking cessation, and medical detoxification for alcohol dependence. Important to note, however, is that these observations provided no accounting for why individuals use substances for nonmedical purposes to begin with or why risk of relapse back to substance use continues after extended periods without use and no discernible remnants of withdrawal. Such unanswered questions led to theoretical notions of underlying biological imbalances that drug abusers were attempting to medicate (self-medication hypotheses) that generally have not been well supported empirically. More importantly, though, is that these unanswered questions opened the door to a series of transformative contributions from the area of learning and conditioning research that led to a more complete understanding of the genesis and maintenance of substance use disorders and remain pillars of contemporary conceptual models.

Drugs Use as Learned Behavior

Reinforcing Effects. Starting in the 1940s, investigators began developing methods for studying voluntary drug consumption in laboratory animals. Initial studies used animals that were first made physically dependent on drugs via experimenter-administered drug exposure. These animals readily learned an arbitrary response that resulted in the delivery of a drug injection. These initial studies were important because they introduced methods that permitted programmatic research on substance use disorders to be conducted in the preclinical biomedical research laboratory. The effects observed were readily accommodated within the conceptual framework outlined above that posited substance use disorders to be largely driven by tolerance and withdrawal. However, pivotal and novel conceptual contributions came soon thereafter when studies using this new methodology showed that laboratory animals who were not physically dependent would similarly learn an arbitrary response when doing so led to injections of commonly abused drugs such as morphine, amphetamine, or cocaine (Deneau et al. 1969; Schuster and Thompson 1969). Importantly, experimental variations of parameters such as the scheduled relationship between the learned response and drug administration in these studies impacted behavior in the same manner as when comparable manipulations were made with laboratory animals responding to present themselves with food, sex, or other basic behavioral consequences with obvious survival value. Scientifically, these observations provided an empirical foundation for concluding that, like food, water, and sex, drugs that humans use and abuse function as unconditioned positive reinforcers and that behaviors directed toward procuring and using these drugs conform to operant conditioning principles; that is, they conform to basic principles of learning. Operant conditioning is a basic form of learning in which the future probability of voluntary behavior is influenced by its consequences.

Several decades of subsequent experimental research has confirmed those conclusions and demonstrated that laboratory animals will voluntarily consume most of the same drugs that humans abuse (Griffiths et al. 1980). Psychomotor stimulants, ethanol, nicotine, opioids, and sedatives function as positive reinforcers and promote voluntary drug consumption in a wide variety of species. Importantly, this body of research also demonstrated that while tolerance, physical dependence, and withdrawal influence patterns of drug consumption in important and orderly ways, these are best considered as consequences of repeated drug use and untoward consequences and not as necessary conditions for the genesis or maintenance of drug use for nonmedical purposes. Moreover, when given unconstrained access to drugs of abuse such as cocaine and opiates, laboratory animals will sometimes consume lethal doses of the drugs or engage in repeated drug use to the exclusion of basic sustenance, demonstrating that even these most extreme characteristics of drug use disorders are not exclusively human phenomena. Initially conducted exclusively at the behavioral level, over time this area of investigation evolved to include highly sophisticated neurobiological and neuropharmacological studies. In the most general terms, such neurobiological research modeled how changes in normal dopamine-based brain reward systems and associated processes during chronic drug use may underpin substance use disorders (e.g., Everitt et al. 2001; Wise and Bozarth 1987). With considerable merit, some posited that substance use disorders can be conceptualized as a form of reinforcement disorder or pathology as can, for example, gambling, eating, and sexual disorders (Bickel et al. 2011a; Higgins et al. 2004).

Conditioned Stimulus Effects. A related and also pivotal observation about repeated drug use involved respondent or classical conditioning, another basic class of learning that occurs when an environmental stimulus (person, place, or thing) reliably predicts presentation of an unconditioned eliciting stimulus, in this case drug administration (Wikler 1980). With repeated pairings, previously neutral environmental stimuli come to acquire conditioned stimulus effects. For example, they become discriminative stimuli for urges to use drugs as well as for drug seeking and use, and they also can elicit a conditioned withdrawal syndrome. This latter capacity provided insight into why individuals with a history of drug dependence often continue to experience strong urges to use months or years after initiating abstinence from drug use. Additionally, rather than promoting tolerance or habituation to drug-related stimuli, chronic drug exposure can come to enhance or sensitize response to drug-related stimuli, a process thought to contribute to the pronounced cravings that are characteristic of substance use disorders (Robinson and Berridge 1993). Importantly, this body of research also elucidated how environmental stimuli that are paired with drug use acquire conditioned reinforcing effects, which work in concert with the unconditioned reinforcing effects of abused drugs to sustain the extraordinary efforts that drug-dependent individuals will often engage in to procure and use drugs. As mentioned above regarding the unconditioned reinforcing effects of drugs, these conditioned eliciting and reinforcing effects of drugs (learned relationships) have also been well documented in humans and nonhumans and have generated sophisticated neuropharmacology studies characterizing the neural substrates and processes underpinning them (e.g., Robbins et al 2008).

Social Factors. Critically important to recognize is that humans are social primates and their drug use typically occurs in social contexts. An important body of research on drug use and dependence as learned behavior addressed the influence of social factors on the behavioral action of drugs. An important advance in this area extended to drug use is the social learning concepts of modeling and imitation. For example, alcoholic individuals imitate the rates and patterns of alcohol ingestion modeled by others drinking in their presence. Another area of research demonstrates that prototypical drugs of abuse share a common effect of facilitating social interaction. For example, when under the acute influence of alcohol or other abused drugs, individuals will forgo opportunities to earn money in order to interact socially with other research volunteers. It is these effects that have led alcohol, for example, to become such a mainstay of formal social activities (dinner parties, weddings, other gatherings). Other studies demonstrated that many of the positive mood effects that abused drugs engender are conditional on them being consumed in social contexts. These studies suggested that in addition to the direct conditioned and unconditioned reinforcing effects of abused drugs, human drug use is also likely maintained via the indirect effects of drug-produced enhancement of social reinforcement.

Drug consumption patterns are also influenced by what users have been told or otherwise learned socially about the expected effects of the drugs that they consume as opposed to the direct pharmacological effects of the substances per se. For example, whether social drinkers or alcoholics consume relatively large amounts of a beverage in a taste-testing procedure is determined in part by whether they are told the beverage contains alcohol rather than by the alcohol content of the drink per se. These findings illustrated a striking influence of nonpharmacological factors on the behavioral effects of drugs that in earlier conceptual models would likely have been attributed to the direct pharmacological effects of the drug. Such interactions are not unique to humans. A programmatic body of research in nonhuman primates, for example, demonstrates a striking interaction between social context, neurobiology, and the reinforcing effects of cocaine (Nader and Banks 2014).

Theoretical and Clinical Implications. This large body of research on the role of learning and conditioning has and continues to have important theoretical and clinical influences. First, it made clear that theories of substance use disorders couched exclusively in terms of tolerance and withdrawal, even those that included conditioned tolerance and withdrawal, were incomplete. Evidence of that advance is discernible, for example, across consecutive editions of the psychoactive substance use disorder syndromes section of the DSM. In DSM-III, the psychoactive substance use disorder syndromes were defined exclusively in terms of tolerance and withdrawal. In DSM-III-R, tolerance and withdrawal were reduced to four of the nine symptoms, and the remaining five addressed behavioral symptoms. In DSM-IV, tolerance and withdrawal were retained and in some ways made more prominent than in DSM-III-R, but they represented only two of the seven symptoms and were explicitly noted to be neither necessary nor sufficient for a diagnosis of dependence. As noted above, tolerance and withdrawal represent only two of the eleven symptoms in the most recent iteration, DSM-V, and are neither necessary nor sufficient for a diagnosis.

In terms of clinical implications, elucidating the critical role of positive reinforcement in the development and maintenance of substance use disorders contributed directly to the development of efficacious clinical interventions such as contingency management interventions and community reinforcement therapy that emphasize explicit use of material and social reinforcement to promote abstinence from drug use and to increase the likelihood of engaging in family, recreational, and vocational activities that are incompatible with a drug-abusing lifestyle (Higgins et al. 2004, 2012). Elucidation of the conditioned stimulus effects of drugs contributed to such mainstream clinical strategies as environmental restructuring to minimize contact with persons, places, and things paired with drug use and exposure strategies to extinguish the conditioned eliciting effects of those stimuli (Carroll and Onken 2005).

Second, this body of evidence contradicted self-medication theories that posited preexisting abnormalities that were ameliorated by drug use as necessary conditions for the genesis of substance use disorders. The literature on drug use as learned behavior provided compelling evidence that repeated substance use was a product of normal learning processes that are an integral part of the evolutionary history of humans and other species. As such, any healthy individual is deemed to possess the necessary biological systems to experience drug-produced reinforcement and associated conditioning processes and hence the potential to develop a substance use disorder given appropriate opportunity and supporting environmental circumstances. In brief, psychopathology or other preexisting conditions appear to be factors that moderate the probability of substance use disorders rather than necessary conditions for them to emerge.

Third, studies on instructions and beliefs as determinants of the behavioral actions of alcohol have provided the impetus for current theoretical and clinical interests in cognitive and expectancy factors in substance use disorders. Humans bring complex histories to drug-using situations, often involving years of direct and vicarious experience with drug use and years assimilating peer influence and other cultural views about reasons for and consequences of substance use. Those rich histories influence substance consumption practices and the consequences of substance use, and they provide therapeutic targets for treatment interventions. Widely used and efficacious cognitive behavioral therapies for substance use disorders, for example, are designed to assist users with restructuring such beliefs about drug use (Carroll and Onken 2005).

Individual Differences in Risk

The learning-based observations described above raised an important conceptual question: if all humans possess the basic prerequisites for initiating and sustaining drug use and drug use for recreational purposes is quite prevalent in most modern societies, why is the prevalence of these disorders not higher? A complete answer to that question is not available, but several decades of research addressing this overarching question have revealed a number of determinants of substance use that have important clinical and policy implications. **Well-Established Observations**. First, cost is an important determinant of drug use by recreational users and dependent individuals. Cost is used broadly here to subsume (1) monetary price; (2) the degree of effort required to search out and obtain substances; (3) the frequency and intensity of aversive events encountered while obtaining, using, and recovering from substance use; and (4) what other reinforcers are forfeited by choosing to allocate time to substance consumption rather than existing alternatives (i.e., opportunity cost).

Second, environmental context is an important determinant of substance use. Human and nonhuman laboratory studies have modeled such influences demonstrating how the presence or absence of alternative sources of nonpharmacological reinforcement, for example, can promote the acquisition and maintenance of drug use. Similarly, individuals lacking skills or resources necessary to find available sources of alternative reinforcement are likely to be at greater risk for substance use disorders. Such skill deficits likely contribute to the strikingly higher prevalence rates of these disorders observed, for example, among individuals with lower socioeconomic status, the unemployed, and those with mental illness (e.g., Higgins and Chilcoat 2009). Also, as noted above, social or peer influence can modulate risk for substance use disorders, likely via modeling of abusive or risky consumption patterns and social reinforcement of deviant behavior. Family context also influences risk for substance use disorders not only via the substance use practices and views of the parents but also via family behavior management practices, interpersonal conflict, and the degree of parent monitoring of child behavior. Finally, neighborhoods and other settings can influence risk through the number of drug-using opportunities afforded to youths and probably through other social learning processes.

Third, genetics significantly influence risk for substance use disorders. How such genetic differences in risk are expressed is not yet well understood, but can be as simple as altering rates at which the abused drug is metabolized. Twin and adoption studies show an association between impulsive personality characteristics and risk for severe forms of early-onset substance abuse. Variations in individual sensitivity to drug-produced reinforcement and related learning factors discussed above have been well demonstrated in animal laboratory studies.

Recognition of the fundamental role of learning and the multiple other contributors to vulnerability to substance use disorders raised challenges to and at times controversy around the disease concept of substance use disorders. The disease model can seem overly simplistic and lacking in face validity, especially when considered as akin to infectious diseases. However, as chronicity came to be accepted as a defining characteristic of substance use disorders (particularly on the severe end of the spectrum) and scientific understanding grew regarding the multiple determinants involved in chronic diseases generally (hypertension, diabetes), debate and controversy around a chronic disease model of substance use disorders have dissipated considerably if not disappeared completely.

Emerging/Novel Observations. Consistent with the association between impulsivity and risk for substance use disorders discussed above in twin and adoption studies, recent behavioral economic research has documented an important association between discounting of delayed reinforcement and substance abuse (Bickel et al. 2007, 2011a). In this area of research, impulsive behavior is operationally defined as choosing a smaller, more immediate over a larger, more delayed reinforcer. Individual preference for immediate versus delayed reinforcement is assessed by allowing individuals to make a series of choices between different amounts of real or hypothetical money, drugs, or other reinforcing events that are available after different temporal delays. A typical question, for example, might be, "Would you prefer $100 today or $200 a month from now?" Cigarette smokers, opioid abusers, problem drinkers, and heterogeneous samples of substance abusers have been demonstrated to exhibit greater preference for smaller, more immediate over larger, more delayed reinforcement compared to samples of non-abusers matched on relevant sociodemographic characteristics. Delay discounting has also been effective in accounting for individual differences in animal models of substance use disorders (Perry and Carroll 2008). Delay discounting has promise for providing insights into differences in risk for substance abuse between individuals and within individuals over time that can be directly related to reinforcement principles.

Studies examining the neurobiology of discounting have revealed relationships that integrate this area of research with a larger and important body of research in cognitive neuroscience focused on self-regulatory processes or executive functions and their role in risk for and recovery from substance use disorders. The observations on discounting demonstrated that choosing smaller, more immediate rewards (i.e., impulsive choices) is associated with increased activity in limbic areas, while those directed toward later, larger rewards are associated with greater activity in the prefrontal cortices. This is important in that the limbic area is associated with basic brain reward centers, while the prefrontal cortex is associated with more complex processes (memory, reward valuation, response inhibition, problem solving) involved in longer-term goal (i.e., reinforcement) seeking and associated self-regulatory skills. Other areas of research on risk factors for substance use disorders also implicate these areas, including studies showing that individuals with cocaine, tobacco, and other substance disorders often have prefrontal cortical deficits relative to those without or who have recovered from those same disorders. This convergence of empirical evidence has led to emerging theories of vulnerability to substance use disorders that posit a tension or competition for control between these more emotion-laden, limbic-based processes versus the more reflective, rationale prefrontal cortex-based processes important to obtaining reinforcement in complex environments involving temporal delays and other complexities (competing brain system models) (e.g., Bickel et al. 2007). These models posit that vulnerability to substance use disorders is underpinned by innate or acquired imbalances between these two systems that allow for greater influence of the limbic-based over the prefrontal-based processes. Such imbalances may result from a hyperactive limbic system that overwhelms a normal prefrontal cortex or alternatively a hypoactive prefrontal cortex that is unable to compete with a relatively more active limbic system. The net results of each are greater control over behavioral choices by limbic-based processes and hence increased vulnerability to the allure of the immediate positive consequences of drug abuse over the more delayed but more substantive sources of reinforcement available from a sober or more moderate lifestyle. Evidence suggesting that chronic exposure to abused drugs can directly degrade the integrity of prefrontal cortices while also inducing enhanced or hyperactivity in limbic systems outlines a feedback loop with the potential to engender a substance use disorder and hence a potentially more developed understanding of the genesis of substance use disorders than was available based on earlier models.

This is among the newer areas of research discussed in this entry, and thus, its impact on the development of clinical interventions is only starting to emerge. Promising observations are being reported, for example, demonstrating that memory training directed at improving executive functions and presumably associated prefrontal cortical activity levels can decrease the excessive delay discounting seen in substance abusers (e.g., Bickel et al. 2011b). To the extent that delay discounting and associated impairments in prefrontal cortical health are associated with the maintenance of substance use disorders, such changes would be expected to increase the likelihood of successful resolution or at least an amelioration in the severity of the drug use problem. This area of research is also offering conceptual insights into other dimensions of substance use disorders not well addressed by the earlier models. For example, the relatively slow pace of development of brain structures and processes underpinning executive functions extending well into early adulthood parallels the increased vulnerability to substance use and associated problems during adolescence and young adulthood. This competing brain system model also offers potential insights into the well-recognized but less well-understood pattern of increased vulnerability to comorbid psychiatric disorders and other health-related risk behaviors among those with substance use disorders compared to the general population.

Acknowledgments The preparation of this entry was supported in part by the Centers of Biomedical Research Excellence grant P20GM103644-01from the National Institute of General Medical Sciences and research grants R01HD075669 from the National Institute of Child Health and Human Development; R01DA030241, R01DA024080, and R01DA012997 from the National Institute on Drug Abuse Grants; and R01DA024080-02S1 from the National Institute on Alcohol Abuse and Alcoholism.

Cross-References

▶ Addiction
▶ Behavioral Economics
▶ Conditioned Reinforcers
▶ Contingency Management
▶ Instrumental Conditioning
▶ Operant
▶ Pharmacodynamic Tolerance
▶ Reinforcement
▶ Reinforcement Disorders
▶ Self-Administration of Drugs
▶ Social Behavior
▶ Tolerance
▶ Withdrawal Syndromes

References

Bickel WK, Miller ML, Yi R, Kowal BP, Lindquist DM, Pitcock JA (2007) Behavioral and neuro-economics of drug addiction: competing neural systems and temporal discounting processes. Drug Alcohol Depend 90S: S85–S91

Bickel WK, Jarmolowicz DP, Mueller ET, Garchalian KM (2011a) The behavioral economics and neuroeconomics of reinforce pathologies: implications for etiology and treatment for addiction. Curr Psychiatry Rep 13:406–415

Bickel WK, Yi R, Landes RD, Hill RD, Baxter C (2011b) Remember the future: working memory training decreases delay discounting among stimulant addicts. Biol Psychiatry 69:260–265

Bickel WK, Mueller ET, Jarmolowicz DP (2013) What is addiction? In: McCrady BS, Epstein EE (eds) Addictions: a comprehensive guidebook, 2nd edn. Oxford University Press, New York, pp 3–16

Carroll KM, Onken LS (2005) Behavioral therapies for drug abuse. Am J Psychiatry 162:1452–1460

Deneau G, Yanagita T, Seevers MH (1969) Self-administration of psychoactive substances by the monkey. Psychopharmacologia 16:30–48

Everitt BJ, Dickenson A, Robbins TW (2001) The neuropsychological basis of addictive behavior. Brain Res Rev 36:129–138

Griffiths RR, Bigelow GE, Henningfield JE (1980) Similarities in animal and human drug taking behavior. In: Mello NK (ed) Advances in substance abuse: behavioral and biological research. JAI Press, Greenwich, pp 1–90

Hassin DS, O'Brien CP, Auriacombe M, Borges G, Bucholz K, Budney A, Compton WM, Crowley T, Ling W, Petry NM, Schukit M, Grant BF (2013) DSM-5 criteria for substance use disorders:

recommendations and rationale. Am J Psychiary 170:834–851
Higgins ST, Chilcoat HD (2009) Women and smoking: an interdisciplinary examination of socioeconomic influences. Drug Alcohol Depend 104(Suppl 1):S1–S5
Higgins ST, Heil SH (2004) Principals of learning in the study and treatment of substance abuse. In: Galanter M, Kleber HD (eds) The American Psychiatric Publishing textbook of substance abuse treatment. American Psychiatric Publishing, Arlington, pp 81–87
Higgins ST, Heil SH, Lussier JP (2004) Clinical implications of reinforcement as a determinant of substance use disorders. Annu Rev Psychol 55:431–461
Higgins ST, Silverman K, Sigmon SC, Naito NA (2012) Incentives and health: an introduction. Prev Med 55:S2–S6
Jellinek EM (1960) The disease concept of alcoholism. Hillhouse Press, New Brunswick
Nader MA, Banks ML (2014) Environmental modulation of drug taking: nonhuman primate models of cocaine abuse and PET neuroimaging. Neuropharmacology 76:510–517
Perry JL, Carroll ME (2008) The role of impulsive behavior in drug abuse. Psychopharmacology (Berl) 200:1–26
Robbins TW, Ersche KD, Everitt BJ (2008) Drug addiction and the memory systems of the brain. Ann NY Acad Sci 1141:1–21
Robinson TE, Berridge KC (1993) The neural basis of drug craving: an incentive-sensitization theory of addiction. Brain Res Rev 18:247–291
Schuster CR, Thompson T (1969) Self administration of and behavioral dependence on drugs. Annu Rev Pharmacol 9:483–502
Wikler A (1980) Opioid dependence. Plenum, New York
Wise RA, Bozarth MA (1987) A psychomotor stimulant theory of addiction. Psychol Rev 94:469–492

Substitutes

Synonyms

Surrogate

Definition

In behavioral economics substitutes are goods that are interchangeable, producing a relationship between the price of one good and the demand for the other – as the price of the first good increases, the demand for the second good increases. For example, increasing the price of an artificial sweetener such as saccharine will increase the demand for sugar.

Cross-References

▶ Behavioral Economics

Suicide

Gustavo Turecki
McGill Group for Suicide Studies, McGill University, Montreal, QC, Canada

Synonyms

Self-destructive behavior; Self-immolation; Suicide completion

Definition

Suicide is the act of taking one's own life voluntarily, usually intentionally. Suicidal behavior is a general term used to refer to suicide and suicide attempts. Under the latter, we refer to the actions taken to end one's life, irrespective of the degree of intentionality, which does not result in death. While suicidal ideation is not a behavior, it is often considered under the category of suicidal behaviors. It refers to the wish to die, including thoughts of actively ending one's life.

Current Concepts and State of Knowledge

Suicide is one of the leading causes of death worldwide, among the top ten causes of death in most of the world and one of the three leading causes of death for people between the ages of 15–34 (WHO 2000). As such, it has been referred to as the "leading cause of unnecessary and premature death." Over the last 45 years, suicide rates have been increasing dramatically, by as

Suicide, Table 1 Distribution (%) of major diagnostic categories found among suicide completers by psychological autopsy studies according to regions of the world (Adapted from Arsenault-Lapierre et al. 2004)

	European (%)	North American (%)	Australian (%)	Asian (%)
Mood disorders	48.5	33.6	32.7	51.3
Substance-related disorders	18.6	40.1	24.1	26.7
Schizophrenia and other psychotic disorders	7.5	4.2	24.3	8.4
Personality disorders	16.8	13.4	17.7	17.7

much as 60 % in some countries. In the USA, this rate is approximately 11 per 100,000, while much of Asia sees a rate exceeding 13 per 100,000. With the notable exception of rural China, suicide is significantly more common among males than females. Western countries, and particularly those in Northern Europe and the ex-Soviet Union, present larger gender effects on suicide rates, while Asian countries tend to have sex ratios closer to one. Age is another important demographic factor that seems to have a large impact on suicide risk. Accordingly, the distribution of suicide risk through the lifespan displays marked age effects, with peaks among youth and elderly age groups.

Suicide methods vary considerably among different countries and world regions and relate to the availability of means to suicide, as well as popular concepts and imagery associated with suicide. As such, in North America, suicide by firearms and hanging are the most common methods, whereas in Asia, suicide by pesticides is the most prevalent. Suicidal behavior has been classified as violent and nonviolent, depending exclusively on the method used. In general, the term nonviolent refers to suicidal behavior by substance intoxication (typically medication overdose) or superficial cuts, which are generally associated with low suicide intent and are frequently of low lethality. All other methods are considered violent.

Suicide is strongly associated with psychopathology (Arsenault-Lapierre et al. 2004). While there is significant regional variability in the percentages found, all studies have consistently shown that most individuals who have died by suicide were affected by mental disorders in the last months of their life. Meta-analyses of studies investigating rates of mental disorders among individuals who died by suicide suggest that up to 90 % of suicides had a history of mental illness. Table 1 lists the most common mental disorders found among suicide completers, according to different world regions. Mood disorders, and major depressive disorder, in particular, are the most common diagnoses among individuals who died by suicide.

Predictors of Suicide

Suicide is a complex behavior, likely the result of several interacting factors. Figure 1 conceptualizes the relationships among key factors believed to play an important role in suicide. While, as mentioned earlier, psychopathology is strongly associated with suicide risk, it is neither sufficient nor specific, as only a fraction of people affected by mental illness will die by suicide.

A positive history of suicide attempts, a history of childhood adversity, certain demographic variables, and issues related to social and medical support have been found to be stronger predictors of suicidal behaviors among psychiatric patients. The risk of suicide completion among clinical populations varies as a function of diagnosis and clinical features. For instance, among patients with major depressive disorder, the risk is conditional on the population of depressed patients, i.e., suicide risk is higher among depressed inpatients and lower among depressed patients from the general population and somewhere in between for depressed outpatients. For the latter group, the percentage of individuals who die by suicide is estimated to be between 2 % and 5 %. Clinical predictors of suicide among patients with major depression include symptom severity (as measured by a requirement of hospitalization), comorbidity with substance-related disorders, high levels of

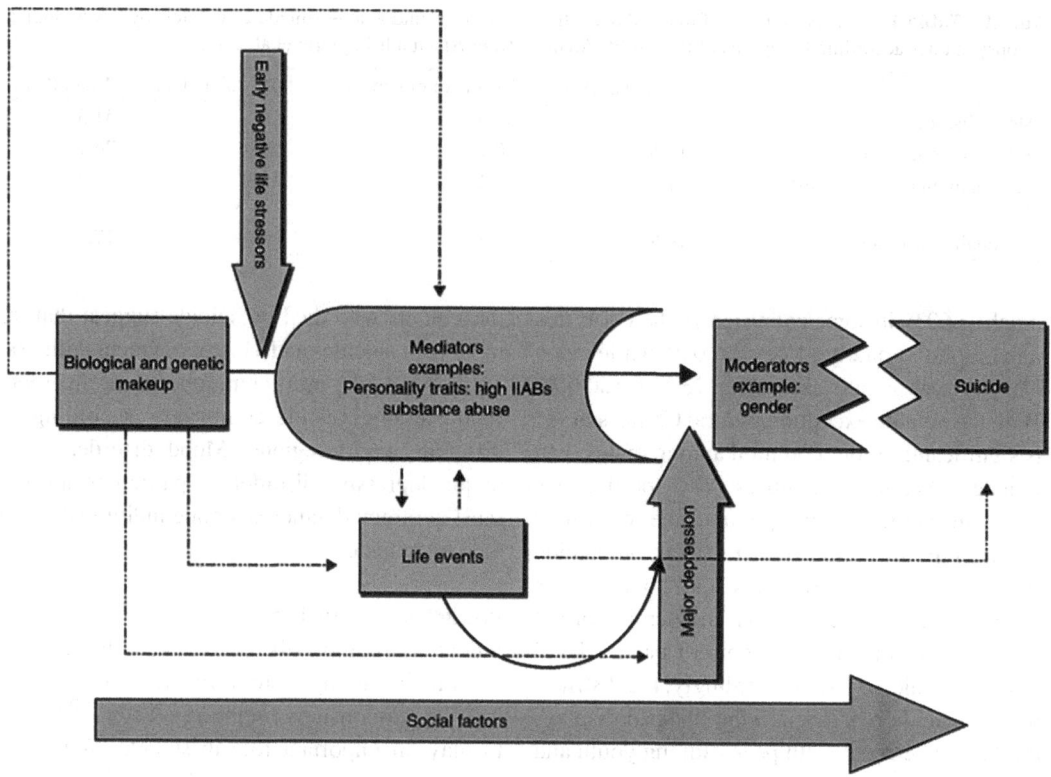

Suicide, Fig. 1 Factors contributing to suicide

hopelessness, and a history of a past suicide attempt.

Over the last decades, it has become increasingly clear that individuals who die by suicide also have constitutional risk factors. However, the relationship between these predisposing factors, which to a certain extent are conferred by the individual's biological makeup, and suicide is not direct, but seems to be mediated and moderated by a number of different factors. Among these factors are clinical and demographic risk factors, such as those listed earlier, history of early life sexual and physical abuse, personality variants such as behavioral traits, and triggering factors such as recent life events and interpersonal stressors.

Epidemiological and clinical studies have consistently suggested that a positive family history of suicide acts as an important risk factor for suicide. As psychopathology also runs in families, a major question has been to what extent suicide aggregates in families independently of psychopathology. Studies have suggested that first-degree relatives of suicide probands have a 4–10 times higher risk of suicidal behavior than relatives of psychiatrically normal controls, and this is once other risk factors and psychopathology are accounted for. Genetic-epidemiological studies suggest that at least part of this familial aggregation is attributable to genetic effects (Turecki 2001).

Several lines of evidence point to the fact that familial transmission of suicidal behavior may be mediated through the transmission of personality traits such as impulsive-aggressive behaviors. Personality traits represent emotional, behavioral, motivational, interpersonal, experiential, and cognitive styles that help us relate to and cope with the world. Clinical and community research suggest links between suicidality and extreme personality profiles. For instance, children who score high on a measure of disruptive behavior, a composite of aggressive, impulsive, and hyperactive behaviors, were found to be more likely to attempt suicide as young adults (Brezo et al. 2008). In general, most clinical and psychological autopsy studies (which use proxy-based interviews to investigate individuals who died by suicide) report elevated measures of impulsivity and aggressive behaviors among suicide attempters and completers as compared to controls, particularly among younger individuals. There is also direct and indirect evidence suggesting that relatives of suicide completers and suicide attempters have higher scores on these personality traits than controls.

Impulsivity may be conceptualized as the inability to resist impulses, which, from the strict phenomenological point of view, refer to explosive and instantaneous, automatic or semiautomatic psychomotor actions that are characterized by their sudden and incoercible nature. A more behavioral definition considers impulsivity as a drive, stimulus, or behavior that occurs without reflection or consideration for the consequences of such behavior. However, studies suggest that it is not the exclusive presence of impulsivity that appears to account for its observed association with suicide. Rather, impulsivity is frequently comorbid with other personality traits, particularly aggressive behaviors. As such, suicides tend to have high levels of aggressive-destructive impulsive traits, generally referred to as impulsive-aggressive behaviors. These have been operationally defined in suicide studies as a tendency to react with animosity or overt hostility, without consideration of possible consequences, when piqued or under stress.

Neurobiology of Suicide

The last decades have seen a growing interest in the understanding of the biological processes underlying suicide. Suicidal behavior has been associated with several neurobiological alterations, particularly in neurotransmission. For close to four decades, molecular studies have considered monoamines as prime suspects in suicidal behavior. Serotonin, in particular, has been the most investigated monoaminergic neurotransmitter system, and several lines of evidence suggest its involvement in the vulnerability to, and process of, suicide (Turecki and Lalovic 2005). Overall, while not always consistent, studies suggest that suicidal behavior is associated with reduced serotonergic neurotransmission. The evidence supporting serotonergic changes associated with suicide comes from studies using different and complementary approaches, including, but not limited to, investigations of cerebrospinal fluid levels of the serotonin metabolite 5-hydroxyindoleacetic acid (5-HIAA), neuroendocrine challenges, and postmortem receptor binding and imaging studies with receptor ligands. The serotonergic alterations seem to be more pronounced in the prefrontal cortex, where there is evidence of a decrease in presynaptic serotonin transporter binding and an upregulation of postsynaptic serotonin receptors. Together, these results imply reduced serotonergic input to this brain region. Among other functions, the prefrontal cortex is involved in the behavioral inhibition and expression of emotions. Reduced serotonergic input to this brain region could result in the impaired inhibition of behaviors such as impulsive aggression, which in turn could increase suicide risk.

Although the most extensively investigated neurotransmitter alterations associated with suicide have been those related to serotonergic changes, other neurotransmitters have also been investigated. Among these neurotransmitters are the noradrenergic, dopaminergic, opioid, glutamatergic, and GABAergic neurotransmitter systems. More recently, promising and consistent results have pointed to the implication of the polyamine system, which is involved in the stress response, depression, and suicide.

Alterations have also been reported in several components of different signal transduction systems, as well as in neurotrophic factors, particularly brain-derived neurotrophic factor (BDNF) and TrkB. These studies suggest that the suicide process may be associated with an altered neuroplastic capacity.

Another interesting avenue investigated in suicide is a possible relationship with lipid metabolism and more specifically with low cholesterol levels. This intriguing association is supported by evidence from studies that have used different designs, including large cohort studies, observational studies in suicidal, impulsive or violent populations, and experimental studies in animals, as well as postmortem brain studies (Golomb 1998). The use of different study designs and the replication of the findings across different populations have added validity to this still polemical association. A number of hypotheses have been advanced in an attempt to understand the possible relationship between cholesterol and suicide, but only a few of these hypotheses have been the subject of experimental testing. Whether this association proves to be true or not, one possible way to understand it is through an evolutionary point of view. It is not unreasonable to think that dietary needs, hence cholesterol levels, should be associated with certain behaviors, such as aggression. Aggressive behaviors may have conferred advantages on early humans who were hunting for food. Such an evolutionary link between cholesterol and aggression, preserved in the link between impulsive aggression and suicide, could underlie the observed relationship between cholesterol and suicide risk. If this is the case, there should also be molecular mechanisms connecting cholesterol levels to molecular correlates of aggression/suicide.

Treatment

There is no specific pharmacological treatment for suicidal behavior. There is evidence, however, that certain pharmacological interventions used for the treatment of the underlying mental disorder are more effective than other ones in reducing suicidal behavior in these patients. Two examples of such treatments are lithium and clozapine. Lithium is a salt used primarily in the maintenance treatment of bipolar disorder. There is substantial evidence indicating that lithium is effective in reducing the risk of suicidal behavior among bipolar patients (Baldessarini et al. 2006). While some preliminary evidence suggests that the antisuicidal effect of lithium may be independent of its mood-stabilizing properties, this issue requires further investigation. Similarly, clozapine, an atypical antipsychotic drug, is believed to have antisuicidal properties (Meltzer 2005). As with lithium, it is unclear if the antisuicidal properties of clozapine are independent of its neuroleptic effects.

Two psychotherapeutic interventions have provided evidence of efficacy in reducing suicidal behavior. Cognitive behavioral therapy applied in 10 sessions designed specifically to prevent suicide attempts has been shown to be effective in primarily depressed adults with recent histories of suicide attempts (Brown et al. 2005). Dialectic behavioral therapy, which is a variant of cognitive behavioral therapy developed specifically to treat patients with borderline personality disorder, has also shown evidence that it reduces suicidal behavior in this population (Linehan et al. 1991).

Cross-References

- ▶ Bipolar Disorder
- ▶ Depression
- ▶ Impulsivity
- ▶ Lithium

References

Arsenault-Lapierre G, Kim C, Turecki G (2004) 3500 cases of suicide: a systematic review. BMC Psychiatry 4:4–37

Baldessarini RJ, Tondo L, Davis P, Pompili M, Goodwin FK, Hennen J (2006) Decreased risk of suicides and attempts during long-term lithium treatment: a meta-analytic review. Bipolar Disord 8:625–639

Brezo J, Barker ED, Paris J, Hebert M, Vitaro F et al (2008) Childhood trajectories of anxiousness and disruptiveness as predictors of suicide attempts. Arch Pediatr Adolesc Med 162:1015–1021

Brown GK, Ten Have T, Henriques GR, Xie SX, Hollander JE, Beck AT (2005) Cognitive therapy for the prevention of suicide attempts: a randomized controlled trial. JAMA 294:563–570

Golomb BA (1998) Cholesterol and violence: is there a connection? Ann Intern Med 128:478–487

Linehan MM, Armstrong HE, Suarez A, Allmon D, Heard HL (1991) Cognitive-behavioral treatment of chronically parasuicidal borderline patients. Arch Gen Psychiatry 48:1060–1064

Meltzer HY (2005) Suicide in schizophrenia, clozapine, and adoption of evidence-based medicine. J Clin Psychiatry 66:530–533

Turecki G (2001) Suicidal behavior: is there a genetic predisposition? Bipolar Disord 3:335–349

Turecki G, Lalovic A (2005) The biology and genetics of suicidality. In: Licinio J, Wong M (eds) Biology of depression: from novel insights to therapeutic strategies. Wiley-CVH, Weinheim, pp 287–305

WHO (2000) Health systems: improving performance. World health report 2000, Geneva

Sulpiride

Definition

Sulpiride is a ▶ typical antipsychotic based on antagonism of postsynaptic D_2 dopamine receptors. Agonism at the gamma-hydroxybutyrate receptor may contribute to its antipsychotic properties. Sulpiride has fewer extrapyramidal side effects, but also reduced antipsychotic potency, as compared with many other ▶ typical antipsychotics. At lower doses than used for antipsychotic treatment, its prominent action is presynaptic dopamine ▶ autoreceptor antagonism, giving rise to antidepressant and stimulating effects. Secondary clinical uses are thus treatment of depression and vertigo. Sulpiride is currently not approved in the USA and Canada. Together with the atypical antipsychotic ▶ sultopride, sulpiride falls under the chemical class of benzamides.

Cross-References

▶ Antipsychotic Drugs
▶ First-Generation Antipsychotics

Sultopride

Definition

Sultopride is an atypical antipsychotic of the same chemical class as ▶ sulpiride (benzamides). Sultopride's action at antipsychotic doses exceeds D_2 dopamine receptor blockade to include antagonism of D_3 and 5-HT_7 receptors. Agonism of the GHB receptor may contribute to its properties. Like sulpiride, low doses of sultopride predominantly block dopaminergic ▶ autoreceptors, facilitating dopamine release. This action, as well as its 5-HT_7 receptor antagonism, likely explains it efficacy at low doses for treating depression. Sultopride is not currently approved in the USA.

Cross-References

▶ Antipsychotic Drugs
▶ First-Generation Antipsychotics
▶ Second- and Third-Generation Antipsychotics

Summary of Product Characteristics

Synonyms

Product information; SPC

Definition

The SPC is the basis of information for healthcare professionals on how to use a medicinal product safely and effectively. The package leaflet (PL) is drawn up in accordance with the SPC. The SPC sets out the agreed position of the medicinal product as distilled during the course of the assessment process. The content cannot be changed except with the approval of the originating competent authority.

Summation Test

Definition

The summation test is one of two widely accepted tests for whether a stimulus functions as a ▶ conditioned inhibitor (the ▶ retardation of acquisition test is the other one). In the summation test, a putative ▶ conditioned inhibitor is presented in compound with a separately trained excitatory conditioned stimulus (CS) – that is, one that evokes a ▶ conditioned response. If the stimulus functions as a conditioned inhibitor, it should decrease conditioned responding evoked by the separately trained excitor relative to controls that receive the excitatory CS alone and the excitor in compound with an alternative stimulus that was not a signal for non-reinforcement. Because alternative accounts for response inhibition remain with just the summation test, the strongest case for a stimulus functioning as a conditioned inhibitor also requires the use of the retardation test.

Cross-References

- ▶ Blocking, Overshadowing, and Related Concepts
- ▶ Classical (Pavlovian) Conditioning
- ▶ Occasion Setting with Drugs
- ▶ Pavlovian Fear Conditioning

Suppressibility

Definition

Many patients with tic disorders describe a capacity to limit tic expression, at least temporarily. Patients will describe that suppression requires concentration and that it is ultimately time limited (see ▶ Tics).

Sustained Attention

Synonyms

Vigilance

Definition

Sustained attention is the capacity to maintain attention over time on repetitive tasks. It is the need for continuous allocation of processing resources that distinguishes vigilance from other forms of attention, such as divided or selective attention. In humans and other animals, under normal conditions, deficits in sustained attention emerge during "boring" and/or routine activities, especially in situations of low event rate.

Cross-References

- ▶ Attention
- ▶ Attention-Deficit and Disruptive Behavior Disorders

Sygen

Synonyms

GM1 ganglioside; H3-α-acetylneuraminosyl-gangliotetragly cosylceramidemonosialoglucoside

Definition

Sygen, a GM1 monosialoganglioside (one of about 60 plasma membrane glycosphingolipids thought to modulate cell signal transduction), was developed as a neuroprotectant for Parkinson's disease and other CNS indications by Fidia. Gangliosides are composed of ceramide-glucose-galactose-N-acetylneuraminic acid. Sygen was withdrawn for association with Guillain-Barre syndrome and lack of benefit in Parkinson's disease. Sygen was not available in the USA, except for compassionate use in spinal cord injury.

Cross-References

- ▶ Anti-Parkinson Drugs
- ▶ Neuroprotectants: Novel Approaches for Dementias
- ▶ Neuroprotection

Sympathectomy

Definition

Lesion of the sympathetic (noradrenergic) branch of the peripheral autonomic nervous system. Experimental sympathectomy is achieved by administration of a systemic toxin, affecting all branches of the system, or by extirpation of individual spinal or cervical ganglia to disconnect (deafferent) specific branches.

Sympathetic Neurons

Definition

These are neurons located in prevertebral and paravertebral ganglia. They are a main component of the so-called autonomic nervous system. These neurons are controlled by preganglionar terminations originating in the spinal cord, and their axons are distributed throughout the skin, blood vessels, glands, heart, and viscera. Upon stimulation, they release the neurotransmitter ▶ noradrenaline (norepinephrine) with the exception of some neurons innervating the sweat glands that release ▶ acetylcholine. The main functions are maintaining the peripheral vascular tone, accelerating the heart frequency, and inhibiting gastrointestinal motility.

Synaptic Consolidation

Definition

Consolidation or synaptic consolidation is defined as the stabilization of new memories over time, typically on the timescale of 4–8 h.

Cross-References

▶ Consolidation and Reconsolidation

Synaptic Plasticity

Johannes Mosbacher
Research & Development, F. Hoffmann – La Roche Ltd, Basel, Switzerland

Synonyms

Homeostatic plasticity; Neuronal plasticity; Short-term plasticity; Synaptic scaling

Definition

In general terms, synaptic plasticity describes a change, persistent or transient, of morphology, composition, or signal transduction efficiency at a neuronal synapse in response to intrinsic or extrinsic signals. Long-term potentiation (LTP) and long-term depression (LTD) likely represent the most extensively studied forms of synaptic plasticity, which itself is the best characterized form of neuronal plasticity, the cellular substrate for learning and memory.

Physiology and psychopharmacological modulation of LTP and LTD are described in detail elsewhere (see Cross-References). Like these special cases, most other forms of synaptic plasticity are induced by an associative coincidence of signals in space and time, in line with the theory of the so-called Hebbian plasticity: a convergence of different second messenger pathways or pre- and postsynaptic activity at the synapse itself. Of note, the term "synaptic" could point to either the origin or the affected target of such signals. However, for many phenomena related to drug effects, this distinction cannot be conclusively drawn because a differentiation of cause and effect in neuronal networks is inherently difficult for drugs acting at multiple sites and over a relatively long period of time.

Forms of synaptic plasticity beyond LTP and LTD comprise short-term plasticity (STP) based on presynaptic transmitter release probability, postsynaptic spine motility, translocation of proteins between extrasynaptic and synaptic sites, epigenetic, posttranscriptional or posttranslational modifications of synaptic proteins, and changes of intra- and extracellular ion concentrations. Some of those changes are described as "▶ Meta-Plasticity" because they have been shown to change the ability of synapses to undergo classical forms of plasticity like LTP or LTD (Abraham 2008). Common to all those changes is their link to an altered function of synaptic transmission. This could be experimentally shown for some but not all of the abovementioned examples, and there are certainly more mechanisms to be uncovered by improved and refined approaches and technologies.

The concept of synaptic plasticity was introduced more than 100 years ago, with W. James, E. Tanzi, E. Lugaro, and D.O. Hebb making milestone contributions in the development of this principle, long before T. Lomo and T.V. Bliss in P. Andersens laboratory could experimentally show in 1973 that specific electric stimuli induce persistent changes in synaptic transmission efficiency, both in vitro and in vivo (see Berlucchi and Buchtel 2009 for a recent review of historical aspects of synaptic plasticity).

For a long time, electrophysiological techniques such as extra- or intracellular recordings remained the gold standard for observing synaptic physiology. It is only recently that optical imaging complemented the functional information with structural data at the level of single synapses in living tissue.

Models of synapses usually show synaptic signal transmission between neurons as a "static element" that can be described by a fixed input–output relation: for a single event of chemical signal transmission between neurons, a presynaptic stimulus leads to a defined postsynaptic response. In the most classical case of synaptic transmission, a presynaptic action potential depolarizes the presynaptic Bouton, which contains the machinery for vesicular release. A rise of intracellular calcium ($[Ca^{2+}]_{int}$) induces the release of vesicle-stored neurotransmitters into the synaptic cleft. The transmitter activates postsynaptic receptors that trigger electric or second messenger signals in the postsynaptic cell. When one or several of the contributing synaptic components change their function, the synapse undergoes plastic changes that can last from milliseconds to days or years. More special cases of synaptic signal transmission, like electrical synapses formed by gap junctions, are not described further, as there are to date still very few studies on their structural or functional plasticity. Figure 1 shows some examples of synaptic plasticity across a large temporal spectrum. In vivo optical imaging of postsynaptic compartments in glutamatergic neurons, the so-called

Synaptic Plasticity, Fig. 1 Types of synaptic plasticity across a timescale spanning several orders of magnitude. Depicted is a nonexhaustive selection of mechanisms underlying synaptic plasticity sorted by their approximate time of expression. A common type of classical LTP and LTD is reflected by altered AMPA receptor-mediated EPSCs. *Cav2* presynaptic, high-voltage-activated calcium channels of the Cav2 family, *PKC* protein kinase type C, *NR2* subunit family 2 of the NMDA receptors, *AChR* acetylcholine receptor, *KCC2* potassium-chloride cotransporter type 2, $[Cl^-]_{int}$ intracellular chloride ion concentration, *Gβ/γ* β/γ subunit of trimeric G-protein, *EPSC* excitatory postsynaptic currents

"▶ Dendritic Spines," suggests that synaptic plasticity is the rule rather than the exception in the lifetime of a synapse.

Impact of Psychoactive Drugs

The diversity of synaptic plasticity implicates that there are numerous potential pharmacological mechanisms that affect synapses and their function. The below-given examples only reflect a selection of the better-studied links between psychopharmaceuticals and synaptic plasticity. As for many other aspects of psychoactive drugs, we have to acknowledge that there are still many gaps in our understanding of their molecular mechanism of action and link to synaptic function. As a general rule – taking into consideration the prerequisite for spatiotemporal coincidence – it is helpful for deciphering such links to focus on "▶ Second Messenger" systems triggered by targets of psychopharmacological drugs and study their effect on synaptic proteins that are involved in the above-described plasticity process.

Linking Dopamine Receptors with NMDA Receptors via DARPP-32

A prominent example of this approach is the link between dopamine and N-methyl-D-aspartate (NMDA) receptors that are key mediators of many forms of LTP. This connection may even provide a molecular basis for the clinical manifestation of both positive and negative symptoms in schizophrenia via the mutual interaction between dopaminergic and glutamatergic signaling networks (Stone et al. 2007). Phosphorylation of the dopamine and cAMP-regulated phosphoprotein (▶ DARPP-32) by either dopamine receptor-induced protein kinase A (PKA) or by NMDA receptor-triggered calcineurin reciprocally affects either receptor system via protein phosphatase 1 (PP1) and modulates the late phase of LTP by influencing the extracellular signal-regulated kinase (ERK) pathway. With its multiple phosphorylation sites, DARPP-32 is an excellent coincidence detector and signal integrator, as many more protein kinases and phosphatases regulate the signaling properties of DARPP-32. Stimulants like amphetamines, nicotine, and caffeine all show impaired efficacy in transgenic mice with ablation or genetically diminished phosphorylation capacity of DARPP-32. Antipsychotics like haloperidol that block dopamine receptors were also shown to affect DARPP-32 phosphorylation in the same direction as the above-described stimulants. The differential behavioral consequences induced by both classes of drugs may arise from cell specificity in DARPP-32 phosphorylation as found in striatonigral versus striatopallidal neurons (Bateup et al. 2008). Several alternative explanations exist. The activation of the immediate early gene necessary for LTP maintenance, Arc/Arg3.1, for example, needs simultaneous signaling of Gαs-coupled receptors like dopamine D_1 receptors and calcium influx through NMDA receptors. It will need further studies to fully understand the interaction between dopaminergic and glutamatergic systems at the synaptic level.

Proteins involved in DARPP-32 signaling, especially PP1 and calcineurin, provide an interesting additional link between receptor function and synaptic plasticity via their modulation of the actin cytoskeleton in postsynaptic compartments.

Antidepressants, Mood Stabilizers, and Antidementia Drugs Modulate the Number of Dendritic Spines in the Hippocampus

The plasticity of synaptic morphology received much attention in recent years because of the technological revolution in genetic labeling and optical imaging. With two-photon, confocal, or evanescent wave imaging, it became possible to investigate single synapses at submicron resolution in unstained tissue. This development significantly accelerated studies on synaptic morphology that formerly required fixation and often electron microscopy. It was also the basis to identify a new downstream effector system of antidepressant drugs and the mood stabilizer lithium in relation to synaptic plasticity: dendritic and postsynaptic morphology. This was first shown in a rodent model for depression, the bulbectomized rat. One of the anatomical consequences of bulbectomy is a decreased hippocampal volume. It could be shown that the ablation of the olfactory bulb, over a period of several weeks, induced a significant reduction in dendritic spines at glutamatergic neurons in the hippocampus. Loss of dendritic spines is thought to be equivalent to a loss of synaptic function. The antidepressant tianeptin was effectively protecting against spine loss in this animal model. Corroborating this finding, electron microscopy studies in ovariectomized rats found a similar increase in synapse number after short-term fluoxetine treatment, and subchronic treatment of imipramine was found to significantly modify synaptic morphology and increase dendritic spine density in hippocampal subregions of healthy adult rats. Increased expression of neurotrophic factors like BDNF may provide a mechanistic link between the monoaminergic system modulated by antidepressants and synaptic morphogens like filamentous actin. Tyrosine phosphorylation of β-catenin by BDNF is known to promote dissociation from cadherin, a major structural component of many synapses. Notably, β-catenin is also downstream of glycogen synthase kinase GSK-3β, the molecular target of mood stabilizers like lithium. Lithium treatment of stressed rats increased dendritic arborization of hippocampal pyramidal cells, also affecting the number of dendritic spines in this brain region (see also Pittenger and Duman 2008).

Dendritic spines can grow and collapse within minutes, and the NMDA receptor was shown to be a necessary trigger for such major though reversible changes in synaptic morphology. Memantine, one of the few available drugs for the treatment of Alzheimer's Disease (AD), acts as a partial NMDA receptor antagonist. In primary hippocampal cultures, memantine was shown to prevent dendritic spine loss and shape changes induced by oligomers of amyloid β (Calabrese et al. 2007), providing a potential mechanism for its therapeutic efficacy in AD patients. Beyond AD, the Fragile X syndrome, the most common inherited cause of mental impairment and the most common known cause of autism, was linked to defects in synaptic plasticity. Lack of the fragile X mental retardation protein (FMRP) induces dysregulation of spine morphogenesis and exaggerated metabotropic glutamate receptor-dependent LTD. FMRP is a synaptic protein-regulating dendritic RNA delivery and translational repression. Based on pharmacological and genetic experiments, current theories see FMRP and the metabotropic glutamate receptor 5 (mGluR5) as counterparts in dendritic protein synthesis. There is preclinical evidence that genetic suppression of mGluR5 or the mGluR5 antagonists MPEP and fenobam can rebalance the system at the physiological level and improve cognitive performance of mice lacking FMRP. This was corroborated by a pilot clinical trial with the mGluR5 antagonist fenobam (Berry-Kravis et al. 2009) and will be followed up by further clinical studies using improved drugs from this new evolving class of psychoactive drugs.

Synaptic plasticity is not a phenomenon restricted to the postsynaptic part of neuronal synapses. In the example of amyloid β-induced spine collapse, presynaptic boutons have been also affected and spontaneous synaptic transmission impaired. Morphological changes of presynaptic structures currently are targets of numerous studies. Functional plasticity of presynaptic proteins is particularly important for LTP at GABAergic synapses and most forms of STP. This short-lasting form of plasticity is dependent on the vesicular release machinery and modulated by a number of mechanisms regulating presynaptic calcium (Zucker and Regehr 2002).

Nicotinic Receptors Affect Short-Term Plasticity by Regulation of Vesicular Transmitter Release

Presynaptic nicotinic acetylcholine receptors (nAChR) are an important trigger for increased presynaptic calcium and have been shown to regulate STP in a number of brain areas. Particularly, the α7 nAChR channel is a target for a number of drug candidates currently in clinical development for the treatment of negative symptoms in "► Schizophrenia" or "► Mild Cognitive Impairment." Those receptors are expressed in many neurons at presynaptic sites while they control as postsynaptic receptors on GABAergic neurons the inhibitory tone in the hippocampus. Presynaptically, there is strong evidence that nAChR control release probability of several neurotransmitters, notably also of dopamine. Earlier studies using tonic application of subtype selective nAChR agonists showed a role for non-α7 AChRs in the regulation of dopamine release. More recently, phasic and short-term activation pattern of synaptosomes revealed a significant role of α7 receptors in the control of the readily releasable pool of dopamine (Turner 2004). The increase in dopamine release upon phasic AChR stimulation was dependent on the calcium-binding protein calmodulin but not on presynaptic high-voltage-activated calcium channels, as was the case for the non-α7 receptor-mediated dopamine release. An increase in the number of vesicles ready to release their neurotransmitter upon stimulation facilitates neurotransmission for the next few synaptic events and thus represents a form of STP. Nicotinic α7 receptors show an agonist-dependent rapid and strong desensitization after activation, and this desensitization likely turns strongly desensitizing agonists such as nicotine into functional antagonists when they are constantly present. Partial α7 receptor agonists like MEM3454 from memory pharmaceuticals may well have a different effect due to their increased activation of α7 receptor-mediated steady-state current. Still, a continuous presence of the drug may impair phasic cholinergic signaling via those receptors. In this respect, allosteric positive α7 receptor modulators are likely to keep phasic signal transmission intact and thus may show a stronger impact on synaptic facilitation at dopaminergic synapses and therapeutic efficacy.

In summary, there is good evidence that psychotherapeutics influence forms of synaptic plasticity beyond LTP and LTD. Whether changes of synaptic structure or function upon treatment with psychoactive drugs are purely coincidental or causally correlated with their therapeutic effect remains to be determined for most cases. Given the relevance of synaptic plasticity for cognitive and emotional processes, this form of neuronal plasticity is no doubt a major contributor to the behavioral effects induced by psychopharmacological drugs.

Cross-References

▶ Extracellular Recording
▶ Intracellular Recording
▶ Nicotinic Agonists and Antagonists
▶ Tonic Signals

References

Abraham WC (2008) Metaplasticity: tuning synapses and networks for plasticity. Nat Rev Neurosci 9:387–399

Bateup HS, Svenningsson P, Kuroiwa M, Gong S, Nishi A, Heintz N, Greengard P (2008) Cell type-specific regulation of DARPP-32 phosphorylation by psychostimulant and antipsychotic drugs. Nat Neurosci 11:932–939

Berlucchi G, Buchtel HA (2009) Neuronal plasticity: historical roots and evolution of meaning. Exp Brain Res 192:307–319

Berry-Kravis E, Hessl D, Coffey S, Hervey C, Schneider A, Yuhas J, Hutchison J, Snape M, Tranfaglia M, Nguyen DV, Hagerman R (2009) A pilot open label, single dose trial of fenobam in adults with fragile X syndrome. J Med Genet 46:266–271

Calabrese B, Shaked GM, Tabarean IV, Braga J, Koo EH, Halpain S (2007) Rapid, concurrent alterations in pre- and postsynaptic structure induced by naturally-secreted amyloid-beta protein. Mol Cell Neurosci 35:183–193

Pittenger C, Duman RS (2008) Stress, depression, and neuroplasticity: a convergence of mechanisms. Neuropsychopharmacology 33:88–109

Stone JM, Morrison PD, Pilowsky LS (2007) Glutamate and dopamine dysregulation in schizophrenia – a synthesis and selective review. J Psychopharmacol 21:440–452

Turner TJ (2004) Nicotine enhancement of dopamine release by a calcium-dependent increase in the size of the readily releasable pool of synaptic vesicles. J Neurosci 24:11328–11336

Zucker RS, Regehr WG (2002) Short-term synaptic plasticity. Annu Rev Physiol 64:355–405

Synaptic Pruning

Definition

A process by which synaptic contacts between neurons, across which neurotransmitters and other chemical messengers are released in support of information processing, are normally reduced or eliminated in postnatal development. The overabundance of metabolically expensive synaptic contacts present in many brain regions in early life is thought to set the stage for processes of competitive elimination, in which those neural connections that carry the most functionally relevant forms of information survive and grow, while those carrying the least are removed. In many brain regions, this process may be particularly robust or entering its finishing stages during adolescent neurodevelopment.

Synaptic Reconsolidation

Definition

Synaptic reconsolidation is defined as the return of a consolidated memory to an unstable state, from which it must restabilize in order to persist. Typically, restabilization, or reconsolidation, takes several hours. Reactivating, for example, recalling or retrieving, existing memory can induce reconsolidation.

Synesthesia

Definition

Synesthesia is the crossing of senses, a phenomenon whereby stimulation of one sensory modality leads to automatic and involuntary experience of stimulation in a second sensory modality, for example, to smell colors and to visualize sounds as colors. Certain individuals are born with this ability and are called synesthetes.

T

Tachykinins

Marc Turiault[1], Caroline Cohen[2] and Guy Griebel[3]
[1]Geneva Business News, Geneva, Switzerland
[2]Therapeutic Strategic Unit Aging, Chilly-Mazarin, France
[3]Translational Sciences Unit, Chilly-Mazarin, France

Definition

Tachykinins are short-chain amino acid neuromodulators found in species ranging from invertebrates to mammals that share the common C-terminal amino acid sequence: Gly-Leu-Met-NH2. The name tachykinin suggests the ability of these molecules to rapidly induce rapidly (tachy, swift) contractions of smooth muscles (kineo, to move).

Pharmacological Properties

History

In 1931, Euler and Gaddum characterized an unidentified substance able to induce rapidly contractions of intestinal tissue. They named it substance P (SP); it was stable in a dry powder form. Substance P remained the only mammalian member of the tachykinin family until identification of neurokinin A (NKA) in 1983. Other mammalian tachykinins have been isolated subsequently neuropeptide K (NPK), neuropeptide γ (NP γ), neurokinin B (NKB), endokinins, and hemokinins. Studies of the distribution pattern of tachykinins show a widespread expression in peripheral tissues where they have various effects such as inducing vasodilatation, hypotension, or contraction of smooth muscle. In the central nervous system, the three prominent mammalian tachykinins SP, NKA, and NKB are widely distributed with different distribution patterns. Maximal NKB concentrations are found in the cortex, whereas SP and NKA share a more similar distribution with a strong expression in the spinal cord and in the nuclei implicated in emotional process (e.g., nucleus accumbens, septum, amygdala). At a cellular level, SP and NKA are mostly co-localized in neurons and interneurons with glutamate, GABA, monoamines, or acetylcholine. One or several tachykinins can be expressed within the same neurons and be co-released with classical neurotransmitters or neuromodulators (Beaujouan et al. 2004). The co-expression of SP and NKA is not surprising since there are three genes that encode for all known mammalian tachykinins. SP, NKA, NPK, and NP γ mRNA are generated by alternative splicing of a unique preprotachykinin-A (PPT-A) gene. NKB is derived from a second gene, the preprotachykinin-B (PPT-B) gene. A third gene, the more recently cloned preprotachykinin-C, is coding for hemokinins and endokinins that are primarily expressed in nonneuronal cells.

In parallel with the discovery of new peptides from the mammalian tachykinin family, three types of receptors have been identified. They belong to the G protein-coupled receptors (GPCR) superfamily containing seven transmembrane domains. The activation of tachykinin receptors leads to a transduction cascade, which in turn activates, among others, phospholipase C, the release of intracellular Ca^{2+}, and the stimulation of neurotransmitter release (Chahl 2006). However, tachykinins are neuromodulators, preferentially released when neurons are strongly activated (or under pathological conditions). Consequently, blockade of their receptors by antagonists may result in effects only when the system is stimulated (Hökfelt et al. 2000). This is of great relevance as it may provide pharmacological targets for therapeutic applications with potentially less pronounced side effects than drugs acting on tonically active modulators such as monoamines. Tachykinin receptors termed NK1, NK2, and NK3 bind with a high-affinity, respectively, SP, NKA, and NKB. Antagonists for these receptors have been suggested to have therapeutic value in a variety of areas, including inflammation, emesis, anxiety, and depression. However, the development of highly selective antagonists was hampered by findings from pharmacological and molecular studies that showed the existence of NK1 receptor isoforms with different affinities for tachykinins and a tissue-specific expression. Furthermore, it was observed that several NK1 receptor antagonists have a greater affinity for the guinea pig and human receptor than for the rat and mouse receptor (Beaujouan et al. 2004). This species heterogeneity, evidenced for NK1 and NK3, but not for NK2 receptors, had a major impact in the development of specific antagonists for these receptors as it required the development of suitable behavioral models in atypical species such as guinea pigs and gerbils to characterize their psychopharmacological properties.

NK1 Receptor Antagonists

The development of highly selective NK1 receptor antagonists was initiated after the discovery of the role of substance P as a key mediator of pain processes. NK1 antagonists were used as tools to specify the topological and functional features of NK1 receptors leading to the idea that SP could be used for the treatment of other pathologies such as emesis, Parkinson's disease, anxiety, or depression. Thus, the use of these antagonists in experimental research on depression and anxiety was based, for instance, on findings showing (1) an expression of SP and NK1 receptor in fear and depression-associated pathways, (2) fear-related behaviors after intracerebroventricular injection of SP and reduced fear following the peripheral administration of NK1 receptor antagonists, and (3) that binding sites for SP are co-localized with those of monoamine transmitters in the human brain.

The antidepressant- and anxiolytic-like effects of MK-0869 (aprepitant), the first NK1 receptor antagonist tested in human, were initially demonstrated in a range of animal models. The further development of NK1 and SP receptor knockout mice confirmed these results as mutant animals displayed an anxiolytic- and antidepressant-like phenotype.

Several randomized, placebo-controlled, double-blind, clinical studies were carried out to measure the safety and efficacy of aprepitant. In an initial clinical phase II trial, aprepitant was shown to display significant antidepressant activity. It was well tolerated and had fewer side effects than the selective serotonin reuptake inhibitor, paroxetine, which was used as a positive control in this study. Unfortunately, this result was never replicated in phase III clinical trials, thereby questioning the idea that NK1 receptor antagonists may be effective antidepressants (Rost et al. 2006). These disappointments have been a major reason why pharmaceutical companies seem to have abandoned the development of drugs targeting NK1 receptors for CNS disorders; to our knowledge there are no drugs targeting this system currently under development.

NK2 Receptor Antagonists

The identification of NK2 receptors in a number of peripheral tissues such as the smooth muscles of the gastrointestinal, the respiratory, and the urinary tracts, along with studies using selective

NK2 receptor antagonists, has led to the idea that this receptor may represent a potential therapeutic target for a wide range of disorders, including irritable bowel syndrome and pulmonary and urinary tract disorders. The demonstration of the existence of NK2 receptors in the brain was made much later. Data obtained from adult brain were not convincing due to nonspecific binding and poor selectivity of ligands and weak expression of the NK2 receptor. In 2001, the demonstration of the existence of central NK2 receptors was made using radiolabeled endogenous NKA in the presence of NK1 and NK3 receptor antagonists to avoid labeling of other tachykinin binding sites. These results were strengthened by detection of NK2 receptor mRNA in human and rat in brain structures affected in mood disorders, such as the prefrontal cortex and the hippocampus. Moreover neurochemical studies have identified a central regulatory role of NK2 receptors on monoaminergic and cholinergic neurotransmission. For instance, in anesthetized rats, the peripheral administration of the NK2 receptor antagonist, saredutant, has no effects on basal norepinephrine levels in the prefrontal cortex but reduces its release elicited by tail pinch. In addition, in non-anesthetized rats, local infusion of saredutant in the septum blocks stress-induced increase of hippocampal acetylcholine release but has no effect on basal conditions (Desvignes et al. 2003). These results underline the ability of NK2 receptor antagonists to regulate neurotransmission only when systems are activated (Steinberg et al. 2001).

Furthermore, recent data suggest an interaction between the tachykinin system and corticotropin-releasing factor (CRF). CRF is a neurohormone known to be involved in the regulation of the stress axis and in the etiology of mood disorders. The injection of saredutant was found to block CRF-induced increase in acetylcholine and norepinephrine release, suggesting that NK2 receptor antagonists counteract, at least in part, the effects of stress on neurotransmission.

At a behavioral level, the activation of central NK2 receptors by intracerebroventricular administration of NKA produces anxiogenic-like effects. Moreover, several NK2 receptor antagonists have been shown to produce anxiolytic-like activity in animal studies. For example, GR159897 and saredutant exhibited anxiolytic-like effects in exploration-based procedures such as the light/dark and the elevated plus-maze tests. It is noteworthy that the anxiolytic-like properties of saredutant were observed across species as evidenced in the mouse defense test battery, the marmoset human intruder test, and the social interaction test in gerbils and rats. In addition, saredutant also exhibited antidepressant-like properties in the rat forced swim test. Moreover, the drug attenuated physical degradation in the mouse chronic mild stress test (Griebel et al. 2001; Louis et al. 2008). In addition to the neurochemical and behavioral effects of NK2 receptor antagonists, molecular and cellular studies showed that chronic treatment with saredutant upregulates cAMP response element-binding protein (CREB) and promotes neurogenesis in the hippocampus after chronic stress exposure in mice. Similar effects are observed with classical antidepressant treatment such as fluoxetine. Together, these results suggested that selective NK2 receptor antagonists could have therapeutic utility for the pharmacological treatment of mood disorders and anxiety. Unfortunately, the low efficacy and poor brain penetration of the NK2 receptor antagonist tested in recent clinical trials did not allow a definitive conclusion on the pertinence of the target. As is the case with NK1 receptor antagonists, there is currently no selective NK2 receptor antagonist in development for a CNS condition.

NK3 Receptor Antagonists

The NK3 receptor has been the least studied of the NK receptor family. The first paper on a selective NK3 receptor antagonist, published in 1995, was the starting point of numerous studies which explored the pharmacology of the NK3 receptor (Emonds-Alt et al. 1995).

In peripheral tissues, similar to other tachykinin receptors, it was found that the NK3 receptor plays a role in smooth muscle contraction. However, NK3 receptors are mostly expressed in the central nervous system with a wide distribution throughout the spinal cord and brain, in particular in limbic areas, well known to play a crucial role in psychiatric disorders. Importantly, the activation of NK3 receptors expressed on dopaminergic neurons by the highly selective agonist senktide leads to an increased dopamine release in the striatum and the prefrontal cortex. This excitatory activity, as well as the stimulation of serotonergic and noradrenergic systems, is blocked by the selective NK3 receptor antagonist osanetant, but not by NK1 or NK2 antagonists (Spooren et al. 2005). Decreasing the dopaminergic activity may be of interest for the treatment of the positive symptoms of schizophrenia since all currently approved antipsychotics share this feature. Unfortunately, the available clinical findings with selective NK3 receptor antagonists in psychiatric diseases, in particular schizophrenia, have not convincingly established that blockade of this neuropeptide receptor may be sufficient to improve these clinical conditions (Griebel and Holsboer 2012). Given these results, it is not surprising that pharmaceutical companies have considerably reduced or abandoned research and development of NK3 receptor ligands for the treatment of CNS disorders.

Cross-References

▶ Elevated Plus Maze
▶ Neurogenesis

References

Beaujouan JC, Torrens Y, Saffroy M, Kemel ML, Glowinski J (2004) A 25 year adventure in the field of tachykinins. Peptides 25:339–357
Chahl LA (2006) Tachykinins and neuropsychiatric disorders. Curr Drug Targets 7(8):993–1003
Desvignes C, Rouquier L, Souilhac J, Mons G, Rodier D, Soubrié P, Steinberg R (2003) Control by tachykinin NK(2) receptors of CRF(1) receptor-mediated activation of hippocampal acetylcholine release in the rat and guinea-pig. Neuropeptides 37:89–97
Emonds-Alt X, Bichon D, Ducoux JP, Heaulme M, Miloux B, Poncelet M, Proietto V, Van Broeck D, Vilain P, Neliat G (1995) SR 142801, the first potent non-peptide antagonist of the tachykinin NK3 receptor. Life Sci 56(1):27–32
Griebel G, Holsboer F (2012) Neuropeptide receptor ligands as drugs for psychiatric diseases: the end of the beginning? Nat Rev Drug Discov 11:462–478
Griebel G, Perrault G, Soubrié P (2001) Effects of SR48968, a selective non-peptide NK2 receptor antagonist on emotional processes in rodents. Psychopharmacology (Berl) 158:241–251
Hökfelt T, Broberger C, Xu ZQ, Sergeyev V, Ubink R, Diez M (2000) Neuropeptides-an overview. Neuropharmacology 39:1337–1356
Louis C, Stemmelin J, Boulay D, Bergis O, Cohen C, Griebel G (2008) Additional evidence for anxiolytic- and antidepressant-like activities of saredutant (SR48968), an antagonist at the neurokinin-2 receptor in various rodent-models. Pharmacol Biochem Behav 89:36–45
Rost K, Fleischer F, Nieber K (2006) Neurokinin 1 receptor antagonists-between hope and disappointment. Med Monatsschr Pharm 29:200–205
Spooren W, Riemer C, Meltzer H (2005) Opinion: NK3 receptor antagonists: the next generation of antipsychotics? Nat Rev Drug Discov 4:967–975
Steinberg R, Alonso R, Griebel G, Bert L, Jung M, Oury-Donat F, Poncelet M, Gueudet C, Desvignes C, Le Fur G, Soubrié P (2001) Selective blockade of neurokinin-2 receptors produces antidepressant-like effects associated with reduced corticotropin-releasing factor function. J Pharmacol Exp Ther 299:449–458

Tachyphylaxis

Definition

Tachyphylaxis is the rapid development of tolerance to the effect of a drug. That is, the response to a drug rapidly decreases after one or a few doses. The term seems to be synonymous with acute tolerance. For example, the hallucinogen LSD produces tachyphylaxis such that if a dose is taken daily, after 3–4 days it produces almost no effect, even if the dose is increased. In the case of LSD, or other classic hallucinogens, tachyphylaxis is thought to be a result of rapid desensitization and internalization of serotonin 5-HT$_{2A}$ receptors.

Tandospirone

Definition

Tandospirone displays anxiolytic and antidepressant properties with a short ▶ half-life. Its main pharmacological mode of action is as a potent 5-HT$_{1A}$ receptor ▶ partial agonist. Its active metabolite displays antagonistic action at α_2-adrenergic receptors. Tandospirone is a member of the piperazine and azapirone chemical classes. It is prescribed mainly in Japan and China. A Western equivalent of the same pharmacological class is buspirone, which is prescribed for ▶ generalized anxiety disorder, ▶ panic attacks, and as augmentation of treatment with serotonin reuptake inhibitors.

Cross-References

▶ Buspirone

Tardive Dyskinesia

Definition

Tardive dyskinesia (TD) is a chronic neuromotor side effect of dopamine-blocking medications, characterized by abnormal involuntary movements of voluntary musculature that are generally slow and that can be irreversible. It can include abnormal rotatory or sinuous movements of the mouth, lips, neck, trunk, hands, arms, and legs. It is identified as a side effect of long-term antipsychotic use; risk factors have been identified (e.g., age, duration of exposure), but there is no specific means of precisely predicting who will develop TD. With the first-generation antipsychotics, prevalence rates in chronically treated patients approximated up to 25 %. Evidence for the newer, "atypical" antipsychotics indicates that none of these agents are without risk of TD, although prevalence rates appear lower.

The precise pathophysiologic mechanisms underlying TD have not been elucidated, although high and sustained levels of dopamine D2 occupancy have been implicated. Many putative pharmacologic treatments have been investigated, although only a few (e.g., tetrabenazine in Canada) have gained an indication for the treatment of TD. TD has proven inconsistent in its response to all treatments, variable over the course of the illness, and is frequently irreversible. Other tardive movements (e.g., tardive dystonia) are also linked to chronic antipsychotic exposure.

Cross-References

▶ Antipsychotic Drugs
▶ First-Generation Antipsychotics
▶ Schizophrenia
▶ Second-Generation Antipsychotics

Taste Reactivity Test

Definition

This is a method that can be used to determine in the rat the motivational valence of a taste without having the test subject actually ingest the fluid. Ralph Norgren and Harvey Grill discovered that rats show distinct orofacial (movement of the mouth and tongue and opening of the mouth, gaping and face washing and wiping) reactions to the intraoral application of sapid solutions. The reactions observed can be characterized as appetitive or aversive, depending on whether the fluid evokes an ingestive or rejective reaction. This test can be used as a method to assess tastes paired with drugs that produce conditioned taste preferences or taste aversions.

Cross-References

▶ Conditioned Taste Aversions
▶ Conditioned Taste Preferences

Temazepam

Definition

Temazepam is an intermediate-acting (half-life 8–20 h) benzodiazepine used to treat insomnia. Side effects include a hangover effect (residual drowsiness) the next morning, dizziness, nausea, and vomiting. Prolonged use of this drug is not recommended because of risks for dependence. Temazepam is also used recreationally usually to dampen the effects of withdrawal associated with other drugs, for example, cocaine and MDMA. The gel capsule form of temazepam (called "jellies") was withdrawn after being associated with intra-arterial injection for the purposes of getting high, which caused severe vascular injury and limb ischemia leading to amputations.

Cross-References

▶ Benzodiazepines
▶ Sedative, Hypnotic, and Anxiolytic Dependence

Temporal Discounting

Synonyms

Delay discounting; Time discounting

Definition

Temporal discounting is the degree to which the subjective value of an outcome decreases as the delay to obtaining that outcome increases.

Cross-References

▶ Behavioral Economics
▶ Impulsivity

Temporal Myopia

Definition

Temporal myopia is a focus on the present or the lack of consideration of delayed outcomes of an action when making a choice.

Cross-References

▶ Behavioral Economics

Teratogenic

Definition

Agents or conditions that are able to interfere with the normal development and growth of an embryo or fetus. Teratogens include radiation, maternal infections, chemicals, and drugs; they can either halt the pregnancy or produce a congenital malformation.

Cross-References

▶ Fetal Alcohol Syndrome
▶ Thalidomide
▶ Valproic Acid

Territorial Aggression

Definition

Territorial aggression is defined as the display of aggressive acts and postures for the purpose of excluding other males from the patrolled and marked locale. This type of aggression may vary with the season.

Tertiary Amine Tricyclic Antidepressants

Definition

Tertiary amine TCAs are molecules composed of a three-ring structure that has two methyl groups on the nitrogen atom of the side chain. This group of TCAs includes ▶ amitriptyline, ▶ imipramine, ▶ clomipramine, ▶ trimipramine, and ▶ doxepin.

Tetrabenazine

Definition

Tetrabenazine is a reversible inhibitor of vesicular monoamine transporters. It reduces uptake of ▶ monoamines such as dopamine into synaptic vesicles causing depletion of monoamine stores. It is approved for the treatment of chorea (abnormal involuntary movements) associated with ▶ Huntington's disease. Extrapyramidal side effects are much less common with tetrabenazine than with typical neuroleptics. However, it can amplify the risk of depression, and suicidal thoughts and behavior already seen in Huntington's disease patients. Other common side effects of tetrabenazine include drowsiness, insomnia, and ▶ akathisia. Tetrabenazine has also been clinically evaluated in the treatment of other hyperkinetic disorders including Gilles de la Tourette's syndrome (Tourette's syndrome). It was also used in experimental psychopharmacology as an amine depletor, enabling the role of endogenous amines in responses to other drugs to be investigated. In this application, it has fallen out of use due to the nonspecific nature of the depletions that it produces.

Cross-References

▶ Amine Depletors
▶ Antipsychotic Drugs
▶ Tic Disorders with Childhood Onset

Thalidomide

Synonyms

Thalomid

Definition

Thalidomide is a sedative agent with anti-inflammatory and anxiolytic effects. The drug was prescribed during the late 1950s and early 1960s to pregnant women as an antiemetic to overcome morning sickness and to improve sleep. Approximately 10,000 children prenatally exposed to thalidomide were born with severe malformations, including the hands and feet attached to abbreviated arms and legs (phocomelia). Thalidomide is now approved by the FDA for the treatment of conditions associated with leprosy, multiple myeloma, and actinic prurigo. Side effects of thalidomide include severe birth defects, sleepiness, drowsiness, constipation, skin rash, severe headaches, stomachaches, peripheral neuropathy, dizziness and nausea, mood swings, and a general sense of illness. The use of thalidomide is forbidden before conception and during pregnancy.

Cross-References

▶ Autism: Animal Models

Therapeutic Drug Monitoring

Pierre Baumann
Centre de Neurosciences psychiatriques,
Département de Psychiatrie, DP-CHUV,
Prilly-Lausanne, Switzerland

Synonyms

Drug plasma level determination for therapy optimization; TDM; Therapeutic drug plasma level monitoring

Definition

The rationale of therapeutic drug monitoring (TDM) is based on the hypothesis that drug (parent compound plus active metabolites) plasma concentrations in the patient reflect better than drug dose in the brain concentrations of the active components. For specific drugs, a «therapeutic window» (therapeutic range) can be defined: it represents a drug plasma concentration range defined by a lower threshold below which there is a high probability for an insufficient clinical response. In contrast, drug plasma concentrations higher than the upper threshold result in an increased risk for adverse effects. Drug plasma concentrations within the "therapeutic window" should therefore be synonymous with optimal clinical response. Hence, TDM is the measurement of drug concentrations in the blood of patients in order to facilitate dose adaptation for optimizing treatment. An update of a previously published consensus guideline by a group of experts (AGNP-TDM group (Arbeitsgemeinschaft für Neuropsychopharmacologie und Pharmacopsychiatrie)) and which should help the clinician to take optimal clinical benefit of TDM in psychiatry is freely available (www.agnp.de; Hiemke et al. 2011).

Pharmacokinetics and Metabolism

The majority of psychiatric patients are treated with psychotropic drugs including antidepressants, mood stabilizers, antipsychotic, or anxiolytic/hypnotic drugs, but complications such as nonresponse, poor tolerance, noncompliance, or drug interactions are frequent. Besides pharmacodynamic, pharmacokinetic factors may also be responsible for unsuccessful treatments, in that in the body and particularly in the brain, inadequate drug concentrations are reached for different reasons, as both environmental and genetic factors control the fate of the drug in the organism.

In order to reach the target organ, the brain, psychotropic drugs are submitted to several steps: absorption, distribution, metabolism, and elimination (ADME). The bioavailability of these drugs is 100 % when intravenously administered. However, after oral administration it generally varies between 20 % and 80 %, depending on the drug, due to a limited absorption from the intestine and to metabolism in the liver ("first-pass effect"). For example, some recently introduced drugs have an absorption $< 10 \%$ (agomelatine, ziprasidone, lurasidone, asenapine), and the administration of the three latter drugs requires special precautions.

Most psychotropic drugs are submitted to metabolism in the liver, in the intestine, and in other organs, by phase I (e.g., cytochrome P-450) and phase II (e.g., UTP-glucuronyltransferase) reactions. An important consequence is the possible formation of active metabolites, mainly after formation by phase I and rarely by phase II enzymes. These products may differ in their pharmacological and pharmacokinetic properties from their parent compound, but they contribute to the overall clinical profile of the drug, and therefore, they have also to be quantified.

Pharmacogenetics

Cytochrome P-450 is the most important enzymatic system implicated in the biotransformation of psychotropic compounds, which are metabolized by one or several isozymes, mainly CYP1A2, CYP2B6, CYP2C9, CYP2C19, CYP2D6, and/or CYP3A4/5/7. Except CYP2D6, these enzymes are inducible and most of them display a genetic polymorphism. This can be defined as the presence of at least two genetically determined variants (e.g., alleles) in a population, and such a variant allele has to occur in at least 1 % of the population. As a consequence, a high, genetically determined variability may exist in the expression of an active enzyme protein, and patients can be categorized according to their phenotype: poor (PM), intermediate (IM), extensive (EM), and ultrarapid (UM) metabolizers. In some UM, gene multiplication is observed, and therefore, the presence of at least three active alleles. These patients are at risk for poor clinical response, as drug levels may never reach concentrations needed for clinical efficacy. In contrast, due to the absence of active alleles, drug plasma concentrations in PM may be within a range leading to adverse effects, already at usual drug doses. IM and EM are characterized by the presence of one or two "normally" active alleles, respectively, but the distinction between these groups is sometimes difficult (Zanger and Schwab 2013; Stingl et al. 2013). This has to be considered as a somewhat simplified presentation of the relationship between the genotype and phenotype. Patients may be genotyped or phenotyped using probes which are specific substrates of the isozyme to be examined (Fuhr et al. 2007). Generally, a single dose of these probes is administered to the patients, and urine or blood is then collected during or after a well-defined period of time for the analysis of the parent compound and its metabolite formed by the enzyme. The metabolic ratio (parent compound/metabolite) informs on the metabolic status, i.e., phenotype, of the patient. Typically, debrisoquin or dextromethorphan are used as probes for phenotyping patients with regard to CYP2D6 activity. Genotyping represents a "trait marker" in comparison to phenotyping ("state marker"), the result of which may be influenced by enzyme inhibiting or inducing comedications.

Numerous psychotropic drugs are substrates, inhibitors, and/or inducers of the transporter molecule P-glycoprotein, which extrudes drugs and other xenobiotics from the intracellular to the extracellular space (Choong et al. 2010; O'Brien et al. 2012). This transmembrane efflux transporter is expressed in many tissues such as the intestine and liver and at the blood–brain barrier. It is encoded by the multidrug resistance 1 gene (*ABCB1*) and belongs to the adenosine triphosphate-binding cassette (ABC) family. It presents a genetic polymorphism and depending on the genotype of the patients, this may have consequences on the transport of drugs in the organism. However, the clinical relevance of genotyping patients for P-glycoprotein needs to be further investigated as in man, the absence of this protein as a consequence of a genetic deficiency has not yet been demonstrated. Moreover, the interaction at the blood–brain barrier between substrates and inhibitors of this transport protein does seemingly not lead to clinically relevant consequences (Kalvass et al. 2013; O'Brien et al. 2012). On the other hand, as only minor modifications of the plasma/brain ratio in drug concentrations are observed, the genetic polymorphism of P-glycoprotein has little influence on the rationale of therapeutic drug monitoring.

In conclusion, TDM may be advantageously combined with pharmacogenetic tests especially regarding cytochrome P-450, as these tests may provide a useful help for diagnosing pharmacokinetic particularities.

Indications for TDM

Pharmacogenetic variability in drug metabolism already demonstrates an indication for TDM. Table 1 summarizes indications for TDM, which is however absolutely mandatory only for a few psychotropic drugs such as lithium. Its clinical usefulness is also widely recognized for antidepressants (Wille et al. 2008) and antipsychotics (Patteet et al. 2012). Indeed, drugs such as lithium, tricyclic antidepressants, and the antipsychotic drug clozapine display a relatively narrow therapeutic range, while it is wide for some antidepressants including selective serotonin reuptake inhibitors (SSRI) (Fig. 1).

One of the main problems in pharmacotherapy is the lack of compliance in patients, as it leads to nonresponse and enhances the risk for hospitalization and rehospitalization. However, there may also be other reasons for insufficient response or adverse effects which lead to interruption of the treatment despite the prescription of doses considered as adequate. This may be a consequence

Therapeutic Drug Monitoring, Table 1 General indications for TDM of psychotropic drugs. Primary clinical relevance of defined therapeutic ranges and dose-related drug concentrations

General indications	Therapeutic range	Dose-related drug plasma concentrations
Suspected noncompliance		x
Drugs, for which TDM is mandatory for safety reasons (e.g., lithium)	x	
Lack of clinical response or insufficient response even at doses considered as adequate	x	
Adverse effects despite the use of generally recommended doses	x	
Suspected drug interactions		x
TDM in pharmacovigilance programs	x	x
Combination treatment with a drug known for its interaction potential, in situations of comorbidities, "augmentation"	x	x
Relapse prevention in long-term treatments, prophylactic treatments	x	x
Recurrence despite good compliance and adequate doses	x	x
Presence of a genetic particularity concerning the metabolism or transport of the drug (genetic deficiency, gene multiplication)		x
Children and adolescents		x
Elderly patients (>65 years)		x
Patients with pharmacokinetically relevant comorbidities (hepatic or renal insufficiency, cardiovascular disease)		x
Forensic psychiatry		x
Problems occurring after switching from an original preparation to a generic form (and vice versa)		x

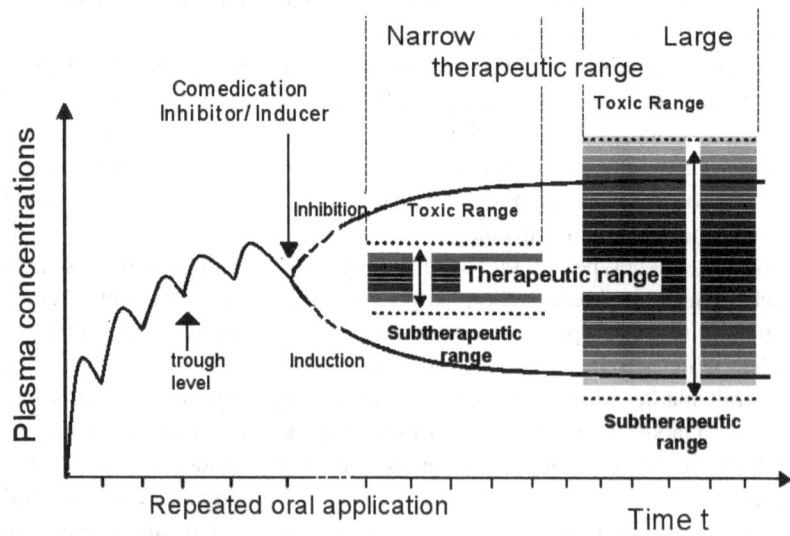

Therapeutic Drug Monitoring, Fig. 1 Pharmacokinetic behavior of drugs after their repeated administration, before and after addition of comedications differing by their interaction potential

of pharmacokinetic interactions leading to an inhibition or induction of the metabolism of the administered drug. Combination treatments are frequent, as many patients suffering from comorbidities need comedications, or in situations of poor response, they are comedicated with augmenting but pharmacokinetically interacting agents. TDM is also recommended in pharmacovigilance programs as it represents an interesting diagnostic tool. Increasingly, the need for long-term treatments is recognized in order to prevent relapse, but despite apparently

Therapeutic Drug Monitoring, Table 2 Recommended target plasma concentration ranges for psychoactive drugs and levels of recommendation for routine monitoring (adapted from Hiemke et al. 2011)

Drug and active metabolite	Recommended therapeutic range (consensus)[a]	Level of recommendation
Antidepressants		
Agomelatine	7–300 ng/ml (1–2 h after 50 mg)	4
Amitriptyline plus nortriptyline	80–200 ng/ml	1
Bupropion plus hydroxybupropion	225–1,500 ng/ml	3
Citalopram	50–110 ng/ml	2
Clomipramine plus norclomipramine	230–450 ng/ml	1
Desipramine	100–300 ng/ml	2
Desvenlafaxine	100–400 ng/ml	2
Dosulepin (= dothiepin)	45–100 ng/ml	2
Doxepin plus nordoxepin	50–150 ng/ml	2
Duloxetine	30–120 ng/ml	2
Escitalopram	15–80 ng/ml	2
Fluoxetine plus norfluoxetine	120–500 ng/ml	2
Fluvoxamine	60–230 ng/ml	2
Imipramine plus desipramine	175–300 ng/ml	1
Maprotiline	75–130 ng/ml	2
Mianserin	15–70 ng/ml	3
Milnacipran	50–110 ng/ml	2
Mirtazapine	30–80 ng/ml	2
Moclobemide	300–1,000 ng/ml	3
Nortriptyline	70–170 ng/ml	1
Paroxetine	30–120 ng/ml	3
Reboxetine	60–350 ng/ml	3
Sertraline	10–50 ng/ml	2
Tranylcypromine	≤50 ng/ml	4
Trazodone	700–1,000 ng/ml	2
Trimipramine	150–300 ng/ml	2
Venlafaxine plus O-desmethylvenlafaxine	100–400 ng/ml	2
Antipsychotics		
Amisulpride	100–320 ng/ml	1
Aripiprazole	150–500 ng/ml	2
Asenapine	2–5 ng/ml	4
Benperidol	1–10 ng/ml	3
Bromperidol	12–15 ng/ml	2
Chlorpromazine	30–300 ng/ml	2
Chlorprothixene	20–300 ng/ml	3
Clozapine	350–600 ng/ml	1
Flupentixol	1–10 ng/ml	2
Fluphenazine	0.5–2 ng/ml	1
Fluspirilen	0.1–2.2 ng/ml	2
Haloperidol	1–10 ng/ml	1
Levomepromazine	30–160 ng/ml	3
Melperone	30–100 ng/ml	3
Olanzapine	20–80 ng/ml	1
Paliperidone	20–60 ng/ml	2

(continued)

Therapeutic Drug Monitoring, Table 2 (continued)

Drug and active metabolite	Recommended therapeutic range (consensus)[a]	Level of recommendation
Perazine	100–230 ng/ml	1
Perphenazine	0.6–2.4 ng/ml	1
Pimozide	15–20 ng/ml	3
Pipamperone	100–400 ng/ml	3
Quetiapine	100–500 ng/ml	2
Risperidone plus 9-hydroxyrisperidone	20–60 ng/ml	2
Sulpiride	200–1,000 ng/ml	2
Thioridazine	100–200 ng/ml	1
Ziprasidone	50–200 ng/ml	2
Zotepine	10–150 ng/ml	3
Zuclopenthixol	4–50 ng/ml	3
Mood stabilizers		
Carbamazepine	6–10 µg/ml	2
Lamotrigine	3–14 µg/ml	2
Lithium	0.5–1.2 mmol/l	1
Valproate	50–100 µg/ml	2
Antidementia drugs		
Donepezil	30–75 ng/ml	2
Galantamine	30–60 ng/ml	3
Memantine	7–150 ng/ml	3
Rivastigmine	Oral: 8–20 ng/ml (1–2 h after dose)	3
	Patch: 5–13 ng/ml (1 h before application of new patch)	
Other drugs		
Levomethadone	250–400 ng/ml	2
Methadone	400–600 ng/ml	2

[a]*Therapeutic ranges* indicate generally trough concentrations of drugs in serum or plasma of patients under steady-state medication

good compliance, recurrence of the illness may occur. The pharmacokinetics of psychotropic drugs may be particular in many subjects belonging to the category of "special populations," such as children and adolescents, elderly patients, and patients suffering from comorbidities which may result in an impaired absorption, distribution, metabolism, or elimination of the drug. Similarly, problems may occur at the level absorption when galenic forms of drugs are changed, e.g., switching from an original preparation to a generic brand, and vice versa. Depending on the indication, either the availability of "therapeutic ranges" or the knowledge of plasma drug concentrations at defined doses may primarily be helpful for the clinician (Table 1).

Table 2 presents a list of recommended plasma concentrations of antidepressants and antipsychotics as defined by the already mentioned group of experts (Hiemke et al. 2011). The levels of recommendation are indicated for each drug, as defined according to the present evidence from studies reported in the literature and to the conclusion from the expert group.

1. *Strongly recommended (for lithium, TDM should be a standard of care)*: Established therapeutic range.

   Clinical significance: At therapeutic plasma concentrations highest probability of response or remission; at "subtherapeutic" plasma concentrations, response rate similar

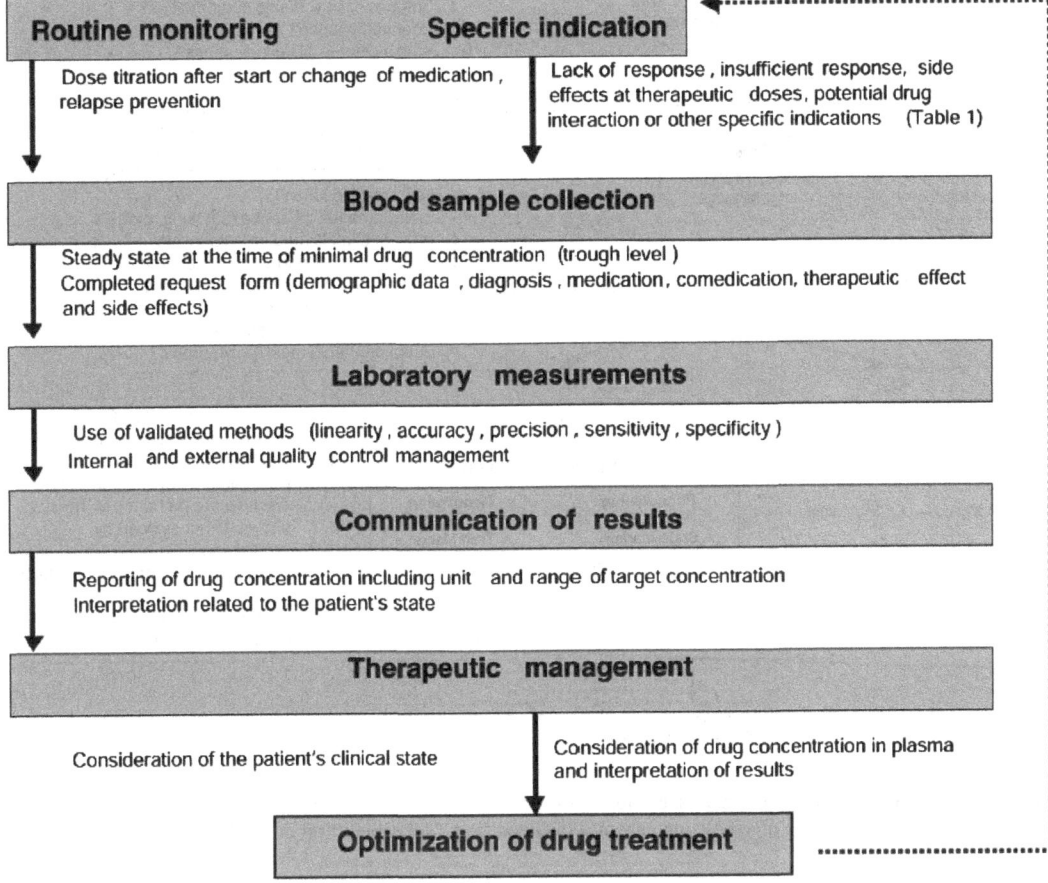

Therapeutic Drug Monitoring, Fig. 2 Schematic presentation of the TDM process for optimization of psychopharmacotherapy

to placebo under acute treatment and risk of relapse under chronic treatment; at "supratherapeutic" plasma concentrations, risk of intolerance or intoxication.

2. *Recommended*: Reported drug concentrations were obtained from plasma concentrations at therapeutically effective doses and related to clinical effects; reports on decreased tolerability or intoxications at "supratherapeutic" plasma concentrations.

   Clinical significance: TDM will increase the probability of response in nonresponders. At "subtherapeutic" plasma concentrations, risk of poor response; at "supratherapeutic" plasma concentrations, risk of intolerance or intoxication.

3. *Useful*: Reported drug concentrations were calculated from plasma concentrations at effective doses obtained from pharmacokinetic studies. Plasma concentrations related to pharmacodynamic effects are either not yet available or based on retrospective analysis of TDM data, single case reports, or nonsystematic clinical experience.

   Clinical significance: TDM can be used to control whether plasma concentrations are plausible for a given dose or clinical improvement may be attained by dose increase in nonresponders who display too low plasma concentrations.

4. *Probably useful*: Plasma concentrations do not correlate with clinical effects due to unique

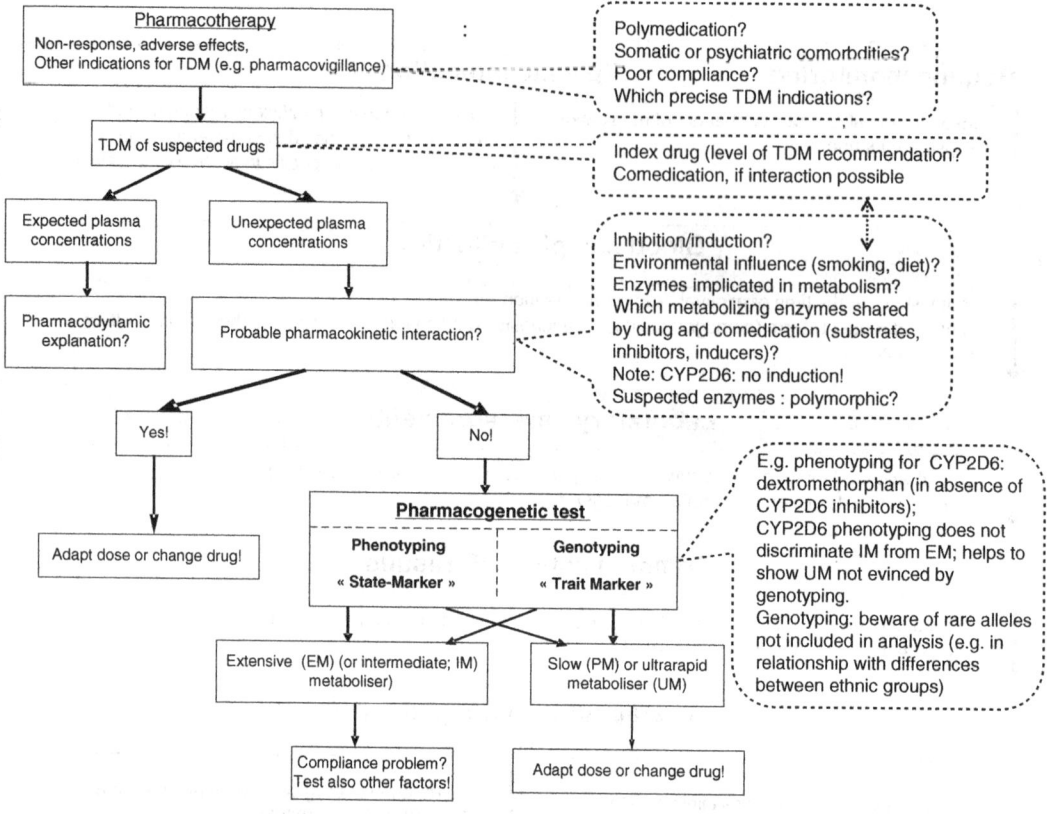

Therapeutic Drug Monitoring, Fig. 3 Decision tree for the combined use of therapeutic drug monitoring (*TDM*) and pharmacogenetic tests of drug metabolizing enzymes (adapted from Jaquenoud Sirot et al. 2006)

pharmacology of the drug, e.g., irreversible blockade of an enzyme, or dosing can be easily guided by clinical symptoms, e.g., sleep induction by a hypnotic drug.

Clinical significance: TDM should be restricted to special indications.

Practical Issues

Figure 2 summarizes the steps which should be considered when TDM is envisaged in a patient. Successful TDM is the result of a fruitful collaboration between the treating physician, the laboratory specialists, and, in special circumstances, the clinical pharmacologist. From a pharmacokinetic point of view, TDM should be carried out in steady-state conditions of both the parent compound and its active metabolite to be co-monitored (Fig. 1). This situation is generally reached within a time period of 4–5 half-lives of the active constituents after treatment is initiated or after adaptation of the dose. Moreover, trough levels (Fig. 1) should be measured, i.e., generally 12–16 h after the last intake of the drug or in the morning before the first dose of the day is administered. In situations of pharmacokinetic interactions, two situations have to be considered separately. In the case an inhibitor of the metabolism of the drug to be monitored is administered, its new steady-state concentrations are only reached after the inhibitor has reached steady-state conditions, when it exerts a maximal inhibiting effect. After comedication with a metabolism inducing drug, maximal induction is only reached after the inhibitor itself reached steady state and after maximal inducing effect, i.e., stimulation of protein synthesis, is obtained (Fig. 1). This takes generally 1–2 weeks. Practically, it is difficult to be compliant with all these recommendations, as many drugs are administered several times per day, and others display very short (e.g., quetiapine: 2–4 h) or very long elimination half-lives, respectively (e.g., fluoxetine (and its active metabolite norfluoxetine): 3–4 days).

Finally, a decision tree about the optimal combination of TDM with pharmacogenetic tests is presented in Fig. 3. For example, TDM helps to demonstrate pharmacokinetic factors contributing to the appearance of adverse effects, while the pharmacogenetic tests allow diagnosing causes of an unexpected pharmacokinetic behavior of the drug leading to a pharmacovigilance problem (Jaquenoud Sirot et al. 2006).

In conclusion, in a period characterized by a lack of novel effective drugs for the treatment of mental diseases, the use of drugs presently available should be ameliorated. TDM of psychotropic drugs, alone or in combination with pharmacogenetic tests, appears to be a useful tool for treatment optimization.

Cross-References

▶ Antidepressants
▶ Antipsychotic Drugs
▶ Blood–Brain Barrier
▶ Drug Interactions
▶ Lithium
▶ Mood Stabilizers
▶ Pharmacogenetics
▶ Pharmacokinetics
▶ Selective Serotonin Reuptake Inhibitors

References

Choong E, Dobrinas M, Carrupt PA, Eap CB (2010) The permeability P-glycoprotein: a focus on enantioselectivity and brain distribution. Expert Opin Drug Metab Toxicol 6(8):953–965

Fuhr U, Jetter A, Kirchheiner J (2007) Appropriate phenotyping procedures for drugs metabolizing enzymes and transporters in humans and their simultaneous use in the "cocktail" approach. Clin Pharmacol Ther 81:270–283

Hiemke C, Baumann P, Bergemann N, Conca A, Dietmaier O, Egberts K et al (2011) AGNP consensus guidelines for therapeutic drug monitoring in psychiatry: update 2011. Pharmacopsychiatry 44(6):195–235

Jaquenoud Sirot E, Van der Velden JW, Rentsch K, Eap CB, Baumann P (2006) Therapeutic drug monitoring and pharmacogenetic tests as tools in pharmacovigilance. Drug Saf 29:735–768

Kalvass JC, Polli JW, Bourdet DL, Feng B, Huang SM, Liu X et al (2013) Why clinical modulation of efflux transport at the human blood–brain barrier is unlikely: the ITC evidence-based position. Clin Pharmacol Ther 94(1):80–94

O'Brien FE, Dinan TG, Griffin BT, Cryan JF (2012) Interactions between antidepressants and P-glycoprotein at the blood–brain barrier: clinical significance of in vitro and in vivo findings. Br J Pharmacol 165(2):289–312

Patteet L, Morrens M, Maudens KE, Niemegeers P, Sabbe B, Neels H (2012) Therapeutic drug monitoring of common antipsychotics. Ther Drug Monit 34(6):629–651

Stingl JC, Brockmoller J, Viviani R (2013) Genetic variability of drug-metabolizing enzymes: the dual impact on psychiatric therapy and regulation of brain function. Mol Psychiatry 18(3):273–287

Wille SMR, Cooreman SG, Neels HM, Lambert WEE (2008) Relevant issues in the monitoring and the toxicology of antidepressants. Crit Rev Clin Lab Sci 45:25–89

Zanger UM, Schwab M (2013) Cytochrome P450 enzymes in drug metabolism: regulation of gene expression, enzyme activities, and impact of genetic variation. Pharmacol Ther 138(1):103–141

Thioridazine

Definition

Thioridazine, like other compounds of its chemical class (▶ phenothiazines), is classified as a ▶ typical antipsychotic, although its pharmacological profile exceeds dopamine D2 receptor blockade to include antagonism at dopamine D1, alpha-1 adrenoceptors, and muscariniccholinoceptors, and it produces a relatively low incidence of extrapyramidal side effects. Cardiotoxicity, retinopathy, and other serious side effects have reduced its use to patients who do not respond to other commonly used antipsychotic compounds.

Cross-References

▶ Antipsychotic Drugs
▶ First-Generation Antipsychotics

Thiothixene

Definition

Thiothixene acts at multiple receptors but mainly by dopamine D2 blockade. It is a first-generation antipsychotic with an ▶ elimination half-life of 34 h and is mainly metabolized by 1A2 CYP450 isoenzymes.

Cross-References

▶ First-Generation Antipsychotics

Thioxanthenes

Definition

A group of drugs that includes key members of the first generation of antipsychotic substances that brought about major changes in the treatment of schizophrenia. Among the widely used drugs in this category are chlorprothixene, flupenthixol, and zuclopenthixol.

Cross-References

▶ Antipsychotic Drugs
▶ First-Generation Antipsychotics

Third-Generation Antipsychotics

Synonyms

Partial dopamine agonists

Definition

All first- and second-generation antipsychotic drugs exert their clinical actions by, among other pharmacologic properties, blocking the dopamine D2 receptor. Aripiprazole, a partial D2 agonist, is the first drug with a different mechanism of action. It is therefore considered to be a third-generation antipsychotic. While opinions differ somewhat as to which pharmacological profile can legitimately be termed third-generation antipsychotics, some investigators have proposed that they can exhibit diverse combinations of antagonism (or partial agonism) at D2 receptors with partial agonism at serotonin 5-HT$_{1A}$ receptors.

Cross-References

▶ Antipsychotic Drugs
▶ Aripiprazole

Threshold

Definition

In psychophysics, the theoretical point at which a sensory stimulus is perceived correctly on at least 50 % of occasions.

Tiagabine

Definition

Tiagabine is an anticonvulsant drug that is used as a third-line adjunct medication for the treatment of partial seizures in adults and children. It acts by selectively inhibiting the ▶ GABA transporter 1 reuptake inhibitor. Thus, it enhances inhibitory ▶ GABAergic neurotransmission and prevents propagation of neuronal impulses and seizures. Tiagabine reduces seizure frequency but does not produce freedom from seizures and has a low tolerability, for example, low anticonvulsant efficacy compared to other third-line antiepileptic drugs. It also has a high rate of side effects including dizziness, lack of energy, nausea, and nonconvulsive status epilepticus. Open-label investigations have suggested that tiagabine may be used as an add-on medication in anxiety disorders, for example, ▶ generalized anxiety disorder, and for the treatment of alcohol dependence. It is thought that tiagabine may reduce both ▶ anxiety and alcohol ▶ craving in withdrawal, possibly via indirect GABA inhibition of the mesolimbic dopamine system. However, its efficacy for these indications has not yet been confirmed in double-blind placebo-controlled studies.

Cross-References

- ▶ Alcohol Abuse and Dependence
- ▶ Anticonvulsants
- ▶ Generalized Anxiety Disorder
- ▶ Placebo Effect
- ▶ Randomized Controlled Trials

Tic Disorders with Childhood Onset

Lawrence David Scahill
Marcus Autism Center, Children's Healthcare of Atlanta, and Department of Pediatrics, Emory University School of Medicine, Atlanta, GA, USA

Synonyms

Tourette syndrome

Definition

Tic disorders of childhood are a group of probably related conditions characterized by the presence of motor tics and/or vocalizations. Motor tics tend to be rapid, jerky movements of the head and neck regions, though other muscle groups may also be involved. Common vocal tics include repetitive throat clearing, grunting, snorting, or shouts. Tic disorders are described as transient or chronic according to the duration. The differential diagnosis is also based on whether both motor and vocal tics are present. The diagnosis of a tic disorder requires that the tics are not due to the effects of a substance (e.g., stimulants) or another medical condition (e.g., Huntington's chorea).

Role of Pharmacotherapy

Tics are common in children, affecting as many as 20 % of the school-age population. In most of these children, the tics are transient and require no intervention. An estimated 1 % of children show a pattern of chronic tics that endure over time. Tourette syndrome, defined by the presence of both motor and vocal tics persisting for at least a year, affects 3–8/1,000 children (Scahill et al. 2005). Even among children with chronic tics, many cases are mild, and medication is not warranted. Children with tic disorders appear to be at higher risk for co-occurring conditions such as attention-deficit hyperactivity disorder (ADHD) and obsessive-compulsive disorder (OCD). Thus, the first step in selecting medication for the treatment of children with tic disorders is to identify the most important source of impairment: tics, ADHD, or OCD. The use of pharmacotherapy in children with tic disorders should always be combined with education for the child and the family about the medication, as well as the natural history of tic disorders.

Diagnostic Categories
The following are the three main diagnostic categories of childhood-onset tic disorders:

1. Provisional tic disorder (motor and/or vocal tics lasting less than 1 year)
2. Persistent (chronic) motor or vocal tic disorder (motor or vocal tics, but not both, lasting for more than a year)
3. Tourette disorder (multiple motor tics and at least one vocal tic, present at some time during the illness, although not necessarily concurrently, lasting for more than a year) (American Psychiatric Association 2013)

Tic Disorders with Childhood Onset, Table 1 Drugs used for the treatment of tics

Drug Typical antipsychotics	Usual starting dose (mg/day)	Usual dose range (mg/day)	Adverse effects
Haloperidol[a]	0.25–0.5	1.5–3.0	Sedation, dyskinesia, dystonia, cognitive dulling Weight gain, social phobia
Pimozide[a]	0.5–1.0	2.0–6.0	
Fluphenazine	1.0–1.5	3.0–6.0	
Atypical antipsychotics			
Risperidone[a]	0.25–0.50	1.5–2.5	Weight gain, sedation, cognitive dulling, social phobia, drooling, dyskinesia, dystonia
Ziprasidone[a]	20	40–60	
Aripiprazole	1.0–2.5	5–10	
Olanzapine	25–50	200–400	
Alpha-2-agonists			
Clonidine[a]	0.025–0.05	0.15–0.3	Sedation, mid-sleep awakening, irritability
Guanfacine	0.5–1.0	1.5–3.0	
Other topiramate[a]	12.5–25	50–125	Sedation, short-term memory problems, headache

[a]Shown to be effective in at least one placebo-controlled trial for tics

All three tic disorders require the onset of tics before the age of 18 years. Parents often report a fluctuating course for the tics (tendency to rise and fall in frequency and intensity over time). By age 10 years, most patients will report a capacity to suppress tics at least for brief periods and the presence of a warning or premonitory urge before the occurrence of some or all of their tics.

There are no diagnostic tests for tic disorders. Except for the tics, the physical and neurologic examinations are usually normal. Neuroimaging procedures (e.g., magnetic resonance imaging) or electroencephalography is rarely indicated. Thus, the assessment of a child referred for a tic disorder includes tracing the onset and course of symptoms, documenting the current severity of motor and phonic tics, inquiring about the presence of premonitory sensations and capacity for tic suppression, establishing the overall impairment caused by the tics, and surveying the treatment approaches implemented to date. Due to their common co-occurrence, ADHD and OCD warrant particular attention in the assessment. The presence of other psychiatric disorders, such as anxiety and depression, also deserves consideration. Because stereotypic movements in children with pervasive developmental disorders can resemble tics, overall development as well as social and language competence should also be part of the assessment (Scahill et al. 2006).

Medications Used in the Treatment of Children with Tic Disorders

Drugs Used to Treat Tics. The most widely prescribed drugs for the treatment of tics include antipsychotic medications and the alpha-2-adrenergic agonists (see Table 1). Other medications such as tiapride (available in Europe), pergolide, botulinum toxin injections, topiramate, and tetrabenazine have also been used for the treatment of tics. Empirical support for these medications varies. For example, botulinum toxin, topiramate, and tiapride have each been shown superior to placebo, tetrabenazine has not. Pergolide, which is a dopamine partial agonist, appears to be effective in reducing tics, but has fallen out of use due to concern about adverse effects. The other dopamine partial agonists have not been rigorously evaluated.

Efficacy. To date, most pharmacological trials for the treatment of tics have been relatively small (ranging from 30 to 60 subjects). The magnitude of effect ranges from 30 % to 50 % reduction in tics, suggesting that even effective medications are not associated with a large reduction in tics.

Tic Disorders with Childhood Onset, Table 2 Drugs used to treat ADHD in children with tics

Drug[a]	Usual starting dose (mg/day)	Usual dose range (mg/day)	Adverse effects
Methylphenidate	12.5–15[b]	25–35[b]	Loss of appetite, insomnia, irritability, mild growth retardation, increase in tics
Clonidine	0.025–0.05	0.15–0.3	Sedation, mid-sleep awakening, irritability
Guanfacine	0.5–1.0	1.5–3.0	Sedation, mid-sleep awakening
Atomoxetine	18–36	36–80	Vomiting, loss of appetite, insomnia

[a]Each of these medications has been shown to be effective in at least one placebo-controlled trial in children with tic disorders and ADHD
[b]Assuming an immediate release product and TID schedule, third dose about half of the morning dose

Tolerability. The antipsychotics are predictably the most effective medications, but are also associated with a large range of adverse effects. The newer atypical antipsychotics offered the promise of decreased risk of motor adverse effects. However, the issue of weight gain and the liabilities associated with obesity have emerged as important concerns.

Clinical Use. Given the fluctuating course of tics, medication for tics should be considered only when there is clear evidence that the tics are causing persistent interference in everyday life. Frequent and forceful tics may cause direct interference with motor activities or speech. In other cases, the child may be distracted by the bombardment of premonitory sensations and efforts to suppress tics. Still other children may suffer social consequences from noticeable tics. Families need education about the fluctuating course of tics. For example, even after achieving satisfactory results with a tic-suppressing medication, tic exacerbations will occur. The impulse to increase the medication dose with the inevitable exacerbations of tics should be tempered by the increased risk of adverse effects. Finally, there is the matter of how long to maintain children on a tic-suppressing medication. To date, no clinical trials have adequately addressed this important clinical question. Follow-up studies have shown that many children will have fewer tics at age 18 years compared to 12 years of age. This replicated finding suggests that it is reasonable to consider a decrease or even discontinuation of a tic-suppressing medication in the mid- to late teenage years. Understandably, adolescents and their parents may be reluctant to undertake such a course of action. Based primarily on clinical consensus, medication withdrawal should be done gradually. In the case of the antipsychotics (even atypicals), abrupt cessation may result in withdrawal dyskinesias and rebound increase in tics. Gradual withdrawal spread over months may also be more acceptable to adolescents and parents.

Drugs Used to Treat ADHD in Children with Tic Disorders. Based on several case reports of tic worsening upon exposure to stimulant medication, the stimulants have been contraindicated for children with ADHD and a tic disorder. However, accumulated evidence over the past 15 years disputes this as an ironclad contraindication. As in children with ADHD who do not have tics, the stimulants may fail. Thus, if stimulant medication is not successful either due to tic worsening, other adverse effects, or lack of efficacy, non-stimulant alternatives warrant consideration. Table 2 presents a summary of information on medications that are commonly used and have been studied in children with a tic disorder and ADHD.

Drugs Used to Treat OCD in Children with Tic Disorders. Children with tic disorders ascertained from clinical settings also have an increased likelihood of OCD compared to general population samples. It has been noted that the obsessive-compulsive symptoms in children with tic disorders differ from those with OCD in the absence of tics. Indeed, a recurring challenge in the assessment of children with tic disorders can be differentiating between tics and compulsive behaviors. For example, some children with OCD express the anxious worry (obsession) that "something bad will happen" if a specific ritual (compulsion) to prevent harm is not completed. By contrast, children with the tic-related form of OCD are more likely to report a feeling or urge to carry out the repetitive behavior and a need to achieve a sense of completion rather than a need to prevent harm. The repetitive behavior in children with Tourette syndrome may be stereotypic in form, such as arranging objects to achieve symmetry or repeatedly picking up and setting down an object on a table.

The first-line treatments for OCD in children are the serotonin reuptake inhibitors (SRIs). Based on available evidence, there is no compelling evidence to support the selection of any one SRI over another. Clomipramine, fluoxetine, sertraline, and fluvoxamine have demonstrated efficacy in children with OCD in placebo-controlled trials. Citalopram, escitalopram, and paroxetine are also used in clinical settings. In general, the approach to treating a child with a tic disorder and OCD is similar to treating children with OCD who do not have a tic disorder. However, several points warrant consideration when selecting a medication for the treatment of children with OCD. First, 30–40 % of children with OCD who are treated with an anti-obsessional medication do not achieve a positive response. In addition, the magnitude of benefit is not large, even for those showing a positive response. Therefore, the benefits of an SRI should not be overstated to children and parents. Second, although compulsive behaviors are common in children with tic disorders, these behaviors may be mild and medication may not be indicated. Third, and a closely related point, there is evidence in adults with OCD that subjects with co-occurring tics are less likely to show a positive response to monotherapy with an anti-obsessional medication. Fourth, although the magnitude of effect appears similar for clomipramine and the more selective SRIs, clomipramine requires periodic electrocardiograms and blood levels, and it is more vulnerable to drug-drug interactions. Thus, clomipramine is generally regarded as a second-line drug for OCD. Fifth, the combination of anti-obsessional medication plus exposure and response prevention therapy appears more effective than medication alone.

Conclusion

Tic disorders of childhood may be chronic, but they are not usually progressive. The severity of tics ranges from mild to severe and tics show a fluctuating course. A substantial percentage of children with tics will show a decline by the end of the second decade of life. In addition to tics, children with tic disorders are at higher risk for ADHD and OCD. The aim of assessment and treatment is to identify and remediate the most important source of impairment in order to promote the child's overall development. Pharmacotherapy has an important role to play in the management of children with tic disorders. In addition to medication, children and families are likely to need education about tics and related problems. Behavioral interventions may be used to augment medication or as an alternative to medication in some cases (Woods et al. 2007). Advocacy and education of school teachers and administrators is often critical in securing appropriate school placement and minimizing the social fallout due to tics.

Cross-References

▶ Antipsychotic Medication
▶ Attention-Deficit/Hyperactivity Disorder
▶ Habit Reversal Therapy
▶ Obsessive–Compulsive Disorders
▶ Suppressibility
▶ Tics, Motor
▶ Tics, Vocal

References

American Psychiatric Association (2013) Diagnostic and statistical manual of mental disorders, 5th edn. American Psychiatric Publishing, Washington, DC

Scahill L, Sukhodolsky D, Williams S, Leckman JF (2005) The public health importance of tics and tic disorders. Adv Neuro 96:240–248

Scahill L, Erenberg G, Berlin CM et al (2006) Contemporary assessment and pharmacotherapy of Tourette syndrome. NeuroRx 3:192–206

Woods DW, Piacentini JC, Walkup JT (2007) Treating Tourette syndrome and tic disorders: a guide for practitioners. Guilford, New York

Tics

Definition

Tics are usually stereotyped rapid, jerky movements or brief vocalizations.

Cross-References

▶ Tic Disorders with Childhood Onset

Tics, Motor

Definition

These may be simple or complex. For example, simple tics may involve single or related muscle groups such as blinking, facial grimacing, or shrugging. Complex tics may include tics occurring in a sequence, such as an arm thrust with near simultaneous head-jerking, or movement such as bending or twirling.

Tics, Vocal

Definition

Vocal tics may be simple or complex. Simple vocal tics may include throat-clearing, grunting, coughing, or snorting. Complex vocal tics may include loud shrieks or blurting out words or parts of words. Verbal tics may involve cursing or other socially inappropriate references, but these tics are not common among patients with Tourette syndrome.

Time Series

Synonyms

Time series plot

Definition

A time series is a series of data points or observations measured at successive time intervals. With respect to ▶ functional magnetic resonance imaging (fMRI) techniques, it refers to a large number of images (e.g., echo planar images) acquired in sequence and at regular time intervals over the duration of an experiment.

Cross-References

▶ Magnetic Resonance Imaging (Functional)

Timing Accuracy

Definition

Timing accuracy refers to the estimated duration of an event. Timing accuracy is estimated by the mean or peak time of the response distribution (see Buhusi and Meck 2005).

References

Buhusi CV, Meck WH (2005) What makes us tick? Functional and neural mechanisms of interval timing. Nat Rev Neurosci 6:755–765

Timing Behavior

Warren H. Meck[1] and Catalin V. Buhusi[2]
[1]Department of Psychology and Neuroscience, Duke University, Durham, NC, USA
[2]Utah State University, North Logan, UT, USA

Synonyms

Interval timing; Temporal conditioning; Temporal processing; Timing

Definition

Timing behavior refers to the capacity of subjects to perceive, estimate, and discriminate time intervals (e.g., durations of and between events); to emit (or avoid emitting) behavioral responses at appropriate time intervals; and, more generally, to modulate their behavior in time (Buhusi and Meck 2005; Gibbon et al. 1997). Timing behavior includes millisecond timing, interval timing, and circadian timing. Millisecond timing refers to the perception, estimation, and discrimination of durations in the subsecond range, crucial for motor control, speech generation and recognition, and music production and perception. Disorders of millisecond timing may result in disorders of motor control (see "▶ Motor Activity and Stereotypy"). Circadian timing (see "▶ Circadian Rhythms") refers to repetition of certain phenomena at about the same time each day as a function of the light/dark cycle. The most studied circadian rhythm is sleep, but other examples include body temperature, blood pressure, production of hormones, and digestive secretions. Here, we discuss the effect of psychoactive drugs on timing behavior in the seconds-to-minutes range ("▶ Interval Timing") which is considered crucial to learning and cognition (Gallistel and Gibbon 2000; Williamson et al. 2008).

Impact of Psychoactive Drugs

Protocols Used to Investigate the Neuropharmacology of Interval Timing

The most reliable approach to investigate the neuropharmacology of interval timing in humans and animals is to use a reproduction protocol, such as the peak-interval (PI) procedure (Matell et al. 2006) although a variety of different timing procedures can be used (see Paule et al. 1999). Idealized data from a PI procedure in two groups of subjects trained to time different durations are shown in Fig. 1. The subject's responses distribute normally around the criterion duration, and the width of this distribution is proportional to the criterion. The way in which the mean and standard deviation of the response distribution covary is typically referred to as the scalar property (Gibbon et al. 1997), a strong form of Weber's law, which is obeyed by most sensory dimensions. The scalar property applies not only to behavioral responses but also to neural activity as measured by electrophysiological recordings as well as the hemodynamic response associated with a subject's active reproduction of a timed interval, but not for passive responses triggered at the untimed interval (Buhusi and Meck 2005).

Estimating the Effect of Psychoactive Drugs

The PI procedure with gaps allows the independent assessment of the effect of psychoactive drugs on timing (see "▶ Timing Accuracy"; "▶ Timing Precision"), as well as their attentional effect, and allows the dissociation of these effects from motor/motivational effects as illustrated in Fig. 2. Generally speaking, motor/motivational effects of drugs are observed as shifts in the amplitude of the response functions along the vertical axis, while effects on timing are observed as leftward or rightward shifts in the response function along the horizontal axis. Timing accuracy is estimated by the peak of the response function in peak trials (Fig. 2 – *left panel*). Timing precision is estimated by the width of the response function in peak trials (Fig. 2 – *left panel*). Attentional effects are estimated by inserting gaps (e.g., retention intervals) in a subset of peak trials (Fig. 2 – *right panel*).

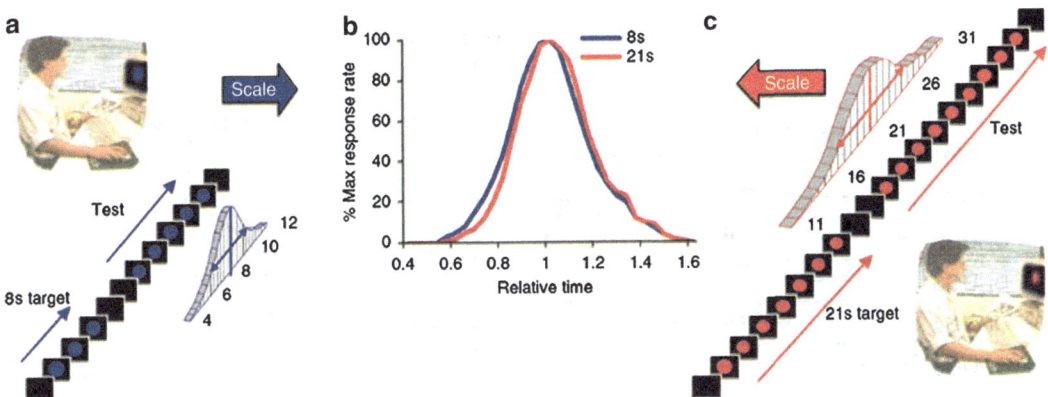

Timing Behavior, Fig. 1 The scalar property is a hallmark of interval timing at both the behavioral and neural levels. In a typical duration reproduction procedure called the PI procedure, subjects receive training trials, during which they are presented with target stimuli of specific durations (here, panel (**a**): 8 s, panel (**c**): 21 s), and test trials, in which the subjects are asked to reproduce the criterion duration. Typically, in peak trials the responses distribute normally around the criterion duration with a width that is proportional to the criterion. When the response distributions are scaled both in amplitude and duration, they superimpose, thus demonstrating the scalar property at the behavioral level (Redrawn from Buhusi and Meck 2005)

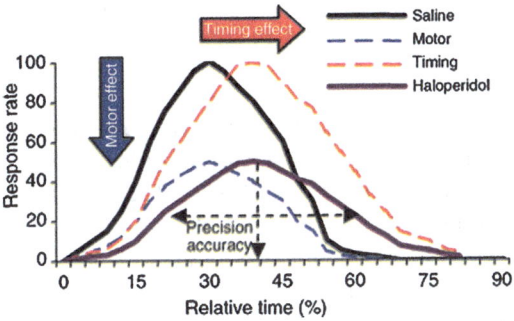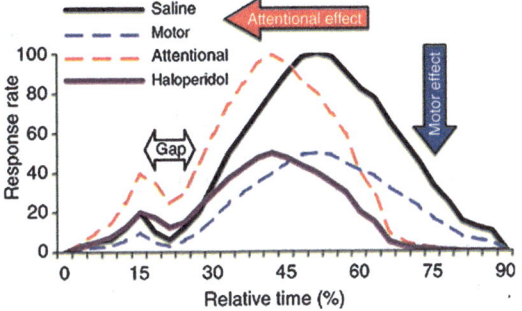

Timing Behavior, Fig. 2 The PI procedure with gaps allows the independent assessment of the effect of a drug on timing accuracy, timing precision, and attention to timing. Here are data from groups of subjects trained to time a 30-s PI procedure, tested under systemic administration of saline and the dopamine D2 antagonist haloperidol (0.04–0.06 mg/kg i.p). Timing accuracy is estimated by the peak response rate, timing precision by the width of the response function, and attentional effect by the delay of the response function in trials with gaps (*right*) relative to trials without gaps (*left*). *Left panel (peak trials)*: Motor/motivational effects are observed as changes on the vertical axis, while timing effects are observed as proportional changes on the horizontal axis. Here, haloperidol has both motor/motivational effect and effect on timing: It slows down timing (rightward shift relative to saline control). *Right panel (gap trials)*: Attentional effects are evaluated by introducing a gap (retention interval) on a subset of peak trials, which delays the response function. Motor/motivational effects are observed as changes on the vertical axis, while attentional effects are observed as absolute changes on the horizontal axis. Here, haloperidol has both a motor/motivational effect and an attentional effect: It reduces the delay following a retention interval, suggesting that subjects are paying more attention to the task (Adapted from Buhusi 2003)

For illustrative purposes, Fig. 2 shows the effect of systemic administration of the dopamine D2 antagonist haloperidol in the PI procedure with gaps. Haloperidol has both motor/motivational and timing effects: It not only reduces the amplitude of the response function (Fig. 2 – *left and right panels*) but also produces an immediate proportional rightward shift in the peak function (e.g., slows down the speed of the clock as shown in Fig. 2 – *left panel*). When retention intervals are inserted in the to-be-timed duration, haloperidol also increases the attention to time

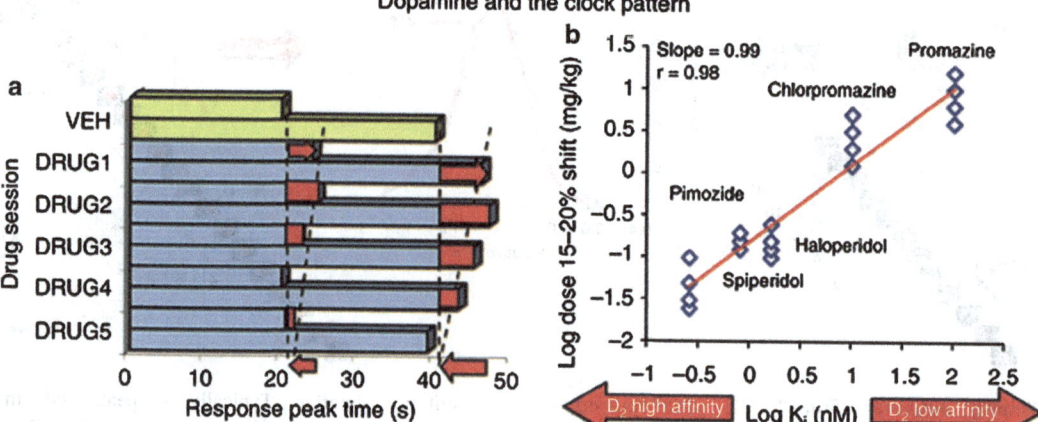

Timing Behavior, Fig. 3 The effects of the dopamine D2 antagonist haloperidol on interval timing are consistent with the slowing down of time accumulation. Panel (**a**): Two groups of subjects were trained to time two durations, using 20-s and 40-s PI procedures. Afterward, in test sessions, subjects received systemic administration of vehicle (VEH), then five consecutive sessions with systemic administration of haloperidol. Acute administration of haloperidol results in a sudden rightward shift that is proportional to the estimated durations, whereas its repeated administration results in a gradual return of the estimated time to the criterion duration. Panel (**b**): The rightward shift of the estimated time is proportional to the affinity of the drug for the D2 dopamine receptor (Adapted from Buhusi and Meck 2005)

("▶ Attentional Effect"): It diminishes the resetting effect of the retention intervals as shown in Fig. 2 (*right panel*) (Buhusi 2003).

Dopaminergic Drugs: The Clock Speed Effect

Dopaminergic drugs selectively affect the subjective speed of an internal clock in a variety of animals (Buhusi and Meck 2005; Meck 1996) (Fig. 3). More specifically, when timing is evaluated in the PI procedure, dopamine agonists produce an immediate leftward shift of response function in proportion to the to-be-timed criterion, indicative of an increase of the speed of an internal clock. In contrast, dopamine antagonists produce an immediate rightward shift of response function in proportion to the to-be-timed criterion, indicative of a decrease in the speed of an internal clock (Fig. 3a) (Meck 1996). Repeated administration of dopaminergic drugs is followed by a characteristic decrease in the effect of the drug (Fig. 3a), which is not interpreted as desensitization but rather as recalibration of the timer. The interruption of the drug regimen results in a rebound in the opposite direction (Meck 1996). Dopamine antagonists were shown to produce a deceleration of the subjective clock speed in proportion to their affinity to dopamine D2 receptor (Fig. 3b) (Meck 1996). The effects of dopamine-like drugs on clock speed are diminished as a function of experience with the timing task (habit formation) prior to drug administration which may reflect a change in the balance between dopamine-glutamate interactions (Williamson et al. 2008).

Cholinergic Drugs: The Memory Storage Effect

When initially administered, cholinergic drugs have no immediate effects on interval timing. However, repeated administration of cholinergic agonists results in a gradual decrease of the estimated duration, while cholinergic antagonists result in a gradual increase of the estimated duration (Fig. 4a). Discontinuing the drug administration results in a gradual return to the baseline. These results are interpreted as alteration of memory storage of duration. More specifically, reference memory error was found to vary in proportion to cholinergic activity in the frontal cortex (Buhusi and Meck 2005; Meck 1996) (Fig. 4b).

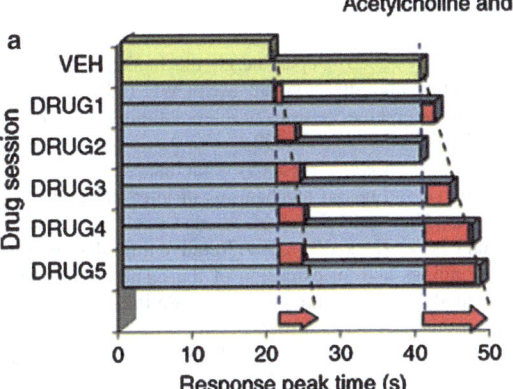

Timing Behavior, Fig. 4 The effects of cholinergic drugs are consistent with effects on reference memory. Panel (**a**) Two groups of subjects were trained to time two durations, using 20-s and 40-s PI procedures. Afterward, in test sessions, subjects received systemic administration of VEH, then five consecutive sessions with systemic administration of the muscarinic cholinergic receptor atropine. Acute administration of atropine results in a gradual scalar (proportional to the timed criterion) rightward shift of the estimated time. Panel (**b**) The effect is correlated with the activity of cholinergic neurons in the frontal cortex as measured by sodium-dependent high-affinity choline uptake (*SDHACU*) (Redrawn from Buhusi and Meck 2005)

Serotonergic Drugs: The Attentional Effect

Time accumulation and attention to the timing task can be dissociated pharmacologically, and they depend on the dopamine and serotonin systems in the cortex and striatum (Buhusi and Meck 2009; Ho et al. 2002). Administration of the indirect dopamine agonist methamphetamine shortens estimated durations and decreases attention to timing (Buhusi 2003). In turn, the dopamine D2 receptor blocker haloperidol lengthens estimated durations but facilitates the maintenance in temporal information in working memory (Buhusi 2003). However, the effect of methamphetamine to shorten estimated durations is blocked by depletion of serotonin, suggesting that an intact serotonergic system is required for interval timing. Indeed, specific stimulation of either 5-HT$_{1A}$ or 5-HT$_{2A}$ receptors shortens estimated durations in qualitatively similar ways, an effect antagonized by specific blockade of these receptors. Interestingly, clozapine – a drug acting on both dopamine and serotonin systems – not only shortens the estimated durations but also facilitates the attention to time ("▶ Attentional Effect") during retention intervals (Buhusi and Meck 2009). Clozapine is reported to have differential effects on dopamine levels in the frontal cortex and striatum: It serves as a dopamine receptor antagonist in the mesolimbic dopaminergic system and as an indirect dopamine agonist in the frontal cortex by its activation of the serotonin 5-HT$_{2A}$ receptors on dopamine neurons. This pattern of pharmacological results is consistent with the hypothesis that the effect of clozapine on time estimation is due to an increase in dopaminergic neurotransmission in frontal cortex, but not in the dorsal striatum. Together, these data suggest that drug effects on time accumulation and attention to time rely on the interaction between the dopaminergic and serotonergic activation in frontal cortex and striatum (Buhusi and Meck 2009; Ho et al. 2002).

Drugs of Addiction

Acute exposure to psychostimulants, e.g., cocaine, methamphetamine, nicotine, etc., produces results compatible with the speeding up of an internal clock (Matell et al. 2006; Meck 1996; Paule et al. 1999). These effects are believed to be linked to the addictive and rewarding properties of psychostimulants (see "▶ Self-Administration of Drugs"; "▶ Addictive Disorder: Animal Models") and are susceptible to the effects of overtraining and habit formation. In addition, repetitive, high-dose exposure to psychostimulants such as methamphetamine produces deficits in time perception, possibly due to neurotoxic effects on the dopaminergic system. Of particular interest is the observation that stimulant-dependent individuals exhibit impaired timing and time perception. Taken together, these findings may help to explain why human subjects that abuse methamphetamine or cocaine show more impulsivity and less self-control in decision making due to greater discounting of delayed rewards (Williamson et al. 2008).

A Model of Drug Action on Interval Timing

The striatal beat frequency (SBF) model of interval timing ascribes a mechanism for detecting event durations to medium spiny neurons within the dorsal striatum (Buhusi and Meck 2005). These striatal neurons have a set of functional properties that place them in an ideal position to detect behaviorally relevant patterns of afferent cortical input. Briefly, the SBF model posits that medium spiny neurons in the dorsal striatum become entrained to fire in response to oscillating, coincident cortical inputs that become active at a previously trained event duration. In the context of this model, the clock effect of dopaminergic drugs is interpreted as reflecting tonic dopamine levels within the striatum, modulating the oscillatory frequency within the cortex through corticostriato-thalamocortical feedback mechanisms. Similarly, the memory effect of cholinergic drugs is interpreted as affecting the synaptic weights of striatal median spiny neurons and tonically active interneurons (Buhusi and Meck 2009; Matell et al. 2006; Williamson et al. 2008).

Cross-References

▶ Addictive Disorder: Animal Models
▶ Attention
▶ Behavioral Economics
▶ Circadian Rhythms
▶ Motor Activity and Stereotypy
▶ Self-Administration of Drugs

References

Buhusi CV (2003) Dopaminergic mechanisms of interval timing and attention. In: Meck WH (ed) Functional and neural mechanisms of interval timing. CRC Press, Boca Raton, pp 317–338

Buhusi CV, Meck WH (2005) What makes us tick? Functional and neural mechanisms of interval timing. Nat Rev Neurosci 6:755–765

Buhusi CV, Meck WH (2009) Relative time sharing: new findings and an extension of the resource allocation model of temporal processing. Philos Trans R Soc B 364(1525):1875–1885

Gallistel CR, Gibbon J (2000) Time, rate, and conditioning. Psychol Rev 107:289–344

Gibbon J, Malapani C, Dale CL, Gallistel CR (1997) Toward a neurobiology of temporal cognition: advances and challenges. Curr Opin Neurobiol 7:170–184

Ho MY, Velazquez-Martinez DN, Bradshaw CM, Szabadi E (2002) 5-Hydroxytryptamine and interval timing behaviour. Pharmacol Biochem Behav 71:773–785

Matell MS, Bateson M, Meck WH (2006) Single-trials analyses demonstrate that increases in clock speed contribute to the methamphetamine-induced horizontal shifts in peak-interval timing functions. Psychopharmacology 188:201–212

Meck WH (1996) Neuropharmacology of timing and time perception. Cogn Brain Res 3:227–242

Paule MG, Meck WH, McMillan DE, McClure GYH, Bateson M, Popke EJ, Chelonis JJ, Hinton SC (1999) The use of timing behaviors in animals and humans to detect drug and/or toxicant effects. Neurotoxicol Teratol 21:491–502

Williamson LL, Cheng RK, Etchegaray M, Meck WH (2008) "Speed" warps time: methamphetamine's interactive roles in drug abuse, habit formation, and the biological clocks of circadian and interval timing. Curr Drug Abuse Rev 1:203–212

Timing Precision

Definition

Timing precision refers to the trial-by-trial errors in the estimated duration. Timing precision is estimated by the width of the response distribution (see Buhusi and Meck 2005).

References

Buhusi CV, Meck WH (2005) What makes us tick? Functional and neural mechanisms of interval timing. Nat Rev Neurosci 6:755–765

Timiperone

Definition

Timiperone is a first-generation ▶ (typical) antipsychotic drug that belongs to the butyrophenone class; it is approved in Japan for the treatment of schizophrenia. It has general properties similar to those of haloperidol, but it has higher affinity (as an antagonist) for D2 and 5-HT$_{2A}$ receptors. It also has modest binding affinity for sigma receptors. It can induce ▶ extrapyramidal motor side effects, drowsiness, and constipation, but it displays low toxicity.

Cross-References

- ▶ Butyrophenones
- ▶ Extrapyramidal Motor Side Effects
- ▶ First-Generation Antipsychotics
- ▶ Haloperidol
- ▶ Schizophrenia

Tolcapone

Synonyms

(3,4-dihydroxy-5-nitrophenyl)(4-methylphenyl); methanone; Tasmar

Definition

Tolcapone is prescribed as an adjunct therapy with levodopa (L-DOPA) for the treatment of ▶ Parkinson's disease. It has been approved by the Food and Drug Administration for use in the United States. Tolcapone inhibits the enzymatic activity of catechol-O-methyltransferase (COMT) in the central and peripheral nervous systems. This action prevents COMT-induced methylation of L-DOPA, thereby increasing the bioavailability of L-DOPA. Tolcapone induces liver toxicity that has limited its use and led to more favorable alternatives. Other side effects of tolcapone use are nausea, vomiting, constipation, dizziness, dyskinesia, and hypotension.

Cross-References

- ▶ COMT Inhibitor

Tolerance

Synonyms

Acquired tolerance; Resistance

Definition

Tolerance is the diminution of a drug effect after its repeated administration. Tolerance is often identified as a shift to the right in a function that describes the ability of a range of doses of a drug to produce a particular response (dose-response curve). Such tolerance is surmountable in the sense that the original drug effect can be restored by increasing the dose at which the substance is administered. Insurmountable tolerance is characterized by a downward shift in the dose-response curve, and it is not sensitive to dose adjustment.

Cross-References

- ▶ Behavioral Tolerance
- ▶ Caffeine
- ▶ Pharmacodynamic Tolerance

Toluene

Synonyms

Methylbenzene, Toluol

Definition

Toluene is a highly volatile liquid at room temperatures with a distinctive sweet odor. It is one of the most widely abused inhalants. Pure toluene is readily available as a household and industrial product, used primary as a paint thinner and remover. Toluene is also a constituent of many other products such as adhesives, inks, paints, sealants, and disinfectants. It is also a component of gasoline. Toluene is typically abused by sniffing directly from the container or by placing the liquid on a rag or paper bag and breathing the vapors. A particularly dangerous practice is placing a toluene-laden plastic bag over one's head. Overdose death from toluene exposure is possible when users become unconscious while still exposed to the vapor. Toluene metabolites such as hippuric acid can be detected in the blood of users; tests for this can be used to diagnose toluene abuse. Neurological problems are often seen in toluene abusers.

Toluene has been the most widely studied abused inhalant in animal and neuropharmacology studies. When given to animals, it produces abuse-related behavioral effects similar to those produced by alcohol and other depressant drugs. Toluene produces very specific cellular effects, primarily on ligand-gated ion channels.

Cross-References

▶ Inhalant Abuse

Tomography

Definition

Tomography is an imaging method that consists of acquiring information about an object, such as the brain, in two-dimensional slices. These slices can then be integrated to form a three-dimensional image of the object.

Cross-References

▶ Positron Emission Tomography (PET) Imaging

Tonic Signals

Definition

Signals induced by constantly present triggers like neurotransmitters at background level or neuropeptides or hormones present in the cerebrospinal fluid.

Topiramate

Definition

Topiramate is an ▶ anticonvulsant drug that is used in the treatment of epilepsy in adults and children. Topiramate has several pharmacodynamic actions that contribute to its anticonvulsant properties including inhibitory effects upon voltage-gated NA^+ and Ca^{2+} channels; inhibitory effects at AMPA and kainate glutamate receptors, in particular the GluR5 ▶ glutamate receptor; and positive or negative modulatory actions at certain subtypes of ▶ $GABA_A$ receptor. Other actions of topiramate that may contribute to its anticonvulsant activity include positive modulation of some voltage-gated K^+ channels, inhibition of carbonic anhydrase, modulation of presynaptic neurotransmitter release, and effects upon intracellular concentrations of ▶ GABA in GABAergic neurons. Topiramate is also approved for the prophylaxis of migraine. Its efficacy as an antimigraine medication is mainly due to inhibitory effects upon AMPA and kainate glutamate receptors and, to a lesser extent, inhibitory effects upon voltage-gated Ca^{2+} channels. Topiramate has a wide therapeutic window, that is, the difference between the minimum effective anticonvulsant dose and that causing motor impairment; however, there are several recognized adverse effects of topiramate including paresthesia (unpleasant tingling in the limbs especially the fingers and toes), metabolic acidosis, kidney stones, and acute cognitive impairment. Like other anticonvulsant medications, topiramate has ▶ teratogenic effects and should not be administered during pregnancy. Other potential indications for topiramate include weight loss (involving complex pharmacodynamic actions at ▶ neuropeptide-Y, corticotrophin-releasing hormone, and ▶ glucocorticoid receptors), diabetes (topiramate promotes glucose transporter activation), and binge-eating, drug addiction, and anxiety-related disorders (most likely via effects at AMPA, kainate, and $GABA_A$ receptors).

Cross-References

▶ Anticonvulsants
▶ Corticotropin-Releasing Factor
▶ Eating and Appetite

Tourette's Disorder

Definition

Tourette's disorder is a neuropsychiatric syndrome that typically presents in young children and is characterized by both motor and vocal tics (abrupt, rapid, repetitive, nonrhythmic, stereotyped, temporally suppressible movement/vocalization). Premonitory urges accompany these tics that can range from simple movements/vocalizations, such as blinking, sniffling, or barking, to the more complex squatting, twirling, or coprolalia (obscene utterances). Tic severity and type fluctuate over time, usually becoming less severe or nonexistent into adulthood. For a diagnosis of Tourette's disorder, multiple motor and vocal tics must have been present for at least 1 year and have had a waxing-waning symptom course.

Modified from the Diagnostic and Statistical Manual of the American Psychiatric Association, Fifth edition (▶ DSM-5) and the NIH-NLM MedlinePlus Encyclopedia (online).

Cross-References

▶ Tic Disorders with Childhood Onset

Toxicity of Drugs

Synonyms

Side effects

Definition

Drug toxicity refers to the level of damage that a compound can cause to an organism. The toxic effects of a drug are dose-dependent and can affect an entire system as in the CNS or a specific organ such as the liver. Drug toxicity usually occurs at doses that exceed the therapeutic efficacy of a drug; however, toxic and therapeutic effects can occur simultaneously. It can be assessed at the behavioral or physiological level. Behaviorally, drug toxicity can be exhibited in a variety of ways, for example, decreases in locomotor activity, loss of motor coordination, and cognitive impairment. Examples of physiological effects include lesions to tissue, neuronal death, and disrupted hormonal cycles.

Trace Amines

Sara Tomlinson[1], Darrell D. Mousseau[2], Glen B. Baker[1] and Ashley D. Radomski[1]
[1]Neurochemical Research Unit and Bebensee Schizophrenia Research Unit, Department of Psychiatry, University of Alberta, Edmonton, AB, Canada
[2]Cell Signalling Laboratory, Department of Psychiatry, University of Saskatchewan, Saskatoon, SK, Canada

Synonyms

Arylalkylamines; Microamines

Definition

Trace amines, which include β-phenylethylamine (PEA), tryptamine (T), phenylethanolamine (PEOH), tyramines (TAs), octopamines (OAs), and synephrine (SYN) [some authors include N,N-dimethyltryptamine (DMT) in this list], are amines related structurally to, but present in the brain at much lower concentrations than, the classical neurotransmitter amines – dopamine (DA), noradrenaline (NA), and 5-hydroxytryptamine (5-HT, serotonin).

Pharmacological Properties

History

The trace amines are so named because of their low absolute concentrations in the brain compared to the classical neurotransmitter amines DA, NA, and 5-HT; they are very similar structurally to these neurotransmitter amines (often only lacking one or both hydroxyl moieties on the benzene ring: Fig. 1), but have much higher turnover rates, and in contrast to their hydroxylated analogs, PEA and T pass the blood–brain barrier readily.

From the 1960s through to the 1990s, there was a great deal of interest in the trace amines in the central nervous system (CNS) as behavioral and pharmacological studies in animals and neurochemical measurements in body fluids from human subjects suggested their involvement in the etiology and pharmacotherapy of a number of psychiatric and neurological disorders, including depression, schizophrenia, phenylketonuria (PKU), Reye's syndrome, Parkinson's disease, attention deficit hyperactivity disorder (ADHD), Tourette's syndrome, epilepsy, and migraine headaches (Baker et al. 1993; Berry 2007).

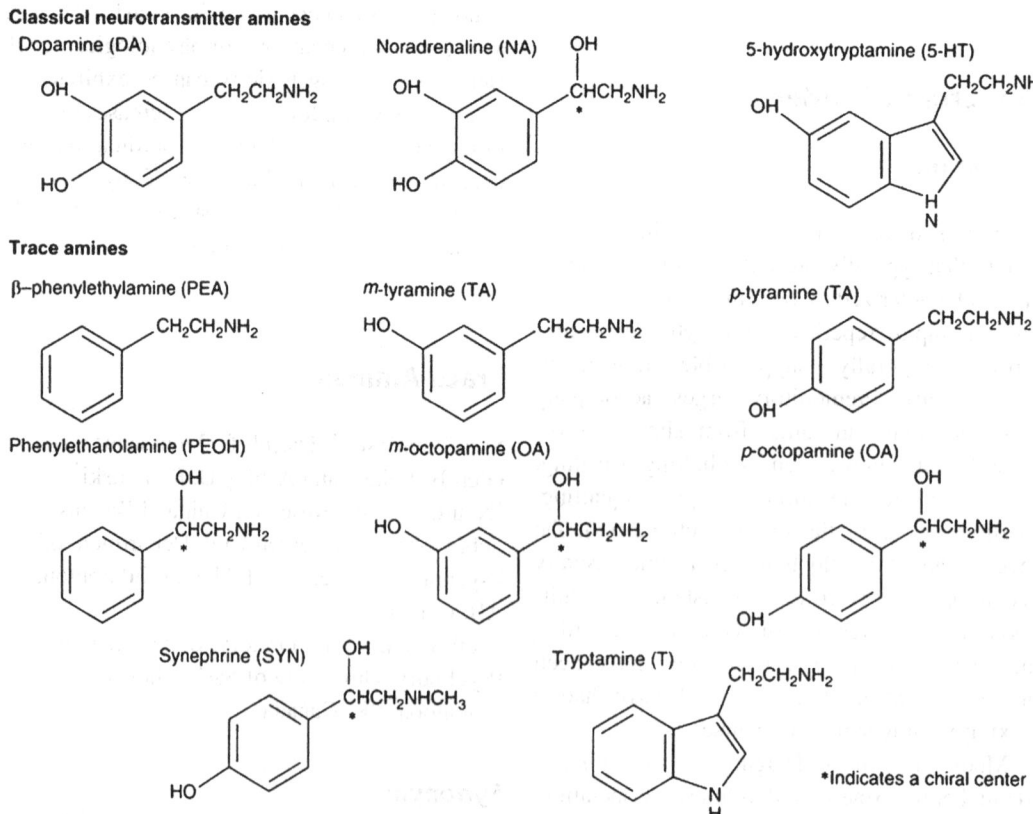

Trace Amines, Fig. 1 Structures of some classical neurotransmitter amines and trace amines

In the 1970s, there was a flurry of activity in trace amine research because of the development of a number of elegant, sensitive analytical techniques which facilitated their measurement in the brain. During the 1980s, binding studies suggested possible receptors for the trace amines, while electrophysiological and behavioral research suggested that these amines might act as neuromodulators for DA, NA, or 5-HT. There was a resurgence of interest in the trace amines in 2001, following reports of the discovery of a novel family of G protein-coupled receptors, some of which appear to be selectively activated by trace amines.

Synthesis, Catabolism, and Localization

The trace amines PEA, T, and TA are synthesized in neuron terminals by decarboxylation of precursor amino acids (phenylalanine, tryptophan, and tyrosine, respectively), catalyzed by the enzyme aromatic L-amino acid decarboxylase (AADC), which is also involved in the decarboxylation of L-3,4-dihyroxyphenylalanine (L-DOPA) and 5-hydroxytrytophan (5-HTP) in the synthesis of the catecholamines (DA, NA) and 5-HT, respectively. However, tyrosine hydroxylase and tryptophan hydroxylase are the rate-limiting enzymes in the synthesis of the catecholamines and 5-HT, while AADC is the major enzyme involved in the synthesis of PEA, T, and TA. Thus, alterations of AADC activity would be expected to have little effect on the levels of DA, NA, and 5-HT in the brain while possibly markedly affecting the levels of trace amines (Berry 2007).

Tyramine is further metabolized to OA and PEA to PEOH by dopamine β-hydroxylase (DBH). Tyramine and OA have been proposed to function as neurotransmitters rather than neuromodulators in invertebrates. Octopamine can be further metabolized to synephrine by phenylethanolamine N-methyltransferase (PMNT) or related methyltransferases. The trace amines are all substrates for monoamine oxidase (MAO). Abnormal levels of the resultant acid metabolites of the trace amines have also been reported in several psychiatric and neurological disorders (Berry 2007; Boulton et al. 1985).

The trace amines are distributed heterogeneously throughout the brain, with the highest concentrations generally reported in the striatum or hypothalamus. Burchett and Hicks (2006) have provided a comprehensive review of their regional brain distribution and localization relative to catecholaminergic and serotoninergic neuronal systems in the brain. Although PEA, T, and TA have been shown to be present in synaptosomes, studies with reserpine and neurotoxins suggest that m- and p-TA may be stored in vesicles, while PEA and T are not (Boulton et al. 1985). PEA and T appear to cross cell membranes passively, but there is some evidence for activity-dependent veratridine (a neurotoxin causing persistent activation of sodium ion channels)-induced release of m- and p-TA from striatal slices.

What Is the Function of the Trace Amines in the CNS?

It has long been known that trace amines such as PEA exert amphetamine-like effects on the CNS when administered to rats or mice. However, the levels required for such effects are well above normal concentrations in brain, and it is thought that other roles for trace amines at their endogenous, physiological concentrations must exist. Researchers found mounting evidence that trace amines may play a neuromodulatory role in the CNS. The trace amines are known to inhibit reuptake of and stimulate release of NA, DA, and/or 5-HT; and in electrophysiological studies, several trace amines have been shown to potentiate the actions of the classical monoamine neurotransmitter amines DA, NA, and/or 5-HT by altering the receptor sensitivity to these neurotransmitter amines, suggesting that the trace amines serve to maintain the activity of the classical monoamine neurotransmitters within defined physiological limits (Berry 2007). PEA has also been reported to stimulate acetylcholine release by activating glutamatergic signaling pathways, and PEA and p-TA have been demonstrated to depress $GABA_B$ receptor-mediated responses in dopaminergic neurons. In the 1980s, specific and saturable binding sites for radiolabeled PEA, T, and p-TA were reported, suggesting that these amines might have a role independent of the classical neurotransmitter amines. Burchett and Hicks (2006) have suggested four kinds of trace amine activity in the CNS: cotransmitters released with the catecholamines or 5-HT, transmitters with their own receptors, false transmitters at catecholamine receptors, and neuromodulators.

Receptors

A resurgence of interest in trace amines in the past few years followed the publication of papers in 2001 on the discovery and cloning of a unique family of G protein-coupled receptors, some of which are selectively activated by trace amines (Borowsky et al. 2001; Bunzow et al. 2001). The mechanisms by which the trace amines activate these receptors are not yet fully defined (Lindemann and Hoener 2005). To date only two members of the family of receptors have been demonstrated to be responsive to trace amines, and endogenous ligands for these receptors other than the trace amines have been proposed; these include O-methyl metabolites of catecholamines, thyronamine metabolites of thyroid hormones, and imidazoline ligands including β-carbolines (Berry 2007; Bunzow et al. 2001; Ianculescu and Scanlan 2010).

The trace amine-associated receptor (TAAR) family consists of three subgroups (TAAR1-4, TAAR5, and TAAR6-9) which are phylogenetically and functionally distinct from other G protein-coupled receptor families and from invertebrate OA and TA receptors (Lindemann and Hoener 2005). Genes for TAARs have been discovered in humans, chimpanzees, rats, and mice. There are marked interspecies differences in the distribution of the TAARs, with more TAARs in rodents than in humans. This variability has led some researchers to suggest that these receptors are linked in an intricate way to species-specific functioning (Berry 2007). In humans, all TAAR genes are located in a narrow region in the locus 6q23.1, which has also been linked to schizophrenia and bipolar disorder. Recent studies on TAAR1 knockout (KO) mice suggest that the TAAR1 is a regulator of dopaminergic neurotransmission and that such mice may represent a useful model for development of drugs for treatment of some positive symptoms of schizophrenia. Studies by Sotnikova et al. (2008) in TAAR1-KO mice, DA transporter (DAT)-KO/TAAR1-KO mice, and TAAR1-deficient/DA-deficient mice suggested that the TAAR1 is involved in tonic inhibitory actions on locomotor activity. Based on these same observations, the authors proposed that blockade of the TAAR1 by antagonists may represent a novel way to enhance the anti-Parkinson effects of L-DOPA. Studies in recent years using volatile amines such as trimethylamine and isoamylamine have shown that these amines are ligands for TAARs, which suggests that some TAARs are olfactory receptors distinct from classical olfactory receptors (Ferrero et al. 2012).

Several amphetamines [amphetamine, MDMA (Ecstasy), DOI, 4-hydroxyamphetamine] are relatively potent agonists at the TAAR1 receptor, as are ergometrine, dihydroergotamine, LSD, so-called hallucinogens [N,N-dimethyltryptamine (DMT), 5-hydroxy-N,N-dimethyltryptamine, and 5-methoxy-N,N-dimethyltryptamine], the anti-Parkinsonian agents bromocriptine and lisuride, and inhibitors of the DA transporter. Interestingly, the trace amine p-TA has been demonstrated to be necessary for sensitization to cocaine in Drosophila. These findings are of interest because it is possible that the TAAR1 may be a mediator of at least some of the effects of these drugs, providing a possible future target for treatment of drug abuse. It is also of interest that several biogenic amine antagonists, including phentolamine, tolazoline, cyproheptadine, metergoline, and chlorpromazine, as well as nomifensine and MPTP, act as agonists at the TAAR1.

Involvement in Psychiatric and Neurological Disorders: Neurochemical Studies

Several studies looking at the levels of trace amines and/or their acid metabolites in body fluids of patients with psychiatric or neurological disorders found potential associations with depression, bipolar disorder, schizophrenia, Reye's syndrome, ADHD, Tourette's syndrome, and PKU (Baker et al. 1993; Berry 2007; Boulton et al. 1985), although these studies are not without controversy. It has been reported that excess DMT in body fluids is related to production of psychoses, these yet findings have been controversial, and more recently it has been suggested that DMT acts on TAARs and that at low levels it is anxiolytic and suppresses the symptoms of psychoses (Jacob and Presti 2005). Increased PEA levels have been reported in mania, while depressed states have been found to be associated with deficits in PEA and the acid metabolites of OA and TA. Associations between paranoid schizophrenia and increased PEA excretion have been proposed as well. Decreased body fluid levels of PEA have been reported in Parkinson's disease. Urine levels of Tryptamine have also been reported to be increased in schizophrenics and to correlate with disease severity, and plasma levels of the p-TA metabolite p-hydroxyphenylacetic acid have been reported to be decreased in schizophrenia. Increased PEA levels in the brain have been reported in PKU. Evidence to date from several research groups suggests decreased urinary PEA in ADHD and Tourette's syndrome; there is also evidence for decreased PEA levels in brain and plasma in ADHD and for decreased urinary levels of m- and p-TA and indole-3-acetic acid (the major metabolite of T) in Tourette's syndrome. Animal studies and limited data in humans suggest that elevated PEA may be associated with an increase in stress and anxiety. High doses of PEA can induce seizures in mice, and this effect can be antagonized by benzodiazepines, suggesting an interaction with the GABA system. Other studies have suggested that PEA modulates glutamatergic and GABAergic systems. It is of interest that the gene for AADC, the rate-limiting enzyme involved in the synthesis of the trace amines, is located in the same region of chromosome 7p that has been suggested as a susceptibility locus for ADHD; 7p has also been linked to nicotine dependence. Elevations of TA and OA have been reported in hepatic encephalopathy and Reye's syndrome.

The effects of drugs used to treat psychiatric illnesses provide further support for the importance of trace amines in physiological and pathological brain function. It is known that monoamine oxidase inhibitor antidepressants such as phenelzine and tranylcypromine cause a much greater increase in levels of trace amines than of classical neurotransmitters such as 5-HT and NA in the brain, and increases in brain levels of PEA have been reported with tricyclic antidepressants and ECT. L-Deprenyl (selegiline) and rasagiline are used in the treatment of Parkinson's disease, and because they are selective inhibitors of MAO-B, they cause a marked increase in PEA levels in the brain relative to other amines. The antipsychotics chlorpromazine, fluphenazine, and haloperidol have been shown in acute studies in rodents to decrease striatal p-TA levels other studies have found that antipsychotics increase the rate of PEA accumulation in the striatum (see Boulton et al. 1985 for studies on these drug effects). Studies with R05166017, a selective TAAR1 agonist, suggest that the TAAR1 is involved in control of several dopamine- and 5-HT-driven behaviors, indicating possible antipsychotic and anxiolytic properties for such agonists (Revel et al. 2011).

Summary

Behavioral, pharmacological, and neurochemical studies in animals as well as investigations in body fluids of humans have long suggested that trace amines such as PEA, T, TA, and OA may be involved in the etiology and/or pharmacotherapy of a number of psychiatric and neurological disorders. There has always been debate about whether the trace amines have a neurotransmitter role. Although there is good evidence that OA may be a neurotransmitter in invertebrates, electrophysiological research has suggested that trace amines act as neuromodulators in the human brain, with their activity related closely to the classical neurotransmitters amines DA, NA, and 5-HT.

There has been a marked resurgence of interest in the trace amines since reports in 2001 of a unique family of G protein-coupled receptors, some of which are selectively activated by trace amines. These receptors, termed TAARs, are helping to explain the possible role of trace amines in the CNS (including their interactions with classical neurotransmitters), the effects of other compounds which may be endogenous ligands at these receptors, and the actions of a number of drugs of abuse, and may prove to be very useful in developing more selective drugs for the treatment of psychiatric and neurological disorders.

Acknowledgments Funding from the following sources is gratefully acknowledged: Canadian Institutes of Health Research [CIHR; MOP86712 (GBB)], Canada Research Chair and Canada Foundation for Innovation programs, the University of Alberta Distinguished University Professor Program (GBB), the Alzheimer Society of Saskatchewan and the Saskatchewan Health Research Foundation (DDM), the Davey Endowment, the Abraham and Freda Berger Memorial Fund, and Donald R. and Nancy Romanow Cranston.

Cross-References

- ▶ Antidepressants
- ▶ Anti-Parkinson Drugs
- ▶ Antipsychotic Drugs
- ▶ Attention-Deficit and Disruptive Behavior Disorders
- ▶ Bipolar Disorder
- ▶ Monoamine Oxidase Inhibitors
- ▶ Schizophrenia

References

Baker GB, Bornstein RA, Yeragani VK (1993) Trace amines and Tourette's syndrome. Neurochem Res 18:951–956

Berry MD (2007) The potential of trace amines and their receptors for treating neurological and psychiatric diseases. Rev Recent Clin Trials 2:3–19

Borowsky B, Adham N, Jones KA, Raddatz R, Artymyshyn R, Ogozalek KL et al (2001) Trace amines: identification of a family of mammalian G protein-coupled receptors. Proc Natl Acad Sci U S A 98:8966–8971

Boulton AA, Bieck PR, Maitre L, Riederer P (eds) (1985) Neuropsychopharmacology of the trace amines: experimental and clinical aspects. Humana Press, Clifton

Bunzow JR, Sonders MS, Arttamangkul S, Harrison LM, Zhang G, Quigley DI et al (2001) Amphetamine, 3,4-methylenedioxymethamphetamine, lysergic acid diethylamide, and metabolites of the catecholamine neurotransmitters are agonists of a rat trace amine receptor. Mol Pharmacol 60:1181–1188

Burchett SA, Hicks TP (2006) The mysterious trace amines: protean neuromodulators of synaptic transmission in mammalian brain. Prog Neurobiol 79:223–246

Ferrero DM, Wacker D, Roque MA, Baldwin MW, Stevens RC, Liberles SD (2012) Agonists for 13 trace amine-associated receptors provide insight into the molecular basis of odor selectivity. ACS Chem Biol 7:1184–1189

Ianculescu AG, Scanlan TS (2010) 3-Iodothyronamine (T(1)AM): a new chapter of thyroid hormone endocrinology? Mol Biosyst 6:1338–1344

Jacob MS, Presti DE (2005) Endogenous psychoactive tryptamines reconsidered: an anxiolytic role for dimethyltryptamine. Med Hypotheses 64:930–970

Lindemann L, Hoener MC (2005) A renaissance in trace amines inspired by a novel GPCR family. Trends Pharmacol Sci 26:274–281

Revel FG, Moreau JL, Gainetdinov RR, Bradaia A, Sotnikova TD, Mory R, Durkin S, Zbinden KG, Norcross R, Meyer CA, Metzler V, Chaboz S, Ozmen L, Trube G, Pouzet B, Bettler B, Caron MG, Wettstein JG, Hoener MC (2011) TAAR1 activation modulates monoaminergic neurotransmission, preventing hyperdopaminergic and hypoglutamatergic activity. Proc Natl Acad Sci U S A 108:8485–8490

Sotnikova TD, Zorina OI, Ghisi V, Caron MG, Gainetdinov RR (2008) Trace amine associated receptor 1 and movement control. Parkinsonism Relat Disord 14:S99–S102

Tracer Dose

Definition

A very small concentration at which a radiotracer is administered, such that the presence of the radiotracer itself does not noticeably influence the pharmacology or pharmacokinetics of the process being imaged. For reversibly binding ligands, the usual convention is that the peak binding of the radioligand does not exceed 5 % of the available receptor pool. Because the equilibrium bound to free ligand ratio is $B/F = B_{max}/(F + K_D)$ and the free ligand associated with 5 % occupancy is approximately 5 % of K_D, this restriction insures that ▶ binding potential will be measured accurately to within 5 %. From the data-fitting perspective, the use of tracer dose also insures that the linearization of the mass action law applied in the compartment model is justified.

Trait

Synonyms

Characteristic

Definition

Set of behavioral propensities that remain stable over time. They are usually measured by self-report questionnaires (e.g., personality or temperament questionnaires). Traits may be attributable to genetic makeup of the individual or an expression of prior experiences.

Trait Independence

Definition

Traits that are independent do not correlate, which indicates that it is not possible to predict an individual's level on trait A from knowledge of the same individual's level on trait B. It is assumed that independent traits have unrelated genetic and ▶ neurobiological foundations; however, the phenotypic manifestations of the traits may interact to alter the magnitude of either or both traits.

Tramadol

Definition

Tramadol is a centrally acting analgesic used to treat mild to moderately severe pain. It is an opiate agonist and analog of ▶ codeine. Notably, tramadol acts quite selectively at the μ-opioid receptor, whereas most other opiate analgesics do not share this property. However, it also acts outside the opioid system, enhancing ▶ serotonin and ▶ norepinephrine neurotransmission. ▶ Serotonin syndrome may occur in patients taking combinations of tramadol and other agents that increase serotonin activity. The relative degree of contribution of each mechanism toward pain control is not fully understood. It is possible that tramadol may have future uses in the treatment of depression, anxiety, and phobias. In contrast to typical opioids, tramadol may enhance immune function. Side effects may include nausea, vomiting, sweating, constipation, and seizure. Tramadol can lead to physical dependence such that rapid cessation leads to a severe ▶ withdrawal syndrome that includes tremors, nervousness, insomnia, and flu-like symptoms.

Cross-References

▶ Analgesics
▶ Opioid Use Disorder and Its Treatment
▶ Opioids
▶ Pain and Psychopharmacology
▶ Physical Dependence
▶ Tolerance

Transcranial Magnetic Stimulation

Kristina G. Gaud, Noah S. Philip and Linda L. Carpenter
Department of Psychiatry and Human Behavior, Butler Hospital, Brown University, Providence, RI, USA

Synonyms

Noninvasive brain stimulation; Repetitive transcranial magnetic stimulation (rTMS); Transcranial magnetic stimulation (TMS)

Definition

Repetitive transcranial magnetic stimulation (rTMS or simply "TMS") is a noninvasive treatment in which magnetic fields are pulsed through a coil placed over a patient's scalp. These magnetic fields induce electrical currents within the cerebral cortex, which are thought to generate antidepressant effects through normalization of brain function.

Current Concepts and State of Knowledge

History

The basic principle of TMS dates back to the work of Michael Faraday in 1839, when he discovered that a magnetic field could interact with an electric circuit to generate an electromotive force; applied to TMS this corresponds to a coil-generated magnetic field, which interacts with brain tissue to produce neuronal depolarization. The idea to harness electromagnetic energy to treat psychiatric disorders first originated in the early 1900s; in 1902, Pollacsek and Beer applied for a patent to treat depression and neuroses with a device that is conceptually similar to today's TMS. Barker and Cain created the first modern TMS device in 1985, for the purpose of stimulating nerve roots in the spine. It was not long before the technology was considered as a possible treatment for depression, and initial case reports were favorable. However, it was not until the 1990s when the technology was systematically examined as a treatment for depression. Currently, the evidence for the clinical efficacy of TMS spans over 30 randomized controlled trials in over 2,000 patients. Data has consistently shown that repetitive stimulation of the dorsolateral prefrontal cortex (DLPFC) can have an antidepressant effect in patients who do not benefit from antidepressant medication (Slotema et al. 2010). In 2008, the Food and Drug Administration (FDA) approved the first device to deliver TMS as an acute treatment for depression; a second device with this indication was FDA cleared in January 2013. TMS is being investigated as a treatment for a multitude of other psychiatric and neurologic disorders, but this chapter will focus on TMS as a treatment for major depressive disorder.

TMS Procedures

In TMS, a coil of copper wire is placed over the scalp, and electrical current is run through the coil in pulse-like fashion, creating a brief but potent pulsed electromagnetic field that can enter the brain without interference from the scalp, skull, or meninges. This magnetic field in turn induces an electrical current in the cerebral cortex, generating action potentials in neurons. Currently available TMS coils generate approximately 1.5 T, comparable to a standard session of magnetic resonance imaging (MRI). Depending on the parameters used, the depth of the electromagnetic field is thought to penetrate between 2 and 6 cm, depending on the coil configuration and the intensity of the pulses (Deng et al. 2013). The pulses can be administered as singles, pairs, or in a repetitive series or "train." Single and paired pulses are used for basic electrophysiology research, whereas repetitive trains (one or more pulses per second) are used therapeutically for psychiatric illnesses.

The intensity of pulses delivered during a treatment session is customized for each patient, calibrated as a function of the individual's "motor threshold" (MT). MT, a measure of global corticospinal excitability, is typically determined by the minimal amount of energy output from the TMS coil needed to produce a visible twitch of the contralateral thumb when a single pulse is delivered over primary motor cortex. For treatment of depression, standard intensity for treatment trains is up to 120 % MT. The point of stimulation of the thumb guides the placement of the coil on the head. The location of coil placement on a patient's head can be determined by several methods, including measurement to a point 5–6 cm anteriorly in a parasagittal plane, relative to the point where the MT was located. The majority of depression studies have stimulated left DLPFC. However, recent findings raise the question of whether focality and laterality are truly critical parameters.

TMS Treatment Parameters

A number of treatment parameters can be manipulated in TMS, including the targeted brain region/laterality, pulse intensity (percent of MT), pulse frequency (number of pulses per second), the number of pulses per train, the duration of inter-train-interval (i.e., pause between pulse trains), the number of pulses per session, and the number of sessions in a course of therapy (Peterchev et al. 2012). A convention has evolved for high-frequency (i.e., 5–10 Hz) stimulation on the left side and low frequency (1 Hz) for right-sided treatment, but it is not based on patient handedness or any specific physiologic rationale. "Slow" TMS is thought to have an inhibitory effect on neuronal activity, whereas "fast" TMS is generally thought to be excitatory. The TMS frequency used in most large published studies is 10 Hz, although evidence exists for positive effects from TMS delivered at other frequencies between 1 and 20 Hz. A standard treatment session consists of 3,000–5,000 pulses. Typically patients have one treatment session per weekday for an acute treatment course of 20–30 sessions over 4–6 weeks. The coils in devices approved for depression are a figure-eight-shaped coil with iron core (NeuroStar®) and the Hesed "H-coil" (Brainsway) with air core and complex windings over a larger area; coil configuration impacts the focality of the field delivered by each pulse.

Mechanism of Action

Similar to other neurostimulation and pharmacologic therapies, the exact mechanism of action of TMS is unclear. Some have hypothesized that, by stimulating action potentials within neurons, TMS increases the release of neurotransmitters into the synaptic cleft, similar to pharmacologic agents. Imaging data using functional MRI provides evidence that stimulation of cortex by TMS can activate deeper structures, such as insula, putamen, hippocampus, and thalamus, both ipsilateral and contralateral to the coil placement, via frontal subcortical neural circuits. Efficacy of TMS stimulation may depend upon polysynaptic connections with the subgenual anterior cingulate cortex (sgACC), an area thought to be a critical region of the brain involved in depression. Several lines of evidence support the hypothesis that TMS may have a circuit-based mechanism of antidepressant action (Fox et al. 2012).

Entrainment of brain electrical activity and resetting of thalamocortical oscillatory rhythms may be a possible mechanism by which TMS restores intrinsic cerebral rhythms and enhances neuroplasticity. Several studies show neuroendocrine correlates of TMS, including increased concentrations of thyroid-stimulating hormone, "normalization" of cortisol secretion as measured by the dexamethasone suppression test, and increased peripheral secretion of neurotrophins such as brain-derived neurotrophic factor (BDNF). Imaging studies also indicate TMS is associated with increased striatal dopamine.

Clinical Application

Two devices have regulatory approval for use in the treatment of major depressive disorder (MDD) in adults. Other TMS devices carry indications for peripheral nerve stimulation or to aid in diagnosis of intracranial masses. TMS for treatment of other psychiatric and neurologic disorders is currently being investigated with promising extant data.

The first large industry-sponsored depression study was a multisite, randomized, double-blind study of TMS as monotherapy in 325 medication-free patients, by O'Reardon et al., in 2007. All patients in the study met criteria for MDD and had failed between one and four antidepressant trials during the current depressive episode. Primary outcome measure was baseline to endpoint change on the Montgomery Asberg Depression Rating Scale (MADRS) at week 4, with several categorical and continuous outcome measures at weeks 4 and 6 examined as secondary outcomes. Results showed trend level differences between active and sham on the week 4 MADRS ($p = 0.057$) and statistically significant differentiation on the Hamilton Depression Rating Scale (HAMD) at weeks 4 ($p = 0.006$) and 6 ($p = 0.005$). Categorical response and remission rates were consistently statistically superior for TMS at week 6. Discontinuation due to side effects was low (4.5 %), and scalp discomfort and pain were the most common adverse events (O'Reardon et al. 2007).

Another pivotal investigation was the OPT-TMS study in 2010 by George et al., an NIH-funded prospective, multisite, randomized, sham-controlled study, in which 190 antidepressant free patients with pharmacoresistant unipolar depression were treated with daily TMS for 3 weeks, followed by up to another 3 weeks of blinded treatment for those who showed at least partial improvement. The active treatment utilized the same stimulation parameters as the industry trial, but extra steps were taken with the sham in this study to produce a similar somatosensory experience as active TMS so patients were unable to guess whether they were receiving active or sham treatment. Primary outcome was remission, which was significantly more likely to be achieved among those who were treated with active TMS (14.1 % vs. 5.1 % sham; $p = 0.02$). The odds of attaining remission were 4.2 times greater with active TMS compared with sham, and lower degrees of pharmacoresistance predicted better outcomes (George et al. 2010).

Supporting the efficacy and safety data generated by research trials, a naturalistic study designed to capture TMS outcomes in real-world practice found similarly positive results (58 % response, 37 % remission using the Clinician Global Impressions Scale) in 307 patients treated with the first commercially available device for depression at 42 clinical practice sites (Carpenter et al. 2012). Patient-rated scales produced response (56.4 % and 41.5 %) and remission (28.7 % and 26.5 %) rates depending on the scale used that aligned with results from research trials. Patients in the naturalistic outcomes study typically continued their previously ineffective antidepressant medications while undergoing treatment with TMS. Interestingly, patients who had failed one antidepressant were just as likely to respond to TMS as patients who had failed multiple antidepressants.

Most recently, the manufacturers of an H-coil device thought to deliver TMS with deeper penetration into cortex conducted a double-blind, randomized controlled multicenter trial ($n = 181$ efficacy sample) resulting in FDA clearance for use of their device to treat adults with major depression (Brainsway FDA 510 K Summary Number K122288) (U.S. Food and Drug Administration; Harel et al. 2012). In this study, treatments (18 Hz trains at 120 % MT, total 1280 pulses per session) were delivered 5 days a week for 4 weeks, followed by a follow-up maintenance phase in which subjects were treated twice weekly for another 12 weeks. "Deep" TMS was associated with a 6.39 point decrease in 21-item HAMD scale, compared to only 3.11 points in the sham group ($p = 0.008$), and efficacy was maintained at 16 weeks. These results were corroborated by statistically significant improvements in Clinical Global Improvement scores at week 5 and maintained through week 16. Similar to clinical trials with the predicate device, no negative cognitive side effects were associated with treatment, and main side effects of active TMS included jaw and stimulation site pain (Bersani et al. 2013).

Side Effects/Risks

The most common side effect encountered with TMS is discomfort at and around the treatment coil. During stimulation trains, a percussive sound is associated with a "tapping" feeling at the spot on a patient's head under the coil where pulsed electromagnetic energy field is maximal. TMS-associated discomfort is usually mild and diminishes or resolves within the first few weeks of daily TMS treatments. The most serious potential adverse effect associated with TMS is seizure induction. When stimulation is delivered within published safety parameters, described by Rossi et al., the risk of seizure during TMS is relatively small, less than 1 in 1,000 patients, which is comparable to treatment with antidepressant medications such as bupropion or tricyclic antidepressants; one seizure was reported in the 229 participants in the H-coil clinical trial. Earplugs or other occlusive earbuds are often worn to protect against hearing loss from loud clicking sounds that may be emitted when the TMS device discharges. Due to risk of inducing hypomania and mania, patients with bipolar disorder have not been included in the large controlled depression trials to date. Typical contraindications to TMS include conditions associated with heightened seizure risk, presence of implanted ferromagnetic metal objects in the head within close proximity to the area of stimulation, and presence of space-occupying intracranial lesions. Careful review of health history and medical status is part of screening for safe treatment with TMS. Several screening instruments have been developed for use in the clinic (Rossi et al. 2009).

Comparison to Other Neurostimulation Techniques and Therapies

In contrast to other approved neurostimulation techniques, TMS is considered noninvasive. While electroconvulsive therapy (ECT) requires anesthesia and seizure induction, TMS is delivered to the awake patient, and no sedation or recovery period is necessary. TMS is available on an outpatient basis, but requires frequent daily trips to a clinic over a period of 4–6 weeks. In contrast to ECT, TMS therapy is not associated with any cognitive side effects. Presently there are insufficient data to evaluate the relative efficacy of TMS and ECT in comparison to one another. ECT is often a treatment of choice for hospitalized patients who are imminently suicidal or have psychotic features, and such patients have not been included in large controlled TMS studies to date (Holtzheimer and Mayberg 2012).

Summary

TMS as a treatment modality is still in the early stages of development. At this point, the last two decades of research have generated data that support TMS as a safe and effective treatment for depression. TMS is particularly appropriate for patients who fail standard antidepressant therapy, do not tolerate medication side effects, or cannot tolerate ECT. Although used in psychiatric clinical practice settings primarily for treatment-resistant depression today, research is underway to optimize TMS technology for customization of parameters and amplification of therapeutic effects, to further improve safety and portability of devices, and to broaden the applications of TMS for treating other neuropsychiatric disorders and conditions, such as schizophrenia, obsessive compulsive disorder, autism, posttraumatic stress disorder, substance abuse, impulse control disorders, tinnitus, neuropathic pain, migraine, tic disorders, movement disorders, and for poststroke rehabilitation.

Cross-References

▶ Antidepressants
▶ Deep Brain Stimulation
▶ Depression
▶ Electroconvulsive Therapy
▶ Vagus Nerve Stimulation

References

Bersani FS, Minichino A et al (2013) This is a 2012 publication (epub 2012/05/09). Deep transcranial magnetic stimulation as a treatment for psychiatric disorders: a comprehensive review. Eur Psychiatry 28(1):30–39, PMID 22559998

Carpenter LL, Janicak PG et al (2012) Transcranial magnetic stimulation (TMS) for major depression: a multisite, naturalistic, observational study of acute treatment outcomes in clinical practice. Depress Anxiety 29(7):587–596, PMID: 22689344

Deng ZD, Lisanby SH, Peterchev AV (2013) Electric field depth-focality tradeoff in transcranial magnetic stimulation: simulation comparison of 50 coil designs. Brain Stimul 6(1):1–13

Fox MD, Buckner RL et al (2012) Efficacy of transcranial magnetic stimulation targets for depression is related to intrinsic functional connectivity with the subgenual cingulate dorsolateral prefrontal cortex. Biol Psychiatry 72:595–603

George MS, Lisanby SH et al (2010) Daily left prefrontal transcranial magnetic stimulation therapy for major depressive disorder: a sham-controlled randomized trial. Arch Gen Psychiatry 67(5):507–516

Harel EV, Rabany L, Deutsch L, Bloch Y, Zangen A, Levkovitz Y (2012) H-coil repetitive transcranial magnetic stimulation for treatment resistant major depressive disorder: an 18-week continuation safety and feasibility study. World J Biol Psychiatry

Holtzheimer PE, Mayberg HS (2012) Neuromodulation for treatment-resistant depression. F1000 Med Rep 4:22

O'Reardon JP, Solvason HB et al (2007) Efficacy and safety of transcranial magnetic stimulation in the acute treatment of major depression: a multisite randomized controlled trial. J Biol Psychiatry 62:1208–1216

Peterchev AV, Wagner TA, Miranda PC et al (2012) Fundamentals of transcranial electric and magnetic stimulation dose: definition, selection, and reporting practices. Brain Stimul 5(4):435–453

Rossi S, Hallett M et al (2009) Safety, ethical consideration, and application guidelines for the use of transcranial magnetic stimulation in clinical practice and research. J Clin Neurophysiol 120. PMID: 19833552

Slotema CW, Blom JD et al (2010) Should we expand the toolbox of psychiatric treatment methods to include repetitive transcranial magnetic stimulation (rTMS)? A meta-analysis of the efficacy of rTMS in psychiatric disorders. J Clin Psychiatry 71:7

U.S. Food and Drug Administration 510(K) Premarket notification "510(k) number K122288 Brainsway deep TMS System" http://www.accessdata.fda.gov/SCRIPTs/cdrh/cfdocs/cfPMN/pmn.cfm?ID=40833

Transcytosis

Definition

A mode of transcellular transport in which a cell entraps extracellular material, mostly macromolecules, in an invagination of the cell membrane to form vesicles, draws them through it, and ejects them on the other side. Membrane interaction with the enclosed material can result from nonspecific or adsorptive mechanisms or receptor-mediated processes. Blood-brain barrier capillaries are an active site for transcytosis by which iron via transferrin transcytosis, insulin, and growth factors are delivered to the brain.

Transgenic Organism

Synonyms

Mutant

Definition

Traditionally, an organism with the addition of foreign DNA, whether from the same species or a different one. More recently the term "transgenic" has been used to refer to any genetically modified organism.

Cross-References

▶ Ethopharmacology
▶ Genetically Modified Animals
▶ Phenotyping of Behavioral Characteristics

Translational Neuroimaging

Definition

Translational neuroimaging refers to the ability to perform imaging methods developed for animals in human subjects. Accordingly, these methods need to fulfill certain criteria such as being noninvasive (MRI) or at least minimal invasive (nuclear imaging techniques) and providing similar readouts, taking size, anatomical, and functional differences into account. In general, structural and functional MRIs are translatable due to their inherent noninvasiveness. Administration of exogenous contrast agents or pharmacological compounds (as performed in fMRI) limit the ability to translate MR methods developed in animals to human subjects. Pharmacokinetics, safety, and ethical approval for human applications need to be clarified before administration.

Cross-References

▶ Functional Magnetic Resonance Imaging
▶ Pharmacological fMRI

Translational Research in Drug Discovery

Thomas Steckler[1] and John Talpos[2]
[1]Pharma R&D Quality & Compliance, Janssen Research and Development, Beerse, Belgium
[2]Neuroscience Therapeutic Area, Janssen Research and Development, Beerse, Belgium

Synonyms

Experimental medicine; Translational medicine; Translational science

Definition

Translational research aims to bring basic preclinical knowledge (from the bench) to clinical practice (to the bedside) by characterizing fundamental mechanisms that play a role in the disease in the laboratory, finding ways to measure this in the human disease state, and developing beneficial healthcare outcomes. It is an iterative process: not only should preclinical findings result in clinical applications, but clinical insights will inform and direct preclinical research. For psychopharmacology, translational research wants to enhance the confidence in a central mechanism that could be of relevance for the treatment of psychiatric or neurological patients by enhancing the predictive value of disease-relevant, preclinical models for the clinic, by refining these models based upon clinical findings (back translation), and by developing innovative human tests based on preclinical findings (translation). It wants to ensure that the rationale for the progression of a psychoactive compound to man (and proper dose selection) is based on a link between target engagement in the brain and pharmacodynamic (PD) response, by integrating preclinical neuroscience, experimental medicine (in particular psychiatry and neurology), and pharmacokinetic/pharmacodynamic (PK/PD) modeling expertise. This should result in greater confidence in a molecular target and a compound having the desired/predicted outcome in patients.

Principles and Role in Psychopharmacology

Over recent years, it has become apparent that the traditional (or classical) approach to develop novel psychoactive drugs to treat patients has not performed well. Many compounds that were identified preclinically failed in the clinic, because they either lacked appropriate efficacy, had poor drug-like PK properties, or were hampered by safety issues. This was because many drugs proceeded into the clinic with a leap of faith, because often preclinical data were poor predictors for clinical effects and clinical measures allowing an early and reliable indication of efficacy did not exist, leading to late attrition in large and expensive patient studies only. Translational research should bridge the gap between preclinical research and clinical research, allowing for an early and quantitative indication of pharmacological activity that predicts the efficacy of a therapeutic intervention in patients (McArthur and Borsini 2008; O'Connell and Roblin 2006). It systematically addresses the questions whether a compound shows target tissue exposure, activity on the molecular target, and modulation of the desired biochemical pathways by looking at PK/PD markers and whether activity in a model system is predictive of clinical outcome, trying to identify suitable surrogate endpoint markers. Besides predicting efficacy, translational research also addresses mechanism-based toxicology to de-risk drug development (of note, idiosyncratic toxicology, unrelated to the mechanism of action of a compound, cannot readily be predicted).

Surrogate Endpoint Markers

A surrogate endpoint marker could be a trait marker, reflecting diagnostic accuracy, or a state marker, being predictive of clinical outcome. In psychopharmacology, surrogate endpoint markers could comprise, for example, genetic or genomic markers, neurochemical markers (e.g., cerebrospinal fluid (CSF) measures), neuroendocrine markers (e.g., reflecting activity of the hypothalamic-pituitary-adrenal (HPA) axis), electrophysiological markers (e.g., electroencephalographic (EEG) measures), molecular imaging data (e.g., functional magnetic resonance imaging (fMRI), measures of brain metabolic activity using deoxyglucose imaging or target occupancy positron emission tomography (PET) imaging to show target engagement), autonomic markers (e.g., measures of heart rate, skin conductance, or body temperature), or types of behavior that can be measured across species. Ideally, analogous or even homologous measures should be used, with the inherent assumption being that they are mediated by the same neural substrates across species. However, a problem inherent to these markers is that often the relationship between changes in marker activity in patients and clinical outcome is not clear. Prepulse inhibition (PPI), for example, can be readily measured and seems to be mediated by the same neural systems across species. PPI deficits have been reported in a number of psychiatric disorders, including schizophrenia. Antipsychotics have been reported to alleviate PPI deficits in schizophrenic patients, but how exactly this translates into clinical improvements remains elusive. Consequently, these markers are currently not accepted as clinical endpoints by health authorities in clinical trials. Accordingly, a marker that is correlated with functional outcome is required and must demonstrate more than just target engagement.

Translation and Back Translation

A translational approach refers to the development of clinical tests based on learning from preclinical experimentation. Current clinical trial endpoints are largely based on patient assessment using rating scales, although notable exceptions do exist, for example, the urine testing for illicit substances in the area of drug dependency, breath CO measures in tobacco smoking studies, or the measurement of CSF amyloid levels in Alzheimer's patients. With the growing emphasis on translational research, ever-increasing efforts are being undertaken to explore alternative, dimensional measures to align clinical research with preclinical findings. A behavioral example in the field of anxiety is fear-potentiated startle, a paradigm that was well characterized in animals and has been translated for use in humans (healthy volunteers and patients) (Nordquist et al. 2008).

Conversely, we can learn from the clinic and aim to back-translate to the preclinical setup. Of note, there are only a few signs and symptoms that are specific for a particular psychiatric disorder, while many symptoms cut across diagnoses. Therefore, preclinical behavioral tests most likely will address a certain diagnostic dimension that may be an important feature of multiple psychiatric disorders (Markou et al. 2009) but are unlikely to be specific for a single disorder.

The degree to which we can back-translate will depend on the species specificity of the diagnostic dimension under investigation. For example, we can readily measure sensorimotor gating in animals (e.g., using PPI), which is impaired in psychiatric disorders such as schizophrenia, and can assess motivational types of behavior that resemble anhedonic symptoms seen in schizophrenia and depression, but we cannot readily measure animal equivalents of some psychiatric symptoms such as hallucinations, delusions, or suicidal thoughts. Nonbehavioral, electrophysiological (e.g., EEG), neurochemical, neuroendocrine, autonomic, or imaging measures also lend themselves well to a (back-)translational approach, for example, the measurement of CSF amyloid levels in animals. Yet, other types of human pathological behavior have not been sufficiently explored in animals to allow conclusions whether preclinical equivalents exist (Markou et al. 2009), especially when going beyond face validity. To fully capitalize on the potential a translational strategy offers, approaches should change not only at the place of the preclinical but also the clinical methodology. In the study of cognition in a clinical setting, a shift is occurring from the use of questionnaires and rating systems to experimental/performance-based measures of ability. This is a great boon to the preclinical researcher who can now attempt to replicate the experimental design used in the clinical setting, something largely impossible without experimental measures. Perhaps the best example of this is the progress made with the Cambridge Neurological Test Battery (▶ CANTAB). CANTAB is an automated battery requiring participants to respond to a computer monitor to solve problems from a battery of tests that are dependent upon different cognitive processes and brain regions. Designed specifically with translation in mind, CANTAB has preclinical nonhuman primate equivalents, not only with a high degree of face but also construct validity. While perhaps not originally envisioned for translation into the rodent, an increasingly large touchscreen-based cognitive test battery is being developed for rats and mice, and many of the tasks originally developed for CANTAB have since been back-translated into this ad hoc rodent battery (Talpos and Steckler 2013). But despite the potential strength of this approach, it is still largely theoretical at this time and it remains to be shown that it will result in preclinical data with improved predictive ability for clinical outcome.

PK/PD Markers

PK/PD markers assess the relationship between the desired or undesired effects of a drug and its effect at the molecular target (PKs, what the body does to the drug; PDs, what the drug does to the body). Possibly the most straightforward marker would be the measurement of the plasma concentration of a drug in relation to its efficacy, either in preclinical species or man. Alternatively, one can study the concentration of a drug or its metabolites in CSF. Preclinically, target occupancy can be measured ex vivo or in vivo in tissue homogenates, brain slices, or other imaging techniques such as μPET, which then can be related to occupancy in man, provided a suitable radioactive PET ligand exists, and the degree of occupancy can be related to the desired or undesired effects of the drug under investigation. In principle, surrogate endpoint markers as those mentioned earlier could also be used as PK/PD markers, provided one can link the PD measures to target occupancy/drug levels. Once this relationship is firmly established preclinically, this information can be used to predict the pharmacologically active dose in man (Nordquist et al. 2008). Of note, occupancy only shows target engagement, and the occupancy of a target required to induce desired or undesired effects in preclinical species can only be used as a guide since different levels of occupancy may be required in patients for a drug to have clinical effects.

The Model System

While it is possible to measure many markers across species (although several exceptions exist – see above), it is equally important to use an appropriate manipulation of the model organism to mimic the perturbation that may occur in a disease state, be it an animal or a healthy volunteer, leading to a model with high construct validity. This seems particularly relevant for the measurement of surrogate endpoint markers.

Psychiatric disorders have traditionally been challenging to model in animals in a manner directly translatable to human research (Markou et al. 2009). First, we lack a good understanding of the pathophysiology of psychiatric disorders, which makes it difficult if not impossible to develop an animal model that incorporates the causative factors leading to disease, that is, our ability to build models based on strong construct validity is limited. This is further complicated by the fact that classification schemes for psychiatric disorders are primarily based on symptom clusters, thereby risking that diseases with different underlying etiologies but related symptoms are grouped within the same category. Consequently, different animal models may be needed to address different etiological factors of a disorder as defined by contemporary classification schemes. Alternatively, different manipulations have to be performed in the same animal to better model the disease. While it seems relatively straightforward to model some psychiatric disorders in animals, for example, drug abuse (Kreek et al. 2009), we do not really know what the ideal model for other psychiatric disorders, such as schizophrenia or depression, should be (Markou et al. 2009).

Complicating this approach is the fact that psychiatric classification schemes have changed over time, sometimes rendering models that were established according to the definition of an older classification scheme inappropriate to depict a disorder as defined in the newer classification. For example, many preclinical anxiety tests were developed before anxiety disorders were split into different categories by the third edition Diagnostics and Statistical Manual of Mental Disorders of the American Psychiatric Association (DSM III), published in 1980. Until recently, we needed models that adhered to the classification scheme of the next edition, DSM IV, published in 1994, and the newly published DSM V (May 2013) has yet additional modifications to the classification schemes that may require new adaptations to our animal models (Steckler et al. 2008).

In healthy volunteers, challenges are for ethical reasons more limited than those that can be performed preclinically. Such challenges, which are also used in preclinical research, can be pharmacological (e.g., NMDA antagonists to model the glutamatergic hypofunction in schizophrenia) or non-pharmacological (e.g., psychological stress exposure to increase stress-related anxiety). However, it should also be noted that not all challenges that are used in man have been successfully back-translated to animals, often because of species-dependent behavioral differences, and often questions about the robustness and statistical power of the healthy volunteer and preclinical models remain to be addressed.

Predictive Validity: Key Requirement for Translational Research

Often, it is this combination of a model system mimicking (some of) the pathophysiology of a psychiatric disorder and the use of a relevant test that holds potential for a translational model, although notable exceptions do exist (e.g., target engagement often can be measured in healthy animals and humans and is very informative for expected target engagement in patients). A stringent requirement for the translational approach is the ability to link the readout from the model system to clinical outcome. Key for that is the demonstration that efficacious treatments in the clinic, be it a drug or other approach, are active in the model (correct positives), while treatments that fail to therapeutically improve the disorder of interest should be inactive on corresponding preclinical measures (correct negatives), that is, it needs to be demonstrated that the model has a high predictive validity. In the absence of a clinically used treatment, one may revert to treatments that have shown efficacy in healthy volunteer models following appropriate challenges to gain more confidence in the predictive validity of an animal model, but this then leaves one with uncertainty about the translational value of the healthy volunteer model for the patient situation, also because the ability to relate such experimental healthy volunteer data to accepted clinical outcome measures is often lacking. Unfortunately, there is a shortage of suitable treatment approaches that can be used for the validation of translational models in psychiatry because suitable treatments for many psychiatric symptoms simply do not exist than can be used as correct positives (Markou et al. 2009). However, there are many examples of "impairments" that cross species, such as the effects of NMDA receptor antagonists, and many brain regions have been shown crucial for related tasks across species. It seems reasonable to suggest that if impairments can be translated across species, then enhancements should translate as well. While correct positives are still necessary to "prove" predictive validity for pro-cognitive effects, certainly proof of concept has been generated with cognitive disruptors.

Advantages and Limitations of the Translational Approach

The major advantages of translational research reside in a more stringent approach to psychopharmacological drug development, trying to bridge more closely between preclinical and clinical measures and using model systems of high construct and predictive validity. It is hoped that this should lead to faster and better informed go/no-go decisions, more efficacious drugs, and consequently reduced attrition rates during late drug development. It is also evident, however, that we still lack good surrogate endpoint markers for certain diagnostic dimensions and also lack clarity on what would constitute the best model systems for certain psychiatric disorders.

Thus, although the approach seems highly credible and promising (Sultana et al. 2007), only time will tell whether translational research will be really more advantageous than classical approaches in reducing attrition rates in drug development (Wehling 2006).

Cross-References

▶ CANTAB
▶ Construct Validity
▶ Deoxyglucose
▶ DSM
▶ Electroencephalography
▶ Ex Vivo

- ▶ Face Validity
- ▶ Hypothalamic-Pituitary-Adrenal Axis
- ▶ In Vivo
- ▶ Pharmacological fMRI
- ▶ Predictive Validity
- ▶ Prepulse Inhibition
- ▶ μPET

References

Kreek MJ, Zhou Y, Butelman ER, Levran O (2009) Opiate and cocaine addiction: from bench to the clinic and back. Curr Opin Pharmacol 9:74–80

Markou A, Chiamulera C, Geyer M, Tricklebank M, Steckler T (2009) Removing obstacles in neuroscience drug discovery: the future path for animal models. Neuropsychopharmacology 34:74–89

McArthur RA, Borsini F (2008) Animal and translational models for CNS drug discovery, vol 1: Psychiatric disorders. Academic, Amsterdam

Nordquist RE, Steckler T, Wettstein JG, Mackie C, Spooren W (2008) Metabotropic glutamate receptor modulation, translational methods, and biomarkers: relationships with anxiety. Psychopharmacology (Berl) 199:389–402

O'Connel D, Roblin D (2006) Translational research in the pharmaceutical industry: from bench to bedside. Drug Discov Today 11:833–838

Steckler T, Stein MB, Holmes A (2008) Developing novel anxiolytics: improving preclinical detection and clinical assessment. In: McArthur RA, Borsini F (eds) Animal and translational models for CNS drug discovery, vol 1: Psychiatric disorders. Academic, Amsterdam, pp 117–132

Sultana SR, Roblin D, O'Connell D (2007) Translational research in the pharmaceutical industry: from theory to reality. Drug Discov Today 12:419–425

Talpos J, Steckler T (2013) Touching on translation. Cell Tissue Res. doi:10.1007/s00441-013-1694-7

Wehling M (2006) Translational medicine: can it really facilitate the transition of research "from bench to bedside"? Eur J Clin Pharmacol 62:91–95

Transmembrane Helix

Definition

Part of an ▶ integral membrane protein that spans the entire cell membrane, with the transmembrane part of the protein coiled as a helix.

Transporter

Definition

A protein whose function is to terminate signaling by a neurotransmitter that has been released into the synaptic cleft. It does so by removing (transporting) the neurotransmitter away from the receptors and back into the nerve terminal that released it. It can then either be destroyed or repackaged for future use.

Tranylcypromine

Definition

Tranylcypromine is a nonselective, irreversible inhibitor of the enzyme monoamine oxidase. It is a cyclopropyl analogue of ▶ amphetamine although its effects and mechanism of action are quite different. It was one of the earliest ▶ antidepressants and was much used prior to the development of more effective and safer medications. Its use is largely limited to cases of depression that have failed to respond to other medications. It increases tissue concentrations of ▶ catecholamine and ▶ indoleamine neurotransmitters by slowing their metabolism; it also elevates the effects of biologically active dietary amines by increasing their absorption from the gastrointestinal tract and by impairing their breakdown. Dangerous interactions occur if foodstuffs high in ▶ monoamines or their precursors are consumed, a common example being the hypertensive crisis caused by the ingestion of tyramine from mature cheese and red wine. Tranylcypromine also potentiates the effects of later antidepressants such as the tricyclics and may produce toxic reactions if taken together.

Cross-References

- ▶ Antidepressants
- ▶ Monoamines
- ▶ Serotonin Syndrome

Trauma- and Stressor-Related Disorders

Joseph Zohar
Division of Psychiatry, Chaim Sheba Medical Center, Tel Hashomer Ramat Gan, Israel

Synonyms

Posttraumatic stress disorder

Definition

Introduction
Conservative figures estimate that between one- and two-thirds of the population in Western countries are exposed to a traumatic event of a magnitude that might eventually lead to posttraumatic stress disorder (PTSD). Of those exposed, 10–20 % develop PTSD, which is a long-lasting debilitating illness with substantial impact on the individual's work, family, social relations, and quality of life (Yehuda et al. 2005; Zohar et al. 2008). The disorder is characterized by the presence of three distinct, but co-occurring, symptom groups. *Reexperiencing* symptoms are intrusions of the traumatic memory in the form of distressing images, flashbacks, nightmares, or dissociative experiences. *Avoidance* symptoms consist of attempts to actively avoid reminders of the traumatic event including persons, places, or things associated with the trauma and/or passive behaviors reflecting emotional numbing and constriction. *Hyperarousal* symptoms include insomnia, irritability, impaired concentration, hypervigilance, and increased startle responses. In order to fulfill the criteria for PTSD, these symptoms must impair social, occupational, or interpersonal function and persist for at least a month following the trauma.

What is unique with regard to the pharmacological treatment after an exposure to trauma is that it is heavily influenced by the time elapsed since the trauma and not only by the severity and nature of the symptoms (which are also judged along a time line) (Nutt et al. 2000; Stein et al. 2006, 2009a, b). Six different conditions might surface following trauma exposure, and each might require different treatment strategies. These include acute stress reaction (ASR), acute stress disorder (ASD), acute PTSD (1–3 months), chronic PTSD (over 3 months), other psychiatric disorders (such as depression, panic disorder, obsessive-compulsive disorder (OCD), psychosis), and specific problems including sleep disorders, sexual dysfunction, aggression-impulsivity problems, addiction, and others (Bandelow et al. 2008; Davidson et al. 2005).

Role of Pharmacotherapy

Acute Stress Reaction (ASR)
This initial reaction to trauma includes symptoms of disorientation, autonomic symptoms of panic and anxiety (e.g., elevated heart rate, sweating), and sometimes amnesia of the traumatic event, lasting for a few days only.

This is considered a normal reaction (up to 70 % of those exposed to a traumatic event meet the criteria for ASR). There is no recommendation for pharmacological intervention for the vast majority of individuals experiencing ASR, as there is no treatment that has been scientifically supported at this stage. However, in specific cases of extreme agitation, medications may be used. The tradition of frequent use of benzodiazepines has recently been criticized, as it might interfere with the normal spontaneous remission which takes place in 80–90 % of cases (Zohar et al. 2008).

Individuals having experienced such a reaction should be advised of the need for monitoring if symptoms do not subside within a few weeks. However, those with past history of depression, anxiety, and substance abuse and those with dissociative reaction, panic-like response, and severe agitation should be followed up more closely, as these are all considered risk factors for developing PTSD (Bandelow et al. 2008; Stein et al. 2006, 2009a; Zohar et al. 2008).

Acute Stress Disorder

ASD is the severe end of ASR and is diagnosed (according to DSM-5) when the prescribed symptoms of PTSD occur, but within 4 weeks of the trauma, and lasting for a minimum of 2 days. The risk of developing PTSD for these individuals is substantially higher and might be up to 50 %. As is the case for ASR, the "non-pharmacologizing" rule is applicable; however, closer monitoring is recommended (Bandelow et al. 2008; Stein et al. 2006, 2009a; Zohar et al. 2008).

Acute PTSD (up to 3 Months)

Although there are no controlled studies for the treatment of acute PTSD, the data derived from treatment of chronic PTSD might be applicable. Paroxetine, fluoxetine, and sertraline are all US Food and Drug Administration (FDA) approved for the treatment of PTSD. The recommendation for treatment with these medications (which are all selective serotonin reuptake inhibitors, SSRIs) is to aim for full therapeutic dose (e.g., paroxetine 20–40 mg/day or sertraline 100–150 mg/day), gradually titrated. If there is no response within 12 weeks, the condition is considered as chronic PTSD with one failed medication trial.

If the response is an improvement to the extent of complete remission of symptoms, it is advisable to continue the treatment for 6–12 months from the time that remission is attained and to taper off the medication very gradually. There are currently no studies supporting the efficacy of treatment lasting more than a year, but as in other anxiety disorders, continuing treatment beyond that is often clinically indicated. A follow-up visit is recommended within 3 months of stopping pharmacological treatment.

In cases in which anxiety and distress symptoms are prominent, adjunctive treatments, including pharmacotherapies, can be used. It is advisable to closely monitor progress after beginning the adjunctive treatment and, if there is no response or numerous side effects, to stop it. Adjunctive treatments often include benzodiazepines, either for treating anxiety and distress that are not responding to an SSRI or to ameliorate such symptoms until the SSRI takes effect. Short-term use is recommended (but, again, not immediately after the trauma). When using benzodiazepines, a high-potency drug with a short half-life may be preferable, such as alprazolam 0.5–1 mg or lorazepam 1 mg, both of which can be given as needed for acute symptoms in multiple doses during the day. If a lengthier (yet still limited) treatment is required, a drug with a long half-life is preferable (e.g., clonazepam 1–3 mg/day, diazepam 15–30 mg/day, or clorazepate 15–45 mg/day). However, as there is high comorbidity of addiction in PTSD, it is very important that the treatment be limited in time, and if tolerance develops to the drug, it should gradually be tapered down. In any case, before recommending benzodiazepines, past drug, alcohol, or benzodiazepine dependence must be ruled out.

Low doses of second-generation antipsychotics, such as olanzapine 2.5–7.5 mg/day, quetiapine 50–200 mg/day, risperidone 1–3 mg/day, and amisulpride 50–200 mg/day, have been examined in preliminary studies. The efficacy of these medications in the treatment of PTSD is yet to be tested in larger, controlled studies (Bandelow et al. 2008; Davidson et al. 2005; Stein et al. 2006, 2009a).

Chronic PTSD (Over 3 Months)

The first-line treatment here is an SSRI, titrated gradually to a full therapeutic dose. Even if remission is achieved, guidelines suggest continuation of treatment for 6–12 months from the time remission is attained and then tapering off the medication very gradually. Although currently there are no studies supporting the efficacy of treatment lasting more than a year, the consensus is to continue treatment beyond that, as in other anxiety disorders. Follow-up visits are recommended at 3, 6, 9, and 12 months after cessation of pharmacological treatment. In cases of partial response (i.e., around 30 % improvement), a dose increase and continuation of the same medication for a further 12 weeks is recommended. In the instance of nonresponse to a treatment trial (less than 50 % decrease in symptom severity) after at least 12 weeks, a treatment change is recommended.

In the case of no response to an SSRI, switching strategies have been proposed. These include switching to another SSRI, switching to a serotonin-norepinephrine reuptake inhibitor (SNRI) at a dose that would ensure dual action (e.g., venlafaxine 225 mg/day and above), or augmentation with an antidepressant with a different mode of action (e.g., desipramine 150–200 mg/day, reboxetine 4–8 mg/day, mirtazapine 15–45 mg/day, mianserin 30–60 mg/day) as an add-on to the SSRI. However, when administering these drug in conjunction with SSRIs (e.g., fluoxetine, fluvoxamine, and paroxetine), drug-drug interactions should be monitored. It should be noted that none of these approaches are supported by controlled trials, and thus, there is no solid evidence to prefer one over another.

Second-line treatment alternatives (after at least two other attempts) include switching to a tricyclic drug (e.g., imipramine 150–300 mg/day) or a monoamine oxidase inhibitor (MAOI) (e.g., phenelzine 30–75 mg). Accepted precautions for the use of an MAOI must be taken (e.g., a washout period of at least 2 weeks after the previous medication and dietary and concomitant medication restrictions).

Nonresponse or "treatment-resistant PTSD" is usually defined as occurring after three full treatment attempts, as described earlier. In treatment-resistant PTSD, treatment combinations such as augmentation with second-generation antipsychotics may be considered (there are reports regarding the use of risperidone 1–3 mg and olanzapine 2.5–10 mg in such cases). Drugs with putative mood-stabilizing properties (e.g., lamotrigine and topiramate) have also been examined in this context. However, as these recommendations are largely based on case series and not controlled studies, they are not well supported. In any case, multiple medications should be avoided wherever possible, with a rule of thumb aiming at no more than two or three concurrent medications (Bandelow et al. 2008; Davidson et al. 2005; Nutt et al. 2000; Stein et al. 2003, 2006, 2009a, b).

Other Psychiatric Conditions

Comorbidity is very common in chronic PTSD patients. Comorbid diagnoses should be made according to the accepted criteria in DSM/ICD. Major depressive disorder (MDD), obsessive-compulsive disorder (OCD), panic disorder, alcohol and substance abuse, psychosis, and bipolar disorder are some of the most frequent comorbidities.

In each of these disorders, the therapeutic approach is to treat both the PTSD and the comorbid disorder. Consequently, the therapeutic approach is the accepted practice for each specific disorder. Priority of treatment will depend on the severity of each disorder and the subjective suffering of the patient or those in his or her immediate environment (e.g., partner, children, relatives) (Nutt et al. 2000; Stein et al. 2009a).

Specific Problems

Sleep Disorders

Complaints of sleep disruption are very common in PTSD, although there is sometimes a large discrepancy between the complaints and sleep laboratory data.

Therapeutic options include prazosin (5–10 mg), currently the only drug to have been tested in double-blind studies and found effective, particularly in cases of nightmares. Other possibilities (which have not been tested specifically) include melatonin 5–10 mg, clothiapine 20–60 mg, or levomepromazine 12.5–50 mg.

Standard hypnotic drugs, such as zolpidem 10–20 mg and zopiclone 7.5–15 mg, have clear benefits, but also disadvantages (e.g., developing tolerance). Therefore, it is not recommended to use these medications for more than 2–3 weeks consecutively. The recommendation is to treat for a defined period wherever possible and not indefinitely.

Antidepressants that cause drowsiness, but are not generally considered "sleeping pills" and are not associated with tolerance, may be of benefit. This group includes mirtazapine 14–45 mg, trazodone 50–200 mg, or amitriptyline 10–25 mg.

Maintenance treatment for sleep disorders should focus on avoiding tolerance during long-term treatment.

If the patient takes a hypnotic drug, treatment should be discontinued after 2–3 weeks. In patients who take the drug for longer duration, it is advisable to stop as soon as an improvement is observed. In both cases, cessation of treatment should be carried out carefully and gradually.

There are reports of an increased prevalence of sleep apnea in PTSD. Therefore, patients at increased risk of sleep apnea (middle-aged, overweight men with short necks, and whose bed partners report snoring) should undergo an appropriate screening and, if needed, a specific treatment tailored for them. Great care should be taken with these patients when prescribing drugs with side effects such as depression of respiration (such as benzodiazepines, nonbenzodiazepine hypnotics, and opioids).

Sexual Dysfunction

It is important to differentiate between problems with libido, arousal, orgasm, or combinations of these. Libido and orgasm problems could be side effects of one of the drug treatments, and if required, an alternative treatment could be tried.

Erectile dysfunction can also be related to side effects of drug treatment. First, an organic cause should be ruled out (e.g., diabetes, dyslipidemia, high blood pressure, hormonal insufficiency). When treating erectile dysfunction, phosphodiesterase inhibitors (e.g., sildenafil [*Viagra*], vardenafil [*Cialis*]) can be effective.

Aggression and Impulsivity Problems

Available treatments include mood stabilizers (e.g., topiramate 50–150 mg/day, lithium 600–1,500 mg/day, or valproic acid 500–2,000 mg/day, monitoring blood levels as appropriate), β-blockers (e.g., propranolol 20–80 mg/day, monitoring pulse and blood pressure), and antipsychotic medications. These approaches are yet to be examined in controlled studies.

Addiction

Alcohol dependence or cannabinoid, heroin, and cocaine addictions are prevalent in PTSD. The co-occurrence of addictions and PTSD is associated with poorer prognosis and increased risk for somatic comorbidity and should therefore be specifically screened for and treated adequately (Bandelow et al. 2008; Davidson et al. 2005; Stein et al. 2009b).

Summary

The pharmacotherapy of PTSD (which is mainly related to SSRIs) is only one component of the multifactorial strategy comprising appropriate treatment for PTSD. Psychoeducation to the patients and their families, including encouragement to resume previous activities as well as refraining from focusing on compensation issues, along with specific psychological intervention (i.e., prolonged exposure) and social and emotional support, are often as important as currently available pharmacological interventions.

Cross-References

▶ Alcohol Abuse and Dependence
▶ Benzodiazepines
▶ Depressive Disorders: Major, Minor, and Mixed
▶ Insomnias
▶ Obsessive–Compulsive Disorders
▶ Panic Disorder
▶ Schizophrenia
▶ SSRIs and Related Compounds

References

Bandelow B, Zohar J, Hollander E, Kasper S, Moller HJ (2008) WFSBP task force on guidelines for anxiety obsessive-compulsive post-traumatic disorders: World Federation of Societies of Biological Psychiatry (WFSBP) guidelines for the pharmacological treatment of anxiety, obsessive-compulsive and post-traumatic stress disorders – first revision. World J Biol Psychiatry 9(4):248–312

Davidson J, Bernik M, Connor KM, Friedman MJ, Jobson KO, Kim Y, Lecrubier Y, Ma H, Njenga F, Stein DJ, Zohar J (2005) A new treatment algorithm for posttraumatic stress disorder. Psychiatr Ann 35(11):887–900

Nutt DJ, Davidson JRT, Zohar J (eds) (2000) Post-traumatic stress disorder: diagnosis, management and treatment. Martin Dunitz, London

Stein DJ, Bandelow B, Hollander E, Nutt DJ, Okasha A, Pollack MH, Swinson RP, Zohar J (2003) World Council of Anxiety. WCA Recommendations for the long-term treatment of posttraumatic stress disorder. CNS Spectr 8(8 suppl 1):31–39

Stein DJ, Ipser JC, Seedat S (2006) Pharmacotherapy for post traumatic stress disorder (PTSD). Cochrane Database Syst Rev 25(1), CD002795

Stein DJ, Cloitre M, Nemeroff CB, Nutt DJ, Seedat S, Shalev AY, Wittchen HU, Zohar J (2009a) Cape Town consensus on posttraumatic stress disorder. CNS Spectr 14(1 Suppl 1):52–58

Stein DJ, Ipser J, McAnda N (2009b) Pharmacotherapy of posttraumatic stress disorder: a review of meta-analyses and treatment guidelines. CNS Spectr 14(1 Suppl 1):25–31

Yehuda R, Bryant R, Marmar C, Zohar J (2005) Pathological responses to terrorism. Neuropsychopharmacology 30(10):1793–1805

Zohar J, Juven-Wetzler A, Myers V, Fostick L (2008) Post-traumatic stress disorder: facts and fiction. Curr Opin Psychiatry 21(1):74–77

Traumatic Brain Injury and Psychopharmacology

Duc A Tran, Saurabha Bhatnagar and Ross Zafonte
Department of Physical Medicine and Rehabilitation, Massachusetts General Hospital, Spaulding Rehabilitation Hospital, Brigham and Women's Hospital, and Harvard Medical School, Charlestown, MA, USA

Synonyms

Concussion; Traumatic disorders of consciousness

Definition

Traumatic brain injury (TBI) is a non-degenerative, non-congenital insult to the brain from an external mechanical force, causing permanent or temporary impairment of cognitive, physical, and psychosocial functions.

Introduction

This entry focuses on the pharmacotherapy of neuropsychiatric disturbances following a TBI. The spectrum of TBI includes concussion, mild/moderate/severe TBI, and disorders of consciousness. Patients with a TBI may develop the neurobehavioral disorders of anxiety, apathy, impaired attention, impaired processing speed, depression, agitation, disinhibition, and psychosis that also manifest in their nontraumatic cohorts. However, it is the unique pathology of the injured brain, with its associated deranged metabolic state, altered consciousness, and impaired cognition, that requires treatment strategies that are distinct from the typical management of the aforementioned neurobehavioral conditions. While this entry focuses on psychopharmacotherapy, it should be borne in mind that the preferred first-line treatment of neurobehavioral disorders in TBI involves non-pharmacologic management strategies that are beyond the scope of this entry. It is also important to acknowledge the limited evidence in this category of therapies and the need for management by experienced physicians.

Heightened Vulnerability to Side Effects

The challenge in treating neurobehavioral disorders in patients with concomitant TBI is their presumed increased sensitivity to the adverse effects of medications (Zasler 2013). Not only do patients manifest side effects at lower doses, but they also have a lower threshold for toxic encephalopathy. Intolerance to a drug may also surface as worsening of neurologic deficits, which is often misinterpreted as failure to respond to therapy. Clinically, these issues are circumvented by starting medications at a low dose and escalating slowly.

Disorders

Hypo-Arousal

In patients with TBI, there is a continuum of wakefulness, ranging from the extreme hypo-aroused state of coma to the hyper-aroused states of posttraumatic agitation and mania. Along this continuum lie the disorders of consciousness, such as coma, unresponsive wakefulness (UR), minimally conscious state (MCS), and posttraumatic confusional state (PTCS). Coma is a state of pathologic unconsciousness in which the eyes remain closed and the patient cannot be aroused. In UR, formally known as vegetative state, awareness of self and environment is presumed to be absent, and there is an inability to interact with others, although the capacity for spontaneous or stimulus-induced arousal is preserved. MCS is a condition of severely altered consciousness in which there is minimal but definite behavioral evidence of self or environmental awareness. In PTCS, patients have emerged from MCS but remain disoriented and amnesic for day-to-day events. Emergence from MCS is signaled by the recovery of interactive communication or functional object use.

The first steps in treating disorders of arousal involve environmental and non-pharmacologic management. The initial pharmacological step involves the withdrawal of offending agents. In the neurocritical care setting, sedatives and paralytics may need to be weaned to allow patients to emerge from hypo-aroused states. Patients requiring seizure prophylaxis may be on highly sedating anticonvulsants, such as phenytoin, phenobarbital, and topiramate. Levetiracetam is one of the least cognitively impairing anticonvulsants. However, long-term use has been associated with depression, irritability, aggression, and psychosis. Other agents that cause hypo-arousal include benzodiazepines, antihistamines, antispastic medications, typical antipsychotics, and atypical antipsychotics.

When treating disorders of consciousness such as coma, UR, MCS, and PTCS, evidence supports the dopaminergic and noncompetitive N-methyl-D-aspartate (NMDA) antagonist amantadine as the first-line treatment (Giacino 2012). This evidence derives from a single large randomized placebo-controlled study and a smaller crossover design study, which demonstrated more rapid rates of recovery. Second-line agents include other dopaminergic facilitators, such as bromocriptine, levodopa, pramipexole, methylphenidate, and atomoxetine. Modafinil has also been reported to be beneficial but is supported by less robust evidence. Zolpidem, the short-acting hypnotic used to treat insomnia, has been demonstrated to paradoxically arouse comatose patients who sustained a severe TBI. A recent study of 84 traumatic and nontraumatic subjects with disorders of consciousness reported a 4.8 % response rate to this medication.

Even after a patient with a TBI emerges from an MCS and becomes more aware of their environment, physicians will find themselves treating fluctuating pathologies that lie along the arousal spectrum. Patients with a TBI will struggle with impaired alertness, initiation, attention, and processing speed. Pharmacologic treatments consist of dopaminergic and noradrenergic medications, such as methylphenidate, which has been shown to improve attention to task after a TBI. Daytime sleepiness has been shown to be attenuated by modafinil; however, studies of fatigue have found mixed results.

Hyperarousal

States of hyperarousal, such as PTCS, lie at the extreme end of the wakefulness spectrum in patients with a TBI. Agitation after injury is expected to appear as patients emerge from their MCS and persist through PTCS. It is imperative to distinguish posttraumatic agitation from restlessness and akathisia, which may not require aggressive pharmacologic treatment. Akathisia may be a transient sequela of a TBI, but as long as safety is attended to, patients should be allowed to pace with supervision. When a patient's agitation becomes aggressive to the point that personal or caregiver safety is compromised and rehabilitation is impeded, pharmacologic management may be required.

Tailored scales, such as the 14-item Agitated Behavior Scale, should be used to objectively adjust the response to pharmacologic treatment (Corrigan 1989). Agitation tends to fluctuate throughout a clinical course and can belie a patient's true trend. One should start pharmacologic treatment by withdrawing potentially offending medications whenever possible. Typical antipsychotics, such as haloperidol, should be avoided, as they have been shown to slow neurologic recovery in animal models. Neuroleptics may produce akathisia, can mimic agitation and even cause paradoxical delusions. Atypical antipsychotics may be required, but their efficacy in reducing aggression in the absence of psychosis has not been clearly established. Benzodiazepines have been reported to produce paradoxical agitation.

The Neurobehavioral Guidelines Working Group performed an evidence-based review of pharmacotherapies for agitation and concluded that available evidence is insufficient to support treatment standards, but does support treatment guidelines and options (Warden 2006). Propranolol and pindolol were recommended as treatment guidelines, since beta-blockers have the largest, although limited, body of data supporting their efficacy. Beta blockers have the added benefit of attenuating the hypermetabolic state that results from TBI with published estimates of up to 200 % above the predicted metabolic values. Selective serotonin reuptake inhibitors (SSRIs), tricyclic antidepressants (TCAs), buspirone, methylphenidate, valproate, and lithium were recommended as second- or third-line treatment options. In a single-center trial, amantadine was shown to improve irritability and aggression in patients with chronic TBI (Hammond et al. 2014).

When choosing the appropriate medication, one should consider the integrity of the anterior and dorsolateral prefrontal circuits. If these are disrupted, nonspecific beta-blockers should be selected to reduce limbic catecholaminergic dysregulation resulting from lesions that have damaged temporolimbic networks. If these circuits are intact, agents should be selected that preferentially improve prefrontal inhibition, such as amantadine, SSRIs, buspirone, and atypical antipsychotics. If anxiety is contributing to the agitation, buspirone should be considered.

If there is a contribution from delirium or if rapid sedation for harmful aggression is required, then membrane-stabilizing anticonvulsants or atypical antipsychotics may be prescribed. If depression plays a large role, treatment with SSRIs should be considered. If sleep disturbance is an issue, appropriate options include trazodone and low-dose mirtazapine. If posttraumatic seizure is a comorbidity, valproate has the added benefit of mood stabilization.

Disinhibition

Disinhibited behavior is a common source of distress for caretakers and families of patients with a TBI because it tends to manifest as inappropriate sexual advances, verbal abuse, and physical aggression. This "frontal release" reflects damage to prefrontal cortical regions. Pharmacotherapy is often adjunctive to environmental modification and redirection. First-line pharmacotherapy involves treatment with SSRIs. Since high doses are often required, citalopram and escitalopram are SSRIs that should be avoided, due to their potential for inciting cardiac complications. Alternatively, anticonvulsants such as carbamazepine or valproate are also effective. Combination treatment with atypical antipsychotics can be useful in select cases, but should be employed with some caution.

Depression

Posttraumatic depression is the most common psychiatric disorder in patients following a TBI, and it should be aggressively screened for. In one study, the prevalence of depression was as high as 52 % within the first year after injury (Fann et al. 2009). Along with increased risk for depression (33 %) in the TBI cohort, there is also an increased incidence of suicidal ideation (21–22 %). Suicide is a multifactorial issue; however, a review of the literature found that 18 % of those with a head injury attempted suicide and were successful three to four times more often than the general population.

Although there are no firm guidelines, first-line pharmacotherapy for persons with a TBI and depression generally are SSRIs. The most studied of those have been sertraline and citalopram.

For agitation with concomitant depression, combination treatment with a TCA, such as amitriptyline or desipramine, can be considered. Methylphenidate can be used in those with associated hypo-arousal or decreased attention and processing speed. Alternatively, bupropion can be used if there is linked sexual dysfunction, but should be considered with caution in those who have a strong risk of seizure.

At times, emotional lability after a TBI from pseudobulbar symptoms, also known as pathological laughing and crying, can mimic depression. Pseudobulbar affect is an involuntary emotional response, with intense episodes of crying and/or laughing that are transient and can occur several times per day. This can occur in 5–11 % of patients in the first year after a TBI. Generally, serotonergic and noradrenergic antidepressants, namely, SSRIs or TCAs, are initially considered as treatment options. Dextromethorphan-quinidine combination therapy was recently approved by the US Food and Drug Administration for the treatment of pseudobulbar affect, and efficacy studies are currently underway.

The mechanical and cellular injury in a TBI may result in significant endocrinologic pathology. Injury to the hypothalamic-pituitary axis can alter mood and confound clinical depression. All patients with a moderate to severe TBI should undergo a neuroendocrine screen at 3 and 12 months post-injury. Those with milder injury should be evaluated for potential risk, with further testing pursued if indicated. Appropriate recognition of a primary endocrine disorder should not be missed when managing depression, especially in patients with a TBI.

Anxiety Following TBI

Posttraumatic depression, as discussed above, is comorbid with posttraumatic anxiety in 73 % of TBI cases. Thirty-eight percent of TBI patients followed up to 1–5 years post-injury developed an anxiety disorder. Most common was generalized anxiety disorder (17 %), followed by posttraumatic stress disorder (14 %), phobias (7 %,) panic disorder (6 %), and obsessive-compulsive disorder (Kim et al. 2007). The anxiety disorders that develop following a TBI become sources of morbidity and are associated with poorer functional outcome.

Premorbid substance abuse, medication withdrawal, and iatrogenic etiologies should be addressed first. Although cognitive behavioral therapy (CBT) is the recommended first-line treatment, adjunctive pharmacotherapy may alleviate symptoms to facilitate the success of CBT. When pharmacological intervention is required, SSRIs are considered first-line medications due to their efficacy and tolerability in the TBI population. However, the latency between initiation and effect of antidepressants may necessitate the use of short-acting benzodiazepines. Unfortunately, benzodiazepines are not ideal for the TBI population. They are sedating, impair memory, delay motor recovery, and exacerbate TBI-induced sleep disorders. This class of medications is associated with substance abuse, dependence, Alzheimer's disease and discontinuation syndrome. Regulating sleep-wake cycles and treating insomnias can be an adjunctive approach to managing anxiety in the TBI population.

Diminished Motivation and Memory Disorders Following TBI

Apathy is the lack of motivation to form an intention, feeling, emotion, or concern. Many patients with a TBI suffer from diminished function of the paralimbic structures that are required to sustain goal-directed thought, emotion, and action. Apathy, abulia, and akinetic mutism denote increasing severities of TBI-related motivational disorders.

When apathy is a result of depression, SSRIs should be considered. When apathy is independent of depression, SSRIs are not only ineffective but may actually exacerbate the disorder. Evidence suggests that the pharmacologic treatment of apathy without depression is only modestly efficacious (Roth et al. 2007). For this condition, agents that augment cerebral catecholamine signaling, such as methylphenidate, dextroamphetamine, amantadine, and bromocriptine, can be considered. One can also choose agents that augment cerebral cholinergic function, such as donepezil, rivastigmine, or galantamine.

Disorders of motivation are among the most difficult neuropsychiatric conditions to manage in patients with a TBI because of both the lack of viable treatments and the barriers they pose to rehabilitation.

Among the most common long-term complaints of those with a TBI are memory disorders, causing real-world functional impairment. Prior studies had considered physostigmine as a potential therapy. Smaller more recent studies of donepezil have suggested a possible benefit, and a large multicenter trial is now underway. A large study that examined the efficacy of rivastigmine failed to show statistical significance when all groups were considered, but a sub-group analysis suggested benefit among those with the most profound memory impairments.

Summary

Psychopharmacologic treatment for TBI-related neurobehavioral disorders requires modification from the norm because of the unique pathology of these conditions. To optimize rehabilitative outcomes and prevent medical complications, the psychopharmacologic treatment of TBI is ideally carried out by clinicians trained and experienced in managing these disorders.

Cross-References

▶ Antipsychotic Drugs
▶ Anxiety
▶ Atypical Antipsychotic Drugs
▶ Benzodiazepines
▶ Cognitive Behavioral Therapy
▶ Depression
▶ Generalized Anxiety Disorder
▶ Methylphenidate and Related Compounds
▶ Posttraumatic Stress Disorder
▶ Selective Serotonin Reuptake Inhibitors
▶ SNRI Antidepressants
▶ Trazodone
▶ Tricyclic Antidepressants

References

Chew E, Zafonte RD (2009) Pharmacological management of neurobehavioral disorders following traumatic brain injury – a state-of-the-art review. J Rehabil Res Dev 46(6):851–878

Corrigan JD (1989) Development of a scale for assessment of agitation following traumatic brain injury. J Clin Exp Neuropsychol 11(2):261–277

Fann JR, Hart T, Schomer KG (2009) Treatment for depression after traumatic brain injury: a systematic review. J Neurotrauma 26(12):2383–2402

Giacino JT, Whyte J, Bagiella E et al (2012) Placebo-controlled trial of amantadine for severe traumatic brain injury. NEJM 366(9):819–826

Hammond FH, Bickett AK, Norton JH, Pershad R (2014) Effectiveness of amantadine hydrochloride in the reduction of chronic traumatic brain injury irritability and aggression. J Head Trauma Rehabil 29(5):391–399

Kim E, Lauterbach EC, Reeve A et al (2007) Neuropsychiatric complications of traumatic brain injury: a critical review of the literature (a report by the ANPA Committee on Research). J Neuropsychiatr Clin Neurosci 19(2):106–127

Roth RM, Flashman LA, McAllister TW (2007) Apathy and its treatment. Curr Treat Options Neurol 9(5):363–370

Warden DL, Gordon B, McAllister TW et al (2006) Guidelines for the pharmacologic treatment of neurobehavioral sequelae of traumatic brain injury. J Neurotrauma 23(10):1468–1501

Whyte J, Nordenbo AM, Kalmar K et al (2013) Medical complications during inpatient rehabilitation among patients with traumatic disorders of consciousness. Arch Phys Med Rehabil 94(10):1877–1883

Whyte J, Rajan R, Rosenbaum A et al (2014) Zolpidem and restoration of consciousness. Am J Phys Med Rehabil 93(2):101–113

Zasler ND, Katz DI, Zafonte RD (2013) Brain injury medicine: principles and practice. Demos Medical, New York

Trazodone

Definition

Trazodone is a triazolopyridine that was developed in Italy as an antidepressant and was much used for that indication, although its use has declined. It was one of the early atypical antidepressants and was developed according to the mental pain hypothesis, which postulated that major depression is associated with a decreased pain threshold. It inhibits the reuptake of ▶ serotonin with lower affinity and selectivity than for typical SSRIs such as ▶ fluoxetine and also acts as an antagonist 5-HT$_{2A}$ and 5-HT$_{2C}$ receptors and a histamine antagonist. Its efficacy is comparable to that of ▶ tricyclic antidepressants, but it has a superior side-effect profile. It has also been prescribed for insomnia but evidence of efficacy is lacking, and it is uncertain whether it has a favorable risk-benefit ratio. Adverse reactions reported include drowsiness and priapism (painful and prolonged penile erection). It is metabolized by the liver enzyme CYP3A4. Inhibition of the enzyme by other substances may result in high blood concentrations of trazodone. Grapefruit juice has such an inhibitory property and its consumption is discouraged when trazodone is prescribed.

Cross-References

▶ Antidepressants

Treatment Response

Definition

The magnitude of clinical change can be expressed as the mean change in baseline ▶ Y-BOCS scores on active drug compared to placebo or as treatment "responders" versus "nonresponders." Though no universally accepted definition of treatment response exists, a CGI-I rating of "improved" or "very much improved" and a decrease in Y-BOCS of 25 % or 35 % are widely used and may separate active from inactive treatments. Thus, an accepted definition of "responder" is an individual who exhibits a Y-BOCS reduction of 35 % after treatment as compared to pretreatment scores. This conservative level of improvement contrasts with the standard 50 % improvement expected for successful treatment for major depressive disorder.

Cross-References

▶ Clinical Global Impressions Scales
▶ Obsessive–Compulsive Disorders
▶ SSRIs and Related Compounds
▶ Yale-Brown Obsessive-Compulsive Scale

Treatment-Resistant Depression

Synonyms

TRD

Definition

Treatment-resistant depression (TRD) is defined as a major depression disorder in patients who fail to achieve remission with standard antidepressant therapies. TRD can be further distinguished from chronic severe depression, as some patients with milder depressive symptoms are still treatment resistant. Present treatments for TRD include various pharmacological combinations, electroconvulsive therapy (ECT), ▶ vagus nerve stimulation (VNS), ▶ transcranial magnetic stimulation (TMS), and ▶ deep brain stimulation (DBS).

Treatment-Resistant Schizophrenia

Synonyms

Refractory schizophrenia

Definition

Schizophrenia is considered as a heterogeneous illness with various trajectories of response and treatment outcome. A minority of individuals demonstrate complete resolution of symptoms; at the other end of the spectrum, approximately 25–30 % of individuals show minimal response to treatment, at least with first-generation antipsychotics. It is these individuals who are designated as refractory or treatment resistant, and there are now established criteria to make this diagnosis, criteria that take into consideration previous treatments (including drug, dose, and duration) as well as outcome.

It is important to distinguish this subpopulation of individuals from "partial responders." This latter group also manifests a suboptimal response, but it is more substantial than what is observed in the treatment-resistant population.

The distinction of treatment resistant has important clinical implications, as clozapine appears superior to all antipsychotics, including other atypicals, in treating this population. It is estimated that a significant response will be seen in approximately 30–50 % of clozapine-treated individuals.

Treatment-resistant schizophrenia can be seen from the earliest stages of schizophrenia, although it can also be observed in individuals who initially appeared responsive to treatment. There are no well-established criteria that can be used clinically to predict which individuals will be refractory or responsive to clozapine.

Cross-References

▶ Second- and Third-Generation Antipsychotics

TR-FRET

Definition

TR-FRET unites TRF (time-resolved fluorescence) and FRET (fluorescence resonance energy transfer).

FRET uses two fluorescent molecules, a donor and an acceptor. Excitation of the donor by an energy source (e.g., a flashlamp) triggers an energy transfer to the acceptor, if the donor and the acceptor are within a given proximity to each other. The acceptor, in turn, emits light at a given wavelength. Because of this energy transfer, molecular interactions between molecules (e.g., a ligand and a receptor) can be assessed by coupling each partner with a fluorescent label and detecting the energy transfer. TRF uses long-lived fluorescent labels combined with time-resolved detection; a delay between excitation and emission detection minimizes prompt fluorescent interferences.

Cross-References

▶ Receptors: Binding Assays

Triazolam

Definition

Triazolam is a short-acting (▶ half-life 1.5–5 h) benzodiazepine that produces sedative effects by potentiating inhibitory GABAergic neurotransmission at ▶ $GABA_A$ receptors. It was a leading hypnotic prescribed worldwide for the treatment of insomnia during the late 1980s. However, a high rate of reports of adverse effects including memory loss, confusion, anxiety, bizarre behavior, and hallucinations, as well as several legal cases, led to the drug's license being suspended in several countries including the UK. It is currently approved in the USA for use in small doses for the treatment of insomnia.

Cross-References

▶ Benzodiazepines
▶ Sedative, Hypnotic, and Anxiolytic Dependence

Trichloroethane

Synonyms

Methyl chloroform; TCE; 1,1,1-Trichloroethane

Definition

Trichloroethane (TCE) is a highly volatile liquid whose vapors are often subject to ▶ inhalant abuse. It is a widely used industrial solvent and is contained in many household products including adhesives, cleaning solutions, inks, paints, correction fluids, and shoe polish. TCE is a source of environmental chlorine, and as such its use was restricted by the Montreal Protocol to protect the ozone layer. Thus, the availability of TCE in products is much diminished. Nonetheless, TCE is an important abused inhalant because many animal and neuropharmacological studies are done with it as a representative halogenated hydrocarbon-based solvent. Evidence suggests that it produced effects similar to alcohol, other depressant drugs, and other abused inhalants.

Cross-References

▶ Inhalant Abuse

Trichostatin A

Synonyms

TSA

Definition

TSA is a chemical agent that has the ability to inhibit the enzymatic activity of class I and II HDACs. Treatment with this compound leads to the opening of chromatin structure due to an increase in histone acetylation. This compound serves as an HDAC inhibitor.

Cross-References

▶ Epigenetics
▶ Histone Deacetylase Inhibitors

Trichotillomania

Michael Bloch
Yale OCD Research Clinic, New Haven, CT, USA

Synonyms

Chronic hairpulling

Definition

Trichotillomania is classified under obsessive-compulsive and related disorders in DSM-5. Trichotillomania is characterized by recurrent hairpulling that causes noticeable hair loss and significant distress or impairment. Additionally, according to DSM-5, trichotillomania requires that an individual should sometimes experience urges prior to pulling and a sense of pleasure, gratification, or relief after the act.

Role of Pharmacotherapy

Trichotillomania has an estimated lifetime prevalence of roughly 0.6 %. However, the estimated lifetime prevalence increases to roughly 3 % when preceding urges and pleasure/relief afterward are removed from the diagnostic criteria. Trichotillomania patients have a high rate of comorbid illnesses, with approximately 82 % of adults who present for treatment experiencing at least one other axis I psychiatric disorder (Woods et al. 2006a). Common comorbidities include depression, anxiety disorders, obsessive-compulsive disorder, substance use disorders, posttraumatic stress disorder, and other body-focused impulse control disorders. Trichotillomania in adults has a strong female predominance and a chronic, waxing, and waning course. Longitudinal studies of adults with trichotillomania have demonstrated little improvement in symptoms over time (Keuthen et al. 2001). Trichotillomania has an average onset age of around 13 years. It should be distinguished from chronic hairpulling in young children. Hairpulling in young children is considered a behavior consistent with developmentally appropriate environmental exploration and is usually self-limited.

A proper assessment in a patient with trichotillomania involves getting a detailed history of their hairpulling. When discussing hairpulling behaviors, it is important to address (1) antecedent cognitions, behaviors, and feelings prior to pulling, (2) the settings in which pulling occurs, (3) body locations from which pulling occurs, and (4) post-pulling behavior. Trichotillomania patients commonly experience emotions such as boredom, tension, and anxiety prior to hairpulling episodes and/or a physical urge to pull. Commonly experienced cognitions prior to hairpulling include beliefs about the inappropriate appearance of certain hairs (gray, coarse, etc.), that hairlines or lengths of hair need to be symmetrical, or that the patient is unattractive or unlovable because of his or her appearance. Common post-pulling behaviors involve biting, rubbing, or eating the hair. Also, discarding of the hairs in fairly stereotyped ways is the norm. In patients who ingest their own hair, trichobezoars (conglomerations of hair and food that form in the gastrointestinal tract) are of particular concern, as they can lead to weight loss, iron deficiency anemia, malabsorption, and even gastrointestinal tract obstruction.

Trichotillomania patients also usually have specific places where they engage in the behavior, i.e., the bedroom or the bathroom. The most common sites of hairpulling are the scalp (73 %), eyebrows (56 %), eyelashes (53 %), pubic region (46 %), and legs (15 %) (Woods et al. 2006a). A physical examination of areas of hairpulling can uncover areas of irritation, follicle damage, and atypical regrowth of hair, all of which are common in patients with trichotillomania. Rating scales, such as the self-report Massachusetts General Hospital Hairpulling Scale and the clinician-rated National Institute of Mental Health Trichotillomania Severity Scale, are useful in measuring the severity of trichotillomania symptoms and tracking changes in symptom severity over time.

Selective serotonin reuptake inhibitors (SSRIs) are the most commonly utilized pharmacological intervention to treat trichotillomania (Woods et al. 2006a) (SSRIs and related compounds). Initial open-label trials showed improvement over time in trichotillomania patients taking SSRIs. However, in four randomized, blinded, placebo-controlled trials, SSRIs have failed to show benefit compared to placebo (Christenson et al. 1991; Dougherty et al. 2006; Streichenwein and Thornby 1995; van Minnen et al. 2003). A meta-analysis that combined the results of the four previously conducted trials similarly failed to show any evidence of an improvement compared to placebo treatment (Bloch et al. 2007). Although there is substantial evidence that pharmacological treatment with SSRIs is no more effective than placebo in the treatment of primary hairpulling in trichotillomania, these medications may still be quite effective in treating comorbid illness in these patients. Depression, anxiety disorders, and posttraumatic stress disorder are all common comorbidities in trichotillomania patients, and there is substantial evidence demonstrating improvements in these conditions with SSRI pharmacotherapy. When SSRI pharmacotherapy is initiated in a trichotillomania patient, the goal of therapy should be to specifically target comorbid illness that is impairing to the patient, as SSRI pharmacotherapy for primary trichotillomania has little evidence of efficacy.

Clomipramine is a tricyclic antidepressant (TCA) that has been extensively studied in the treatment of trichotillomania (antidepressants). Initial results from a 10-week, randomized, double-blind, crossover study of 13 women that compared clomipramine, a serotonergically potent TCA, to desipramine, a noradrenergically potent TCA, demonstrated substantial improvement in trichotillomania patients treated with clomipramine (Swedo et al. 1989). However, most of the patients in this trial experienced a relapse in their symptoms after longer-term treatment with clomipramine. Another small, 9-week, randomized, placebo-controlled parallel-group study showed some increased improvement of trichotillomania symptoms with clomipramine compared to placebo, but not to the level of statistical significance (Ninan et al. 2000). Clomipramine was very poorly tolerated in this study, with 40 % of subjects dropping out early due to side effects. Common side effects associated with clomipramine include weight gain, anticholinergic symptoms, and sedation. The meta-analysis of randomized trials with clomipramine suggests modest short-term benefits when compared to control conditions. However, evidence suggests that benefits from clomipramine are short-lived.

A substantial number of case reports and uncontrolled trials have suggested the possible efficacy of both typical and atypical antipsychotics in trichotillomania (antipsychotic drugs). Additionally, case reports suggest the possible efficacy of glutamate-modulating agents such as riluzole, N-acetylcysteine, and topiramate. Naltrexone, an opioid antagonist used to treat urge-related disorders such as alcoholism and pathological gambling, has shown some evidence of efficacy in uncontrolled trials. Skepticism is warranted when viewing the likely efficacy of any treatment of trichotillomania studied in an unblinded, uncontrolled fashion. Many patients with trichotillomania improve over the short-term following initial presentation regardless of the treatment administered. Greater familiarity and psychoeducation about the disorder, meeting other individuals with the disorder, supportive clinicians, and engaging in any active intervention against the disorder are all powerful forces toward clinical improvement. These aspects of treatment, along with the waxing and waning course of the disorder, make it difficult to assess efficacy in uncontrolled trials.

Role of Non-pharmacological Therapies

Habit reversal therapy (HRT) is a behavioral therapy designed for the treatment of trichotillomania and tics. HRT is a manualized, behavioral technique that is administered over a period of 2–3 months, with a maintenance period for relapse prevention. HRT involves several different components: self-monitoring, awareness training, stimulus control, and competing response training. The self-monitoring component of HRT requires patients to keep records of their hairpulling. Awareness training works to increase patient awareness of hairpulling behaviors and of high-risk situations that increase the risk of hairpulling. Stimulus control employs interventions designed to decrease the opportunities to pull and to intervene or prevent pulling behaviors. Competing response training involves teaching a patient to engage in a behavior that is physically incompatible with pulling for a set period of time when they feel the urge to pull. In HRT, patients are only permitted to pull after they have engaged in the competing response behaviors.

In three randomized, parallel-group studies, HRT demonstrated superior efficacy compared to wait list or placebo controls (Ninan et al. 2000; van Minnen et al. 2003; Woods et al. 2006b). HRT also demonstrated superiority to pharmacotherapy with fluoxetine and clomipramine in two of these randomized trials (Ninan et al. 2000; van Minnen et al. 2003). A recent meta-analysis demonstrated that in randomized, controlled trials, HRT shows superior effect sizes compared to pharmacological agents for trichotillomania (Bloch et al. 2007). Figure 1 depicts the relative effect sizes of HRT compared to pharmacotherapies for the treatment of trichotillomania. Further trials are needed to demonstrate that HRT will maintain efficacy when compared to control conditions that account for the nonspecific aspects of therapy (i.e., emotional support and psychoeducation).

Support groups (e.g., the Trichotillomania Learning Center in Santa Cruz, California, www.trich.org) can also be very helpful in treating trichotillomania. Such groups can provide treatment referrals and support to individuals experiencing the condition. Often, hearing other individuals' stories, coping mechanisms, and strategies can help the individual develop

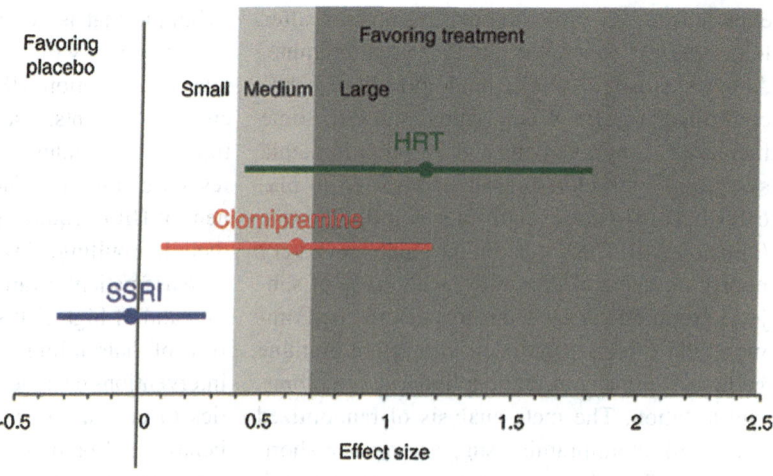

Trichotillomania, Fig. 1 Relative efficacy of treatments in trichotillomania: effect sizes of habit reversal therapy (*HRT*), clomipramine, and SSRI. *Circles* represent the point estimate and lines the 95 % confidence intervals for effect sizes for each intervention (Adapted with permission from Bloch et al. 2007)

more effective personal approaches for adjusting to and managing the disorder.

Conclusion

Trichotillomania can cause substantial impairment for individuals who suffer from it. Individuals may have significant dermatological and medical problems as a direct result of hairpulling. Substantial psychosocial impairment may occur due to the resulting hair loss. Effective behavioral treatments for trichotillomania, such as HRT, have been developed over recent years. Access to skilled therapists practicing HRT remains a major challenge for trichotillomania treatment. Currently, expert HRT treatment for trichotillomania is available at only a few academic centers. No pharmacological agents have convincingly demonstrated long-term efficacy in the treatment of trichotillomania. Novel pharmacological treatments are urgently needed given the problem of access to effective behavioral treatments for most patients and given the substantial proportion of patients for whom this behavioral technique is not tolerable or who still do not experience adequate symptom relief.

Cross-References

▶ Antidepressants
▶ Antipsychotic Drugs
▶ Impulse Control Disorders

References

Bloch MH, Landeros-Weisenberger A et al (2007) Systematic review: pharmacological and behavioral treatment for trichotillomania. Biol Psychiatry 62(8):839–846

Christenson GA, Mackenzie TB et al (1991) A placebo-controlled, double-blind crossover study of fluoxetine in trichotillomania. Am J Psychiatry 148(11):1566–1571

Dougherty DD, Loh R et al (2006) Single modality versus dual modality treatment for trichotillomania: sertraline, behavioral therapy, or both? J Clin Psychiatry 67(7):1086–1092

Keuthen NJ, Fraim C et al (2001) Longitudinal follow-up of naturalistic treatment outcome in patients with trichotillomania. J Clin Psychiatry 62(2):101–107

Ninan PT, Rothbaum BO et al (2000) A placebo-controlled trial of cognitive-behavioral therapy and clomipramine in trichotillomania. J Clin Psychiatry 61(1):47–50

Streichenwein SM, Thornby JI (1995) A long-term, double-blind, placebo-controlled crossover trial of the efficacy of fluoxetine for trichotillomania. Am J Psychiatry 152(8):1192–1196

Swedo SE, Leonard HL et al (1989) A double-blind comparison of clomipramine and desipramine in the treatment of trichotillomania (hair pulling). N Engl J Med 321(8):497–501

van Minnen A, Hoogduin KA et al (2003) Treatment of trichotillomania with behavioral therapy or fluoxetine: a randomized, waiting-list controlled study. Arch Gen Psychiatry 60(5):517–522

Woods DW, Flessner CA et al (2006a) The trichotillomania impact project (TIP): exploring phenomenology, functional impairment, and treatment utilization. J Clin Psychiatry 67(12):1877–1888

Woods DW, Wetterneck CT et al (2006b) A controlled evaluation of acceptance and commitment therapy plus habit reversal for trichotillomania. Behav Res Ther 44:639–656

Tricyclic Antidepressants

Synonyms

TCAs

Definition

Class of antidepressants first developed in the 1950s and named after their chemical structure. They are primarily used to treat depression, anxiety disorders, and pain.

Cross-References

- Amitriptyline
- Antidepressants
- Imipramine
- NARI Antidepressants
- Nortriptyline
- Secondary Amine Tricyclic Antidepressants
- Tertiary Amine Tricyclic Antidepressants

Trier Social Stress Test

Clemens Kirschbaum
Department of Psychology, Dresden University of Technology, Dresden, Germany

Synonyms

TSST

Definition

The Trier Social Stress Test (Kirschbaum et al. 1993) is a protocol for the induction of moderate to intense psychosocial stress under laboratory conditions. It comprises a 3 min anticipatory period, a 5 min public speaking task, and a 5 min mental arithmetic task, all in front of an evaluative panel of two adults. The TSST can be employed in adolescents and adults of all ages; a slightly modified version is used when studying children from the age of 7 years and older (TSST-C). Today, the TSST is the most widely used psychosocial stress protocol in laboratory studies of human subjects and patient populations.

Principles and Role in Psychopharmacology

Stress and Biological Responses to Threat

Psychosocial stress is a major burden for individual and societies alike. The direct and indirect costs of stress amount to billions of euros and US dollars in many countries because of the fact that approximately 50 % of all days of work absence can be attributed to consequences of stress exposure. While such financial effects burden economies significantly, the individual suffering in response to traumatic or chronic psychosocial stress threatens physical and psychological health. In 2001, the World Health Organization (WHO) therefore listed stress as "one of the most significant health problems of the twenty-first century."

In trying to understand the mechanisms through which stress affects well-being and what measures need to be taken in order to avoid adverse health outcomes, research protocols are required for a reliable induction of stress under laboratory conditions. In animals, different footshock or restraint stress paradigms are widely used. Only relatively few laboratories, however, employ more complex stress protocols that mimic stress exposures in modern human societies.

In human stress research, several paradigms and protocols exist for induction of moderate, acute stress responses. Most of these procedures, such as the *cold pressor test* or the *Stroop test*, evoke sympathetic nervous system responses of small or moderate magnitudes. The other most important physiological system, which conveys the brain's response to threatening and detrimental stimulation to the organs, remains unaffected by such protocols. The hypothalamic-pituitary-adrenal (HPA) axis requires more intense stimulation and threat to the ego before a significant increase in corticotropin-releasing hormone (CRH), adrenocorticotropic hormone (ACTH), or cortisol occurs.

Deliberate stress activation of the human HPA axis has proven to be difficult under ethically acceptable laboratory conditions. Only variable and inconsistent cortisol responses had been obtained with computerized or several other interactive tasks. As revealed in a meta-analysis of 208 laboratory stress studies (Dickerson and Kemeny 2004), a combination of social evaluative threat and uncontrollability is required for a reliable and strong ACTH and cortisol release. Social evaluative threat and uncontrollability are key components of the TSST.

The TSST Protocol

The standard TSST protocol requires at least two different rooms and a total of three lab members as a minimum. Upon arrival at the laboratory, subjects are guided to a standard office room for a first rest period of 30–60 min. Then, the experimenter takes them to a second room for the actual stress task. This room is sparsely equipped with two tables and three chairs, a video camera, and a microphone. Behind the larger table, two members of an *evaluation panel* are seated, dressed in white lab coats. The test subject is asked to stand at the microphone facing the seated panel approximately 2 m away. At this point, the experimenter gives instructions for the TSST to the subject, while the panel members remain silent. The subject is told that he or she should imagine having applied for an open position at a company or an institution. Out of many other applications, he or she was invited to present himself or herself before an evaluation panel to convince them that he or she is the best candidate for the job based on his or her personal characteristics. The first of two tasks, therefore, is to give an oral presentation about his or her personal strengths and positive aspects. The subjects are told that a video will be recorded for later analysis of nonverbal signs of stress and a voice frequency analysis. Furthermore, the two panel members are described as trained in the detection of verbal and nonverbal stress signals. In order to prepare this speech, the subjects are given 3 min of preparation time, which they spend sitting at the small desk in the same room. Paper and pencil are provided for sketching the talk; however, the notes cannot be used later in the oral presentation. The experimenter then leaves the subject alone with the two evaluation panel members in the room.

After 3 min, the experimenter asks the test subject to step up to the microphone and to begin with the free speech about himself or herself in a clear, loud voice. Most test subjects finish their talk after about 2–3 min of speech time. One of the panel members then asks them to continue their speech, since there is still time left. When the subject halts a second time before the first 5 min are over, the two panel members look at the subject with a neutral facial expression and do not speak for 20 s. Thereafter, they begin to ask personal questions (e.g., "Do you have friends?"). After 5 min exactly, the panel members stop the speech and proceed with a second task. They ask the subjects to perform mental arithmetic, subtracting an odd number from a larger number. Depending on the age of the subject, the difficulty of the problem is adjusted (e.g., for healthy young adults, they have to serially subtract 13, starting at 2011). The subjects are told that upon each error, the panel member will ask them to start anew from the initial number. After 5 min of mental arithmetic, the experimenter enters the room again and takes the subject back to the first room, where he or she rests for another 20–120 min depending on the psychological or biological measures taken.

As a mandatory part of every TSST (unless a study on habituation of stress responses is conducted), the subjects have to be fully debriefed after the last psychological or biological measure has been obtained. It has proven helpful to have the two panel members come to the tested subject then (without the white lab coats now) and introduce themselves. They should explicitly explain that the test protocol required them to be nonresponsive and cold in the interaction with the subject.

Trier Social Stress Test, Table 1 Biological parameters responsive to TSST exposure

Endocrine	Immunological	Other
ACTH	Neutrophils	Coagulation factors
Cortisol	Eosinophils	Hemoconcentration
Epinephrine	Basophils	Heart rate
Norepinephrine	Lymphocytes	Heart rate variability
Prolactin	Tumor necrosis factor alpha	Blood pressure
Growth hormone	Interleukin-6	Amylase activity
Testosterone	Interleukin-1 receptor antagonist	MAO-A inhibitory activity
DHEA		NFκB
		Electrodermal activity
		Prefrontal cortex activity
		Amygdala activity

Small variations of the standard TSST protocol are required when testing retired adults or children and adolescents between ages 7 and 16 years. The variations pertain to the topic of the free speech task only. Test subjects of advanced age will be presented with a fabricated advertisement which calls for people willing to donate their time for working in a nonprofit organization (e.g., helping with experiments with elderly subjects in a psychology department). The experimenter asks the subjects to talk to the evaluation panel as a job applicant, presenting his or her personal strengths and positive characteristics. When testing children, the so-called TSST-C (Buske-Kirschbaum et al. 1997) protocol is used. Here, the experimenter reads the beginning of the story to the child and asks him or her to continue and finish the story in as interesting and suspenseful a fashion as possible. All other aspects of the standard TSST protocol are also used in the elderly and young study populations.

Responses to the TSST

A wide range of psychological and biological response parameters have been studied in the past 15 years of TSST research (Kudielka et al. 2007). The TSST typically induces moderate to large subjective and physiological responses that peak 1–30 min after cessation of the stressful procedure. Approximately 80–85 % of all subjects tested show a substantial increase in the respective parameters from the resting (baseline) period to peak values. Self-reported negative mood changes and moderate increases in anxiety ratings are typical. Endocrine, immunological, and cardiovascular parameters increase by 50–300 % over baseline. Table 1 provides a partial list of biological parameters which are significantly changed in response to the TSST.

Repeated Exposure to the TSST

In many instances, a repeated stress exposure is the ideal study design for the investigation of specific treatment effects. For example, the potency of an anxiolytic drug could be tested or the efficacy of a psychotherapeutic intervention studied using a prepost challenge study design. The TSST can be used for such purposes. Careful attention should be paid, however, as to whether the crucial TSST response readout parameter shows habituation effects. When parameters controlled by the sympathetic nervous system are the main readout variables, the standard TSST protocol can be used repeatedly for the same subjects. ACTH or cortisol responses, however, decline upon the second TSST exposure (Kirschbaum et al. 1995). HPA axis habituation can be circumvented by changing the test setting (novel rooms/labs, panel members, and experimenters) for each TSST.

Using the TSST as a Group Stress Protocol

Researchers have begun to evaluate a modified TSST protocol for use with groups of 2–6 individuals at the same time (Childs et al. 2006). This allows for investigations of complex interactions between members of a certain peer group or serves as an economical way of stressing larger numbers of individuals within a brief period of time.

A "Placebo" Version of the TSST

When stressing individuals with the TSST, the researcher might want to differentiate between the specific effects of the social evaluative stress ("distress") and the effort component involved (orthostatic responses, speech-induced physiological changes, etc.). A simple no-intervention control session cannot provide such information. It is therefore necessary to contrast the TSST-induced subjective and biological responses with responses observed under similar setting and effort conditions, but without a distress component. Such a "placebo" version of the TSST has been recently published (Het et al. 2009).

References

Buske-Kirschbaum A, Jobst S, Wustmans A, Kirschbaum C, Rauh W, Hellhammer D (1997) Attenuated free cortisol response to psychosocial stress in children with atopic dermatitis. Psychosom Med 59:419–426

Childs E, Vicini LM, De Wit H (2006) Responses to the Trier Social Stress Test (TSST) in single versus grouped participants. Psychophysiology 43:366–371

Dickerson SS, Kemeny ME (2004) Acute stressors and cortisol responses: a theoretical integration and synthesis of laboratory research. Psychol Bull 130:355–391

Het S, Rohleder N, Schoofs D, Kirschbaum C, Wolf OT (2009) Neuroendocrine and psychometric evaluation of a placebo version of the 'Trier Social Stress Test'. Psychoneuroendocrinology 34:1075–86. doi:10.1016/j. psyneuen. 8 Feb 2009

Kirschbaum C, Pirke KM, Hellhammer DH (1993) The 'Trier Social Stress Test' – a tool for investigating psychobiology stress responses in a laboratory setting. Neuropsychobiology 28:76–81

Kirschbaum C, Pruessner JC, Stone AA, Federenko I, Gaab J, Lintz D, Schommer N, Hellhammer DH (1995) Persistent high cortisol responses to repeated psychological stress in a subpopulation of healthy men. Psychosom Med 57:468–474

Kudielka BM, Hellhammer DH, Kirschbaum C (2007) Ten years of research with the Trier Social Stress Test - revisited. In: Harmon-Jones E, Winkielman P (eds) Social neuroscience: integrating biological and psychological explanations of social behavior. Guilford Press, New York, pp 56–83

Trifluoperazine

Definition

Trifluoperazine acts at multiple receptors but mainly as an antagonist at dopamine D2 receptors. It is a ▶ phenothiazine, a first-generation antipsychotic with a plasma ▶ half-life of around 13 h. The active 7-hydroxy metabolite has a half-life of 10 h.

Cross-References

▶ First-Generation Antipsychotics

Trihexyphenidyl

Synonyms

(RS)-1-cyclohexyl-1-phenyl-3-(1-piperidyl)propan-1-ol; Apo-Trihex; Artane; Pacitane; Parkin

Definition

Trihexyphenidyl is an anticholinergic drug indicated in the treatment of ▶ Parkinson's disease either prescribed as a monotherapy or an adjunct therapy with levodopa. It is also used to treat extrapyramidal motor symptoms induced by antipsychotic drug treatment. Trihexyphenidyl functions as an M_1 muscarinic acetylcholine receptor antagonist. Anxiety, confusion, drowsiness, blurred vision, dry mouth, headache, and constipation are side effects of trihexyphenidyl use. Trihexyphenidyl has been approved by the Food and Drug Administration for use in the United States.

Cross-References

▶ Anticholinergic Side Effects
▶ Antimuscarinic/Anticholinergic Agent
▶ Scopolamine

Trimipramine

Definition

Trimipramine is a ▶ tricyclic antidepressant with atypical pharmacological properties. Trimipramine produces moderate enhancement of noradrenergic (NA), serotonergic (5-HT), and dopaminergic (DA) transmission by inhibiting reuptake of the neurotransmitters. It is also a potent antagonist of 5-HT$_2$ and α_1 adrenergic receptors. In addition to its antidepressant properties, trimipramine also has sedative-hypnotic and anxiolytic effects. Furthermore, due to its moderate effectiveness as a D2 dopamine receptor antagonist, trimipramine has been evaluated as an atypical antipsychotic with some limited clinical effectiveness.

Clinically observable antidepressant effects of trimipramine usually become noticeable after 10 days to 4 weeks of treatment. Trimipramine has strong anticholinergic and antiadrenergic side effects, including dry mouth, sweating, hypotension, and arrhythmias, as well as other effects such as fatigue, dizziness, drowsiness, confusion (especially in the elderly), anxiety, nightmares, increased appetite, and, more rarely, hypomania, seizures, delirium, and hematological and hepatic problems.

As with other antidepressants, abrupt termination of therapy triggers withdrawal symptoms, including rebound depression, anxiety, insomnia, or mania.

Trophic Factors

Definition

The original definition of trophic (from the Greek) is "to nourish." The definition has been expanded to the growth and sustenance of cells and tissues. In biology, the term "factor" refers to substances (of known or unknown molecular identity) displaying specific actions on specific tissues or cells. The term "neurotrophic factor" is presently applied to a large number of well-characterized substances with specific actions over cells (neurons and glia) of the peripheral and central nervous systems. They are classified in families according to similar molecular characteristics and related biological actions.

Tryptamines

Definition

Tryptamine is the general chemical name for a substance that possesses a bicyclic indole nucleus with an ethylamine moiety attached to the indole 3-position. Within this broad definition, tryptamines may be substituted on the indole ring itself, on the ethylamine side chain, or on the amino group. Molecules with any or all of these types of substitutions would still be generically referred to as tryptamines. Examples are the neurotransmitter serotonin, which is 5-hydroxytryptamine (5-HT), or the naturally occurring ▶ hallucinogen N,N-dimethyltryptamine. The neuromodulator melatonin is 5-methoxy N-methyltryptamine, also a substituted tryptamine.

TrkB

Definition

The receptor for brain-derived neurotrophic factor (BDNF). It is a receptor protein tyrosine kinase.

Tryptophan

Definition

Tryptophan is one of the eight essential amino acids required by humans; these substances cannot be synthesized by the organism and must therefore be supplied in the diet. Tryptophan is metabolized via the rate-limiting enzyme tryptophan hydroxylase to 5-hydroxytryptophan (5-HTP), which is in turn converted into the major neurotransmitter serotonin (5-hydroxytryptamine; 5-HT). ▶ Serotonin is broken down to the inactive metabolite 5-hydroxyindoleacetic acid (5-HIAA), but can also be metabolized further to the neurohormone ▶ melatonin. Tryptophan is sold as a dietary supplement and has been investigated as a mild ▶ hypnotic and as an adjunctive agent to enhance the efficacy of ▶ antidepressant drugs; studies of such uses have been inconclusive. In the early 1990s, tryptophan was implicated in the pathogenesis of eosinophilia-myalgia syndrome (EMS), a potentially disabling autoimmune reaction; subsequent investigation suggested that impurities introduced during a flawed manufacturing process were responsible, although some authorities have held that tryptophan itself is also implicated.

Cross-References

▶ Melatonin
▶ Serotonin

Tryptophan Depletion

Simon N. Young
Department of Psychiatry, McGill University, Montréal, QC, Canada

Synonyms

Dietary tryptophan depletion; Rapid tryptophan depletion

Definition

Acute tryptophan depletion (ATD) is a technique designed to investigate directly the effects of low serotonin levels in humans. Participants in ATD studies ingest an amino acid mixture that is devoid of tryptophan. This lowers brain tryptophan and, therefore, brain serotonin synthesis.

Principles and Role in Psychopharmacology

Principles of ATD

Ingestion of an amino acid mixture containing all the essential amino acids except for tryptophan induces protein synthesis. As tryptophan is incorporated into protein its level in the blood and tissues declines. Because brain tryptophan hydroxylase, the rate-limiting enzyme on the pathway from tryptophan to serotonin (5-hydroxytryptamine, 5-HT), is not normally saturated with tryptophan, this results in a decline in serotonin synthesis. ATD was originally developed to investigate the symptoms associated with low brain serotonin levels. It has also been used in conjunction with drugs (and nonpharmacological treatments) to see effects of the drug that are reversed by ATD. Reversal of the effect suggests that normal serotonin function is an important component of the drug effect. The assumption behind the ATD method is that decreased synthesis of serotonin results in decreased release of serotonin from neurons, i.e., decreased serotonin function. This assumption remains untested but presumably the extent to which changes in serotonin synthesis will change release of serotonin depends in part on the firing rate of serotonin neurons, which may vary depending on the exact experimental conditions used. Therefore, the question of whether changes in serotonin synthesis lead to changes in serotonin function is too simplistic. The relevant issue is whether there are circumstances in which it is possible to demonstrate that ATD leads to changes in aspect of brain function that seem to be mediated by serotonin. Furthermore, because the coupling between serotonin synthesis and serotonin function is not necessarily the same in humans and experimental animals, studies on the effect of ATD on extracellular levels of serotonin in rodents may not be relevant to human studies. These issues are discussed in detail in a review (Young 2013).

Methodology of ATD

The principles, methodology, and results of ATD studies have been discussed in a number of reviews (Hood et al. 2005; Neumeister 2003; Young and Leyton 2002). The most commonly used amino acid mixture is that of Young et al. (1985), which is based on the amino acid content of human milk (and therefore presumably optimized for human consumption) and contains 100 g of amino acids. The control mixture is usually the same except for the addition of 2.3 g L-tryptophan. Both control and active mixtures contain no glutamate, glutamine, aspartate, or asparagine to reduce the size of the mixture and to avoid any potential adverse effects of the ingestion of large quantities of these free amino acids. As they are not essential amino acids, this should not alter the ability of the tryptophan deficient mixture to lower tryptophan levels. Some researchers have used mixtures that omit more of the nonessential amino acids, while some have used 50 g of amino acids in the same proportion as in the 100 g mixture. The 102.3 g control amino acid mixture will tend to lower brain tryptophan levels somewhat, like most proteins, because it has lower levels of tryptophan than of the other large neutral amino acids that tend to inhibit tryptophan uptake into brain. While this could be seen as an advantage, as it is a conservative control, some researchers have used control mixture with higher tryptophan levels. There is little evidence from direct comparisons of different methods on the advantages and disadvantages of the various depletion and control mixtures.

Plasma tryptophan reaches its lowest level in about 5 h after amino acid mixture ingestion and experimental measures are usually performed 4–7 h after mixture ingestion. Depletions of plasma tryptophan are normally in the range 70–90 %, with even greater lowering of serotonin synthesis as measured by positron emission tomography. Depletion of tryptophan is usually greater in women than in men. How large the depletion in serotonin needs to be to induce changes in mood is not clear, but the decline in plasma tryptophan probably needs to be more than 50 % (Booij et al. 2002).

Effects of ATD

The majority of ATD studies have noted effects on mood, behavior, and cognition in drug-free participants. ATD has a wide range of effects on mood, depending on the characteristics of the participant undergoing the procedure. In general, effects are greater in those with a greater susceptibility to depression. For example, healthy never depressed males usually show no change in mood, but healthy never depressed males with a family history of depression show a modest lowering of mood. In keeping with the greater incidence of depression in women than in men, healthy never depressed women sometimes (but not always) show a modest lowering of mood. Formerly, depressed patients off medication occasionally show a dramatic lowering of mood into the clinical range that resolves quickly at the end of the study once brain serotonin synthesis is normalized. Effects in such patients are more likely to be seen with recurrent depressive episodes, female gender, and previous suicidal ideation or attempts. There is some evidence that supports the idea that the magnitude of the effect on mood after ATD may be a predictor of future depression, but more research is needed on this topic. Overall, studies in drug-free individuals suggest that low serotonin can contribute directly to lowered mood, but that it is not sufficient by itself to lower mood. The other neurochemical factors contributing to a susceptibility to lowered mood after ATD are not known.

Effects of ATD on anxiety in health individuals, when seen, are smaller than those on mood. ATD has no effect on untreated patients with panic disorder and has minimal effect on the response to panicogenic treatments. In keeping with the large animal literature showing that low serotonin increases aggression, ATD has been shown in several studies to increase the response in laboratory tests of both aggression and impulsivity. Results of the effect of ATD on cognition have not always been consistent, but there is some consistency in the findings that ATD has adverse effects on aspects of memory, particularly memory consolidation, while it can improve focused attention. Effects on cognition have been seen in the absence of effects on mood.

ATD has been used to investigate the role of serotonin in a variety of antidepressant treatments. In newly recovered depressed patients who are on serotonin reuptake inhibitors, the majority of patients showed a marked lowering of mood, similar to that when they were depressed, which resolves when their serotonin synthesis is normalized. No such effect is seen in patients on norepinephrine reuptake inhibitors. These results suggest that, in the short-term, maintenance of serotonin function is necessary component of the antidepressant action of serotonin reuptake inhibitors, but not norepinephrine reuptake inhibitors. In a small number of studies, ATD also reversed the antidepressant effect of bright light therapy and monoamine oxidase inhibitors, but not that of tricyclic antidepressants, lithium, electroconvulsive therapy, and cognitive therapy. While ATD reverses the therapeutic effect of serotonin reuptake inhibitors in patients with depression, results in anxiety disorders are not as clear-cut. ATD can enhance the response to panicogenic agents in patients with panic disorder who are on serotonin reuptake inhibitors and reverse the therapeutic effect in patients with social anxiety disorder. It does not cause greater obsessive or compulsive symptoms in patients with obsessive-compulsive disorder (OCD), who have responded to treatment with a serotonin reuptake inhibitor. On the other hand, such patients may show lowered mood or enhanced anxiety, raising the possibility that the therapeutic effect of serotonin reuptake inhibitors in OCD may be due in part to effects on mood and anxiety rather than directly on the core symptoms of OCD.

Other studies have used ATD to look at the role of serotonin in the action of a variety of different drugs. The number of these studies is small and results must be considered tentative. ATD has little effect on symptoms in patients with schizophrenia treated with neuroleptics. It decreased the cocaine-induced high and desire for cocaine suggesting that serotonin release has a role in the action of cocaine. It blocked the analgesic effect of morphine in response to the cold pressor test in participants, who had not taken morphine previously. However, the relevance of this to a role for serotonin in the clinical analgesic effects of morphine are not clear. Finally, ATD decreases the release of prolactin in response to the serotonin releaser, fenfluramine, suggesting that fenfluramine releases primarily newly synthesized serotonin.

Ethical Aspects of ATD

Given that low serotonin is associated with depression, aggression, and suicide, participants in ATD studies need to be monitored carefully. At the end of an ATD study, serotonin levels of participants should be restored to normal as quickly as possible. This is best done by administration of tryptophan but has been done more often by allowing participants to eat high protein foods. In a study of ATD in recovered depressed patients, some of whom reexperienced depressed mood, the participants had a positive opinion of their participation in the study. Common views were that participation in the study had changed their perception of their illness and that the study had helped them to gain more insight and facilitated acceptance of being vulnerable and had made them realize how much they had improved during treatment (Booij et al. 2005).

Limitations of ATD

From the practical point of view, one of the limitations of the ATD method is the unpalatability of the amino acid mixtures, which often results in nausea and sometimes vomiting. Various strategies have been adopted to reduce these symptoms, including adding strong flavors such as chocolate to the amino acid mixtures and putting the more unpalatable amino acids in capsules, but they do not eliminate the problem. From the theoretical point of view, a limitation is that relating any effect of ATD to lowered serotonin function involves two main assumptions. First, the effect seen is due to decreased release of serotonin, and second, other mechanisms do not contribute to the effect. Amino acid imbalance can alter aspects of brain function. However, the effects of ATD are unlikely to be due just to nonspecific effects of amino acid imbalance as (1) acute phenylalanine/tyrosine depletion causes somewhat different effects from ATD and (2) in one study the effects on mood seen with ATD did not occur when a positive control mixture deficient in lysine was used (Klaassen et al. 1999). ATD will tend to lower the levels of tryptophan metabolites in addition to serotonin. These include tryptamine, melatonin, kynurenic acid, and quinolinic acid. While a contribution of these compounds to the effect of ATD cannot be ruled out, given that the effects of ATD are consistent with other research on the role of serotonin, effects of the other compounds are unlikely (Young 2013). Finally, an important limitation in extrapolating from ATD studies to the role of low serotonin in psychopathology, or to the role of serotonin in drug effects, is the short duration of the effect of ATD on serotonin. Longer-term depletion of serotonin through repeated administration of amino acid mixtures is not advisable due to the effects such a strategy would have on protein synthesis and amino acid metabolism.

Cross-References

▶ Aggression
▶ Amine Depletors
▶ Aminergic Hypotheses for Depression
▶ Antidepressants
▶ Emotion and Mood
▶ Impulsivity
▶ Obsessive–Compulsive Disorders
▶ Panic Disorder
▶ Social Anxiety Disorder
▶ SSRIs and Related Compounds

References

Booij L, Van der Does W, Benkelfat C, Bremner JD, Cowen PJ, Fava M, Gillin C, Leyton M, Moore P, Smith KA, Van der Kloot WA (2002) Predictors of mood response to acute tryptophan depletion: a reanalysis. Neuropsychopharmacology 27:852–861

Booij L, van der Does AJW, Haffmans PMJ, Spinhoven P, McNally RJ (2005) Acute tryptophan depletion as a model of depressive relapse: behavioural specificity and ethical considerations. Br J Psychiatry 187:148–154

Hood SD, Bell CJ, Nutt DJ (2005) Acute tryptophan depletion. Part I: rationale and methodology. Aust N Z J Psychiatry 39:558–564

Klaassen T, Riedel WJ, Deutz NEP, van Someren A, van Praag HM (1999) Specificity of the tryptophan depletion method. Psychopharmacology (Berl) 141:279–286

Neumeister A (2003) Tryptophan depletion, serotonin, and depression: where do we stand? Psychopharmacol Bull 37:99–115

Young SN (2013) Acute tryptophan depletion in humans: a review of theoretical, practical and ethical aspects. J Psychiatry Neurosci 38:388–397, online at http://www.cma.ca/multimedia/staticContent/HTML/N0/l2/jpn/vol-38/issue-2/pdf/pg120209.pdf

Young SN, Leyton M (2002) The role of serotonin in human mood and social interaction: insight from altered tryptophan levels. Pharmacol Biochem Behav 71:857–865

Young SN, Smith SE, Pihl RO, Ervin FR (1985) Tryptophan depletion causes a rapid lowering of mood in normal males. Psychopharmacology (Berl) 87:173–177

Two-Dimensional Gel Electrophoresis

Synonyms

2-DE; 2-D Electrophoresis; Gel electrophoresis

Definition

Two-dimensional gel electrophoresis is based on the separation of proteins in two dimensions based on differences in isoelectric point (pI) and size (molecular weight, Mw). The first dimension, isoelectric focusing (IEF), is usually performed using individual DryStrips with immobilized pH gradients (IPG). A number of different pH ranges are available from wide-range strips to more narrow-range drystrips. The second dimension in 2-DE is a traditional sodium dodecyl sulfate-polyacrylamide electrophoresis (SDS-PAGE) separation. A wide variety of protein detection methods are compatible with 2-DE such as Coomassie brilliant blue, silver staining, and fluorescent staining.

Cross-References

- Electrospray Ionization
- Mass Spectrometry
- Matrix-Assisted Laser Desorption Ionization
- Posttranslational Modification
- Proteomics

Typical Antipsychotics

Synonyms

Classical neuroleptics; Conventional antipsychotics; First-generation antipsychotics

Definition

Typical antipsychotics are a class of antipsychotic drugs first developed in the 1950s to treat schizophrenia. Typical antipsychotics may also be used for the treatment of acute ▶ mania, agitation, and other conditions.

Cross-References

- Antipsychotic Drugs
- Bipolar Disorder in Children

U

Ubiquitin-Proteasome System

Synonyms

UPS

Definition

A specialized system in cells for regulating their protein composition. Degraded, misfolded, or aberrant proteins are tagged by ubiquitylation (attachment of ubiquitin to a substrate protein) for removal by digestion at the cytoplasmic 20/26S proteasome enzyme complex.

Ultrasonic

Definition

Refers to sound energies having a frequency above the human hearing range. The highest frequency that the human ear can detect is approximately 20,000 cycles per second (20,000 Hz).

Attachment bonds signaled by ultrasonic vocalizations have been proposed to underlie a variety of social relationships, for example, parent-infant, filial, and pair (male-female) bond formation. These are all typically characterized by preferential proximity seeking and all involve a response to separation. These forms of attachment appear to be common to many species including humans, suggesting that the neural basis can be investigated in animal models. Emerging evidence suggests that the biology of attachment in its many forms may also be similar across species.

Ultrasonic Vocalizations

Paul B. S. Clarke and Jennifer Wright
Department of Pharmacology and Therapeutics, McGill University, Montréal, QC, Canada

Synonyms

Ultrasonic call; Ultrasonic vocalization

Definition

Ultrasonic vocalizations (USVs) occur at sound frequencies above the normal limit of human hearing, i.e., 20 kHz. Rodent ultrasonic vocalizations are generated by the larynx and are abolished by surgical denervation of this organ.

© Springer-Verlag Berlin Heidelberg 2015
I.P. Stolerman, L.H. Price (eds.), *Encyclopedia of Psychopharmacology*,
DOI 10.1007/978-3-642-36172-2

Current Concepts and State of Knowledge

Background and Technical Aspects

Many animals vocalize in order to convey signals related to agonistic behavior, territory, courtship, food, predators, etc. However, only a relatively small number of animal species are known to emit vocalizations composed of pure ultrasound. These include various species of frogs, cetaceans, bats, rodents, and at least one primate species.

The idea that rodents might communicate via ultrasound was first mooted in the nineteenth century, but rat and mouse ultrasounds were not detected until the 1950s. The terms "ultrasonic vocalization" (USV) and "ultrasonic call" are most often used interchangeably; other terms such as "chirp" or "song" are also occasionally used. Rodent USVs have been reported in a wide variety of behavioral settings and have been postulated to serve several functions, including social communication, affective expression, echolocation, and thermoregulation (Brudzynski 2010).

The interpretability of USV data depends critically on how the ultrasounds are recorded. Many pre-2000 studies featured heterodyne and frequency-division bat detectors; these devices greatly distort the acoustic signal, thereby limiting the analysis of the acoustic features of the USVs. Thus, many early USV investigators could only report on the number of calls emitted. Detailed analysis of USVs requires instead a "broadband" detection system using high sampling rates (e.g., 250 kHz), exceeding the Nyquist rate; this approach allows all the critical acoustic information to be retained. When the broadband signal is Fourier transformed, USVs are clearly revealed in the resulting frequency versus time spectrogram. Commercial software is now available which not only generates spectrograms but also reports a wide range of acoustic parameters for each individual call (e.g., frequency, duration, bandwidth).

Which Kinds of USV-Related Parameters Are Useful?

Most calls produced by laboratory rats and mice comprise only a single peak sound frequency at any moment in time. However, this peak frequency generally fluctuates throughout the call, such that individual USVs display varying degrees of pitch modulation. Signal analysis yields numerous acoustic features, including mean intensity, mean peak (i.e., most intense) frequency, frequency range (i.e., bandwidth), duration, and spectrographic shape ("subtype," e.g., upward ramp, inverted U, step, trill). It is not known which acoustic features are most salient to rodents nor whether each call subtype has a particular behavioral function (see below). However, numerous studies have shown that experimental conditions can modulate several USV parameters and the propensity to emit certain USV subtypes. Although some calls may occur in isolation, others are emitted in bouts with a distinct temporal structure that may also potentially carry meaning.

Current psychopharmacological research on USVs does not exploit the vast amount of data that analysis of these calls can generate. For example, many studies report call rate (e.g., calls per minute) as the sole dependent variable. Changes in other acoustic parameters (e.g., peak frequency, duration) are sometimes documented, whereas call intensity is rarely useful since it is typically confounded with distance from, and orientation toward, the microphone. The call "profile" describes the relative proportion of call subtypes emitted in a given situation. Call profiles can be altered by drugs (e.g., amphetamine) and behavioral manipulations (e.g., social contact), independent of changes in the overall call rate. Mouse vocalizations are often extensively subtyped, providing detailed call profile information. In contrast, most studies of adult rat vocalizations continue to rely on a simplified (2- or 3-way) classification such as 22 kHz versus flat 50 kHz versus frequency-modulated (FM) 50 kHz (see below).

Rat Ultrasonic Vocalizations

Rat USV Categories

Laboratory rats emit ultrasonic calls from the first week of life. Three main call types are recognized: pup calls, adult 22-kHz calls, and adult

50-kHz calls. Importantly, a considerable amount of heterogeneity exists within each of these call classes. Rat pup calls can occur between 30 and 65 kHz, but are nevertheless often termed "40-kHz calls." Rat pups emit ultrasonic calls in response to stressful stimuli, notably maternal separation (Litvin et al. 2007) (see "▶ Distress Vocalization"); hypothermia appears to be at least partly responsible. These separation calls readily induce maternal approach. In total, at least nine subtypes of pup calls have been recognized spectrographically.

The adult rat 22-kHz call type appears to express distress or alarm (see "▶ Distress Vocalization"), and not surprisingly, these calls tend to be salient: loud and long (typically 0.3–1 s). Collectively, 22-kHz calls usually fall within a 20–26-kHz frequency range; depending on the behavioral context, individual calls may occur at a single frequency or else occupy a frequency range of up to about 6 kHz. At least ten subtypes of 22-kHz calls have been reported, based on spectrographic analysis; recent evidence indicates a possible subclass of unusually short 22-kHz calls lasting less than 100 ms.

Calls of the so-called "50-kHz" type collectively encompass a wide frequency range (30–90 kHz) and are softer and much shorter (typically 10–100 ms) than the 22-kHz calls described above. Some individual calls are "flat" (i.e., constant frequency), but most are frequency modulated (i.e., FM), often fluctuating over 25 kHz or more. Until recently, only three or four subtypes of 50-kHz calls had been recognized in the literature, e.g., trills, harmonic, and step calls. However, a detailed spectrographic analysis has revealed much greater call diversity, with 14 identified 50-kHz call subtypes. This analysis was performed under restricted conditions (i.e., single vs. pair tested, drug-free vs. after amphetamine), and so it is quite possible that additional call subtypes await identification.

Impact of Drugs on Rat Pup Isolation USVs

Rat pup isolation calls are inhibited by several drug classes. For example, they are suppressed by benzodiazepines, SSRIs, and 5-HT1A agonists and increased by the anxiogenic drugs pentylenetetrazole and FG 7142. Although this initial drug profile appeared predictive of efficacy in generalized anxiety disorder, other drug effects (e.g., alpha- and beta-adrenergic agonists) now appear inconsistent with this interpretation. Furthermore, evidence suggests that isolation calls are not directly triggered by maternal separation per se but rather by the resultant hypothermia; these calls are dramatically reduced when isolated pups are tested in temperatures resembling those of the nest. Even so, not all drugs that inhibit isolation calls increase body temperature.

Impact of Drugs on Adult Rat 22-kHz USVs

Footshock-induced 22-kHz USV calls have been extensively evaluated as a potential screen for anxiolytic drugs. These calls are inhibited by acute systemic administration of benzodiazepine anxiolytics (albeit at near-sedative doses), by SSRIs, and less effectively by SNRIs. 5-HT1A receptor agonists markedly reduce calling associated with footshock and other aversive stimuli; agonists at 5-HT1B and 5-HT2A receptors are also inhibitory. Alpha-adrenergic drugs exert complex effects on 22-kHz calling, whereas beta antagonists are ineffective. Other drug classes have been shown to affect emission of 22-kHz calls by adult rats. Dopamine D2 receptor agonists potently inhibit footshock-induced 22-kHz calls, likely through activation of inhibitory autoreceptors. Surprisingly, however, dopaminergic D1 and D2 antagonists tend not to be inhibitory except at cataleptic doses. Morphine preferentially suppresses 22-kHz calls emitted in anticipation of electric shock; it also inhibits 22-kHz calls emitted by an intruder that is threatened by an aggressive resident male rat. While 22-kHz calls have been used to screen for potential anxiety-modulating drugs, anxiety is also a common feature of withdrawal from psychostimulants, sedatives, and opiates. Rats withdrawn from these drugs emit 22-kHz USVs, particularly when they are provoked with an aversive air puff (Covington and Miczek 2003).

Impact of Drugs on Adult Rat 50-kHz USVs

The notion that 50-kHz calls may indicate positive affect (discussed below) has motivated several studies of the euphorigenic D-amphetamine. This drug robustly increases the 50-kHz call rate after acute systemic or intra-accumbens administration. Cocaine also has a stimulatory effect on the call rate, as shown after systemic administration. Amphetamine and cocaine produce analogous shifts in the 50-kHz call profile, promoting trill calls at the expense of flat calls. These psychostimulant effects on call rate and profile were countered by D1 and D2 dopaminergic antagonists, even in some cases where motor impairment was highly unlikely. Neither the DA transporter blocker GBR 12909 nor the D1-like or D2-like receptor-selective agonists (given alone or in combination) promote calling; these negative findings suggest that tonic stimulation of DA receptors may not suffice to evoke these calls. In contrast, tests with antagonists indicate that dopaminergic signaling via D1, D2, and D3 DA receptor subtypes may be required for 50-kHz call emission.

Noradrenergic drugs also have a profound effect on amphetamine-induced and spontaneous 50-kHz calls. Both the alpha1 receptor antagonist prazosin and the alpha2 agonist clonidine exerted profound depressant effects on the rate of calling, even at doses unlikely to produce appreciable sedation. The suppressant effect of clonidine is thought to have occurred through autoreceptor-mediated inhibition of (nor)adrenergic transmission. The beta receptor antagonist propranolol, in contrast, left the call rate unchanged but normalized the amphetamine call *profile*; subsequent drug tests tended to implicate simultaneous blockade of beta1 and beta2 receptors in the CNS. While inhibition of (nor)adrenergic transmission can profoundly affect 50-kHz calling, the converse does not hold: administration of a noradrenaline transporter blocker (nisoxetine), either alone or with GBR 12909, failed to stimulate calling.

Morphine tends to inhibit 50-kHz USVs after acute systemic administration in drug-naïve rats. Such an effect has been observed using conventional testing cages and also during resident-intruder interactions. In unstressed rats, morphine has been shown to suppress calling even at a low dose (1 mg/kg) which stimulated locomotor activity. The acute depressant effect of morphine makes it difficult to detect potential effects on the call profile, although in one study morphine selectively increased the emission of "downward ramp" calls in the resident-intruder test. With repeated daily administration, the call-suppressant effect of morphine tends to wane and may possibly be replaced by a stimulant effect over an extended treatment period, although this remains to be confirmed. Antagonism of mu opioid receptors (by naloxone or naltrexone) results in little or no change in spontaneous 50-kHz calling, according to several studies. However, recent evidence indicates that calls which are made in anticipation of food delivery may be markedly suppressed.

Several other drugs have been evaluated for their acute, unconditioned effects on 50-kHz call rate, notably methylphenidate, MDMA, nicotine, 5-HT1A partial agonists, diazepam, and ethanol. Effects of these drugs were typically either undetectable, small, or variable across studies. Effects of repeated drug administration have been little investigated to date.

Interindividual and Genetic Differences in Rat 50-kHz Calling

A striking feature of the 50-kHz call literature is the existence of large and stable interindividual differences among adult rats (Wohr and Schwarting 2010). Amphetamine- and tickling-induced callings represent two documented examples. Treated with amphetamine, rats differ consistently not only in terms of call rate but also call profile and in the acoustic properties of individual call subtypes; an initial report suggests that rats that call more after amphetamine administration tend to form stronger conditioned place preferences to this drug. Individual differences in tickling-evoked 50-kHz calling have been correlated, either positively or negatively, with several neurochemical and behavioral measures including adult hippocampal neurogenesis, exploration, and sucrose preference. Interindividual differences in tickling-evoked 50-kHz calling have formed the basis for selective breeding. It has been suggested that the high-calling line may be more stress resilient, whereas the low-calling line may serve as a rat model of social deficits seen in autism.

Mouse Ultrasonic Vocalizations

Mouse USV Types and Subtypes

Mice, like rats, emit USVs from an early age. Pups emit several kinds of calls in response to diverse stimuli (Ehret 2005), including isolation calls at around 65 and 94 kHz (see "▶ Distress Vocalization"). Adult mouse calls are spectrally diverse, with peak frequencies typically between 60 and 100 kHz (Lahvis et al. 2011); multiple call classification schemes are currently in use, each with as many as ten or so distinguishable categories. Several adult mouse calls resemble particular 50-kHz call subtypes seen in adult rats (e.g., flat, short, upward and downward ramp). However, mice notably lack an analogue of the rat's 22-kHz alarm call and 50-kHz trill call. As in rats, the relative proportion of different call subtypes varies according to the behavioral context (Lahvis et al. 2011).

Mouse USVs and Genetic Manipulations

Numerous gene modifications interfere with USV emission in mice. Here, pup isolation calls have been the focus, with impairments reported after knockout of gene encoding: receptors (e.g., dopamine D2, oxytocin, vasopressin 1b), intracellular signaling molecules (e.g., CD38, EPAC2), and cell adhesion and scaffolding proteins (e.g., neuroligins, cadherin-6, SHANK3). Some examples of impaired ultrasonic calling stem from mutations with direct relevance to human neurodevelopmental disorders including autism (multiple genes), fragile X syndrome (fragile X mental retardation 1), and impaired language development (FoxP2).

Impact of Drugs on Mouse USVs

A wide range of drugs and other chemicals has been tested for their effects on mouse vocalizations. Alterations in pup isolation calls have been shown after gestational exposure to toxins (e.g., aluminum, trihalomethanes, organophosphate insecticide, cocaine). Pup calls were also affected by early postnatal administration of various growth factors and drugs. The latter included those targeting transmission via GABAergic, glutamatergic, 5-HT, muscarinic cholinergic, dopaminergic, or tachykinin receptors. Ultrasonic calling is altered in adult mice subjected to stressors, by antidepressant and anxiolytic drugs and by analgesic treatment after painful formalin injection. In addition, administration of sex steroids to adult gonadectomized mice has demonstrated an activational effect on courtship vocalizations.

Behavioral and Emotional Significance of Rodent Ultrasonic Vocalizations

Since their discovery, ultrasonic vocalizations have been proposed to subserve numerous functions (Wohr and Schwarting 2010). These include communicating hypothermia, arousal state, positive affect, submission, appeasement, frustration, reward anticipation, and affiliative intent, as well as coordinating circadian rhythms within the colony. Studies employing devocalization, deafening, or ultrasound playback have shown that rodent ultrasonic calls can modulate, and in some cases initiate, a range of affiliative, sexual, agonistic, and maternal behavior. At least two types of calls serve a defined communicative function: pup isolation calls induce retrieval behavior in the mother, whereas adult rat 22-kHz calls can serve as alarm calls warning conspecifics of potential danger (e.g., a predator). The finding that 22-kHz USVs occur in other contexts as well (see below) suggests that a given call may serve multiple purposes; alternatively, there may be sufficient acoustic heterogeneity between calls to ensure functional specificity. Adult rats and mice possess a rich repertoire of ultrasonic calls, and to better understand the potential meaning of individual calls will require careful consideration of rigorously defined call categories. Detailed analysis of this sort has shown, for example, that ultrasonic call profiles vary with the behavioral context (Wohr and Schwarting 2010; Lahvis et al. 2011).

Adult Rat 22-kHz USVs as an Indicator of Distress or Alarm

Adult rat 22-kHz USVs frequently occur in response to aversive or startling events (Schwarting and Wohr 2012). Examples include the presence of a feline predator, inescapable footshock, acoustic or physical startle stimuli, social defeat, and subsequent threat in the resident-intruder test, drug withdrawal, and unexpected termination of rewarding brain stimulation. Stimuli or contexts that predict aversive electroshock also typically promote 22-kHz USVs, showing that anxious anticipation itself can trigger calling. However, adult rat 22-kHz USVs do not always coincide with anxiety, fear, or other dysphoric states. For example, 22-kHz calls have also been reported in response to a safety cue that predicts a respite from electric shock; a possibly related finding is that predator-induced 22-kHz calls occur more readily when adult rats have access to a place of relative safety. In addition, 22-kHz calls are not consistently seen in pain states, and they are emitted by male rats after ejaculation. These post-ejaculation calls serve to alleviate postcoital brain hyperthermia, while also signaling that the animal is sexually refractive. These calls also appear to differ spectrally from calls emitted after social defeat or electrical tail shock.

Adult Rat 50-kHz USVs as an Indicator of Positive Affect

Whereas 22-kHz calls express distress or alarm, it is widely held that adult rat 50-kHz calls express positive affect (i.e., hedonia). For example, rat 50-kHz calls have been proposed as an antecedent of playful laughter in children (Panksepp and Burgdorf 2003). This general idea has become so influential in the field that it is now quite commonly stated, without qualification, that adult rat 50-kHz USVs represent positive affect. Proponents of this view argue for a strong association between 50-kHz USVs and reward-related events. For example, 50-kHz calls have been detected in a wide variety of natural appetitive contexts such as juvenile play, social reunion, and sexual behavior. In addition, a number of discrete appetitive stimuli also elicit 50-kHz calls; such stimuli include tickling by a human experimenter, systemic administration of amphetamine or cocaine, and drug-paired sensory cues. Rats are also reported to emit 50-kHz calls in *anticipation* of rewards, notably access to food, play, tickling, a sexually receptive partner, or cocaine.

If 50-kHz USVs reflect positive affect, they should be suppressed by aversive stimuli. Such an effect was initially reported, albeit based on the less reliable heterodyne recording method (see above). However, 50-kHz calls can also occur in presumed aversive contexts: upon brief isolation from a cage mate, during morphine withdrawal, or during CO_2 exposure. In addition, 50-kHz calls are emitted by intruder rats encountering an aggressive resident rat, and these calls actually escalated across consecutive "threat" episodes following repeated attacks (Tornatzky and Miczek 1994). In more general terms, the effects of aversive stimuli have frequently been studied using recording equipment incapable of detecting 50-kHz calls; in some other cases, investigators have deliberately restricted their analyses to 22-kHz calls.

Recently, the USV-positive affect hypothesis has been refined so that only *frequency-modulated* (i.e., FM) 50-kHz calls are proposed to convey hedonia. This revised proposal is based mainly on two sets of observations. First, FM calls were much more prevalent than flat calls in situations where hedonia might be expected: before ejaculation, during rough-and-tumble play, during tickling, and after administration of amphetamine. Second, FM but not flat calls were reduced by dopamine-depleting lesions of the forebrain. This revised hypothesis is also problematic, however. First, it is controversial whether hedonia itself is dopamine dependent, either in animals or humans. Second, morphine produces positive subjective effects in human subjects (even when opioid naïve), yet it tends to inhibit 50-kHz calling in adult rats; this depressant effect has been seen even at doses that were demonstrably nonsedative and rewarding. Importantly, morphine failed to promote even the trill call subtype, which is preferentially elevated by the euphoriants amphetamine and cocaine.

Conclusion

Animals cannot directly tell us about their emotional experience, a reality that once led to the Radical Behaviorism of B. F. Skinner. Today, however, there is growing interest in "affective neuroscience," wherein some degree of anthropomorphism is considered desirable; this growing scientific subdiscipline is underpinned by clear evolutionary links between emotional expression in humans and other mammals. For example, facial expressions are now being used to gauge the emotional component of pain in mice, and "hedonia" has been operationalized in terms of facial reactions to sweet tastes in rats. USVs appear to offer a potentially rich window into emotion and may ultimately provide a more informative and versatile output. It is widely thought that CNS drug discovery should strive for greater interspecies consilience by adopting animal tests that are more "humanlike." Rodent ultrasonic vocalizations appear to offer such a possibility.

Cross-References

▶ Aggression
▶ Amphetamine
▶ Animal Models for Psychiatric States
▶ Anxiety: Animal Models
▶ Anxiogenic
▶ Anxiolytics
▶ Autism: Animal Models
▶ Cocaine
▶ Distress Vocalization
▶ Dopamine
▶ Emotion and Mood
▶ Fear
▶ Knockout/Knockin
▶ Morphine
▶ Nucleus Accumbens
▶ Psychostimulants
▶ Selective Breeding
▶ Spectrograms
▶ Withdrawal Syndromes

References

Brudzynski SM (2010) Handbook of mammalian vocalization: an integrative neuroscience approach. Academic, London

Covington HE III, Miczek KA (2003) Vocalizations during withdrawal from opiates and cocaine: possible expressions of affective distress. Eur J Pharmacol 467:1–13

Ehret G (2005) Infant rodent ultrasounds – a gate to the understanding of sound communication. Behav Genet 35:19–29

Lahvis GP, Alleva E, Scattoni ML (2011) Translating mouse vocalizations: prosody and frequency modulation. Genes Brain Behav 10:4–16

Litvin Y, Blanchard DC, Blanchard RJ (2007) Rat 22 kHz ultrasonic vocalizations as alarm cries. Behav Brain Res 182:166–172

Panksepp J, Burgdorf J (2003) "Laughing" rats and the evolutionary antecedents of human joy? Physiol Behav 79:533–547

Schwarting RK, Wohr M (2012) On the relationships between ultrasonic calling and anxiety-related behavior in rats. Braz J Med Biol Res 45:337–348

Tornatzky W, Miczek KA (1994) Behavioral and autonomic responses to intermittent social stress: differential protection by clonidine and metoprolol. Psychopharmacology (Berl) 116:346–356

Wohr M, Schwarting RKW (2010) Rodent ultrasonic communication and its relevance for models of neuropsychiatric disorders. e-Neuroforum 4:71–80

Unblocking

Synonyms

Posttrial surprise

Definition

The liberation of attention for new learning that is seen when an unconditioned stimulus is either more or less than predicted. This is an effect seen in classical (Pavlovian) conditioning and a constraint on the general importance of temporal coincidence as a determinant of new learning.

Cross-References

▶ Classical (Pavlovian) Conditioning

Unconditioned Response

Synonyms

UCR; UR

Definition

A response that occurs to a stimulus without previous experience (training), e.g., flinching in response to electrical shock.

Unconditioned Stimulus

Synonyms

US

Definition

An unconditioned stimulus, for example, electrical shock, elicits a characteristic behavioral response without previous learning (i.e., an unconditioned response).

Unit Price

Definition

The unit price is the cost per benefit unit of a commodity.

Cross-References

▶ Behavioral Economics

Uptake

Definition

In neuropharmacology uptake and reuptake refer to the reabsorption of a neurotransmitter by a transport mechanism located on a presynaptic neuron or another cell, after it has performed its function of transmitting the signal. Reuptake is necessary for normal functioning because it regulates how long a signal lasts, allowing the recycling of neurotransmitters and regulating its normal levels in the releasing cell. The uptake process can be experimentally assessed as the incorporation of radiolabeled neurotransmitter by the target cell.

Urge

Synonyms

Craving

Definition

Urge is an involuntary impulse promoting engagement in a given activity.

Cross-References

▶ Impulse Control Disorders

Utilitarianism

Definition

The philosophical doctrine that an action is right if it promotes happiness and the greatest happiness of the greatest number should be its guiding principle.

V

Vaccination

Synonyms

Active immunization

Definition

In psychopharmacology vaccination refers to the administration of an immunogen (e.g., vaccine) to the subjects being studied in order to stimulate the immune system to produce drug-specific antibodies. Primary advantages of vaccination are that it requires relatively few administrations to achieve high serum antibody concentrations that can persist for several months and it is relatively inexpensive. The main disadvantages are the delay (1–2 months) to reaching effective serum antibody concentrations and the inability to precisely control those concentrations. Marked variability in antibody concentrations is observed between subjects, with some not achieving required levels.

Vaccines and Drug-Specific Antibodies

Mark G. LeSage[1] and Paul R. Pentel[2]
[1]Department of Medicine, Minneapolis Medical Research Foundation and University of Minnesota, Minneapolis, MN, USA
[2]Department of Medicine, Hennepin County Medical Center, Minneapolis, MN, USA

Synonyms

Immunotherapy

Definition

In the context of psychopharmacology, vaccines consist of a target drug chemically linked or otherwise combined with additional components required to initiate an immune response. Addiction vaccines stimulate production of drug-specific antibodies that can bind the target drug and alter its pharmacokinetics.

Pharmacological Properties

Background

Vaccines and drug-specific antibodies are being developed for the treatment of drug abuse. Most pharmacotherapies for drug abuse target the neuropharmacological processes that mediate a drug's dependence-related behavioral effects by binding to associated receptors in the CNS (▶ Nicotine Dependence and Its Treatment, ▶ Opioid Use Disorder and Its Treatment). In contrast, vaccines target the drug itself rather than the brain. Vaccines elicit the production of drug-specific antibodies that bind drug in blood and reduce its distribution into the brain. By acting outside the brain, a primary advantage of vaccines is that they lack the CNS side effects associated with CNS receptor-based pharmacotherapies that can alter normal neural function.

Immunologic interventions such as ▶ Vaccination or ▶ Passive Immunization were first suggested as a pharmacological approach to treating drug abuse 40 years ago, when it was shown that immunization against heroin could reduce the self-administration of the drug in monkeys (Bonese et al. 1974). Since then, vaccines and drug-specific antibodies that can modify the ▶ Pharmacokinetics and behavioral effects of a wide range of drugs of abuse in animals have been developed, and vaccines against cocaine and nicotine have entered clinical trials (Kosten and Owens 2005; Fahim et al 2013).

Mechanisms of Action

Drugs of abuse act as reinforcers, and the pharmacokinetic properties of drugs are key determinants of their reinforcing effects (▶ Operant Behavior in Animals, ▶ Pharmacokinetics). The concentration of drug in the brain and the rate at which drug enters the brain are directly related to the strength of its subjective and reinforcing effects. In addition, the rate of elimination of drug is inversely related to the duration of its subjective and reinforcing effects. Vaccines and drug-specific antibodies target these key pharmacokinetic processes.

The antibodies produced by vaccination against a given drug contain binding sites for that drug. In a vaccinated subject, these drug-specific antibodies are present in blood and extracellular fluid but excluded from the brain because they are too large to cross the blood-brain barrier. When a vaccinated subject receives the drug, a substantial fraction of the drug is bound to antibody, sequestered in blood, and prevented from entering the brain. The binding of drug in serum also attenuates the normally rapid rise in brain concentrations of drug. In addition, the binding of drug by antibody may make it less available for elimination and prolong its elimination half-life (LeSage et al. 2008). Slower elimination may be a beneficial effect in some circumstances, as it has been associated with reduced cigarette smoking and enhanced smoking cessation rates.

The efficacy of a vaccine depends upon three key variables: the serum concentration, affinity, and specificity of the elicited antibodies. Because higher ratios of antibody to drug result in greater binding of drug in serum, vaccines must elicit high serum concentrations of antibody to be maximally effective. Higher affinity (strength of binding to drug) increases the bound fraction of drug. Specificity refers to the extent to which the antibodies bind the drug in preference to other compounds. Greater specificity reduces competition from other compounds (e.g., metabolites, endogenous compounds) for antibody binding sites and reduces the likelihood of adverse side effects. However, in cases where the metabolites of a drug contribute to its behavioral effects (e.g., methamphetamine, heroin), antibodies that cross-react with those metabolites would be advantageous.

Although vaccines and receptor antagonists can produce similar attenuation of a drug's behavioral effects, they should not be considered analogous. Receptor antagonists block the binding of endogenous compounds (e.g., dopamine, acetylcholine) to receptors, while antibodies do not. In addition, drug-specific antibodies may slow distribution to brain or increase the elimination half-life of some drugs, while receptor antagonists do not.

Formulation

Because drugs are typically too small to elicit an immune response, they are rendered immunogenic by covalently linking (conjugating) the drug to a carrier protein to form the complete immunogen. Carrier proteins are foreign (nonhuman) proteins but virus-like particles and disrupted adenovirus have also been used (Fahim et al. 2013; Maoz et al. 2013). Such conjugate vaccines are typically combined with an adjuvant such as alum to enhance the immune response. Alternatively, drug and other vaccine components can be linked to synthetic structures or nanoparticles such as liposomes. All human trials to date have used conjugate vaccines.

Administration

Initial vaccination schedules in both animals and humans typically involve two to six injections at 2–4 week intervals. Periodic booster doses are then needed to maintain satisfactory antibody levels since exposure to the drug by itself, which is not the complete immunogen, does not elicit a booster response (Hatsukami et al. 2005; Fahim et al. 2013).

Animal Models

Pharmacokinetics: The effects of immunologic interventions on drug pharmacokinetics are generally similar across different drugs (Kosten and Owens 2005; LeSage et al. 2008). In rodents, vaccines or drug-specific antibodies produce increases in serum drug concentrations due to the binding and retention of drug in serum and reduce brain concentrations of cocaine, phencyclidine, methamphetamine, morphine, and nicotine following an acute administration of clinically relevant doses. Importantly, some studies have shown that the early distribution of an acute dose to the brain continues to be reduced or slowed during chronic drug administration, even when total drug exposure exceeds the binding capacity of the antibodies (Kosten and Owens 2005; LeSage et al. 2008). This finding suggests that the initial subjective and reinforcing effects of a drug could still be suppressed in the context of ongoing drug use. Finally, immunization can markedly prolong the elimination half-life of some drugs because antibody binding restricts elimination to the unbound fraction of drug. For example, immunization can produce a two- to sixfold increase in the half-life of nicotine and morphine in rats and rabbits, respectively (LeSage et al. 2008).

Behavioral effects: Animal studies have shown that vaccines and drug-specific antibodies can significantly reduce the behavioral effects of the aforementioned drugs in a wide range of behavioral assays, including seizure induction, acute locomotor activation, development of locomotor sensitization, responding under simple operant schedules of food delivery, and drug discrimination. Active or passive immunization against cocaine, methamphetamine, nicotine, and heroin has also been shown to attenuate the acquisition, maintenance, and reinstatement of self-administration of these drugs, which are key animal models of drug use and relapse in humans (▶ Reinstatement of Drug Self-administration, ▶ Self-administration of Drugs, ▶ Addictive Disorder: Animal Models, ▶ Operant Behavior in Animals) (Kosten and Owens 2005; LeSage et al. 2008). Immunization can also decrease the ability of drug to relieve withdrawal symptoms that occur upon the cessation of drug exposure, yet it has not been found to elicit withdrawal signs when initiated during chronic drug administration (▶ Withdrawal Syndromes) (LeSage et al. 2008; Lindblom et al. 2005).

Clinical Trials

Immunogenicity and safety: Clinical trials have been conducted with four nicotine conjugate vaccines (NicVax, NicQb, TA-NIC, and Niccine) and one cocaine conjugate vaccine (TA-CD) (Fahim et al. 2013) (▶ Randomized Controlled Trials). Phase I/II trials have shown that these vaccines elicit drug-specific antibodies in a dose-dependent fashion, with marked variability between participants. All vaccines have been well tolerated with no serious adverse events. Antibody levels decrease slowly after the last vaccine injection of the initial immunization period (e.g., 50 % decline over 6–8 weeks) but increase again when a booster dose is administered.

Pharmacokinetics: Proof of concept that immunization with a nicotine vaccine increases plasma nicotine concentrations and reduces nicotine concentrations in brain (using nicotinic receptor binding as a surrogate measure) has been demonstrated in humans (Esterlis et al. 2013). In addition, immunization significantly reduces nicotine clearance in humans, as it does in rats.

Efficacy: Although not designed to rigorously assess the efficacy of vaccination, the phase I/II trials mentioned earlier included drug use as either a primary or secondary end point to provide the preliminary assessment of efficacy. A phase II study of NicVax showed a higher smoking abstinence rate in the highest vaccine dose group, and both this and a phase II trial of NicQb showed higher abstinence rates in the one-third of subjects with the highest serum antibody levels (Fahim et al. 2013). Although reported only briefly in a review article, two phase III trials of NicVax vaccine did not confirm overall efficacy (defined as greater continuous abstinence from month 8 to month 12 after the target quit date (Fahim et al. 2013). Effects of NicVax on secondary end points in these trials (e.g., point-prevalence abstinence, cigarette consumption, smoking satisfaction, level of dependence) have not yet been reported. Importantly, neither withdrawal symptoms nor compensatory increases in drug use to surmount the effect of vaccination have been reported in any trial (Hatsukami et al. 2005).

Conclusion

Taken together, both preclinical and clinical studies suggest that vaccines could have utility in the treatment of drug addiction. However, the lack of control over antibody levels and large variability between subjects are the primary limitation of all current vaccines. Because achieving the highest antibody levels possible will be essential to maximizing the efficacy of vaccination, methods of boosting immunogenicity need to be developed in order to address this issue. Toward this end, vaccine formulations that contain novel haptens, different adjuvants and/or carrier proteins, multivalent vaccines (simultaneous administration of two or more distinct drug-protein conjugates), passive immunization with drug-specific monoclonal antibodies, and viral vector mediated transfer of antibody producing genes are being explored (Keyler et al. 2008; Kosten and Owens 2005, Hicks et al. 2012). Although not explicitly tested, vaccines would not be expected to directly reduce drug craving or withdrawal, since these occur when drug levels are already low or absent. Combining vaccination with medications that address craving and withdrawal might therefore enhance their efficacy.

Cross-References

▶ Addictive Disorder: Animal Models
▶ Drug Discrimination
▶ Nicotine Dependence and Its Treatment
▶ Operant Behavior in Animals
▶ Opioid Use Disorder and Its Treatment
▶ Passive Immunization
▶ Pharmacokinetics
▶ Randomized Controlled Trials
▶ Reinstatement of Drug Self-administration
▶ Self-administration of Drugs
▶ Vaccination
▶ Withdrawal Syndromes

References

Bonese KF, Wainer BH, Fitch FW, Rothberg RM, Schuster CR (1974) Changes in heroin self-administration by a rhesus monkey after morphine immunization. Nature 252:708–710

Esterlis I, Hannestad JO, Perkins E, Bois F, D'Souza DC, Tyndale RF, Seibyl JP, Hatsukami DM, Cosgrove KP, O'Malley SS (2013) Effect of a nicotine vaccine on nicotine binding to β2*-nicotinic acetylcholine receptors in vivo in human tobacco smokers. Am J Psychiatry 170:399–407

Fahim RE, Kessler PD, Kalnik MW (2013) Therapeutic vaccines against tobacco addiction. Expert Rev Vaccines 12:333–342

Hatsukami DK, Rennard S, Jorenby D, Fiore M, Koopmeiners J, de Vos A, Horwith G, Pentel PR (2005) Safety and immunogenicity of a nicotine conjugate vaccine in current smokers. Clin Pharmacol Ther 78:456–467

Hicks MJ, Rosenberg JB, De BP, Pagovich OE, Young CN, Qiu J, Kaminsky SM, Hackett NR, Worgall S, Janda KD, Davisson RL, Crystal RG (2012) AAV-directed persistent expression of a gene encoding anti-nicotine antibody for smoking cessation. Sci Transl Med 140:1–7

Keyler DE, Roiko SA, Earley CA, Murtaugh MP, Pentel PR (2008) Enhanced immunogenicity of a bivalent nicotine vaccine. Int Immunopharmacol 8:1589–1594

Kosten T, Owens SM (2005) Immunotherapy for the treatment of drug abuse. Pharmacol Ther 108:76–85

LeSage MG, Keyler DE, Pentel PR (2008) Current status of immunologic approaches to treating tobacco dependence: vaccines and nicotine-specific antibodies. In: Rapaka RS, Sadee W (eds) Drug addiction: from basic research to therapy. Springer, New York, pp 455–475

Lindblom N, de Villiers SH, Semenova S, Kalayanov G, Gordon S, Schilstrom B, Johansson AM, Markou A, Svensson TH (2005) Active immunization against nicotine blocks the reward facilitating effects of nicotine and partially prevents nicotine withdrawal in the rat as measured by dopamine output in the nucleus accumbens, brain reward thresholds and somatic signs. Naunyn Schmiedebergs Arch Pharmacol 372:182–194

Maoz A, Hicks MJ, Vallabhjosula S, Synan M, Kothari PJ, Dyke JP, Ballon DJ, Kaminsky SM, De BP, Rosenberg JB, Martinez D, Koob GF, Janda KD, Crystal RG (2013) Adenovirus capsid-based anti-cocaine vaccine prevents cocaine from binding to the nonhuman primate CNS dopamine transporter. Neuropsychopharm 38:2170–2178

Vagus Nerve Stimulation

Synonyms

VNS

Definition

A technique for activating brain neurons through the stimulation of afferent fibers of the left vagus nerve by electrical impulses from a small generator implanted in the upper chest. Introduced in the 1990s to treat epilepsy, VNS is clinically available and may be beneficial in severe depression that is resistant to pharmacotherapy. In addition to the risks of implantation surgery, the most common side effects are vocal changes and hoarseness.

Valproic Acid

Synonyms

Convulex; Depakene; Depakine; Epival; Stavzor; Valproate

Definition

A simple branched-chain carboxylic acid that possesses antiseizure and mood-stabilizing properties. It is rapidly absorbed after oral administration and exhibits 90 % binding to plasma proteins, hepatic metabolism, and a half-life of approximately 15 h.

Valproic acid is used for the treatment of convulsions, migraines, and acute manic or mixed episodes associated with psychiatric disorders. Valproate exerts its effects by increasing the concentration of gamma-aminobutyric acid (GABA) through the inhibition of enzymes catabolizing GABA: transferase and succinic aldehyde dehydrogenase. Valproic acid is also a histone deacetylase inhibitor, a property that might be responsible for its teratogenic effects. The side effects due to valproic acid include liver injury, pancreatitis, abnormal bleeding, birth defects, drowsiness, dizziness, nausea, vomiting, indigestion, diarrhea, weight loss, and tremors. If taken during pregnancy, there is a risk of harm to the offspring.

Cross-References

▶ Anticonvulsants
▶ Autism: Animal Models
▶ Mood Stabilizers

Values-Based Medicine

Definition

The theory and practice of effective healthcare decision-making for situations in which legitimately different, and hence potentially conflicting, value perspectives are in play.

Varenicline

Definition

Varenicline is a nicotinic ▶ partial agonist developed specifically as a smoking cessation agent.

As a partial agonist, it reduces ▶ cravings for and decreases the pleasurable effects of cigarettes and other tobacco products and also attenuates effects of nicotine obtained by smoking. Through these mechanisms, it can assist some patients to quit smoking, where its efficacy appears greater than that for nicotine replacement therapies. Its primary action as a ▶ partial agonist is on the α4β2 subtype of the nicotinic acetylcholine receptor. In addition, it acts on α3β4 and weakly on α3β2 and α6-containing receptors. A full agonist effect has also been shown on α7-receptors. Nausea occurs commonly in people taking varenicline. Other less-common side effects include headache, difficulty in sleeping, and abnormal dreams. Rare side effects reported by people taking varenicline compared to placebo include change in taste, vomiting, abdominal pain, flatulence, and constipation. There has been concern about more serious neuropsychiatric symptoms such as suicidal ideation.

Cross-References

▶ Nicotine
▶ Nicotine Dependence and Its Treatment
▶ Nicotinic Agonists and Antagonists

Vascular Dementia

Synonyms

Multi-infarct dementia

Definition

Vascular dementia is the second most common form of dementia after Alzheimer's disease (AD) in older adults. The term refers to a group of syndromes caused by different mechanisms all resulting in vascular lesions in the brain. Early detection and accurate diagnosis are important, as vascular dementia is at least partially preventable.

Venlafaxine

Synonyms

Venlafaxine hydrochloride

Definition

Venlafaxine was the first SNRI developed and was marketed as Effexor in the USA, the European Union, and elsewhere in the 1990s. Because it has been available longer than other members in its class and in a greater number of countries, more clinical data have accumulated to describe its efficacy, tolerability, and safety. In the USA, it is approved for major depressive disorder in adults as well as for the treatment of certain anxiety disorders including generalized anxiety disorder, social anxiety disorder, and panic disorder. Venlafaxine is available in two different formulations: an immediate release tablet and an extended release capsule.

Ventral Tegmental Area

Synonyms

VTA

Definition

The ventral region of the midbrain, where ▶ dopamine (DA) cell bodies that project to limbic structures and cortical areas are located.

Ventromedial Prefrontal Cortex

Definition

Ventromedial prefrontal cortex is a part of the ventral part of the brain that plays a role in decision making and the processing of fear and risk taking.

Verbal and Nonverbal Learning in Humans

Cross-References

▶ Impulse Control Disorders

Verbal and Nonverbal Learning in Humans

Joachim Liepert
Department of Neurorehabilitation, Kliniken Schmieder, Allensbach, Germany

Synonyms

Associative learning; Motor learning; Novel word learning

Definition

Learning is the acquisition of a new ability. The most relevant areas are knowledge, behavior, and skills. In this entry, nonverbal learning mainly focuses on motor functions and how this learning can be influenced by psychoactive drugs in health and disease. Verbal learning explicitly focuses on the use of language and how our ability to learn and recall new words is modified by psychoactive drugs.

Impact of Psychoactive Drugs

One major aspect of learning is the ability to encode and recall new material or a new function. A neurobiological correlate of learning and memory is long-term potentiation (LTP). In LTP, synaptic plasticity is evoked by repeated and synchronized firing of pre- and postsynaptic neurons. Thus, all substances that induce or enhance LTP might support or augment learning capacities. Several lines of evidence suggest that glutamate metabolism and in particular *N*-methyl-D-aspartate (NMDA) agonists facilitate LTP induction.

Animal experiments have also indicated other neurotransmitters that are presumably involved in learning. For example, in rats, the acquisition of motor skills is improved by noradrenergic substances. Lesions of dopaminergic neurons can produce cognitive deficits and impair attentional processes which are relevant for learning abilities.

In some learning paradigms, e.g., the serial reaction time task, it is possible to follow the process of learning and to detect the development of implicit and explicit knowledge. In most paradigms, however, it is rather tested whether learning had occurred and whether the application of a psychoactive drug had been able to speed up the learning process or the amount of what had been learned.

Nonverbal Learning in Humans

Dopaminergic Drugs

Healthy Subjects The application of a single dose of levodopa (100 mg) was able to improve a training-induced motor memory. The task consisted of practicing of thumb movements in a different direction than thumb movements evoked by transcranial magnetic stimulation (TMS). After the training period, TMS induced thumb movements in the new, practiced direction. In young subjects, the encoding process of this movement was accelerated by levodopa. Elderly subjects performed worse in this task but improved their motor memory encoding by the intake of levodopa to levels present in younger subjects (Flöel et al. 2005a; Figs. 1 and 2). Presumably, older subjects have a subclinical dopaminergic deficit per se. The application of 300 mg levodopa induced a small, but significant improvement of motor functions even without training. This effect was only seen in elderly, but not in young subjects. In a study that used a tactile coactivation protocol to induce non-associative learning in the somatosensory system, tactile two-point discrimination was only improved after placebo, but not after levodopa application, suggesting that a potential beneficial effect of levodopa in the motor system is not necessarily generalized to other systems.

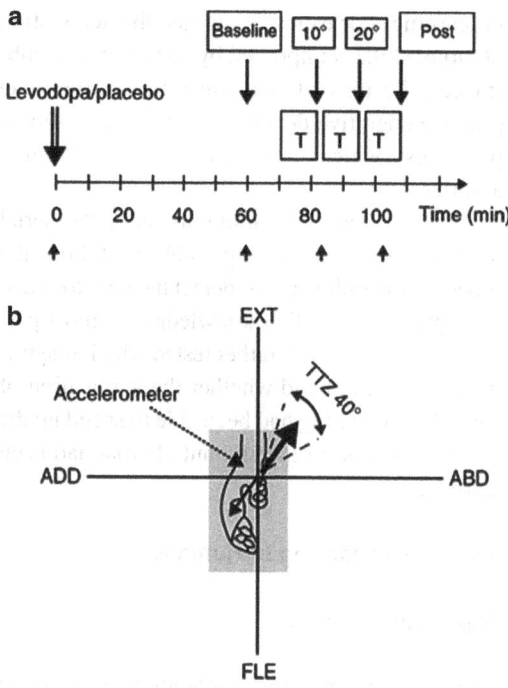

Verbal and Nonverbal Learning in Humans, Fig. 1 (a) Experimental design. L-Dopa/placebo was administered at time 0 in each session, followed by determination of transcranial magnetic stimulation (*TMS*)-evoked thumb movement directions at baseline (60 min after drug intake) and after 10, 20, and 30 (post) min of training (*downward arrows*). Training consisted of three blocks of brisk thumb movements performed at 1 Hz in the direction opposite to the baseline TMS-evoked thumb movement direction (*T*). Fatigue, attention towards the training task, blood pressure, and heart rate were assessed four times during the experiment (*upward arrows*). (b) Diagram showing measurement of thumb movements with an accelerometer positioned on the distal interphalangeal joint (rectangle on the thumb). Baseline TMS-evoked thumb movements in this example fell in a flexion-adduction direction (*thin solid arrow*). Training voluntary thumb motions were performed in the opposite direction (extension-abduction, *thick solid arrow*). At the end of the training period, we measured the percentage of TMS-evoked thumb movements falling in the training target zone (*TTZ*), the end point measure of the study (From Flöel et al. 2005a)

Patient Groups In stroke patients, several studies using multiple or single doses of levodopa have been published. Results are inconsistent. In one study with repeated intake of 100 mg levodopa per day for 3 weeks, patients improved in their motor functions significantly stronger than the placebo group. This higher level of performance persisted even after termination of the drug intake period (Scheidtmann et al. 2001). However, a more recent placebo-controlled study with application of 100 mg levodopa per day for 2 weeks did not demonstrate a superiority of levodopa. Single doses of levodopa were effective in the encoding of a motor memory and in improving procedural motor learning (serial reaction time task) (Flöel et al. 2005b) but did not modulate more clinical aspects of motor functions (dexterity and strength) (Liepert 2008).

Undoubtedly, levodopa reduces motor symptoms in patients with Parkinson's disease (PD). Astonishingly, its effects on cognitive functions and learning abilities are questionable and have even been shown to be detrimental. For example, PD patients treated with levodopa deteriorated in visual memory functions and motor sequence learning. It was suggested that levodopa effects depend on task demands and basal dopamine levels in distinct parts of the striatum.

Methylphenidate produces an increase in dopamine signaling through multiple actions. A trial with 21 subacute stroke patients indicated that the combination of methylphenidate with physical therapy over a period of 3 weeks improved motor functions and decreased depression. In patients with traumatic brain injury, the drug improved the speed of mental processing and had some effect on tests of attention and motor performance.

Noradrenergic Drugs

Healthy Subjects The noradrenaline reuptake inhibitor reboxetine improved motor skill acquisition in a velocity-depending motor task (Plewnia et al. 2004).

D-Amphetamine modulates not only noradrenergic but also dopaminergic and serotonergic neurotransmission. In a paradigm with training of voluntary thumb movements, D-amphetamine facilitated use-dependent plasticity. However, in another study, the administration of D-amphetamine did not improve sensory functions in a tactile frequency discrimination training. There is some evidence that D-amphetamine exerts its effects in cognitive rather than motor networks.

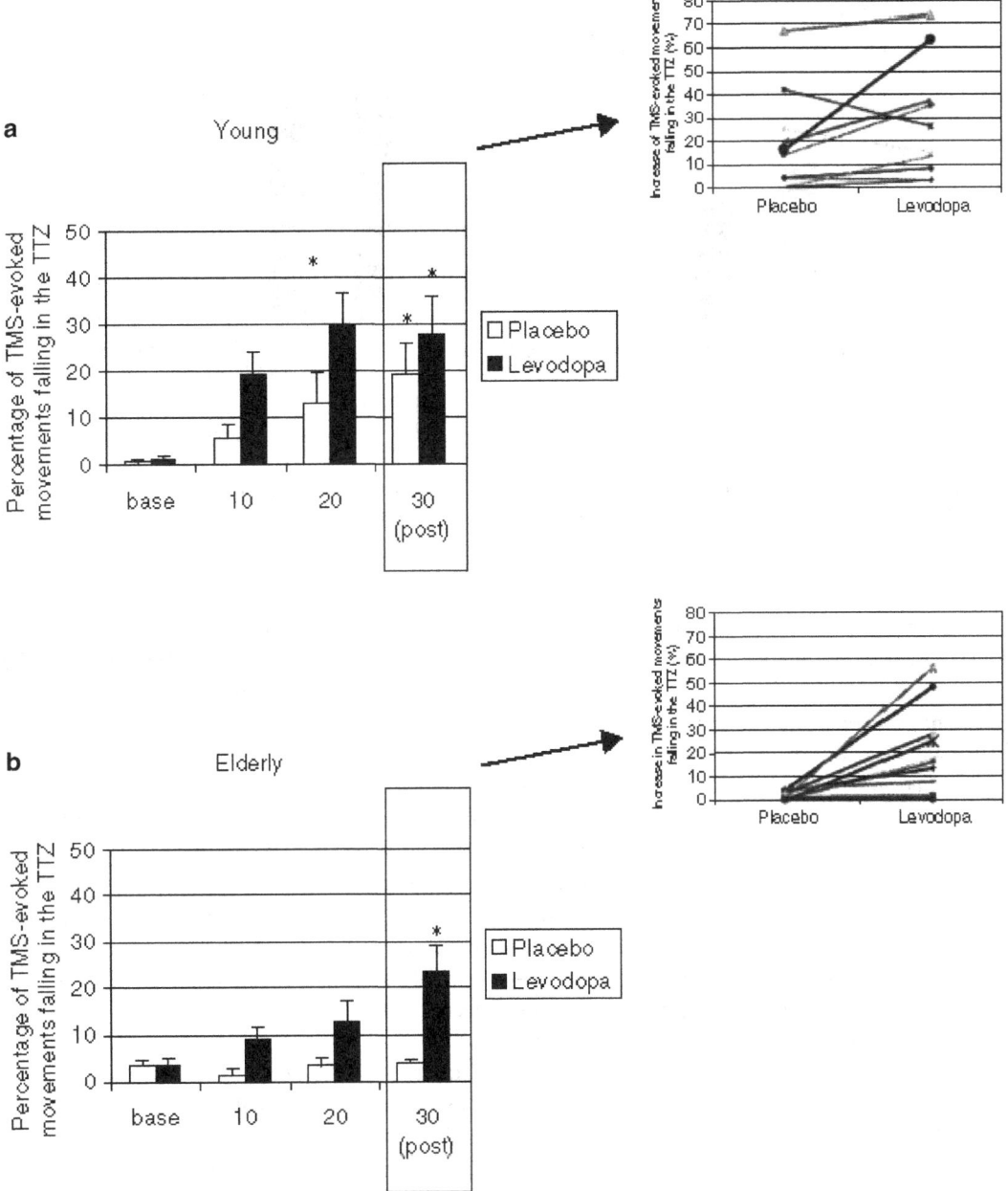

Verbal and Nonverbal Learning in Humans, Fig. 2 Percentage of transcranial magnetic stimulation (*TMS*)-evoked thumb movements falling in the training target zone (*TTZ*) in young (**a**) and elderly (**b**) healthy volunteers. In young subjects (**a**), training under placebo led to a progressive increase in TMS-evoked thumb movements falling in the TTZ that became significant after 30 min (**a**, 30 min [post], *white bar*). L-Dopa + training accelerated the development of this form of plasticity, which became significant after only 10 min of training (**a**, 10 min, *black bar*). In elderly subjects (**b**), consistent with previous results, training under placebo did not induce changes in TMS-evoked thumb movements falling in the TTZ (**b**, 30 min [post], *white bar*). L-Dopa + training substantially enhanced the response to motor training, which became significant after 30 min (**b**, 30 min [post], *black bar*), and that was comparable in magnitude with that identified in younger subjects under placebo and L-dopa (**a**, 30 min [post], *black and white bars*). Note that this effect was evident in five of the seven elderly subjects tested (*inset*). To illustrate the percentage change in the training + L-dopa versus the training + placebo condition, we summarize the mean change for each subject in each condition in the above insets (**a**, young; **b**, elderly). *$p < 0.05$ (From Flöel et al. 2005a)

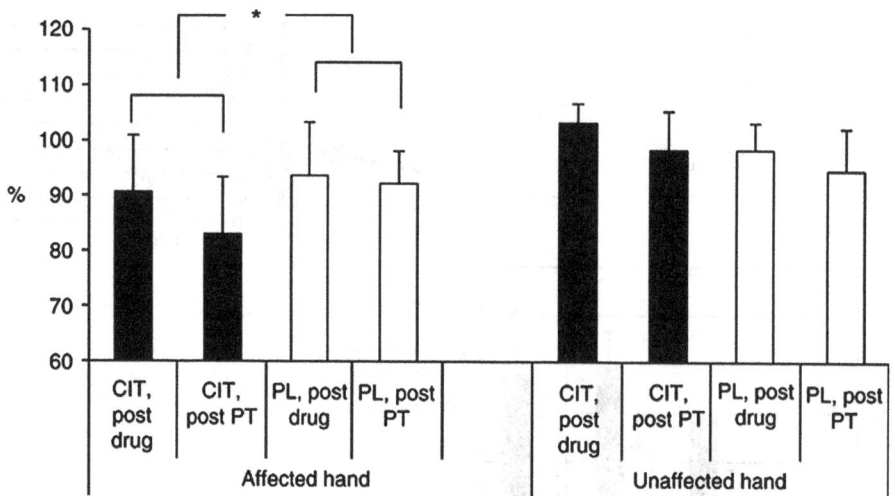

Verbal and Nonverbal Learning in Humans, Fig. 3 Nine-Hole Peg Test results, expressed in percentage of the values obtained prior to drug ingestion. *CIT* citalopram; *PL* placebo; *PT* physiotherapy. Error bars indicate standard deviations. *$p < 0.05$ (From Zittel et al. 2008)

For example, D-amphetamine improved performance of selective attention tasks.

Patient Groups In chronic stroke patients, a single dose of reboxetine improved tapping speed and grip strength in the paretic but not in the non-affected hand.

Studies with D-amphetamine in stroke patients are inconclusive. In two studies, drug effects on motor recovery were found. In subsequently performed studies ($n = 6$), these positive results could not be replicated. A Cochrane review summarized that it is currently impossible to draw any definite conclusions about the potential role of D-amphetamine in motor rehabilitation (Martinsson et al. 2007).

Serotonergic Substances

Healthy Subjects There are no studies with serotonergic drugs for enhancement of learning in healthy subjects available.

Patient Groups The serotonin reuptake inhibitor fluoxetine improved walking and activities of daily living in a group of subacute stroke patients. In eight chronic stroke patients, a single dose of fluoxetine improved motor functions of the affected hand. In another study, the more selective serotonin reuptake inhibitor citalopram was used in chronic stroke patients. A single dose of citalopram was able to improve dexterity but not strength in the paretic hand. The effect was pronounced after 1 h of physiotherapy aimed at improving hand function (Zittel et al. 2008; Fig. 3). Taking these three studies together, results seem promising but one has to consider that only a very limited number of subjects have been studied so far.

Cholinergic Substances

This group mainly includes drugs that act as inhibitors of the enzyme acetylcholinesterase. This increases the amount of acetylcholine. Such drugs have been successfully tested in dementia and are recommended for patients with mild to moderate Alzheimer's disease.

Healthy Subjects In elderly healthy subjects, rivastigmine improved motor learning and visuospatial processes. In contrast the anticholinergic agent biperiden impaired these abilities. In another study donepezil was applied in elderly subjects and improved their memory performance. In addition, rapid eye movement (REM) sleep was enhanced, and the positive correlation between memory function and REM sleep duration suggested an interrelationship between these two factors.

Patient Groups There is only limited evidence regarding the beneficial effects of acetylcholinesterase inhibitors in stroke patients. One case study and one open-label pilot study suggest that these drugs may enhance the recovery of motor functions, in particular in cognitively impaired stroke patients. One study combining the application of donepezil with constraint-induced movement therapy only showed a trend towards stronger improvement in the drug-treated group.

Modafinil
This wake-promoting drug is approved for application in patients with abnormal sleepiness, e.g., narcolepsy and sleep disorders of shift workers. Some studies also suggest that it is helpful in patients with multiple sclerosis suffering from fatigue. A recent study demonstrated that modafinil blocks dopamine transporters and increases dopamine in the human brain.

Healthy Subjects Studies in healthy subjects are inconclusive. One study reported a drug-related improvement of several cognitive abilities including digit span, visual pattern recognition memory, spatial planning, and stop-signal reaction time. Another study suggested an effectiveness in monotonous working-memory tasks, and in a third study, modafinil was found to be equal to placebo regarding reaction time, dexterity, and the d2 test.

Patient Groups In patients with Huntington's disease, modafinil increased alertness but did not improve cognitive functions. In contrast, deleterious effects on visual recognition and working memory were observed. In schizophrenic patients, improvements of executive functions and attention were found.

Verbal Learning

Dopaminergic Substances

Healthy Subjects Single doses of 100 mg levodopa improved verbal learning in young healthy subjects. Learning occurred faster and more successful, and the long-term retention of novel word learning was better than in the placebo-treated control group (Knecht et al. 2004; Fig. 4). A comparison between the application of levodopa, D-amphetamine, and placebo indicated a similar efficacy of levodopa and D-amphetamine and the superiority of both drugs compared to placebo. In contrast to beneficial effects of levodopa, the dopamine-receptor agonist pergolide impaired novel word learning. The authors suggested that this finding can be explained by the tonic dopaminergic effects produced by pergolide. Due to a much shorter half-life, levodopa rather exerts a phasic dopaminergic stimulation. Phasic stimulations might be more effective for associative learning.

Patient Groups A very recent placebo-controlled study in stroke patients indicated that levodopa induced a greater improvement of verbal fluency and repetition than placebo. This positive effect was particularly obvious in patients with frontal lesions.

Bromocriptine, a dopamine-receptor agonist, has yielded inconclusive results. In one study, reading comprehension, repetition, dictation, and verbal latency improved in chronic stroke patients with nonfluent aphasia. However, in another study, bromocriptine was not superior to placebo in stroke patients with nonfluent aphasia.

Some preliminary evidence coming from an open-label study with four patients indicated that amantadine might be effective in nonfluent speech.

Noradrenergic Substances

Healthy Subjects Several studies have demonstrated that D-amphetamine facilitates verbal memory performance, improves information

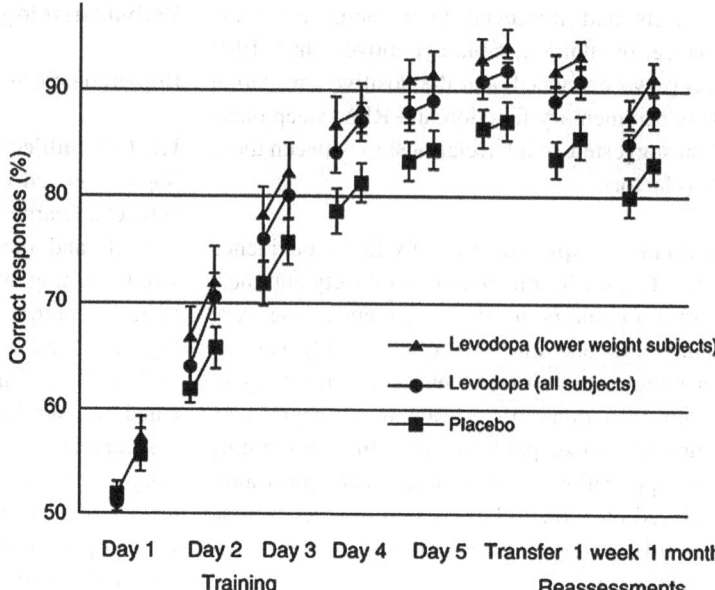

Verbal and Nonverbal Learning in Humans, Fig. 4 Success in novel word learning in subjects receiving placebo or L-dopa (mean values with standard errors of the means for two daily sessions with 200 trials each). In addition, also shown are results in the subgroup ($n = 10$) receiving relatively higher doses of L-dopa because their body weights were below the group median. Scores for the reassessments and transfer sessions also are displayed (From Knecht et al. 2004)

processing, and enhances new word learning. These effects were independent of drug-induced increases of blood pressure and heart rate. It seems that there is an individual optimal D-amphetamine dose, since an inverted U-shaped relationship was observed between the amount of D-amphetamine and working-memory processing efficiency. This suggests that high doses of D-amphetamine can even impair memory processing. Another factor that influences the results of D-amphetamine application is the cognitive performance at baseline. The drug was effective in individuals with a low working-memory capacity, but deteriorated performance in subjects with high working-memory capacity at baseline.

Patient Groups Two studies performed in stroke patients with aphasia suggested a beneficial effect of D-amphetamine. However, when patients were reevaluated 6 months later, the effect was gone.

Cholinergic Substances

Healthy Subjects There are no studies that explicitly investigate the effect of cholinergic transmission on verbal learning.

Patient Groups In one study with 26 stroke patients, donepezil reduced the severity of aphasia and improved picture naming to a stronger degree than placebo.

Piracetam

The mode of action is unknown, but there is some evidence that piracetam enhances glucose utilization and cellular metabolism in the brain.

Healthy Subjects There are no studies available that investigated piracetam effects on verbal learning.

Patient Groups Placebo-controlled trials in subacute stroke patients ($n = 203$) indicated that the application of 4.8 mg piracetam daily reduced aphasic symptoms as evaluated by the Aachener Aphasie Test (Greener et al. 2001). Treatment duration was at least 6 weeks. A positron emission tomography (PET) study demonstrated increased activity in speech-relevant brain areas as the left transverse temporal gyrus, Wernicke's area, and Broca's area in the piracetam group, but not in the placebo group.

Conclusions

Currently, evidence is limited regarding a positive effect of psychoactive drugs that enhance neurotransmission or support cell metabolism. In healthy subjects, the most convincing and most consistent studies were found for the improvement of cognitive function and verbal learning by D-amphetamine and levodopa. Results obtained in elderly healthy subjects might suggest that aging per se is associated with a subclinical reduction of neurotransmission and that substitution of dopaminergic, cholinergic (and potentially also serotonergic) neurotransmission can be more beneficial than in young subjects.

In patients, acquisition of new abilities or reacquisition of old abilities is more difficult. We do not know enough about the consequences of brain lesions, e.g., stroke or traumatic brain injury, on neurotransmission. Therefore, it is still unpredictable who might benefit from which type of enhancement of neurotransmission. Hopefully, brain imaging techniques will help to develop hypothesis-driven concepts of specific neuropharmacological interventions. Until then, one can recommend piracetam for treatment of poststroke aphasia and levodopa for recovery of motor functions.

Cross-References

▶ Acetylcholinesterase and Cognitive Enhancement
▶ Attention
▶ Cognitive Enhancers: Role of the Glutamate System
▶ Declarative and Nondeclarative Memory
▶ Long-Term Potentiation and Memory
▶ Neuroprotectants: Novel Approaches for Dementias
▶ Pharmaceutical Cognitive Enhancers: Neuroscience and Society
▶ Psychomotor Performance in Humans
▶ Psychostimulants
▶ Short-Term and Working Memory in Humans
▶ Synaptic Plasticity

References

Flöel A, Breitenstein C, Hummel F, Celnik P, Gingert C, Sawaki L, Knecht S, Cohen LG (2005a) Dopaminergic influences on formation of a motor memory. Ann Neurol 58(1):121–130

Flöel A et al (2005b) Dopaminergic effects on encoding of a motor memory in chronic stroke. Neurology 65(3):472–474

Greener J, Enderby P, Whurr R (2001) Pharmacological treatment for aphasia following stroke. Cochrane Database Syst Rev CD000424

Knecht S, Breitenstein C, Bushuven S, Wailke S, Kamping S, Flöel A, Zwitserlood P, Ringelstein EB (2004) Levodopa: faster and better word learning in normal humans. Ann Neurol 56(1):20–26

Liepert J (2008) Pharmacotherapy in restorative neurology. Curr Opin Neurol 21(6):639–643

Martinsson L, Hårdemark H, Eksborg S (2007) Amphetamines for improving recovery after stroke. Cochrane Database Syst Rev CD002090

Plewnia C, Hoppe J, Cohen LG, Gerloff C (2004) Improved motor skill acquisition after selective stimulation of central norepinephrine. Neurology 62:2124–2126

Scheidtmann K, Fries W, Muller F, Koenig E (2001) Effect of levodopa in combination with physiotherapy on functional motor recovery after stroke: a prospective, randomized, double-blind study. Lancet 358:787–790

Zittel S, Weiller C, Liepert J (2008) Citalopram improves dexterity in chronic stroke patients. Neurorehabil Neural Repair 22(3):311–314

Verum

Definition

Literally, "the truth" – a drug with a known and established effect on psychomotor function – which when included as one of the treatment conditions in a study of the effects of an unknown drug serves to demonstrate the sensitivity of the selected psychometric tests to drug-induced changes.

Violence

Definition

The term "violence" is most frequently applied in reference to pathological forms of aggression in humans. It refers to behavior that exceeds the species-normative levels in terms of intensity, duration, and frequency that are typically associated with tissue damage.

Cross-References

▶ Aggression
▶ Aggressive Behavior: Clinical Aspects

Visuospatial Process

Definition

Any process in which vision is needed to explore or pay attention to the external space surrounding the subject.

Voltage Clamp

Definition

Voltage-clamp recording is an ▶ intracellular recording modality where the experimenter controls the voltage across the cell membrane by injecting current as to maintain it at a fixed "command" voltage level. The injected current is the mirror of the ionic current flowing across the membrane at any one time and thus allows for its direct measurement.

Cross-References

▶ Intracellular Recording

Voltage-Gated Calcium Channels

Synonyms

Voltage-dependent calcium channels; VGCC; VDCC

Definition

Voltage-gated calcium channels are a group of ion channels that permit the influx of calcium under conditions of changes in voltage (depolarization of the membrane). These channels can be particularly important as the influx of calcium into the presynaptic cell may be required for the release of transmitter substance into the synaptic cleft. Chronic exposure to ethanol can cause an increase in the number of these channels (i.e., upregulation). Increasing the number of these channels can lead to an increased or prolonged duration of transmitter release, which may in turn lead to an increased hyperexcitability in the neural circuit.

Voltage-Gated Potassium Channels

Definition

Voltage-gated K_V channels are a large family of membrane ion channels that are selectively permeable to potassium ions. Activation of these channels by changes in the voltage across cell membranes produces membrane ▶ hyperpolarization which controls action potential repolarization, frequency, and firing patterns.

Volume of Distribution

Synonyms

Apparent volume of distribution; V_D

Definition

The volume of distribution (V_D) is the fictitious (theoretical) volume, expressed in liter or in liter per kilogram, in which a drug would have been distributed by supposing that its concentration is homogeneous, that is, the average tissue concentration is identical to that of the plasma.

It is expressed as V_D = dose/C0 (initial concentration). For example, after intravenous injection of 100 mg of a drug whose initial concentration, C0, in plasma is 10 mg/L, the V_D is of 10 L.

Cross-References

- ▶ Bioavailability
- ▶ Elimination Half-Life
- ▶ Pharmacokinetics

Volume Transmission

Definition

Volume transmission in the brain usually refers to actions of ▶ neurotransmitters or ▶ neuropeptides at distances far from their release sites in cells or synapses. It is established that some peptides such as β-endorphin and ▶ enkephalins can act on neurons at distances much greater than a synapse, but the actual range of influence is still unknown.

Wakefulness

Jaime M. Monti and Héctor Jantos
Department of Pharmacology and Therapeutics,
School of Medicine Clinics Hospital,
Montevideo, Uruguay

Synonyms

Behavioral and electroencephalographic arousal; Waking state

Definition

Wakefulness is regulated by several neurotransmitter systems. Strategies aimed at determining their role in the regulation of wakefulness have included electrophysiological, genetic, neurochemical, and neuropharmacological approaches.

Wake-Promoting Neurotransmitter Systems

The central nervous system structures involved in the promotion of the waking state are located in the brainstem, hypothalamus, and basal forebrain. The nuclei found in the brainstem contain serotonin (dorsal raphe nucleus, median raphe nucleus), noradrenaline (locus coeruleus), dopamine (ventral tegmental area, substantia nigra pars compacta, ventral periaqueductal gray matter), and acetylcholine (laterodorsal and pedunculopontine tegmental nuclei). The structures located in the hypothalamus contain histamine (tuberomammillary nucleus) and orexin (posterior lateral hypothalamus around the fornix). The cholinergic and glutamatergic neurons of the basal forebrain involved in the regulation of wakefulness are located predominantly in the diagonal band, substantia innominata, and medial septal area.

The serotonergic, noradrenergic, histaminergic, and cholinergic neurons that participate in the regulation of wakefulness give rise to mainly ascending projections to (1) the thalamus (dorsal route), which in turn project to the cerebral cortex, and (2) the basal forebrain (ventral route) where cells in turn project to the cerebral cortex and hippocampus. The dopamine-containing cells of the ventral tegmental area and substantia nigra pars compacta project to the basal ganglia and the prefrontal cortex, while these corresponding to the ventral periaqueductal gray matter project predominantly to the basal forebrain and midline thalamus. Furthermore, the orexin-containing neurons carry projections to the entire forebrain and brainstem arousal systems (Watson et al. 2010; Brown et al. 2012). Isolated activation of each of the arousal systems already provokes wakefulness. However, under normal conditions they all participate in the occurrence of behavioral and electroencephalographic arousal. This depends, in part, on the interconnections of most wakefulness-promoting structures. In addition, the systems that promote wakefulness inhibit neural structures located in the brainstem and hypothalamus that promote and/or induce non-rapid eye movement sleep (NREMS) and REM sleep (REMS) (España and Scammell 2011).

Neurotransmitters Involved in the Occurrence of Wakefulness

Serotonin

Serotonergic Receptors
The 5-HT1A and 5-HT1B receptor subtypes are linked to the inhibition of adenylate cyclase, and their activation evokes a membrane hyperpolarization. The actions of the 5-HT2A, 5-HT2B, and 5-HT2C receptor subtypes are mediated by the activation of phospholipase C, with a resulting depolarization of the host cell. The 5-HT3 receptor directly activates a 5-HT-gated channel which leads to the depolarization of monoaminergic, aminoacidergic, and cholinergic cells. The primary signal transduction pathway of 5-HT6 and 5-HT7 receptors is the stimulation of adenylate cyclase which results in the depolarization of the follower neurons.

Electrophysiological Approach
During quiet wakefulness, serotonergic neurons fire in a slow and regular fashion; during active wakefulness, neuronal activity shows a significant increase. As the animal enters NREMS, the firing rate decreases to 50 % of the quiet waking state; during REMS, there is a further decrease or even a cessation of neuronal activity.

Genetic Approach
Mutant mice that do not express 5-HT1A or 5-HT1B receptor exhibit greater amounts of REMS than their wild-type counterparts, while 5-HT2A and 5-HT2C receptor knockout mice show a significant increase of wakefulness and a reduction of NREMS. Although the basis for this apparent discrepancy is unclear, it has been posed that the greater amounts of wakefulness in 5-HT2A and 5-HT2C receptor knockout mice could be related to the increase of catecholaminergic neurotransmission involving mainly the noradrenergic and dopaminergic systems. Furthermore, REMS is significantly reduced in 5-HT7 receptor knockout mice.

Neuropharmacological Approach
System and/or intracerebroventricular injection of full agonists of postsynaptic 5-HT1A, 5-HT1B, 5-HT2B, 5-HT2C, 5-HT3, 5-HT6, and 5-HT7 receptor increases wakefulness and reduces NREMS and REMS irrespective of their operational characteristics. Direct infusion of 5-HT1A receptor agonists into the dorsal raphe nucleus significantly enhances REMS in the rat. In contrast, microinjection of agonists corresponding to 5-HT1B, 5-HT2A/2C, 5-HT3, 5-HT6, and 5-HT7 receptor subtypes induces the opposite effect (Monti 2011a).

In conclusion, all serotonin receptor agonists studied to date share the ability to promote wakefulness and to reduce sleep when given systematically or by the intracerebroventricular route.

Noradrenaline

Noradrenergic Receptors
The noradrenergic receptors have been divided into three classes with their corresponding subtypes: $\alpha 1$, β, and $\alpha 2$. The $\alpha 1$ receptor is related to the enzyme phospholipase C and acts by increasing the Ca^{2+} concentration in the target cell. The primary signal transduction pathway of the β receptor is the stimulation of cyclic adenosine monophosphate (AMP) synthesis. In contrast, the $\alpha 2$ receptor is linked to the inhibition of cyclic AMP synthesis.

Electrophysiological Approach
The locus coeruleus noradrenergic neurons fire actively during wakefulness, less during NREMS, and virtually cease activity during REMS.

Genetic Approach
Sleep variables have been studied in dopamine β-hydroxylase-deficient mice. The mutant mice

showed an increase of REMS during the dark period. Moreover, NREMS latency was significantly shorter following a mild stress or the systemic administration of amphetamine in the dopamine β-hydroxylase-knockout mice.

Neurochemical Approach

The progressive cooling of the locus coeruleus or the local administration of the neurotoxin 6-hydroxydopamine resulted in an increase of NREMS and REMS in the cat and/or the rat. Similar effects on REMS have been reported after systemic administration of the neurotoxin DSP-4 that selectively lesions the noradrenergic neurons. However, the 6-OHDA- and DSP-4-related effects persist for a few days only because of the inherent redundancy of the noradrenergic system.

Neuropharmacological Approach

The injection of noradrenaline reuptake inhibitors (amphetamine, cocaine) is followed by an increase of noradrenaline levels at central sites. As a result, wakefulness is augmented and sleep is reduced. Microinjection of the selective α2 receptor agonists clonidine and dexmedetomidine into the locus coeruleus decreases the firing rate of noradrenergic cells and extracellular levels of the neurotransmitter. Furthermore, systemic, intracerebroventricular, or local administration into the locus coeruleus of clonidine produces profound sedation, whereas the α2 antagonist yohimbine prevents this effect. Intraperitoneal injection of methoxamine, a selective α1 receptor agonist, significantly increases W and reduces sleep; these effects are antagonized by the α1 antagonist prazosin. A number of basal forebrain structures including the medial septal area, medial preoptic area, and substantia innominata receive an input from locus coeruleus noradrenergic neurons. In this respect, microinjection of the α1- and β-receptor agonists phenylephrine and isoproterenol, respectively, into the medial septal area or medial preoptic area is followed by an increase of electroencephalographic/behavioral indices of wakefulness.

It can be concluded that the increased availability of noradrenaline in the central nervous system augments wakefulness and reduces NREMS and REMS. The increase of wakefulness would be partly dependent on the activation of wakefulness-promoting neurons occurring in the basal forebrain. Noradrenaline would participate in the occurrence of wakefulness during situations that require high attention or activation of the sympathetic nervous system (Berridge et al. 2012).

Dopamine

Dopaminergic Receptors

Two distinct groups of dopaminergic receptors, D1- and D-2-like receptors, have been described. The D1 subfamily includes the D1 and D5 receptors which are located at postsynaptic sites and linked to the activation of adenylate cyclase. Dopaminergic receptors of the D2 subfamily are predominantly coupled to the inhibition of adenylate cyclase. The D2 receptor is the predominant D2-like subtype in the brain. It has been characterized at postsynaptic sites and also on cell bodies and dendrites in the ventral tegmental area and substantia nigra pars compacta, where it functions as an autoreceptor. D3 and D4 receptors are found at much lower levels and more narrowly distributed in the central nervous system compared with the D2 receptor.

Electrophysiological Approach

Opposite to what occurs in serotonergic, noradrenergic, and histaminergic neurons, dopaminergic cells in the ventral tegmental area and substantia nigra pars compacta show a change in the temporal pattern rather than the firing rate during the sleep-wake cycle. The change in temporal pattern manifests as an increase in burst firing during wakefulness and REMS. This results in an increase of dopamine release in the prefrontal cortex and nucleus accumbens. The firing pattern of ventral periaqueductal gray matter dopaminergic neurons during the behavioral state has not been quantified to date.

Genetic Approach

The evidence gathered using a genetic approach tends to indicate that the effect of dopamine on behavioral arousal depends on the spectrum of receptors involved. Accordingly, behavioral arousal is impaired in D1- and D2-receptor-deficient mice, whereas it is facilitated in mutant mice lacking functional D3 receptors.

Neurochemical Approach
Systemic or intracerebral administration of the neurotoxin 1-methyl-4-phenyl-1,2,5,6-tetrahydropyridine (MPTP) was found to have a marked preferential neurotoxic effect on dopamine-containing neurons of the ventral tegmental area and substantia nigra pars compacta, although it also affects serotonergic and noradrenergic cells. The neurotoxin produces a syndrome similar to Parkinson's disease and suppresses REMS in the cat and NREMS and REMS in the rat. Microinjection of 6-OHDA into the ventral periaqueductal gray matter has been shown to lesion tyrosine hydroxylase-immunoreactive neurons and to significantly increase sleep and reduce wakefulness.

Neuropharmacological Approach
Systemic administration of the selective dopamine D1 receptor agonist SKF 38393 increases wakefulness and reduces NREMS and REMS in laboratory animals. Opposite effects have been described after injection of the D1 antagonist SCH 23390. Systemic or intracerebroventricular administration of (−)3-PPP, a selective DA D2 autoreceptor agonist, suppresses locomotor activity and induces behavioral and electroencephalographic sleep in the rat. The effects of systemic administration of the dopamine D2 receptor agonists apomorphine (D1 > D2), bromocriptine (D2 > D1), or quinpirole (D2 > D3) have been compared with those produced by the D2 antagonists haloperidol or YM-09151-2. Apomorphine, bromocriptine, and quinpirole produced biphasic effects, such that low doses decreased wakefulness and augmented sleep whereas large doses induced the opposite effects in the rat. A relatively large dose of haloperidol has been shown to reduce wakefulness and to increase NREMS. On the other hand, YM-09151-2 increased light sleep and suppressed REMS. Haloperidol and YM-09151-2 prevented the effects of apomorphine, bromocriptine, and quinpirole on the sleep-wake cycle, respectively. Administration of a small dose of the D3-preferring dopaminergic agonist pramipexole induced an increase of NREMS and REMS whereas wakefulness was reduced in the rat. In contrast, a greater dose of pramipexole induced the opposite effects. Systemic administration of a relatively large dose of the selective dopamine D4 receptor antagonist has been shown to reduce total sleep time and to increase wakefulness in laboratory animals. The effects of dopamine D4 receptor agonists on sleep and wakefulness have not been determined to date (Monti and Jantos 2008).

Histamine

Histaminergic Receptors
Presently available evidence tends to indicate that histamine-related functions in the central nervous system are regulated at postsynaptic sites by the H1 and the H2 receptors. On the other hand, the H3 receptor shows the features of a presynaptic autoreceptor mediating the synthesis and release of histamine and of a presynaptic heteroreceptor, controlling the release of several neurotransmitters. Recently, a fourth receptor (H4) has been localized in the central nervous system. However, its physiological role has not been characterized yet. The H1 receptor is related to the enzyme phospholipase C and acts primarily by increasing the Ca^{2+} concentration in the target cell. Histamine acting through the H2 receptor activates a stimulatory G protein, which in turn stimulates adenylate cyclase. The H3 receptor is negatively coupled to adenylate cyclase and its stimulation induces a decrease of cAMP.

Electrophysiological Approach
Histaminergic neurons display a slow and regular discharge during quiet waking. When the animal is moving (active waking), the mean discharge rate increases significantly. As the animal enters NREMS, the mean discharge rate shows a progressive decrease. During deep NREMS and REMS, all the neurons become silent.

Genetic Approach
The sleep-wake cycle and cortical electroencephalogram have been examined in both

histidine decarboxylase and H1 receptor knockout mice. Compared to wild-type mice, histidine decarboxylase-knockout mice show an increase of REMS. H1 receptor knockout mice have fewer brief awakening episodes and a shorter NREMS latency.

Neurochemical Approach
Systemic injection of α-fluoromethylhistidine, a highly specific, irreversible inhibitor of histidine decarboxylase, induced a significant reduction of wakefulness while sleep was increased.

Neuropharmacological Approach
Intracerebroventricular infusion of histamine or the relatively selective H1 receptor agonist 2-thiazolylethylamine in the conscious animal has been shown to elicit a marked increase of wakefulness while sleep is reduced. In the opposite, systemic administration of the H1-antagonists mepyramine, diphenhydramine, and promethazine decreases wakefulness and REMS and augments NREMS. The H3 receptor agonist BP 2.04 given by oral route produced a significant increase of NREMS in the rat. In contrast, the H3 antagonist carboperamide increased wakefulness and reduced sleep.

As a whole, histamine would participate in attention and psychomotor performance and promote motivated behavior (Monti 2011b).

Acetylcholine

Cholinergic Receptors
Acetylcholine acts on two different types of receptors referred to as nicotinic and muscarinic receptors. Central nervous system nicotinic receptors have two protein subunits α and β with 2α- and 3β-subunits forming the ion channel. To date, five muscarinic receptors (M1-M5) have been described, and all of them are coupled to G proteins. The M1 receptor is involved in most of the central nervous system muscarinic excitatory effects of acetylcholine, and its stimulation causes hydrolysis of polyphosphoinositides and mobilization of intracellular Ca^{2+}. The M2 receptor behaves as an inhibitory presynaptic autoreceptor, and its activation leads to the opening of K^+ channels and the closing of Ca^{2+} channels.

Neuropharmacological Approach
Acetylcholine contributes to wakefulness-related activation and to REMS promotion. Neurons of the basal forebrain, including the diagonal band of Broca, substantia innominata, and medial septal area, provide the cholinergic innervation of the cerebral cortex and the hippocampus. In addition, cells located in the laterodorsal and pedunculopontine tegmental nuclei release acetylcholine in the subcoeruleus/sublaterodorsal nucleus and the thalamus. Activation of basal forebrain cholinergic neurons projecting to the cerebral cortex and hippocampus induces the occurrence of wakefulness through actions on nicotinic and muscarinic M1 receptors; cholinergic wakefulness-on cells occurring in the dorsal tegmentum contributes in this respect. On the other hand, depolarization of laterodorsal and pedunculopontine tegmental nuclei cholinergic REM on neurons leads to the appearance of REMS.

A series of pharmacological studies have provided further evidence for the involvement of acetylcholine in the regulation of wakefulness and REMS. Thus, central administration of the choline uptake inhibitor hemicholinium-3 reduces wakefulness and suppresses REMS in laboratory animals. Moreover, systemic administration of the muscarinic receptor agonists pilocarpine and arecoline evokes a characteristic cortical arousal which is reduced by atropine, while microinjection of the broad-spectrum cholinergic agonist carbachol or the anticholinesterase agents physostigmine and neostigmine into the medial pontine reticular formation induced a long-lasting REMS-like state in the cat (Jones 2005).

Orexin

Orexin Receptors
The orexin-A and orexin-B neuropeptides are derived from a common prepro-orexin gene and synthesized by neurons located in the lateral and perifornical hypothalamic areas. Two orexin receptors have been characterized in the central nervous

system, namely, OX1 receptor that shows higher affinity for orexin-A than orexin-B, and OX2 receptor that exhibits similar affinity for both neuropeptides. The binding of orexin to OX1 activates intracellular signaling pathways by coupling to G_q, while OX2 is known to couple to G_q or G_i/G_o. Orexin-induced depolarization of postsynaptic neurons is related to the inhibition of K^+ channels and the increase in intracellular Ca^{2+}. In contrast, activation of presynaptic nerve terminal has been found to increase the release of γ-aminobutyric acid and glutamate.

Genetic Approach
Orexin-knockout mice show severe sleepiness and are unable to maintain long periods of wakefulness during the dark phase. Patients with a diagnosis of narcolepsy have also severe sleepiness and, in addition, REMS episodes can occur at any time of the day.

Electrophysiological Approach
Immunohistochemically identified orexinergic neurons have been recorded during the sleep-wake cycle in the rat. Orexinergic neurons are active during wakefulness and remain silent during NREMS and REMS.

Neuropharmacological Approach
Administration of orexin into the central nervous system increases wakefulness while suppressing NREMS and REMS, which has been related to the activation of arousal systems located in the brainstem, hypothalamus, and basal forebrain. The effects are similar to those seen with optogenetic stimulation of orexinergic cells. Opposite, the orexin receptor antagonist almorexant increases NREMS and REMS in laboratory animals (Sinton 2011).

Summary

Wake-promoting structures are located in the brainstem, hypothalamus, and basal forebrain and include monoamines-, acetylcholine-, and orexin-containing neurons. Under normal conditions they all participate in the occurrence of wakefulness. In addition, the wake-promoting systems inhibit neural structures that induce NREMS and REMS. Systemic or intracerebroventricular administration of serotonergic, noradrenergic, dopaminergic, histaminergic, cholinergic, and orexinergic agonists consistently increase wakefulness and reduce sleep.

References

Berridge CW, Schmeichel BE, España RA (2012) Noradrenergic modulation of wakefulness/arousal. Sleep Med Rev 16:187–197

Brown RE, Basheer R, McKenna JT, Strecker RE, McCarley RW (2012) Control of sleep and wakefulness. Physiol Rev 92:1087–1187

España RA, Scammell TE (2011) Sleep neurobiology from a clinical perspective. Sleep 34:845–858

Jones BE (2005) From waking to sleeping: neuronal and chemical substrates. Trends Pharmacol Sci 26:578–586

Monti JM (2011a) Serotonin control of sleep-wake behavior. Sleep Med Rev 15:269–281

Monti JM (2011b) The role of tuberomammillary nucleus histaminergic neurons, an of their receptors, in the regulation of sleep and waking. In: Mallick BN, Pandi-Perumal SR, McCarley RW, Morrison AR (eds) REM sleep: regulation and function. Cambridge University Press, Cambridge, pp 223–233

Monti JM, Jantos H (2008) The roles of dopamine and serotonin, and of their receptors in regulating sleep and waking. Prog Brain Res 172:625–646

Sinton CM (2011) Orexin/hypocretin plays a role in the response to physiological disequilibrium. Sleep Med Rev 15:197–207

Watson CJ, Baghdoyan HA, Lydic R (2010) Neuropharmacology of sleep and waking. Sleep Med Clin 5:513–528

Water Maze

Synonyms

Morris water maze

Definition

The water maze task, originally developed by Richard G. Morris, is one of the most commonly used paradigms to assess spatial learning and memory in rodents. In the typical paradigm, a rat or mouse is placed into a pool of opaque water that contains an escape platform hidden below the water surface. Distal visual cues are placed around the pool that the animal uses to guide its search for the hidden platform. When released, the rat swims around the pool in search of an exit, while various parameters are recorded, including the time spent in each quadrant of the pool, the time taken to reach the platform (latency), and the total distance travelled. The rat's escape from the water reinforces its attempts to quickly find the platform, and on subsequent trials, the rat is able to locate the platform more rapidly. Rats and mice typically learn to find the platform in an efficient manner after relatively few trials. In the classic version that assesses long-term spatial memory, the location of the platform remains in one location over the duration of testing. A short-term memory variation has also been used, whereby the location of the platform changes each training day and animals display substantial reductions in escape latencies and path lengths from the first to subsequent trials. A control version of the task uses a visible platform, so that animals can solve the task without the use of distal spatial cues.

Cross-References

▶ Rodent Tests of Cognition

Whole-Cell Recording

Definition

Whole-cell recording is a form of patch-clamp recording where the membrane patch covering the tip of the recording electrode after tight seal formation is broken to gain electrical access to the cell's interior. This configuration allows for the recording of a cell's membrane potential, in current clamp, or of membrane currents originating from the entire cell, in voltage clamp, hence the name.

Cross-References

▶ Intracellular Recording

Wisconsin Card Sorting Test

Synonyms

Set-shifting test

Definition

In this test, subjects are presented with a series of multidimensional stimuli, which they have to sort into piles, based on one of the stimulus dimensions, for example, color, number, and shape. After subjects have learned to sort according to one of the dimensions, the rule is changed, so that now subjects have to shift their attention and start sorting according to a different stimulus dimension. Adequate performance depends on accurate working memory representations of the currently relevant dimension as well as recently chosen stimuli.

Cross-References

▶ Behavioral Flexibility: Attentional Shifting, Rule Switching and Response Reversal

Wisconsin General Test Apparatus

Definition

A behavioral testing apparatus for monkeys that includes a stimulus tray containing three food wells and a one-way screen for experimenter observation. Objects or plaques can be placed over the food wells, and any combination of wells may be baited. The monkey may or may not be given the opportunity to see the food wells being baited. A screen can be placed between the monkey and the food wells if required. A variety of learning and memory tests can be administered in this apparatus.

Withdrawal Syndromes

Martine Cador, Stéphanie Caillé and Luis Stinus
Team Neuropsychopharmacology of addiction,
UMR CNRS 5287, University of Bordeaux,
Bordeaux, France

Synonyms

Abstinence syndrome; Drug abstinence; Drug discontinuation

Definition

A withdrawal syndrome that can be characterized both at the behavioral and the physiological level occurs when the use of a psychoactive substance that has been taken, usually for a prolonged period of time and/or in high doses, is either stopped or reduced. The onset and course of the withdrawal syndrome are time limited and related to the type of substance and dose taken previously. A withdrawal syndrome is one of the indicators of a drug-dependent state, which is very often paralleled by drug tolerance, both being due to adaptations within the body and the brain.

Impact of Psychoactive Drugs

Withdrawal in Drug Addiction

Withdrawal is part of the drug addiction process and is a recurring feature of drug addiction because of the aversive state it engenders (negative reinforcement theory). The symptoms defining withdrawal syndromes are rather well characterized in humans and appear in a time-dependent fashion (Jaffe 1990) depending on the drug being used (heroin, cocaine, cannabinoids, nicotine, or alcohol). During the initial phase following drug discontinuation, a common feature for all drug abuse is that the individual is in a dysphoric state. No physical signs are observed at this stage though a strong motivation to use the drug again to relieve this discomfort is experienced. It has been suggested that drug withdrawal induces a motivational state that contributes to the maintenance of drug-taking behavior (Koob et al. 1989). If the drug is not provided, then as a continuum to the affective component of the withdrawal, several somatic symptoms of variable intensity emerge. In humans, the signs range from discomfort, anxiety, decreased appetite, sleep disorder, nightmares, irritability, nervousness, restlessness, sweating, aggressiveness, anger, shakiness, stomach pain, diarrhea, and weight loss.

Animal models of drug dependence and drug withdrawal have been developed to study both the behavioral and the neurobiological targets of withdrawal. Usually animals are rendered dependent on a drug through either continuous exposure (osmotic minipumps, drug pellets) or repeated exposure (noncontingent systemic administration or self-administered injections) to drugs such as morphine, heroin, cocaine, amphetamine, nicotine, alcohol, tetrahydrocannabinol (THC), or their analogs. A *spontaneous withdrawal* is observed when drug exposure is discontinued and the substance is slowly cleansed from the body. The signs of withdrawal appear quite slowly and become stronger as time passes, progressing from a negative affective state to the expression of somatic signs. A *precipitated withdrawal* is obtained through the administration of an antagonist of the receptors on which the drug acts (for instance, naloxone for opiate dependence, mecamylamine for nicotine dependence, and CB1 receptor antagonist for THC dependence) though it should be mentioned that some cross-precipitation of withdrawal may occur (e.g., naloxone is able to induce both THC and nicotine withdrawal symptoms). When precipitated, the signs of withdrawal appear very rapidly and quite intensely due to the abrupt blockade of the receptors. Both affective and somatic components of withdrawal can be reproduced through the use of different doses of the antagonist (low doses induce only a negative affective state without somatic signs, whereas high doses induce somatic signs as well) (see Fig. 1 for an example of the different aspects of opiate withdrawal, depending on the dose of naloxone used).

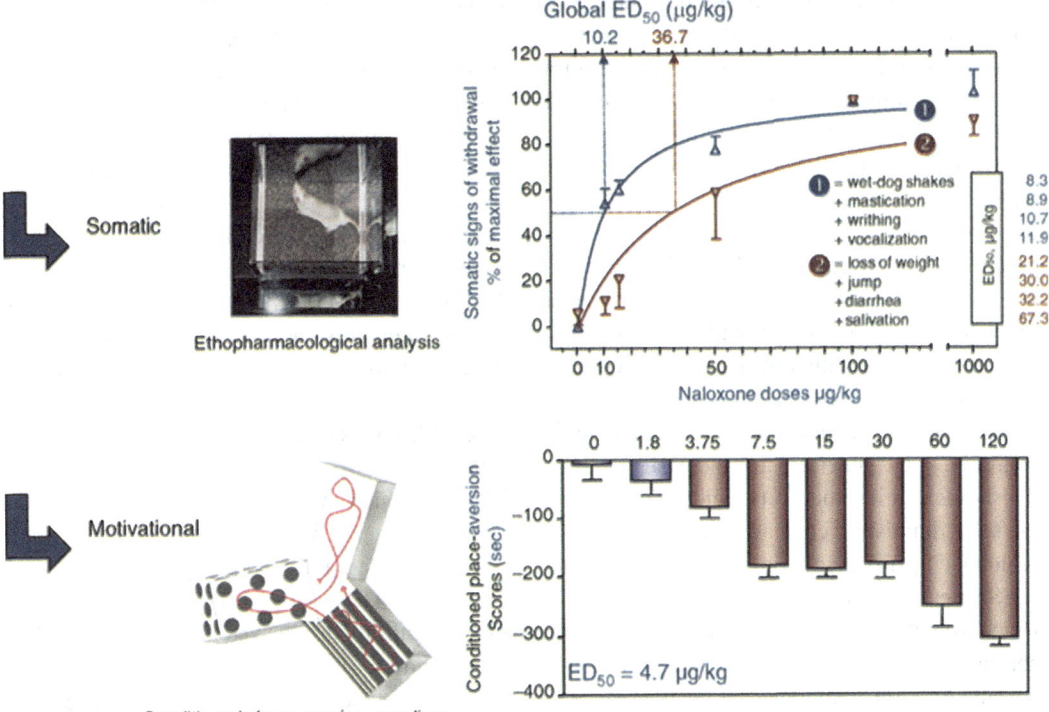

Withdrawal Syndromes, Fig. 1 Global behavioral analysis of naloxone-precipitated opiate withdrawal. The *upper panel* (*curves*) shows the expression of two groups of somatic signs of withdrawal (means ± SEM) in response to naloxone injections at the following doses: 0, 10, 15, 50, 100, and 1,000 μg/kg. The first group is represented by the blue curve that fits the means of the signs showing an ED_{50} below 20 μg/kg (body shake, mastication, writhing, and vocalization). The second group is represented by the *red curve* that fits the means of the signs showing an ED_{50} above 20 μg/kg (weight loss, jumping, diarrhea, and salivation). The global ED_{50} values corresponding to the first and the second group of somatic signs were, respectively, 10.2 and 36.7 μg/kg. The *lower panel* (*histogram*) shows the expression of naloxone-induced conditioned place aversion using the following doses of naloxone during the conditioning phase: 0, 1.8, 3.75, 7.5, 15, 30, and 120 μg/kg. Each bar represents the aversion score (means ± SEM) for one arm of an unbiased Y-maze, which had been previously paired with a naloxone injection during the conditioning phase (*brown histograms* illustrate significant aversions) (Adapted from Frenois et al. 2005)

Affective Component of Drug Withdrawal: The negative affective component of withdrawal can be seen in various rodent behavioral models, including conditioned place aversion, conditioned suppression of operant response to food, increased threshold for intracranial self-stimulation, and anxiety measures. In the conditioned place aversion (CPA) model, the negative affective state of withdrawal is repeatedly paired via classical conditioning with a specific environment. Subsequently, rats will avoid this environment and thus display conditioned place aversion, showing that the general negative affective state generated by withdrawal has been transferred to environmental stimuli (Frenois et al. 2005). Such conditioned stimuli have been shown to influence an animal's motivation toward the drug later on. The conditioned suppression model takes advantage of the interruption of an operant response (i.e., pressing a lever in order to obtain food). In this model, withdrawal is induced in dependent rats which are engaged in an operant response and paired with the presentation of a stimulus (sound or odor, for instance).

The aversive state generated by acute precipitated withdrawal then interrupts the operant response. Thereafter, the simple presentation of this conditioned stimulus will also be able to interfere with the operant response. In the intracranial self-stimulation (ICSS) model, animals learn to electrically self-stimulate a specific brain area – the medial forebrain bundle – until they reach a threshold of stable self-stimulation. In dependent rats, the induction of withdrawal produces an increase in *ICSS* thresholds, which is equivalent to a reduction in the reward sensitivity to the electrical stimulation of this brain area, indicating a withdrawal-induced dysphoric state. When withdrawal is paired with environmental stimuli, such stimuli are then able by themselves to generate an increase in ICSS thresholds. For instance, rats which were allowed to self-administer cocaine intravenously for prolonged periods of time showed elevated ICSS thresholds, the magnitude and duration of which were proportional to the amount of cocaine consumed. Anxiety and anhedonia are also components of early withdrawal: a decrease in time spent in the open arms of a plus maze or in the periphery of an open field has been repeatedly reported, as well as a decrease in the intake of natural rewards such as sucrose.

Physical Component of Drug Withdrawal: The rating of the many physical signs of drug withdrawal in animals is based on the ethological quantification of a broad range of behaviors and somatic symptoms that may be very specific to the drug used. These rated behaviors may vary in intensity and quality depending on the drug used. The major signs in rodents are weight loss, jumping, writhing, body shake, diarrhea, teeth chattering, mastication, swallowing, salivation, chromodacryorrhea, ptosis, chewing, vocalization, eye twitches, rhinorrhea, scratching, facial rubbing, hunched posture, ataxia, irritability, abnormal motor responses, anxiety-like responses, decreased reward sensitivity, and seizures. For instance, opiate withdrawal symptoms are quantified by using the different scales found in the literature. According to the scale of Gellert and Holtzman (1978), some of these indices are quantified in a graded manner whereas others are quantal (either present or absent). Another method, the ethoscore described by Espejo et al. (1994), also makes it possible to obtain a global score for the appearance of somatic withdrawal signs but is based only on the frequency of chewing and loss of weight. The scale of Gellert-Holtzman presents an asymptotic progression relative to increasing doses of naloxone (ceiling effect, see Fig. 1), whereas the ethoscore progresses linearly. Nicotine withdrawal symptoms are rated according to a scale designed by Malin et al. (1992). No scale exists for psychostimulants, as somatic symptoms from psychostimulants have not yet been reported.

Neural Circuits and Neurotransmitters Involved

The expression of a withdrawal syndrome is the consequence of neuroadaptations, which have been engendered by continuous/repeated drug exposure. These neuroadaptations can take place within the brain system on which the drug is acting directly (within system adaptations) and lead, for instance, to decreased or increased sensitivity and/or number of receptors, and increased or decreased release of the neurotransmitter, which activates these receptors endogenously. These neuroadaptations can also take place in systems different from the one on which the drug is acting directly (between-system adaptations), which are often counteracting the drug effects. These within- and between-system adaptations might develop with very different kinetics and may follow one after the other. Some of them are directly responsible for the onset of the behavioral manifestations of withdrawal, but others can be very long lasting and persist long after the drug is cleared and the overt signs of withdrawal have disappeared (Koob et al. 1989).

Using the different behavioral paradigms mentioned above, the neural substrates of withdrawal syndromes have been studied using different approaches such as brain site-specific injections of drug antagonist, c-fos imaging, and knockout mice (KO) for specific receptors. A brief synthesis of the findings related to the neurobiological substrates of the withdrawal syndrome is given below.

Brain Structures Involved: One of the major findings related to drug withdrawal is the dissociation between the neural systems involved in the motivational and somatic components of withdrawal. In terms of brain structures, it has been shown that limbic structures encompassing the nucleus accumbens, amygdala, BNST, and prefrontal cortex are involved in the negative affective part of withdrawal from drugs like opiates, cocaine, alcohol, and nicotine, whereas other structures such as the locus coeruleus and the periaqueductal gray (PAG) are involved in the somatic aspect of withdrawal. Using intracerebral injections of either opiate receptor antagonists (naloxone or methynaloxonium), nicotinic receptor antagonists (mecamylamine and dihydro-beta-erythroidine), or cannabinoid receptor antagonists (SR141716A or AM251), negative affective states are produced and expressed, by either conditioned place aversion, increased anxiety, or increased threshold for ICSS (reviews, Koob et al. 1992 for opiates; Kenny and Markou 2001 for nicotine; Tanda and Goldberg 2003 for cannabinoids). However, when the antagonists were injected in other structures such as the locus coeruleus (for opiates), somatic signs emerged even with low doses (Koob et al. 1992). These data were corroborated by c-fos imaging data showing that with low doses of antagonists, limbic structures were activated (among them the central nucleus of the amygdala has been shown to be activated with all drug withdrawal), whereas with higher doses leading to the expression of somatic signs, structures such as the locus coeruleus or PAG are activated (e.g., see Frenois et al. 2005 for opiate withdrawal).

Neurotransmitters and Receptors Involved: KO mice have been used for different types of receptors, showing that opiate mu-receptor KO mice do not show opiate withdrawal and cannabinoid CB1 receptor KO mice do not show THC withdrawal. Regarding nicotinic receptor KO mice, there is some evidence that some receptor subtypes are involved in the motivational component ($\alpha 4\beta 2$), while others are involved in the physical component ($\alpha 5\beta 4$) of withdrawal. However, available data are conflicting and there is at present no consensus over the role of receptor subtypes in nicotine withdrawal. Among neurotransmitters, corticotropin-releasing factors (CRF) and dopamine have been repeatedly involved in the negative affective state associated with withdrawal. Peripheral administration of a CRF antagonist blocks the conditioned place aversion produced by either opiate withdrawal, nicotine withdrawal, or alcohol withdrawal. This effect takes place principally at the level of the central nucleus of the amygdale (CEA) where local CRF receptor antagonist administration (principally CRF1 receptor) diminishes the negative affective state of opiate, nicotine, and alcohol withdrawal, using CPA, ICSS, or anxiety measures (Koob 2008). Thus, intra-amygdala injection of CRF precipitates a negative affective state in opiate-, nicotine-, and alcohol-dependent rats. In rodents, during nicotine, alcohol, opiate, and cannabinoid withdrawal, CRF neurotransmission is increased (review Koob 2008). This CRF neurotransmission involvement which has for long been related to the stress system and the stress response may participate in the anxiogenic aspect of withdrawal.

Dopamine release is decreased in the nucleus accumbens during withdrawal from cocaine, opiate, nicotine, alcohol, and cannabinoid. This reduced release of dopamine may explain the decreased sensitivity to rewards, as demonstrated using ICSS or sweet solution consumption. However, this might be oversimple as, for example, a lesion of dopamine neurons does not modify the affective and somatic aspects of withdrawal (Caillé et al. 2003).

Besides dopamine, other monoamines have also been involved. It has been suggested that noradrenergic functional antagonists can block some aspects of ethanol or nicotine withdrawal (Koob 2008 for alcohol). However, a noradrenergic lesion did not block opiate withdrawal, whereas clonidine (alpha2 agonist) abolished the negative affective state of opiate withdrawal and some (but not all) aspects of somatic withdrawal.

Other neurotransmitters have also been involved, though data are still incomplete. For instance, an increase in glutamatergic neurotransmission has been associated with opiate withdrawal and dynorphin neurotransmission has also been proposed to mediate the negative state of drug withdrawal (Koob 2008).

It is clear that our understanding of the behavioral and neurobiological effects of drug withdrawal is far from complete, though it is progressing rapidly. Using behavioral and neurobiological screening, systematic comparisons among withdrawal from different drugs should bring important information about potential therapeutic targets that can ameliorate withdrawal states and maintain abstinence.

Cross-References

- ▶ Addiction
- ▶ Addictive Disorder: Animal Models
- ▶ Alcohol
- ▶ Anxiety
- ▶ Cannabinoids
- ▶ Cocaine
- ▶ Conditioned Place Preference and Aversion
- ▶ Dopamine
- ▶ Elevated Plus Maze
- ▶ Liquid Diet for Administering Alcohol
- ▶ Morphine
- ▶ Nicotine
- ▶ Noradrenaline
- ▶ Open-Field Test
- ▶ Opioids
- ▶ Physical Dependence
- ▶ Serotonin

References

Caillé S, Rodriguez-Arias M, Minarro J, Espejo EF, Cador M, Stinus L (2003) Changes in dopaminergic neurotransmission do not alter somatic or motivational opiate withdrawal-induced symptoms in rats. Behav Neurosci 117:995–1005

Espejo EF, Stinus L, Cador M, Mir D (1994) Effects of morphine and naloxone on behaviour in the hot plate test: an ethopharmacological study in the rat. Psychopharmacology (Berl) 113(3–4):500–510

Frenois F, Le Moine C, Cador M (2005) The motivational component of withdrawal in opiate addiction: role of associative learning and aversive memory in opiate addiction from a behavioral, anatomical and functional perspective. Rev Neurosci 16(3):255–276

Gellert VF, Holtzman SG (1978) Development and maintenance of morphine tolerance and dependence in the rat by scheduled access to morphine drinking solutions. J Pharmacol Exp Ther 205:536–546

Jaffe JH (1990) Drug addiction and drug abuse. Pergamon Press, New York

Kenny PJ, Markou A (2001) Neurobiology of the nicotine withdrawal syndrome. Pharmacol Biochem Behav 70(4):531–549

Koob GF (2008) A role for brain stress systems in addiction. Neuron 59:11–34

Koob GF, Stinus L, Le Moal M, Bloom FE (1989) Opponent process theory of motivation: neurobiological evidence from studies of opiate dependence. Neurosci Biobehav Rev 13(2–3):135–140 (Review)

Koob GF, Maldonado R, Stinus L (1992) Neural substrates of opiate withdrawal. Trends Neurosci 15(5):186–191 (Review)

Malin DH, Lake JR, Newlin-Maultsby P, Roberts LK, Lanier JG, Carter VA, Cunningham JS, Wilson OB (1992) Rodent model of nicotine abstinence syndrome. Pharmacol Biochem Behav 43(3):779–784

Tanda G, Goldberg SR (2003) Cannabinoids: reward, dependence, and underlying neurochemical mechanisms – a review of recent preclinical data. Psychopharmacology (Berl) 169(2):115–134 (Review)

Working Memory

Definition

A theoretical construct within cognitive psychology that refers to the structures and processes used for temporarily storing and manipulating information.

Cross-References

- ▶ Short-Term and Working Memory in Animals
- ▶ Short-Term and Working Memory in Humans

Y

Yale-Brown Obsessive-Compulsive Scale

Synonyms

Y-BOCS

Definition

The 10-item Y-BOCS scale, which evaluates the severity of obsessions and compulsions separately, is the standard scale used in treatment outcome studies of obsessive-compulsive disorder (OCD) (Goodman et al. 1989a, b). The rater measures obsessions and compulsions in terms of the time occupied, how much they interfere with functioning, the patient's degree of distress, and his or her attempts to resist symptoms and the ability to control them successfully. To assess the patient's symptoms, the clinician may wish to use the Y-BOCS symptom checklist, which lists 40 obsessions and 29 compulsions. The 18-item Obsessive-Compulsive Inventory is a shorter alternative. The Y-BOCS scale and checklist along with instructions for their use are available at http://apple.cmu.edu.tw/_u901039/Rating-Scale-YBOCS.pdf. In clinical practice, some clinicians measure symptom change by simply asking the patient to estimate the amount of time taken by symptoms in an average day in the past week.

Cross-References

▶ Compulsions
▶ Obsessions

References

Goodman WK, Price LH, Rasmussen SA, Mazure C, Fleischmann RL, Hill CL, Henninger GR, Charney DS (1989a) The Yale-Brown obsessive compulsive scale: part I. Development, use, and reliability. Arch Gen Psychiatry 46:1006–1012

Goodman WK, Price LH, Rasmussen SA, Mazure C, Delgado P, Heninger GR, Charney DS (1989b) The Yale-Brown obsessive compulsive scale: part II. Validity. Arch Gen Psychiatry 46:1012–1016

Yohimbine

Definition

Yohimbine is a naturally occurring alkaloid. By blocking presynaptic α2-adrenoceptors (autoreceptors), it increases noradrenaline release and produces sympathomimetic effects in some organs. It has stimulating properties and is purported to have aphrodisiac qualities.

Cross-References

▶ Sexual Behavior
▶ Sexual Disorders
▶ Short-Term and Working Memory in Animals

Z

Zaleplon

Definition

Zaleplon is a non-benzodiazepine sedative-hypnotic that potentiates gamma-amino-butyric acid (► GABA) neurotransmission by binding to the α-subunit of the ► $GABA_A$ receptor complex, with some selectivity for the α1-subunit. Zaleplon is an effective treatment for insomnia and has an extremely brief duration of action that makes it particularly suitable for treating difficulties in falling asleep (it decreases latency of sleep onset). This drug does not substantially alter the quality of sleep and has little or no residual hangover effects, and the recovery from sedation is more rapid than with other ► hypnotics. Zaleplon has a similar side-effect profile to benzodiazepines and other non-benzodiazepine ► hypnotics, although the adverse effects on cognition and psychomotor function are reduced because of the relative selectivity for the ω_1 receptor and the short half-life. Tolerance and dependence may develop with long-term use.

Cross-References

► Benzodiazepines
► Hypnotics
► Non-benzodiazepine Hypnotics

Zero-Order Elimination Kinetics

Definition

Sometimes a drug is absorbed at essentially a constant rate, called zero-order absorption. Zero-order kinetics occur when a constant amount of drug is eliminated per unit time and the rate is independent of the concentration of the drug.

Cross-References

► Bioavailability
► Elimination Half-Life
► Pharmacokinetics

Ziprasidone

Definition

Ziprasidone acts at multiple receptors but mainly as a potent $5HT_{2A}$ receptor antagonist. It is a second-generation antipsychotic with a benzothiazolyl-piperazine structure. The ► half-life is 4–10 h and it is metabolized by various CYP450 isoenzymes, most potently by 3A4. It has to be taken with food. Ziprasidone also

I.P. Stolerman, L.H. Price (eds.), *Encyclopedia of Psychopharmacology*,
DOI 10.1007/978-3-642-36172-2

inhibits noradrenaline and serotonin reuptake, but the clinical relevance of these pharmacological effects has not yet been explored. It is also available as an acute intramuscular preparation. Among the newer antipsychotics, ziprasidone has a lower propensity to induce metabolic side effects.

Cross-References

▶ Second-Generation Antipsychotics

Zolpidem

Definition

Zolpidem is a non-benzodiazepine sedative-hypnotic belonging to the class of imidazopyridines. It potentiates gamma-amino butyric acid (▶ GABA) transmission by binding to the α-subunit of the ▶ $GABA_A$ receptor complex with some selectivity for the α1-subunit. It is commonly prescribed for short-term treatment of insomnia which should not exceed 2 weeks' duration as the likelihood of ▶ tolerance, ▶ dependence, and rebound ▶ withdrawal symptoms increases with prolonged use. Like benzodiazepines, it also has anxiolytic, muscle relaxant, and anticonvulsant properties, but these are very weak and require higher doses, which increase the severity of side effects. Clinical reports suggest that zolpidem may also be used to dramatically improve the condition of patients with some brain injuries. The side-effect profile of zolpidem is similar to that of other sedative-hypnotics and additionally includes symptoms such as hallucinations and delusions. Additionally, some patients report sleepwalking, migraine, subjective feelings of intoxication, manic reactions, and panic attacks. Tolerance and dependence may develop with long-term use.

Cross-References

▶ Benzodiazepines
▶ Hypnotics
▶ Non-benzodiazepine Hypnotics

Zonisamide

Definition

Zonisamide is a sulfonamide ▶ anticonvulsant. It is a ▶ GABA agonist that also reduces ▶ glutamate function and has a ▶ half-life of 63 h. It has been studied as an adjunctive treatment in various ▶ neuropsychiatric disorders ranging from migraine to ▶ bipolar disorder without leading to conclusive recommendations for its use beyond epilepsy.

Zopiclone

Definition

Zopiclone is a non-benzodiazepine sedative-hypnotic acting as an agonist at the benzodiazepine binding site on gamma-aminobutyric acid ($GABA_A$) receptors. It is used for short-term treatment of insomnia, to improve both the initiation and maintenance of sleep. Common side effects include bitter metallic taste, disruption of REM sleep, drowsiness, nausea and vomiting, irritability, confusion, depression, and a lack of coordination. More severe side effects including headache, hallucinations, nightmares, and amnesia have also been reported but appear to be rare. Tolerance and dependence may develop with long-term use.

Cross-References

▶ Benzodiazepines
▶ Hypnotics
▶ Non-benzodiazepine Hypnotics

Zotepine

Definition

Zotepine is a second-generation antipsychotic that acts as a potent dopamine D2 and 5HT$_{2A}$ antagonist. Chemically, it is in the dibenzodiazepine class. It has a plasma half-life of 21 h and is metabolized by 1A2 and 3A4 CYP450 isoenzymes. It also inhibits noradrenaline reuptake but the clinical relevance of this effect has not yet been explored.

Cross-References

▶ Second-Generation Antipsychotics

Zuclopenthixol

Definition

Zuclopenthixol is a multi-receptor blocking ▶ thioxanthene ▶ antipsychotic of the first generation. Its plasma ▶ half-life amounts to 12–28 h, and it is mainly metabolized by 2D6 CYP450 isoenzymes. It is also available in a short-acting intramuscular depot preparation that has an ▶ elimination half-life of 48–72 h.

Cross-References

▶ First-Generation Antipsychotics

List of Synonyms

A

Abeta
▶ Amyloid-Beta

Abridged Somatization
▶ Illness Anxiety Disorder, Somatic Symptom Disorder, and Body Dysmorphic Disorder

Abstinence Syndrome
▶ Withdrawal Syndromes

Abuse Potential
▶ Abuse Liability
▶ Abuse Liability Evaluation

Acquired Tolerance
▶ Tolerance

Action Inhibition
▶ Behavioral Inhibition

Active Immunization
▶ Vaccination

Activities of Living Scales
▶ Functional Scales

Acute Brain Failure
▶ Delirium

Acute Brain Syndrome
▶ Delirium

Acute Confusional State
▶ Delirium

Acute Tyrosine/Phenylalanine Depletion
▶ Phenylalanine and Tyrosine Depletion

AD-5423
▶ Blonanserin

Adaptability
▶ Behavioral Flexibility: Attentional Shifting, Rule Switching, and Response Reversal

Add on Therapy
▶ Drug Interactions

Addiction Stroop Test
▶ Attentional Bias to Drug Cues

Adenine-9-β-D-Ribofuranoside
▶ Adenosine

Adenine Riboside
▶ Adenosine

ADH
▶ Acetaldehyde
▶ Alcohol Dehydrogenase
▶ Arginine-Vasopressin

ADHD
▶ Attention-Deficit/Hyperactivity Disorder

Adjunctive Behavior
▶ Schedule-Induced Polydipsia

Adjunctive Drinking
▶ Schedule-Induced Polydipsia

Adjustability
▶ Elasticity

Adolescent Neurodevelopmental Vulnerability
▶ Adolescence and Responses to Drugs

Adult Neurogenesis
▶ Neurogenesis

Adventuresome
▶ Risk Taking

Adverse Reaction
▶ Adverse Effect

Adverse Side Effect
▶ Adverse Effect

AEA
▶ N-Arachidonoylethanolamine

Affect
▶ Emotion and Mood

Affective Disorders
▶ Mood Disorders

Aggressive Behavior
▶ Aggression
▶ Aggressive Behavior: Clinical Aspects

Aggressiveness
▶ Aggression
▶ Aggressive Behavior: Clinical Aspects

AIF
▶ Apoptosis-Inducing Factor

Alarm Calls
▶ Distress Vocalization

Alcohol Addiction
▶ Alcohol Abuse and Dependence

Alcohol (Ethanol) Choice Tests
▶ Alcohol Preference Tests

Alcohol (Ethanol) Drinking Preference Tests
▶ Alcohol Preference Tests

Alcohol (Ethanol) Reward Tests
▶ Alcohol Preference Tests

Alcohol Use Disorder
▶ Alcohol Abuse and Dependence

Alcohol Withdrawal Symptoms
▶ Alcohol Withdrawal-Related Anxiety

ALDH
▶ Alcohol Dehydrogenase

Allotropy
▶ Genetic Polymorphism

Allowable
▶ Legal Aspects of Psychopharmacology

(RS)-1-[2-(Allyloxy)Phenoxy]-3-(Isopropylamino)Propan-2-ol
▶ Oxprenolol

(S)-Alpha-Hydrazino-3,4-Dihydroxy-α-Methyl-Bensemonopropanoic Acid Monohydrate
▶ Carbidopa

Alpha Waves
▶ Function of Slow and Fast Alpha Waves

Amentia
▶ Autism Spectrum Disorders and Intellectual Disability

2-Amino-3-Hydroxy-N'-[(2,3,4-Trihydroxyphenyl) Methyl] Propanehydrazide
▶ Benserazide

(2R,3R,4R,5R)-2-(6-Aminopurin-9-yl)-5-(Hydroxymethyl)Oxolane-3,4-diol
▶ Adenosine

Amnestic Compounds
▶ Inhibition of Memory

AMPAR
▶ AMPA Receptor

Amperometry
▶ Electrochemical Techniques

Amphipathic
▶ Amphiphilic

Amylobarbitone
▶ Amobarbital

Anabolic Steroids
▶ Sex Hormones

Analgesia Tests
▶ Antinociception Test Methods

Anandamide
▶ N-Arachidonoylethanolamine
▶ Cannabinoids and Endocannabinoids

Anesthesia
▶ General Anesthesia

Animal Model with Construct Validity
▶ Simulation Models

Animal Models of Psychopathology
▶ Animal Models for Psychiatric States

Animal Tests of Anxiety
▶ Anxiety: Animal Models

Anorectics
▶ Appetite Suppressants

Anorexia
▶ Eating Disorder: Anorexia Nervosa

Antianxiety Agents
▶ Punishment Procedures

Antianxiety Drugs
▶ Anxiolytics

Antianxiety Medication
▶ Anxiolytics
▶ Minor Tranquilizer

Anti-Cholinesterases
▶ Acetylcholinesterase and Cognitive Enhancement

Anticipatory Food Seeking
▶ Goal Tracking

Anticipatory Goal Seeking
▶ Goal Tracking

Antidiuretic Hormone
▶ Arginine-Vasopressin

Antiepileptics
▶ Anticonvulsants

Antihistamines
▶ Histaminic Agonists and Antagonists

Antipsychotic Drugs
▶ Classification of Psychoactive Drugs

Antipsychotic-Induced Movement Disorders
▶ Movement Disorders Induced by Medications

Antisense DNA
▶ Antisense Oligonucleotides

Anxiety Neurosis
▶ Generalized Anxiety Disorder

Anxiety or Mixed States
▶ Adjustment Disorders

Anxiety-Reducing Drugs
▶ Anxiolytics

Anxiolytic Dependence
▶ Sedative, Hypnotic, and Anxiolytic Dependence

APOE
▶ Apolipoprotein E

Apparent Volume of Distribution
▶ Volume of Distribution

Appetitive Conditioning
▶ Conditioned Taste Preferences

Approach-Avoidance
▶ Punishment Procedures

Approach Response
▶ Appetitive Responses

Approval and Marketing of Psychotropic Drugs
▶ Ethical Issues in Human Psychopharmacology

ARC
▶ Addiction Research Center

ARCI
▶ Addiction Research Center Inventory

ARND
▶ Alcohol-Related Neurodevelopmental Disorder

Aropax
▶ Paroxetine

Arousal Disorders
▶ Parasomnias

Arylalkylamines
▶ Trace Amines

Assessing Brain Function
▶ Rating Scales and Diagnostic Schemata

Associated Depression in Schizophrenia
▶ Postpsychotic Depressive Disorder of Schizophrenia

Associative Learning
▶ Classical (Pavlovian) Conditioning
▶ Verbal and Nonverbal Learning in Humans

At-Risk Mental State
▶ Prepsychotic States and Prodromal Symptoms

Atomoxetine Hydrochloride
▶ Atomoxetine

Atypical Autism
▶ Pervasive Developmental Disorder Not Otherwise Specified

AUC
▶ Area Under the Curve
▶ Distribution
▶ Pharmacokinetics

AUD
▶ Alcohol Abuse and Dependence

Aurorix
▶ Moclobemide

Autistic Disorder
▶ Autism

Autonomous
▶ Independents

Aversive Drug Effects
▶ Conditioned Taste Aversions

Avoidance of Feared Places and Situations: Anticipatory Anxiety
▶ Agoraphobia

Avoidant Personality Disorder
▶ Social Anxiety Disorder

AVP
▶ Arginine-Vasopressin

Aβ
▶ Amyloid-Beta

B

B_{avail}
▶ B_{max}

Barbitone
▶ Barbital

Barbiturate Derivatives
▶ Barbiturates

BDNF
▶ Brain-Derived Neurotrophic Factor

Bed Nucleus of the Stria Terminalis
▶ BNST

Behavioral Addictions
▶ Impulse Control Disorders

Behavioral Augmentation
▶ Sensitization to Drugs

Behavioral Characterization
▶ Phenotyping of Behavioral Characteristics

Behavioral Facilitation
▶ Sensitization to Drugs

Behavioral Flexibility
▶ Behavioral Inhibition

Behavioral Models of Psychopathology
▶ Animal Models for Psychiatric States

Behavioral Phenotyping
▶ Phenotyping of Behavioral Characteristics

Behavioral Tests
▶ Primate Models of Cognition

Benzedrine
▶ Amphetamine

Benzodiazepine Dependence
▶ Sedative, Hypnotic, and Anxiolytic Dependence

Benztropine
▶ Benzatropine

Beta Amyloid
▶ Amyloid-Beta

Beta-Adrenoceptor Blocking Drugs
▶ Beta-Adrenoceptor Antagonists

Beta-Blockers
▶ Beta-Adrenoceptor Antagonists

Biased Agonism
▶ Functional Selectivity

BIM-23014
▶ Lanreotide

Binge Eating
▶ Binge

Binge-Eating Disorder
▶ Eating and Appetite

Biologic Rhythms and Medicine
▶ Circadian Rhythms

Biological Clock and Drugs
▶ Circadian Rhythms

Biological Clock and Pharmacology
▶ Circadian Rhythms

Biological Half-Life
▶ Elimination Half-Life
▶ Half-Life

Biotransformation
▶ Metabolism

Bipolar Illness
▶ Bipolar Disorder

Blood Oxygenation Level-Dependent Contrast
▶ BOLD Contrast

BMI
▶ Body Mass Index

BOLD
▶ Cerebral Perfusion

BP
▶ Binding Potential

BP_1
▶ Binding Potential

BP_2
▶ Binding Potential

BP_F
▶ Binding Potential

BP_{ND}
▶ Binding Potential

BP_P
▶ Binding Potential

Brain Atrophy
▶ Neuroprotectants: Novel Approaches for Dementias

Brain Imaging
▶ Neuroimaging

Brain Mapping
▶ Neuroimaging

Brain Microdialysis
▶ Microdialysis

Brain Reward Systems
▶ Mesotelencephalic Dopamine Reward Systems

Breaking Point
▶ Breakpoint

Brexiaceae
▶ Celastraceae

2-Bromo-Alpha-Ergocryptine
▶ Bromocriptine

2-Bromo-12'-Hydroxy-2'(1-Methylethyl)-5'-(2-Methylpropyl)Ergotoman-3',6',18-Trione
▶ Bromocriptine

BSR
▶ Brain Stimulation Reward

C

Caffeine Abstinence
▶ Caffeine Withdrawal Syndrome

Caffeine 'Jitters'
▶ Caffeine Intoxication

Calcium Acetylhomotaurinate
▶ Acamprosate

CAM
▶ Confusion Assessment Method

Camazepam
▶ Benzodiazepines

Cambridge Neuropsychological Test Automated Battery
▶ CANTAB

Campral
▶ Acamprosate

Cannabis Addiction
▶ Cannabis Use Disorders

Canotiaceae
▶ Celastraceae

Capgras Syndrome
▶ Delusional Disorder

Catalytic RNA Molecule
▶ Ribozyme

Catathrenia
▶ Parasomnias

Catecholamine Depletion
▶ Amine Depletors

Catecholamine Hypothesis
▶ Aminergic Hypotheses for Depression

Catecholamine Toxins
▶ Neurotoxins

CB-154
▶ Bromocriptine

CBF
▶ Cerebral Perfusion

CBIT
▶ Habit Reversal Therapy

CBT
▶ Cognitive Behavioral Therapy

CBV
▶ Cerebral Perfusion

Cell Type-Specific Knockout
▶ Conditional Knockout

Cerebral Blood Flow
▶ Cerebral Perfusion

Cerebral Blood Volume
▶ Cerebral Perfusion

Cerebrovascular Accident
▶ Stroke
▶ Clinical Global Impressions Scales

Characteristic
▶ Trait

Chat
▶ Khat

ChAT
▶ Choline Acetyltransferase

Childhood Depression
▶ Depressive Disorders in Children

Childhood-Onset Obsessive–Compulsive Disorder
▶ Obsessive-Compulsive Disorders in Childhood

Childhood-Onset Schizophrenia
▶ Pediatric Schizophrenia

Childhood Schizophrenia
▶ Autism Spectrum Disorders and Intellectual Disability

Chingithamnaceae
▶ Celastraceae

ChIP
▶ Chromatin Immunoprecipitation

Chlormethiazole
▶ Clomethiazole

Chronic Hairpulling
▶ Trichotillomania

Chronic Low-Grade Depression
▶ Dysthymic Disorder (Persistent Depressive Disorder)

Chronomedicine
▶ Circadian Rhythms

Chronopharmacology
▶ Circadian Rhythms

CL
▶ Clearance

Classical Anticonvulsants
▶ First-Generation Anticonvulsants

Classical Antipsychotics
▶ First-Generation Antipsychotics

Classical Fear Conditioning
▶ Pavlovian Fear Conditioning

Classical Neuroleptics
▶ First-Generation Antipsychotics
▶ Typical Antipsychotics

Clinical Depression
▶ Depressive Disorders: Major, Minor, and Mixed

Clinical Outcome
▶ Efficacy

CM
▶ Contingency Management
▶ Contingency Management in Drug Dependence

Coca Paste
▶ Cocaine

Cocaine Abuse
▶ Cocaine Dependence

Cocaine Addiction
▶ Cocaine Dependence

Cocaine Hydrochloride
▶ Cocaine

Cocaine Hydrochloride Abuse
▶ Cocaine Dependence

Cocaine Hydrochloride Addiction
▶ Cocaine Dependence

Coffee Nerves
▶ Caffeine Intoxication

Cognitive Control
▶ Executive Functions

Cognitive Deficit
▶ Cognitive Impairment

Cognitive Flexibility
▶ Behavioral Flexibility: Attentional Shifting, Rule Switching, and Response Reversal

Companion
▶ Complements

Comprehensive Behavioral Intervention for Tics or Trichotillomania
▶ Habit Reversal Therapy

Compulsive Acts
▶ Compulsions

Compulsive Gambling
▶ Pathological Gambling

Compulsive Rituals
▶ Compulsions

Concept Formation Test
▶ Wisconsin Card Sorting Test

Conditioned Avoidance Response
▶ Active Avoidance

Conditioned Catalepsy
▶ Context-Dependent Catalepsy

Conditioned Emotional Response
▶ Pavlovian Fear Conditioning
▶ Punishment Procedures

Conditioned Flavor Preferences
▶ Conditioned Taste Preferences

Conditioned Freezing
▶ Pavlovian Fear Conditioning

Conditioned Inhibition
▶ Conditioned Inhibitor

Conditioned Locomotion
▶ Conditioned Activity

Conditioned Locomotor Activity
▶ Conditioned Activity

Conditioned Reinforcement
▶ Appetitive Responses

Conditioned Taste Avoidance
▶ Conditioned Taste Aversions

Confusional Arousals
▶ Parasomnias

Consciously Accessible and Nonconsciously Accessible Memory
▶ Declarative and Nondeclarative Memory

Constitutional
▶ Legal Aspects of Psychopharmacology

Constitutive Knockout
▶ Knockout/Knockin

Context and Cued Conditioning
▶ Pavlovian Fear Conditioning

Context-Specific Drug Effects
▶ Conditioned Drug Effects

Contingent Tolerance
▶ Behavioral Tolerance

Continuous Reinforcement
▶ Fixed Ratio

Controlled Clinical Trial
▶ Phase II Clinical Trial
▶ Phase III Clinical Trial

Controlled Clinical Trials
▶ Randomized Controlled Trials

Conventional Antipsychotics
▶ First-Generation Antipsychotics
▶ Typical Antipsychotics

Conventional Neuroleptics
▶ First-Generation Antipsychotics

Convulex
▶ Valproic Acid

Cortical or Brain Waves
▶ Electroencephalography

Corticotropin-Releasing Hormone
▶ Corticotropin-Releasing Factor

Cost Utility of the Latest Antipsychotic Drugs in Schizophrenia Study
▶ CUtLASS

CPP
▶ Conditioned Place Preference and Aversion

CR
▶ Conditioned Response

CRF
▶ Corticotropin-Releasing Factor

CRH
▶ Corticotropin-Releasing Factor

Crying
▶ Distress Vocalization

CS
▶ Conditioned Stimulus

CSA
▶ Controlled Substances Act

5-CSRT
▶ Five-Choice Serial Reaction Time Task

5-CSRTT
▶ Five-Choice Serial Reaction Time Task

CST
▶ Cortistatin

CST14
▶ Cortistatin

CST17
▶ Cortistatin

CST29
▶ Cortistatin

Cue (in psychology)
▶ Discriminative Stimulus
▶ Drug Cues
▶ Stimulus Generalization

Cue Competition
▶ Overshadowing

Cueing Properties of Drugs
▶ Drug Discrimination

Cumulative Distance Moved
▶ Distance Traveled

Current Mood
▶ Affective State
▶ Decision Making

Cyclic-Adenosine Monophosphate
▶ cAMP

CYP
▶ Cytochrome P450

CYP450
▶ Cytochrome P450

Cytotec
▶ Misoprostol

D

Δ^9-Tetrahydrocannabinol
▶ Cannabinoids
▶ Cannabinoids and Endocannabinoids

DA
▶ Dopamine

DAT
▶ Dopamine Transporter

Davedax
▶ Reboxetine

DBS
▶ Deep Brain Stimulation

2-DE
▶ Two-Dimensional Gel Electrophoresis

De Clerambault's Syndrome
▶ Delusional Disorder

Decomposition
▶ Spectrograms

Deconvolution
▶ Spectrograms

Defensive Responses
▶ Defensive Behaviors

Defensive Threat and Attack
▶ Defensive Behaviors

Deficit Symptoms Syndrome
▶ Negative Symptoms Syndrome

Degraded Contingency
▶ Relative Validity

Delorazepam
▶ Benzodiazepines

Delta Activity
▶ Function of Delta Waves

Delusion of Interpretation
▶ Delusional Disorder

Delusions of Infestation
▶ Delusional Disorder

Delusional Halitosis
▶ Delusional Disorder

Delusional Parasitosis
▶ Delusional Disorder

Dementia Praecox
▶ Schizophrenia

Dementia with Lewy Bodies
▶ Lewy Body Dementia

Demerol
▶ Pethidine

Demoralization Syndrome
▶ Chronic Disappointment Reaction

Depakine
▶ Valproic Acid

Depression Medications
▶ Antidepressants: Recent Developments

Depression NOS
▶ Postpsychotic Depressive Disorder of Schizophrenia

Depression Superimposed on Residual Schizophrenia
▶ Postpsychotic Depressive Disorder of Schizophrenia

Dermatophobia
▶ Delusional Disorder

Desoxyephedrine
▶ Methamphetamine

Desvenlafaxine Succinate
▶ Desvenlafaxine

Developmental Neurotoxins
▶ Perinatal Exposure to Drugs

Dexmethylphenidate
▶ Methylphenidate and Related Compounds

DF-118
▶ Dihydrocodeine

2-DG
▶ Deoxyglucose

DHC
▶ Dihydrocodeine

Diacetylmorphine
▶ Diamorphine

Dietary Tryptophan Depletion
▶ Tryptophan Depletion

Diethyl Barbituric Acid
▶ Barbiturates

Dihydrocodeinone
▶ Hydrocodone

Dihydromorphinone
▶ Hydromorphone

1,3-Dihydro-7-Nitro-5-Phenyl-2H-1,4-Benzodiazepin-2-One
▶ Nitrazepam

L-3,4-Dihydroxyphenylalanine
▶ Levodopa, L-DOPA

N-[(3-Dimethylamino)Propyl]-N-[(Ethylamino)Carbonyl]-6-(2-Propenyl)-Ergoline-8b-Carboxamide
▶ Cabergoline

N,N Dimethyl-2[α-(o-Tolyl)Benzyloxy] Ethylamine HCl or Citrate
▶ Orphenadrine

Dimorphone
▶ Hydromorphone

2-[(Diphenylmethyl)Sulfinyl]acetamide
▶ Modafinil

Disability Scales
▶ Functional Scales

Discriminative Cue
▶ Discriminative Stimulus
▶ Stimulus Generalization

Disruptive Behavior Disorders
▶ Externalizing Disorders

Dissociated Learning
▶ State Dependence of Memory

DLB
▶ Lewy Body Dementia

DMTS
▶ Delayed Match-to-Sample Test

DNA-Marking
▶ Epigenetics

Dopamine- and Cyclic-AMP-Regulated 32 KDa Phosphoprotein
▶ DARPP-32

Dopamine Reward Systems
▶ Mesotelencephalic Dopamine Reward Systems

Double-Masked Studies
▶ Double-Blinded Study

DRS
▶ Delirium Rating Scale

Drug
▶ Medicine

Drug Abstinence
▶ Withdrawal Syndromes

Drug Addiction
▶ Dependence
▶ Opioid Use Disorder and Its Treatment
▶ Substance Use Disorders: A Theoretical Framework

Drug as Cues
▶ Drug Discrimination
▶ Occasion Setting with Drugs

Drug-Associated Stimuli
▶ Drug Cues

Drug Augmentation
▶ Drug Interactions

Drug Combinations
▶ Drug Interactions

Drug Discontinuation
▶ Withdrawal Syndromes

Drug Discriminative Stimulus
▶ Drug Discrimination
▶ Occasion Setting with Drugs

Drug Effects
▶ Environmental Enrichment and Drug Action
▶ Rate-Dependency Theory

Drug Facilitator
▶ Occasion Setting with Drugs

Drug Licensing
▶ Licensing and Regulation of Medicines in the UK

Drug Modulator
▶ Occasion Setting with Drugs

Drug Occasion Setter
▶ Occasion Setting with Drugs

Drug Plasma Level Determination for Therapy Optimization
▶ Therapeutic Drug Monitoring

Drug Preferences
▶ Drug Taste Preference Conditioning

Drug Seeking
▶ Self-Administration of Drugs

Drug Stimuli
▶ Drug Cues
▶ Drug Discrimination

Drug Taking
▶ Self-Administration of Drugs

Drug-Induced Motor Syndromes
▶ Movement Disorders Induced by Medications

Drugs as States
▶ State Dependence of Memory

Duloxetine Hydrochloride
▶ Duloxetine

Dysthymia
▶ Dysthymic Disorder (Persistent Depressive Disorder)

Dystrobrevin-Binding Protein 1
▶ Dysbindin

E

E
▶ Methylenedioxymethamphetamine (MDMA)

EAAT
▶ Glutamate and EAA Transporters

Early-Onset Schizophrenia
▶ Pediatric Schizophrenia

EBM
▶ Evidence-Based Medicine

Eccie
▶ Methylenedioxymethamphetamine (MDMA)

Ecological Genetics
▶ Ecogenetics

Ecstasy
▶ Methylenedioxymethamphetamine (MDMA)

ECT
▶ Electroconvulsive Therapy

ED
▶ Extradimensional

Edronax
▶ Reboxetine

EEG
▶ Electroencephalography

Effectiveness of Antipsychotic Drugs in First-Episode Schizophrenia and Schizophreniform Disorder Study
▶ EUFEST

Ekbom Syndrome
▶ Delusional Disorder

Eldepryl
▶ Selegiline

Electroencephalogy
▶ Electroencephalography

2-D Electrophoresis
▶ Two-Dimensional Gel Electrophoresis

Elevated Prolactin (Luteotropic Hormone)
▶ Hyperprolactinemia

Elevated X-Maze
▶ Elevated Plus Maze

Elimination
▶ Excretion

Emotional State
▶ Affective State
▶ Decision Making

Empathogen
▶ Entactogen

Emsam Patch
▶ Selegiline

Enantiomers
▶ Stereoisomers

Enantiomorphs
▶ Stereoisomers

Endo-3-(Diphenylmethoxy)-8-Methyl-8-Azabicyclo[3.2.1]Octane Methanesulfonate
▶ Benzatropine

Energy Intake
▶ Eating and Appetite

Entheogen
▶ Hallucinogen Abuse and Dependence

Entheogens
▶ Hallucinogens
▶ Ritual Uses of Psychoactive Drugs

EOS
▶ Pediatric Schizophrenia

Epival
▶ Valproic Acid

EPS
▶ Extrapyramidal Motor Side Effects
▶ Movement Disorders Induced by Medications

Erotomania (Also Known as De Clerambault's Syndrome)
▶ Delusional Disorder

ERP
▶ Cognitive Behavioral Therapy
▶ QT Interval

ERP Components
▶ Event-Related Potential Components

ERPs
▶ Event-Related Potential

ES
▶ Electrospray Ionization

ESI
▶ Electrospray Ionization

Eszopiclone
▶ Benzodiazepines

Ethanal
▶ Acetaldehyde

Ethanol
▶ Alcohol
▶ Alcohol Abuse and Dependence

Ethics of Placebo Prescription
▶ Ethical Issues in Human Psychopharmacology

Ethics of Prescription
▶ Ethical Issues in Human Psychopharmacology

Ethyl-Alcohol
▶ Alcohol

EtOH
▶ Alcohol

Euonymaceae
▶ Celastraceae

Excitatory Amino Acid Transmitters
▶ Excitatory Amino Acids and their Antagonists
▶ Glutamate

Excitatory Neurotransmitters
▶ Excitatory Amino Acids and their Antagonists

Excitatory Postsynaptic Potentials
▶ EPSPs and IPSPs

Exocentric
▶ Allocentric

Expectations
▶ Expectancies and Their Influence on Drug Effects

Experimental Animal Models of Attention Deficit/Hyperactivity Disorder
▶ Attention Deficit Hyperactivity Disorders: Animal Models

Experimental Conflict
▶ Punishment Procedures

Experimental Drug Dependence
▶ Addictive Disorder: Animal Models

Experimental Medicine
▶ Translational Research in Drug Discovery

Experimental Models of Anxiety
▶ Anxiety: Animal Models

Explicitly and Implicitly Assessed Memory
▶ Declarative and Nondeclarative Memory

Exploding Head Syndrome
▶ Parasomnias

Exploratory Behavior
▶ Motor Activity and Stereotypy

Exposure and Response Prevention
▶ Cognitive Behavioral Therapy

Eye Tracking
▶ Eye Movement Tasks

Eyes Closed Occipital Waves
► Function of Slow and Fast Alpha Waves

E-α-Cyano-N,N-Dietyl-3,4-Dihydroxy-5-Nitrocinnamamide
► Entacapone

F

FAAH
► Fatty Acid Amide Hydrolase

FAS
► Fetal Alcohol Syndrome

Fast-Scan Cyclic Voltammetry
► Electrochemical Techniques

Fat Soluble
► Lipophilicity

FCE-21336
► Cabergoline

Fear of Being Alone
► Agoraphobia

Fear of Crowded Areas
► Agoraphobia

Fear of Public Places
► Agoraphobia

Fear-Potentiated Startle
► Pavlovian Fear Conditioning

Fear Reaction to Somatic Anxiety Symptoms
► Agoraphobia

Federal Medical Center, Lexington
► Narcotics Prison Farm

Feeling
► Emotion and Mood

FFT
► Fast Fourier Transformation

Final Ratio
► Breakpoint

Financial Incentives
► Contingency Management in Drug Dependence

Flexibility
► Elasticity

Flicker Fusion Rate
► Critical Flicker Fusion Frequency

Flicker Fusion Threshold
► Critical Flicker Fusion Frequency

Fluid
► Liquid Diet for Administering Alcohol

Flumazepil
► Flumazenil

fMRI
► Magnetic Resonance Imaging (Functional)

Force Platform
► Force-Plate Actometer

Forced Swimming Test
► Behavioral Despair

Forward Genetics – Random Mutagenesis
► Forward Genetics/Reverse Genetics

Fourier Analysis of Time Series
► Power Spectral Analysis

Fourier Spectrum
► Spectrograms

Fourier Transform
► Power Spectral Analysis

Free-Floating Anxiety
► Generalized Anxiety Disorder

Frequency Estimation
► Power Spectral Analysis

Frequency of Oscillation
▶ Rhythmicity

Frequency Spectrum
▶ Spectrograms

Functional Somatic Syndromes
▶ Illness Anxiety Disorder, Somatic Symptom Disorder, and Body Dysmorphic Disorder

G

GAD
▶ Generalized Anxiety Disorder

Gambling Addiction
▶ Pathological Gambling

Gamma Camera
▶ Gamma Camera System

Gamma-Hydroxybutyrate
▶ Oxybate

Gat
▶ Khat

Gel Electrophoresis
▶ Two-Dimensional Gel Electrophoresis

Gender Differences
▶ Sex Differences in Drug Effects

Genetically Engineered Animal
▶ Genetically Modified Animals

Genetically Modified Organism
▶ Genetically Modified Animals

Genomic Imprinting
▶ Imprinted Genes

Geocentric
▶ Allocentric

GHB
▶ Oxybate

Global Knockout
▶ Knockout/Knockin

GMO
▶ Genetically Modified Animals

GnRH
▶ Gonadotropin-Releasing Hormone

Gradient-Echo EPI
▶ Echo-Planar Imaging

Gradient-Echo MRI
▶ Gradient-Echo Images

Gradient-Recalled-Echo (GRE) Planar Images
▶ Gradient-Echo Images

H

Halazepam
▶ Benzodiazepines

Hallucinogen
▶ Hallucinogen Abuse and Dependence

Hallucinogen Intoxication
▶ Hallucinogen Abuse and Dependence

Haloxazepam
▶ Benzodiazepines

Haploid Genotype
▶ Haplotype

HATs
▶ Histone Acetyltransferases

HD
▶ Huntington's Disease

HDACs
▶ Histone Deacetylase

HDACs Inhibitors
▶ Histone Deacetylase Inhibitors

Health Anxiety
▶ Illness Anxiety Disorder, Somatic Symptom Disorder, and Body Dysmorphic Disorder

Hedonic Reactions
▶ Taste Reactivity Test

Heightened Illness Concern
▶ Illness Anxiety Disorder, Somatic Symptom Disorder, and Body Dysmorphic Disorder
▶ Blood–Brain Barrier

Herbal
▶ Herbal Medicinal Product

Herbal Medicine
▶ Herbal Medicinal Product

Herbal Remedy
▶ Herbal Medicinal Product

Heroin
▶ Diamorphine

Heroin Addiction
▶ Opioid Use Disorder and Its Treatment

High-Performance Liquid Chromatography
▶ High-Pressure Liquid Chromatography

High-Speed Chronoamperometry
▶ Electrochemical Techniques

Higher-Order Cognitive Processing
▶ Executive Functions

Hippocampal EEG Domain
▶ RSA

Hippocrateaceae
▶ Celastraceae

Histamine Receptor Agonists and Antagonists
▶ Histaminic Agonists and Antagonists

HPA
▶ Hypothalamic-Pituitary-Adrenal Axis

HPLC
▶ High-Pressure Liquid Chromatography

HPPD
▶ Hallucinogen Persisting Perception Disorder

HRT
▶ Habit Reversal Therapy

5-HT
▶ Serotonin

5-HT$_{1A}$ Receptor
▶ Antidepressants
▶ 5-HT$_{2A}$ Receptor
▶ Serotonin

Hug Drug
▶ Methylenedioxymethamphetamine (MDMA)

Human Medicinal Product
▶ Medicine

Hunger-Mimetics
▶ Appetite Stimulants

14-Hydroxy-Dihydromorphinone
▶ Oxymorphone

3-Hydroxy-L-Tyrosine
▶ Levodopa, L-DOPA

9-Hydroxyrisperidone
▶ Paliperidone

5-Hydroxytryptamine
▶ Serotonin

Hyoscine
▶ Scopolamine

Hyperkinesias
▶ Hyperactivity

Hyperkinetic Child Syndrome
▶ Attention-Deficit and Disruptive Behavior Disorders
▶ Attention Deficit Hyperactivity Disorder

Hyperlocomotion
▶ Hyperactivity

Hyperphagics
▶ Appetite Stimulants

Hypnotic Dependence
▶ Sedative, Hypnotic, and Anxiolytic Dependence

Hypochondria
▶ Illness Anxiety Disorder, Somatic Symptom Disorder, and Body Dysmorphic Disorder

Hypophagics
▶ Appetite Suppressants

Hypovase®
▶ Prazosin

I

ID
▶ Intradimensional

Illness Anxiety
▶ Illness Anxiety Disorder, Somatic Symptom Disorder, and Body Dysmorphic Disorder

Immunotherapy
▶ Vaccines and Drug-Specific Antibodies

Immunotoxins
▶ Neurotoxins

Impairment of Functioning Scales
▶ Impairment of Functioning; Measurement Scales

Impulse Control
▶ Behavioral Inhibition

Impulsive–Compulsive Gambling
▶ Pathological Gambling

IMS
▶ Imaging Mass Spectrometry

Inability to Delay Gratification
▶ Impulsivity

Inability to Sleep
▶ Insomnias

Incentive Learning
▶ Reward-Related Incentive Learning

Incentive Motivation
▶ Appetitive Responses
▶ Conditioned Taste Preferences

Incentive Properties of Drug Cues
▶ Attentional Bias to Drug Cues

Indoleamine Hypothesis
▶ Aminergic Hypotheses for Depression

Infantile Autism
▶ Autism Spectrum Disorders and Intellectual Disability

Inhibitory Postsynaptic Potentials
▶ EPSPs and IPSPs

Institutional Animal Care and Use Committee
▶ IACUC

Instrumental Aversive Conditioning
▶ Operant Behavior in Animals
▶ Operant Conditioning

Instrumental Behavior
▶ Operant Behavior in Animals

Instrumental Performance
▶ Operant Behavior in Animals

Intellectual Disability
▶ Autism Spectrum Disorders and Intellectual Disability
▶ Mild Cognitive Impairment

Intensive Care Unit Psychosis
▶ Delirium

Intracerebral Microdialysis
▶ Microdialysis

Inventories
▶ Rating Scales and Diagnostic Schemata

Inventors Delusion
▶ Delusional Disorder

Inventors' Psychosis or "Inventors" Delusion ("Erfinderwahn" or "Erfindungswahn")
▶ Delusional Disorder

Iontophoresis
▶ Microiontophoresis and Related Methods

IPT
▶ Interpersonal Psychotherapy

IRBS
▶ Insulin-Resistant Brain State

Isolation Calls
▶ Distress Vocalization

Isomers
▶ Stereoisomers

J

Justifications
▶ Ethical Issues in Animal Psychopharmacology
▶ Ethical Issues in Human Psychopharmacology

K

Kanner Syndrome
▶ Autism Spectrum Disorders and Intellectual Disability

Kat
▶ Khat

Kath
▶ Khat

K_D
▶ Equilibrium Dissociation Constant

K_D^{-1}
▶ Affinity

Kel
▶ First-Order Elimination Kinetics

Ketazolam
▶ Benzodiazepines

Kinetics
▶ Pharmacokinetics

Koro
▶ Delusional Disorder

L

LAAM
▶ L-Alpha-Acetyl-Methadol

Labeled Ligand Binding
▶ Receptors: Binding Assays

Labeled Ligand Concentration Binding Isotherm
▶ Receptors: Binding Assays
▶ Saturation Binding Curve

Laboratory Animal Models of Minimal Brain Dysfunction
▶ Attention Deficit Hyperactivity Disorders: Animal Models

Laser Desorption Ionization
▶ Matrix-Assisted Laser Desorption Ionization

Late Luteal Phase Dysphoric Disorder
▶ Premenstrual Dysphoric Mood Disorder

Lawful
▶ Legal Aspects of Psychopharmacology

L-Deprenyl
▶ Antidepressants
▶ Selegiline

Legitimate
▶ Legal Aspects of Psychopharmacology

Lep^{db}/Lep^{db} mouse
▶ db/db Mouse

Lep^{ob}/*Lep^{ob}* mouse
▶ ob/ob Mouse

Levodopa-L-Di Ortho Phenylalanine
▶ Anti-Parkinson Drugs

Levo-Duboisine
▶ Scopolamine

Lexington Narcotics Farm
▶ Narcotics Prison Farm

Ligand Binding
▶ Receptors: Binding Assays

Ligand-Directed Trafficking
▶ Functional Selectivity

Lipid Soluble
▶ Lipohilic

Lithium Salts
▶ Lithium

Liver Toxicity
▶ Hepatotoxicity

Locomotor Activity
▶ Motor Activity and Stereotypy

Lofepramine
▶ Antidepressants
▶ Tricyclic Antidepressants

Long-Lasting Synaptic Depression
▶ Long-Term Depression and Memory

Long-Term Depression
▶ Long-Term Depression and Memory
▶ Synaptic Plasticity

Long-Term Treatments for Bipolar Disorder
▶ Mood Stabilizers

Love Drug
▶ Methylenedioxymethamphetamine (MDMA)
▶ Oxytocin

Love Hormone
▶ Methylenedioxymethamphetamine (MDMA)
▶ Oxytocin

LQTS
▶ Long QT Syndrome

LSD
▶ Hallucinogen Abuse and Dependence
▶ Hallucinogens

LTD
▶ Long-Term Depression and Memory

LTP
▶ Long-Term Potentiation and Memory

LY127809
▶ Pergolide Mesylate

LY139603
▶ Atomoxetine

Lysergic and Diethylamide
▶ Hallucinogen Abuse and Dependence
▶ Hallucinogens

M

Magnesium Pemoline
▶ Pemoline

Magnetic Resonance
▶ Nuclear Magnetic Resonance

Mairungi
▶ Khat

MALDI
▶ Matrix-Assisted Laser Desorption Ionization

Malleability
▶ Elasticity

Malonylurea
▶ Barbiturates

Manerix
▶ Moclobemide

Manic-Depressive Illness
▶ Bipolar Disorder
▶ Bipolar Disorder in Children

Marijuana Abuse
▶ Cannabis Use Disorders

Marijuana Addiction
▶ Cannabis Use Disorders

Marijuana Dependence
▶ Cannabis Use Disorders

Marplan
▶ Isocarboxazid

Mass Spectrometry Imaging
▶ Mass Spectrometry

Mass Spectroscopy
▶ Mass Spectrometry

Mating Behavior
▶ Sexual Behavior

MDAS
▶ Memorial Delirium Assessment Scale

MDMA
▶ Methylenedioxymethamphetamine (MDMA)

MEA
▶ Glutamate Microelectrode Arrays
▶ Microelectrode Arrays

Measures
▶ Rating Scales and Diagnostic Schemata

Measure of Drug Activity
▶ Potency

Measurement and Treatment Research to Improve Cognition in Schizophrenia
▶ MATRICS

Measurement of Biological Effect Resulting from Interaction with the Receptor
▶ Receptors: Functional Assays

Measurement of Neuronal Activity
▶ Extracellular Recording

Measurement of Receptor Signaling
▶ Receptors: Functional Assays

Medical Herbalism
▶ Herbal Remedies

Medically Unexplained Symptoms (MUS)
▶ Illness Anxiety Disorder, Somatic Symptom Disorder, and Body Dysmorphic Disorder

Medicines Control
▶ Licensing and Regulation of Medicines in the UK

Medicines Regulation
▶ Licensing and Regulation of Medicines in the UK

Megalomania
▶ Delusional Disorder

Membrane Potential Recording
▶ Current Clamp

Membrane Voltage
▶ Membrane Potential

Memory Consolidation
▶ Consolidation and Reconsolidation
▶ Protein Synthesis and Memory

Mental Deficiency
▶ Autism Spectrum Disorders and Intellectual Disability

Memory Dysfunction
▶ Dementias and Other Amnestic Disorders

Memory-Impairing Drugs
▶ Inhibition of Memory

Memory Impairment
▶ Dementias and Other Amnestic Disorders

Memory: Information Storage, Learning and Memory
▶ Long-Term Depression and Memory

Memory Persistence
▶ Protein Synthesis and Memory

Memory Restabilization
▶ Consolidation and Reconsolidation

Memory Stabilization
▶ Consolidation and Reconsolidation

Memory Storage
▶ Consolidation and Reconsolidation
▶ Protein Synthesis and Memory

Mental Disorders
▶ Neuropsychiatric Disorders

Mental Retardation
▶ Autism Spectrum Disorders and Intellectual Disability

MEOS
▶ Microsomal Ethanol-Oxidizing System

Meperidine
▶ Opioids
▶ Pethidine

Merital®
▶ Nomifensine

Metabolic Encephalopathy
▶ Delirium

Metabolic Toxins
▶ Neurotoxins

Metabonomics
▶ Metabolomics

Metabotropic Glutamate Receptors 2 and 3
▶ Group II Metabotropic Glutamate Receptor

Metergoline
▶ Methergoline

Methotrimeprazine
▶ Levomepromazine

Methylamphetamine
▶ Methamphetamine

N-Methylamphetamine
▶ Methamphetamine

Methylbenzene, Toluol
▶ Toluene

Methyl (1R,2R,3S,5S)-3-(Benzoyloxy)-8-Methyl-8-Azabicyclo[3.2.1] Octane-2-Carboxylate
▶ Cocaine

Methylbenzylpropynylamine
▶ Pargyline

Methyl Chloroform
▶ Trichloroethane

3,4 Methylenedioxymethamphetamine
▶ Methylenedioxymethamphetamine (MDMA)

Methyl-Lorazepam
▶ Lormetazepam

(3R)-N-Methyl-3-(2-Methylphenoxy)-3-Phenyl-Propan-1-Amine
▶ Atomoxetine

Methylmorphine
▶ Codeine

Methylphenoxy-Benzene Propanamine
▶ Atomoxetine

N-Methyl-N-2-Propynylbenzylamine
▶ Pargyline

(–)-1-Methyl-2-(3-Pyridyl)Pyrrolidine
▶ Nicotine

8 β[(Methylthio)Methyl]-6-Propylergoline Monomethanesulfonate
▶ Pergolide Mesylate

mGluR
▶ Metabotropic Glutamate Receptor

Michael Kohlhaas' Syndrome
▶ Delusional Disorder

Microamines
▶ Trace Amines

Microelectrophoresis
▶ Microiontophoresis and Related Methods

Milnacipran Hydrochloride
▶ Milnacipran

Miltown
▶ Meprobamate

Minimal Brain Damage
▶ Attention-Deficit and Disruptive Behavior Disorders

Minimal Cerebral Dysfunction
▶ Attention-Deficit and Disruptive Behavior Disorders

Minipress®
▶ Prazosin

Minor Cerebral Dysfunction
▶ Attention-Deficit and Disruptive Behavior Disorders

Miraa
▶ Khat

Mismatch Field
▶ Mismatch Negativity

MK486
▶ Carbidopa

MMN
▶ Mismatch Negativity

Model Organisms of Hyperkinetic Syndrome
▶ Attention Deficit Hyperactivity Disorders: Animal Models

Mogadon
▶ Nitrazepam

Monoamine Depletion
▶ Amine Depletors

Monoamine Hypotheses
▶ Aminergic Hypotheses for Depression
▶ Aminergic Hypotheses for Schizophrenia

Morals
▶ Ethical Issues in Animal Psychopharmacology
▶ Ethical Issues in Human Psychopharmacology

Morphine-Like Compounds
▶ Mu-Opioid Agonists

Morphogenesis
▶ Ontogeny

Morris Water Navigation Task
▶ Morris Water Maze

Mosaic
▶ Chimera

Motivational Valence
▶ Taste Reactivity Test

Motor Activity: Repetitious Behavior
▶ Motor Activity and Stereotypy

Motor Inhibition
▶ Behavioral Inhibition

Motor Learning
▶ Verbal and Nonverbal Learning in Humans

Movement Disorder
▶ Tic Disorders with Childhood Onset
▶ Tics

m-PD
▶ Meta-Phenylenediamine

mPFC
▶ Medial Prefrontal Cortex

mRNA Splice Variants
► Alternative Splicing

MSI
► Imaging Mass Spectrometry

Multi-Infarct Dementia
► Vascular Dementia

Multiple-Unit Spiking Activity
► Multiunit Activity

Murungu
► Khat

Muscarinic Agonists
► Muscarinic Cholinergic Receptor Agonists and Antagonists

Muscarinic Antagonists
► Muscarinic Cholinergic Receptor Agonists and Antagonists
► Scopolamine

Mutant
► Transgenic Organism

Mutant Animal
► Genetically Modified Animals

N

NAcc
► Nucleus Accumbens

nAChR
► Nicotinic Receptor

NAD
► Nicotinamide Adenine Dinucleotide

Nanospray
► Electrospray Ionization

Narcotic Analgesics
► Opioid Analgesics
► Opioids

Narcotics
► Opioid Analgesics
► Opioids

Nardil
► Phenelzine

Negative Antagonists
► Inverse Agonists

Neostriatum
► Striatum

Nerve Impulse
► Action Potentials

Nerve Spikes
► Action Potentials

NET
► Norepinephrine Transporter

Neuronal Network Oscillations
► Electroencephalography

Neural Plasticity
► Molecular Mechanisms of Learning & Memory

Neurobehavioral Teratogens
► Perinatal Exposure to Drugs
► Teratogens

Neurocognitive Dysfunction
► Cognitive Impairment

Neurokinin 1
► Substance P

Neurokinin 2
► Neurokinin A

Neurokinin 3
► Neurokinin B

Neurokinin-α
► Neurokinin A

Neuromedin K
► Neurokinin B
► Tachykinins

Neuromedin L
► Neurokinin A
► Tachykinins

Neuronal Plasticity
► Synaptic Plasticity

Neuroprotectant
► Neuroprotective Agent

Neuroprotective Drug
► Neuroprotective Agent

Neuropsychopharmacology
► Psychopharmacology

Neurotic Depression
► Dysthymic Disorder (Persistent Depressive Disorder)

Newer Anticonvulsants
► Second-Generation Anticonvulsants

NGF
► Nerve Growth Factor

NIDA IRP
► Addiction Research Center

Nightmares
► Parasomnias

NK2
► Neurokinin A

NK3
► Neurokinin B

NKA
► Neurokinin A
► Tachykinins

NKB
► Neurokinin B
► Tachykinins

NMDA Receptors
► N-Methyl-D-Aspartate Receptor

NMS
► Neuroleptic Malignant Syndrome

NNH
► Number Needed to Harm

NNT
► Number Needed to Treat

Non-Invasive Neuro-Imaging
► Magnetic Resonance Imaging (Functional)

Nonspecific Memory Impairment
► Delay-Independent Deficit

Nordazepam
► Benzodiazepines

Norebox
► Reboxetine

Novel Antipsychotics
► Second- and Third-Generation Antipsychotics

Novel Word Learning
► Verbal and Nonverbal Learning in Humans

Novelty Preference Test
► Novel Object Recognition Test

NREM
► Parasomnias

NREM Sleep
► Non-Rapid Eye Movement Sleep

NSAIDS
► Nonsteroidal Anti-Inflammatory Drugs

O

OB Protein
▶ Leptin

Obsessive Ruminations
▶ Obsessions

Obsessive–Compulsive Neurosis
▶ Obsessive–Compulsive Disorders

OCD
▶ Obsessive-Compulsive Disorders in Childhood
▶ Obsessive–Compulsive Disorders

Oculomotor Tasks
▶ Eye Movement Tasks

Oestrogens
▶ Estrogens

6-OHDA
▶ 6-Hydroxydopamine

Old Antipsychotics
▶ First-Generation Antipsychotics

Old Neuroleptics
▶ First-Generation Antipsychotics

Older Anticonvulsants
▶ First-Generation Anticonvulsants

Oligophrenia
▶ Autism Spectrum Disorders and Intellectual Disability

Olton Maze
▶ Radial Arm Maze

One-/two-Way Active Avoidance
▶ Active Avoidance

Ontogenesis
▶ Ontogeny

Opioid Addiction
▶ Opioid Use Disorder and Its Treatment

Opposite Effect
▶ Effect Inversion

Optical Isomers
▶ Stereoisomers

Oral Self-Administration
▶ Drug Taste Preference Conditioning

Orexigens
▶ Appetite Stimulants

Orexins
▶ Hypocretins

Othello Syndrome
▶ Delusional Disorder

Ovaries or Testes
▶ Gonads

Overconsumption
▶ Hyperphagia

2-Oxopyrrolidin-1-Acetamide
▶ Piracetam

P

P3
▶ P300

P450
▶ Cytochrome P450

Pain-Relievers
▶ Analgesics

Paracetamol
▶ Analgesics

Paracodeine
▶ Dihydrocodeine

Paramagnetism
▶ Paramagnetic

Paranoia
▶ Delusional Disorder

Paranoid Delusions
▶ Delusional Disorder

Paranoid Psychosis
▶ Delusional Disorder

Partial Reinforcement
▶ Relative Validity

Paxil
▶ Paroxetine

PCP
▶ Phencyclidine

PD
▶ Parkinson's Disease

PDD NOS
▶ Pervasive Developmental Disorder Not Otherwise Specified

PDE Inhibitors
▶ Phosphodiesterase Inhibitors

Pediatric-Onset Obsessive–Compulsive Disorder
▶ Obsessive-Compulsive Disorders in Childhood

Pentobarbitone
▶ Pentobarbital

Peptidomics of the Brain
▶ Neuropeptidomics

Peri-Adolescent Psychopharmacology
▶ Adolescence and Responses to Drugs

Periciazine
▶ Pericyazine

Persecutory Delusions
▶ Delusional Disorder

PET
▶ Positron Emission Tomography (PET) Imaging

PET Imaging
▶ Positron Emission Tomography (PET) Imaging

Pharmaco-Ethology
▶ Ethopharmacology

Pharmacokinetics Study
▶ Phase I Clinical Trial
▶ Pharmacokinetics

Phasic Neuronal Firing
▶ Phasic Neurotransmission

Phasic Neurotransmitter Release
▶ Phasic Neurotransmission

Phenobarbitone
▶ Phenobarbital

Philosophical Basis for Using Animals
▶ Ethical Issues in Animal Psychopharmacology

phMRI
▶ Pharmacological fMRI

Phobic Anxiety
▶ Agoraphobia

Physiological Cell Death
▶ Apoptosis

Physiological Dependence
▶ Physical Dependence

Phytotherapy
▶ Herbal Remedies

Pinazepam
▶ Benzodiazepines

Pitocin®
▶ Oxytocin

PK
▶ Pharmacokinetics

PKI
▶ Kinase Inhibitors

PKU
▶ Phenylketonuria

Place Conditioning
▶ Conditioned Place Preference and Aversion

Place Fields
▶ Place Cells

Place Learning
▶ Spatial Learning in Animals

Placebo Response
▶ Placebo Effect

Plant Toxins
▶ Neurotoxins

Plasticity
▶ Elasticity
▶ Synaptic Plasticity

Pleomorphism
▶ Genetic Polymorphism

Pliancy
▶ Elasticity

Plus Maze Test
▶ Elevated Plus Maze

Point of Subjective Equality
▶ Indifference Point

Poisons
▶ Neurotoxins

Polyphagia
▶ Hyperphagia

Porsolt Test
▶ Behavioral Despair

Positive and Negative Syndrome Scale (for Schizophrenia)
▶ PANSS

Postnatal Neurogenesis
▶ Neurogenesis

Postpsychotic Depression
▶ Postpsychotic Depressive Disorder of Schizophrenia

Post-translational Amino Acid Modification
▶ Posttranslational Modification

Post-translational Protein Modification
▶ Posttranslational Modification

Post-trial Surprise
▶ Unblocking

Potentiation
▶ Drug Interactions

Power Spectrum
▶ Spectrograms

Pragmatic Outcome
▶ Effectiveness

Prefrontal Lobe
▶ Prefrontal Cortex

Premenstrual Syndrome
▶ Premenstrual Dysphoric Mood Disorder

Premenstrual Tension
▶ Premenstrual Dysphoric Mood Disorder

Preparatory Behavior
▶ Appetitive Responses

Pre-Psychotic Prodrome
▶ Prepsychotic States and Prodromal Symptoms

Price
▶ Behavioral Economics
▶ Intracranial Self-Stimulation

Product Information
▶ Summary of Product Characteristics

Programmed Cell Death
► Apoptosis

Prolift
► Reboxetine

Propericiazine
► Pericyazine

Protein Array
► Protein Microarray

Protein-Binding Microarray
► Protein Microarray

Protein Kinase Inhibitors
► Kinase Inhibitors

Prozac
► Fluoxetine

Pseudo-parkinsonism
► Akinesia

Psychedelics
► Hallucinogens
► Hallucinogen Abuse and Dependence
► Ritual Uses of Psychoactive Drugs

Psychiatric Disorder
► Neuropsychiatric Disorders

Psychomotor Function
► Psychomotor Performance in Humans

Psychomotor Stimulants
► Indirect-Acting Dopamine Agonists
► Psychostimulants

Psychophysics
► Psychophysiological Methods

Psychose Passionelle
► Delusional Disorder

Psychostimulant Dependence
► Psychostimulant Use Disorder

Psychotomimetics
► Hallucinogens

PTM
► Posttranslational Modification

PTSD
► Posttraumatic Stress Disorder

Punished Behavior
► Punishment Procedures

Pure Insertion
► Cognitive Subtraction

Purposive Behavior
► Operant Behavior in Animals

Q

Q'at
► Khat

Qat
► Khat

QOL
► Quality of Life

Quality of Life Scales
► Functional Scales

Quantified (Wake) EEG
► qEEG

Querulous Paranoia
► Delusional Disorder

R

Radial Maze
► Radial Arm Maze

Radioactive Tracer
► Radiopharmaceutical

Radiolabeled Probe
► Radiopharmaceutical

Radioligand
► Radiopharmaceutical

Radioligand Binding Assays
► Receptors: Binding Assays

Radiotracer
► Radiopharmaceutical

Radiotracer Imaging
► Positron Emission Tomography (PET) Imaging

Rapid Eye Movement Disorder
► Parasomnias

Rapid Tryptophan Depletion
► Tryptophan Depletion

Rat or Mouse Models
► Rodent Tests of Cognition

Rate of Responding
► Rate-Dependency Theory

RBD
► Parasomnias
► Recurrent Brief Depressive Disorder
► REM Sleep Behavior Disorder

Reactive Depressions
► Adjustment Disorders

Receptor Activator
► Agonist

Receptor Inhibitor
► Antagonist

Recurrent Brief Depression
► Recurrent Brief Depressive Disorder

Refractory Schizophrenia
► Treatment-Resistant Schizophrenia

Region-Specific Knockout
► Conditional Knockout

Reinforcer Efficacy
► Subjective Value

Reinstatement of Drug Seeking
► Reinstatement of Drug Self-Administration

Reinstatement of Drug-Taking Behavior
► Reinstatement of Drug Self-Administration

REM Sleep
► Parasomnias
► Rapid Eye Movement Sleep

Remeron
► Mirtazapine

Repetitive Behavior
► Motor Activity and Stereotypy

Repetitive Thoughts
► Obsessions

Reporting Ligand
► Radiopharmaceutical

Reproductive Behavior
► Sexual Behavior

Resistance
► Tolerance

Respondent Conditioning
► Classical (Pavlovian) Conditioning
► Pavlovian Conditioning
► Pavlovian Fear Conditioning

Response Conflict Task
► Go/No-Go Task

Response Inhibition or Behavioral Inhibition
► Impulsivity

Response Inhibition Task
► Stop-Signal Task

Responsiveness
► Behavioral Flexibility: Attentional Shifting, Rule Switching, and Response Reversal

Restlessness
► Hyperactivity

Retardation Test
▶ Retardation of Acquisition Test

Reverse Genetics – Targeted Mutagenesis
▶ Forward Genetics/Reverse Genetics

Reverse Tolerance
▶ Sensitization
▶ Sensitization to Drugs

ReVia
▶ Naltrexone

9-β-D-Ribofuranosyladenine
▶ Adenosine

Rights
▶ Ethical Issues in Animal Psychopharmacology

Ritalin
▶ Methylphenidate and Related Compounds

RNAi
▶ RNA Interference

Ro 4-4602
▶ Benserazide

Ro-15-1788
▶ Flumazenil

Rodent Behavioral Test Paradigms or Procedures
▶ Animal Models for Psychiatric States
▶ Rodent Tests of Cognition

Rodent Models of Attention, Memory, and Learning
▶ Rodent Tests of Cognition

Rodent Models of Autism
▶ Autism: Animal Models

Rodent Models of Cognition
▶ Rodent Tests of Cognition

ROS
▶ Reactive Oxygen Species

S

Saccades
▶ Eye Movement Tasks

Salaciaceae
▶ Celastraceae

Sandostatin
▶ Octreotide

Secobarbitone
▶ Secobarbital

Secondary Depression in Schizophrenia
▶ Postpsychotic Depressive Disorder of Schizophrenia

Secondary Messenger Molecule
▶ Second Messenger

Secondary Reinforcer
▶ Conditioned Reinforcers

Secondary Reward
▶ Appetitive Responses

Sedative
▶ Minor Tranquilizer

Sedative Dependence
▶ Sedative, Hypnotic, and Anxiolytic Dependence

Sedative–Hypnotics
▶ Barbiturates
▶ Benzodiazepines
▶ Meprobamate
▶ Non-Benzodiazepine Hypnotics

Selective Noradrenergic Reuptake Inhibitors
▶ NARI Antidepressants

Selective Serotonin Reuptake Inhibitors
▶ SSRIs and Related Compounds

Self-Control Failure
▶ Impulsivity

Self-Destructive Behavior
▶ Suicide

Self-Immolation
▶ Suicide

Self-Injection
▶ Self-Administration of Drugs

Sensitive Delusion of Reference
▶ Delusional Disorder

Sensorimotor Gating
▶ Prepulse Inhibition

Sensory Gating
▶ Prepulse Inhibition

Separated
▶ Independents

Separation Calls
▶ Distress Vocalization

Serazide
▶ Benserazide

Serotonin Antidepressants
▶ SSRIs and Related Compounds

Serotonin-Norepinephrine Reuptake Inhibitors
▶ SNRI Antidepressants

Seroxat
▶ Paroxetine

Set-Shifting Test
▶ Wisconsin Card Sorting Test

Sexsomnia
▶ Parasomnias

Sexual Dimorphism
▶ Sex Differences in Drug Effects

Sharp Microelectrode Recording
▶ Intracellular Recording

Short Interfering RNA
▶ Antisense Oligonucleotides
▶ siRNA

Short-Term Plasticity
▶ Synaptic Plasticity

Shyness
▶ Social Anxiety Disorder

Side Effects
▶ Toxicity of Drugs

Signal Transduction Pathways
▶ Signaling Cascades

Signals
▶ Discriminative Stimulus
▶ Drug Cues

Silencing RNA
▶ siRNA

Simulations of Psychopathology
▶ Animal Models for Psychiatric States

Sinemet
▶ Anti-Parkinson Drugs

Single Nucleotide
▶ Single-Nucleotide Polymorphism

Single-photon Emission Computed Tomography
▶ SPECT Imaging

Single-photon Emission Tomography (SPET)
▶ SPECT Imaging

Siphonodontaceae
▶ Celastraceae

Sirolimus
▶ Rapamycin

Site-Specific Knockout
▶ Conditional Knockout

Skilled Performance
▶ Psychomotor Performance in Humans

Sleep Enuresis
▶ Parasomnias

Sleep Paralysis
▶ Parasomnias

Sleep-related Dissociative Disorder
▶ Parasomnias

Sleep-related Eating
▶ Parasomnias

Sleep Terrors
▶ Parasomnias

Sleeplessness
▶ Insomnias

Sleepwalking
▶ Parasomnias

Slow Waves
▶ Function of Delta Waves

Small Interfering RNA
▶ siRNA

Smart Drugs
▶ Pharmaceutical Cognitive Enhancers: Neuroscience and Society

Smooth Pursuit
▶ Eye Movement Tasks

SMS201-995
▶ Octreotide

SNP
▶ Single-Nucleotide Polymorphism

Social Challenge
▶ Social Stress

Social Phobia
▶ Social Anxiety Disorder

Sodium 4-Hydroxybutyrate
▶ Oxybate

Sodium 5-Ethyl-5-(1-Methylbutyl) Barbiturate
▶ Pentobarbital

Solvent Abuse
▶ Inhalant Abuse

Solvex
▶ Reboxetine

Somatization
▶ Illness Anxiety Disorder, Somatic Symptom Disorder, and Body Dysmorphic Disorder

Somatotropin Release-Inhibiting Factor
▶ Somatostatin

Somatuline
▶ Lanreotide

Source Analysis
▶ Source Localization Techniques

SP
▶ Substance P

Spatial Navigation
▶ Spatial Learning in Animals

SPC
▶ Summary of Product Characteristics

Special K
▶ Ketamine

SPET
▶ SPECT Imaging

Spin-Echo EPI
▶ Echo-Planar Imaging

Spindles
▶ Function of Slow and Fast Alpha Waves

Spontaneous Activity
▶ Motor Activity and Stereotypy

SR141716
▶ Rimonabant

SRIF
► Somatostatin

SROMs
► Slow-Release Morphines

SSRIs
► Selective Serotonin Reuptake Inhibitors

SSRT
► Stop-Signal Task

sst_1
► Somatostatin Receptors

sst_2
► Somatostatin Receptors

sst_3
► Somatostatin Receptors

sst_4
► Somatostatin Receptors

sst_5
► Somatostatin Receptors

Stackhousiaceae
► Celastraceae

Stage R Sleep
► Rapid Eye Movement Sleep

Standardized Mean Difference
► Effect Size

Startle Modulation
► Prepulse Inhibition

Startle Response
► Startle

State-Dependent Learning
► State Dependence of Memory

State-Dependent Retrieval
► State Dependence of Memory
► State-Dependent Learning and Memory

Stavzor
► Valproic Acid

Stereotypy
► Motor Activity and Stereotypy
► Stereotypical and Repetitive Behavior

Stimulants
► Psychostimulants

Stimulus Enrichment
► Environmental Enrichment and Drug Action

Stimulus Pre-exposure Effect
► Latent Inhibition

Stimulus Trafficking
► Functional Selectivity

Strattera™
► Atomoxetine

Stress-Related Mood Disorders
► Adjustment Disorders

Stressful Stimuli
► Stress: Influence on Relapse to Substance Use

Stretched-Attend Posture
► Risk Assessment

Substance Abuse Disorder
► Opioid Use Disorder and Its Treatment

Substance K
► Neurokinin A
► Tachykinins

Substance Use Disorders: A Theoretical Framework
► Substance Abuse

Substitution
► Cross-Dependence

Subthreshold Diagnoses
► Adjustment Disorders

Subtraction Method
▶ Cognitive Subtraction

Suicide Completion
▶ Suicide

Supervisory Attentional System
▶ Executive Functions

Suppressed Behavior
▶ Punishment Procedures

Surface Field Potentials
▶ Electroencephalography

Surrogate
▶ Substitutes

Switch in Preference
▶ Preference Reversal

Syntocinon®
▶ Oxytocin

T

Tacrine
▶ Acetylcholinesterase and Cognitive Enhancement

Tail Suspension Test
▶ Behavioral Despair

Taste Aversion Learning
▶ Conditioned Taste Aversions
▶ Long-Delay Learning

Tastes as Conditioned Stimuli for Drugs
▶ Conditioned Taste Aversions
▶ Conditioned Taste Preferences

TCAs
▶ Secondary Amine Tricyclic Antidepressants
▶ Tertiary Amine Tricyclic Antidepressants
▶ Tricyclic Antidepressants

TCE
▶ Trichloroethane

TDM
▶ Therapeutic Drug Monitoring

Temporal Conditioning
▶ Timing Behavior

Temporal Knockout
▶ Inducible Knockout

Temporal Processing
▶ Timing Behavior

Teratogenetic
▶ Teratogenic

Terminal Half-Life
▶ Elimination Half-Life

Tests
▶ Rating Scales and Diagnostic Schemata

Tetracyclic Antidepressants
▶ NARI Antidepressants

Tetraethylthiuram Disulphide
▶ Disulfiram

Tetrazepam
▶ Benzodiazepines

Thalamocortical 10 Hz Rhythm
▶ Function of Slow and Fast Alpha Waves

Thalomid
▶ Thalidomide

The European Union First Episode Schizophrenia Trial
▶ EUFEST

Therapeutic Benefit
▶ Efficacy

Therapeutic Drug Plasma Level Monitoring
▶ Therapeutic Drug Monitoring

Theta Rhythm
▶ RSA

Thrill Seeking
▶ Risk Taking

Thrombocytes
▶ Platelets

Thymoleptics
▶ Antidepressants

Tight Seal Recording
▶ Patch Clamp

Time Discounting
▶ Temporal Discounting

Time Series Plot
▶ Time Series

Time-Specific Knockout
▶ Inducible Knockout

Timing
▶ Timing Behavior

TMS
▶ Transcranial Magnetic Stimulation

Tobacco Dependence
▶ Nicotine Dependence and Its Treatment

Tomoxetine
▶ Atomoxetine

Tourette Syndrome
▶ Tic Disorders with Childhood Onset

Toxicants
▶ Neurotoxins

Traditional Anticonvulsants
▶ First-Generation Anticonvulsants

Traditional Antipsychotics
▶ First-Generation Antipsychotics

Traditional Neuroleptics
▶ First-Generation Antipsychotics

Traffic Safety and Medicines
▶ Driving and Flying Under the Influence of Drugs

Tranquilizer
▶ Minor Tranquilizer

Transgenic Animal
▶ Genetically Modified Animals

Translational Medicine
▶ Translational Research in Drug Discovery

Translational Science
▶ Translational Research in Drug Discovery

Tranxene
▶ Clorazepate

TRD
▶ Treatment-Resistant Depression

1,1,1-Trichloroethane
▶ Trichloroethane

Tricyclo[3.3.1.1^{2,7}]Decan-1-Amine
▶ Amantadine

Triggers
▶ Drug Cues

1,3,7-Trimethylpurine-2,6-Dione
▶ Caffeine

Trimethylxanthine
▶ Caffeine

1,3,7-Trimethylxanthine
▶ Caffeine

Tripping
▶ Hallucinogen Abuse and Dependence

TSA
▶ Trichostatin A

Tschat
▶ Khat

TSST
▶ Trier Social Stress Test

Typical Neuroleptics
▶ First-Generation Antipsychotics

U

Unipolar Depression
▶ Depressive Disorders: Major, Minor, and Mixed

Unrelated
▶ Behavioral Economics
▶ Independents

Unsatisfactory Sleep
▶ Insomnias

UPS
▶ Ubiquitin-Proteasome System

UR
▶ Unconditioned Response

Urge-Driven Behaviors
▶ Impulse Control Disorders

US
▶ Unconditioned Stimulus

Usefulness
▶ Effectiveness

V

V3″
▶ Binding Potential

Vasoflex®
▶ Prazosin

Vasopressin
▶ Arginine-Vasopressin

VD
▶ Volume of Distribution

VDCC
▶ Voltage-Gated Calcium Channels

Venlafaxine Hydrochloride
▶ Venlafaxine

Venoms
▶ Neurotoxins

VEOS
▶ Pediatric Schizophrenia

Very Early Onset Schizophrenia (VEOS)
▶ Pediatric Schizophrenia

Vestra
▶ Reboxetine

VGCC
▶ Voltage-Gated Calcium Channels

Vigilance
▶ Attention
▶ Sustained Attention

Vitamin R
▶ Methylphenidate and Related Compounds

Vivitrol (Injectable)
▶ Naltrexone

VNS
▶ Vagus Nerve Stimulation

Volatile Nitrites
▶ Nitrites

Volatile Substance Abuse
▶ Inhalant Abuse

Voltage-Dependent Calcium Channels
▶ Voltage-Gated Calcium Channels

Voltammetry
▶ Electrochemical Techniques

Voucher-Based Reinforcement Therapy
▶ Contingency Management in Drug Dependence

VTA
▶ Ventral Tegmental Area

W

Weakening of Synaptic Connections
▶ Long-Term Depression and Memory

Weight Control Drugs
▶ Appetite Suppressants

Weight Management Drugs
▶ Appetite Suppressants

Welfare
▶ Ethical Issues in Animal Psychopharmacology

X

XTC
▶ Methylenedioxymethamphetamine (MDMA)

Y

Y-516
▶ Mosapramine

Y-BOCS
▶ Yale-Brown Obsessive-Compulsive Scale

Z

Z-Drugs
▶ Benzodiazepines
▶ Non-Benzodiazepine Hypnotics

Zelapar
▶ Selegiline

Zispin
▶ Mirtazapine

9783642361715VOL02